This new edition of a successful textbook describes and explains in a refreshingly clear way the origin and evolution of plants as revealed by the fossil record. It summarizes paleobotanical information relevant to our present understanding of the relationships among the major plant groups, extant and extinct. As in the first edition, the text is profusely illustrated with line drawings and halftones. For those students with little knowledge of plant structure and morphology, there is a brief résumé of those features of extant plants that will be needed to gain a better understanding of the fossil record. Summarizing charts are also used to help students visualize the interpretative material. For this edition new material on the evolution of the angiosperms has been added, and there is a new chapter dealing with the paleoecology of ancient plants. The text has also been extensively updated to include new information on the methodology of cladistics.

Paleobotany and the evolution of plants

Paleobotany and the evolution of plants

Second edition

Wilson N. Stewart

Professor Emeritus, Botany
University of Alberta
Edmonton, Alberta

Gar W. Rothwell

Professor of Environmental and Plant Biology
Ohio University
Athens, Ohio

CAMBRIDGE
UNIVERSITY PRESS

Published in South Asia by

Cambridge University Press India Pvt. Ltd.
Cambridge House
4381/4, Ansari Road, Daryaganj
New Delhi 110002
India

First South Asian Edition 2005
Reprinted 2013

ISBN-13 978-81-7596-295-8 (Paperback)

This edition is for sale in South Asia only, not for export elsewhere.

This edition of *"Wilson N. Stewart, Gar W. Rothwell / Paleobotany and the evolution of plants"*, is published by arrangement with Cambridge University Press, The Edinburgh Building, Shaftesbury Road, Cambridge CB2 2RU, UK.

Published by Manas Saikia for Cambridge University Press India Pvt. Ltd. and printed and bound at Raj Press, R-3 Inderpuri, New Delhi - 110012

In memory of
James Morton Schopf,
1911–1978

Contents

Preface

Although this new edition of *Paleobotany and the Evolution of Plants* represents an extensive revision of old topics found in the first edition as well as the inclusion of many new topics, the objective here continues to be unchanged. It is a textbook for upper- and graduate-level students interested in the origin and evolution of extant and extinct groups of plants as revealed by the fossil record. In some areas the coverage is fairly comprehensive and may require the use of the special references that appear in the body of the text and review articles and books that give more complete bibliographies. From past experiences as teachers of paleobotany, we know that some students are deficient in basic information about plant morphology and anatomy. To help overcome these deficiencies there are sections at the beginning of most chapters that provide a brief résumé of the main organographic and anatomical characteristics of extant representatives that should give some insight as to what one might expect to find among the extinct members.

As in the first edition there is much illustrative material. The number of photographs of actual fossils has been increased. The quantity of interpretative line drawings and charts has also been increased to aid in better visualizing the often fragmentary remains of the plant fossils. Among the new line drawings are reconstructions of "whole" plants that are now more or less completely understood. Reconstruction is one of the ultimate objectives of most paleobotanists in search of the origins and relationships of the plants they work with.

All of the summarizing charts have been updated and reflect new ideas about the relationships and temporal distributions of the major groups of plants depicted. These charts should be used with caution, however, because of the incomplete nature of the fossil record. It is possible to get a wrong impression of the relative abundance of plants in a major group at a given time as well as a false idea of the extent of the group through geological time. Because of the way they are constructed these same charts can be misleading when the degrees of relationships among the groups are considered. In spite of these pitfalls the continued use of the "bubble" charts has been encouraged as a useful summarizing tool for students and teachers.

Teachers will find the new edition to be considerably more provocative in content because of the introduction of new concepts to such topics as the origin and subsequent evolution of eukaryotes, the early evolution of ferns, the origin of seeds, and the origin and early evolution of gymnosperms and angiosperms. All are controversial. These and other new concepts are placed in juxtaposition to the older, more traditional interpretations. In this way it is hoped to provide students and teachers an opportunity to more critically evaluate our ideas about plant origins and subsequent evolution.

The unifying theme provided by elements of the telome theory still pervades the early chapters on vascular plants. This so-called gradualistic concept of plant evolution has been reevaluated in light of "punctuated equilibrium" as an explanation for the gaps in the fossil record.

The study of paleoecology, which was conspicuous because of its absence in the first edition, represents a relatively new topic that provides another theme that appears at several places in this edition – from chapters describing the environments in which Precambrian organisms lived and evolved to the environments that supported the origin and subsequent evolution of the flowering plants. To develop this theme more fully, Chapter 12, a new chapter, is devoted to the paleoecology of Pennsylvanian coal swamps.

Paleobotanists, as do many other biologists interested in systematics, continue with the often frustrating process of establishing "natural" systems of classification. In an effort to avoid subjectivity in this process, the methodology of cladistics has been adopted by some during the past ten years. The cladistic method and its results provide yet another more or less pervasive theme. It starts with a brief explanation of the methodology in Chapter 3 and closes in Chapter 30 with a cladistic analysis explaining the origin of angiosperms.

Perhaps the major deficiency of the first edition was the "light" treatment given to the fossil record of the most obvious group of extant plants – the angiosperms. Reacting to the advice of our many colleagues we have expanded the space for this important group by adding a new chapter, Chapter 31, which covers the paleobotany of flowering plants from the mid-Cretaceous through the Eocene. It is here that the paleoecology of the controversial Cretaceous-Tertiary (K/T) boundary is also considered.

The preparation of this revised edition of *Paleobotany and the Evolution of Plants* would not have come about without the contributions of a score or more friends, colleagues, students, and family. To the most durable of all, through more than four years of preparation of this revision, we owe a debt of gratitude to our wives who encouraged us, sustained us, performed secretarial and editing functions, and put up with our, at times, bad humor. Our thanks to these sublime people is so profound that we can find no adequate way of conveying our feelings for their continued help. Even with their help we would not have been able to complete our task without the unstinting help of many others who have freely given us advice on content, editorial assistance, and illustrative materials as well as insights into new concepts and ideas developed in the text. This is particularly true of Peter Crane, who has been exceptionally supportive of our work, and of Gene Mapes, who performed the monumental task of proofreading the text. We also wish to extend special thanks to Ruth Stockey for her critical reading of Chapters 30 and 31, to Andrew Knoll for his extensive review of Chapter 4, and to Sara Stubblefield for her analysis and suggested revisions of Chapter 5. In addition to reviewing the chapters, all of these colleagues supplied quantities of illustrative materials.

To all who have been exceedingly generous in supplying photographs, drawings, charts, and permissions, we give special thanks. They are D. L. Dilcher, W. A. DiMichele, J. A. Doyle, E. M. Friis, R. M. Kosanke, S. R. Manchester, K. J. Niklas, M. Nishida, R. A. Peppers, T. L. Phillips, K. B. Pigg, S. A. J. Pocock, J. W. Schopf, A. C. Scott, B. M. Stidd, T. Tanai, E. L. Taylor, T. N. Taylor, G. R. Upchurch, J. Watson, and J. A. Wolfe.

In spite of our efforts and the expertise of all who have helped us, there will be mistakes for which we assume full responsibility.

W. N. S.
G. W. R.

Preface to first edition

This book is a summation of the paleobotanical information germane to our present understanding of the origins and relationships of major groups of plants, extant as well as extinct. It comprises the information I have accumulated during the past 30 years, some obtained through my own efforts, but most provided by others – many of whom I have taught, worked with in research laboratories, and toiled with in the field. My main contribution has been one of synthesis and interpretation of information with the emphasis on paleobotanical discoveries made during the last three decades, the most productive period in the history of our science.

The presentation of the information closely follows the format of lecture notes I have developed since 1949, when I initiated course work and research in paleobotany at the University of Illinois, Urbana. Some will find my style of writing somewhat prolix and personal because of a conscious effort on my part to reflect a presentation one might expect in a more informal classroom environment. My hope in doing this is to make the text more interesting, if not more readable, and to reflect some of my own enthusiasm for paleobotany.

To the student

Beyond providing the pertinent information and interpretations, the purpose of this book is to stimulate the interest of students in paleobotany. Some students take a paleobotany course with a minimal understanding of plant structure and morphology. To help students with such a marginal background in these subjects, I have included at the beginning of many chapters a brief résumé of those features of extant plants that will be needed to gain a better understanding of the fossil record. Even with this assistance some students will encounter unfamiliar terms. To obviate this difficulty, an explanation of each new term is given either parenthetically or in the text that immediately follows. I have attempted to keep the documentation in the text to a minimum, but citation of some research articles cannot be avoided. However, references to review articles and reference books, where the interested student can find more complete bibliographies, have taken precedence.

Because fossil plants are usually fragmentary, students find interpretive drawings to be a great aid in undertstanding the parts that are preserved and described. Thus, to clarify and amplify the descriptions in the text, I have employed large numbers of line drawings and diagrams. To make sure that we do not lose sight of what the fossils actually look like I have included a number of authenticating photographs. Summarizing charts also are used to help students visualize the interpretive material. A word of caution is required, however, about using charts where plant distributions in geological time and their relationships are shown. Because the fossil record is incomplete, such charts can give a distorted impression of the relative abundance of a plant group at a given time in the geological column, as well as a false idea of the extent of the group through time. The same kinds of charts, because of the way they are constructed, can also give a wrong impression of the degrees of relationships of the groups being treated. If, however, we are aware of these problems, then the chart becomes a useful summarizing tool where, in a general way, degrees of relationships of plant groups can be shown superimposed on a background of geological time.

To the teacher

The teacher will find that the text includes more than an adequate amount of material for a two-semester course in paleobotany. Although the subject matter is highly integrated, one-semester courses can be arranged by using Chapters 1 through 18, which will cover the introductory material and the origin and evolution of free-sporing cryptograms. A second unit, including Chapters 1 through 3 and 19 through 28, will provide coverage of the introductory material and the origin and evolution of seed plants. The arrangement of Chapters 4 through 27 reflects the order of the taxa in the system of classification presented in Chapter 3.

There is minimal coverage of such topics as paleoecology, paleoclimatology, paleogeography, and palynology as it applies to stratigraphy. The main thrust of the text is plant evolution as revealed by what we presently know of the fossil record. In this connection free use has been made of hypotheses, theories, and concepts in the interpretation of this record. For example, I have found the telome theory to be an exceedingly useful teaching tool providing a pervasive theme that can aid students in understanding the origin and evolution of vascular plants. The use of this theory provides an opportunity for students and teachers to use fossil evidence to test the hypotheses incorporated in this far-reaching, unifying concept.

Because this book is intended for students, no effort has been made to make it a compendium of species, genera, families, or even orders of fossil plants. My motivation prompting the selection of the included taxa has been to choose those that best illustrate the idea under consideration. As a result of this selective process, for which I assume full responsibility, many researches describing taxa that might be considered important have been omitted.

A teacher can hope that some students will become sufficiently interested in paleobotany to want to continue with additional reading or a special problem. In anticipation of this possibility I have tried at various places in the text to indicate topics that might be suitable for further study or research.

To my colleagues

Those who know my attitudes toward our science will recall that I early adopted a philosophy that required students working with an abundance of well-preserved fossils to interpret their findings in the light of known plants, extant and extinct. There are those who, for good reason, maintain that reporting the facts is all that should be undertaken and that interpretations resulting in hypotheses, theories, or concepts should be rejected. I have not been able to adopt this approach because, for me, it would make our science dull and static. There would be no opportunity to test old hypotheses or to formulate new ones that stimulate more researches leading to rejection or adoption of speculative ideas and in this way make the study of fossil plants a dynamic science. Right or wrong, I have opted for the point of view that paleobotany can be a dynamic and interesting subject. I can only hope that this is reflected in the presentation of this book. In adopting this philosophy, one is obliged to have a thorough knowledge of the factual material that provides the foundations on which speculations are based. Every effort has been made in writing this book not to misrepresent the factual material or to overstate that which is speculative.

To those who have helped

The magnitude of the task of writing this book was not appreciated at its inception some years ago. Had I known the amount of work involved, there is reason to believe that the project never would have been undertaken. It would have been impossible from the start without the help, encouragement, and even persuasion of many students, friends, and colleagues.

Among those listed are those who have critically read sections of the manuscript and who made many helpful suggestions. There also are many who have unstintingly provided photographs and other illustrative materials, and yet others who have provided me with insight into new concepts and ideas which have been developed in the text. The following are those whose contributions in the above categories are acknowledged with heartfelt thanks: Henry N. Andrews, Harlan P. Banks, James F. Basinger, Robert W. Baxter, Sheila D. Brack-Hanes, William G. Chaloner, William L. Crepet, Theodore Delevoryas, David L. Dilcher, William A. DiMichele, Jeffrey B. Doran, James A. Doyle, Donald A. Eggert, Patricia G. Gensel, Charles W. Good, Leo J. Hickey, Hans J. Hofmann, Francis M. Hueber, James R. Jennings, Andrew H. Knoll, Gilbert Leisman, Geoffery Lyon, J. Frank McAlpine, Michael A. Millay, Charles N. Miller, Karl J. Niklas, Russel A. Peppers, Hermann W. Pfefferkorn, Tom L. Phillips, Gar W. Rothwell, Albert S. Rouffa, Stephen E. Scheckler, James M. Schopf, J. William Schopf, Jack Simon, Benton M. Stidd, Ruth A. Stockey, and Thomas N. Taylor. Of those listed above, I owe a special debt of gratitude to Dr. James M. Schopf (1911–1978), who first stimulated and directed my interest in paleobotany.

As I am sadly inexpert with the typewriter, the physical preparation of the manuscript would have been impossible without the dedication of Mary Liz, my wife, who patiently transcribed more than 1,000 pages of my nearly illegible longhand into the working typescript. The removal of typos and improvement in grammar, punctuation, and spelling were undertaken by Diana Horton, whose suggestions and encouragement provided a much-needed impetus. After this initial editing the final typing was done by members of the office staff of the Department of Botany, University of Alberta. I am grateful to these proficient people who, in spite of a heavy work load for the department, always found time to work on the manuscript, a task that went on for two and one-half years.

In spite of my own efforts and the expertise of all who helped this book on its way to completion there will be mistakes, for which I must accept full responsibility.

W. N. S.

1

Introduction

What is a fossil?

The word fossil has been defined in many different ways, some of which are cumbersome and lack precision. One that seems to be satisfactory, because it is simple and brief, defines a fossil as "any evidence of prehistoric life." The catchword prehistoric makes our definition highly arbitrary and excludes those fossil-like objects that have accumulated on Earth since the onset of recorded history, approximately 6,000 years ago. In spite of this limitation, you will find that our definition gives us plenty of latitude as to what can be categorized as a fossil. We are allowed to include not only all direct evidence of prehistoric life, but all indirect evidence as well. The state of preservation has very little to do with determining whether an object is a fossil even though we generally think of fossils as being objects turned to stone. To illustrate, a park ranger reported finding the remains of a coyote in the outflow of a geyser. He left the carcass undisturbed and found, when he returned some four years later, that the remaining coyote bones had already become silicified in the same way as the bones of dinosaurs. No one would argue that a dinosaur bone is a fossil, yet the coyote bones, according to our definition, are not fossils even though they are petrified.

The dinosaur bone is direct evidence of a once-living organism that roamed Earth some 100 m.y. (million years) ago. Often, however, the evidence of dinosaurs in the fossil record is indirect. The preserved footprints that the animal left in the mud, the fecal material (coprolites), or even the highly polished pebbles (gastrolliths) that presumably aided in digestion, all provide indirect evidence that dinosaurs existed, and as such these objects also should be considered fossils. It is not unusual for paleontologists to find prehistoric worm borings or even chemical substances (chemical fossils) that are the products of the metabolism of certain prehistoric algae and bacteria. All of these and other kinds of fossils give us clues about the organisms that inhabited Earth.

How are fossils used?

Some assemblages of fossils can be used to identify stratigraphic units of Earth's surface. In this way they serve the geologist as index fossils. For plants this is particularly true of such reproductive units as spores and pollen grains. When a palynologist analyzes an assemblage of these microscopic units, he frequently is able precisely to identify the relative age and position of the rock or stratigraphic unit containing the spores and pollen (Kosanke, 1988a, b). Using this kind of information when studying drill cores, for example, it is possible to help those searching for new deposits of fossil fuels to make new discoveries or avoid costly mistakes. The same index fossils can be used to correlate sedimentary rocks often occurring at great distances from one another.

A good index fossil is one that is abundant and is easily identified, with a wide horizontal distribution, but a relatively short vertical range of approximately 1 million years. Until recently plant megafossils rarely have been used as index fossils because they are relatively scarce and often difficult to identify. Even though they may have wide geographical distributions in a variety of sedimentary rocks, they often have long vertical ranges of millions of years. However, using assemblages of megafossils as indices Phillips (1980) and other investigators have been able to characterize fairly restricted stratigraphic units. These studies are often complemented by palynological information obtained from the rock units containing the megafossils. Other than the practical applications, palynologists have made good use of spores and pollen to help us in understanding the origins and relationships of major groups of plants, especially gymnosperms and angiosperms.

Relative duration of Eras	ERA	PERIOD	EPOCH	Characteristic organisms and some major geological events	Duration (m.y.)	Began (m.y.) ago
CENOZOIC	CENOZOIC	QUATERNARY	RECENT		Last 5,000 years	
MESOZOIC			PLEISTO-CENE	Several advances and retreats of continental ice sheets, accompanied by some elevation of continents and drainage of major synclines. Redistribution of floras reflecting advances and retreats of glaciers. Woolly mammoth and giant bison. Appearance of modern man.	2.5	2.5
PALEOZOIC		TERTIARY	PLIOCENE	Elevation of the Andes and general continued uplift of continents bringing about climatic changes in temperate latitudes that allowed spread of grasslands caused local restriction or extinction of some species. Mastodons, camels, horses, and cats.	4.5	7
			MIOCENE	Marked worldwide continental uplifts and a major orogeny resulting in rise of the Alps. Climatic cooling and restriction of broad-leafed evergreens to lower latitudes. Establishment of present-day forest associations. Rise and rapid evolution of grazing mammals and apelike creatures.	19	26
			OLIGOCENE	Mild temperate climates and widespread occurrence of now relic taxa (*Metasequoia, Cercidiphyllum*) in higher latitudes. Dryer climates in southwestern North America. Saber-toothed cats, true cats, dogs, rodents, and great rhinoceroses.	12	38
			EOCENE	Subtropical climates with heavy rainfall supported distinctive forests in northern and southern latitudes. Grasslands limited. Many genera of angiosperms became extinct with new, more modern types appearing. All modern orders of mammals present; first horses.	16	54
			PALEOCENE	Trend from temperate climates to mild subtropical. Some seasonal variations. Continued inundation of embayments and synclines. Angiosperms having affinities with Magnoliaceae, Lauraceae, Juglandacae. Floras having more in common with Upper Cretaceous than modern ones. First lemurs; some modern groups of birds.	11	65
	MESOZOIC	CRETACEOUS	UPPER	Climate tending to be uniform, temperate because of extensive inundation of continents. Elevation of Rocky Mountains. Angiosperms rise to dominance in Upper Cretaceous. First recognition of angiosperm pollen, leaves, and flowers in Lower Cretaceous. Monocots and dicots present. Modern groups of insects; first pouched and placental mammals; extinction of giant land and marine reptiles.	76	141
			LOWER			
		JURASSIC	UPPER	Uniform, mild climates from North to South Poles. Highly similar plant communities composed of ginkgos, conifers, ferns, cycads, and cycadeoids. Some pteridosperms. Rise of higher insects and birds. Dinosaurs abundant.	54	195
			MIDDLE			
			LOWER			
		TRIASSIC	UPPER	Arid to semiarid, savanna-type climates. Rise of the cycadophytes, Ginkgoales. Diversification of conifers and ferns. Decline of glossopterids. First mammals; rise of the dinosaurs.	30	225
			MIDDLE			
			LOWER			

Era	Period	Epoch	Characteristics	Duration	Boundary
PALEOZOIC	PERMIAN	UPPER / LOWER	Uplift in Appalachian geosyncline. Cooler and drier climates with extensive glaciation in Southern Hemisphere. Rise of Northern Hemisphere. Voltziales and diversification of Southern Hemisphere glossopterids. Extinction of arborescent lycopsids and sphenopsids. Diversification of reptiles.	55	280
	CARBONIFEROUS: PENNSYLVANIAN	UPPER / MIDDLE / LOWER	Uniformly warm, humid climates. Further intrusion of epicontinental seas associated with the formation of the great Carboniferous swamps. Mosses, lycopods, sphenopsids, ferns, seed ferns, and cordaites. Origin of conifers, reptiles, diversification of amphibians. Insects abundant.	45	325
	CARBONIFEROUS: MISSISSIPPIAN	UPPER / LOWER	Warm, equable climate. Continued inundation of low-lying continents and their synclines. Primitive ferns, seed ferns, arborescent lycopods, and calamites associated with extensive swamps in lowlands. Spread of amphibians, sharks, and bony fish. Insects evolved wings.	20	345
	DEVONIAN	UPPER / MIDDLE / LOWER	Heavy rainfall and aridity. Extensive inundation of continents to form seas in areas of major synclines. Diversification of vascular plants; all major groups except flowering plants present by end of Devonian. Heterospory and the seed habit in early seed ferns. Liverworts, fungi. Diversification of fishes. Origin of amphibians.	50	395
	SILURIAN	UPPER / LOWER	Mild climates. Low-lying continents with some epicontinental flooding. First vascular plants in mid-Silurian (probably among the first land plants). Scorpions and millipedes, first air-breathing animals. Brachiopods, corals, and eurypterids.	40	435
	ORDOVICIAN	UPPER / LOWER	Warm, mild climates. Warm epicontinental seas with an abundance of green and red algae. First vertebrates and a great variety of marine invertebrates. Graptolites, nautiloids, cystoids in maximum abundance.	65	500
	CAMBRIAN	UPPER / MIDDLE / LOWER	Climate warm and equable. Warm, epicontinental seas supporting an abundance of cyanophytes, green and red algae. Abundance of marine invertebrates; first trilobites and foraminifers.	70	570
PRECAMBRIAN			Climates during parts of the Precambrian were warm enough to support the growth of cyanophytes, red algae and bacteria and possibly green algae. Questionable invertebrate fossils.	4,130	4,700

(Left column of chart: PRECAMBRIAN)

Chart 1.1. Geological column and time scale. (Based on information obtained from Banks, 1970; Gifford & Foster, 1989; Harland, Smith, and Wilcock, 1964; Stirton, 1959; and van Eysinga, 1975.)

In their structure and distribution, extant organisms often reflect the composition of the environments in which they grow and reproduce. We make the assumption that organisms of the past, in the same ways as those of the present, became adapted to their environments. If the assumption is correct, then we can interpret the presence of growth rings in a fragment of petrified wood as evidence of seasonal variations in the availability of water usually accompanied by temperature changes in the paleoenvironment. Conversely, fossil woods lacking evidence of growth rings reflect an equable environment with a more or less continuous supply of water and a more uniform temperature regimen. Thick cuticles and sunken stomata of fossil leaves also suggest lack of available water, while roots and stems with spongy tissues indicate an aquatic or swamp paleoenvironment. Information of this kind gleaned from the morphology and anatomy of fossil plants provides part of the basis of paleoecology and paleoclimatology. Other sources of information that tell us about paleoenvironments are derived from studies of the sedimentary materials in which the fossils occur and the way the fossils are preserved. Studies of this kind provide us with a picture of the paleoenvironments into which plants living at a particular time radiated and evolved. We are, however, cautioned by paleoecologists not to base our interpretations of the interactions between extinct plants and their paleoenvironments on our observations of the interactions of extant plants to present-day environments. In other words, what we observe occurring today is not necessarily the key to the past (DiMichele & Wing, 1987).

Paleofloristics, the study of assemblages of fossil plants in time and space, also gives us clues as to climates of the past in restricted as well as wide areas of Earth's surface. For example, a fossil flora may comprise elements of a tropical rain forest, a warm temperate vegetation, or a boreal forest of cool temperate regions. These same studies can be used to provide insight into past distributions of plant populations and their migrations in response to changes in the ancient environments. Studies revealing the successions of plants at a given time and place in the geological column are becoming popular with paleoecologists, especially those interested in Carboniferous, Tertiary, and Quaternary floras. On a worldwide scale, paleofloristic studies have provided evidence of plate tectonics and the drifting of continents. In short, the paleoecologist can study many of the same aspects of paleoenvironments that are studied by ecologists interested in extant organisms. The paleoecologist has the advantage, however, of being able to study plant successions through long periods of time.

The successional changes of organisms through geological time also provide the fundamental basis for those who wish to study evolution of plants and animals, studies that many consider the most rewarding when deciphering the fossil record. Thus, the purpose of this book is to give an overview of the geological record of those organisms commonly called plants interpreted in light of phylogeny. Although paleobotanists have answered many questions about evolutionary relationships among plants, there are many that await answers to be provided by present and future generations of paleobotanists.

The emphasis when studying relationships of ancient plants usually focuses, by necessity, on the morphology and anatomy of one or two isolated plant fragments. Recently, however, studies utilizing large quantities of well-preserved material have allowed insight into developmental stages (ontogeny) of the life cycles of many fossil plants (Delevoryas, 1964). Several examples of such ontogenetic studies appear at appropriate places in later chapters.

Knowledge of fossil plants and making collections of them have provided many amateur paleontologists with an interesting and permanent hobby. In our experience, many amateurs have brought valuable specimens to our laboratories, as well as giving information about highly productive localities. Moreover, it is impossible to ignore the aesthetic attributes of a polished piece of agatized wood or the intricate patterns of fossilized leaves on a slab of light gray shale. Fossil plants are beautiful; their beauty is only enhanced by a better understanding.

Chart 1.1 shows some of the levels of plant and animal evolution related to major geological events and stratigraphic units. It is fair to say that most paleobotanists are more concerned with the relative stratigraphic positions of the fossils they study than their absolute ages. It is more important for us to determine how a fossil specimen relates, in a temporal and spatial context, to other specimens either extinct or extant, with the ultimate objective of unraveling patterns of plant evolution.

From the late 1940s on, paleobotanists have made significant contributions to our understanding of the antiquity of algae and fungi; the origin and subsequent evolution of early vascular plants; the evolution of the seed habit; the origin and evolution of primary and secondary vascular systems, especially in gymnosperms; and the evolution of foliar and reproductive structures of gymnosperms and angiosperms. In addition to these contributions, a noticeable change in emphasis has occurred in the last 15 years – from researches that are structurally oriented to those that take into account the paleoenvironments in which plants of the past evolved, and those that explore changes in the patterns of plant diversity. More attention has been focused on the interactions of plants and animals in the paleoenvironments, most notably in the area of pollination biology. The progressive increase in the total diversity of land plants has been further documented, with additional emphasis on the decrease in diversity of some major groups and the extinction of others (Niklas, Tiffney, & Knoll, 1985; Traverse, 1988). The cladistic method of analysis has been adopted by paleobotanists in an effort to better understand the relationships among major groups of plants both extinct and extant. The results of these analyses have produced some surprises and controversies that are considered in later chapters. Our understanding of the biology of the Precambrian continues to increase at a phenomenal rate as does the revelation of angiosperm radiations in the Lower Cretaceous. Where possible the morphologist has turned his attention to the task of "whole plant" reconstructions in an effort to better provide bases for comparative studies of extant and extinct plants and the floras they comprise. These are but a few of the more important advances that are very well summarized by Andrews & Mamay (1955), Arnold (1968), Delevoryas (1969), Andrews (1974), Beck (1976), Dilcher (1979), Knoll & Rothwell (1981), Gould (1981), Smoot & Taylor (1985), and Banks (1987).

Some guiding principles

If he were with us today F. O. Bower (1935), one of the twentieth century's greatest plant morphologists, would be content with the progress made by paleobotanists in the last 40 years and the application of one of his guiding principles: "an upward outlook" in plant morphology. According to Bower, "An upward outlook is in itself a practical application of any evolutionary view. This seems like a platitude today; but it is not difficult to quote from current literature where it has been neglected. It is only in the 'Nature Philosophy' of a former century that organisms lower in the scale have been interpreted in terms of the higher." Although this was written in 1935, some plant morphologists who have been labeled "angiosperm-centered" still resist this guiding principle. To paraphrase Delevoryas (1969), neontologists fall into the trap of accepting dogmas formalized by the exclusive study of extant plants and then look "backward" for fossil evidence supporting these dogmas. This is not to say, however, that we should ignore the information provided by the study of extant plants when trying to determine their origins. Rather, it is important to use all available information from every source.

Bower also recognized that plant parts do not spring fully formed with all of those characteristics that would make it possible to assign them to one organ or another. In the course of evolution the characteristics of organs have gradually appeared so that a structure that we would call a branching system at an early stage of evolution would be called a leaf at a later stage. Categories, such as branch and leaf, for example, are human devices for handling and communicating information. "Organs need not conform *ab initio* to one or to another. The lower in the [evolutionary] scale the object studied the less the degree of conformity may be expected to be." Thus, paleobotanists in their search for the "beginnings" should expect the unexpected, such as branches evolving integuments and cupules, plants with gymnospermous vegetative structure and fernlike reproduction, or seed ferns that share characteristics with conifers. These and many other discoveries that deviate from the "norm" established by studies of extant plants bring to mind the astute observation made by Lamarck more than a century-and-a-half ago: "Nature exceeds on all sides the limits we so gratuitously impose on her." This applies to taxonomic as well as structural categories and raises the problem of nomenclature and classification. For example, what do you call a photosynthetic organ that is at an intermediate stage of evolution between a branching system and a frond? How do you classify those extinct plants that do not "fit" into existing categories? What system of classification should we use? Should we establish a separate system based on evidence from the fossil record? These are but a few of the many questions that confront paleobotanists, some of which are difficult to answer.

References

Andrews, H. N. (1974). Twenty-five years of botany (paleobotany 1947-1972). *Annals of the Missouri Botanical Garden*, **61**, 179–202.

Andrews, H. N., & Mamay, S. H. (1955). Some recent advances in morphological paleobotany. *Phytomorphology*, **5**, 372–93.

Arnold, C. A. (1968). Current trends in paleobotany. *Earth Science Review*, **4**, 283–309.

Banks, H. P. (1970). *Evolution and Plants of the Past.* Belmont, Calif.: Wadsworth.

Banks, H. P. (1987). Comparative morphology and the rise of paleobotany. *Review of Palaeobotany and Palynology*, **50**, 13–29.

Beck, C. B. (1976). Current status of the Progymnospermopsida. *Review of Palaeobotany and Palynology*, **21**, 5–23.

Bower, F. O. (1935). *Primitive Land Plants.* London: Macmillan, p. 611. Reprint. New York: Macmillan (Hafner Press), 1959.

Delevoryas, T. (1964). Ontogenetic studies of fossil plants. *Phytomorphology*, **14**, 299–314.

Delevoryas, T. (1969). Paleobotany, phylogeny, and a natural system of classification. *Taxon*, **18**, 204–12.

Dilcher, D. L. (1979). Early angiosperm reproduction: an introductory report. *Review of Paleobotany and Palynology*, **27**, 291–328.

DiMichele, W. A., & Wing, S. L. (1987). Preface: Methods and applications of plant paleoecology, notes for a short course. Paleobotanical Section, Botanical Society of America.

van Eysinga, F. W. B. (1975). *Geological Time Table.* Amsterdam: Elsevier.

Gifford, E. M., & Foster, A. S. (1989). *Comparative Morphology of Vascular Plants*, 3rd ed. San Francisco: Freeman.

Gould, R. (1981). Paleobotany is blooming: 1970–1979, a review. *Alcheringa*, **5,** 49–70.

Harland, W. B., Smith, A. G., & Wilcock, B. (1964). The Phanerozoic time scale. *Quarterly Journal of the Geological Society, London*, **120**(S), 260–2.

Knoll, A. H., & Rothwell, G. W. (1981). Paleobotany: perspectives in 1980. *Paleobiology*, **7,** 7–35.

Kosanke, R. M. (1988a). Palynological analyses of Upper Pennsylvanian coal beds and adjacent strata from the proposed Pennsylvanian system stratotype in West Virginia. *U.S. Geological Survey Professional Paper*, **1486**.

Kosanke, R. M. (1988b). Palynological studies of Middle Pennsylvanian coal beds of the proposed Pennsylvanian System stratotype in West Virginia. *U.S. Geological Survey Professional Paper*, **1455**.

Niklas, K. J., Tiffney, B. A., & Knoll, A. H. (1985). Patterns in vascular land plant diversification: an analysis at the species level. In *Phanerozoic Diversity Patterns: Profiles in Macroevolution*, ed. J. W. Valentine. Princeton: Princeton University Press.

Phillips, T. L. (1980). Stratigraphic and geographic occurrences of permineralized coal swamp plants – Upper Carboniferous of North America and Europe. In *Biostratigraphy of Fossil Plants*, eds. D. L. Dilcher & T. N. Taylor. Stroudsburg, Pa.: Dowden, Hutchinson, & Ross.

Smoot, E. L., & Taylor, T. N. (1985). Paleobotany: recent developments and future research directions. *Palaeogeography, Palaeoclimatology, Palaeoecology*, **50,** 149–62.

Stirton, R. A. (1959). *Time, Life, and man*. London: Chapman & Hall.

Traverse, A. (1988). Plant evolution dances to a different beat. Plant and animal evolutionary mechanisms compared. *Historical Biology*, **1,** 277–301.

2

Plant fossils: preservation, preparation, and age determination

The vast majority of plant fossils are preserved in sedimentary rocks. In general, sedimentary rocks are formed in an environment where deposition, as opposed to erosion, has occurred. Deposition usually takes place when rock particles of various sizes accumulate in a body of water. The particles, which range in size from colloidal to pebbles and boulders, are derived from the erosion of igneous, metamorphic, or other sedimentary rocks by the well-known mechanical actions of wind, water, freezing and thawing, erupting volcanoes, and movement of glaciers. When converted into rock (lithified), the sedimentary materials produce strata that, depending on particle size, range from fine-grained siltstones and shales to coarse sandstones and conglomerates.

Rocks that contain fossils

Among the sedimentary rocks of Earth's crust we often find extensive deposits of limestone and dolomite. Deposition of these usually occurs in marine environments. Some geologists and paleontologists are of the opinion that their formation is dependent on the activity of living organisms. In some cases to be discussed later, it is clear that primitive organisms such as some cyanophytes and bacteria are responsible for converting soluble carbonates in water into insoluble limestone. There is still reason to believe, however, that some of the limestones and dolomites have been precipitated by purely physical and chemical means. This is certainly true when soluble carbonates, silicates, iron, and limonitic and phosphatic compounds are deposited in plant parts to "turn them into stone." This kind of sedimentation is of great importance to paleobotanists and will be discussed more fully later in this chapter.

Coal, a combustible rock, can also be classified as sedimentary in which the sediment is derived from plants. By special techniques in which bituminous and brown coals can be broken down or made into thin sections, we can recognize such vegetative structures as wood fragments, pieces of bark, cuticles from leaf surfaces, and colonies of algal cells. Reproductive units in the form of spores or pollen are almost universally present in bituminous coals. Because bituminous coals are so abundant and of such great economic importance they have been the object of study by palynologists, whose contributions to paleobotany are indicated at several places in later chapters.

Other interesting, indeed unique, forms of sedimentary rocks in which plant and animal fossils abound are diatomite and amber. Diatomite is a rock formed from the cell walls of a group of unicellular algae called diatoms. They occur in great abundance in freshwater and saltwater environments. When diatoms die, their highly resistant silica cell walls are deposited on the bottoms of lakes, seas, and oceans where, after the passage of time and subsequent consolidation, they form a white, lightweight, sedimentary rock: diatomite. Diatomite may, in turn, contain the well-preserved remains of fossilized plants and animals. Thus, we have the rather unusual situation where there is a sedimentary matrix composed of fossils in which fossils are embedded.

Amber has attracted much attention through the last few centuries as a soft, semiprecious stone of rare beauty. It is common knowledge that these sedimentary materials are plant resins that have undergone chemical changes during the process of fossilization (Langenheim, 1964, 1969). These changes are not completely understood. Before fossilization, the sticky, flypaperlike resins trapped insects, old flower parts, windblown pollen grains, fungal spores, and other plant and animal debris. The fossil amber with the entombed plant or animal parts (Fig. 2.1A,B) are prized equally by lapidaries and paleontologists. The preservation of these Tertiary fossils within fossils is often exquisite and has given us insight into the evolution of insects and the flowering plants they pollinated.

Environments for fossilization

Without further explanation one might get the idea that fossils will occur wherever sedimentary rocks have been formed. This is clearly not the case. Relative to the abundance of sedimentary rocks on Earth, those that contain plant fossils are very rare indeed. This implies that plant fossils are formed under very special environmental conditions. It is not sufficient to have a body of water, a source of inorganic sediments, and plant or animal remains to make fossils. We all know that when an organism dies under normal conditions in nature, that organism decays. This happens because of the activity of bacteria and fungi, especially those that are aerobic. These decomposers break down the complex organic materials that constitute the dead organism into simple compounds that can be recycled into new living systems. If we conceive of an environ-

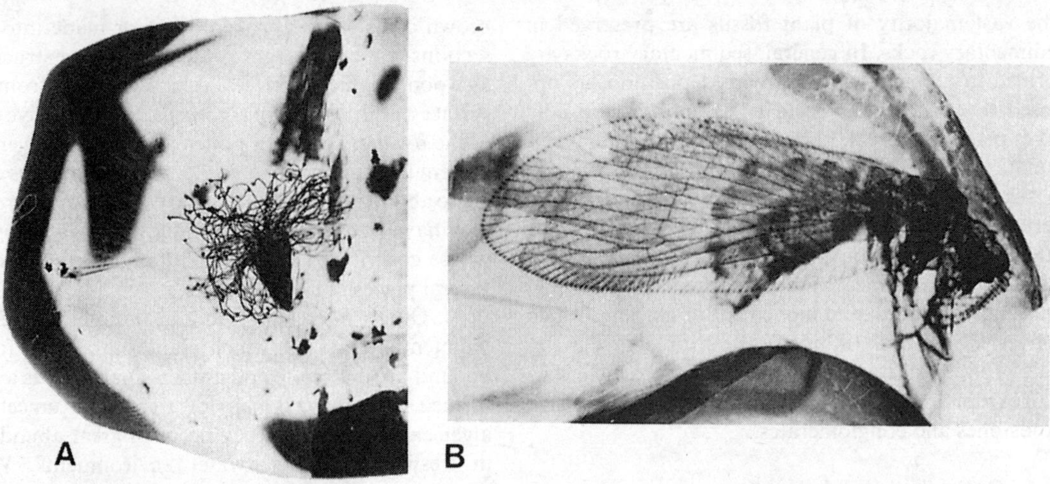

Figure 2.1. **A.** Dominican amber containing a mimosalike flower with numerous filamentous stamens. Probably Oligocene. (Courtesy of Dr. F. M. Hueber.) **B.** An insect belonging to the Neuroptera in Upper Cretaceous amber. From Canadian Amber Collection. (Courtesy of Dr. J. F. McAlpine.) **C.** Conifer wood (*Taxaceoxylon*) preserved in bitumen of the Cretaceous oil sands, Alberta, Canada. The section shows perfect preservation of pits on the radial walls of wood cells. (From Roy, 1972.) **D.** Silicified petrifaction of a fern petiole (*Dennstaedtiopsis*) with exceptional preservation of the cortical cells. (Photograph by Dr. J. F. Basinger.)

ment where decomposition is retarded, where plant or animal remains can remain undisturbed in a body of water, and where there is a source of sediment, then we have the makings of a place where preservation might go on. We usually think of swamp environments as such places. Here, the pools of water in lagoons and slowly moving streams provide a low-energy environment where deposited plant parts can remain undisturbed and intact. Aerobic decomposition will go on, but in a swamp metabolic wastes of decomposition generally cause an increase in acidity, which in turn limits the rate of activity of decomposers. If the plant parts fall into water that is fairly deep – say more than a meter – then the activity of aerobic decomposition is severely restricted. That anaerobic organisms contribute little to the decay process is attested to by the parts of trees beautifully preserved for many years in deep, quiet bodies of water. Finally, there must be a source of sedimentary material in a soluble or particulate form. In the swamp environment there can be soluble silicates and carbonates, iron compounds, or other minerals. Soluble silicates may originate from the waters of hot springs and volcanoes. It has been suggested that soluble carbonates may have been supplied by marine environments, a mixed source (both marine and nonmarine), or an exclusively nonmarine source such as an aquifer or other groundwater supply. In Figure 2.2, Scott & Rex illustrate models of the principal ways in which Carboniferous coal balls were formed. The environmental factors that favored their formation are those given here that promoted permineralization (petrifaction) of uncompacted peat in a coal swamp that encroached upon the peat beds where fossilization took place (Evans & Amos, 1961; Scott & Rex, 1985). A lowering of the pH (increase in acidity) could trigger the deposition of compounds such as SiO_2 into the plant parts, petrifying them. Or evaporation of water from the swamps containing the plant debris could have caused the formation of a saturated solution of soluble minerals followed by their precipitation when supersaturation was reached.

Plant parts can be preserved as compressions in sedimentary rock if there is an abundance of such particulate, sedimentary material as silt, clay, or fine sand available in the environment. As the plant parts accumulate in the body of water they may become covered with the sediment and entombed in the subsequently formed rock.

Underground plant parts such as roots and rhizomes are frequently found preserved in what is called "stigmarian" underclay, which is found immediately beneath coal seams of Carboniferous age. The underclay represents the substrate that supported the plants in the swamp where they grew.

According to our definition, those prehistoric remains of plants that qualify as fossils but are found in other than ancient swamp environments are unaltered twigs, seeds, and fruits that have been retrieved from the sediments in the floors of desert caves. The high temperatures and low humidity of the desert environment plus the protection of the cave prevents decomposition.

Refrigeration of mammoths trapped in snow and ice fields of Siberia during the late Pleistocene is a spectacular example of yet another way in which organisms have been preserved for thousands of years. Their unaltered remains have been unfrozen and edible meat obtained. Tillites deposited at the time of glacial retreat during the Pleistocene often contain unaltered branches and trunks of trees. The most spectacular of all of the deposits in this category are large tree trunks found at the base of the Cretaceous oil sands of northern Alberta. These 100 m.y. old fossils are literally preserved in a black, gooey oil. It is possible to cut the fossil logs with a saw and drive nails into them! The cells of the unaltered wood are so perfectly preserved (Fig. 2.1C) that paleobotanists have been able to identify the genus to which the trees belong. This mode of preservation in an oil-saturated environment is the same as the preservation of animal remains in the world-famous Pleistocene Rancho La Brea tar pits of California.

Modes of preservation and techniques

In presenting the previous account of ancient environments that favored preservation it is necessary to allude to some categories of fossilization. We have mentioned, for example, petrified and compressed plant remains. We should, however, have a much more complete understanding of these and other modes of preservation. James M. Schopf (1975) prepared a scholarly and comprehensive account on this subject that you should read. What follows is a condensation of parts of that publication.

According to his account, Schopf recognizes four distinct modes of preservation. One is cellular permineralization, which includes all of these specimens that, in past accounts, have been called petrifaction fossils. Permineralization occurs when the soluble silicates, carbonates, iron compounds, and so on infiltrate cells and the spaces between them. The precipitation of these compounds forms a rock matrix supporting the plant tissues. The process is analogous to the embedding of plant or animal tissues in paraffin when preparing them for sectioning and microscopic examination. A few examples of extensively studied fossils that show silicified permineralization are the Devonian Rhynie cherts, Precambrian Gunflint cherts, Triassic woods from the Petrified Forest of Arizona, and the giant Tertiary *Sequoia* trunks, some of which stand upright and in place on Specimen Ridge in Yellowstone Park. Not so widely known, but even more spectacular in their preservation, are the silicified specimens (Fig. 2.1D) described by Basinger & Rothwell (1978) from Eocene deposits in British Columbia.

Figure 2.2. Models for the formation of coal balls in Pennsylvanian age coal swamps. Explanation of key: (1) coal ball, (2) mixed coal ball, (3) coal seam, (4) interseam sediments, (5) marine shale, (6) roof nodule, (7) uncompacted peat, (8) goniatites, (9) pathway of the carbonate-rich permineralizing water. **A.** Marine-influenced coal balls. Carbonate precipitated directly from water of transgressing sea. **B.** Mixed coal-ball formation. Carbonate and mud rollers deposited in the coal swamp from the sea by storm activity. The resulting coal balls present a mixture of marine animals and peat-forming plants. **C.** Nonmarine coal-ball formation. Carbonate-rich permineralizing water derived from a groundwater supply such as an aquifer. (From Scott & Rex, 1985, with permission.)

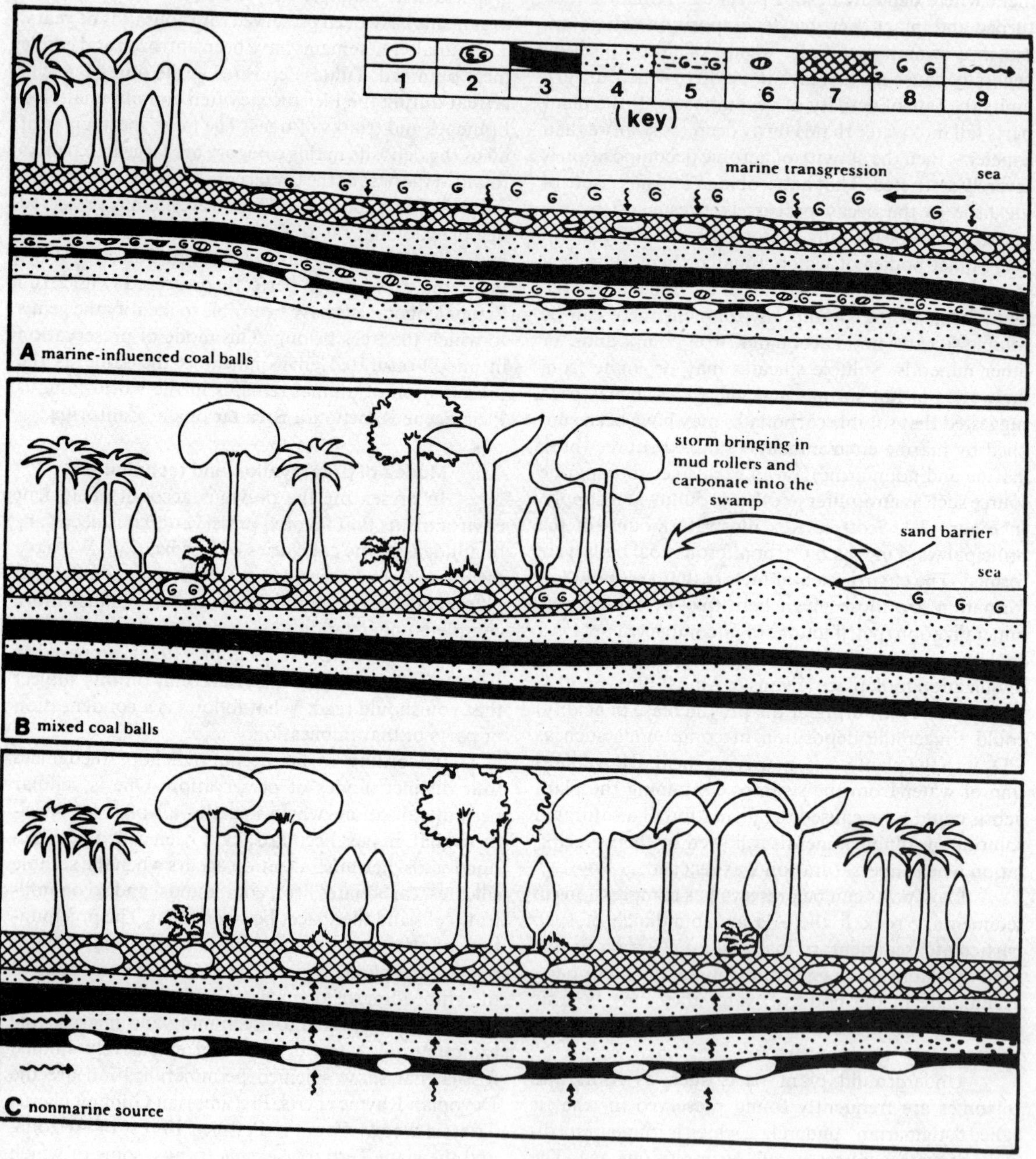

Vegetative and reproductive remains of flowering plants, conifers, ferns, bryophytes, and fungi have been found in a variety nearly equal to that described from silicified specimens obtained from the Tertiary Deccan Intertrappean series of India. New localities in the Transantarctic Mountains of Antarctica have recently also provided exquisitely preserved specimens of vascular plants, bryophytes, and fungi (Schopf, 1970; Smoot, Taylor, & Delevoryas, 1985; Taylor & Taylor, 1989). These permineralized fossils are Triassic and Permian in age and represent some of the first material from these periods to reveal the internal structure of plants that are usually preserved as impression/compression fossils.

Figure 2.3. Pile of coal balls in a strip mine near Wilmington, Illinois. The coal balls occur in the #4 coal member, which is Middle Pennsylvanian in age. (Courtesy of the Illinois State Geological Survey.)

Although silicified permineralized fossils have played a role of tremendous importance in paleobotanical studies. Even more extensive studies have been completed in the last four decades using calcareous specimens, commonly called coal balls. Unfortunately, coal-ball specimens are restricted in occurrence to Carboniferous rocks where they are found associated with seams of bituminous coal. Large collections of coal balls have been obtained from stream banks and coal mines (Fig. 2.3) in central and southern Illinois, Indiana, Kentucky, Ohio, Kansas, and Iowa in the United States as well as in Great Britain, Belgium, France, the Donetz basin, and other localities in Europe (Phillips, 1980).

When first discovered and described, the name coal ball seemed appropriate because the specimens retrieved from coal seams were spherical or ovoid limestone rocks. More recently irregular masses of coal-ball material have been found extending from top to bottom through coal seams 1 meter or more

Figure 2.4. **A.** Cellulose acetate peel from prepared surface of a large coal ball. **B.** Interpretive drawing of plant fragments shown in the panel. Pennsylvanian. (Courtesy of the Illinois State Geological Survey.)

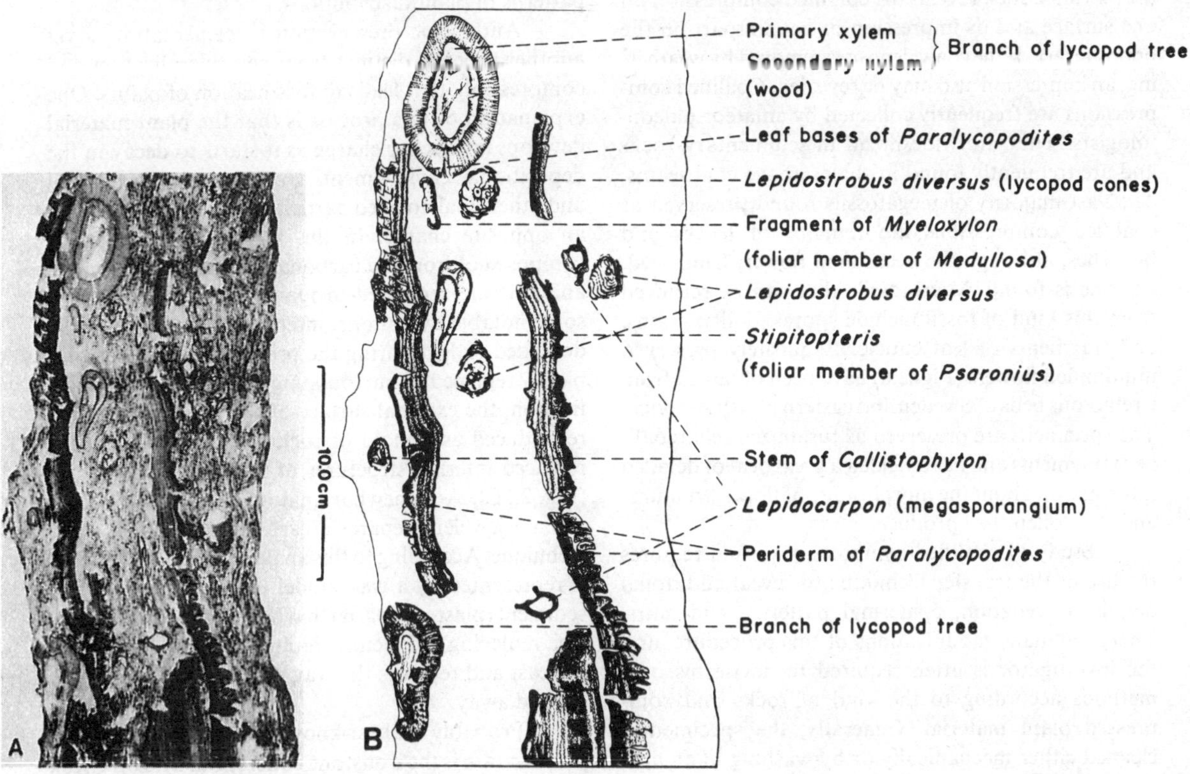

thick and having a lateral extent 10 times as great. These fossil-containing limestone masses (Fig. 2.4A,B) are usually found when they are uncovered by drag lines or by shovels operating in strip mines. The hard limestone can ruin loading and sorting machines and has earned it the local name "bastard limestone" from unappreciative miners. By blasting, using bulldozers, and hand collecting with pick, sledge, and bull point, tons of specimens have been obtained. The importance of these massive collections in revealing the ontogeny and phylogeny of Carboniferous plants will be obvious to you in later chapters.

Although coal balls are among the most intensively studied and abundant of the permineralized specimens, relatively small pyritic, limonitic, and phosphatic specimens have also provided material of great importance, especially in recent studies of early vascular plants recovered from Devonian rocks. We never cease to be intrigued by the wealth of information that our colleagues, who specialize in the study of these plant parts, can glean from such seemingly insignificant and fragmentary specimens.

Another category of fossil is the coalified compression (Fig. 2.5A). Following deposition in the sedimentary environment, the cell walls of the plant part soften and then collapse. After a loss of gas, moisture, and other soluble materials, because of pressure exerted by accumulated sediments and water, the residues are altered and consolidated to form a black, coaly deposit. When the sediments have become lithified, splitting of the stratified rock reveals the coalified compression on one surface and its impression or counterpart on the opposite face. If the coaly deposit is removed by weathering, an impression also may be revealed. Coalified compressions are frequently collected by amateur paleontologists. They are widespread in sedimentary rocks and are frequently found in shales above coal seams. The vast majority of megafossils found preserved as coalified compressions are remains of leaves and branches, although the occasional flower, fruit, seed, or cone is found. Microfossils that can be retrieved from this kind of fossil include spores, pollen grains, and fragments of leaf cuticle. Exquisitely preserved mummified flowers (Fig. 2.6) have been obtained from Cretaceous beds of Sweden and eastern North America. The specimens are preserved as fusainized (charcoalified) fragments and have yielded a great deal of detailed information about the morphology of floral structures and the pollen they produce.

Study of coalified compressions often requires the use of the transfer technique to reveal additional details of venation, epidermal patterns, and hairs. There are many modifications of this procedure, and the investigator is often required to devise his own methods according to the kind of rocks and compressed plant material. Generally, the specimen is cleaned either mechanically or by washing in an acid to remove rock particles and to slightly etch the surface. A plastic film (either in the form of a liquid or an acetate sheet softened in acetone) is applied to the prepared surface of the fossil. After hardening, the film is loosened from the rock surface. If the technique has worked, the coalified material will adhere to the film and be "transferred" from the rock surface. Subsequent treatment may involve the use of strong oxidizing agents to make the coalified compression more transparent for microscopic study. Figure 2.7B is a beautiful example of a successful transfer of a leaf showing details of the leaf structure that otherwise would never have been seen.

Paleobotanists rely on coalified compressions to learn about leaf form and venation patterns (Fig. 2.7A–C), as well as details of epidermal cells and stomatal characteristics (Fig. 2.7D). These are important features used in classification of extinct plants as well as in establishing evolutionary pathways, especially among flowering plants (Hickey & Doyle, 1977). More traditional floristic studies have been prepared utilizing coalified compressions. Many studies of this kind have been produced since the middle of the last century. Although their value has been questioned by some paleobotanists, recent studies (Chandrasekharam, 1974; Christophel, 1976) provide a sound basis for making comparisons with other floras when working on problems of stratigraphy. In addition, we learn more about climates of the past, other aspects of physical and biological environments, as well as changing patterns in plant distributions.

Authigenic preservation or cementation is yet another process, distinct from permineralization and compression, that leads to fossilization of plants. One explanation of this process is that the plant material develops an electric charge as it starts to decay in the depositional environment. Here it attracts colloidal and other small ionized particles of sediment that have an opposite charge. In this way sediments usually composed of iron and carbonate minerals accumulate and become cemented around the plant part. With some notable exceptions, internal structure is usually degraded or lost during the process and can be completely replaced by surrounding sediments. After lithification, the external surface of the plant is faithfully reproduced as a mold of rock (Fig. 2.5B), while the replaced internal structures of the plant form a cast (Fig. 2.5C). A somewhat analogous situation occurs when a jeweler prepares a cast using the "lost wax" technique. According to this analogy, the plant material is represented by a wax model that is embedded in a sediment (plaster of paris) that hardens to form a mold. The replacing sediment is molten metal, which forms the cast and replaces the wax model as it is melted or burned away.

Probably the best-known example of authigenic preservation is the ironstone concretion. Mazon Creek

Figure 2.5. Some types of plant fossils. **A.** Coalified compression of *Metasequoia occidentalis* on volcanic shale. Upper Cretaceous, Alberta, Canada. (Photograph by Dr. A. Chandrasekharam.) **B.** Mold of a bark fragment of *Lepidodendron*, a Carboniferous lycopod. **C.** Pith cast of a calamite stem. Carboniferous. **D,E.** Mazon Creek concretions containing leaf parts of Middle Pennsylvanian seed ferns preserved by the authigenic process. (Photographs by J. A. Wollin.)

concretions (Fig. 2.5D,E) from the Braidwood area of Illinois have been extensively collected and widely distributed. These contain plant or animal parts whose preservation is exquisite. In leaf fossils the venation pattern and epidermal hairs are often preserved while the outline of the leaf is exactly reproduced. On rare occasions some infiltration and permineralization can occur, thus preserving the internal structure as well as external features. Other exceptional specimens combine authigenic preservation to form an internal cast covered by a coalified compression of the external tissues.

Some plants, notably certain algae, precipitate hard coats of limestone or silica. These resistant parts are preserved without being changed by oxidation or physical factors. Such "hard part" preservation is called duripartic. So-called coralline algae belonging to the red and green algal groups, as well as some cyanophytes, are examples of plants whose calcareous hard parts are preserved. Diatoms with their indestructible silicified cell walls provide us with another example of duripartic preservation.

The degree of preservation is dependent on several factors. For coalified compressions the plant parts initially may be transported in rapidly moving water along with coarse sediments. These factors of the environment tend to macerate the parts prior to their deposition and fossilization. A rule of thumb might be that the finer the sediment the better the preservation. It follows that well-preserved coalified compressions are usually found in siltstone, shale, or fine-grained

Figure 2.6. Mummified flower bud, preserved as fusainized (charcoalified) fragment showing calyx and corolla of the saxifragalean, *Silvianthemum*. (Photograph courtesy of Dr. E. M. Friis.)

sandstone, while fragmentary, poorly preserved compressions occur in coarse sandstone and conglomerate.

Another factor may be the amount of decay of the plant tissues prior to preservation. Some plant tissues and structures are highly resistant to biological oxidation and mechanical breakdown. Without knowing anything of plant anatomy, one might guess that cells from wood and some bark tissues would be more resistant to degradation than soft tissues. Conducting cells of the phloem as well as parenchyma cells of pith and cortex are thin-walled cells that tend to break down prior to fossilization. With this information in mind, it is not surprising to see in a permineralized specimen that phloem tissues, pith, and parts of the cortex are often missing while bark and the woody portion of the conducting system remain. Cuticles of leaves and exines of spores and most pollen grains are highly resistant to degradation. For this reason we find them well preserved in brown and bituminous coals and the matrices of other sedimentary rocks. A rather unusual example of the preservation of cuticular material from leaves of ancient plants is found in "paper coal" (Neavel & Guennel, 1960). In the coal seam they describe, the upper 10 cm is composed entirely of layer after layer of leaf cuticles that can be taken directly from the coal for microscopic study. Usually a maceration technique, employing a strong oxidizing reagent followed by a strong basic solution, is used to release cuticles, spores (Fig. 2.8A), and pollen grains from the sedimentary matrix. Numerous accounts giving the details of maceration techniques have been published. You are referred to the *Handbook of Paleontological Techniques* by Kummel & Raup (1965) for a comprehensive description.

Many of us have studied specimens of fossil plants as a hobby and profession for several decades, yet there is still a thrill when examining a specimen that shows perfection in its preservation. The cross section of the permineralized fern stem, 6 cm in diameter (Fig. 2.8B), is just such a specimen. It was retrieved from a Carboniferous coal seam in southern Illinois and is roughly 280 m.y. old. The illustration shows that nearly all tissues are intact except for those adjacent to the conducting strands. It is doubtful whether such perfection could be achieved when sectioning a modern tree fern stem of equal size, using the techniques of embedding in paraffin or plastic, sectioning, and staining.

Visitors to paleobotanical laboratories are unusually curious about the method used in making such large thin sections of permineralized specimens. In the early days, thin sections were made by hand-grinding a slab of the specimen to the point where it was thin enough to transmit light and could be examined under the microscope. This technique requires considerable expertise. It also is slow and wasteful of material. It was replaced several decades ago by the peel technique that avoids the difficulties encountered in hand-grinding.

The peel technique was brought to its present state of perfection by Joy, Willis, & Lacey (1956). These investigators introduced the use of preformed cellulose acetate film in the preparation of sections (Fig. 2.4A; 2.8B), an improvement that makes possible rapid preparation of serial sections of permineralized fossil specimens.

The success of this wonderfully simple technique depends on the composition of the specimen. We already know that the matrix is usually an inorganic substance such as limestone or a silicon compound. The plant material is not replaced by the mineral, but retains its organic composition, including carbon compounds. If the cut and smoothed surface of the specimen is dipped into an appropriate acid of predetermined concentration, the acid will dissolve the matrix but not the plant material. Dissolving the matrix causes the organic portions (cell walls and cell contents) of the plant to stand in relief from the acid-treated (etched) surface of the matrix. The thickness of the section is determined by the concentration of the acid and the length of time the etching process goes on. Good sections as thin as 20 µm can be obtained. The etched surface is washed, dried, and flooded with acetone. Then a film of clear cellulose acetate is placed in the acetone that softens and partially dissolves the acetate film. As the acetone evaporates, the film hardens around those plant parts in relief on the surface. When we "peel" or pull the hardened film from the surface of the specimen, we find that the plant material has "transferred" from the specimen and is now embedded

Figure 2.7. Products of the transfer technique. **A.** Compression of a *Cercidiphyllum* leaf prior to preparation of the transfer. Paleocene. Alberta, Canada. **B.** Transfer made from a *Cercidiphyllum* leaf. **C.** Portion of transfer enlarged to show details of vein endings. **D.** Highly magnified portion of leaf cuticle prepared from a transfer showing details of the stomata (*Paleorubiaceophyllum eocenium* var. *amplum*). Eocene. (**A–C** from Chandrasekharam, 1974; **D** from Roth & Dilcher, 1979.)

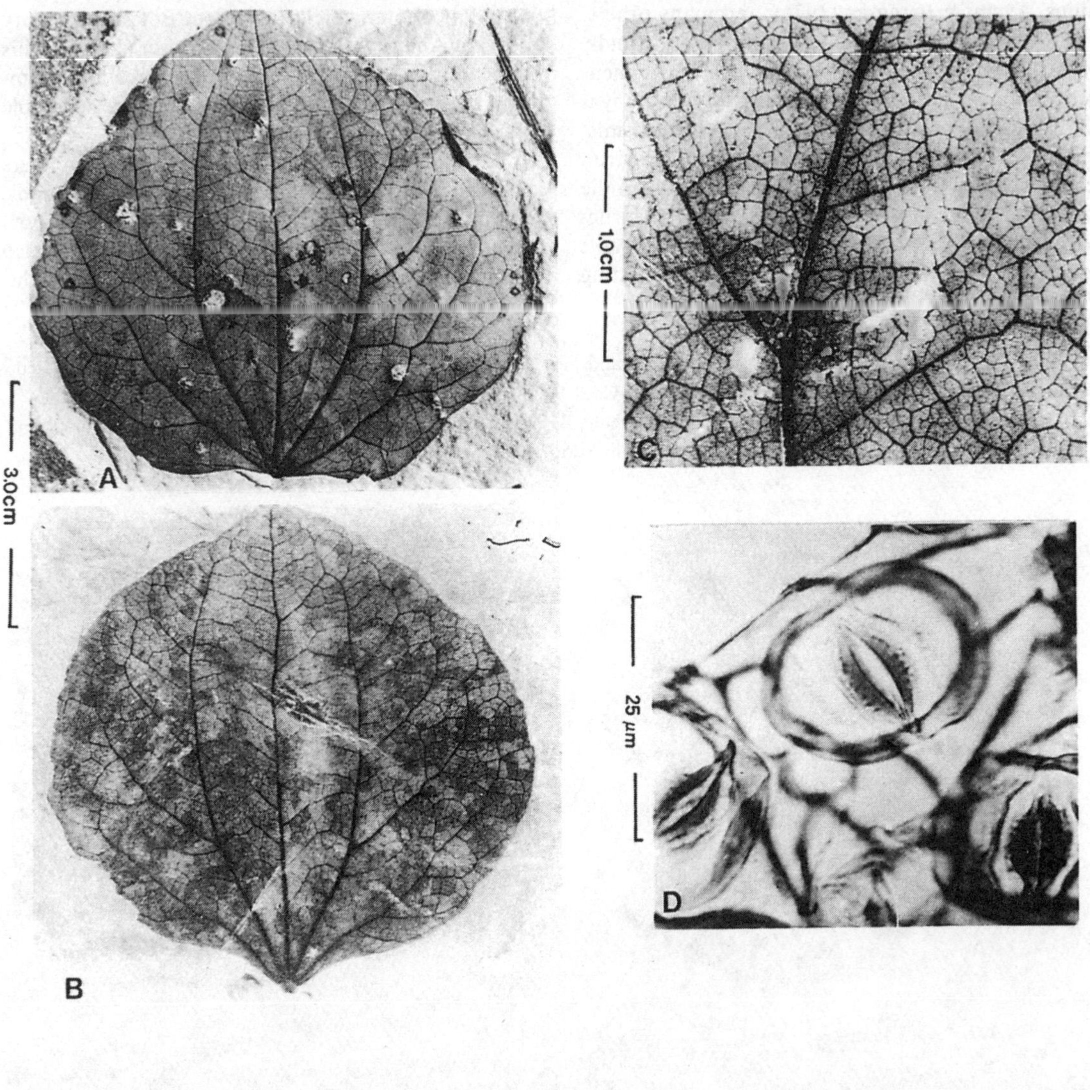

in the transparent plastic film. By smoothing and etching the rock surface again, another peel can be made, and the process can be repeated over and over so that serial sections result. With enough serial sections made at proper intervals, a reconstruction of the internal structure of the plant part is possible, and reconstruction is a paleobotanist's principal occupation. Often it is desirable to study the delicate walls of parenchyma cells and phloem of permineralized specimens. When observed in preparations prepared by the standard peel technique, carbonate residues and other defects tend to produce artifacts that have been erroneously interpreted as fine structures in the walls of these cells. Holmes & Lopez (1986) have developed a disappearing peel technique, which eliminates these artifacts and the possibility of making such mistaken interpretations.

Even though the peel technique has been described repeatedly in books, pamphlets, and research articles, it is so important to paleobotanical studies that this brief account is required. More information spelling out the details of this indispensable technique can be found in articles by Stewart & Taylor (1965), Phillips, Avcin, & Berggren (1976), and many others.

Slides for microscopic examination are usually prepared from the peel, transfer, and other specimens described above. Although optical light microscopy is still most widely used (Figs. 2.8A; 2.9A), the transmission electron microscope (TEM) and scanning electron microscope (SEM) now play an important role in the observation of such fossil plant remains as wall structures of wood cells, spores (Fig. 2.9B,C), pollen grains (Fig. 2.9D), and epidermal cell patterns including stomata. Recent studies of Devonian plants using high-energy x-rays show structures as small as 50 μm embedded in the rock matrix. These and many other technological advances have given paleobotanists the tools needed to explore the fine structures of fossil plants in ways not possible three decades ago.

Because of the fragmentary nature of plant megafossils and their occurrence in sedimentary rocks, paleobotanists have often found it difficult to develop careful and efficient collecting techniques for the field. For paleoecologists, such techniques are especially important if a meaningful sample of a plant-fossil assemblage is to be obtained that can be subjected to subsequent quantitative and statistical analyses. For quantitative studies of megafossils in coal balls, Phillips, Kunz, & Mickish (1977) developed a sampling technique, which has been used in recent extensive studies of Carboniferous peat swamps. This technique has been improved upon by Pryor (1988), who has developed in much more time-efficient method. Techniques for sampling megafossil assemblages in clastic sediments (e.g., siltstone, shale, sandstone, volcanic ash) and their analysis are the main topics of a comprehensive account prepared by the Paleobotanical Section of the Botanical Society (1987) of America. The article by Spicer (1987) in this publication will give you an excellent, readable summary of the sampling techniques that are available to the paleoecologist.

Modern paleobotanical studies utilize computers as a basic research tool with a wide range of applications. In addition to performing statistical analyses of fossil assemblages for paleoecological studies, they are also being used to recreate the major trends in plant evolu-

Figure 2.8. Specimens prepared by maceration and peel techniques. **A.** Megaspore of the *Triletes* type macerated from a "spore coal" occurring in the Michigan Coal Basin, Pennsylvanian. Photographed using reflected light. Actual specimen approximately 500 μm in diameter. **B.** Transverse section of a tree fern stem. Specimen made by the peel technique. Middle Pennsylvanian. (From Stidd, 1971.)

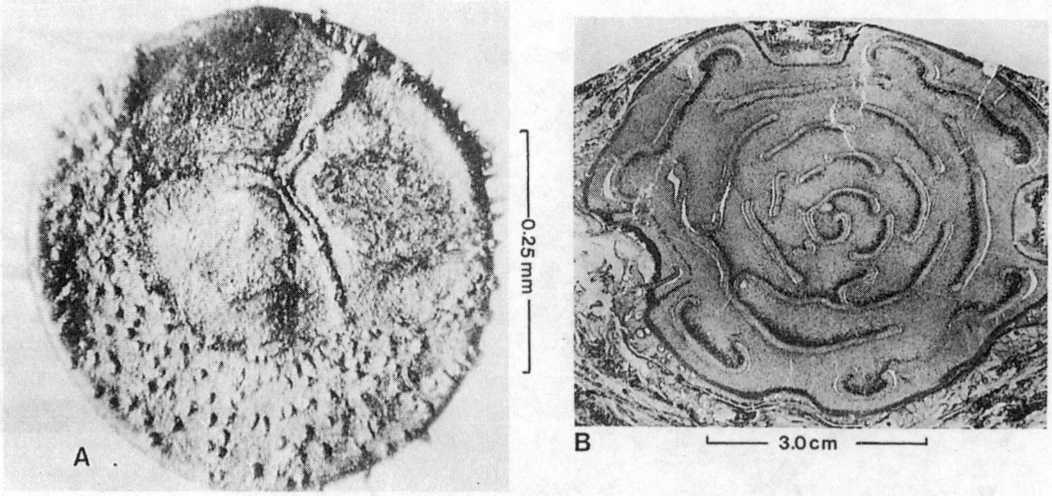

tion. This is because computers can handle large quantities of information and efficiently perform the required repeated calculations (Niklas, 1986).

The perfection of preservation

When we examine the cell walls of permineralized specimens, the preservation is often so good that all of the pits, membranes around pits, organization of secondary wall thickenings, and primary wall layers can be seen in detail. Being able to see even a few of these details helps a paleobotanist identify a plant or plant part, whether it be the thick-walled wood cells of a permineralized conifer stem (Ramanujam, 1972) or the thin-walled cells of phloem of an extinct sphenopsid (Eggert & Gaunt, 1973).

In view of their delicate nature, it seemed unlikely that well-preserved apical meristems could be found. True, there are early reports of permineralized meristems, but none of these included a detailed investigation of the cellular organization of the apex. In 1961 Melchior & Hall discovered a calamite stem tip with a well-defined apical cell and its derivatives. This was an important find, for it gave some support to the idea that the histological study of developmental stages of fossilized plants might be possible. Later, Good (1971) found a calamite twig with an apical cell and was able to tell us that this Carboniferous plant produced the cells of its leafy shoots from four faces of the apical cell. Some exceptional material of another Carboniferous member of the Sphenopsida – *Sphenophyllum* – showed that the apical meristem of its branches also are terminated by apical cells (Good &

Figure 2.9. Microscopy. **A.** Prepollen grain of *Parasporites* photographed using optics and transmitted light. Compare with Figure 2.8A, where reflected light was used. **B.** Photomicrograph taken with scanning electron microscope (SEM). Compare the appearance of this specimen of *Parasporites* with that at **A. C.** Broken sporoderm of *Parasporites* photographed with SEM at high magnification showing surface features and internal structure. **D.** Sectioned sporoderm of *Monoletes* photographed using transmission electron microscope (TEM). Note the alveolar structure. **A–D:** prepollen of Pennsylvanian seed ferns. (**A–C** from Millay, Eggert, & Dennis, 1978; D from Taylor, 1978.)

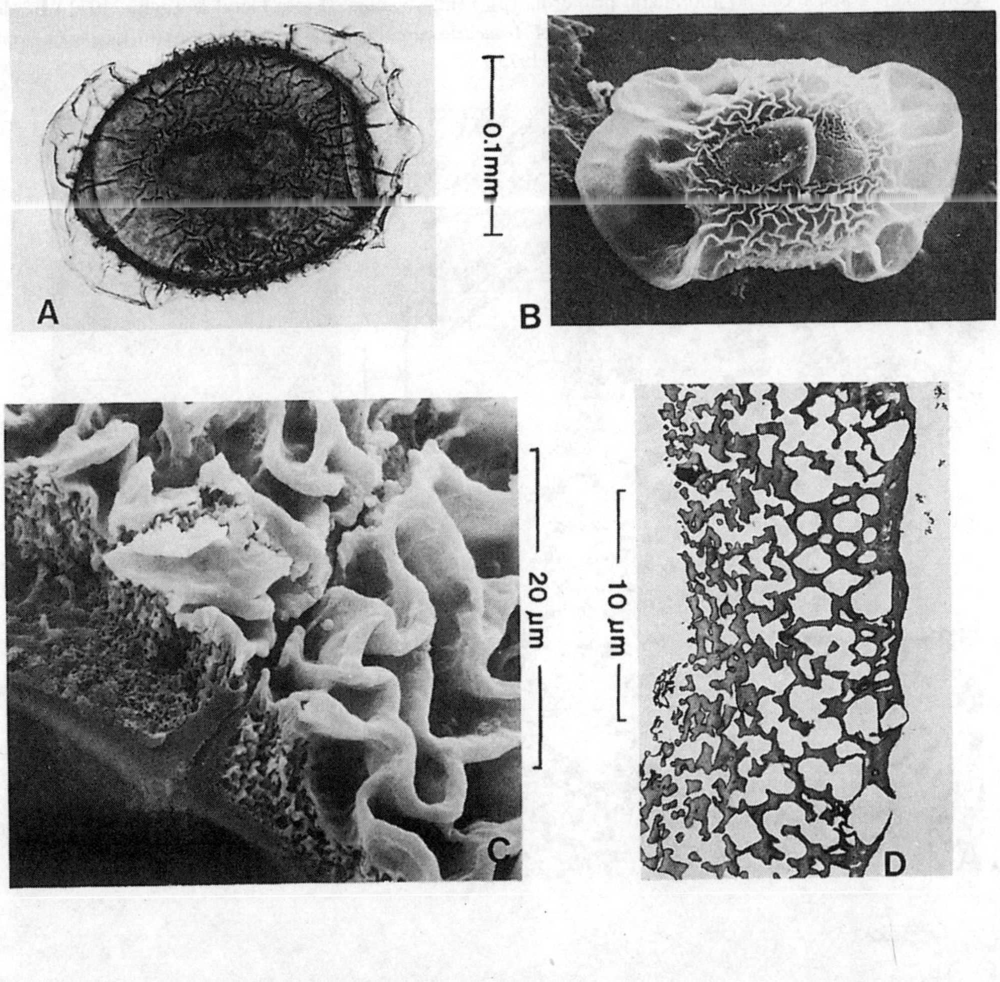

Taylor, 1972). The preservation of one specimen is so perfect that these authors were able to determine the relationship of the leaf primordia of the shoot to the cells of its apical meristem (Fig. 2.10A).

Other tissues having the same delicate construction as apical meristems are those of vascular plants bearing the gamete-forming sex organs (gametophytes). There are many accounts of permineralized gametophytes belonging to Carboniferous lycopods and pteridosperms. One that is most interesting is a male gametophyte of a pteridosperm (seed fern) where one can see a pollen grain with a branched pollen tube (Rothwell, 1972). Pollen grains from the pollen sacs of other pteridosperms show stages in the development of their delicate microgametophytes (Fig. 2.11A) (Millay & Eggert, 1974; Millay & Taylor, 1976). Another spectacular find in yet another pteridosperm pollen grain shows cells (Fig. 2.11C) resembling sperm cells of a modern cycad (Stewart, 1951). This specimen was found in the pollen chamber of a pteridosperm seed that also contained remains of a female gametophyte, including the "endosperm" and an egg or zygote (Fig. 2.11B) within an archegonium.

Megagametophytes with beautifully preserved archegonia in many stages of development have been described and illustrated (Figs. 2.11D,E) by Galtier (1970), Brack (1970), and Brack-Hanes (1978). Paleobotanists always get excited when they discover a fragile embryo preserved in a seed (Fig. 2.10B). Numerous such discoveries have been made that help us to understand early development of the sporophyte in ancient seed plants, especially conifers (Stockey, 1978) and other gymnosperms (Smoot & Taylor, 1987).

Up to this point we have been concerned with the preservation of tissues and cells. Now we take the next step, one that takes us into the cell. When Darrah (1938) illustrated a fossil cellular megagametophyte showing what were claimed to be nuclei in each of the cells, there was considerable skepticism because paleobotanists doubted that cytological detail could be preserved. Since that time, however, so many descriptions of fossil cytoplasm (Fig. 2.12B), nuclei, nucleoli, starch grains, and chloroplasts have been reported that now there is no doubt that internal cell structures

Figure 2.10. Preservation of meristems. **A.** Median longitudinal section of apical meristem of *Sphenophyllum*. Note wedge-shaped apical cell (a) and lateral primordia (p). Pennsylvanian. (From Good & Taylor, 1972.) **B.** Median longitudinal section of a dicotyledonous embryo of *Araucaria mirabilis*. Jurassic. Cotyledons (c), megagametophyte (m), hypocotyl (h), epicotyl (e). (From Stockey, 1975.)

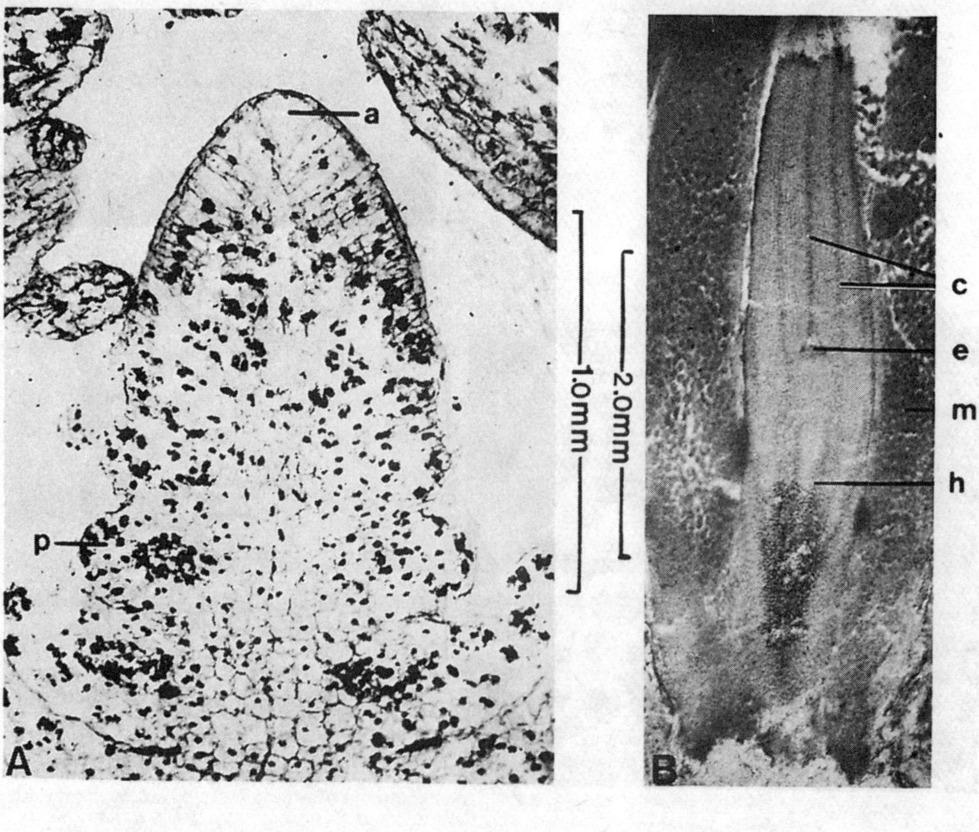

are preserved. One of the most convincing descriptions and illustrations of fossil nuclei is shown in Figure 2.12A (Millay & Eggert, 1974). If confirmed, the most spectacular of all is the discovery by Brack-Hanes & Vaughn (1978) who found what they interpret to be nuclei and mitotic chromosomes (Figs. 2.12C–H) in developmental stages of microgametophytes of a Pennsylvanian lycopod cone.

The possibility that TEM could be used to study the fine structure of fossilized cell contents was explored by Taylor & Millay (1977), who examined the cytoplasmic structure of gametophytic cells. Further, studies by Niklas & Brown (1981), Niklas (1982a), and others, using coalified leaf compressions from Miocene deposits, have demonstrated conclusively that in well-preserved specimens one can observe such fine structures as cytoplasmic membranes within the cell wall (Fig. 2.13A) and those that form the grana of chloroplasts (Fig. 2.13B). Structures resembling mitochondria have also been seen.

With such excellent preservation of the detailed structure of cells and their contents, it is little wonder that questions about the process and rapidity of fossilization are asked. After all, in an aerobic environment, pollen tubes, meristematic tissues, sperm cells, nuclei, and so forth do not last long after being detached from the parent plant. From this we are led to the conclusion that infiltration and lithification probably occurred rather suddenly in the proper environment. Experiments were conducted by Oehler & Schopf (1971) to answer questions about the rate and the process of permineralization. In designing their experiments, they

Figure 2.11. Preservation of gametophytic structures. **A.** Prepollen grain, *Vesicaspora,* containing a three-celled microgametophyte. Middle Pennsylvanian. (From Millay & Eggert, 1974). **B.** Longitudinal section of apical portion of a pteridosperm ovule (*Pachytesta*) showing megagametophyte (m), two archegonia (a), one containing a protoplast. **C.** Microgametophyte of *Pachytesta* with two cells similar in size and organization to sperms of modern cycads. **B,C:** Upper Pennsylvanian. (From Stewart, 1951.) **D,E.** Archegonia of the megagametophyte of *Flemingites schopfii.* Pennsylvanian. **D.** Longitudinal section showing neck cells (n), position of egg (e), and cells of the venter (v). **E.** Transverse section of archegonium showing neck canal (c) and four tiers of neck cells (n). (From Brack-Hanes, 1978.)

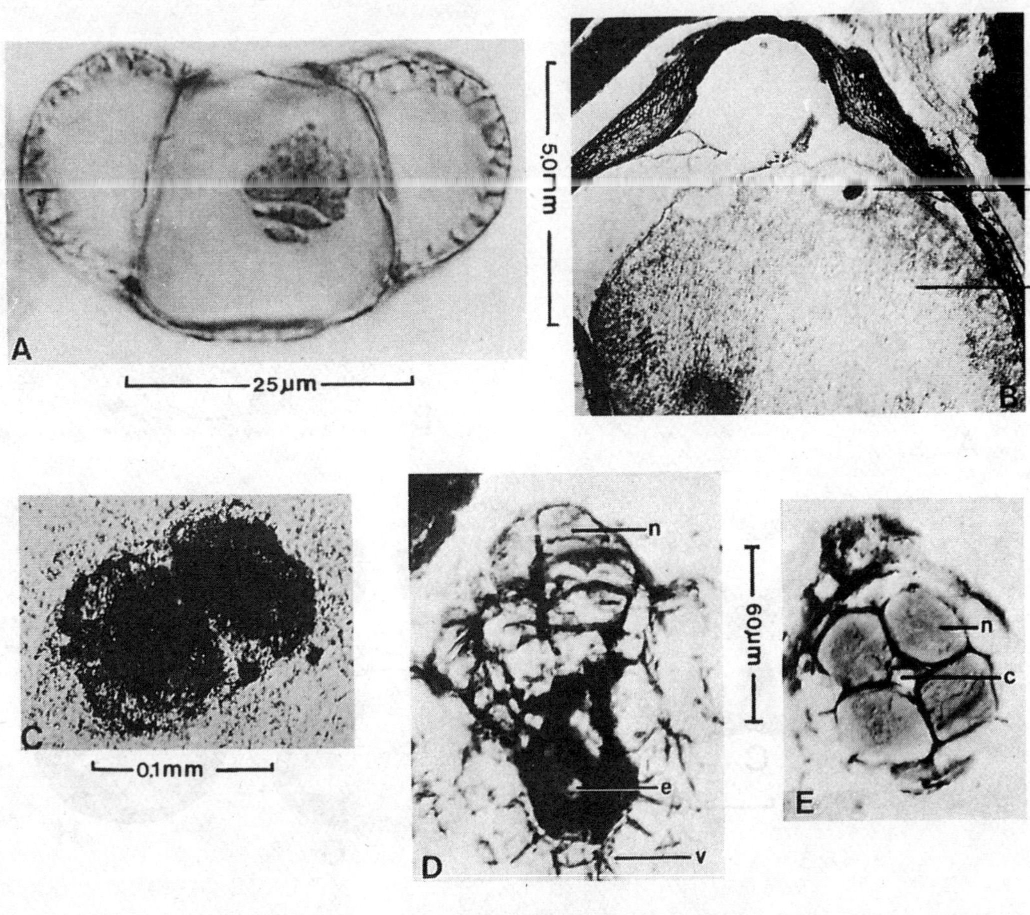

attempted to re-create a Precambrian environment analogous to a place where cyanophytes might have grown approximately 2 billion years ago. Preparing the model required water and a source of soluble silicates which can be converted into a colloidal state. The colloidal silicates must infiltrate the living cells, and then convert into a solid gel. The latter provides the material that changes into a microcrystalline quartz, forming the "stone" matrix in which the "fossilized" cyanophytes are entombed. To get the experimental model working required adjustments in the pH to produce the solid gel from the soluble silicates as well as moderate increases in temperature to 150°C and pressures between 1,000 and 3,000 bars to obtain the quartz. Good artificial permineralization of the cyanophytes was obtained in experiments lasting 2 to 4 weeks. These figures tend to support our earlier assumption that permineralization in past environments could have been a rapid process, indeed, and did not take years to accomplish.

A difficult but potentially important area of investigation being developed is the study of "biochemical" fossils. Using actual fossil leaves from Miocene deposits, Niklas & Giannasi (1977) isolated organic compounds called flavonoids. These compounds abound in extant plants. Other chemical analyses of Tertiary leaves have led to the identification of chlorophylls, aromatic acids, steroids, and branched hydrocarbons (Niklas & Brown, 1981). Substances related to cellulose, lignin, cutin, and sporopollenin also have been identified in much older fossils (Niklas, 1981). Although the presence of amino acids and the degradation products of chlorophylls have been reported indigenous to Precambrian cherts containing the remains of microorganisms, more current work (Smith, Schopf, & Kaplan, 1970) shows that these substances are prob-

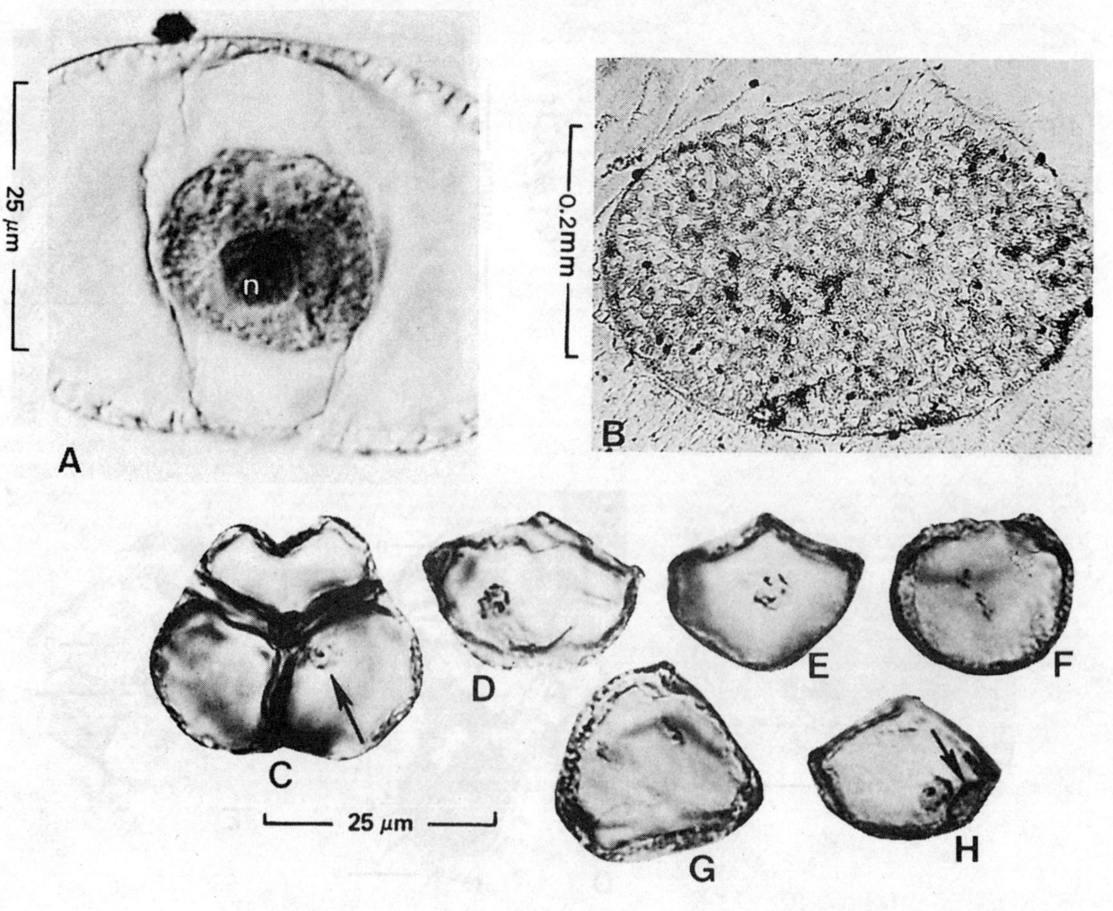

Figure 2.12. Preservation of cytological structures. **A.** Prepollen grain of *Vesicaspora* with a well-defined nucleus (n) embedded in shrunken cytoplasm. Middle Pennsylvanian. (From Millay & Eggert, 1974.) **B.** Cytoplasm belonging to egg or zygote in the archegonium of a pteridosperm ovule. Upper Pennsylvanian. (From Stewart, 1951.) **C–H.** Microspores of *Flemingites schopfii* showing stages in nuclear division (meiosis) and microgametophyte development. **C.** Microspore tetrad; **D,E.** Prophase. **F.** Metaphase. **G.** Anaphase. **H.** Two-celled microgametophyte. Arrow points to cell next to wall. Pennsylvanian. (From Brack-Hanes, 1978.)

ably of recent origin and thus may represent contaminants. With further study and improvement in techniques, however, we can expect paleobotanical biochemists to characterize those chemical components of plant groups that can be used as yet another source of evidence indicating phylogenies (Niklas, 1982b).

Some methods of age determination

One of the most frequently recurring questions about fossils, plant or animal, concerns their antiquity. Prior to the advent of radiometric methods used in making absolute age determinations of rocks associated with fossils, it was difficult to give an answer except to say that one fossil was older, the same age as, or younger than another. If you look at the geological time scale (Chart 1.1), however, you will see that the onset of various units of geological time (eras, periods, and epochs) is given in absolute terms. For example, the chart indicates that the Devonian period started approximately 395 m.y. ago and lasted for some 50 m.y., or until the beginning of the Mississippian period. The ages given are not precise. All of those who are concerned with radiometric dating realize that there is a percentage of error.

The technique of radiometric dating was developed early in this century and was based on an important discovery made in the laboratories at Cambridge University, England. It was demonstrated that there are radioactive isotopes of certain elements that decay at a constant rate, irrespective of heat, pressure, or any other factor in the environment. Radioactive isotopes that have been extensively used in making age determinations are uranium-238, uranium-236, thorium-232, and potassium-40. Uranium and thorium are found most frequently in an igneous rock called pegmatite, while potassium-40 is a component of some sedimentary rocks. The sanadine and biotite fractions of the volcanic rock bentonite are important sources of potassium-40 used in potassium-argon dating of associated plant fossils.

When the radioisotopes of uranium and thorium disintegrate, they ultimately produce a stable form of lead, helium, and heat. Thus, ^{236}U after passing through 14 intermediate radioactive stages will yield ^{206}Pb plus 8H and heat. The rate at which 1 gram of ^{236}U decays into ^{206}Pb is 1/7,600,000 gram of lead in 1 year. Knowing this constant rate of decay, it is clear that the ratio of lead to the remaining uranium can be used to determine the age of the rock. Another way of portraying the same idea is to express the rate of decay in terms of the half-life of the radioscope. Uranium-238 has a half-life of 4.50×10^9 or 4,500 m.y. If the ^{206}Pb in pegmatite is the product of the decay of ^{238}U from the time the pegmatite was formed, then the ratio of ^{238}U to ^{206}Pb, when related to the half-life of ^{238}U, will give us an indication of the age of the rock. Igneous rocks such as pegmatite do not contain plant fossils, so the age of fossiliferous sedimentary rocks is inferred where dated igneous pegmatites have intruded into the sedimentary strata. It is this inference that can be the greatest source of error. Using the techniques of radiometric dating and other sources of evidence, the age of Earth is estimated at 4,700 m.y. The oldest sedimentary rocks, the ones we are interested in, are believed to be 3,750 m.y. old.

Figure 2.13. Preservation of ultrastructure. **A.** TEM photograph showing double membranes in cell of a Miocene leaf. **B.** TEM photograph showing detail of grana comprising a chloroplast in a Miocene leaf. Unlabeled bar scales = 100 nm. (**A,B** from Niklas & Brown, 1981.)

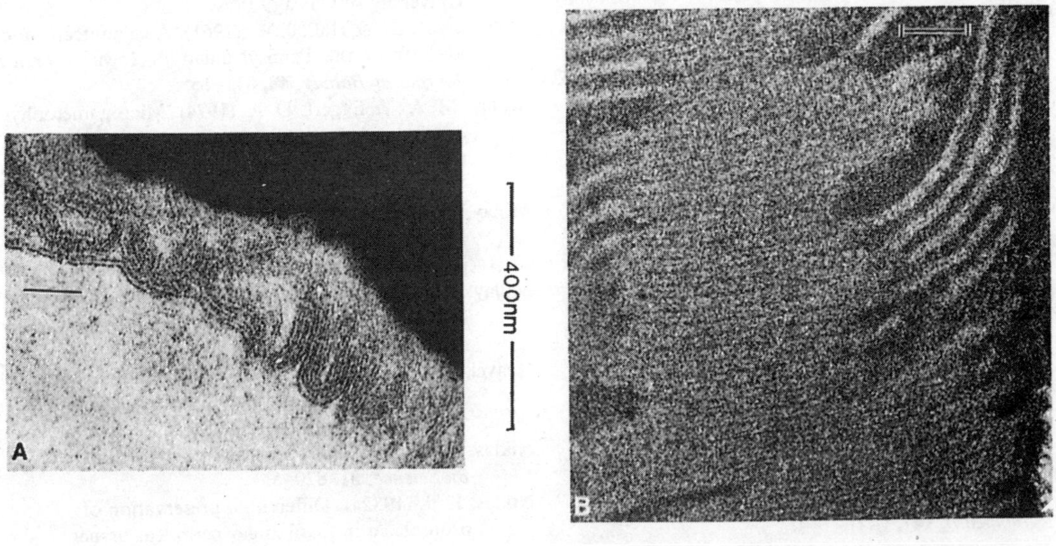

Paleobotanists feel more comfortable in making absolute age determinations of the sedimentary rocks containing fossils by utilization of the potassium–argon techniques because the potassium-containing minerals are a part of the depositional materials in which the specimens are embedded. When ^{40}K decays it gives rise to ^{40}Ca and ^{40}A. Argon-40, an inert gas, makes up 12 percent of the end product of decay; calcium-40 makes up 88 percent. Measurements of the known amount of ^{40}A derived from a unit of ^{40}K in a specific amount of time will give a fairly accurate dating of sedimentary rocks.

The techniques involving radioactive uranium, thorium, and potassium are widely used in dating rocks of considerable antiquity. There reliability, however, is in question when used on rocks of less than 100,000 years.

Radiocarbon dating techniques are used, in part, when determining the age of relatively recent fossils from the Pleistocene. The technique is one employing the isotope carbon-14 and was developed by W. F. Libby (1955) and his colleagues at the University of Chicago. Carbon-14 is formed when nitrogen atoms are bombarded by cosmic rays in the upper atmosphere. It is assumed that the rate of ^{14}C formation has been constant in the past so that a predictable number of ^{14}C atoms have been produced and become mixed evenly through living matter, past and present. The ^{14}C becomes incorporated in living systems by the process of photosynthesis. In this process, ^{14}C is fixed when CO_2 is taken from the atmosphere and utilized in carbohydrate manufacture by the green plant. Further mixing is accomplished in animals when they, in turn, ingest the ^{14}C-containing carbohydrates. Incorporation and mixing ceases when the individual plant or animal dies and the accumulated ^{14}C decays. The rate of decay in terms of the half-life of ^{14}C is $5{,}568 \pm 30$ years. The short half-life of ^{14}C limits its usefulness to approximately 40,000 to 50,000 years. The degree of accuracy is within 2 to 3 percent.

References

Basinger, J. F., & Rothwell, G. W. (1978). Anatomically preserved plants from the Middle Eocene (Allenby Formation) of British Columbia. *Canadian Journal of Botany*, **55**, 1984–90.

Brack, S. D. (1970). On a new structurally preserved arborescent lycopod fructification from the lower Pennsylvanian of North America. *American Journal of Botany*, **57**, 317–30.

Brack-Hanes, S. D. (1978). On the megagametophytes of two lepidodendracean cones. *Botanical Gazette*, **139**, 140–6.

Brack-Hanes, S. D., & Vaughn, J. C. (1978). Evidence of paleozoic chromosomes from lycopod microgametophytes, *Science*, **200**, 1383–5.

Chandrasekharam, A. (1974). Megafossil flora from the Genesee locality, Alberta, Canada, *Palaeontographica B*, **147**, 1–41.

Christophel, D. C. (1976). Fossil floras of the Smoky Tower Locality, Alberta, Canada. *Palaeontographica B*, **157**, 1–43.

Darrah, W. C. (1938). A remarkable fossil *Selaginella* with preserved female gametophyte. *Botanical Museum Leaflets, Harvard University*, **6**, 113–36.

Eggert, D. A., & Gaunt, D. D. (1973). Phloem of *Sphenophyllum*. *American Journal of Botany*, **60**, 755–70.

Evans, W. D., & Amos, D. H. (1961). An example of the origin of coal balls. *Proceedings of the Geologists' Association*, **72**, 445–54.

Friis, E. M. (1990). *Silvianthemum sucicum* gen. et sp. nor. a new saxifragatean flower from the late Cretaceous of Sweden. *Biologiske Skriften*, **36**, 1–35.

Galtier, J. (1970). Observations nouvelles sur le gametophyte femelle des lapidodendress. *Comptes Rendus des Séances de l'Académie des Sciences, Paris*, **271**, 1495–7.

Good, C. W. (1971). The ontogeny of Carboniferous articulates: calamite leaves and twigs. *Palaeontographica, B*, **133**, 137–58.

Good, C. W., & Taylor, T. N. (1972). The ontogeny of Carboniferous articulates: the apex of *Sphenophyllum*. *American Journal of Botany*, **59**, 617–26.

Hickey, L. J., & Doyle, J. A. (1977). Early Cretaceous fossil evidence for angiosperm evolution. *Botanical Review*, **43**, 3–104.

Holmes, J., & Lopez, J. (1986). The disappearing peel technique: an improved method for studying permineralized plant tissues. *Palaeontology*, **29**, 66–70

Joy, K. W., Willis, A. J., & Lacey, W. S. (1956). A rapid cellulose peel technique in paleobotany. *Annals of Botany* (N.S.) **20**, 635–7.

Kummel, B., & Raup, D. (1965). *Handbook of Paleontological Techniques*. San Francisco: Freeman.

Langenheim, J. H. (1964). Present status of botanical studies of ambers, *Botanical Museum Leaflets, Harvard University*, **20**, 225–87.

Langenheim, J. H. (1969). Amber: a botanical inquiry. *Science*, **163**, 1157–69.

Libby, W. F. (1955). *Radiocarbon Dating*, 2nd ed. Chicago: University of Chicago Press.

Melchior, R. C., & Hall, J. W. (1961). A calamitean shoot apex from the Pennsylvanian of Iowa. *American Journal of Botany*, **48**, 811–15.

Millay, M. A., & Eggert, D. A. (1974). Microgametophyte development in the Paleozoic seed fern family Callistophytaceae. *American Journal of Botany*, **61**, 1067–75.

Millay, M. A., Eggert, D. A., & Dennis, R. L. (1978). Morphology and ultrastructure of four Pennsylvanian prepollen types, *Micropaleontology*, **24**, 305–15.

Millay, M. A., & Taylor, T. N. (1976). Evolutionary trends in fossil gymnosperm pollen. *Review of Palaeobotany and Palynology*, **21**, 65–91.

Neavel, R. C., & Guennel, G. K. (1960). Indiana paper coal: composition and deposition. *Journal of Sedimentary Petrology*, **30**, 241–8.

Niklas, K. J. (1981). The chemistry of fossil plants. *BioScience*, **31**, 820–5.

Niklas, K. J. (1982a). Differential preservation of protoplasm in fossil angiosperm leaf tissues.

American Journal of Botany, **69**, 325–34.

Niklas, K. J. (1982b). Chemical diversification and evolution of plants as inferred from paleobiochemical studies. In *Biochemical aspects of evolutionary biology*, ed. M. H. Nitecki. Chicago: University of Chicago Press, pp. 29–91.

Niklas, K. J. (1986). Computer-simulated plant evolution. *Nature*, **254**, 78–86.

Niklas, K. J., & Brown, R. M. (1981). Ultrastructural and paleobiochemical correlations among fossil leaf tissues from the St. Maries River (Clarkia) area, Northern Idaho, USA. *American Journal of Botany*, **68**, 332–41.

Niklas, K. H., & Giannasi, D. E. (1977). Flavonoids and other chemical constituents of fossil Miocene *Zelkova (Ulmaceae)*. *Science*, **196**, 877–8.

Oehler, J. H., & Schopf, J. W. (1971). Artificial microfossils: experimental studies of permineralization of blue-green algae in silica. *Science*, **174**, 1229–31.

Phillips, T. L. (1980). Stratigraphic and geographic occurrences of permineralized coal swamp plants – Upper Carboniferous of North America and Europe. In *Biostratigraphy of Fossil Plants*, eds. D. L. Dilcher & T. N. Taylor. Stroudsburg, Pa.: Dowden, Hutchinson & Ross, pp. 25–92.

Phillips, T. L., Avcin, M. J., & Berggren, D. (1976). Fossil peat from the Illinois Basin: a guide to the study of coal balls of Pennsylvanian age. *Educational Ser. II Illinois State Geological Survey, Urbana.*

Phillips, T. L., Kunz, A. B., & Mickish, D. J. (1977). Paleobotany of permineralized peat (coalballs) from the Herrin (No. 6) of the Illinois Basin. In *Interdisciplinary Studies of Peat and Coal Origins*, eds. P. N. Given & A. D. Cohen. Geological Society of America, Microform Publication, 18–49.

Pryor, J. S. (1988). Sampling methods for quantitative analysis of coal-ball plants. *Palaeogeography, Palaeoclimatology, Palaeoecology*, **63**, 313–26.

Ramanujam, C. G. K. (1972). Fossil coniferous woods from the Oldman Formation (Upper Cretaceous) of Alberta. *Canadian Journal of Botany*, **50**, 595–602.

Roth, J. L., & Dilcher, D. L. (1979). Investigations of angiosperms from the Eocene of North America: stipulate leaves of the Rubiaceae including a probable polyploid population, *American Journal of Botany*, **66**, 1194–1207.

Rothwell, G. W. (1972). Evidence of pollen tubes in Paleozoic pteridosperms. *Science*, **175**, 722–74.

Roy, S. K. (1972) Fossil wood of Taxaceae from the McMurry formation (Lower Cretaceous) of Alberta, Canada, *Canadian Journal of Botany*, **50**, 349–52.

Schopf, J. M. (1970). Petrified peat from a Permian coal bed in Antarctica, *Science*, **169**, 274–277.

Schopf, J. M. (1975). Modes of fossil preservation. *Review of Palaeobotany and Palynology*, **20**, 27–53.

Scott, A. C., & Rex, G. (1985). The formation and significance of Carboniferous coal balls. *Philosophical Transactions of the Royal Society of London, B*, **311**, 123–37.

Smith, J. W., Schopf, J. W., & Kaplan, I. R. (1970). Extractable organic matter in Precambrian cherts. *Geochimics et Cosmochimica Acta*, **34**, 649–75.

Smoot, E. L., & Taylor, T. N. (1987). Structurally preserved fossil plants from Antarctica III. Permian seeds. *American Journal of Botany*, **74**, 904–13.

Smoot, E. L., Taylor, T. N. & Delevoryas, T. (1985). Structurally preserved fossil plants from Antarctica. I. *Antarcticycas*, gen. nov., a triassic cycad stem from the Beardmore Glacier area. *American Journal of Botany*, **72**, 1410–1423.

Spicer, E. L. (1987). Quantitative sampling of plant assemblages. In *Methods and Applications of Plant Paleoecology – Notes for a Short Course*, eds. W. A. DiMichele & S. L. Wing. Paleobotanical Section, Botanical Society of America.

Stewart, W. N. (1951). A new *Pachytesta* from the Berryville locality of southeastern Illinois. *American Midland Naturalist*, **46**, 717–42.

Stewart, W. N., & Taylor, T. N. (1965). The peel technique. In *Handbook of Paleontological Techniques*, eds. B. Kummel & D. Raup. San Francisco: Freeman, pp. 224–32.

Stidd, B. M. (1971). Morphology and anatomy of the frond of *Psaronius*. *Palaeontographica, B*, **134**, 87–123.

Stockey, R. A. (1975). Seeds and embryos of *Araucaria mirabilis*. *American Journal of Botany*, **62**, 856–68.

Stockey, R. A. (1978). Reproductive biology of Cerro Cuadrado fossil conifers: ontogeny and reproductive strategies in *Araucaria mirabilis* (Spegazzini) Windhausen. *Palaeontographica, B*, **166**, 1–16.

Taylor, T. N., (1978). The ultrastructure and reproductive significance of *Monoletes* (Pteridospermales) pollen. *Canadian Journal of Botany*, **56**, 3105–18.

Taylor, T. N., & Millay, M. A. (1977). Structurally preserved fossil cell contents. *Transactions of the American Microscopical Society*, **93**, 390–93.

Taylor, E. L., & Taylor, T. N. (1989). Structurally preserved Permian and Triassic floras from Antarctica. In *Antarctic paleobiology: its role in reconstruction of Gondwana*, eds. T. N. Taylor & E. L. Taylor. Berlin, Springer-Verlag.

3

The fossil record: systematics, reconstruction, and nomenclature

Unfortunately botanists do not agree as to what should be included in the plant kingdom (Plantae). No one would disagree that mosses, ferns, pine trees, grasses, and similar organisms are plants. They have cellulose in their cell walls. Biochemically they are characterized by similar photosynthetic pigments (chlorophylls *a* and *b*) and food stored as true starch. They are multicellular, embryo-forming organisms with life cycles that we associate with the condition of being a plant, a condition that will be explained later. But many organisms do not fit these criteria at all. Among them are primitive unicellular organisms; bacteria and cyanophytes (blue-green algae), which, unlike other organisms, lack true nuclei. For this reason they are said to be prokaryotic. Because they are so different and show little relationship to other organisms, these two groups are placed in a separate kingdom, the Monera. Similarly, the fungi have been removed from the plant kingdom because current evidence indicates that the mushrooms and their relatives did not evolve from plant ancestors. So yet another kingdom, Fungi, was set up. Finally, a poorly defined kingdom, Protista, was established some time ago for certain primitive organisms, some of which show characteristics of both plants and animals. If we add the animal kingdom (Animalia) to our list, we have a total of five kingdoms. In this book we are concerned with the remains of organisms belonging to the Monera, Fungi, Protista, and of course Plantae. From this point on, the word plant will be used in its broad, traditional meaning to include organisms from these four groups.

Natural systems of classification

Botanists are confronted with the monumental challenge of establishing a natural system of plant classification. This is one that attempts to express in its organization the true evolutionary relationships among plants. If one were to argue that there is such a thing as a truly natural system of classification, it would be the same as saying that botanists know everything about living and extinct plants – further, that everyone agrees as to what characteristics are important in determining plant relationships. Thus, it is not surprising to find that there are many different ways of classifying plants according to a natural system. In our opinion, the system used in this book is the most nearly natural system presently available because it defines taxonomic groups by shared derived characters and takes into account the maximum amount of evidence from the fossil record that reflects evolutionary relationships among both extinct and extant plants. Such a system is referred to as a phylogenetic system, as opposed to a phenetic system that reflects degrees of similarity, usually among modern plants.

During the past 30 years efforts to construct phylogenetic systems of classification have received support from the development of a more precise and objective methodology that is commonly termed cladistics. The two major advantages of the cladistic approach are (1) only those characters that are appropriate to each study are used, and (2) all of the data and methods on which a classification is based are presented in detail. Cladistic analyses were initiated independently by the zoologist W. Henning (1950) and the botanist W. H. Wagner (1952), and developed further by a large number of workers during the 1970s and 1980s. A brief explanation of the cladistic methodology will provide the basic information needed to understand systematic interpretations presented here. The interested student may want to consult Hill & Crane (1982), Crane (1988), or Doyle & Donoghue (1987) for recent examples of cladistic treatments of vascular plants.

Cladists attempt to identify evolutionary lineages, or clades, and to separate the latter from groups or organisms that have independently obtained a similar level of evolutionary advancement. These groups may be referred to as grades of evolution, or simply as grades. Cladists recognize essentially three categories of groups of organisms. These are monophyletic, paraphyletic, and polyphyletic groups (Chart 3.1). Monophyletic groups consist of an ancestor and all of its descendants, whereas paraphyletic groups consist of an ancestor and only some of its descendants. In constrast, polyphyletic groups of organisms do not have an immediate or direct common ancestor that belongs to the group.

Cladists begin a study by focusing their attention on all the characteristics of each organism under investigation. They apply uniform methods for interpreting the characters and for deciding which are appropriate for assessing relationships among the organisms under study. The results of a cladistic study are presented in the form of a diagram that is termed a cladogram (Chart 3.1). In each cladistic study, a list of all of the characters used and an explanation of the methods used are presented along with the cladogram so that future workers can assess the evidence upon which a classification is based. The cladogram can, therefore, be viewed as a testable hypothesis of the relationships of the plants being considered.

If a similar character in two plants has been inherited from a common ancestor, the characters are said to be homologous, but if they are the result of

Chart 3.1. Cladogram showing the relationships among six hypothetical species (**A–D**). Numbered marks along the axes indicate positions where character-state changes occur. Species **A** is the common ancestor to species **B–F,** and together species **A–F** constitute a monophyletic group. Species **A–D** represent a paraphyletic group because not all of the descendants of the common ancestor **A** are included, and species **B–D** represent a polyphyletic group because the group does not include an immediate common ancestor (**A**).

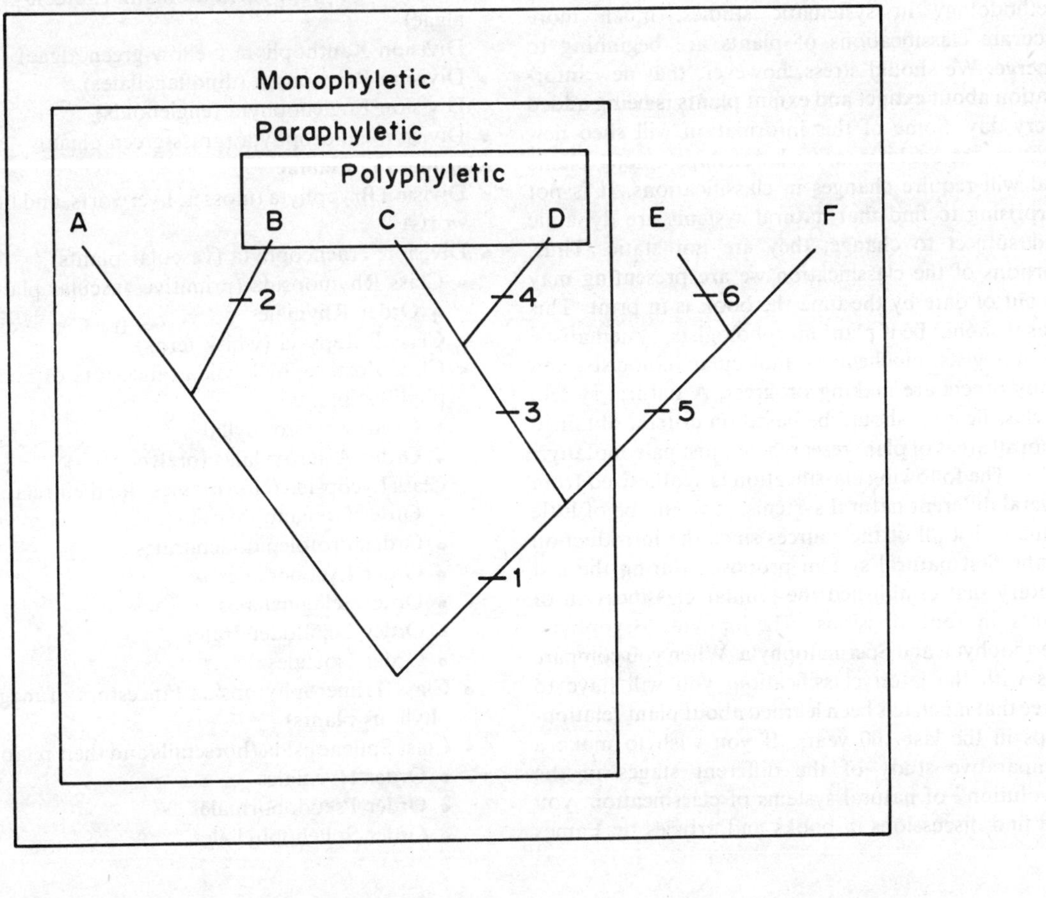

independent evolutionary events (i.e., parallel or convergent evolution), the characters are considered to be analogous. Only homologous characters are used by cladists to assess phylogenetic relationships. When evaluating contrasting states of a given character, cladists apply uniform methods to determine which of the character states represents the ancestral or plesiomorphous condition and which represent a derived or apomorphous condition. Shared ancestral characters are termed symplesiomorphies, whereas shared derived characters are called synapomorphies. Only synapomorphies are used to interpret the interrelationships among the group under investigation. For example, having water-conducting cells with differentially thickened and lignified walls (tracheary elements) is a synapomorphy for the vascular plants, and an important character for including a species in the Division Tracheophyta. However, within the Division Tracheophyta, tracheary elements are a symplesiomorphy and therefore of no value for interpreting relationships among vascular plants.

Once the appropriate characters and character states have been determined for all of the species under consideration, the minimum number of character-state changes required to account for the interrelationships of the species are assessed. This is usually carried out with the aid of computer programs to generate one or more cladograms. By employing the cladistic methodology in systematic studies, much more accurate classifications of plants are beginning to emerge. We should stress, however, that new information about extinct and extant plants is being added every day. Some of this information will shed new light on the evolutionary relationships among plants and will require changes in classifications. It is not surprising to find that natural systems are dynamic and subject to change; they are not static. Thus, portions of the classification we are presenting may be out of date by the time the book is in print. This is as it should be if plant morphologists, systematists, physiologists, biochemists, molecular biologists, and many others are making progress. A natural system of classification should be based on criteria obtained from all areas of plant research, not just paleobotany.

The following classification is synthesized from several different natural systems. It would be of little value to list all of the sources since the introduction of the first natural system proposed during the last century that established the familar classification of plants in four divisions: Thallophyta, Bryophyta, Pteridophyta, and Spermatophyta. When you compare this with the latest classification, you will have to agree that much has been learned about plant relationships in the last 100 years. If you wish to make a comparative study of the different stages in the "evolution" of natural systems of classification, you will find discussions in books and articles by Eames (1936); Tippo (1942); Smith (1955); Cronquist (1960); Lawrence (1963); Scagel et al. (1967); Banks (1968), Foster, & Gifford (1974); Raven et al., (1976); Bold, Alexopoulos, & Delevoryas (1980); Knoll & Rothwell (1981); and Meyen (1987). These are arranged in chronological order so that the most modern classifications appear in the later references. With some modifications, the classification presented here is a synthesis of the classifications presented by these authors. (The sequence of the divisions, classes, and orders presented in the text is the same as presented in the classification given here. Further amplification of the classification within each of the orders will be undertaken in the appropriate chapters. Those groups marked by a bullet have a fossil record that is treated in this book.)

Kingdom I. Monera
- • Division Cyanophyta (cyanobacteria)
- • Division Schizomycota (bacteria)

Kingdom II. Fungi
- Division Gymnomycota (slime molds)
- • Division Mastigomycota (phycomycetes)
- • Division Amastigomycota (true fungi)

Kingdom III. Protista
- • Division Phaeophyta (brown algae)
- • Division Rhodophyta (red algae)
- • Division Chrysophyta (diatoms and golden-brown algae)
- Division Xanthophyta (yellow-green algae)
- • Division Pyrrophyta (dinoflagellates)
- Division Euglenophyta (euglenoids)
- • Division Chlorophyta (grass-green algae)

Kingdom IV. Plantae
- • Division Bryophyta (mosses, liverworts, and hornworts)
- • Division Tracheophyta (vascular plants)
  - • Class Rhyniopsida (primitive vascular plants)
    - • Order Rhyniales
  - Class Psilopsida (whisk ferns)
  - • Class Zosterophyllopsida (ancestors of microphyllous plants)
    - • Order Zosterophyllales
    - • Order Asteroxylales (prelycopods)
  - Class Lycopsida (club mosses and their relatives)
    - • Order Drepanophycales
    - • Order Protolepidodendrales
    - • Order Lycopodiales
    - • Order Selaginellales
    - • Order Lepidodendrales
    - • Order Isoetales
  - • Class Trimerophytopsida (ancestors of megaphyllous plants)
  - • Class Sphenopsida (horsetails and their relatives)
    - • Order Hyeniales
    - • Order Pseudoborniales
    - • Order Sphenophyllales

- Order Equisetales

Class Filicopsida (ferns and their relatives)

- Order Cladoxylales
- Order Stauropteridales
- Order Zygopteridales
- Order Ophioglossales
- Order Marattiales
- Order Filicales
- Order Salviniales
- Order Marsileales

Class Progymnospermopsida (ancestors of gymnosperms)

- Order Aneurophytales
- Order Archaeopteridales
- Order Protopityales

Class Gymnospermopsida (plants with naked seeds)

- Order Pteridospermales
- Order Cycadales
- Order Cycadeoidales
- Order Caytoniales
- Order Glossopteridales
- Order Pentoxylales
- Order Czekanowskiales
- Order Gnetales
- Order Ginkgoales
- Order Cordaitales
- Order Voltziales
- Order Coniferales
- Order Taxales

Subdivision Angiospermophytina (flowering plants)

- Class Magnoliopsida (dicotyledons)
- Class Liliopsida (monocotyledons)

Kingdom V. Animalia

In the hierarchy of major taxonomic categories, the division in a natural system of classification is the most inclusive of related organisms. Although some attempt has been made to show a degree of relationship at the kingdom level, this is certainly not the case for the heterogeneous Protista, which includes many divisions of organisms that are believed to have had independent evolutionary origins from unknown unicellular ancestors. Because of this, we can say that the Protista is a highly polyphyletic group. In contrast, the kingdom Plantae with its two divisions, Bryophyta and Tracheophyta, can be considered monophyletic. In this case biologists agree that there is overwhelming evidence indicating an origin of these green land plants from tetrasporine members of the grass-green algae. The tetrasporine line is characterized by photosynthetic organisms with vegetative cells that divide in the monmotile condition and have uninucleate daughter cells. All Chlorophyta (grass-green algae) have a similar biochemistry of pigments and stored food, as well as cellulose cell walls. For these and other reasons, some investigators think the relationship is so close that the Chlorophyta should be included in the plant kingdom rather than the Protista. At the level of division, the Tracheophyta is another striking example of monophylesis. As the name implies, all Tracheophyta form cells in the vascular (conducting) tissue called tracheids. Also, in the life cycle of such plants, the sporophyte phase is predominant and independent. Sporophytes of tracheophytes – for example, ferns, gymnosperms, and flowering plants – are comprised of roots, stems, leaves, and accessory reproductive parts. Of equal importance, however, in establishing a monophyletic concept for the Tracheophyta is the paleobotanical evidence that vascular plants have evolved from a primitive type known to have existed from mid-Silurian well into the Devonian. The revelation of the primitive type and its subsequent evolution into other groups of vascular plants is one of several high points in the investigation of plant evolution. As for the Tracheophyta, the fossil record has contributed extensively in establishing the natural relationships within the major groups that comprise it.

Discoveries from the fossil record have made a minimal contribution to the natural classification of nonvascular plants. On the other hand, fossil evidence has had a tremendous impact on the classification of vascular plants. This can be demonstrated when we compare an early classification by Eichler to that of Eames (1936) (see Chart 3.2). In Eichler's classification, vascular plants are placed in two divisions, Pteridophyta and Spermatophyta. This tells us that the mode of reproduction was considered the most important characteristic. As a result, free-sporing vascular plants, those that shed their spores, were placed in the Pteridophyta, a group distinct from the Spermatophyta (seed plants). This idea prevailed until 1936 when Eames took into account the extinct pteridosperms, a group confirmed by Oliver & Scott (1904) 32 years earlier. The pteridosperms are vascular plants with many vegetative characteristics of ferns, but surprisingly, they bore seeds. Eames and others interpreted this and other evidence as an indication that the basis for separating the two divisions was not valid, especially when the presence or absence of seeds is the essential criterion.

By 1936 the renowned work of Kidston & Lang (1917–21) had been available for some time. They showed that there existed in Devonian times a flora of primitive vascular plants that they called the Psilophytales. In light of this and other discoveries from fossil and extant plants, it became apparent that vascular plants came from a primitive ancestral stock, the psilophytes. To reflect these discoveries, a single division, Tracheophyta, for all vascular plants was proposed to replace the two divisions of Eichler. Further evidence provided by the extinct pteridosperms indicated that ferns and seed plants were more closely related than ferns to other free-sporing vascular plants. Knowing this, Eames adopted the subdivision Pteropsida, proposed earlier by Jeffrey (1917) to include the ferns, gymnosperms, and flowering plants. The other free-sporing vascular plants, psilophytes, club mosses, and horsetails had been traced as an independent group extending into the Paleozoic quite separate from the ferns. Eames used the subdivisions Psilopsida, Lycopsida, and Sphenopsida to indicate this apparent lack of close relationship (Chart 3.2).

After its proposal this classification was quickly adopted (Tippo, 1942) and made its way into textbooks where it persisted for nearly 30 years. During this time paleobotanists were concentrating on fossil plants of the Silurian, Devonian, and Carboniferous. The results of their studies brought about many more significant changes which are reflected in the classification at the beginning of this section. If you compare the part of this classification dealing with vascular plants with those of Eichler & Eames, you can easily identify those major changes that have occurred since 1940. Evidence for these changes will be presented in the appropriate chapters.

Problems of reconstruction

It would help to understand some of the special problems confronting paleobotanists when reconstructing and naming plants if we were to examine a collection of Carboniferous fossils. The collection might be in the form of permineralized specimens from the renowned Berryville locality of Illinois. Peels made from these specimens show a wealth of exceedingly well-preserved plant fragments. Microscopic examination only amplifies the fragmentary nature and the wide variety of plant organs present. There are roots of various kinds, and a plethora of leaf types, many with attached sporangia. Stem fragments are present in bewildering variety; there are little seeds and large seeds distributed in the matrix along with pollen-containing reproductive structures of numerous shape, and so on. Our problem is: Which part belongs to what plant? Paleobotanists have worked in earnest reconstructing the flora at the Berryville locality since 1950. It has taken numerous investigators 40 years of painstaking, often laborious research to accomplish this, and the job is still not complete.

If we undertake the reconstruction of a whole organism from its parts, we must keep in mind that the description of an organism includes its developmental stages, both vegetative and reproductive, in addition to the mature vegetative aspect of the plant with which we are most familiar. When a paleobotanist realizes the magnitude of the task of reconstruction, he is a little more than envious of those plant systematists who deal with modern, intact, developing organisms going through relatively easily observed life cycles.

There are three common forms of evidence used by paleobotanists in making reconstructions. They include evidence from actual attachment of parts; evidence for attachment inferred by similarity of anatomical features; and evidence for attachment inferred by frequency with which parts are associated, or by the total absence of parts.

Evidence from actual attachment is obviously the most reliable and desirable. Let us take the example of the gymnospermous Devonian *Callixylon* and the associated fernlike foliage *Archaeopteris* (see Chapter 21). These two genera were believed to belong to unrelated groups until a fortuitous find by C. B. Beck (1960) showed the two in organic connection. This discovery was the basis for establishing the Progymnospermopsida and consequently a completely new approach to understanding gymnosperm origin and evolution.

Chart 3.2. A comparison of two vascular plant classifications prior to 1940.

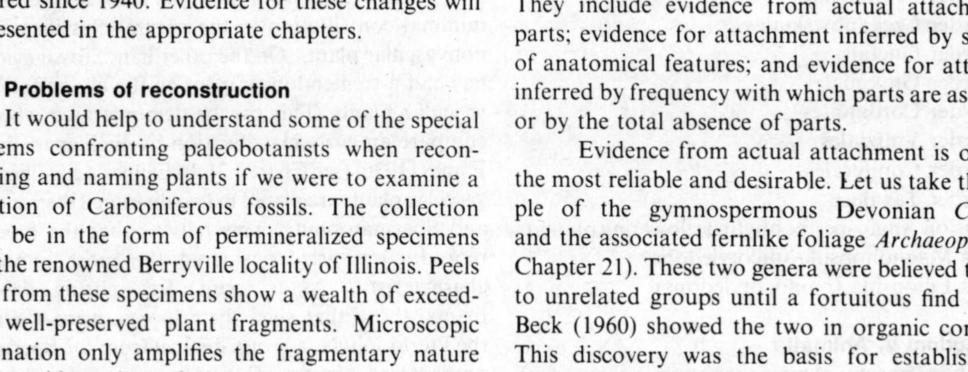

Attachment of parts may be inferred from a more esoteric form of evidence. Such evidence is derived from morphologically different organs that have similar unique anatomical characteristics. A classical example of this kind of evidence was reported by Oliver & Scott (1904), who confirmed the discovery of extinct pteridosperms. When studying the cupule surrounding the Carboniferous seed *Lagenostoma lomaxi*, Oliver & Scott observed microscopic capitate glands (see Chapter 21) attached to the surface of the cupule. Glands having the same structure had been found on leaves and stems of *Lyginopteris oldhamia*. Using this peculiar anatomical similarity, the researchers were able to infer organic connection between foliage-bearing stems and cupulate seeds. A quarter-century later, a compression foliage specimen of *Sphenopteris hoeninghausi* was discovered with gland-bearing cupules in attachment. The investigator was able to show that the foliage was of the same borne by *Lyginopteris* stems, and final proof for organic connection was assured.

Attachment of plant parts inferred by frequency of association is a risky approach to reconstruction. After many years of working with the Carboniferous Berryville flora, certain associations have become apparent. When we find the pteridosperm stem *Medullosa noei*, we frequently find the foliage of *Alethopteris*, the seeds of *Pachytesta illinoensis*, and the pollen-containing structure of *Bernaultia* (= *Dolerotheca*) *formosa* in the same assemblage (see Chapter 21). Unfortunately, these parts are not actually attached. When part or all of this assemblage is repeatedly found, it is tempting to conclude that all are organs belonging to the same species of pteridosperm. The *Bernaultia formosa* fructification with its *Monoletes* (= *Schopfipollenites*) type of pollen grains is known to be attached to *Alethopteris*-type foliage (Ramanujam, Rothwell, & Stewart, 1974). In 1954, Stewart observed *Monoletes* in pollen chambers of at least 10 different seeds of *Pachytesta illinoensis*. Because of this association, it again is tempting to conclude that *Pachytesta* and *Bernaultia* are congeneric and that these types of reproductive structures were borne by *Medullosa* stems with their *Alethopteris*-type foliage. The danger of inferring such a relationship lies in the fact that pollen or spores from unrelated plants can contaminate pollen chambers of different seeds. However, when the same pollen type is repeatedly found in pollen chambers of a specific seed, then the chance that there is a relationship between the seed and the pollen-producing organ is greatly enhanced. In this case, attachment of parts inferred by association has validity.

The absence of certain plant parts in an assemblage can also be used to infer connection between parts of plants in that assemblage. During the many years of investigation of the Berryville flora, occasional specimens of roots borne by stigmarian axes were found. Stigmarian systems are composed of roots attached to a root-bearing axis (rhizomorph), and they are known to be the underground absorbing and anchoring systems for *Diapharodendron*, *Lepidodendron*, and *Lepidophloios* stem types (see Chapter 11). The *Stigmaria* roots from the Berryville locality appeared to be slightly different from those of the common *Stigmaria ficoides*, but there was no way of telling for sure to what lycopod stem type they might belong. Subsequent investigation (Delevoryas, 1957) showed the arborescent lycopod *Sigillaria* to be a part of the Berryville flora. After 10 years of investigation, no other remains of arborescent lycopods except *Sigillaria* had been found. With this observation in mind the obvious question was asked: Could it be that these *Stigmaria*-type roots were, in reality, the roots attached to the rhizomorphs of *Sigillaria*? It followed that the roots could not be parts of a stigmarian system of *Lepidodendron*, *Diapharodendron*, or *Lepidophloios* because these three genera proved to be totally lacking in the Berryville flora. Investigations were undertaken to answer the question with the result that the rhizomorph of a *Stigmaria* was found. A fortunate cut through a coal ball showed the actual attachment of the rhizomorph to *Sigillaria approximata* stems (Eggert, 1972). Thus, one more step in the monumental task of reconstruction was accomplished, first by inference and later by finding actual attachment.

Although the process is a slow one, floras are gradually being reconstructed, as are the environments that sustained them. The ultimate aim, however, of reconstruction of ancient plants and floras is to provide an accurate basis for comparison with other extinct and extant plants. These comparisons can result in the revelation of patterns of plant evolution. If problems of stratigraphy and paleoecology also can be solved, so much the better.

Problems of nomenclature

How much easier the paleobotanist's work would be if plants were preserved in their entirety. The process of reconstruction would be rapid. Difficulties encountered in naming them would be greatly alleviated. As you read the preceding paragraphs on reconstruction you may have noticed that detached plant organs are given a binomial (genus and species name) in the same way as modern plants. It is not the intent to go into a discussion of the advantages of the binomial system of nomenclature or the International Rules of Botanical Nomenclature that regulate it. Suffice it to say that the same rules apply to naming fossil and extant plants. Arnold (1947) presents an excellent summary of those rules that apply.

Because we have already alluded to the Carboniferous genus *Lepidodendron* (see Chapter 11), let us use it as a basis for developing our ideas about naming fossil plant parts. The first valid description of *Lepidodendron* is to be found in the important publications of Sternberg in 1820. This date has been chosen as the starting point for paleobotanical nomenclature in the same way that Linnaeus's *Species Plantarum*, published in 1753, was chosen as the starting point for the nomenclature of modern vascular plants. Originally the name *Lepidodendron* was used to describe bark fragments from arborescent lycopods with characteristic scars on the surface (Fig. 2.5B). The bark fragments were preserved as casts or molds that lacked internal structure. Later, specimens were found that revealed internal structure as well. Leaves, underground rhizomorphs with roots, and cones containing spores and sporangia also were found attached. With the passage of time, there has been a gradual expansion of the characteristics of the genus to include all parts, so that today when we think of *Lepidodendron* we envisage the whole organism, not just a bark fragment.

The living *Lepidodendron* in its Carboniferous swamp environment continually shed its leaves, spores, and cones. Eventually it died and its branches and trunks became fragmented. Many of these scattered parts were preserved to be recovered some 280 m.y. later by today's paleobotanists. As the fossilized organs are discoverd and described, investigators are not always sure about the relationship of one part to another. For example, the leaves of *Lepidodendron* are grasslike and of varying lengths, but so are the leaves of *Sigillaria* and *Bothrodendron*. In the compressed state it is impossible to tell them apart, so that noncommital generic name *Cyperites* (= *Lepidophylloides*) is applied to these detached organs. Accordingly, *Cyperites* should be considered a form or artificial genus. The reason for this is that the genetic relationships of *Cyperites* are not well enough established so that the genus can be assigned to a family. Using the present system of classification, *Cyperites* can be assigned to any one of three different families: Lepidodendraceae, Sigillariaceae, or Bothrodendraceae. All are included in the order Lepidodendrales. Following the same reasoning, *Stigmaria*, the underground anchoring and absorbing system for all members of the Lepidodendrales, and the spore genus *Triletes* (Fig. 2.8A) also are form genera. To put it another way, a form genus is one that cannot reliably be assigned to a single family. It may be assigned, however, to an order or some higher taxonomic category. Some form genera serve no other purpose than a way of reporting fragmented fossil material. Such documentation is, however, as much a necessity for fossil plant remains as for modern plants.

Although there is no provision in the rules of nomenclature for the use of organ genera, paleobotanists find it a useful category. When the relationships among different types of fossils (roots, stem parts, leaves, and reproductive structures) are well established and can be assigned to the same family, the genera can be called organ genera. This idea can be explained by using certain genera belonging to the pteridosperm family Medullosaceae. As noted in an earlier section on methods of reconstruction, we have accumulated evidence indicating that the stem genus *Medullosa*, the leaf genus *Alethopteris*, the pollen-producing organ *Bernaultia*, and the seed genus *Pachytesta* are genetically related. To express this, all are assigned to the same family, Medullosaceae. Thus, all can be considered organ genera. It has been noted that an organ genus is not fundamentally different from what is generally recognized as a natural genus, the generic category for living plants where the whole life cycle is known, including vegetative, reproductive, and developmental stages.

Let us suppose that we have completed our investigations of the medullosan pteridosperms so that there is no doubt that the various organ genera listed above, plus a few more, give us an exact idea of what the whole organism was like. When this stage of reconstruction is reached, the investigator may, using the rule of priority, select the earliest (after 1820) validly published generic name applied to any one of its parts and use this organ genus as the generic name for the whole organism. At this point one could say that the organ genus is the equivalent of a natural genus. It so happens that the stem genus *Medullosa* was the first of all the related genera to be validly published and is also the logical one to use when describing the whole plant. This is not always the case, however, and the rule of priority can require the use of some seemingly illogical generic names for the whole organism. The case in point is illustrated by the selection of the generic name for another pteridosperm, *Calymmatotheca*. The stem genus for the assemblage of organ genera is *Lyginopteris*, but this name was not published until after valid publication of *Calymmatotheca*. The incongruity is that the name *Calymmatotheca* refers to the cupules borne by the organism, and to many it seems inappropriate for a whole plant. To avoid such incongruities, authors naming a newly reconstructed whole plant often choose a new binomial.

Dissociated fossil specimens that are obviously equivalent as far as their preserved parts are concerned (e.g., the seed genus *Pachytesta*) are usually designated as species of that genus. It follows that if there is reasonable doubt that a fossil seed is indeed a *Pachytesta*, it should be designated by a different generic name. The latter procedure may be used to indicate that different genera, although closely related, evolved slightly different characteristics at different geological times. In some instances, however, the characteristics of the fossil are identical with those of modern plants. For example, it has been shown that Eocene petrified wood of the conifer *Metasequoia* is difficult to distinguish from the wood of modern *Metasequoia*. Some taxonomists insist that in spite of this similarity the fossil genus should have a different name. To indicate this, the generic names of fossil wood types are often given the suffix *oxylon*. *Pinoxylon* and *Taxodioxylon* are examples. Other paleobotanists do not follow this practice. If they find a Cretaceous leaf that has characteristics of a modern *Platanus*, they continue to use the same generic name with a different species designation in the binomial. Obviously there can be confusion, and this results because there are no rules governing these very special problems paleobotanists are confronted with when naming fossils.

Finally, it is important to remember that when a paleobotanist applies a binomial to a detached plant organ or fragment, the existence of a whole plant is implicit in the name he gives (Schopf, 1978). Thus, we can assume that when Sternberg first assigned the name *Lepidodendron obovatum* to a bark fragment in 1820 he was fully aware that this was only a part of a plant species whose other characteristics remained to be discovered.

References

Arnold, C. A. (1947), *An Introduction to Paleobotany*. New York: McGraw-Hill.

Banks, H. P. (1968). The early history of land plants. In *Evolution and Environment*, ed. E. T. Drake. New Haven: Yale University Press.

Beck, C. B. (1960). Connection between *Archaeopteris* and *Callixylon*. *Science*, **131,** 1524–5.

Bold, H. C., Alexopoulos, C. J., & Deleroryas, T. (1980). *Morphology of Plants and Fungi*. New York: Harper & Row.

Crane, P. R. (1988). Major clades and the relationships in the "higher" gymnosperms. In *The Origin and Evolution of Gymnosperms*, ed. C. B. Beck. New York: Columbia University Press.

Cronquist, A. (1960). The divisions and classes of plants. *Botanical Review*, **26,** 425–82.

Delevoryas, T. (1957). Anatomy of *Sigillaria approximata*. *American Journal of Botany*, **44,** 654–60.

Doyle, J. A., & Donoghue, J. J. (1987). The origin of angiosperms: A cladistic approach. In *The Origins of Angiosperms and Their Biological Consequences*, eds. E. M. Friis, W. G. Chaloner & P. R. Crane. Cambridge: Cambridge University Press.

Eames, A. J. (1936). *Morphology of Vascular Plants*. New York: McGraw-Hill.

Eggert, D. A. (1972). Petrified *Stigmaria* of sigillarian origin from North America. *Review of Palaeobotany and Palynology*, **14,** 85–99.

Foster, A. S., & Gifford, E. M. (1974). *Comparative Morphology of Vascular Plants*, 2nd ed. San Francisco: Freeman.

Hennig, W. (1950). *Grundzuge einer Theorie der phylogenetischen Systematik*. Berlin, Deutscher Zentralverlag.

Hill, C. R., & Crane, P. R. (1982). Evolutionary cladistics and the origin of angiosperms. In *Problems of Phylogenetic Reconstruction*, eds., K. A. Joysey & A. E. Friday, Systematics Association Special Vol. 21: Oxford: Oxford University Press.

Jeffrey, A. C. (1917). *The Anatomy of Woody Plants*. Chicago: University of Chicago Press.

Kidston, R., & Lang, W. H. (1917–21). On Old Red Sandstone plants showing structure from the Rhynie chert bed, Aberdeenshire. Parts I–V. *Transactions of the Royal Society of Edinburgh*.

Knoll, A. H., & Rothwell, G. W. (1981). Paleobotany: perspectives in 1980. *Paleobiology*, **7,** 7–35.

Lawrence, G. H. M. (1963). *Taxonomy of Vascular Plants*. New York: Macmillan.

Mayen, S. V. (1987). *Fundamentals of Paleobotany*. London: Chapman and Hall.

Oliver, F. W., & Scott, D. H. (1904). On the structure of the Paleozoic seed *Lagenostoma lomaxi*, with a statement of the evidence upon which it is referred to *Lyginodendron*. *Philosophical Transactions of the Royal Society of London*, **197,** 193–247.

Ramanujam, C. G. K., Rothwell, G. W., & Stewart, W. N. (1974). Probable attachment of the *Dolerotheca* campanulum to a *Myeloxylon–Alethopteris*-type frond. *American Journal of Botany*, **61,** 1056–66.

Raven, P. H., Evert, R. F., & Curtis, H. (1976). *Biology of Plants*, 2nd ed. New York: Worth.

Scagel, R. F., Bandoni, R. J., Rouse, G. E., Schofield, W. B., Stein, J. R., & Taylor, T. M. C. (1967). *An Evolutionary Survey of the Plant Kingdom*. Belmont, Calif.: Wadsworth.

Schopf, J. M. (1978). Unstated requirements in nomenclature for fossils. *Taxon*, **27,** 485–99.

Smith, G. M. (1955). *Cryptogamic Botany*, 2nd ed. New York: McGraw-Hill, Vols. 1 and 2.

Stewart, W. N. (1954). The structure and affinities of *Pachytesta illinoense* comb. nov. *American Journal of Botany*, **41,** 500–8.

Tippo, O. (1942). A classification of the plant kingdom. *Chronica Botanica*, **7,** 203–6.

Wagner, W. H. (1952). The fern genus *Diellia*: Structure, affinities, and taxonomy. *University of California Publication, Botany*, **26,** 683–711.

4

Life in the Precambrian

In his book *Plant Life Through the Ages,* published in 1931, A. C. Seward described the Precambrian "as an age of algae – algae with doubtful credentials." His description reflects the questionable nature of the evidence for life forms in the Precambrian available at that time. In 1954, nearly 30 years later, the landmark publication by Tyler & Barghoorn appeared, and this provided us with the first definitive evidence on life forms in the Precambrian. Here, the authors describe in a general way an assemblage of fossil microorganisms from the carbonaceous algal member of the Gunflint chert. A detailed account of the organisms comprising the assemblage appeared in 1965, this prepared by Barghoorn & Tyler. If any doubts remained in the minds of paleontologists about the presence of microorganisms in the Precambrian, this publication completely converted the skeptics. In the relatively short time since 1965, rapid advances have been made, some showing that early Precambrian rocks contain remains of small, unicellular life forms, and that later in the Precambrian more complex organisms appeared. To accommodate these basic biological discoveries in a geological time frame, an older geologically founded stratigraphic classification was adopted by Precambrian paleontologists. This classification subdivides the Precambrian Era into three eons, the Proterozoic, Archaeozoic or Archean, and Hadean.

In the time interval between the present and the appearance of Barghoorn & Tyler's 1965 publication, those interested in Precambrian paleobiology have discovered that there are three principal sources of evidence for early evolution of life on Earth. These are from stromatolites, microfossils, and geochemical analyses (Knoll, 1989).

Stromatolites and clastics

A fascinating display of present-day stromatolites in various stages of development has been discovered on the mud flats (Fig. 4.1) along the west coast of Australia at Shark Bay. Here there are extensive mats of microorganisms that grow out over the surface of the substrate of the intertidal zone. At least 28 species of algae, mostly cyanobacteria and green algae, are known to comprise these extant mats. Some of the filamentous cyanophytes produce copious quantities of mucilaginous material that give the living mats the consistency of soft rubber. When sectioned, the mats show a laminated structure produced by layers of sand that are incorporated into the growing mats. In addition, many members of the biota, especially cyanophyte components growing at the surface of the mats, produce layers of carbonates so characteristic of stromatolites. The mats vary in morphology depending on their location in the intertidal environment. Some are flat or low-domed where they are protected from wave action. Others growing where waves and currents are more pronounced tend to be columnar or cabbage-shaped.

Observations of Precambrian stromatolites (Fig. 4.2) led C. D. Walcott (1914) to the conclusion that, like modern stromatolites, the Precambrian forms were of algal origin. After microscopic examination of thin sections of Precambrian cherts from stromatolites, J. W. Gruner (1925) claimed to have found filaments of microorganisms that he interpreted to be sheathed bacteria. From these early beginnings, studies have shown that stromatolites occur throughout geological time and have a widespread geographical distribution. The Precambrian stromatolitic reefs of the Big Belt Mountains of Montana provide us with a good example. Here the limestone stromatolites cover an area of more than 23,000 square kilometers. Most stromatolites, especially those occurring in Archaeozoic sediments, are devoid of fossil microorganisms. However, because they are believed by most investigators to have been formed by photoautrophic organisms, they are generally accepted as evidence of early biological activity. It is estimated that of the dozen or so Archaeozoic localities that do show such evidence, only three have supplied material containing possible microfossils. The remaining areas are represented by barren stromatolites (Knoll, 1985a).

Figure 4.1. Present-day stromatolites exposed at low tide, Shark Bay, Western Australia. Formed by green algae and cyanophytes. (Photograph courtesy of H. J. Hofmann.)

Small spheroidal stromatolitic structures called oncolites are known from Proterozoic sediments. Some show evidence of microbiotas that, in the presence of strong currents of water, coated and bound particles of detritus together. These sand grain-sized particles may be silicified.

Stromatolites often display morphological characteristics in their size, shape configuration of the laminae, special relationships, internal microscopic structures, and so on. These and other features have been used to distinguish genera and species of stromatolites such as *Collenia columnaris* and *Cryptozoön proliferum*. This practice has led to many nomenclatural difficulties and has not been universally adopted by biologists. For a comprehensive account, the interested student is referred to the book *Stromatolites* (Walter, 1976).

Clastics

The emphasis placed on stromatolites as a source of Precambrian microbiotas limited the study of the microorganisms that comprise them to a few marine environments. Expanding their search for new biotas, some investigators have turned their attention

Figure 4.2. Early Archean stromatolite showing characteristic laminations, Bulawayan Group, South Africa 3,200 m.y. old. (From Schopf et al., 1971.)

to the study of clastics (shales, siltstones, and mudstones) in the Precambrian. The search has been productive, so that now there are more than 2,800 localities known from the Proterozoic that have yielded microbiotas. These represent a much wider variety of depositional environments than those where only stromatolites were formed.

The antiquity of life

There is general agreement among micropaleontologists that microorganisms existed as early as 3,500 m.y. ago (Knoll, 1983a, 1989) in the early Archean (Archaeozoic). Much of the evidence supporting this great antiquity for life forms comes from studies of the Onverwacht, Bulawayan, and Fig Tree groups of South Africa (Knoll & Barghoorn, 1977; Byerly, Lower, & Walsh, 1986), Warrawoona Group of Western Australia (Schopf & Packer, 1987). Stromatolites occur as components of all of these groups, but none is known to contain microfossils. Nonetheless, the stromatolites are accepted by most investigators as representing trace fossils of microbial assemblages (Knoll, 1989). That this may be the case is supported by the frequent association of stromatolites and carbonaceous cherts containing spheroidal microstructures that have the characteristics of prokaryotic microorganisms (Fig. 4.3A). Although the biogenicity of the spheroids is difficult to establish, paleontologists have formulated a set of criteria, based on morphological characteristics, that makes it possible to evaluate their biogenicity (Knoll, 1989). These criteria are (1) a nearly normal size frequency of spheroids with a mean diameter of 2.5 μm; (2) a clear evidence of binary division (Fig. 4.3A); (3) a sedimentary environment similar to that of more recent known microfossils; (4) and features produced by degradation and diagenesis such as flattening and wrinkling of membranes around carbonaceous residues. Cherts that contain spheroids having the characteristics of microorganisms listed above occur in the Onverwacht and Fig Tree groups of South Africa. These early Archean rocks are 3,200–3,500 m.y. old (Barghoorn & Schopf, 1966; Schopf & Barghoorn, 1967; Knoll & Barghoorn, 1977). Because of their small size and binary method of division, these ancient organisms may have been prokaryotic bacteria, or unicellular cyanophytes. It is impossible, however, to determine whether or not they were heterotrophic or autotrophic. The antiquity of these putative organisms is reflected in their generic names, e.g., *Archaeosphaeroides* and *Eobacterium*.

Less equivocal evidence comes from the Towers Formation and the Apex Basalt of the Warrawoona Group, Western Australia (Schopf & Packer, 1987). The rocks containing microfossils from the Towers Formation (Fig. 4.3B) include colonies of coccoidal cells surrounded by lamellated sheaths enveloping cells that are similar in size to extant, colony-forming

Figure 4.3. Early Archean organisms. **A.** Putative unicellular spheroids similar to *Archaeosphaeroides*, showing a late stage of division in two focal planes. Unlabeled bar scales = 10 μm. **B–C.** Microfossils from the Warrawoona Group, Australia. **B.** Colony of chroococcalean cells with sheaths. Arrow indicates sheath. **C.** Oscillatorealeanlike filament with discoidal medial cells. (From Schopf & Packer, 1987.) (**A–C** courtesy of Dr. J. W. Schopf.)

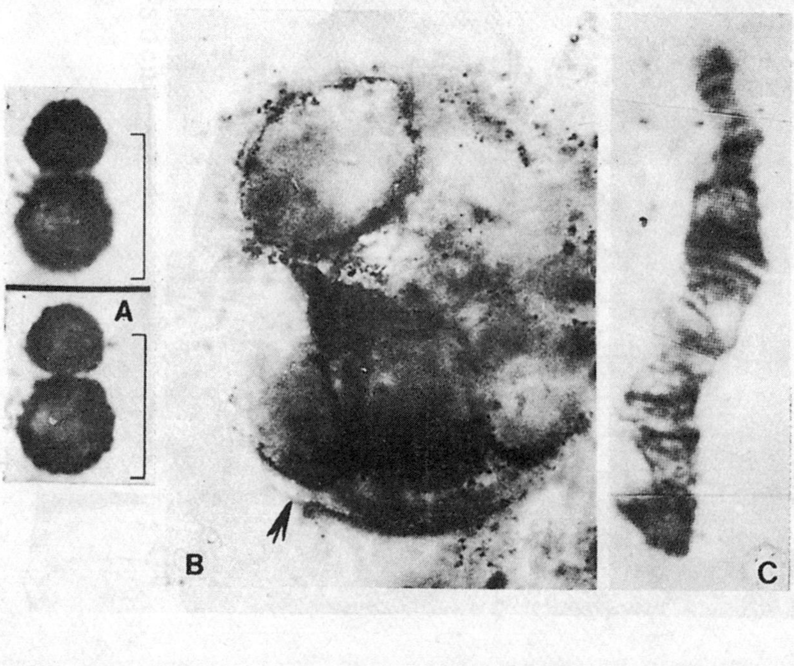

chroococcalean cyanophytes (e.g., *Chroococcus,* Fig. 4.5I). In addition to the spheroids, microbial filaments (Fig. 4.3C) have been found in cherts of the Apex Basalt. The filaments appear to have cross walls and resemble trichomes of fossil and extant prokaryotes. Organisms with similar morphologies are common among the oscillatorians (Fig. 4.5B) as well as the nonoxygen-producing beggiatoaceans and chloroflexaceans. Stromatolites also occur in rocks of the Warrawoona Group, and it is tempting to conclude that they were produced by ancient cyanobacteria and bacteria in much the same way and in a sedimentary environment similar to that found in the hot springs of Yellowstone Park. Here thermophilic bacteria and cyanophytes form mats and deposit layers of carbonates in a way similar to those formed at Shark Bay, Australia.

Organisms of the Proterozoic

The Gunflint chert

The organisms of the Gunflint chert described by Tyler & Barghoorn (1954), and Barghoorn & Tyler (1965) were discovered in a black, carbonaceous chert that came from the Gunflint Formation of western Ontario. Age determinations obtained from adjacent dated Early Proterozoic magnatites indicate an approximate age of 2,000 m.y. In their 1965 article, the authors give a comprehensive account of the assemblage of microorganisms. Some are very different from any previously described microorganisms, while others can be recognized and compared with extant organisms. It seemed surprising to some of us that there were any familiar forms to be found among organisms that existed that long ago. One that is especially striking and common is *Animikiea* (Fig. 4.4A). In size and cell organization, it is nearly identical to the modern *Lyngbya*-type cyanophyte (Fig. 4.5B) even to the presence of a conspicuous sheath enclosing the filament of cells. *Gunflintia* (Fig. 4.4B) also is a common component found in most sections of the chert. Some authors suggest that in their contents and organization these filaments resemble chemosynthetic iron-forming bacteria. Others see a resemblance to the sheaths of degraded filamentous cyanobacteria. Unicells having the shape and size range of extant eubacteria spores are scattered throughout the chert. These have been placed in the genus *Huroniospora* (Fig. 4.4C). *Eoastrion* (Fig. 4.4D) is an infrequent component of the Gunflint chert. The suggestion has been made that *Eoastrion* is related to extant metal-oxidizing, budding bacteria.

Figure 4.4. Suite of organisms from the Proterozoic Gunflint Formation. **A.** *Animkiea.* **B.** *Gunflintia.* **C.** *Huroniospora.* **D.** *Eoastrion.* (Drawn from photographs by Barghoorn & Tyler, 1965.)

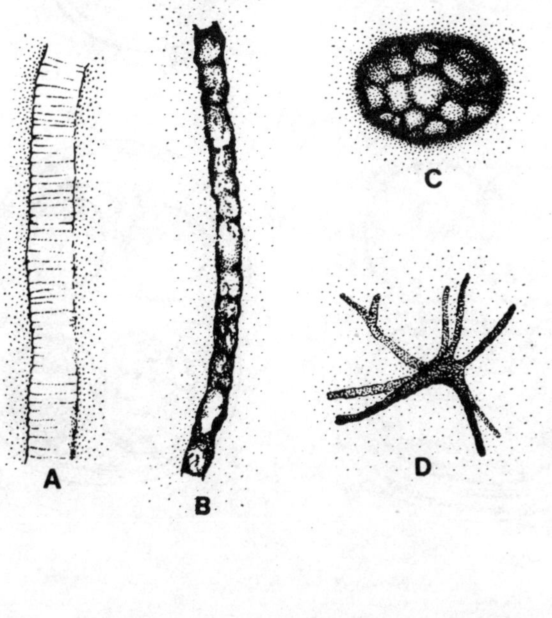

The Duck Creek dolomite

Unlike the Gunflint chert assemblages, which come from the stromatolitic members of the Gunflint Formation, the microorganisms from the Lower Proterozoic Duck Creek dolomite of Western Australia (Knoll, Strother, & Ross, 1988) are found in or associated with silicified carbonaceous clasts and clots, which the authors interpret as representing the remains of "a subtidal, mud-dwelling microbenthic community." The assemblage of microorganisms discovered here is quite similar to that of the Gunflint chert both in age (approximately 2,000 m.y. old) and microbial composition. In addition to *Huroniospora, Gunflintia, Eoastrion,* and *Siphonophycus,* large specimens of *Oscillatoriopsis* were recovered. These spectacular fossils have filaments that may exceed 700 μm in length and 63 μm in width. Until the discovery of these microbial giants, it was thought that the small size of Archean and Lower Proterozoic microorganisms (1–15 μm) was an expression of an early level of evolution. Now it seems that the apparent small size of these ancient microorganisms is a reflection of restricted sampling and not necessarily a limitation controlled by evolution.

Bitter Springs cherts

The Bitter Springs cherts of Central Australia are about 850 m.y. old. This places them in the lower part of Late Proterozoic. The chert is in the form of algal stromatolites that were apparently formed in a lacustrine environment by a diverse assemblage of cyanophytes and other forms that cannot be assigned with certainty. The preservation of the microfossils is so superb that it is possible to relate them to extant oscillatorian and chroococcalean cyanobacteria. The description and nomenclature of the biota comprising the Bitter Springs cherts was initiated by Schopf

(1968) and Schopf & Blacic (1971). It should be noted here that the nomenclature proposed by these authors is followed in the text, even though the interpretations of the morphology and relationships of the organisms may be at variance with what they proposed. Only a few of the approximately 25 genera are considered here.

Of the oscillatorian cyanophytes, the fossil *Palaeolyngbya* (Fig. 4.5A) is very similar to a modern *Lyngbya* (Fig. 4.5B) with its gelatinous tubular sheath around an unbranched trichome composed of disc-shaped cells and hemispherical terminal cells. Other oscillatoria-like fossils (Fig. 4.5D) lack the conspicuous sheath and have trichomes with attenuated apices

Figure 4.5. Suite of extant and Precambrian microorganisms. All highly magnified. **A.** *Palaeolyngbya*. Proterozoic. Compare with extant *Lyngbya* (**B**). **C.** *Microcoleus*, extant cyanophyte. **D.** *Cephalophytarion*. Proterozoic. Compare with *Microcoleus* (C). **E.** *Glenobotrydion*. Proterozoic. Cell containing putative nuclear structure. **F.** *Gloeodiniopsis*. Proterozoic. Note lamellated sheath similar to that of extant *Chroococcus* (I) and *Gloeocapsa* (J). **G.** *Eozygion*, two-celled colony with lamellated sheath. Proterozoic. **H.** *Eozygion*, dividing cell. Proterozoic. **I.** *Chroococcus*. Extant colonial cyanophyte. **J.** *Gloeocapsa*. Extant colonial cyanophyte. (All Proterozoic genera are from the Bitter Springs chert and are drawn from photographs in Schopf, 1968, and Schopf & Blacic, 1971.)

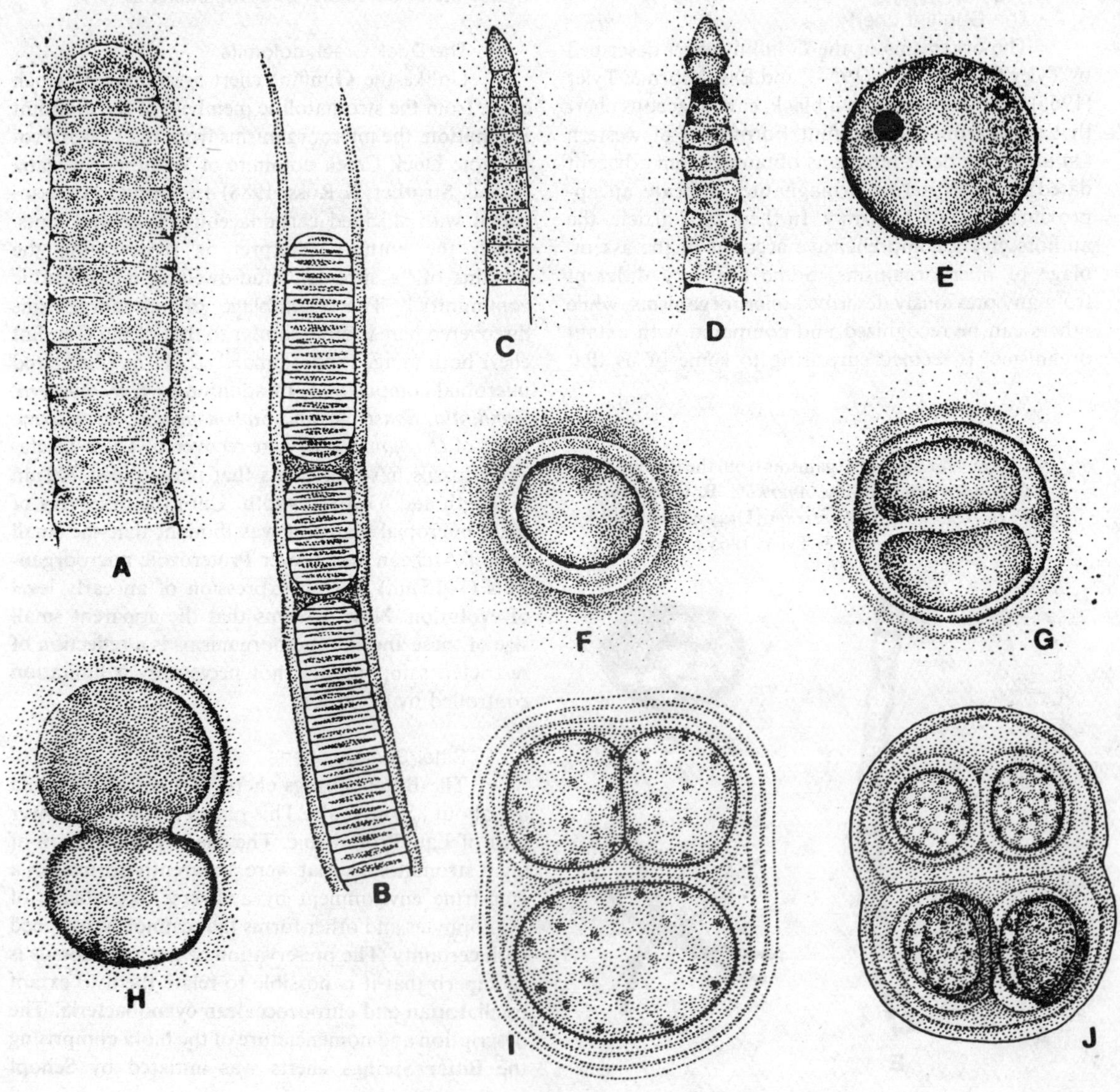

as in *Cephalophytarion*. The extant mat-building blue-green alga *Microcoleus* (Fig. 4.5C) compares, favorably with *Cephalophytarion*. Knoll (1981, 1985b) points out, however, that there are more than a dozen morphotypes of fossil blue-green algal trichomes, many of which may represent stages in degradation and diagenesis of one or two taxa. For example, the attenuated apices of *Cephalophytarion* may represent a morphotype produced in this way. Thus, the comparison with *Microcoleus* may not be valid.

Figure 4.6. Evidences of cell division in chroococcalean cyanobacteria. Proterozoic. **A.** Daughter cells of *Caryosphaeroides*, each containing a nucleuslike structure. (From Schopf, 1968.) **B.** *Eozygion*, completion of cell division. **C.** *Gloeodiniopsis*, two-celled colony. **A–C** from Bitter Springs. (Photographs for **B & C** from Knoll, 1985b.)

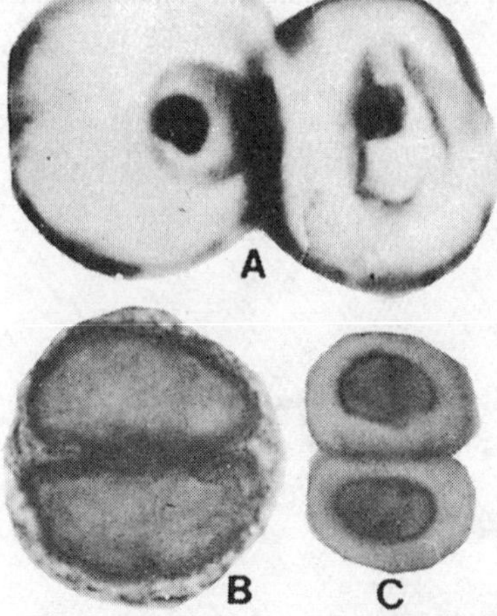

It is generally agreed that the filamentous genera of cyanophytes represent the mat builders that produced stromatolites. The chroococcalean components of the Bitter Springs cherts were probably part of the phytoplankton in the water where the stromatolites were forming. These coccoid forms, which include *Caryosphaeroides* (Fig. 4.6A), *Glenobotrydion* (Figs. 4.5E; 4.7A), *Eozygion* (Figs. 4.5G,H; 4.6B), and *Gloeodiniopsis* (Figs. 4.5F; 4.6C), are very similar, with the possible exception of a "spot" in each cell of species of *Glenobotrydion* and *Caryosphaeroides* (Figs. 4.5E; 4.6A). The spots have been variously interpreted as eyespots, pyrenoids, starch grains, and a nucleus all of which are structures found in cells of eukaryotic organisms.

The unicellular genera, *Eozygion* and *Gloeodiniopsis*, which lack internal structures, have conspicuous sheaths and are abundant in the chert. Their cells often show stages in division (Figs. 4.5G,H; 4.6B,C) and colony formation (Fig. 4.7B). In their morphology they are highly similar to extant cyanophytes belonging to the genera *Chroococcus* (Fig. 4.5I) and *Gloeocapsa* (Fig. 4.5J).

Nonseptate tubelike structures resembling the hyphae of fungi belonging to the Oomycetes and Zygomycetes have been placed in the genus *Eomycetopsis* (Schopf, 1968). Because cellular remains of trichomes of cyanobacteria have been found in the

Figure 4.7. **A.** Cluster of planktonic unicells of the *Glenobotrydion* type showing "spot" cells. Bitter Springs. **B.** Colonies of pleurocapsalean cyanobacteria. Draken Conglomerate Formation. **A,B:** Proterozoic. (Photographs from Knoll, 1983b, 1985a.)

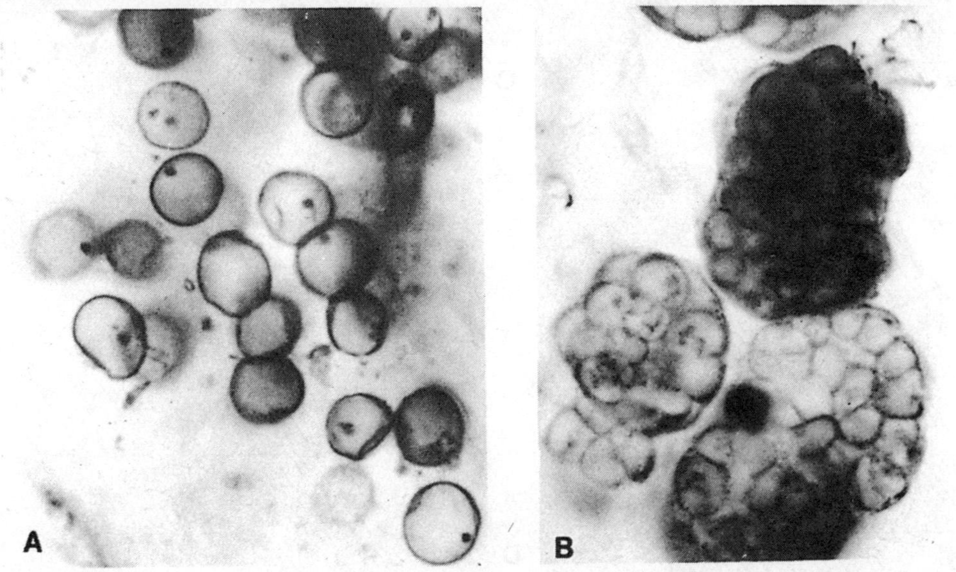

tubes, it is suggested that they represent the empty sheaths of filamentous blue-greens and not fungal hyphae (Knoll, 1981).

The marine shales that are a part of the lower Bitter Springs Formation are now known to contain acritarchs (Fig. 4.8B). These unicellular structures appear to be carbonaceous spheres often with conspicuous ornamentation and are interpreted to be reproductive cysts of planktonic algae (Knoll, 1983a). Their importance will become apparent later in the chapter.

The Svanbergfjellet shale

Of the nearly 200 fossil-producing localities now known to occur in Late Proterozoic rocks, the Svanbergfjellet shales of Spitsbergen Island provided the most taxonomically varied assemblages (Butterfield, Knoll, & Swett, 1988) known for the Late Proterozoic. An age of 700–800 m.y. is indicated for the fossil-containing green-to-black siliciclastic shales.

There are the usual remains of oscillatorian and chroococcalean cyanophytes so common in Bitter Springs cherts and some older fossiliferous horizons

Figure 4.8. Diversity of microorganisms in Late Proterozoic. **A.** Probable heterotrophic bacteria rods. **B.** Large ornamented acritarch within a vesicle. **C.** Multicellular alga similar to extant cladophoralean green algae. **D.** Vesicle similar to germinating zoospore of the extant *Vaucheria*. From Svanbergfjellet Formation. (From Butterfield, Knoll, & Swett, 1988.)

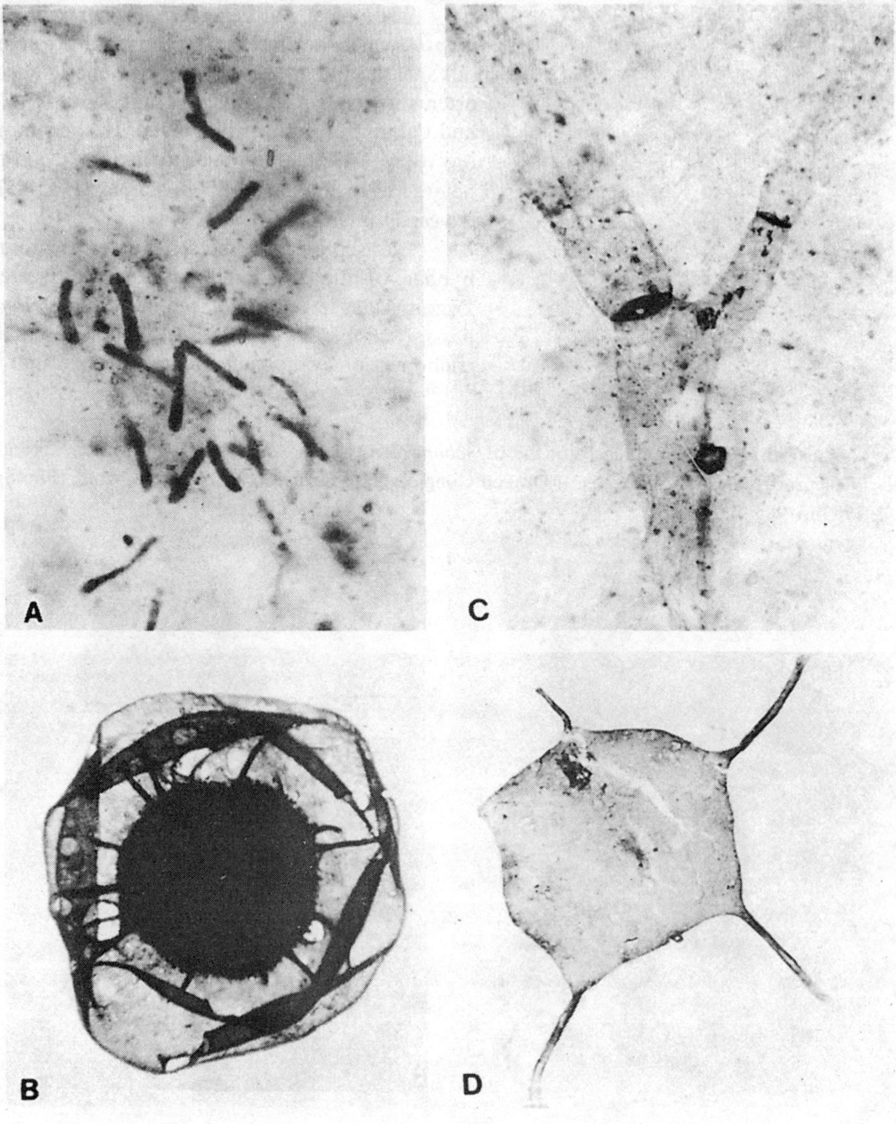

of the Late Proterozoic. Geochemical evidence indicates that heterotrophic bacteria were a part of the microbiotas of Proterozoic communities, but there is very little visible evidence for their presence. In the Svanbergfjellet shale there are dense colonies of small (1–1.5 × 5–19 μm) rods (Fig. 4.8A) that on the bases of shape and size are thought to be bacterial heterotrophs. Other than the prokaryotes, there are several fossils that represent a diverse assemblage of eukaryotes. In addition to acritarchs (Fig. 4.8B), of which eight taxa have been identified, perhaps more unexpected was the discovery of fossils (Fig. 4.8C) that compare favorably with multicellular cladophoralean green algae of which the common extant *Cladophora* is a representative. Heterotrichous members of the Chaetophorales may be represented by complex branching structures that consist of well-developed rhizoids. Other enigmatic fossils are constructed of vesicles bearing two to seven filamentous appendages. It is suggested that these may represent the germinating zoospores of an organism similar to the extant xanthophyte, *Vaucheria* (Fig. 4.8D).

In this survey of microorganisms, from early Archean rocks to those of the Late Proterozoic, it is apparent from what we now know of the fossil record, that a shift has occurred from small, simple, if not primitive, microorganisms of the Archean to biotas consisting of larger prokaryotic cyanobacteria and a large variety of eukaryotic algae.

Major evolutionary events

From the foregoing account of the Archean and Proterozoic microorganisms two major events that had a profound effect on the evolution of their biotas become evident. One is the change in a biotic influence (oxygen production) on the physical environment that ultimately affected the evolution of microbiotas (Tiffney, 1985). The second is the evolution of eukaryotic organisms from prokaryotic ancestors (Knoll, 1983a, 1989). Here, the interactions of biological environments with organisms play an increasingly important role in the subsequent patterns of evolution (Knoll, 1985b).

During the early Archean the synthesis, survival, and reproduction of the first organic molecules depended on the materials and energy obtained from the atmosphere and surface in a reducing environment. Subsequently, communities of heterotrophic microorganisms evolved that were later augmented by autotrophic organisms, probably prokaryotic cyanobacteria that produced free atmospheric O_2 as a function of their photosynthetic activity. It now seems reasonable to suggest that limited O_2 production by (< 1% PAL) prokaryotes had been initiated as early as 3,500 m.y. ago. Geochemical evidence supports the idea that initially the limited supply of free O_2 was used up, for the most part, in the oxidation of organic matter, ferrous iron, and sulfides. Deposits of oxidized sedimentary iron have been found in Archean formations as old as 3,800 m.y., which suggests an even earlier origin for photosynthetic prokaryotes. The highly active and unstable environments of the Archean Eon, which were the products of massive tectonic activity, severely limited the habitats where early autotrophic prokaryotes could grow and reproduce. Not until the Early Proterozoic (2,500 m.y. ago), when tectonic activity had subsided, were stable near-shore environments established permitting a rapid increase of O_2-producing autotrophs. Thus, for the first time O_2 production was able to exceed O_2 consumption. By this time, enough O_2 production had occurred to supply an ozone layer, which acted as an early filter of harmful ultraviolet rays from the sun. Another important consequence of the increase in free O_2 in the atmosphere would be the evolution of aerobic respirers (Knoll, 1983a; Tiffney, 1985), both autotrophs and heterotrophs. This increase in diversity in the Early Proterozoic may have increased the rate of evolution. One result was probably the origin of eukaryotes through endosymbiosis (Knoll, 1983a).

The endosymbiotic theory states that the mitochondria and plastids of eukaryotic cells originated as free-living aerobic bacteria and cyanobacteria, which entered into an endosymbiotic relationship with a host organism. The host organism is thought to have been an anaerobic prokaryote or primitive aerobe. The nucleus is hypothesized to be the result of invaginations of the plasma membrane of the host cell that moved the genetic material to the cell's center, where it was surrounded by an irregular membrane system. A second theory visualizes the origin of plastids, mitochondria, and nuclei to have occurred totally within the evolving cells (Knoll, 1983a). Irrespective of which is the correct theory, the evolution of eukaryotes set the stage for multicellular organisms, mitosis, meiosis, and life cycles with sexual reproduction. One would expect that the resulting increase in diversity would speed up the rate of evolution as organisms adapted to many new environments.

Biochemical analyses

Earlier we gave a list of morphological criteria by which the biogenicity of spheroids could be evaluated. Equally important, however, in determining whether or not putative Archean life forms were actually living is to understand their chemical composition. The biogenicity of the spheroids has been questioned, because it is known that inorganic spheroids having some characteristics of *Archaeosphaeroides* frequently occur in sedimentary rocks. Even though their organic nature has been confirmed, this does not necessarily mean that spheroids once were living organisms. Similar spheroids of organic composition have been produced abiotically in laboratory experiments designed to simulate the events that may have produced similar spheroidal elements in carbonaceous meteorites. What must be done is to distinguish between carbon that has been fixed in the form of CO_2 by once-living photosynthetic systems and carbon of inorganic origin. It now seems that this can be done with carbonaceous sediments as old as 3,500 m.y. When carbon isotopes are identified in the sediments and analyzed, ^{12}C and ^{13}C appear in a ratio comparable to carbonaceous materials of known biological origin, and this is distinct from that of carbonaceous material of nonbiological origin (Schopf, 1975).

One of the problems related to the recovery of organic substances in rocks of any age is the possibility that such soluble compounds as amino acids, hydrocarbons, fatty acids, and many others may have infiltrated the rocks long after lithification. However, the substance kerogen ($^{13}C/^{12}C$), which is used as proof of biogenicity when found in carbonaceous cherts and other rocks containing suspected microorganisms, is an insoluble substance. Since it is insoluble, kerogen is believed to be indigenous to the rock where it is found and could not have permeated the rock matrix at a later time to become a contaminant.

Biochemical analyses of Precambrian carbonaceous cherts and actual plant fossil remains from many different places in the geological column have shown small concentrations of two biologically important hydrocarbons, pristane and phytane. These compounds are known to be the products of degradation of the chlorophyll molecule, the organic compound necessary for photosynthesis. For the reasons that these may be contaminants or produced by other nonphotosynthetic organisms, one cannot accept their presence in carbonaceous cherts as proof of photosynthesis. However, as Schopf (1968) points out, the early appearance of photosynthesis would account for the abundance of carbonaceous cherts in the Precambrian.

Eukaryotes in the Precambrian

As far as can be determined with any degree of certainty, unicellular prokaryotic Monera were the only life forms to inhabit the seas of ancient Earth prior to the late Precambrian, or about 1,300 m.y. ago. About that time, according to Schopf & Blacic (1971) and Schopf & Oehler (1976), unicellular eukaryotic life forms had evolved from autotrophic, prokaryotic ancestors. Because the appearance of the nucleus is such an important event in evolution, its presumed presence in organisms of the Bitter Springs biota has attracted much attention. Some have suggested, however, that the organellelike body (Fig. 4.5E; 4.6A) is a pyrenoid, not a nucleus. A pyrenoid is a protein-containing body around which starch accumulates in the cell. It occurs in cells of many eukaryotic algae, but not in the prokaryotic Cyanophyta. Others have suggested that the nucleus-like structure is nothing more than the degraded protoplast of a fossilized cell that bears a superficial similarity to a nucleus or a pyrenoid (Knoll & Barghoorn, 1975; Knoll & Golubic, 1979). Their evidence is based on observations of moribund, modern prokaryotic algae. After making observations and measurements of more than 1,000 cells containing protoplasmic remains, Schopf also came to the conclusion that Bitter Springs cells do contain remnants of degraded cytoplasm. This material, however, can be distinguished from the organelle-like bodies that number one per cell, and in size and shape are like nuclei or pyrenoids of unicellular green algae. As noted earlier, one should not be surprised if such organelles were preserved in the cells of such exquisitely permineralized fossils.

If we agree that eukaryotic organisms had evolved by the time the Bitter Springs biota was extant some 900 m.y. ago, then it follows that mitosis must have occurred. This would set the stage for life cycles in which sexual reproduction, necessarily accompanied by meiosis, could evolve. To support this possibility, Schopf points to the diverse organisms that appear to be eukaryotic in the Bitter Springs material, followed by an ever-increasing diversity through the late Precambrian into the Cambrian. Specifically, he describes cells that occur in a tetrahedral arrangement (Fig. 4.9A) along with isolated cells that show what appear to be triradiate ornamentations on the surface (Fig. 4.9B). The significance of these finds lies in the fact that the product of meiosis in primitive plants is a tetrad or quartet of cells called meiospores. When these cells mature, they develop a superficial, triradiate ornamentation that becomes evident when the cells of the quartet separate. This evidence by itself cannot be taken as proof that meiosis was occurring some 850 m.y. ago because it is known that certain modern algae can form vegetative colonies and groups of cells

Figure 4.9. Putative evidence for life cycles in eukaryotic algae. **A.** *Eotetrahedrion,* tetrad of cells. **B.** *Eotetrahedrion,* single cell with triradiate configuration. (Drawn from Schopf & Blacic, 1971.)

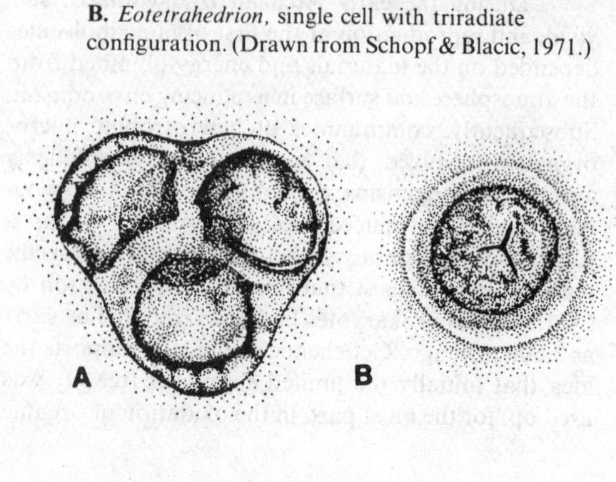

in which the cells have a tetrahedral arrangement. This is true of many colonial members of the Cyanophyta (Fig. 4.5J). After a careful reexamination of cells showing triradiate ornamentations, Knoll & Golubic (1979) have come to the conclusion that these structures are folds produced by the collapsing contents of the cells and that the tetrahedral arrangement of the cells can be equated with colonies as they are formed by the modern as well as fossil cyanophyte genera.

To some micropaleontologists (Knoll, 1983a) the occurrence of acritarchs (Fig. 4.8B) in fossil microbiotas is considered to provide unequivocal evidence of eukaryotes. Most Proterozoic acritarchs are, in their external morphology and large size, similar to their younger Phanerozoic counterparts. These are generally accepted as reproductive cysts of eukaryotic planktonic algae. New discoveries of ornamented acritarchs in Spitsbergen and Arctic Canada extend their fossil record back to at least 1,000 m.y. Other less convincing spheroidal microfossils having the size range of eukaryotic acritarchs occur in the approximately 1,400-m.y.-old Belt Supergroup, Montana. Added to the evidence for eukaryotes in the Precambrian are fossil biotas, described earlier in the chapter, comprised in part of filamentous green and yellow-green algal remains. These occur in Late Proterozoic beds of approximately the same age as the Bitter Springs cherts. When all the morphological and geochemical evidence is considered pro and con, Knoll (1989) believes that eukaryotes can be traced back through time at least 1,500 m.y.

Paleoecology

Once having been established in the Middle Proterozoic, populations of eukaryotic aerobes were added to the preexisting populations of prokaryotic bacteria and cyanophytes. From the Early Proterozoic onward tectonic activity decreased, resulting in stable continental shelf environments into which the rapidly diversifying microbiotas soon radiated. There are many factors that regulated their distribution laterally into the various environmental niches; some were factors of the physical environment, and others were the products of natural selection acting on the ever-increasing variability among the eukaryotes. It is worth noting that the omnipresent prokaryotes in the form of bacteria and cyanophytes had undergone little change in their morphologies since their first appearance in the Archean. This apparent conservatism or stasis may be partially explained as a result of their early adaptation to a variety of environments, which have continued to exist from that time to the present. It is also possible that their lack of life cycles incorporating sexual reproduction severely limited the diversity that could be acted upon by natural selection. As Knoll (1985b) has pointed out, however, our present interpretations of rates of evolution over long periods of geological time are dependent on comparisons of "benchmark" microbiotas that represent vertical distributions in time. These usually represent a limited sample of environments in which the microbiotas grew and may not reflect the true diversity of organisms that were actually present. Thus, the true variety of microorganisms present in a given environment will be revealed only by studies that show their lateral extent within limited units of geological time.

To get some idea of the lateral diversity of Late Proterozoic supratidal, intertidal, subtidal, and planktonic environments and the microbiotas they contained, Knoll (1985b) selected microfossil assemblages from 17 formations of the Late Proterozoic ranging from 900–570 m.y. in age.

Geochemical evidence indicates that the oceans and atmosphere of the Late Proterozoic were similar in chemical composition to what exists today. Thus, it can be assumed that differences imposed by chemical composition of seawater would not be a factor in determining the composition of biotas in different continental-shelf environments. Wave action, changes in tide levels, inshore currents, etc., however, can have a profound effect on the biota of a given environment. These and numerous other physical and biological factors can affect not only the composition of a biota but the degree of preservation during fossilization. The study of these factors is called taphonomy (Gastaldo, 1987) and starts with necrology, which can include the death of an individual or the physiological or traumatic alteration of lost parts. This is followed by biostratinomy where the parts are transported by wind, water, and mass movements to the site of deposition and burial. Further alteration may be activated by diagenesis, which involves chemical reactions in the depositional environment followed by lithification. Taphonomic processes can bring about the complete elimination of some parts of a biota in a given environment. By transportation, microorganisms can be moved (allochthonous) from their original environment to another. In the process the organisms may undergo physical changes, which may lead to alterations of their morphology and biological composition. Burial leading to preservation in the sedimentary environment followed by lithification can produce additional postmortem changes that also can be misinterpreted as the result of an evolutionary process. for this reason, it is very important when interpreting plant fossils to attempt to distinguish, by studying the environment in which the fossils occur, changes that are caused by the environment as opposed to those that are the products of evolution.

Taking the effects of taphonomy into consideration, plus evidence from microfossils, stromatolites, oncolites, geochemical studies, and inferences from extant biotas, Knoll (1985b) produced a model (Chart 4.1) based on 17 Late Paleozoic assemblages which include nonmarine, supratidal-to-intertidal, subtidal benthic, and planktonic environments. What follows is a summary of this model.

Nonmarine environments

The observation that extant cyanophytes occur on rocks and desert soils that receive a minimal amount of water has prompted the suggestion that similar organisms resistant to desiccation could have lived in similar Late Proterozoic environments. Some paleosols obtained from the Proterozoic contain organic matter, which leads to the conclusion that some kind of organisms lived there long before the arrival of land vascular plants. Microbial mats are formed in present-day saline lakes, hot springs, and freshwater lakes that have unusual chemistries. Stromatolites present in nonmarine lacustrine environments have been reported from Archean rocks as old as 2,800 m.y. There is now reason to believe that the Bitter Springs cherts and stromatolites were also formed in an ancient lake environment. This and other evidence based on the fine structure of trace stromatolites leads to the reasonable conclusion that microbial mat-forming communities were abundant on Late Proterozoic terrestrial substrates associated with shallow water, environments where there was little or no current or wave action, and only slow sedimentation.

Supertidal-to-intertidal environments

Supertidal-to-intertidal environments were studied by Knoll (1985b) from a dozen assemblages of Late Proterozoic age (Chart 4.1). A summary of the information obtained shows that the mat communities of the Late Proterozoic localities compare favorably with similar communities inhabiting present-day supratidal and intertidal coastal regions, e.g., Shark Bay of Western Australia. Supratidal communities are derived from the microbiota of the intertidal environment. These may include nonstromatolitic, surface-dwelling cyanophytes and bacterialike organisms that produce a laminated precipitate. In areas protected from wave action, flat laminated mats may occur.

Intertidal benthic environments

In intertidal zones several different mat-building communities may occur, each with a characteristic biota and stromatolitic structure. For example, at the Narssârssuk Formation of Greenland there are a half-dozen different types of flat-laminated and low-domal stromatolites. One distinctive type of stromatolite is pustular in form and is dominated by a colonial, coccoid cyanophyte, *Eoentophysalis*. Intertidal stromatolites having the same structure occur at Shark Bay and are formed by microorganisms having the same characteristics as *Eoentophysalis*. Intertidal mat assemblages from the Ryssö Formation of Svalbard, Norway, are dominated by filamentous sheaths in which a mixture of chroococcalean cyanophytes is incorporated. These are similar to the extant *Chroococcus* (Fig. 4.5I). The mat-building filaments are thought to be oscillatorian cyanophytes similar to the modern mat-forming species of *Phormidium*, *Lyngbya* (Fig. 4.5B), and *Microcoleus* (Fig. 4.5C). It

Chart 4.1. Summary of the environmental distribution of Late Proterozoic microfossils showing the paleoecological distribution of 17 assemblages from selected formations. See text for elaboration. (From Knoll, 1985b.)

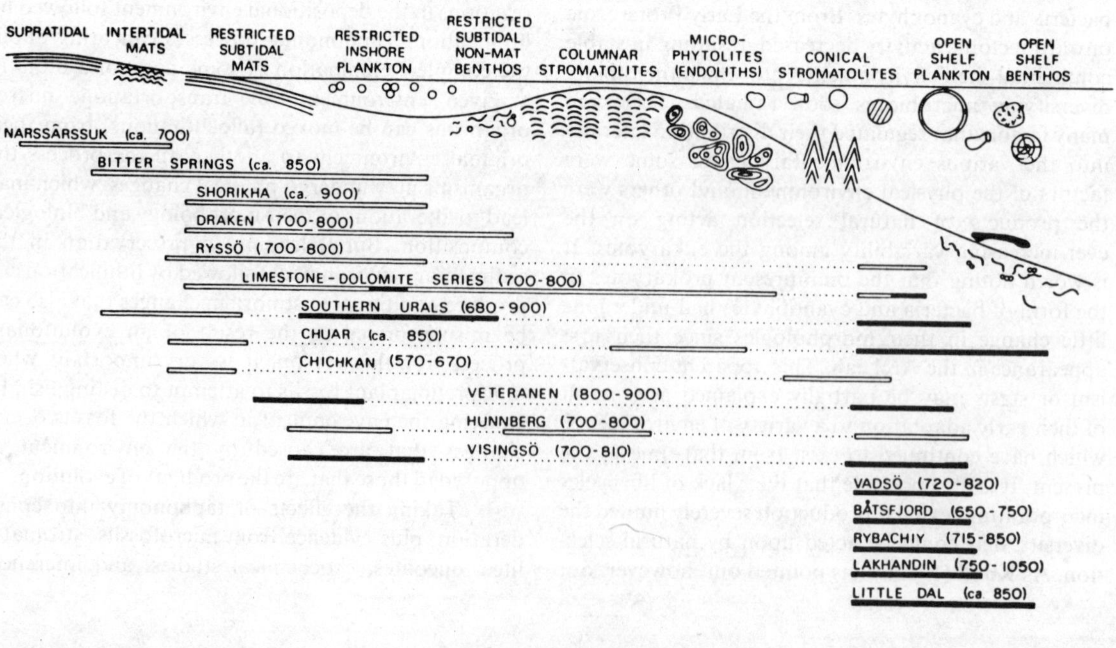

seems reasonable to assume that the distribution of the Late Proterozoic mat builders in the tidal flats was determined by environmental factors such as salinity, substrate composition, and the tidal gradient. There is no question that components of mat builders were redistributed by fragmentation and transportation to different parts of the environment as occurs in modern mat-building communities. The various communities of mat builders are not necessarily distinct entities within a given environment but may overlap as a result of environmental change produced by transgression or regression of the seas. There are many other variables that must be considered when attempting to reconstruct the paleoecology of Proterozoic communities.

In those communities protected from direct wave action, the intertidal mats graded seaward into deeper water where they were totally submerged. In this subtidal environment the mats formed flat, laminated structures (Chart 4.1). Three distinctive mat communities have been recognized in the Bitter Springs Formation and the Old Draken Conglomerate Formation of Svalbard. The organisms primarily responsible for the formation of the mat communities appear to be sheath-forming oscillatorians. Based on their distribution within the rock, some of the fossils are thought to be parts of the mat-dwelling benthos. Included with the mats are unicellular chroococcalean and cyanophycean bacteria. Also occurring as components of the benthos are unicells, either solitary or in aggregates forming large populations in the microbiota. These may occur throughout the mats or in nonmat sediments. The unicellular components of the community were probably a part of the plankton that settled to the bottom where they became a part of the benthic environment. The spheroidal unicells are difficult to classify, but may represent remains of cyanobacteria, planktonic algae, or cysts of heterotrophic protists.

In shallow marine environments on the seaward side of the subtidal benthic zone exposed to strong currents and some wave action, the flat, laminated mats give way to columnar stromatolites (Chart 4.1). Farther seaward and below the level of wave action, the columnar stromatolites give way to coniform stromatolites. Studies of the microstructure of these stromatolites provide evidence suggesting that there were several distinct communities and habitats. There is very limited evidence of microfossils in these stromatolites.

Subtidal benthos containing microorganisms is often preserved in silicified lagoonal muds, chert nodules that form a part of subtidal carbonates, mud stones, and shales that were formed in lagoonal or quiet offshore environments. Sheaths and trichomes of oscillatorian filaments and a variety of small spheroidal and colonial unicells have been removed from carbonaceous shales representing lagoonal or other quiet offshore environments. Shales from the 850 m.y. old Chuar Group of the Grand Canyon, Arizona, have yielded large organic tubes (40–130 μm in diameter), which have been interpreted as filaments of green algae. Other groups of approximately the same age have produced fossils from shales that demonstrate the presence of multicellular eukaryotic algae.

Microphytolites in the form of oncolites and other similar structures provide further evidence for life in the Late Proterozoic. These small grain-sized structures often occur in open-shelf environments exposed to strong wave action. Oncolites contain large populations of sheaths that are morphologically similar to oscillatorian sheaths of flat stromatolites. Colonies of cells that resemble chroococcalean cyanobacteria also occur in the laminae and spaces between oncolites.

Late Proterozoic plankton

Thus far the evidence that cyanobacteria and other prokaryotes dominated the shallow marine environments is unequivocal. However, study of shales and siltstones representing open-shelf environments (Chart 4.1) show that eukaryotic algae and heterotrophs were the primary constituents of these Late Proterozoic planktonic assemblages. Especially conspicuous by their size and number are acritarchs in the form of large (30–70 μm in diameter) vesicles with a variety of surface ornamentations. That many of these are the reproductive cysts of eukaryotic algae seems established. Also present in some assemblages are distinctive vase-shaped microfossils (50–200 + μm in length). Their walls are rigid and composed of organic material. These microfossils are fairly uniform in shape, with a rounded base and apical pore, and in their general morphology they are similar to some simple chitinozoans. The probability is good that they represent heterotrophic protists that lived on other heterotrophic and autotrophic components of the plankton and functioned as heterotrophic "micrograzers" of the Late Proterozoic ecosystems. If these conclusions are correct, then they represent the first evidence of invertebrate metazoans (Knoll, 1985c).

Patterns of evolution

A. H. Knoll (1985a, 1989) has prepared schematic representations of the major patterns of evolution in the Precambrian (Chart 4.2). Most of the patterns have been presented briefly in this chapter with evaluation of the supporting evidence. An attempt has been made to summarize some of the Precambrian environments in which evolution of microbiotas occurred, with a brief consideration of taphonomic factors and their effect on evolutionary interpretations. The effects of limited sampling on phylogenetic interpretations also have been considered. Taking into account these and other factors, we can see (Chart 4.2) that once established early archaeobacteria and eubacteria, representing major groups of anaerobic prokaryotes, soon were widespread. These have survived to the present as a predominant part of Earth's heterotrophic biota. From these ancestral organisms and after the appearance of the first photoautotrophs, free atmospheric O_2 started to accumulate. Subsequently, several lines of photoautotrophic aerobic prokaryotes (cyanophytes) appeared at different times in the geological record, adding to the existing populations of anaerobic prokaryotes. With the passage of time from the late Archean to the Middle Proterozoic, cyanophytes with differing morphologies appeared, most of which have survived to the present as chroococcalean, oscillatorian, and other representatives of the Cyanophyta. Although little can be determined about the rates of evolution of their biological systems, it is apparent that their morphotypes have remained fairly constant through the Precambrian to the present.

The early evidence of unicellular protists, e.g., acritarchs and related structures, first appears at about 1,500 m.y. ago, increasing dramatically in the Phanerozoic Eon. By endosymbiosis it is hypothesized that plastids and mitochondria were introduced into prokaryotic unicells more-or-less concomitantly with the evolution of the nucleus within the cell. In this way the evidence suggests that unicellular, eukaryotic protists evolved. As outlined in classification in Chapter 3, the Kingdom Protista includes unicellular and multicellular, eukaryotic algae. As you will see, we believe that members of the Kingdom Plantae evolved from certain Chlorophyta (grass-green algae), which are photoautotrophic members of the Protista. Evidence for the first appearance of the Chlorophyta is found in the Late Proterozoic, and this "paves the

Chart 4.2. Representation of major patterns of evolution in the Precambrian. See text for clarification. (Modified from Knoll, 1985a.)

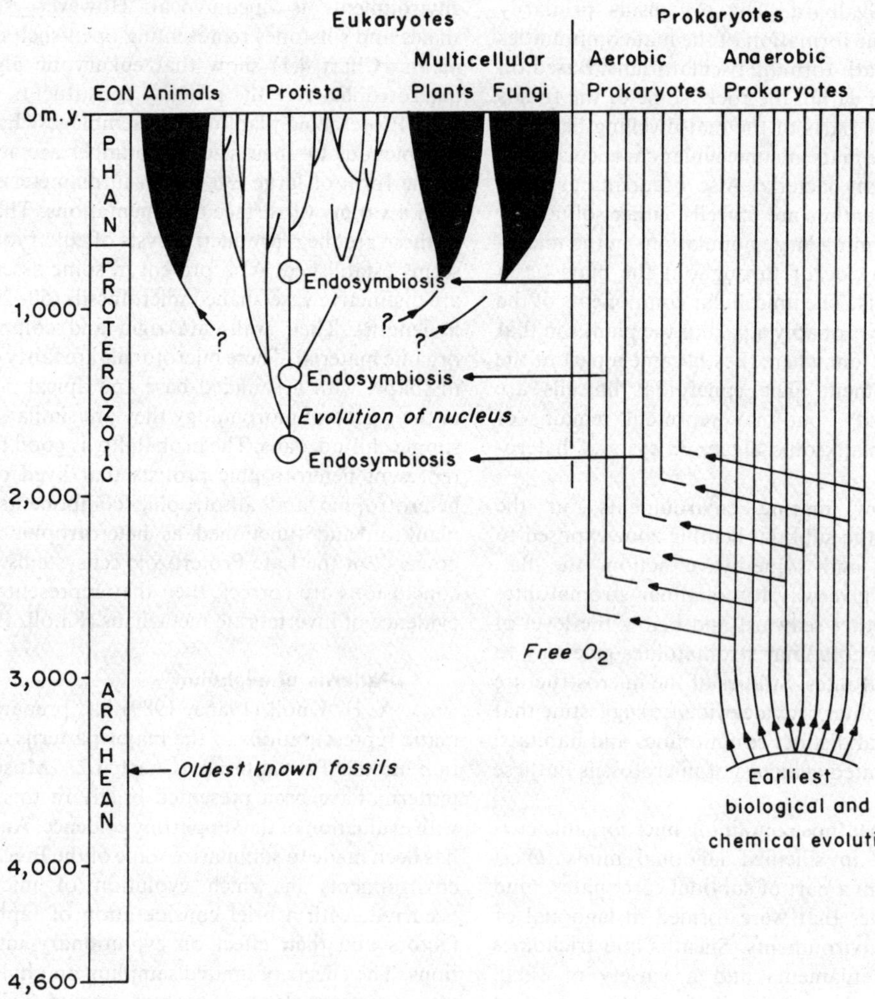

way" for consideration of the algal groups of the Lower Paleozoic.

References

Barghoorn, E. S., & Schopf, J. W. (1966). Microorganisms three billion years old from Precambrian of South Africa. *Science*, **152,** 758–63.

Barghoorn, E. S., & Tyler, S. A. (1965). Microorganisms from the Gunflint chert. *Science*, **147,** 563–77.

Butterfield, N. J., Knoll, A. H., & Stewart, K. (1988). Exceptional preservation of fossils in an Upper Proterozoic shale. *Nature*, **334,** 424–7.

Byerly, G. R., Lower, D. R., & Walsh, M. M. (1986). Stromatolites from the 3,300–3,500 Myr Swaziland Supergruop, Barberton Land, South Africa. *Nature*, **319,** 489–91.

Gastaldo, R. A. (1987). A conspectus of phytotaphonomy. In *Methods and Applications of Plant Paleoecology, Notes for a Short Course*, ed. W. A. DiMichele & S. L. Wing. Paleobotanical Section, Botanical Society of America.

Gruner, J. W. (1925). Discovery of life in the Archean. *Journal of Geology*, **33,** 146–8.

Knoll, A. H. (1981). Paleoecology of Late Precambrian microbial assemblages. In *Paleobotany, Paleoecology, and Evolution*, ed. K. J. Niklas. New York: Praeger.

Knoll, A. H. (1983a). Biological interactions and Precambrian eukaryotes. In *Biotic Interactions in Recent and Fossil Benthic Communities*, eds. M. J. S. Tevesz & P. L. McCall. New York: Plenum.

Knoll, A. H. (1983b). African and Precambrian biological evolution. *Bothalia*, **14,** 329–36.

Knoll, A. H. (1985a). Patterns of evolution in Archean and Proterozoic Eons. *Paleobiology*, **11,** 53–64.

Knoll, A. H. (1985b). The distribution and evolution of microbial life in the Late Proterozoic Era. *Annual Review of Microbiology*, **39,** 391–417.

Knoll, A. H. (1985c). Exceptional preservation of silicified microfossils from 700–800 m.y. old Draken Conglomerate Formation, Spitsbergen. *Philosophical Transactions of the Royal Society of London, B*, **311,** 111–22.

Knoll, A. H. (1989). Precambrian evolution of prokaryotes and protists. In *Paleobiology, a synthesis*, eds. Briggs, D. E. G. & P. T. Crowther. Oxford: Blackwell Sci. Pub.

Knoll, A. H. & E. S. Barghoorn (1975). Precambrian eukaryotic organisms: A reassessment of the evidence. *Science*, **190,** 52–54.

Knoll, A. H., & Barghoorn, E. S. (1977). Archean microfossils showing cell division from the Swaziland System of South Africa. *Sciences*, **168,** 369–9.

Knoll, A. H., & Golubic, S. (1979). Anatomy and taphonomy of a Precambrian algal stromatolite. *Precambrian Research*, **10,** 115–51.

Knoll, A. H., Strother, P. K., & Ross, S. (1988). Distribution and diagenesis of microfossils from the Lower Proterozoic Duck Creek dolomite, Western Australia. *Precambrian Research*, **38,** 257–79.

Schopf, J. W. (1967). *McGraw-Hill Yearbook: Science and Technology*. New York: McGraw-Hill.

Schopf, J. W. (1968). Microflora or the Bitter Springs Formation, late Precambrian, central Australia. *Journal of Paleontology*, **42,** 651–88.

Schopf, J. W. (1975). The age of microscopic life. *Endeavour*, **34,** 51–8.

Schopf, J. W., & Barghoorn, E. S. (1967). Algal-like fossils from the early Precambrian of South Africa, *Science*, **156,** 508–12.

Schopf, J. W., & Blacic, J. M. (1971). New microorganisms from the Bitter Springs Formation (late Precambrian) of north-central Amadeus Basin, Australia. *Journal of Paleontology*, **45,** 925–60.

Schopf, J. W., & Oehler, D. Z. (1976). How old are the eukaryotes? *Science*, **193,** 47–49.

Schopf, J. W., Oehler, D. Z., Horodyski, R. J., & Kvenvolden, K. A. (1971). Biogenicity and significance of the oldest known stromatolites. *Journal of Paleontology*, **45,** 477–83.

Schopf, J. W., & Packer, B. M. (1987). Early Archean (3.3 billion to 3.5 billion-year-old) microfossils from the Warrawoona Group, Australia. *Science*, **237,** 70–3.

Seward, A. C. (1931). *Plant Life Through the Ages*. Cambridge: Cambridge University Press.

Tiffney, B. N. (1985). Geological factors and the evolution of plants. In *Geological Factors and the Evolution of Plants*, ed. B. H. Tiffney. New Haven: Yale University Press.

Tyler, S. A., & Barghoorn, E. S. (1954). Occurrence of structurally preserved plants in Precambrian rocks of the Canadian shield. *Science*, **119,** 606–8.

Walcott, C. D. (1914). Precambrian Algonkian algal flora. *Smithsonian Miscellaneous Collections*, **64,** 77–156.

Walter, M. R. (ed.) (1976). *Stromatolites*. Amsterdam: Elsevier.

5

Diversification of the fungi

As more information has accumulated about organisms traditionally placed in the Kingdom Fungi, it has become clear to most mycologists that three divisions of these heterotrophs can be recognized. They are Gymnomycota (slime molds), Mastigomycota (flagellate fungi), and the Amastigomycota (nonflagellated fungi). The classes Oomycetes and Chytridomycetes are placed with the Mastigomycota; the Zygomycetes, Ascomycetes, and Basidiomycetes are classes of the Amastigomycota (Bold, Alexopoulos, & Delevoryas, 1980).

Recent counts (Graham, 1962; Tiffney & Barghoorn, 1974) place the number of species of fossil fungi at approximately 500, these distributed among 250 genera. The vast majority of the genera come from Cretaceous and Tertiary rocks. Many are exactly like modern fungi even with respect to their life cycles (Dilcher, 1965). For an updated and more complete account of Mesozoic and Cenozoic fungal diversity see Stubblefield & Taylor (1988).

Earliest evidence of fungi

Chart 4.2 shows us that students of Precambrian life forms believe that fungi comprised a part of the Late Proterozoic biota (Knoll, 1985). In their accounts of the microorganisms of the 2,000-m.y.-old Gunflint chert, Tyler & Barghoorn (1954) and Barghoorn & Tyler (1965) described filamentous fossils lacking cross walls, which they suggested might be hyphae of Oomycetes. The possibility remains, however, that these are the remains of empty sheaths belonging to cyanobacteria as seems to have been the case for the filamentous remains found by Schopf (1968) in the 800-m.y.-old Bitter Springs chert. An ascuslike microfossil (Fig. 5.1A) was described by Schopf & Barghoorn (1969) from the Skillogalee Dolomite of South Australia. This 900- to 1,050-m.y.-old structure has since been reinterpreted as the sporangium of an Oomycete (Sherwood-Pike & Gray, 1985). It is clear from the above that the Precambrian record of the fungi is a poor one. This can be accounted for because of the lack of good morphological features that can be used to distinguish members of the Kingdom Fungi from other components of these ancient biotas. It seems reasonable to assume, however, that saprophytic decomposers and parasites similar to aquatic Oomycetes were present by the end of the Proterozoic Eon. During the subsequent Cambrian and Ordovician of the Lower Paleozoic the fossil record of fungi is wanting, and it is not until the late Silurian Period that microfossils have been found with morphologies similar to those of certain Ascomycetes (Sherwood-Pike & Gray, 1985). Assuming that the interpretations of these authors are correct, then there is a correlation between the time when early land plants first appeared and the appearance of terrestrial fungi, such as the Ascomycetes, in the Silurian. It has been suggested that Ascomycetes and Basidiomycetes having their beginnings in the aquatic environments of the Cambrian and Ordovician seas became saprophytes and parasites of the algae that occurred there. As land plants evolved from green algal ancestors during the Silurian, heterotrophic fungi living on the algae continued as saprophytes and parasites of the evolving land plants. In this way one can explain how aquatic fungi were able to achieve the terrestrial environment as "passengers of autotrophic vehicles." If this is the way it happened, then one can conclude that the host-parasite relationship and saprophytism were well established on land plants by late Silurian times.

The idea of a mutualistic relationship between fungi and evolving land plants has been developed by Pirozynski & Malloch (1975). They suggest that an endomycorrhizal relationship existed between Zygomycetes and vascular plants and that this symbiotic association was essential to the evolution of terrestrial floras. In their scenario, the authors suggest that the fungal partner increased the water- and mineral-absorbing capacity through its hyphal system in a poor terrestrial environment, while the plant provided food in the form carbohydrated and protection from excessive drying and ultraviolet radiation. Evidence for endomycorrhizal associations in the fossil record is given later in this chapter.

More recent fungi

The fossil evidence for fungi prior to the Upper Paleozoic is fragmentary at best, and much of it is equivocal. By the end of the Carboniferous, however, there is well-founded evidence for the presence of all the major classes of the Kingdom Fungi. These include Chytridomycetes, Oomycetes, Zygomycetes, Ascomycetes, and Basidiomycetes. From time to time all of us who work with Carboniferous coal balls have encountered poorly preserved plant parts that appear to have decayed. In the broken cells or in areas where the tissue has distintegrated we often encounter clusters of well-preserved bodies resembling spores. Because they are never associated with hyphae or sporangia, Baxter (1975) suggests that these may be chytrids (Fig. 5.2A). More convincing chytridlike cells have been found in tissues of seeds, megaspores, and pollen grains (Millay & Taylor, 1978). The evidence that these bodies are chytrid "swarmsporangia" comes from the many included protoplasts, which are interpreted as zoospores and an exit pore in the wall of the sporangium. More recently Illman (1984) has provided a convincing description of zoosporangia within the spores of *Horneophyton*. The latter is a vascular plant occurring in the Rhynie chert beds of Scotland. These beds are close to the Siegenian/Emsian boundary of the Lower Devonian. According to the author, the fossils are quite similar to extant hyphochytrids, which belong to a small group of marine fungi placed with the Chytridomycetes.

Figure 5.1. **A.** Oomycetelike structure from Skillogalee Dolomite. Precambrian. (Drawn from Schopf & Barghoorn, 1969.) **B.** *Palaeomyces* branching, nonseptate hyphae terminated by enlarged vesicles. **C.** Thick-walled resting spore of *Palaeomyces* terminating a hypha. **B,C:** Devonian. (Drawn from Kidston & Lang, 1921.) **D.** Fungal oogonium containing an oosphere (o). Cell at (a) a possible antheridium. Pennsylvanian. (Drawn from Stidd & Cosentino, 1975.) **E.** *Palaeancistrus,* evidence for Basidiomycetes with clamp connections in the Pennsylvanian. (Drawn from Dennis, 1970.)

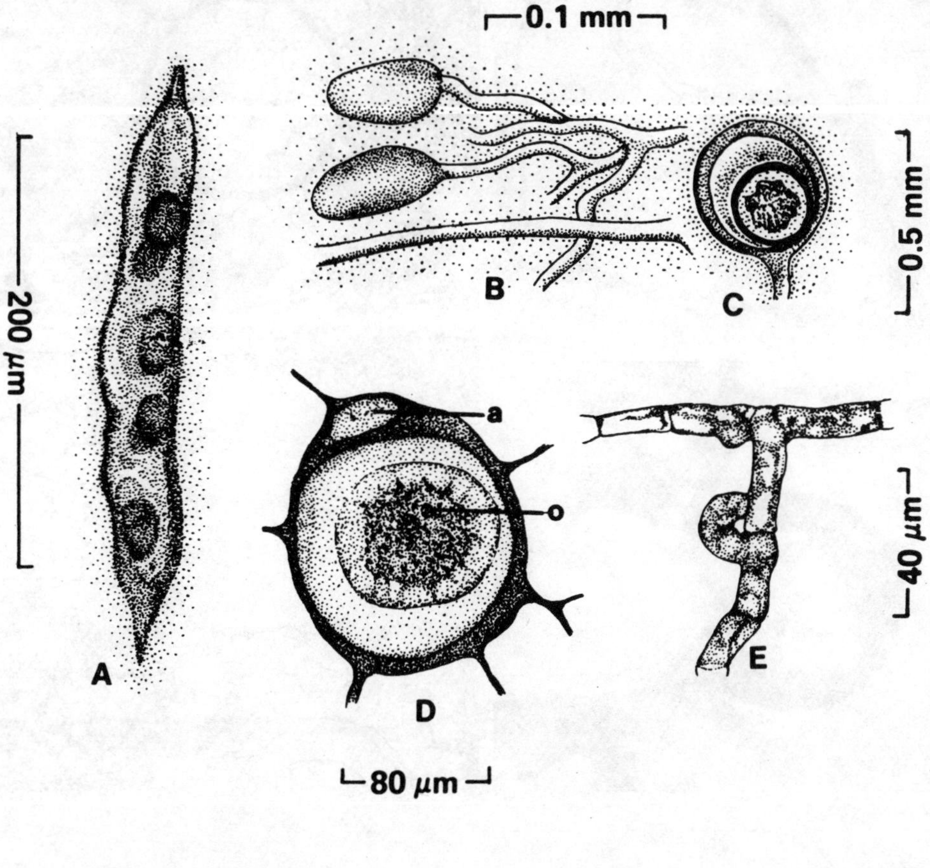

Mycelia (a collective term for many hyphae) composed of nonseptate hyphae belonging to Oomycetes have been found in the Lower Devonian Rhynie chert (Kidston & Lang, 1921). These beautifully preserved hyphae (Fig. 5.1B) often bear reproductive structures, resting spores or chladospores characteristic of Oomycetes (Fig. 5.1C). Kidston & Lang assigned these fungi to the noncommittal genus *Palaeomyces*. Pennsylvanian (Upper Carboniferous) coal balls have provided several specimens of branching, nonseptate hyphae (Fig. 5.2B). The possibility that these are hyphae of Oomycetes is suggested by

Figure 5.2. Some Pennsylvanian age fungi. **A.** Chytrid-like cell with exit pore (p). **B.** Branching, nonseptate hyphae in wood cell of a calamite. **C.** *Protoascon,* the probable fruiting structure of an oomycete. **D–F:** Cleistothecia of Ascomycetes. **D.** *Dubiocarpon* containing asci. **E.** *Traquairia* containing asci. **F.** *Palaesclerotium*. (**A–C, E** from Baxter, 1975; D,E from Stubblefield & Taylor, 1988; F from Rothwell, 1972.)

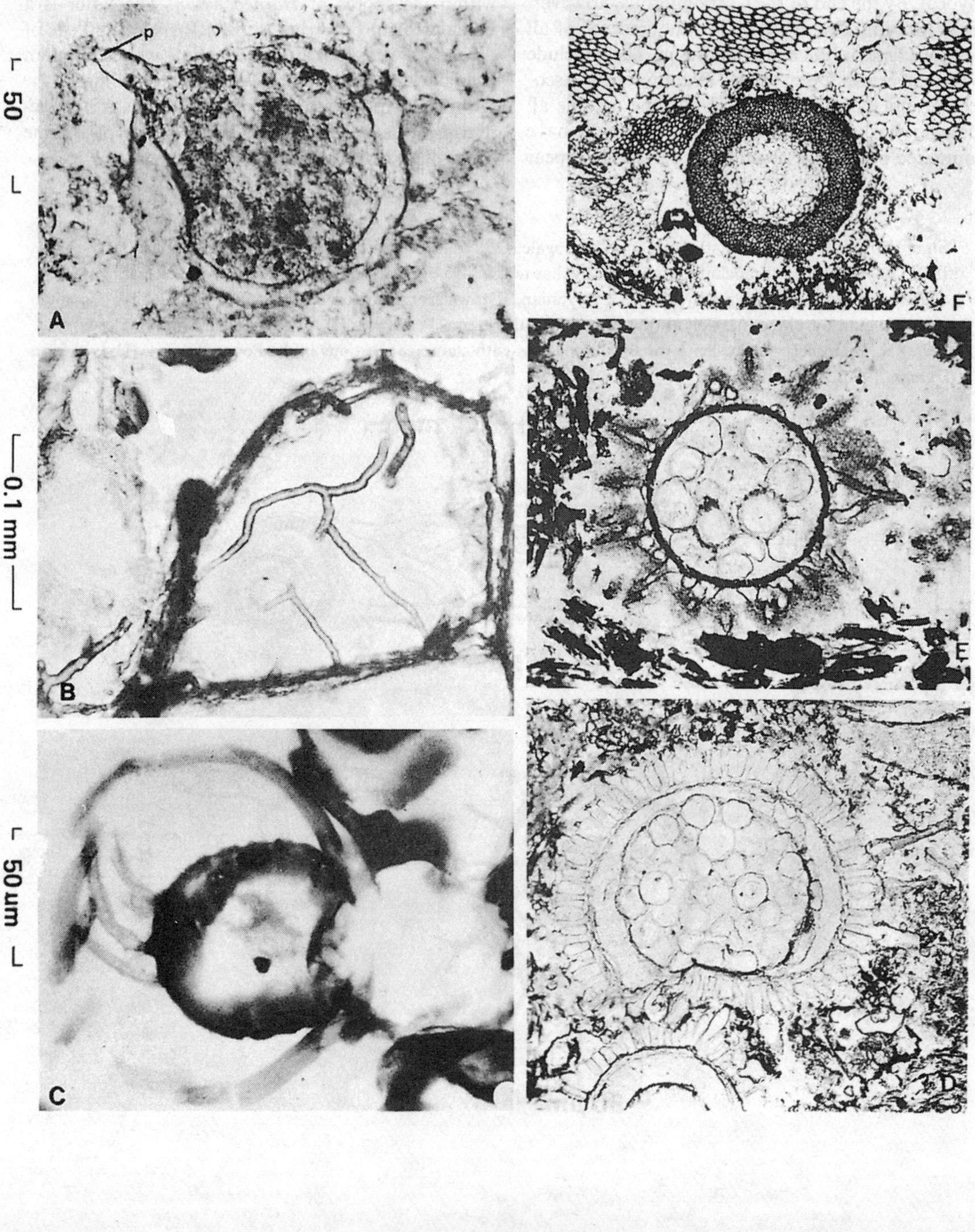

Stidd & Cosentino (1975), who discovered oogonia (female cells with eggs) and antheridia (male gamete-producing cells) in the cells of an ovule, *Nucellangium* (Fig. 5.1D). These fungal reproductive structures are similar to the extant parasite *Albugo* of the Oomycetes. Another structure (Fig. 5.2C), probably belonging to the Oomycetes, shows what appears to be a terminal oogonium with attached antheridial appendages. Originally this was thought to be the fruiting structure of an ascomycete, and thus the generic name *Protoascon*.

Well-preserved endogonaceous fungi (those living within host tissues) have been described by Wagner & Taylor (1981) in cortical tissues of roots and other underground organs of Pennsylvanian age plants. The fossils consist of chlamydospores with a multilayered wall attached to a hyphal stalk (Fig. 5.1C). These special spores are highly similar to those of the extant fungus, *Glomus*. The latter is a member of the Zygomycetes and is of particular interest because it produces vesicular-arbuscular endomycorrhizae (VAM) in the host cells. These are branched, bushlike structures that apparently function in the absorption of nutrients. Although VAMs were not found associated with the Pennsylvanian chlamydospores, beautifully preserved specimens from the Triassic of Antarctica were discovered (Stubblefield, Taylor, & Trappe, 1987) in host cells of roots that may have belonged to cycads (Fig. 5.5A). Other fossils from the same Triassic beds show terminal chlamydospores (Fig. 5.5B), which are very similar to the modern zygomycete *Sclerocystis* (Stubblefield, Taylor, & Seymour, 1987).

In coal-ball material from the Pennsylvanian, mycelia occur that are composed of branching hyphae with septa (Fig. 5.3). When we see fungal structures of this kind we usually associate them with Ascomycetes or Basidiomycetes. Fruiting structures resembling cleistothecia (small spherical reproductive bodies that produce asci and ascospores) of Ascomycetes (Fig. 5.4) have been found by Davis & Leisman (1962). They describe *Sporocarpon, Dubiocarpon* (Fig. 5.2D), and *Mycocarpon*. Of these *Dubiocarpon* shows appendages and surrounding hyphae that strongly suggest the fruiting structures of an extant powdery mildew (Erisyphales). Another cleistothecium of the genus *Traquaira* (Fig. 5.2E) has recently been reinvestigated and found to contain up to 30 asci. Scanning electron microscope studies of these and other genera, e.g., *Coleocarpon* (Stubblefield et al., 1983) have confirmed that these structures are indeed cleistothecia of Ascomycetes containing ascospores (Fig. 5.3B). *Endochaetophora,* a parathecium-like reproductive structure (an ascocarp with an ostiole), has been found in Triassic age permineralized deposits from Antarctica (White & Taylor, 1988). Even though asci and ascospores were not found and the wall structure of the perithecia differs from that of extant ascomycetes, the authors suggest a relationship with that group.

When it was originally described (Rothwell,

Figure 5.3. Pennsylvanian. **A.** Branching septate hyphae in parenchyma cells. **B.** Ascus of *Coleocarpon* with ascospores. (**A** from Stubblefield, 1987; **B** from Stubblefield & Taylor, 1988.)

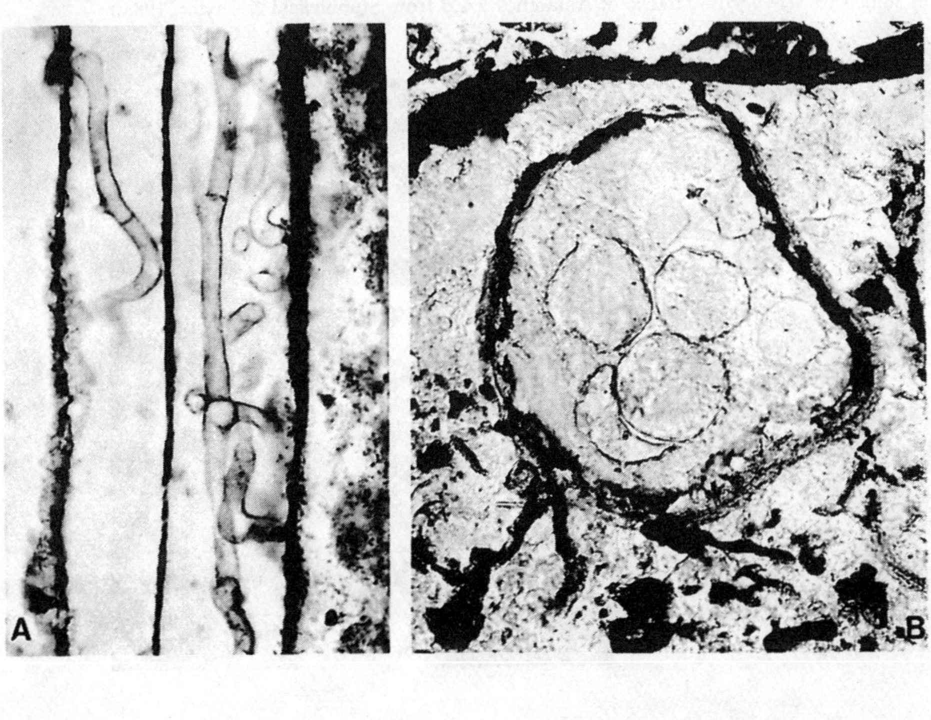

1972), *Palaeosclerotium* (Fig. 5.2F) was believed to be a sclerotium. These tiny spherical bodies are composed of differentiated layers of vegetative hyphae. They seem to function as resting stages in life cycles of several Ascomycetes and Basidiomycetes. A reinvestigation of these Pennsylvanian structures by Dennis (1976) shows very clearly that these were reproductive. They contained asci with four to eight ascospores and are similar in most respects to Ascomycete cleistothecia. Most remarkable are the characteristics of the vegetative hyphae of the cleistothecium wall. The outer layer of a loosely woven mycelium has hyphae with clamp connections (Fig. 5.1E). The inner pseudoparenchymatous mycelium is composed of hyphae that show dolipores. These are complex pores in the cross walls. Both clamp connections and dolipores are found in the dikaryotic hyphae of Basidiomycetes. Assuming that the interpretation of the evidence is correct, Dennis has demonstrated an unexpected and important combination of Ascomycete and Basidiomycete characteristics in *Palaeosclerotium*. Mycologists are generally agreed that Ascomycetes and Basidiomycetes are closely related and that Basidiomycetes evolved form Ascomycetes. As one might anticipate, with such an unexpected discovery, questions have been raised by mycologists (McLaughlin, 1976; Singer, 1977) about the interpretation of the material. They point out that dolipores and clamp connections are not exclusive characteristics of Basidiomycetes and that possibly more than one fungus is present in the speciman. That *Palaeosclerotium* could represent more than one organism is suggested by Dennis (1976), and thus the interpretation that the fossil possesses features of both Ascomycetes and Basidiomycetes should be viewed with caution. Further research is required.

In an earlier investigation Dennis (1970) described and named *Palaeancistrus martinii*, also characterized by septate hyphae and clamp connections. The

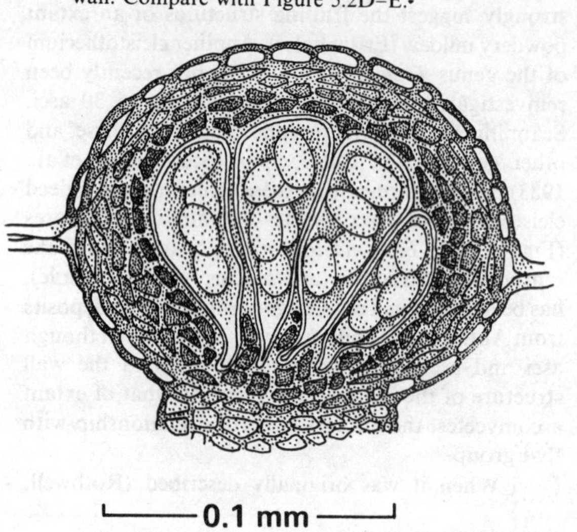

Figure 5.4. Section through the cleistothecium of extant *Erysiphe* showing asci containing ascospores and bases of appendages extending from ascocarp wall. Compare with Figure 5.2D–E.

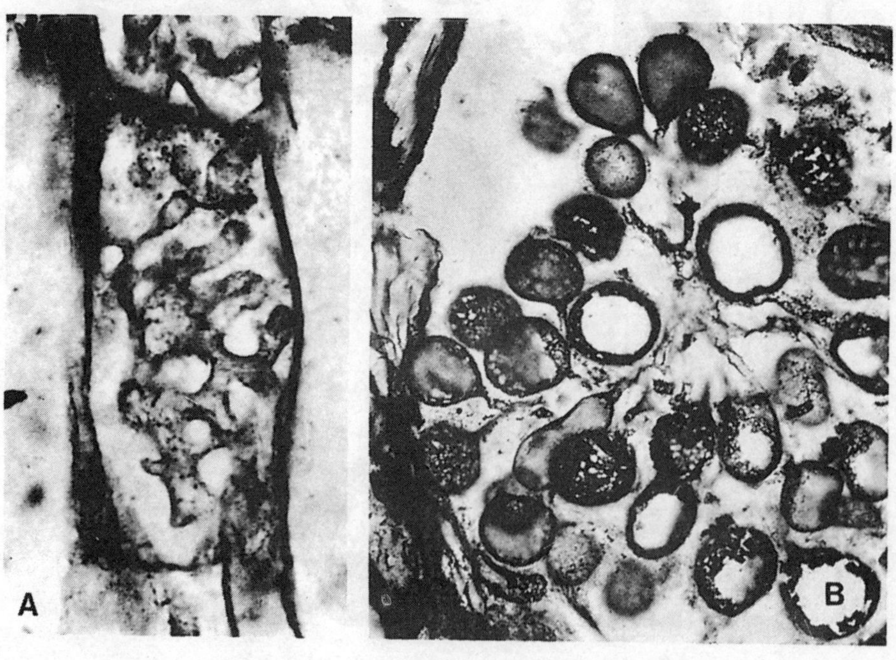

Figure 5.5. **A.** Root cell of Triassic plant containing highly branched arbuscule. **B.** Cluster of chlamydospore-like structures similar to *Sclerocytis*. Triassic of Antarctica. (**A,B** from Stubblefield & Taylor, 1988.)

specimens were found in the wood of a Pennsylvanian vascular plant, *Zygopteris*. This was the first report of a basidiomycete in rocks older than those of the Mesozoic. Clearly it places the origin of this group much earlier in geological time than was anticipated by most mycologists, who believed that Basidiomycetes were much more recent in their origin.

Rare finds of other Basidiomycetes have been reported in the literature from time to time that confirm their presence in the fossil record from the Pennsylvanian to the present. Much of this evidence is in the form of spores that have the characteristics of parasitic Basidiomycetes. Occasionally the fruiting structures of fleshy Basidiomycetes are reported; *Geasterites* from the Tertiary and *Fomes* (Fig. 5.6A), a bracket fungus of probable Pleistocene age (Andrews & Lenz, 1947), are examples.

Fungi as indicators of paleoclimates

Perhaps the most elegant fossil fungal material ever to be described appears in research by Dilcher (1965). While working with compression fossils of Eocene angiosperm leaves, he discovered beautifully preserved epiphyllous fungi (Fig. 5.6B). Many of these show development stages of their vegetative and reproductive structures (Fig. 5.6C). Because their life cycles are known, Dilcher was able to assign many species to families of the Ascomycetes. Another interesting outcome of this study of epiphyllous fungi shows us that of the 150 genera described, the vast majority have appeared since the Cretaceous in association with the evolution of the angiosperm hosts. The genera to which the fungi belong are presently found in humid, tropical, and subtropical environments. This supports the idea that during early Eocene times, the region in Tennessee where the fossils are found was characterized by a similar climate. From this example it is clear that with well-preserved material we can learn much about climates of the past as well as rates of evolution and factors regulating the appearance of major groups of the Kingdom Fungi.

Fungal phylogeny

At best, conclusions about the evolutionary original (s) and relationships of the major groups of fungi are speculative and often based on equivocal interpretations of the fossil evidence. Taking the evidence from studies of extant fungi, one fairly widely accepted hypothesis that is supported by the fossil record suggests that the Mastigomycota (Chytridomycetes and Oomycetes) represent lineages quite independent of the Amastigomycota (Ascomycetes and Basidiomycetes). The Zygomycetes, as indicated by their biochemistry, ultrastructure, and life histories, seem to be closer to the higher fungi (Basidiomycetes and Ascomycetes). Verification of Oomycetes in the Precambrian would fit well with the concept that the members of this class represent primitive fungi that became parasitic on green algae during the Cambrian and Ordovician. These and aquatic Zygomycetes formed a mycorrhizal association with evolving terrestrial plants during the Silurian and Devonian, an association that has prevailed in tracheophytes to the present.

Because of the similarities in their biochemistry, life cycles, and absence of flagellated cells, many mycologists believe that the higher fungi owe their origin(s) to the red algae, which are known to have been a part of the Precambrian, Cambrian, and Ordovician

Figure 5.6. **A.** *Fomes idahoensis*. Tertiary. Upper picture shows underside of bracket with many pores. Lower picture depicts the bracket fungus seen from above. (From Brown, 1940.) **B.** Hyphae of the epiphyllous fungus *Asterina* on leaf of *Sapindus*. **C.** The stroma of the fungus *Callimothallus on Sapindus*. **B,C:** Eocene. (From Dilcher, 1965.)

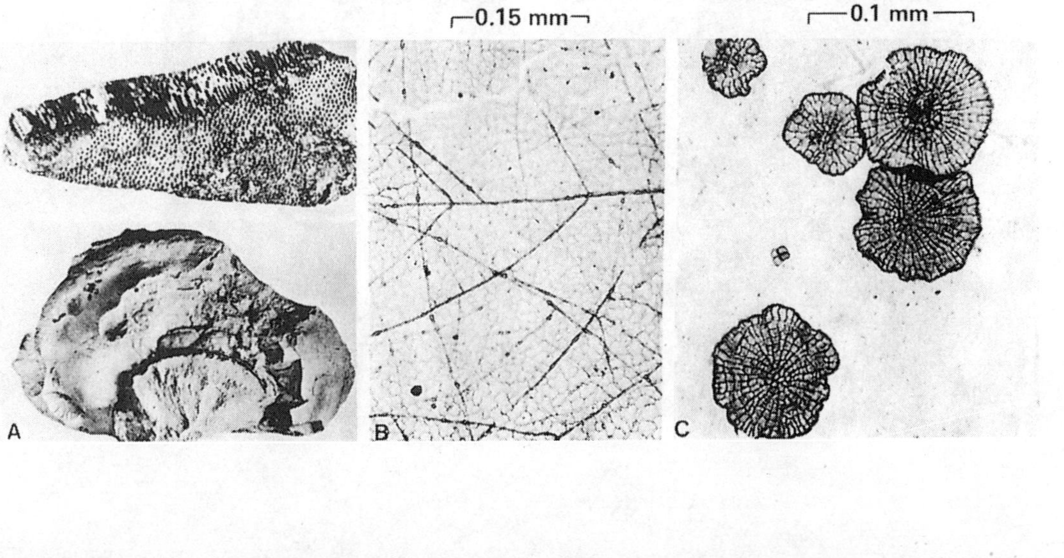

biota. Others argue that the higher fungi have evolved from ancient Zygomycetes which have a first appearance in the Devonian. Based on similarities of their life cycles, apparent homologies of reproductive structures, and vegetative morphologies, it is believed further that Basidiomycetes evolved from Ascomycetes. Until the recent discovery of probable Ascomycetes from the Silurian (Sherwood-Pike, & Gray, 1985), both Ascomycetes and Basidiomycetes were known to occur in the Carboniferous. The new information tends to support the idea that Basidiomycetes could have evolved from Silurian ancestors. One hypothesis accounting for the origin of a terrestrial habitat for Ascomycetes is that they formed a loose lichenlike association with blue-green algae, which would already have been adapted to terrestrial conditions prior to the colonization of land vascular plants. Implied here is an evolutionary sequence starting with a primitive symbiotic lichenlike organism in the terrestrial environment, to a saprophytic Ascomycete decomposing organic remains of early land plants, becoming parasitic on higher plants and finally forming obligate lichens in which Ascomycetes are the major fungal components. This scenario, which fits what fossil evidence we have for the origin of higher fungi, is but one of many that have been developed. See Sherwood-Pike & Gray (1985) for comparisons among the various theories for ancestry of the fungi.

From the Pennsylvanian to the Cretaceous there is little evidence that there was rapid evolution of the fungi. After the appearance and rapid evolution of flowering plants in the Upper Cretaceous and Tertiary, major radiation of the fungi occurred resulting in a fungal biota that is essentially similar to that of the present. There is little doubt that this rapid radiation was the result of coevolution of fungi with the rapidly evolving flowering plants (Pirozynski & Malloch, 1975).

Interactions of fungi

Paleoecological studies of the fungi clearly must be directed to understanding their interactions with the biotic environment provided by plants and animals. For example, we already have been exposed to the concept of mutualism that involves endomycorrhizal symbiosis. Although mutualism of this kind has been postulated as occurring in Devonian land plants of the *Rhynia* type, proving that such an association exists is difficult. There is an obvious lack of information about symbiotic metabolic interactions between fungus and host even though the morphology of nonseptate hyphae in the host cells suggests a mutualistic relationship. In some cases, structures that are reported to be hyphae with coils, vesicles, or arbuscules characteristic of mycorrhiza turn out to be nonfungal in origin. In spite of a much earlier origin for an endomycorrhizal association postulated by Pirozynski & Malloch, the most convincing evidence for such an association comes from the discovery of the Triassic permineralized arbuscules (Fig. 5.5A) (Stubblefield, Taylor, & Trappe, 1987).

Little attention has been given to the study

Figure 5.7. **A.** *Araucarioxylon*, transverse section of woody stem with many pockets of decay. Triassic. (From Stubblefield & Taylor, 1986, with permission.) **B.** Fungal hyphae in tracheids of *Callixylon*. (From Stubblefield et al., 1985, with permission.)

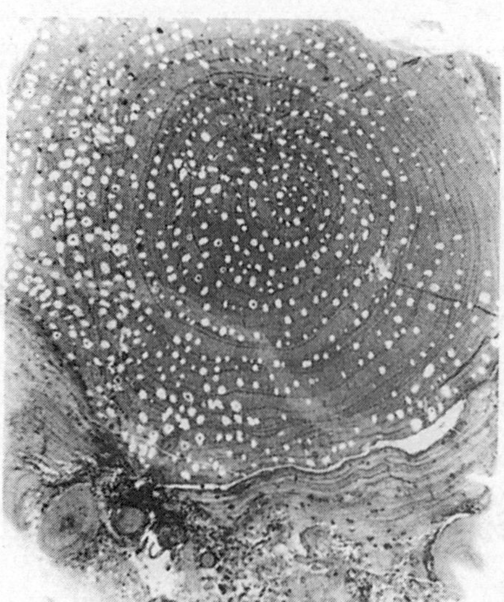

A

B

of saprophytic fungi in the fossil record, although it is generally accepted that these saprophytes played the same role in decomposition in the past as they do today. Attention to the role of saprophytic fungi of the past has been centered on two studies. Pockets of decay-producing fungi were found in the secondary wood of *Callixylon newberryi* from the Upper Devonian (Stubblefield, Taylor, & Beck, 1985). The permineralized specimens showed branched septate hyphae throughout the tracheids of the secondary xylem (Fig. 5.7B). Here walls of the tracheids affected by the fungal hyphae showed grooves, oval cavities, and general deterioration similar to that found in white rots of present-day woody plants caused by basidiomycetes. The branched, septate hyphae from the Devonian fossil could belong to a basidiomycete or ascomycete. A second example of fossil wood-decaying fungi has been found in permineralized specimens of a conifer wood, *Araucarioxylon* (Fig. 5.7A) from Permian-Triassic age rocks of Antarctica (Stubblefield & Taylor, 1986). Again symptoms of decay were found in the tracheids where more observations of the stages of disintegration of their walls were possible. The branched septate hyphae of the fungus displayed clamp connections of the typical basidiomycete found in white rot decay.

Trying to identify parasitic fungi in the fossil record is highly problematical, especially in the absence of symptoms being displayed by the host plant. As has been pointed out, the fossil fungal hyphae seen in plant tissues could represent those of a saprophyte or those involved in a symbiotic relationship. Only a few examples of possible parasitism are known from the fossil record. One is found in a gymnospermous cone *Lasiostrobous* from the Pennsylvanian (Stubblefield et al., 1984). In addition to the fungal hyphae, there are structures produced that are interpreted to be wall appositions and abundant resinous material forming vesicles. In extant plants these features are known to be responses to fungal parasites. Because of their similarity to the extant Oomycete *Albugo*, the fungal oogonium and antheridium described by Stidd & Cosentino (1975) (Fig. 5.2A) from a Middle Pennsylvanian ovule are throught to represent fossil parasites. Epiphyllous fungi abundant in the Tertiary (Dilcher, 1965) represent another parasitic association common to extant plants.

In addition to these studies that emphasize the ecological role that ancient fungi played in their interactions with evolving plants, other areas of research have been initiated that concentrate on their interactions with animals and the abiotic environment. As Stubblefield & Taylor (1988) emphasize, these new areas of research could be very productive in increasing our understanding of paleoecological mycology.

Summary

From the evidence provided by morphological paleomycological studies it is becoming increasingly clear that every class of the fungi (Chitridomycetes, Oomycetes, Zygomycetes, Ascomycetes, and Basidiomycetes) was present by the close of the Carboniferous. Further, the morphologies of the genera existing earlier in geological time are very similar to those of fungi appearing in the fossil record at least until the Cretaceous or Tertiary. Then, in conjunction with the evolution of angiosperms, the fungi appear to have evolved and radiated into new environments pioneered by the flowering plants. Not only does it appear, based on morphological characteristics, that the fungi went through a period of relative stasis prior to the Tertiary, it also seems that their biochemical activities as evidenced by their interactions as parasites are comparable with extant fungi.

References

Andrews, H. N., & Lenz, L. W. (1947). Fossil polypores from Idaho. *Annals of the Missouri Botanical Garden*, **34**, 113–14.

Barghoorn, E. S., & Tyler, S. A. (1965). Microorganisms from the Gunflint Chert. *Science*, **147**, 563–77.

Baxter, R. W. (1975). Fossil fungi from American Pennsylvanian coal balls. *University of Kansas Paleontological Contributions*, **77**, 1–6.

Bold, H. C., Alexopoulos, C. J., & Delevoryas, T. (1980). *Morphology of Plants and Fungi*, 4th ed. New York: Harper & Row.

Brown, R. W. (1940). A bracket fungus from late Tertiary of southwestern Idaho. *Journal of the Washington Academy of Science*, **30**, 422–4.

Davis, B., & Leisman, G. A. (1962). Further observations on *Sporocarpon* and allied genera. *Bulletin of the Torrey Botanical Club*, **89**, 97–109.

Dennis, R. L. (1970). A Middle Pennsylvanian Basidiomycete mycelium with clamp connections. *Mycologia*, **62**, 578–84.

Dennis, R. L. (1976). *Palaeosclerotium*, a Pennsylvanian-age fungus combining features of modern ascomycetes and basidiomycetes. *Mycologia*, **62**, 578–84.

Dilcher, D. L. (1965). Epiphyllous fungi from Eocene deposits in western Tennessee, U.S.A. *Palaeontographica, B*, **116**, 1–54.

Graham, A. (1962). The role of fungal spores in palynology. *Journal of Paleontology*, **36**, 60–8.

Illman, W. I. (1984). Zoosporic fungal bodies in spores of the Devonian fossil vascular plant, *Horneophyton*. *Mycologia*, **76**, 545–7.

Kidston, R., & Lang, H. W. (1921). On old Red Sandstone plants showing structure, from the Rhynie chert bed, Aberdeenshire. Part V. The Thallophyta occurring in the peatbed. *Transactions of the Royal Society of Edinburgh*, **52**, 855–902.

Knoll, A. H. (1985). Patterns of evolution in the Archean and Proterozoic Eons. *Paleobiology*, **11**, 53–64.

McLaughlin, F. J. (1976). On *Plaeosclerotium* as a link between ascomycetes and basidiomycetes. *Science*, **193**, 602.

Millay, M. A., & Taylor T. N. (1978). Chytridlike fossils of Pennsylvanian age. *Science*, **200,** 1147–49.

Pirozynski, K. A., & Malloch, D. W. (1975). The origin of land plants: a matter of mycotrophism. *Bio Systems*, **6,** 153–64.

Rothwell, G. W. (1972). *Palaeosclorotium pusillium* gen. et sp. nov., a fossil eumycete from the Pennsylvanian of Illinois. *Canadian Journal of Botany*, **50,** 2353–6.

Schopf, J. W. (1968). Microflora of the Bitter Springs Formation, Late Precambrian, Central Australia. *Journal of Paleontology*, **42,** 651–88.

Schopf, J. W., & Barghoorn, E. S. (1969). Microorganisms from the late Precambrian of South Australia. *Journal of Paleontology*, **43,** 111–18.

Sherwood-Pike, M. A., & Gray, J. (1985). Silurian fungal remains: probable records of the Class Ascomycetes. *Lethaia*, **18,** 1–20.

Singer, R. (1977). An interpretation of *Palaeosclerotium*, *Mycologia*, **69,** 850–4.

Stidd, B. M., & Cosentino, K. (1975). *Albugo*like oogonia from the American Carboniferous. *Science*, **190,** 1092–3.

Stubblefield, S. P., & Taylor, T. N. (1986). Wood decay in silicified gymnosperms from Antarctica. *Botanical Gazette*, **147,** 116–25.

Stubblefield, S. P., & Taylor, T. N. (1988). Recent advances in palaeomycology. Tansley Review No. 12. *New Phytologist*, **108,** 3–25.

Stubblefield, S. P., Taylor, T. N., & Beck, C. B. (1985). Studies of Paleozoic Fungi V. Wood-decaying fungi in *Callixylon newberryi* from the Upper Devonian. *American Journal of Botany*, **72,** 1765–74.

Stubblefield, S. P., Taylor, T. N., Miller, C. E., & Cole, G. T. (1983). Studies of Carboniferous Fungi. II. The structure and organization of *Mycocarpon, Sporocarpon, Dubiocarpon* and *Coleocarpon* (Ascomycotina). *American Journal of Botany*, **70,** 1482–98.

Stubblefield, S. P., Taylor, T. N., Miller, C. E., & Cole, G. T. (1984). Studies of Paleozoic Fungi. III. Fungal parasitism in a Pennsylvanian gymnosperm. *American Journal of Botany*, **71,** 1275–82.

Stubblefield, S. P., Taylor, T. N., & Seymour, R. L. (1987). A possible endogonaceous fungus form the Triassic of Antarctica. *Mycologia*, **79,** 905–6.

Stubblefield, S. P., Taylor, T. N., & Trappe, J. M. (1987). Vesicular-arbuscular mycorrhizae from the Triassic of Antarctica. *American Journal of Botany*, **74,** 1904–11.

Tiffney, B. J., & Barghoorn, E. S. (1974). *The fossil record of fungi.* Occasional Papers of the Farlow Herbarium of Cryptogamic Botany. Cambridge: Harvard University Press.

Tyler, A. S., & Barghoorn, E. S. (1954). Occurrence of structurally preserved plants in Pre-Cambrian rocks of the Canadian shield. *Science*, **119,** 606–8.

Wagner, C. A., & Taylor, T. N. (1981). Evidence for endomycorrhizae in Pennsylvanian age plant fossils. *Science*, **212,** 562–3.

White, J. F., & Taylor, T. N. (1988). Triassic fungus from Antarctica with possible ascomycetous affinities. *American Journal of Botany*, **75,** 1495–500.

6

Diversification among the algae and related plants

Although there is some evidence of an increase in algal diversity during the Precambrian (Schopf, 1970; Butterfield, Knoll, & Swett, 1988), it is not until we turn our attention to the fossils of the Lower Paleozoic (the Cambrian, Ordovician, and Silurian periods) that we find a wide variety of algal morphologies. As you might expect, evidence abounds in the Lower Paleozoic for the unicellular and colonial cyanophytes so characteristic of the Precambrian. In addition, however, we find the first solid evidence of multicellularity in plant architecture. This represents an event of the greatest significance in the evolution of the green land plants that was to follow. To further emphasize the importance of this event, remember that many algae and fungi and all embryo-forming plants (bryophytes and tracheophytes) are multicellular organisms.

Among extant plants the condition of multicellularity is believed to occur among eukaryotes where the protoplasts of contiguous cells are interconnected by protoplasmic strands called plasmodesmata. It should be noted that prokaryotes – bacteria and cyanophytes – do not have plasmodesmata. This suggests that the level of multicellularity was not reached until the end of the Precambrian or beginning of the Paleozoic when we have our first unequivocal evidence of eukaryotic organisms.

The best explanation for the diversity that resulted in the evolution of multicellularity is mutation in combination with sexual reproduction (nuclear fusion accompanied by meiosis) to produce an increased variety of organisms to be acted on by natural selection. At the beginning of the Paleozoic there is ample evidence that both multicellular plants and animals had evolved life cycles that included some form of sexual reproduction.

Geologists tell us that during the Lower Paleozoic the continents were relatively low-lying, with extensive geosynclines filled with water. These formed shallow seas in which the diversification of plants and animals could take place. There is considerable evidence suggesting that during the Cambrian the seas were clear and the water fairly warm, an ideal environment for the growth and reproduction of autotrophic organisms. For the most part we find remains of algae belonging to those Cyanophyta, Rhodophyta, and Chlorophyta that precipitate $CaCO_3$ and form stromatolites, lime-encrusted parts, or deposits of limestone in which the tissues become embedded as the alga grows.

The Cyanophyta (cyanobacteria)

With few exceptions, evidence of the Cyanophyta during the Lower Paleozoic is derived from reef formations composed of stromatolites (Fig. 6.1). Fine examples are found in the Cambrian algal reefs of New York (Goldering, 1938). Algal limestones having the characteristics of stromatolites occur from early Precambrian to the present. They provide us with both direct and indirect evidence of the presence of blue-green algae throughout this vast extent of geological time. The oldest known noncalcareous representative thought to be a cyanophyte is *Gloeocapsamorpha* (Fig. 6.2E). As its name suggests, these fossils consist of irregular colonies of cells embedded in an ensheathing matrix similar to a modern *Gloeocapsa* colony with its gelatinous sheath. These specimens are from the Ordovician, but several collections of *Gloeocapsamorpha* from the Silurian have been reported. Permineralized specimens of cyanophytes have been described by Croft & George (1959) from the Devonian Rhynie chert. The specimens show the same excellence of preservation found in the Precambrian Bitter Springs cherts. The gelatinous sheaths are preserved, recognizable heterocysts and akinetes occur in the filaments, and prostrate and erect systems of filaments can be seen. This combination

Figure 6.1. Cambrian algal barrier reef composed of *Cryptozoön* stromatolites. (Courtesy of the New York State Museum. In Goldering, 1938.)

Figure 6.2. **A.** Tangential section of *Solenopora compacta* showing pseudoparenchymatous structure. Ordovician. **B.** Vertical section of *S. compacta*. (**A,B** redrawn from Seward, 1898.) **C.** *Epiphyton fruticosum* sporangia terminating short lateral branches. Cambrian. **D.** *Epiphyton* habit showing conspicuous dichotomous branching. (**C,D** redrawn from Johnson, 1966.) **E.** *Gloeocapsamorpha*, irregular colonies embedded in a lamellated sheathing matrix. Ordovician. (Redrawn from Pia, 1927.)

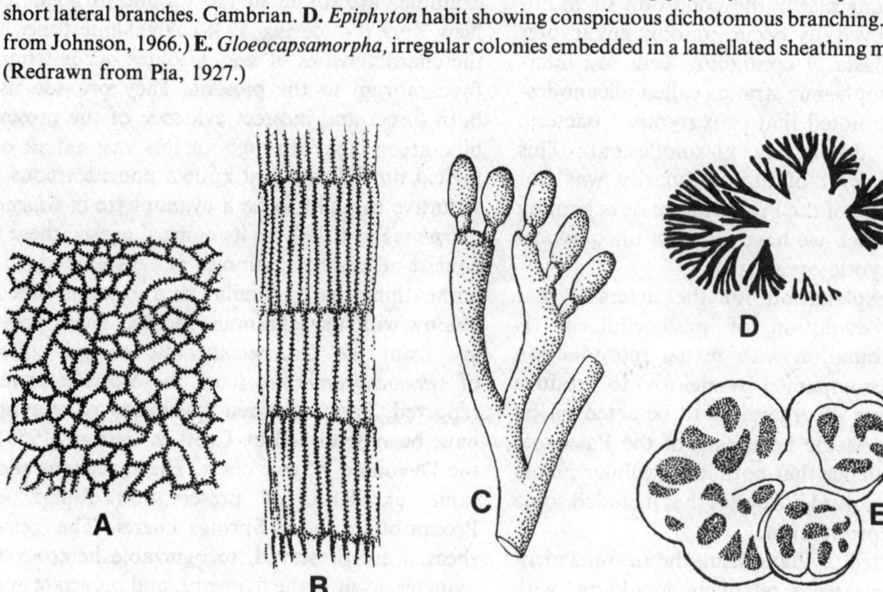

of characteristics is found among the modern members of the Stigonemataceae, an advanced family of the Cyanophyta. Coexisting with these representatives of Stigonemataceae in the Rhynie chert, Edwards & Lyon (1983) describe species of a coccoid algal form, *Rhyniococcus,* which they believe can be assigned to the Chroococcaceae of the Cyanophyta. The small cell size, the arrangement of the cells in a sheet, and uniform spacing of the *R. uniformus* cells are very similar to the extant cyanophyte *Merismopedia.* From evidence of this kind we might conclude that the cyanophytes had approximated their maximum degree of diversification and specialization by the Lower Paleozoic.

Comparisons of the biochemistry and life cycles of extant red algae (Rhodophyta) and Cyanophyta lead us to suspect that the two groups are in some way related. The only chlorophyll found in both groups is chlorophyll *a,* and the phycobilins are the same. Neither has significant cellulose in its cell walls nor do they form the flagellated cells so characteristic of other groups of algae. The major differences between the two groups are seen in the multicellular organization of most red algae and the eukaryotic nature of their cells. Cyanophytes are unicellular or colonial and are prokaryotic.

The Rhodophyta

The fossil record of the Rhodophyta is questionably extended into the late Precambrian (Schopf, 1970). There is no doubt, however, that red algae abounded in the warm seas of the Lower Paleozoic along with cyanophytes. Since the Lower Paleozoic some 600 m.y. ago, remains of the Rhodophyta form a fossil record that is more or less continuous to the present. Comprehensive accounts of this record were prepared by J. H. Johnson (1961) in his book *Limestone-Building Algae and Algal Limestones,* and by Wray (1977).

Of the various members of the Rhodophyta, those that precipitate limestome (Fig. 6.3A) are the ones best preserved in the fossil record. Red algae belonging to the Corallinaceae, Solenoporaceae, and Gymnocodiaceae have their tissues to some degree immersed in calcium carbonate. This material forms a matrix in which the cell and tissue structures are preserved, making it easier to study the fossil forms.

Among the fossil remains of the Rhodophyta, two genera found in the Cambrian are worth mentioning. One is *Solenopora* (Fig. 6.3C), of the Solenoporaceae, that forms irregular nodular masses of limestone. Vertical sections of *Solenopora* show slightly radiating filaments of cells (Fig. 6.2B). In cross section the cells are polygonal in outline (Fig. 6.2A). The arrangement of the filaments in the matrix is nearly identical with extant multicellular coralline red algae, leaving little doubt about the relationships of *Solenopora. Solenopora* disappears from the fossil record in the Jurassic.

Epiphyton (Fig. 6.2C, D) is one of the commonest algal forms found in Cambrian rocks. This bushy, highly branched form has been placed in the family Epiphytaceae. Unfortunately, its position with the Rhodophyta is questioned. Unlike *Solenopora,* the internal structure is poorly preserved if preserved at all. Other than the Rhodophyta, this dichotomously branched organism has been assigned to either the Chlorophyta or Cyanophyta (Wray, 1977). This apparent indecision as to its relationships graphically illustrates the point that most coalified compressions exhibiting an algal form are of little value to paleobotanists.

Fossil coralline red algae that have vegetative and reproductive structures indistinguishable from extant genera are *Lithothamnium* and *Lithoporella,* extending back into the Jurassic, and *Lithophyllum,* whose record extends from the present into the Upper Cretaceous.

This kind of evidence, and there is much more, assures us that the Rhodophyta is a group with a long fossil record that probably was initiated at least by late Precambrian times when they coexisted with the Cyanophyta. This evidence, added to the evidence of their biochemical and morphological similarities, strongly suggests that the relationship of the Rhodophyta is with the Cyanophyta. With such abundant and beautifully preserved Precambrian fossils available to us, it is not too much to hope that more definite answers will be forthcoming.

The Chlorophyta

The fossil record of the Chlorophyta (grass-green algae) is of particular interest to us because it is generally accepted that the ancestors of green land plants belonged to this division. We have a good idea of when the first land plants appeared. Thus, it is important to see if the Chlorophyta were present prior to the time that land was "invaded" by green plants. Have the grass-green algae been around long enough to be the ancestors? We already know (Chapter 4) that Schopf believes there is a possibility that the Chlorophyta were extant during the late Precambrian, long before the first known record of green land plants. This possibility has been supported by Butterfield, Knoll, & Swett (1988), who have found well-preserved algal remains in a Late Proterozoic shale 700 to 800 m.y. old. In their morphologies the fossil algae are comparable to certain cladophoralean and chaetophoralean green algae. Before we examine unquestionable evidence for ancient Chlorophyta we have to know something about a few extant representatives. Similar to the Cyanophyta and Rhodophyta, there are limestone-forming Chlorophyta. The modern representatives are found in the families Codiaceae and Dasycladaceae of the order Siphonales. The order is characterized by species that form vegetative filaments lacking cross walls. They are nonseptate coenocytes that, in some genera, secrete limestone. As the limestone is formed the older coenocytes are immersed in the calcareous matrix. When the filaments die a system of branching tubules remains in the limestone. Members of Codiaceae are characterized by coenocytic filaments that are interwoven with one another to form a plant body of definite shape. *Halimeda* (Fig. 6.4B), a common component of algal reefs, is an example. The filaments of Dasycladaceae are arranged in the form of a central axis with radiating branches that may divide. A simple extant member of this family is *Acetabularia* (Fig. 6.4E) with its main axis and a single whorl of radiating branches at the apex. All parts are encrusted in a secreted layer of lime.

It is interesting that lime-secreting Chlorophyta show these two basic morphologies as early as the Cambrian (Pia, 1927; Johnson, 1966). There is abundant evidence that they were a conspicuous part of the algal floras of the warm Ordovician seas and

Figure 6.3. **A.** *Porolithon* sp., a coralline red alga forming compact heads of closely packed branches. Extant. **B.** *Parka decipiens,* a Middle Devonian plant with a thallus similar to that of extant green alga *Coleochaete*. (From Niklas, 1976c.) **C.** *Solenopora* sp., a small nodule from the Ordovician. **D.** *Botrycoccus* sp., colonies from thin sections of boghead coal. Pennsylvanian. (Photograph by Dr. R. Kosanke.)

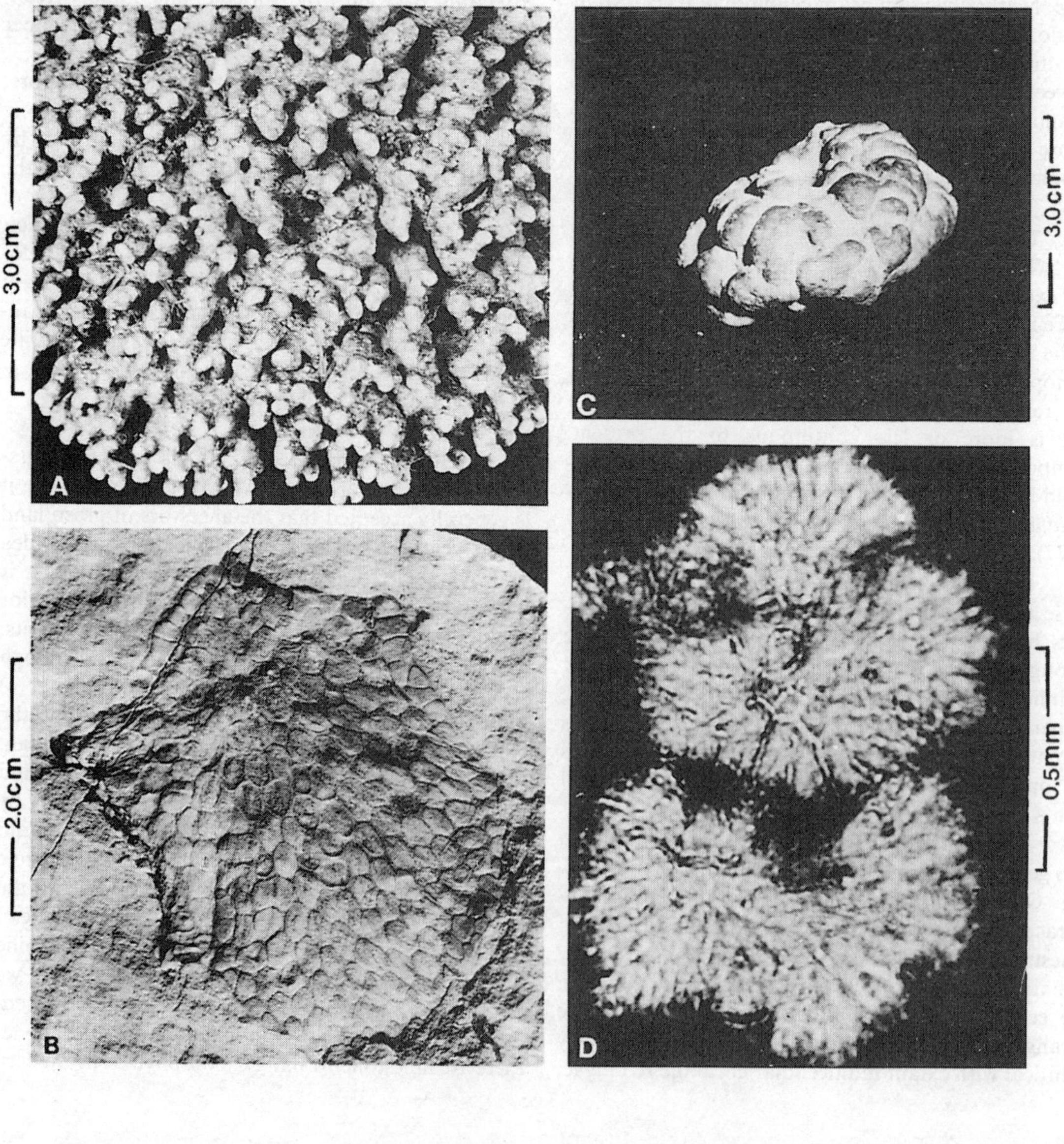

Figure 6.4. Some extant Chlorophyta with fossil records. **A.** *Closterium*, a freshwater desmid. **B.** *Halimeda*, a marine alga of algal reefs. **C.** *Chara*, a lime-encrusted freshwater green alga. Habit drawing. **D.** *Chara*, enlarged portion of thallus showing nucule (n) and corticating filaments (cf). **E.** *Acetabularia*, a marine green alga of the warm seas.

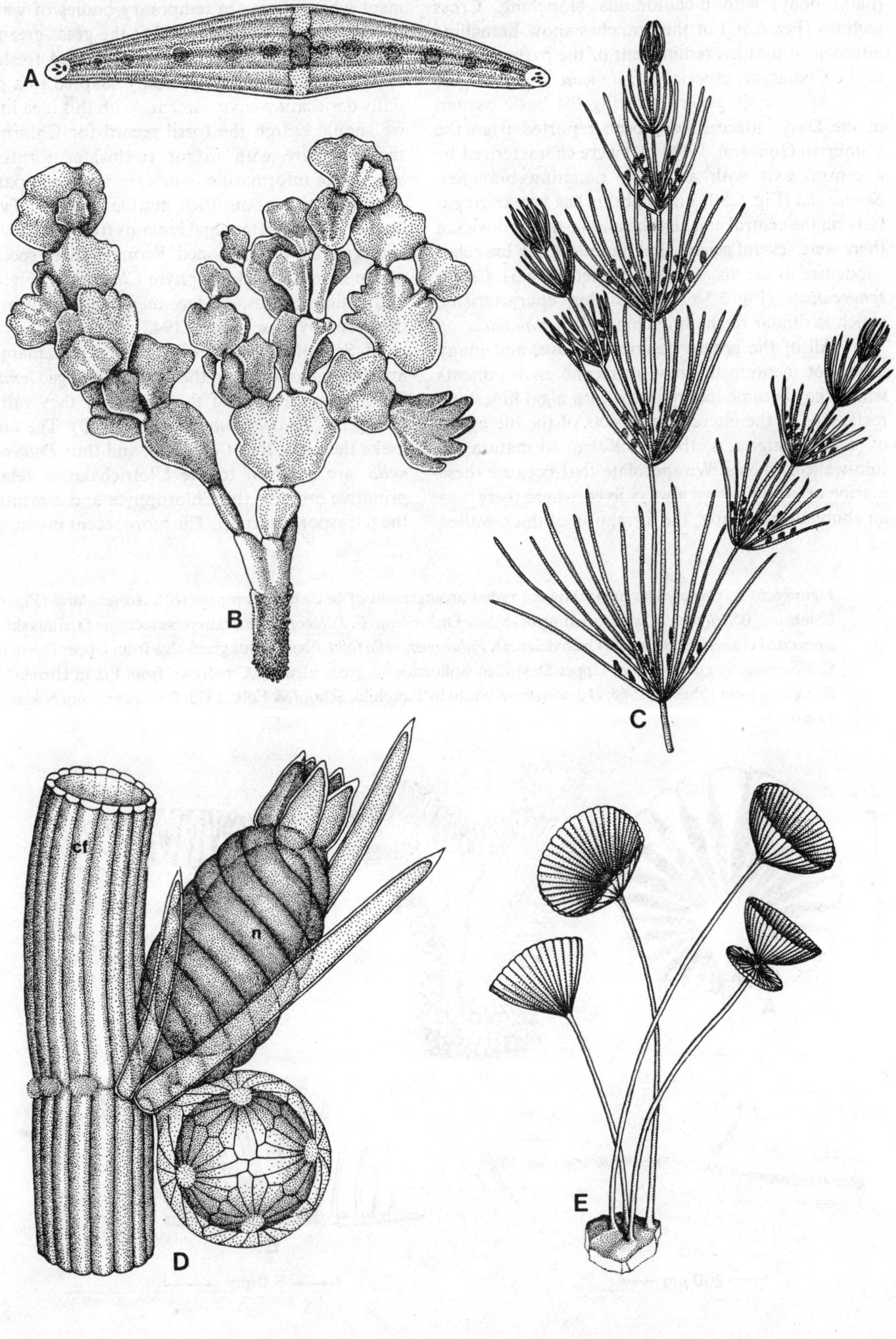

that they have been in continuous existence to the present.

Palaeoporella from the late Cambrian is a limestone-building alga having an articulated thallus (plant body) with dichotomous branching. Cross sections (Fig. 6.5C) of the branches show branching intertwined tubules, reminiscent of the pattern found in the Codiaceae, especially *Halimeda.*

At least six genera showing the basic pattern of the Dasycladaceae have been reported from the Cambrian (Johnson, 1966). All were characterized by a central axis with attached, radiating branches. *Seletonella* (Fig. 6.5B) had its branches borne irregularly on the central axis. By the end of the Ordovician there were several genera that had radiating branches condensed at the distal end of the central axis. *Coelosphaeridium* (Fig. 6.5A) shows this characteristic, which is similar to the branching of *Acetabularia.*

All of the genera described above, and many more not mentioned, grew in marine environments where they became incorporated into algal limestone reefs. One of the interesting aspects of the life cycles of marine algae is their uniform formation of thin-walled zygotes. We speculate that because these marine organisms have always lived where there was an abundance of water, the formation of thick-walled zygotes would be of little advantage. The evolution of thick-walled zygotes in freshwater algae is, however, of great selective advantage. The formation of such a structure adapts these organisms to an environment where there are temporary bodies of water. It makes sense if we presume that the grass-green algal ancestors of green land plants inhabited freshwater environments and were partially adapted to a potentially desiccating environment. With this idea in mind we should search the fossil record for Chlorophyta that compare with extant freshwater forms. The amount of information is meager when compared to what is known about their marine relatives. Two accounts of freshwater algal remains from the Devonian cherts should be mentioned. Permineralized specimens of the unicellular chlorophyte *Closterium* (Fig. 6.4A) and a filamentous, *Oedogonium*-like organism were described by Baschnagel (1942). More recently Fairchild, Schopf, & Folk (1973) found algal remains that are strikingly similar to the freshwater alga *Geminella.* To convey the idea of this similarity they call their new genus *Palaeogeminella* (Fig. 6.5D). The authors make the point that *Geminella,* and thus *Palaeogeminella,* are assigned to the Ulotrichales, a relatively primitive order of the Chlorophyta and a member of the tetrasporine group. The more recent investigation

Figure 6.5. **A.** *Coelosphaeridium* showing radial arrangement of branches. Compare with *Acetabularia* (Fig. 6.4E). Cambrian. **B.** *Seletonella,* with radiating branches. Ordovician. **C.** *Palaeoporella,* transverse section of thallus showing embedded branching filaments. Ordovician. **D.** *Palaeogeminella folki,* filamentous green alga from Upper Devonian. **E.** *Courvoisiella ctenomorpha,* Upper Devonian siphonaceous green alga. (**A,C** redrawn from Pia in Hirmer, 1927; **B** redrawn from Johnson, 1966; **D** drawn from photo by Fairchild, Schopf, & Folk, 1973; **E** redrawn from Niklas, 1976b.)

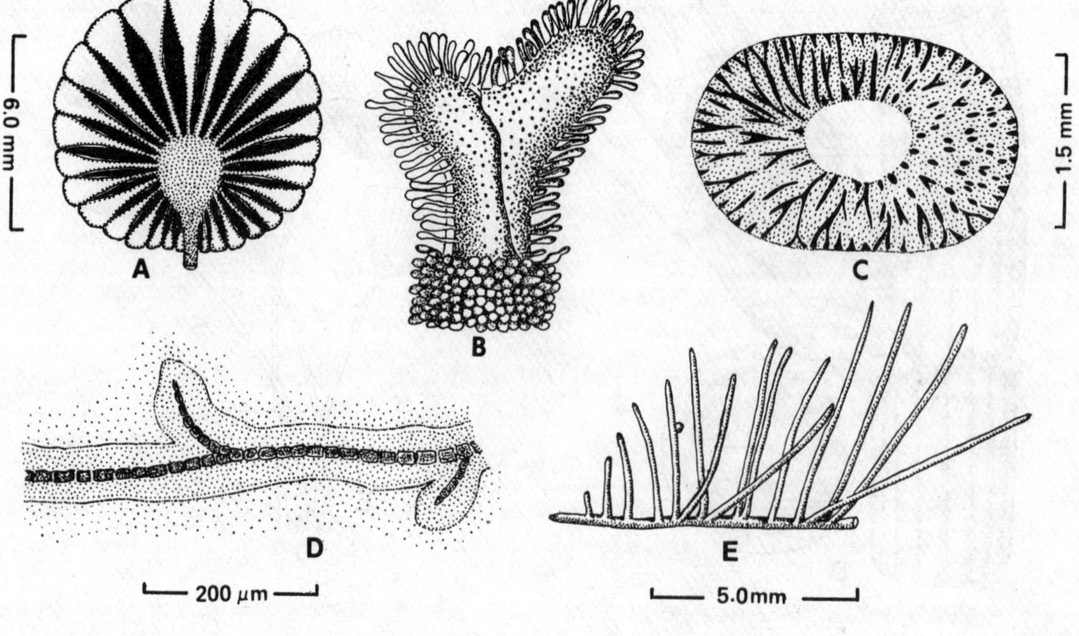

of the Rhynie chert by Edwards & Lyon (1983) confirms the presence of ulotrichalean Chlorophyta. Their new genus *Rhynchertia* compares favorably with the extant *Ulothrix cylindricum.* A second new genus, *Mackiella* does not fit well with most orders of the Chlorophyta, but is tentatively assigned to the Ulotrichales because of the similar length-to-width ratios of its cells. These fossils would be of tremendous significance if they had been found in the Lower Paleozoic prior to the mid-Silurian where we find the first unequivocal evidence of green land plants. We can be optimistic, however, that studies of Lower Paleozoic and late Precambrian cherts deposited in shallow freshwater environments will reveal them. This would give us a proper geological sequence and "bridge the gap" between the tetrasporine Chlorophyta and green land plants.

The Charophyceae also have received attention as possible ancestors of primitive green land plants, especially bryophytes. These freshwater algae, commonly called stoneworts (Fig. 6.4C), are considered by some to be related to the Chlorophyta because of their similar biochemistry and flagellated cells. Others place this small group in a separate division, the Charophyta. Indeed, these organisms are very different in the structure of their vegetative and reproductive parts when compared to other grass-green algae. As their common name implies, some species of stoneworts secrete a surface layer of limestone that makes the thallus brittle and harsh to the touch. The plant has an axial filament with whorls of branches borne at distinct nodes. In *Chara,* corticating filaments usually surround the axial filament. The reproductive structures (Fig. 6.4D) are complex and it is the female reproductive structure, the nucule, that is of interest to paleontologists. The nucule comprises an egg in a unicellular oogonium surrounded by five sinistrally spiralled corticating filaments. The latter can be partially or completely calcified.

Casts of nucules of Charophyceae abound in the fossil record and are called gyrogonites (Fig. 6.6A). Their geological distribution and evolution has been intensively studied by Grambast (1964) and others, who have shown that the fossilized reproductive structures can be useful index fossils for nonmarine strata. According to Grambast, primitive charophyte nucules of the Devonian had an irregular number of vertical corticating filaments (Fig. 6.6C,D). By the Middle Devonian the number of filaments was stabilized at five. This was accompanied by the introduction of a sinistral spiral to the filaments and a pore at the apex of the nucule. This basic type appeared in the Carboniferous period, with numerous variations in size, shape, and ornamentation (Fig. 6.6E,F) of nucules making their appearance from the Carboniferous to the present. Vegetative remains believed to be those

Figure 6.6. Reproductive structures (gyrogonites) of charophytes. **A.** *Gyrogonites* sp., cast of nucule. Eocene. **B.** *Lagynophora,* nucules attached to node and surrounding vegetative branches. Paleocene. **C,D.** *Eochara* sp., lateral and apical views. Devonian. **E.** *Atopochara.* Cretaceous. **F.** *Harrisichara.* Eocene. (**B** redrawn from Pia in Hirmer, 1927; **A,C–F** redrawn from Grambast, 1964.)

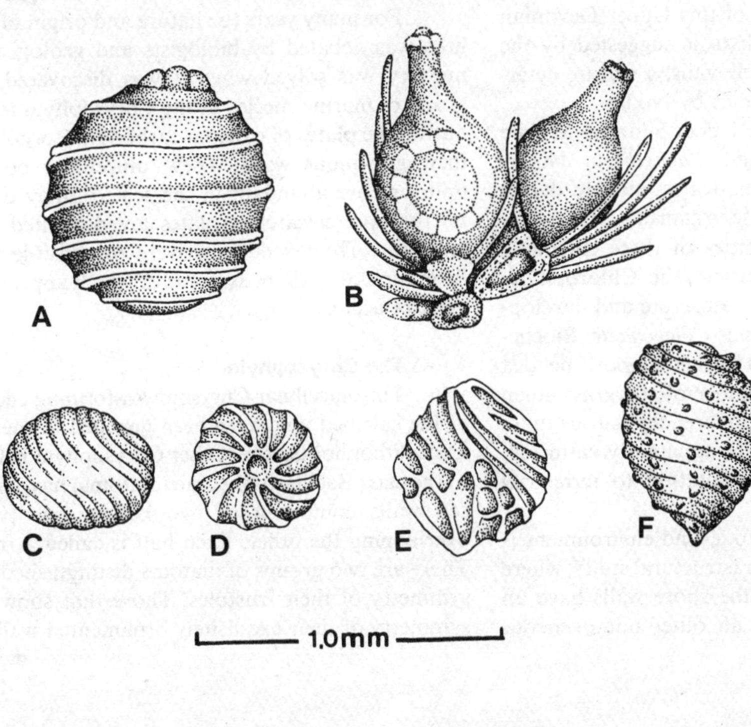

of charophytes were first described by Kidston & Lang (1921) from the Rhynie chert. Because the fragments lacked the corticating filaments of *Chara* and were like *Nitella* in this respect, these authors assigned a provisional generic name, *Palaeonitella*. Edwards & Lyon (1983) have confirmed the occurrence of the *Nitella*-like fragments in these Lower Devonian cherts. Tertiary deposits frequently contain fragments of *Chara* vegetative parts. Occasionally compressed reproductive remains as represented by the Paleocene genus *Lagynophora* (Fig. 6.6B) are found.

Other genera assigned to the Chlorophyta that deserve more than passing attention as organisms with chemical and morphological characteristics one might expect to find in ancestors of land plants are *Botrycoccus* (Fig. 6.3D), *Courvoisiella* (Fig. 6.5E), and *Parka* (Fig. 6.3B). It has been suggested that *Botrycoccus* and *Gloeocapsamorpha* are congeneric. Both form irregular colonies of cells embedded in a conspicuous matrix rich in hydrocarbons (Traverse, 1955). *Botrycoccus,* which is an important genus of extant marine and freshwater algal floras, is known to occur in rocks of Ordovician age onward. It has been discovered to be the principal constituent of boghead coals and torbanites and may be the chief source of hydrocarbons in the highly important oil shales. The chemical nature of these hydrocarbons has been investigated by Niklas (1976a), who has compared these compounds with those of other nonvascular and vascular plants from the Devonian. He concludes that some of the plants he investigated show a degree of parallelism in their chemical adaptations to land environments.

Courvoisiella, a genus described by Niklas (1976b), is thought to belong to the Siphonales (Eusiphonales). The affinity of this Upper Devonian organism with the Chlorophyta is suggested by the presence of cellulose in its cell walls, a feature determined from biochemical studies by Niklas.

In a novel study of the Upper Silurian–Middle Devonian genera *Parka* and *Pachytheca,* Niklas (1976c) made use of mathematical models to depict the ontogenetic stages of these organisms. His results show that the possible affinities of these enigmatic nonvascular plants also are with the Chlorophyta, especially those with a thallus structure and development similar to the extant genus *Coleochaete*. Biochemical studies (Niklas, 1976d) further support the idea that *Parka* is an ancient member of the grass-green algae. The nonvascular pseudoparenchymatous thalli of *Parka* and *Pachytheca* were apparently cutinized, again suggesting an early adaptation to terrestrial environments.

Adaptation of *Parka* to a land environment is also suggested by a recent ultrastructural study, where Hemsley (1986) found that the spore walls have an inner laminated region and an outer homogeneous region that is similar to some bryophyte spores. As in some liverworts, *Parka* spores show no evidence of a trilete mark. At the present time we do not know if this latter feature indicates that the spores were produced by mitosis, or if they merely separated from the meiotic tetrad early in development.

Hystrichospheres and coccoliths

Many unicellular organisms belonging to various groups of algae can form cells with walls that are resistant to chemical or mechanical breakdown. Among these are the highly ornamented thick-walled zygotes of freshwater Chlorophyta called desmids. Their zygotes have been reported from the Jurassic and Cretaceous. In size and ornamentation they are strikingly similar to objects described by micropaleontologists as hystrichospheres (Fig. 6.7D). Another conjecture about the nature of hystrichospheres is that they represent cysts (aplanospores) of those Pyrrophyta called dinoflagellates. The thick-walled cysts may be highly ornamented with stout spines, a conspicuous characteristic of hystrichospheres. These cystlike bodies have been reported from the Lower Paleozoic, but are much more abundant in the Mesozoic, especially the Cretaceous.

The vegetative cells of the armored dinoflagellates have tough cellulose walls deposited in discrete plates (Fig. 6.7B). These too are found in abundance in Cretaceous rocks. Over 100 genera of fossil dinoflagellates have been described, mostly Mesozoic and Tertiary in age. The earliest to be described comes from the Ordovician. Fossil hystrichospheres and dinoflagellates occur most commonly in marine sediments. They have been used to study and identify nearshore marine environments of Cretaceous age.

For many years the nature and origin of coccoliths was debated by biologists and geologists. The mystery was solved when it was discovered that a group of marine species of the Chrysophyta form the buttonlike plates of calcium carbonate (coccoliths) in their gelatinous walls. These unicellular coccolithophorids are abundant in the sea. As they die their microscopic calcareous plates are deposited on the sea floor. These sediments were responsible for the formation of chalk beds, some of which appropriately are Cretaceous in age.

The Chrysophyta

The unicellular Chrysophyta-forming coccoliths of the calcified type have been grouped in the family Coccolithophoridaceae. Other Chrysophyta belonging to the class Bacillariophyceae (diatoms) have silicified cell walls composed of two halves, one partially overlapping the other. Each half is called a frustule. There are two groups of diatoms distinguished by the symmetry of their frustules. Those that show radial symmetry of their exquisitely ornamented wall structures are called centric diatoms and are placed in the order Centrales; those showing bilateral symmetry are commonly called pennate diatoms and are assigned to the Pennales. Centric diatoms are usually marine, while pennate forms occur more frequently in freshwater. During certain times of the year diatoms produce a prodigious number of offspring. As they die their frustules accumulate on the bottom of the lake or sea where they form a diatomaceous ooze. The accumulation of diatomaceous material has gone on as far back in geological time as the Cretaceous. The marine deposits of fossil diatoms may become partially lithified to form a soft, lightweight rock, diatomite. Deposits of diatomaceous earth may vary in thickness from a few meters to as much as 1,000 meters. Subterranean formations of fossil diatoms are often indicators of oil and gas and are of interest to micropaleontologists searching for new supplies of fossil fuels.

Just how many species of fossil diatoms have been described is difficult to say. A fairly recent count indicates 70 fossil genera. Most of these are very similar to or identical with modern genera (Fig. 6.7A), even those that come from the Lower Cretaceous. Students of diatoms believe that certain marine centric diatoms are more primitive and that pennate forms were derived from them. Unfortunately, both types are present in the Cretaceous. There is some debate about the antiquity of diatoms. Questionably they are reported from Paleozoic rocks. Usually accepted as unequivocal evidence for the oldest known diatom is *Pyxidicula* (Fig. 6.7C) from the Jurassic.

The Phaeophyta

There is no need to take up much time or space explaining the inadequate fossil record for the Phaeophyta (brown algae). The problem already alluded to lies in the state of preservation, plus the fact that compression fossils having the form of extant brown algae also have counterparts in form among thallose red algae. Fry & Banks (1955), unlike some investigators, make this point in describing a new genus *Drydenia* from the Devonian (Fig. 6.8). The first inclination is to classify the plant as a small specimen of the brown alga *Laminaria* with its obvious holdfast, short stipe, and blade. This morphology, however, can be found in species of *Porphyra, Gigartina,* and *Dilsea.* All of these are Rhodophyta. Some supposed algal

Figure 6.7. **A.** *Cymbella* sp., a pennate diatom with a fossil record extending from the Cretaceous to the present. **B.** *Peridinium* sp., a dinoflagellate from the Cretaceous. **C.** *Pyxidicula bollensis* showing the two frustules and markings of a diatom. Jurassic. **D.** *Xanthidium* sp., a hystrichosphere from the Cretaceous. A dinoflagellate cyst or desmid zygote. **E.** *Prototaxites southworthii.* Tangential section showing tubules of two sizes. Lower Devonian, (**B,C** redrawn from Pia in Hirmer, 1927; **D,E** redrawn from Delevoryas, 1962.)

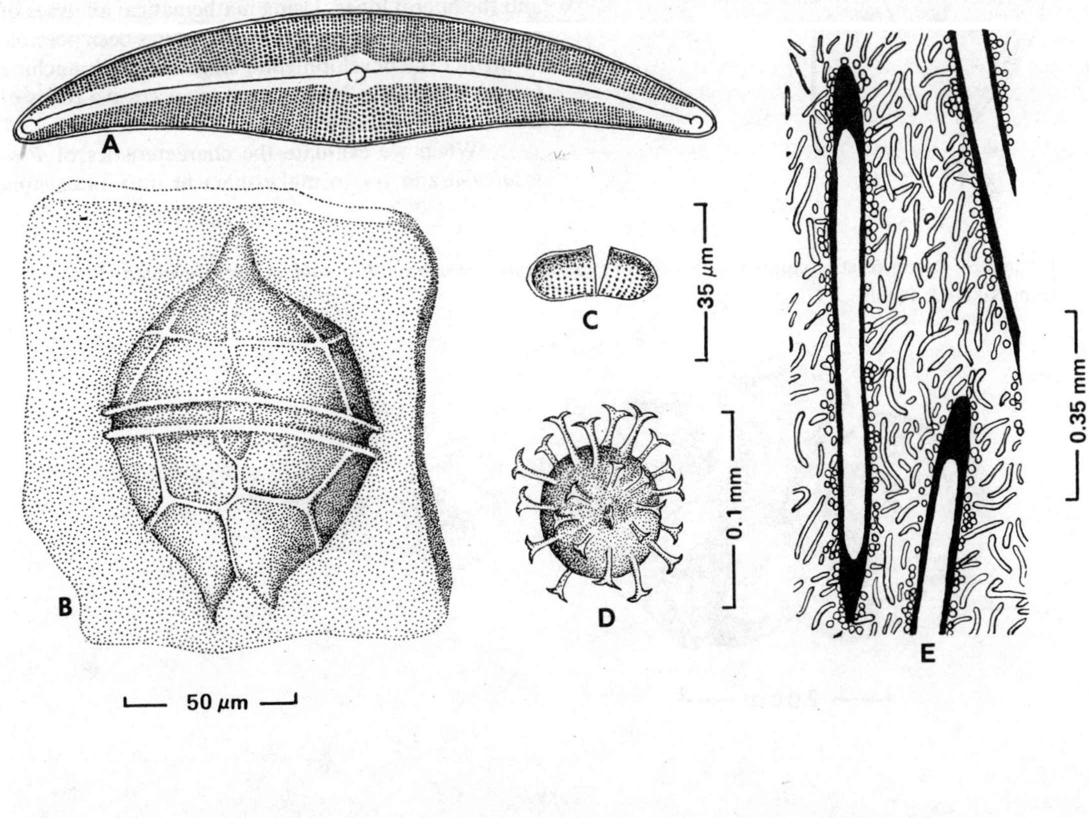

remains that superficially resemble brown algae turn out to be worm burrows or animal trails.

Some enigmatic nonvascular plants

Some authors believe there is a relationship between the Nematophytales, *Protosalvinia,* and Phaeophyta. *Protosalvinia* and genera of the Nematophytales are found in Devonian strata. The coalified compressions of *Protosalvinia* (Fig. 6.10D,E) do bear a superficial resemblance to the dichotomous tips of the marine brown alga *Fucus.* The internal structure of stemlike parts of species belonging to the Nematophytales is reminiscent of the pseudoparenchymatous organization found in the thalli of brown algae belonging to the kelps. However, careful analysis of these fossils shows that these similarities may be more apparent than real.

Figure 6.8. *Drydenia* sp., a putative brown alga from the Upper Devonian. (Drawn from photo by Fry & Banks, 1955.)

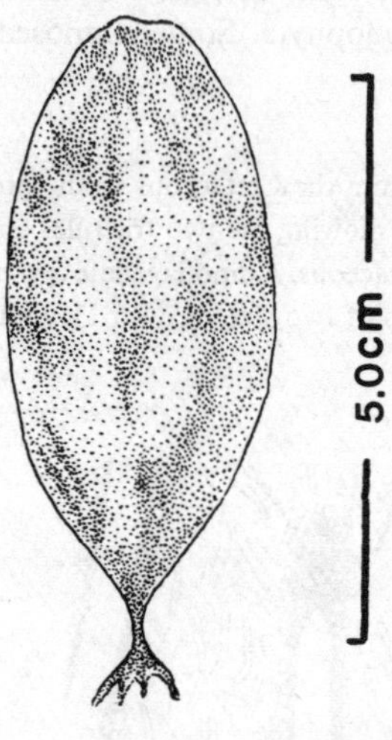

Protosalvinia (Fig. 6.9), is commonly found in shales of the Upper Devonian. The composition of the shale suggests a shoreline (littoral) environment where these unbranched and dichotomously branched portions of plants were present in great abundance. The principal investigators (Arnold, 1954; Phillips, Niklas, & Andrews, 1972; Niklas & Phillips, 1976; Schopf, 1978; Gray & Boucot, 1979) describe some highly interesting features of *Protosalvinia* that make it even more difficult to assign the genus to an extant group. Those who study *Protosalvinia* have been puzzled by the truncated appearance of the branches. It was obvious that they were attached to something, but what? Niklas & Phillips were fortunate in finding an ill-defined dorsiventral parenchymatous thallus with branches attached (Fig. 6.9). Careful examination of the erect, dichotomizing, and unbranched tips revealed hypodermal, conceptaclelike cavities (Fig. 6.10A,B) often containing a single tetrahedral quartet of thick-walled spores resembling meiospores (Fig. 6.10C). These were found in a cuplike depression at the apex of the unbranched forms and in an elongated, apical groove for those plants that are dichotomously branched. Isolated spores have been chemically analyzed resulting in the discovery of sporopollenin, the substance universally present in the walls of spores and pollen of embryo-forming land plants. Chemical analyses of *Protosalvinia* thalli indicate that cutin and ligninlike compounds also were present. Sterile specimens often show an apical dome flanked by the two lobes of a dichotomizing axis. The dome appears to have been meristematic, contributing to the length of the branch and the lateral lobes. Using mathematical analyses of a large population of *Protosalvinia* it has been possible to predict the developmental stages of the branching forms. These have been related to successive zones of the littoral environment.

When we evaluate the characteristics of *Protosalvinia* and try to make them fit into an existing

Figure 6.9. Reconstruction combining features of *Protosalvinia ravenna* and *P. arnoldii.* Upper Devonian. (Redrawn from Niklas & Phillips, 1976.)

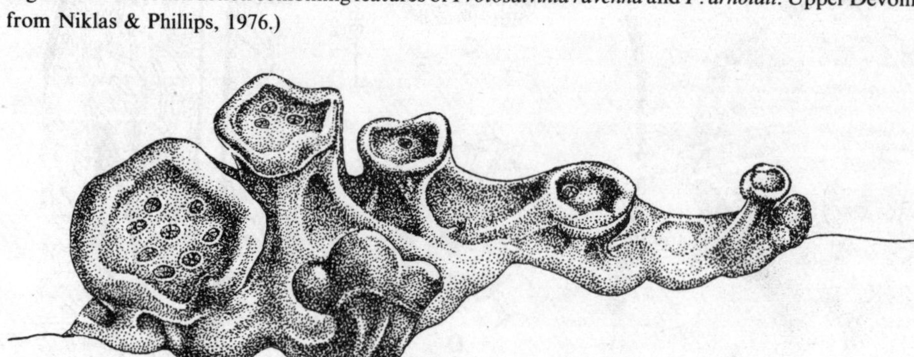

group, we are confronted with a dead end. Attempts have been made to relate the genus to the liverworts as well as the fucoid brown algae (Schopf, 1978). In both cases the fit is uncomfortable. It is tempting to try to relate *Protoslavinia* to a *Riccia*-like thalloid liverwort because both have a dorsiventral parenchymatous plant body, and both dichotomize and have a meristematic zone at the base of the dichotomy. The thick-walled meiospores of *Riccia* are produced in an exceedingly simple sporophyte, a sacklike structure embedded in the tissues of the thallus in a way that is reminiscent of the conceptacles of *Protosalvinia.* Or could the upright cylindrical spore-bearing branches of *Protosalvinia* be homologous with the sporophyte of the hornwort *Anthoceros*? This has been suggested by Arnold (1954). It would be interesting to know how the upright branches of *Protosalvinia* are attached to the prostrate thallus. Is this a sporophyte attached to a gametophyte?

Now that we are speculating, the relationship of *Protosalvinia* with the fucoid Phaeophyta is just as interesting and logical. The Fucales, the group to which *Fucus* and its relatives belong, have life cycles where the plant we see is a multicellular, partially parenchymatous sporophyte. In *Fucus* the thallus is dichotomously branched. Conceptacles are formed in the swollen tips of the branches. Inside the conceptacles the plant produces a number of fertile and sterile branched filaments. In female plants, the fertile units in a conceptacle form meiospores that, when released, function as eggs. There is no gametophyte. For some members of the Fucales the number of functional spores is four, as in *Protosalvinia. Fucus* meiospores, however, are thin-walled, a characteristic of most algae. If we interpret the life cycles of *Protosalvinia* to be like that of the Fucales, then *Protosalvinia* is a diploid sporophyte, and the tetrads of thick-walled haploid meiospores in their conceptacles would have functioned as eggs.

Because considerable emphasis has been attri-

Figure 6.10. **A.** Upper surface of *Protosalvinia* thallus showing conceptacles (arrow) with tetrads of spores. **B.** Conceptacles enlarged. **C.** *P. arnoldii* tetrad of spores. **D.** *P. arnoldii* portion of thallus showing terminal dichotomy. **E.** *P. furcata.* A–E Upper Devonian. (A–E from Niklas & Phillips, 1976.)

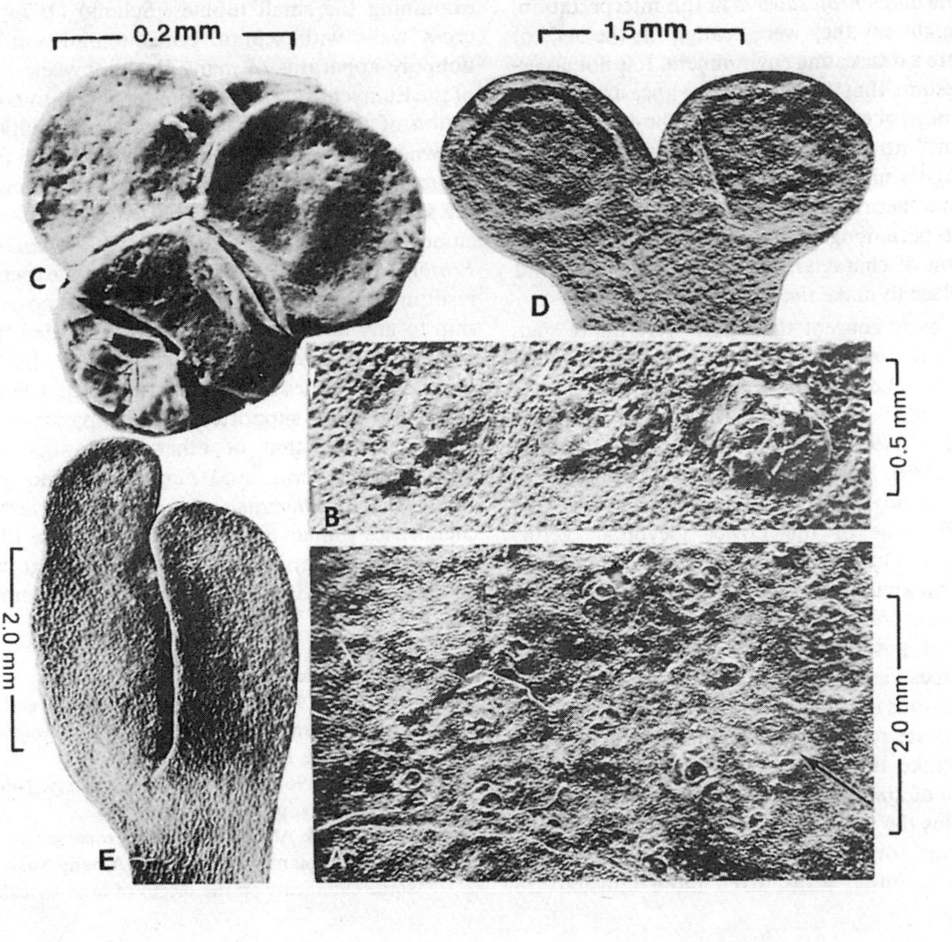

buted to the nature of the tetrads of *Protosalvinia* spores, Taylor & Taylor (1987) studied the ultrastructure of their walls and compared it with the walls of eggs and tetraspores of extant brown and red algae. Although they found little or no similarity in the fine structure in the walls of the respective reproductive structures, they were able to determine a suture beneath the proximal trilete mark of the *Protosalvinia* spores. This observation demonstrates rather conclusively that they are meiospores. Although the wall structure of *Protosalvinia* spores is unlike any presently known for any other plants, fossil or living, Taylor & Taylor emphasize the durable nature of the spores as being like that found exclusively in land plants.

You may think the foregoing is a lengthy way to illustrate what we consider to be an important point about plant evolution. It must be clear that *Protosalvinia* does not fit well into any group of plants known to us. Perhaps we should not excpect it to. *Protosalvinia* grew on mud flats in an environment transitional between water and dry land. If our concept about the origin of land plants from aquatic ancestors is valid and we employ the "upward outlook," we should expect to find organisms that show a combination of characteristics, some of aquatic organisms and some showing adaptations to a terrestrial environment. Such organisms as *Protosalvinia* fit this interpretation, and one might say they were "caught in the act" of evolving into a desiccating environment. It is not necessary to presume that *Protosalvinia* is ancestral to one or more groups of extant land plants. The evolutionary "experiment" attempted by *Protosalvinia* may have failed, being terminated by extinction. It seems highly probable that there must have been many such failures, with one type among many evolving the necessary combination of characteristics at the right time and the right place to make the "experiment" a success.

The basic concept that there were many such attempts in the evolution of land plants from algal ancestors has been expressed by several authors (Leclercq, 1954; Andrews & Alt, 1956; Andrews, 1959; Axelrod, 1959; Krassilov, 1981; Taylor, 1988; Chaloner, 1989). Not only *Protosalvinia* but members of the Nematophytales seem to support this idea. This is especially true of the Lower Devonian genus *Nematothallus.* These fragmentary thalloid compression fossils exhibit an unusual set of characteristics in their internal structure. It appears that the thalli are composed of a system of tubes of two sizes. The smaller tubes intertwine to produce a pseudoparenchymatous tissue like that of the Codiaceae. The larger tubes are reinforced on the inside with closely spaced ringlike thickenings that give the tubes the appearance of annular tracheids. Toward the surface of the thallus the tubes are more compact so that the surface layer forms an epidermis with thickened, apparently cutinized walls. Even more unusual are the thick-walled spores, sometimes in tetrahedral quartets, that occur inside the larger tubes. Here again we find a strange combination of characteristics, some belonging to the algae, some to land plants. Andrews's (1961) evaluation of *Nematothallus* is most appropriate: "If one can put aside for a moment the traditional concept of a primitive vascular plant in which we seek root, stem, and leaf and a spore-bearing organ, we may see in *Nematothallus* a uniquely unspecialized plant in the early stages of becoming established in the land environment." Here Andrews is practicing the "upward outlook" in plant morphology.

In spite of its name, which suggests gymnosperm affinities, *Prototaxites* has turned out to be a colossal fossil enigma. The specimens are often trunklike objects. Large ones are 1 meter in diameter and up to 2 meters long. They were first described by Sir William Dawson in 1859 when he was investigating plant fossils of the Lower Devonian. He thought they were trunks of an ancient yew (*Taxus*), and this explains the misnomer. Subsequent investigation of the enigmatic plant shows that the trunks have essentially the same vegetative structures as *Nematothallus* with two sizes of tubules; those that are small and intertwining and those that are large, more or less vertical, and thick-walled (Fig. 6.7E). While examining the small tubules, Schmid (1976) found cross walls with septate pores reminiscent of the dolipore apparatus of many Basidiomycetes, a class of the Eumycota. The pseudoparenchymatous organization of the cells is like that of some multicellular brown and red algae and of fungi. The large tubes of *Prototaxites* have no counterparts. Schmid comes to the same conclusion about *Prototaxites* that we have emphasized for *Protosalvinia* and *Nematothallus: Prototaxites* is another evolutionary "experiment" resulting in extinct organisms that bear no relationship to any other group of plants. The idea that the evolutionary "experiment" represented by *Prototaxites* was one leading to adaptation in a terrestrial environment is supported by comparisons of its chemistry with that of other fossil algal genera, *Protosalvinia, Parka,* and *Pachytheca,* and the vascular plant *Taeniocrada.* All of these genera have chemistries similar to *Prototaxites* (Niklas, 1976e) – chemistries that are associated with plants growing in or presumably adapting to terrestrial environments.

References

Andrews, H. N. (1959). Evolutionary trends in early vascular plants. *Cold Springs Harbor Symposia on Quantitative Biology*, **24**, 217–34.

Andrews, H. N. (1961). *Studies in Paleobotany*. New York: Wiley, p. 51.

Andrews, H. N., & Alt, K. S. (1956). *Crocalophyton* a new fossil plant from the New Albany Shale and some comments on the origin of land vascular plants. *Annals of the Missouri Botanical Garden,* **43,** 355–78.

Arnold, C. A. (1954). Fossil sporocarps of the genus *Protosalvinia* Dawson, with special reference to *P. furcata* (Dawson) comb. nov. *Svensk Botanisk Tidskrift,* **48,** 292–300.

Axelrod, D. I. (1959). Evolution of the psilophyte paleo flora. *Evolution* **13,** 264–75.

Bashnagel, R. A. (1942). Some microfossils from the Onondaga chert of central New York. *Buffalo Society of Natural Science Bulletin,* **17,** 1–8.

Butterfield, N. J., Knoll, A. H., & Swett, K. (1988). Exceptional preservation of fossils in an Upper Proterozoic shale. *Nature,* **334,** 424–7.

Chaloner, W. G. (1988). Early land plants – the saga of a great conquest. In *Proceedings of the XIV International Botanical Congress,* eds. W. Greuter & B. Zimmer. Königstein: Koeltz Scientific Books.

Croft, W. N., & George, E. A. (1959). Blue-green algae from the Middle Devonian of Rhynie, Aberdeenshire. *British Museum Geological Bulletin,* **3,** 341–53.

Dawson, J. W. (1859). On fossil plants from the Devonian rocks of Canada. *Quarterly Journal of the Geological Society of London,* **35,** 477–88.

Delevoryas, T. (1962). *Morphology and Evolution of Fossil Plants.* New York: Holt, Rinehart, and Winston.

Edwards, E. S., & Lyon, A. G. (1983). Algae from the Rhynie Chert. In *Contributions to Palaeobotany,* eds. M. W. Dick & D. Edwards. London: Academic Press, pp. 37–55.

Fairchild, T. R., Schopf, J. W., & Folk, R. L. (1973). Filamentous algal microfossils from the Caballos Novaculite, Devonian of Texas, *Journal of Paleontology,* **47,** 946–52.

Fry, W. L., & Banks, H. P. (1955). Three new genera of algae from the Upper Devonian of New York. *Journal of Paleontology,* **29,** 37–44.

Goldering, W. (1938). Algal barrier reefs in the lower Ozarkian of New York. *New York State Museum Bulletin,* **315,** 9–73.

Grambast, L. (1964). Precisions nouvelles sur la phylogenie des charophytes. *Natura Monspeliensia, Botanical Ser.,* **16,** 71–7.

Gray, J., & Boucot, A. J. (1979). The Devonian land plant *Protosalvinia, Lethaia,* **12,** 57–63.

Hemsley, A. R. (1989). The ultrastructure of the spores of the Devonian plant *Parka decipiens. Annals of Botany,* **64,** 349–67.

Johnson, J. H. (1961). *Limestone-Building Algae and Algal Limestones.* Boulder, Colo.: Johnson.

Johnson, J. H. (1966). A review of the Cambrian algae. *Quarterly of the Colorado School of Mines,* **61,** 1–161.

Kidston, R., & Lang, W. H. (1921). On Old Red Sandstone plants showing structure from the Rhynie Chert Bed, Aberdeenshire, Part V. The thallophytea occurring in the peat-bed: the succession of plants throughout a vertical section of the bed and conditions of accumulation and preservation of the deposit. *Transactions of the Royal Society of Edinburgh,* **52,** 855–902.

Krassilov, V. (1981). *Orestovia* and the origin of vascular plants. *Lethaia,* **14,** 235–50.

Leclercq, S. (1954). Are the Psilophytales a starting or a resulting point? *Svensk Botanisk Tidskrift,* **48,** 301–15.

Niklas, K. J. (1976a). Chemical examinations of some nonvascular Paleozoic plants. *Brittonia,* **28,** 113–37.

Niklas, K. J. (1976b). Morphological and chemical examination of *Courvoisiella ctenomorpha* gen. and sp. nov., a siphonous alga from the Upper Devonian, West Virginia, U.S.A. *Review of Palaeobotany and Palynology,* **21,** 181–203.

Niklas, K. J. (1976c). Morphological and ontogenetic reconstruction of *Parka decipiens* Fleming and *Pachytheca* Hooker from the Lower Old Red Sandstone, Scotland. *Transactions of the Royal Society of Edinburgh,* **69,** 483–99.

Niklas, K. J. (1976d). The chemotaxonomy of *Parka decipiens* from the Lower Old Red Sandstone, Scotland (U.K.). *Review of Palaeobotany and Palynology,* **21,** 205–17.

Niklas, K. J. (1976e). Chemotaxonomy of *Prototaxites* and evidence for possible terrestrial adaptation. *Review of Palaeobotany and Palynology,* **22,** 1–17.

Niklas, K. J., & Phillips, T. L. (1976). Morphology of *Protosalvinia* from the Upper Devonian of Ohio and Kentucky. *American Journal of Botany,* **63,** 9–29.

Phillips, T. L., Niklas, K. J., & Andrews, H. N. (1972). Morphology and vertical distribution of *Protosalvinia* (*Foerstia*) from the New Albany Shale (Upper Devonian). *Review of Palaeobotany and Palynology,* **14,** 171–96.

Pia, J. (1927). I Abteilung: Thallophyta. In *Handbuch der Paläobotanik,* ed. M. Hirmer. Munich and Berlin: Oldenbourg.

Schmid, R. (1976) Septal pores in *Protoaxites,* an enigmatic plant. *Science.* **191,** 287–8.

Schopf, J. M. (1978). *Forestia* and recent interpretations of early, vascular land plants. *Lethaia,* **11,** 139–43.

Schopf, J. W. (1970). Precambrian microorganisms and evolutionary events prior to the origin of vascular plants. *Biological Reviews,* **45,** 319–52.

Seward, A. C. (1898). *Fossil Plants.* Vol. 1. Cambridge: Cambridge University Press. Reprint. New York: Macmillan (Hafner Press), 1969.

Taylor, T. N. (1988). The origin of land plants: some answers, more questions. *Taxon,* **37,** 805–33.

Taylor, W. A., & Taylor, T. N. (1987) Spore wall ultrastructure of *Protosalvinia. American Journal of Botany,* **74,** 437–43.

Traverse, A. (1955). Occurrence of the oil-forming alga *Botrycoccus* in lignites and other Tertiary sediments. *Micropaleontology,* **1,** 343–50.

Wray, J. L. (1977). Late Paleozoic calcareous red algae. In *Fossil Algae: Recent Results and Developments,* ed. E. Flügel. Berlin: Springer, pp. 167–76.

7

How the land turned green: speculation

We are about to embark on an admittedly "backward outlook" in our search for the ancestors of green land plants. We will evaluate the characteristics of these most important of all organisms and then speculate what their ancestors were like. Delevoryas (1969) and others have warned against the pitfalls of this kind of an approach, but as we shall see, we have the advantage of evidence from the fossil record to temper our interpretations.

The organisms on which we focus our attention are those belonging to the divisions Bryophyta and Tracheophyta. The importance of the latter group cannot be overemphasized, especially when we realize that all green land plants that are of economic importance are tracheophytes. We can state unequivocally that without green land plants there would be no land animals, including humans. It follows that land plants evolved prior to land animals, providing them with food, shelter, and an environment where initially they could grow and reproduce with a minimal amount of competition.

All biologists agree, and an overwhelming amount of evidence supports the idea, that land-dwelling organisms evolved from aquatic ancestors. As we have learned, all organisms known in the Precambrian, Cambrian, and Ordovician lived in aquatic environments. The evolutionary conquest of land probably occurred sometime between the late Cambrian and early Silurian. Prior to this event, the land was barren, and its desiccating environment was a major obstacle to its colonization by land plants. If the plants were photosynthetic, they were confronted with yet another problem, that of obtaining carbon from the terrestrial environment for the manufacture of their food in the form of carbohydrates. In an aquatic environment, plants are literally bathed in a solution containing carbonates from which carbon can be extracted. How did primitive land plants adapt to overcome the problems imposed by the terrestrial environment – to overcome drying out and obtain carbon? When did this occur? Where? What kinds of aquatic organisms were their probable ancestors? These are but a few of the questions that are of great importance to those who study plant evolution.

The Chlorophyta as ancestors of green land plants

In an earlier chapter we alluded to the idea generally agreed to by botanists that green land plants evolved from ancestors belonging to the Chlorophyta. Their biochemical characteristics (the presence of chlorophylls *a* and *b*, true starch, and cellulose in their cell walls) point like an evolutionary signpost to green land plants. We know that green algae were present in the Cambrian and may have been a part of the late Precambrian floras. Biochemical evidence suggests that photosynthetic organisms were present some 3,400 m.y. ago, but these were probably autotrophic monerans. The first solid biochemical evidence of the Chlorophyta comes from the Upper Devonian in the form of the siphonaceous algae *Courvoisiella* (Niklas, 1976a). By isolating portions of the thallus from the matrix and submitting these to chemical analysis, Niklas was able to demonstrate cellulose in the filaments. Perhaps similar chemical analyses of presumed Lower Paleozoic Chlorophyta will be possible.

For the moment we have to be satisfied with evidence from comparative studies of extant Chlorophyta and their fossil record (Chapter 5). When we examine the Chlorophyta we discover that there are several evolutionary trends or lines within the division. Only one of these need be considered. This is the tetrasporine line. As explained earlier, we see that when cells of organisms belonging to the group divide they do so in the nonmotile condition and produce cells that are uninucleate. This is precisely what happens in multicellular green land plants. If there is a division of labor and cells of the alga divide in three or more planes, then a multicellular parenchymatous thallus will be produced. Some genera of the tetrasporine group, notably *Fritschiella* (Fig. 7.1A), exhibit this condition. *Fritschiella* has attracted much attention as being a grass-green alga that might represent a condition ancestral to land plants because it has prostrate, rhizoid-bearing, and erect portions to its thallus. Among the algae this is called the heterotrichous condition. This morphology may be reflected in the organization of green land plants where there may be a prostrate or underground parenchymatous portion as well as branched, erect parts.

Figure 7.1. **A.** *Fritschiella* sp. showing prostrate and erect portions of the multicellular thallus. **B.** Section through the upper portion of a *Marchantia* thallus showing relationship of a pore (p) to the underlying air chamber containing the photosynthetic cells. **C.** Stoma with paired guard cells from the sporophyte of *Anthoceros*. **D.** Simple embryo (e) of *Riccia* in the venter (v) of an archegonium. **E.** Archegonium of *Riccia* in longitudinal section. Egg (e); venter (v); neck cells (n). **F.** Longitudinal section of an antheridium of *Sphaerocarpos* sp. **G.** Tetrad of thick-walled spores, *Sphaerocarpos*. **H.** Diagrammatic representation of the longitudinal section of *Anthoceros* sporophyte attached to gametophyte (g) by a foot (f). All extant.

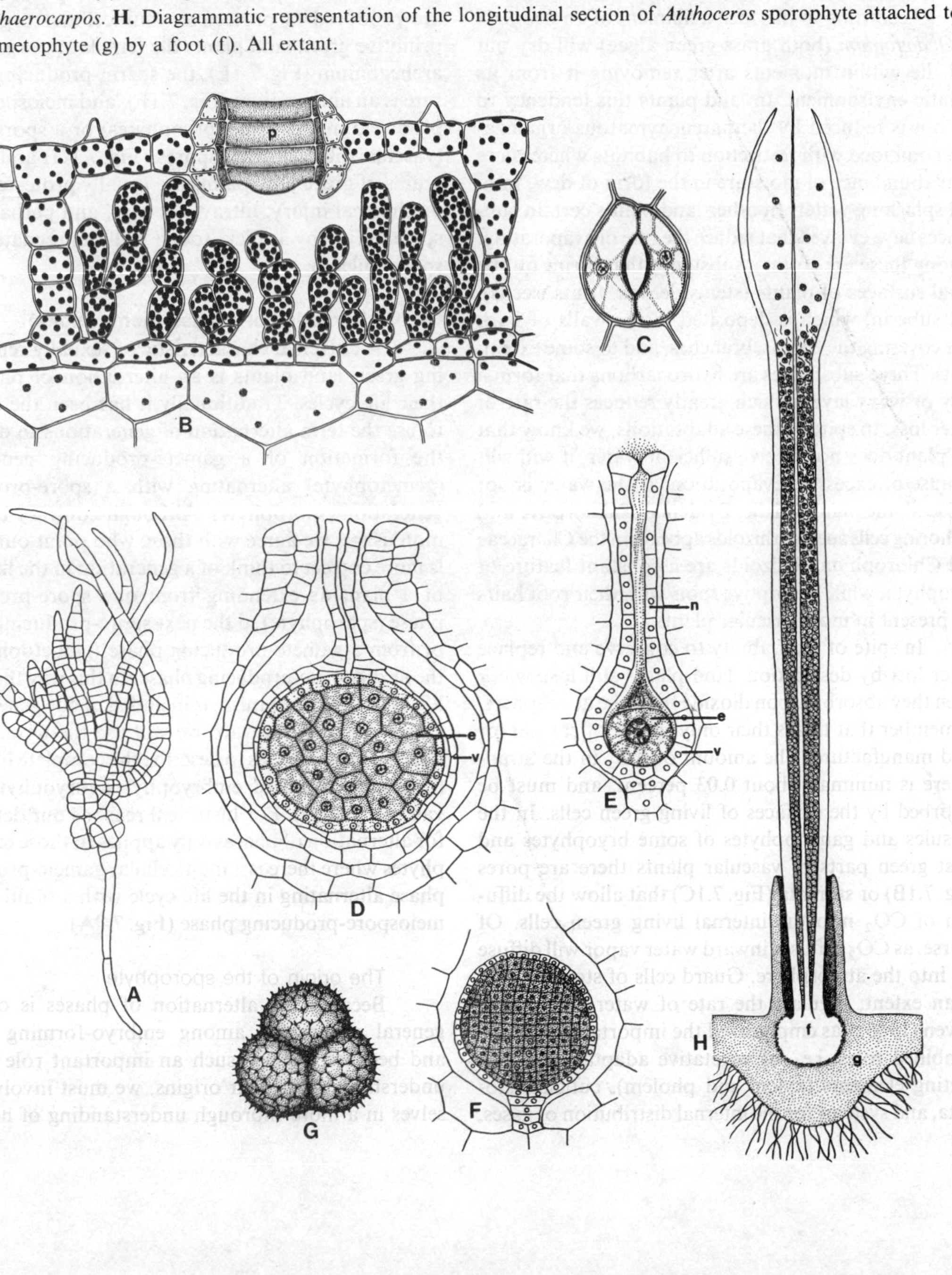

Vegetative adaptations

A parenchymatous organization is the only one amenable to a prolonged existence in the desiccating terrestrial environment. Not only does it provide a plant body where tissues can differentiate, but it also tends to reduce the relative rate of water loss. So if we compute the surface area of a filamentous green alga and that of a green land plant and compare the percentage of surface from which evaporation can take place in an aerial environment, it must be clear that the alga presents a relatively greater surface, per number of cells, to desiccation than the parenchymatous land plant. In absolute terms, however, the land plant, because of its greater surface area, will lose much more water.

According to this idea a filament of an *Ulothrix* or *Oedogonium* (both grass-green algae) will dry out and die within moments after removing it from its aquatic environment. In land plants this tendency to dry out is reduced by the parenchymatous organization combined with restriction to habitats where there is an abundance of moisture in the form of dew, rain, and splashing water. In other land plants certain substances have evolved that reduce the rate of evaporation. Among these are cutin, a substance that forms on the aerial surfaces of plants (stems, leaves, fruits, seeds), and suberin, which is deposited in the walls of bark cells covering the trunks, branches, and to some extent, roots. These substances are hydrocarbons that form a fatty or waxy layer, which greatly reduces the rate of water loss. In spite of these adaptations, we know that if a plant does not receive sufficient water, it will wilt because of excessive evaporation. If the water is not replaced, the plant will die. Evidence of absorptive and anchoring cells such as rhizoids appears in the Characeae and Chlorophyta. Rhizoids are a constant feature of Bryophyta, while absorptive roots with their root hairs are present in most vascular plants.

In spite of their ability to conserve and replace water lost by desiccation, land plants still lose water when they absorb carbon dioxide from the atmosphere. Remember that this is their only source of carbon for food manufacture. The amount of CO_2 in the atmosphere is minimal, about 0.03 percent, and must be absorbed by the surfaces of living green cells. In the capsules and gametophytes of some bryophytes and most green parts of vascular plants there are pores (Fig. 7.1B) or stomata (Fig. 7.1C) that allow the diffusion of CO_2 into the internal living green cells. Of course, as CO_2 diffuses inward water vapor will diffuse out into the atmosphere. Guard cells of stomata can, to an extent, regulate the rate of water vapor loss. Raven (1986) has emphasized the importance that the combined roles, i.e., of vegetative adaptations, conducting elements (xylem and pholem), cuticles, stomata, and systems for the internal distribution of gases, all play in promoting the survival of green land plants in a desiccating environment.

Reproductive adaptations

When green land plants reproduce, we find that their developing reproductive cells (eggs, sperm, and spores) also are protected from the effects of evaporation. With the exception of the Characeae, Chlorophyta have unicellular reproductive cells. In some, oogamous reproduction occurs and the zygote is retained in the oogonium for a period of time. This tendency to retain the zygote is amplified in bryophytes and vascular plants where it is retained in a multicellular reproductive organ to develop into a well-protected multicellular embryo (Fig. 7.1D). Sperms and meiospores likewise are produced in protective, multicellular organs. In primitive green land plants the female sex organ is an archegonium (Fig. 7.1E), the sperm-producing structure is an antheridium (Fig. 7.1F), and meiospores are produced in a capsule (bryophytes) or a sporangium (vascular plants). Meiospores (Fig. 7.1G) and pollen grains of green land plants are usually protected from mechanical injury, ultraviolet rays, and perhaps also some drying by a thick, tough wall impregnated with sporopollenin.

Alternation of phases (generations)

A universal characteristic of sexually reproducing green land plants is an alternation of phases in their life cycles. Traditionally it has been the custom to use the term alternation of generations to describe the formation of a gamete-producing generation (gemetophyte) alternating with a spore-producing generation (sporophyte). Although contrary to common usage, we agree with those who point out that it is more correct to think of a generation in the life cycle of a plant as extending from one spore-producing phase (sparophyte) to the next spore-producing phase or from a gamete-producing phase (gametophyte) to the next gamete-producing phase. In this way the whole life cycle of the organism is included in one generation. From this point onward we will use the terminology alternation of phases to describe the complete life cycle (one generation) of embryophytes (bryophytes and vascular plants). For historical reasons our definition for alternation of phases only applies to those embryophytes where there is a multicellular gamete-producing phase alternating in the life cycle with a multicellular meiospore-producing phase (Fig. 7.2A).

The origin of the sporophyte

Because the alternation of phases is of such general occurrence among embryo-forming plants and because it plays such an important role in our understanding of their origins, we must involve ourselves in a more thorough understanding of how the process evolved. Both Bower (1935) and Allen (1937) have excellent discussions of this topic, while Smith (1955) has a good summary. The basic premise assumes that the gamete-forming phase is primitive. It evolved first. This is well substantiated by evidence from many of the tetrasporine Chlorophyta. In the life cycles of these algae syngamy occurs in water, and a diploid zygote is formed. Meiosis takes place when the zygote germinates and haploid meiospores are produced. These develop into new haploid gamete-producing organisms. This type of life cycle is called haplontic (Fig. 7.2B).

According to the interpolation (antithetic) theory, the sporophyte phase is a new, multicellular structure that has been interposed between two successive gametophyte phases to produce a diplohaplontic life cycle (Fig. 7.2A,C). This is accomplished when the zygote germinates and divides mitotically to produce a number of diploid cells: a rudimentary sporophyte. This phase is terminated by meiosis in the diploid cells and the formation of meiospores. In this example the gametophyte and sporophyte phases are different in morphology; they are heteromorphic (Fig. 7.2C). There is little evidence among the algae that the sporophyte evolved by gradual intercalation. Life cycles of the Bryophyta, however, can be interpreted as showing step-by-step intercalation of a multicellular sporophyte phase. Thus, those bryophytes with simple sporophytes, as in *Riccia* (Fig. 7.1D), are considered primitive, and those with complex sporophytes – for example, *Anthoceros* (Fig. 7.1H) – are advanced. According to this theory, complexity has been achieved by "progressive sterilization of potentially sporogenous tissue." This is a rather pompous way of saying that as the sporophyte increased in size there were relatively fewer cells to produce meiospores when compared to the number of vegetative cells. If we adopt the interpolation theory for the origin of the sporophyte in bryophytes, it seems to provide us with a satisfactory explanation of the life cycles of mosses and liverworts where the sporophyte phase is dependent on a predominant and independent gametophyte. This dependency of the sporophyte is explained by the fact that the zygote germinates by mitosis within the archegonium to form the embryo. The embryo gets its nourishment from the photosynthetic gametophyte. In essence the sporophyte is parasitic on the gametophyte from the outset. As evolution of the sporophyte continued, more sterile tissues were produced, some of which became photosynthetic, thus giving the sporophyte a degree of independence, as is the case of *Anthoceros*. If we extend the idea of increasing independence of the sporophyte to tracheo-

Figure 7.2. **A.** Diplohaplontic life cycle of a hypothetical alga with isomorphic sporophyte and gametophyte phases. **B.** Haplontic life cycle of a hypothetical alga. **C.** Diplohaplontic life cycle of a hypothetical bryophyte with alternation of heteromorphic phases. **A–C:** All stippled structures outlined in heavy black lines are diploid portions of the sporophyte phases; all other parts are haploid structures of the gametophyte. Zygote (z); oogonium (o); archegonium (ar); antheridium (an); sperm (s); zoospores (zs); spores (sp).

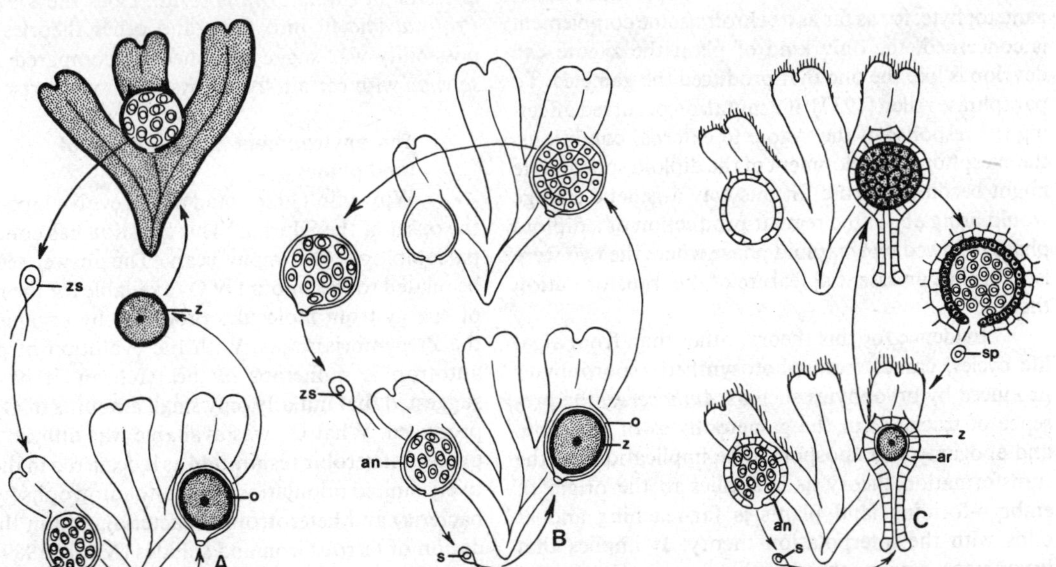

phytes, then we can explain why the sporophyte phase of vascular plants has become predominant and independent and differs in appearance from the gametophyte.

For many years Bower (1935) and many others thought this to be a satisfactory explanation of the evolutionary origin of the sporophyte phase in embryo-forming land plants. It explained how in evolution to a terrestrial environment the gametophyte could transport the sporophyte and protect it from the vicissitudes of land. Once adapted to the land environment the sporophyte gradually achieved independence by evolving photosynthetic and vascular tissues. In this way, it was believed, tracheophytes evolved from bryophyte ancestors.

Some years after the adoption of the interpolation theory and after much more had been learned about the life cycles of algae, it became clear that there was a second possible interpretation for the origin of the sporophyte phase, for which Bower proposed the name transformation (homologous) theory. This theory is based on the observation that many algae belonging to the Chlorophyta, Rhodophyta, and Phaeophyta have alternation of free-living isomorphic phases (Fig. 7.2A). Here we see that the gametophyte and sporophyte phases are identical in appearance. They differ only in the fact that the gametophyte is haploid and produces gametes and the diploid sporophyte produces meiospores. Advocates of the transformation theory believe that the sporophyte phase appeared suddenly, caused by a delay in meiosis from the time of zygote germination until a whole new diploid plant was formed. It is reasonable to assume that if exposed to the same external conditions the new diploid phase would have the same appearance as the gametophyte, for as far as its chromosome complement is concerned, the only kind of plant the zygote can develop is like the one that produced the gametes. To paraphrase Allen (1937), if a mutation occurred affecting the response of the zygote to external conditions, the exceptional development of the diploid sporophyte might become the rule. In this way a genetic change would bring about the regular production of a diploid phase followed by a haploid phase where the two were isomorphic, an essential feature of the transformation theory.

Evidence for this theory, other than from algal life cycles, comes from photosynthetic sporophytes produced by bryophytes such as *Anthoceros*, the presence of tracheids in the gametophytes of *Psilotum*, and apogamy and apospory. The implications of the transformation theory as it applies to the origin of embryo-forming land plants is far-reaching and at odds with the interpolation theory. It implies that bryophytes having photosynthetic sporophytes and gametophytes are primitive, not advanced; that vascular plants did not evolve from bryophytic ancestors, but independently from the Chlorophyta; and that the primitive type of land vascular plant should have alternation of isomorphic, free-living phases. Subsequent evolution in a desiccating land environment would produce an elaborated, free-living sporophyte and a reduced gametophyte (Fig. 7.5B). The reason for the emphasis on the progressive evolution of the sporophyte is clear. Sexual reproduction by the gametophyte phase of algae is dependent on the presence of free water. In transmigration to a land environment, access to water for sexual reproduction would be occasional at best, so the emphasis on the reproductive process shifted to that phase not requiring the aquatic medium. This was the sporophyte where small, well-protected reproductive structures – meiospores – were formed that could be widely distributed in the land environment. Here they could remain dormant for long periods until environmental conditions favored their growth. The production of large numbers of meiospores would also favor survival of the organism. To support the production of numerous meiospores the sporophyte must be photosynthetic and be able to supply quantities of water. It must produce its meiospores in a position on the plant favorable to their widest dispersal. This would be on the erect aerial parts of the sporophyte exposed to air currents or other dispersal mechanisms.

From the information presented thus far you will begin to understand how important evidence from the fossil record can be in solving the problem of the origin of green land plants from algal ancestors. Can we reconstruct the morphology and life cycle of the grass-green algal ancestor? Did bryophytes and vascular plants have independent origins among the algae, or did vascular plants evolve from some bryophytic ancestor not unlike *Anthoceros?* Does the Devonian *Protosalvinia* fit into these and other theories? This possibility was suggested when we compared *Protosalvinia* with certain liverworts and hornworts.

The environment and evolution of land plants

Why didn't green land plants evolve long before the onset of the Silurian? This question has concerned paleobiologists for many years. The answer seems to be related to the amount of O_2 available for the release of energy from molecules produced by organisms in the Precambrian seas. With the evolution of photoautotrophic monerans in the Archean, it has been suggested that initially only small amounts of O_2 were produced. What O_2 was available was utilized in the process of aerobic respiration as it occurred in the relatively limited populations of photoautotrophs (cyanobacteria) and heterotrophs (bacteria), and in the oxidation of ferrous iron and sulfides (Knoll, 1989). It is estimated that the amount of atmospheric O_2 at the end of the Early Proterozoic was probably less than 2 percent of the present 21 percent. The amount of O_2

increased in the Late Proterozoic as tectonic activity declined producing new, sizeable, and stable nearshore environments (Knoll, 1983, 1989). Here O_2-producing autotrophs grew, reproduced, and evolved. They increased in diversity and in numbers, and this allowed for the production of O_2 that exceeded, for the first time, the rate of O_2 consumption. In this way an O_2-rich atmosphere developed (Tiffney, 1985).

In addition to allowing the more efficient use of energy-rich molecules produced by photosynthetic organisms growing in the seas, the increase of O_2 levels in the atmosphere permitted the formation of ozone (O_3). It is well known that, at present, there is a layer of ozone in our outer atmosphere that acts as a filter removing most of the destructive ultraviolet rays originating from the sun. A sufficient accumulation of O_2 was required to produce an effective ozone layer that allowed eukaryotic organisms to survive near the surface of water and on land. Such a layer was probably in place by the early Silurian approximately 400 m.y. ago when the first evidence of multicellular green land plants appears.

Thus, we must conclude that a minimal concentration of O_2 in the atmosphere and a protective ozone layer were preconditions of the environment for the subsequent evolution of land plants and animals. Many other environmental factors, however, were of importance in the evolutionary process. If, for example, we agree that *Protosalvinia* is indeed a plant "caught in the act" of transmigration to land, we can gain some additional insight as to what the environment would be like. The fine-grained carbonaceous shales that contain the plants tell us that this was a low-energy shoreline (littoral) environment. Prior to lithification the sediments formed extensive mud flats in a marine or brackish-water estuary.

Streams and rivers flowing into an estuary would supply a source of freshwater containing a higher concentration of inorganic nutrients favoring plant growth. The encroachment of evolving autotrophs into the terrestrial environment and their subsequent death and decomposition must have provided organic nutrients and substrates for plants growing on land and in nearshore habitats, resulting in more favorable environments for the production of new, rapidly radiating populations. In short, for the first time the C, H, O, N cycles in nature included green land plants and their products, as well as providing the environment in which a terrestrial food chain could be established. Adaptation of evolving land plants also could have been affected by local epirogenic movements of Earth's crust causing cyclic emergence and subsidence in the littoral zone. In this way the various combinations of characteristics of the littoral inhabitants would be exposed to more prolonged periods of drying than those permitted by tidal action. Nearshore populations of these evolving plants would tend to be stabilized against the physical action of water by speeding the weathering of rocks, contributing to the formation of soil and slowing the rate of its erosion.

From time to time *Coleochaete* has received much attention as representing a possible prototype of parenchymatous green land plants (Graham, 1984, 1985). This possibility was further amplified when Pickett-Heaps (1975) discovered that cytokinesis (cell division) in this genus and a few other Chlorophyta, e.g., *Spirogyra* and the stonewort *Chara,* was similar to cytokinesis in bryophytes and vascular plants. Here, cell division takes place by the formation of a phragmoplast (cell plate). This structure is initiated at the onset of telophase of nuclear division when fibrils, like those of the mitotic spindle, are formed between daughter nuclei. The fibrils consist of bundles of microtubules that are oriented at right angles to the plane of cell division. In the vast majority of green algae, a phycoplast is formed during cytokinesis where the microtubules are oriented in the plane of cell division. The discovery of a phragmoplast among a few Chlorophyta only amplifies the possible origin of green land plants from green algal ancestors.

Vegetatively *Coleochaete* (Fig. 7.4A) consists of a flat plate of cells attached to the substrate. The plate of cells is comprised of short radiating files of cells appressed laterally to form a monostromatic, pseudoparenchymatous thallus. When it reproduces, internal cells of the thallus may form sperms or zoospores. Some marginal cells metamorphose into egg cells (oogonia), which subsequently are enveloped by further growth of the vegetative filaments. After fertilization, the zygote formed by the union of egg and sperm, remains in the oogonium for some time where it germinates in place by meiotic divisions. No other green alga is known to exhibit this characteristic. The propensity to retain the zygote, which then germinates in the female reproductive structure, is a characteristic shared with all embryophytes (bryophytes and vascular plants).

In Chapter 6 we reported the elegant work of Niklas (1976b), who used mathematical models to demonstrate that ontogenetic stages of *Coleochaete* and the Upper Silurian to Middle Devonian genus *Parka* (Fig. 6.3B) are very similar. The sporangia of *Parka* have been tentatively homologized with the oogonia of *Coleochaete* (Taylor, 1988). The sporangia of the fossil are located on the upper surface of the thallus (Fig. 7.4B) and contain many dense clusters of spore-like cells (Fig. 7.4C). Taylor suggests that the so-called spores could be thick-walled zygotes.

In addition to the biochemical similarities between *Parka* and *Coleochaete* already described by Niklas (1976c), Delwiche, Graham, & Thomson (1989) have found ligninlike compounds in *Coleochaete* and the bryophyte *Anthoceros* (Fig. 8.1D). They also identified a highly resistant material, sporopollenin, in the zygote

Figure 7.3. **A.** Hypothetical primitive vascular plant. (Redrawn from Lignier, 1908.) **B.** Diagram of hypothetical algal ancestor of green land plants (bryophytes and tracheophytes) with diplohaplontic isomorphic phases. (Redrawn in Stewart, 1964, from Zimmermann, 1952.)

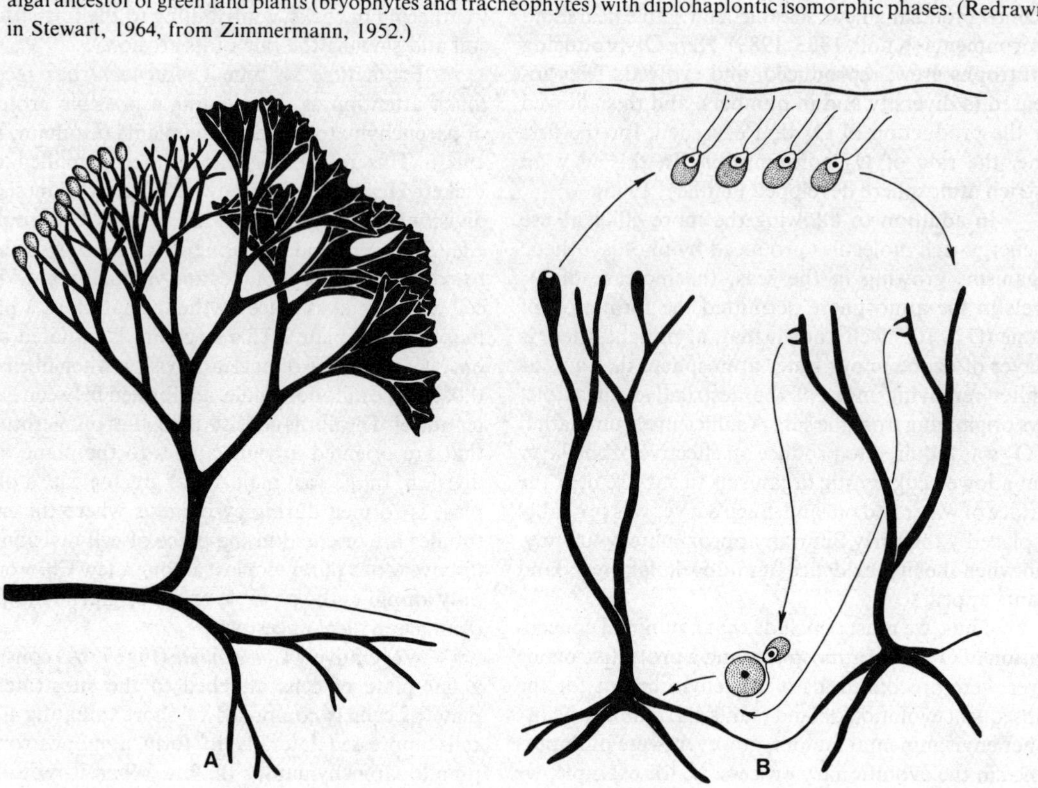

Figure 7.4. **A.** Vegetative thallus of *Coleochaete* (extant). **B.** Reconstruction of *Parka*, upper surface showing location of sporangium-like reproductive structures. **C.** *Parka* reproductive structures containing thick-walled sporelike bodies. **B,C:** Upper Silurian – Lower Devonian. (**B,C** from Taylor, 1988.)

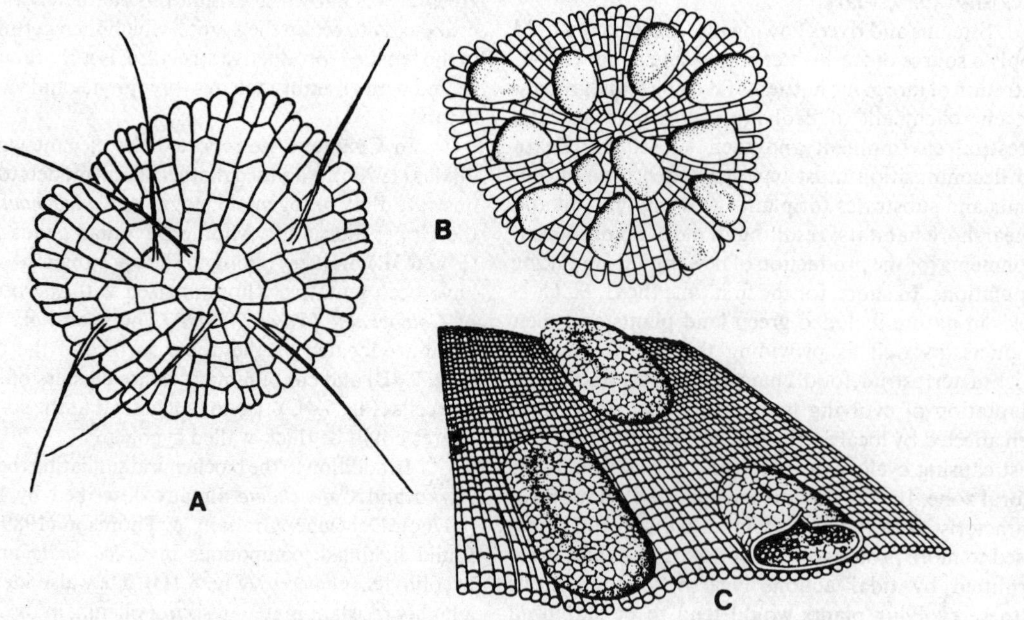

walls of *Coleochaete,* a substance found by Niklas (1976b) to occur in the regions of the reproductive organs of *Parka* and the zygotes of extant *Chara.* Because ligninlike compounds and sporopollenin are components of most embryophytes, Delwiche et al. support the conclusion arrived at by cladistic analyses, that *Coleochaete* is a sister taxon to embryophytes. More specifically, they interpret the evidence as indicating the possible descent of *Anthoceros* from a *Coleochaete*-like ancestor.

Based on cytological evidence, biochemical similarities, similar vegetative morphologies, and the tendency to retain the zygote in the oogonium until after germination, many morphologists have come to favor an origin of green land plants from a *Coleochaete* prototype. We are reminded by Taylor (1988), however, that at least 400 m.y. have elapsed without additional fossil evidence to bridge the gap between the Silurian–Devonian *Parka* and the extant freshwater alga, *Choleochaete*. And, as Chaloner (1988) points out, there is no evidence from the fossil record to link present-day *Coleochaete* with any of the earliest vascular land plants. With these observations, these authors make it clear that we should exercise considerable caution in arriving at conclusions about the origin of green land plants when using such a "backward outlook."

Hypothetical primitive land plants

Using speculations of the kind presented in this chapter and available evidence from extant and fossil plants, Lignier (1908) and Zimmermann (1952, 1965) synthesized hypothetical ancestors of green land plants. Lignier suggested that the land plants evolved from Chlorophyta with a dichotomizing parenchymatous thallus (Fig. 7.3A). As transmigration to land occurred a basal prostrate portion entered the soil to function in anchorage and absorption. Erect parts retained the photosynthetic function with some portions of the branching system becoming flattened and filled in with photosynthetic tissues to form rudimentary leaves. Some aerial portions with terminal sporangia retained the primitive three-dimensional dichotomous branching system. Although Lignier was aware of fossil evidence

Figure 7.5. **A.** Diagrammatic life cycle of a hypothetical bryophyte showing sporophyte dependent on gametophyte. **B.** Diagrammatic life cycle of a hypothetical tracheophyte with predominant sporophyte and reduced gametophyte. **(A,B redrawn in Stewart, 1964, from Zimmermann, 1952.)**

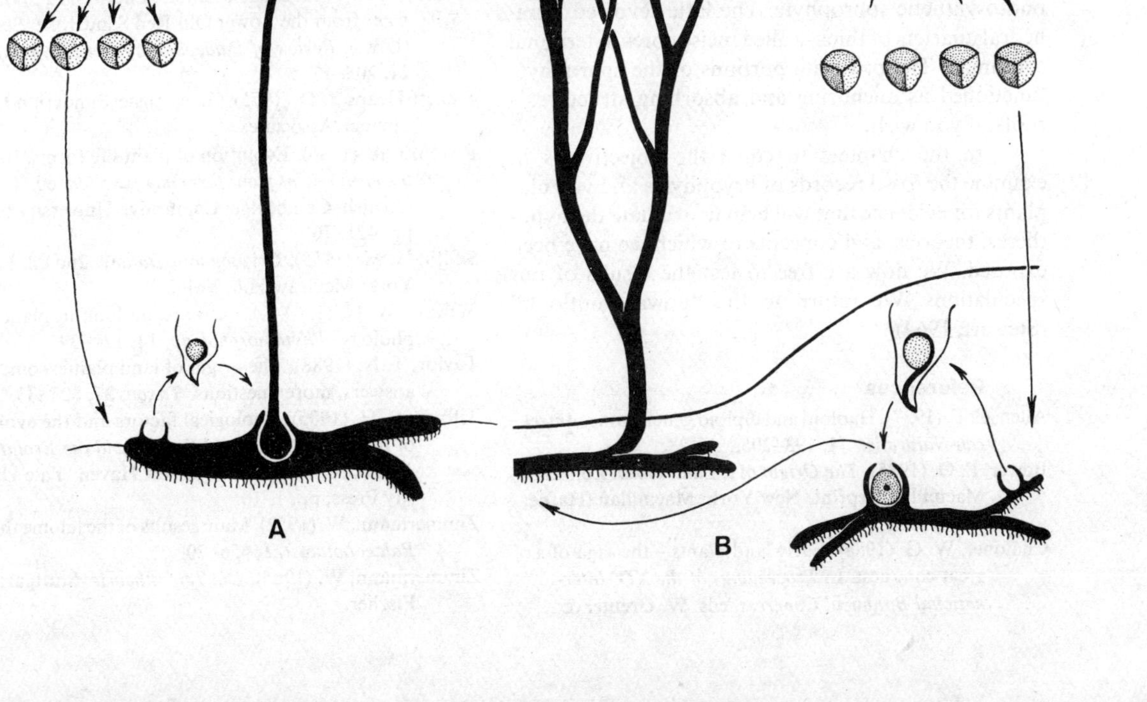

showing primitive vascular plants (psilopsids and lycopsids), his reconstruction of a hypothetical green land plant was prophetic of evidence that was to be discovered at a later date.

By 1938 Zimmermann had formulated his telome theory, but its foundations already had been laid by a host of evolutionarily oriented botanists. Much new information explaining life cycles of algae and bryophytes was available. The classical works of Kidston & Lang (1921) describing primitive vascular plants in the Devonian had appeared. In other words the telome theory is a synthesis of information accumulated prior to 1938. Like Lignier, Zimmermann believed that green land plants evolved from grass-green algae with parenchymatous three-dimensional dichotomizing axes. These axes were heterotrichous, having both prostrate and erect portions. Zimmermann further speculated that the algal ancestor had a diplohaplontic life cycle with isomorphic alternating phases (Fig. 7.3B). He adopted the ideas that the sporophyte bore its sporangia in the terminal position, that the meiospores produced by the sporangia were free-swimming, and that the gametophyte had oogamous sexual reproduction with syngamy occurring in water. From this basic hypothetical type, Zimmermann conceived two main lines of evolution, one leading to the bryophytes, the other to vascular plants. In adopting the transformation theory for the origin of the sporophyte, phase he believed that those bryophytes (Fig. 7.5A) with complex and partially independent sporophytes (for example *Anthoceros*) are primitive and "caught-in-the-act" of becoming reduced and dependent on the photosynthetic gametophyte.

For the evolutionary origin of vascular land plants from the hypothetical algal ancestor, Zimmermann shows us (Fig. 7.5B) that the oogamous gametophyte phase is reduced when compared with the photosynthetic sporophyte. The latter evolved tetrahedral quartets of thick-walled meisopores in terminal sporangia. The prostrate portions of the sporophyte functioned as anchoring and absorbing structures – roots, if you wish.

In the chapters to come the objective is to examine the fossil records of bryophytes and vascular plants for evidence that will help us evaluate the hypotheses, theories, and concepts to which we have been exposed. We now are free to test the results of our speculations. We return to the "upward outlook" (Stewart, 1964).

References

Allen, C. E. (1937). Haploid and diploid generations. *American Naturalist,* **71,** 193–205.

Bower, F. O. (1935). *The Origin of a Land Flora.* London: Macmillan Reprint. New York: Macmillan (Hafner Press) 1959.

Chaloner, W. G. (1988). Early land plants – the saga of a great conquest. In *Proceedings of the XIV International Botanical Congress,* eds. W. Greuter & B. Zimmer. Königstein: Koeltz Scientific Books.

Delevoryas, T. (1969). Paleobotany, phylogeny, and a natural system of classification. *Taxon,* **18,** 204–12.

Delwiche, C. F., Grahm, L. E., & Thomson, N. (1989). Lignin-like compounds and sporopollenin in *Coleochaete,* an algal model for land plant ancestry. *Science,* **245,** 399–401.

Graham, L. E. (1984). *Coleochaete* and the origin of land plants. *American Journal of Botany,* **71,** 603–8.

Graham, L. E. (1985). The origin of the life cycle of land plants. *American Scientist,* **75,** 178–86.

Kidston, R., & Lang, W. H. (1921). On Old Red Sandstone plants showing structure from the Rhynie chert bed, Aberdeenshire. Part V. The Thallophyta occurring in the peat bed: the succession of plants throughout a vertical section of the bed and the conditions of accumulation and preservation of the deposit. *Transactions of the Royal Society of Edinburgh,* **52,** 855–902.

Knoll, A. H. (1983). Tectonics, productivity, and ecosystems on early earth. In *Proceedings Third North American Paleontological Conference,* **2,** 307–11.

Knoll, A. H. (1989). Precambrian evolution of prokaryotes and protists. Article prepared for the *Encyclopedic Dictionary of Palaeobiology.*

Lignier, O. (1908). Essai sur l'èvolution morphologiqué de règne végétal. *Compte Rendu de l'Association Français pour l'Avance du Science* (Clermont-Ferrand), **37,** 530–42.

Niklas, K. J. (1976a). Morphological and chemical examinations of *Courvoisiella ctenomorpha* gen. and sp. nov.: a siphonous alga from the Upper Devonian, West Virginia, U.S.A. *Review of Palaeobotany and Palynology,* **21,** 187–203.

Niklas, K. J. (1976b). Morphological and ontogenetic reconstruction of *Parka decipiens* Fleming and *Pacytheca* Hooker from the Lower Old Red Sandstone, Scotland. *Transactions of the Royal Society of Edinburgh,* **69,** 483–99.

Niklas, K. J. (1976c). The chemotaxonomy of *Parka decipiens* from the Lower Old Red Sandstone, Scotland (U.K.). *Review of Palaeobotany and Palynology,* **21,** 205–17.

Pickett-Heaps, J. D. (1975). Green algae. Sunderland, Mass.: Sinauer Associates.

Raven, J. A. (1986). Evolution of plant life forms. In *On the economy of plant form and function,* ed. T. J. Givnish, Cambridge: Cambridge University Press, pp. 421–76.

Smith, G. M. (1955). *Cryptogamic Botany,* 2nd ed. New York: McGraw-Hill, Vol. 2.

Stewart, W. N. (1964). An upward outlook in plant morphology. *Phytomorphology,* **14,** 120–34.

Taylor, T. N. (1988). The origin of land plants: some answers, more questions. *Taxon,* **37,** 805–33.

Tiffney, B. H. (1985). Geological factors and the evolution of plants. In *Geological Factors and the Evolution of Plants,* ed. B. H. Tiffney. New Haven: Yale University Press, pp. 1–10.

Zimmermann, W. (1952). Main results of the telome theory. *Palaeobotanist,* **1,** 456–70.

Zimmermann, W. (1965). *Die Telomtheorie.* Stuttgart: Fischer.

8

How the land turned green: Bryophyta

The division Bryophyta (liverworts, mosses, and hornworts) is the only division that has been retained as a natural group since it was established long before the turn of the twentieth century. Some bryologists, however, question the implication that the classes comprising the Bryophyta had a common ancestry and are monophyletic. Without going into their reasons, Bold, Alexopoulos, & Delevoryas (1980) and Crandall-Stotler (1980) believe that there are at least three independent lines of bryophytes and that this is best reflected by establishing three divisions – the Bryophyta (mosses), Hepatophyta (liverworts), and Anthocerotophyta (hornworts). Studies using cladistic techniques tend to substantiate this polyphyletic origin of bryophytes. However, before we look at the fossil record to see if it can help in determining the monophyletic or polyphyletic nature of these plants, we should get some idea of what characteristics bryophytes have in common. First, they are all embryo-forming plants. It has been customary to distinguish them from embryo-forming tracheophytes by their lack of vascular tissue. Many bryologists do not accept this distinction and make the valid point that many bryophytes have conducting tissues that are similar in structure, organization, and function to vascular tissues of the Tracheophyta. They suggest the possibility that primitive vascular plants could have been the ancestors of bryophytes, that, by reduction, the vascular tissue has either been lost or modified. If we accept this hypothesis, this leaves us with just one good criterion to distinguish bryophytes from vascular plants. This has to do with their life cycles where there is an alternation of heteromorphic phases in which the sporophyte is dependent on the dominant independent gametophyte (Fig. 7.5A).

For our purposes we need to know the general characteristics of the major groups of the Bryophyta. Most members of the classes Hepaticeae (liverworts) and Anthocerotae (hornworts) have green gametophytes that are dorsiventral. Dichotomous branching of the gametophyte is common but not universal. Some liverworts belong to the general category of thalloid liverworts. *Marchantia* (Fig. 8.1A) and *Riccia* (Fig. 8.1B) are examples. Other liverworts, *Porella* (Fig. 8.1C) and *Fossombronia* (Fig. 8.1E), show varying degrees of differentiation into stemlike and leaflike parts. *Anthoceros* (Fig. 8.1D) has a gametophyte that superficially resembles that of a thalloid liverwort but lacks internal differentiation. Unlike other bryophytes, the hornshaped sporophytes of *Anthoceros* are photosynthetic at maturity and capable of indeterminant growth. These are features that will be of importance to us later.

The class Musci comprises the true mosses (Eubrya) represented by *Polytrichum* (Fig. 8.1F) and bog mosses (Sphagnobrya) with a single extant genus, *Sphagnum* (Fig. 8.1G). *Polytrichum* and *Sphagnum* show the upright, radially symmetrical gametophores characteristic of the class. The gametophores, which arise from a protonema, generally have spirally arranged leaves. Leaves of true mosses are differentiated into a median costa (midrib) and lamina, a feature that distinguishes their leaves from those of leafy liverworts and bog mosses. In bog mosses the leaf cells are of two types, a reticulum of slender photosynthetic cells separated by islands of large colorless water storage cells.

Sporophytes of true mosses are terminal or lateral in position on branches of the gametophores. They have complex meiospore-producing capsules connected by a long seta to a foot embedded in the gametophytic tissue. The sporophyte of *Sphagnum* is much less conspicuous. It is characterized by a columella (a dome of sterile tissue that extends into the capsule of the sporophyte). We have chosen to describe these extant bryophytes and emphasize

Figure 8.1. A variety of extant bryophytes. **A.** Female plant of *Marchantia*, a thallus liverwort. **B.** *Riccia natans*. **C.** *Porella*, a leafy liverwort. **D.** *Anthoceros* gametophyte with three sporophytes. **E.** *Fossombronia* showing partial differentiation into leaflike and stemlike parts. **F.** The common moss *Polytrichum* with sporophyte attached. Note well-differentiated stemlike and leaflike parts. **G.** *Sphagnum*, the bog-moss with sporophytes at the tips of gametophytic branches.

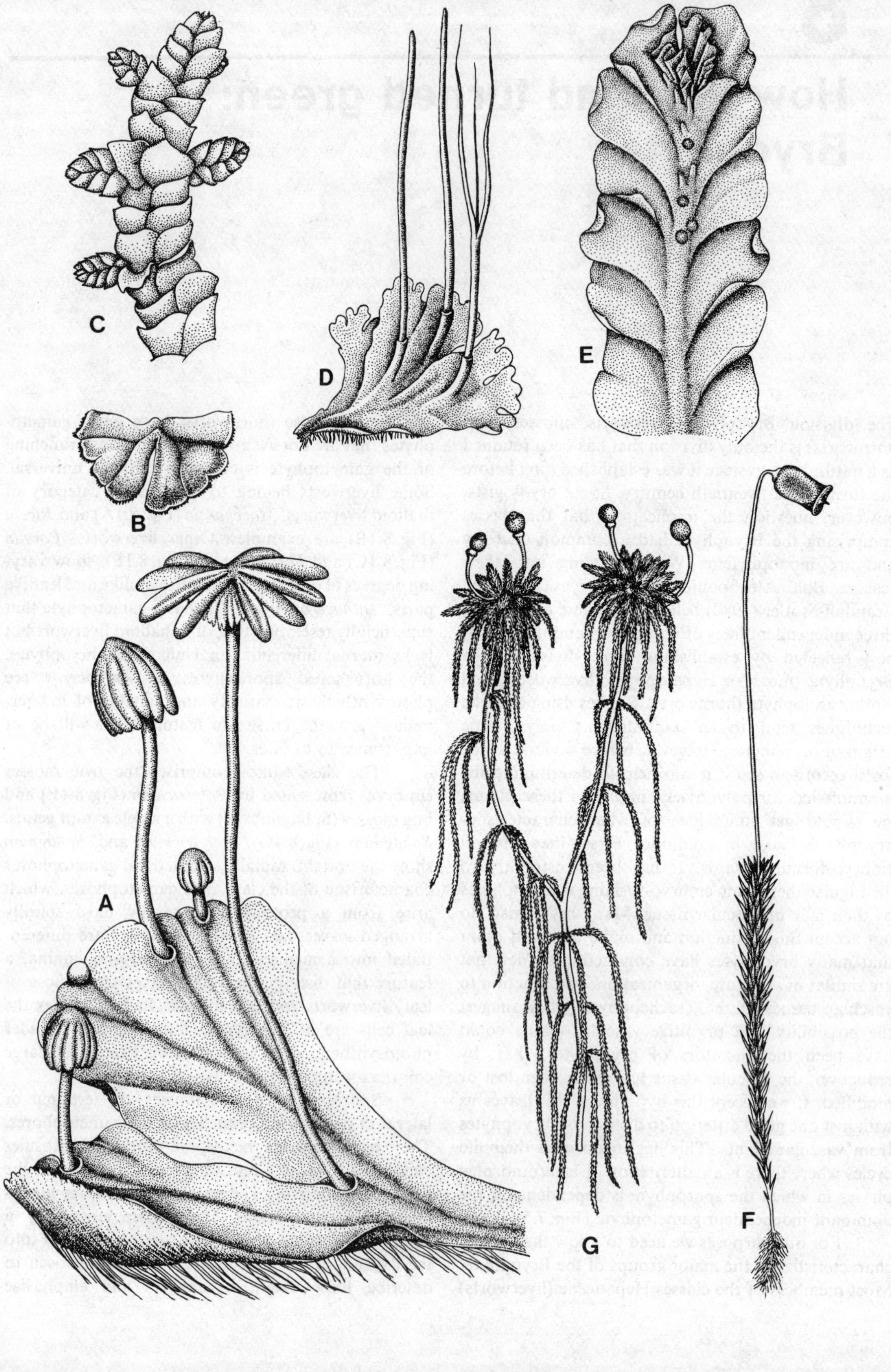

certain characteristics because they are the ones that are important in the interpretation of the bryophyte fossil record.

There are some puzzling aspects of this fossil record. It is usually maintained that we should expect a poor record because of the delicate nature of many bryophytes, especially liverworts. Yet some of the oldest and best-preserved are fossil liverworts. Our colleagues, after many years of studying coal-ball material with superb preservation, have noticed the apparent lack of bryophyte remains. One would think that the Carboniferous swamp environment, which supported such a wealth of vascular plants, would have been a favorable environment for liverwort-like bryophytes.

During the last five decades many additions have been made to the bryophyte record. By far the most comprehensive account of this record was prepared by Jovet-Ast (1967). A critical review of Cenozoic and Mesozoic bryophytes was synthesized by Steere (1946), while Savicz-Ljubitzkaja & Abramov (1959) and Lacey (1969) added to our general information about the evolutionary history and paleoecology of bryophytes. Other, more specific contributions will be noted as we proceed.

Liverworts

The oldest unequivocal fossil bryophyte is a small ribbonlike thalloid liverwort, *Pallavicinites devonicus* (Hueber, 1961; Schuster, 1966). The fossils are compressions found in rocks of Upper Devonian age (Fig. 8.2A). Specimens obtained by maceration show a surprising amount of cellular detail, including cells of a dichotomizing apex (Fig. 8.2B). The characteristics displayed by the fossils enabled Hueber to suggest affinities with the Jungermanniales–Anacrogynae (thallus liverworts). There is an unusually high degree of similarity between the extant genus *Pallavicinia* and *Pallavicinites devonicus*. It is safe to say that many paleobotanists were amazed at this similarity between these Devonian fossils and extant liverworts. Even more beautifully preserved specimens of anacrogynous liverworts were discovered by Walton (1925, 1928) from the Carboniferous. One is a leafy form, *Hepaticites kidstoni* (Fig. 8.3G), which is likened to the extant anacrogenous *Treubia*. *Thallites willsi* (Fig. 8.3E,F) has a dorsiventral thallus that dichotomizes. It looks like a modern *Metzgeria*. In the absence of conclusive evidence of their exact affinities, however, it is best to consider *Hepaticites* and *Thallites* to be form genera. There are at least ten accounts of Mesozoic *Thallites* and eight of *Hepaticites*. The report of *Metzgerites* from the Triassic only confirms that the anacrogenous liverworts have a more or less continuous fossil record initiated in the Devonian.

If recent interpretations of the fossil members of the Jungermanniales–Acrogynae (leafy liverworts) are correct, this group did not make its appearance much before the Tertiary. Specimens resembling extant leafy liverworts are assigned either to the form genus *Jungermannites* or to extant genera. Interestingly, some specimens of leafy liverworts and mosses have been found preserved in the famous Oligocene Baltic amber. These are assigned to modern genera.

Critical examination and evaluation of those fossils placed in the Marchantiales has been accomplished by Steere (1946) and Lundblad (1954). All questionable specimens have been excluded so that the earliest specimen showing unequivocal relationships with the Marchantiales is *Hepaticites cyathodoides* from the Middle Triassic. Other than the presence of an undifferentiated, dorsiventral dichotomizing thallus (Fig. 8.3H), pores with air chambers (Fig. 8.3D,I), and small scales on the ventral surface

Figure 8.2. **A.** *Pallavicinites devonicus*, fragment of thallus showing differentiation of axis with laminalike portion. **B.** *P. devonicus*, apex of thallus showing dichotomy and cellular detail. Upper Devonian. (From Hueber, 1961.)

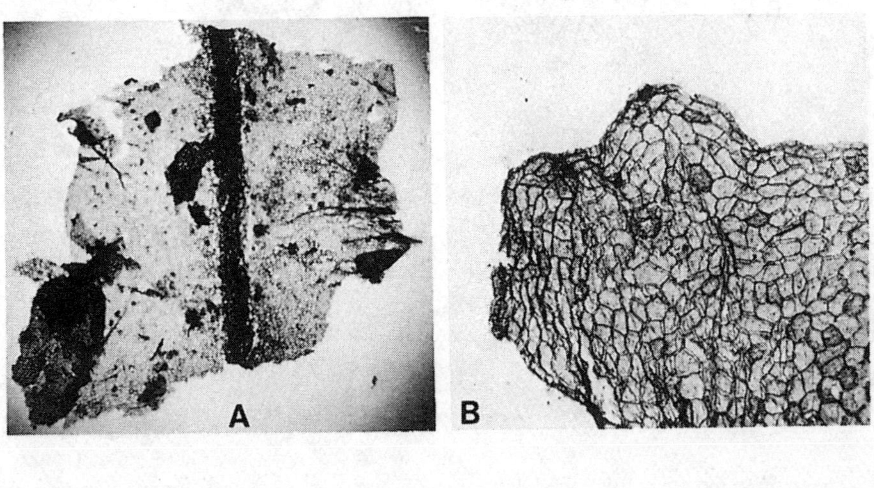

are also accepted as characteristics that allow assignment to the Marchantiales. A fine flora of liverworts has been described by Lundblad (1954) from the Jurassic–Triassic rocks of Sweden. Among these is the genus *Ricciopsis* (Fig. 8.3C) that bears a striking resemblance to the modern *Riccia* (Fig. 8.1B). The most complete specimen of a *Marchantia*-like fossil is the Eocene *Marchantites sezannensis*. In addition to the characteristics described above, this specimen bore antheridial gametophores similar to an extant *Marchantia*. As in the Jungermanniales, many Cenozoic members of the Marchantiales are like extant genera. You may have noticed that there is a predictable pattern where many genera of Tertiary floras bear a marked resemblance to extant ones. This has been repeated in every group of land plants we have encountered including the fungi. Further, much of the diversification of species first appears in the Upper Mesozoic and Cenozoic.

The enigma, Naiadita

As we have gone from group to group, we have emphasized one or two investigations that stand out among the others because of beautifully preserved material that reveals most of the stages in the life cycle of the ancient organism. The work by Harris (1938) on *Naiadita lanceolata* (Fig. 8.3A) is in this category. This Triassic genus is unquestionably a liverwort, but its unusual combination of characteristics makes it impossible to assign to a family let alone a genus of extant Hepaticae. An affinity with the Sphaerocarpales is suggested, especially the extant genus *Riella*. Like *Riella, Naiadita* has unicellular rhizoids, sessile archegonia with a venter wall one cell layer thick, and leaf cells of a similar shape. The gametophores were probably erect with spirally arranged leaves like those of most mosses. The leaves, however, lack a costa so characteristic of mosses. In their placement, archegonia are lateral on the gametophore axis and apparently replaced the leaves. Some archegonia are enclosed in a "perianth" of several leaflike lobes. Even sporophytes (Fig. 8.3B) were observed by Harris at the tips of a short lateral branch of the gametophore. The sporophyte with its poorly developed bulbous foot and globose sporangium is similar to that of *Corsinia*, an extant member

Figure 8.3. **A.** Habit drawing of *Naiadita lanceolata*. Triassic. **B.** *N. lanceolata*, sporophyte(s) at tip of branch. **(A,B** redrawn from Harris, 1938.) **C.** *Ricciopsis florinii*. (Compare with Figure 8.1B.) Triassic. (Redrawn from Lundblad, (1954.) **D.** *Marchantiolites porosus*. Highly magnified pore in upper surface of thallus. Triassic. (Redrawn from Lundblad, 1954.) **E.** *Thallites willsi*. Fragment of thallus showing dichotomous branching. **F.** *T. willsi*. Enlarged tip of thallus showing position of apical cells (a). **E,F:** Upper Carboniferous. **(E,F** drawn from photographs by Walton, 1925.) **G.** *Hepaticites kidstoni*. Lower surface of thallus showing differentiation into leaflike parts as in laefy liverworts, Figure 8.1C. Upper Carboniferous. (Drawn from photograph by Walton, 1925.) **H.** *Marchantites gracilis*, dichotomizing thallus. **I.** *M. gracilis*, upper surface magnified to show pores. (Redrawn from Troll, 1927.)

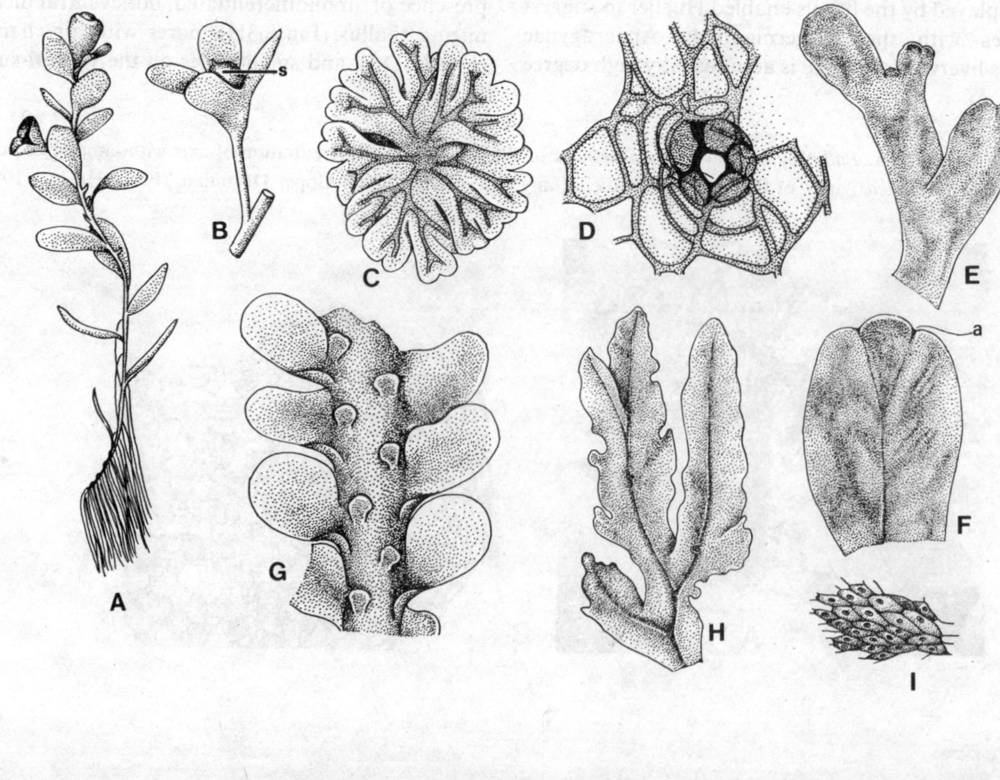

of the Marchantiales. *Naiadita* is another fossil enigma. It is tempting to think of this genus as one that synthesizes characteristics of the Musci and Hepaticae. Because both Hepaticae and Musci are known to have occurred long before Triassic times it seems less than likely that *Naiadita* represents an ancestor that gave rise to these two groups. It is fair to say that this genus raises more questions than it answers about bryophyte evolution.

Spore types from the Lower Carboniferous that look like those of some Spaerocarpales have been found as early as the Upper Cretaceous (Jarzen, 1979).

Mosses

Fossil bryophytes with a distinctive leaf cell pattern like that of bog mosses have been reported from the Permian of Russia. Unlike *Sphagnum* leaves found later in the fossil record and those of extant

Figure 8.4. **A.** *Aulacomnium heterostichoides*. Habit. Eocene. (From Janssens, Horton, & Basinger, 1979.) **B.** *Ditrichites* sp. Eocene. (From Chandrasekharam, 1974.)

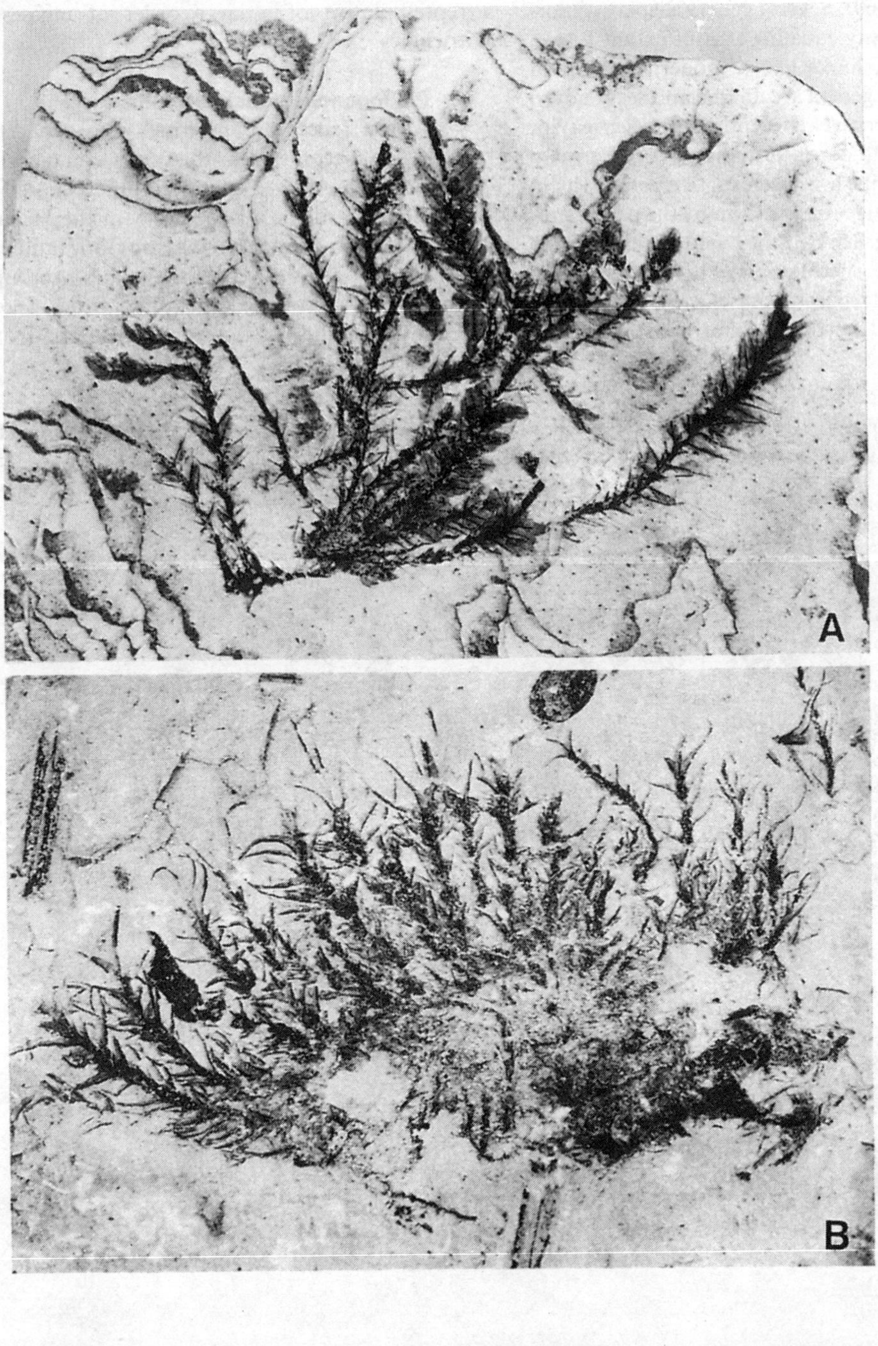

Sphagnum, the leaves have midribs and the occasional lateral vein. There are three genera, one of which is *Protosphagnum* placed in the order Protosphagnales. Since there is no more recent fossil record of this order, and unequivocal remains of Sphagnales do not appear prior to the lower Jurassic, it has been suggested that remains of the Protosphagnales were precursors of today's bog mosses.

The oldest known fossil validated by the experts as a true moss (Eubyra) is the Carboniferous *Musites polytrichaceus*. It was described by Renault & Zeiller in 1888. Certainly the most beautifully preserved and thus convincing fossil mosses of pre-Tertiary age are those described from Permian permineralized specimens (Neuberg, 1960). Several genera display a morphological complexity equaling that of extant forms. *Intia*, for example, looks like a modern *Mnium* or *Bryum*, while other genera are unique and seem not to have extant counterparts. Recently Smoot & Taylor (1986) have added to our information about Permian mosses. They describe a new genus, *Merceria*, which, based on anatomical features shown by the permineralized specimen (Fig. 8.5A,B) they were able to assign to the Bryales. The specimen, which came from the late Permian of Antarctica, represents the oldest anatomically preserved bryophyte known from Gondwanaland. By the early Cenozoic most mosses can be referred to extant genera, with increasing numbers of species again appearing in the Tertiary and Quaternary. An excellent example of a Tertiary (Eocene) moss, that is generically identical to an extant genus, is *Aulacomnium heterostichoides* (Janssens, Horton, & Basinger, 1979). With the exception of a few minor leaf characteristics this fossil moss (Fig. 8.4A) has the characteristics of the modern *Aulacomnium heterostichum*. It is worth noting that the Northern Hemisphere moss flora is very much like the present one. Although the details of the leaves are not well preserved and no sporophytes are available, the Paleocene moss (Fig. 8.4B) *Ditrichites* is probably a representative of a large order of mosses, the Ditrichales.

Phylogenetic considerations

As we reflect on the fossil record of known bryophytes it becomes clear that the two major groups (Musci and Hepaticae) were well differentiated long before the end of the Paleozoic, with the Hepaticae appearing before the Musci. Watson (1967) and others believe that only the Jungermanniales–Acrongeny, Anthocerotales, and one other small order were not represented in the fossil record by the end of the

Figure 8.5. *Merceria angustica*. **A.** Transverse section of axis showing differentiated central column and surrounding cortex. **B.** Leaf in transverse section showing one-cell-thick lamina and midrib. Permian. (From Smoot & Taylor, 1986.)

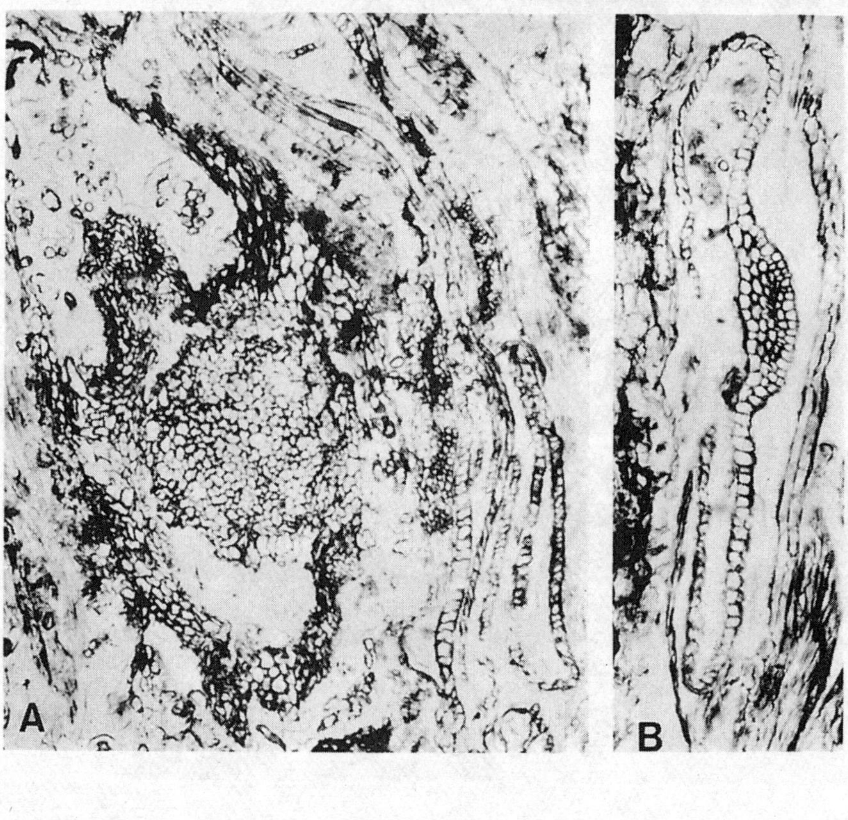

Permian. With this information in mind and the obvious lack of any fossil evidence linking the Musci and Hepaticae, we are obliged to adopt the interpretations of Steere (1958), Lacey (1969), & Crandall-Stotler (1980) that the Bryophyta are polyphyletic in origin. This origin must be placed at least as far back as the Upper Devonian, and we can assume that evolution of liverworts occurred earlier, with the mosses diverging at a later time. Thus far there is no evidence supporting this supposition. No primitive bryophyte type that can be considered ancestral to both groups has yet been recognized. We still must depend on students of extant bryophytes to determine the relationships among mosses, liverworts, and hornworts, if any. Taylor (1988) has summarized the various often-conflicting articles on the phylogenies of bryophytes based on extant representatives. This and the lack of definitive fossil record has prompted Smith (1986) to conclude that it is impossible to determine whether bryophytes are monophyletic or polyphyletic.

I hope you have not forgotten *Protosalvinia*. We included this Upper Devonian genus with the algae because of the initial conclusions by Niklas Phillips (1976) that "the Protosalviniales are considered a modified diplobiontic, polystichous algal group belonging to the Heterogeneratae." Later in the discussion, after an exhaustive analysis of the algal and bryophytelike characteristics of *Protosalvinia*, their final conclusion is that "the morphology and chemistry of *Protosalvinia* suggest a level of complexity intermediate between that of the algae and bryophytes. Since the reproductive organization and proposed life cycle of *Protosalvinia* exclude taxonomic assignment to bryophytes, the genus is considered an alga showing convergent evolution with that of the land plant habit." We would like to hazard the opinion that *Protosalvinia* is just as much a bryophyte as an alga. But this is not the point. As emphasized earlier, *Protosalvinia* has a set of characteristics that defies placing it with any known taxon. Its importance lies in the fact that it gives us a glimpse of one way that

Figure 8.6. **A.** *Sporogonites exurberans,* reconstruction, Lower Devonian. (Redrawn from Andrews, 1960.) **B.** *Andreaea* sp. Diagram longitudinal section of sporophyte with columella overarched by sporogenous tissue. Extant.

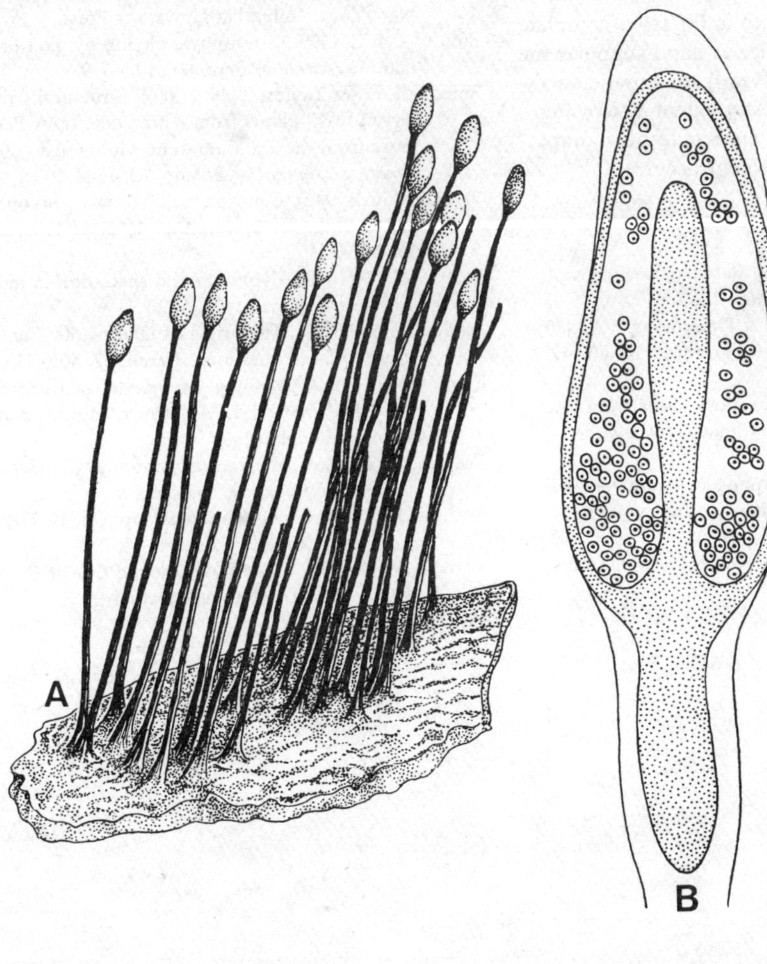

aquatic plants partially adapted to a terrestrial environment. It gives us insight as to how land plants evolved at an earlier time in the history of Earth.

When we examine the morphology of the Lower Devonian *Sporogonites* (Fig. 8.6A) (Andrews, 1960), it is clear that there is a basic similarity between it and *Protosalvinia*. As in *Protosalvinia*, Andrews believes that *Sporogonites* had a prostrate thalluslike structure with erect branches bearing sporangia that contained thick-walled spores. The details of the spore-bearing structures differ, but the prostrate–erect growth form is repeated. As we intimated earlier, it would be an exciting discovery if the prostrate portion turned out to be a gametophyte with an attached sporophyte, a pattern repeated in many liverworts including *Anthoceros*. Andrews examined many specimens of *Sporogonites* without finding any evidence of conducting tissue. The terminal sporangium has a multicellular wall and a columella projecting into the center covered by an overarching mass of spores. Columellate sporangia of this type occur among the Bryophyta (*Anthoceros*, *Sphagnum*, and *Andreaea*, Fig. 8.6B) as well as *Horneophyton*, a Devonian vascular plant. Because he could find no vascular tissue, Andrews tentatively assigns *Sporogonites* to the Bryophyta.

Even though the picture is far from complete it seems to us that the evidence at hand supports the concept that bryophytes evolved from aquatic ancestors. *Protosalvinia* and to a lesser extent *Sporogonites* provide us with concrete evidence as to how adaptation to the terrestrial environment occurred.

References

Andrews, H. N. (1960). Notes on Belgium specimens of *Sporogonites*. *Palaeobotanist*, **7**, 85–9.

Bold, H. C., Alexopoulos, C. J., & Delevoryas, T. (1980). *Morphology of Plants and Fungi*, 4th ed. New York: Harper & Row.

Chandrasekharam, A. (1974). Megafossil flora from the Genesee locality, Alberta, Canada. *Palaeontographica B*, **147**, 1–41.

Crandall-Stotler, B. (1980). Morphogenetic designs and a theory of bryophyte origins and divergence. *BioScience*, **30**, 580–85.

Harris, T. M. (1938). *The British Rhaetic Flora*. London: British Museum Press.

Hueber, F. M. (1961). *Hepaticites devonicus*, a new fossil liverwort from the Devonian of New York. *Annals of the Missouri Botanical Garden*, **48**, 125–32.

Janssens, J. A. P., Horton, D. G., & Basinger, J. F. (1979). *Aulacomnium heterostichoides* sp. nov., an Eocene moss from southcentral British Columbia. *Candian Journal of Botany*, **57**, 2150–61.

Jarzen, D. M. (1979). Spore morphology and the occurrence of *Phaeoceros* spores in the Cretaceous of North America. *Pollen et Spores*, **21**, 211–31.

Jovet-Ast, S. (1967). Bryophyta. In *Traité de Paléobotanique*, ed. E. Boureau. Paris: Masson.

Lacey, W. A. (1969). Fossil bryophytes. *Biological Reviews*, **44**, 189–205.

Lundblad, B. (1954). Contributions to the geological history of the Hepaticae. I. Fossil Marchantiales from the Rhaetic–Liassic coal mines of Skromberge, Sweden. *Svensk Botanisk Tidskrift*, **48**, 381–417.

Neuberg, M. F. (1960). Leafy mosses from the Permian of Angaraland. *Trudy Geological Institute*, **19**, 1–104.

Niklas, K. J., & Phillips, T. L. (1976). Morphology of *Protosalvinia* from the Upper Devonian of Ohio and Kentucky. *American Journal of Botany*, **63**, 9–29.

Savicz-Ljubitzkaja, L., & Abramov, I. I. (1959). Geological annals of the Bryophyta. *Review of Bryology and Lichenology*, **28**, 330–42.

Schuster, R. (1966). *The Hepaticae and Anthocerotae of North America East of the Hundredth Meridian*. New York: Columbia University Press.

Smith, A. J. E. (1986). Bryophyte phylogeny: fact or fiction? *Journal of Bryology*, **14**, 83–9.

Smoot, E. L., & Taylor, T. N. (1986). Structurally preserved fossil plants from Antarctica: II. A Permian moss from the Transantarctic Mountains. *American Journal of Botany*, **73**, 1683–91.

Steere, W. C. (1946). Cenozoic and Mesozoic bryophytes of North America. *American Midland Naturalist*, **36**, 298–324.

Steere, W. C. (1958). Evolution and speciation in mosses. *American Naturalist*, **92**, 5–20.

Taylor, T. N. (1988). The origin of land plants: Some answers, more questions. *Taxon*, **37**, 805–33.

Troll, W. (1927). 2 Abteilung: Bryophyta. In *Handbuch der Paläobotanik*, ed. M. Hirmer. Munich and Berlin: Oldenbourg.

Walton, J. (1925). Carboniferous Bryophyta. I. Hepaticae. *Annals of Botany*, **39**, 563–72.

Walton, J. (1928). Carboniferous Bryophyta. II. Hepaticae and Musci. *Annals of Botany*, **42**, 707–16.

Watson, E. V. (1967). *The Structure and Life of Bryophytes*, 2nd ed. London: Hutchinson.

9

How the land turned green: vascular plants, primitive types

The origin of vascular plants has been the theme of many symposia (Stewart, 1960), the topic of numerous research papers, and the subject of many chapters in other books. It is fair to say that in its degree of importance, the origin of vascular plants is at least equal to that accorded the origin of vertebrates. We already have considered a classification of vascular plants (Chapter 3) based on the idea that they had a monophyletic origin. This, however, is not universally accepted. There are many who look for a polyphyletic origin of tracheophytes (Andrews, 1959; Axelrod, 1959; Banks, 1968; Bold, Alexopoulos, & Delevoryas, 1980; Leclercq, 1954; Scagel et al., 1969) from different, unrelated ancestral types. As a result of the differing interpretations of the evidence relating to vascular plant origins, two opposing points of view – monophyletic vs. polyphyletic – have developed during the past quarter-century. It is the examination of this evidence, pro and con, that provides us with the main theme of this and parts of succeeding chapters.

When we examine the characteristics of extant vascular plants as Delevoryas (1962) has done, we find several compelling reasons suggesting that the origin of these plants is from a single ancestral type. We already know that the biochemistry of vascular plants is highly similar. They have chlorophylls *a* and *b*, store true starch, and have cellulose in their cell walls. Another striking similarity is found in the arrangement of their vascular tissues and the sequence in which their tracheary cells develop. A common configuration appears in the protostele. Here the xylem cells form a solid core with phloem around the outside (Fig. 9.1A). The protostelic arrangement is found in roots of most vascular plants and in the aerial axes and stems of many, but not all psilopsids, lecopods, sphenopsids, and ferns. The protostele appears first in the early stages of development of many vascular plants. These may give way later to medullated steles (steles with a pith) (Fig. 9.1B). Even more striking is the order in which the xylem cells of the stele develop (Fig. 9.1C). Usually the first to differentiate are annular elements followed by helical (spiral) and scalariform types. This sequence is found in all groups of vascular plants.

Life cycles of vascular plants are basically similar. They have alteration of phases with a predominant, independent, sporophyte – the structure developing vascular tissue. The archegonia of free-sporing vascular plants (psilopsids, lycopsids, sphenopsids, and ferns) are alike in structure and development.

If vascular plants were similar in only one of the ways described, then the argument that the similarity was the product of convergent or parallel evolution might be acceptable. This seems much less plausible, however, in view of the many ways that tracheophytes are alike. The cliché that vascular plants are more alike than they are different may well apply here, and tends to support the idea of their monophyletic origin.

Evidence of the first vascular plants

The geological column has been sifted and scrutinized for evidence of the earliest vascular plants. For reasons to be given later we should look for this evidence in the Lower Silurian, although there are fragments of fossils from the Cambrian that are claimed to belong to tracheophytes. Thick-walled spores showing trilete markings and other ornamentations have been reported from Cambrian rocks of India (Ghosh & Bose, 1949–50) and Russia (Naumova, 1949). The initial interpretation suggested that these were meiospores of primitive vascular plants. If proven, this would place their oldest known remains in the Lower Cambrian, much earlier in geological time than one might expect. As is often the case with finds of this kind, a healthy skepticism developed.

There is always the possibility of contamination, and we must remember that primitive vascular plants are not the only plants that form thick-walled meiospores with trilete markings. After a detailed study of meiospore types produced by bryophytes, Knox (1939) concluded, "Except where fossil spores are found in organic union with recognizable parent material, however, there can be no certainty as to their relationships." This conclusion applies equally to vascular plants.

Much significance has been attached to the discovery of *Aldanophyton* by Kryshtofovich (1953). The fossils are coalified compressions–impressions from the Cambrian of Siberia. They have the general appearance of small lycopods with spirally arranged leaves. The demonstration of vascular tissue, however, was not provided, and the relationships of these fragmentary remains are highly equivocal (Stewart, 1960). At present, there is no sound evidence that vascular plants occurred in the Cambrian.

Numerous other plantlike fragments have been isolated from Ordovician and Lower Silurian rocks. Among these remains are what appear to be spore tetrads. In addition to dyads and tetrads of cells, isolated thick-walled spores with trilete ornamentation have been reported from the Landoverian of the Lower Silurian (Pratt, Phillips, & Dennison, 1978). Some suggest that the tetrads represent the products of meiosis and that the spores with trilete markings could belong to land plants, possibly vascular plants. The relationships indicated by these plant fragments have been debated vigorously for more than a decade. Could they be remains of algae, bryophytic plants, vascular plants, or all of these? This same question has been applied to microfossils that look like cells with spiral and reticulate wall thickenings simulating tracheary elements, and to sheets of cells and cuticular fragments showing epidermal cell patterns. Banks, 1975b; Pratt et al., 1977; Gray & Boucot, 1977; and others who have made recent studies of these assemblages of fragments have come to the conclusion that the structures do not represent the remains of ancient vascular plants, but are parts of other kinds of organisms. The spores, for example, are like those of the nonvascular plant *Nematothallus*. You may recall that this genus is composed of branching tubes with secondary wall thickenings that have the appearance of tracheids. This same genus is known to have had a cuticle. All of these parts could be formed by this or some other nonvascular plants, such as algae, bryophytes, or transitional nonvascular plants – for example, *Protosalvinia* (Gray & Boucot, 1977).

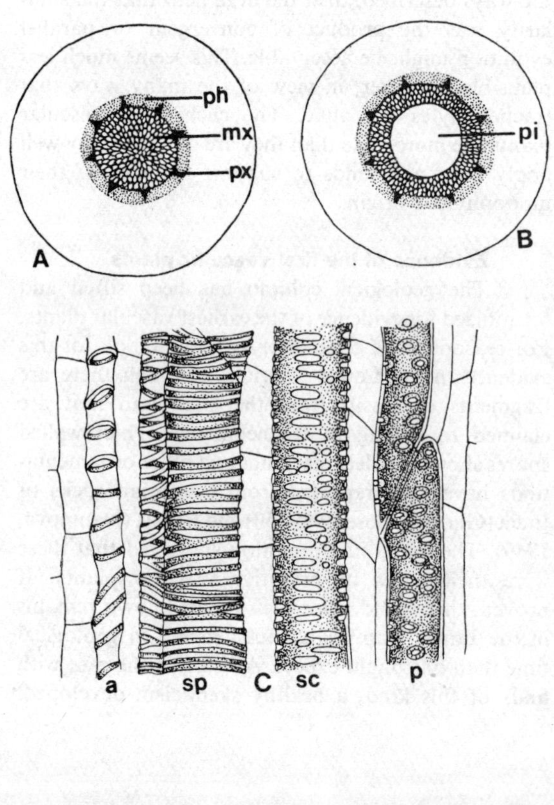

Figure 9.1. **A.** Diagram of exarch protostele. Phloem (ph); metaxylem (mx); protoxylem (px). **B.** An exarch siphonostele. Pith (pi). **C.** Longitudinal section of primary xylem of *Aristolochia* showing sequence of development from left to right. Tracheids with annular thickenings (a); with spiral (helical) thickenings (sp); with scalariform thickenings (sc); with circular, bordered pits (p). (Redrawn from Esau, 1965.)

For a comprehensive review and extended bibliography pertaining to Silurian and Devonian fossils of land plants and their precursors you are referred to Chaloner (1988); Chaloner & Sheerin (1979); Gensel & Andrews (1984); and Taylor (1982, 1988).

The first plants with vascular tissue

For a number of years *Baragwanathia* (Chapter 10) was accepted as the oldest known vascular plant. It was first described by Lang & Cookson (1935) from Upper Silurian beds of Victoria, Australia. In this instance the investigators were able to demonstrate the presence of the all important vascular tissue in the form of a stellate protostele with annular tracheids. The spirally arranged leaves are supplied with a single simple bundle of vascular tissue as in the leaves of the Lycopsida. In addition, Lang & Cookson found meiospore-containing sporangia in the adaxial position at the base of a leaf, yet another characteristic of lycopods. There is no question where the relationships of *Baragwanathia* lie. There is, however, a question about the Upper Silurian age assigned to the fossils (Edwards, Bassett, & Rogerson, 1979). A restudy of the invertebrate index fossil *Monograptus*, used to determine the age of the rocks, indicates that they are Lower Devonian. The question of age is still not settled to everyone's satisfaction. For example,

Garratt (1981) has discovered additional *Baragwanathia* specimens in strata that lie below the Lower Devonian beds containing the specimens first described by Lang & Cookson. Based on identification of invertebrate fossils, these beds are placed in the Upper Silurian. Well-preserved specimens of *Baragwanathia* have also been found in Lower Devonian (Emsian) beds in Ontario, Canada (Hueber, 1983). The age of these beds has been verified.

Although *Baragwanathia* seems to be a fairly complex vascular plant, there are others that are exceedingly simple in structure occurring in the same beds and in older strata in other parts of the world. One of these that is well characterized is *Cooksonia* (Fig. 9.2C). This plant, first described by Lang (1937) from the Lower Devonian, is a small plant only a few centimeters high. It is dichotomously branched, without appendages (lateral branches or leaves) of any kind. The slender, naked axes are terminated by sporangia (Fig. 9.2B) that are short and wide – almost reniform (kidney-shaped). Spores isolated from the sporangia are smooth and have the trilete mark of meiospores. The plants were isosporous. These are the salient features of the class Rhyniopsida.

Although Lang demonstrated the presence of annular tracheids in vegetative axes of *Cooksonia*, to date, vascular tissue has not been found in fertile, sporangium-bearing axes. Because one cannot be sure that the fertile specimens were actually parts of a *Cooksonia* with vascular tissue, there has been some hesitancy to accept the idea that the two came from the same kind of plant (Taylor, 1988). Since Lang's initial description of *Cooksonia* in 1937, however, there have been many accounts of fertile and sterile specimens assigned to this genus on the basis of size and general morphology alone. Some of these are mid-Silurian in age (Obrhel, 1962; Banks, 1970; Edwards, 1980). The specimens described by Obrhel come from the base of the Pridolian stage of the Upper Silurian of Bohemia. Edwards (1970), while working on coalified compressions from the Emsian (Lower Devonian) of South Wales, encountered some rather complete specimens of *Cooksonia*, one of which shows four orders of branching in the dichotomous system. In the same beds she found another simple plant, *Steganotheca* (Fig. 9.2A), a bushy, dichotomously branched plant 5.0 cm high with terminal sporangia. The sporangia are truncated at the apex and taper to the base. A central strand can be seen in the compressed branches suggesting the presence of a vascular system. Specimens similar to *Steganotheca* known to be Upper Silurian were found in strata below the Downtonian.

Continuing the search for early vascular plants, Shute & Edwards (1989) reexamined permineralized

Figure 9.2. **A.** *Steganotheca striata*. Fertile specimen. Upper Silurian. **B.** *Cooksonia caledonica*. Reniform, terminal sporangia, Lower Devonian. **C.** *C. caledonica*. Fertile specimen. Lower Devonian. (**A–C** redrawn from Edwards, 1970.)

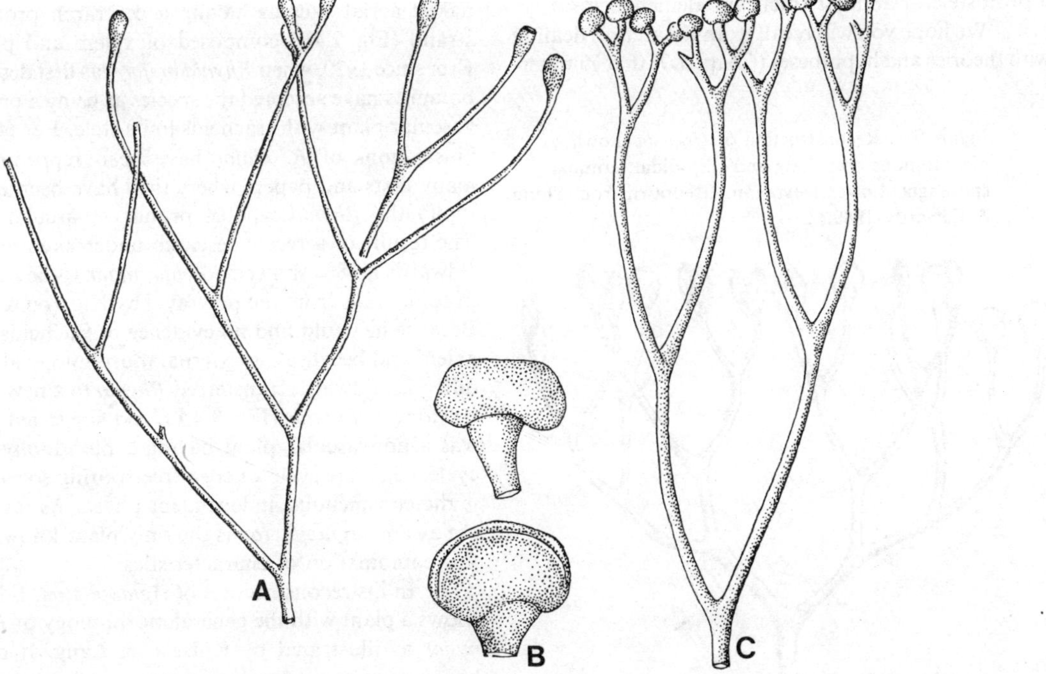

plants from the Lower Devonian (Siegenian) beds of South Wales. These had previously been identified as specimens of *Cooksonia*. As in this genus the fossils consisted of a simple dichotomous branching system of naked axes. Their sporangia, however, are longer and wider than those of *Cooksonia* and possess a more complex wall structure. For this reason these plants were placed in a new genus, *Uskiella* (Fig. 9.3). In addition to characterizing the spores and wall structure of its sporangia, Shute & Edwards found evidence of annular tracheids in the axes of *Uskiella*. The authors make the point that this genus combines a primitive organization in its vegetative parts and structurally advanced sporangia.

In an earlier study, Edwards & Feehan (1980) report finding remains of *Cooksonia* sporangia in outcrops classified as mid-Silurian (Wenlockian) in Ireland. If one believes that *Cooksonia* is a vascular plant, this discovery assumes considerable significance in placing the origin of vascular plants and colonization of land by these plants at about 420 m.y. This is in keeping with the conclusions of Chaloner (1967), who, after an extensive study of Devonian spores, believes that his evidence favors a Silurian origin of land plants followed by a slow rate of diversification, rather than a Cambrian of Precambrian origin as suggested by others. With the evidence from megafossils it is safe to conclude that the sporophytes of these first land vascular plants were extremely simple in organization, having three-dimensional dichotomizing naked axes with terminal sporangia. From what little is known of their vascular systems it has been determined that some possessed a protostele of annular tracheary elements.

We hope you will recall from the section dealing with theories and hypotheses (Chapter 7) that Zimmermann and Lignier speculated about the morphology of vascular plant ancestors, a morphology strikingly similar to *Cooksonia, Uskiella* and *Steganotheca*. The question remains: Are the characteristics of these two genera those of the primitive type for all vascular plants, or are there additional morphologies that can be interpreted as primitive?

Figure 9.3. Reconstruction of *Uskiella* showing dichotomous branching and ellipsoidal terminal sporangia. Lower Devonian. (Redrawn from Shute & Edwards, 1989.)

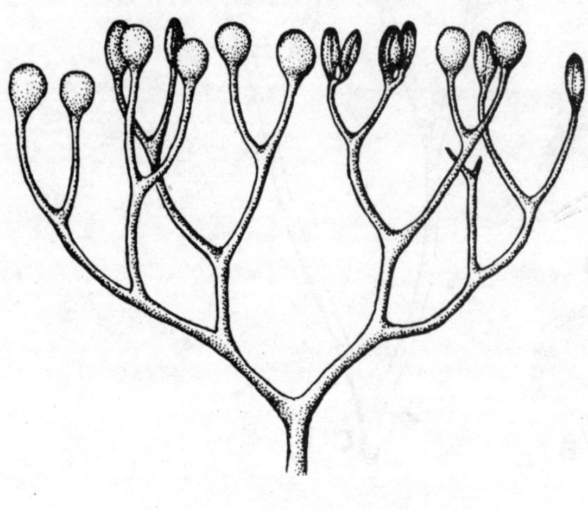

Rhyniopsida from the Rhynie chert

In our search for the best-preserved early vascular plants our attention, sooner or later, will focus on the famous Rhynie chert beds of Scotland. The age of the beds is believed to be Lower Devonian (Siegenian/Emsian). The plant remains that we are interested in are permineralized in a matrix of volcanic chert and from their often vertical orientation it seems they were preserved in situ in a swampy peat bed.

Of the many superbly preserved plant remains found in the chert and the most completely known members of the class Rhyniopsida are the type genera *Rhynia* and *Horneophyton*. These are among the many plants first described by Kidston & Lang in a series of papers published between 1917 and 1921.

In their description of the flora of the chert beds, Kidston & Lang illustrated and gave the characteristics of what they believed to be a primitive vascular plant similar in many ways to members of the old order Psilophytales. The plant, which they called *Rhynia major*, was reconstructed as a plant approximately 50 cm high with prostrate axes bearing rhizoids and erect dichotomizing axes, some of which were terminated by ellipsoidal sporangia (Fig. 9.4D) filled with thick-walled isospores (Fig. 9.4C). They described the naked aerial axes as having a centrarch protostelic strand (Fig. 9.4B) composed of xylem and phloem. Ever since 1920, when *Rhynia major* was first described, botanists have accepted the species as being a primitive vascular plant with tracheids in its stele. For 60 years illustrations of *R. major* have been reproduced in many texts and papers where they have been used to typify the *Rhynia* type of primitive vascular plant. The results of a recent research undertaken by D. S. Edwards (1986) showed *Rhynia major* to be different in some ways from the portrayal by Kidston & Lang. Because he could find no evidence of tracheids in the xylem and because the external morphology also was different, Edwards transferred *Rhynia* to a new genus, *Aglaophyton major* (Fig. 9.4A), and suggested that it was a nonvascular plant having a pteridophytic life cycle, i.e., a life cycle where a free-sporing sporophyte is the conspicuous, independent phase. As far as we are aware, *Aglaophyton* is the only plant known with this combination of characteristics.

In his reconstruction of *Aglaophyton*, Edwards shows a plant with the general morphology of *Rhynia major* as illustrated by Kidston & Lang. It differs, however, in being smaller, approximately 18 cm high, and having erect dichotomizing branches that parted at a rather wide angle. The erect branches come from the prostrate rhizoid-bearing axes that show a rather unusual arched configuration. A careful examination of the thick-walled cells comprising the central portion of the conducting strand of *Aglaophyton* failed to show the characteristic secondary wall thickenings of tracheophytes (Fig. 9.1C). Instead these cells appear to be similar to the hydroids (water-conducting cells) found in the conducting strands of gametophytes and sporophytes of many moss plants.

In the same beds with *Aglaophyton* (= *Rhynia major*) Kidston & Lang found a second species of *Rhynia*, *R. gwynne-vaughanii* (Fig. 9.4E), which they estimated to be about 18 cm high. It too had prostrate rhizomelike parts and erect, naked dichotomizing axes presumably terminated by sporangia (Edwards, 1980) containing isospores. A distinctive feature of *R. gwynne-vaughanii* is its lateral branches believed by Kidston & Lang to be adventitious. When they were shed, these branches left scars on the aerial axes.

Adaptive structures

Earlier we speculated about the ways in which sporophytes of early land plants adapted to a terrestrial environment. Because of their excellent preservation we can see many characteristics present in *Rhynia gwynne-vaughanii* that promoted its survival on land.

Figure 9.4. **A.** Reconstruction of *Aglaophyton major*. **B.** *A. major*, transverse section of an aerial axis showing differentiation of tissues; conducting strand (cs). **C.** *A. major*, spores in a tetrad. **D.** *A. major*, terminal eusporangia enlarged. **E.** *R. gwynne-vaughanii*, reconstructed. **F.** *R. gwynne-vaughanii* stoma and guard cells. All Lower Devonian (Siegenian–Emsian) from Rhynie chert. (**A** redrawn from Edwards, 1986; **B–D** redrawn from Kidston & Lang, 1917, 1921.)

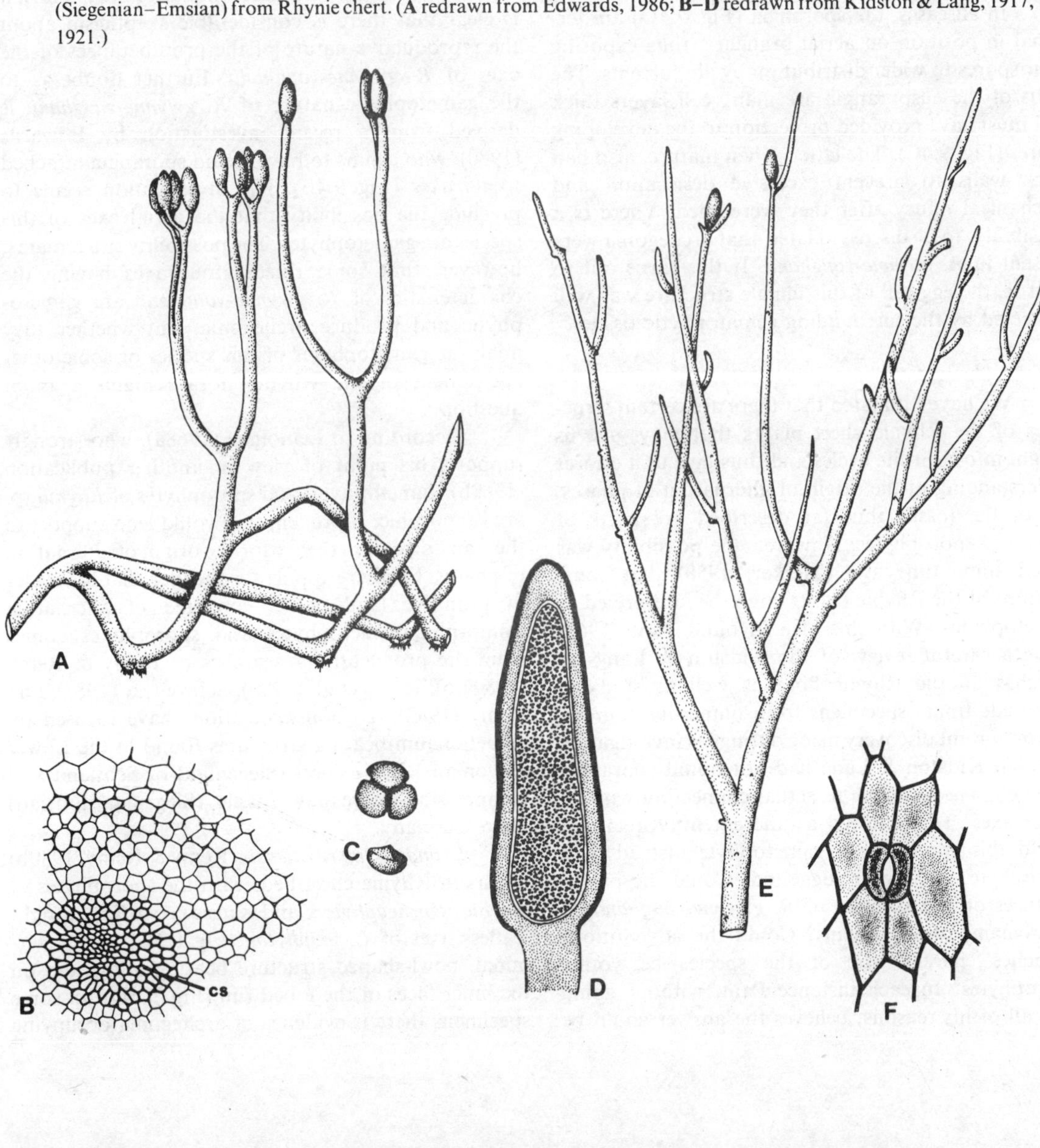

Cross sections of its axes show a parenchymatous organization differentiated into an epidermis, cortex, and stele. The protostele is equipped with supporting and conducting annular tracheids with smaller tracheids in the center. This xylem tissue is surrounded by a layer of phloemlike cells. Although there is as yet no way of telling for sure, the surfaces of aerial branches are covered with a layer that has the appearance of a cuticle. Again on the aerial branches, there are stomata and guard cells among the cells of the epidermis (Fig. 9.4F). This is a good indication that these axes were photosynthetic. We know that stomata allow the inward diffusion of CO_2 from the atmosphere and that the guard cells regulate the outward diffusion of H_2O vapor. To replace the water inevitably lost, the prostrate rhizomes or corms are equipped with large numbers of rhizoids. These are nothing more than root-hairlike extensions of the epidermis increasing the absorbing surface of the basal portions of the plant manyfold.

In all cases, the sporangia (Fig. 9.4D) are terminal in position on aerial branches, thus exposing their spores to wider distribution by air currents. The walls of the eusporangia are many cell layers thick and must have provided protection to the developing spores (Fig. 9.4C). The latter, when mature, also had thick walls to prevent excessive desiccation and mechanical injury after they were shed. There is a possibility, to be discussed later, that archegonia were present in *R. gwynne-vaughanii.* If this turns out to be true, the egg cell of this female structure was well protected by the surrounding gematophytic tissues.

A life cycle?

We have intimated that there are certain structures of the Rhynie chert plants that may give us insight into their life cycles and thus give us a clearer understanding of the origin of alternation of phases. All of the fossils thus far described are parts of presumed sporophytes. However, the possibility was raised some time ago (Merker, 1958) that some portions of the Rhynie plants might be interpreted as gametophytes. With this idea in mind Pant (1962) made a careful review of the Kidston & Lang researches on the Rhynie flora as well as studying many additional specimens from numerous sources. The important discovery made during his investigation was that Kidston & Lang had not found sporangia of *R. gwynne-vaughanii* in actual connection with its aerial axes. This allowed a different interpretation. Could this species be a gametophyte instead of a sporophyte, as Merker suggested? Could the protuberances on the surfaces of *R. gwynne-vaughanii* be the remains of sex organs? Could the adventitious branches characteristic of the species be young sporophytes? In each instance Pant, without going into all of his reasons, believes the answer could be yes. Pant has confirmed that the tissues comprising the embedded portions of the adventitious branches are quite distinct from the tissues of the axes bearing them and that they have the appearance of emerging embryos. He further suggests that absence of vascular tissue connecting the axial branches of *R. gwynne-vaughanii* with its lateral adventitious branches is another piece of evidence showing that the latter are embryo stages of a new sporophyte that will differentiate its own vascular tissue at a later time. Pant received considerable support from Lemoigne (1986a), who described and figured structures claimed to be archegonia of *R. gwynne-vaughanii.* Although these structures are convincing with their neck cells, canal, and basal egg, there is a possibility that the "neck cells" are guard cells in section view and the "canal" is a substomatal air space. Bierhorst (1971) compares the archegoniumlike structures with hydathodes or secretory structures of some kind. Others suggest that the structures resemble insect-induced injury. Thus, it is clear that there is considerable skepticism about the reproductive nature of the protuberances of the axes of *R. gwynne-vaughanii.* Further doubt as to the gametophytic nature of *R. gwynne-vaughanii* is derived from a recent investigation by Edwards (1980), who claims to have found sporangia attached to its axes (Fig. 9.4E). His investigation seems to preclude the possibility that the aerial axes of this species are gametophytes. The possibility still remains, however, that some rhizomatous axes having the characteristics of *R. gwynne-vaughanii* are gametophytic and produce archegonia, but whether they might be gametophytes of this species or some other component of the Rhynie flora remains a moot question.

According to Lemoigne (1968a), who strongly supports his point of view in another publication (1968b), gametophytes and sporophytes of *Rhynia* sp. are isomorphic. If proven, this would lend support to the transformation theory for the origin of alternation of phases. Further support for the idea that life cycles of primitive land plants consisted of alternating isomorphic gametophytes and sporophytes comes from the provocative researches of Remy & Remy (1980a,b), Remy et al. (1980), Schweitzer (1981), and Remy (1982). Their investigations have focused on gametangium-bearing structures found in the Lower Devonian Rhynie chert (Siegenian) of Scotland and compression-impression fossils (Siegenian–Emsian) from Germany.

Lyonophyton rhyniensis (Remy & Remy, 1980b) occurs in Rhynie chert beds with the sporophytes of *Rhynia, Horneophyton,* and *Asteroxylon.* The upright leafless axes of *Lyonophyton* (Fig. 9.5C) have a terminal, bowl-shaped structure bearing antheridia on the inner faces of the lobed rim (Fig. 9.5D). In some specimens there is evidence of archegonia occupying a more central position within the bowl. In size and structure the terete axes of the gametophores are reminiscent of *A. major* sporophytes with a central column of conducting tissue and a simple parenchymatous cortex covered by an epidermis with stomata and a cuticle.

In size and general morphology the compression-impression fossils of *Sciadophyton* are similar to *Lyonophyton*. Although they are believed to be the remains of sporophytes, it was suggested five decades ago that the branches of *Sciadophyton* with their terminal, bowl-shaped structures might be gametophores bearing antheridia and archegonia. Detailed studies by Remy et al. (1980), and Schweitzer (1981), have confirmed the gametophytic nature of *Sciadophyton* (Fig. 9.5A,B) where gametophoric axes radiate from a central corm. Occasionally these terete axes branch dichotomously. Recognizable tracheids have been found in the central strand of tissue of these gametophytic organs. Based on these observations and studies of development stages of *Sciadophyton, Taeniocrada langii,* and *Zosterophyllum rhenanum,* Schweitzer (1981) has proposed the following life cycle for these primitive vascular plants: Spore germination produced a prostrate gametophytic thallus composed of dichotomously branched, terete axes. These were probably photosynthetic and contained vascular tissue. As development proceeded, some of the axes grew upright to become gametophores terminated by bowl-shaped discs bearing antheridia and archegonia as in *Sciadophyton*. Fertilization initiated production of young sporophytes having dichotomously branched, photosynthetic axes of the *Taeniocrada* and *Zosterophyllum* type, also with vascular tissue. As these sporophytes matured they produced erect branches with sporangia.

Such a life cycle strongly supports the ideas of Pant (1962), Lemoigne (1968a,b), and others that some primitive vascular plants of the Lower Devonian had alternation of phases where gametophyte and sporophyte were isomorphic, both being photosynthetic and containing vascular tissue.

Horneophyton and alternation of phases

Horneophyton (Fig. 9.6A) has been the focal point of much speculation about the origin of alternation of phases. You may recall (Chapter 7) that there are two theories explaining the origin of the sporophyte phase – the interpolation and transformation theories. These theories relate directly to our understanding of the origins of green land plants (bryophytes and tracheophytes). To understand this rather complex topic we have to know something of the structure of *Horneophyton* and the bryophyte *Anthoceros* (Fig. 7.1H).

Horneophyton was first described by Kidston

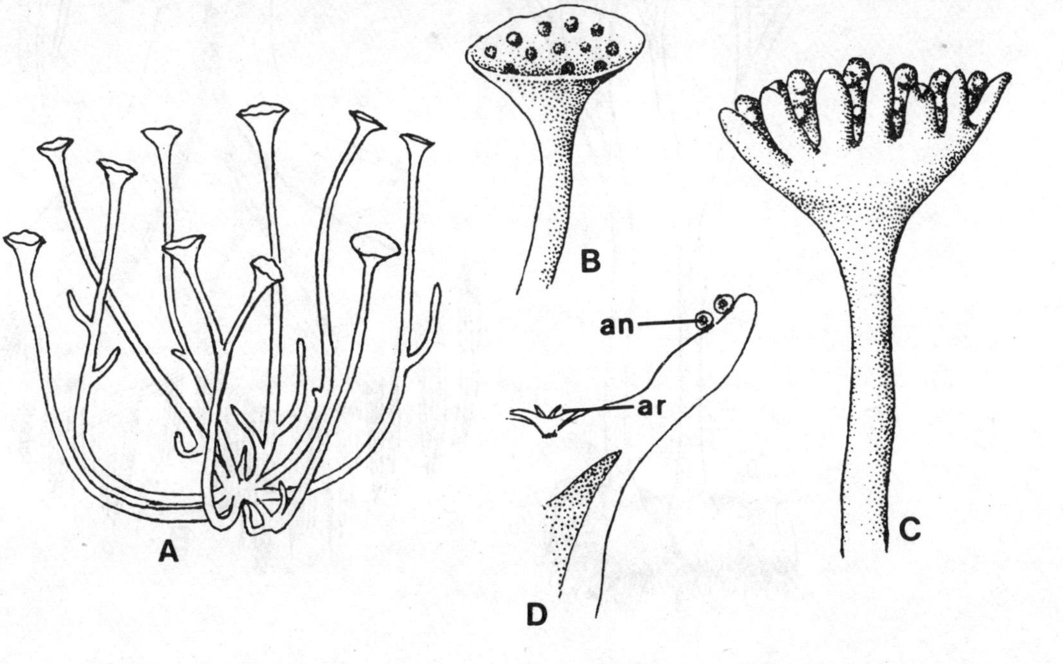

Figure 9.5. Putative gametophores of Devonian vascular plants. **A.** *Sciadophyton* with gametophores arising from a central disc. **B.** *Sciadophyton*, enlarged tip of gametophore with antheridia and archegonia. **C.** *Lyonophyton* gametophore cup showing antheridia on inner surface at tips of fingers. **D.** Longitudinal section of gametophore cup with archegonia (ar) at base of cup and antheridia (an) at tips of lobes. (**A** redrawn from Taylor, 1988; **B–D** redrawn from Remy & Remy, 1980a, 1980b.)

& Lang (1920) from the Lower Devonian (Siegenian–Emsian boundary) Rhynie chert beds. They provided the name *Hornea*, a generic name that later was found (Barghoorn & Darrah, 1938) to have been used for a flowering plant, thus requiring the name change. In spite of its presumed importance in understanding the evolution of green land plants, very little new research was reported until 1974 when a paper by Eggert appeared describing an unusual and significant interpretatoin of the sporangial structure.

The sporophyte of *Horneophyton* was approximately 15 to 20 cm high. The naked aerial axes branched dichotomously several times, each dichotomy producing a pair of branches at right angles to the branches immediately below. At the base of each aerial axis was a presumably subterranean cormlike structure (Fig. 9.6B) covered with a multitude of unicellular rhizoids (Fig. 9.6C). The epidermis of the aerial branching system possessed scattered stomata and guard cells.

Internally, the aerial axes have what appears to be a terete protostele that, according to Kidston &

Figure 9.6. **A.** *Horneophyton lignieri,* reconstruction, **B.** *H. lignieri,* longitudinal section, basal corm. **C.** *H. lignieri,* detail of rhizoids. **D.** *H. lignieri,* sporangia. All Lower Devonian (Siegenian–Emsian). (**A,D** redrawn from Eggert, 1974; **B,C** redrawn from Kidston & Lang, 1920.)

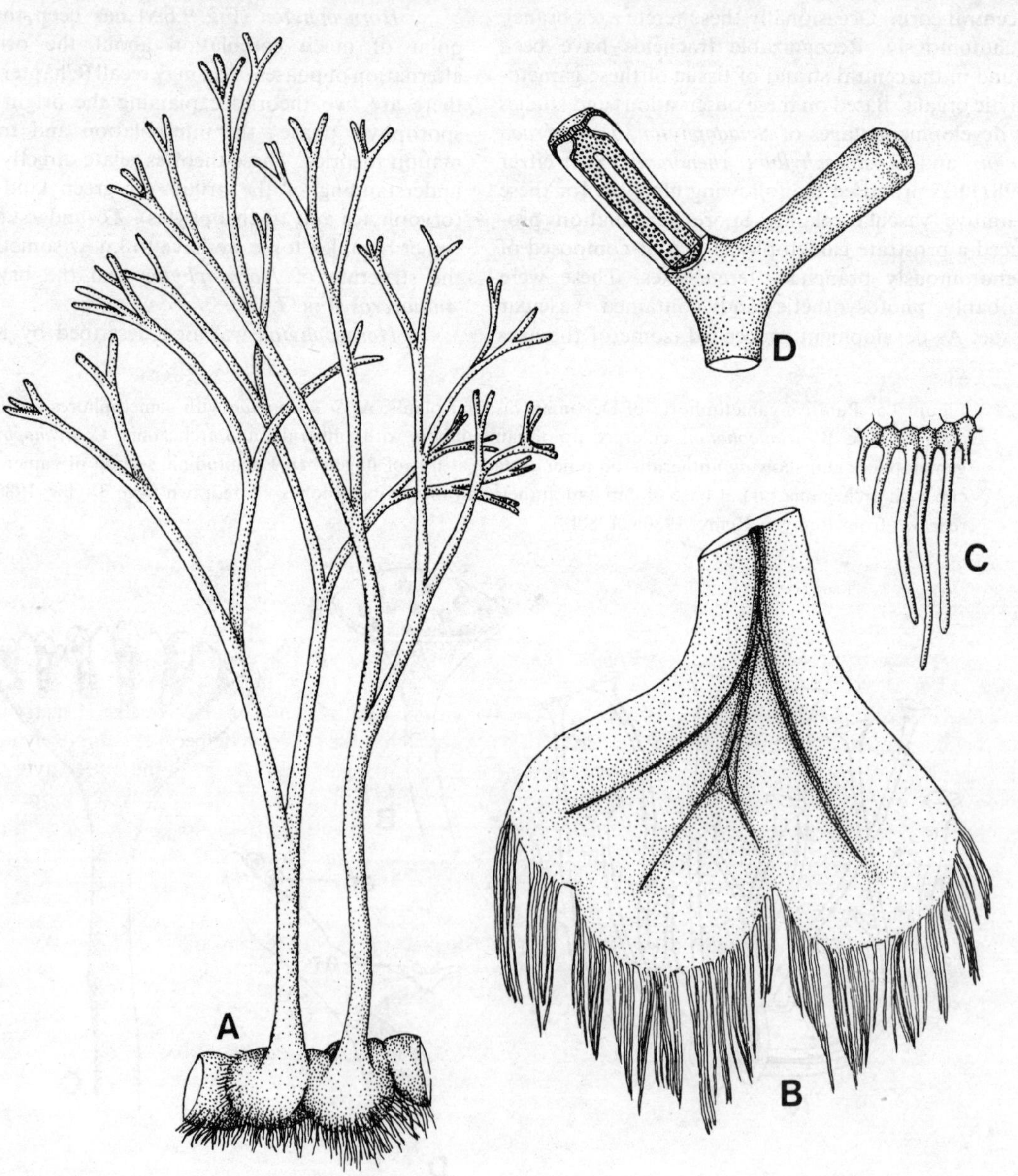

Lang, has smaller, centrally placed protoxylem cells surrounded by larger cells of metaxylem. As in *Rhynia gwynne-vaughanii* the arrangement of the primary xylem of the stele is centrarch. A presumed layer of phloem completely surrounds the xylem. Toward the cormlike base the vascular strand of the aerial branches disappears to be replaced by dark, thick-walled parenchyma cells that extend into the center of the corm (Fig. 9.6B).

According to Kidston & Lang's original description, the sporangia of *Horneophyton* were terminal and eusporangiate. Although usually borne singly at the tips of the aerial branches, occasional branching sporangia were described. The sporogenous layer, composed of thick-walled isospores with conspicuous trilete ornamentations, overarched a central columella that projected from the base of the sporangium upward into the sporangium cavity (Fig. 9.6D). Dehiscence of the sporangium was by means of an apical slit.

From the studies of serial sections, Eggert (1974) was able to demonstrate that sporangia of *Horneophyton* are usually dichotomously branched and may have as many as four branches per sporangium. The columella extending upward from the base of a branched sporangium also branches dichotomously and is surrounded by spores that overarch the tips of the columella that supply each of the sporangial branches.

At the turn of the century, D. H. Campbell originated and vigorously supported the idea that primitive vascular plants had evolved from bryophytic ancestors of the *Anthoceros* type. The discovery of *Rhynia* and *Horneophyton* by Kidston & Lang seemed to provide the fossil evidence supporting this idea. G. M. Smith (1955), an associate of Campbell's, was the chief advocate of the theory that the fossils bridged the gap between the two groups. Prior to evaluating the theory, however, we should review some of the salient characteristics of the *Anthoceros* type that made this theory so appealing.

The gametophytes of *Anthoceros* (Fig. 8.1D) are usually found in damp, shaded environments where they form rosettes of dorsiventral dichotomously branched thalli. These dark green gametophytes bear unicellular rhizoids on the ventral surface where one can find stomatalike slits that open into cavities in the gametophytic tissue. Sex organs are deeply embedded in the tissues of the dorsal surface. The position of the archegonia is, of course, related to the position of the development of sporophytes.

Like the gametophyte, the mature sporophyte (Fig. 7.1H) is photosynthetic. It has stomata with guard cells in the epidermis, and internally it is differentiated into a centrally placed slender columella that, early in development, is overarched by spore-forming tissue. In the closely related genus *Dendroceros*, the outer layer of elongated cells belonging to the columella have weakly developed spiral thickenings suggesting tracheids (Proskauer, 1960). Similar thickenings may be found in pseudoelaters of *Anthoceros* intermixed with the spores. Growth of the sporophyte is indeterminate by means of a basal intercalary meristem. Here the cells divide and add to the upper, spore-producing portion of the sporophyte. Below the meristem is a globose foot that is embedded in the tissues of the gametophyte. Cells of the foot in contact with gametophytic tissues form rhizoidlike extensions that increase the absorptive surface of the sporophyte.

It has been demonstrated that photosynthetic sporophytes excised from gametophytes and placed on nutrient agar or soil may remain alive for months but show only limited growth.

When Campbell, Smith, and others evaluated these and other characteristics of the anthocerotean-type sporophyte they interpreted it as one that, in an evolutionary sense, is "caught in the act" of becoming independent of the gametophyte. This, plus its indeterminate growth, strongly suggested the condition found in the independent mature sporophytes of vascular plants. Later the discovery of the tracheidlike cells in the columella of *Dendroceros* enhanced this idea. With this explanation it is easy to see how such fossil plants as *Horneophyton* and *Sporogonites* (Fig. 8.6A) with their columellate sporangia were interpreted as being transitional between the *Anthoceros*-type ancestor (Fig. 9.7A) and vascular plants. In the case of *Horneophyton* one can homologize its cormlike base bearing rhizoids with the foot of the *Anthoceros*-type and its rhizoidlike projections. The columellate branching sporangia can be interpreted as an intermediate step in which the *Anthoceros*-type sporophyte branches dichotomously and where the intercalary meristem has moved toward the apices of the branches. This closely approximates the hypothetical step shown by Smith (Fig. 9.7C).

Accepting this explanation for the origin of vascular plants from *Anthoceros*-like ancestry presupposes that the sporophyte became completely independent of the gametophyte when the sporophyte reached maturity and the sporophyte evolved a more complex structure (roots, stems, and leaves) containing well-developed vascular tissue, and terminal meristems (Fig. 9.7D). This interpretation favors the gradual step-by-step interpolation, into the heteromorphic life cycles of the ancestral plants, of a predominant, free-living, vascularized sporophyte. This is the essence of the interpolation theory (Chapter 7) – an attractive theory.

From the fossil record, it seems that plants similar to *Horneophyton* and to a lesser degree *Sporogonites* provide the intermediates between bryophytes and vascular plants. Is it necessary, however, to assume a "progressive" type of evolution from a simple dependent primitive sporophyte to a complex independent sporophyte? There are those who think the "evolutionary signposts" can be read in the opposite direction (Haskell, 1949); that the bryophytes represented by the *Anthoceros* type have evolved from primitive vascular plants of the *Horneophyton* type by reduction. This hypothesis assumes that the free-living branching sporophyte of the *Horneophyton* type became semiparasitic on its gametophyte. We do not know what that gametophyte looked like, but it has been suggested that *Sporogonites* and *Protosalvinia* might have had prostrate thalloid structures comparable to the gametophytes of the *Anthoceros* type. Further evolution along this line would lead to sporophytes more or less dependent on gametophytes, a characteristic of most bryophytes. When we consider the morphology of *Horneophyton*, we see it is quite different from *Rhynia gwynne-vaughanii* in having branched sporangia with apical dehiscence. Further, *Horneophyton* has the unusual cormlike base not found in *Rhynia*, which has a rhizomatous underground branching system. *Horneophyton* is much less like *Rhynia* than we suspected. Because of this it seems plausible that within the Rhyniopsida we can see a divergence in the evolution of vascular plants: those represented by the *Rhynia* type that provided the ancestors of vascular plants with independent free-living sporophytes and those of the *Horneophyton* type leading to the evolution of primitive bryophytes with their dependent sporophytes and independent gametophytes. Remember, bryologists have demonstrated that many bryophytes have functional conducting cells that in some have the appearance of tracheary elements. Thus, the absence of vascular tissue can no longer be used to distinguish bryophytes from vascular plants. Only their life cycles can be considered.

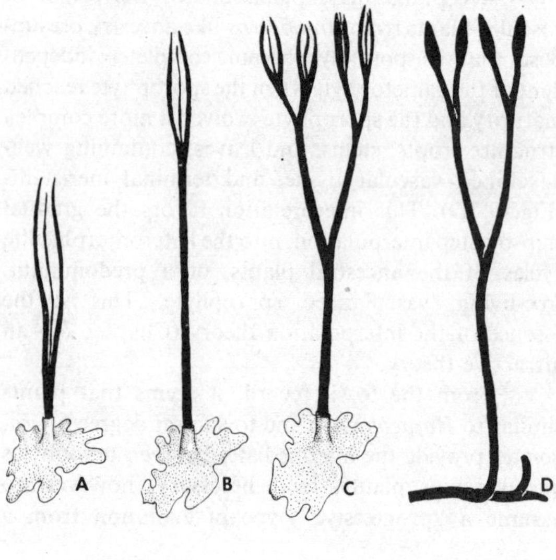

Figure 9.7. Hypothetical steps in the evolution of *Rhynia*-type sporophyte from an anthocerotan ancestor. **A.** *Anthoceros*-type ancestor. **B.** Localization of sporogenous tissue at distal end of sporophyte. **C.** *Horneophyton*-like stage introducing dichotomous branching of sporophyte. **D.** *Rhynia*-like vascular plant independent of gametophyte. Gametophyte, stippled; sporophyte, black. (Adapted from Smith, 1955.)

If the gametophyte and sporophyte evolved from ancestors with isomorphic alternating phases, there should be some evidence that the two phases were alike. In *Anthoceros*, however, where gametophyte and sporophyte can be studied, we see that both phases are photosynthetic, and both have structures that can be interpreted as stomata with guard cells. One investigator suggests that evolution of the gametophyte has been from an upright organism, with a terete axis similar to the sporophyte, to a prostrate thalloid structure. This idea fits well with the discovery of *Pallavicinites* from the Upper Devonian, which places those bryophytes with thalloid gametophytes later in geological time than *Lyonophyton* and *Sciadophyton* from the Lower Devonian, with their upright, highly differentiated gametophytes.

Based on the new evidence provided by Remy & Remy (1980a,b), Remy et al. (1980), Schweitzer (1981), and Remy (1982), we must favor the idea that some primitive Lower Devonian vascular plants had life cycles with alternation of isomorphic gametophyte and sporophyte phases. According to Remy (1982), some gametophytic characteristics of *Lyonophyton* and *Sciadophyton* and sporophytic characteristics of *Aglarphyton major* and *Horneophyton* are like those of bryophytes. For example, *Lyonophyton* is similar to a bryophyte in the general morphology of its gametophore, the structure of its conducting cells, and antheridial features. Bryophytic characteristics of sporophytes of the *Rhynia* type (e.g., *A. major*) are conducting cells interpreted by some authors to be hydroids, rhizoid-bearing "rhizomes," and columellate sporangia as in *Horneophyton*. Tracheary elements known to be in the sporophytes of the *Rhynia* type and in gametophytes of *Sciadophyton* tell us that the evolution of vascular tissue in both sporophyte and gametophyte had been accomplished by the Lower Devonian. Remy (1982) interprets the evidence given above as indicating a plexus of Lower Devonian green land plants with bryophyte and tracheophyte characteristics, and with independent phases, from which vascular plants with independent sporophytes and bryophytes with dependent sporophytes evolved. This idea favors the monophyletic interpretation adopted by Zimmermann (1952, 1959) in the telome theory for the origin of vascular plants and bryophytes, and later supported by Taylor (1988).

This leaves us with a third possibility, which

also embraces the transformation theory. It differs from the hypothesis of Haskell in assuming that Bryophyta and Tracheophyta had separate origins among green algal ancestors. They were in this sense polyphyletic. Accordingly, the bryophytes are thought to be an evolutionary dead end, with vascular plants evolving independently from the primitive type of vascular plant represented by the Rhyniopsida.

Psilopsida and the Rhyniopsida

While we are still concerned with the evidence relating to the life cycles of primitive vascular plants, we should turn our attention to the extant genus *Psilotum* (Fig. 9.8A) class Psilopsida. Until recently this simple plant and *Tmesipteris* were believed to be living relatives of the Rhyniopsida. A superficial examination of *Psilotum* reveals a shrubby plant with three-dimensional dichotomous branching, a plant that has a low degree of organ differentiation in which roots and vascularized leaves are absent and where sporangia are borne terminally on short lateral axes. You will recognize these as characteristics of the Rhyniopsida except for the sporangia of the latter that are terminal on axial (main) portions of the plant. Internally the structure of *Psilotum* can be interpreted as primitive. The underground axis is a solid protostele, becoming medulated in the aerial portions. Development of the primary xylem is centripetal (from the periphery to the center) so that protoxylem develops first to the outside of the metaxylem. This produces the exarch arrangement (Fig. 9.1A) of primary xylem that is conceded to be a primitive organization. Protoxylem cells are annular or helical; the larger centripetal metaxylem cells have scalariform or circular bordered pits. Like the Rhyniopsida, sporangia of *Psilotum* are eusporangiate and produce large numbers of isospores; all are primitive characteristics.

After germination the isospores give rise to gametophytes that may grow on the trunks of tree ferns, in rock crevices filled with humus, or in subterranean environments. The unusual aspect of the gametophyte is its similarity in appearance and structure to the underground brown axes of the sporophyte. Both bear rhizoids, are radially symmetrical, and dichotomize. Perhaps most significant is the

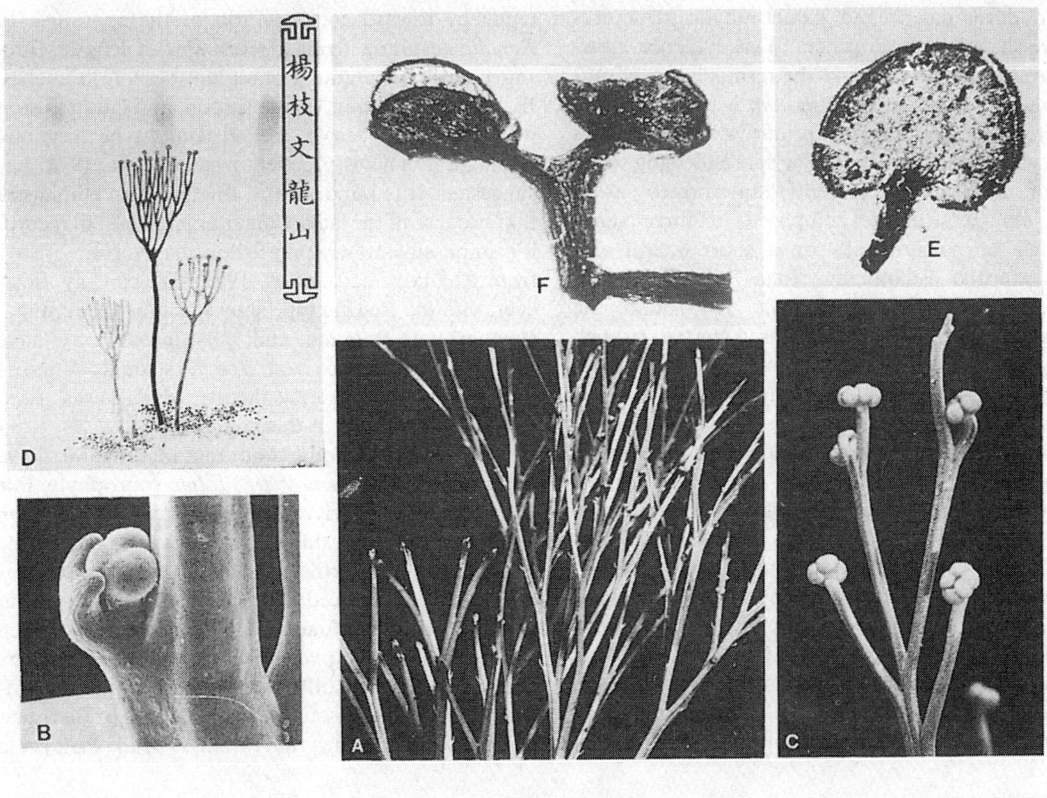

Figure 9.8. **A.** *Psilotum nudum,* habit. Extant. Note three-dimensional branching, small enations, and synangia. **B.** *P. Nudum,* normal synangium subtended by bifurcate "foliar" unit. **C.** *P. nudum,* appendageless form with terminal synangia. **D.** Illustration of an appendageless *Psilotum* ("Bunryu-zan" cultivar) in a nineteenth-century Japanese book. **E.** *Renalia hueberi,* single reniform sporangium. Lower Devonian. **F.** *hueberi,* sporangium showing transverse dehiscence. (**A–D** from Rouffa, 1971, 1973, 1978; **E,F** from Gensel, 1976.)

occurrence of vascular tissue in some gametophytes (Bierhorst, 1953). The annular and scalariform tracheids are centrally located in the gametophytic tissues and have been surrounded by phloem and an endodermis. Vascular tissue is not known to occur in the gametophytes of any other extant vascular plants. As you might conclude, the similarity in structure of *Psilotum* sporophytes and gametophytes supports the transformation theory for the origin of alternation of phases (Chapter 7). Further, it is clear that the life cycle of *Psilotum* would compare favorably with that of the *Rhynia* type if it is demonstrated that the *Rhynia* type has an alternation of isomorphic phases in which gametophyte and sporophyte are endowed with vascular tissue, a possibility suggested by Remy (1982).

Because of similarities in their appearance and structure it was suggested long ago that the Psilopsida (*Psilotum* and *Tmesipteris*) were extant relatives of the Silurian–Devonian Rhyniopsida. There are, however, some problems in arriving at this conclusion. The first is the complete absence of a fossil record of rhyniophytes after the Middle Devonian or 360 m.y. ago. One might expect to find some evidence of ancestors of extant Psilopsida in the fossil record since the Middle Devonian if the two groups are actually related. In a series of investigations into the structure and life cycle of *Psilotum*, Bierhorst (1971) has come to the conclusion that *Psilotum* is closely related to extant *Stromatopteris*, a genus belonging to a group of leptosporangiate ferns. He reports similarities in embryo development and gametophytic structures that are shared by both genera. This evidence, however, overlooks the fact that the fern is leptosporangiate and has roots, while *Psilotum* is eusporangiate and is rootless. The latter are primitive characteristics shared with the Rhyniopsida. The spore-forming structure of *Psilotum* is presently interpreted as a synangium – a structure composed of three fused sporangia borne terminally on a short lateral axis associated with a dichotomous "foliar" unit (Fig. 9.8B). There is good evidence (Rouffa, 1973) that the synangium represents a condensed fertile branch system with terminal sporangia associated with "foliar" units. The evidence comes from Rouffa's studies of wild clones and cultivars of *Psilotum nudum* that produced synangia terminally on aerial axes lacking the sterile "foliar" units (Fig. 9.8C). Specimens collected in the wild as well as those grown under controlled intensities and durations of light showed these characteristics. An illustration (Fig. 9.8D) taken from a nineteenth-century Japanese book on *Psilotum* shows cultivars of "Bunryu-zan" with their three-dimensional naked dichotomizing axes terminated by synangia. The resemblance of these cultivars to members of the Rhyniopsida is apparent. In fact, Rouffa (1978) strongly supports the idea that the fertile axes with their terminal synangia were evolutionarily derived from ancestral forms similar to members of the Rhyniopsida. Some support for this interpretation has recently been provided by the investigations of Gensel (1976), who described *Renalia*, a new Lower Devonian rhyniophyte (Fig 9.9A).

Renalia is estimated to have attained a height of at least 20 cm. The main axes of the plant divide more or less dichotomously, and these may bear smaller dichotomous branches terminated by round-to-reniform sporangia (Fig. 9.8E). The sporangia dehisce along their distal margins (Fig. 9.8F) and often contain trilete isospores. A centrally located vascular strand is present both in the axes and their branches. Spiral and scalariform tracheids have been isolated from the vascular system, but the arrangement of the vascular tissues has not been determined from the compression fossils.

As noted by Gensel, the sporangia are similar in form to those of *Cooksonia*. Unlike most species of *Cooksonia* and *Uskiella*, which show dichotomous branching in all parts of their aerial branching systems, *Renalia* has both dichotomous branching in the distal parts of the branching system and pseudomonopodial branching (at the point of branching one member of the branch pair is larger than the other, producing a main axis with a smaller lateral) of the main axes. Of particular interest, however, are the lateral, dichotomizing branches with terminal sporangia. When comparing with other rhyniophytes such as *Cooksonia*, *Uskiella*, and *Rhynia*, one can see a tendency toward condensation of the sporangia in *Renalia* resulting from a reduction in length (foreshortening) of lateral sporangium-bearing branches. By further reduction, condensation, and lateral fusion it is easy to derive the *Psilotum*-type terminal synangium. This is precisely what Rouffa (1978) has suggested. It is important at this point to emphasize that because of the lack of an intervening fossil record, it is impossible to say that Psilopsida evolved directly from Rhyniopsida (Gensel, 1977). We can say, however, that the *Renalia* type shows us how the terminal synangia of *Psilotum* and *Tmesipteris* may have evolved. The evidence from *Renalia* strongly supports one aspect of the telome theory, which is discussed in some detail later in this chapter.

It should be pointed out that investigators have other interpretations of the *Psilotum* sporophyte. For example, one conceives the aerial axes of *Psilotum* to be reduced leafy shoots bearing sterile and fertile appendages. The fertile appendage, according to this concept, is a reduced bifid sporophyll bearing an adaxially inserted synangium. This conclusion, which is supported by Kaplan (1977), is not confirmed by the researches of Rouffa and interpretations of evidence from the fossil record. Although Bierhorst (1971) accepts the idea that synangia of the Psilopsida are terminal, he hypothesizes that the aerial axes of an ancestral form became flattened to produce an appendicular axis homologous with the fertile frond of the fern *Stromatopteris*. When we consider these and other equally divergent interpretations of the *Psilotum* sporophyte, it becomes abundantly clear that we need to know much more about the fossil record of the class Psilopsida.

Questionable Rhyniopsida

Other than the genera discussed thus far, there remain only a few other Rhyniopsida. Their assignment to the class is questionable because their remains are incomplete (Banks, 1975a; Taylor, 1988). One of these is the genus *Taeniocrada* (Fig. 9.9B), a fairly wide-ranging fossil with a record extending through the Lower Devonian into the Upper Devonian. Characteristic of other rhyniophytes, *Taeniocrada* had naked, dichotomously branched axes. Rather than being terete, the branches were ribbonlike with a central vascular strand. Sporangia were terminal on more or less dichotomous recurved branches. In this characteristic, *Taeniocrada* is unlike other rhyniophytes. The abundance of the fossils and their disposition in the rock matrix suggest an aquatic environment for the genus.

Hedeia (Lang & Cookson, 1935) and *Yarravia* (Cookson, 1935) are Lower Devonian fragments consisting of short axes with terminal sporangia. The sporangia of *Hedeia* (Fig. 9.9C) are similar in size and shape to those of *Rhynia* but are grouped closer to one another at the tips of naked dichotomous branches. *Yarravia* (Fig. 9.9D) has its five to six sporangia fused laterally forming a terminal synangium. In light of our interpretations of terminal synangia in *Psilotum* and how they evolved, it is most interesting to find that synangia had evolved during Lower Devonian times.

Figure 9.9. **A.** *Renalia hueberi,* reconstructed part of branching axis with terminal sporangia. Lower Devonian. **B.** *Taeniocrada* sp. Lower Devonian. **C.** *Hedeia corymbosa,* an excellent example of a fertile telome truss. Lower Devonian. **D.** *Yarravia oblonga,* synangium. An excellent example of syngenesis of sporangia. Lower Devonian. (A redrawn from Gensel, 1976; **B** redrawn from Banks, 1968; **C,D** redrawn from Lang & Cookson, 1935.)

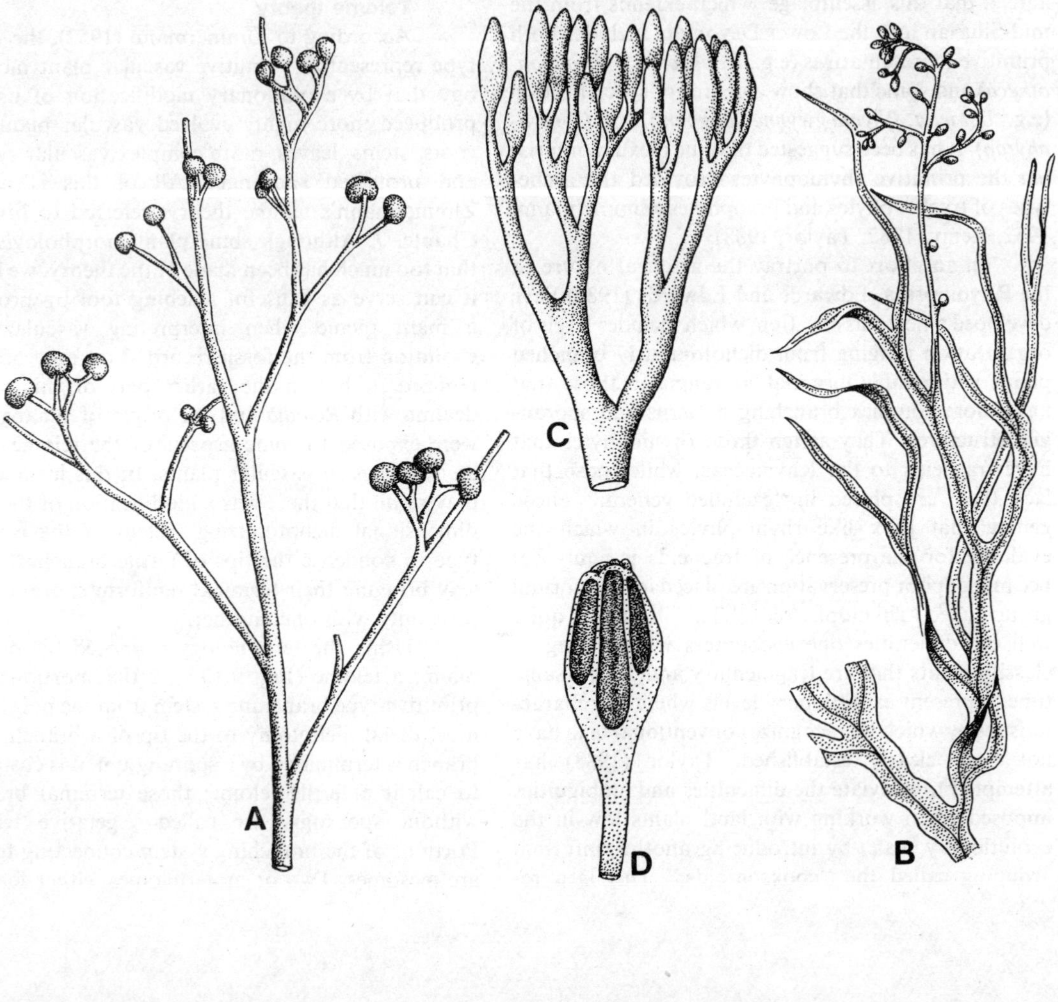

Other genera such as *Hicklingia, Dutoitea,* and *Eogaspesiea* are compression fossils consisting of slender naked dichotomizing axes with what appear to be terminal sporangia. *Hostinella* is a form genus for fragments of naked dichotomizing axes.

So far much of the subject matter of this chapter is speculative. The main point, however, that should be clear, and most botanists agree, is that the Rhyniopsida is represented by a group of the most primitive land plants thus far discovered. They appear in the mid-Silurian prior to any other land plants, and in this respect they are considered to represent a plexus of organisms from which vascular plants and perhaps bryophytes have evolved.

Classifications and the *Rhynia* type

It has become apparent to paleobotanists during the last decade that the class Rhyniopsida is a heterogeneous assemblage of land plants. We have learned that its members may show some diversity in their branching patterns, conducting systems, and sporangial shape and dehiscence. They show uniformity, however, (1) in having aerial axes that are naked (leafless), (2) usually in having three-dimensional dichotomizing axes, (3) in having sporangia that terminate these axes, and (4) in producing isospores. It is agreed that this assemblage, which extends from the mid-Silurian into the Lower Devonian, includes both primitive representatives (e.g., *Cooksonia* and *Steganotheca*) and some that show a degree of specialization (e.g., *Uskiella, Rhynia gwynne-vaughanii,* and *Horneophyton*). It has been suggested that the plexus comprising the primitive rhyniophytes provided the archetypes of tracheophytes and bryophytes (Zimmermann, 1952; Remy, 1982; Taylor, 1988).

In an effort to portray the artificial nature of the Rhyniopsida, Edwards and Edwards (1986) have developed a new classification, which includes levels of organization ranging from dichotomously branched plants with simple terminal sporangia to those that have more complex branching patterns and sporangial structures. They assign those rhyniophytes that have tracheids to the Rhyniaceae, while those that lack them are placed in "excluded genera." Those genera that look like rhyniophytes in which the evidence for the presence of tracheids is equivocal because of poor preservation are placed in an informal group called "rhyniophytoids." This illustrates quite well the difficulties one encounters when trying to classify plants that are fragmentary and, at the same time, represent evolutionary levels where the characteristics by which we recognize conventional taxa have not been clearly established. Taylor (1988) has attempted to alleviate the difficulties and ambiguities imposed when working with land plants low in the evolutionary scale, by introducing another informal grouping called the "cooksonioids." This is a replacement for the "rhyniophytoids" of the Edwards & Edwards (1986) classification.

With the recognition that, like the Rhyniopsida, the "rhyniophytoid" and "cooksonioid" groups are artificial, it seems that using the old terminology *Rhynia* type is a much less ambiguous way of treating the rhyniophytes than the systems proposed above. Here we recognize a level of evolution which includes those rhyniophytes that have naked leafless axes, branching systems that are three dimensional and dichotomous, and that have terminal sporangia containing isospores. It makes little difference whether the *Rhynia* type lacks conducting tissue, has a bryophytic system with hydroids, or has a vascular system with tracheids. If our hypothesis is correct, we can expect variability of this type to occur among members of the plexus from which land plants evolved. In the chapters to come, we shall see many other primitive vascular plants that provide ample evidence of having originated from archetypes of the *Rhynia* type. Until the time arrives when sufficient evidence has accumulated to make possible the separation of the Rhyniopsida into natural, recognizable taxa, as Banks (1975a) accomplished with the Psilophytales, we will continue to refer to the *Rhynia* type when it seems appropriate.

Telome theory

According to Zimmermann (1952), the *Rhynia* type represents a primitive vascular plant morphology that by evolutionary modification of its parts produced more highly evolved vascular plants with roots, stems, leaves, more complex vascular systems, and protected sporangia. All of this is part of Zimmermann's telome theory, referred to briefly in Chapter 7. Although some plant morphologists feel that too much has been made of the theory, we believe it can serve as a useful teaching tool by providing a main theme when interpreting vascular plant evolution from the fossil record. You may not have realized it, but in the earlier part of this chapter dealing with *Renalia* and the origin of synangia you were exposed to some aspects of the telome theory as it applies to vascular plants. In this instance you may recall that there was a modification of the three-dimensional dichotomizing system of the primitive type to condense the tips of fertile branches, in this way bringing their terminal reniform sporangia into proximity with one another.

Using the terminology proposed by Zimmermann, a telome (Fig. 9.10A) is that portion of the primitive-type branching system from the point of the most distal dichotomy to the tip of a branch. If the branch is terminated by a sporangium, it is customary to call it a fertile telome; those terminal branches without sporangia are called vegetative telomes. Portions of the branching system connecting telomes are mesomes. Two or more telomes, either fertile or

Figure 9.10. The telome concept. **A.** Primitive vascular plant of the *Rhynia* type. Telome truss (tt); sterile telome (st); fertile telome (ft); mesome (m). **B–D.** Evolutionary process of overtopping. Stippled axes represent those parts of the dichotomous axes that are overtopped to produce pseudomonopodial branching. **E,F.** Evolutionary process of plantation. Note fertile and sterile telome trusses. **G,H.** Evolutionary process of webbing to form a lamina. (A–H redrawn and modified from Zimmermann, 1959.)

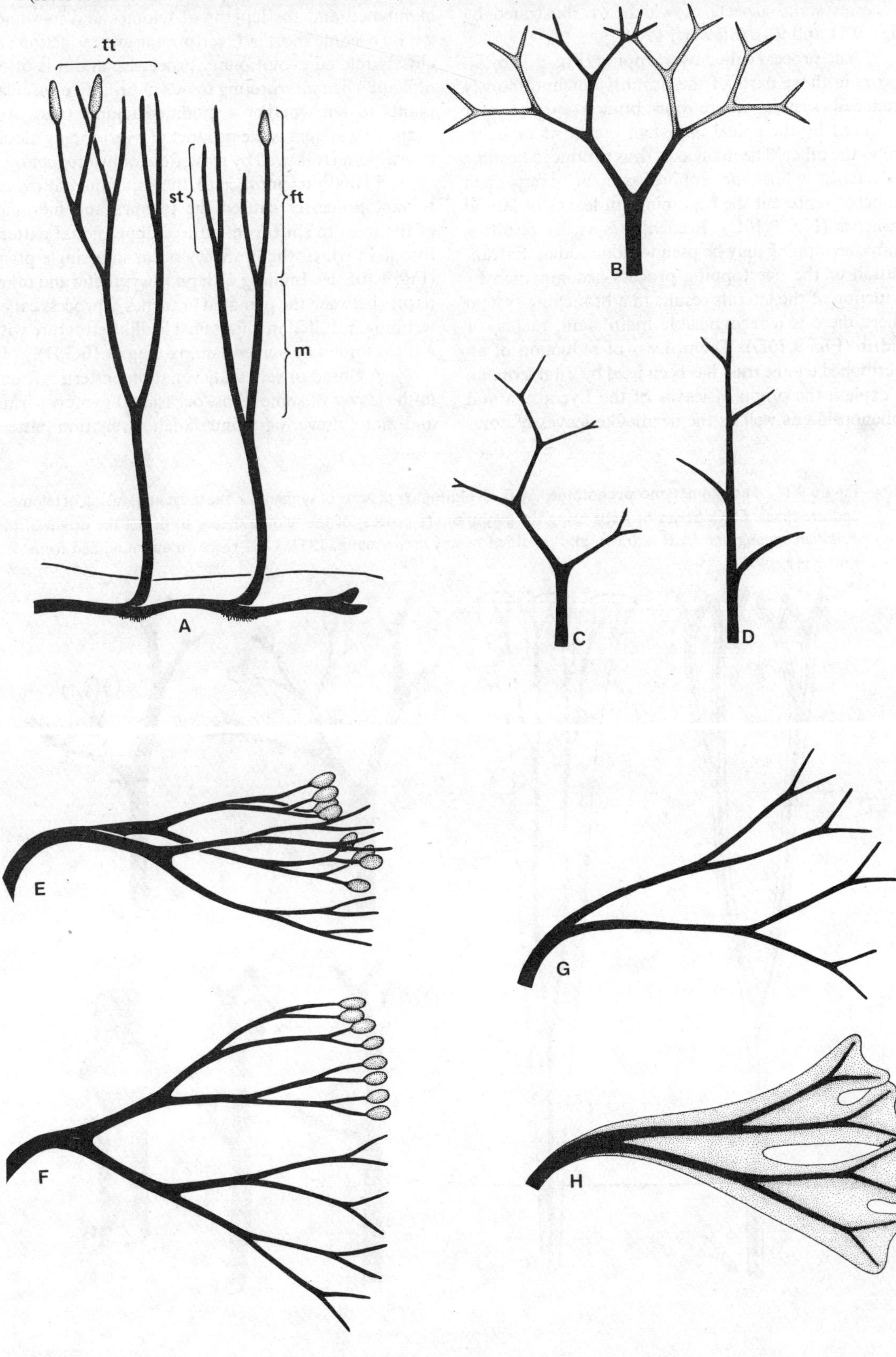

vegetative, connected by mesomes, comprise a telome truss. According to the theory, telomes or telome trusses have been subjected to certain evolutionary processes some aspects of which are illustrated by Figs. 9.11 and 9.12 (Stewart, 1964).

The process called overtopping (Fig. 9.10B–D) occurs in those parts of the primitive dichotomously branched system where one branch of the pair produced by the apical meristem outgrows or overgrows the other. The main axes thus produced become the stems, while the subordinate or overtopped branches represent the beginnings of leaves or lateral branches (Fig. 9.10C). Branching systems resulting from overtopping may be pseudomonopodial. Extrapolation of the overtopping process accompanied by reduction of the laterals results in a branching system where there is a recognizable main stem, rachis, or midrib (Fig. 9.10D). The process of reduction of an overtopped telome truss has been used by Zimmermann to explain the origin of leaves of the Lycopsida and Sphenopsida as well as the needlelike leaves of conifers. Reduction would result if the activity of the terminal meristem of each telome of the truss was suppressed so that branching would be decreased or eliminated and the lengths of telomes and mesomes would become shorter. Overtopping and reduction are closely related evolutionary processes, and it is often difficult when interpreting fossils of primitive vascular plants to tell whether a modified telome truss of a branching system is the product of overtopping alone, overtopping followed by reduction, or just reduction.

Planation, another of the fundamental evolutionary processes caused the telomes and mesomes of the truss to shift from a three-dimensional pattern to one in which the branches occur in a single plane (Fig. 9.10E,F). Infilling with photosynthetic and other tissues between the planated branches, a process called webbing, resulted in a flattened leaflike structure with a dichotomously veined lamina (Fig. 9.10G,H).

A closed or reticulate venation pattern is found in the leaves of some ferns, occasional gymnosperms, and many flowering plants. Such a venation pattern

Figure 9.11. The telome concept continued. **A,B.** Evolutionary process of syngenesis, the tangential fusion of telomes and mesomes. **C–G.** Series of steps using the evolutionary process of the telome theory to depict the origin of the *Psilotum* synangium. (**A,B** redrawn and modified from Zimmermann, 1959; **C–G** redrawn and modified from Emberger, 1944.)

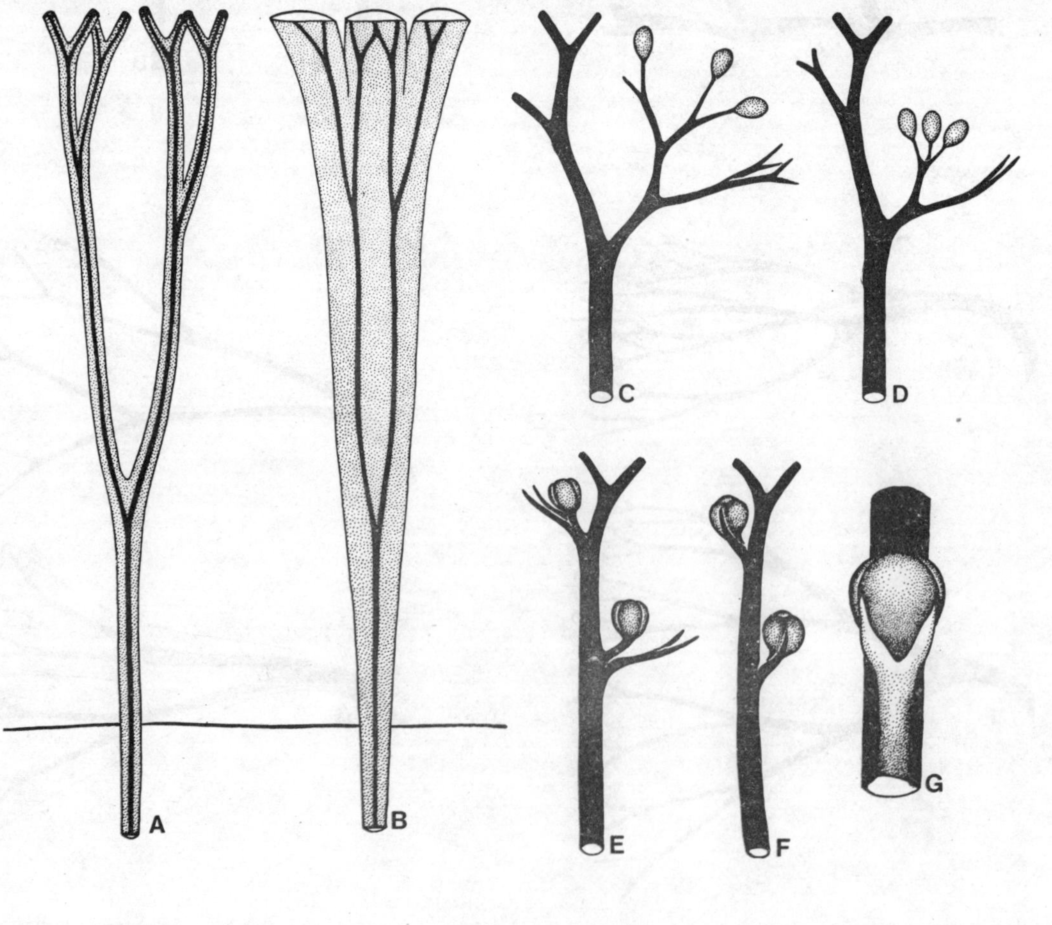

resulted from syngenesis or lateral fusion of the dichotomizing veins of the primitive leaf to form the anastomoses characteristic of venation patterns of the reticulate type. The evolutionary process of syngenesis also is applied by Zimmermann to the tangential fusion of mesomes and telomes, either fertile or vegetative (Fig. 9.11A,B). Such tangential fusion resulted in the formation of synangia (Fig. 9.11C–G) when fertile trusses with their terminal sporangia became reduced. Complex anastomosing vascular systems (Fig. 9.11B) in stems were produced as a result of syngenesis of vegetative telomes and mesomes.

Another evolutionary process recognized by Zimmermann is recurvation (Fig. 9.12). The idea to be conveyed is that fertile or vegetative telomes may bend inward toward an axis. An example would be the inward-projecting sporangia on a sporangiophore of *Equisetum*.

Now let us see how many of the elementary evolutionary processes (overtopping, reduction, planation, webbing, syngenesis, and recurvation) apply to the evolution of a synangium of the *Psilotum* type (consult Fig. 9.11C–G). Starting with the primitive *Rhynia* type with its fertile and vegetative telomes and using *Renalia* (Fig. 9.9A) as fossil evidence representing the way in which synangia evolved, we can see that overtopping occurred to produce a pseudomonopodial branching system with laterals consisting of three-dimensional dichotomously branched fertile and sterile telome trusses (Fig. 9.11C). Judging from the proximity of the sporangia in condensed clusters (Fig. 9.11D), it appears that some reduction has occurred. Continued reduction of the fertile telome truss would cause the sporangia to come to lie side by side, allowing syngenesis to occur (Fig. 9.11E) and in this way form a synangium. According to the theory, the bifid vegetative structure normally subtending a synangium of *Psilotum* (Fig. 9.11F,G) is the product of reduction of a vegetative portion of a telome truss associated with the fertile telomes. Presently there is no evidence from the fossil record of psilophytes supporting this origin of the bifid structure. In other groups, the Sphenopsida and Lycopsida, however, we will see fossil evidence supporting the origin of terminal sporangia with subtending leaves from dichotomously branched telomic systems.

Figure 9.12. Evolutionary process of recurvation of fertile telomes so that the terminal sporangia are directed toward the axis. (Redrawn from Zimmermann, 1959.)

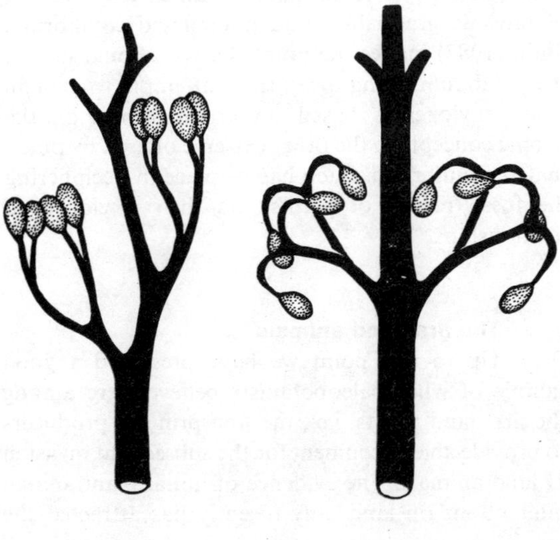

The concept of hologeny

As other authors have pointed out, it is wise to emphasize that even though we often state that such a mature structure as a telome truss "gave rise to or resulted in" an evolutionary more complex or specialized mature plant organ, it is implicit that the new organ has come about by changes in the genetic makeup of the ancestors. Further, that natural selection acting on the new genotypes has brought about gradual modifications that were expressed in the ontogenies of mature structures developed from apical meristems and embryos. The idea just expressed is depicted by Zimmermann (1959) as part of the telome theory. He calls it a hologenetic interpretation of phylogeny. Appropriately, the illustration (Fig. 9.13) shows his interpretation of the evolution of the synangium according to hologeny.

After Zimmermann had completed the synthesis of the telome concept, evidence accumulated that cast considerable doubt on the idea that the *Rhynia* type, which he selected as representing the ancestral-type vascular plant, was truly primitive. Presently, based on evidence much of which has been presented in this chapter, we think now it is safe to say that the *Rhynia* type is represented by a complex of primitive vascular plants from which several major groups of the division Tracheophyta have evolved, either directly or indirectly. These are the classes Trimerophytopsida, Sphenopsida, Filicopsida, and Progymnospermopsida. You may have noticed that the classes Zosterophyllopsida and Lycopsida are not included in the list. The reasons for this will be one of the main topics of the next chapter.

Telomes and heterochrony

When the first edition of this book was initiated in 1977, evolutionary theory was influenced to a high degree by what is called the modern synthesis. Zimmermann's telome theory provides us with a good example of the gradualistic assumptions that are a part of this synthesis. All students of biology, and many who are not, are acquainted with Darwin's concept of evolution, which incorporated the idea of gradual change. Accordingly, a given species would "descend with modification" from another. This he believed came about by natural selection operating on small inherited differences, and in this way the amount of variability would be increased. Darwin had no idea of the mechanisms involved in producing the variability other than the possible influence of the environment. Today we know that these minor variations (micromutations) are the product of changes in the genetic makeup of the species. Natural selection operating on the variants is thought to eliminate those individuals unsuited to the environment in which the population exists. Those that were better suited (adapted) became the progenitors of new species. Recently, however, many studies have questioned the idea that successful variants always arise by this process (Wight, 1987).

Figure 9.13. The concept of hologeny depicting the evolution of the synangium from a fertile telome truss. (Redrawn from Zimmermann, 1959.)

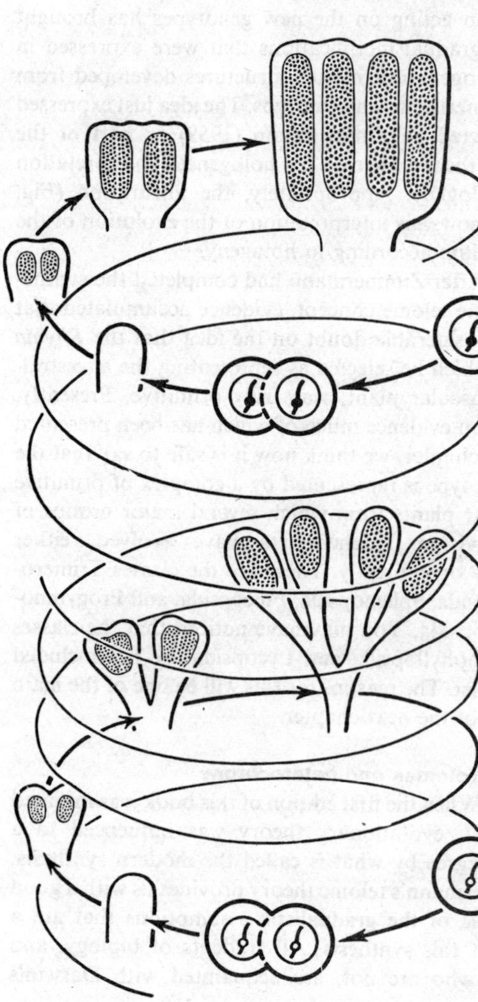

When a paleobotanist looks at the fossil record in an effort to determine the degrees of relationship of plants at various taxonomic levels, he is more often than not frustrated by the lack of evidence supporting the concept of gradual evolutionary change. Often there are abrupt gaps in the phylogenetic story, which are often filled with hypothetical organisms. The hope, of course, is that with more complete information, these discontinuities will be occupied by intermediate forms discovered at a later time. What we see emerging, however, is a pattern of evolution that is punctuated by gaps that might be explained not only by the incomplete nature of the fossil record, but by the process of evolutionary change themselves. In this category are those processes that have been categorized as macromutations, i.e., those changes that can have a dramatic effect in altering the development of plant characteristics (Rothwell, 1987). Among the types of developmental processes that can produce dramatic "jumps" or saltations of this kind are progenesis, neoteny, and postdisplacement, all of which lead to the development of normally juvenile characteristics in adult organisms. Other processes such as hypermorphosis, acceleration, and predisplacement give rise to characteristics developed beyond those of the progenitor adults. All of these refer to changes in the relative rate and timing of the development of characteristics already present in progenitors. This is called heterochrony, examples of which appear in later chapters and are discussed at length by Rothwell (1987).

Thus far, we have not attempted to do more than demonstrate that, within the last decade, new ideas grouped under the heading of punctuated equilibrium have been introduced that are at variance with the much older ideas of evolutionary gradualism. The new concepts offer alternate explanations for the apparent discontinuities in the fossil record that in the past have been explained as failures in fossilization or discoveries yet to be made. In an excellent review of phyletic gradualism and punctuated equilibrium Stidd (1987) makes the point that we should not let the pendulum swing away from attempts to explain plant phylogenies based on gradualism (e.g., the telome concept) to the other extreme offered by punctuated equilibrium. Each has its place in deciphering the fossil record of plants and their phylogenetic relationships.

The first land animals

Up to this point we have presented a good sample of what paleobotanists believe were among the first land plants, i.e., the first primary producers to provide an environment for the subsequent invasion of land animals. The evidence of initial plant/animal interactions on land only recently has attracted the attention of paleontologists (Scott, 1984; Scott, Chaloner, & Paterson, 1985). They focus our attention once more on the Lower Devonian Rhynie chert beds of Scotland where there is evidence of interaction between arthropods (Fig. 9.14) and certain primitive land plants described earlier in this chapter. The evidence consists of *Rhynia* specimens that show penetrating wounds, which in turn suggest that sucking insects may have fed on the plant's sap. These wounds are very similar in appearance to the archegoniumlike structure of *R. gwynne-vaughanii*. Earlier studies by Kevan, Chaloner, & Savile (1975) revealed an assemblage of arthropods (Fig. 9.14) which includes herbivorous myriapods (millipedes), carnivorous arachnida (spiders and scorpions), a collembolan, and unidentified chitinous jaws. The arachnid was discovered inside an empty *Rhynia* sporangium where it is suggested that it was feeding on spores or using the sporangium for shelter and as a place to feed on other small insects.

While macerating a fine-grained mudstone from the Middle Devonian near Gilboa, New York, Shear et al. (1984) obtained more fossilized remains of arthropods that occur as unaltered cuticular material. The specimens provide good detail and greater variety than other Devonian-age localities. These authors believe that these specimens include parts of arachnids, centipedes, a complete mite, and a compound eye of an insect.

Other than indicating an early invasion of a land flora by arthropods in Devonian times, the presence of phytophagous animals interacting with plants indicates the initiation of a producer/consumer food chain in a terrestrial environment. The implications of this event in terms of the subsequent evolution of terrestrial organisms is of the greatest biological magnitude.

References

Andrews, H. N. (1959). Evolutionary trends in early vascular plants. *Cold Springs Harbor Symposium on Quantitative Biology*, **24**, 217–34.

Axelrod, D. I. (1959). Evolution of the psilophyte paleo flora. *Evolution*, **13**, 264–75.

Banks, H. P. (1968). The early history of land plants. In *Evolution and Environment*, ed. E. T. Drake. New Haven: Yale University Press.

Banks, H. P. (1970). Occurrence of *Cooksonia*, the oldest vascular land plant macrofossil in Upper Silurian of New York state. *Journal of the Indian Botanical Society*, **50A**, 227–35.

Banks, H. P. (1975a). Reclassification of psilophyta. *Taxon*, **24**, 401–13.

Figure 9.14. Arthropods from the Lower Devonian. **A.** Trigonotarbid arachnid. **B.** Myriapod. **C.** Mite. **D.** Collembolan. (From Scott, 1984.)

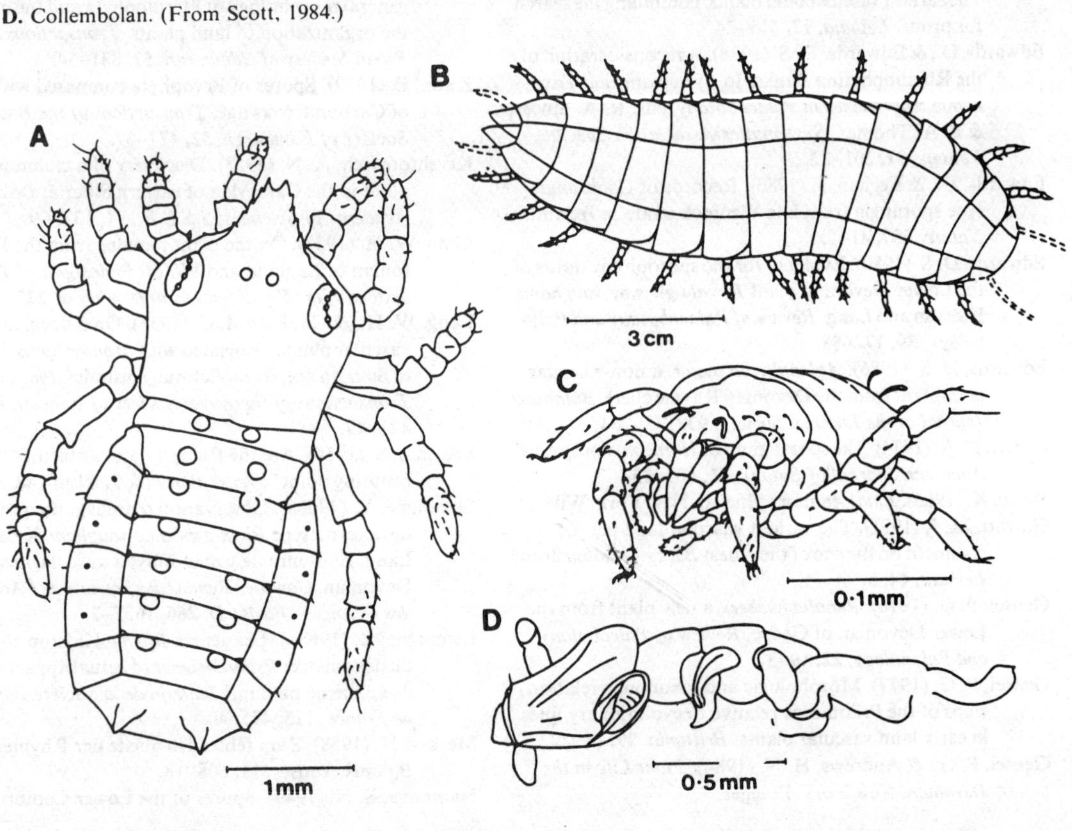

Banks, H. P. (1975b). The oldest vascular plants: a note of caution. *Review of Palaeobotany and Palynology*, **20**, 13–25.

Barghoorn, E. S., & Darrah, W. C. (1938). *Horneophyton*, a necessary change of name for *Hornea*. *Harvard University Botanical Museum Leaflets*, **6**, 142–4.

Bierhorst, D. W. (1953). Structure and development of the gametophyte of *Psilotum nudum*. *American Journal of Botany*, **40**, 649–58.

Bierhorst, D. W. (1971). *Morphology of Vascular Plants*. New York: Macmillan.

Bold, H. C., Alexopoulos, C. J., & Delevoryas, T. (1980). *Morphology of Plants and Fungi*, 4th ed. New York: Harper & Row.

Chaloner, W. G. (1967). Spores and land plant evolution. *Review of Palaeobotany and Palynology*, **1**, 89–93.

Chaloner, W. G. (1988). Early land plants – the saga of a great conquest. In *Proceedings of the XIV International Botanical Congress*, eds. W. Greuter & B. Zimmer. Königstein: Koeltz.

Chaloner, W. G., & Sheerin, A. (1979). Devonian macrofloras. In *The Devonian System*, eds. M. R. House, C. T. Scrutton, & M. G. Bassett. *Special Papers in Palaeontology*, **23**, 145–61.

Delevoryas, T. (1962). *Morphology and Evolution of Fossil Plants*. New York: Holt, Rinehart and Winston.

Edwards, D. (1970). Fertile Rhyniophytina from the Lower Devonian of Britain. *Palaeontology*, **13**, 451–61.

Edwards, D. (1980). Early land floras. In *The Terrestrial Environment and the Origin of Land Vertebrates*, ed. A. L. Panchen. London: Academic Press, pp. 55–85.

Edwards, D., Bassett, M. G., & Rogerson, C. W. (1979). The earliest vascular land plants: continuing the search for proof. *Lethaia*, **12**, 313–24.

Edwards, D., & Edwards, D. S. (1986). A reconsideration of the Rhyniophytina Banks. In *Systematic and Taxonomic Approaches in Palaeobotany*, eds. R. A. Spicer, & B. A. Thomas. *Systematics Association Special Volume*, **31**, 201–22.

Edwards, D., & Feehan, J. (1980). Records of *Cooksonia*-type sporangia from late Wenlock strata in Ireland. *Nature*, **287**, 41–2.

Edwards, D. S. (1980). Evidence for the sporophytic status of the Lower Devonian plant *Rhynia gwynne-vaughanii* Kidston and Lang. *Review of Palaeobotany and Palynology*, **29**, 177–88.

Edwards, D. S. (1986). *Aglaophyton major*, a non-vascular land-plant from the Devonian Rhynie chert. *Botanical Journal of the Linnean Society*, **93**, 173–204.

Eggert, D. A. (1974). The sporangium of *Hornephyton lignieri*. *American Journal of Botany*, **61**, 405–13.

Essau, K. (1965). *Plant Anatomy*, 2nd ed. New York: Wiley.

Garratt, M. J. (1981). The earliest vascular land plants: comment on the age of the oldest *Baragwanathia* flora. *Lethaia*, **14**, 8.

Gensel, P. G. (1976). *Renalia hueberi*, a new plant from the Lower Devonian of Gaspé. *Review of Paleobotany and Palynology*, **22**, 19–37.

Gensel, P. G. (1977). Morphologic and taxonomic relationships of the Psilotaceae relative to evolutionary lines in early land vascular plants. *Brittonia*, **29**, 14–29.

Gensel, P. G., & Andrews, H. N. (1984). *Plant Life in the Devonian*. New York: Praeger.

Ghosh, A. K., & Bose, A. (1949–50). Microfossils from the Cambrian strata, Salt Range, Punjab. *Transactions of the Bose Research Institute*, **18**, 71–8.

Gray, J., & Boucot, A. J. (1977). Early vascular land plants: proof and conjecture. *Lethaia*, **10**, 145–74.

Haskell, G. (1949). Some evolutionary problems concerning the Bryophyta. *The Bryologist*, **52**, 49–57.

Hueber, F. M. (1983). A new species of *Baragwanathia* from the Sextant Formation (Emsian) northern Ontario, Canada. In *Contributions to Palaeobotany*, eds. M. W. Dick, & D. Edwards. *Botanical Journal of the Linnean Society*, **86**, 57–79.

Kaplan, D. R. (1977). Morphological status of the shoot systems of Psilotaceae. *Brittonia*, **29**, 30–53.

Kevan, P. G., Chaloner, W. G., & Savile, D. B. O. (1975). Interrelationships of early terrestrial arthropods and plants. *Palaeontology*, **18**, 391–417.

Kidston, R., & Lang, W. H. (1917). On Old Red Sandstone plants showing structure from the Rhynie chert bed, Aberdeenshire. Part I. *Rhynia gwynne-vaughanii*, Kidston and Lang. *Transactions of the Royal Society of Edinburgh*, **51**, 761–84.

Kidston, R., & Lang, W. H. (1920). On Old Red Sandstone plants showing structure from the Rhynie chert bed, Aberdeenshire. Part II. Additional notes on *Rhynia major* n.sp. and *Hornea Lignieri* n.g., n.sp. *Transactions of the Royal Society of Edinburgh*, **52**, 603–27.

Kidston, R., & Lang, W. H. (1921). On Old Red Sandstone plants showing structure from the Rhynie chert bed, Aberdeenshire, Part IV. Restorations of the vascular cryptogams, and discussion of their bearing on the general morphology of Pteridophyta and the origin of the organization of land plants. *Transactions of the Royal Society of Edinburgh*, **52**, 831–54.

Knox, E. (1939). Spores of Bryophyta compared with those of Carboniferous age. *Transactions of the Botanical Society of Edinburgh*, **32**, 477–87.

Kryshtofovich, A. N. (1953). Discovery of a clubmoss-like plant in the Cambrian of eastern Siberia. *Doklady Academy of Sciences, S.S.S.R.*, **91**, 1377–9.

Lang, W. H. (1937). On the plant remains from the Downtonian of England and Wales. *Philosophical Transactions of the Royal Society of London, B*, **227**, 245–91.

Lang, W. H., & Cookson, I. C. (1935). On a flora, including vascular plants associated with *Monograptus*, in rocks of Silurian age, from Victoria, Australia. *Philosophical Transactions of the Royal Society of London, B*, **224**, 421–49.

Leclercq, S. (1954). Are the Psilophytales a starting or a resulting point? *Svensk Botanisk Tidskrift*, **48**, 301–15.

Lemoigne, Y. (1968a). Observation d'archégones portés par des axes du type *Rhynia gwynne-vaughanii* Kidston et Lang. Existence de gametophytes vascularises au Devonian. *Comptes Rendus des Séances de l'Académie des Sciences (Paris) D*, **266**, 1655–7.

Lemoigne, Y. (1968b). Les genres *Rhynia* Kidston et Lang du dévonian et *Psilotum* Seward actual appartiennent ils au même phylum? *Bulletin de la Société Botanique de France*, **115**, 425–40.

Merker, H. (1958). Zum fehlenden gliede der Rhynien flora. *Botanik Notiser*, **11**, 608–18.

Naumova, S. N. (1949). Spores of the Lower Cambrian.

Geological Bulletin, Academy of Science, S.S.S.R., **4,** 49–56.

Obrhel, J. (1962). Die flora der Přidolischichten (Budňany-Stufe) des mittelböhmischen Silurs, *Geologie*, **11,** 83–97.

Pant, D. D. (1962). The gametophyte of the Psilophytales. *Proceedings of the Summer School of Botany – Darjeeling*, eds. P. Maheshwari, B. M. Johri, & I. K. Vasil, pp. 276–301.

Pratt, L. M., Phillips, T. L., & Dennison, J. M. (1978). Evidence of non-vascular land plants from the early Silurian (Llandoverian) of Virginia, U.S.A. *Review of Palaeobotany and Palynology*, **25,** 121–49.

Proskauer, J. (1960). Studies in the Anthocerotales, VI. *Phytomorphology*, **10,** 1–19.

Remy, W. (1982). Lower Devonian gametophytes: relation to phylogeny of land plants. *Science*, **215,** 1625–27.

Remy, W., & Remy, R. (1980a). Devonian gametophytes with anatomically preserved gametangia. *Science*, **208,** 295–6.

Remy, W., & Remy, R. (1980b). *Lyonophyton rhyniensis* nov. gen. et nov. sp., ein Gametophyte aus dem Chert von Rhynie (Unterdevon, Schottland). *Argumenta Palaeobotanica*, **6,** 37–72.

Remy, W., Remy, R., Hass, H., Schultka, St., & Franzmeyer, F. (1980). *Sciadophyton* Steinmann–ein Gametophyt aus dem Siegen. *Argumenta Palaeobotanica*, **6,** 73–94.

Rothwell, G. W. (1987). The role of development in plant phylogeny: a paleobotanical perspective. *Review of Palaeobotany and Palynology*, **50,** 97–114.

Rouffa, A. S. (1971). An appendageless *Psilotum*: Introduction to aerial shoot morphology. *American Fern Journal*, **61,** 75–86.

Rouffa, A. S. (1973). Natural origin of the appendageless *Psilotum*. *American Fern Journal*. **63,** 145–6.

Rouffa, A. S. (1978). On phenotypic expression, morphogenetic pattern, and synangium evolution in *Psilotum*, *American Journal of Botany*, **65,** 692–713.

Scagel, R. F., Bandoni, R. J., Rouse, G. E., Schofield, W. B., Stein, J. R., & Taylor, T. M. C. (1969). *Plant Diversity: An Evolutionary Approach*, 2nd. ed. Belmont, Calif.: Wadsworth.

Schweitzer, H. (1981). Der Generationswechsel rheinischer Psilophyten. *Bonner Paläobotanaishe Mitteilungen*, **8,** 1–19.

Scott, A. C. (1984). The early history of life on land. *Journal of Biological Education*, **18,** 207–19.

Scott, A. C., Chaloner, W. G., & Paterson, S. (1985). Evidence of pteridophyte–arthropod interactions in the fossil record. *Proceedings of the Royal Society of Edinburgh*, **86b,** 133–40.

Shear, W. A., Bonamo, P. M., Grierson, J. D., Rolfe, W. D. I., Smith, E. L., & Norton, R. A. (1984). Early land animals in North America: Evidence from Devonian age arthropods from Gilboa, New York. *Science*, **224,** 492–94.

Shute, C. H., & Edwards, D. (1989). A new rhyniopsid with novel sporangium organization from the Lower Devonian of South Wales. *Botanical Journal of the Linnean Society*, **100,** 111–37.

Smith, G. M. (1955). *Cryptogamic Botany*, 2nd. ed. New York: McGraw-Hill, Vol. 2.

Stewart, W. N. (1960). More about the origin of vascular plants. *Plant Science Bulletin*, **6,** 1–4.

Stewart, W. N. (1964). An upward outlook in plant morphology. *Phytomorphology*, **14,** 120–34.

Stidd, B. M. (1987). Telomes, theory change, and the evolution of vascular plants. *Review of Palaeobotany and Palynology*, **50,** 115–26.

Taylor, T. N. (1982). The origin of land plants: a paleobotanical perspective. *Taxon*, **31,** 155–77.

Taylor, T. N. (1988). The origin of land plants: some answers, more questions. *Taxon*, **37,** 805–33.

Wight, D. C. (1987). Non-adaptive change in early land plant evolution. *Paleobiology*, **13,** 208–14.

Zimmermann, W. (1952). Main results of the "Telome Theory." *The Paleobotanist*, Birbal Sahni Memorial Volume, 456–70.

Zimmermann, W. (1959). *Die Phylogenie der Pflanzen*, 2nd. ed. Stuttgart: Fischer.

10

The evolution of microphylls and adaxial sporangia

Although they are of little economic importance and they form only a minor part of our extant floras, the eight living genera of class Lycopsida have attracted more than their share of attention from those interested in plant evolution. Most taxonomists recognize three orders that comprise the extant lycopods; the Lycopodiales, Selaginellales, and Isoetales. Originally there were only two genera, *Lycopodium* (ground pine) (Fig. 10.1A) and *Phyloglossum* assigned to the Lycopodiales. Subsequently, after the reevaluation of the 400 species of *Lycopodium,* four new genera including the genus *Huperzia* (Fig. 10.2A) were created. *Selaginella* (Fig. 10.2B) remains as the sole living genus of the Selaginellales. Until recently two extant genera were included in the Isoetales, *Isoetes* (Fig. 10.1B) and *Stylites* (Fig. 10.1C). As a result of recent investigations (Karrfalt, 1984), *Stylites* has been transferred to *Isoetes* so that only the one genus, *Isoetes*, remains in the order. All of these plants are relatively small herbaceous organisms in contrast to some of the fossil representatives, which were arborescent.

Some characteristics of the lycopods

The combination of characteristics – microphylls and adaxial sporangia – is usually used to distinguish members of the Lycopsida from other free-sporing vascular plants. The term microphyll ("little leaf") is a misnomer since it is known that some extinct lycopods had large grasslike leaves that attained a length of more than 78 cm (Kosanke, 1979). Obviously size is not a good criterion by which to distinguish these foliar structures. With few exceptions (Wagner, Beitel, & Wagner, 1982), however, microphylls have a single unbranched vascular strand that arises from the vascular cylinder of the stem without leaving a leaf gap (Fig. 10.3A). A leaf gap occurs in the stems of plants (many Filicopsida) where there is a vascular cylinder with a pith (siphonostele) (Fig. 10.3B). Immediately above the point of departure of the leaf trace from the siphonostele there is a parenchyma-filled interruption (gap) in the cylinder of vascular tissue. At a higher level the gap is closed by the vascular tissue of the siphonostele.

As can be seen in Fig. 10.3C,D, the sporangia of the Lycopsida do not appear to be terminal on aerial branches of the sporophyte as in the Rhyniopsida and Psilopsida. Instead, they are associated with the upper (adaxial) surface of the leaf (sporophyll), either in the leaf axil or near it. One of our objectives is to find answers to questions about the evolutionary origins of microphylls and the adaxial position of sporangia on sporophylls. Another objective is to sample the early diversification of the lycopods. To do this we have to return to our "upward outlook" and search for their beginnings.

Zosterophyllopsida

As far as has been determined to date, fossil remains that can be interpreted as early ancestors of the Lycopsida first appear in the Lower Devonian (Gedinnian). The principal genus, *Zosterophyllum* (Fig. 10.4A), was first described by Penhallow in 1892. Lang (1927) reinvestigated the original as well as new specimens and gave us our first comprehensive account of the genus. Because of its primitive characteristics, the genus was placed in the old order Psilophytales. It was not until 1968 that Banks placed *Zosterophyllum* and other genera having similar characteristics in a new subdivision, the Zosterophyllophytina (our class Zosterophyllopsida), separate from other fossil psilophytes. According to Banks, there are three characteristics that, in combination, set the zosterophylls apart from other psilophytes. The first of these is the lateral position of sporangia in terminal portions of aerial axes. Second, the sporangia are globose or reniform and dehisce transversely along the distal edge (Fig. 10.4B). Third, in cross section the xylem strand of the aerial axes is a relatively massive elliptical to terete exarch protostele.

Zosterophyllum is usually depicted as a plant with a tufted growth habit growing to a height up to 15 cm. Its naked somewhat flattened axes have dichotomous and a curious H-shaped branching (Fig. 10.4C). Internally, the exarch protostele is composed of annular and scalariform tracheids. It is terete in vegetative axes and elliptical in fertile ones. Cuticle, epidermal cells, and stomata have been found on the upper aerial branches. The absence of cuticle and stomata on the lower portions of the branching system suggests that *Zosterophyllum* grew in a marshy if not aquatic habitat.

Sporangia of *Zosterophyllum* terminate short lateral branches, which are radially arranged on the distal portions of the aerial axes. Edwards (1969) was able to show that the sporangia dehisced along the distal convex surface by means of a special mechanism and that the sporangial wall was composed of several layers of cells, thus verifying its eusporangiate structure. Isospores were isolated from the sporangia and these proved to be the *Retusotriletes* type.

Similar to *Zosterophyllum,* the genera *Gosslingia* (Fig. 10.4D) and *Rebuchia* (Fig. 10.4E) have more or less smooth dichotomously and pseudomonopodial branched axes. The vascular system, known for *Gosslingia,* is an elliptical exarch protostele composed of spiral and reticulate tracheids. Reniform to globose sporangia are scattered irregularly along the sides of the axes. Those of *Rebuchia* (Hueber, 1971a) terminate short, rather stout branches. The latter are arranged in two rows that extend to the tips of the axes forming a bilaterally symmetrical spike. Both *Gosslingia* and *Rebuchia* have sporangia like those of other zosterophylls in form and position of dehiscence along the distal convex margin. *Rebuchia* appears in strata that are Lower Devonian (Emsian) in age while *Gosslingia* is somewhat older and has been found in the Siegenian–Emsian boundary of the Lower Devonian.

Figure 10.1. **A.** *Lycopodium annotinum,* a species with cones terminating aerial axes. **B.** *Isoetes melanapoda,* habit. **C.** *Isoetes* (= *Stylites*) *andicola.* (Drawn from specimens provided by E. Karrfalt.) **D.** *Isoetes* sp., adaxial surface, base of megasporophyll: ligule (l), megasporangium (sp).

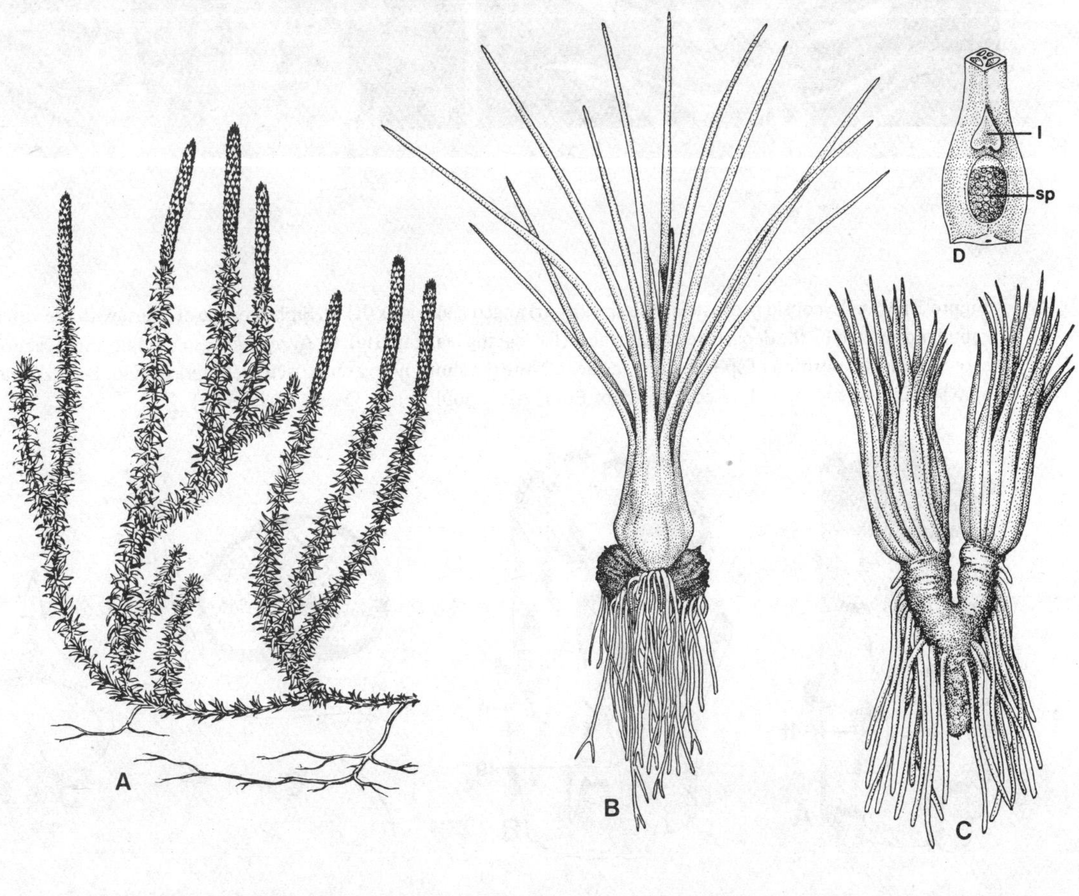

Figure 10.2. **A.** *Huperzia* (= *Lycopodium*) *selago*, a species with alternating zones of sporophylls and vegetative leaves. Note the adaxial position of the sporangia and spiral arrangement of leaves. **B.** *Selaginella grandis* with cones. Note dimorphic nature of vegetative leaves. (**A,B** courtesy of the General Biological Supply House, Chicago.)

Figure 10.3. **A.** Lycopsid protostele with spirally arranged leaf traces (lt). **B.** Siphonostele of a fern with the vascular cylinder dissected by the departure of leaf traces (lt), leaving leaf gaps (lg). **C.** *Lycopodium* sp., longitudinal section showing adaxial position of sporangium on a leaf. Note the short, nonvascularized archesporial pad. **D.** *L. clavatum*, adaxial view of sporophyll. (Redrawn from Bold, Alexopoulos, and Delevoryas, 1980.)

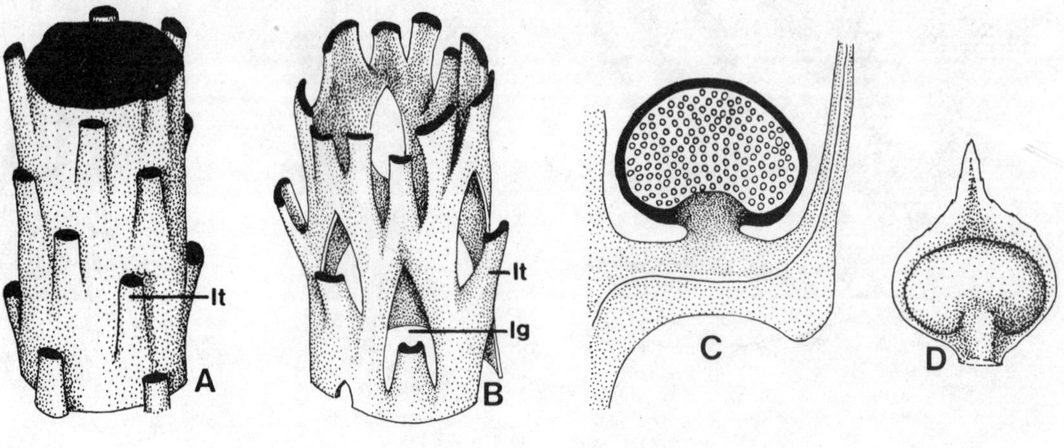

Bathurstia, Crenaticaulis, and *Serrulacaulis* (Fig. 10.4F) have axes that are dichotomously branched. In *Crenaticaulis* (Fig. 10.4G) some branching is pseudomonopodial. Internally, its vascular system is an elliptical strand of exarch primary xylem. *Bathurstia* and *Serrulacaulis* have branch tips that are circinate (coiled). Unlike the rhyniophytes and zosterophylls described thus far, the branching axes of *Bathurstia, Crenaticaulis,* and *Serrulacaulis* have toothlike enations (small nonvascular projections) arranged in vertical rows on oppostie sides of the axes. Sporangia of the three genera are reniform and terminate short branches. The sporangium-bearing branches occur in two rows on the distal axes. In *Serrulacaulis* the two rows are on one side of the axis, and the short branches with terminal sporangia alternate with one another. *Crenaticaulis* and *Bathurstia* have their sporangium-bearing branches opposite one another in rows on opposite sides of the main axes. Their sporangia are more or less reniform and dehisce transversely along the distal margins, a characteristic of the zosterophylls. Of the three genera, *Bathurstia* is the oldest, occurring in rocks of Siegenian (Lower Devonian) strata. *Crenaticaulis* is Emsian (Middle Devonian), while *Serrulacaulis* has been found in Givetian Frasnian (Upper Devonian) rocks.

Similar to other zosterophylls, *Sawdonia* (Fig. 10.5A) is wide-ranging in the Devonian. Hueber (1971a) reported the genus from the Siegenian strata in which he found *Rebuchia* and *Bathurstia*. The

Figure 10.4. **A.** *Zosterophyllum rhenanum,* reconstruction. (Redrawn from Kräusel & Weyland, 1935.) **B.** *Z. australianum,* purse-shaped sporangia terminating short lateral branches. **C.** *Z. australianum* showing rhizome with H-shaped branching. All Lower Devonian. (**B,C** drawn from photographs by Cookson, 1935.) **D.** *Gosslingia breconensis.* Lower to Middle Devonian. (Redrawn from Edwards, 1970.) **E.** *Rebuchia ovata,* reconstruction of sterile and fertile parts. Lower Devonian. (Redrawn from Hueber, 1972.) **F.** *Serrulacaulis furcatus,* reconstruction. Upper Devonian. (From Hueber & Banks, 1979.) **G.** *Crenaticaulis verruculosus,* reconstruction showing two rows of crenulations and purselike sporangia. Uppermost Lower Devonian. (Redrawn from Gensel, 1975.)

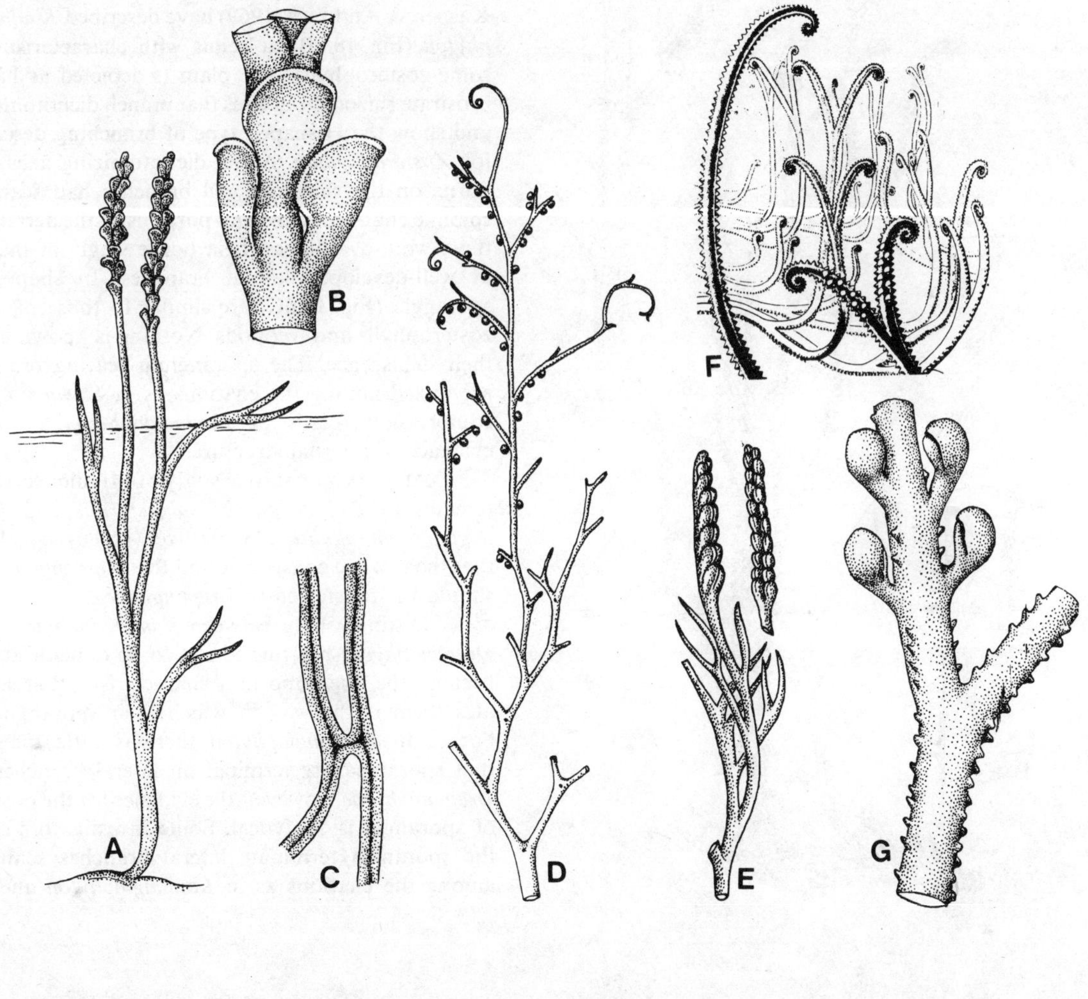

greatest number of finds of *Sawdonia ornata*, however, are reported from the Emsian stage. The plant is pseudomonopodially branched with circinate tips. The presence of irregularly disposed tapered and pointed spines on the axes distinguishes *Sawdonia* from other genera of zosterophylls. Sporangia of *Sawdonia* are described by Hueber (1971b) to be globose and borne singly at the tips of short lateral branches (Fig. 10.5B). The lateral fertile branches are borne among the spinelike enations and tend to be concentrated toward the apices of the plant where they form loose spikes (Fig. 10.5C). The arrangement of the sporangium-bearing branches in the spike is irregular to subopposite. As in other zosterophylls,

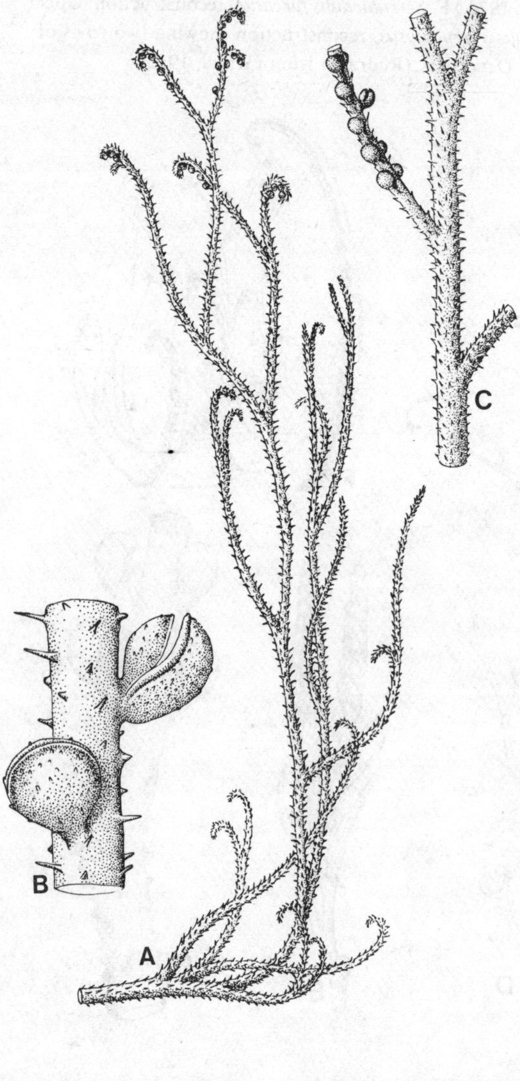

Figure 10.5. **A.** *Sawdonia ornata*, reconstruction. Lower Devonian. (Redrawn from Ananiev & Stepanov, 1968.) **B.** *S. acanthotheca*, reconstruction of purselike sporangia. Note transverse dehiscence. **C.** *S. acanthotheca*, portion of branching system reconstructed. **B,C:** Middle Devonian. (**B,C** redrawn from Gensel, 1975.)

the dehiscence of the sporangia occurred by splitting along the convex margins to form a two-part purselike structure. Other characteristics of *Sawdonia* are similar to those already described for other genera of the class.

Without going into all of the convolutions of nomenclature, it is of interest that what is now named *Sawdonia ornata* (Hueber, 1971b) was first described, in part, as *Psilophyton princeps* by Sir William Dawson (1859). The binomial *Psilophyton princeps* used by Dawson has been retained for a member of the Trimerophytopsida with naked or spiny axes bearing clusters of terminal sporangia (Chapter 13).

Discalis longistipa (Hao Shou-Gang, 1989) is a well-preserved representative of the zosterophylls from the Lower Devonian of Yunnan, China. In many respects this new genus (Fig. 10.6A) is quite similar to *Sawdonia* (Fig. 10.5A) in the shape, size, and dehiscence of its sporangia; the morphology of its spines; and in its spore type (*Retusotriletes*). Characteristics that distinguish *D. longistipa* from other zosterophylls are the long stalks (up to 5.0 mm) on which the terminal sporangia are borne, and the abundant spines that are scattered on all parts of the plant, including the walls of the sporangia.

From Lower Devonian outcrops, Gensel, Kasper, & Andrews (1969) have described *Kaulangiophyton* (Fig. 10.6B), a genus with characteristics of some zosterophylls. The plant is depicted as having prostrate rhizomatous axes that branch dichotomously and show the H-shaped type of branching described for *Zosterophyllum*. Aerial dichotomizing axes were borne on the rhizomes. All branches had scattered spinlike enations. On distal portions of the aerial axes there were ovoid sporangia borne singly at the tips of well-developed lateral branches. In shape, the sporangia (Fig. 10.6C) are similar to those of other zosterophylls and lycopods. Nothing is known about their dehiscence. The sporangium-bearing branches are mixed among the enations as in *Sawdonia*. The compression fossils of *Kaulangiophyton* show no evidence of internal structure.

After a comparative study of Middle Devonian remains of *Drepanophycus spinaeformis* and *Kaulangiophyton akantha*, Schweitzer (1980) concludes that the two are conspecific and that *Kaulangiophyton* should be transferred to *Drepanophycus*.

Distinguishing between *Kaulangiophyton* and *Drepanophycus* has turned out to be difficult at best because the one important characteristic that separates them is the way in which their sporangia are borne. In *Kaulangiophyton* there is little question that sporangia are terminal on lateral branches. In *Drepanophycus*, however, the evidence for the position of sporangia is equivocal. Some investigators claim the sporangia terminate lateral branches scattered among the enations as in *Kaulangiophyton* and the

Zosterophyllopsida. Others describe the stalked sporangia as occurring on the adaxial surface of sporophylls, a characteristic of the Lycopsida. Because of the absence of definitive evidence as to sporangial position, Gensel & Andrews (1984) suggested that the specimens of *Drepanophycus* described by Schweitzer (1980) could just as well be specimens of *Kaulangiophyton*. Until definitive evidence demonstrating the position of sporangia in *Drepanophycus* is provided, we will continue to recognize *Kaulangiophyton* and *Drepanophycus* as distinct genera.

According to Gensel et al., *Kaulangiophyton* is similar to *Asteroxylon* (Fig. 10.7A), described by Kidston & Lang (1920) from the famous Rhynie chert beds of the late Lower Devonian. These exquisite permineralized fossils revealed a robust lycopodlike plant with upright pseudomonopodial axes more than half a meter high bearing dichotomizing laterals. In contrast to the naked rhizomatous portions of the plant, the aerial axes were covered with crowded spirally arranged spinelike enations approximately 5 mm long and provided with stomata. When Kidston & Lang described *Asteroxylon*, they thought that sporangia were terminal on dichotomously branched, naked axes. A reinvestigation of *Asteroxylon* by Lyon (1964) showed (Fig. 10.7D), however, that the sporangia are borne at the tips of short vascularized branches scattered among the enations of the aerial axes (Fig. 10.7B), a feature of *Sawdonia* and *Kaulangiophyton*. The sporangia are reniform and dehisced along the convex distal margin. As reconstructed, *Asteroxylon* bears a superficial resemblance to the extant *Huperzia selago* (Fig. 10.2A).

Internally, *Asteroxylon* has a massive actinostele (Fig. 10.7C) (a protostele that is stellate when seen in cross section) with four or more lobes in the aerial axes. Smaller protoxylem cells are slightly immersed in metaxylem near the tips of the lobes. Metaxylem cells are usually well preserved and show spiral thickenings in their walls. Of considerable interest are traces of vascular tissue that depart from the tips of the lobes and terminate at the bases of enations. Phloem may surround the xylem, but is best developed in the embayments between the xylem lobes. The cortex of the aerial axes is differentiated into several zones. The distinctive middle cortex consists of vertical radiating plates of cells separated by large intercellular spaces. Aerating tissue of this type is usually found in plants growing in aquatic or marshy environments.

The naked rhizomatous portions of the plant branched by repeated dichotomies to produce a system of vegetative telomes that penetrated the substrate. Here we are provided with a good example of an early stage in the evolution of anchoring and absorbing systems from vegetative telome trusses. There is little

Figure 10.6. **A.** *Discalis longistipa*, reconstruction of branch bearing sporangia and spines. **B.** *Kaulangiophyton akantha*, restoration showing H-shaped branching and fertile axes. **C.** *K. akantha*, segment of fertile branch with sporangia and enations. **A–C:** Lower Devonian. (**A** redrawn from Hao Shou-Gang, 1989; **B,C** redrawn from Gensel, Kasper, & Andrews, 1969.)

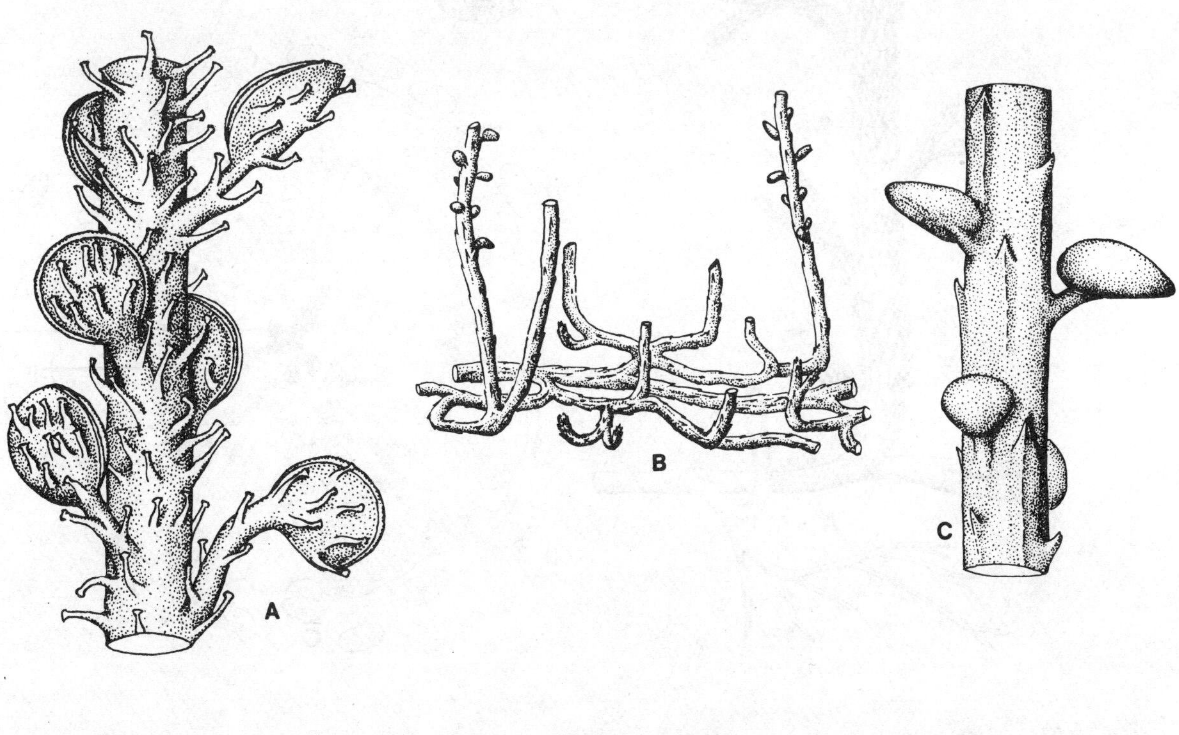

doubt in the minds of paleobotanists that this is the way that root systems evolved.

Evolution of terminal sporangia of zosterophylls

When summarizing the morphologies of the Zosterophyllopsida we find that they share many primitive characteristics with the Rhyniopsida. Both are herbaceous plants with dichotomous to pseudomonopodial branching. Both are protostelic, differing only in the maturation of their primary xylem. Both have sporangia that are terminal on branches and are homosporous. For these and other reasons it can be argued that zosterophylls shared a common ancestry with rhyniophytes.

Problems arise, however, in the minds of many paleobotanists when the lateral position of the sporangium-bearing branches is considered. Banks (1968) uses this feature as one of the important characteristics distinguishing Zosterophyllopsida from Rhyniopsida.

Figure 10.7. **A.** *Asteroxylon mackiei,* restoration of aerial and rootlike axes. (Redrawn from Kidston & Lang, 1920.) **B.** *A. mackiei,* reconstruction of portion of fertile axis showing purselike sporangia with transverse dehiscence. (Based on photograph of model, Royal Scottish Museum, Edinburgh.) **C.** Transverse section aerial axis of *A. mackiei.* Foliar structures (f); leaf traces (lt); xylem of actinostele (x). **D.** Diagrammatic reconstruction, longitudinal section, portion of *A. mackiei* axis with sporangia. Foliar structure (f); leaf trace (lt); sporangium (s). (Drawing based on photograph by Lyon, 1964.) **A–D:** From Rhynie chert (Lower Devonian, Siegenian–Emsian).

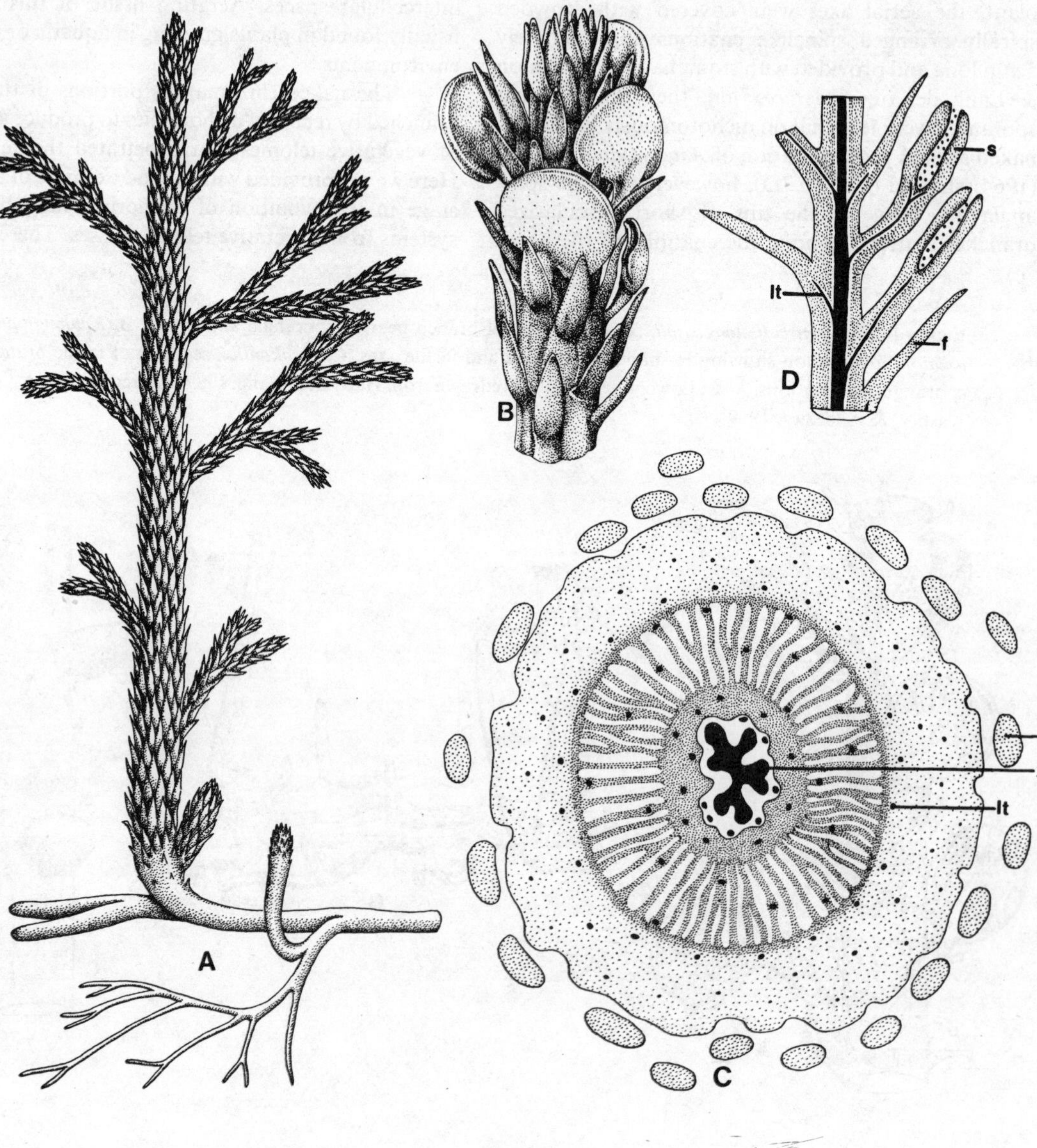

We have already been exposed to the idea, with evidence, that synangium-bearing branches may be the product of reduction. The fossil record as revealed by *Renalia* (Gensel, 1976) of the Rhyniopsida, shows us how by overtopping, accompanied by reduction of fertile telomes, lateral branches with a terminal sporangium could have evolved. The fact that most zosterophylls discovered thus far have sporangia at the tips of short branches supports the idea that they evolved as a group from an ancestral type with terminal sporangia on fertile telomes. This concept is further supported by *Discalis* (Fig. 10.6A), where the lateral sporangium-bearing branches are better developed than on other zosterophylls and seem to be intermediate between *Renalia* and *Sawdonia* in this characteristic.

Although a dehiscence mechanism has not been reported for the sporangia of *Cooksonia*, some specimens figured by Edwards (1970) had apparently dehisced along the distal convex margins, which is similar to the position of the dehiscence mechanism for *Renalia* and the Zosterophyllopsida. The shapes of the sporangia of the rhyniophytes *Cooksonia* and *Renalia* are globose to reniform, as are those of zosterophylls. These similarities in sporangial characteristics add support to the idea that zosterophylls evolved from a primitive *Rhynia* type similar to *Renalia* (Niklas & Banks, 1990).

The other alternative to explain the position of sporangia at the tips of lateral branches is the elevation of a localized area of sporogenous tissue on the tip of a lateral nonvascularized emergence. We are not aware of any fossil or other evidence from vascular plants that suggests this could have been the case.

Irrespective of how the terminal position of sporangia on short lateral branches evolved, we should keep in mind that it was an important step in positioning the developing sporangia so that they could be protected by enations and microphylls. This leads us to two questions. (1) How did the microphyll evolve? and (2) How did the sporangium achieve the adaxial to axillary position on the microphyll? Before trying to answer the questions, let us review the evidence at hand and examine additional evidence from fossil representatives of the class Lycopsida.

Microphylls: Enations or branches?

In *Psilotum* the terminal synangia are normally subtended by a bifid structure that has been interpreted as foliar by some investigators. Rouffa (1978) has shown us, however, that in his cultivars and clones of *Psilotum* the bifid structures are not associated with synangia, but have the appearance and development of modified branches. Such an interpretation applies to *Tmesipteris* as well. In most species of *Psilotum*, however, the bifid structures lack any evidence of vascularization and in this respect are similar to the spirally arranged vegetative enations on the main axes. In *P. complanatum*, however, a trace of vascular tissue from the vascular system of the axis traverses the cortex and ends at the base of the "foliar" structure. Does this reflect the evolution of an enation that is "caught in the act" of becoming vascularized, or is it a reduced axis that is losing its vascular tissue, as is suggested by the telome theory? Do we have both sterile "foliar" units derived from enations and bifid "foliar" units associated with sporangium-bearing branches derived from reduced branches on the same plant? Or are all of the "foliar" appendages, both sterile and fertile, derived from enations?

Nonvascularized enations occur on the axes of many zosterophylls. In some they are blunt, toothlike structures in vertical rows (*Bathurstia, Crenaticaulis,* and *Serrulacaulis*) quite different from those of *Sawdonia, Discalis,* and *Kaulangiophyton,* which have spinelike enations scattered on prostrate and aerial axes and from *Asteroxylon,* which has spirally arranged enations. Although none is known to have a vascular trace, it is pertinent that in *Asteroxylon* traces of vascular tissue do extend from the protostele of the axes to the bases of their enations (Fig. 10.7D). Could this condition represent the initiation of vascular tissue in a microphyll? Certainly, even without vascular tissue, enations increased those surfaces exposed to light and performing photosynthesis. We know because of the presence of stomata that the enations of *Asteroxylon* were functional foliar units even though they lacked vascular tissue.

Classification of transitional forms

Thus far we have raised more questions than we have answered about the evolution of microphylls and adaxial sporangia in lycopods. We have been exposed, however, to a group of primitive vascular plants (*Sawdonia, Kaulangiophyton, Discalis,* and *Asteroxylon*) that, according to Banks (1968) and others, are transitional between the Zosterophyllopsida and Lycopsida. They are like the zosterophylls in having nonvascularized enations and terminal sporangia on lateral branches without evident relationship to the enations. They are like the Lycopsida, however, in having reniform to globose sporangia that dehisce along the convex distal margin and, when preserved, they have massive exarch protosteles.

Because they do seem to be transitional, the classification of the four genera becomes difficult. It is our opinion, however, that because they lack vascular tissue in thier enations and their sporangia are not associated with foliar units, *Sawdonia, Kaulangiophyton,* and *Asteroxylon* should be placed in an order of the Zosterophyllopsida – the Asteroxylales (Bierhorst, 1971). Plants have dichotomous or pseudomonopodial branching axes more or less covered with spinelike enations borne randomly. Sporangia are reniform to globose, borne singly at the tips of short lateral branches. Sporangium-bearing branches are more or less grouped on more distal portions of aerial axes among enations, but without evident association with enations. Dehiscence of sporangia is along the convex distal margin. All appear to be hornosporous. Where internal structure is preserved, an exarch protostele is present.

In their 1990 paper Niklas & Banks established a similar group of transitional genera for which they adopted the name "prelycopods" (Gensel & Andrews, 1984). They envisage this group as originating from those zosterophylls that have radially arranged spines, and stalked sporangia, which may have a radial or bilateral arrangement. Of primary importance, however, is the nonterminate nature of the aerial axes, i.e., there is a distal extension of a vegetative axis beyond the more proximal fertile regions. The vegetative extension may end in a circinate tip (Fig. 10.5A). *Sawdonia* and possibly *Discalis* (Fig. 10.6A) are given as examples of zosterophylls with these characteristics. By changes in the morphogenetic activities of their terminal meristems, the prelycopods (e.g., *Asteroxylon, Discalis,* and *Drepanophycus*) with nonterminate axes evolved radial symmetry. Further readjustments instituted by the terminal meristems gave rise to the Lycopsida with adaxial sporangia on spirally arranged microphylls.

You may have noted that Niklas & Banks have used the genus *Drepanophycus* in place of *Kaulangiophyton,* presumably on the assumption that the two are congeneric. In the opinion of several paleobotanists this remains to be demonstrated. They suggest that the only characteristic that keeps *Drepanophycus* with the Zosterophyllopsida is the presence of stalked lateral sporangia scattered among the microphylls (Schweitzer, 1980). If, on the other hand, sporangial position of *Drepanophycus* is on the adaxial surface of the microphyll as suggested by some (Kräusel & Weyland, 1935), then this genus is best placed in its own order with the Lycopsida. Irrespective of the final solution to sporangial position in the genera discussed, it is generally agreed that they form a transitional group between the Zosterophyllopsida and Lycopsida, more specifically the Lycopodiales.

Other zosterophylls with terminate radial fertile axes (e.g., *Zosterophyllum*) and those with nonterminate, bilaterally symmetrical fertile axes (e.g., *Gossligia*) are thought to represent phylogenetic dead ends (Niklas & Banks, 1990).

Herbaceous Lycopsida of the Devonian

With some exceptions, the Devonian lycopods to be described share characteristics with the order Lycopodiales. Plants belonging to this order are herbaceous, have dichotomous branching microphylls that are usually spirally arranged, and sporangia borne singly on the adaxial surface or in the axils of microphylls. As far as we know, the plants were homosporous and barring one exception are without ligules. (A ligule is an enation with a well-defined base embedded in the adaxial surface of the microphyll.) Genera of the Selaginellales and Isoetales are examples of extant plants that are ligulate (Fig. 10.1D).

Even though they are very similar to extant and other fossil Lycopodiales, the Devonian herbaceous lycopods have been placed in separate orders. One of these is the Drepanophycales whose type genus, *Drepanophycus* (Fig. 10.8A), occurs in Lower, Middle, and Upper Devonian strata (Banks & Grierson, 1968). At one time, the genus was thought to be a good index fossil for rocks of Lower Devonian age.

The robust aerial axes of *Drepanophycus* dichotomize and presumably arose from horizontal rhizomes. The upright axes attained a diameter of up to 4 cm and they bore stout falcate, spinelike microphylls more or less irregularly scattered to spirally arranged on the axes. An exarch, lobed protostele that supplies traces of vascular tissue to the microphylls has been detected in *D. spinaeformis*. The oval to reniform sporangia are borne singly on microphylls, which are either scattered or condensed into zones on the aerial axes. Each sporangium appears to be borne at the tip of a small branch on the adaxial surface midway between the base and tip of the leaf. Some reconstructions, however, indicate that the sporangia were borne on or near the leaf tips while others show the sporangia terminating short branches scattered among the leaves.

For a complete report on species of *Drepanophycus* and other Devonian lycopods the papers by Grierson & Banks (1963) and Bonamo, Banks, & Grierson (1988) are recommended.

Ever since it was first described by Lang & Cookson (1935), *Baragwanathia* (Fig. 10.8C), the second genus of the Drepanophycales, has been the object of considerable discussion and speculation. For example, paleobotanists have, for some time, noticed the remarkable similarity among fossil herbaceous lycopods and extant representatives of *Lycopodium* and *Huperzia* (Fig. 10.2A). Although *Drepanophycus spinaeformis* (Fig. 10.8A) looks like a *Lycopodium*, the resemblance of *Baragwanathia longifolia* (Fig. 10.8C) to *Huperzia* is even more pronounced. The similarities include axes clothed in spirally arranged unbranched microphylls occurring in zones alternating with zones of sporophylls having the same morphology as the vegetative leaves. The sporangia of both are in the axils of the sporophylls, and spores isolated from the sporangia of *Baragwanathia* indicate that it was homosporous as are all members of the Lycopodiales. Internally, there is an actinostele composed of annular tracheids, which give rise to leaf traces that supply the vegetative leaves and sporophylls with a single trace. Beginning with *Baragwanathia*, whose verified fossil record first appears in the early Devonian (Hueber, 1983), there is a more-or-less continuous record of the Lycopodiales to the present. This apparent stasis in the rate of evolution within the order for some 370 m. y. has been the object of considerable interest to paleobotanists. Although Hueber and others have placed *Baragwanathia* with the Drepanophycales, it seems appropriate in view of the uncertainty about the classification of prelycopods to place *Baragwanathia* with the Lycopodiales.

Because *Baragwanathia* has the characteristics of an extant *Lycopodium*, it is difficult to believe that such a relatively highly evolved Lower Devonian plant could have had ancestors of the primitive *Rhynia* type, which were either younger than or coexisted with *Baragwanathia*. Clearly, this casts much doubt on the idea that rhyniophytes were the ancestors of lycopods, suggesting instead an independent origin for each group. Based, however, on their presently known position in the fossil record and their structure, we believe it has been demonstrated rather conclusively that zosterophylls bridge the gap between mid-Silurian rhyniophytes and Lower Devonian lycopods and do not represent a group that evolved independently of the Rhyniopsida. In taking this position and admitting that the evidence is far from being unequivocal, we support the idea that vascular plants are monophyletic, having their origins with the mid-Silurian complex of *Rhynia*-type plants.

Archaeosigillaria has been reported from Lower Devonian to the Upper Carboniferous. This stratigraphically wide-ranging genus is known from compressions of dichotomizing axes that reached a diameter up to 5 cm. The spirally arranged microphylls were simple and had a flattened lamina. Although shorter (5 to 6 mm long), they are similar to the leaves of *Baragwanathia*. On older stems where outer stem tissues have been lost, leaf bases are hexagonal with a leaf trace scar in the upper part. As in most Devonian lycopods where internal structure is known, the vascular system of aerial axes is an exarch actinostele where scalariform and reticulate tracheids are to be found. Structures that are interpreted as sporangia have been found associated with the adaxial surfaces of unmodified leaves.

The next lycopods to be considered comprise the genera of the Protolepidodendrales. The type genus Protolepidodendron (Fig. 10.8B), like *Baragwanathia* and *Drepanophycus*, bears a striking resemblance in gross morphology and anatomy to an extant *Lycopodium*. Unlike these genera, however, the spirally arranged leaves of *Protolepidodendron* and some other related genera branch dichotomously (Fig. 10.9E). On compression fossils the basis of the decurrent leaves is enlarged to form an elongated obovate cushion. The actual point of leaf attachment was above the midpoint of the cushion. Apparently the leaves did not abscise and as a consequence did not produce the elaborate leaf cushion scars so characteristic of their arborescent relatives, the Lepidodendrales. Internally, one species of *Protolepidodendron* (Grierson & Banks, 1963) is known to have an exarch actinostele (Fig. 10.9F) composed of annular scalariform and reticulate tracheids.

There is considerable uncertainty about the identity of *Protolepidodendron* when vegetative leaf-bearing axes are encountered. The type species, *P.*

Figure 10.8. **A.** *Drepanophycus spinaeformis*. Lower Devonian. (Redrawn from Kräusel & Weyland, 1933.) **B.** *Protolepidodendron scharyanum*. Lower Devonian. (Redrawn from Kräusel & Weyland, 1932.) **C.** *Baragwanathia longifolia*, shoot with axillary sporangia. Lower Devonian. (Drawn from photograph by Lang & Cookson, 1935.)

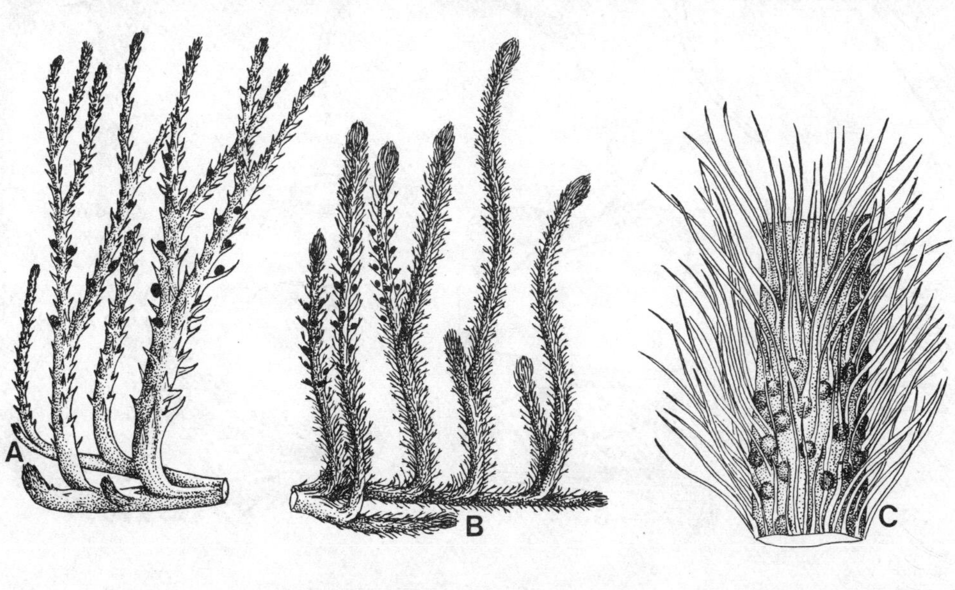

scharyanum (Fig. 10.9E), is defined as having spirally arranged leaves forking at or near the distal end. On careful investigation, however, many specimens were found to have leaves that forked more than once near the midpoint and to belong to other genera. Equal difficulty has been encountered in identifying fertile specimens of *Protolepidodendron*. It is generally agreed that *P. scharyanum* bore an oval-shaped sporangia broadly attached to the adaxial surface of each sporophyll (Fig. 10.9E). The sporophylls, like those of some other lycopsids, are morphologically similar to vegetative leaves. A line of dehiscence parallel to the long axis of the sporangium appears to be present on the upper surface. Based on the number of sporangia per sporophyll and the number of times it forked, the species *P. wahnbachense* was established (Schweitzer, 1980). Subsequent investigations (Fairon-Demaret, 1978, 1979) of fertile specimens of this species resulted in the transfer of *Protolepidodendron wahnbachense* to the new genus *Estinnophyton* (Fig. 10.9D) because they bore two pairs of sporangia per sporophyll rather than the single sporangium characteristic of other lycopods. The sporophylls of *E. gracile* fork just above the midpoint, and two pairs of recurved, stalked sporangia are borne just below the dichotomy. *Estinnophyton wahnbachense* turned out to be similar to *E. gracile* except that the sporophyll forked twice to form four tips (Fig. 10.9D). There are no organized cones or strobili. Instead the sporophylls occur occasionally on the aerial axes.

Other members of the Protolepidodendrales that have microphylls with branched tips are *Sugambrophyton* (Lower Devonian), *Colpodexylon* (Middle and Upper Devonian), and *Leclercqia* (Middle Devonian). All three genera have a gross morphology similar to that of *Protolepidodendron*, where there is a

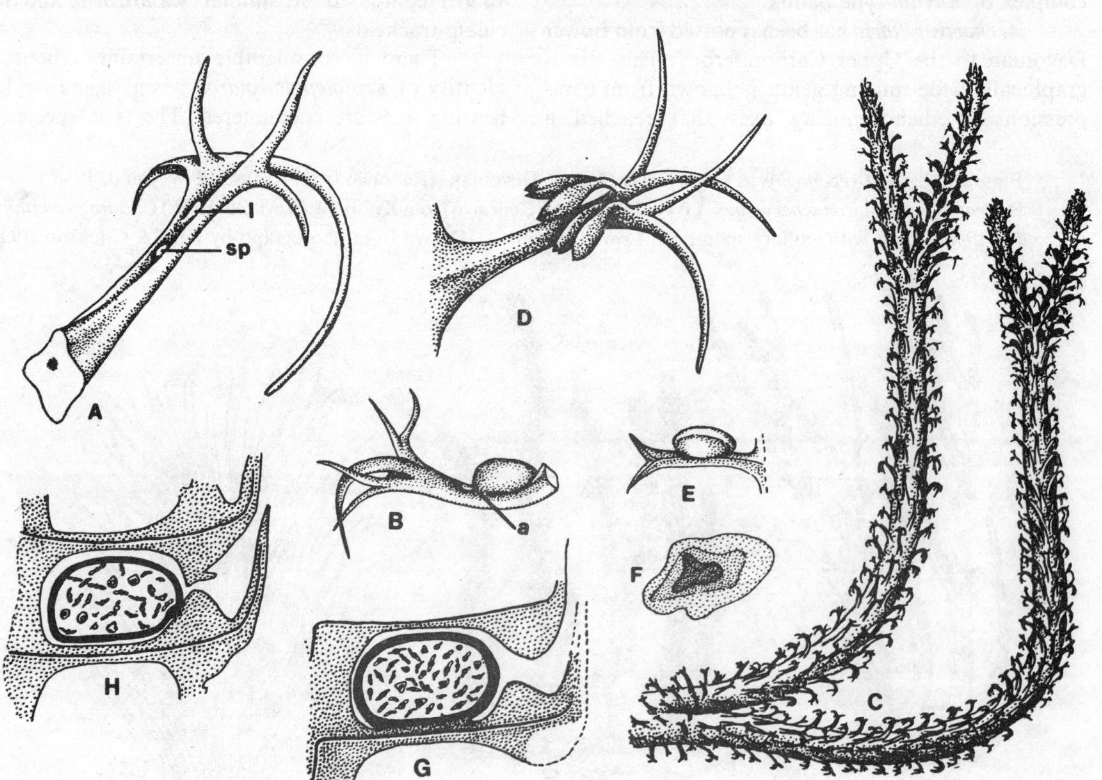

Figure 10.9. **A.** *Leclercqia complexa* sporophyll showing positions of ligule (l) and sporangial pad (sp). **B.** *L. complexa* single sporophyll with attached sporangium (a). **C.** Reconstruction of *L. complexa* with sterile and fertile leaves. Middle Devonian. (A–C based on Banks, Bonamo, & Grierson, 1972; Bonamo, Banks, & Grierson, 1988). **D.** Sporophyll of *Estinnophyton wahnbachense* reconstructed. Lower Devonian. (Redrawn from Fairon-Demeret, 1979). **E.** *Protolepidodendron scharyanum*, bifurcate sporophyll with a single adaxial sporangium. Lower Devonian. **F.** *P. scharyanum*, transverse section of axis with three-ribbed protostele. (E,F redrawn from Kräusel & Weyland, 1932). **G.** Longitudinal section of *Spencerites moorei*. Note broad attachment of sporangium to adaxial surface of sporophyll. **H.** *S. insignis*, longitudinal section. Note attachment of sporandium to sporophyll by a short stalk. G,H: Carboniferous (G,H redrawn from Leisman & Stidd, 1967).

rhizome-bearing erect dichotomous axis covered with small leaves. In *Sugambrophyton* and *Colpodexylon* the tips of some leaves are trifurcate. In *Colpodexylon* (Banks, 1944) the trifurcate leaves are arranged in pseudowhorls. Internally there is an exarch-to-mesarch actinostele composed of scalariform, scalariform reticulate, and pitted tracheids. The reniform sporangia are adaxial on unmodified leaves.

Of all Protolepidodendrales, *Leclercqia* (Fig. 10.9C) is by far the most completely known (Banks, Bonamo, & Grierson, 1972; Bonamo, Banks, & Grierson, 1988) because of the extraordinary preservation of its compressed parts. In their reconstruction, they indicate that even the detailed structure of *Leclercqia* compares favorably with that of *Protolepidodendron* (Grierson & Banks, 1963). The structure of the vascular strand, an exarch actinostele with 14 to 18 protoxylem ridges, is the same. Protoxylem is composed of annular and spiral elements, while the centripetal metaxylem has scalariform tracheids as well as those with oval bordered pits. The globose to ellipsoidal sporangia of *Leclerqia* (Fig. 10.9B) are borne singly on the adaxial surface of unmodified microphylls and are situated some distance from the leaf axil. The long axis of the sporangium and its dehiscence mechanism is parallel to the long axis of the sporophyll. Of particular interest is the way the sporangium is attached to the sporophyll by a pad connecting the distal end of the sporangium to the subtending leaf. Except for the point of attachment, the rest of the adaxial surface of the sporangium is free from the sporophyll. The sporophylls tend to be grouped into zones alternating with vegetative leaves, as in *Huperzia*.

Leclercqia differs from all other genera of the order in having the distal portion of the leaf divided into as many as seven parts (Bonamo & Grierson, 1981), with five being the common number (Fig. 10.9A). Leaves with five parts have a central unbranched segment and two lateral segments that dichotomize near the base. Those with seven parts have an extra dichotomy near the tip of a lateral or central segment. The extra dichotomies occur in a plane at right angles to the dichotomy below. Leaf traces arise from the protoxylem ridges of the axes, and at a higher level single traces enter each leaf base as a vein that extends to near the tip of the central segment. Lateral veins arising from the midvein supply the bases of the lateral segments.

While reinvestigating material of *Leclercqia,* using scanning electron microscopy (SEM), Grierson & Bonamo (1979) discovered a ligule (Fig. 10.9A) on the adaxial surface of both vegetative and fertile leaves. What makes this unusual is that *Leclercqia* is surely homosporous (scores of sporangia containing spores have been investigated), and thus is the first known ligulate homosporous lycopod. It may be that other members of the Protolepidodendrales are ligulate, but their relatively poor preservation does not permit the structure to be observed.

Spencerites insignis is an example of a Carboniferous cone in which the sporangium is attached distally by a short stalk on the lamina of the sporophyll (Fig. 10.9H). There is no indication of a ligule, and because of the fragmentary nature of the permineralized fossils one cannot determine whether the cones were homosporous or heterosporous. The spores of *S. insignis* are distinctive and of the *Spencersporites* type. Another cone, *S. moorei* (Leisman & Stidd, 1967), has the sporangium attached through most of its length to the adaxial surface of the sporophyll (Fig. 10.9G).

The record of homosporous eligulate lycopods that can be assigned to the order Lycopodiales is meager at best. The difficulty in establishing the fossil record for this group lies with the problem of identification. Seldom is the preservation adequate to determine whether a *Lycopodium*-like fossil is homosporous and without ligules. Further complicating the identification is the possibility that the specimens are distal branches of arborescent lycopods or conifers. Only four or five specimens with cones and believed to be lycopodiums have been reported. Our information about Mesozoic Lycopodiales was recently augmented by Skog (1986), who discovered that a Lower Cretaceous fossil thought to be a fern actually had the characteristics of the homosporous Lycopodiales. The new genus, *Tanydorus* (Fig. 10.10), has a strobilus that is similar to that of the extant *Lycopodium subulatum,* a pendant epiphyte of the neotropics. *Lycopodites oosensis* described by Kraüsel & Weyland (1937) is an Upper Devonian plant. The specimen has dichotomous branches clothed in short spirally arranged leaves. The adaxial sporangia are borne on unmodified leaves in zones alternating with the vegetative leaves. In its general morphology, the extant *Huperzia* again is suggested.

Herbaceous heterosporous lycopods

Other compression fossils of herbaceous lycopods with cones have been found in Carboniferous strata and assigned to the form genus *Lycopodites* (Fig. 10.11). Some of these are heterophyllous (leaves of two sizes) with leaves in four rows. Plants with these characteristics may be related to the Selaginellales. If cones are present and it can be demonstrated that the plant was heterosporous and had ligules (Fig. 10.12B), then the affinities are with *Selaginella* (Fig. 10.2B). *Selaginella harrisiana* (Fig. 10.12A) Townrow (1968), from the Permian beds of New South Wales, is a fossil with all of the characteristics of a modern *Selaginella.* The cones have ligulate megasporophylls with four spores per megasporangium. Microsporophylls with microspores were found in the

Figure 10.10. Partial restoration of *Tanydorus* with strobilus (st) and sporangia. Lower Cretaceous (Redrawn from Skog, 1986.)

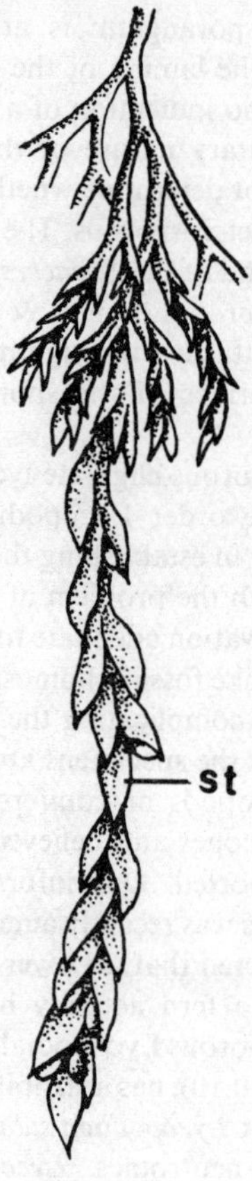

Figure 10.11. *Lycopodites meeki* in Pennsylvanian concretion. (Courtesy Illinois State Museum, Springfield.)

distal portions of the cone. *Selaginellites hallei* is from Jurassic (Rhaetic) coal beds of Sweden. As described by Lundblad (1950), this species was herbaceous and heterosporous. Ligules were not observed.

Miadesmia (Fig. 10.13) is a name originally used for small Upper Carboniferous (Westphalian A) ligulate sporophylls each containing a single megasporangium. These unique structures have an elaborate distal lamina with flaps of tissue on the outer surface that nearly enclose the megasporangium. There is a single functional megaspore that fills the megasporangium. This is in contrast to the selaginellas where there are usually four functional megaspores.

Based on the association of *Miadesmia* with certain vegetative axes, it is thought that the plant bearing these reproductive units had delicate dichotomizing axes bearing spirally arranged leaves and had a gross morphology similar to *Selaginella fraipontii*. Like this species, axes assigned to *Miadesmia* have exarch actinosteles with annular protoxylem elements occupying the tips of the radiating arms. The centripetal metaxylem cells are large scalariform tracheids with fimbrils.

A summary with evolutionary overtones

Are the conspicuous enations on the aerial axes of some zosterophylls and the microphylls of the Lycopsida reduced branches or have they evolved as emergence from the surface of axes? The idea that they may have evolved as emergences was developed by F. O. Bower (1935). According to this concept (Fig. 10.14A) microphylls were initiated, in an evolutionary sense, as nonvascularized spinelike outgrowths on the axes of primitive vascular plants. Examples are found among the zosterophylls (*Discalis, Sawdonia,* and *Kaulangiophyton*). The enations of these plants are not arranged in a definable pattern and are unlike *Asteroxylon* with its spirally arranged enations. Enations of *Asteroxylon* show the next step (Fig. 10.7D), where a trace of vascular tissue arises from the protostele of the axis and terminates at the base of the enation. With an increase in size of the enation, vascular tissue was supplied, by extension of the trace into the lamina of the enation. In this way a microphyll with a single leaf trace evolved as exemplified by *Drepanophycus* and *Baragwanathia*. It has been suggested that by further elaboration of the microphyll, a structure with branched tips as found in *Estinnophyton, Protolepidodendron, Colpodexylon, Sugambrophyton,* and *Leclercqia* was produced.

The branching tips of the microphylls can be interpreted, however, as a modification of a telome truss by overtopping and reduction (Fig. 10.14B), an interpretation first supported by Bonamo & Grierson (1981) and later rejected by Bonamo, Bauks, &

Grierson (1988). The linear, unbranched microphyll was derived by further reduction of the telome truss. Extrapolation of the idea of reduction could produce a nonvascularized enation as in *Asteroxylon* and the zosterophylls. In other words, the evolutionary "series" depicted for the enation theory can be read in reverse if the telome theory is applied.

Generally speaking, we find those zosterophylls with naked axes or with scattered enations appearing either before or at about the same time in the Lower Devonian as those lycospods with spirally arranged leaves. The Lower Devonian *Drepanophycus* is said to have its microphylls scattered irregularly or in a detectable spiral. One might expect to find this intermediate condition exhibited by a plant of this age if the enation theory is applied. If the telome theory is used, then we must conclude that Upper Devonian lycopods with spirally arranged microphylls are primitive and Lower Devonian zosterophylls with scattered enations are the product of reduction. Based on stratigraphy, if for no other reason, this is a bad "fit."

Figure 10.12. **A.** Reconstruction of *Selaginella harrisiana* and its parts. Note leaf with ligule; microsporophyll (mi); megasporophyll (me) and ligule (1). Permian. (Redrawn from Townrow, 1968.) **B.** Diagram, longitudinal section through cone of *Selaginella* showing mega- and microsporangia and ligules; megasporangium (me); microsporangium (mi); ligule (1).

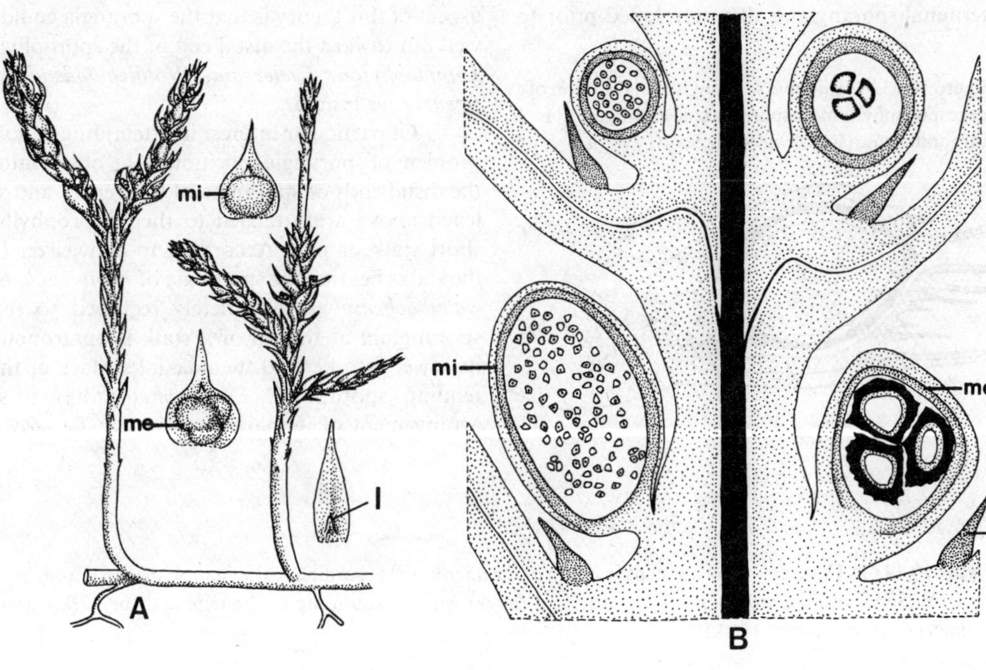

If we accept the interpretation of Gensel (1975), Banks (1968), Bonamo et al. (1988), and others that the microphylls of all lycopsids evolved from enations, then we are confronted with the question of how the sporangium-bearing lateral branches (stalks) occupied the axillary position on the microphyll. Prior to attempting the answer, we think it is pertinent to keep in mind that in becoming adapted to terrestrial environments, plants of all kinds have evolved an infinite variety of mechanisms to reduce desiccation of their developing reproductive structures. Certainly the evolution of the adaxial position of the sporagium on a microphyll, in close juxtaposition to other sporophylls, and in a spiral would be of advantage to land plants, especially those radiating into drier environments. In *Asteroxylon* we know that the short-stalked sporangia are scattered among the closely placed enations. These must have afforded some protection to the terminal sporangia as they developed prior to spore dispersal. It would not require a major morphogenetic change in the shoot apex to alter development that would result in a shift of the fertile branch from a point adjacent to a microphyll to its axil (Fig. 10.15A,B). (Bonamo et al., 1988; Hao Shou-Gang, 1989; Niklas & Banks, 1990). When attempting to explain the adaxial position of the sporangium to or beyond the midpoint of the sporophyll as in the Protolepidodendrales, additional morphogenetic changes altering development are required.

Another simpler explanation for the axillary or adaxial juxtaposition of a sporangium-bearing branch and subtending microphyll is provided by the telome theory. By overtopping and reduction of a telome truss comprised of both fertile and vegetative telomes, a branch with one or more terminal sporangia could become associated with the adaxial surface of the subtending foliar unit (Fig. 10.16A–D). The attractive aspect of this theory is that the sporangia could occur well out toward the distal end of the sporophyll as in *Estinnophyton, Leclercqia, Protolepidodendron,* and *Spencerites insignis.*

Of particular interest in attempting to solve the problem of sporangial position is the observation that the distal ends of sporangia of the genera and species listed above are attached to their sporophylls by a short stalk or pad. According to Schweitzer (1980), the stalks bearing the sporangia of *Protolepidodendron wahnbachense* are completely recurved so that the sporangium at the tip of a stalk is anatropous with its lower face next to the adaxial surface of the subtending sporophyll. *Estinnophyton* has a similar arrangement of its paired sporangia. *Leclercqia* and

Figure 10.13. *Miadesmia* sp., longitudinal section of megasporophyll, megasporangium, and ligule (l). Carboniferous. (Redrawn from Scott, 1908.)

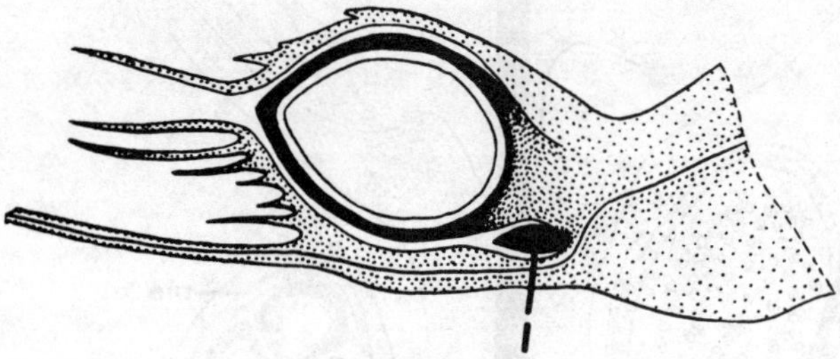

Figure 10.14. **A.** Diagrams showing steps in origin of the microphyll according to the enation theory. (From Stewart, 1964.) **B.** Series of diagrams showing the origin of the microphyll according to the telome theory. (Based on Zimmermann in Stewart, 1964.)

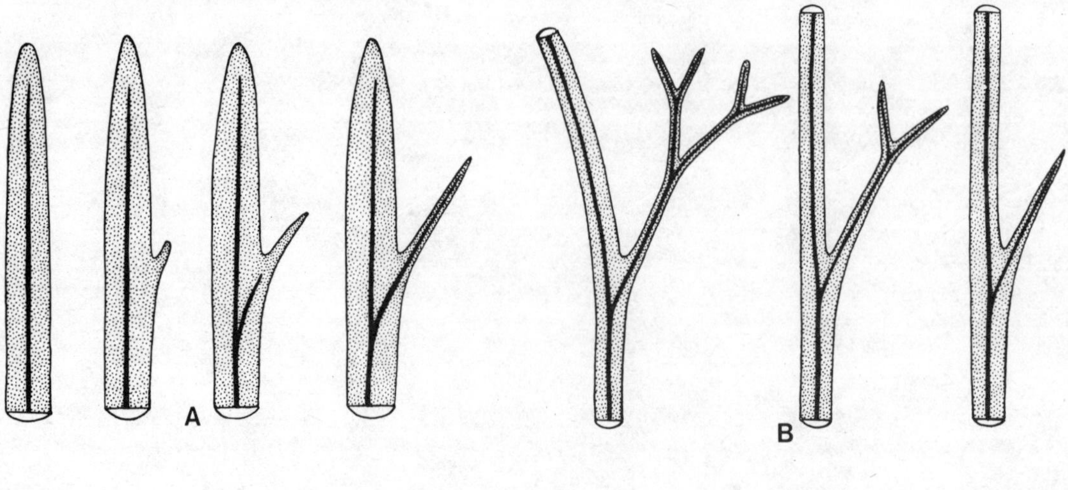

Figure 10.15. Hypothetical steps leading to the juxtaposition of an adaxial sporangium and enation. **A.** Stalked sporangium above enation axil as in *Asteroxylon*. Note that trace of vascular tissue ends at base of enation, but the stalk bearing the sporangium has a vascular bundle. **B.** Sporangium has "moved" to the axillary position as in *Drepanophycus, Baragwanathia*, and some extant lycopods. Note extension of vascular trace into the microphyll. **C.** Further distal "movement" of sporangium onto the adaxial surface of the microphyll as in members of the Protolepidodendrales.

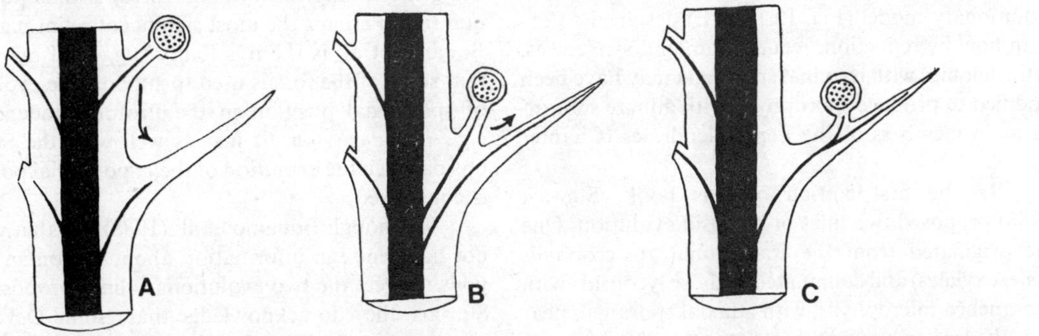

Figure 10.16. Evolution of sporangial position in the Protolepidodendrales according to the telome theory. **A.** Hypothetical telome truss composed of fertile and sterile segments. **B.** Reflexed fertile telomes (*Estinnophyton*). **C.** Reduction in number of sporangia. **D.** Reduction in number of sporangia to one. Reduction of sterile telomes to form a sporophyll with a bifurcate tip (*Protolepidodendron scharyanum*). The telome truss and its derivatives have been rotated by 90° better to show their organization.

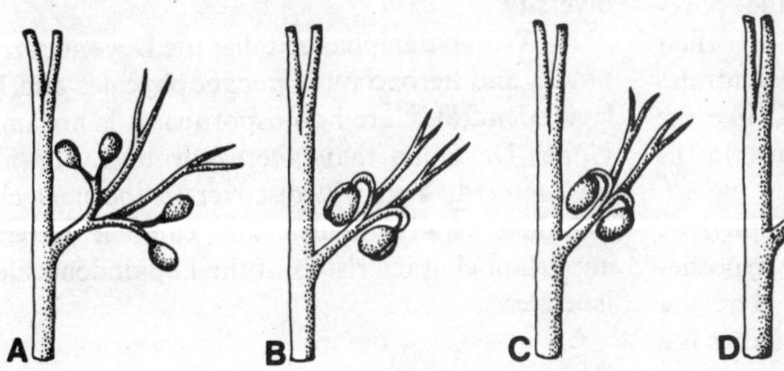

Figure 10.17. Model for the evolution of sporangial position leading to that of the Lepidodendrales. **A.** Upright fertile telome with subtending sterile, bifurcate sterile telome. Hypothetical. **B.** Recurvation of fertile telome (*Estinnophyton wahnbachensis*). **C.** Reduction of fertile telome and syngenesis with sterile telome (*Protolepidodendron scharyanum*). **D.** Further syngenesis and reduction resulting in a stalk or pad (p) connecting the distal end of the sporangium with the midpoint of the adaxial surface of the sporophyll (*Leclercqia, Spencerites insignis*). **E.** Complete loss of sporangial stalk accompanied by adnation of sporangium to sporophyll. Fusion and some webbing of sterile telomes to form a sporophyll with a lamina (*S. moorei; Lepidostrobus*).

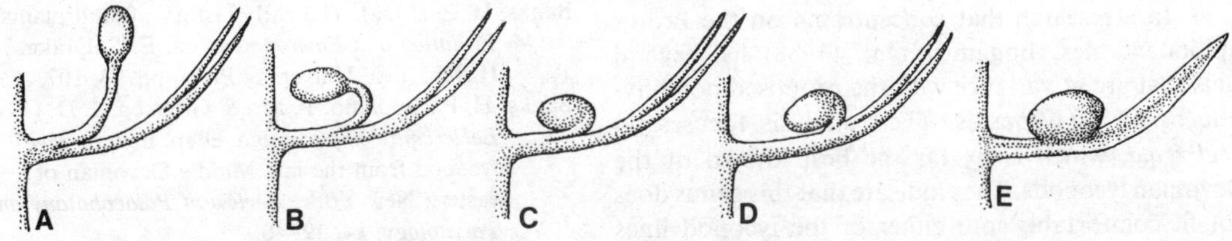

Spencerites insignis have sporangia attached by their distal ends to a pad of tissue that probably represents a reduced stalk. Fusion between the sporangium and the adaxial surface of the sporophyll occurs in *Spencerites moorei* (Leisman & Stidd, 1967) so that the whole lower face of the sporangium is adnate with the sporophyll. What is presented above serves as an evolutionary model (Fig. 10.17A–E) designed to explain how by reduction, recurvation, and syngenesis, fertile telomes with terminal sporangia may have been modified to produce sporophylls with adnate sporangia as in members of the Lepidodendrales (Chapter 11).

In the first edition of this book, Stewart (1983) proposed two lines of lycopsid evolution. One line originated from the transitional zosterophylls (Asteroxylales) and culminated in those lycopsids with unbranched microphylls, with adaxial sporangia near or at the leaf axil, and with sporangia with transverse distal dehiscence. Members of the Drepanophycales, which first appeared in the Lower Devonian, and the Lycopodiales, which appeared later, are two orders that are claimed by some to exhibit these characteristics. The second line originated from a *Rhynia* type similar to *Renalia* and included those lycopsids with fertile and vegetative microphylls that branch dichotomously one or more times, have adaxial sporangia that occur well out toward the distal end of the sporophyll, and have longitudinal dehiscence of their sporangia. Most genera of the Protolepidodendrales fit these criteria. The Protolepidodendrales, like the Drepanophycales, make their first appearance in the Lower Devonian.

Since the introduction of this polyphyletic origin for the Lycopsida a new monophyletic hypothesis (Niklas & Banks, 1990) has appeared. Here the *Rhynia* type represented by *Renalia* is accepted as the ancestral type for zosterophylls. Out of the zosterophylls through the prelycopods, a single line evolved from which all other lycopods radiated. This included the Devonian members of the Lycopodiales and Protolepidodendrales. As the authors of this concept note, it provides a more parsimonious monophyletic derivation of lycopods. At the same time it precludes the interpretation that the branched foliar units of the Protolepidodendrales represent reduced telome trusses. Instead, they are thought to be elaborated from simple unbranched enations similar to those of the prelycopods.

In a research that concentrated on the Protolepidodendrales, Bonamo et al. (1988) list several facts that are at variance with the proposed polyphyletic origin of lycopsids. Their analysis focuses on *Leclercqia,* which is by far the best known of the Devonian lycopods. They indicate that this genus does not fit comfortably into either of the lycopod lines proposed by Stewart (1983). Their evidence suggests that the elaborately branched leaves of *Leclercqia* were derived from less branched ones and probably evolved from ancestors with simple microphylls and not from telome trusses. The evolution of sporangial position for the Protolepidodendrales as proposed by using the telome theory (Fig. 10.16) is thus put into question. Perhaps the most significant point made by Bonamo et al. is the possibility suggested by others that some of the fossils used to support the evolution of sporangial position in the Protolepidodendrales e.g., *Estinnophyton,* fit just as well with the Sphenopsida when the evolution of their sporangial position is considered.

Although Bonamo et al. (1988) feel that we do not have enough information about Devonian lycopods to erect the two evolutionary lines proposed by Stewart, they do acknowledge that "some Devonian lycopods evolved in the direction of arborescent, ligulate, heterosporous Lepidodendrales and others remained herbaceous, eligulate, and homosporous from Devonian to present." There is general agreement that lycopods were derived from zosterophylls through the transitional (prelycopod) members (Niklas & Banks, 1990). All agree that there was considerable diversity among early lycopods; however, there is a lack of agreement about how to interpret this diversity.

We have emphasized that the Devonian zosterophylls, and herbaceous Drepanophycales and Protolepidodendrales, are homosporous. It is not until the Upper Devonian that heterosporous Lycopsida are encountered. You will discover in the next chapter that heterospory, ligules, and cambial activity are important characteristics of the Lepidodendrales and Isoetales.

References

Ananiev, A. R., & Stepanov, S. A. (1968). Finds of sporogenous organs of *Psilophyton princeps* Dawson emend. Halle in the Lower Devonian of the South-Minusinsk hollow. *Treatises Tomsk Order Red Banner of Labor State University, Geological Series,* **202,** 30–46.

Banks, H. P. (1944). A new Devonian lycopod genus from southeastern New York. *American Journal of Botany,* **31,** 650–9.

Banks, H. P. (1968). The early history of land plants. In *Evolution and Environment,* ed. E. T. Drake. New Haven: Yale University Press, pp. 73–107.

Banks, H. P., Bonamo, P. M., & Grierson, J. D. (1972). *Leclercqia complexa* gen. et sp. nov., a new lycopod from the late Middle Devonian of eastern New York. *Review of Palaeobotany and Palynology,* **14,** 19–40.

Banks, H. P., & Grierson, J. D. (1968). *Drepanophycus spinaeformis* Göppert in the early Upper Devonian of New York State. *Palaeontographica B.*, **123,** 113–20.

Bierhorst, D. W. (1971). *Morphology of Vascular Plants.* New York: Macmillan.

Bold, H. C., Alexopoulos, C. J., & Delevoryas, T. (1980). *Morphology of Plants and Fungi,* 4th ed. New York: Harper & Row.

Bonamo, P. M., Banks, H. P., & Grierson, J. D. (1988). *Leclercqia, Haskinsia,* and the role of leaves in definition of Devonian lycopod genera. *Botanical Gazette,* **149,** 222–39.

Bonamo, P. M., & Grierson, J. D. (1981). Leaf variation in *Leclercqia complexa* and its possible significance. *Botanical Society of America Miscellaneous Series,* **160,** 42.

Bower, F. O. (1935). *Primitive Land Plants.* London: Macmillan. Reprint, New York: Macmillan (Hafner Press), 1959.

Cookson, I. C. (1935). On plant-remains from the Silurian of Victoria, Australia, that extend and connect floras hitherto described. *Philosophical Transactions of the Royal Society of London, B,* **225,** 127–48.

Dowson, J. W. (1859). On fossil plants from the Devonian rocks of Canada. *Quarterly Journal, Geological Society of London,* **15,** 477–88.

Edwards, D. (1969). Further observations on *Zostero-phyllum llanoveranum* from the Lower Devonian of South Wales. *American Journal of Botany,* **56,** 201–10.

Edwards, D. (1970). Fertile Rhyniophytina from the Lower Devonian. *Palaeontology,* **13,** 451–61.

Fairon-Demaret, M. (1978). *Estinnophyton gracile* gen. et sp. nov., a new name for specimens previously determined *Protolepidodendron wahnbachense* Kräusel and Wayland from the Siegenian of Belgium. *Académie Royale de Belgique., Bulletin de la Classede Sciences Series* 5, **64,** 597–610.

Fairon-Demeret, M. (1979). *Estinnophyton wahnbachense* (Kräusel et Weyland) comb. nov., une plante remarquable du Siegenien d'Allemagen. *Review of Palaeobotany and Palynology,* **28,** 145–60.

Gensel, P. G. (1975). A new species of *Sawdonia* with notes on the origin of microphylls and lateral sporangia. *American Journal of Botany,* **136,** 50–62.

Gensel, P. G. (1976). *Renalia hueberi,* a new plant from the Lower Devonian of Gaspé. *Review of Palaeobotany and Palynology,* **22,** 19–37.

Gensel, P. G., & Andrews, H. N. (1984). *Plant Life in the Devonian.* New York: Praeger.

Gensel, P. G., Kasper, A., & Andrews, H. N. (1969). *Kaulangiophyton,* a new genus of plants from the Devonian of Maine. *Bulletin of the Torrey Botanical Club,* **96,** 265–76.

Grierson, J. D., & Banks, H. P. (1963). Lycopods of the Devonian of New York State. *Paleontographica Americana,* **4,** No. 31, 220–95.

Grierson, J. D., & Bonamo, P. M. (1979). *Leclercqia complexa*: earliest ligulate lycopod (Middle Devonian). *American Journal of Botany,* **66,** 474–6.

Hao Shou-Gang (1989). A new zosterophyll from the Lower Devonian (Siegenian) of Yunnan, China. *Review of Palaeobotany and Palynology,* **57,** 155–71.

Hueber, F. M. (1971a). Early Devonian land plants from Bathurst Island, District of Franklin. *Geological Survey of Canada,* Paper 71–28, pp. 1–17.

Hueber, F. M. (1971b). *Sawdonia ornata*: a new name for *Psilophyton princeps* var. *ornatum. Taxon,* **20,** 641–2.

Hueber, F. M. (1972). *Rebuchia ovata,* its vegetative morphology and classification with the Zosterophyllophytina, *Review of Palaeobotany and Palynology,* **14,** 113–27.

Hueber, F. M. (1983). A new species of *Baragwanathia* from the Sextant Formation (Emsian) Northern Ontario, Canada. In *Contributions to Paleobotany,* eds. M. W. Dick & D. Edwards. *Botanical Journal of the Linnean Society,* **86,** 57–79. London: Academic Press.

Hueber, R. M., & Banks, H. P. (1979). *Serrulacaulis furcatus* gen. et sp. nov., a new zosterophyll from the Lower Devonian of New York State. *Review of Palaeobotany and Palynology,* **28,** 169–89.

Karrfalt, E. (1984). The origin and early development of the root-producing meristem of *Isoetes andicola* L. D. Gomez. *Botanical Gazette,* **145,** 372–7.

Kidston, R., & Lang, W. H. (1920). On Old Red Sandstone plants showing structure from the Rhynie chert bed, Aberdeenshire. Part III. *Asteroxylon mackiei,* Kidston and Lang. *Transactions of the Royal Society of Edinburg,* **52,** 643–80.

Kosanke, R. M (1979). A long-leaved specimen of *Lepidodendron. Geological Society of America, Bulletin I,* **90,** 431–4.

Kräusel, R., & Weyland, H. (1932). Pflanzenreste aus dem Devon, IV. *Protolepidodendron* Krejci. *Senckenbergiana,* **14,** 391–403.

Kräusel, R., & Weyland, H. (1933). Die Flora des bohmischen Mitteldevons. *Paleontographica B,* **78,** 1–46.

Kräusel, R., & Weyland, H. (1935). Neue Pflanzenfunde im rheinischen Unterdevon. *Palaeontographica B,* **80,** 170–90.

Kräusel, R., & Weyland, H. (1937). Plazenreste aus dem Devon, X. Zwei Planzenfunde im Oberdevon der Eifel. *Senckenbergiana,* **19,** 338–55.

Lang, W. H. (1927). Contributions to the study of the Old Red Sandstone flora of Scotland. VI. On *Zosterophyllum myretonianum,* Penh, and some other plant remains from the Carmyville Beds of the Lower Old Red Sandstone. *Transactions of the Royal Society of Edinburgh,* **55,** 443–52.

Lang, W. H., & Cookson, I. C. (1935). On a flora, including vascular land plants, associated with *Monograptus,* in rocks of Silurian age, from Victoria, Australia. *Philosophical Transactions of the Royal Society of London, B,* **224,** 421–49.

Leisman, G. A., & Stidd, B. M. (1967). Further occurrences of *Spencerites* from the Middle Pennsylvanian of Kansas and Illinois. *American Journal of Botany,* **54,** 316–23.

Lundblad, B. (1950). On a fossil *Selaginella* from the

Rhaetic of Hyllinge, Scania. *Svensk Botanisk Tidskrift*, **44**, 447–87.

Lyon, A. G. (1964). The probable fertile region of *Asteroxylon mackiei* K. and L. *Nature*, **203**, 1082–3.

Niklas K. J., & Banks, H. P. (1990). A reevaluation of the Zosterophyllophytina with comments on the origin of lycopods. *American Journal of Botany*, **77**, 274–83.

Rouffa, A. S. (1978). On phenotypic expression, morphogenetic pattern, and synangium evolution in *Psilotum*. *American Journal of Botany*, **65**, 692–713.

Schweitzer, H. J. (1980). Uber *Drepanophycus spinaeformis* Goeppert. *Bonner Palaeobot. Mitteil. no.* **7**, 29 pp.

Scott, D. H. (1908). *Studies in Fossil Botany*. London: Black, Part 2.

Skog, J. E. (1986). The supposed fern *Onychiopsis psilotoides* from the English Wealden (Lower Cretaceous) reinterpreted as a lycopod. *Canadian Journal of Botany*, **64**, 1453–66.

Stewart, W. N. (1964). An upward outlook in plant morphology. *Phytomorphology*, **14**, 120–34.

Stewart, W. N. (1983). *Paleobotany and the Evolution of Plants*, 1st ed. New York: Cambridge University Press.

Townrow, J. A. (1968). a fossil *Selaginella* from the Permian of New South Wales. *Journal of the Linnean Society* (Botany), **61**, 13–23.

Wagner, W. H., Beitel, J. M., & Wagner, R. S. (1982). Complex venation patterns in the leaves of *Selaginella*: megaphyll-like leaves in lycophytes. *Science*, **218**, 793–4.

11

The isoetalean clade

The reconstruction of a Carboniferous swamp in the Field Museum of Natural History (Fig. 11.1) must be seen by all who are interested in paleobotany. Approximately 3 million man-hours of research and work went into the manufacture and assembly of this highly realistic diorama. As in most reconstructions of Carboniferous floras, the arborescent lycopods (Lepidodendrales) predominate. We can see trunk fragments, abscised leaves, and cones scattered in the foreground. The bases of the upright trees can be seen attached to the supporting and anchoring stigmarian axes, while dichotomizing branches of young trees are shown with their grasslike leaves. The features of the bark are rendered in exquisite detail and show the spirally arranged leaf cushions so characteristic of these plants. In the background and towering above all other plants of the swamp forest are the three-dimensional dichotomizing branching systems of these magnificent trees.

When we envisage such an arborescent growth habit, we usually associate it with cambial activity and the production of abundant secondary tissues, an important evolutionary innovation. Plants having this characteristic obviously can achieve greater size, thus presenting their leaves to maximum light energy and their spores shed from cones to wider distribution by air currents and gravity.

Transitional forms

The fossil record, as we have examined it thus far, makes it clear that plants with the herbaceous growth habit were the first to evolve on land. Difficulty is encountered, however, in determining for sure just when arborescent lycopsids with cambial activity and secondary tissues first appeared. Fragments of fossils that have dimensions suggesting an arborescent habit are found in the Middle and Upper Devonian, but these are casts and molds that show nothing of internal structure. From their external morphologies some of these are believed to represent plants transitional between the Protolepidodendrales of the Devonian and Lower Carboniferous and the Lepidodendrales that formed a predominant part of the Upper Carboniferous landscape.

Other than being herbaceous and dichotomously branched plants that were probably homosporous, you may recall (Chapter 10) that Protolepidodendrales had microphylls arranged in spirals or pseudowhorls. At least one genus, *Leclercqia,* had ligules. Another characteristic of the order is exhibited by their microphylls, which did not abscise. By contrast, leaves of the Lepidodendrales, which are expanded at the base, did abscise, revealing a conspicuous leaf cushion on the stem surface (Fig. 11.2A). When well preserved, a leaf scar with its vascular bundle scar, parichnos (aerating tissue) scars, and a ligule pit can be identified on the surface of the leaf cushion. With these characteristics in mind we can proceed to examine three genera, *Valmeyerodendron, Cyclostigma,* and *Lepidosigillaria,* which exhibit various combinations of characteristics that seem to be transitional.

Valmeyerodendron (Fig. 11.2B), described by Jennings (1972), was found in rocks of Lower Carboniferous (Mississippian) age. In his reconstruction of the plant he shows a three-dimensional dichotomizing aerial axis about 60 cm high. The main axis is approximately 3 cm in diameter and marked on its surface with widely spaced leaf cushions that are spirally arranged. As with the lepidodendrids, the leaf cushions each have a leaf and vascular bundle scar. They lack, however, a ligule pit or parichnos scars. The distal dichotomizing branches were clothed in spirally arranged leaves that are triangular in shape and not more than 1.5 cm long. In size and leaf characteristics, *Valmeyerodendron* recalls members of the Protolepidodendrales. In its leaf abscission and leaf cushions it is

Figure 11.1. Reconstruction of a Carboniferous swamp. (Photograph courtesy of the Field Museum of Natural History, Chicago.)

like the Lepidodendrales. Nothing is known of its internal structure or reproduction.

An Upper Devonian genus *Cyclostigma* has been interpreted as another transitional form closely allied with the Carboniferous lepidodendrids. *Cyclostigma kiltorkense* is perhaps the best-known species. It was described by Chaloner & Boureau (1967) as an arborescent plant attaining a height of 8 m and a diameter of at least 30 cm. Its dichotomizing branches suggest the branching system of a *Lepidodendron*. Like many members of this genus, the leaves were long and grasslike, some up to 15 cm in length. On the smaller branches, the leaves were arranged in a shallow spiral and their bases were expanded. When the leaves abscised they left conspicuous leaf cushions on the surfaces of the older branches. Similar to *Valmeyerodendron*, a leaf cushion of *Cyclostigma* has a leaf and vascular bundle scar, and there is no evidence of a ligule pit or parichnos scars. Aerial axes of *Cyclostigma* have been found divided at the base to form short and blunt lobes of a rhizomorph bearing roots or root scars.

Figure 11.2. **A.** *Lepidodendron* sp., surface of branch with attached leaves and leaf cushions where leaves have been shed. Carboniferous. (Redrawn from Walton, 1935.) **B.** *Valmeyerodendron triangularifolium*, reconstruction. Mississippian. (From Jennings, 1972.)

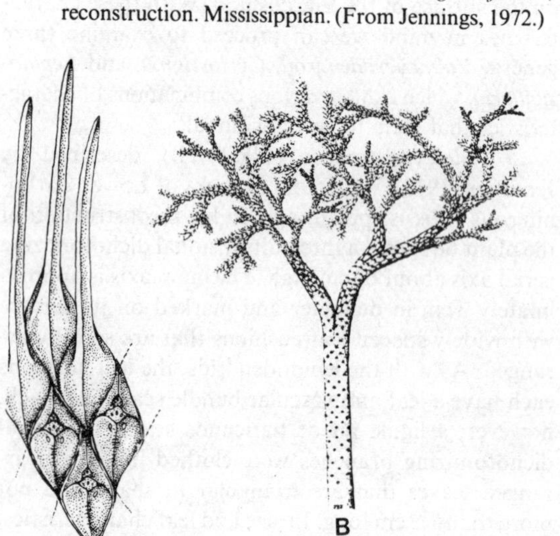

Cones of *Cyclostigma* were borne at the tips of aerial branches (Chaloner, 1968) and in at least one specimen Chaloner & Boureau were able to demonstrate megaspores in megasporangia. A pair of permineralized lepidendroid cones borne at the tips of a forked branch were described by Chitaley & McGregor (1988) from the Upper Devonian. These heterosporous cones, which they called *Bisporangiostrobus*, are similar to those of *Cyclostigma*. These finds are of importance because they demonstrate that heterospory had evolved in ancestors of the arborescent Lycopsida by Upper Devonian times.

A plant that attracted much attention as an early arborescent lycopod was found in Upper Devonian rocks near the village of Naples in western New York State. As described by Arnold (1947), the compressed trunk of the specimen was 5 m long, 38 cm in diameter at its swollen base, tapering to 12 cm at the upper end. Although nothing is known of its branching, several reconstructions of the Naples tree show the plant with dichotomous recurved branches clothed in short, spirally arranged microphylls. A few leaves were found attached, and these are not more than 3 cm long with broad bases fastened to leaf cushions. It is apparent that the leaves of this plant abscised as in the lepidodendrids. The leaf cushions have leaf, vascular bundle, and parichnos scars as well as a ligule pit. The leaf cushion arrangement noted by Arnold (1939) varies from the base where they are aligned in vertical rows, then change to a steep spiral approximately 1 m above. This is of particular interest because the vertical arrangement is characteristic of Carboniferous sigillarioid arborescent lycopods, while the spiral arrangement is found in the lepidodendrids. The implication that the two groups evolved from a common ancestor having the characteristics of the Naples tree is obvious. Apparently roots of the plant were borne on a rounded cormlike base.

Originally the Naples tree was assigned to *Archaeosigillaria;* however, for reasons well documented by Grierson & Banks (1963), it has been transferred to *Lepidosigillaria.* Nothing is known about the internal structure or reproduction of this genus.

Although there are several other genera that show sets of characteristics transitional between the herbaceous and arborescent lycopods, the three genera discussed provide us with sufficient evidence supporting an origin of arborescent ligulate heterosporous lycopods that abscised their leaves, from herbaceous eligulate homosporous ancestors that retained their leaves. Having among the ancestors such plants as *Leclercqia* and *Cyclostigma,* both of which show combinations of characteristics not found in more recently evolved major groups of fossil and extant lycopods, only emphasizes an expected genetic plasticity of the ancestral group. Because of this, transitional forms are often difficult to assign to established groups. As a result we find that *Lepidosigillaria, Cyclostigma, Valmeyerodendron,* and other genera have been placed in various families of the Protolepidodendrales or Lepidodendrales depending on what characteristics the individual investigator believes to be important.

Lepidodendrales of the Carboniferous

The Lepidodendrales predominated in the Upper Carboniferous swamps of Europe and North America, and for many years we considered most species of the Lepidodendraceae to belong to the genus *Lepidodendron* (Fig. 11.3A). In recent comprehensive studies, however, DiMichele (1979a, 1979b, 1980, 1981, 1983, 1985) has dramatically expanded our understanding of the family and has transferred most of the permineralized species to *Lepidophloios, Paralycopodites,* and the new genus *Diaphorodendron.* Because *Lepidodendron* (Fig. 11.3A) and the other genera are similar in their gross form, we will begin our discussion with the general features of the order. From a cursory examination of the reconstruction an arborescent lycopod (Fig. 11.3A) it is apparent that it had bipolar growth. This simply means that these plants had a main axis that grew and branched at both ends. The branches of the aerial part formed a three-dimensional system of dichotomous or pseudomonopodial branches with spirally arranged leaves and terminal cones. The basal end also branched dichotomously to form the anchoring and water-absorbing system comprised of rhizomorphs bearing spirally arranged roots.

Figure 11.3. **A.** Reconstruction of *Lepidodendron* sp. **B.** Leaf cushion of *L. aculeatum.* Ligule pit (l); vascular bundle (vb); parichnos (p); infrafoliar parichnos (ip). Carboniferous.

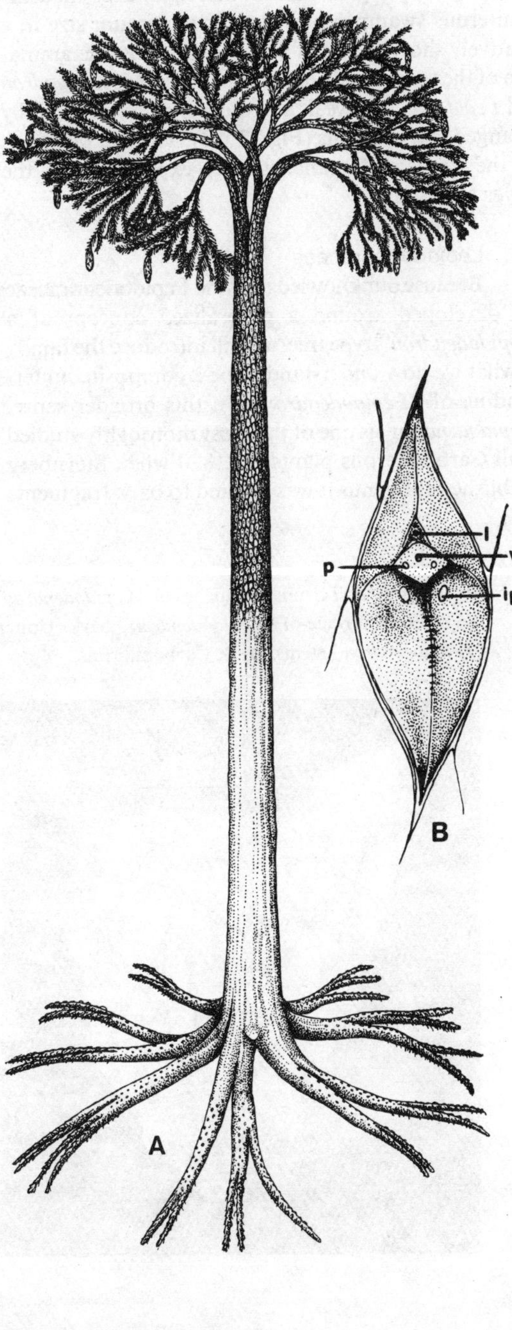

Even by modern standards of size, these arborescent lycopods were large trees. Casts and compressions of trunks belonging to *Lepidodendron* and *Lepidophloios* are frequently retrieved from Carboniferous coal mines. One specimen measured 35 m up to the first branching at the distal end, and the branched portion continued for another 6 m. Reasonable estimates suggest that these trees grew to a height of 54 m. At the base, trunks attached to branching rhizomorphs are known to be more than 1 m in diameter. The rhizomorphs extended into the substrate of the swamp as much as 12 m. There is reason to believe, because of uniform environmental conditions favoring continuous and rapid growth, that these giants of the Carboniferous swamp achieved their maximum size in a relatively short period of a few years. Closer examination of the trunks and aerial branches of *Lepidodendron* and *Lepidophloios* reveals an exquisite pattern of spirally arranged leaf cushions (Fig. 11.4B). The leaf cushions on the trunk and branches were exposed when the leaves abscised.

Lepidodendraceae

Because our knowledge of the Lepidodendraceae has developed around a generalized concept of a "*Lepidodendron*"-type tree, we will introduce the family by what we now understand to be a composite understanding of "*Lepidodendron.*" In this broader sense, "*Lepidodendron*" is one of the most thoroughly studied of all Carboniferous plants. In 1820 when Sternberg established the genus it was applied to bark fragments of arborescent lycopods (Fig. 11.4B) showing spirally arranged leaf cushions where the rhomboidal cushions are greater in their vertical than in their transverse dimensions (Fig. 11.3B). A leaf scar with a vascular bundle scar and flanking parichnos scars is usually situated just above the middle of the cushion. The ligule pit, in turn, is situated just above the leaf scar, while below and to either side of the leaf scar are two additional parichnos scars. Transverse wrinkles are sometimes present along a median ridge on the lower attenuated portion of the cushion. The individual cushions may be adjacent or separated by grooves or bands of bark tissue. Their degree of separation seems to have been correlated with aging and increasing in diameter of the trunk. On the oldest parts of large trees the outer cortex with the cushions was sloughed off, exposing subsurface stem features (Fig. 11.4B).

Since the genus *Lepidodendron* was first established, more than 100 species have been described. We can conclude that many of these, however, are nothing more than different developmental stages having leaf cushions belonging to a single species. For example, it has been shown that the small leaf cushions on distal branches of a mature tree (Fig. 11.4A) did not increase in size even after leaf abscission. By contrast, first-formed growth stages of "*Lepidodendron*" axes had large leaf cushions from the start and were not formed by the subsequent enlargement of small leaf cushions. This has been determined by the histological examination of a great number of permineralized stem specimens that, based on characteristics other

Figure 11.4. **A.** Terminal branches of "*Lepidodendron*" sp. with short, awl-shaped leaves. **B.** Lower center, mold of the external surface of "*Lepidodendron*" bark. Upper three specimens, molds of various stages in decortication of "*Lepidodendron*" stems. **A, B:** Carboniferous.

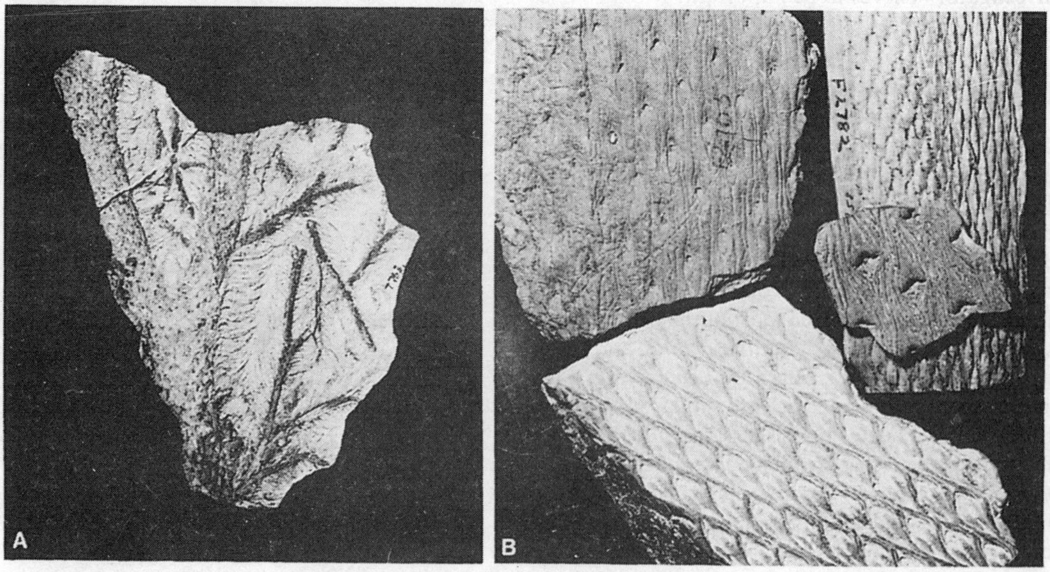

than those of leaf cushions, can be assigned to the same species. Such an informative specimen as the one described by Kosanke (1979) shows what is believed to be a young "*Lepidodendron*" axis 1.03 m long and 10 cm in diameter. The attached leaves are more than 78 cm long. Where the leaves have been shed, large rhomboidal leaf cushions are revealed. If the interpretation of the specimen is correct, then we can conclude that the large leaf cushions did not develop from small ones after leaf abscission, but that they had reached their ultimate size by the time the leaf was mature prior to being shed. Thus, on a mature "*Lepidodendron*" we find large leaf cushions on the main axis, smaller ones on the distal portions of the branched aerial axes, with further decrease in size toward the ends of the branches. Concomitant with the decrease in size of leaf cushions from the base of the main axis to the apices of the ultimate branches we can conclude that there was a decrease in size of "*Lepidodendron*" leaves from those many centimeters long and with greatly expanded bases to those only a few centimeters long with relatively narrow bases (Chaloner & Meyer-Berthaud, 1983).

Internal structure: leaf-bearing axes

To more easily understand variations of internal structure that reflect the developmental stages in the stem and branches of the lepidodendrids, we should familiarize ourselves with major structures and tissues found when examining the cross section of a branch of a mature tree (Fig. 11.5A). It is important to have some

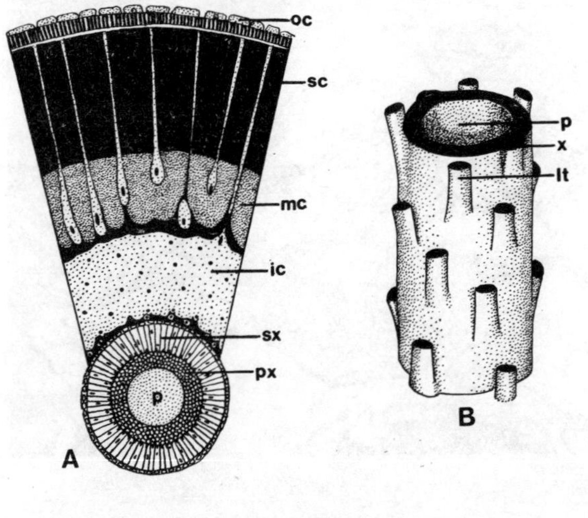

Figure 11.5. **A.** Diagram, transverse section stem of "*Lepidodendron.*" Outer cortex (oc); secondary cortex (sc); middle cortex (mc); inner cortex (ic); secondary xylem (sx); primary xylem (px); pith (p). **B.** Stereodiagram of medullated lycopsid stele with spirally arranged leaf traces (lt). Note absence of leaf gaps. Position of pith (p); xylem (x).

idea of the level or position in the branching system from which the section was taken. For reasons explained later, we have selected a section that is somewhere between the older, basal portions of the trunk and the distal, leaf-bearing branches of a mature tree.

Sections from this general region usually have a medullated protostele (Fig. 11.5B) where the primary xylem immediately outside the pith is exarch. By far the greatest part of the primary xylem is composed of metaxylem tracheids that are scalariform and have fimbrils traversing the apertures between adjacent wall thickenings (Fig. 11.6C). The small protoxylem tracheids form vertical ridges at the periphery of the primary xylem cylinder. In cross section, the ridges appear as undulations or small radiating teeth of protoxylem, producing what has been called a corona. Where protoxylem cells are well preserved they appear to be annular and spiral tracheids.

Leaf traces supplying the microphylls arise at a steep angle from the protoxylem ridges in a spiral sequence. No leaf gaps are formed (Fig. 11.5B). At a higher level the leaf traces bend outward and traverse the intervening tissues horizontally and terminate in leaf scars. Prior to abscission of the leaf, the leaf trace continues as a single vein into the attached microphyll.

At the level of the branching system from which our transverse section (Fig. 11.6A) is taken we can expect to find some secondary xylem surrounding the primary xylem. The scalariform tracheids of which it is composed are radially arranged but smaller than those of the metaxylem (Fig. 11.6B). They also have fimbrils (Fig. 13.6C). According to Eggert & Kanemoto (1977) the secondary xylem was produced by a unifacial vascular cambium. Unlike the vascular cambium of arborescent seed plants, the vascular cambium of a lepidodendrid did not produce secondary phloem. In their beautifully preserved specimens, Eggert & Kanemoto were able to show a continuous band of cortical parenchyma immediately outside the secondary xylem. Departing leaf traces and strands of primary phloem are embedded in this tissue. It has now been demonstrated conclusively that cells of the cortical parenchyma can assume a meristematic function to produce cells of secondary cortex or the so-called periderm. In some species of lepidodendrids the meristem producing the secondary cortex seems to form a continuous ring, and it has been reported to be two or three cells wide. In other species, multiple concentrically placed cortical meristems have been described.

Irrespective of whether there is a single continuous or series of discontinuous meristems responsible for the formation of secondary cortex, one of the distinctive features of arborescent lycopod stems is the production of copious amounts of this tissue. In the region from which our stem section was taken, as much as 50 percent of the volume of the axis may be composed of secondary cortex, the percentage increasing toward the base of the tree where it may be as much as 40 cm thick and decreasing toward the tips of branches to a point where it is absent in the ultimate twigs.

The secondary cortex is a complex tissue that varies in its composition depending on its stage of development and, to some extent, on the species. The most obvious portion of the secondary cortex is composed of fiber cells arranged in radial files superficially resembling the radial files of cork cells produced by a cork cambium in bark tissues of arborescent seed plants. Scattered through this tissue are areas of thin-walled cells that may have had a secretory function. Other thin-walled cells called "chambered" cells have been reported to occur among the fiber cells.

Functionally, the secondary cortex must have provided support for stems and branches with their leaves and cones. The amount of support provided by the relatively small amount of xylem tissue would not have been sufficient for trees of such magnitude. There is a good possibility that the secondary cortex also functioned to reduce the rate of desiccation from the aerial parts of old trees where the outer tissues had sloughed off. It is here that the secondary cortex is developed to the greatest thickness.

Intercalation of the secondary cortex has the effect of separating the primary cortex into two parts, an inner part around the vascular cylinder and an outer part bearing the leaf cushion in such a way as to show the pit containing a ligule (Fig. 11.6D). Further examination of the sectioned leaf cushions can reveal branched channels filled with aerenchyma (parenchyma tissue with large intercellular spaces) that terminate on the surface of the leaf cushion in the form of the parichnos scars. These may have functioned as a means of gas exchange in much the same way as lenticels in the bark of arborescent seed plants.

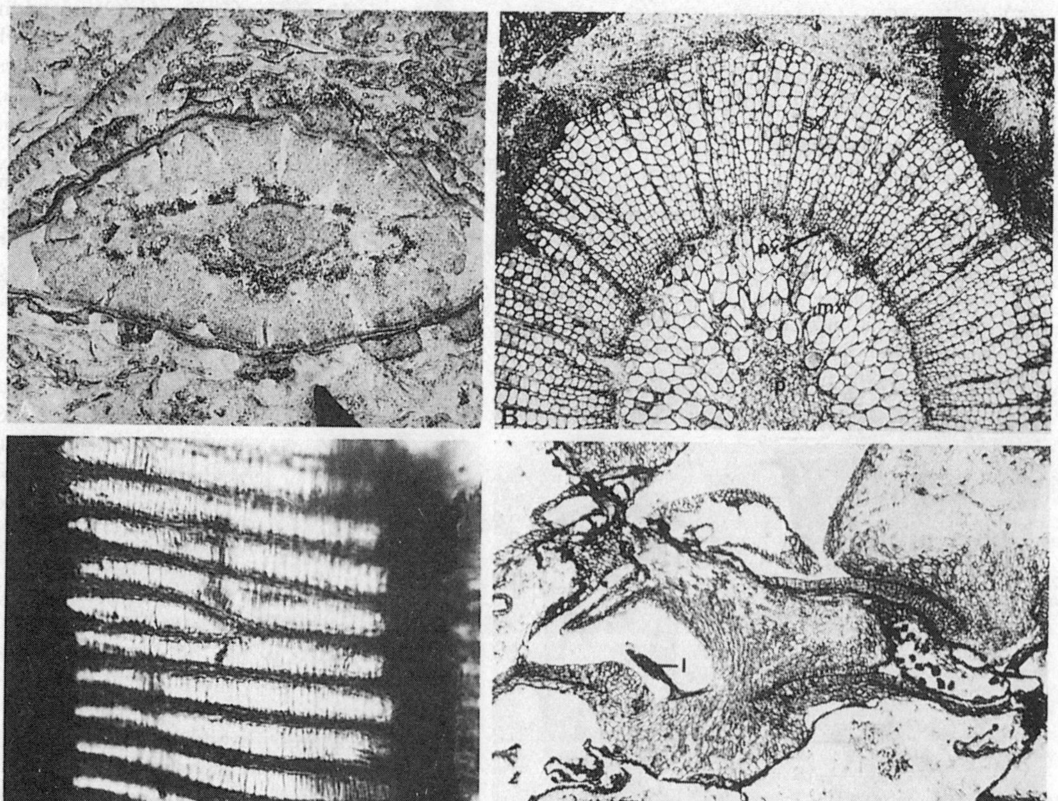

Figure 11.6. **A.** Transverse section of a small stem of "*Lepidodendron*." Compare with Figure 11.5A. **B.** Transverse section of stele of "*Lepidodendron*" enlarged. Pith (p); metaxylem (mx); secondary xylem (sx); protoxylem (px). **C.** "*Lepidodendron*" scalariform tracheid of secondary xylem highly magnified to show fimbrils traversing the apertures. **D.** Longitudinal section leaf cushion of "*Lepidodendron*" showing ligule (l) in ligule pit. **A–D:** Carboniferous.

Lepidodendron

Since the more precise characterization of *Lepidodendron* by DiMichele (1983), we now recognize the genus to consist of the permineralized species *L. hickii*, the compression species *L. aculeatum*, and similar forms. *Lepidodendron* was a large tree with a prominent trunk and a spreading crown that produced cones at the tips of the branches (Fig. 11.3A). The leaf cushions remained on the plant to near the base of the trunk and varied in shape at differrent levels in the branching system, but were generally longer than broad (Fig. 11.3B). The leaf scar of *Lepidodendron* is located near the center of the cushion, with a prominent ligule pit and two areas below the leaf scar known as infrafoliar parichnos (Fig. 11.3B). In cross sections of the stele, the primary xylem consists of a solid zone of tracheids with relatively inconspicuous protoxylem strands at the periphery. Two types of monosporangiate cones (those that produce megasporangia and those that produce microsporangia), which we will discuss in more detail later, occur in association with the vegetative remains. The cones are known as *Achlamydocarpon takhtajanii*, and microspores are of the *Lycospora* type.

Paralycopodites

Paralycopodites is known from the single species, *P. brevifolius*, that occurred in swamps throughout much of the Carboniferous (DiMichele, 1980). It was a small tree with a columnar trunk and dichotomizing, deciduous lateral branches that bore terminal cones. Leaf cushions are rhomboidal and similar to the compression genus *Ulodendron*. The leaves remained attached to the lateral branches and some regions of the trunk. Below the crown, the leaf bases may have become detached along with a thick zone of periderm. The stele is similar to *Lepidodendron*, but with even less conspicuous protoxylem. The cortex is similar to other genera of the Lepidodendraceae, but with no secretory contents or sclerenchyma. Of particular significance to the diversity within the family, *Paralycopodites* is the only genus of arborescent lycopods that is thought to have produced bisporangiate cones (those that produce both mega sporangia and microsporangia). We will discuss these cones later in the chapter as the genus *Flemingites* (Brack-Hanes & Thomas, 1983).

Diaphorodendron

Four species that previously had been assigned to *Lepidodendron* are now recognized as the distinct genus *Diaphorodendron* (DiMichele, 1985), (Fig. 11.7). As in *Lepidodendron* there are monosporangiate cones, but in *Diaphorodendron* the cones are borne on lateral branches near the tips of vegetative branches. *Diaphorodendron vasculare* and *D. phillipsii* are trees that range from 8 to 10 m tall, with deciduous lateral branches. *Diaphorodendron scleroticum* was larger, up to 20 m tall, with lateral branches that also may have been deciduous. In contrast, *D. dicentricum* had a columnar trunk and a crown of forking branches. Leaf cushions of *Diaphorodendron* are difficult to tell apart. In large stems with a pith, the tracheids of the primary xylem are separated into wedges by xylem parenchyma, and the protoxylem forms an inconspicuous, uniformly thick ring at the periphery. The cones associated with *Diaphorodendron* are known as *Achlamydocarpon varius*, and the microspores are of the *Capposporites* type.

Lepidophloios

Several species of *Lepidophloios* occur in sediments that range from the Mississippian through the Middle Pennsylvanian of North America and equivalent strata of Europe. The best-known species *L. halli* is found in Lower and Middle Pennsylvanian deposits of North America (DiMichele, 1979b). It formed a tree 10 to 20 m tall, with a crown of dichotomizing branches that were larger and more erect than those of *Lepidodendron*. Cones were monosporangiate and occurred on two to four rows of lateral branches, or on terminal branches of other species. Leaf cushions of *Lepidophloios* (Figs. 11.8, 11.9A), unlike those of the other genera of the Lepidodendraceae, are broader than they are high. The leaf scar is in the lower half of the leaf cushion instead of above the midpoint. Internally, most species of *Lepidophloios* have well-developed ridges of protoxylem around the metaxylem of the vascular cylinder. In transverse section these ridges form a corona of toothlike projections resembling a cog wheel (Fig. 11.9B). The megasporangiate cones of *Lepidophloios* are known as *Lepidocarpon*, while the microsporangiate cones are assigned to *Lepidostrobus*.

Ontogeny of the vascular system

In presenting the description of the cross section of a lepidodendrid stem, we have tried to make it clear that we can interpret what we see as a reflection of an ontogenetic stage of the plant. The objection to this is that we obviously cannot directly relate a particular level of development in a mature axis to the meristem that produced it. It is true that our ontogenetic interpretations as they relate to the activities of meristems must, because of the nature of the material, be inferred. This, however, does not detract from the validity of our interpretations.

To accomplish ontogenetic studies of conspecific plant parts requires large numbers of well-preserved specimens each of which may represent only a fragment of the whole. By accumulating and observing these many fragments, one can establish a spectrum of pieces whose structural characteristics exhibit a degree of continuity. The first comprehensive study of this kind for the arborescent lycopods was produced by Eggert (1961). In this monumental work, the author was able to construct the changes occurring in the vascular

Figure 11.7. Reconstructions of species of *Diaphorodendron*. **A.** *D. vasculare*. **B.** *D. scleroticum*. **C.** *D. phillipsii*. **D.** *D. dicentricum*. Pennsylvanian. (From DiMichele, 1981, with permission.)

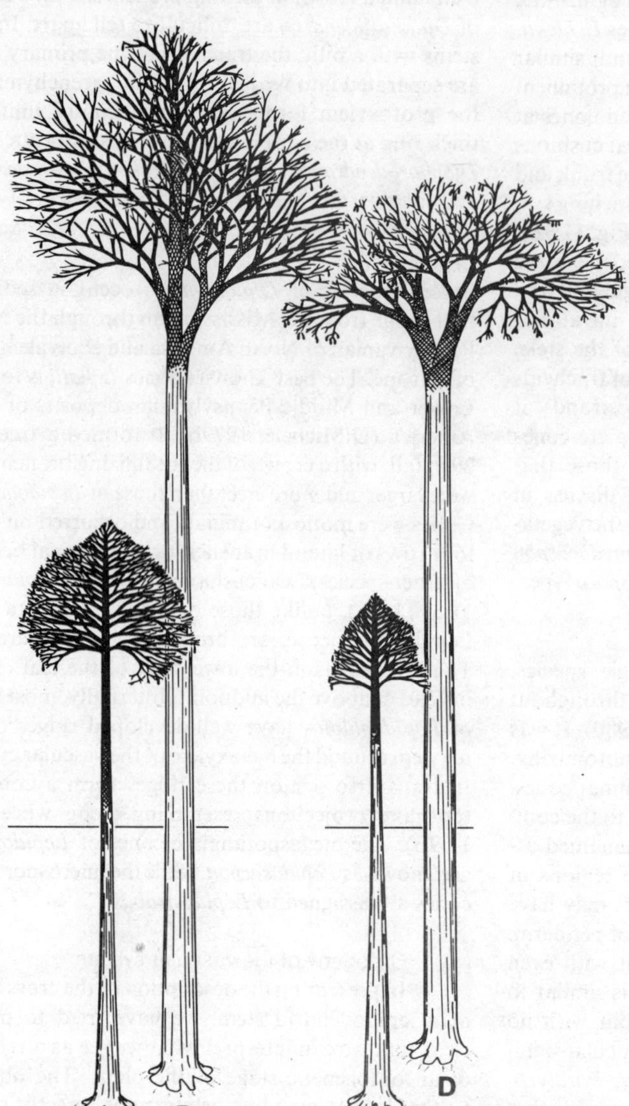

Figure 11.8. Detail, single leaf cushion of *Lepidophloios*. Ligule pit (lp); leaf scar (ls). Carboniferous.

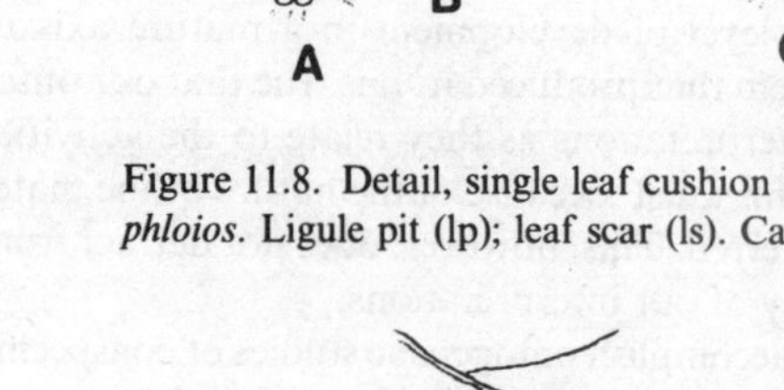

system of an arborescent lycopod extending from the base of the stem to the tip of an aerial branch.

Williamson (1895) first suggested that such an approach to the study of arborescent lycopods was possible by using permineralized (coal-ball) specimens. From his researches, he developed the idea that those stem specimens of arborescent lycopods having the largest primary vascular cylinder around a pith represented the basal, first-formed portion of the aerial axis.

Walton (1935) realized that another interpretation was possible after examining some unusual permineralized specimens of *Lepidophloios* trunks from

Arran, Scotland. Prior to preservation, the soft tissues of a trunk apparently decayed allowing the stelar vascular structures to fragment and drop into a cavity formed by the surrounding secondary cortex at the base of the trunk. In a single cross section of one specimen Walton observed four fragments of the vascular system of the plant, each fragment representing a different level in the trunk. Critical observation showed that the primary vascular system differed from piece to piece. One was an exarch protostele, and the other three exarch medullated steles. One of the latter had a small pith surrounded by primary xylem; the other two showed increased medullation (amount of parenchyma) and size of the pith. All four fragments were surrounded by a thick zone of secondary xylem. In other specimens of the Arran fossils, Walton noted that those vascular segments with large medullated steles usually had smaller amounts of secondary xylem. These observations allowed him to conclude that fragments showing the maximum amount of secondary xylem should represent the vascular system in the oldest, basal portion of the axis. Some specimens showing abundant development of secondary xylem were also protostelic. He also noted that as the primary xylem cylinder increased in size and became medullated, there was an accompanying decrease in the amount of secondary xylem. These he interpreted as ontogenetic stages formed later in development and thus having come from younger, more distal portions of the axes. Such an interpretation was in keeping with the idea developed by F. O. Bower that the oldest, first-developed portions of primitive vascular plants were protostelic. Figure 11.10A shows the reconstruction of the vascular system of *Lepidophloios* with a basal protostele as determined from the Arran fossils.

Eggert (1961) was able to apply Walton's findings to *Diaphorodendron scleroticum* (Fig. 11.7**B**), a common Pennsylvanian (Upper Carboniferous) arborescent lycopod of North America. Eggert confirmed that the primary vascular system at the base of a mature tree was indeed protostelic; that the primary cylinder became medullated, with the resulting pith increasing in size at higher levels in the trunk. This aspect of development, where there is progressive enlargement and increase in complexity of the primary vascular system, has been termed epidogenesis.

Where dichotomous branching occurred in the aerial axes, it is apparent that the resulting branches were smaller than the axis from which they were formed. Thus, with each successive dichotomy, there was an ever-decreasing size of branches. Andrews & Murdy (1958) suggested that this progressive diminution in size ultimately led to termination of growth and that arborescent lycopods had a determinate growth pattern. If this interpretation is correct, we can conclude that with each successive dichotomy the terminal meristems also decreased in size until they reached such small dimensions that further dichotomies were not possible. We might say the apical meristems producing the branches were "used up." Such diminution in size occurs in the aerial branching systems of extant Lycopsida.

Accompanying the successive decrease in size of the aerial branches, Eggert was able to demonstrate a decrease in size and complexity of the primary vascu-

Figure 11.9. **A.** Compression fragment of *Lepidophloios* bark. **B.** Highly magnified transverse section of a *Lepidophloios* stele Note the corona of protoxylem (px) ribs radiating from the continuous band of metaxylem (mx). **A, B:** Carboniferous. (**B** from DiMichele, 1979a.)

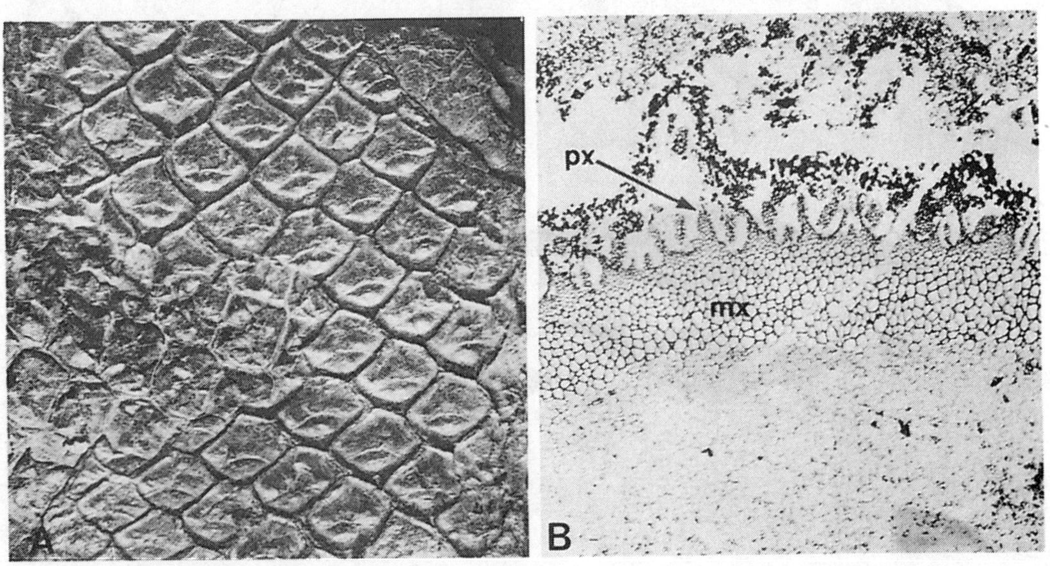

Figure 11.10. **A.** Diagram, longitudinal section vascular system of *Lepidophloios* based on Walton (1935). Primary xylem, solid black; pith, stippled, secondary xylem, hatched; outer tissues, white. **B.** Diagram of vascular system aerial axes of an arborescent lycopod. Levels a–e indicate position of transverse sections a–e. Primary xylem, solid black; pith, white; secondary xylem, hatched. (From Eggert, 1961.) **C.** Common configurations of *Cyperites* in transverse section. **D.** Optical view of stomata from epidermis in stomatal furrow, *Cyperites* **E.** Transverse section, *Cyperites*. Vein (v); stomatal furrow (sf). (**C–E** modified from Reed, 1941.)

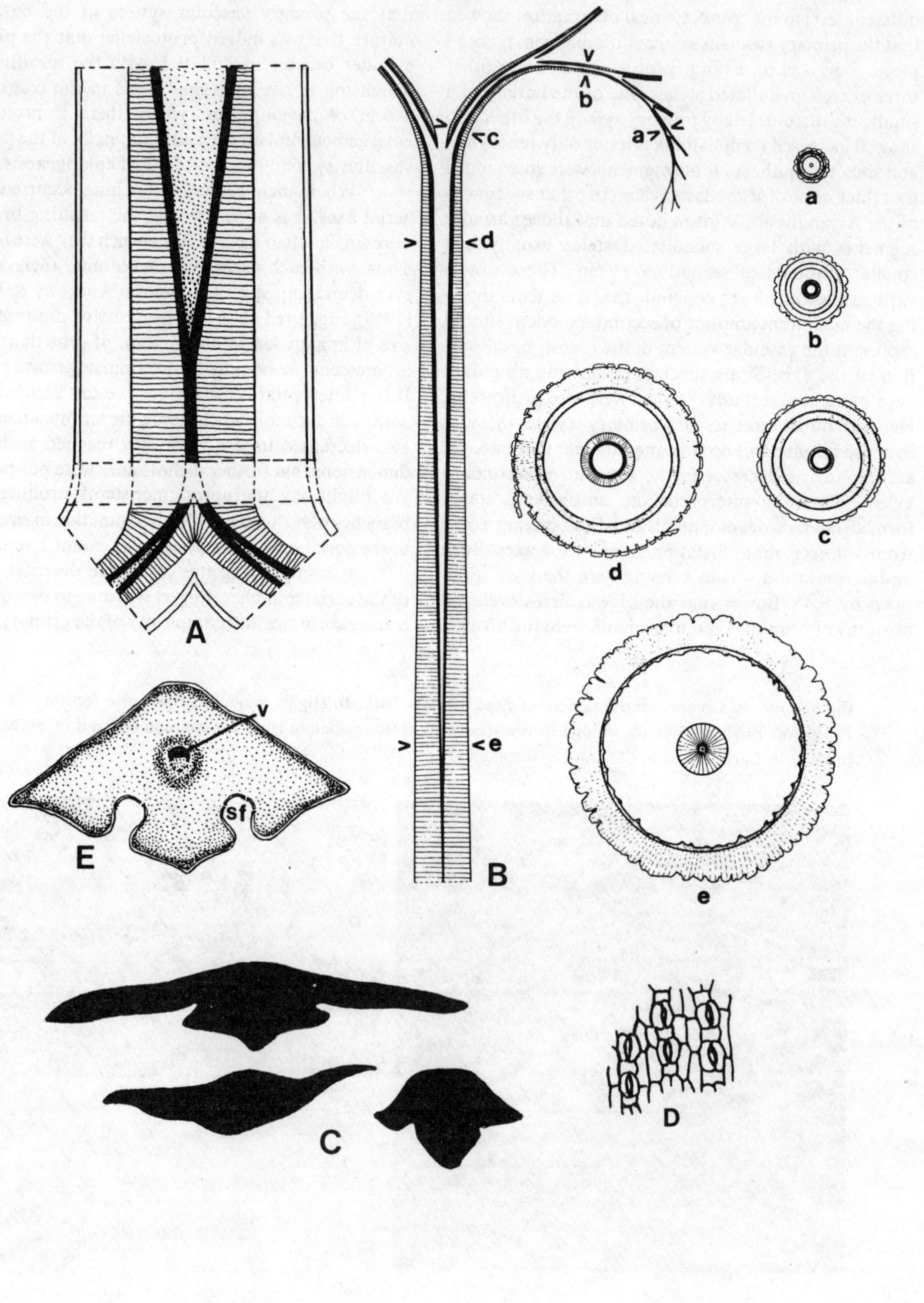

lar system (Fig. 11.10B). This is reflected in a decrease in size of the stele so that in penultimate branches, the vascular cylinder is only slightly medullated while in the ultimate branches only a tiny protostele remains. Concomitant with the diminution of the primary vascular system was a decrease in the amount of secondary xylem and secondary cortex. This type of development, which involves the production of progressively smaller and simpler anatomy, is called apoxogenesis.

Prior to studies that revealed the spectrum of ontogenetic variability within a species of fossil plants, it was not unusual, but understandable, that investigators established several species for fragments of a plant organ that have turned out to be developmental stages of a single species. From the example presented above, we now know that a basal fragment of a mature *Diaphorodendron scleroticum* stem may exhibit a very thick secondary cortex without leaf cushions at the surface and that there will be well-developed wood around a protostele. These characteristics are quite different from those of another stem fragment where there is little or no secondary cortex, leaf cushions are present, there is no wood, and the primary vascular system is a medullated protostele. If you were confronted with these two stem fragments and nothing more, the inclination would be to interpret the variability exhibited by the two specimens as representing differences between species of *Diaphorodendron*. Careful examination of many stem fragments might reveal, however, that the variability actually reflects the developmental stages within a species. Paleobotanists are making every effort to distinguish these two kinds of variability, and many studies have been done or are ongoing in an effort to unravel the problem of how to interpret the variability presented by fossils of related plant parts.

Leaves

The external form of lepidodendrid leaves, even when attached to stem fragments, cannot be used as a reliable species characteristic. The reason is that leaf length varies for an individual plant and is correlated with its developmental stage (Chaloner & Meyer-Berthaud, 1983). A specific example of this can be seen in a specimen of *Lepidodendron obovatum* where Nemejc (1947) demonstrated conclusively that leaves on distal branches were shorter than those on older proximal branches.

Leaves on the small ultimate and penultimate branches of mature specimens of lepidodendraleans are awl-shaped and may be slightly recurved at the tips (Fig. 11.2A). On larger branches of mature trees or on unbranched axes of developing plants where they may persist, the leaves are longer, more grasslike (Fig. 11.11), and with conspicuously expanded bases. From our experience in examining the compression fossils of abscised leaves of lepidodendrids, it seems that all are long and grasslike. The absence of detached awl-shaped leaves suggests that they were not shed. These observations seem to support the idea that the leaves were longer on older parts of aerial axes, and as they became nonfunctional they were abscised. The shorter leaves formed later on small terminal branches remained functional and attached until the tree died or was destroyed.

In the detached, compressed state, one cannot be sure whether a lepidodendrid-type leaf belongs to *Diaphorodendron, Lepidodendron, Lepidophloios, Sigillaria,* or some other arborescent lycopod. For this reason a form genus, *Cyperites,* has been adopted (Rex, 1983).

Just as there is variability in leaf size for a given species, so there is variability in leaf shape from the base to the apex (Fig. 11.10C). At the expanded base covering the leaf cushion the cross section of a leaf tends to be flatly rhomboidal. Distally the configuration changes to angular rhomboidal or nearly triangular. In the past these developmental states were believed to represent different species.

Except at the base, transverse sections of attached lepidodendrid leaves (Fig. 11.10E) exhibit two conspicuous furrows on the abaxial (lower) surface. The furrows contain several rows of stomata parallel to the long axis of the leaf (Fig. 11.10D). Except for tissues to the inside of the furrows, there is a well-developed hypodermis of thick-walled cells that have pointed ends characteristic of fibers. The thickness of the hypodermis varies in different parts of the same leaf. To the inside of the hypodermis are thin-walled

Figure 11.11. *Cyperites* in large Mazon Creek concretion. Pennsylvanian.

cells of mesophyll (presumably the photosynthetic tissue), which surround the single sheathed vein. The sheath is composed in part of transfusion cells that have the general shape of parenchyma cells, but with reticulate secondary wall thickenings reminiscent of tracheids. Thin-walled cells surround the single strand of scalariform xylem cells. Some of the thin-walled cells that occur below the xylem strand are believed to be phloem.

Anchoring and water-absorbing structures

It is not an overstatement of the facts to say that the anchoring and water-absorbing structures of *Lepidodendron, Lepidophloios,* and other arborescent lycopods are unusual. As described earlier, the root-bearing system at the base branches dichotomously or pseudomonopodially similar to the leaf-bearing aerial branching systems at the distal end of the main axis. For the lepidodendraceae the basic morphology of the mature tree is a main axis, which branches dichotomously at both ends (Fig. 11.3A). The root-bearing branches that penetrated the substrate of the swamp are called rhizomorphs and, as will be explained, they have many features in common with the aerial axis that rests on the basal pad formed by the rhizomorph. Many casts of in situ stumps of lepidodendrid trunks with the basal rhizomorph attached (Fig. 11.12A) have been discovered, but perhaps the most famous are those of Victoria Park in Glasgow.

At the base, the main axis of the rhizomorph formed four branches as a product of two successive and closely spaced dichotomies. As these axes continued outward growth into the substrate, they perpetuated the dichotomizing branching pattern in the horizontal plane. Roots borne on the rhizomorphs are found in the younger, distal portions. Here the roots, like the leaves of the aerial branches, have a spiral arrangement. The positions of roots on older, more proximal portions of rhizomorphs are marked by spirally arranged scars of roots that have abscised! Although differing in function, it is apparent that the water-absorbing roots of lepidodendrids have characteristics in common with the photosynthetic leaves. This raises the intriguing suggestion that ancestors of the Lepidodendrales with bipolar growth were endowed with appendages having similar structure and arrangement at each end of the dichotomously branched main axis. Those appendages penetrating the substrate evolved the functions of roots, while aerial appendages became leaves.

Detached rhizomorphs and their roots are placed in the form genus *Stigmaria.* In the detached condition, casts and molds of *Stigmaria* (Fig. 11.12B) are frequently found in the unstratified shale that represents the underclay (fire clay) of the original swamp. A seam of coal representing the decayed and compressed vegetation on the floor of the swamp usually lies on top of the stigmarian underclay. It has been suggested that

Figure 11.12. **A.** Exceptional cast, *Stigmaria ficoides* from near Bradford, Yorkshire. The distance across the system is 6 m. (From Frankenberg & Eggert, 1969.) **B.** Cast, segment of *S. ficoides* showing spirally arranged root scars. **A, B:** Carboniferous.

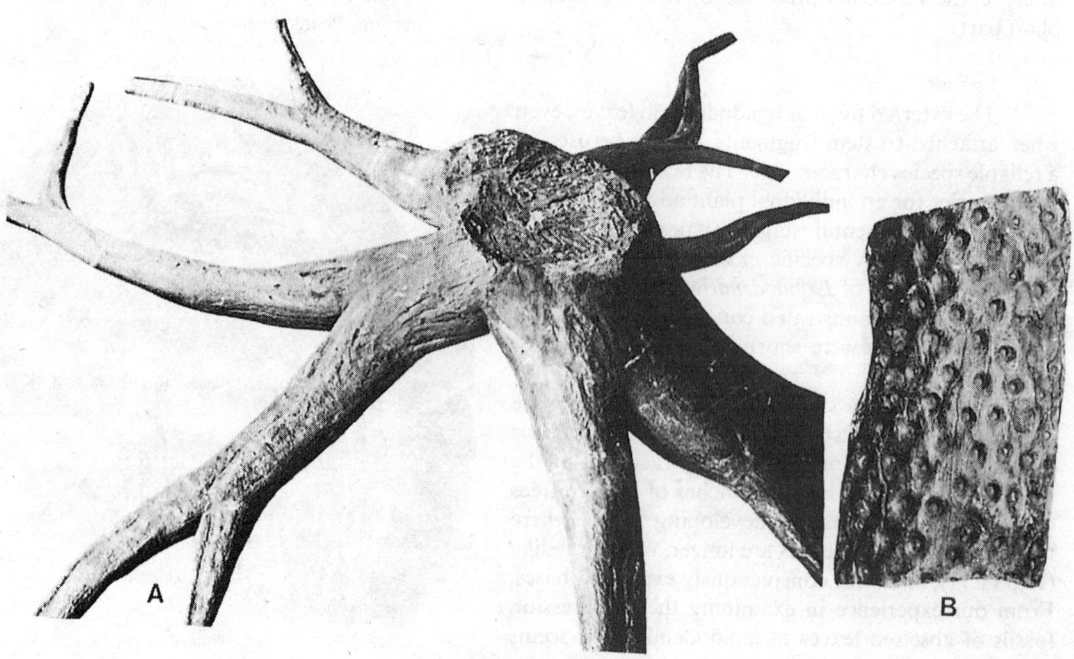

Figure 11.13. **A.** Transverse section showing anatomy of *Stigmaria ficoides*. (Reproduced with permission of Dr. R. Stockey.) **B.** Tangential section of *S. ficoides* wood showing root traces (rt) associated with rays (r). (From Frankenberg & Eggert, 1969.)

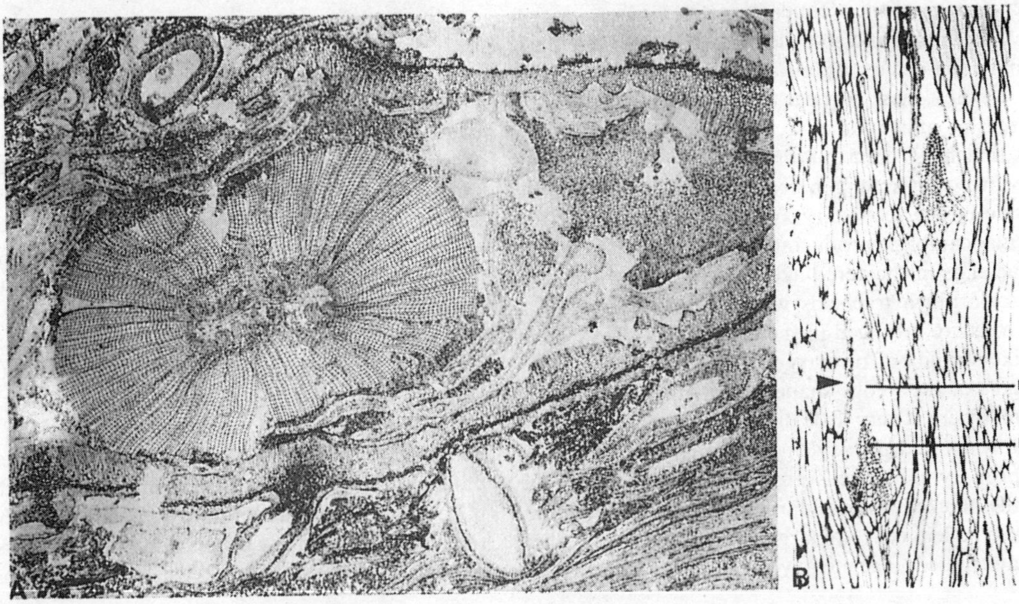

the fossilized rhizomorphs are in situ and were produced by the arborescent lycopods that grew in the ancient swamp and are now part of the coal seam.

By far the commonest species of *Stigmaria* is *S. ficoides*. Several comprehensive accounts of the structure of this species have appeared in the last 100 years. The earliest account by Williamson (1887) is remarkable in its accuracy. That of Frankenberg & Eggert (1969) is the most profusely illustrated account of *S. ficoides* ever prepared.

The spirally arranged scars, which indicate the points where the roots were attached to the rhizomorph, are shallow dishlike pits 3 to 7 mm in diameter. Near the center of each is a raised vascular bundle scar. The root scars are partially embedded in and are a part of the outer primary cortical tissues of the rhizomorph. In old rhizomorphs, which may be as much as 80 cm in diameter, the outer cortex with its root scars has been sloughed off. In the cross section of a root-bearing rhozomorph (Figs. 11.13A; 11.14) we can see that the outer cortex surrounds an extensive and complex layer of secondary cortex (periderm). The way in which the secondary cortex developed has been thoroughly treated by Frankenberg & Eggert. Suffice it to say that although a more or less continuous lateral meristem can be identified as initiating development of the secondary cortex, it is obvious from the construction of outer tissues that many cells retained their capacity to divide and produce short irregular files of cells with thin walls. Cells of these files show indications that they

Figure 11.14. Stereodiagram of *Stigmaria ficoides*. Secondary cortex (sc); middle cortex (mc); primary xylem (px); secondary xylem (sx); cambium (c); root trace (rt). (Redrawn from Stewart, 1947, in Chaloner & Boureau, 1967.)

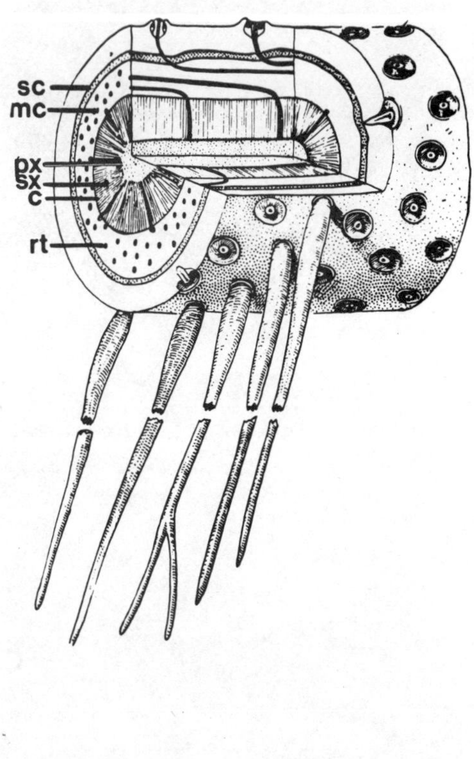

divided not only tangentially but radially as well. The tissues of secondary cortex produced in this way no doubt compensated for increases in diameter of developing rhizomorphs. Secretory ducts form a ring in the outer portion of the secondary cortex.

The inner and so-called middle cortex are both primary and occupy a zone around the vascular cylinder. Usually these cortical tissues are not preserved. There is no indication of secondary phloem, although cells that have been interpreted as those of primary phloem occur among parenchyma cells just to the outside of the secondary xylem. It appears that the vascular cambium producing the secondary xylem was unifacial as in the aerial axes. Secondary xylem cells are scalariform tracheids with fimbrils (Fig. 11.6C). The cells are arranged in the usual radial files. Small barlike strands of primary xylem occur at the inner margins of the wedge-shaped segments of secondary xylem. There is very little distinction between cells of protoxylem and metaxylem. The most centripetal of the primary xylem cells are smallest and may have spirally thickened secondary walls. To the outside, primary xylem cells are larger and scalariform. They are interpreted to be metaxylem. If these observations are correct, then the primary xylem is endarch, a condition at variance with the exarch primary xylem of the aerial axes. At the position in the rhizomorph represented by our transverse section we can expect to find a pith or pith cavity.

From a developmental point of view, Eggert (1961) has shown that in the younger, distal portions of mature stigmarian rhizomorphs there is a decrease in size of the pith accompanied by a decrease in the amount of wood and secondary cortex. As in the stem, much of the decrease found in *Stigmaria* results from the axes distal to a fork being significantly smaller than the axis from which they were produced (Fig. 11.12A). This pattern generally repeats the phenomenon of apoxogenesis observed for the aerial axes of the plant. Along with the decrease in size of the rhizomorphs, there is a decrease in number and size of the attached roots.

Unlike the stem, the primary xylem of the rhizomorph does not form a continuous uninterrupted ring of tissue around the pith. Instead, massive parenchyma-filled rays continuous with the parenchyma of the pith interrupt the primary xylem and then traverse the secondary xylem. Root traces originating from margins

Figure 11.15. *Stigmaria ficoides* root anatomy. **A.** Transverse section of free root with excentrically placed bundle. **B.** Transverse section of monarch, collateral bundle in free root. (**A–C** from Frankenberg & Eggert, 1969.) **C.** Longitudinal section of root base showing dark abcission layer.

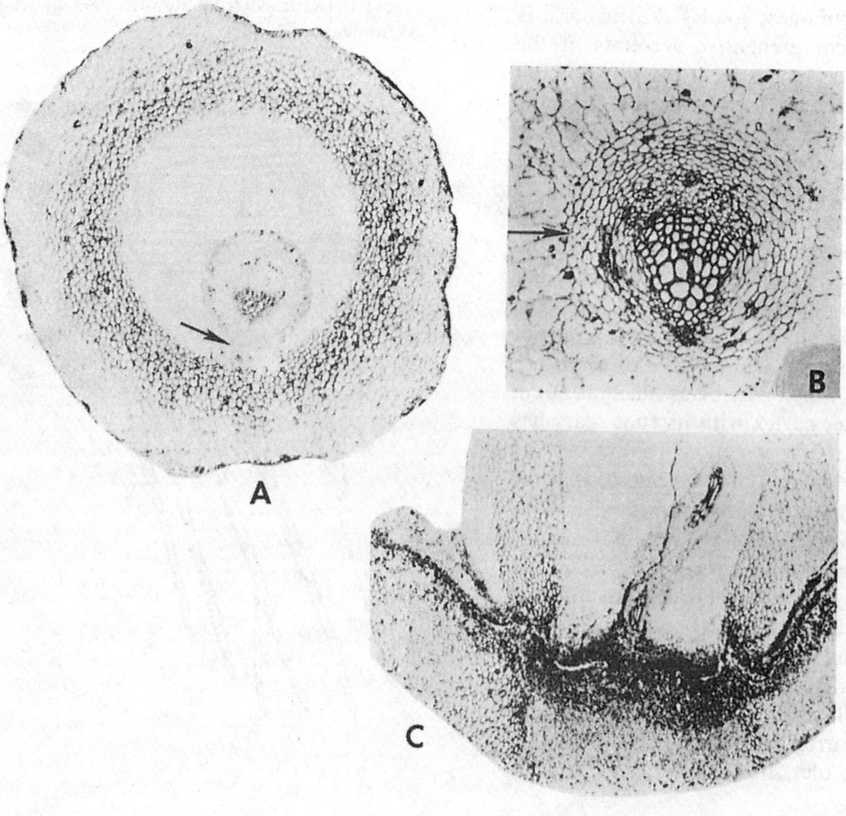

of the barlike strands or primary xylem turn outward at right angles to the rhizomorph axis as they pass through the wood. The root traces are produced in a spiral sequence and a single trace is associated with each ray (Fig. 11.13B). In a tangential section of the wood one can see the spirally arranged rays each with its root trace. This configuration is characteristic of the wood of *Stigmaria,* and it explains why in cross section the wood appears to be in wedge-shaped segments separated by wide rays.

After traversing the secondary wood, the root traces bend at a steep angle distally, pass through the surrounding primary and secondary cortical tissues, then turn outward again and into the bases of roots. If the root has abscised, the root trace forms the centrally placed vascular bundle scar. Longitudinal sections through the bases of attached roots (Fig. 11.15C) show the abscission tissue.

In free roots, the root trace strand can be seen in cross section (Fig. 11.15B) as a monarch collateral bundle, one in which protoxylem, metaxylem, and a cap of phloem are arranged in this centripetal sequence (Stewart, 1940). The root trace, surrounded by an inner cortex, may be centrally or slightly excentrically placed within a large cavity (Fig. 11.15A), which in turn is delimited by an outer cortex. The cavity is formed by the breakdown of middle cortex cells at the point of attachment of the root to the rhizomorph. In certain species of *Stigmaria* there is a parenchymatous connective between the inner cortex and outer cortex. The roots of *Stigmaria* are known to dichotomize. Nothing is known of the developmental origin of the root or if root hairs or a root cap were formed.

Although specimens interpreted to be apices of *Stigmaria* rhizomorphs have been described from time to time, perhaps the most instructive specimen thus far discovered is a mold/cast from the Middle Pennsylvanian (Fig. 11.16A). Near the tip, the specimen tapers rather abruptly and terminates in a circular rim (Fig. 11.16B). Within the rim (Fig. 11.16B) there is a discontinuous groove that circumscribes a concave plug at the tip of the rhizomorph. Rothwell (1984) suggests that the plug formed a protective cap for the rhizomorph as it penetrated the substrate. Proximal to the circular rim there are rootlet scars arranged in a helix. Considerable significance has been attributed to the morphology of this rhizomorph apex (Rothwell & Erwin, 1985) as explained later in the chapter.

Reproductive morphology

Two characteristics come to mind when we initiate consideration of the reproductive biology of the Lepidodendraceae. We know that (1) they were heterosporous and (2) their sporophylls were organized into well-defined cones (Fig. 11.17A,B). We also know that both of these characteristics had evolved in the arborescent lycopod *Cyclostigma* by Upper Devonian times. In *Cyclostigma* the cones are bisporangiate (Chaloner, 1968) with approximately 24 megaspores per megasporangium. The closely related Carboniferous genus *Pinakodendron* is also bisporangiate, but its sporophylls are not organized into a cone. Instead they are borne on major branches, and the sporangia are adaxial on unmodified leaves. Each megasporangium contains one tetrad (four) of megaspores. From this and evidence derived from other genera not discussed, it is plausible that ancestors of the arborescent lycopods first developed heterospory and that the bisporangiate condition was followed by the condensation of sporophylls into cones on terminal branches.

Phillips (1979), in a superb article summarizing

Figure 11.16. Cast, tip of branch of *Stigmaria* rhizomorph. **A.** Side view showing spirally arranged rootlet scars (r). **B.** View of apex showing plug (p) within the circular rim (cr). Pennsylvanian. (From Rothwell, 1984.)

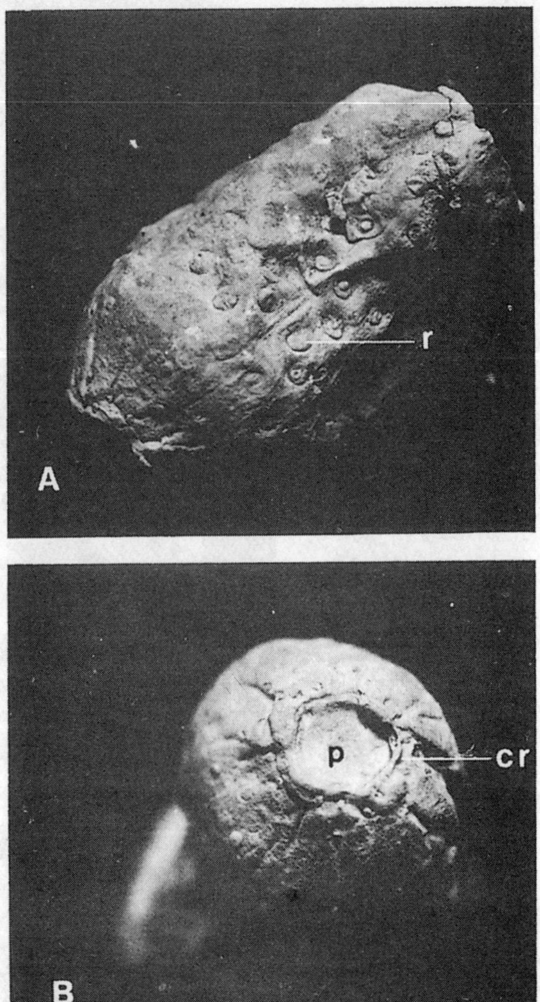

the reproductive morphology of arborescent lycopods, supports the idea that the bisporangiate *Flemingites* type cone, similar to that of *Cyclostigma,* represents the type from which diverse modifications of heterospory evolved to be expressed later in Carboniferous arborescent lycopods. You will recall from the description of the various genera of the Lepidodendraceae, that *Flemingites* is considered to be the cone of *Paralycopodites*. In addition to the bisporangiate free-sporing *Flemingites* type organization, Phillips recognizes three additional groups that show extensive modifications of megaspore number, megasporangium, and megasporophylls. Cones belonging to those genera displaying one or more of these modifications may be monosporangiate.

Flemingites schopfii (Fig. 11.17C) described by Brack (1970) is a good example of the bisporangiate free-sporing cone type. The permineralized material

Figure 11.17. **A.** *Lepidostrobus oldhamius,* a large cone. Actual specimen at least 32 cm long. **B.** *Lepidostrobus* sp., concretion specimen broken to show cone in transverse section. Distal lamina (dl); sporangial pad (s). **C.** *Flemingites schopfii,* longitudinal section with megasporangia (me) and microsporangia (mi). (From Brack, 1970.) **D.** *F. oldhamium,* transverse section of cone showing cone axis and traces to sporophylls. (Courtesy Dr. D. H. Vitt.) **E.** *F. schopfii,* archegonium with four tiers of neck cells and egg chamber (e). (From Brack-Hanes, 1978.) **F.** *Lepidostrobophyllum* sp., isolated sporophylls of different sizes in Mazon Creek concretions. **A–F:** Pennsylvanian.)

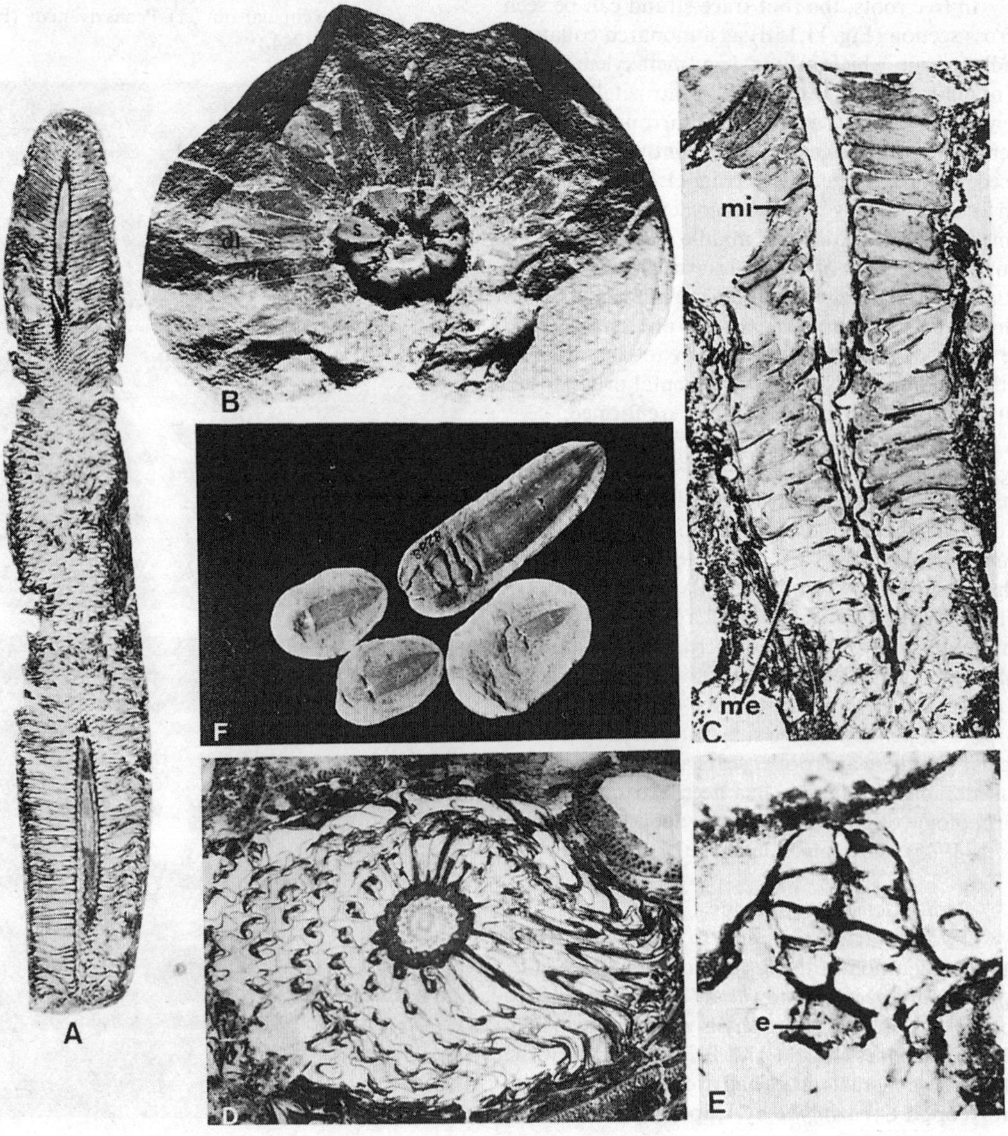

she describes comes from the Lower Pennsylvanian of North America and displays most of the characteristics of the genus. The largest incomplete cone fragment was 8 cm long and 1.3 cm in diameter. Cones assigned to *Lepidostrobus oldhamius* (Fig. 11.17A) produced only microspores and are known to have been at least 35 cm long and 4 cm in diameter (Balbach, 1967). Most, however, seem to be 8 to 20 cm in length.

Cone anatomy
The anatomy of the cone axis recalls the primary structures of a lepidodendrid twig where there is an exarch siphonostele and spirally arranged sporophyll traces formed from the tips of the protoxylem points (Fig. 11.17D). In the cone, sporophyll traces depart at a steep angle, pass through the intervening cortical tissues into the conspicuous outer cortex where they are surrounded by parichnos tissue, and then continue into the sporophyll.

Being bisporangiate, *Flemingites schopfii* has both micro- and megasporophylls attached to a cone axis. They are compactly arranged in a tight spiral that exhibits a variable phyllotaxy. Each sporophyll (Fig. 11.18A) comprises a horizontal pedicel and an upturned distal lamina that is awl-shaped. Sporophylls of the lower part of the cone have megasporangia; microsporangia are on the upper sporophylls, an arrangement repeated in the cones of *Selaginella* (Fig. 10.12B) and *Selaginellites*. Micro- and megasporangia are radially elongated and attached to the adaxial

Figure 11.18. **A.** *Flemingites* sp., a single sporophyll showing pedicels, distal lamina, and sporangium. **B.** *Lepidocarpon*, lateral lamina enclosing sporangium except for "micropylar" slit (ms). **C.** *Achlamydocarpon* sp. Note absence of lateral lamina. (A–C redrawn from Phillips, 1979.) **D.** Embryo, *Lepidophloios*. Stem (s); rhizomorph axis (r). **E.** Embryo, *Bothrodendrostrobus*. First leaf (l); first root (r). (D, E redrawn from Stubblefield & Rothwell, 1981.)

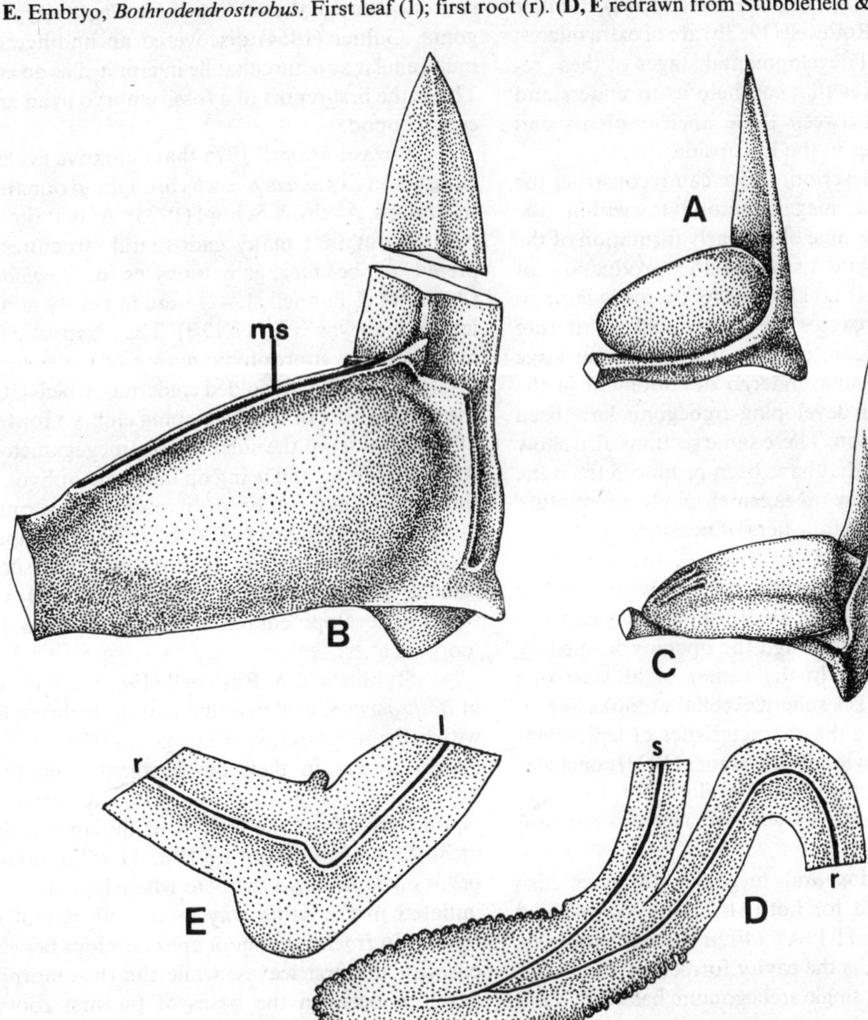

surface of the pedicel. Each sporangium is fused with the subtending pedicel by about four-fifths its length. The ligule is placed in a pit at the distal end of the sporangium. A subarchesporial pad with trabeculae extends into the sporangial cavity from the base of the sporangium. The sporangium wall is only one cell layer thick. The microspores are 20 to 30 μm in diameter and compare with the dispersed spore genus *Lycospora*. There are 12 to 29 megaspores per sporangium. These vary from 700 to 1250 μm and compare to spores of the *Lagenicula* section of the dispersed spore genus *Triletes*.

Megagametophytes

The *Flemingites* cones described by Brack (1970) and Brack-Hanes (same author) (1978) proved to be exceptional because of beautifully preserved megagametophytes in the megaspores. Several reports of these structures have appeared in the literature from time to time since the turn of the century. The descriptions of megagametophytes by Brack-Hanes (1978), Galtier (1970), and Pigg & Rothwell (1983b) are of extra interest because they reveal developmental stages of these reproductive structures that can help us to understand the relationships between these ancient plants and their extant relatives in the Lycopsida.

From their descriptions we can reconstruct the development of a megagametophyte within the megaspore from the time of the early formation of the cellular gametophytic tissue to the production of mature archegonia (Fig. 11.17E). As the megagametophyte enlarges, it causes the spore wall to rupture along the triradiate suture. At the same time large numbers of archegonia undergo development in the exposed tissues. Ten developing archegonia have been counted in one section. These same sections also show clusters of rhizoids that have been produced from the exposed surface of the megagametophyte and mature archegonia that have four tiers of neck cells with four cells per tier (Fig. 2.11D,E). Cells of the outermost tier are rounded and protrude from the surface of the megagametophyte. Galtier (1970) shows megagametophytes that protrude through the opening formed by the triradiate suture. In the venter of at least one archegonium there is a spherical cell that looks like an egg. After reviewing the characteristics of lepidodendrid megagametophytes, Brack-Hanes (1978) concludes that in their structure and development they are most like megagametophytes of *Isoetes* and not *Selaginella*.

Cellular endosporal megagametophytes also have been observed for both *Achlamydocarpon* and *Lepidocarpon* (Fig. 11.19A). Often the gametophytic tissue completely fills the cavity formed by the megaspore membrane. A single archegonium has been found by Phillips (1979), deeply sunken in the gametophyte tissue just beneath the trilete suture of the functional megaspore. The suture opened as the megagametophyte developed to expose the archegonium for fertilization.

Microgametophytes

More recent, but no less remarkable, than the description of megagametophytes is the discovery by Brack-Hanes & Vaughn (1978) of specimens showing early developmental stages of lepidodendrid microgametophytes (Fig. 2.12C–H). By using transmission and scanning electron microscopy, it was possible for the investigators to observe two-celled stages. Some cells have nuclei and what the authors claim to be division figures. Comparison of the fossil microgametophytes with those of extant lycopsids suggests that there is a similarity to those of *Selaginella*.

Embryos

With evidence of microgametophytes and megagametophytes in bisporangiate cones one might expect that embryos should be found. After making serial sections through the venters of *Flemingites*-type archegonia, Galtier (1964) discovered an undifferentiated multicellular structure that he interpreted as an embryo. This is the first report of a fossil embryo in an arborescent lycopod.

It was not until 1975 that definitive evidence of embryos in *Lepidocarpon* was brought to our attention by Phillips, Avcin, & Schopf (1975). At that time it was pointed out that many endosporal structures interpreted to be megagametophytes of *Lepidocarpon* (Andrews & Pannell, 1942) were in reality nonvascularized embryos (Fig. 11.19B). The diagnostic feature indicating the sporophytic nature of the tissue is the distinctive isoclinally folded epidermis, which increased the surface area of the developing embryo for transfer of nutrients from the surrounding megagametophyte. After absorbing and using up the gametophytic tissue, the nonvascular embryos remained ungerminated. Subsequently, Phillips (1979) found and described vascularized embryos (Fig. 11.19A) with an isoclinally folded epidermis. These embryos showed various stages of development into aerial stem and basal rhizomorphic axes.

Stubblefield & Rothwell (1981) in their studies of *Bothrodendrostrobus* found that the embryos formed within the megaspores of this genus (Fig. 11.19C) are quite different in their development from those of *Lepidocarpon* (Phillips et al., 1975). In *Lepidocarpon*, bipolar growth of the embryo is initiated early by a dichotomy of the shoot tip (Fig. 11.18D), unlike embryos of *Bothrodendrostrobus* where bipolar growth is initiated in the same way as in embryos of extant *Isoetes*. In *Isoetes* the shoot apex develops between the bases of the first leaves, while the rhizomorph apex develops between the bases of the first roots (Fig. 11.18E). These observations prompted Stubblefield and Rothwell to suggest that there are two basic types of embryo development among the Lepidodendrales. However, further evidence is presented later in this chapter that these differences in embryogeny represent alternate developmental pathways within a single lineage.

We think you will agree, as do most paleobotanists, that these are exciting discoveries. They help us immeasurably in understanding the life cycles of these intriguing plants and also give us renewed hope that we will be better able to understand development and thus evolution in other groups of extinct plants.

Evolution of megasporophylls, megasporangia, and megaspores

It has been noted many times that, depending on the species, the number of megaspores per megasporangium in *Flemingites* varies considerably. In *F. noei*, for example, there are hundreds of megaspores per megasporangium, while in *F. foliaceus* only four megaspores are present. These represent the extremes. Other species of *Flemingites* have intermediate numbers of megaspores. Demonstrated here is the recognized evolutionary tendency for the reduction of megaspore

Figure 11.19. **A.** *Lepidocarpon lomaxi*, longitudinal section showing vascularized embryo (e) embedded in tissue of megagametophyte (mg). **B.** *L. lomaxi*, transverse section of avascular embryo showing folded epidermis. (**A, B** from Phillips, 1979.) **C.** Embryo of *Bothrodendrostrobus mundus* in megaspore. Arrow points to vascular tissue. (From Stubblefield & Rothwell, 1981.) **A–C:** Pennsylvanian.

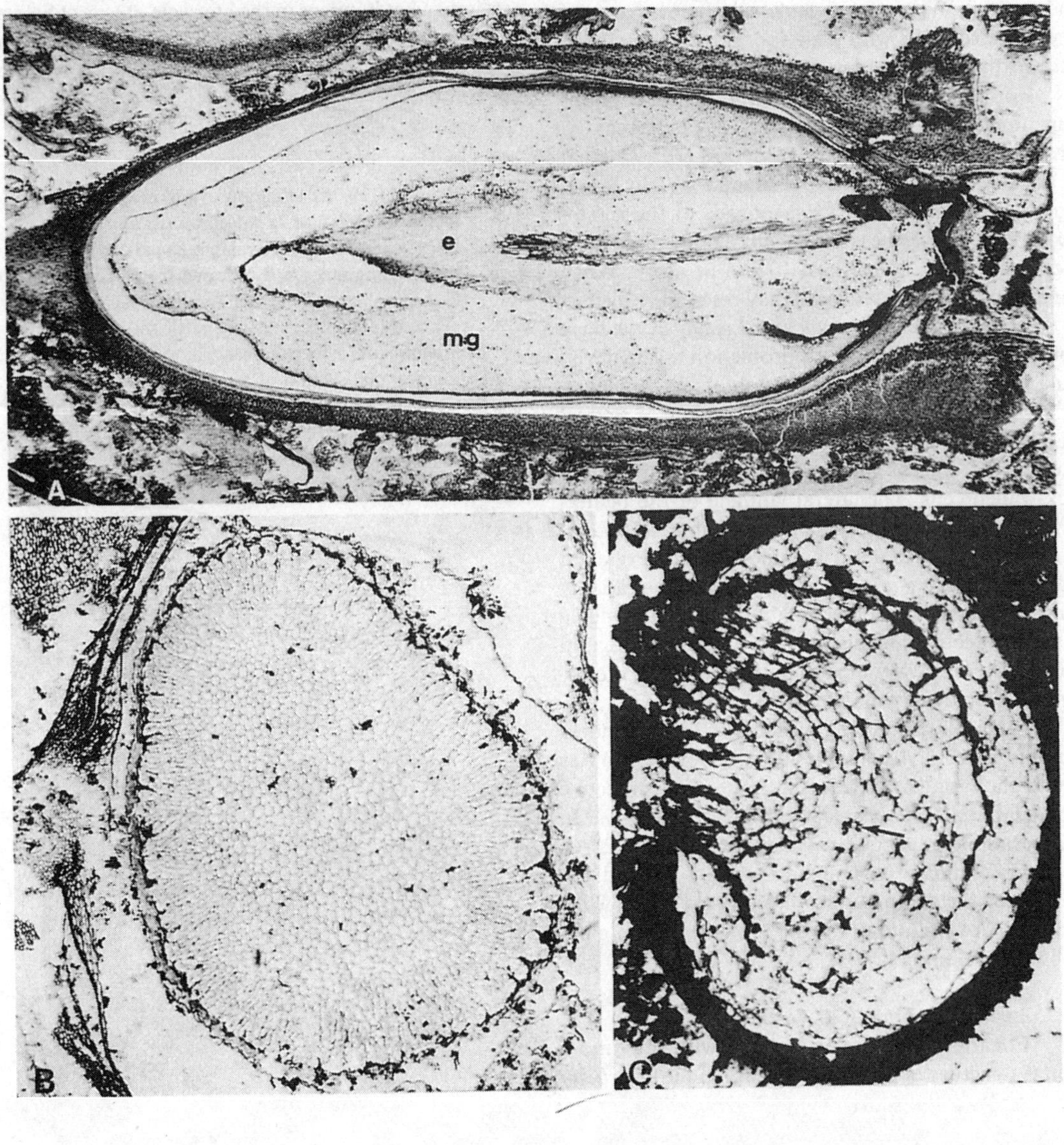

number from numerous functional megaspores to a single functional megaspore. For the Lepidodendraceae, this tendency is epitomized by the genera *Achlamydocarpon* (Schumaker-Lambry, 1966; Leisman & Phillips, 1979) and *Lepidocarpon* (Hoskins & Cross, 1941). The cones of both are monosporangiate. An *Achlamydocarpon takhtajanii* cone type has been found attached to a *Lepidodendron* stem, while *Lepidocarpon*-type cones were probably attached to *Lepidophloios*. Unlike the sporophylls of bisporangiate *Flemingites* the megasporophylls of *Achlamydocarpon* (Fig. 11.18C) and *Lepidocarpon* (Fig. 11.18B) were shed from the cones axis while it was attached to the tree. This no doubt aided in the dispersal of these bulky reproductive structures. In addition to a massive adaxial micro- or megasporangium, the sporophyll is composed of a pedicel with a more or less developed lateral lamina. The distal portion of the sporophyll is expanded and at right angles to the pedicel. This distal lamina may be 4 to 5 cm long. When found detached, these large compressed sporophylls with their sporangia are assigned to the form genus *Lepidostrobophyllum* (Fig. 11.17F).

For both genera, the megasporangium contains a functional megaspore that fills the sporangial cavity. Three abortive megaspores of the tetrad are located at the distal end of the megasporangium. When found dispersed, the functional megaspore is of the *Cystosporites* type. These, the largest of all lycopod megaspore types, may attain a length of 10 mm or more. When found in the compressed state they resemble raisins.

Both *Achlamydocarpon* and *Lepidocarpon* have evolved structures that afford protection to the developing megaspore and its gametophyte. In *Achlamydocarpon* the wall of the megasporangium is several cells thick and shows differentiation into two or three tissue layers, one or more of which is composed of thick-walled cells. The lateral laminae of the pedicel are only slightly developed. This is in contrast to *Lepidocarpon* where the lateral laminae are well developed and envelop the megasporangium except for a radially elongated adaxial slit. Functionally, the lateral laminae form a protective integument. The megasporangium wall is several cells thick, but the cells are mostly thin-walled parenchyma.

After studying large numbers of lepidodendrid cone and sporophyll compressions, Abbott (1963) was able to show stages in the evolution of the protective lateral laminae so conspicuous in *Lepidocarpon*. In one species of her genus *Lepidocarpopsis* (Fig. 11.20A), the lateral laminae extend horizontally beyond the margins of the sporangium. In another (Fig. 11.20B) they fold upward along the sides of the sporangium without enveloping it.

Lepidocarpon: Is it an ovule?

For those who are acquainted with the fundamental structure of an ovule, it is apparent that *Lepidocarpon* is well on the way, evolutionarily speaking, to fulfilling the characteristics of this structure. When we examine a transverse section through the megasporangium and pedicel of *Lepidocarpon* (Fig. 11.20C) and compare it with an ovule of the gymnosperm type (Fig. 11.20D), the similarity is apparent. Both are integumented, both have an endosporic megagametophyte with archegonia and food storage tissue, and both have what appears to be a micropyle for the entrance of the microgametophytes. This last point of similarity is more apparent than real, for in *Lepidocarpon* we know that the "micropyle" is really an elongated, radially oriented slit and not a small hole giving access to the enclosed megasporangium. Further, it has been demonstrated by Ramanujam & Stewart (1969) and Phillips (1979) that the megasporangium dehisced to expose the archegonia on the surface of the megagametophyte. This would allow flagellated sperms direct access to the female reproductive structure. This is precisely what happens in the life cycle of extant Selaginellales and Isoetales. Thus, the discovery of dehiscent megasporangia in the arborescent lycopods was not unexpected.

Figure 11.20. **A.** Diagram, transverse section of megasporophyll of *Lepidocarpopsis lanceolatus*. **B.** *L. semialata* with lateral lamina partially investing the sporangium. (**A, B** redrawn from Aboot, 1963.) **C.** *Lepidocarpon lomaxi*, diagram, transverse section of megasporophyll. "Micropyle" (m); nucellus (n); integument (i); megagametophyte (mg). **D.** Diagram, longitudinal section of a gymnosperm ovule. Compare with *Lepidocarpon* (Fig. 11.20C). Micropyle (m); nucellus (n); integument (i); megagametophyte (mg). **A–C:** Pennsylvanian.

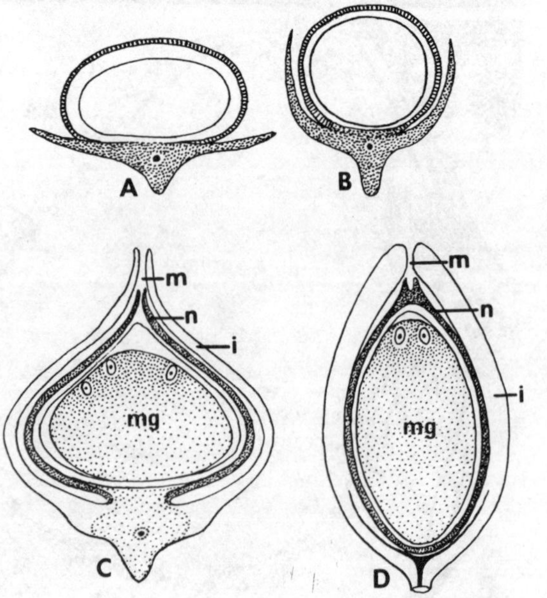

The dehiscence of the megasporangium in heterosporous lycopods is the important characteristic that distinguishes them from seed plants where the megasporangium of the ovule is indehiscent. In the latter, the egg-containing structure (archegonium in gymnosperms) is separated from the external environment by the megasporangium. In other words, there is no direct access to the archegonium by the sperms. There must be a pollen tube to convey the sperms through the megasporangium in proximity to an egg. If one considers these to be important criteria, then we can define an ovule as an integumented indehiscent megasporangium in which the single functional megaspore develops into a megagametophyte. You can see that *Lepidocarpon* meets all of the criteria of an ovule except one. Certainly the arborescent lycopods with *Lepidocarpon*-type reproductive units have come a long way toward evolving the seed habit. They did not quite make it, however, prior to their extinction by the onset of the Permian.

From the information we have at hand, we can begin to see examples of parallel evolution in both related and unrelated groups of plants. Obviously heterospory and the resulting ovulelike structures of the arborescent lycopods have evolved quite independently of the evolution of the ovule of seed plants. It is worth repeating that the protective integument around the megasporangium of the *Lepidocarpon*-type reproductive unit is foliar in origin, while the integument of seed plants appears to have been derived from a telome truss. In *Achlamydocarpon* the megasporangium has a very thick wall for protection of the female gametophyte. The point of this summary is to emphasize once more that protection of reproductive structures is accomplished among land plants, whether they be mosses or mimosas, by the evolutionary modification of available structures, and these can vary remarkably even among closely related organisms.

Sigillariaceae of the Carboniferous

Another prominent tree of the Upper Carboniferous swamp interspersed among the Lepidodendraceae was the arborescent lycopod *Sigillaria* (Fig. 11.21A). Although clearly related to the Lepidodendraceae, *Sigillaria* differs in several important ways that will become apparent.

Initially the genus *Sigillaria* was established by Brongniart in 1822 for bark fragments in which the spirally arranged leaf cushions are in vertical rows (Fig. 11.21B). Later, as reconstruction of the plant became possible, *Sigillaria* was depicted as a tree form with a main aerial axis unbranched or with one or two dichotomies at the distal end and attached at its base to *Stigmaria*-type rhizophores. Casts of some *Sigillaria* trunks are at least 30 m long and over 1 m in diameter at the slightly swollen base. They, too, were large trees.

The grasslike leaves of *Sigillaria* are, on the average, longer than those of lepidodendraceans. *Sigillaria* leaves up to 1 m in length are not uncommon. As is characteristic of lepidodendrids, *Sigillaria* leaves on developmentally older parts of the aerial axis abscised leaving leaf cushions (Fig. 11.21B). These have hexagonal, round, or oval leaf scars each with a vascular bundle scar flanked by two conspicuous parichnos scars all of which are placed at or above the midpoint

Figure 11.21. **A.** Restoration, *Sigillaria* sp. Leaves deleted in area of cones. **B.** Leaf cushions of the *S. elegans* type. Note vascular bundle scar between two parichnos scars. (Redrawn from Hirmer, 1927.) Carboniferous.

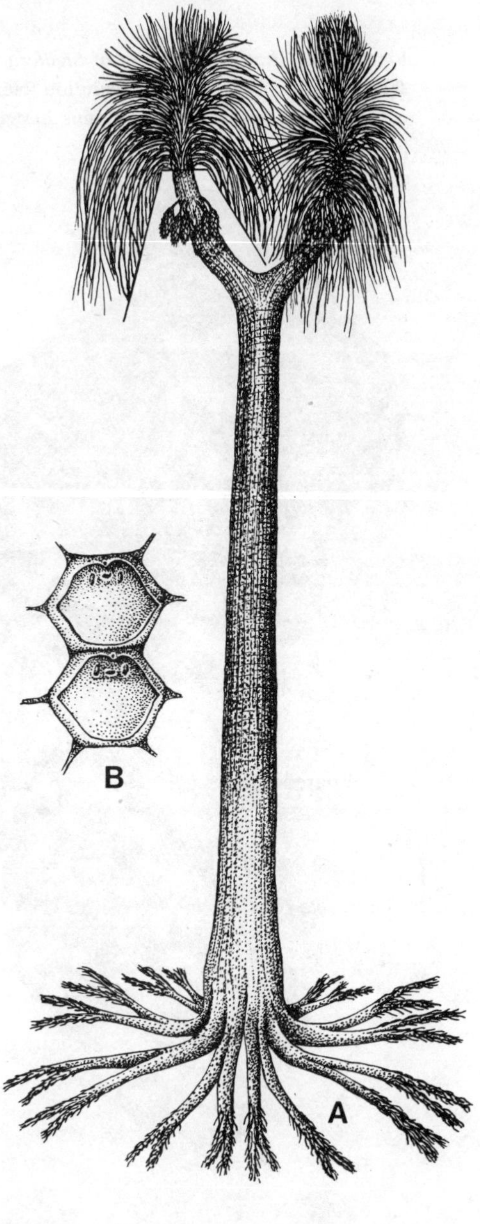

of the leaf scar. A ligule pit is situated directly above the leaf scar.

Those species of *Sigillaria* where the leaf cushions are in discrete vertical rows separated by bands of cortical tissue are placed in the subgenus *Eusigillaria*. Others lacking the separating bandlike ribs are placed in the subgenus *Subsigillaria* (Fig. 11.22A). It is now known that both stem surface configurations can occur on the same plant.

The cones of *Sigillaria*, like those of *Lepidophloios*, were borne on short lateral axes (peduncles) among the leaves (Fig. 11.21A, 11.23E). As the tree grew, it produced successive zones where cones were produced alternating with zones without cones.

Internal structure: aerial stem and rhizomorph

Until fairly recently very little was known about the internal structure of stems and rhizomorphs of *Sigillaria*. Although fragmentary permineralized specimens of *Sigillaria* have been reported from the Pennsylvanian of North America (Pigg, 1983; Feng & Rothwell, 1989), as far as we are aware, the material described by Delevoryas (1957) provides the best detail of the internal structure of *Sigillaria* aerial axes. In the permineralized stem fragments, secondary cortex, leaf bases, and vascular cylinders were found.

The beautifully preserved vascular cylinders are medullated protosteles with a large pith composd en-

Figure 11.22. **A.** Cast, fragment of *Sigillaria* of the subgenus *Subsigillaria*. **B.** Fragment of decorticated *Sigillaria* showing a *Syringodendron* configuration. Note that leaf cushions were in vertical rows and separated by bands of cortical tissue, typical of the subgenus *Eusigillaria*. **A, B:** Pennsylvanian.

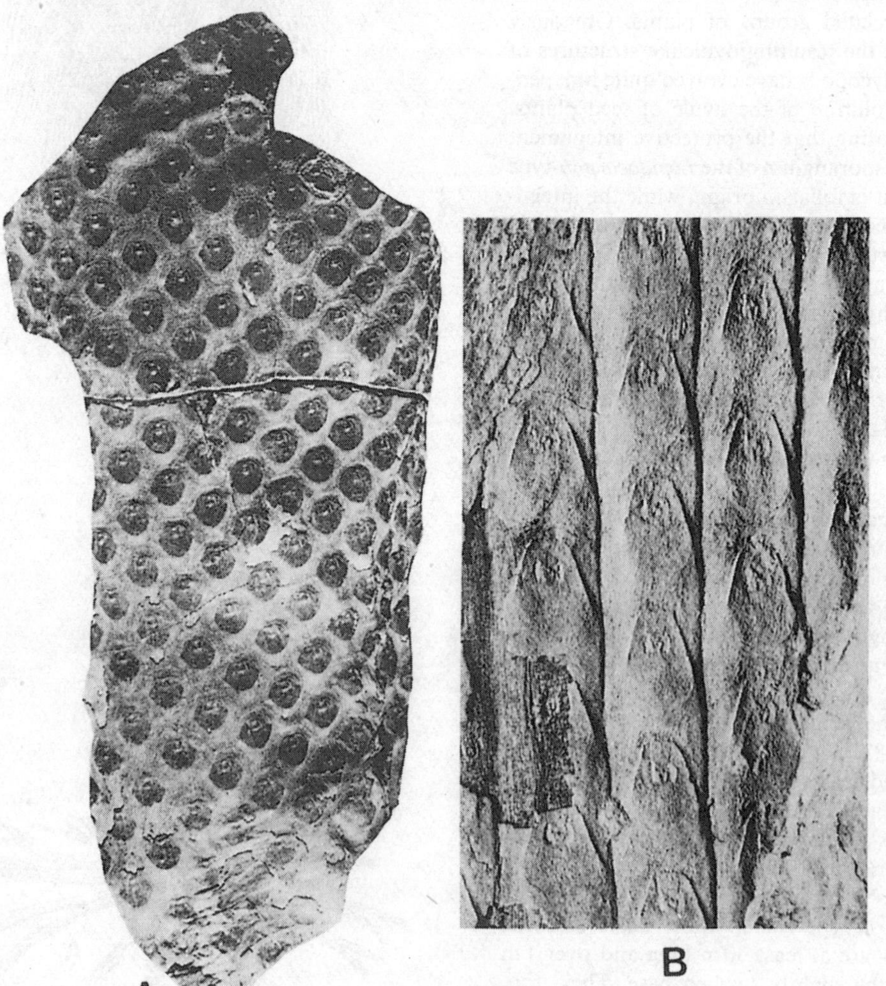

A B

tirely of parenchyma cells (Fig. 11.23B). Around the pith is a continuous cylinder of exarch primary xylem. The outer face of the primary xylem cylinder is marked by vertical ridges and furrows that, in cross section, give an undulating appearance to the corona of the primary xylem cylinder (Fig. 11.23C). The outermost protoxylem cells of the primary xylem usually have spiral secondary walls. The centripetal metaxylem tracheids have scalariform thickenings with fimbrils. Characteristics of the secondary xylem are the same as those of other lepidodendrids.

If the thickness of the secondary cortex is a criterion, then the specimens described by Delevoryas must have come from fairly old trunks. The thick

Figure 11.23. **A.** Bark of *Sigillaria approximata* sectioned to show banded appearance. **B.** Transverse section of stele of *S. approximata.* Note the pith (p) of the medullated stele with a continuous ring of primary xylem (px) surrounded by secondary xylem (sx). **C.** Enlarged portion of stele in B. Note undulating corona formed by primary xylem with departing leaf traces (lt). **D.** Tangential section of *S. approximata* bark showing dumbbell-shaped parichnos strands. **E.** Tangential section through outer bark of *S. approximata* showing anatomy and arrangement of leaf bases with intermixed cone peduncles (p). (**A–E** from Delevoryas, 1957.) All Pennsylvanian.

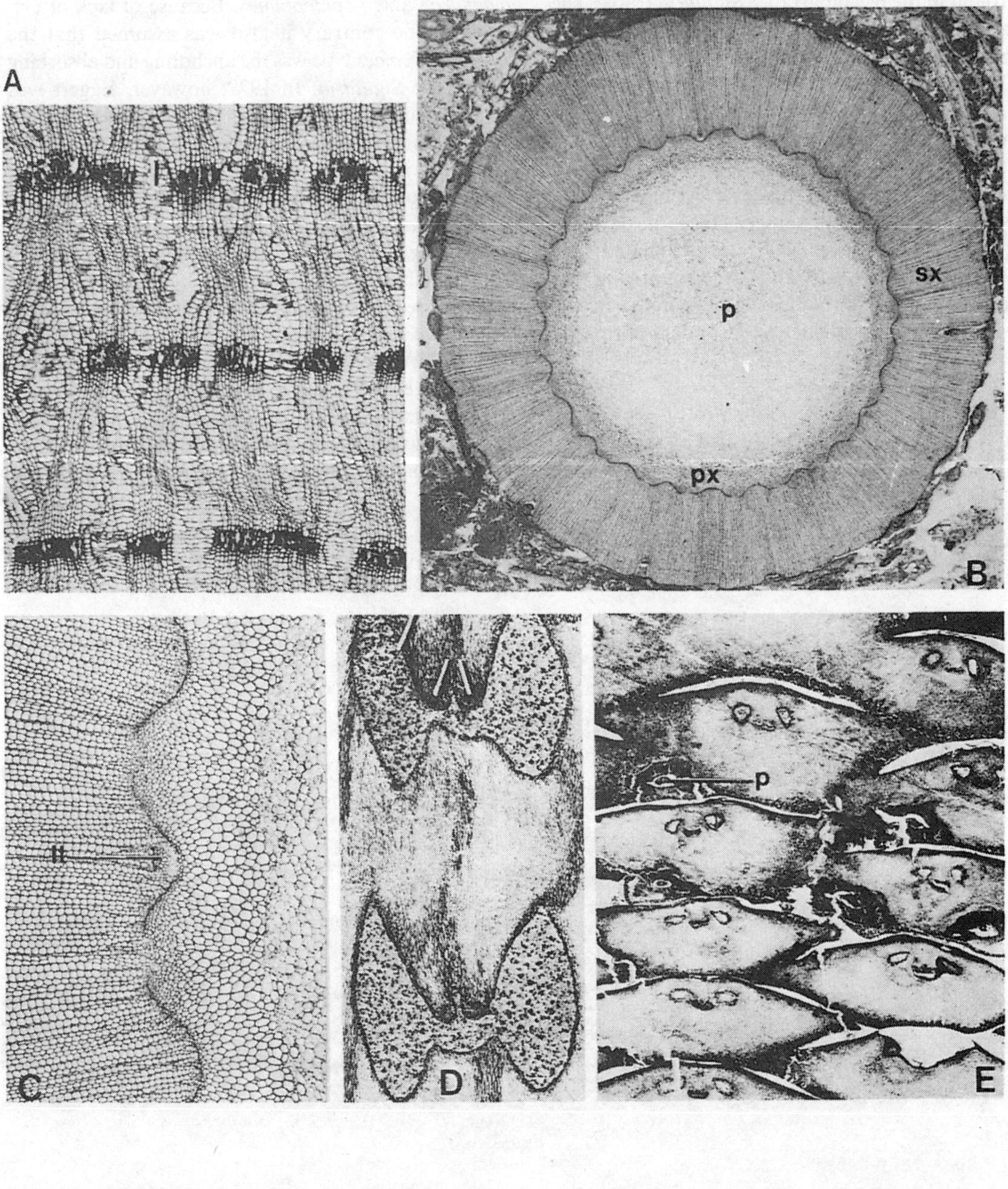

secondary cortex (Fig. 11.23A) is characterized by a banded appearance produced by zones of dark secretory tissues alternating with tissues composed of thin-walled cells that are radially aligned. Some of these show evidence of continued tangential and radial divisions. Tangential sections through the secondary cortex show vertical rows of dumbbell-shaped parichnos strands (Fig. 11.23D) that terminate as scars on the leaf cushions. When seen on decorticated compressed stems of *Sigillaria* they look like rows of "rabbit tracks" and show the *Syringodendron* configuration (Fig. 11.22B). Leaf traces are associated with the parichnos tissue.

In 1961, the same year that Eggert published his work on the ontogeny of *Lepidodendron* and *Lepidophloios*, Lemoigne described the developmental stages to be found in the basal part of a *Sigillaria* trunk. The results of his work, like those of Eggert, confirmed the occurrence of epidogenesis described by Walton (1935) long before. In his study of *Sigillaria brardii* (Fig. 11.25A), Lemoigne (1961) found that the basal portion of a mature trunk contained a protostele surrounded by a thick layer of secondary xylem. The secondary cortex was also thickest at this level. At higher levels representing successively younger stages in development, the protostele gives way to a medullated stele that increases in size distally. Accompanying the development of the pith there is a decrease in the amount of secondary xylem and secondary cortex.

Leaf traces and leaves

Leaf traces originate from protoxylem along the sides or base of the furrows formed by the primary vascular cylinder. According to Lemoigne, the trace is double at first and then fuses to form a single trace prior to entering a leaf base. In some species of *Sigillaria*, however, the trace apparently remains double both in the stem and leaf. Thus, leaves of *Sigillaria* may have one or two veins. When found detached and compressed, the long grasslike leaves showing two parallel veins are assigned to *Sigillariophyllum*. If the internal structure is preserved and sections through the leaves show two veins, then the leaf is placed in the genus *Sigillariopsis*. Unless found attached, compressed leaves of *Sigillaria* with a single vein are assigned to the form genus *Cyperites*.

Rhizomorphs and roots

It was suspected for many years that *Stigmaria ficoides* represented the rhizomorphic system of *Lepidodendron* and *Lepidophloios*. Because of lack of evidence to the contrary it also was assumed that the *Stigmaria ficoides* type was the anchoring and absorbing system for *Sigillaria*. In 1972, however, Eggert was able to demonstrate that the *Stigmaria* attached to *Sigillaria approximata*-type stems (Delevoryas, 1957) was not *S. ficoides*. As Eggert points out, at localities where the only arborescent lycopod stem remains are of the *Lepidodendron–Lepidophloios* type, the only detached *Stigmaria* present is *S. ficoides*. At the locality from which Eggert's specimens were obtained, the only arborescent lycopod stem remains were of *Sigillaria approximata* and the associated *Stigmaria* axes (Fig. 11.24A) were distinctly different from *S. ficoides*. They differed in having a small pith, distinct primary xylem, and a secondary cortex similar to that of *Sigillaria approximata* where there are several bands of secretory tissues. The rays in the secondary wood are

Figure 11.24. **A.** Transverse section stele of *Stigmaria* of the sigillarian type. **B.** *Stigmaria* roots of the sigillarian type. Note well-defined connective (c). (**A, B** from Eggert, 1972.) **A, B:** Pennsylvanian.

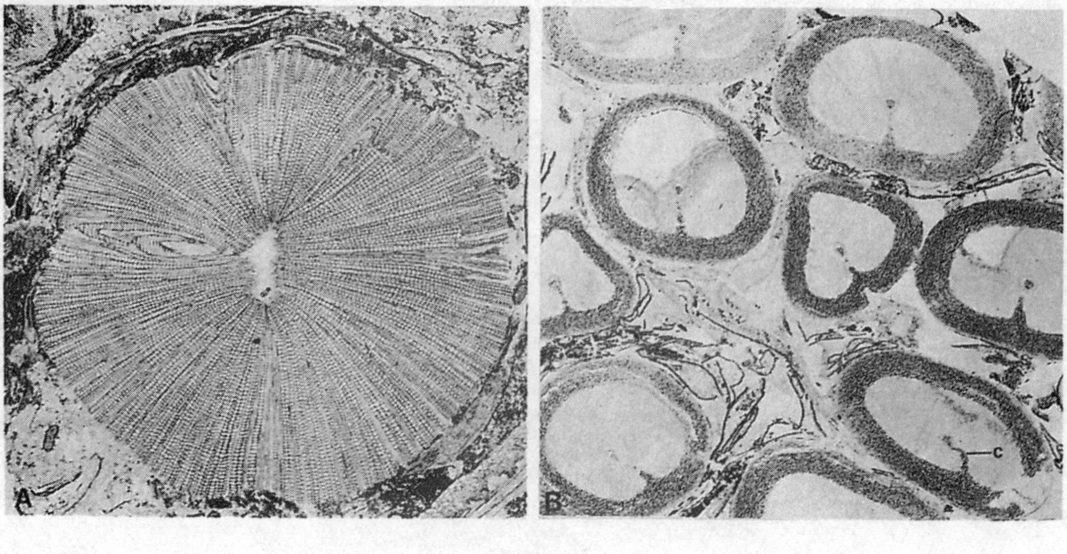

distinctive, being relatively wider, shorter, and less abundant than those of *Stigmaria ficoides*. The free roots (Fig. 11.24B) are generally smaller than those of *S. ficoides* and always have a distinct connective with traces of vascular tissue. Roots of this type were placed in the species *S. bacupensis* by Leclercq (1930). Finally, actual attachment of the new *Stigmaria* to *Sigillaria approximata* was demonstrated. As far as we are aware, this is the first and only time such connection has been discovered in permineralized specimens.

Reproductive morphology

Perhaps the best known of all cone types attributed to *Sigillaria* is *Mazocarpon oedipternum* (Fig. 11.25C) described by J. M. Schopf (1941) from the Upper Pennsylvanian of Illinois. In his researches on *Sigillaria approximata,* Delevoryas (1957) was able to demonstrate that this species of *Mazocarpon* was the cone type of *S. approximata.* He observed bases of peduncles among the leaf cushions on the stem with an internal anatomy exactly the same as the anatomy of the base of the cone axis. The distribution of the peduncles, which are restricted to zones along the stem, provided additional evidence for the lateral position of cones on the stem of *Sigillaria.*

Mazocarpon is known to be monosporangiate, i.e., with separate microsporangiate and megasporangiate cones. This is similar to the condition for *Lepidocarpon* and *Achlamydocarpon,* but differs from *Flemingites* with its bisporangiate cones. The cones of *Mazocarpon* are 10 cm or more in length and about 12 mm in diameter. The sporophylls are attached in whorls or in a low spiral. It appears that the entire cone was shed at maturity, unlike *Lepidocarpon* and *Achlamydocarpon* which shed their sporophylls.

Individual sporophylls of *Mazocarpon oedipternum* (Fig. 11.25B) are somewhat boat-shaped. The pedicel has a slightly expanded lateral lamina; the distal lamina is short and barely covers the distal end of the sporangium. A distinctive feature of the sporophyll is the conspicuous heel that projects downward from the outer end of the pedicel.

Each *Mazocarpon oedipternum* megasporangium (Fig. 11.25C) contains eight dish-shaped megaspores arranged around a columellate dome of tissue that extends into the sporangial cavity from the base of the sporangium. Twelve megaspores are found in the megasporangium of *M. pettycurense*. The megaspores have their concave surfaces so tightly appressed to the tissue of the columella that parts of it adhere to the spores after they are released by decay of the surrounding tissues. The individual megaspores of *M. oedipternum* are up to 2 mm in diameter. They are easily recognized in the dispersed state as the *Laevigatisporites glabratus* type. The columella in the microsporangia is poorly developed; otherwise they have the same characteristics as megasporangia. The microspores, examined in situ, are similar to dispersed spores of the *Crassispora kosankei* type.

Megagametophytes of *M. oedipternum*, each with a single archegonium, were reported by Schopf (1941) in eight different megaspores. Structures resembling embryos in the 8- to 12-celled stage were also observed. Similar structures in *M. bensonii* have since been recognized as archegonial neck cells (Pigg, 1983).

Monosporangiate cones known from compressions, but having the same general characteristics as *Mazocarpon*, are assigned to the genus *Sigillariostrobus*. These cones have been found with peduncles bearing *Sigillaria*-type leaf cushions. Other than being monosporangiate, *Sigillariostrobus* differs from *Flemingites* in having sporophylls with a distal lamina that is acicular (sharply pointed). These are quite distinct from the more obtusely pointed distal laminae of

Figure 11.25. **A.** Diagram, vascular system of a sigillarian. Transverse sections a–e correspond to levels a–e indicated on reconstruction. (Redrawn from Lemoigne, 1966.) **B.** Single megasporophyll of *Mazocarpon oedipternum*. (Redrawn from Phillips, 1979.) **C.** Megasporophylls of *M. oedipternum* sectioned to show arrangement of megaspores around the columellate dome (c). (Redrawn from Schopf, 1941.) All Carboniferous.

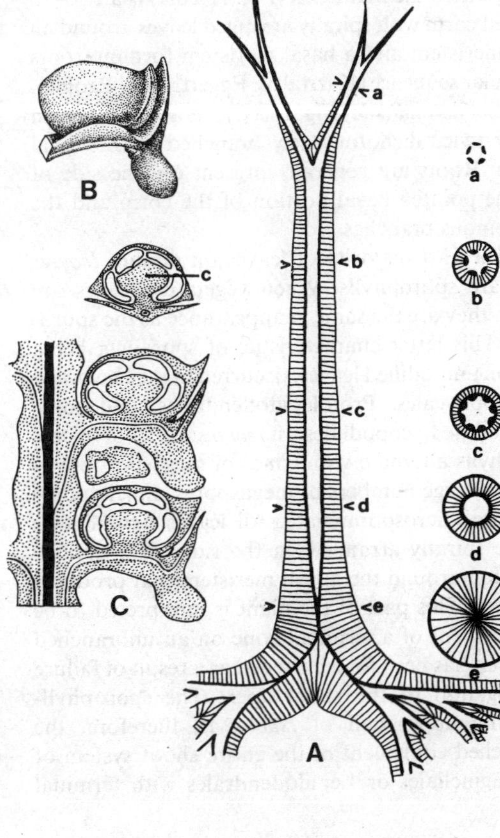

Flemingites sporophylls. Often there is evidence of two veins in the *Sigillariostrobus* sporophyll.

Origin of the Isoetales

In our presentation of fossil Lycopsida it was necessary for obvious reasons to limit the discussion to only a few genera and species. The motivations regulating our selection were twofold: (1) to present information about those extinct lycopods that are most thoroughly understood, and (2) to provide students with the basic information required to comprehend the evolutionary origins of extant plants.

In this latter context, we need to turn our attention to the origin of the extant Isoetales with a single genus, *Isoetes* (Fig. 10.1B,C). Paleobotanists are generally agreed that *Isoetes* represents the last living remnant of a long line whose origin is among the Upper Devonian or Lower Carboniferous lycopods, a line that probably had its origin among an ancestral group that gave rise to the Lycopodiales, Selaginellales, and Lepidodendrales. (See Chapter 10.)

To understand the evidence pro and con for this interpretation it is necessary to know something of the salient characteristics of *Isoetes*. It is a relatively small herbaceous plant that grows in marshy or aquatic environments and is widespread in parts of the world having cool climates. In its general appearance *Isoetes* looks like a sedge and is often overlooked by inexperienced collectors. The genus exhibits a bipolar growth pattern where the main axis (Fig. 11.26A) is a two- to six-lobed corm with spirally arranged leaves around an apical meristem and a basal meristem forming roots in a regular sequence (Karrfalt & Eggert, 1978; Paolillo, 1982). *Isoetes andicola* (Fig. 10.1C) has a pointed corm once or twice dichotomously branched in the distal portion. Roots are generally present on one side of both the pointed basal portion of the corm and the dichotomous branches.

The vast majority of leaves on mature *Isoetes* plants are sporophylls. When vegetative leaves are present, they are the same in appearance as the sporophylls. This latter characteristic, of sporangia being borne on unmodified leaves, occurred in the Devonian Drepanophycales, Protolepidodendrales, and some species of the Lycopodiales. In *Isoetes*, zones of microsporophylls alternate with zones of megasporophylls that have large numbers of megaspores. Inner sporophylls are microsporangiate. All leaves have ligules and are spirally arranged on the sides of a cuplike depression around the apical meristem that produces the leaves. This part of the plant is interpreted to be the equivalent of a terminal cone on an unbranched shoot that has been foreshortened as a result of failure of elongation of the shoot apex. The sporophyll-producing upper half of *Isoetes* is, therefore, the unbranched equivalent of the entire shoot system of the Selaginellales or Lepidodendrales with terminal bisporangiate cones.

Examining the base of the *Isoetes* corm (Fig. 11.26A), we see free roots arranged in a regular sequence, with the youngest roots appearing along a central furrow that divides the base of the corm into two lobes. Furrow development in *Isoetes* has been studied by Karrfalt & Eggert (1977a), and we now understand the origin of multilobed forms. In three-lobed species the furrow branches at one end to become triradiate; in four-lobed ones the furrow forks at both ends to produce a furrow system that consists of a central segment from each end of which two peripheral furrows extend. *Isoetes* with larger number of basal lobes have additional forks at the periphery of the furrow system, with each peripheral furrow separating two of the lobes (Jennings, Karrfalt, & Rothwell, 1983). In cross section, the free roots have an excentrically positioned monarch collateral bundle surrounded by an inner cortex. In orientation of the bundle, general configuration, development, and arrangement, roots of *Isoetes* are like those of *Stigmaria* (Stewart, 1947). This similarity is so striking that it was noticed many decades ago. Consequently, the relationship of *Isoetes* to Carboniferous arborescent lycopods was proposed and has been debated ever since. It has been repeatedly suggested that the basal meristem of the *Isoetes* corm represents the highly reduced, foreshortened rhizomorph of *Stigmaria*-type axes. This is an old interpretation questioned by Paollilo (1982), but supported by the work of Karrfalt & Eggert (1977a, 1977b, 1978), Karrfalt (1982, 1984b), Jennings, Karrfalt, & Rothwell (1983), Rothwell, (1984), and Rothwell & Erwin (1985).

An examination of the internal structure of the corm of *Isoetes* reveals some additional enigmas. Longitudinal sections (Fig. 11.26A) show that there is a protostele, the upper part of which is radially symmetrical and derived, along with leaves and their traces, from the apical meristem. The lower portion is bilaterally symmetrical, two to six lobed, and added to by the basal meristem from which roots and their root traces arise. A layer of tissue with radially arranged cells surrounds the xylem of the protostele. This is called the "prismatic" layer, and its formation is related to a most unusual lateral meristem. For many years the lateral meristem of *Isoetes* has been described as an anomalous cambium that forms cells of the prismatic layer on the inside and secondary cortical cells on the outside. After a detailed study of the meristematic regions in *Isoetes* corms, Karrfalt (1984b) has concluded that its lateral meristem is analogous to the primary thickening meristem found in the stems of many monocotyledonous angiosperms. Here a discrete meristem cuts off files of cells that are often radially arranged as are the cells of the prismatic layer of *Isoetes*. The prismatic layer is comprised of parenchyma cells that occur in layers alternating with cells that have the characteristics of phloem and xylem. The

Figure 11.26. **A.** Stereodiagram of the corm of *Isoetes*. Plane a at right angles to Plane b. Primary xylem (Prim. x); primary stele (Prim. st); prismatic zone (Pris. 2nd). (From Stewart, 1947.) **B.** *Pleuromeia sternbergi*. Triassic. Reconstruction. (After Hirmer, 1933, in Andrews, 1961.) **C.** *Nathorstiana arborea*. Lower Cretaceous. Reconstruction. (After Mägdefrau, 1957, in Boureau, 1967.)

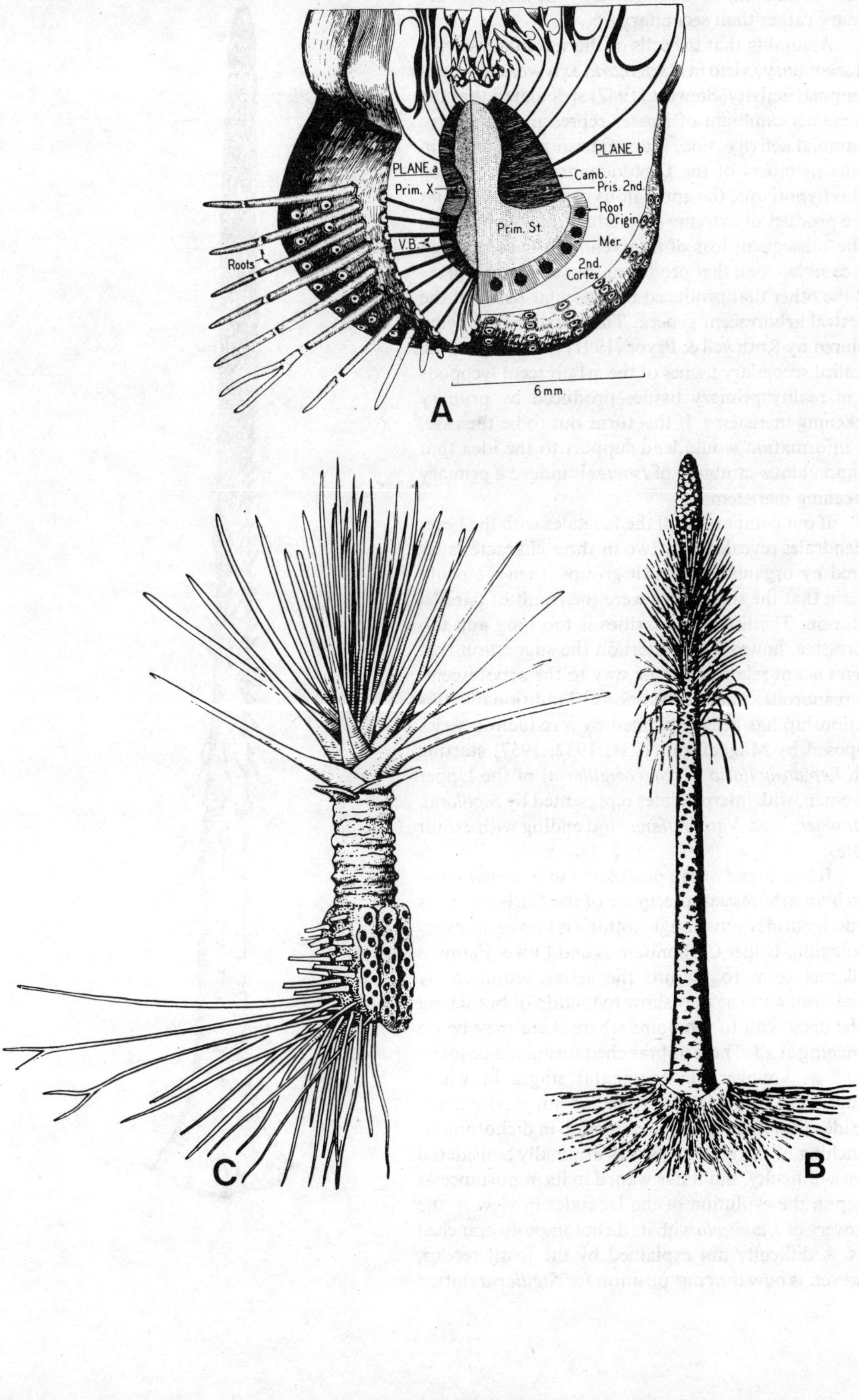

bulk of the food-storing, lobed corm of *Isoetes* is produced by the primary thickening meristem. The latter forms new cortical cells, which continue to divide to the exterior of the meristem. According to this interpretation, all of the tissues formed by the meristem are primary rather than secondary.

Assuming that the cells of the secondary cortex and secondary xylem in a *Stigmaria* axis were products of cambial activity, Stewart (1947) speculated that the anomalous cambium of *Isoetes* represented a vestige of cambial activity, since it appears so prominently in extinct members of the Lepidodendrales. According to this hypothesis, the anomalous cambium of *Isoetes* is the product of extreme size reduction which results in the subsequent loss of intervening tissues between two cambia – one that produced the secondary cortex and the other that produced the vascular tissue of the ancestral arborescent genera. The possibility is being explored by Rothwell & Pryor (1991) that some of the so-called secondary tissues of the arborescent lycopods are in reality primary tissues produced by primary thickening meristems. If this turns out to be the case, this information would lend support to the idea that the anomalous cambium of *Isoetes* is indeed a primary thickening meristem.

If our comparison of the Isoetales with the Lepidodendrales revealed only two or three characteristics shared by organisms of both groups, then we could explain that the similarities were the result of parallel evolution. The list of similarities is too long and too impressive, however, to entertain the suggestion that *Isoetes* is not related in some way to the arborescent, heterosporous, ligulate lycopods. Traditionally this relationship has been explained by a reduction series proposed by Mägdefrau, (1931; 1932; 1957) starting with *Lepidosigillaria* (*Archaeosigillaria*) of the Upper Devonian, with intermediates represented by *Sigillaria, Pleuromeia,* and *Nathorstiana,* and ending with extant *Isoetes*.

It has already been postulated that *Lepidosigillaria* is an arborescent precursor of the Carboniferous lepidodendrids even though nothing is known of its reproduction. Upper Carboniferous and Lower Permian sigillarias seem to fit into the series proposed by Mägdefrau because they show reduction of branching at the distal end to the point where there may be no branching at all. These unbranched forms can be interpreted as younger developmental stages in which branching of the apical meristem had not yet occurred. The idea that there has been reduction in dichotomous branching of the aerial axis was originally considered to be a difficulty, but it has waned in its importance as a step in the evolution of the Isoetales in view of the discovery of *I. andicola* with its dichotomously branched axes. A difficulty not explained by the fossil record, however, is how the cone position for *Sigillaria* shifted

Figure 11.27. Reconstruction of *Chaloneria* approximately 1 m high. Note the slightly lobed cormose base. Pennsylvanian. (Redrawn from Pigg & Rothwell, 1983a.)

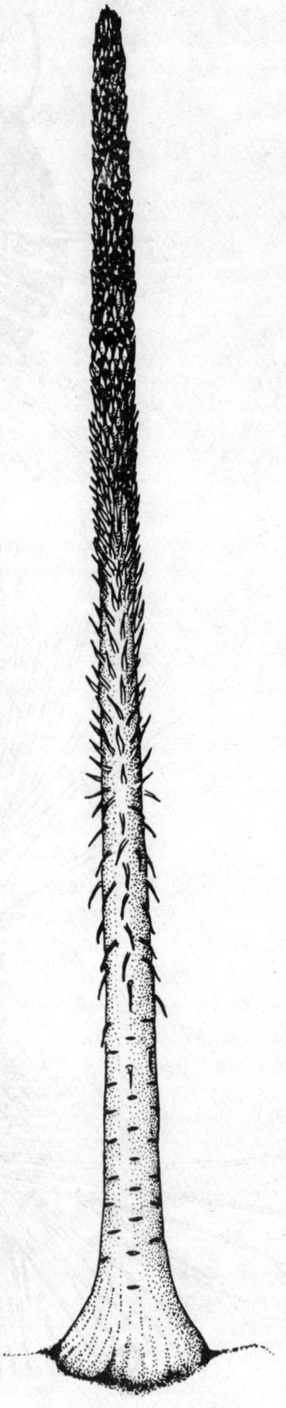

from the lateral position to the terminal one found in *Isoetes*.

Chaloneria

Although not included in the series proposed by Mägdefrau, the more recently described genus *Chaloneria* (Pigg & Rothwell, 1979; 1983a; 1983b) has helped considerably in our understanding of the clade comprised of the rhizomorphic lycopsids. This Upper Pennsylvanian plant was found in coal balls in the Appalachian Basin. In gross morphology *Chaloneria* (Fig. 11.27) appears as an unbranched plant up to 2 m high with a stem as much as 10 cm in diameter. The stem is clothed in small, spirally arranged leaves with ligules. When they were broken from the stem, they left irregular leaf scars unlike most lepidodendrids where conspicuous leaf cushions are produced. Also unlike the Lepidodendrales, the base of *Chaloneria* is a rounded, slightly lobed cormose structure. Vascular traces found in the cortical tissues of the base suggest that the roots of the *Stigmaria–Isoetes* type were attached. It is clear that the rooting structure of *Chaloneria* is quite different from the extensive, dichotomously branched rhizomorph of *Stigmaria*. Within the proximal portion of the stem there is an exarch protostele, which at higher levels becomes medullated. Secondary wood is abundant in the basal portions of the stem. This decreases in amount toward the apex. The distal vegetative portion of the stem gives way to fertile regions (Fig. 11.27) where leaflike sporophylls become smaller, more abundant, and more condensed toward the stem apex.

Compression fossils of fertile portions of the *Chaloneria* type are assigned to *Polysporia*. Permineralized fertile specimens of *Chaloneria periodica* (DiMichele et al., 1979) have zones of vegetative leaves alternating with sporophylls (Fig. 11.28). Megasporophylls with sporangia containing megaspores of the *Valvisisporites* dispersed type were described by Pigg & Rothwell (1983b). Some megaspores show various stages in development of megagametophytes that are highly similar to those of *Isoetes*. The microspores of *Chaloneria* are of the *Endosporites* dispersed type.

In many ways the Upper Pennsylvanian *Chaloneria* seems to bridge the gap between the Upper Devonian treelike lycopsids with their cormose rhizomorphs and the cormose Triassic *Pleuromeia*.

The terminal position of the cone is a conspicuous feature of the next genus in the series, *Pleuromeia* (Fig. 11.26B). Remains of this plant occur as casts and compressions obtained from rocks of Triassic age of Australia, Europe, South America, Japan, and Russia (Retallack, 1975). In general appearance *Pleauromeia* looks like *Chaloneria*, about 2 m high. It bore spirally arranged linear leaves that abscised, leaving leaf cushions exposed on the lower parts of the trunk. The base of the trunk is a four-lobed structure, clearly the product of furrow branching like that of *Isoetes* (Jennings, Karrfalt, & Rothwell, 1983). On the surface of the lobes there are curved rows of scars that mark the position of abscised roots. Attached roots have been found and are known to dichotomize.

A single cone of the detached *Cyclostrobus* type was borne terminally on the main axes of *Pleuromeia*. The cones are bisporangiate with spirally arranged megasporophylls borne on the lower part and microsporophylls above. The individual sporophylls are dorsiventrally flattened, each with a sporangium that covers most of the adaxial surface.

Nathorstiana (Fig. 11.26C), the next organism in Mägdefrau's series, is an *Isoetes*-like plant from the Lower Cretaceous. The plant, with its tuft of sedgelike leaves, was about 10 cm high. There has been no success in distinguishing sporophylls. The basal portion of the stem axis is a rhizomorph with approximately 12 vertical ribs. The surfaces of the ribs have root scars with a rhizotaxy similar to that of *Paurodendron fraipontii*.

After its proposal, this series was widely accepted as the best documented example of evolutionary reduction known to have occurred in the plant kingdom. As a result of new discoveries, however, questions have been raised that challenge what seemed to be a well-established series. The logic of the stratigraphic sequence reflected by the series was first questioned by Bock (1962), who reported that plants similar to extant *Isoetes*, placed in the genus *Isoetites*, occurred in beds of Triassic age. Thus, direct derivation of *Isoetes* from Cretaceous-age *Nathorstiana* through Triassic *Pleuromeia* was put into question.

Figure 11.28. Portion of *Chaloneria* aerial axis showing spirally arranged leaves alternating with sporophylls. Pennsylvanian. (From DiMichele, Mahaffy, & Phillips, 1979.)

Paurodendron

Further insight into the relationships among lycophytes of the isoetalean clade has been obtained from the study of *Paurodendron* (Fig. 11.29). One way we can learn about the structure of this diminutive plant is by following the chronology of its reconstruction from the time the first organ was described to the time a whole plant emerged. As far as we are aware, the investigator who first described a part of *Paurodendron* was Dr. S. Leclercq (1924). She observed cross sections of petrifaction fossils containing small axes (Fig. 11.29C) with a conspicuous exarch actinostele having 5 to 10 radiating protoxylem points. This, plus the fact that the cells of the cortex were thick-walled, suggested to Dr. Leclercq that this was the stem of an ancient fern, *Botryopteris*. She named the new fossil *B. fraipontii*. Similar material was described by Darrah (1941) under the name *B. radiata*. In 1954, Fry published a comprehensive study of the "fernlike" stems of the *B. fraipontii* type. During his investigation he discovered new characteristics of the stems that convinced him that these were axes of ancient lycopods and not ferns at all. He found delicate linear microphylls spirally arranged on the stems. The leaves bore axillary ligules, and the metaxylem cells were scalariform and had fimbrils (Cichan, Taylor, & Smoot, 1981). (Fimbrils are striations that extend across the apertures between the scalariform thickenings of the tracheids). These were thought to be characteristic of the tracheids of arborescent lycopods. Realizing that the stems were not those of ferns, Fry established the genus *Paurodendron* for these plant organs, and correctly retained the specific epithet *fraipontii*. Thus the name *Paurodendron fraipontii* was established in 1954.

Twelve years later, in 1966, Phillips & Leisman found the stems of *P. fraipontii* attached to a root-bearing rhizomorph (Fig.11.29A), which they thought were similar to several extant species of *Selaginella*. Examination of the petrified rhizomorphs showed a protostele surrounded by secondary xylem, the product of a vascular cambium. The radial alignment of cells in the outer cortex of the rhizomorph suggested that

Figure 11.29. **A.** *Paurodendron fraipontii,* restoration of root-bearing rhizomorph and shoot. **B.** *P. fraipontii,* transverse section of root. Note monarch collateral vascular bundle. **A, D:** Middle Pennsylvanian. (**A** from Phillips & Leisman, 1966; **B** drawn from photograph from Phillips & Leisman, 1966) **C.** Transverse section of shoot axis of *P. fraipontii.* Pennsylvanian. (Drawn from photograph in Fry, 1954.)

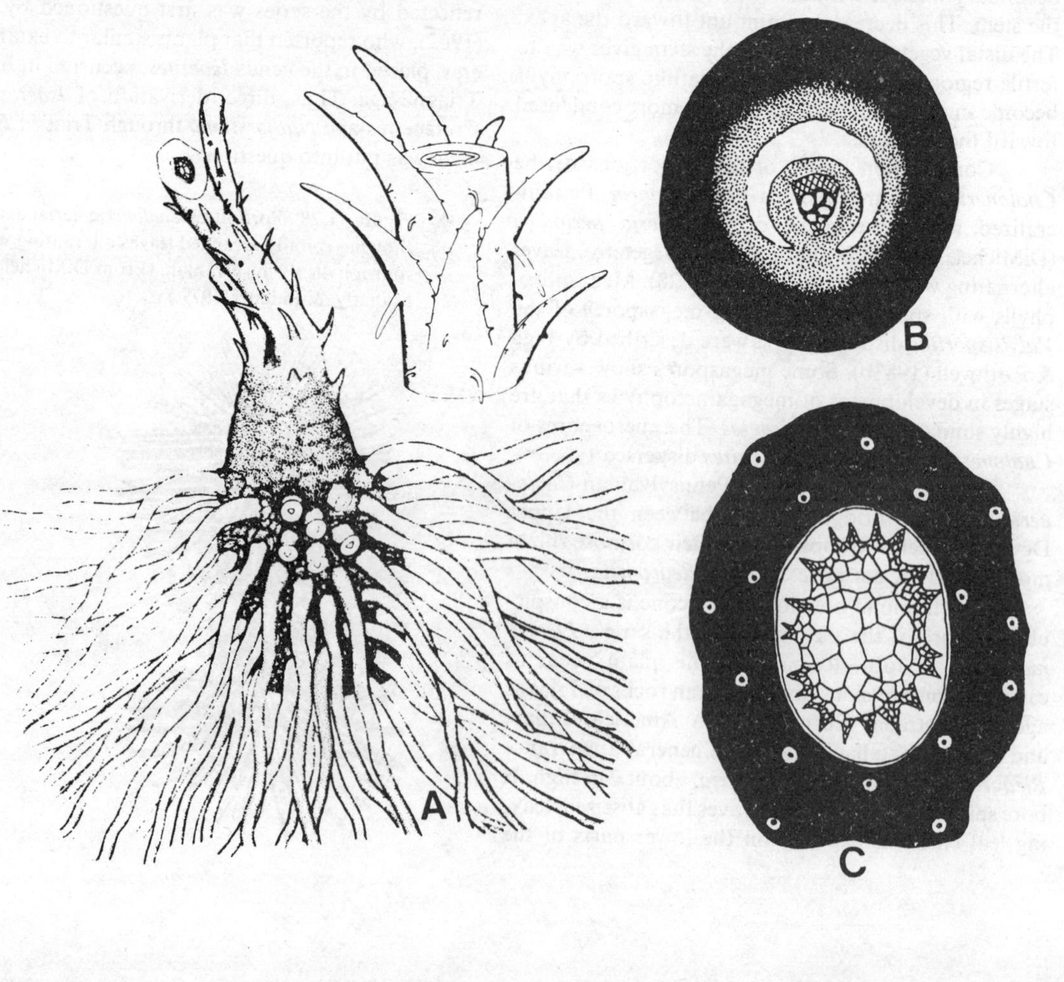

it too was the product of cambial activity. Aerial branches having characteristics of *Paurodendron* (Fry, 1954) were found attached to the upper end of the erect rhizomorph.

The roots borne by the rhizomorph were spirally arranged, a rather unusual root arrangement found in several extinct arborescent lycopods. The roots themselves are different in structure when compared to those of most plants. Instead of a vascular system that is a bilaterally or radially symmetrical protostele, Phillips & Leisman found a monarch collateral bundle centrally placed in each root (Fig. 11.29B). As you will recall this un-rootlike characteristic is also present in *Isoetes* and extinct arborescent lycopods.

Still working with permineralized specimens (coal balls) from the Middle Pennsylvanian, Schlanker & Leisman (1969) were able to demonstrate a connection between the Carboniferous lycopod cone type *Selaginellites crassicinctus,* which is heterosporous and ligulate, and *Paurodendron fraipontii.* In addition to actual attachment, the authors were able to infer connection because of histological similarity of the vascular systems of the two parts. Because the entire plant as reconstructed by Schlanker & Leisman was similar to a *Selaginella,* it was their opinion that *Paurodendron* should be transferred to that genus. The story of the reconstruction was continued by Rothwell & Erwin (1985) when they described a specimen of the rooting organ of *Paurodendron* (= *Selaginella*) *fraipontii* which confirmed what had been suspected (Karrfalt, 1981; 1984b), i.e., that the rooting organs of *Paurodendron* were fundamentally different in their development from the adventitious roots of *Selaginella.* This suggested that the relationships of *Paurodendron* were with the Isoetales and not the Selaginellales. For this reason, the generic name *Paurodendron* as established by Fry (1954) has been retained.

This rather detailed historical account of the reconstruction of *P. fraipontii* is presented for several reasons, not the least of which is to reveal the structure of the most completely known early members of the Isoetales. Another reason is to guide you through the intricacies of paleobotanical nomenclature that are required in the reconstruction of whole plants. Perhaps the most important reason for this presentation is, however, to emphasize again the time that may be required by paleobotanists to accomplish the reconstruction of a single species of extinct plants. The elapsed time from Leclercq's 1924 paper to Rothwell & Erwin's 1985 paper is 61 years!

Oxroadia

A new genus, *Oxroadia,* having many of the vegetative and reproductive features of *Paurodendron,* was first described by Alvin (1965) from the Lower Carboniferous of Scotland. Subsequent studies (Long, 1971; Bateman, 1988) not only confirm the relationship between these two genera, but give additional insight into the evolution of rootstocks in the isoetalean clade. With attention to the rootstock (rhizomorph) of *Oxroadia,* Bateman has shown that it is a dichotomously branched structure attached to the base of the stem. The vascular architecture indicates that the rhizomorph underwent one to three dichotomies in successive planes at right angles. The rhizomorph is nearly radially symmetrical and is similar to the rhizomorph of *Paurodendron* in its anatomy. Relatively large roots with monarch traces are emitted from the base of the stem and the lobed rhizomorph.

It has been argued that *Oxroadia* rhizomorphs resemble a compact, diminutive *Stigmaria* and it is an intermediate between *Stigmaria* and *Isoetes.* In view of new interpretations by Rothwell & Erwin (1985), who believe that the radially symmetrical rhizomorph of *Paurodendron* is homologous with that of *Stigmaria,* it now is tempting to conclude that the radially symmetrical *Paurodendron*-like rhozomorphs of *Oxroadia* with dichotomous branching represent a stage in a rhizomorphic reduction series. In this series the lepidodendraleans with their extensive, dichotomously branched rhizomorphs with radially arranged rootstocks are thought to be primitive. *Oxroadia* occupies an intermediate position in which the rhizomorph is reduced but has dichotomously branched rootstocks that show internal radial arrangement. Further reduction is evidenced by *Paurodendron* with its unbranched rhizomorphic rootstock.

In the minds of many investigators, the magnitude of evolutionary reduction implied by the derivation of Isoetales from Lepidodendrales seems insuperable. This applies especially to the reduction of stigmarian axes into the basal rhizomorph of *Isoetes.* Based on his observations of *Protostigmaria* (Fig. 11.29), Jennings (1975) advanced the theory that the evolutionary line terminating in extant Isoetales may have originated with Upper Devonian arborescent lycopods where the basal rhizomorph is not developed into horizontally extended stigmarian rhizomorphs. For example, the base of *Lepidosigillaria* is depicted as a rounded cormlike structure with roots, and that of *Cyclostigma* as a rhizomorph with short blunt arms bearing their roots.

In a more detailed description of the Lower Mississippian *Protostigmaria,* Jennings, Karrfalt, & Rothwell (1983) determined that it represents the lobed and furrowed rhizomorph of the branched lycopod tree, *Lepidodendropsis.* The smallest rhizomorph has two lobes (Fig. 11.30) that are separated by a linear furrow as is characteristic of two-lobed *Isoetes* plants. Larger specimens are up to 40 × 28 cm, with as many as 13 lobes that are separated by a system of linear furrows that fork successively toward the periphery as do multilobed *Isoetes* plants. Rows of root scars like those of *Stigmaria* extend between the furrows of *Pro-*

tostigmaria in the same way that rows of roots extend away from the furrows of *Isoetes*. If the *Isoetes*-type rhizomorph originated from an arborescent ancestor with lobed root-bearing rhizomorphs, then it becomes unnecessary to account for its origin by reduction of a stigmarian system. However, such an origin is complicated by the fact that the *Isoetes*-like rhizomorphs of *Lepidosigillaria* and *Protostigmaria* from the Upper Devonian and Lower Mississippian both predate the earliest fossil record of *Stigmaria*, which appears later in the Mississippian.

The discovery of a *Paurodendron* rhizomorph with an apex (Rothwell & Erwin, 1985) further clarified the homologies between the genus *Stigmaria* and other rhizomorph-producing lycopods (Karrfalt, 1984a). The authors clearly show that the apex of the *Paurodendron* rhizomorph apex (Fig. 11.31) is equivalent in structure to the apex of *Stigmaria*. In addition to reflecting the radial symmetry of both organs with their spirally arranged roots, the apex of the *Paurodendron* rhizomorph, like that of *Stigmaria*, is covered by an apical plug circumscribed by a circular groove and a parenchymatous rim from which young rootlets of the *Stigmaria* type arise. From these and other observations, it appears that the rooting structures of *Paurodendron, Oxroadia, and Stigmaria* are homologous. It is further postulated that the circular groove of the *Paurodendron* rooting structure with its rootlet-forming rim is the homologue of the rootlet-producing, linear groove found in the bilaterally symmetrical, often-lobed corms of *Cyclostigma, Protostigmaria, Chaloneria, Pleuromeia, Nathorstiana, Isoetes*, and other genera showing this characteristic (Karrfalt, 1984a). In conclusion Rothwell & Erwin interpret these rhizomorphic lycopsids as representatives of a distinct clade of lycophytes. Thus, rather than visualizing the rhizomorphs of the Lepidodendrales and Isoetales as belonging to two different lines, as has been suggested earlier, they can now be recognized merely as growth variations in the same clade, with extant *Isoetes* representing the last highly reduced relative of a group of rhizomorphic antecedents first known to appear in the Upper Devonian.

It is important to recall at this point that the Upper Devonian arborescent lycopod *Cyclostigma* had a lobed basal rhizomorph. It is quite plausible that the Upper Carboniferous *Chaloneria* described by Pigg & Rothwell (1983a) represents a transitional form between the *Cyclostigma* type and the herbaceous Isoetales. If this is an acceptable interpretation, then the authors have succeeded in narrowing the time gap between the Upper Devonian arborescent lycopods and *Isoetites*, which makes its first known appearance in the Triassic. This interpretation also implies a reduction in size from treelike Devonian ancestors and not from the Upper Carboniferous *Sigillaria* as proposed by Mägdefrau. Further, the evidence demonstrates

Figure 11.30. *Protostigmaria eggertiana*. Mississippian. (Redrawn from Jennings, 1975.)

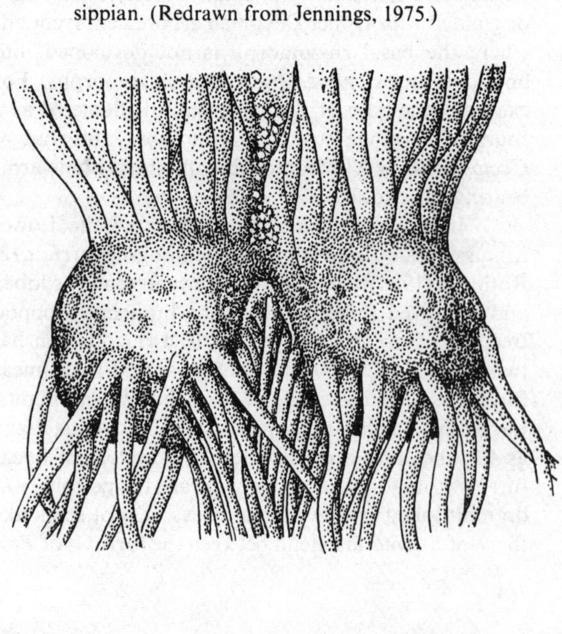

Figure 11.31. *Paurodendron fraipontii* rhizomorph. Pennsylvanian. Stereodiagram of sectioned rhizomorph. Parenchymatous plug (p); rootlet scar (rs); rootlet origin (ro); primary xylem (px); secondary xylem (sx). (Redrawn Rothwell & Erwin, 1985.)

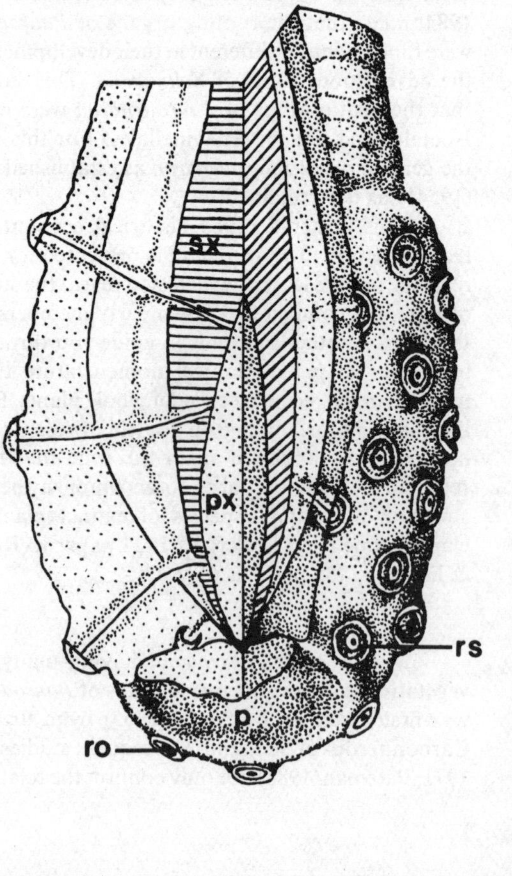

rather conclusively, as suggested earlier by Jennings (1975), that the bilobed rhizomorph of *Isoetes* might not have evolved by reduction from the horizontal dichotomous stigmarian system of arborescent lycopods. Instead, it now appears that the lobed cormlike rhizomorph is a characteristic appearing in the Upper Devonian, which was transmitted to the Carboniferous *Chaloneria* of the Isoetales. If the Isoetales evolved from lepidodendrids with cormlike rhizomorphs, it follows that the aerial parts must have become further reduced resulting in some limitation of branching, foreshortening of the axes, and reduction of the cones as postulated by Mägdefrau.

The function of this rather detailed explanation of research into the origin of the Isoetales is to give you an example of how old concepts and theories are gradually replaced by new ones as new evidence is revealed from the fossil record and extant plants. It is a slow and difficult process, but progress is being made.

Summary of evolutionary origins and relationships

As we reflect on the evolutionary origin of the Lycopsida, it seems fairly certain that they had their beginnings among the mid-Silurian–Lower Devonian Rhyniopsida (Chart 11.1). If we accept the *Rhynia* type as representing the primitive type of vascular plant, then we can logically start with mid-Silurian–Lower Devonian *Cooksonia* with its naked dichotomizing axes, terminated by globose to reniform sporangia resembling those of lycopods.

The weak link in our search for the evolutionary origin of the lycopods occurs in the derivation of the Lower to Middle Devonian Zosterophyllopsida from the Rhyniopsida. It was suggested that the Lower Devonian *Renalia* may show us how the sporangia of the zosterophylls achieved their terminal position on lateral branches from the primitive terminal position at the tips of dichotomous branches as in the Rhyniopsida. In addition to having a shape similar to the sporangia of *Cooksonia,* zosterophyll sporangia have dehiscence mechanisms that are similar to those of herbaceous lycopods. *Zosterophyllum* has smooth axes, like those of the Rhyniopsida, that branch dichotomously or pseudomonopodially. Internally, the vascular system is an elliptical, exarch protostele, a condition reflected by the herbaceous lycopods with their exarch actinosteles.

Other than *Zosterophyllum* and *Gosslingia,* genera of the zosterophylls have their axes ornamented with rows of teeth as in *Crenaticaulis,* covered with sparsely scattered spinelike enations (*Sawdonia, Discalis,* and *Kaulangiophyton*) or with condensed spirally arranged enations, a feature of *Asteroxylon.* These genera, also including *Drepanophycus,* are considered by some to be prelycopods forming a transitional group between the Zosterophyllopsida and Lycopsida. However, the first unequivocal evidence we have of a lycopsid with spirally arranged leaves and adaxial sporangia comes from the late Silurian–early Devonian *Baragwanathia.* This genus gives us our first glimpse of the herbaceous homosporous Lycopodiales. The Isoetales are represented by a clade whose origins seem to be with Upper Devonian lycopods with cormose or lobed rhizomorphs, e.g., *Lepidosigillaria* and *Cyclostigma.* This morphology appears in the Carboniferous genera *Oxroadia, Paurodendron,* and *Chaloneria.* The rooting structures of these plants bear spirally arranged roots with an internal structure similar to that of extant *Isoetes.* Recent studies have shown the radial rhizomorph of *Paurodendron* to be the homologue of the much-branched rhizomorph of *Stigmaria.* This observation suggests that the Lepidodendrales with their rhizomorphs of the *Stigmaria* type are part of the isoetalean clade. This part of the clade comprised of the arborescent, ligulate, heterosporous lycopsids (e.g., *Lepidodendron, Diaphorododendron, Lepidophloios, Sigillaria,* and others) became extinct in the Permian. The Triassic *Pleuromeia,* with its cormose rhizomorph, resembles *Chaloneria* in gross morphology and extends the clade into the Mesozoic along with *Nathorstiana* and *Isoetites,* which are very similar to present-day *Isoetes.*

Members of the lepidodendrid group evolved a great variety of modifications of their heterosporous reproductive structures, including the reduction in number of functional megaspores to one and various adaptations of protection megaspores and megagametophytes that approximate that seed habit.

An herbaceous line that parallels the lepidodendrids in being ligulate, and heterosporous, showing a reduction in megaspore number and protection of the megaspores is that of the Selaginellales. This order probably evolved from heterosporous ligulate ancestors derived from the Protolepidodendrales, but did not make its appearance until the Upper Carboniferous. Protection of the megasporangium is found in the diminutive integumented *Miadesmia,* which is, in all probability, closely related to the Selaginellales and shows considerable similarity to *Lepidocarpon* in the way the megasporangium is covered by a foliar lamina.

From the overview provided by this imcomplete summary, it must be apparent that we have dealt with several different aspects of evolution. Perhaps the most conspicuous is that kind of evolution that results in elaboration and/or specialization of parts, and often referred to as progressive evolution. In this category, the evolution of the medullated stele from a protostele and the appearance of cambial activity, both of which increase the strength of arborescent lycopod stems, are examples of an evolutionary response to natural selection resulting in more specialized structures.

Reduction, another aspect of evolution, results

Chart 11.1. Suggested origins and relationships of the Zosterophyllopsida, Rhyniopsida, and major groups of the Lycopsida and their temporal distributions. Drep. = Drepanophycales; Ast. = Asteroxylales; Zost. = Zosterophyllopsida; Proto. = Protolepidodendrales; Selag. = Selaginellales.

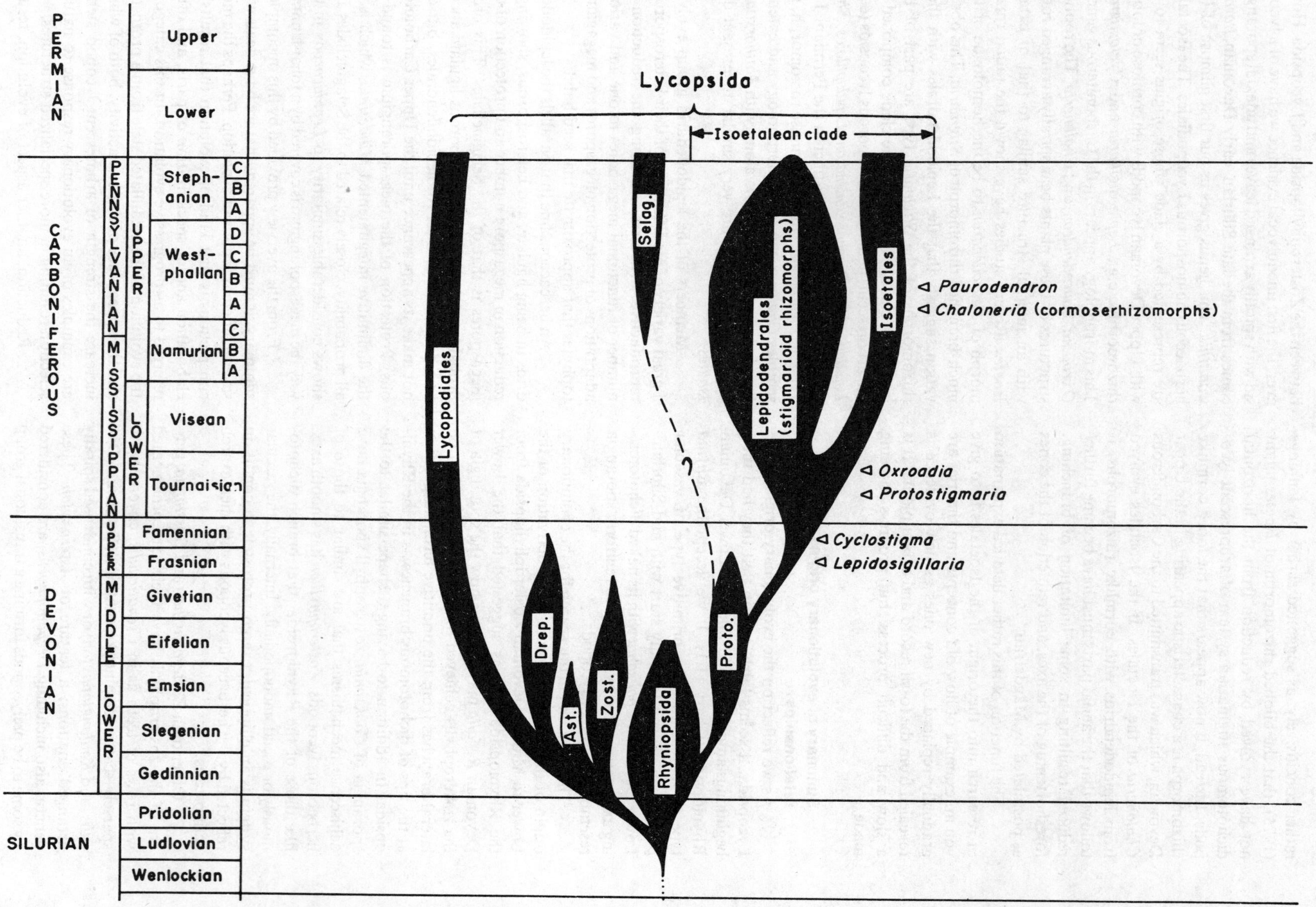

in a decrease in size, number, or complexity. Reduction of megaspore number in the Lycopsida is an example. Another, where decrease in size and complexity results, is found in the evolution of the diminutive Isoetales.

Parallel or convergent evolution occurs when similar characteristics evolve in two or more distantly or unrelated groups. As an example, let us remind you again that evolution of heterospory occurred in the herbaceous Selaginellales independently of heterospory in the distantly related arborescent Lepidodendrales, and that the cones of the Lycopodiales, Selaginellales, and Lepidodendrales also evolved independently. In the chapters that follow we will see many expressions of these and other aspects of plant evolution.

References

Abbott, M. L. (1963). Lycopod fructifications from the upper Freeport (No. 7) coal in southeastern Ohio. *Palaeontographica B*, **112**, 93–118.

Alvin, K. L. (1985). A new fertile lycopod from the Lower Carboniferous of Scotland. *Palaeontology*, **8**, 281–93.

Andrews, H. N. (1961). *Studies in Paleobotany*. New York: Wiley.

Andrews, H. N, & Murdy, W. H. (1958). *Lepidophloios* and ontogeny in arborescent lycopods. *American Journal of Botany*, **45**, 522–60.

Andrews, H. N., & Pannell, E. (1942). Contributions to our knowledge of American Carboniferous flora. II. *Lepidocarpon*. *Annals of the Missouri Botanical Garden*, **29**, 19–28.

Arnold, C. A. (1939). Observations on fossil plants from the Devonian of eastern North America. IV. Plant remains from the Catskill Delta deposits of northern Pennsylvania and southern New York. *Contributions from the Museum of Paleontology, University of Michigan*, **5**, 271–313.

Arnold, C. A. (1947). *An Introduction to Paleobotany*. New York: McGraw-Hall, pp. 92–3.

Balbach, M. K. (1967). Paleozoic lycopsid fructifications. III. Conspecificity of British and North American *Lepidostrobus* petrifactions. *American Journal of Botany*, **54**, 867–75.

Bateman, R. M. (1988). Palaeobotany and palaeoenvironments of Lower Carboniferous floras of two volcanigenic terrains in the Scottish Midland Valley. Unpublished thesis, University of London.

Bock, W. (1962). A study of fossil *Isoetes*. *Journal of Paleontology*, **36**, 53–9.

Brack, S. D. (1970). On a new structurally preserved arborescent lycopsid fructification from the Lower Pennsylvanian of North America. *American Journal of Botany*, **57**, 317–30.

Brack-Hanes, S. D. (1978). On the megagametophytes of two lepidodendracean cones. *Botanical Gazette*, **139**, 140–6.

Brack-Hanes, S. D., & Thomas, B. A. (1983). A re-examination of *Lepidostrobus* Brongniart. *Botanical Journal of the Linnean Society*, **86**, 125–33.

Backs-Hanes, S. D., & Vaughn, J. C. (1978). Evidence of Paleozoic chromosomes from lycopod microgametophytes. *Science*, **200**, 1383–5.

Chaloner, W. G. (1968). The cone of *Cyclostigma kiltorkense* Houghton, from the Upper Devonian of Ireland. *Journal of the Linnean Society (Botany)*, **61**, 25–36.

Chaloner, W. G., & Boureau, E. (1967). Lycophyta. In *Traité de Paléobotanique*, ed. E. Boureau. Paris: Masson, Vol. 2.

Chaloner, W. G., & Meyer-Berthaud, B. (1983). Leaf and stem growth in the Lepidodendrales. *Botanical Journal of the Linnean Society*, **86**, 135–48.

Chitaley, S., & McGregor, D. C. (1988). *Bisporangiostrobus harrisii* gen. et sp. nov., an eligulate lycopsid cone with *Duosporites* megaspores and *Geminospora* microspores from the Upper Devonian of Pennsylvania, U.S.A. *Palaeontographica, B*, **210**, 127–49.

Cichan, M. A., Taylor, T. N., & Smoot, E. L. (1981). The application of scanning electron microscopy in characterization of Carboniferous lycopod wood. *Scanning Electron Microscopy*, **3**, 197–201.

Darrah, W. C. (1941). The coenopterid ferns in American coal balls. *American Midland Naturalist*, **25**, 233–69.

Delevoryas, T. (1957). Anatomy of *Sigillaria approximata*. *American Journal of Botany*, **44**, 654–60.

DiMichele, M. A. (1979a). Arborescent lycopods of Pennsylvanian age coals; *Lepidophloios*. *Palaeontographica B*, **171**, 57–77.

DiMichele, W. A. (1979b). Arborescent lycopods of Pennsylvanian age coals: *Lepidodendron dicentrum* c. Felix. *Palaeontographica B*, **171**, 122–36.

DiMichele, W. A. (1980). *Paralycopodites* Morey & Morey, from the Carboniferous of Euramerica. A reassessment of generic affinities and evolution of *"Lepidodendron" brevifolium* Williamson. *American Journal of Botany*, **67**, 1466–76.

DiMichele, W. A. (1981). Arborescent lycopods of Pennsylvanian age coals: Lepidodendron, with description of a new species. *Palaeontographica B*, **175**, 85–125.

DiMichele, W. A. (1983). *Lepidodendron hickii* and generic delimitations in Carboniferous lepidodendrid lycopods. *Systematic Botany*, **8**, 317–33.

DiMichele, W. A. (1985). *Diaphorodendron*, gen. nov., a segregate from *Lepidodendron* (Pennsylvanian age). *Systematic Botany*, **10**, 453–8.

DiMichele, W. A., Mahaffy, J. F., & Phillips, T. L. (1979). Lycopods of Pennsylvanian age coals: *Polysporia*. *Canadian Journal of Botany*, **57**, 1740–53.

Eggert, D. A. (1961). The ontogeny of Carboniferous arborescent Lycoposida. *Palaeontographica B*, **108**, 43–92.

Eggert, D. A. (1972). Petrified *Stigmaria* of sigillarian origin from North America. *Review of Palaeobotany and Palynology*, **14**, 85–99.

Eggert, D. A., & Kanemoto, N. Y. (1977). Stem phloem of a Middle Pennsylvanian *Lepidodendron*. *Botanical Gazette*, **138**, 102–11.

Feng B.-C., & Rothwell, G. W. (1989). Microsporangiate cones of *Mazocarpon bensonii* (Lycopsida) from the Upper Pennsylvanian of the Appalachian basin. *Review of Palaeobotany and Palynology*, **57**, 289–97.

Frankenberg, J. M., & Eggert, D. A. (1969). Petrified

Stigmaria from North America. I. *Stigmaria ficoides*, the underground portions of Lepidodendraceae. *Palaeontographica B*, **128** 1–47.

Fry, W. (1954). A study of the Carboniferous lycopod, *Paurodendron*, gen. nov. *American Journal of Botany*, **41,** 415–28.

Galtier, J. (1964). Sur le gamétophyte femelle des lépidodendracees. *Compte Rendus des Séances de l'Académie des Sciences, Paris*, **258,** 2625–8.

Galtier, J. (1970). Observations nouvelles sur le gaméophyte femelle des lépidodendracees. *Compte Rendus des Séances de l'Académie des Sciences, Paris*, **271,** 1495–7.

Grierson, J. D., & Banks, H. P. (1963). Lycopods of the Devonian of New York State. Palaeontographica Americana, **4,** 220–79.

Hirmer, M. (1927). *Handbuch der Paläobotanik*. Munich and Berlin; Oldenbourg.

Hirmer, M. (1933). Rekonstruktion von *Pleuromeia sternbergi* Corda. nebst bemerkungen zur Morphologie der Lycopodiales. *Palaeontographica B*, **78,** 47–56.

Hoskins, J. H., & Cross, A. T. (1941). A consideration of the structure of *Lepidocarpon* Scott based on a new strobilus from Iowa. *American Midland Naturalist*, **25,** 523–47.

Jennings, J. R. (1972). A new lycopod genus from the Salem limestone (Mississippian) of Illinois. *Palaeontographica B*, **137,** 72–84.

Jennings, J. R. (1975). *Protostigmaria*, a new plant organ from the Lower Mississippian of Virginia. *Paleontology*, **18,** 19–24.

Jennings, J. R., Karrafalt, E. E., & Rothwell, G. W. (1983). Structure and affinities of *Protostigmaria eggertiana*. *American Journal of Botany*, **70,** 963–74.

Karrfalt, E. E. (1981). The comparative and developmental morphology of the root system of *Selaginella salaginoides* (L.) Link. *American Journal of Botany*, **68,** 244–53.

Karrfalt, E. E. (1982). Secondary development in the cortex of *Isoetes*. *Botanical Gazette*, **143,** 439–45.

Karrfalt, E. E. (1984a). Further observations on *Nathorstiana* (Isoetaceae). *American Journal of Botany*, **71,** 1023–30.

Karrfalt, E. E. (1984b). The origin and early development of the root-producing meristem of *Isoetes andicola* L. D. Gomez. *Botanical Gazette*, **145,** 372–7.

Karrfalt, E. E., & Eggert, D. A. (1977a). The comparative morphology and development of *Isoetes* L. I. Lobe and furrow development in *I. tuckermanii* A. Br. *Botanical Gazette*, **138,** 236–47.

Karrfalt, E. E., & Eggert, D. A. (1977b). The comparative morphology and development of *Isoetes* L. II. Branching of the base of the corm in *I. tuckermanii* A. Br. and *I. nuttallii* A. Br. *Botanical Gazette*, **138,** 357–68.

Karrfalt, E. E., & Eggert, D. A. (1978). The comparative morphology and development of *Isoetes* L. III. The sequence of root initiation in three- and four-lobed plants of *I. tuckermanii* A. Br. and *I. nuttallii* A. Br. *Botanical Gazette*, **139,** 271–83.

Kosanke, R. M. (1979). A long-leaved specimen of *Lepidodendron*. *Geological Society of America Bulletin, Part I*, **90,** 431–4.

Leclercq, S. (1924). Observations nouvelles sur la structure anatomique de quelques végétaux du houiller Belgique. *L'Académie Royale de Belgique*, **5–9,** 352–4.

Leclercq, S. (1930). A monograph of *Stigmaria bacupensis*. *Annals of Botany*, **44,** 31–54.

Leisman, G. A., & Phillips, T. L. (1979). Megasporangiate and microsporangiate cones of *Achlamydocarpon varius* from the Middle Pennsylvanian. *Palaeontographica B*, **168,** 100–28.

Lemoigne, Y. (1961). Études analytiques et comparées des structures des Sigillaires. *Annales des Sciences Naturelles Botanique*, **7,** 473–578.

Lemoigne, Y. (1966). Les tissus vasculaires et leur histogénese chez les Lépidophytales arborescentes du Paléozoique. *Annales des Sciences Naturelles, Botanique, et Biologie Végétal*, **7,** 445–74.

Long, A. G. (1971). A new interpretation of *Lepidodendron calamopsoides* Long and *Oxroadia gracilis* Alvin. *Transactions of the Royal Society of Edinburgh*, **68,** 491–506.

Mägdefrau, K. (1931). Zur Morphologie und phylogenetischen Bedeutung der fossilen Pflanzengattung *Pleuromeia*. *Beiheften zum Botanischen Centralblatt*, **48,** 119–40.

Mägdefrau, K. (1932). Ueber *Nathorstiana* eine Isoetaceae aus dem Nekom von Quedlinburg a. *Beiheften zum Botanischen Centralblatt*, **49,** 706–18.

Mägdefrau, K. (1957). *Palaobiologie der Pflanzen*. Jena: Fischer.

Nemejc, F. (1947). The Lepidodendraceae of the coal districts of Central Bohemia. *Sbornik Narodniho Musea v Praze*, **38,** 45–87.

Paollilo, D. J. (1982). Meristems and evolution: developmental correspondence among rhizomorphs of the lycopsids. *American Journal of Botany*, **69,** 1032–42.

Phillips, T. L. (1979). Reproduction of heterosporous arborescent lycopods in the Mississippian-Pennsylvanian of Euramerica. *Review of Palaeobotany and Palynology*, **27,** 239–89.

Phillips, T. L., Avcin, M. J., & Schopf, J. M. (1975). Gemetophytes and young sporophyte development in *Lepidocarpon* (abstract). Corvallis, Ore.: Botanical Society of America, p. 23.

Phillips, T. L., & Leisman, G. A. (1966). *Paurodendron*, a rhizomorphic lycopod. *American Journal of Botany*, **53,** 1086–100.

Pigg, K. B. (1983). The morphology and reproductive biology of the sigillarian cone *Mazocarpon*. *Botanical Gazette*, **144,** 600–13.

Pigg, K. B., & Rothwell, G. W. (1979). Stem-shoot transition of an Upper Pennsylvanian woody lycopsid. *American Journal of Botany*, **66,** 914–24.

Pigg, K.B., & Rothwell, G. W. (1983a). *Chaloneria* gen. nov.: heterosporous lycophytes from the Pennsylvanian of North America. *Botanical Gazette*, **144,** 132–47.

Pigg, K. B., & Rothwell, G. W. (1983b). Megagametophyte development in the Chaloneriaceae fam. nov., permineralized Paleozoic Isoetales (Lycopsida). *Botanical Gazette*, **144,** 295–302.

Ramanujam, C. G. K., & Stewart, W. N. (1969). A *Lepidocarpon* cone tip from the Pennsylvanian of Illinois. *Palaeontographica B*, **127,** 159–67.

Reed F. D. (1941). Coal flora studies: Lepidodendrales. *Botanical Gazette*, **102,** 663–83.

Retallack, G. (1975). The life and times of a Triassic lycopod. *Alcheringa*, **1**, 3–29.

Rex, G. M. (1983). The compression state of preservation of Carboniferous lepidodendrid leaves. *Review of Palaeobotany and Palynology*, **39**, 65–85.

Rothwell, G. M. (1984). The apex of *Stigmaria* (Lycopsida), rooting organ of Lepidodendrales. *American Journal of Botany*, **71**, 1031–4.

Rothwell, G. W., & Erwin, D. M. (1985). The rhizomorph apex of *Paurodendron*, implications for homologies among the rooting organs of the Lycopsida. *American Journal of Botany*, **72**, 86–98.

Rothwell, G. W., & Pryor, J. S. (1991). Developmental dynamics of arborescent lycophytes – apical and lateral growth in *Stigmaria ficoides*. *American Journal of Botany*, **78**, 1740–5.

Schlanker, C. M., & Leisman, G. A. (1969). The herbaceous Carboniferous lycopod *Selaginella fraipontii*, comb. nov. *American Journal of Botany*, **130**, 35–41.

Schopf, J. M. (1941). Contributions to Pennsylvanian paleobotany: *Mazocarpon oedipternum* sp. nov. and sigillarian relationships. *Illinois Geological Survey, Report of Investigations*, **75**, 1–40.

Schumaker-Lambry, J. (1966). Étude d'un cone de Lepidocarpaceae du houiller belge: *Achlamydocarpon belgicum* gen. et sp. nov. *Académie Royale de Belgique, Sciences*, **17**, 6–27.

Stewart, W. N. (1940). Phloem histology in stigmarian appendages. *Transactions of the Illinois Academy of Sciences*, **33**, 54–7.

Stewart, W. N. (1947). A comparative study of stigmarian appendages and *Isoetes* roots. *American Journal of Botany*, **34**, 315–24.

Stubblefield, S. P., & Rothwell, G. W. (1981). Embryogeny and reproductive biology of *Bothrodendrostrobus mundus* (Lycopsida). *American Journal of Botany*, **68**, 625–34.

Walton, J. (1935). Scottish Lower Carboniferous plants: the fossil hollow trees of Arran and their branches. (*Lepidophloios wunschianus* Carruthers). *Transactions of the Royal Society of Edinburgh*, **58**, 313–37.

Williamson, W. C. (1887). A monograph on the morphology and histology of *Stigmaria ficoides*. *The Palaeontographical Society*, **40**, 1–62.

Williamson, W. C. (1895). On the light thrown upon the question of the growth and development of the Carboniferous arborescent lepidodendra by study of the details of their organization. *Memoirs of the Proceedings of the Manchester Literary and Philosophical Society*, **39**, 31–65.

12

Paleoecology of the Pennsylvanian coal swamps

One of the most rewarding encounters experienced by paleobotanists and their students interested in floras of the Pennsylvanian Period is to see, for the first time, one of the highly realistic reconstructions of the flora of a Pennsylvanian coal swamp. The Field Museum of Natural History in Chicago has one of several life-size dioramas (Figs. 11.1; 24.1) depicting a coal swamp with startling realism. In the previous chapter we have surveyed one of the important groups of plants that comprise a major element of the coal-swamp flora. These are the arborescent lycopsids that dominated the forest. On further examination we can see that members of at least three other groups can be identified. These are ferns, seed ferns (pteridosperms), and horsetails (sphenopsids). The origin and evolution of these and other groups provide the subject material of later chapters. Right now we are interested in understanding something about the environments of Pennsylvanian coal swamps and their influence on the distribution and evolution of the plants that inhabited them.

At about the time the manuscript for the first edition of this book was being prepared, paleobotanists had just started to ask questions about environments that supported plant communities in the past, to wonder about what influences climatic and edaphic factors had, and to speculate to what extent the changes in communities could be attributed to evolution of new species. To do this required, among other things, interpreting the effects of taphonomic factors on plant remains at the time of their deposition, an aspect that seldom was considered prior to the 1980s. Before that time, most paleobotanists when making field collections were concerned primarily with collecting the best specimens without regard to the substrate or stratigraphy in the area of deposition and other associated geological features. Today publications describing fossil plants regularly give the geological setting in addition to paleoecological interpretations.

From earlier reports indicating that there were temporal changes in coal-swamp vegetation, it seemed that comprehensive studies of Pennsylvanian coal-swamp communities might provide a means of learning more about their structure, organization, and distribution. From such studies one might be able to provide inferences about the effects of environmental factors (edaphic and climatic) in producing change as distinct from the products of speciation. Previous studies (Peppers & Phillips, 1973; Phillips et al., 1974) indicate that the information required for an undertaking of such magnitude could be obtained from coal, permineralized peat (coal balls), and compression/impression fossil assemblages found in clastic substrates associated with the coal swamps. Quantitative analyses of the vegetation contained in these rocks revealed by palynological and morphological studies should provide the information from which inferences about community composition could be deduced. Thus it should be possible to make comparisons among communities of the same age but different geographical localities (horizontal comparisons) and those of different stratigraphic positions (vertical comparisons).

Palynological studies

Palynology is, in itself, a specialized field of paleobotany. It has been known for a long time that the spores, pollen grains, and other microfossils (palynomorphs) of many plants are highly resistant to mechanical breakdown and chemical decomposition. Their resistance is related to a substance called sporopollenin in the outer coats (sporoderms) of spores and pollen grains. The chemical nature of this highly resistant material is not thoroughly understood. Palynomorphs are often preserved in sedimentary rocks where no other plant fossils occur. The vast majority of these microfossils are found in a dispersed state having been distributed by air currents and water, often great distances from the parent plants. Dispersed spores provide one of the primary sources of evidence for stratigraphic studies (Kosanke, 1950; 1988a; 1988b) where the primary objective is to establish correlations between rock units often composed of coal or clastic materials. For many years dispersed palynomorphs have been described and characterized by palynologists according to size, shape, sculpturing of the walls, presence or absence of pores, sutures, and so on (Figs. 12.1A–H). The various combinations of these characteristics makes it possible to distinguish types of palynomorphs and the assemblages they comprise. It is known that assemblages of palynomorph types vary according to their position in the geological column. It is obvious that these variations reflect changes in the populations of plants that produced them. Such changes may be

Figure 12.1. Some dispersed spore types found in sporangia of coal swamp plants. **A.** *Lycospora* from microsporangia of *Lepidodendron* and *Lepidophloios*. **B.** *Cappasporites* (= *Crassispora*) from microsporangia of species of *Diaphorodendron*. **C.** *Punctatisporites* and **D.** *Laevigatosporites* found in sporangia of the *Psaronius*-type tree fern. **E.** *Endosporites* recovered from sporangia of heterosporous isoetalean lycopods. **F.** *Calamospora* occurs in calamite- and sphenophyll-type cones. **G.** *Florinites* from microsporangiate structures of cordaites. **H.** *Vesicaspora*, a pollen type of a callistophyte pteridosperm. Pennsylvanian (Photographs courtesy of Dr. R. M. Kosanke and Dr. R. A. Peppers.)

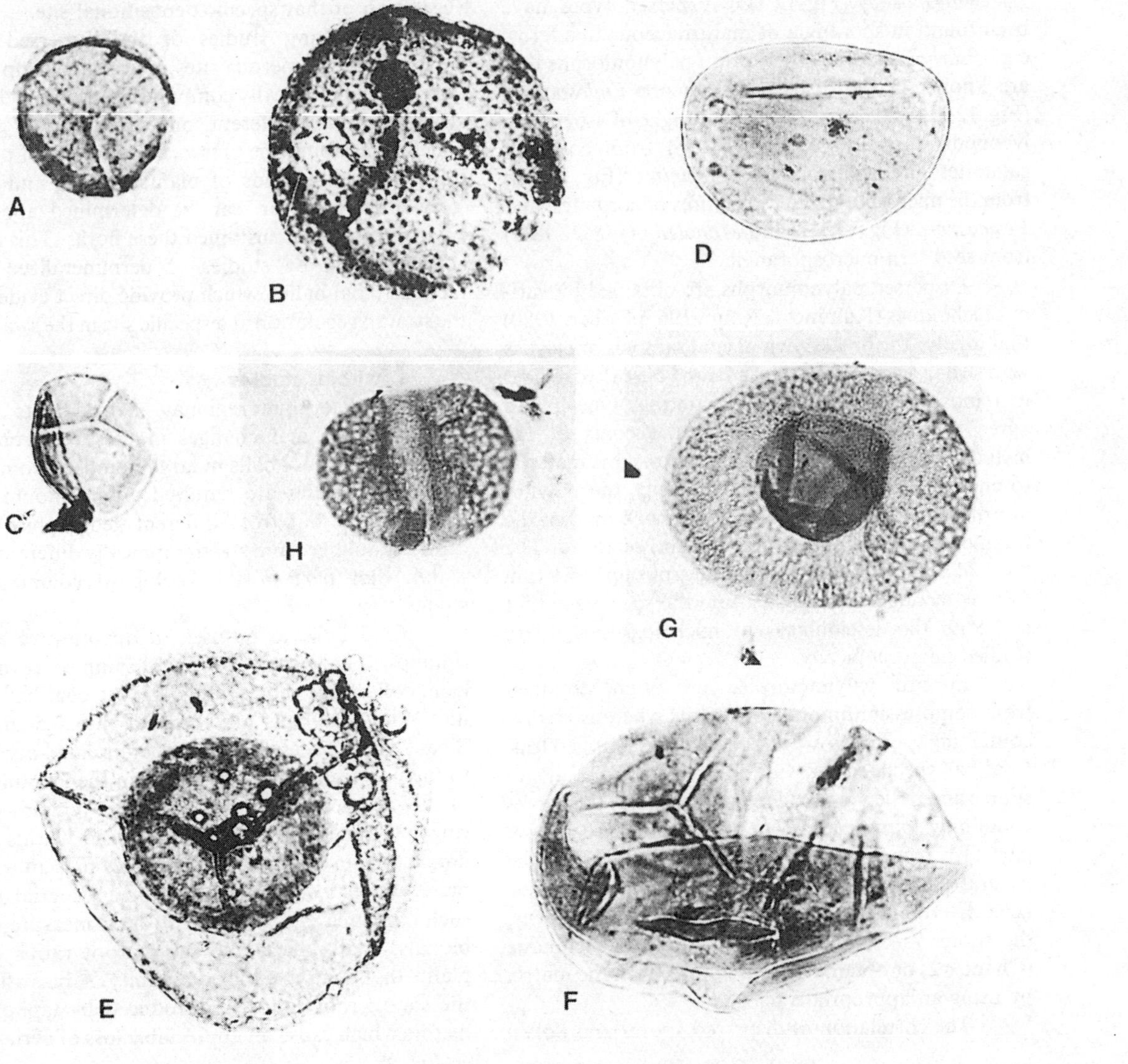

the result of alterations of the environment, the evolution of new species, the migration of existing species into the population, and extinctions. As we will discover later in this and other chapters, distinguishing between environmentally and evolutionarily produced variants is not an easy task. Geologists are primarily interested in using palynomorphs to construct stratotypes for stratigraphic units that can be used for age determinations and correlations.

Although the vast majority of palynomorphs occur in the dispersed state, many have been found in situ within the sporangia of the parent plants. Thus, it is possible to correlate dispersed spores and pollen-grain types with the plants that produced them. For example, the dispersed spore *Lycospora* (Fig. 12.1A) has been found in the microsporangia of *Lepidodendron* and *Lepidophloios,* while the *Capposporites* (Fig. 12.1B) dispersed type occurs in microsporangia of various species of *Diaphorodendron.* Similarly, isospores of the *Punctatisporites* (Fig. 12.1C) and *Laevigatosporites* (Fig. 12.1D) dispersed types have been found in sporangia of marattiaceous tree ferns, e.g., *Psaronius.* Some other in situ palynomorphs that are known in the dispersed state are *Endosporites* (Fig. 12.1E) recovered from sporangia of isoetalean lycopods; *Calamospora* (Fig. 12.1F) from cones of calamites and sphenophylls; *Florinites* (Fig. 12.1G) from the microsporangiate structures of cordaites; and *Vesicaspora* (Fig. 12.1H) and *Monoletes* (Fig. 22.13D) from seed fern microsporangia.

Dispersed palynomorphs are obtained by various techniques (Kummel & Raup, 1965; Doher, 1980) that involve the breakdown of coal samples or clastics with strong acids followed by a strong base (hydroxide) to remove the opaque humic materials. One has to agree that the sporopollenin in the coats of the included palynomorphs is indeed a resistant material to endure such drastic treatment. Once the palynomorphs have been released from the rock matrix, the fragmented sediments are washed and screened. The mesh of the screen allows palynomorphs up to 210 μm to pass through. These constitute the specimens that make up the assemblages of microfossils that are studied microscopically.

In situ palynomorphs are often obtained from compression/impression fossils where isospore-containing sporangia are attached to an identified leaf. Coal-ball specimens provide another source of in situ spores and pollen grains. Here, once again, the spore-containing sporangia may be attached to a vegetative part of a recognizable whole plant. In this way the natural affinities of many dispersed spore types have been determined. In the case of coal-ball specimens, the spores are released by using the peel technique (Chapter 2) or obtaining them in toto from the matrix by using an appropriate acid.

The correlation of dispersed spores and pollen with their parent plants has made it possible for coal palynologists to establish extensive records of plant distributions during the Pennsylvanian Period. Unlike coal balls, however, which provide in situ evidence of the vegetation of a coal swamp, dispersed palynomorphs can only tell us that one or more plants, for example lepidodendrids, were present in the general region of the swamp. At the time the lepidodendrid shed its spores, the spores may have been transported by air currents or water and deposited in a different environment some distance away. The spores, because of their small size and large numbers, may be transported great distances by relatively small air currents, while larger and fewer spores produced by smaller plants at the same location would be carried only short distances and would not appear in distant depositional sites. For these and other reasons, it must be apparent that one cannot use the dispersed spore content of a locality in a coal swamp to determine the number and kinds of plants that made up the vegetation of that specific depositional site.

By making studies of the dispersed spore content from numerous sites in a coal swamp, some that are geographically comparable and some that are stratigraphically different, one can arrive at general conclusions about the presence or absence of populations of certain kinds of plants in time and space. Very little, however, can be determined about the environments that sustained these floras. This can be achieved only by studies of permineralized petrifactions (coal balls), which provide direct evidence of the swamp vegetation at a specific site in the swamp.

Coal-ball studies

To determine regional environments within a coal swamp and changes in these environments through time, coal balls in large numbers from many different localities are required. Some should be of the same age but from different geographical sites; others should be from stratigraphically different levels within that part of the geological column being studied.

Coal balls to be used in quantitative studies (Phillips & DiMichele, 1981) are sampled at random from collections representing specific coal-ball localities. The coal balls are then cut into 2.5-cm-thick slices. The surfaces of the slices are treated according to the peel technique, which aids in the identification of the included plant remains. Those remains occurring in the peels are identified within cm^2 grids according to the taxa, organs, and tissues present and the preservational state. The identified material within each cm^2 area is considered to be a measure of the biomass that reflects the shoot: root ratios of the plants that comprise the community. A bias affecting the shoot: root ratios is introduced by taphonomic factors which cause an appreciable loss of aerial plant parts during preservation. This tends to lower the ratios from approximately 4:1 to 1:1. Another source of error in the ratios is often introduced by the subsequent penetration of root systems of plants that were not part of the community being studied.

Although compression/impression fossils are abundant in clastic substrates associated with coal swamps, less emphasis is placed on the specific composition of the diverse floras displayed by these fossils. The plant parts occurring in these clastic substrates were transported prior to deposition and thus were influenced to a greater degree by taphonomic factors than the autochthonous (in situ) deposits of vegetation, which subsequently were converted into the coal balls (permineralized peat) of the coal swamps (Spicer, R. A., 1980; Gastaldo, R. A., 1987).

Using the techniques outlined above to study their massive collections of coal balls and coals containing dispersed spores, Phillips, Peppers & DiMichele (1985) have prepared an extensive analysis of the changes in coal-swamp vegetation for the Westphalian and early Stephanian of western Europe and the Pennsylvanian Period of the Appalachian and Midcontinent regions of the United States (Chart 12.1).

There can be little doubt that the formation of the ancient supercontinent Pangaea and the paleocontinent of Laurussia had a marked effect on temperature, atmospheric moisture levels, changing sea levels, and the emergence of new land masses over long periods of time. All of these factors had a marked influence on the Pennsylvanian (Late Carboniferous) tropical coal swamps of Euramerica. Cyclic climatic oscillations plus the tectonic activity related to the repositioning and emergence of land masses during the Upper Paleozoic combined to create an ever-changing mosaic of land and swamps that contracted and expanded to produce regional areas of deposition and erosion (Phillips et al., 1985). The cyclic nature of deposition and erosion in the major coal basins of the eastern half of the United States (Chart 12.1) was clarified by Wanless, et al. in 1963.

Changes in coal-swamp vegetation

Major changes in coal-swamp vegetation were first recorded for the Upper Pennsylvanian in studies of compression/impression floras, clastic sediments, and coal deposits by White & Thiessen (1913); Kosanke (1947); Schemel (1957); Winslow (1959), and Read & Mamay (1964). The information provided by these authors indicated a rapid decrease of lepidodendrids in the Upper Pennsylvanian communities with an abundance of the large tree fern *Psaronius* apparently replacing them. It was noticed that the fronds produced by the tree ferns tended to be villous (hairy), a common adaptation of the leaves of plants living in dry environments. In addition, a possible dry period during the Lower Pennsylvanian was inferred by Peppers (1979) from palynological studies. Using evidence from palynology and coal-ball studies, Peppers & Phillips (1973) and Phillips, et al. (1974) confirmed the onset of lepidodendracean extinction in the late Middle Pennsylvanian (Westphalian D) during a dry interval that was followed by a third dry period at the onset of the Permian (Chart 12.1).

It appears that the composition of floras of the Pennsylvanian tropical coal swamps was strongly influenced by the amount of fresh water that was available (DiMichele, Phillips, & Peppers, 1985). As these authors point out, the flora present should serve as an indication of the degree of wetness or dryness that prevailed during the existence of the swamp. Thus, changes in the vegetation of a coal swamp should reflect, over long periods of time during the Pennsylvanian Period, climatic and edaphic trends that regulated the availability of water. Although coal swamps of the major coal basins were influenced by the availability of fresh water, many coastal brackish-water swamps sustained extensive floras, which were influenced by major tectonic events resulting in widespread marine transgressions and regressions. These in turn produced fluctuations in the sea levels that regulated the amount of water available to the coastal coal swamps.

After an analysis of coal resource abundance, corroborated by palynological and coal-ball studies, it has become clear that the Pennsylvanian Period can be divided into five climatic intervals (DiMichele et al., 1985), (Chart 12.1). (1) Early Pennsylvanian (Westphalian A), a time of moderate, probably seasonal wetness. (2) Middle Pennsylvanian (Westphalian B), a decline in the amount of available moisture followed by a gradual increase in the amount of water available. (3) Late Middle Pennsylvanian (Westphalian D), a possible reduction in the seasonal distribution of increased amounts of moisture culminated in the most favorable conditions for growth of coal-swamp vegetation. (4) Middle and Late Pennsylvanian (late Westphalian D to Stephanian A), a sharp decline in the amount of fresh water available in lowlands and swamps. (5) Upper half of Pennsylvanian, a return of climatic or tectonic conditions that favored growth of coal-swamp floras; this was followed by the onset of the third drier interval that continued into the Permian.

Coal-swamp vegetation – environmental inferences

As Chart 12.1 illustrates, the vast majority of paralic coal swamps of the Early and Middle Pennsylvanian Period in the eastern half of the United States were dominated by lycopod trees (Phillips et al., 1985). Genera occurring in abundance in these swamps included *Lepidophloios* (Fig. 11.10), *Lepidodendron* (Fig. 11.3), *Diaphorodendron* (Fig. 11.7), and

Chart 12.1. Generalized patterns of abundance of Pennsylvanian Period coal-wamp vegetation in the United States with a relative wetness curve indicating the three most important times and amounts of changes. The data for the diagram are derived from the distribution of microfossils and megafossils. (Modified from Phillips, Peppers, & DiMichele, 1985.)

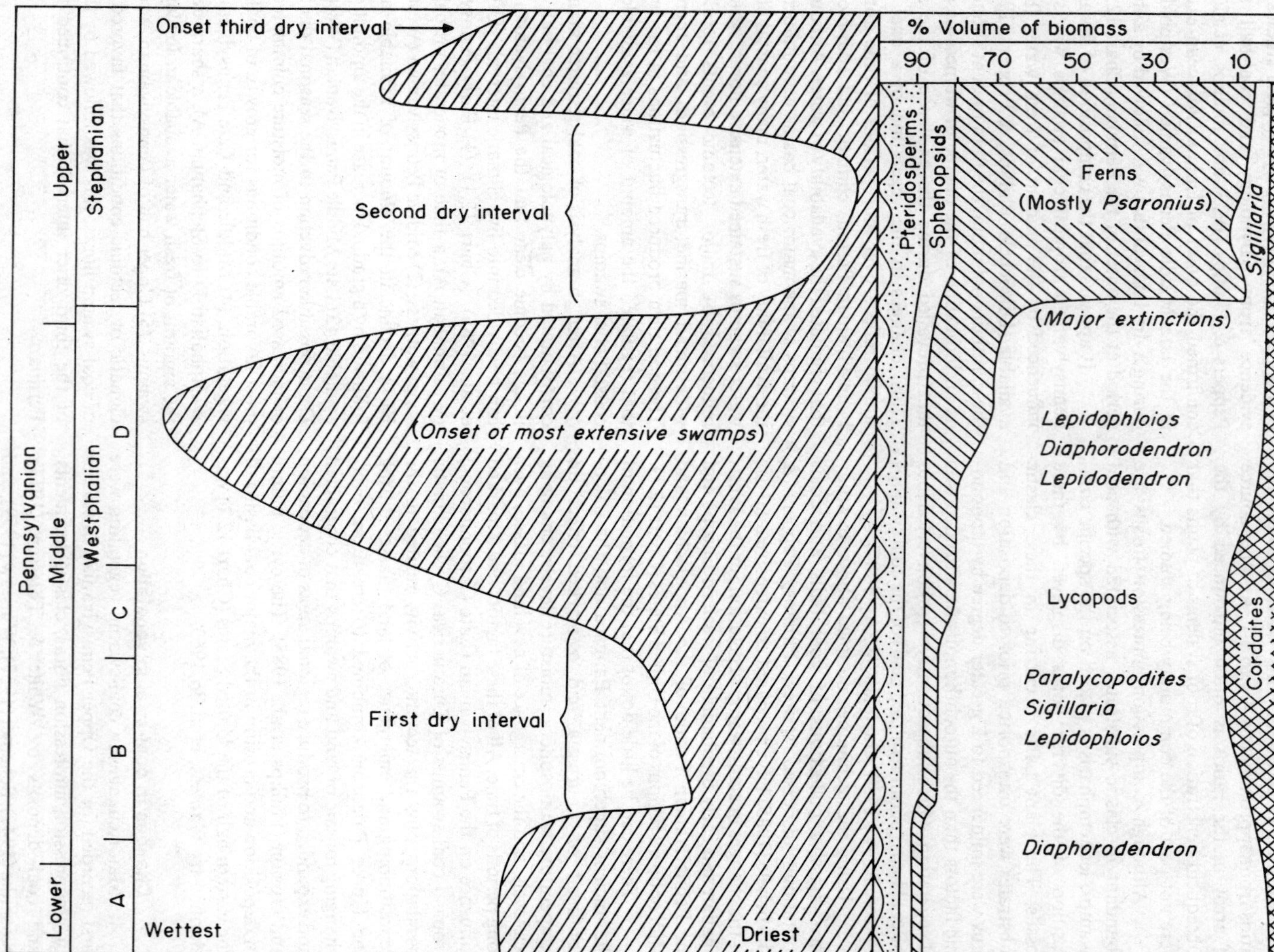

Paralycopodites. All became extinct during the Middle to Late Pennsylvanian. Although they did not predominate, two cordaites genera *Mesoxylon* and *Cordaixylon* (Chapter 28), the tree fern *Psaronius* (Chapter 18), and the seed fern *Medullosa* (Chapter 23) were important parts of the vegetation.

As noted earlier, the major environmental factor regulating the distribution of coal-swamp plants was the amount of available water. In the wettest swamps with standing water, *Lepidophloios* predominated. Apparently some species of *Diaphorodendron* were better adapted to somewhat less wet environments and tolerated slightly brackish conditions in the swamp, while *Lepidodendron* seems to have been somewhat restricted to wet environments where nutrient levels were higher. This genus and *Sigillaria* were common components of the higher nutrient, wet clastic environments during the Early and Middle Pennsylvanian. *Sigillaria,* however, was the least important of the lycopod trees in coal swamps except in the Late Pennsylvanian. *Paralycopodites* occurred in a spectrum of environments varying from clastic to coal-swamp conditions. Unlike other lycopod tree genera, *Paralycopodites* was not a dominant part of the Early Pennsylvanian but did come into dominance during the lower Middle Pennsylvanian. The appearance of abundant *Paralycopodites* at this time is taken as an indication of the end of the first drier interval (Chart 12.1). During the Lower–Middle Pennsylvanian, the relative importance of the lycopod-tree genera changed, with several changes occurring within the Middle Pennsylvanian (DiMichele et al., 1985; Phillips et al., 1985). All of the lycopod-tree genera except for *Sigillaria* became extinct during the early part of the second drier interval (Chart 12.1). Only *Sigillaria* continued through the Late Pennsylvanian into the Permian (Tidwell, 1988).

Although not corroborated by the record of megafossil remains, palynological studies indicate that during the first dry interval of the Early Pennsylvanian, tree ferns were present in regions outside the coal swamps. Coal balls show that the tree fern *Psaronius* occasionally occurred in the coal swamps, but this genus did not comprise an important part of the Early Pennsylvanian flora as indicated by dispersed-spore studies. It was not until just above the Middle-Late Pennsylvanian boundary that there was a dramatic shift to tree fern dominance. This followed the extinction of the arborescent lycopods at the onset of the second dry interval (Chart 12.1). It has been suggested (DiMichele et al., 1985) that these marattialean ferns evolved a "broad ecological amplitude," from the time of their appearance in Early Pennsylvanian coal swamps to the late Middle Pennsylvanian where there is evidence that they were adapted to wetland areas and possibly to lowlands surrounding the coal swamps. It follows, if this assumption is correct, that by the end of the Pennsylvanian and the initiation of the third drying trend that extended into the Permian, there may have been tree-fern populations that were adapted to survival in relatively dry environments as well as in regional wetland clastic swamps.

The cordaites (Chapter 28) seem to have had the same broad ecological amplitude exhibited by the tree ferns. These gymnosperms became important elements of the Pennsylvanian during the first drier interval and later during the time of increased brackishness and subsequent increased wetness of the late Middle Pennsylvanian (Chart 12.1). In the coal basins of the eastern half of the United States, the cordaites were the only gymnosperms to achieve dominance in the coal swamps. Their rise and decline occurred primarily in the Middle Pennsylvanian. The cordaite *Mesoxylon* probably succeeded in migrating into the coal swamps during the first drier interval from surrounding lowlands because of an ability to grow under conditions of moisture stress (DiMichele et al., 1985). Later in the Middle Pennsylvanian (Westphalian D) when wet and brackish water conditions returned to the expanding coal swamps, some cordaites (*Cordaixylon*) became an important part of the coal-swamp assemblages. This is probably related to a tolerance to brackish conditions by some members of the population.

The seed fern *Medullosa* (Fig. 23.13) did not become abundant until the middle of the Middle Pennsylvanian, at about the same time that *Psaronius* started to increase in numbers in coal swamps. Compression/impression fossils from the first drier period of the lower Middle Pennsylvanian suggest that species of *Medullosa* did not migrate into the coal swamps from the mesic lowland environments they occupied. With increasing wetness in the late Middle Pennsylvanian, *Medullosa* increased in abundance in the coal swamps (DiMichele et al., 1985). This increase correlates with an increase in abundance of the tree fern *Psaronius* and the lycopod *Lepidophloios,* an indicator of wet coal swamps. In this environment, *Medullosa* probably grew in parts of the swamp that were drier and higher in available nutrients. In the Upper Pennsylvanian these seed ferns were second in abundance to the marattialean tree ferns, and, like the ferns, the seed ferns continued on into the Permian.

The sphenopsids are represented in coal swamps and areas of clastic substrates by the shrub or vinelike *Sphenophyllum* (Fig. 15.1) and the arborescent calamites (Fig. 16.3A). Both were widespread and occurred in the Pennsylvanian swamps. Calamites have been depicted as growing in habitats along the shores of lakes and streams in disturbed substrates that may have been covered with water from time to time. Like the extant horsetail *Equisetum,* calamites may have favored localized areas of disturbed substrates where they could readily reproduce vegetatively from its rhizomes. Although calamites form a part of the coal-swamp biomass, it appears that most were located outside the coal swamps in lowland clastic substrate areas. Because of its small size, *Sphenophyllum* contributed very little to the coal-swamp biomass in the Pennsylvanian. It may have favored a habitat in swamps with little or no standing water as well as lowland clastic substrate areas outside the swamps. Both calamites and *Sphenophyllum* may have migrated into the swamps during the periods of drying.

Summary

In summarizing the major changes in the vegetation of Pennsylvanian coal swamps, DiMichele et al. (1985) focus our attention on climate changes initiated by the two intervals of drying during the Middle and Late Pennsylvanian. These climatic changes (1) induced the expansion of subdominant parts of the flora to dominant levels within the swamps; (2) provided environments favoring the migration of plants from the surrounding lowlands into the swamps; (3) initiated environments unfavorable to the growth of some species causing their extinction; and (4) created changing environments in the coal swamps that imposed selective pressures that resulted in the evolution of new species. Changes introduced by the evolution of new species are difficult to document and distinguish from changes introduced by migration of species into the swamps from lowlands where stages in their evolution are not recorded.

The alterations in the climate and accompanying changes in availability of water lead to the following changes in the coal-swamp floras during the Pennsylvanian Period in the eastern half of the United States. (1) The dominance of lycopod trees and herbaceous forms during the Early Pennsylvanian; (2) the decline of lycopods and an increase of tree ferns, calamites, and cordaites during the early Middle Pennsylvanian; (3) a resurgence of the lycopod trees to dominance in the Middle to upper Middle Pennsylvanian prior to the Westphalian-Stephanian boundary with cordaites and ferns as secondary elements; (4) late Middle Pennsylvanian still dominated by lycopods with ferns and seed ferns in abundance; (5) the abrupt termination of *Lycospora*-producing lycopod trees at the Middle–Late Pennsylvanian boundary; and (6) the dominance of tree ferns, during the Upper Pennsylvanian in most floras with *Sigillaria* and *Chaloneria* most abundant in others. Seed ferns and calamites were subordinate (Phillips et al., 1974).

Based on palynological and megafossil studies, the similarity of the floras of the Westphalian of Europe and Pennsylvanian of the midcontinent and eastern half of the United States has been interpreted as additional evidence that the two were once parts of a continuous land mass. With this in mind, it is not surprising that further investigations have demonstrated that the major times incorporating changes in coal-swamp climate and vegetation are more-or-less synchronous not only between western and eastern coal regions of the United States, but to some extent between the United States and Europe (Phillips et al., 1985; Galtier & Phillips, 1979).

Late Pennsylvanian into the Permian

The increased aridity dominating the Late Pennsylvanian and continuing into the Permian is correlated with the continued change in position of continental masses, extensive Gondwanan glaciation that affected sea levels and salinity on a worldwide basis, regional orogenic movements that created new highlands contributing to the deposition of clastics, and restrictions in global circulation patterns (Mapes & Gastaldo, 1986). The changes in climatic conditions induced by these major physical events spelled an end to the coal swamps of the Pennsylvanian Period. The onset of the drying trend into the Permian may have provided an opportunity for the subsequent evolution and migration of plants that were physiologically adapted to drier environments. Among these were the walchian conifers (Fig. 12.2), which represent the only major group of plants not found in the Pennsylvanian coal swamps. Compression/impression remains of conifers first appear in clastic sediments (Scott, 1974; Scott & Chaloner, 1983) of the Westphalian B in Europe and deposits of similar age in North America. Their absence from the coal-swamp floras of the Pennsylvanian seems to be related to their physiological adaptation to the drier upland environments some distance from the swamps. Here they evolved and migrated to other areas, which in turn became drier as the physical aspects of the environment changed. Glimpses of these upland plants are found in clastic sediments where the plant remains show the obvious effects of fragmentation and other taphonomic factors from the place the plant was growing, for example on or near a stream bank, to the final depositional site some distance from the original community. Although many specimens of the walchian conifer flora have been found in clastic substrates in the Middle and Late Pennsylvanian, they did not become an important component of these floras until the upper part of the Late Pennsylvanian and Lower Permian.

Several Upper Pennsylvanian compression/impression floras that reflected the plant communities of the drier upland sediments of the early Lower Permian have been reported. A transitional flora slightly below the Pennsylvanian–Permian boundary has been described by Tidwell (1988). The flora is comprised of the arborescent lycopod *Sigillaria;*

Sphenophyllum and *Calamites* of the sphenophytes; foliage probably from marattialean ferns; seed-fern foliage; and remains of the gymnosperm, *Cordaites.* All of the above are constituents of the swamp environments of the earlier Pennsylvanian. That upland elements were not included is evidenced by the absence of walchian conifers. It should be noted that this swamp flora is represented by plants belonging to five of the major groups of vascular plants that continued on into the Permian, i.e., the lycopods, sphenopsids, ferns, seed ferns, and cordaitean gymnosperms.

In their descriptions of the Upper Pennsylvanian Hamilton, Kansas flora, which is close to the Pennsylvanian–Permian boundary, Rothwell & Mapes (1988) and Mapes & Rothwell (1988) had the advantage of having fossils with internal structures as well as compression/impression remains and spores. It became apparent that the flora consisted of an assemblage representing at least two different environments, those from coal swamps and those from an upland conifer community which were transported to the depositional site. The upland conifer community of the Hamilton flora is represented by walchians and cordaiteans which may have formed a canopy with seed ferns forming an understory. When the Hamilton flora is compared with other Paleozoic conifer deposits, it is clear that gymnosperms including walchians, cordaiteans, seed ferns, calamites, and the lycopod *Sigillaria* make up the principal parts of their assemblage. The absence of marattialean ferns was expected since these plants appear to have been restricted to wetter environments of the Pennsylvanian.

Figure 12.2. Walchian conifer branch from Pennsylvanian–Permian boundary. (From Mapes & Rothwell, 1988.)

The absence of the marattialean tree ferns from the Upper Pennsylvanian megafossil flora of the Hartford Limestone further suggests that this flora grew in drier habitats (Leisman, Gillespie, & Mapes, 1988) not favorable to the ferns. The fragmentary nature of the fossils as well as the limestone matrix indicated that they were transported some distance prior to reaching the depositional site. The most numerous components of this assemblage were gymnosperms of the walchian complex and the Cordaitales. As in other assemblages, *Sigillaria,* calamites, and seed-fern remains were present.

In this and other chapters, we have attempted to provide some insight into the ecological factors – climatic and edaphic – as well as taphonomic factors that affect our interpretations of the fossil record. As more paleoecological studies are completed, using all sources of evidence available to investigators, it becomes apparent that our conclusions about evolutionary relationships, especially lineages of plants, must be tempered by taking into account all aspects of the environment before we arrive at conclusions about evolutionary pathways.

References

DiMichele, W. A.., Phillips, T. L., & Peppers, R. A. (1985). The influence of climate and depositional environment on the distribution and evolution of Pennsylvanian coal-swamp plants. In *Geological Factors and the Evolution of Plants,* ed. B. H. Tiffney. New Haven: Yale University Press.

Doher, L. I. (1980). Palynomorph preparation procedures currently used in the paleontology and stratigraphy laboratories, U.S. Geological Survey: *U.S. Geological Survey Circular,* **830,** 29 pp.

Galtier, J., & Phillips, T. L. (1979). Swamp vegetation from Grand Croix (Stephanian) and Autun (Autunian), France and comparisons with coal-ball peats of the Illinois Basin. In *Paleontology, Paleoecology, Paleogeography,* eds. J. T. Dutro, Jr., & H. W. Pfefferkorn. Urbana, Illinois. Ninth International Congress of Carboniferous Stratigraphy and Geology, Compte Rendu, **5**.

Gastaldo, R. A. (1987). A conspectus of phytotaphony. In *Methods and Applications of Plant Paleoecology, Notes for a Short Course,* eds. W. A. DiMichele &

S. L. Wing. Paleobotanical Section, Botanical Society of America.

Kosanke, R. M. (1947). Plant microfossils in correlation of coal beds. *Journal of Geology*, **55**, 280–4.

Kosanke, R. M. (1950). Pennsylvanian spores of Illinois and their use in correlation. *Illinois Geological Survey Bulletin*, **74**, 128 pp.

Kosanke, R. M. (1988a). Palynological analyses of Upper Pennsylvanian coal beds and adjacent strata from the proposed Pennsylvanian System stratotype in West Virginia. *U.S. Geological Survey Professional Paper*, **1486**, 24. pp.

Kosanke, R. M. (1988b). Palynological studies of Middle Pennsylvanian coal beds of the proposed Pennsylvanian System stratotype in West Virginia. *U.S. Geological Survey Professional Paper*, **1455**, 73 pp.

Kummel, B., & Raup, D. (1965). *Handbook of Paleontological Techniques*. San Francisco: Freeman.

Leisman, G. A., Gillespie, W. H., & Mapes, G. (1988). Plant megafossils from the Hartford Limestone (Virgilian-Upper Pennsylvanian) near Hamilton, Kansas. In *Regional Geology and Paleontology of Upper Paleozoic Hamilton Quarry Area in Southeastern Kansas*, eds, G. Mapes & R. H. Mapes. Guidebook 6, Kansas Geological Survey. Lawrence, Kansas.

Mapes, G., & Gastaldo, R. A. (1986). Late Paleozoic non-peat accumulating floras. In *Land Plants, Notes for a Short Course*. T. W. Broadhead. Paleobotanical Society.

Mapes, G., & Rothwell, G. W. (1988). Diversity among Hamilton conifers. In *Regional Paleontology of Upper Paleozoic Hamilton Quarry Area in Southeastern Kansas*, eds. G. Mapes & R. H. Mapes. Guidebook 6, Kansas Geological Survey. Lawrence, Kansas.

Peppers, R. A. (1979). Development of coal-forming floras during the early part of the Pennsylvanian in the Illinois Basin. In *Depositional and Structural History of the Pennsylvanian System of the Illinois Basin*, Part 2, eds. J. E. Palmer & R. R. Dutcher. Urbana: Illinois State Geological Survey.

Peppers, R. A., & Phillips, T. L. (1973). Pennsylvanian coal-swamp floras in the Illinois Basin. Abstract, *1972 Annual Meeting, Geological Society of America*, **4**, 624–25.

Phillips, T. L., & DiMichele, W. A. (1981). Paleoecology of Middle Pennsylvanian age coal swamps in southern Illinois – Herrin coal member at Sahara Mine No. 6, In *Paleobotany, Paleoecology, and Evolution*, ed. K. J. Niklas. New York: Praeger.

Phillips, T. L., Peppers, R. A., Avcin, M. J., & Laughnan, P. F. (1974). Fossil plants and coal: Patterns of change in Pennsylvanian coal swamps of the Illinois Basin. *Science*, **184**, 1367–9.

Phillips, T. L., Peppers, R. A., & DiMichele, W. A. (1985). Stratigraphic and interregional changes in Pennsylvanian coal-swamp vegetation: Environmental inferences. *International Journal of Coal Geology*, **5**, 43–109.

Read, C. B., & Mamay, S, H. (1964). Upper Paleozoic floral zones and floral provinces of the United States. *United States Geological Survey Professional Paper*, **454-k**, 35 pp.

Rothwell, G. W., & Mapes, G. (1988). Vegetation of a Paleozoic conifer community. In *Regional Geology and Paleontology of Upper Paleozoic Hamilton Quarry Area in Southeastern Kansas*, eds. G. Mapes & R. H. Mapes. Guidebook 6, Kansas Geological Survey. Lawrence, Kansas.

Schemel, M. P. (1957). Small spore assemblages of mid-Pennsylvanian coals of West Virginia and adjacent areas. *Ph.D. Dissertation in Geology, West Virginia University*, Morgantown, W. Va. 188 pp.

Scott, A. C. (1974). The earliest conifers. *Nature*, **251**, 707–8.

Scott, A. C., & Chaloner, W. G. (1983). The earliest fossil conifer from the Westphalian B of Yorkshire. *Proceedings of the Royal Society of London*, **B220**, 163–82.

Spicer, R. A. (1980). The importance of depositional sorting to biostratigraphy of plant megafossils. In *Biostratigraphy of Fossil Plants*, eds. D. L. Dilcher & T. N. Taylor. Stroudsburg, Pa. Dowden, Hutchinson & Ross.

Tidwell, W. D. (1988). A new Upper Pennsylvanian or Lower Permian flora from southeastern Utah. *Brigham Young University – Geology*, **35**, 33–56.

Wanless, H. R., Tubb, J. B., Jr., Gednetz, D. E., & Weiner, J. L. (1963). Mapping sedimentary environments of Pennsylvanian cycles. *Bulletin of the Geological Society of America*, **74**, 437–486.

White, D., & Thiessen, R. (1913). The origin of coal. *U.S. Bureau of Mines Bulletin 38*. Washington, D. C.: U.S. Government Printing Office, 390 pp.

Winslow, M. R. (1959). Upper Mississippian and Pennsylvanian megaspores and other plant microfossils from Illinois. *Illinois State Geological Survey Bulletin*, **86**, 135 pp.

13

More diversity in the Devonian: the Trimerophytopsida

In our search for the beginnings of the Lycopsida (Chapters 9–11), we have had a glimpse of the diverse Devonian flora represented by the members of the classes Rhyniopsida, Zosterophyllopsida, and the Drepanophycales and Protolepidodendrales of the class Lycopsida. From the 1950s onward it has become progressively apparent that the amount of heterogeneity in the Devonian was much more extensive than was previously thought. Much credit for deciphering Devonian vascular plant diversity must be given to Andrews, Banks, Chaloner, and their many students. We already know that much of the supporting evidence for a diverse flora has come from the study of macrofossils. Using microfossils, however, in the form of dispersed vascular plant spores, Chaloner (1967) was able to show progressive increase in spore size, ornamentation, and speciation from the Upper Silurian to the end of the Devonian. These findings obviously support the evidence derived from the study of macrofossils, which show us unequivocally that those vascular plants producing the spores originated in the Silurian, evolved rapidly in the Lower Devonian, and increased in diversity to the end of the Upper Devonian.

A part of this diversity that we have not yet examined is represented by the Trimerophytopsida, a class of the division Tracheophyta. The Trimerophytopsida was first proposed in 1968 by H. P. Banks as a subdivision (Trimerophytina) for those Devonian vascular plants with a main axis that branched pseudomonopodially and had laterals that in turn branched repeatedly either dichotomously or trichotomously. Some of the ultimate branches terminated in clusters of paired fusiform sporangia. The sporangia dehisced by means of a longitudinal slit. All known species were homosporous. Internally, the trimerophytes are characterized by a relatively large centrarch to mesarch protosteles. Where internal structure can be observed, an outer cortex with a hypodermis is present.

A superficial examination of trimerophyte characteristics reveals that many are shared with the Rhyniopsida. Both exhibit three-dimensional dichotomizing branching patterns. With exceptions, the trimerophytes are also like the rhyniophytes in having naked axes with terminal sporangia that contain isospores. Where internal structure is known, both groups appear to be protostelic.

The Rhyniopsida differ from the Trimerophytopsida in having sparsely branched dichotomous systems with more or less solitary sporangia at the tips of the branches. With the exception of *Horneophyton* and *Renalia,* sporangia of rhyniophytes show no modification for dehiscense. Their protosteles, like those of trimerophytes, are centrarch, but they are much smaller relative to the axis. There is apparently no hypodermis in the cortex of rhyniophytes.

In spite of these differences it is clear to paleobotanists that there is a close relationship between the two groups, a relationship that will become well substantiated before the end of the chapter. The Rhyniopsida, you may recall, appear in the mid-Silurian and continue into the Middle Devonian. The Trimerophytopsida first appear in the geological column in the Lower Devonian and continue on into the Upper Devonian.

Trimerophyton

The type genus of the Trimerophytopsida, *Trimerophyton* (Fig. 13.1A), was established by Hopping (1956) and based on *Psilophyton robustius* described earlier by Dawson (1871) from the Lower Devonian of the Gaspé of Canada. These were compression–impression fossils that show little internal structure. The external morphology, however, is quite clear. The largest of the plant fragments is 9.5 cm long and about 1 cm in diameter. It consists of a main axis with spirally arranged lateral branches. These trifurcate almost immediately. The three branches produced in this way are of equal size and each of these divides again into three. This time the branches are unequal in size. Two smaller ascending branches arise from the upper side of the central larger branch produced by the unequal trifurcation. The smaller branches dichotomize twice and the resulting ultimate branches are terminated by clusters of sporangia. The third and largest branch dichotomizes three times before the sporangium bearing branches are produced. The sporangia are fusiform and 4 to 5 mm in length. Nothing is recorded of a dehiscence mechanism. However, smooth spores with triradiate markings were found in the sporangia. They averaged about 50 μm in diameter and apparently are isopores.

Psilophyton

To date nine well-delineated species of *Psilophyton* have been described, and most of these occur in Devonian strata of Emsian (Lower Devonian), Eifelian (Middle Devonian), and early Upper Devonian. An excellent review of the general characteristics of the species of *Psilophyton* appears in a paper by Doran, Gensel, & Andrews (1978), and a book by Gensel & Andrews (1984).

The first description of a *Psilophyton* was made by Dawson (1859) and based on material he obtained from the Gaspé coast. In 1870, using additional information, he published a reconstruction of *P. princeps*. This figure became a classic, and you may have seen it repeatedly reproduced in botany and geology texts. More recent investigations by Hueber & Banks (1967) of the original materials used by Dawson show us that Dawson took some liberties in making his reconstruction. He actually combined dissociated fragments of three different specimens. It was not until many years later that it was discovered that Dawson had not proved actual connection among these parts. In 1964, Hueber was able to show that the spiny axes, described by Dawson as *P. princeps* var. *ornatum*, did not bear the terminal sporangia previously described for *P. princeps*, but had lateral sporangia instead. For these specimens, Hueber established a new genus, *Sawdonia*, which has been discussed as a member of the Zosterophyllopsida (Chapter 10). It was necessary, because Dawson had included both spiny and naked axes with terminal sporangia in *P. princeps* without any evidence of connection between the two, to redefine the genus and select a new type specimen.

Hueber & Banks (1967) and Banks, Leclercq, & Hueber (1975) emended the genus *Psilophyton* so that at present it is characterized as follows: "Stems branch dichotomously and laterally [pseudomonopodially], the latter in an irregular, close spiral in zones of vegetative branching; fertile branches, in zones alternating with vegetative zones, more distant and alternate, distichous [in two rows]; lateral branches themselves branch either laterally [pseudomonopodially] or dichotomously with successive dichotomies each formed at right angles to the other [three-dimensional, dichotomous branching]; axes naked or spinous, ridged or unridged, compressions may be marked by punctiform scars; sporangia elongate–elliptical, pendulous, paired, and borne terminally on repeatedly dichotomized lateral branches, total number per lateral branch large; dehiscence longitudinal along facing surfaces of paired sporangia; xylem a solid strand, centrarch; outer cortex collenchymatous except in areas of substomatal chambers."

Among the specimens of *Psilophyton princeps* (Fig. 13.1B) collected and examined by Hueber (1968), several had spiny axes with attached terminal sporangia. The spines of these specimens were cuplike and truncate at their tips. In compression fossils the spines look like short pegs. Hueber noted that his specimens had the same characteristics as one illustrated by Dawson in 1871. Dawson, however, failed to note that spines were present. It is this specimen with its terminal pendulous sporangia that Hueber has designated as the Neotype (generitype) of *Psilophyton*.

Branching patterns and anatomy

Psilophyton dapsile (Kasper, Andrews, & Forbes, 1974) comes as close to the growth habit of the Rhyniopsida as any member of the Trimerophytopsida. Like the rhyniophytes, this species (Fig. 13.1C) is only a few centimeters tall. Its three-dimensional dichotomizing axes are differentiated into fertile and sterile telome trusses. There is some evidence of overtopping in certain parts of the branching system to produce a weakly pseudomonopodial system in the basal parts of the plant.

The pseudomonopodial aspect of the branching pattern is emphasized by *Psilophyton microspinosum* and *P. forbesii* (Fig. 13.1D), where a main axis has evolved with overtopped laterals that have the form of three-dimensional dichotomously branched fertile and sterile telome trusses. Like rhyniophytes, *P. dapsile* and *P. forbesii* (Andrews, Kasper, & Mencher, 1968; Gensel, 1979) have smooth, naked axes while *P. microspinosum*, as its name implies, has small, spinelike enations scattered on the surface of the main axes.

The effects of overtopping and reduction are more pronounced in *P. dawsonii* (Banks, Leclercq, & Hueber, 1975), where there are short spirally arranged lateral telome trusses (Fig. 13.2A) that can be interpreted as precursors of megaphylls. We will have occasion to return to this idea at a later time.

Although Dawson was able to describe something of the internal structure of *Psilophyton* from the compression fossils available to him, internal structure is best seen in some beautifully permineralized specimens of *P. dawsonii* (Banks et al., 1975). By making serial

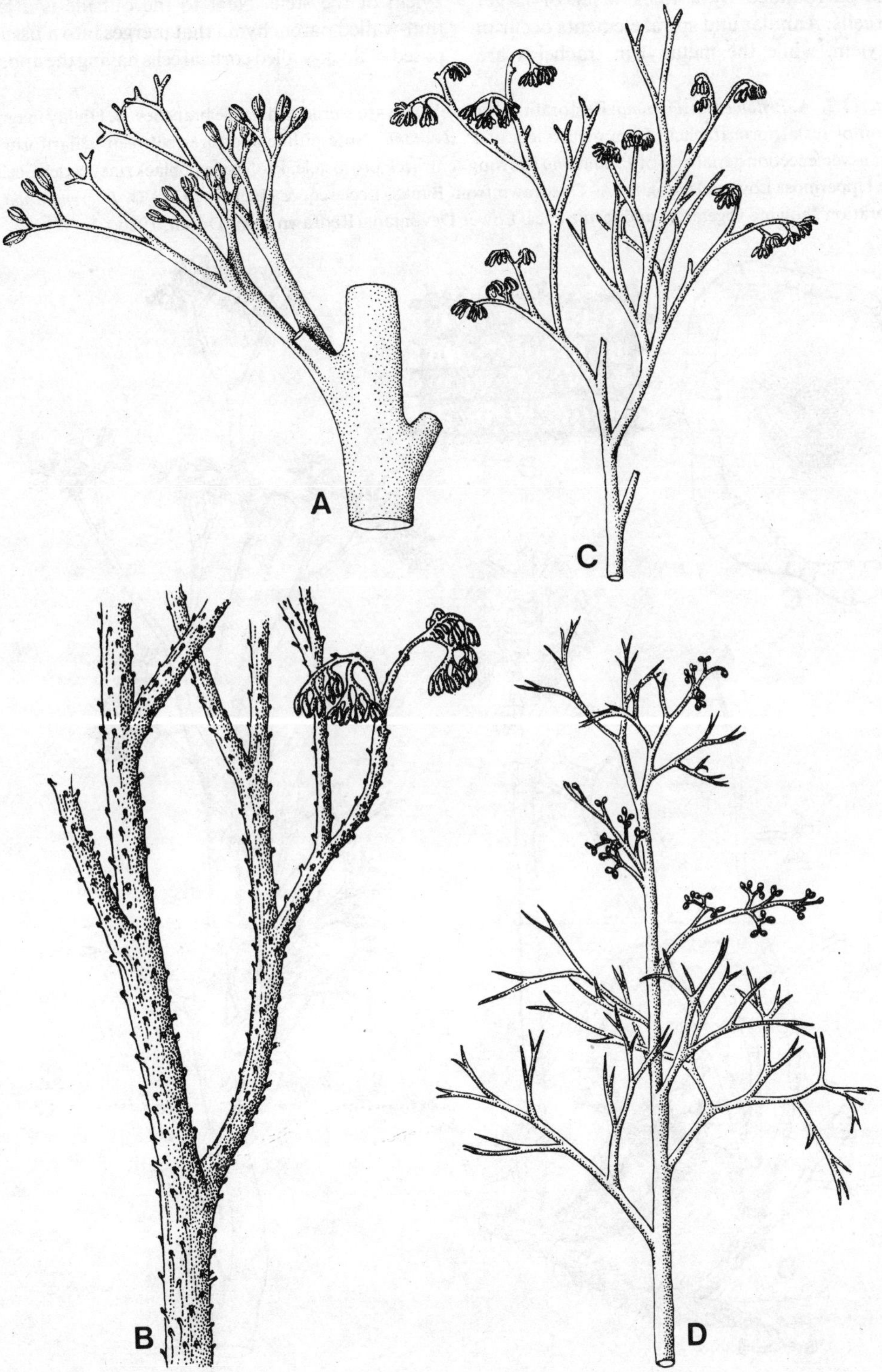

Figure 13.1. **A.** *Trimerophyton robustius.* Segment of axis showing "trifurcations" of fertile lateral branches. One of the three branches is deleted better to show branching. Lower Devonian. (Redrawn from Hopping, 1956.) **B.** *Psilophyton princeps.* Restoration showing peglike spines and pendulous sporangia. Upper Emsian or Lower Eifelian. (Redrawn from Hueber, 1968.) **C.** *P. dapsile.* A *Rhynia*-like plant with clusters of terminal sporangia. Early Middle Devonian. (Redrawn from reconstruction by Kasper, Andrews, & Forbes, 1974.) **D.** *P. forbesii.* Note pseudomonopodial branching. Uppermost Lower Devonian. (Redrawn from Andrews, Kasper, & Mencher, 1968.)

sections of the branching axes, Banks et al. were able to demonstrate a vascular system that extends from the base of the specimen into ultimate fertile and sterile branches. The vascular system is composed of a terete (haplostelic) protostele (Fig. 13.2C) that is quite massive in the basal portions. Cells of protoxylem are centrally located and surrounded by a thick layer of larger metaxylem cells. Annular and spiral elements occur in the protoxylem, while the metaxylem tracheids are scalariform with rows of circular pits between the scalariform bars (Fig. 13.2B). Cells of the metaxylem are often aligned in radial rows and have the appearance of secondary xylem, a feature of some importance in later descriptions of vascular plant structure. A layer of cells having the position of phloem lies around the xylem of the stele. Next to the outside is a layer of thin-walled parenchyma that merges into a tissue composed of thick-walled cortical cells having the appearance

Figure 13.2. **A.** *Psilophyton dawsonii.* Restoration of aerial axis with sterile and fertile branches. **B.** Highly magnified portion of scalariform tracheid from metaxylem of *P. dawsonii.* Note pitlike apertures between scalariform bars. **C.** Transverse section aerial axis of *P. dawsonii* showing centrarch protostele. Protoxylem, black; metaxylem, cellular. **A–C:** Uppermost Lower Devonian. (**A–C** redrawn from Banks, Leclercq, & Hueber, 1975.) **D.** *P. crenulatum.* Restoration showing vegetative and fertile axes. Lower Devonian. (Redrawn from Doran, 1980.)

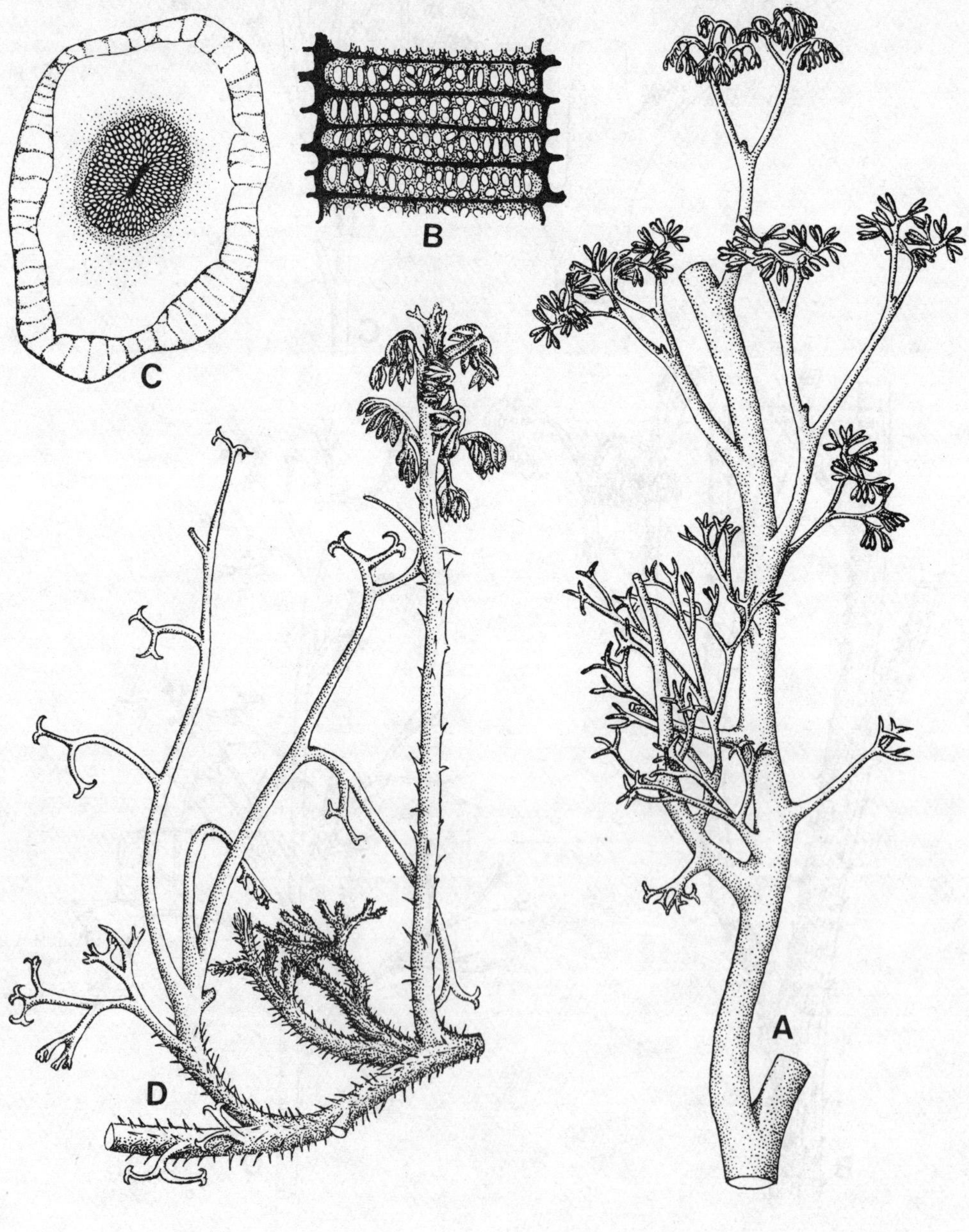

of collenchyma. This is the hypodermis. Stomata have been found in the epidermis covering the axes.

Fusiform sporangia occur in pairs at the tips of fertile branches where they form clusters of many sporangia. Dehiscence is longitudinal with the openings of the two sporangia of a branch facing one another. Spores isolated from sporangia are isospores of the dispersed *Retusotriletes* type.

Although *Psilophyton dawsonii* shows the effects of evolutionary overtopping and reduction, these processes are even more pronounced in the Emsian *P. crenulatum* (Doran, 1980). The specimens studied by

Figure 13.3. **A–D:** *Psilophyton crenulatum*, Lower Devonian. **A.** Aerial branches, some naked, some with enations. Note recurved tips of naked axes. **B.** Aerial shoots with large enations, some showing dichotomies. **C.** Aerial shoot with possible axillary branching. **D.** SEM photograph of dehisced sporangia terminating fertile branches. (**A–D** from Doran, 1980.)

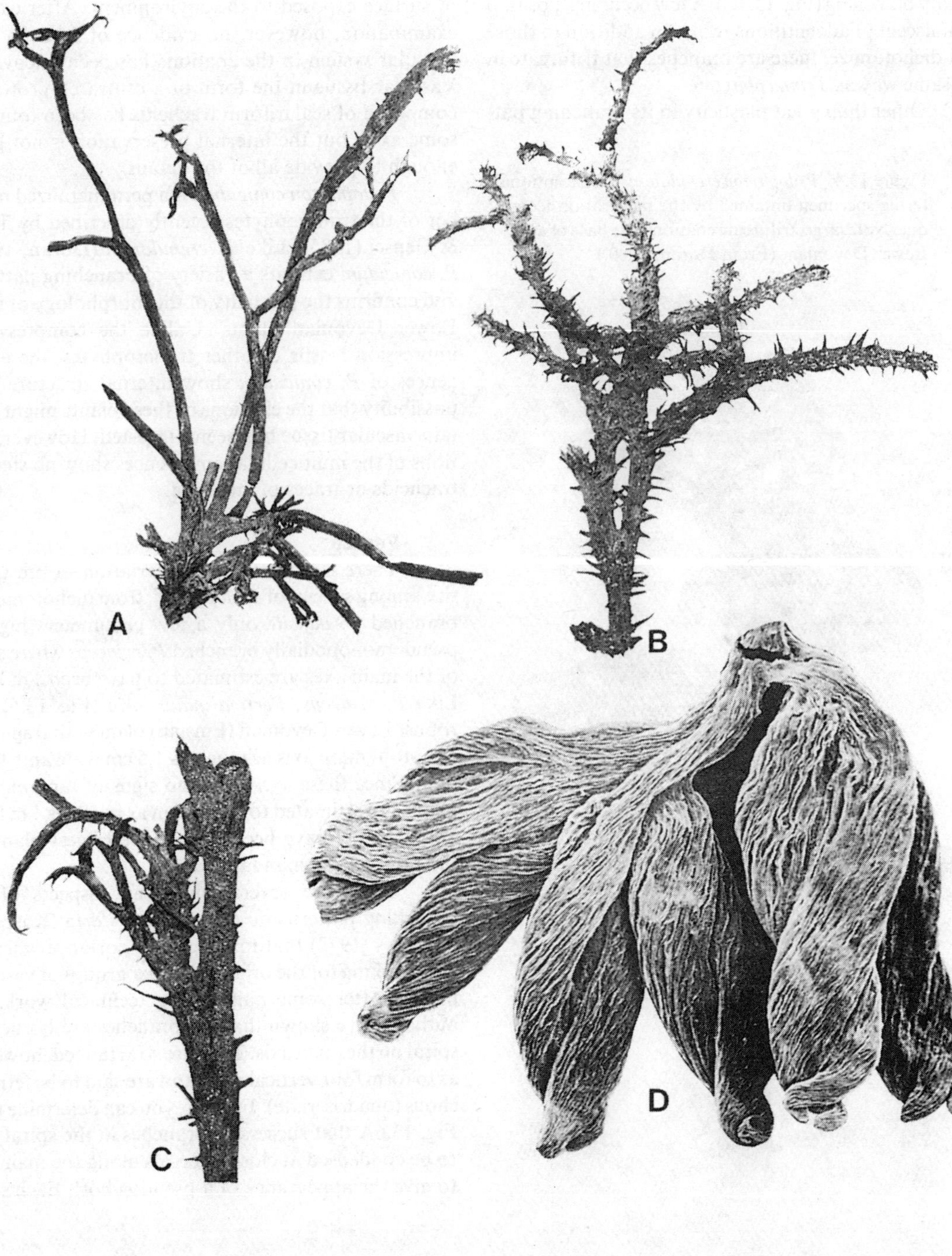

this investigator are compression fossils that are truly remarkable in their completeness and delineation of external morphology (Fig. 13.2D). Every degree of overtopping and reduction of the dichotomizing sterile telome trusses can be observed. There are main axes with dichotomizing laterals whose ultimate branches are recurved (Fig. 13.3A). These may be spirally arranged while other branches appear to be alternate and decussate. Often many branches are produced over a short distance on the main axis to give bushy appearance (Fig. 13.3B). Some branches appear to be the beginnings of leaves while others have an arrangement suggesting axillary branching (Fig. 13.3C). A few occur in a position reminiscent of adventitious roots. In addition to those that dichotomize, there are branches that trifurcate in the same way as *Trimerophyton*.

Other than great plasticity in its branching patterns, *P. crenulatum* has the most elaborate enations (Fig. 13.3B) of any Devonian plant we know of. They may be up to 6 mm long. Many are simple, but some dichotomize or trifurcate. Often they are condensed at shoot tips, while others are scattered on older portions of the dichotomizing axes so that they have the general appearance of lycopod rhizomes. A few dichotomizing axes are naked, especially those with recurved tips or bearing large clusters (up to 128) of banana-shaped sporangia (Figs. 13.3D, 13.4). There is a possibility that the spines served as photosynthetic or water absorbing structures by increasing the amount of surface exposed to the environment. After careful examination, however, no evidence of stomata or a vascular system in the enations has been uncovered. Vascular tissue in the form of a centrarch protostele composed of scalariform tracheids has been found in some axes, but the internal preservation is not good enough to provide all of the details.

Figure 13.4. *Psilophyton crenulatum*. An exceptional fertile specimen obtained by the maceration technique. Note large, trifurcate enation near base of axis. Lower Devonian. (From Doran, 1980.)

Psilophyton coniculum is a permineralized member of the trimerophytes recently described by Trant & Gensel (1985). Like *P. crenulatum* (Doran, 1980), *P. coniculum* exhibits a variety of branching patterns and confirms the plasticity of the morphology of these Lower Devonian plants. Unlike the compression–impression fossils of other trimerophytes, the emergences of *P. coniculum* show internal structure. The possibility that the enations of these plants might contain vascular tissue has been suggested. However, sections of the multicellular emergences show no signs of tracheids or traces of any kind.

Pertica

There is a considerable variation in presumed size among species of *Psilophyton*, from dichotomously branched *P. dapsile* only a few centimeters high to pseudomonopodially branched *P. princeps* where some of the main axes are estimated to have been 1 m long. Like *P. princeps, Pertica quadrifaria* (Fig. 13.5) is a robust Lower Devonian (Emsian) plant with fragments of a stout main axis as much as 1.5 cm wide and 45 cm long. Since these axes show no signs of tapering, the plants are estimated to have been as much as 1 m high. Perticas may have been among the largest plants of the Lower Devonian landscape.

There are several significant aspects of the branching patterns of *Pertica quadrifaria* (Kasper & Andrews, 1972) that may be of importance when we start looking for the origins of other groups of vascular plants. After some painstaking technical work, the authors have shown that the branches are borne in a spiral on the main axis. They are so arranged, however, as to form four vertical rows and are said to be tetrastichous (quadriseriate). Further, you can determine from Fig. 13.6A that successive branches in the spiral tend to be condensed at close intervals along the main axis to give the appearance of a pseudowhorl. Each set of four branches is separated from the next set by an unbranched portion of the axis.

All lateral branches are naked, essentially three-dimensional fertile or sterile telome trusses that dichotomize repeatedly (Fig. 13.6B). Fertile telomes are terminated by dense spherical clusters of tiny sporangia with scores of sporangia per cluster (Fig. 13.6C). Nothing is known of the spore type or method of dehiscence. The compression fossils failed to reveal any internal structure.

Figure 13.5. *Pertica quadrifaria.* Reconstruction. Plants may have been as much as 1 m high. Lower Devonian (From Kasper & Andrews, 1972.)

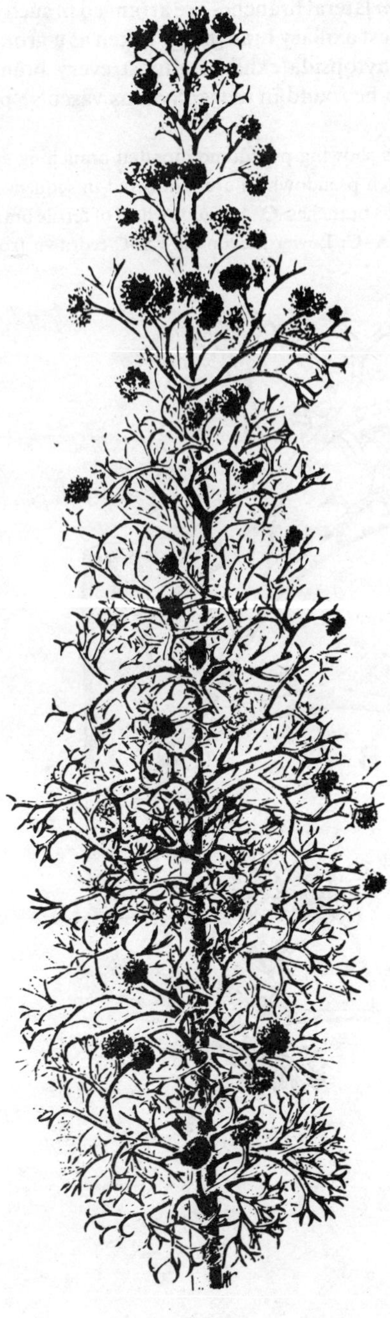

Since the discovery of *Pertica quadrifaria,* two other species have been described that give further insight into the Trimerophytopsida. Axes of *P. dalhousii* are described by Doran, Gensel, & Andrews (1978) to be as much as 3 m long. The lateral sterile telome trusses are borne in a spiral sequence on the main axes. They are three-dimensional and dichotomize. A few trichotomies of the type found in *Trimerophyton* are reported for *P. dalhousii and P. varia.* In *P. varia* (Granoff, Gensel, & Andrews, 1976), the lateral trusses tend to be in subopposite pairs with successive pairs decussate (at right angles to one another). Unlike other species of *Pertica,* the sterile trusses of *P. varia* are overtopped and show a degree of pseudomonopodial branching.

The tips of the dichotomously branched fertile trusses are terminated by pairs of fusiform sporangia. These are aggregated into clusters of up to a dozen sporangia with longitudinal dehiscence. Spores isolated from the sporangia resemble those of the dispersed-spore genus *Apiculiretuspisora.* They are small isospores, 56 to 90 μm in diameter.

Summary of evolutionary trends

One might wonder why so much attention has been given to the Trimerphytopsida. We have considered a majority of those species belonging to the class that are well documented. The reason for this rather detailed presentation is in recognition of the fact that the trimerophytes occupy a cardinal position in the "mainstream" of evolution of all those vascular plant groups that make their appearance in the Devonian and are yet to be considered. These are the Sphenopsida, Cladoxylales, and Coenopteridales of the Filicopsida, and the Progymnospermopsida.

There are so many obvious points of similarity between the Rhyniopsida and Trimerophytopsida that it prompted Banks et al. (1975) to conclude, "The trimerophytes can only be regarded as an elaboration of the rhyniophyte level of evolution resulting from an interplay of overtopping and elaboration in both anatomy and morphology." Some of the characteristics of the primitive rhyniophytes reflected in trimerophyte structure are terminal sporangia in which the sporangium wall is composed of several layers of differentiated cells similar to the axis on which they are borne; vascular tissue that extends into the bases of the sporangia; and branching systems that are three-dimensional, usually dichotomous, and invested with a protostelic vascular system. The gap, if there is one, between the rhyniophytes and trimerophytes is bridged by Trimerophytopsida of the *Psilophyton dapsile* type. In its size and sparse three-dimensional forked branching and naked axes, this species is nothing more than a rhyniophyte. The only characteristic that requires placing it in the Trimerophytopsida is the presence of paired elliptical sporangia in clusters. That this species was contemporaneous with the rhyniophytes of the Lower Devonian may have some significance if we interpret *P. dapsile* as a transitional form.

From this starting point and considering the characteristics of the Trimerophytopsida as a plexus from which other groups of vascular plants have evolved, we can see several well-defined evolutionary trends resulting from modification of the primitive type. That the evolutionary processes of overtopping and reduction have been effective in altering the primitive dichotomous system is evidenced by the evolution of trimerophytes with pseudomonopodial branching. The ultimate expression of this trend is in *Pertica* with its main axis and relatively small lateral telome trusses. The disposition of the laterals on the main axes varies from an irregular arrangement to those that are spiral (*Psilophyton forbesii* and *Pertica dalhousii*) and yet others where some branches are alternate (*Psilophyton princeps*), alternate decussate (*Pertica varia*), distichous (*P. dawsonii*), probably tristichous (*Trimerophyton*), and even tetrastichous (*Pertica quadrifaria*). In *P. quadrifaria*, we see what may be interpreted as a step in the evolution of the whorled branching pattern. Here successive branches of the spiral (four of them) are condensed and separated by a long segment of the axis from the next set of branches. In *Psilophyton crenulatum* lateral branches are arranged in such a way as to suggest axillary branching. Taken as a group, the Trimerophytopsida exhibits almost every branching pattern to be found in megaphyllous vascular plants.

Figure 13.6. **A.** *Pertica quadrifaria.* Diagram, portion of aerial axis showing pseudomonopodial branching system with three pseudowhorls of second-order branches. Branches of each pseudowhorl are numbered in sequence. **B.** *P. quadrifaria.* Sterile truss showing three-dimensional dichotomous branches. **C.** A pseudowhorl of fertile branches of *P. quadrifaria* showing clusters of terminal fusiform sporangia. **A–C:** Lower Devonian. (**A–C** redrawn from Kasper & Andrews, 1972.)

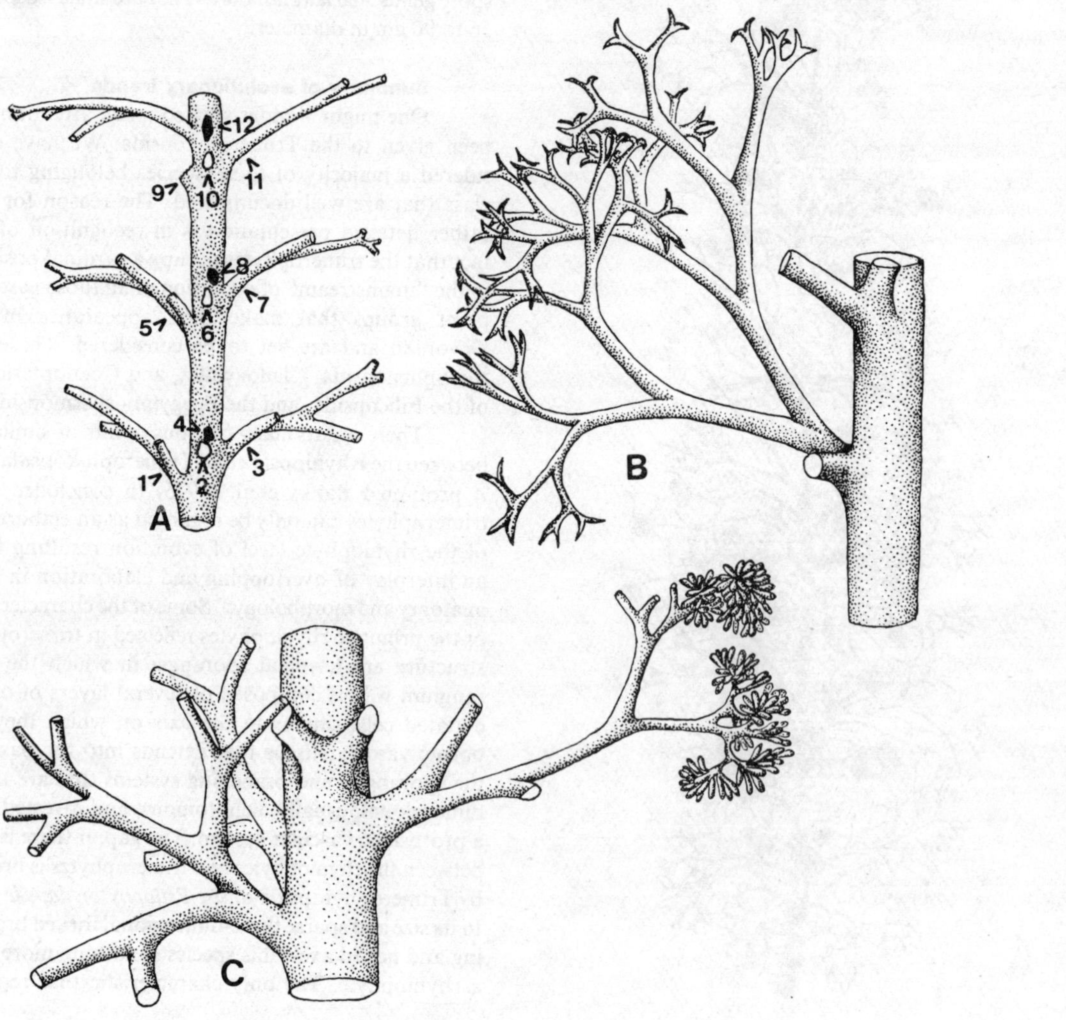

This is a remarkable example of an "evolutionary burst" preparing the way for further diversity found in later evolving megaphyllous plants.

Further modification of sterile laterals to form small, compact dichotomizing telome trusses (*Psilophyton dawsonii* and *P. crenulatum*) foreshadows the later appearance of planated and webbed megaphylls. By their position some of these smaller laterals suggest the possibility of being adventitious roots.

Accompanying the onset of evolutionary modification of the dichotomous branching system was the development of a robust growth form (*Psilophyton princeps* and species of *Pertica*). Although nothing is known of the internal structure of the perticas, their main axes are large enough in diameter, 1.5 cm, and long enough, over 3 m, to require the additional support that is usually supplied by secondary tissues. Further investigation is required here.

As in the Zosterophyllopsida some members of the Trimerophytopsida (*Psilophyton microspinosum*, *P. charientos*, and *P. crenulatum*) have evolved enations. In *P. crenulatum*, these have become more elaborate than any known for a Devonian plant. Just what their significance is in the evolution of plant organs such as microphylls or rootlike structures remains to be determined.

In contrast with the rapid evolution of their vegetative structures, the reproductive branches with their terminal, fusiform sporangia seem to have been conservative. In all cases, the branches bearing the sporangia are dichotomous and three-dimensional. The sporangia are usually paired and borne in clusters (Fig. 13.3D). Dehiscence, where it is known, is always longitudinal and on facing sides of sporangia pairs.

With the above summarizing overview at hand, we are ready to undertake the investigation of origins of other major groups of vascular plants.

References

Andrews, H. N., Kasper, A. E., & Mencher, E. (1968). *Psilophyton forbesii*, a new Devonian plant from northern Maine. *Bulletin of the Torrey Botanical Club*, **95**, 1–11.

Banks, H. P. (1968). The early history of land plants. In *Evolution and Environment*, ed. E. T. Drake. New Haven: Yale University Press, pp. 73–107.

Banks, H. P., Leclercq, S., & Hueber, F. M. (1975). Anatomy and morphology of *Psilophyton dawsonii* sp. n. from the late Lower Devonian of Quebec (Gaspé), and Ontario, Canada. *Palaeontographica Americana*, **8**, 77–126.

Chaloner, W. B. (1967). Spores and land plant evolution. *Review of Palaeobotany and Palynology*, **1**, 83–94.

Dawson, J. W. (1859). On fossil plants from Devonian rocks of Canada. *Quarterly Journal of the Geological Society, London*, **15**, 477–88.

Dawson, J. W. (1871). The fossil plants of the Devonian and Upper Silurian formations of Canada. *Geological Survey of Canada*, **1**, 1–92.

Doran, J.B. (1980). A new species of *Psilophyton* from the Lower Devonian of northern New Brunswick. *Canadian Journal of Botany*, **58**, 2241–62.

Doran, J. B., Gensel, P. G., & Andrews, H. N. (1978). New occurrences of trimerophytes from the Devonian of eastern Canada. *Canadian Journal of Botany*, **56**, 3052–68.

Gensel, P. G. (1979). Two *Psilophyton* species of eastern Canada with a discussion of morphological variation within the genus. *Palaeontographica B*, **168**, 81–9.

Gensel, P. G., & Andrews, H. N. (1984). *Plant Life in the Devonian*. New York: Praeger.

Granoff, J. A., Gensel, P. G., & Andrews, H. N. (1976). A new species of *Pertica* from the Devonian of eastern Canada. *Palaeonotographica, B*, **155**, 119–28.

Hopping, C. A. (1956). On a specimen of "*Psilophyton robustius*" Dawson from the Lower Devonian of Canada. *Proceeding of the Royal Society of Edinburgh*, **66**, 10–28.

Hueber, F. M. (1964). The psilophytes and their relationships to the origin of ferns. *Memoires of the Torrey Botanical Club*, **21**, 5–9.

Hueber, F. M. (1968). *Psilophyton:* the genus and the concept. In *International Symposium on the Devonian System, Calgary*, ed. D. H. Oswald. Calgary: Alberta Society of Petroleum Geologist, Vol. 1.

Hueber, F. M., & Banks, H. P. (1967). *Psilophyton princeps:* the search for organic connection. *Taxon*, **16**, 81–5.

Kasper, A. E., & Andrews, H. N. (1972). *Pertica*, a new genus of Devonian plants from northern Maine. *American Journal of Botany*, **59**, 897–911.

Kasper, A. E., Andrews, H. N., & Forbes, W. H. (1974). New fertile species of *Psilophyton* from the Devonian of Maine. *American Journal of Botany*, **61**, 339–59.

Trant, C. A., & Gensel, P. G. (1985). Branching in *Psilophyton:* A new genus from the Lower Devonian of New Brunswick, Canada. *American Journal of Botany*, **72**, 1256–73.

14

The origin of the Sphenopsida

To better understand the origin and subsequent evolution of the Sphenopsida, it is important to have in mind certain salient characteristics of their morphology and anatomy. Those characteristics that clearly delineate the class Sphenopsida are exemplified by the genus *Equisetum* (Fig. 14.1), the only extant member of the class.

Equisetum: general morphology, organography, and internal structure

No attempt is made here to provide all the details of the structure and life cycle of a horsetail. There are many fine textbooks in which this is accomplished. All that we need to consider are those characteristics germane to our interpretations of extinct members of the group.

Although there is some variability in size from small plants not more than 10 cm high to some robust species 10 m or more high, we usually think of *Equisetum* as being an herbaceous plant without secondary tissues. Irrespective of the species, horsetails have a distinctive jointed appearance reminiscent of a bamboo stem with its well-defined nodes and internodes. At the nodes of *Equisetum* stems there are verticels (whorls) of small leaves, which are more or less fused to form a leaf sheath. Unlike many vascular plants where branches are produced in the axils of leaves (gymnosperms, angiosperms, and some ferns), branches of *Equisetum* alternate with the leaves at the node (Fig. 14.2B). This is an important characteristic of most Sphenopsida. The shoot internodes are marked with vertical ribs that reflect the position of the internal primary vascular system. The aerial shoots arise from horizontal, underground rhizomes that share the characteristics of the shoots. In addition, rhizomes produce adventitious roots at nodes along with branches and a leaf sheath.

The spores of these free-sporing plants are produced in groups of sporangia at the end of a peltate vascularized branch (sporangiophore) (Fig. 14.2C). Whorls of sporangiophores are organized into terminal cones on the aerial shoots (Fig. 14.2D). The isospores in a sporangium are unique in having an outer layer deposited on the spore wall in the form of paddlelike bands called elaters (Fig. 14.2E,F). These aid in spore dispersal at the time the sporangia dehisce along an inward-facing longitudinal slit.

The most conspicuous features of a cross section through an internode of a mature *Equisetum* shoot (Fig. 14.2A) are the canals and cavities. We see a central pith cavity, a ring of protoxylem canals (carinal canals), and an outer ring of cortical canals (vallecular canals). These are formed by a highly specialized mode of growth. As in other vascular plants, new leaves and branches of *Equisetum* are produced by the apical meristem, but most of the length of the stem results from the activity of intercalary meristems that are located just above each node. When a stem is initiated, the apical meristem produces a large number of tightly packed whorls of leaves. Later, the intercalary meristems become active. This causes the extremely rapid internodal elongation that tears apart the mature protoxylem elements to produce the carinal canals. This dramatic elongation also leaves empty spaces in the pith and cortex (vallecular canals). The characteristics of the carinal canals (Fig. 14.2G) are of particular interest when it comes to determining origins and relationships of sphenopsids. Careful observation of the canals shows stretched and distorted annular and spiral wall thickenings from the ruptured protoxylem cells. Metaxylem occurs in two strands that partially flank the canal to the outside. Phloem differentiates between the two metaxylem strands. On the radius and to the outside of each vascular bundle, there is a rib composed of thick-walled cells. Stomata occur in the furrows between the ribs, and they open into substomatal cavities surrounded by photosynthetic tissues.

The vascular bundles remain unbranched until they reach the level of a node. Here, each trifurcates (Fig. 14.2B), the middle branch of the trifurcation becoming the vein of a leaf. The two lateral members of the trifurcate bundle each join with a lateral strand from an adjacent trifurcate bundle. These anastomosing strands become the vascular bundles of the next internode. This has the effect of producing a vascular bundle pattern where the bundles of one internode alternate with those of internodes above and below.

Figure 14.1. *Equisetum arvense*. Note fertile and sterile shoots with distinct verticles of appendages at nodes. Extant. (Photograph courtesy of the Field Museum of Natural History, Chicago.)

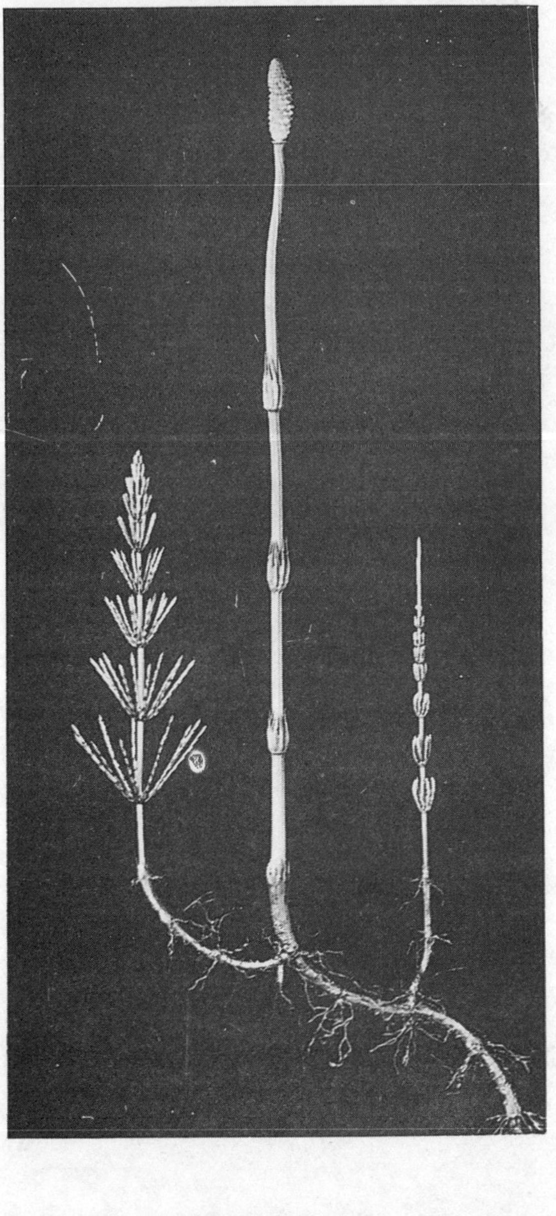

This is another important characteristic we will use in our evaluation of sphenopsid fossils.

A plate of pith tissue with a ring of intercalary metaxylem around it occurs at the node. There are no vallecular or carinal canals at this level. In developing shoots there is an intercalary meristem at the node, and in addition to the primary tissues of the main shoot, it produces the primordia of branches that alternate with the leaves.

The stele of *Equisetum* has been called a siphonostele or a eustele of the *Equisetum* type. We will see, however, in terms of its probable evolutionary origin, that it is neither of these, but had an independent origin. Like a eustele it is composed of a ring of branching sympodia (axial vascular bundles) that give rise to leaf traces. The latter, however, are not associated with leaf gaps, because these do not occur in the Sphenopsida. Based on structural features, the leaves of sphenopsids are microphylls, but their evolutionary origin is quite different from those Lycopsida whose microphylls apparently evolved from enations.

The first Sphenopsida

Even before 1930, when Zimmermann established the group, it was customary to think of the Hyeniales as Middle Devonian ancestors of the Sphenophyllales and Equisetales. In 1973 a new potential precursor, *Ibyka* (Fig. 14.3A), was described from the Givetian (Middle Devonian) by Skog & Banks, and this genus has recently been placed in the new order Iridopteridales by Stein (1982). Skog & Banks interpret *Ibyka* as transitional between the Trimerophytopsida and the somewhat more highly evolved Hyeniales. To understand their conclusion we have to know something of the characteristics of *Ibyka,* which were determined by the study of compression fossils where segments of the axes were permineralized with pyrite and limonite.

The largest specimen of *Ibyka* is 55 cm long with branches that extend as much as 30 cm on either side. Since the main axis of this specimen showed no signs of tapering, the authors assumed that the plant was much larger. In their branching pattern, specimens of *Ibyka* have a pseudomonopodial axis with spirally arranged overtopped laterals of a first and second order. Ultimate three-dimensional dichotomizing laterals are spirally arranged on all orders of branches. These reduced telome trusses are either sterile with recurved tips and represent the precursors of leaves or are fertile with small obovoid sporangia terminating the ultimate branches. It is of interest to note that often the ultimate appendages are in such a close spiral that they appear to be in whorls, a feature of the trimerophyte *Pertica quadrifaria.*

Because portions of the axes and appendages of *Ibyka* are permineralized, it is possible to say

Figure 14.2. **A.** Diagram transverse section of *Equisetum* internode. Pith cavity (p); carinal canal (c); vallecular canal (v). Extant. B. Vascular system of *Equisetum* shoot expanded into one plane. Node (n); leaf trace (lt); branch trace (bt); leaf (l). Note that branch traces and leaf traces alternate at a node. (Redrawn from Foster & Gifford, 1974.) **C.** A single sporangiophore of *E. maximum*. **D.** *E. arvense*, fertile branch terminated by cone with whorls of sporangiophores. **E.** *Equisetum* spore with coiled elaters. **F.** *Equisetum* spore with expanded elaters. (**C–F** from Eames, 1936.) **G.** *Equisetum* sp., protoxylem canal in transverse section. Note thick-walled protoxylem cells. (Redrawr from Bierhorst, 1971.)

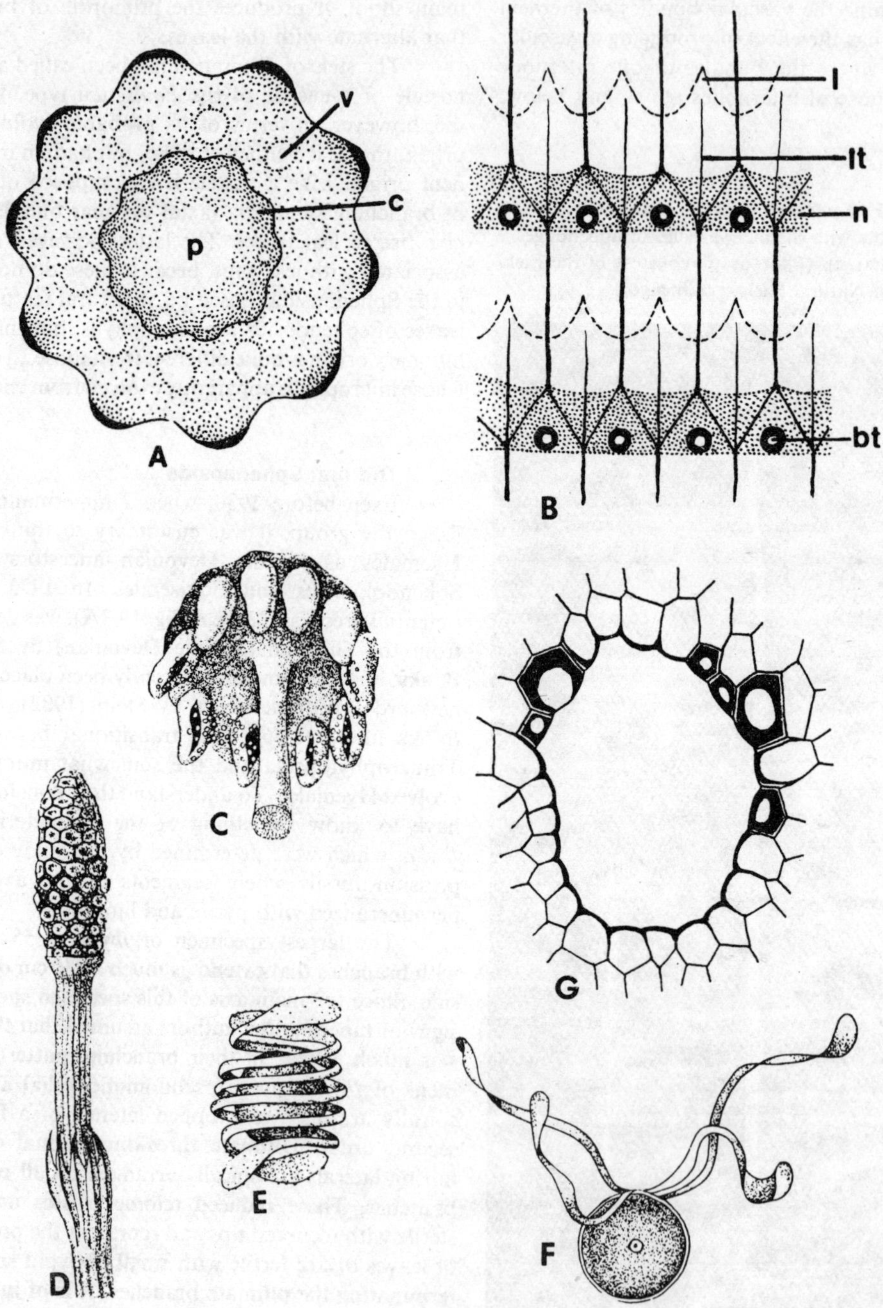

something about their internal structure. Stem axes contain a primary vascular system that is an actinostele with five or six arms and primary xylem that appears to be mesarch. Of particular interest are the lacunae near the tips of the arms that were formed by the disintegration of protoxylem cells. Although these lacunae were interpreted as having formed in the same way that carinal canals are formed in *Equisetum*, there is no evidence that *Ibyka* underwent the dramatic internodal elongation that characterizes *Equisetum*. The presence of lacunae in the axes of *Ibyka* is an important characteristic that indicates what the relationship of *Ibyka* is to the Sphenopsida. There are fragments of annular protoxylem cells in the protoxylem canals of *Ibyka*. The metaxylem cells that make up the bulk of the actinostele are mostly scalariform tracheids. A few of these have thickenings between the scalariform bars that delineate circular pits.

Terete traces arise from the tips of the arms of the actinostele in a simple spiral and then enter the dichotomized, ultimate sterile trusses. The fact that terete traces are known to supply megaphylls of more highly evolved plants supports the contention that the spirally arranged sterile telome trusses of *Ibyka* with their terete traces are, indeed, leaf precursors.

Other genera placed in the Iridopteridales by Stein (1982) include *Arachnoxylon*, *Iridopteris*, and *Asteropteris*. They have actinosteles with protoxylem canals near the tips of the arms, and most produce whorled laterals. However, the fertile parts of these genera have not been discovered, and there is no evidence in the Iridopteridales of the specialized mode of intercalary growth that produces the carinal canals in *Equisetum*. For these reasons, the relationships of the Iridopteridales to *Equisetum* remain uncertain.

Another Devonian genus of questionable affinities is *Protohyenia* (Fig. 14.3B). It is tentatively placed with the Sphenopsida as a transitional form because of its pseudomonopodial axes bearing dichotomizing laterals that tend to be in verticels. Most of these are terminated by erect, oval sporangia. There is some indication of a vascular system that is an actinostele with three or more arms. Although the specimens are about 6 cm high and in the size range of *Hyenia*, it has been suggested that *Protohyenia* may represent the branches of *Pseudosporochnus*, a Devonian tree form.

At present the three genera, *Ibyka, Protohyenia*, and *Arachnoxylon*, are believed by Skog & Banks (1973) to be transitional between trimerophytes and the Hyeniales.

Hyeniales

Traditionally, two genera are recognized as belonging to the Hyeniales. They are *Hyenia* and *Calamophyton*, which are usually found in rocks of late Middle Devonian (Givetian) age from Europe.

Figure 14.3. **A.** *Ibyka amphikoma*, reconstruction of habit and sterile and fertile branches. Middle Devonian. (From Skog & Banks, 1973.) **B.** *Protohyenia janovii*. Devonian. (Based on reconstruction by Ananiev, 1957.)

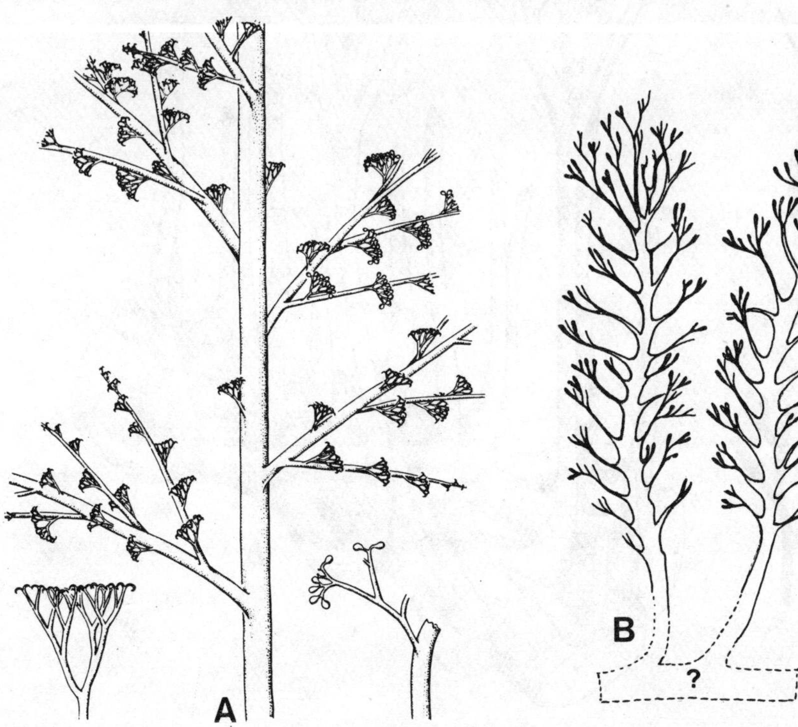

Reinvestigation of the original as well as new material of these genera and the putative fern *Cladoxylon scoparium* (Leclercq, 1940, 1961; Leclercq & Schweitzer, 1965; Schweitzer, 1972, 1973; Schweitzer & Giesen, 1980) has led these authors to the conclusion that *Hyenia* and *Calamophyton* have affinities with the Cladoxylales of the class Filicopsida. The accumulated evidence for such a relationship with the order is convincing. That the Cladoxylales are closely related to other members of the Filicopsida is, however, equivocal. It seems to us that those plants placed in the order Hyeniales are best construed as a transitional group not far removed from their trimerophyte ancestors and sharing characteristics with the sphenopsids, Cladoxylales, and some lycopods. The reasons for this are presented later.

Hyenia

In the most recent account of its growth habit, *Hyenia elegans* (Fig. 14.4A) is depicted by Schweitzer (1972) as having stout dichotomizing rhizomes bearing upright aerial shoots that are spirally arranged. Usually the shoots are repeatedly branched pseudomonopodially or dichotomously toward the distal end where they form a bushy crown. The branches of the crown bear lateral sterile and fertile ultimate appendages. The sterile appendages (Fig. 14.4B), which are borne on branches separate from fertile ones, branch dichot-

Figure 14.4. **A.** Reconstruction of *Hyenia elegans* with sets of fertile and sterile branches. Note digitate branching. (Modified from Schweitzer, 1972.) **B.** Enlarged leaves of *H. elegans*. **C.** Sporangiophore enlarged. **A–C:** Middle Devonian. (**B**,**C** redrawn from Leclercq, 1940.)

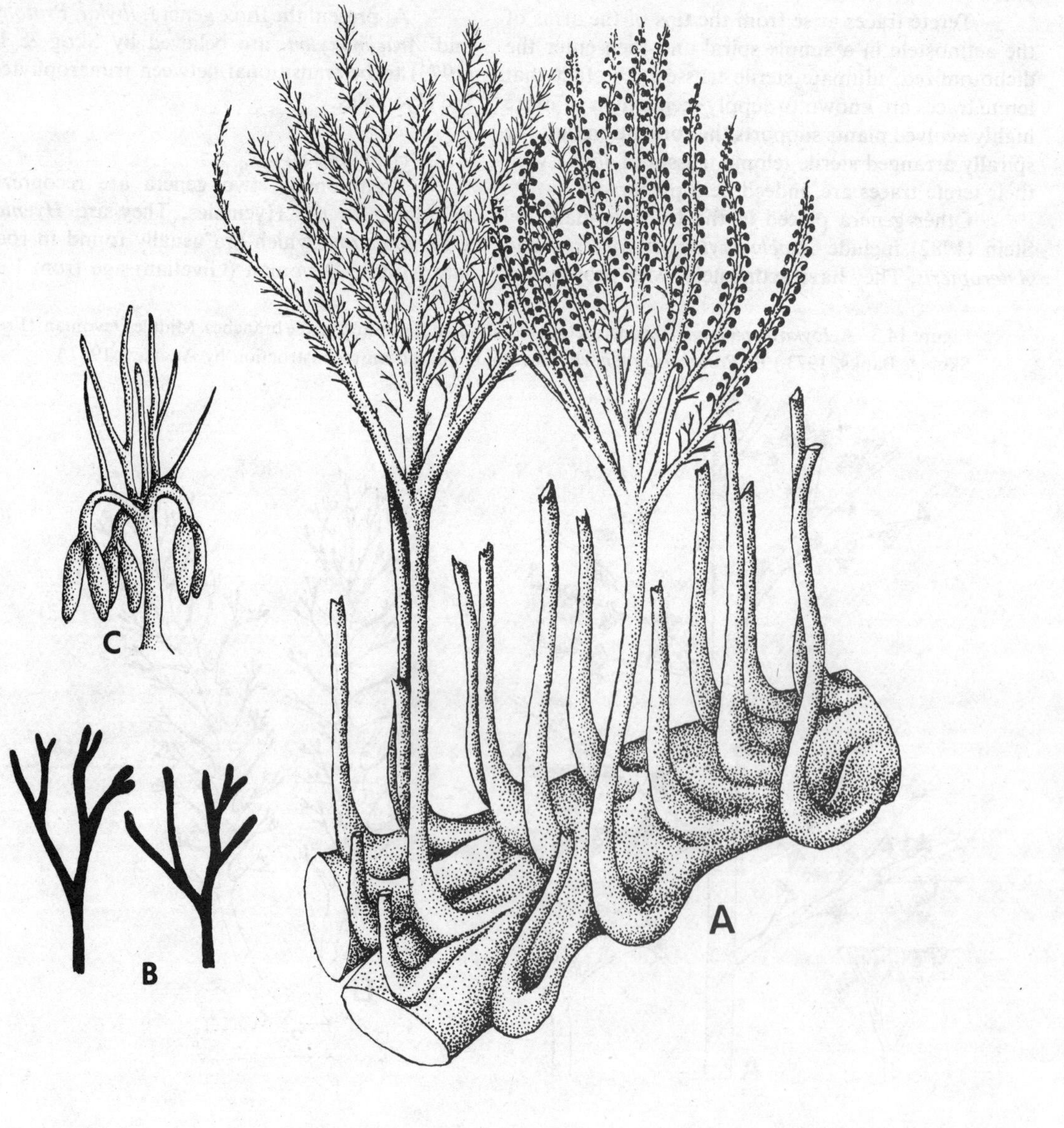

omously one or two times to form telome trusses that unquestionably functioned as leaves. These foliar units vary in arrangement on the sterile shoots from a shallow spiral to pseudoverticillate (almost in whorls).

According to the original reconstruction of *Hyenia elegans* by Kräusel & Weyland (1932), the fertile appendages are shown to be dichotomous sporangiophores with recurved tips, each bearing two to four sporangia. These were borne in pseudowhorls on the fertile shoots where they formed a loose spike. By meticulous degagment (the painstaking process of uncovering plant parts in a matrix of rock with needles), Leclercq (1961) was able to show that the sporangiophores of *H. elegans* (Fig. 14.4C) were more elaborate structures than originally described by Kräusel & Weyland. Instead of two to four sporangia per sporangiophore, Leclercq, found that the fertile units are borne in pairs on the fertile shoots. Each sporangiophore of the pair is terminated in a filiform sterile projection and has three lateral recurved branches each of which is terminated by a pair of fusiform sporangia. Thus, a sporangiophore of *H. elegans* has six sporangia each of which is dehisced by a longitudinal slit.

Figure 14.5. **A.** *Calamophyton bicephalum.* Habit restoration illustrating digitate branching with one fertile and one sterile branch. **B.** Sterile appendages of *C. bicephalum.* **C.** Fertile appendages. **A–C:** Middle Devonian. (**A–C** redrawn and modified from Leclercq & Andrews, 1960.) **D.** Restoration of upper portion *Pseudobornia ursina.* (After Schweitzer, 1967.) **E.** Node of *P. ursina* with whorl of leaves. Upper Devonian. (Redrawn from Nathorst in Hirmer, 1927.)

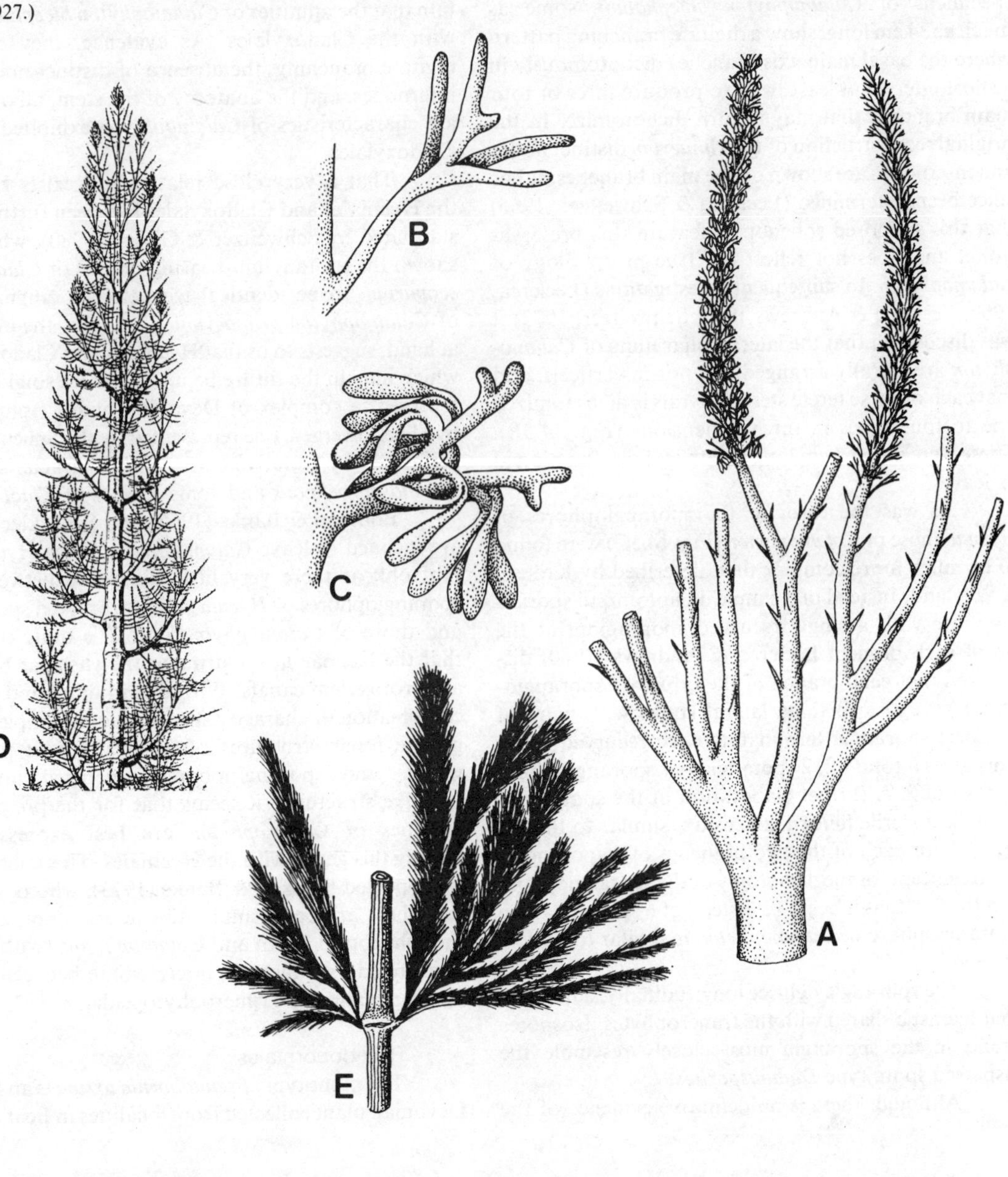

Unfortunately, very little is known about the vascular system of *Hyenia* that might help in determining its relationships. If we depend on external morphology, there are four characteristics that suggest that the genus belonged to a group that was ancestral to the Sphenopsida. These are (1) small sterile telome trusses of determinate growth that functioned as leaves, (2) sporangiophores with recurved tips bearing terminal sporangia, (3) aggregations of sporangiophores into lax conelike structure, and (4) a tendency for fertile and sterile appendages to be arranged in a shallow spiral or pseudowhorls as might be expected to occur in ancestors of the Sphenopsida.

Calamophyton

The external morphology of *Calamophyton* (Fig. 14.5A) is quite similar to that of *Hyenia*. Compressed specimens of *Calamophyton bicephalum*, some as much as 34 cm long, show a digitate branching pattern where the basal main axis branches dichotomously in a short interval at least twice to produce three or four main branches that may in turn dichotomize. In the original reconstruction of *C. primaevum* distinct nodes and internodes are shown on the main branches. It has since been determined (Leclercq & Schweitzer, 1965) that this supposed sphenopsid feature was preservational and does not reflect the true morphology of *Calamophyton*. In subsequent investigations (Leclercq, 1960; Leclercq & Andrews 1960; Schweitzer, 1973) it was discovered that the lateral foliar units of *Calamophyton* are spirally arranged and not in verticels, and that each of these terete sterile laterals is dichotomized one to four times in three dimensions (Fig. 14.5B). These were about 1 cm long and must have functioned as leaves.

As was the case for the sporangiophores of *Hyenia*, those of *Calamophyton* (Fig. 14.5C) were found to be much more complex than described by Kräusel & Weyland. Instead of a simple dichotomized sporangiophore with a single recurved sporangium at the tip of each branch, Leclercq & Andrews (1960) discovered that each branch of the bifurcate sporangiophore gave rise to three lateral stalks with pairs of fusiform sporangia terminating each recurved stalk. This gives a total of 12 sporangia per sporangiophore instead of 2. A further elaboration of the sporangiophore is a sterile filiform extension, similar to that of *Hyenia*, for each of the two branches of a sporangiophore. Except for the dichotomy of the sporangiophore and the fact that it bears 12 instead of 6 sporangia, the sporangiophore of *Calamophyton* is similar to that of *Hyenia*.

The sporangia dehisce longitudinally, a familiar characteristic shared with the trimerophytes. Isospores found in the sporangia most closely resemble the dispersed spore type *Diabolisporites*.

Although there is no definitive evidence of the configuration of vascular tissue in *Hyenia*, fragments of small stems of *Calamophyton bicephalum* were discovered by Leclercq & Schweitzer (1965) to contain the remains of the vascular system. It is best described as multifasicular, comprised of many bundles some of which are terete; others are straight or U- or V-shaped. Near the periphery of the stem, in tips of the radiating arms of the U- or V-shaped strands, it is possible to identify the position of the mesarch primary xylem by a lacuna. The nature of the lacuna is problematical. Is it a protoxylem canal as demonstrated for *Ibyka?* Was it filled with parenchyma that has disintegrated and is thus homologous with the peripheral loop of certain ferns? Or, was it distinct from both of these? In the final analysis, it is this characteristic that has to be determined in classifying *Calamophyton*. Leclercq & Schweitzer (1965) maintain that the affinities of *Calamophyton bicephalum* are with the Cladoxylales. As evidence, they cite the digitate branching, the absence of distinct nodes and internodes, and the anatomy of the stem, all of which are characteristics of *Calamophyton* exhibited by the Cladoxylales.

That a very close relationship exists between the Hyeniales and Cladoxylales has been further substantiated by Schweitzer & Giesen (1980), who have shown the sporangium-bearing organs of *Cladoxylon scoparium* to be identical with the sporangiophores of *Calamophyton primaevum*. This, plus the evidence at hand, suggests to us that Hyeniales and Cladoxylales, which may in the future be united into a single order, represent a complex of Devonian plants from which two lines diverged; one represented by the Sphenopsida, the other by putative ferns with characteristics of *Pseudosporochnous* and some species of *Cladoxylon*.

Bonamo & Banks (1965) and Scheckler (1974) are inclined to leave *Calamophyton* in the Hyeniales. Not only is there very little difference between the sporangiophores of *Hyenia* with its recurved sporangia and those of *Calamophyton*, but it is quite possible that the lacunae in the arms of the vascular bundles are protoxylem canals. With this in mind and with a combination of characteristics, including simple determinate foliar structures often arranged in pseudowhorls and sporangiophores organized into lax conelike structures, it seems that for the present the affinities of *Calamophyton* are best expressed by leaving this genus with the Hyeniales. This conclusion is supported by Skog & Banks (1973), who point out that there are basic similarities in anatomy between the sphenopsid *Ibyka* and *Calamophyton*, with *Ibyka* interpreted as somewhat intermediate between *Calamophyton* and the Trimerophytopsida.

Pseudoborniales

The monotypic *Pseudobornia ursina* is an Upper Devonian plant collected from localities in Bear Island

and Alaska. Sufficient material has been studied (Schweitzer, 1967) to give us a clear picture of *Pseudobornia* (Fig. 14.5D) as an unequivocal member of the Sphenopsida. Larger specimens indicate that *P. ursina* achieved a tree form comprising a central trunk 15 to 20 m high with distinct nodes and internodes. Basal portions of the trunk have been found that are 60 cm in diameter and that have internodes of 80 cm. The nodes of the trunk bore one or two branches some of which were 3 m long. These first-order branches show some evidence of longitudinal ribbing that extends through the nodes, a feature of another Upper Devonian sphenopsid, *Archaeocalamites*. First-order branches of *Pseudobornia* have second-order branches attached mostly in a decussate manner. These, in turn, bear ultimate branches, usually in a distichous arrangement.

The leaves (Fig. 14.5E) are produced in distinct whorls on the ultimate branches. There are about four leaves per whorl, each of which dichotomizes two or three times near the point of attachment. Each of the "leaflets" of the dichotomized leaf consists of a lamina that is pinnately dissected along the margin.

The fertile units are produced at the distal ends of first-order branches at the upper end of the trunk. They consist of whorls of bracts and sporangiophores forming a conelike structure. The sporangiophores are divided into two segments with approximately 30 sporangia borne on their distally recurved tips. Schweitzer (1967) supports the idea that the affinities of this group are with the Equisetales, while others suggest that, because of the structure of the fertile units, the relationships are with the Sphenophyllales. It is also possible that *Pseudobornia* represents another divergent line of the Sphenopsida that ended in extinction.

Origin of the sphenopsid stele

In the first part of this chapter, we described the stele in the stem of *Equisetum*. It should be apparent that the fossil Sphenopsida thus far described do not have siphonosteles or eusteles. Instead, all are protostelic with radiating arms (actinosteles and plectosteles), when seen in transverse section. The question raised by this observation is: How did the sphenopsid stele, present in the Calamitales and Equistales, evolve from the ancestral actinostele? There are several Devonian genera whose stem anatomy is known that seem to provide some answers. These genera have actinosteles that are basically similar to, but more elaborate than, those of *Ibyka* and *Arachnoxylon*. Among these are *Asteropteris* (Fig. 14.6A) and *Langoxylon*, where the well-developed radiating arms are joined in the center of the stele by relatively smaller amounts of metaxylem. The tips of the arms of *Asteropteris* have conspicuous protoxylem lacunae arranged in a ring as in *Arachnoxylon*.

From this actinostelic starting point, Arnold (1952) suggested that a phylogenetic dissection occurred by failure of the central metaxylem cells to differentiate. This would produce a multifasicular protostele of the *Calamophyton–Cladoxylon* type (Fig. 14.6B). If you agree to this part of the hypothesis, then it would not be surprising to find a genus (*Calamophyton*) with multifasicular xylem strands and a ring of protoxylem canals as it occurs among many Sphenopsida.

This idea is supported by Bierhorst (1971), who suggests that the next step in evolution of the sphenopsid stele would be the complete loss of almost all of the centripetal metaxylem in the stele of the *Calamophyton* type. This would leave a central pith surrounded by a ring of mesarch bundles each with a protoxylem canal (Fig. 14.6C). Long ago it was observed that the bundles in an *Equisetum* stem tend to be mesarch, a condition reflected in the early differentiation of its primary xylem. In mature stems the centripetal xylem usually is missing. Figure 14.6 diagrams the main steps in the hypothetical evolutionary origin of the sphenopsid stele from protostelic ancestors. We will be exposed to additional evidence supporting these ideas when we come to the Equisetales and the Progymnospermopsida.

Figure 14.6. Model depicting stages in the evolution of an equisetalean stele. **A.** Actinostele with protoxylem canals at tips of radiating arms as in *Asteropteris*. Metaxylem, black. **B.** Plectostele produced by reduction of metaxylem development as in *Calamophyton*. **C.** Restriction of metaxylem to the area of the protoxylem canals to produce a stele in *Equisetum*.

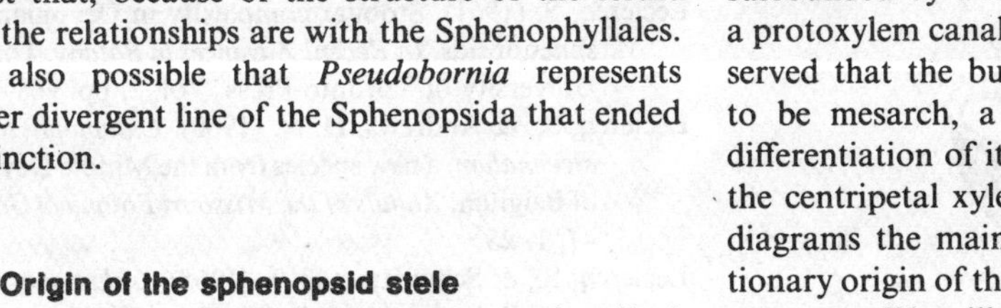

Origin of the sporangiophore

You may recall from an earlier chapter that recurvation was described as one of the evolutionary processes operative on the primitive telomic branching system (Fig. 9.12). The genera *Protohyenia* (Fig. 14.3B), *Arctophyton* (Fig. 14.7A), and *Hyenia* (Fig. 14.4C) provide us with evidence of how recurvation in conjunction with overtopping and reduction of fertile telomes (Fig. 14.7B,C) produced sporangiophores typical of the Sphenopsida. The sporangia in the tips of the fertile telome trusses of *Protohyenia* and *Ibyka* tend to be erect, while those of *Arctophyton, Hyenia,* and *Calamophyton* show the effects of recurvation of the fertile tips of the truss. Overtopping within the truss is apparent in *Protohyenia* and *Arctophyton,* while the sporangiophores of *Hyenia* show further reduction of the sporangiophore axes bringing the sporangia into close proximity. All of the branches of the sporangiophores of *Arctophyton* terminate in recurved sporangia (Fig. 14.7A), but those of *Hyenia* and *Calamophyton* also produce several erect sterile branches (Figs. 14.4C; 14.5C). This later morphology is remarkably similar to both the multisporangiate sporophylls of the lycopod genera *Protolepidodendron* and *Estinnophyton* (Fig. 10.9D) and to the fertile structures of some sphenophyllaleans that are discussed in the next chapter (Fig. 15.6A). The possible evolutionary implications of these similarities are summarized at the end of Chapter 16 (Chart 16.1) with the completion of the Sphenopsida.

Figure 14.7. **A.** *Arctophyton gracile.* Reconstruction of sporangiophore. Upper Devonian. (Redrawn from Schweitzer, 1968.) **B.** Hypothetical stage in the evolution of the sporangiophore showing recurvation. Compare with *Arctophyton* at **A. C.** Formation of peltate sporangiophores of the sphenopsid type. (**B,C** based on Zimmermann, 1952.)

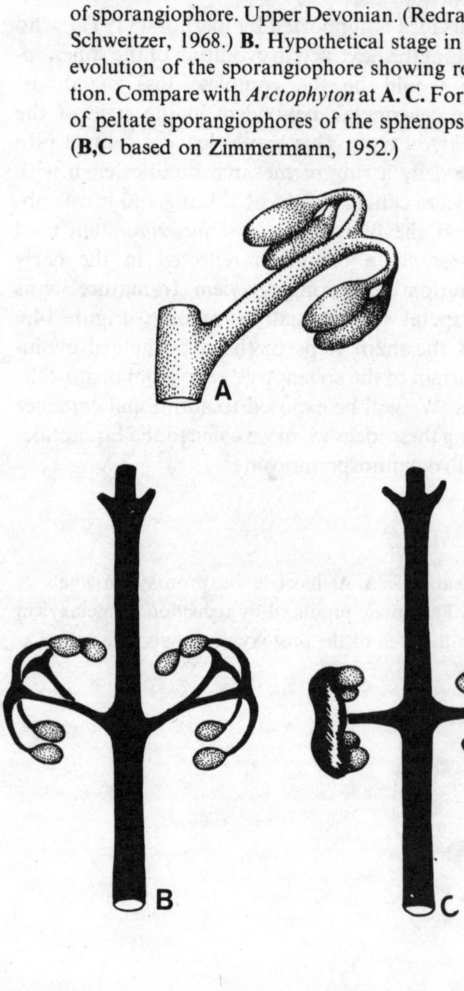

References

Ananiev, A. P. (1957). New Lower Devonian fossil plants from Torgachine of the southeast of western Siberia. *Akademia Nauk USSR* (*Botany*), **42,** 691–702.

Arnold, C. A. (1952). Observation on fossil plants from the Devonian of eastern North America. VI. *Xenocladia medullosina* Arnold. *Contributions of the Museum of Paleontology, the University of Michigan,* **9,** 297–309.

Bierhorst, D. W. (1971). *Morphology of Vascular Plants.* New York: Macmillan.

Bonamo, P. M., & Banks, H. P. (1965). *Calamophyton* in the Devonian of New York state. *American Journal of Botany,* **53,** 778–91.

Eames, A. J. (1936). *Morphology of Vascular Plants.* New York: McGraw-Hill.

Foster, A. S., & Gifford, E. M. (1974). *Comparative Morphology of Vascular Plants,* 2nd ed. San Francisco: Freeman.

Hirmer, M. (1927). *Handbuch der Paläobotanik.* Munich and Berlin: Oldenbourg.

Kräusel, R., & Weyland, H. (1932). Pflanzenreste aus dem Devon III. Ueber *Hyenia* Nath. *Senckenbergiana,* **14,** 275–79.

Leclercq, S. (1940). Contribution à l'étude de la flore du Dévonian de Belgique. *Mémoires, L'Académie Royal, de Belgique, Sciences,* **12,** 3–65.

Leclercq, S. (1961). Stobilar complexity in Devonian sphenopsids. In *Recent Advances in Botany.* Toronto: University of Toronto Press. Vol. 2, pp. 968–71.

Leclercq, S., & Andrews, H. N. (1960). *Calamophyton bicephalum,* a new species from the Middle Devonian of Belgium. *Annals of the Missouri Botanical Garden,* **47,** 1–23.

Leclercq, S., & Schweitzer, H. J. (1965). *Calamophyton* is not a Sphenopsid, *Bulletin, L'Académie Royal de Belgique, Sciences,* **51,** 1395–1403.

Scheckler, S. E. (1974). Systematic characters of Devonian ferns. *Annals of the Missouri Botanical Garden,* **61,** 462–73.

Schweitzer, H. J. (1967). Die Oberdevon-flora der Bäreninsel. 1. *Pseudobornia ursina* Nathorst. Palaeontographica *B,* **120,** 116–37.

Schweitzer, H. J. (1968). Pflanzenreste aus dem Devon Nord-Westspitzbergens. *Palaeontographica B,* **123,** 43–75.

Schweitzer, H. J. (1972). Die Mittledevon-flora von Lindlar (Rheinland). 3. Filicinae-*Hyenia elegans* Kräusel & Weyland. *Palaeontographica B,* **137,** 154–75.

Schweitzer, H. J. (1973). Die Mitteldevon-flora von Lindler (Rheinland). 4. Filiciane-*Calamophyton primaevum* Kräusel & Weyland. *Palaeontographica B,* **140,** 117–50.

Schweitzer, H. J., & Giesen, P. (1980). Ueber *Taeniophyton inopinatum, Protolycopodites devonicus,* und

Cladoxylon scoparium aus dem Mitteldevon von Wuppertal. Palaeontographica B, **173,** 1–25.

Skog, J. E., & Banks, H. P. (1973). *Ibyka amphikoma* gen. et sp. n., a new protoarticulate precursor from the late Middle Devonian of New York state. *American Journal of Batany,* **60,** 366–80.

Stein, W. E., Jr. (1982). *Iridopteris eriensis* from the Middle Devonian of North America, with systematics of apparently related taxa. *Botanical Gazette,* **143,** 401–16.

Zimmermann, W. (1952). The main results of the telome theory. *Palaeobotanist,* **1,** 456–70.

15

Unique and extinct: the Upper Paleozoic sphenophylls

If we return to the reconstruction of the Carboniferous swamp (Fig. 11.1), a cursory inspection will reveal that part of the understory vegetation consisted of members of the Sphenophyllales. As we study the fossils of this group, we will see that they do share some characteristics with extant *Equisetum* in their verticels of leaves at distinct nodes and the presence of protoxylem canals in their stems. In other characteristics, however, the Sphenophyllales are quite unlike other members of the class Sphenopsida.

Sphenophyllales

As originally conceived by Brongniart in 1828, the genus *Sphenophyllum* (Fig. 15.1) applied to compression fossils of an herbaceous Carboniferous plant with whorls of wedge-shaped leaves attached to distinctly jointed stems. Over the years the genus has been expanded to include remains of branches, roots, and cones, which have been found in organic connection. Our knowledge of the internal structure of *Sphenophyllum* has been greatly amplified by beautifully preserved permineralized specimens.

Although there are reports of *Sphenophyllum* in the Upper Devonian, the genus did not reach its zenith until the Upper Carboniferous, only to disappear by the end of the Permian.

General morphology

The exact habit of the *Sphenophyllum* is not known. In general, however, present evidence (Eggert & Gaunt, 1973; Batenburg, 1977) indicates a plant with a scrambling habit where slender aerial axes, not more than 7 mm in diameter but several meters long, grew up from the floor of the swamp to be supported by the surrounding vegetation. It is quite possible that the aerial shoots were borne on a rhizome. The shoots are known to branch dichotomously in the distal portions and to bear lateral branches, usually one per node. However, several branches per node have been reported. The branches tend to alternate with leaves at each node. In *Sphenophyllum emarginatum* a single branch is borne at every third node and all of the branches are produced in a spiral.

The number of leaves per verticel is usually 6 or 9 and is related to the trimerous internal structure of the stem. In some cases, where the leaves are small, up to 18 leaves may be present at a node. Even the largest leaves of *Sphenophyllum* are usually less than 2 cm long. They are supplied at the constricted base with one or two veins that dichotomize two to six times (Fig. 15.2A,B) before reaching the distal extremities of the leaf. The leaves of *Sphenophyllum* vary in external morphology from those that have the appearance of a planated sterile telome truss without webbing (Fig. 15.2A) to those which are completely webbed leaves that have margins with pointed or rounded teeth.

Studies have been done (Abbott, 1958) that graphically illustrate that the number of leaves per verticel and leaf morphology cannot be used to distinguish species of *Sphenophyllum*. This is well documented by Batenburg (1977) for *S. emarginatum*. When reconstructing this plant he found a whole spectrum of leaf morphologies (Fig. 15.2B) ranging from simple leaves with a single unbranched vein, reminiscent of a microphyll, to planated and webbed leaves where the vein dichotomizes up to four times. The later leaf type was found on lateral branches and more basal portions of aerial shoots. There were six such leaves per verticel. Distally, the leaves become smaller and simpler in structure and increased in number to nine per verticel. The margins of some leaves have pointed teeth while others are rounded.

Sphenophyllum emarginatum provides us with an excellent example of leaf polymorphism (heterophylly) and helps emphasize the importance of having large collections of fossils to aid in circumscribing the variability that occurs within a species. Generally, today's paleobotanists are loath to describe a species of fossil plant based on one or two fragments.

Vegetative anatomy

The ontogenetic studies of fossils thus far presented do not attempt to postulate the organization of the shoot apex and subsequent development leading to tissue differentiation. Only in rare cases where the shoot apex is preserved is this possible. Such is the case for *Sphenophyllum*, where Good & Taylor (1972) found an apical cell and its derivatives (Fig. 2.10A). The apical cell has the shape of a tetrahedron with a triangular outer face and three internal cutting faces. Cells derived from the inner cutting faces contribute to the procambium, leaf primordia, and intercalary meristems in much the same way as in *Equisetum*. In its shape and cutting faces the apical cell compares favourably with that of the arborescent sphenopsids and *Equisetum*, an observation that further supports the close relationships of these organisms.

Figure 15.1. Restoration of *Sphenophyllum emarginatum*. Carboniferous. (Photograph courtesy of the Field Museum of Natural History, Chicago.)

Because there are developmental stages to be seen in the anatomy of *Sphenophyllum* twigs and branches, as in all other vascular plants, we have to stipulate that the anatomy described below comes from the internodal region of a mature axis with secondary growth. In transverse section, we can see (Fig. 15.3A) that the primary xylem is a three-ribbed exarch protostele. This kind of primary xylem development recalls that of zosterophylls and lycopods. Except for roots, however, it is an uncommon arrangement in other vascular plants. The strands of protoxylem, one of which is located at the tip of each of the three radiating arms of the actinostele, are of particular interest. In mature stems, cells of the protoxylem occasionally break down to leave a lacuna. This, you may recall, is an important characteristic of the Sphenopsida. The small exarch cells of protoxylem are annular, helical, or scalariform tracheids. The larger centripetal metaxylem tracheids may bear multiseriate bordered pits.

Sphenophyllum is provided with a bifacial vascular cambium (Eggert & Gaunt, 1973), the kind we are familiar with in the woody stems of gymnosperms and

Figure 15.2. **A.** Leaves of various species of *Sphenophyllum* arranged to show degrees of webbing. (a) *S. myriophyllum;* (b) *S. cuneifolium;* (c) *S. majus;* (d) *S. cuneifolium;* (e) *S. emarginatum;* (f) *S. verticillatum.* (Redrawn from Abbott, 1958.) **B.** *Sphenophyllum emarginatum.* Series of leaves to show variability of leaf morphologies within the species. Carboniferous. (Redrawn from Batenburg, 1977.)

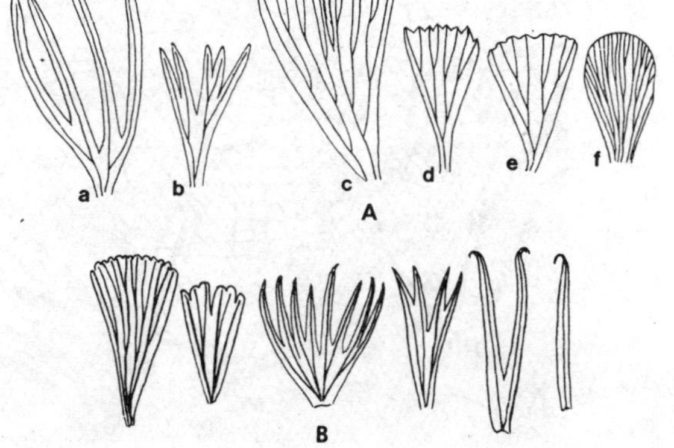

angiosperms. The vascular cambium of *Sphenophyllum* first develops in the embayments between the arms of the actinostele and is completed later around the tips of the arms. Cichan & Taylor (1982) and Cichan (1985) have demonstrated that the secondary xylem produced by the vascular cambium is unique. The large radially arranged tracheids have multiseriate bordered pits on their tangential and radial walls (Fig. 15.3B). They are extremely long cells, and some investigators believe them to be vessel elements. Between the corners of the tracheids are vertical strands of parenchyma and in some species (*S. plurifoliatum*) there are short horizontal rays that connect with the vertical strands (Fig. 15.4A). The secondary xylem opposite the protoxylem is less well developed and composed of smaller tracheids than in other parts. Although primary phloem has been identified in younger stems, in older axes small amounts of secondary phloem and secondary ray tissue can be found. Again in older stems, the primary tissues of the cortex are mostly lost because of the activity of a phellogen and the production of a periderm. The phellogen is a lateral meristem that originates in the parenchyma of the cortex where it forms a continuous ring around the stele. It gives rise to a phellem tissue composed of radial files of cells that are similar in their appearance to the cork cells of modern seed plants. As the stem increases in diameter, the periderm is ruptured to form segments that probably sloughed off, much as bark fragments slough off from the branches of modern trees. A new phellogen was produced from inner parenchyma tissues of the stem to form a new layer of periderm replacing the old outer bark layers.

Leaf traces diverge singly or in pairs from the protoxylem at the tips of the radiating arms. As they pass horizontally to the leaf base they may divide dichotomously or remain unbranched. If branched, a pair of leaf traces may enter the leaf base; if unbranched, there is a single leaf trace.

The relationship of branch traces to leaf traces is variable. In compression–impression fossils, branch traces appear to alternate with leaf traces. In studies using serial sections through the nodes of permineralized *Sphenophyllum* stems, Baxter (1972) found the alternate arrangement, while Phillips (1959) and Good (1973) confirmed that branches can originate from the protoxylem above a leaf trace in the axillary position. In yet others where the leaf traces diverge in the stem, the branch trace departs above, but between leaf traces so that leaves and branches appear to alternate with one another at the node.

The internal structure of a *Sphenophyllum* leaf (Fig. 15.4B) varies with the level at which the section is taken. At about the midpoint of a leaf there may be several veins more or less evenly spaced except

Figure 15.3. **A.** Transverse section of an exceptional specimen of *Sphenophyllum plurifoliatum*. Triarch protostele (p); secondary xylem (sx); secondary phloem (ph); vascular cambium (c); periderm (pe). Middle Pennsylvanian. (From Eggert & Gaunt, 1973.) **B.** *S. plurifoliatum*. Radial section of tracheary element of secondary xylem showing pits on radial walls, horizontal rays, and vertical parenchyma. Middle Pennsylvanian.

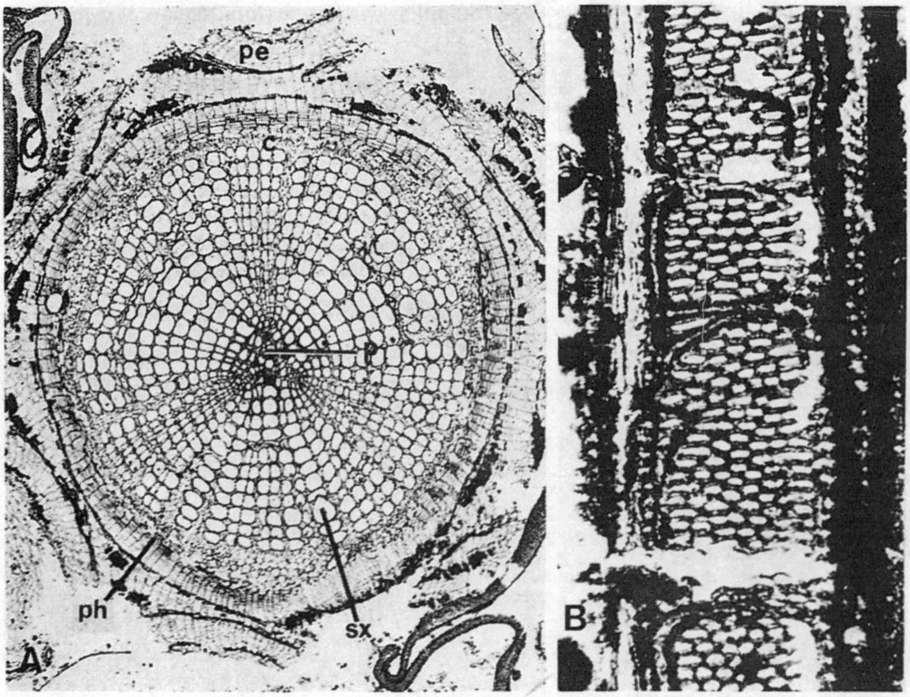

near a dichotomy. Each vein (Fig. 15.4C) is a concentric bundle with a centrally placed group of protoxylem cells surrounded by a layer of thin-walled cells interpreted as phloem. Around this vascular bundle, according to Good (1973), is a layer of thin-walled cells with dark contents that make up a melasmatic layer (a layer of storage cells). The mesophyll between adjacent veins is composed of thin-walled parenchyma with large intercellular spaces. The epidermal cells, with their sinuous wall outlines, are found not only on leaves, but cone bracts, sporangiophores, and sporangia of the Sphenophyllales (Reed, 1949; Baxter, 1950; Leisman, 1964). The sunken stomata are abaxial. In the distal portion of the leaf they are confined to two rows, one on each side of abaxial furrows.

Sphenophyllum roots (Fig. 15.4D) were adventitious and produced at the nodes, often with leaves. This suggests that as the stems of *Sphenophyllum* increased in length the basal portions assumed a horizontal position on the floor of the swamp. Although the roots are much smaller than the stems that bore them, they have essentially the same internal structure except for the primary xylem, which is usually diarch instead of triarch.

Reproductive morphology

Based on its internal vegetative structure *Sphenophyllum* exhibits a high degree of uniformity. Thus, it is surprising, as one reviews the literature, to discover numerous species of detached cones that are believed to belong to stems of the *Sphenophyllum* type. Many compression–impression fossils of *Sphenophyllum* axes with attached cones have been described (Boureau, 1964). Attachment also has been confirmed by Good (1978), who found permineralized cones attached to *Sphenophyllum* stems with leaves.

The principal organ genus for detached cones of the Sphenophyllales is *Bowmanites*. Some authors (Boureau, 1964), however, have adopted the name *Sphenophyllostachys*. As Hoskins & Cross (1943) and others have pointed out, *Bowmanites* has priority over *Sphenophyllostachys;* therefore, these authors placed all cones believed to be referable to the Sphenophyllales in the genus *Bowmanites*. At that time they recognized 18 species. Although *Bowmanites* is still recognized as the principal genus, at least two other genera described since 1943, *Peltastrobus* and *Sphenostrobus,* will be considered.

Most species of cones assigned to the Sphenophyllales consist of whorls of bracts that are usually fused tangentially for part of their length to form a shallow disclike structure. Immediately above the whorl of bracts and partially fused to their adaxial surface are spongiophores with terminally borne sporangia. As we shall see, the number of bracts per whorl, the number of sporangiophores, and the orientation of their terminal sporangia varies even within a single cone (Good, 1978). However, features of the spores (including the presence of a distinctive perispore) are characteristic of most species.

Figure 15.4. **A.** Transverse section enlarged portion of secondary wood, *S. plurifoliatum*. Tracheid (t); vertical parenchyma (v); horizontal ray (hr). **B.** *Sphenophyllum* leaf sectioned. Note I beam arrangement of sclerotic tissue associated with vascular bundles. **C.** Highly magnified portion of sectioned leaf showing distribution of tissues composed of sclereids. **D.** Transverse section of *Sphenophyllum* root. (From Scott, 1920.) All Carboniferous.

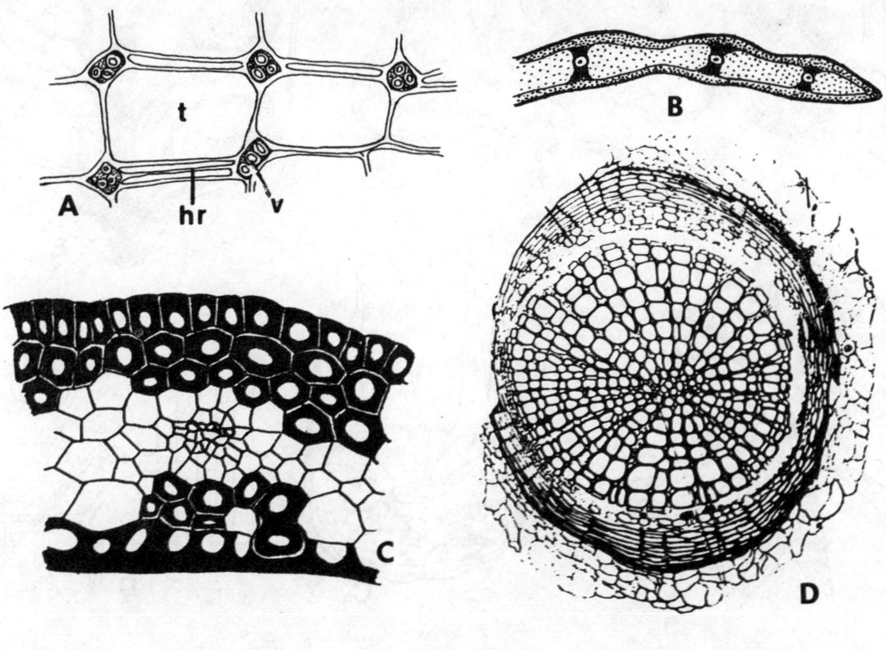

Bowmanites

Bowmanites dawsoni (Fig. 15.5A, B) exemplifies the general characteristics of *Sphenophyllum* cones, as well as some of the variability in their structure. The vascular strand of the cone axis is an exarch actinostele, which may be triarch as in the stem or with additional radiating arms conforming to the hexarch condition. In all those specimens of *B. dawsoni* where the vascular system can be determined, the vascular supply to the bracts and sporangiophores originates from the tips of the arms of the protostele. In these specimens of *B. dawsoni* described by Taylor (1969), where the vascular strand of the cone axis is hexarch, a massive vascular bundle departs at the nodal level from each of the six arms. This trifurcates immediately in the horizontal plane within the cone axis. Each of the three branches produced in this way trichotomizes and supplies the vascular bundles for one bract and two sporangiophores. The bract trace departs horizontally from the axis and goes directly into the bract as an unbranched vein. The two sporangiophore traces ascend at an angle, and each supplies one of the two sporangiophores associated with the bract. One sporangiophore is shorter than the other, so that two concentric rings of sporangia are produced in the whorl. All sporangia are anatropous (the distal end is directed toward the main axis). Theoretically, with a hexarch stele in the cone axis, there should be 18 bracts per whorl and 36 sporangiophores, 2 associated with each bract. Actually the bract number is variable (14 to 22), as is the number of sporangiosphores. In some specimens of *B. dawsoni* there are 3 sporangiophores per bract, while others have a 1:1 ratio.

Of particular interest from an evolutionary point of view is the vascular system of *Bowmanites bifurcatus* (Fig. 15.5C) as worked out by Andrews & Mamay (1951). In these small but well-preserved permineralized cone fragments, it was possible to determine the origin and branching of the vascular supply of the bract–sporangiophore complex at a node. The complex consists of a whorl of 6 bracts fused to form a basal disc. The tip of each bract is bifurcate, resulting in 12 free tips per whorl. A single sporangiophore arises from the adaxial basal part of each bract. At the distal

Figure 15.5. **A.** *Bowmanites dawsoni.* Diagram of longitudinal section of portion of cone. **B.** *B. dawsoni.* Transverse section of cone showing arrangement and position of sporangiophores. Carboniferous. (**A,B** redrawn from Hirmer, 1927.) **C.** Reconstruction of node of the cone *Bowmanites bifurcatus.* Upper Pennsylvanian. **D.** Representation of the vascular system of a bract–sporangiophore complex. Note that it is a three-dimensional dichotomous branching system. (**C,D** from Andrews & Mamay, 1951, in Boureau, 1964.) **E.** *Sphenostrobus thompsonii.* Longitudinal section through portion of cone. This is structurally the simplest of sphenophyll cones. (Redrawn from Levittan & Barghoorn, 1948.) **F.** Transverse section, *Peltastrobus reedae.* (Diagram redrawn from Leisman & Graves, 1964.) **E,F:** Middle Pennsylvanian.

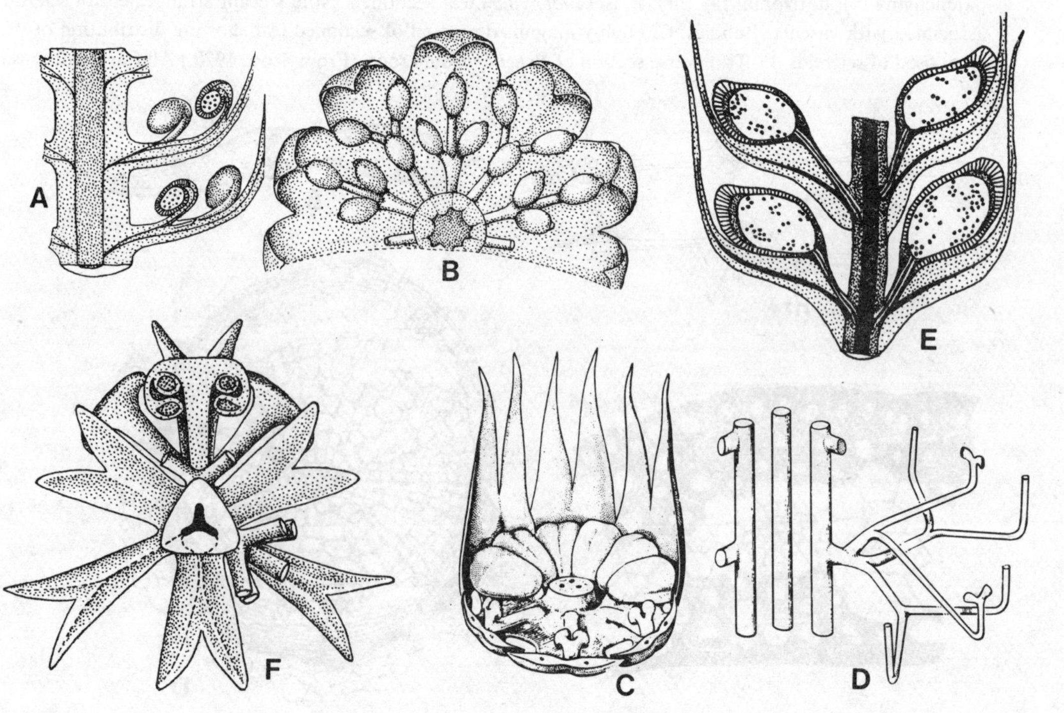

end, the sporangiophore dischotomizes and each of the two branches bears a single anatropous sporangium. So there are 12 sporangia in the whorl.

The vascularization of a unit (Fig. 15.5D), consisting of one bifurcate bract with its two sporangia, starts with a horizontal vascular bundle that arises from one of three vascular strands in the cone axes. The bundle dichotomizes almost immediately in the vertical plane to form an upper and lower branch. The upper branch dichotomizes again, this time in the horizontal plane. One of the two branches resulting from the dichotomy supplies each of the sporangiophores. The lower branch dichotomizes twice in the horizontal plane. Two of the four ultimate branches thus produced provide the vascular supply of one bifurcate bract. When one examines Figure 15.5D showing the branching of the vascular system of *B. bifurcatus*, it is clear that it reflects a three-dimensional branching system composed of fertile and sterile units.

Sphenostrobus

Cones assigned to the genus *Sphenostrobus* Levittan & Barghoorn (1948) and Good (1978) are, in their organization, among the simplest of sphenophyll cones (Fig. 15.5E). There are 8 to 16 bracts per whorl with a single axillary sporangiophore bearing one terminal sporangium per bract. In some species the sporangium appears to be orthotropous (an erect sporangium), while in others there is a tendency for the sporangium to be anatropous as in *Bowmanites*. The isospores are of the *Calamospora* and *Vestispora* dispersed spore types.

Peltastrobus

As is often the case, discoveries are made of fossil plant parts that do not fit our preconceived ideas of those characteristics that should define a certain plant organ. *Peltastrobus* (Baxter, 1950; 1972) is a good example of a sphenopsid cone genus with characteristics quite different from those of other cone genera. If it were not for the distinct nodes and internodes, and the whorled arrangement of the appendages, one might wonder about its inclusion in the Sphenophyllales.

According to the description provided by Leisman & Graves (1964), the cone (Fig. 15.5F) consists of a whorl of six units at each node. Three are bracts that alternate with three units composed of five sporangiophores each subtended by a bract. The bracts tend to be terete to slightly flattened and dichotomize at their tips. The five axillary sporangiophores are arranged so that two are ascending, one is at right angles to the cone axis, and two are descending. Unlike *Bowmanites* and *Sphenostrobus*, *Peltastrobus* has peltate sporangiophores. Its sporangia are in two concentric rings on the adaxial surface of an expanded peltate disc. There are approximately 16 sporangia per sporangiophore. The spores are of the monolete type.

Evolutionary considerations

Cones

Some paleobotanists believe that the Sphenophyllales evolved from Middle Devonian ancestors belonging to the Hyeniales. Much of the evidence for this interpretation hinges on the characteristics of an Upper Devonian (Famennian) fossil called *Eviostachya* (Fig. 15.6A), described in detail by Leclercq (1957). The compressed and permineralized specimens represent small cones on a peduncle with a whorl of six bracts below the fertile region. A triarch protostele not unlike that of *Sphenophyllum* was found in the cone axis. This gives off traces from the tips of the arms that supply whorls of six sporangiophores. Each sporangiophore

Figure 15.6. **A.** *Eviostachya* sp. Upper Devonian. (Redrawn from Leclercq, 1957.) **B.** *Bowmanites fertilis*. Carboniferous. (Redrawn from Leclercq, 1935.)

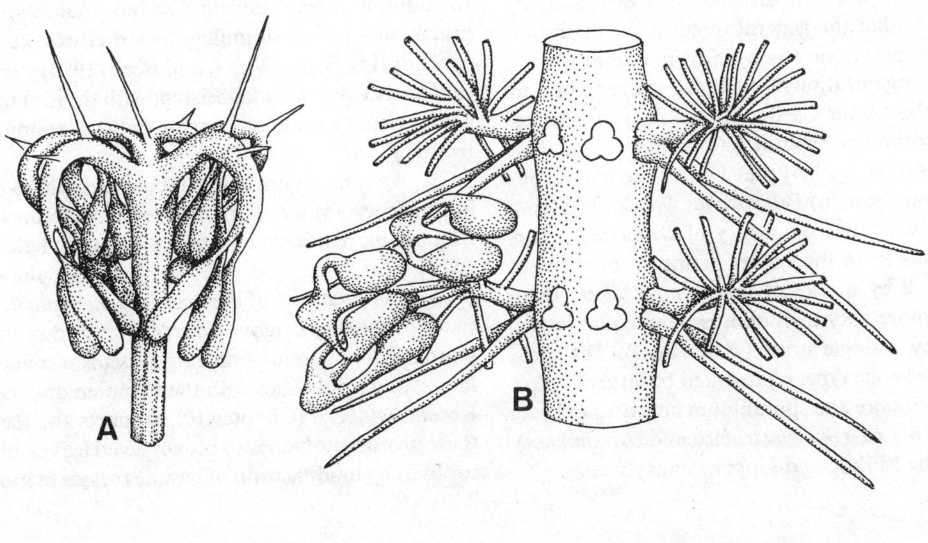

bears both sporangia and sterile processes (Fig. 15.6B). There is no evidence, however, of leafy bracts being associated with the sporangiophores. The similarity in the organization of a sporangiophore of *Eviostachya*, with its trichotomizing and then dichotomizing recurved branches each terminated by a cluster of three sporangia, recalls the complex sporangiophores of *Hyenia* and *Calamophyton*. Surprisingly, however, the sterile processes and recurved sporangia of *Eviostachya* sporangiophores are equally reminiscent of sporophylls of the protolepidodendralean lycopod *Estinnophyton* (Fig. 10.9D).

Just how the verticels of bracts became associated with sporangiophores in the Sphenophyllales is hypothetical. As stressed by Stein, Wight, & Beck (1984), there is no compelling evidence that the bracts and sporangiophores of sphenophyllalean cones are the homologs of those of other sphenopsids. Could bracts associated with sporangiophores have evolved by sterilization of an abaxial segment of a sporangiophore of the *Eviostachya* type? Some support for this idea is found in the Lower Carboniferous *Bowmanites fertilis* (Fig. 15.6B). Whorls of six units are borne at successive nodes of this cone fragment. Each unit consists of a lower pair of terete branches, the homologs of bracts, and an axillary sporangiophore. The latter bears a distal cluster of about 16 branches each of which terminates in a peltate structure with two sporangia. The complex structure that we see here can be interpreted as a stage in the sterilization of a fertile telome truss to produce bracts that subtend a sporangiophore, or as a stage in the reduction of sterile and fertile telome trusses accompanied by recurvation as suggested by Zimmermann's telome theory (1959). It must be emphasized, however, that no matter which of the two interpretations turns out to be the correct one, foliar units (bracts) and branches (sporangiophores) have an evolutionary origin from dichotomous branching systems as evidenced by the vascular system in the cone of *Bowmanites bifurcatus* and other species.

All of this fits well with the idea expressed by Taylor (1969) that the general trend in the evolution of the sphenopsid cone has been from complexity to simplicity in organization of the bract–sporangiophore units, with the Lower Carboniferous *B. fertilis* representing the primitive type. It has been speculated by Leisman & Graves (1964) that Upper Carboniferous (Middle Pennsylvanian) *Peltastrobus* was derived from an ancestor with the complexity of *B. fertilis*. Intermediate forms from the Lower Pennsylvanian might be represented by species of *Bowmanites* where there are two or more sporangiophores of unequal length subtended by a single bract (*B. dawsoni*). The most highly evolved cone type, represented by species where the sporangiophore and sporangium number per bract is reduced to one (*Sphenostrobus* and *B. simplex*), appears in the Middle and Upper Pennsylvanian.

Leaves

According to our definition, a microphyll is a leaf usually with a single vein arising from the axial vascular system without forming a leaf gap. In *Sphenophyllum* where the axial vascular system is a protostele there can be no leaf gaps, and in this latter characteristic Sphenophyllales are similar to many Lycopsida. It has been demonstrated repeatedly that some leaves of *Sphenophyllum* have a single unbranched vein, but most have a single vein that branches dichotomously in the leaf lamina. It is quite clear, and paleobotanists agree, that this kind of leaf probably evolved by planation and webbing (Fig. 9.10) from the three-dimensionally branched sterile telome trusses of the type found on the axes of Hyeniales (*Hyenia* and *Calamophyton*) and possibly the leaves of the Protolepidodendrales. Thus, the foliar organs of the Sphenophyllales, which by definition could be called microphylls, have a different evolutionary origin from those of other Lycopsida (Drepanophycales), where microphylls apparently evolved from enations. As we shall see, the leaves of other Sphenopsida, like those of the Sphenophyllales, have evolved from and are thus homologous with branching systems.

Possible relationships

In a recent phylogenetic analysis, Stein, Wight, & Beck (1984) have identified the characteristics used by most authors to assess the relationships of the Sphenopsida. These include whorled appendages, the occurrence of sporangiophores, stelar type, leaf morphology, secondary tissues, and features of the spores. This careful study supports the proposal that sphenopsids may be derived from the trimerophytes through a form such as *Ibyka* (Fig. 14.3A), but the possible roles of the other Iridopteridales, the Hyeniales, and the Cladoxylales are uncertain. Also, this study leads us to question whether the Sphenophyllales are as closely related to the other articulates as is commonly believed. In addition to emphasizing that sporangiophores and bracts may not be homologous structures among all sphenopsids, Stein, Wight, and Beck (1984) stress that the fossil evidence is inconsistent with the idea that leaf arrangement and stelar form are both synapomorphies for the group.

We cannot conclude this chapter without repeating a suggestion we often have made in lectures about the possible relationships of the Sphenophyllales to the Lycopsida. When one takes into account the exarch protostelic anatomy of the axes of *Sphenophyllum* and the evolutionary origin of their leaves and sporangiophores from telomic branching systems, we find some interesting similarities with the Protolepidodendrales. Recent evidence (Chapter 10) supports the idea that these protostelic lycopods also evolved leaves and sporophylls by modification of telome trusses in much the same way as the Sphenophyllales. This suggests the possibility that these two orders diverged from a common ancestor, with retention of the spiral phyllotaxy in the Protolepidodendrales and subsequent evolution of verticillate appendages, perispore, and a vascular cambium producing unique wood in the Sphenophyllales.

References

Abbott, M. L. (1958). The American species of *Asterophyllites, Annularia,* and *Sphenophyllum. Bulletin of American Paleontology,* **38,** 289–390.

Andrews, H. N., & Mamay, S. H. (1951). A new American species of *Bowmanites. Botanical Gazette,* **113,** 158–65.

Batenburg, L. H. (1977). The *Sphenophyllum* species in the Carboniferous flora of Holz (Westphalian D, Saar Basin, Germany). *Review of Palaeobotany* and *Palynology,* **24,** 69–99.

Baxter, R. W. (1950). *Peltastrobus reedae:* a new sphenopsid cone from the Pennsylvanian of Indiana. *Botanical Gazette,* **112,** 174–82.

Baxter, R. W. (1972). A comparative study of nodal anatomy in *Peltastrobus reedae* and *Sphenophyllum plurifoliatum. Review of Paleobotany* and *Palynology,* **14,** 41–7.

Boureau, E. (1964). *Traité de Paléobotanique.* Paris: Masson, Vol. III.

Cichan, M. A. (1985). Vascular cambium and wood development in Carboniferous plants. *Sphenophyllum plurifolium* Williamson and Scott (Sphenophyllales). *Botanical Gazette,* **146,** 395–403.

Cichan, M. A., & Taylor, T. N. (1982). Vascular cambium development in *Sphenophyllum:* A Carboniferous arthrophyte, *International Association of Wood Anatomists Bulletin* n.s., **3,** 155–60.

Eggert, D. A., & Gaunt, D. D. (1973). Phloem of *Sphenophyllum. American Journal of Botany,* **60,** 755–70.

Good, C. W. (1973). Studies of *Sphenophyllum* shoots: species delimination within the taxon *Sphenophyllum. American Journal of Botany,* **60,** 929–39.

Good, C. W. (1978). Taxonomic characteristics of sphenophyllalean cones. *American Journal of Botany,* **65,** 86–97.

Good, C. W., & Taylor, T. N. (1972). The ontogeny of Carboniferous articulates: the apex of *Sphenophyllum. American Journal of Botany,* **59,** 617–26.

Hirmer, M. (1927). *Handbuch de Paläobotanik.* Munich and Berlin: Oldenbourg.

Hoskins, J. H., & Cross, A. T. (1943). Monograph of the Paleozoic genus *Bowmanites* (Sphenophyllales). *American Midland Naturalist,* **30,** 113–63.

Leclercq, S. (1935). Sur un epi fructifére de Sphenophyllale, Premier Partie. *Annals de la Societé Géologique de Belgique,* **58,** 183–94.

Leclercq, S. (1957). Etude d'une fructification de Sphenopsida un structure conservée de Devonian Superieur. *Mémoires, l'Académie Royale de Belgique, Sciences,* **14,** 3–39.

Leisman, G. A. (1964). *Mesidiophyton paulus* gen. et sp. nov., a new herbaceous sphenophyll. *Palaeontographica B.* **114,** 135–46.

Leisman, G. A., & Graves, C. (1964). The structure of the fossil sphenopsid cone *Peltastrobus reedae. The American Midland Naturalist,* **72,** 446–37.

Levittan, E. D., & Barghoorn, E. S. (1948). *Sphenostrobus thompsonii:* A new genus of the Sphenophyllales. *American Journal of Botany,* **35,** 350–58.

Phillips, T. L. (1959). A new sphenophyllalean shoot system from the Pennsylvanian. *Annals of the Missouri Botanical Garden,* **46,** 1–17.

Reed, F. D. (1949). Notes on the anatomy of two Carboniferous plants *Sphenophyllum* and *Psaronius. Botanical Gazette,* **110,** 501–10.

Scott, D. H. (1920). *Studies in Fossil Botany,* 3rd ed. London: Black.

Stein, W. E., Jr., Wight, D. C., & Beck, C. B. (1984). Possible alternatives for the origin of Sphenopsida. *Systematic Botany,* **9,** 102–18.

Taylor, T. N. (1969). On the structure of *Bowmanites dawsoni* from the lower Pennsylvanian of North America. Palaeontographica *B,* **125,** 65–72.

Zimmermann, W. (1959). *Die Phylogenie der Pflanzen,* 2nd ed. Stuttgart: Fischer.

16

The origin of the horsetails

The genus *Equisetum* with 25 species and the only extant representative of the horsetails could hardly be considered an important part of our modern flora. The evolutionary origin of *Equisetum* has, however, been the object of much discussion and speculation. We are fortunate now in having the additional comprehensive works of several investigators at our disposal that make it possible to remove many of our ideas about the origin of *Equisetum* from the realm of speculation.

Equisetales

In textbook treatments of the Sphenopsida, it is customary to recognize three major orders: the Sphenophyllales, which we have already considered (Chapter 15), the Equisetales, and the Calamitales. For reasons that will become apparent, it is best to group the plants belonging to the Calamitales with the Equisetales (Boureau, 1964). Some authors, including Good (1975), recommend placing the arborescent calamites in the family Equisetaceae. However, the calamites are arborescent and as such are extinct; therefore, we prefer to follow Boureau's classification and place them in two families, the Calamitaceae and the Archaeocalamitaceae.

Archaeocalamitaceae: transitional sphenopsids

Upper Devonian–Lower Permian *Archaeocalamites* (Fig. 16.1A) is known from permineralized specimens, compression–impression fossils, and pith casts. The latter were formed when the pith cavity of the stem axis filled with sediment and lithified. All tissues external to the cast soon disintegrated, and some were replaced by sedimentary material which then became a mold around the cast. Molds and casts formed in this way reflect the arrangement of the primary vascular system of the axis as well as the position of leaves and branches at the nodes. In the pith casts of *Archaeocalamites* and other calamites, the primary vascular system is represented by the furrows on the surface of the cast.

From the pith casts of *Archaeocalamites*, we can determine that the plant was arborescent and that the primary vascular system did not alternate at the nodes as occurs in *Equisetum*. Compression fossils tell us that there were verticels of sterile appendages (Fig. 16.1B) branching up to four times and these probably functioned as leaves (Fig. 16.1C). Some were up to 10 cm long and show no sign of fusion at their bases.

Permineralized specimens of stem fragments show abundant secondary xylem with narrow rays. The radial walls of the tracheids are marked with one to three rows of circular pits. Of particular interest is the ring of carinal canals to the inside of the secondary xylem. These, as you will recall from *Equisetum*, mark the position of the primary vascular system. As in *Equisetum*, the primary vascular system of *Archaeocalamites* is a specialized stele with a pith cavity. This is the first time we have encountered this stelar type among the extinct Sphenopsida, and it seems to be an important characteristic distinguishing *Archaeocalamites* from *Hyenia* and relating it to the Carboniferous calamites.

Compressed and permineralized cones of *Pothocites* have been found associated with the vegetative material of *Archaeocalamites*. The cone axis bears many whorls of sporangiophores with an occasional whorl of sterile bracts having the form of *Archaeocalamites* leaves. Some cones have been found without whorls of bracts, and in this respect they are like those of *Equisetum*. Each sporangiophore dichotomizes twice to form four recurved branches each of which is terminated by a sporangium. Rather than being peltate, the sporangiophore is cruciate in appearance. Spores of the cone appear to have elaters.

Because of the similarity of its anatomy to that of *Calamites*, the genus *Archaeocalamites* has attracted attention as a possible precursor of the Carboniferous

Figure 16.1. **A.** Pith cast of *Archaeocalamites*. **B.** *A radiatus*. Reconstruction of shoot with sterile appendages. **C.** *A. radiatus* foliar appendage enlarged. All Lower Carboniferous. (**B,C** redrawn from Scott, 1920.)

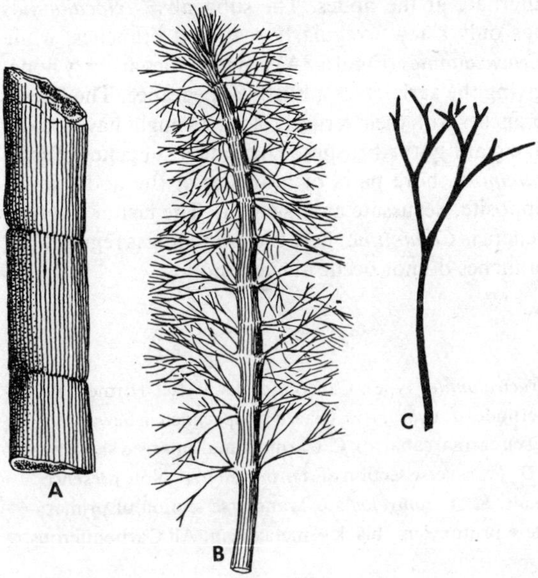

arborescent sphenopsids. A derivation of *Archaeocalamites* from the Hyeniales is "incontestable" according to Boureau (1964). His conclusion is based primarily upon the similarity of their sterile foliar appendages and their sporangiophores with terminal pendulous sporangia. Skog & Banks (1973) support the idea expressed by Boureau that *Archaeocalamites* is a transitional genus between the Hyeniales and Calamitaceae. This conclusion, however, would be difficult to justify if we agree that the Hyeniales are related to the Filicopsida and not the Sphenopsida (Schweitzer, 1972), or that the relationships of the Hyeniales are still unclear (Stein, Wight, & Beck, 1984). If, on the other hand, we think of the Hyeniales and Cladoxylales as representing a complex of Middle and Upper Devonian plants with a synthesis of sphenopsid and fernlike characteristics, then it is simple to visualize them as a group from which several lines have diverged, including the Archaeocalamitaceae of the Equisetales and some Upper Devonian–Lower Carboniferous Cladoxylales.

Calamitaceae: general morphology

The generic name *Calamites* was first proposed by Suckow in 1784. It was made official by Brongniart in 1828, who recognized that there was a close affinity between this genus and *Equisetum*. Originally, *Calamites* was applied to fragments of pith casts (Fig. 16.2B), but later the genus was expanded to include plants with roots, stems, leaves, and cones. It was apparent

Figure 16.2. **A.** Transverse section of an exceptional specimen of *Arthropitys*. Pennsylvanian. (Courtesy of the Illinois State Geological Survey, photograph by Dr. T. L. Phillips.) **B.** Mold of a calamite pith cast showing node (n); branch trace scar (b); and vertical, internodal ribbing. Pennsylvanian.

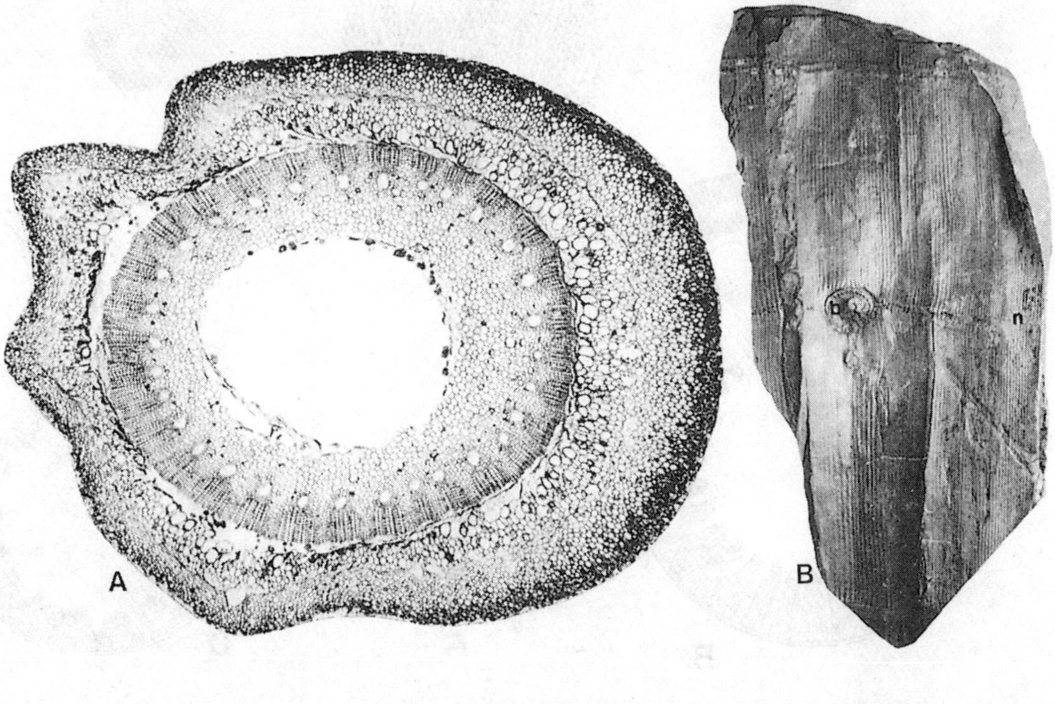

from the beginning that *Calamites* was an arborescent sphenopsid that attained a height of up to 10 m. From studies of pith casts and compressions the bizarre branching patterns and general morphology have been worked out. In all reconstructions of calamites (Fig. 16.3A) the aerial axes are shown arising from the nodes of a stout underground rhizome. The rhizome had adventitious roots also at the nodes. All branches and leaves supported by the aerial axes are in verticels with branches alternating with leaves at the nodes.

Based on the characteristics shown by pith casts, five subgenera of *Calamites* have been established that together pretty well depict the variety in the gross morphology of these unusual plants. The subgenus *Mesocalamites* is used for those middle and late Carboniferous calamites that have casts similar to those of *Archaeocalamites* except that some of the primary vascular strands alternate at the nodes. It should be noted that all other *Calamites* species are like *Equisetum* in having a primary vascular system in which all strands alternate at the nodes. The subgenus *Stylocalamites* has only a few irregularly scattered branches, while *Crucicalamites* (Fig. 16.3A) has branches at every node, giving the aerial axes a bushy appearance. The lateral branches with their verticels of leaves might have looked like giant bottle brushes. The fourth subgenus, *Diplocalamites,* bore pairs of branches at the nodes in an opposite, decussate arrangement. The last of the subgenera is *Calamitina;* here the branching is regular, but branches do not occur at every node.

Figure 16.3. **A.** Reconstruction of *Calamites carinatus* (*Crucicalamites* type). Carboniferous. (After Hirmer, 1927, in Takhtajan, 1956.) **B.** Diagram transverse section of an internode of *Arthropitys*. (Based on specimen in Seward, 1898.) Pith cavity (p); secondary xylem (sx); vallecular canal (v); carinal canal (c). **C.** Carinal canal enlarged showing protoxylem elements in canal and metaxylem around canal. **D.** Transverse section of *Astromyelon* sp. Note presence of pith (p); cortical lacunae (l); and absence of carinal canals. **E.** *Astromyelon* sp. Transverse section of primary xylem strand enlarged showing exarch arrangement. Circle = protoxylem; black = metaxylem. All Carboniferous.

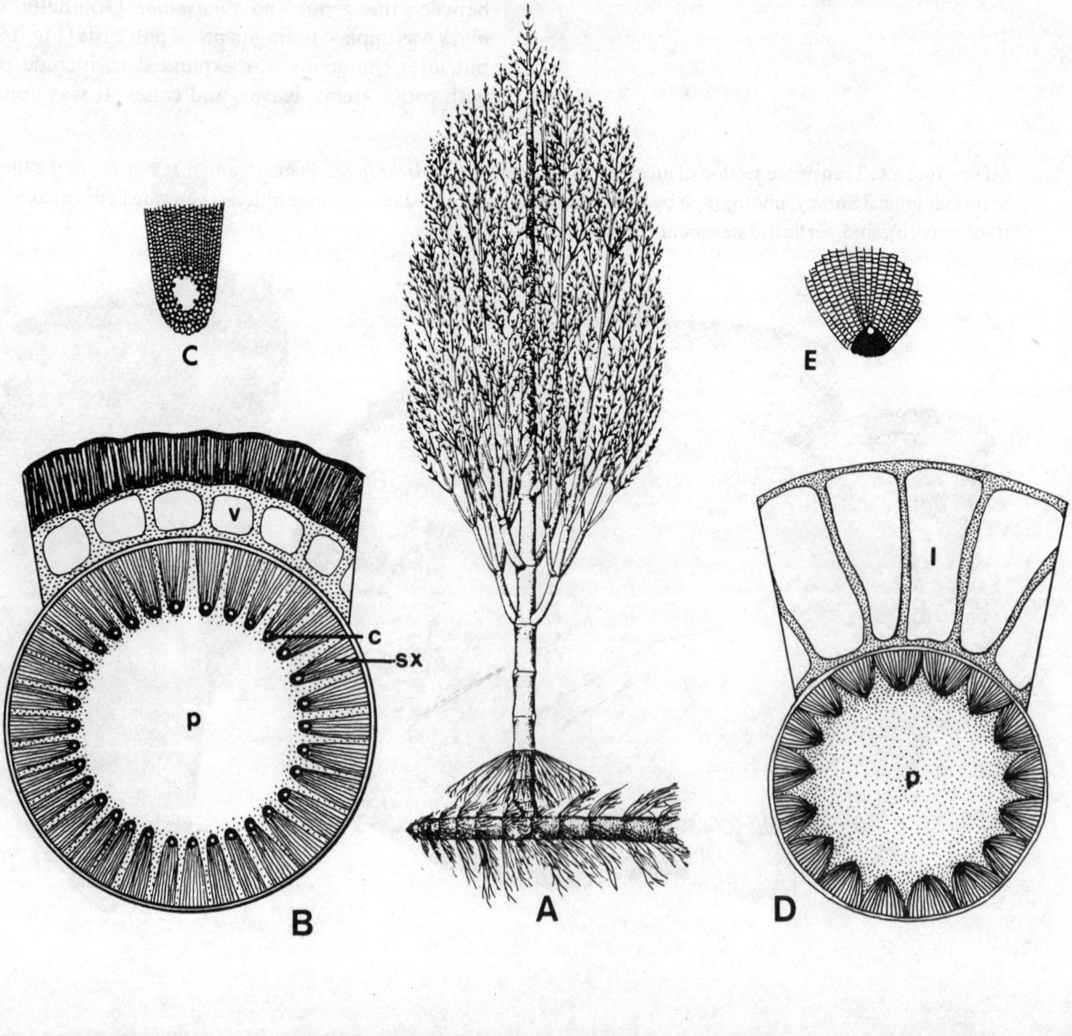

Leaf compressions
The commonest representatives of detached calamite leaves, found as compression–impression fossils, belong to the genera *Annularia* (Fig. 16.4A) and *Asterophyllites* (Fig. 16.4B,C). The numerous species of these two genera have been treated fully by Abbott (1958) and Boureau (1964). The whorls of leaves of *Annularia* are disposed in a plane that is oblique to the branch that bore them. When compressed, the leaves form beautiful stellate patterns at each node that appear to be in the plane of the axis. The linear leaves, characteristic of microphylls, are simple and supplied with a single unbranched vein. They may be fused at their bases to form an inconspicuous collar around the branch at the node. The number of leaves per node varies with the species. There may be as few as 8 leaves and as many as 20 per verticel. The shape and length of leaves also are characteristics used to differentiate species. The amount of leaf polymorphism of a species does not seem to be as great as in species of *Sphenophyllum*.

The leaves of *Asterophyllites* are almost needle-like and are borne in a plane at right angles to the branch. They too vary in length (from 5 mm to 4 cm) and number. There may be up to 40 leaves at a node. In every instance, the leaves of a whorl turn up so that they overlap the leaves of the verticel above.

When found detached, compressed roots, presumably belonging to *Calamites*, are assigned to the form genus *Pinnularia*. If attached, these branched adventitious organs are found at the nodes of the rhizomes or the basal portions of the aerial axes.

Stems: internal anatomy
When describing the internal structure of a calamite stem, we will be concentrating on a mature stem with abundant secondary tissues (Fig. 16.3B). Stems showing this characteristic are frequently found in coal balls, where they are permineralized. Specimens of this kind have not been assigned to the genus *Calamites* because it has proven impossible to demonstrate precisely which permineralized axes correspond to what

Figure 16.4. **A.** *Annularia radiata.* **B.** A superb branching specimen of *Asterophyllites.* **C.** Leaf-bearing shoot of *Asterophyllites* enlarged. **A–C:** Pennsylvanian. (**A,C** from Janssen, 1957, courtesy of the Illinois State Museum, Springfield; **B** courtesy of Dr. R. Peppers.)

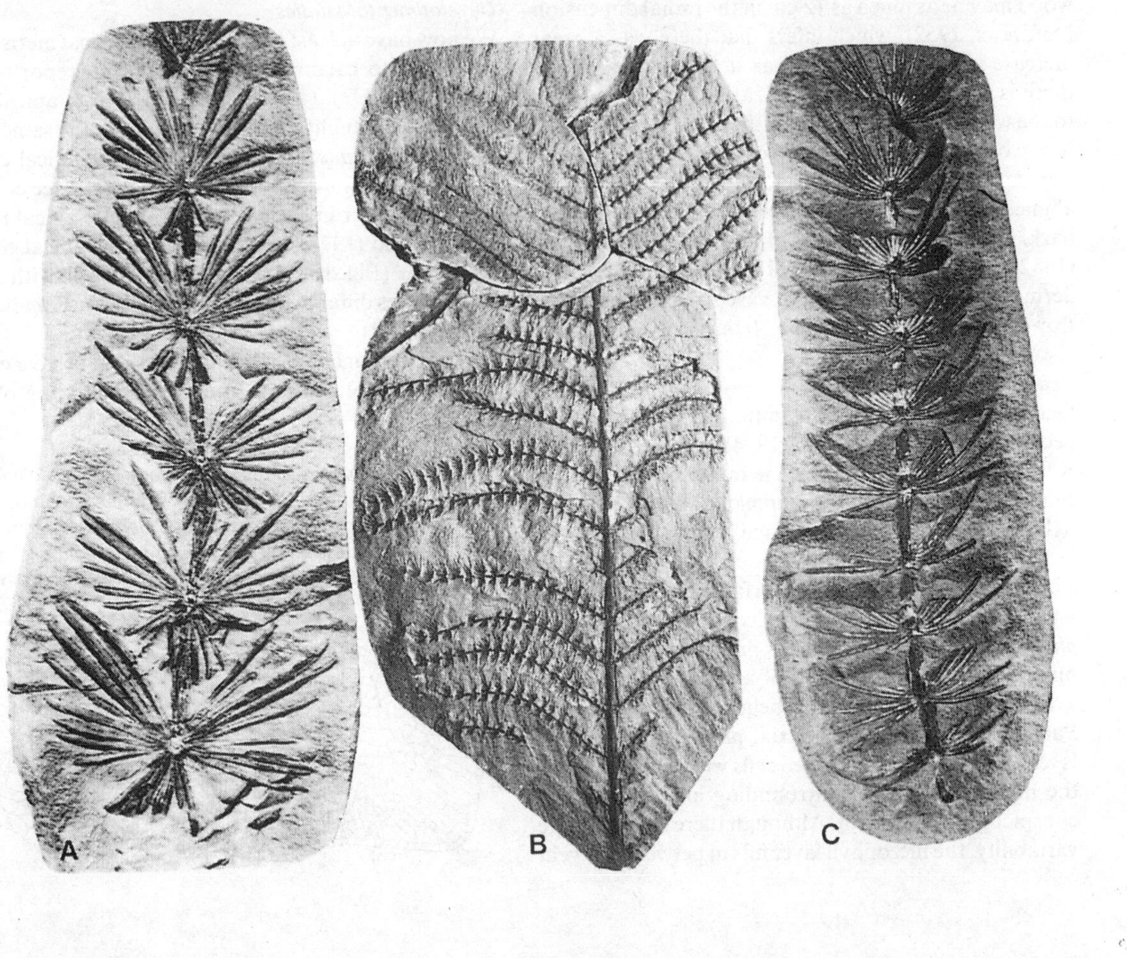

specific *Calamites* pith cast or compression. As a result, permineralized stems and roots have been given different generic names. The most common permineralized stem genus is *Arthropitys* (Fig. 16.2A), which when seen in transverse section through the internode shows a conspicuous pith cavity. A few pith cells may be preserved at the periphery of the cavity where they are associated with a ring of primary vascular strands whose position can be determined by the ring of carinal canals (Fig. 16.3C). Each strand is a sympodium of a stele that gives rise to alternating leaf and branch traces at the nodes in the same way as *Equisetum*. Development of the primary xylem is said to be endarch; however, there is a possibility that some centripetal metaxylem is present and that they are actually mesarch. Annular, spiral (helical), and scalariform tracheids have been observed in the primary vascular strands.

The secondary wood of *Arthropitys* consists of tracheids and rays. The rays are of two types: wide interfascicular rays that radiate from between two adjacent primary xylem points, and narrow rays in the wood segments. Tracheids of the wood segments have pits on their radial walls that vary from scalariform to circular bordered. The vascular cambium producing the secondary wood was unifacial. No secondary phloem was produced (Wilson & Eggert, 1984).

In some large *Arthropitys* stems the secondary wood may be as much as 12 cm in the radial dimension (Andrews, 1952), which infers that there was a great increase in its circumference as it grew. As a result, there is considerable disorganization of cells exterior to the woody cylinder. Some authors claim that tissues to the outside compare with the secondary cortex of a *Lepidodendron*. Others find no indication of radially aligned cells that would suggest a protective layer of bark. This has been confirmed by Cichan and Taylor (1982), who recently described a well-developed periderm produced by successive vascular cambia in the Lower/Middle Pennsylvanian *Arthropitys deltoides*.

Leaf anatomy

Detailed descriptions of calamite leaf anatomy have been provided by Anderson (1954) and Good (1971). A leaf type that is commonly encountered attached to calamite twigs is *Calamites rectangularis*, found in coal balls of Upper Pennsylvanian age. In cross section the leaves are rectangular to five-sided (Fig. 16.5A) with a median ridge on the adaxial surface. They are at least 4.0 mm long, and in shape they compare with *Asterophyllites*. The single leaf vein is probably a concentric bundle with a central strand of xylem surrounded by a layer of thin-walled cells believed to be phloem. Partially enveloping the abaxial part of the vein is a layer of more or less developed cells with dark contents, the melasmatic tissue. Surrounding all of these is a conspicuous vein sheath. Although there may be some variability, the mesophyll layer fills in between the vein sheath and the epidermis. The radiating cells of the mesophyll are palisade parenchyma. Stomata with their guard cells are scattered on all surfaces and usually are oriented parallel to the long axis of the leaf. Leaves of the *Calamites rectangularis* type are believed to belong to *Arthropitys communis* var. *septata* stems.

Root anatomy

Isolated fragments of permineralized calamite roots are usually referred to the genus *Astromyelon* (Fig. 16.3D). A parenchymatous pith is present in all but the smallest roots. The primary vasculature consists of a ring of exarch bundles (Fig. 16.3E) that lie near the periphery of the pith. Unlike the strands of primary xylem in the stem, they do not anastomose. Branch roots arise between adjacent strands of the primary vascular system, while the main roots are adventitious and arise at nodes of stem axes. A notable feature of root anatomy is the absence of carinal canals. Secondary xylem develops in wedge-shaped masses to the outside of the primary xylem strands. Pitting of tracheary elements in *Astromyelon* is the same as in *Arthropitys*. Phloem and endodermis have been reported, and in some exceptional specimens a middle cortex with large lacunae that look like vallecular canals in stems of *Equisetum* have been found.

Developmental studies

We now have at least two accounts of apical meristems belonging to calamite shoots. The first, reported by Melchior & Hall (1961), describes a large apical cell from which daughter cells are derived in the same way as in extant *Equisetum arvense*. Here the apical cell is a four-sided inverted pyramidal cell with three cutting faces. In a later investigation of calamite apical meristems, Good (1971) determined that the apical cell of *Calamites* (Fig. 16.5B) is a five-sided cell with four cutting faces differing from the apical cells of *Equisetum* in this respect.

Both Eggert (1962) and Good (1971) have made significant contributions to our understanding of the

Figure 16.5. **A.** Transverse section, leaf of *Calamites rectangularis* showing vein (v); vein sheath (s); palisade parenchyma (p); **B.** *Calamites* sp. longitudinal section of apical cell and its derivatives. **A,B:** Lower Pennsylvanian. (**A,B** redrawn from Good, 1971.)

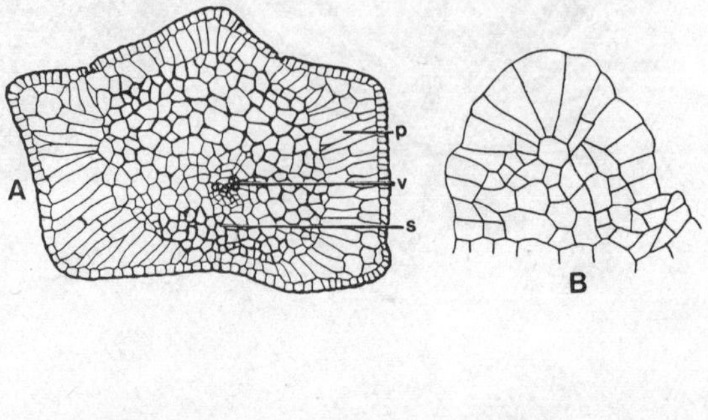

ontogeny of arborescent Sphenopsida. For example, it has been demonstrated that the growth habit of a calamite was determinate, as in the lepidodendrids and shoots of *Equisetum*. After examining numerous calamite roots and shoots, Eggert was able to reconstruct the primary vascular system of the aerial parts of a calamite (Fig. 16.6). In the basal portion of a mature aerial axis, the stele is represented by a small pith cavity and a limited number of sympodia. These in turn are surrounded by the maximum development of secondary xylem. Acropetally from the basal portion, the aerial axes underwent epidogenesis, that is, a progressive development marked by an increase in number of sympodia and an increase in the size of the pith cavity. From the basal position upward there was a gradual decrease in the amount of secondary wood so that in the distal portions of the branching system secondary xylem is totally lacking.

Distal to the region of epidogenesis, there is a gradual decrease in size of the pith cavity and of number and size of the sympodia. Apoxogenesis continues to

Figure 16.6. Diagrammatic reconstruction of vascular system of a calamite shoot and root. (From Eggert, 1962.)

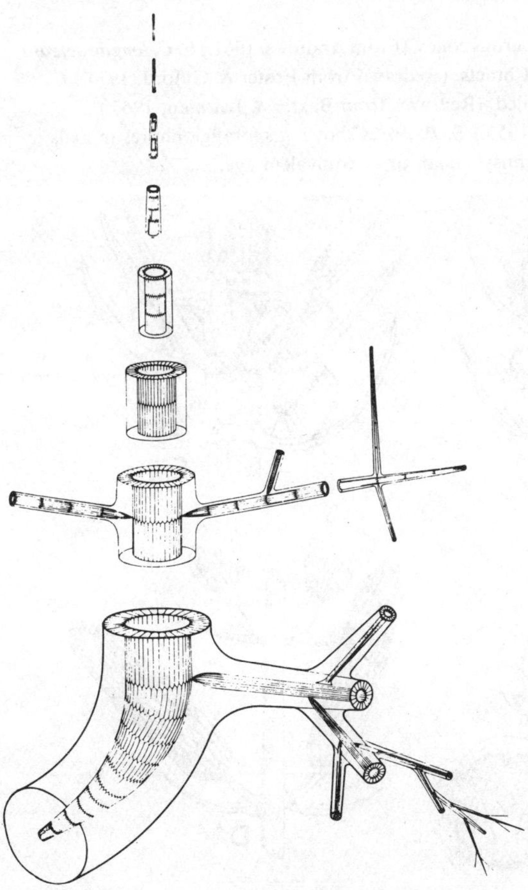

the determinate tips of ultimate branches. In transverse section the small twigs of calamites, with the beginnings of a pith cavity, about eight sympodia in the ring, and no secondary xylem, look exactly like *Equisetum* stems even to the presence of incipient cortical lacunae.

Roots of the *Astromyelon* type have a primary vasculature that decreases in size and number of bundles with each successively produced order of branching (Fig. 16.6). This also is accompanied by a decrease in the amount of secondary xylem.

Cones and spores

Cones recognized as belonging to the Calamitaceae are easily distinguished in compression–impression and petrifaction fossils because of their alternating verticels of bracts and peltate to pendant sporangiophores. There are numerous accounts of cones of this type attached by short peduncles to the nodes of calamite branches (Fig. 16.7A). Some of the larger cones are up to 4 cm in diameter and 12 cm long. Others are diminutive, not more than 2.6 cm long and 4 mm wide.

As was the case for the Sphenophyllales, we are confronted with a large number of different cone types all belonging to the Calamitaceae. Only a few of these, which best illustrate the evolutionary trends displayed by these organ genera, are discussed. *Pothocites* from the Upper Devonian and early Lower Carboniferous already has been briefly described. We will concentrate our attention on *Calamostachys, Palaeostachya,* and *Mazostachys* from the Middle and Upper Pennsylvanian (Upper Carboniferous).

Calamostachys

The *Calamostachys*-type cone is characterized by its verticels of sporangiophores, which are attached at right angles midway between successive verticels of sterile bracts (Fig. 16.7B). The sporangiophores are peltate, each with four sporangia that face the cone axis. Depending on the species, there are 6 to 18 sporangiophores per whorl and 10 to 45 bracts. The bracts of a whorl are fused laterally to form expanded discs with free tips. The degree of fusion varies with the species.

One of the most common and best-known permineralized calamite cones from European and American Carboniferous deposits is *Calamostachys*. Good (1975), in one of the most comprehensive treatments of the calamite cones and spores ever prepared, has reviewed the structure of *C. binneyana*. Although a cellular pith has been described in the cone axes of European material, this feature is lacking in American specimens. The stele consists of a ring of 7 to 12 vascular bundles each with a carinal canal. The shape of the stele is reported to be either triangular or quadrangular in transverse section. Secondary xylem has been observed in internodes of some specimens. The vascular bundles of the cone axis alternate at the nodes and supply the traces to the whorls of bracts in the same way as in calamite stems. A bract trace arises at right angles to the cone axis and passes directly into the bract as an unbranched vein. Midway between successive whorls of bract traces, the sporangiophore traces also arise in a whorl from the vascular supply of the cone axis. Like a bract trace, a trace to a sporangiophore is given off at right angles to the cone axis and passes directly into the sporangiophore continuing on into the distal portion. There are usually twice as many bracts per whorl as sporangiophores.

As in most calamite cones, the sporangium wall is one cell layer thick at maturity. The rectangular cells have walls with peglike thickenings that project into the lumens. These projections are unique and distinguish calamite sporangia from those of all other vascular plants. *Calamostachys binneyana* is homosporous. It has been confirmed by Good (1975) that the isospores of *C. binneyana* bear three circinate coiled elaters. Spores of this type found in the dispersed condition are assigned to *Elaterites* (Fig. 16.7C). The elaters were easily detached from the spore body, and without them the spores correspond to the dispersed spore genus *Calamospora.*

Palaeostachya

Similar to *Calamostachys,* specimens of *Palaeostachya* (Fig. 16.7D) have been recorded from the Carboniferous of Europe and North America. The *Palaeostachya*-type cone is characterized by the insertion of its sporangiophores in the axils of bracts (Fig. 16.7E) at an angle of approximately 45°. In many ways the structure of *Palaeostachya* resembles that of *Calamostachys.* Other than the position of the sporangiophores, however, there are additional differences that are exemplified by *P. andrewsii* (Baxter, 1955; Good, 1975) and *P. decacnema* (Delevoryas, 1955). Of particular interest is the path of the sporangiophore trace in *P. andrewsii* (Fig. 16.8A). After its departure from a node of the cone axis, the vascular bundle supplying a sporangiophore ascends through the tissues of the internode above and then descends to enter the axillary sporangiophore. In *P. decacnema,* a sporangiophore trace arising from a node ascends at an oblique angle and enters the sporangiophore directly. Again emphasizing variability within a cone species, Good found *P. decacnema* to have 14 to 24 bracts per whorl and 6 to 13 sporangiophores in a verticel. The ratio of bracts to sporangiophores is approximately 2:1. The same kind of variability was found in *P. andrewsii* with 18 to 30

Figure 16.7. **A.** *Calamostachys ludwigi,* branch bearing numerous cones. (From Andrews, 1961.) **B.** *C. longibracteata* showing position of sporangiophores relative to whorls of bracts. (Redrawn from Foster & Gifford, 1974.) **C.** Specimens of *Elaterites triferens,* elaters expanded and coiled. (Redrawn from Baxter & Leisman, 1967.) **D.** *Palaeostachya decanema.* (Redrawn from Delevoryas, 1955.) **E.** *P. ovalis* showing sporangiophores in axils of bracts. (Redrawn from Foster & Gifford, 1974.) All Pennsylvanian or of equivalent age.

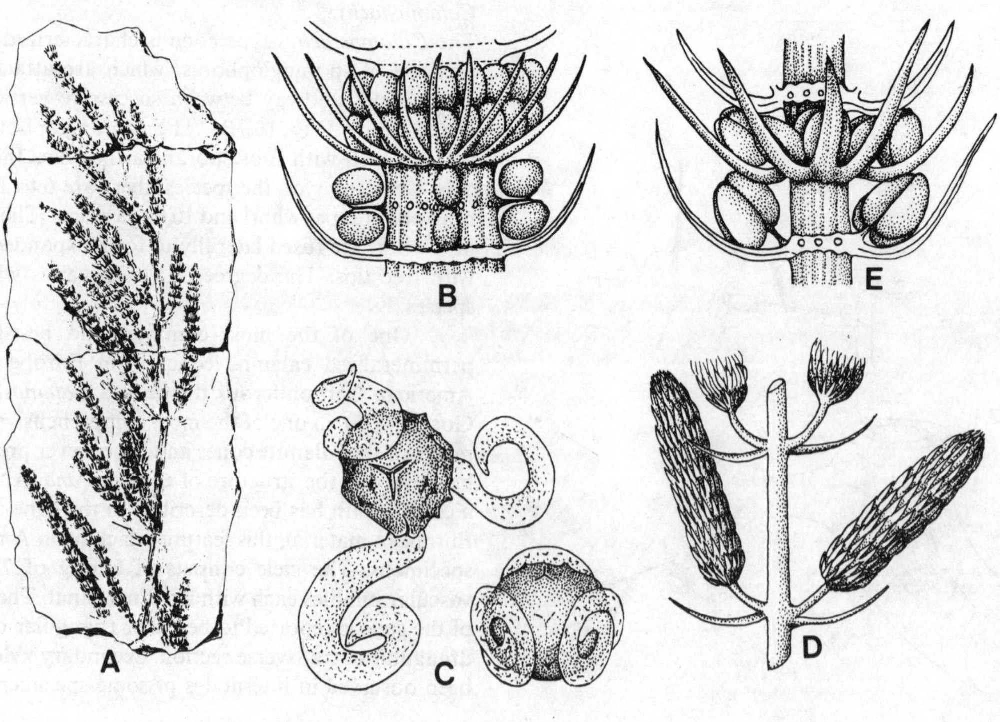

bracts in a verticel and 12 to 20 sporangiophores. Of particular interest is the discovery that *P. andrewsii* is heterosporous. The microspores, which range from 56 to 110 μm, have elaters. The larger megaspores, 235 to 345 μm, lack these structures and have the appearance of dispersed spores of the *Calamospora* type.

Mazostachys

Unlike *Calamostachys* and *Palaeostachya,* we find that the sporangiophores of *Mazostachys* (Kosanke, 1955) are borne in a whorl just below the verticel of bracts of the node directly above (Fig. 16.8B). The bract to sporangiophore ratio is 2:1, with a whorl of 12 bracts subtended by a whorl of 6 sporangiophores. Instead of being peltate with four sporangia, *Mazostachys* has sporangiophores with two pendant sporangia. The vascular trace to a sporangiophore descends from the node directly above prior to bending outward into the sporangiophore. In general organization, *Mazostachys* resembles the *Cingularia*-type cone known from compression specimens of Europe. In *Cingularia* there are four pendant sporangia instead of two, and sporangiophores are flat and bifurcate at their tips.

Calamocarpon

Heterospory among the Calamitaceae seems to have reached its maximum expression in the cones of *Calamocarpon* (Baxter, 1963; Good, 1975). In general morphology and internal structure, *Calamocarpon* is similar to *Calamostachys* with its whorls of sporangiophores attached at right angles midway between adjacent verticels of bracts. The bracts and sporangiophores are produced in a 1:1 ratio, with a variable number of appendages per whorl. There are four sporangia per sporangiophore directed toward the cone axis. In each megasporangium (Fig. 16.8C), which is 2 to 3 mm long, there is a single functional megaspore surrounded by sterile tissue and a uniseriate epidermis. Megametophytes retained inside the megasporangia have been discovered, and the evidence suggests that the megasporangium and its contents were shed as a unit from the cone. Microspores of the *Elaterites* type have been found by Good & Taylor (1974) in cones of *Calamocarpon.*

Heterospory and cone evolution

Heterospory in the Calamitaceae was demonstrated before the turn of the century when Williamson & Scott (1894) found spores of the *Calamospora* type in sporangia of a *Calamostachys* that were three times as large as spores of the same type in other sporangia of the same cone. Such a condition was discovered by Arnold (1958) in *C. americana.* These and other similar observations have provided ample evidence for the evolution of heterospory, which culminated in *Calamocarpon,* where there presumably was a reduction in megaspore number to one functional unit. Although not reaching the level of specialization represented by the integumentary structures in heterosporous lepidodendrons, it is obvious that the evolution of heterospory in the Calamitaceae closely parallels that of the arborescent lycopods. It has been apparent for some

Figure 16.8. **A.** *Palaeostachya andrewsii.* Diagram of longitudinal section to show vascularization of sporangiophores and bracts. (Redrawn from Baxter, 1955.) **B.** *Mazostachys pendulata.* Diagram showing relationships of sporangiophores to bracts. (Redrawn from Kosanke, 1955.) **C.** *Calamocarpon insignis,* single megasporangium. (Redrawn from photograph in Baxter, 1963.) All Middle Pennsylvanian.

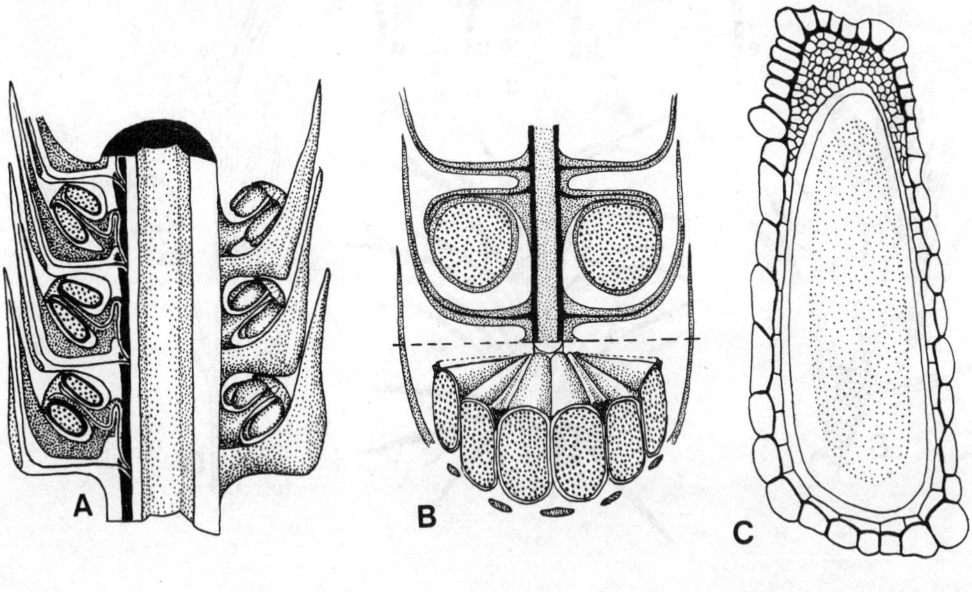

time that the evolution of heterospory, followed by reduction in megaspore number and retention of the megagametophyte, is an evolutionary trend encountered in many major groups of vascular plants and as such was and still is important to their survival.

Where did the whorls of bracts alternating with whorls of sporangiophores come from in calamite cones? Which cone type is primitive? How did the different sporangiophore positions in the cone evolve? Of the various theories that attempt to answer these questions, that proposed by Kosanke (1955) has the advantage of taking into consideration the stratigraphic positions of the cone genera. All seem to agree that the Calamitaceae evolved from a plexus represented by the Archaeocalamitaceae of the Upper Devonian and Lower Carboniferous and that the cone type *Pothocites* represents the primitive condition. With the exception of irregularly interspersed whorls of leaflike structures in some specimens, cones of *Pothocites* are characterized by the absence of whorls of bracts that alternate with whorls of sporangiophores. Because alternating whorls of bracts and sporangiophores are constant characteristics of cones of the Calamitaceae, we must have some explanation of how this might have come about. The presence of occasional whorls of foliage leaves in the cones of *Pothocites* suggests that they became modified into bracts, which were intercalated between adjacent whorls of sporangiophores (Brown, 1926). Another explanation is based on the idea that sporangiophores and sterile appendages (bracts) are homologous and that the alternating whorls of sterile bracts evolved from the *Pothocites* type by failure of sporangia to develop on alternate appendage whorls (Page, 1972).

Leaf evolution

It has been implied repeatedly that leaves of the Calamitaceae are reduced from sterile telome trusses of the type found among the Hyeniales. Leaves of *Archaeocalamites* that dichotomize three or four times in three dimensions support this idea. In the Calamitaceae, Boureau (1964) illustrates a series (Fig. 16.9A) starting with the leaves of the Lower Carboniferous *Dichophyllites*, which are planated and dichotomized twice. Further reduction of the leaf, so that a single dichotomy remains, is exemplified by *Sphenastero-*

Figure 16.9. **A.** Leaf evolution among the calamites; (a) *Archaeocalamites;* (b) *Dichophyllites;* (c) *Sphenastrophyllites;* (d) *Autophyllites.* (Redrawn from Boureau, 1964.) **B.** *Koretrophyllites vulgaris,* fertile branch showing clusters of sporangiophores alternating with whorls of bracts. Lower Carboniferous. **C.** *K. multicostatus,* shoot with sterile appendages. **D.** *Neocalamites* shoot with cone. Triassic. (**B–D** redrawn from Boureau, 1964.)

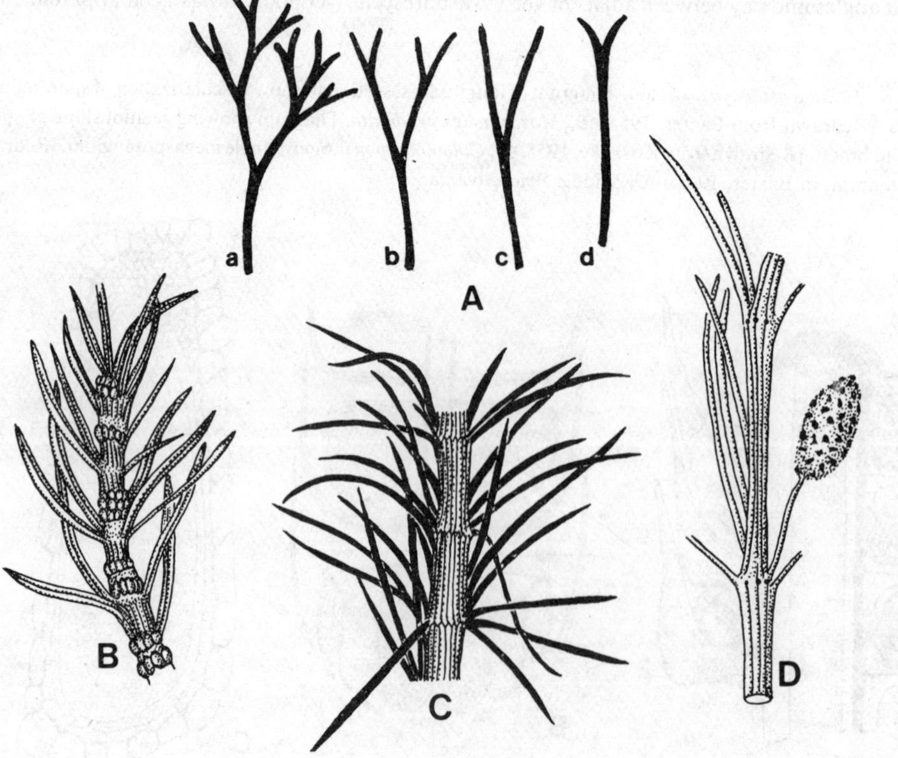

phyllites of the Upper Carboniferous. *Autophyllites* of the Middle and Upper Carboniferous has leaves that dichotomize once near their tips and are laterally fused at their bases. *Annularia* and *Asterophyllites* are unbranched and also show some basal fusion into a collarlike structure around the node. This series is representative of the way sterile telome trusses have been evolutionarily modified to produce the microphyll-like leaves of the Calamitaceae. The evolution of calamite leaves seems to parallel that of the Sphenophyllales, except for the extensive webbing of the planated telome truss in the leaves of *Sphenophyllum*.

The origin of Equisetum

In the presentation of the Calamitaceae, frequent reference has been made to the ways in which members of this family are similar to or differ from the Equisetaceae. Having studied the Calamitaceae, we can better appreciate these characteristics.

The Equisetaceae differ from the Calamitaceae in (1) lack of any indication of secondary tissues and (2) cones that are normally bractless. Equisetaceae are like the Calamitaceae in (1) gross morphology (nodes with whorls of appendages where leaves and branches alternate at the nodes); (2) rhizomatous habit; (3) whorls of microphyll-like leaves that are more or less fused to form a collar or sheath around the node; (4) internal structure (a stele made up of sympodia that alternate at the nodes, carinal canals, and an internodal pith cavity); (5) apical cells that give rise to aerial shoots; (6) developmental stages of aerial axes that show epidogenesis and apoxogenesis; (7) cones with whorls of pelate sporangiophores; and (8) spores with elaters. This is an impressive if incomplete list of similarities. However, a careful analysis has not been conducted to determine which if any of the features represent shared derived characters (synapomorphies). Nevertheless, there is little doubt about the relationships of the equisetums and calamites. The question has been raised whether *Equisetum* evolved directly from Carboniferous Calamitaceae by loss of cambial activity in their stems and whorls of bracts in their cones or from a more ancient plexus of Devonian sphenopsid ancestors without cambial activity and lacking whorls of bracts. These last two characteristics are satisfied by the Middle Devonian Hyeniales. Even though known to have secondary wood, the Archaeocalamitaceae with their more or less bractless cones have attracted much attention as representing the plexus from which the Equisetaceae and Calamitaceae diverged in the Upper Devonian and Lower Carboniferous (Boureau, 1964; Bierhorst, 1971; Page, 1972). If this is an acceptable idea, then there should be some evidence that *Equisetum* coexisted with the calamites during the Carboniferous. Two compression–impression fossils of cones resembling those of *Equisetum* have been described. In his analysis of these specimens, Good (1975) concludes that if the specimens are cones of sphenopsids they could be interpreted just as well as bractless calamite cones.

Boureau believes that *Koretrophyllites* of the Sorocaulaceae bridges the gap between the Archaeocalamitaceae and Equisetaceae. *Koretrophyllites* (Fig. 16.9C) is known from compression–impression fossils extending from the Mississippian through the Permian. The attractive feature of the genus is that it was apparently herbaceous and had fructifications of the *Archaeocalamites* type (*Pothocites*) where several whorls of sporangiophores can occur between whorls of vegetative leaves (Fig. 16.9B). Their arrangement is superficially similar to that which occurs in abnormal cones of *Equisetum*, where whorls of vegetative leaves occur among the sporangiophores of otherwise normal cones. A review of *Koretrophyllites* material indicates that in its vegetative characteristics it is very similar to the widely distributed *Phyllotheca* (Fig. 16.10B). In both genera the leaves are longer than the internodes and are fused at their bases. The fructifications of *Phyllotheca*, however, are quite different from those of *Koretrophyllites*. There is some reason to question the validity of *Koretrophyllites* and, according to Good (1975), much additional information and confirmation is needed before the genus can be considered as a direct ancestor of *Equisetum*.

The fossil record of *Equisetum* and *Equisetites* extends from mid-Permian through the Pleistocene. The genus *Equisetites* is usually used for casts or compression–impression fossils that display well-defined leaf sheaths like those of *Equisetum* and have bractless cones with peltate sporangiophores. *Equisetum* is the name frequently used for Cenozoic fossils that cannot be distinguished from their modern counterpart.

For the reason that the Calamitaceae and Equisetaceae share so many characteristics in common, Good strongly supports the idea that *Equisetum* is a direct descendant of the arborescent calamites. If this is a correct interpretation, and the bulk of the evidence favors it, then one must account for the differences between the two groups. Although *Equisetum* is characterized by its reduced leaves and leaf sheaths, it is now known that leaves of calamites are not free at the node, but fused by their bases. The absence of cambial activity, although real enough in extant *Equisetum*, may be more apparent than real in some Triassic and Jurassic representatives of *Equisetites*. Some specimens from these Mesozoic deposits are truly large, measuring as much as 8 to 14 cm in diameter. Although there is no way of proving it from the casts and compressions, it seems likely that they were endowed with the additional supporting tissue produced by a vascular cambium. By the beginning of the Cenozoic, only species of *Equisetum* that are small appear.

Two sources of evidence suggest that there was a gradual change from the regular bracteate condition found in calamite cones to the nonbracteate condition of *Equisetum*. From the fossil record an upper Triassic *Equisetites* cone has been described (Daugherty & Stagner, 1941) that shows five areas, each consisting of several whorls of sporangiophores separated by a whorl of *Equisetum*-like leaves. The sporangiophores are similar in structure to those of *Equisetum*. Fossils of this type can be interpreted as representing a stage in the loss of whorls of bracts from the calamite cone type.

That the potential for the formation of whorls of sterile appendages is still present in cones of modern equisetums is apparent in teratological forms in which there is a spectrum of change from sterile bracts to sporangiophores through several nodes. Some intermediate whorls include unfused leaves, while others consist of leaflike sporangiophores. Evidence from this kind of morphological plasticity suggests that leaves and sporangiophores may be homologous and that whorls of sterile bracts were lost when sterile appendages were replaced by fertile ones to produce the non-bracteate cone of *Equisetum*.

Other Equisetales

Our attention has been concentrated on those sphenopsids that seem to have a more or less direct bearing on the evolutionary origins of the principal groups of the order. There are other sphenopsids, however, that are so widespread and are such common compression–impression fossils of Mesozoic rocks that they cannot be overlooked.

One of these is the genus *Neocalamites* (Fig. 16.9D) commonly found in Triassic and Lower Jurassic rocks from Greenland to South Africa, throughout Asia and Europe. In gross morphology the axes of *Neocalamites* suggest a small calamite with vascular strands that do not alternate at the nodes. In some specimens as many as 80 to 100 slender leaves were found attached to a node. The *Asterophyllites*-like leaves are up to 20 cm in length, but are not fused at their bases. Bractless cones of the *Equisetum* type were borne singly at the end of a slender stalk inserted at the node. Good (1975) has suggested that *Neocalamites* might be considered an intermediate between the Calamitaceae and Equisetaceae.

Figure 16.10. **A.** *Schizoneura paradoxa*. Reconstruction. Triassic. **B.** *Phyllotheca equisetoides*. Permian. (**A,B** from Boureau, 1964.)

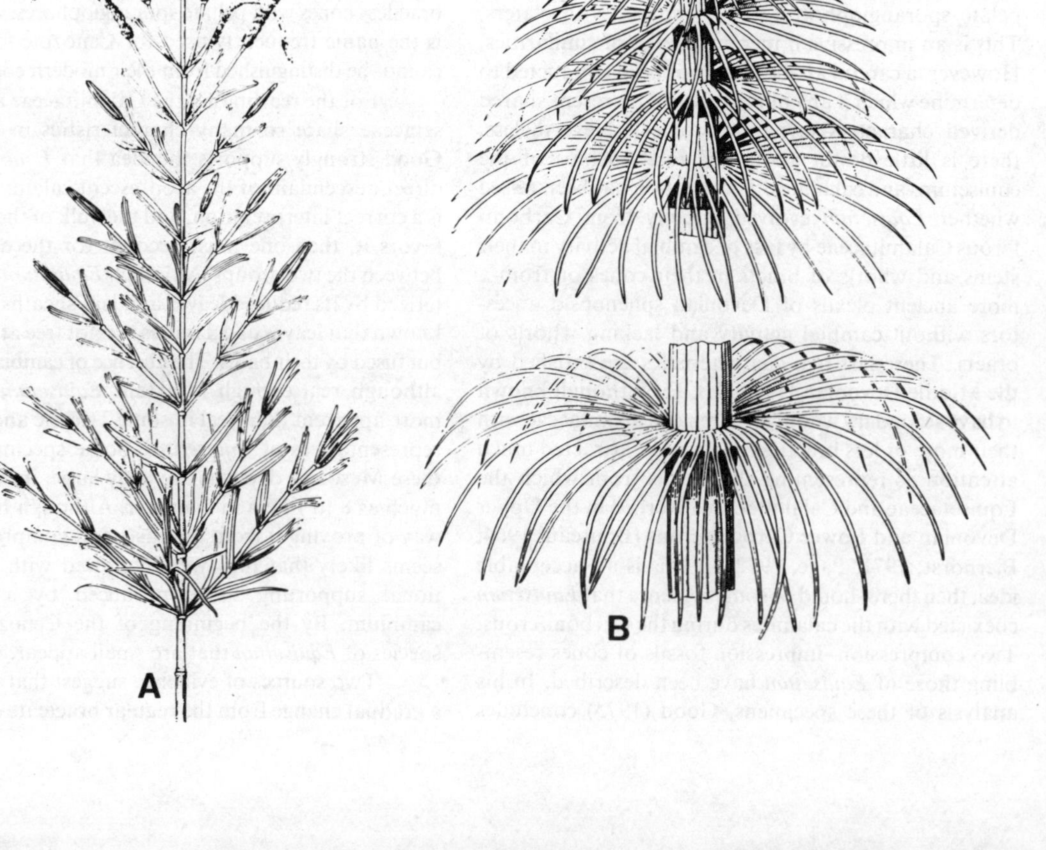

Schizoneura (Fig. 16.10A), once thought to be a monocot, is another widely distributed sphenopsid and is a part of the Cathaysian flora of China and the Glossopteris flora of India as well as European and South African Triassic and Jurassic floras. In Asia, *Schizoneura* is a conspicuous part of the Permian–Carboniferous floras. The plants grew up to 2 m high with diameters of 2 cm. At each node they bore two flat many-veined leaves. Their cones, which were slender and catkinlike, were produced at the ends of slender peduncles. Each apparently bractless cone consisted of whorls of sporangiophores.

Phyllotheca (Fig. 16.10B) is another important constituent of the Permian–Carboniferous *Glossopteris* flora of India and Africa. Like *Schizoneura,* it is almost worldwide in distribution and is found in strata of the Upper Paleozoic into the Cretaceous. One of the conspicuous characteristics of this herbaceous plant is the lateral fusion of its leaf bases into a disclike sheath from which the free tips of the leaves radiate. Verticels of sporangiophores alternate with verticels of bracts, as described for *Koretrophyllites*. The sporangiophores, usually six per whorl, are branched twice dichotomously with a group of four peltate sporangia at the tip of each branch.

The sphenopsid genera belonging to the Equisetales described above give us only a glimpse of the variety of forms that have been described from the Mesozoic and Cenozoic. For a full presentation the student is referred to the comprehensive and well-illustrated work of Boureau in *Traité de Paléobotanique,* Volume III.

Summary of evolutionary origins and relationships

With the evidence now available, it has become possible to trace the beginnings of the Sphenopsida to the Rhyniopsida of the Upper Silurian and Lower and Middle Devonian (Chart 16.1). Investigators with a well-founded knowledge of these ancient vascular plants agree that the Trimerophytopsida, which reached their zenith by the end of the Lower and beginning of Middle Devonian times, evolved from the Rhyniopsida. The more simply constructed members of the trimerophytes (for example, *Psilophyton dapsile*) are little more than rhyniophytes with clusters of paired sporangia. Otherwise, they have naked dichotomous branching systems. In those species where internal structure is known (*Psilophyton dawsonii*), the protostelic condition characteristic of the rhyniophytes is found. If there is one overall summarizing statement that we can make for the Trimerophytopsida, it would be that this was a rapidly evolving group exhibiting a high degree of morphological plasticity, not only among its genera but within species as well (*Psilophyton crenulatum*). With this in mind, it is little wonder that the Trimerophytopsida have assumed an important place in our understanding of vascular plant evolution. The evidence tells us that they represent a plexus of vascular plants from which several groups diverged including the Sphenopsida. We see in the Middle Devonian (Givetian) Iridopteridales just such a divergent group that can be interpreted as transitional between the trimerophytes and the Middle Devonian Hyeniales. This is evidenced by *Ibyka,* which like the Hyeniales has dichotomous fertile branches, dichotomously branched leaf precursors, and what appears to be protoxylem canals in a protostele. In other characteristics *Ibyka* is like *Pertica* of the Trimerophytopsida.

Until recently most paleobotanists believed that two groups had emerged from the Hyeniales by the Upper Devonian, one the Sphenophyllales, the other the Equisetales. However, the phylogenetic study of Stein, Wight, & Beck (1984) has cast considerable doubt on the homologies of some characters, and demonstrated incompatibilities among other characters upon which these interpretations have been made. More surprisingly, previously unrecognized similarities among the Protolepidodendrales, Hyeniales, and Sphenophyllales that we have emphasized in Chapters 14 and 15 suggest that the Sphenopsida may represent a polyphyletic group. *Eviostachya* of the Upper Devonian (Famennian) has been proposed as a transitional genus leading to the Sphenophyllales, but it is unclear whether the transition is from the Hyeniales or the Lycopodiales. Its stelar structure is a triarch protostele with protoxylem canals, and its sporangiophores are similar both to those of the Hyeniales and the sporophylls of the Protolepidodendrales in arrangement and structure. Thus, plants with an organization resembling that of the Sphenophyllales (*Eviostachya* and *Pseudobornia*) were present in the Upper Devonian. The Sphenophyllales did not, however, reach their maximum development until the Upper Carboniferous (Middle Pennsylvanian) and then declined to become extinct by early Permian times.

The Equisetales are represented first in the fossil record by the Upper Devonian–Lower Carboniferous Archaeocalamitaceae. These arborescent plants have whorls of *Hyenia*-like leaves and cones that are reminiscent of *Hyenia* and *Calamophyton,* where there are pseudowhorls of sporangiophores without whorls of bracts. All other characteristics are shared with the arborescent Calamitaceae, which evolved directly from the Archaeocalamitaceae. The first evidence of the Calamitaceae was found in the Lower Carboniferous (Mississippian), where pith casts assigned to *Mesocalamites* were discovered. Like the Sphenophyllales, the Calamitaceae did not reach their zenith until the Upper Carboniferous. Their extinction as arborescent sphenopsids was complete by the mid-Permian.

Depending on our evaluation of the evidence, we can derive the Equisetaceae directly from the arborescent calamites at some time in the Carboniferous

Chart 16.1. Suggested origin and relationships of major groups of the Sphenopsida and their distribution in geological time. Trim. = Trimerophytopsida; Id. = Iridopteridales; Hye. = Hyeniales; Arc. = Archaeocalamitaceae; Ps. = Pseudoboniales; Neo. = Neocalamitaceae; Proto. = Protolepidodendrales.

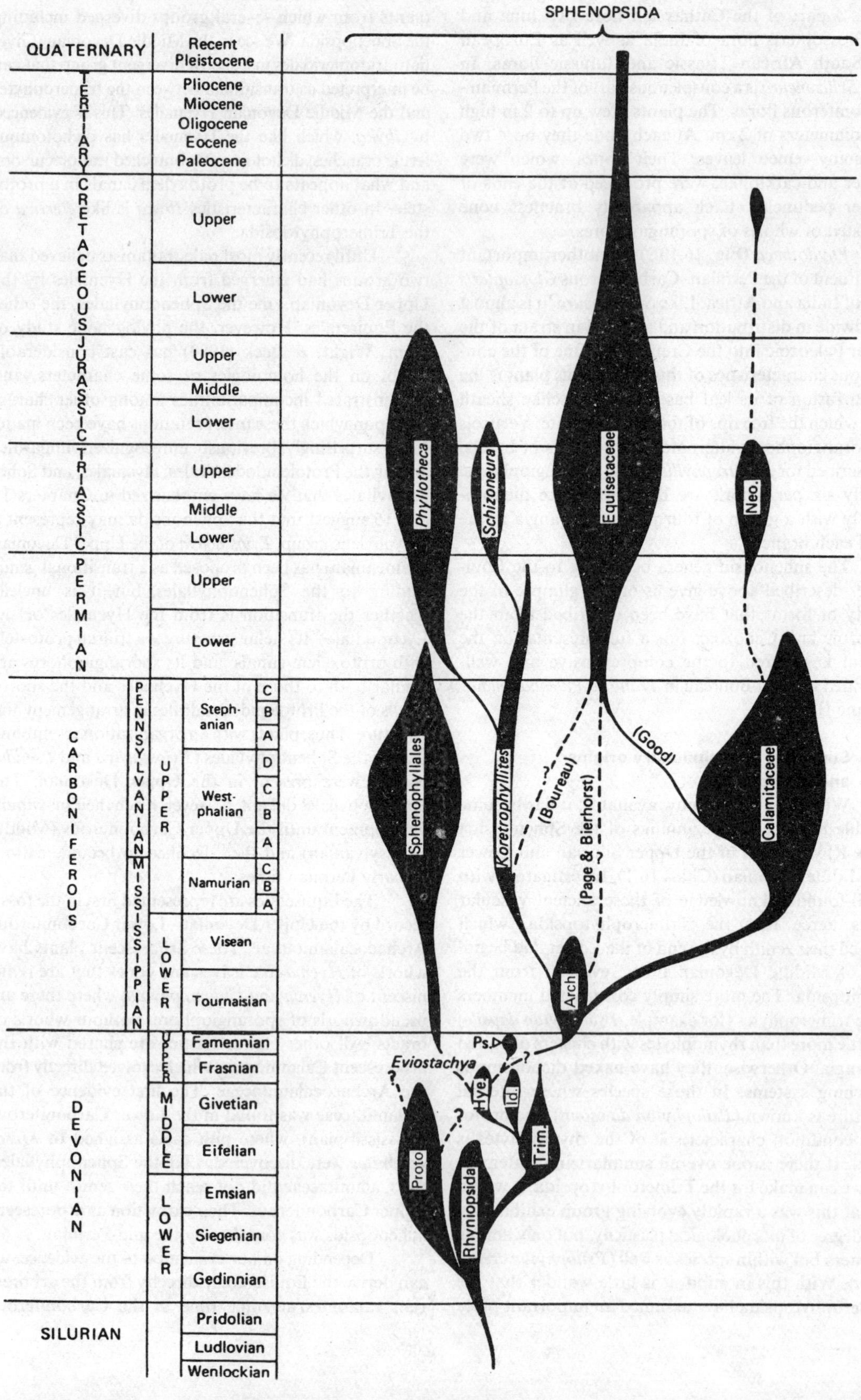

(Good, 1975) by reduction in size, formation of a leaf sheath, and loss of cambial activity and verticels of sterile bracts in the cones. According to Boureau (1964) and others, the evidence suggests an origin of the Equisetaceae from the Upper Devonian Archaeocalamitaceae either directly or through *Koretrophyllites*, which appears in the Lower Carboniferous and extends through the Permian. This is an attractive scheme because *Koretrophyllites* had an herbaceous habit and its fructifications were apparently bractless from the start. This would leave the evolution of the leaf sheath and loss of cambial activity to be explained. Unfortunately, nothing is known of the internal structure or spores of *Koretrophyllites*, and some investigators have questioned the interpretation of the material. Based on the quantities of evidence given earlier in this chapter, we can safely conclude that *Equisetites* and thus *Equisetum* evolved from Carboniferous calamite ancestors with *Elaterites* spore types.

References

Abbott, M. L. (1958). The American species of *Asterophyllites, Annularia,* and *Sphenophyllum. Bulletin of American Paleontology,* **38,** 598–611.

Anderson, B. R. (1954). A study of American petrified *Calamites. Annals of the Missouri Botanical Garden,* **41,** 395–418.

Andrews, H. N. (1952). Some American petrified calamitean stems. *Annals of the Missouri Botanical Garden,* **39,** 189–218.

Andrews, H. N. (1961). *Studies in Paleobotany.* New York: Wiley.

Arnold, C. A. (1958). Petrified cones of the genus *Calamostachys* from the Carboniferous of Illinois. *Contributions from the Museum of Paleontology, University of Michigan,* **14,** 149–65.

Baxter, R. W. (1955). *Palaeostachya andrewsii,* a new species of calamitean cone from the American Carboniferous. *American Journal of Botany,* **42,** 342–51.

Baxter, R. W. (1963). *Calamocarpon insignis,* a new genus of heterosporous petrified calamitean cones from the American Carboniferous. *American Journal of Botany,* **50,** 469–76.

Baxter, R. W., & Leisman, G. A. (1967). A Pennsylvanian calamitean cone with *Elaterites triferens* spores. *American Journal of Botany,* **54,** 748–54.

Bierhorst, D. W. (1971). *Morphology of Vascular Plants.* New York: Macmillan.

Boureau, E. (1964). *Traité de Paléobotanique,* Paris: Masson, Vol. III.

Brown, I. M. P. (1926). A new theory of the morphology of the calamarian cone. *Annals of Botany,* **41,** 301–20.

Cichan, M. A., & Taylor, T. N. (1982) Vascular cambium development in *Sphenophyllum:* A Carboniferous arthrophyte. *International Association of Wood Anatomists Bulletin* n.s., **3,** 155–60.

Daugherty, L. H., & Stagner, H. R. (1941). The upper Triassic flora of Arizona with a discussion of its geological occurrence. *Carnegie Institute of Washington Publications,* **526,** 1–34.

Delevoryas, T. (1955). *A Palaeostachya* from the Pennsylvanian of Kansas. *American Journal of Botany,* **42,** 481–8.

Eggert, D. A. (1962). The ontogeny of Carboniferous arborescent Sphenopsida. *Palaeontographica B,* **110,** 99–127.

Foster, A. S., & Gifford, E. M. (1974). *Comparative Morphology of Vascular Plants,* 2nd ed. San Francisco: Freeman.

Good, C. W. (1971). The ontogeny of Carboniferous articulates: calamite leaves and twigs. *Palaeontographica B,* **113,** 137–58.

Good, C. W. (1975) Pennsylvanian-age calamitean cones, elater-bearing spores, and associated vegetative organs. *Palaeontographica B,* **153,** 28–99.

Good, C. W., & Taylor, T. N. (1974). The establishment of *Elaterites triferens* spores in *Calamocarpon insignis* microsporangia. *Transactions of the American Microscopical Society,* **93,** 148–51.

Hirmer, M. (1927). *Handbuch der Paläobotanik.* Munich and Berlin: Oldenbourg.

Janssen, R. E. (1957). *Leaves and Stems from Fossil Forests,* 2nd ed. Springfield: Illinois State Museum.

Kosanke, R. M. (1955). *Mazostachys*—a new calamite fructification. *State Geological Survey of Illinois, Report of Investigations,* **180,** 7–23.

Melchior, R. C., & Hall, J. W. (1961). A calamitean shoot apex from the Pennsylvanian of Iowa. *American Journal of Botany,* **48,** 811–15.

Page, C. N. (1972). An interpretation of the morphology and evolution of the cone shoot of *Equisetum. Journal of the Linnean Society of London (Botany),* **65,** 359–97.

Schweitzer, H. J. (1972). Die Mitteldevon-flora von Lindlar (Rheinland). 3. Filicinae–*Hyenia elegans* Kräusel & Weyland, *Palaeontographica, B,* **137,** 154–75.

Scott, D. H. (1920). *Studies in Fossil Botany,* 3rd ed. London: Black, Vol. I.

Seward, A. C. (1898). *Fossil Plants,* Vol. I. Cambridge: Cambridge University Press. Reprint. New York: Macmillan (Hafner Press).

Skog, J. E., & Banks, H. P. (1973). *Ibyka amphikoma,* gen. et sp. n., a new protoarticulate precursor from the late Middle Devonian of New York State. *American Journal of Botany,* **60,** 366–80.

Stein, W. E., Wight, D. C., & Beck, C. B. (1984). Possible alternatives for the origin of Sphenopsida. *Systematic Botany,* **9,** 102–18.

Takhtajan, A. L. (1956). *Telomophya.* Moscow: Academiae Scientiarum.

Williamson, W. C., & Scott, D. H. (1894). Further observations on the organization of the fossil plants of the coal-measures. Part I. *Calamites, Calamostachys,* and *Sphenophyllum. Philosophical Transactions of the Royal Society, London, B,* **185,** 863–960.

Wilson, M. L., & Eggert, D. A. (1984). Root phloem of fossil tree-sized arthrophytes. *Botanical Gazette,* **135,** 319–28.

17

Putative ferns of the Paleozoic

Some characteristics of ferns

In gross morphology, the general impression of a fern sporophyte (Fig. 17.1A) is a photosynthetic plant with large compound leaves (fronds) arising from a rhizome with adventitious roots. The fronds develop at the tips of the rhizome from croziers or fiddleheads. As the fronds develop from the unfolding croziers, we see that the leaflets (pinnae) expand into a single plane along the sides of a rachis. Each pinna, in turn, may be subdivided into pinnules and these may be further subdivided. Thus, compound leaves of ferns may be once to many times pinnate. In the foliar lamina there is usually an open venation where veins may be dichotomized toward the margin (Fig. 17.2A). Of course, there are ferns with entire simple leaves, some with very small leaves (*Azolla*), some with closed or reticulate venation (Fig. 17.2B), and many other deviations from the generalized type. All ferns are said to be megaphyllous because their leaves, like those of the Sphenophyllales, have been derived from planated, webbed, and overtopped branching systems. In most ferns, except the protostelic forms, the leaf trace is associated with a leaf gap in their siphonostelic stems (Fig. 17.2C). The leaf trace is usually C-shaped with adaxial curvature.

With few exceptions ferns lack secondary tissues, but often have abundant sclerenchyma (a tissue made up of thick-walled cells) in their stems. Their primary xylem is commonly mesarch.

When sporangia of ferns reach maturity they dehisce and shed their spores. In this characteristic they are free-sporing, as are all other vascular plants discussed thus far. Fern sporangia are usually abaxial on leaves and in clusters called sori (Fig. 17.3), but some are marginal. Some ferns are eusporangiate (Fig. 17.2D) and have rather massive, thick-walled sporangia characteristic of many such primitive vascular plants as rhyniophytes, lycopods, sphenopsida, and trimerophytes. Others are leptosporangiate (Fig. 17.2E) with delicate sporangium walls, one cell layer thick, and highly specialized dehiscence mechanisms. Most ferns are homosporous, but a few aquatic ferns are heterosporous. The vast majority of ferns have exosporic, free-living, photosynthetic gametophytes. It is important to keep in mind that there is no single characteristic by which we can define a plant as being a fern. We need a combination of characteristics, most of which are set forth above.

The classification of ferns

You will recall from Chapter 3 that we are attempting to use a phylogenetic classification, where natural or monophyletic groups are recognized on the basis of shared, derived features (synapomorphies). Ferns do not fit this definition. If you think of all of the free-sporing pteridophytes that have vegetative shoot systems consisting of stems that bear leaves, and then remove those groups that can be recognized by the presence of synapomorphies (lycopsids and sphenopsids), nearly everything that remains is classified as a fern. Remember that the ancient rhyniopsids, zosterophylls, trimerophytes, iridopterids, and the living psilophytes do not have well-differentiated stems and leaves, so they should not be included in our considerations.

In terms of the features discussed above, ferns (Filicopsida) have the following essential characteristics by which they may be recognized. They are free-sporing pteridophytes (homosporous or heterosporous) that have (1) large complex leaves derived from modified branching systems, (2) mesarch steles, (3) usually no secondary growth, and (4) sporangia attached at the tips or at the margin of the pinnules, or to the abaxial surface of the leaves. However, all of these characters could also be inferred for the probable ancestors of ferns, the trimerophytes we

Figure 17.1. **A.** *Polystichum lonchitis.* Habit of plant showing circinate vernation, fronds with abaxial sori, and adventitious roots. **B.** *Osmunda regalis* frond. **A,B:** Extant. (**A,B** from Takhtajan, 1956.)

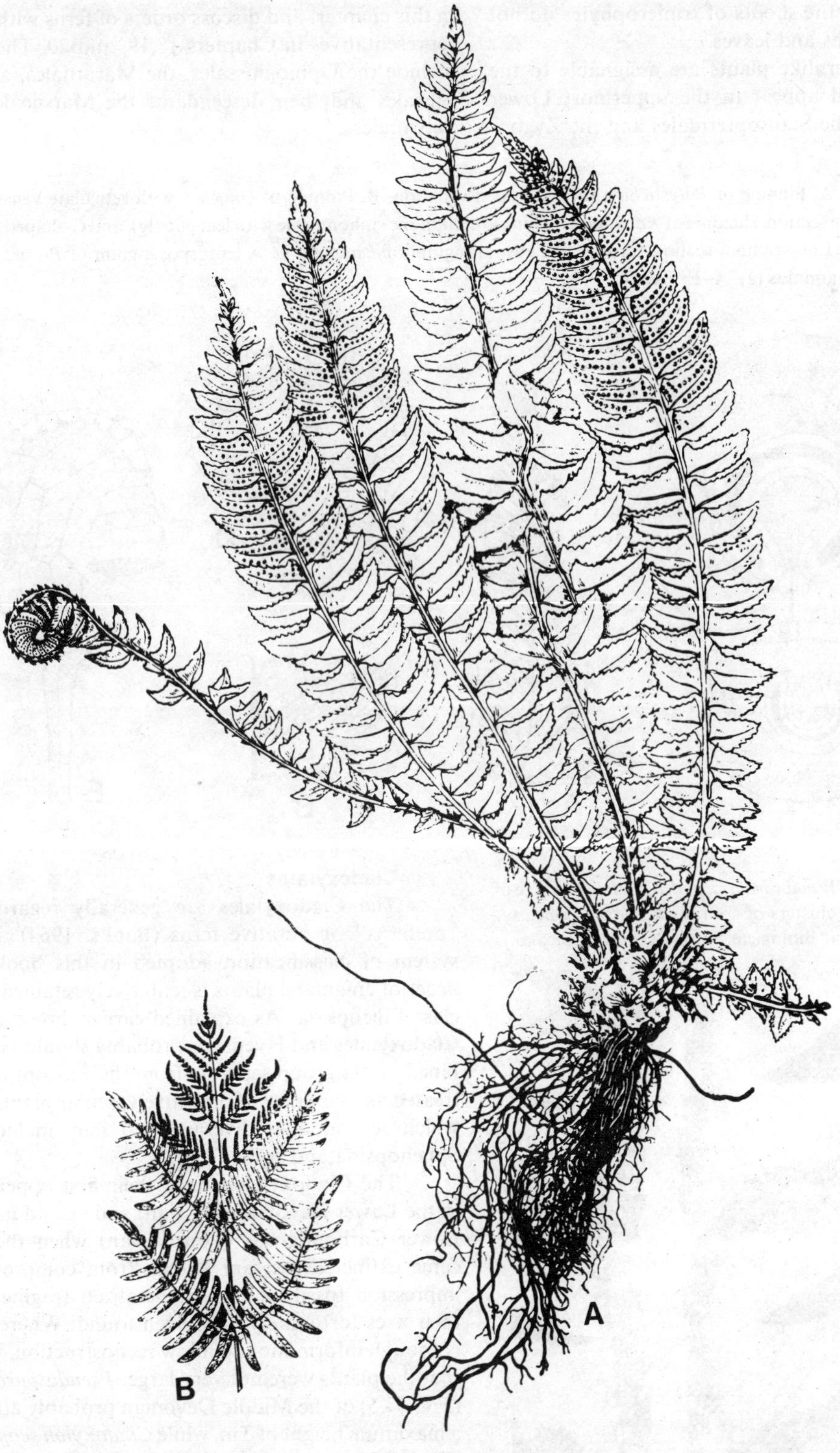

discussed in Chapter 13. Therefore, the ferns are a polyphyletic assemblage. Ferns differ from trimerophytes primarily by having well-differentiated stems and leaves, while the shoots of trimerophytes do not have distinct stems and leaves.

The first fernlike plants are assignable to the Cladoxylales, and appear in the uppermost Lower Devonian, with the Stauropteridales and the Zygopteridales first occurring in Upper Devonian sediments. All three orders became extinct before the end of the Paleozoic. We will consider these orders in this chapter, and discuss orders of ferns with living representatives in Chapters 18,19, and 20. The latter include the Ophioglossales, the Marattiales, and the Filicales and their descendants the Marsileales and Salviniales.

Figure 17.2. **A.** Pinnule of *Polystichum* showing open venation. **B.** Pinnule of *Onoclea,* with reticulate venation. **C.** Transverse section, rhizome of *Adiantum* showing amphiphloic siphonostele with leaf gap (lg), and C-shaped leaf trace (lt). **D.** Longitudinal section through the eusporangium of *Psilotum.* **E.** A leptosporangium of *Pteris.* Note conspicuous annulus (a). **A–E:** Extant.

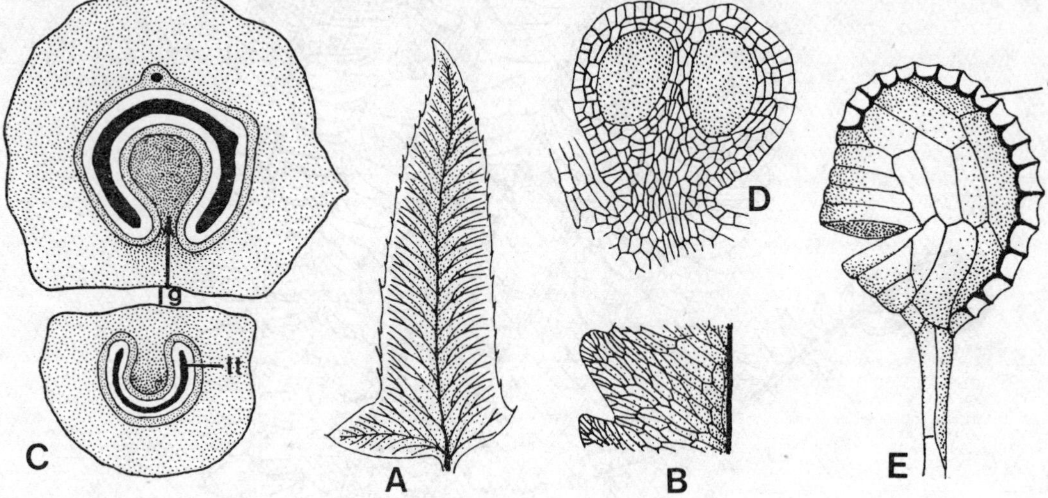

Figure 17.3. *Woodwardia virginica,* abaxial surface of leaf showing clusters of sporangia (sori). (Courtesy of the General Biological Supply House, Chicago.)

Cladoxylales

The Cladoxylales are generally regarded as "preferns" or putative ferns (Banks, 1964). In the system of classification adopted in this book, this order of enigmatic plants is tentatively retained in the class Filicopsida. As explained earlier, however, the Cladoxylales and Hyeniales probably should be combined into a group separate from the Filicopsida and treated as a complex of primitive vascular plants from which several lines diverged that may include the Sphenopsida and some ferns.

The Cladoxylales made their first appearance in the Lower Devonian (Emsian) and extend into the Lower Carboniferous (Tournaisian) when they became extinct. Most are known from compression–impression fossils, but permineralized fragments of their axes fortunately have been found. Where there is enough information to allow reconstruction, we see that the plants were not very large. *Pseudosporochnus* (Fig. 17.5) of the Middle Devonian probably attained a maximum height of 3 m, while *Cladoxylon scoparium* (Fig. 17.7A) was not more than 35 cm tall. Both genera show branching of the distal parts that is characteristically digitate. Fertile and sterile appendages were borne in a spiral on the main branches.

Internally, the axes of the Cladoxylales exhibit a multifascicular condition characterized by many anastomosing, often vermiform vascular segments each of which may have what appears to be secondary xylem around the outside. The primary xylem is mesarch with the protoxylem strands associated, in most cases, with peripheral loops located near the tips of radiating arms of the vascular segments (Fig. 17.7C). By definition (Leclercq, 1970), a peripheral loop (Fig. 17.4) is a rod of parenchyma around which protoxylem tracheids occur. If the delicate cells of the parenchyma strand disintegrate, as is usually the case, a cavity is formed that has the appearance of a protoxylem canal characteristic of most Sphenopsida. Peripheral loops, on the other hand, are commonly found in certain primitive fernlike plants. Difficulties in differentiating between protoxylem canals and peripheral loops have caused some problems in assigning several Devonian genera to their natural groups (Scheckler, 1974, 1975). Such is the case for *Ibyka, Calamophyton,* and perhaps some Cladoxylales.

Pseudosporochnus

Prior to the researches of Leclercq & Banks (1962), *Pseudosporochnus* (Fig. 17.5) was thought to be a robust member of the psilophytes, which says something about the primitive nature of some of its characteristics. Leclercq & Lele (1968) were able to show, however, that the affinities of *P. nodosus* from the Middle Devonian were with the Cladoxylales. This was based primarily on the internal structure of its stems and branches, where the vascular system was found to be composed of the typical anastomosing segments of vascular strands with what appear to be peripheral loops. From the compression–impression fossils they were able to determine what was believed to be a root system that, although not actually attached, had certain anatomical characteristics shared with stems and branches of *P. nodosus* from the same beds.

Figure 17.4. Peripheral loop of *Cladoxylon dawsonii.* Upper Devonian. Note parenchyma cells filling the loop. (Based on a photograph in Scheckler, 1975.)

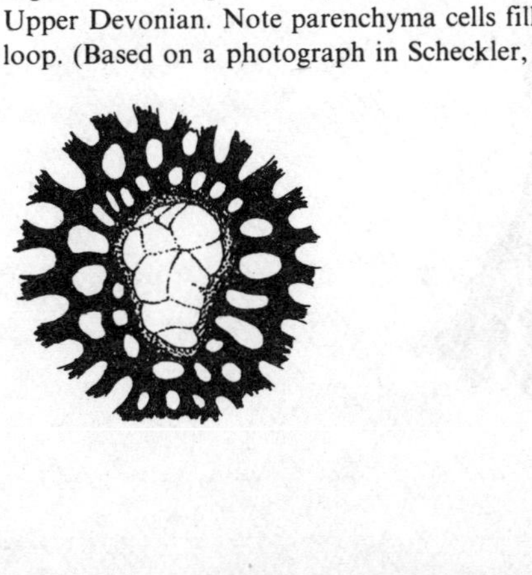

The branches arise from the distal end of the main axis in a digitate manner and branch dichotomously two or three times to produce branches of the first order (Fig. 17.6). These in turn bear spirally arranged planted terete branches of the second order. Branches of the third order were borne laterally and more or less opposite (Fig. 17.6B) on second-order branches. Third-order branches dichotomized two or three times in one plane. They are terete and some were terminated by erect fusiform sporangia borne in pairs (Fig. 17.6A).

Cladoxylon

The best-known species of *Cladoxylon* is *C. scoparium* (Fig. 17.7A), also from the Middle Devonian (Givetian). In its general morphology it is very similar to *Pseudosporochnus nodosus* with its digitate branching. Internally, however, the anastomosing vascular segments (Fig. 17.7B,C) are more vermiform when seen in transverse section. A recent reinvestigation (Schweitzer & Giesen, 1980) of the spirally arranged second-order branches of *C. scoparium* shows that the fertile units are, in reality, sporangiophores similar to those of *Calamophyton primaevum.* These same authors also discovered that those planated, webbed branches, originally described as fertile units with marginal sporangia, are "the terminations of several sterile organs superimposed in a lattice formation." Other sterile, presumably foliar structures (Fig. 17.7D) branch dichotomously two to four times and those on the lower portions of the main axes show indications of planation and some webbing. The results of the investigations by Schweitzer & Giesen (1980) and Schweitzer (1972) show that *Cladoxylon scoparium, Calamophyton bicephalum,* and *Hyenia elegans* are similar to ferns. Contrary to the conclusions of Schweitzer (1972), however, it is the opinion of your authors that the relationship of these species to the Filicopsida is still a moot question.

Origin and relationship of the Cladoxylales

The Cladoxylales first appear at about the same time as the Lower (Emsian to Givetian) of the Middle Devonian Trimerophytopsida from which they probably evolved. Like the trimerophytes, Middle Devonian *Pseudosporochnus nodosus* and *Cladoxylon scoparium* branch dichotomously to pseudomonopodially and have sporangia borne on the tips of fertile telome trusses. In *C. scoparium* the fertile telome truss is similar to a sphenopsid sporangiophore with recurved tips bearing the terminal sporangia. The sporangia dihisce longitudinally as in the trimerophytes. In both groups, the protoxylem is submerged in metaxylem and is centrarch to mesarch in the Trimerophytopsida and mesarch in all Cladoxylales thus far examined.

We started this short chapter by raising the question: Are the Cladoxylales ferns? Now that we have some idea of the characteristics of both groups let us try to answer the question. In *Pseudosporochnus nodosus* there is some indication of the early evolution of a fernlike frond. Like some other primitive fernlike plants, the Cladoxylales have what appear to be peripheral loops associated with mesarch primary xylem. Ferns are almost always without secondary xylem, and if the interpretation is correct that the aligned xylem cells in the stems of the Cladoxylales are really metaxylem, then the Cladoxylales are like ferns in lacking secondary xylem. Although marginally fernlike in some respects, the Cladoxylales are different in many other characteristics.

When it was believed that the Cladoxylales were polystelic and had secondary xylem around the individual vascular segments, affinity with the medullosan pteridosperms was suggested. Unlike the ferns, some Cladoxylales are eustelic, and not siphonostelic. In this characteristic, and in the way their leaf traces arise and supply leaf bases, they do resemble the medullosan pteridosperms. But here the similarity seems to end.

The possibility still exists that some genera assigned to the Cladoxylales have protoxylem lacunae that are homologous with carinal canals, and that when more is known of their general morphology and

Figure 17.5. Reconstruction of *Pseudosporochnus nodosus*. Middle Devonian. (From Leclercq & Banks, 1962.)

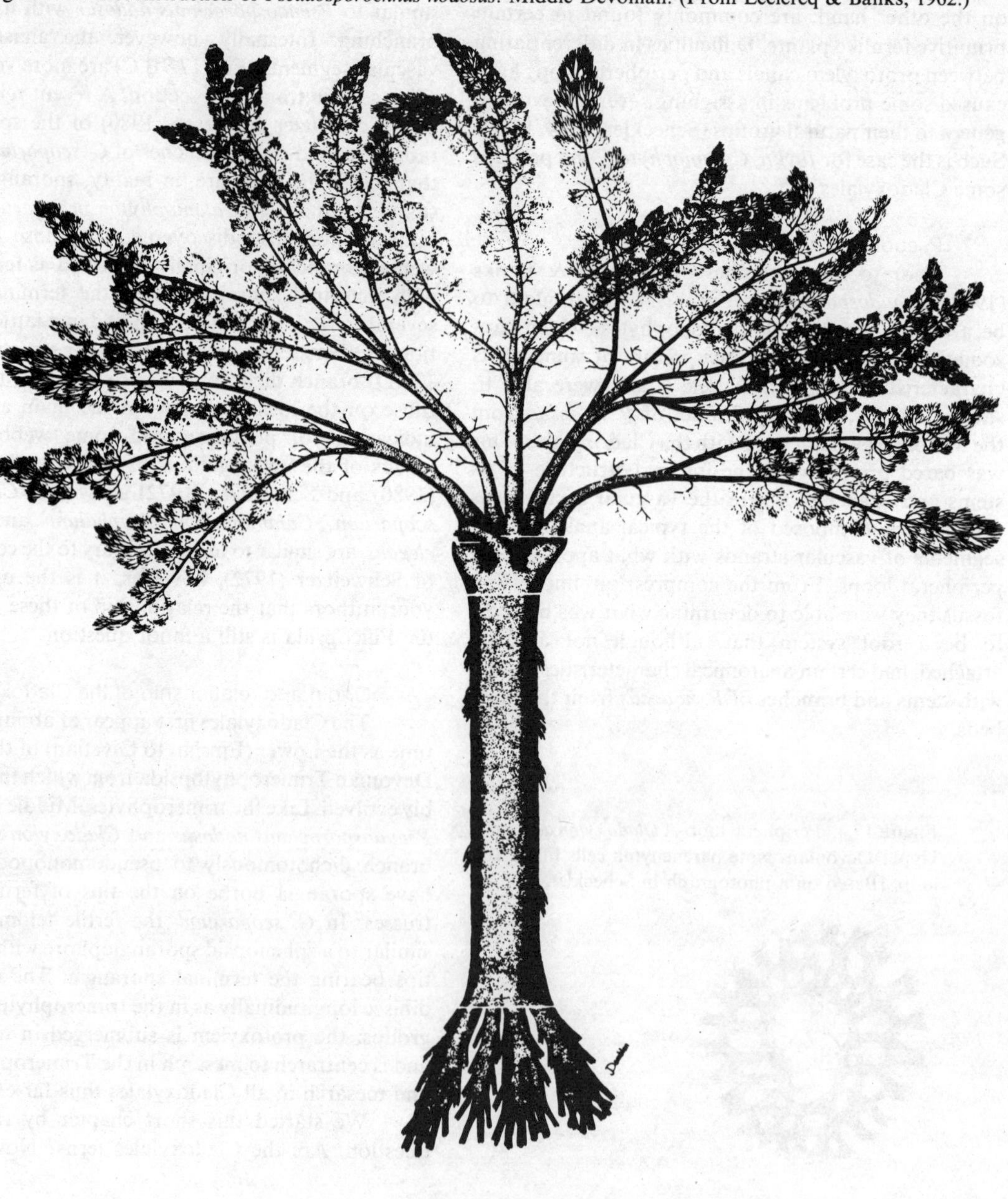

the nature of the lacunae they will join the Hyeniales as a precursor of the Sphenopsida.

Thus, the Cladoxylales can be visualized as a group of plants having a unique combination of characteristics shared with the trimerophytes, sphenopsids, ferns, and even the seed plants, and as a group that became extinct at the end of the Lower Carboniferous without having furnished the ancestors of any other vascular plants. It seems highly probable that the Cladoxylales, as we presently know them, can be added to our list of plant groups that represent unsuccessful evolutionary "experiments" that ended in extinction.

Stauropteridales

The Stauropteridales ranges from uppermost Devonian to the Upper Carboniferous, and is represented by three genera, *Stauropteris, Rowleya,* and *Gillespiea.*

The genus *Stauropteris* is known from both the Lower and Upper Carboniferous in the form of permineralized specimens. One of the more complete descriptions of *Stauropteris* is provided by Surange (1952) for *S. burntislandica* (Fig. 17.8A). The plants were relatively small and bushy without apparent fronds or leaves with laminae. In many specimens, the branching pattern of *S. burntislandica* is quadriseriate in all orders, from the largest axes to the smallest. The pairs of branches that are borne alternately on opposite sides of all axes arise from dichotomies. The ultimate branches are often terminated by sporangia. Pairs of vascularized aphlebiae are borne at every level of branching. The aphlebiae of *S. burntislandica* show considerable morphological variation in form and method of attachment.

When seen in cross section the vascular supply of the largest first-order axes consists of four mesarch strands of primary xylem, two of which divide to give rise to the vascular supply of the paired second-order branches (Fig. 17.8B). The vascular strands of the latter may be tetrarch protosteles. At a higher level and on the opposite side, a second pair of traces is produced. This alternating pattern is repeated through several orders of branching to produce the quadriseriate branches. One interpretation suggests that this unusual branching system represents an early stage in the evolution of a frond from a three-dimensional dichotomous branching system.

A variation of the branching pattern for species of *Stauropteris* was discovered by Cichan & Taylor (1982) in specimens from the Lower–Middle Pennsyl-

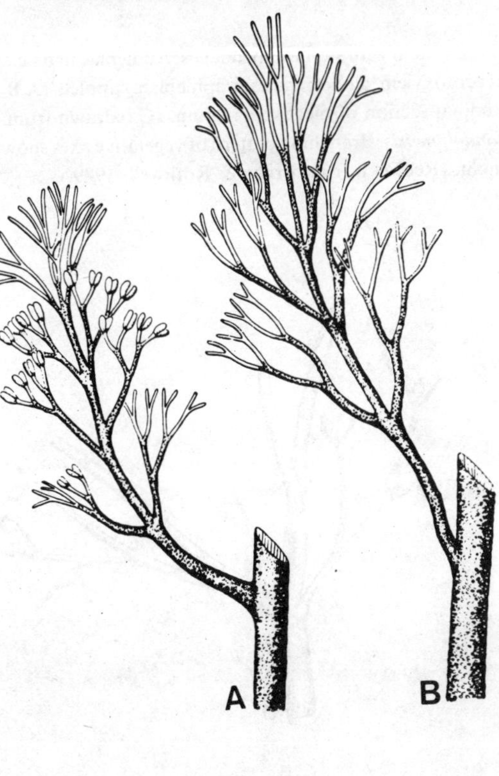

Figure 17.6. **A.** Fertile branch of *Pseudosporochnus.* **B.** Branch with sterile appendages. *P. nodosus.* Middle Devonian. (**A,B** from Leclercq & Banks, 1962.)

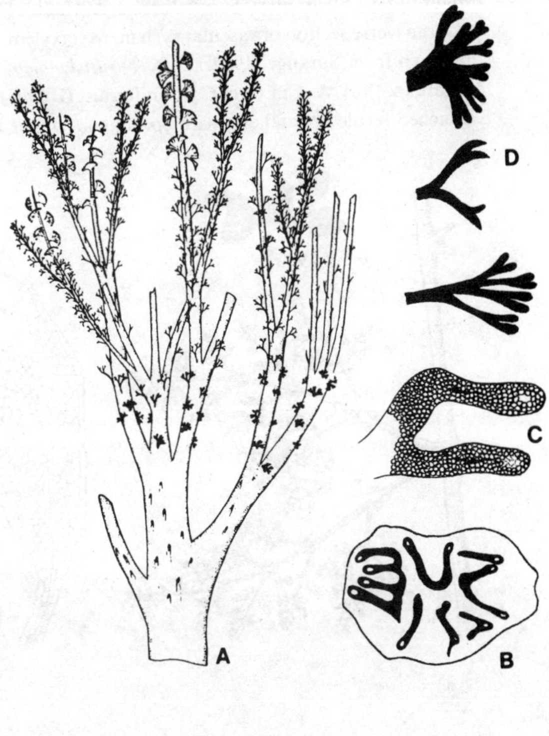

Figure 17.7. **A.** Reconstruction of *Cladoxylon scoparium.* (From Delevoryas, 1962.) **B.** *C. scoparium* transverse section showing vermiform vascular segments. **C.** Enlarged tips of vascular segments showing peripheral loops (?). **D.** *C. scoparium,* sterile appendages. **A–D**: Middle Devonian. (**B–D** redrawn from Leclercq, 1970.)

vanian of Kentucky. As the name *S. biseriata* implies, the branching pattern is distichous and two-ranked in all orders of branches.

A most unusual and unique feature of *Stauropteris burntislandica* is the fact that it is heterosporous! Its eusporangiate megasporangia (Fig. 17.8C) are terminal on ultimate branches. Two functional and two abortive megaspores have been found in the distal half of the spindle-shaped sporangium. The lower half of the sporangium is parenchymatous and is supplied with a trace of vascular tissue. There is no annulus. Terminal eusporangiate microsporangia have been found on another species of *Stauropteris*.

Although the structure of an *S. burntislandica* megasporangium is more primitive than the sporangium of any other fern, heterospory has evolved in this species. If we agree that heterospory represents a relatively advanced level of evolution, then we are confronted with an excellent example of different rates of evolution in different characteristics of a single species. We also add to our list another example of parallel evolution, as it applies to heterospory.

Rowleya trifurcata, (Long, 1976) from the Carboniferous (Westphalian A) of England, with its tetrach protostele, is similar to *Stauropteris*. As the specific epithet implies, however, the axis bears lateral branches in threes resulting from a trifurcation unlike *Stauropteris* with its pairs of branches. Small terete branches, interpreted as pairs of leaves similar to those of *Psilotum,* are produced at the tips of lateral branches.

Gillespiea randolphensis is a recently described (Erwin & Rothwell, 1989) addition to the Stauropteridaceae that occurs as both compressed and permineralized axes, and extends the range of the order back to the Upper Devonian. The basic branching pattern and internal anatomy of *Gillespiea* are like *Stauropteris burntislandica,* but branching is less frequent and regular so that plant probably produced a ground cover of tangled axes. In *Gillespiea* (Fig. 17.8D) delicate branching systems with lateral and terminal sporangia replace the aphlebiae found in other stauropterid genera. The pointed megasporangia bear one or two trilete megaspores.

The occurrence of both homosporous and heterosporous species in the Stauropteridales shows that there is a wide diversity of reproductive biology in the order. Most unexpectedly, the heterosporous stauropterid species of the Upper Devonian and Lower Carboniferous predate the homosporous species of the Upper Carboniferous. Because it is extremely unlikely that the homosporous forms were derived from the heterosporous species, there is reason to suspect that the order includes at least two separate evolutionary lineages (clades) that share quadraseriate branching and simple, mesarch xylem anatomy.

Zygopteridales

Representatives of the Zygopteridales first appear in the Upper Devonian, increasing in abundance in the Lower Carboniferous of Scotland and France, the Mississippian and Pennsylvanian of North America,

Figure 17.8. *Stauropteris burntislandica*. **A.** Habit of branching system with sterile appendages. Aphlebiae in black. **B.** Transverse section of vascular system. Metaxylem, black; protoxylem, circles; traces to aphlebiae, stippled. (**A,B** redrawn from Surange, 1952.) **C.** *S. burntislandica,* longitudinal section of megasporangium. (**C** redrawn from Andrews, 1961) **A–C:** Lower Carboniferous. **D.** *Gillespiea randolphensis*. Branching pattern of vegetative axes showing attached fertile appendages with sporangia. Upper Devonian. (Redrawn from Erwin & Rothwell, 1989.)

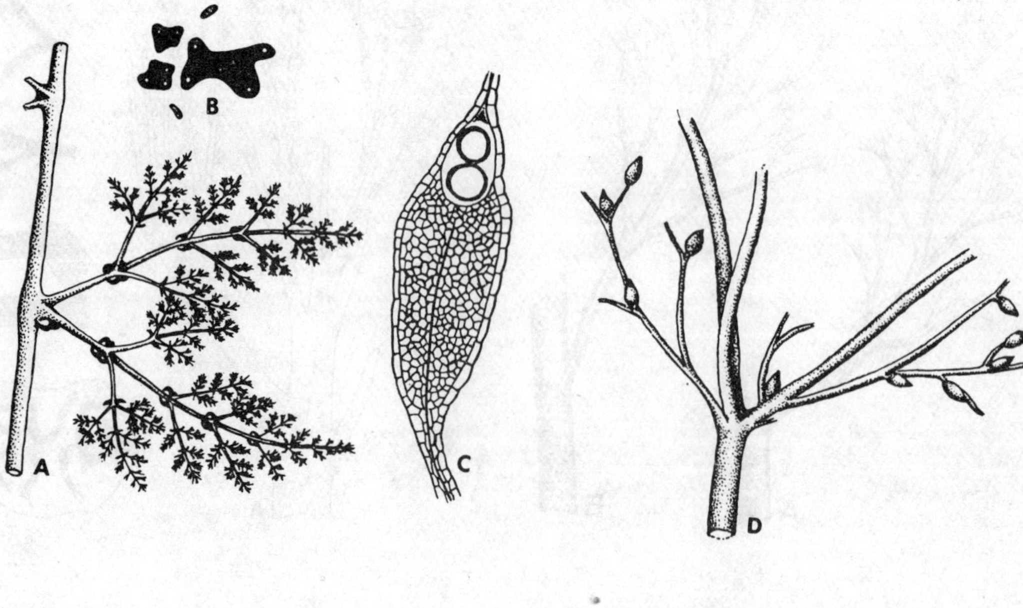

and finally disappearing in the Lower Permian. They are characterized in their vegetative frond morphology by quadraseriate branching like that of the Stauropteridales, and in their vegetative anatomy by the presence of a clepsydroid trace and secondary xylem in their frond axes. The occurrence of secondary xylem in the Zygopteridales is a most unusual specialization for ferns. The clepsydroid configuration applies to the trace of primary xylem that is somewhat hourglass-shaped in transverse section (Fig. 17.9) and has a peripheral loop at each end. The bilaterally symmetrical clepsydroid strand gives off pinna traces laterally from the area of the peripheral loops (Fig. 17.10).

Figure 17.9. Stereodiagrams of frond vascular systems. **A.** *Rhacophyton zygopteroides*. **B.** *Metaclepsydropsis duplex*. **C.** *Diplolabis roemeri*. **D.** *Etapteris scottii*. (**A–D** from Phillips, 1974.)

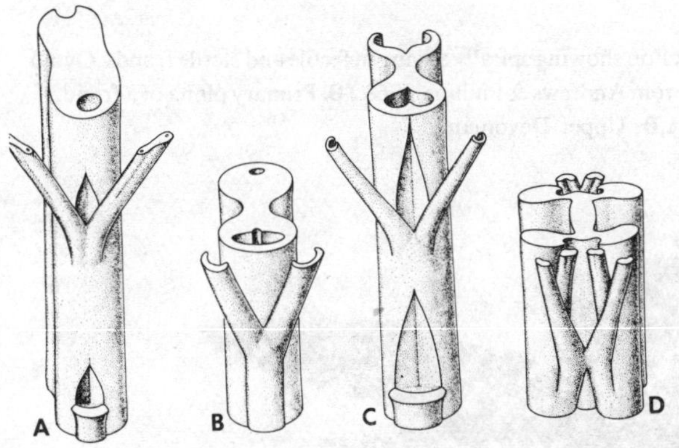

There are two trace morphologies found among frond axes assigned to the Zygopteridales. These are used to distinguish genera assigned to the subgroups Clepsydroideae and Etapteroideae of the family Zygopteridaceae (Phillips, 1974). The clepsydroid type is found in axes where the peripheral loops remain more or less closed at the point of pinna trace departure (Fig. 17.9B). The etapteroid type occurs where the peripheral loop is temporary as in *Diplolabis* (Fig. 17.9C) or permanently open as in the H-shaped phyllophores of *Etapteris* (Fig. 17.9D).

Zygopteridaceae: Devonian genera

Unfortunately the fossil record of Devonian Zygopteridaceae is fragmentary at best, except for the Upper Devonian *Rhacophyton* (Leclercq, 1951; Andrews & Phillips, 1968; Cornet, Phillips, & Andrews, 1976; Dittrich, Matten, & Phillips, 1983), which is well known from compression–impression and permineralized specimens. In reconstruction, *Rhacophyton ceratangium* (Fig. 17.11A) is shown to be about 1.5 m high with an erect to semierect stem

Figure 17.10. Transverse section, frond of *Metaclepsydropsis duplex*. Pinna traces (pt); clepsydroid strand trace (cs); peripheral loop (pl). Carboniferous. (From Phillips, 1974.)

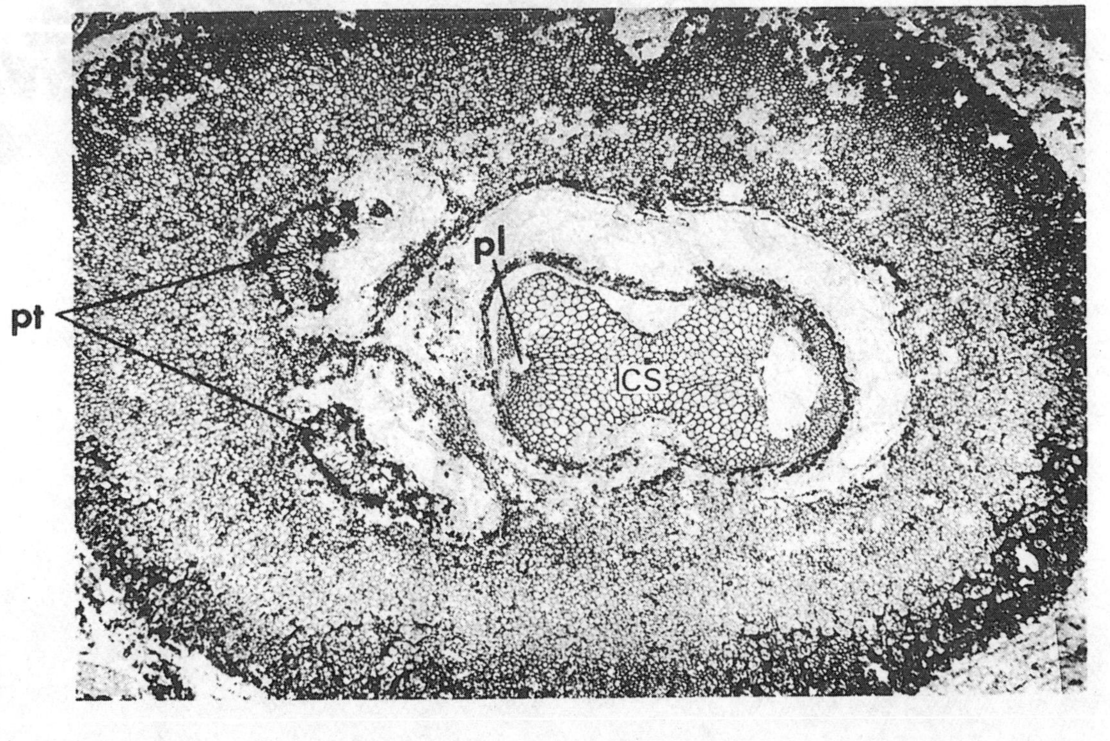

axis bearing crowded, fertile, and sterile fronds in two rows of paired axes. The fronds exhibit two basic types of branching patterns: one in which two lateral rows of alternate pinnae are produced to form a planated biseriate structure similar to the frond of a modern fern, the other which produces a three-dimensional quadriseriate branching pattern that has no modern counterpart. Here, pairs of pinnae arise in four instead of two rows. Generally those fronds with quadriseriate branching bear fertile and sterile pinnae, while those that are planated and biseriate are completely sterile.

The pinna of a biseriate frond of *Rhacophyton zygopteroides* has a clepsydroid xylem strand that gives off alternating lateral pinna traces. Quadriseriate fronds have clepsydroid strands that give off lateral traces, but each of these divides dichotomously to give rise to a pair of pinna traces supplying the pairs of pinnae that alternate on opposite sides of the rachis. The pinna traces also are clepsydroid and give off alternating lateral traces to ultimate branching segments. The sterile frond of *Rhacophyton* is bipinnate, and the ultimate dichotomously branched segments (Fig. 17.11B) are probably the homologs of pinnules. Some of these are three-dimensional in their branching, while others show degrees of planation.

The fertile appendages of the three-dimensional quadriseriate frond occur in pairs at the base of the bifurcate pinnae (Fig. 17.11A). The appendages are composed of three-dimensional branches that dichoto-

Figure 17.11. *Rhacophyton ceratangium*. **A.** Reconstruction showing spirally arranged fertile and sterile fronds. Ovoid structures represent positions of fertile complexes (f). (From Andrews & Phillips, 1968.) **B.** Primary pinna of a frond. (Redrawn from Cornet, Phillips, & Andrews, 1976.) **A,B:** Upper Devonian.

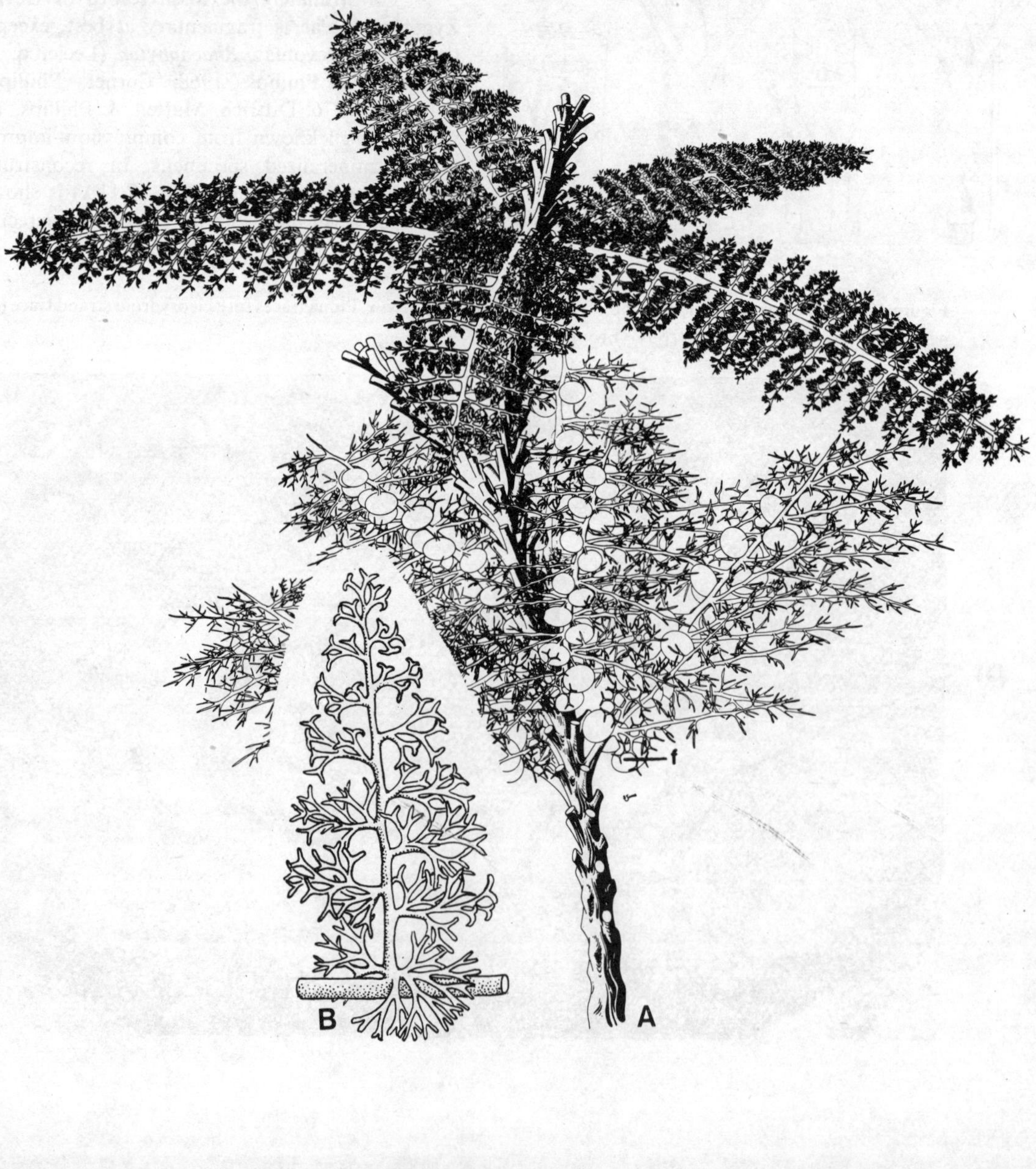

mize several times, then bear pinnate appendages with terminal fusiform sporangia with long tips. Spores of the dispersed *Perotriletes* type have been isolated from the sporangia. In the sterile frond, the basal pair of fertile branches is replaced by sterile aphlebialike branches.

The discovery of secondary xylem in *R. ceratangium* (Andrews & Phillips, 1968) suggests to these researchers that the affinities of *Rhacophyton* are close to the progymnosperm-gymnosperm line. In the later 1976 paper with Cornet, they assign *Rhacophyton* to the old order Protopteridales with the comment, "The Protopteridales may serve as a focal point for fern-like plants which may have progymnosperm characters but are apparently not in a direct line of gymnosperm evolution." The unsettled position of *Rhacophyton* in a system of classification is further emphasized by Taylor (1981), who places the genus in a separate class, the Rhacophytopsida.

Various permineralized stem fragments have been reported from the Upper Devonian, where the clepsydroid configuration has been found in the second-order branches. *Asteropteris* and *Stenokoleos* are examples. Other fragments of axes with clepsydroid xylem strands occur in the Upper Devonian and Lower Carboniferous. These are assigned to the genus *Clepsydropsis* (Galtier, 1966). Some, however, are probably fragments of *Rhacophyton*.

Zygopteridaceae: Carboniferous and Permian genera

Zygopteris

The reconstruction of *Zygopteris primaria* by Sahni (1932) is an excellent example of paleobotanical serendipity. Prior to Sahni's discoveries, the plant known today as *Zygopteris* was known only by its separate permineralized parts. The original specimen was a portion of a trunk of a small tree from Permian deposits of France. This was cut up and the pieces distributed to museums on the continent and England. The specimen, originally described by Cotta in 1832, contained petioles of the *Etapteris* type (Fig. 17.12A). In 1912 Scott independently described the genus *Botrychioxylon* from a different specimen than the one described by Cotta. While visiting a museum in Berlin in 1929, Dr. Sahni had, as he put it, "a pleasant surprise" when he discovered a beautiful petrifaction of *Z. primaria* with its etapteroid petioles attached to a *Botrychioxylon*-type stem. Sahni recognized this unusual specimen as the basal part of the one described by Cotta. In due time and after a good deal of paleobotanical detective work, Sahni reassembled the pieces of the original specimen, which had been scattered in five different repositories! Sahni concluded that if the specimen as it was first collected had been described in its entirety, *Botrychioxylon* and the form genus *Etapteris* would never have been established. Today it is accepted that *Zygopteris* and *Botrychioxylon* are congeneric.

Where the internal structure is well enough preserved to be seen, the vascular system of *Zygopteris* stems, like other Zygopteridales, is a protostele. In *Z. illinoiensis* (Fig. 17.12B), described by Baxter (1952), from the Middle Pennsylvanian of the United States, the stem stele is a protostele of mixed protoxylem and parenchyma. The protostele is surrounded by a conspicuous zone of radially aligned tracheids with multiseriate scalariform pits on the walls. Baxter considers this tissue to be aligned metaxylem; others believe it was produced by a cambium (Dennis, 1974). Petioles

Figure 17.12. **A.** *Etapteris scottii,* transverse section of petiole with pinna traces (pt). **B.** *Zygopteris,* transverse section of stem stele showing secondary xylem. **A,B:** Pennsylvanian. (From Phillips, 1974.)

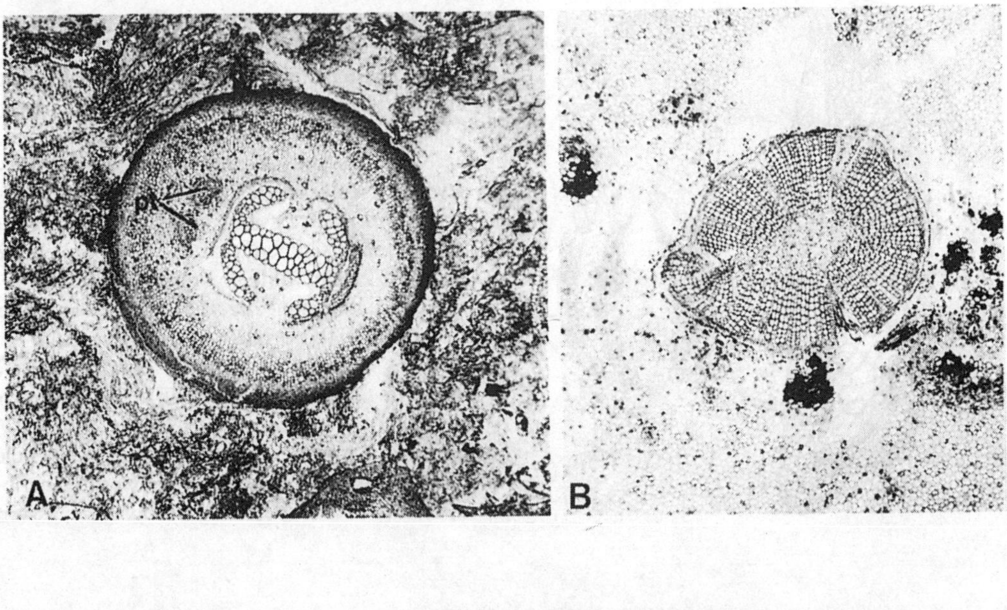

of the *Etapteris* type arise in a spiral from the stems of *Zygopteris*. After their emission from the stem axis the trace assumes the characteristic H-shape with open peripheral loops on the lateral faces. Pinna trace emission is initiated along the margins of the peripheral loop by two traces, one from each side. The two traces fuse to form a bar opposite the peripheral loop. The bar then divides to form two pinna traces (Fig. 17.12A). This process is repeated on the opposite side of the petiole trace so that the pairs of pinna traces alternate. The quadriseriate branching pattern is produced in this way. Pinnule traces arise in two rows from the pinna traces to form a more or less planated frond with pinnules that are thought to be planated, and webbed aphlebialike structures with dichotomizing veins. Such structures have been found on the stems and lower portions of the fronds.

Some fertile zygopterid fronds have huge aggregations of densely packed banana-shaped sporangia (Fig. 17.13A) associated with reduced pinnules (Millay & Rothwell, 1983). These are assigned to *Biscalitheca musata* (Mamay, 1957). *Biscalitheca* clusters are as much as 3 cm in diameter and consist of highly branched stalks with about seven terminal sporangia (Fig. 17.13B) (Phillips & Andrews, 1968). In another genus, *Corynepteris*, rosettes of 4–10 sporangia occur on the abaxial surface of the pinnules (Galtier & Scott, 1979). Galtier & Scott (1985) examined the sporangia of zygopterid ferns from the Upper Devonian through the Lower Permian and summarized the changes that have occurred through time. They concluded that the earliest representatives, including *Rhacophyton*, had no annulus, but that annulate forms evolved by the Lower Carboniferous (Tournaisian). All of the genera described as sporangia of *Zygopteris* have conspicuous annuli developed in the lateral walls. The output of one sporangium is estimated to have been approximately 8,000 isospores. This large output of small spores is similar to eusporangiate ferns. The wall of the sporangium varies in numbers of wall layers from one to several. In having a sporangium wall that may be a single layer thick and an annulus, many zygopterids are similar to the leptosporangiate filicaleans we will cover in Chapters 18 and 19. However, these features appear to have evolved independently in more than one lineage of plants, and it is likely that zygopterids and filicaleans had origins independent from the trimerophytes.

Figure 17.13. *Biscalitheca musata*. **A.** Detail of single sporangium. **B.** Single sorus showing positions of annuli (a) in black. **A–B:** Upper Pennsylvanian. (**A–B** redrawn from Phillips & Andrews, 1968.)

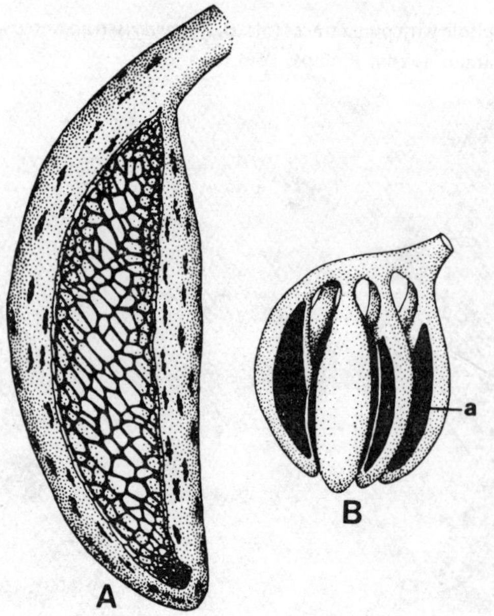

Origin of the megaphyll

All of those who have worked firsthand with material of *Pseudosporochnus nodosus* and *Cladoxylon scoparium* agree that these plants show early stages in the evolution of megaphylls (Fig. 17.14) from sterile and fertile telome trusses in the way hypothesized by Lignier (1908) and later by Zimmermann in his telome theory. In *P. nodosus*, we see planation of terete axes in both second- and third-order branches. Branches of the third order are disposed on opposite sides of second-order branches to produce a bilaterally symmetrical planated foliar system that one can easily visualize as a fern "prefrond." In *C. scoparium* we see what appear to be spirally arranged simple megaphylls "caught in the act" of evolving from overtopped and reduced telome trusses, which have become planated and partially webbed. As we proceed, we will find many beautiful examples of stages in the evolution of both compound and simple leaves from primitive systems.

Stelar evolution

The term polystely traditionally has been used in the literature that deals with the anatomy of the Cladoxylales and other groups of plants. It is usually used in a descriptive sense to indicate the presence of several vascular segments each surrounded by secondary xylem in stems and branches. It is implied that each vascular segment represents a complete stele with primary xylem, phloem, and associated secondary xylem. Phylogenetically, there are two interpretations that explain the origin of several vascular segments in stems of primitive vascular plants. One is derived from the telome theory, which accounts for origin of polystely by syngenesis of telomes (Fig. 9.10A,B) to produce a syntelome with two or more steles. The second explanation, supported by ontogenetic studies of certain extant species of *Lycopodium*, assumes that the basal, oldest, and first-formed part of the plant was protostelic, that as development continued and the axis went through stages of epidogenesis there were changes in the pattern of the primary tissues from haplostelic, actinostelic, and then plectostelic. A plectostele (Fig. 17.15A) is one in which the primary xylem and phloem, when seen in transverse section, appear as bands, some of which are straight or U- or V-shaped. These bands anastomose within the stem. *Lycopodium lucidulum* is an example of an extant plant that shows the development of a plectostele from a protostele.

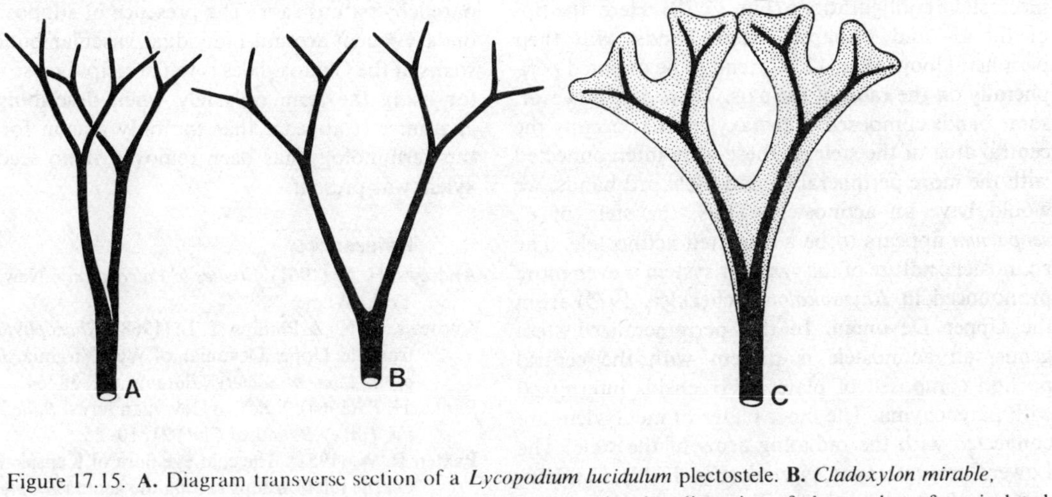

Figure 17.14. Stages in the evolution of the megaphyll according to the telome concept. **A.** Primitive dichotomous axis. **B.** Planation of axis. **C.** Webbing to form a foliar unit. Compare these diagrams with Fig. 17.6. **A,B** and 17.7.**C.** (Based on Zimmermann, 1959.)

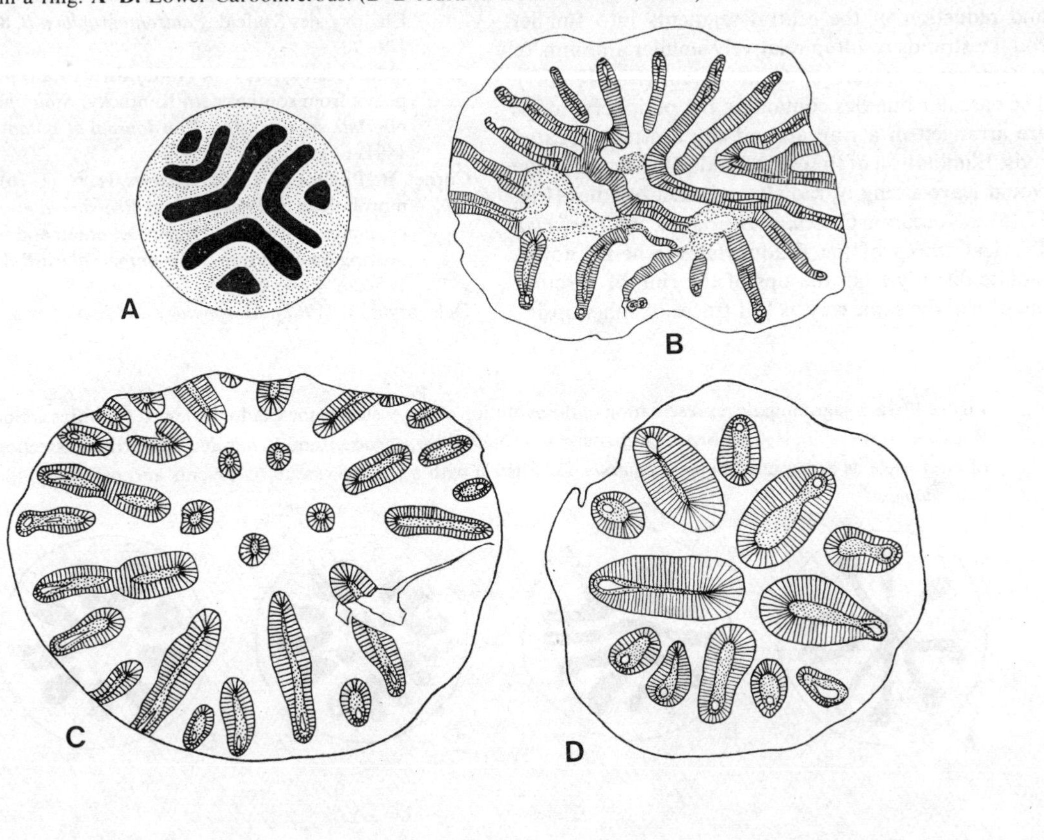

Figure 17.15. **A.** Diagram transverse section of a *Lycopodium lucidulum* plectostele. **B.** *Cladoxylon mirable*, plectostele. Note position of peripheral loops. **C.** *C. taeniatum*, showing dissection of plectostele to form isolated vascular segments in center of axis. **D.** *C. dubium*. Note arrangement of vascular segments and their peripheral loops in a ring. **A–D:** Lower Carboniferous. (**B–D** redrawn from Bertrand, 1935.)

The cross-sectional configuration of a *Lycopodium* plectostele bears a striking resemblance to the polystelic stems and branches of the Cladoxylales. The development of the plectostele from an actinostele (Fig. 17.16) offers the best explanation for the phyletic origin of the "polystele" in the Cladoxylales. In other words, "ontogeny recapitulates phylogeny."

Taken collectively, we can reconstruct stages in the evolution of the polystelic Cladoxylales. In Middle Devonian *Cladoxylon scoparium,* transverse sections taken at several levels of the stem show perfectly good plectostelic configurations (Fig. 17.7B). Here, the tips of the U- and V-shaped xylem bands with their peripheral loops (Fig. 17.7C) tend to be oriented peripherally on the radii of the axis. There are, however, some bands composed of metaxylem that occupy the central area of the stele. If these were interconnected with the more peripheral U- and V-shaped bands, we would have an actinostele. Thus, the stele of *C. scoparium* appears to be a dissected actinostele. The actinostelic nature of the vascular system is even more pronounced in *Rhymokalon* (Scheckler, 1975) from the Upper Devonian. In this permineralized stem genus, an actinostele is present with the central portion composed of plates of tracheids intermixed with parenchyma. The inner plates of metaxylem are connected with the radiating arms of the stele. The Lower Carboniferous *C. mirable* (Fig. 17.15), *C. solmsi,* and *C. radiatum* have the free tips of the radiating arms more or less evenly spaced and extended so that the peripheral loops included in the tips form a ring. Breakup and reduction of the central segments into smaller rodlike strands result in relatively smaller amounts of metaxylem, a feature of *C. taeniatum* (Fig. 17.15C). The vascular bundles containing the peripheral loops are arranged in a ring around the centrally located rods. Elimination of the central rods of vascular tissue would leave a ring of radially oriented bundles (Fig. 17.15D) as occurs in *C. dubium, Syncardia,* and *Voelkelia.* The leaf traces of the Cladoxylales, where known, radiate directly from the tips of the ring of vascular bundles in the same way as leaf traces of sphenopsids and some gymnosperms with their ring of sympodia around a pith. Thus, there seems to be no obstacle in designating the Cladoxylales as a group of vascular plants showing us how the eustele evolved (Fig. 17.16).

A tissue that resembles secondary xylem in having radially aligned xylem cells is found around the primary xylem bundles of some species of *Cladoxylon* (Fig. 17.15B,C), *Voelkelia,* and *Rhymokalon.* According to Scheckler (1975) and Leclercq (1970), these cells that look like secondary xylem are, in reality, aligned cells of metaxylem as evidenced by the absence of parenchymatous rays. The presence of supposed secondary xylem around individual vascular bundles in stems of the Cladoxylales is the principal reason given for using the term polystely when describing their anatomy. It appears that the only reason for using this terminology has been removed if no secondary xylem was present.

References

Andrews, H. N. (1961). *Studies in Paleobotany.* New York: Wiley.

Andrews, H. N., & Phillips, T. L. (1968). *Rhacophyton* from the Upper Devonian of West Virginia. *Journal of the Linnean Society (Botany),* **61,** 38–64.

Banks, H. P. (1964). Putative Devonian ferns. *Bulletin of the Torrey Botanical Club,* **21,** 10–25.

Baxter, R. W. (1952). The coal-age flora of Kansas. II. On the relationships among the genera *Etapteris, Scleropteris,* and *Botrychioxylon. American Journal of Botany,* **39,** 263–74.

Bertrand, P. (1935). Contributions a'l'étude des Cladoxylales Saafeld. *Palaeontographica B,* **80,** 101–70.

Cichan, M. A., & Taylor, T. N. (1982). Structurally preserved plants from southeastern Kentucky: *Stauropteris biseriata* sp. nov. *American Journal of Botany,* **69,** 1491–6.

Cornet, B., Phillips, T. L., & Andrews, H. N. (1976). The morphology and variation in *Rhacophyton ceratangium* from the Upper Devonian and its bearing on evolution. *Palaeontographica B,* **158,** 105–29.

Delevoryas, T. (1962). *Morphology and Evolution of Fossil*

Figure 17.16. Diagrammatic representation of the evolution of the eustele in the Cladoxylales. **A.** Primitive actinostele. **B.** Dissection of actinostele to form a plectostele as in *Cladoxylon scoparium* and *C. mirable.* **C.** Further dissection of plectostele as exemplified by *C. taeniatum.* **D.** Eustele with a ring of vascular segments and peripheral loops as in *C. dubium.*

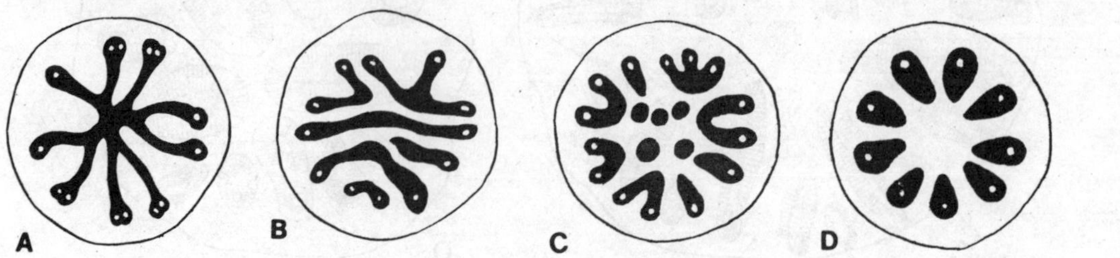

Plants. New York: Holt, Rinehart, & Winston.

Dennis, R. L. (1974). Studies of Paleozoic ferns: *Zygopteris* from the Middle and Late Pennsylvanian of the United States. *Palaeontographica B*, **148,** 95–136.

Dittrich, H. S., Matten, L. C., & Phillips, T. L. (1983). Anatomy of *Rhacophyton ceratangium* from the Upper Devonian (Famennian) of West Virginia. *Review of Palaeobotany and Palynology*, **40,** 127–47.

Erwin, D. M., & Rothwell, G. W. (1989). *Gillespiea randolphensis* gen. et sp. nov. (Stauropteridales), from the Upper Devonian of West Virginia. *Canadian Journal of Botany*, **67,** 3063–77.

Galtier, J. (1966). Observations nouvelles sur le genre *Clepsydropsis. Naturalia Monspeliensia, Botanique*, **17,** 111–132.

Galtier J., & Scott, A. C. (1979). Studies of Palaeozoic ferns: On the genus *Corynepteris*. A redescription of the type and some other European species. *Palaeontographica B*, **170,** 81–125.

Galtier, J., & Scott, A. C. (1985). Diversification of early ferns. *Proceedings of the Royal Society of Edinburgh*, **86B,** 289–301.

Leclercq, S. (1951). Etude morphologique et anatomique d'une fougere due Devonien Supérieur. *Annals de la Société Géologique de Belgique*, **9,** 5–50.

Leclercq, S. (1970). Classe des Cladoxylopsida Pichi-Sermoli, 1959. In *Traité de Paléobotanique*, ed. E. Boureau. Paris: Masson. Vol. IV(1).

Leclercq, S., & Banks, H. P. (1962). *Pseudosporochnus nodosus* sp. nov., a Middle Devonian plant with cladoxylalean affinities. *Palaeontographica B*, **110,** 1–34.

Leclercq, S., & Lele, K. M. (1968). Further investigations on the vascular system of *Pseudosporochnus nodosus* Leclercq et Banks. *Palaeontographica B*, **123,** 97–112.

Lignier, O. (1908). Essai sur l'évolution morphologique de regne végétal. *Compte Rendu de l'Association Français pour l'Avance du Science. Clermont-Ferrand*, **37,** 530–42.

Long, A. C. (1976). *Rowleya trifurcata* gen. et sp. nov., a simple petrified vascular plant from the Lower Coal Measures (Westphalian A) of Lancashire. *Transactions of the Royal Society of Edinburgh*, **69,** 467–81.

Mamay, S. H. (1957). *Biscalitheca*, a new genus of Pennsylvanian coenopterids, based on fructifications. *American Journal of Botany*, **44,** 229–39.

Millay, M. A., & Rothwell, G. W. (1983). Fertile pinnae of *Biscalitheca* (Zygopteridales) from the Upper Pennsylvanian of the Appalachian Basin. *Botanical Gazette*, **144,** 589–99.

Phillips, T. L. (1974). Evolution of vegetative morphology in coenopterid ferns. *Annals of the Missouri Botanical Garden*, **61,** 427–61.

Phillips, T. L., & Andrews, H. N. (1968). *Biscalitheca* (Coenopteridales) from the Upper Pennsylvanian of Illinois. *Palaeontology*, **11,** 104–115.

Sahni, B. (1932). On the structure of *Zygopteris primaria* (Cotta) and on the relations between the genera *Zygopteris, Etapteris,* and *Botrychioxylon. Philosophical Transactions of the Royal Society of London B*, **222,** 29–45.

Scheckler, S. E. (1974). Systematic characters of Devonian ferns. *Annals of the Missouri Botanical Garden*, **61,** 462–73.

Scheckler, S. E. (1975). *Rhymokalon*, a new plant with cladoxylalean anatomy from the Upper Devonian of New York State. *Canadian Journal of Botany*, **53,** 25–38.

Schweitzer, H. J. (1972). Die Mitteldevon-flora Von Lindlar. 3. Filicinae-*Hyenia elegans* Krausel & Weyland. *Paleontographica B*, **137,** 154–75.

Schweitzer, H. J., & Giesen, P. (1980). Ueber *Taeniophyton inopinatum, Protolycopodites devonicus,* und *Cladoxylon scoparium* aus dem Mitteldevon von Wuppertal. *Palaeontographica B*, **173,** 1–25.

Surange, K. R. (1952). The morphology of *Stauropteris burntislandica* P. Bertrand and its megasporangium *Bensonites fusiformis* R. Scott. *Philosophical Transactions of the Royal Society of London B*, **237,** 73–91.

Takhtajan, A. L. (1956). *Telomophyta*. Moscow: Academiae Scientiarum.

Taylor, T. N. (1981). *Paleobotany: An Introduction to Fossil Plant Biology*. New York: McGraw-Hill.

Zimmermann, W. (1959). *Die Phylogenie der Pflanzen*, 2nd ed. Stuttgart: Fisher.

18

The emergence of the Marattiales and Ophioglossales

In the previous chapter we discovered that ferns and fernlike plants had several separate origins from the trimerophyte complex. Among the distinct groups of ferns with living representatives, the Marattiales and Ophioglossales are homosporous and eusporangiate, i.e., their sporangia develop from several cells. At maturity eusporangia are large with relatively thick walls and a large number of spores, sometimes over 1,000. Unlike the Zygopteridales, and the leptosporangiate filicaleans that we will examine in Chapters 19 and 20, eusporangia do not have an annulus for dehiscence. Rather, the sporangia of marattialeans and ophioglossaleans dehisce by a longitudinal slit or apical pore.

Until recently, the Ophioglossales had no creditable megafossil record. In contrast, the marattialeans appeared as an important part of the Upper Carboniferous flora, where they coexisted with the Zygopteridales and Filicales.

Marattiales

The six extant genera of the Marattiales are distributed in tropical areas. They produce tuberous upright, unbranched stems that may attain a diameter of 60 cm. Unlike many ferns, there is very little sclerenchyma present in their stems. The vascular system is a polycyclic dictyostele, a complex type of siphonostele with two or more cylinders of vascular tissue and overlapping leaf gaps (Fig. 10.3B). A pair of large stipules that clasp the stems are produced at the base of each leaf. The adventitious roots of these plants are large and fleshy, often associated with the stipules. Rather than showing a diarch protostele, a construction found with monotonous regularity in the adventitious roots of the Filicales, their roots have an aerenchymatous cortex around an actinostele with numerous radiating arms. The leaves are usually large, up to 4 m, and pinnately compound (Fig. 18.1A). With one exception (Fig. 18.1C) the venation is of the open dichotomous type (Fig. 18.1B). Their eusporangia are abaxial and fused into synangia associated with the veins. In *Christensenia* (Fig. 18.1E,F), the sporangia are laterally fused into a ringlike synangium, while *Marattia* (Fig. 18.1D,G) has clam-shaped linear synangia.

Tree ferns of the Upper Paleozoic

Until recently the magnificent upper Paleozoic marattiaceous tree ferns, represented by the genus *Psaronius* (Fig. 18.2), seem to have sprung fully formed in the Carboniferous swamps as though from Medusa's head. There are only a few scattered, poorly preserved specimens that attest to their presence in the Mississippian; however, by the Lower Pennsylvanian onward into the Permian, fossils of these plants abound. Originally the genus *Psaronius* was established by Cotta in 1832 for stem fragments of tree ferns with polycyclic dictyosteles. As time went on the various parts were found attached, and the concept of the genus changed to include the whole treelike plant. In general habit, these plants compare favorably with modern tree ferns. They grew as high as 8 m, with a crown of large pinnately compound leaves arranged in a spiral or in vertical rows. As the leaves abscised, they left petiole base scars that usually show a C-shaped trace of vascular tissue that is adaxially oriented. This leaf trace configuration is characteristic of some Marattiales and many Filicales. The lower portions of the trunk were clothed in an accumulation of adventitious roots that caused the base of the trunk to appear buttressed. Fertile fronds bore synangia on the abaxial surface and associated with the veins. The sporangia of the synangia are of the eusporangiate type.

Figure 18.1. **A.** *Angiopteris evecta.* Habit of plant about 2 m high. (From Takhtajan, 1956.) **B.** *Marattia fraxinea,* pinna showing open venation and distribution of synangia on abaxial surface. **C.** *Christensenia aesculifolia,* pinna showing circular synangia on abaxial surface. **D,G.** *Marattia fraxinea* linear synangia; **D** sectioned **E,F.** *C. aesculifolia* circular synangia; **E** sectioned. **A–G:** Extant. (**B–G** redrawn from Eames, 1936.)

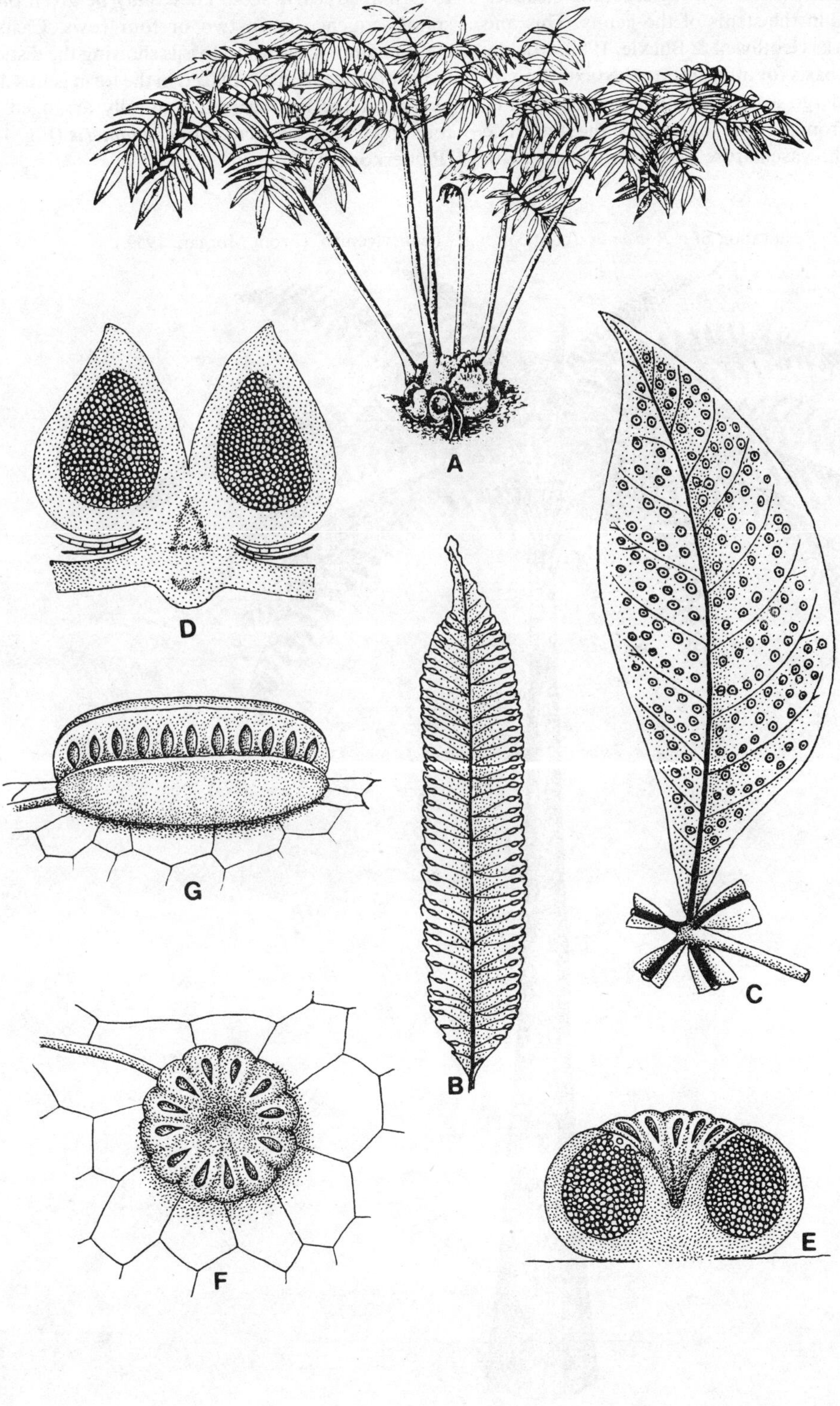

Internal structure of stems

Dr. Jeanne Morgan (1959) was the first to make sense out of the seemingly incomprehensible vascular system of a *Psaronius* stem. Her monograph of the American species is a monument to our understanding of developmental stages in the stems of the genus. This and more recent works (Rothwell & Blickle, 1982; Mickle, 1984) provide a basis for interpreting the taxonomy of species that undergo a tremendous amount of antogenetic change from the base to the tip of the stem. The complexity of the vasculature of *Psaronius* is seen in transverse sections (Fig. 18.3A) taken from the distal end of a mature tree. Here we see the full expression of the polycyclic dictyostele with as many as six inner cycles of vascular tissue and an outer cycle giving rise to C-shaped leaf traces. These may be given off in a spiral sequence or in two or four rows. Casts and compression–impression fossils showing the distichous frond arrangement are placed in the form genus *Megaphyton;* if the leaf scars are spirally arranged or in four rows, they are placed in *Caulopteris* (Fig. 18.3B) (Pfefferkorn, 1976).

Figure 18.2. Restoration of a *Psaronius* about 3 m high. Pennsylvanian. (From Morgan, 1959.)

The strands of cauline vascular tissue are composed of scalariform metaxylem with annular and helical protoxylem cells on the inner face. The primary xylem is endarch. There is a layer of phloem around each of the vascular segments, but no endodermis. In a given transverse section the vascular segments have the appearance of meristeles as in the stems of some extant ferns. In a *Psaronius* stem, however, the vascular strands anastomose and it can be demonstrated that they have developed from a simple siphonostele (Fig. 18.4A) (Morgan, 1959; Stidd & Phillips, 1968; Stidd, 1974) in the basal portion of the stem. Additional cycles of meristeles are formed to the inside of the innermost cycle of the siphonostele. The new central cycle is first represented in the basal part of the stem by a protostelic strand that assumes the configuration of a simple siphonostele at a higher level. Higher yet, the siphonostele becomes dissected by overlapping leaf gaps and becomes an amphiploic dictyostele. New cycles are added distally so that six or more cycles of the polycyclic dictyostele are found in the apical portions of a large *Psaronius* stem (Fig. 18.4B). The vascular supply to a leaf base originates with the innermost cycle. In its upward and outward course it fuses and separates from successively more peripheral cycles until it finally bends outward from the outermost cycle into a leaf base. New leaf traces are added distally. Thus, in the basal portion of the stem, there may be a single leaf trace; at the apex there may be as many as six. From this description we can visualize the vascular supply of a mature *Psaronius* stem as comprising a series of inverted cones fitting inside of the one below, each being protostelic at the base and giving rise to a siphonostele dissected by overlapping leaf gaps at higher levels. The basal and thus the outermost cone of vascular tissue extends from the base of the plant to the apex. The innermost and last-formed cone would extend for a short distance in the distal portion of the stem axis.

The development of the stem of *Psaronius* is epidogenic as in modern tree ferns. Based on a comparison with the vascular supply of a modern tree fern, we can guess that the basalmost part of a *Psaronius* stem was protostelic, with a few spirally or distichously arranged C-shaped leaf traces that are adaxially oriented. This is precisely what Phillips & Andrews (1966) described in their genus *Catenopteris* (Fig. 18.4C). Although *Catenopteris* may not represent the basal part of a developing *Psaronius*, it certainly provides us with a model of what this part of a young *Psaronius* stem should look like.

Development of the complex *Psaronius* stem from a small protostelic or siphonostelic base also suggests that the stem may have evolved in a similar fashion. Support for this suggestion was provided by DiMichele & Phillips (1977), who discovered a siphonostelic species in Upper Mississippian deposits of the Illinois Basin. The stem of *Psaronius simplicicaulis* was up to 6.5 cm in diameter, producing a small, upright tree with two rows of alternate leaf traces and a mantle of adventitious roots that are similar to those of more recent species of the genus. It is of interest to note that *P. simplicicaulis* was found in clastic rather than peat-swamp deposits, which supports the proposal that marattiaceous ferns probably originated outside of the peat-swamp environment (Phillips, Peppers, & DiMichele, 1985).

Roots

Traces supplying the adventitious roots arise horizontally from the vascular segments of the stem. All of the roots are exarch, polyarch, and actinostelic. In older

Figure 18.3. **A.** *Psaronius blicklei,* transverse section of stem with many cycles of vascular tissue. (From Morgan, 1959.) **B.** *Caulopteris* sp., cast of *Psaronius*-type stem fragment with several rows of spirally arranged leaf scars. **A,B:** Pennsylvanian.

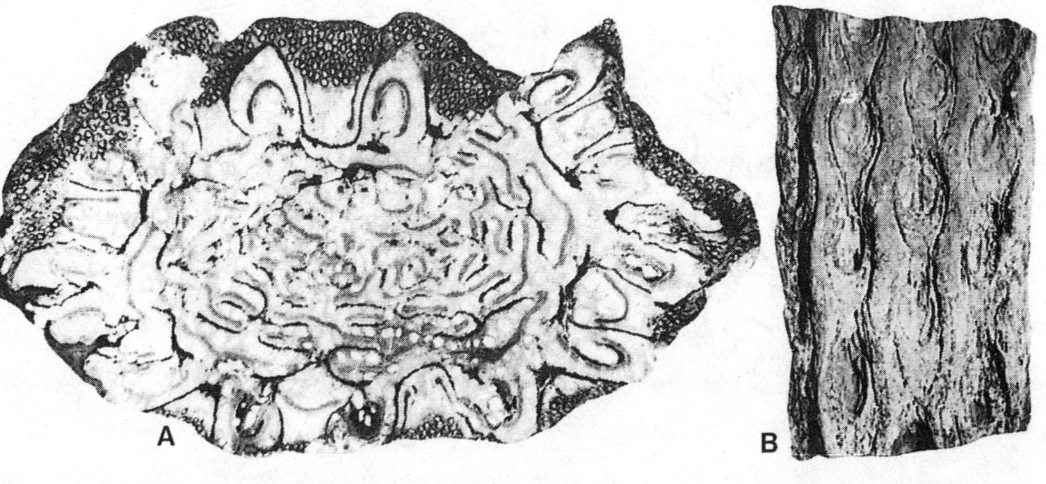

basal parts of stems, a mantle of roots developed around the outside that can be separated into an inner compact zone and an outer zone of free roots (Fig. 18.5A). The latter are amazingly similar to the adventitious roots of extant Marattiales. The compact inner zone appears first near the stem apex and continues to the base. The roots in this zone (Fig. 18.5B) have a thickened sclerenchymatous cortex and are submerged in an interstitial tissue, part of which develops in the same way as the secondary cortex in axes of lepidodendrids. The secondary cortex of *Psaronius* develops from the cortex of stem and roots (Ehret & Phillips, 1977). In the absence of secondary xylem it is clear that the inner root zone provided the additional support for these large tree ferns. This is just one more example of the supportive mechanisms evolved by land plants, other than those related to xylem tissues.

In their downward course, the roots branch with some degree of frequency often penetrating the already established root mantle. Here, Ehret & Phillips (1977) discovered well-preserved root tips (Fig. 18.5C), which display a multilayered root cap, an apical meristem with up to four apical initials, and differentiated procambial strands.

Figure 18.4. **A.** Series of transverse sections (a–i) of *Psaronius* stems showing increasing complexity of vascular system from base to apex. Note increase in number of cycles and leaf traces. **B.** Restoration of vascular system of *Psaronius* stem, in transverse section. Leaf trace (lt). (**A,B** from Morgan, 1959.) **C.** Stereodiagram, *Catenopteris simplex*. (From Phillips & Andrews, 1966.) **D.** Stereodiagram fragment of *Psaronius* stem showing origin of leaf traces with *Stipitopteris*-type configurations. (From Stidd, 1971.) **A–D:** Pennsylvanian.

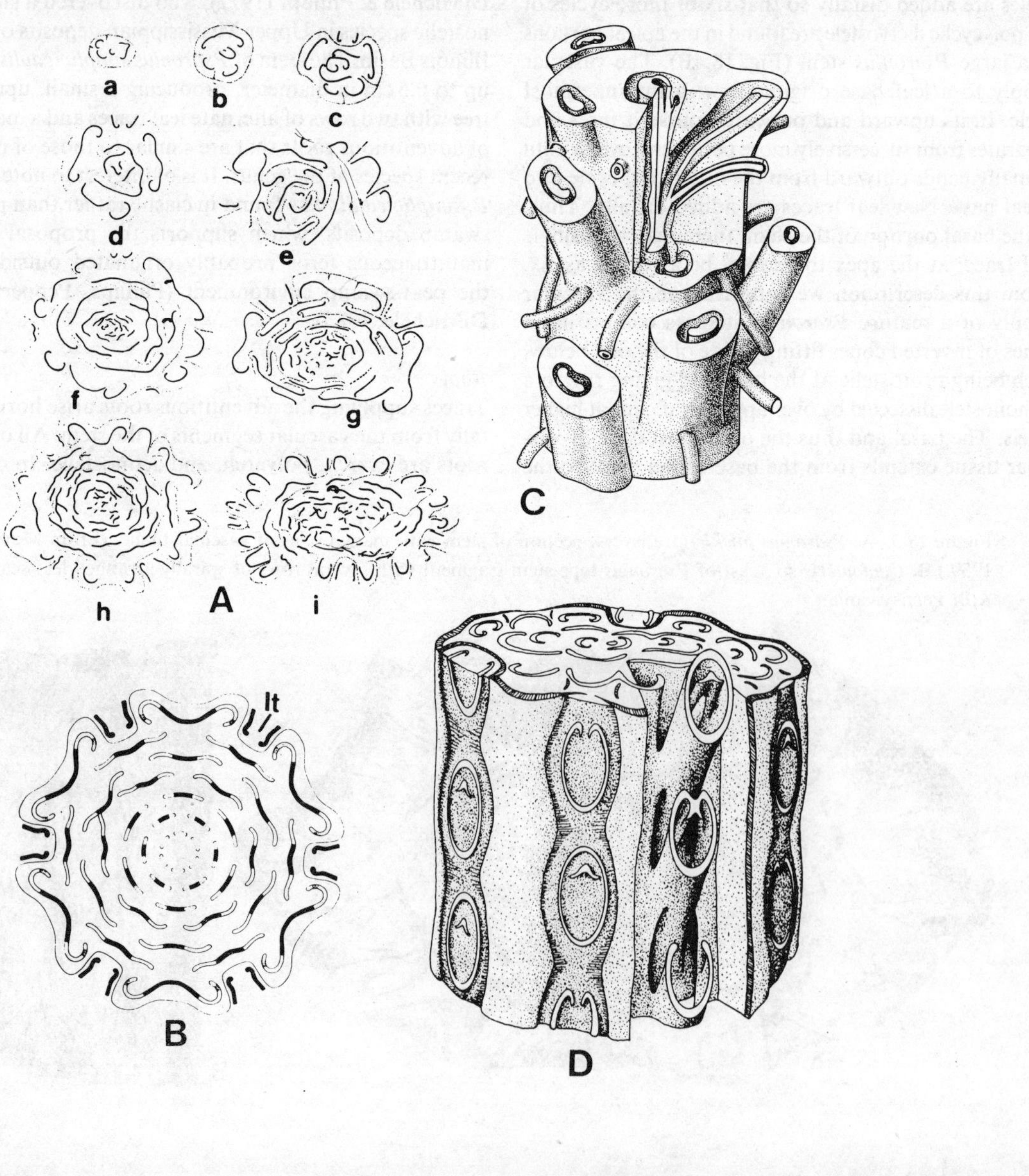

Fronds

The fronds of *Psaronius* developed from croziers at the apex of the stem (Fig. 18.6A). These expanded into planated, bilaterally symmetrical leaves of the *Pecopteris* type. Some compression–impression fossils of this form genus are more than 1.5 m long, and some estimate their total length to be in the vicinity of 3 m. These large fronds are unipinnate to quadripinnate with pinnae and pinnules borne alternately on each side of a pinna rachis. Toward the tip of the frond the pinnules become pinnatifed, that is, reduced in size and fused by their lateral margins. The pinnules exhibit an open-type, often dichotomous venation. Clusters of sporangia in the form of synangia occur on the abaxial surfaces of pinnules below a vein (Fig. 18.6B). Stidd (1971), in a beautifully illustrated article, has worked out the vascularization in the petiole and rachises of the fronds. As with other aspects of *Psaronius* vascularization, the emission of pinna traces from the rachis is complex, resulting in a *Stipitopteris* configura-

Figure 18.5. **A.** Transverse section, *Psaronius* free root. **B.** Transverse section, *Psaronius* inner root zone. Note thick-walled cortex (c) of each root and interstitial tissue (i). **C.** Median longitudinal section, *Psaronius* root tip. Root cap (rc); apical meristem (am). **A–C:** Pennsylvanian. (**A,C** from Ehret & Phillips, 1977; **B** from Morgan, 1959.)

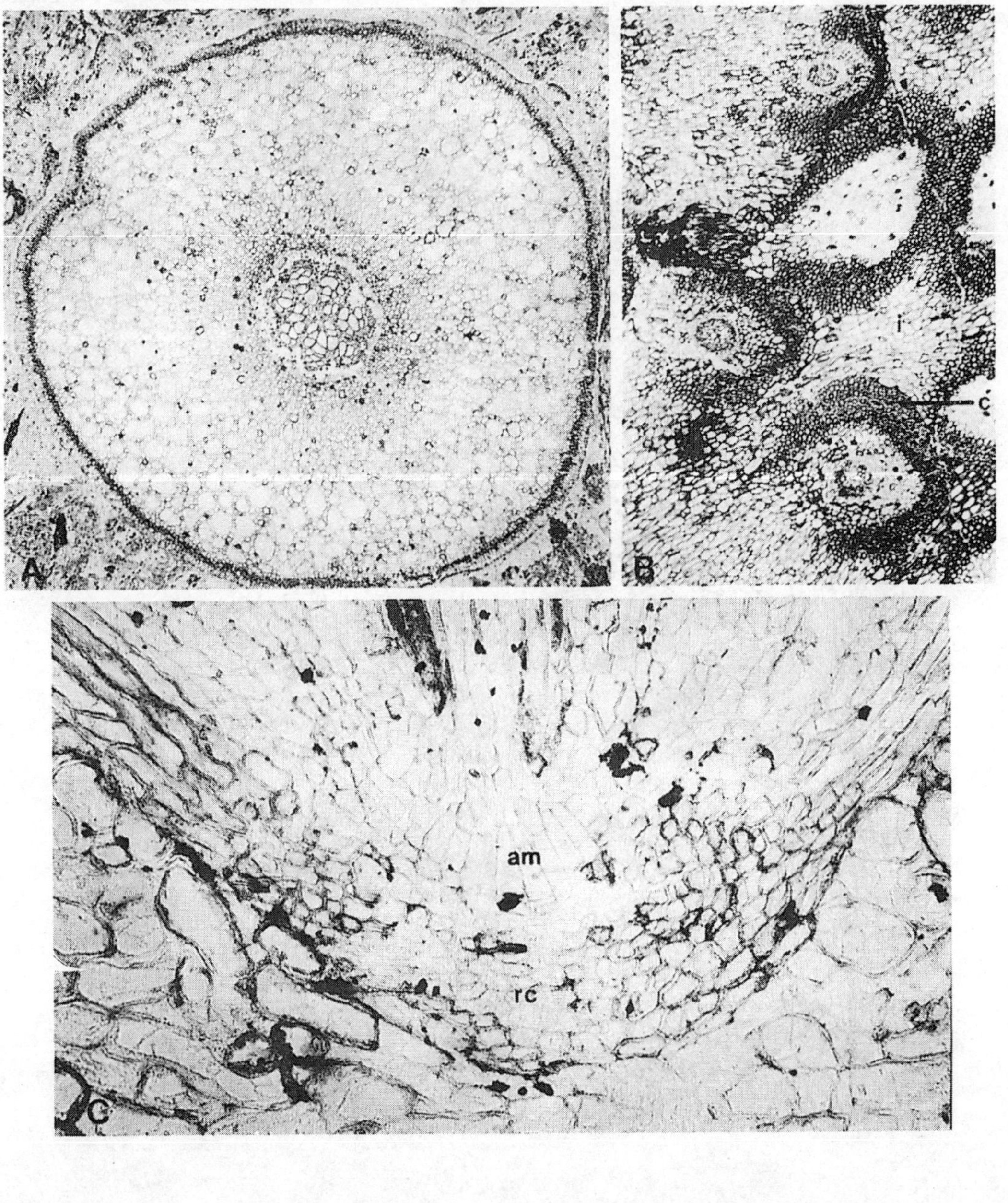

tion (Fig. 18.4D) where the C-shaped trace of the petiole may close adaxially after the production of lateral pinna traces and an internal W-shaped trace.

No effort has been made here to describe the multitude of form genera applied to compression–impression fossils of leaves, stem fragments, and roots unless these can be assigned with a degree of certainty to a parent plant. It has been demonstrated that certain species of *Pecopteris* were fertile and bore sporangia and synangia of the marattialean type (Fig. 18.7B). These can be related to stems of *Psaronius*. When found with synangia, the fertile frond parts have been assigned to separate genera and species. For example, the *Pecopteris miltoni* (Fig. 18.7A) group, a sterile foliage type known from compression–impression fossils, bore *Asterotheca*-type synangia. One way of citing this condition is to give the generic name for the fertile frond parts, followed by the one describing the sterile characteristics. Thus, the name *Asterotheca* (ex *Pecopteris*) *miltoni* tells us what kind of fertile structure one can expect to find on *Pecopteris miltoni* foliage. We know that *Pecopteris*-type foliage was also borne by some filicalean and zygopterid ferns and seed plants, which accounts for the fact that it is a form genus.

Sporangia and synangia

Many of the genera applied to fertile frond portions are known from permineralized specimens as well as

Figure 18.6. **A.** Restoration of *Psaronius* fronds attached to shoot apex. Frond estimated to be 1 to 2 m long. **B.** Portion of frond with *Scolecopteris*-type synangia, enlarged. Pennsylvanian. (From Stidd, 1971.)

compression–impression fossils. These too have been given generic and species names. It is now known that permineralized synangia placed in the genus *Scolecopteris* are equivalent to the compression fossils of *Asterotheca* (Stubblefield, 1984). Because permineralized specimens show much more detail, we will briefly consider a few of their characteristics.

Scolecopteris (Fig. 18.7C) is one of the most common of all synangiate fructifications found in North American coal balls. The abaxial synangia are each composed of a ring consisting of four to five laterally fused sporangia on a stalk. The sporangia may be organized around a centrally located hollow or attached to a parenchymatous extension of the stalk. The wall of a sporangium is several cells thick and without an annulus. Dehiscence, however, is uniformly along the midline of the inward-facing sporangial walls. The synangia usually occur on the abaxial surface of the pinnule, one on each side of the pinnule midrib. Isospores taken from the sporangia are frequently of the *Punctatisporites* dispersed spore type.

In evaluating the morphology of a *Scolecopteris* synangium (Fig. 18.8A), we see that it reflects the morphology of fructifications of other, more primitive groups such as the Trimerophytopsida, where the sporangia are borne in clusters, often two at the tip of a single branch, and dehisce longitudinally. As in *Scolecopteris*, the line of dehiscence in the cluster is along the inward-facing walls of adjacent sporangia.

Clusters of sporangia at the margin of a foliar unit are found in the genus *Chorionopteris* (Fig. 18.8B). Here the cluster terminates a vein at the leaf margin and is an excellent example of a fertile telome truss

Figure 18.7. **A.** Pinna of *Pecopteris miltoni*. (From Janssen, 1957; courtesy of the Illinois State Museum, Springfield.) **B.** *Stellotheca ornata*, synangia on pinnules of the *Pecopteris* type. (From Pfefferkorn, Peppers, & Phillips, 1971; courtesy of Illinois State Geological Survey.) **C.** Sectioned pinnule with *Scolecopteris*-type synangia. **A–C:** Pennsylvanian.

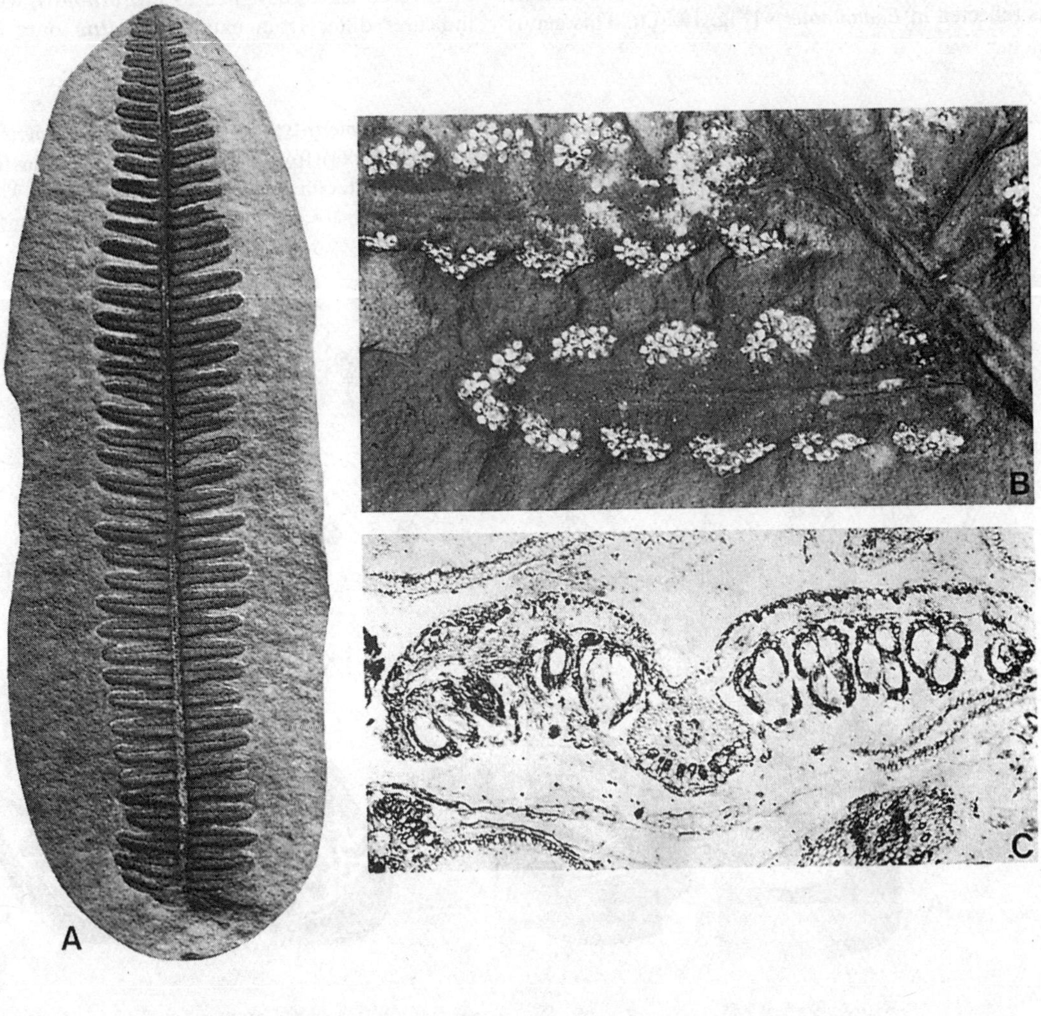

that has become planated, overtopped, and webbed to produce a fertile megaphyll. The sporangia form a synangium in which the sporangia are fused at their bases (Fig. 18.8B). According to Mamay (1950) and Millay (1977), the *Chorionopteris*-type synangium probably represents the primitive type from which sporangia of the Carboniferous marattialeans evolved. By "phyletic slide" the stalked circular synangium was "moved" from the marginal (terminal) position to the abaxial surface of the leaf by overgrowth of the lamina (Fig. 18.8C). *Chorionopteris* helps us interpret how the abaxial synangia of ferns evolved. However, because the relationships of this genus are still in doubt, it offers little information about the phylogenetic origin of the Marattiales.

Several general evolutionary trends are depicted by Mamay (1950) and Stidd (1974) as emerging from the primitive *Chorionopteris*-type fructification. One is the loss of the stalk to produce a sessile circular synangium of the kind found in *Acitheca*. Another involves the complete lateral fusion of the sporangia in the circular synangium as it occurs in *Ptychocarpus*. Yet another trend is the evolution of the linear synangium by lateral coalescence of circular synangia as reflected in *Eoangiopteris* (Fig. 18.8D). This genus (Millay, 1978) has its synangia abaxially placed under lateral veins of pinnules. The sporangia are borne on an elongate pad of tissue with their bases embedded in and fused with the pad. The distal portions of the sporangia are free, and their dehiscence is along the midline of the inner sporangial wall. The isospores are similar to the dispersed spore type, *Verrucosisporites*. The morphology and structure of the *Eoangiopteris* synangium are nearly identical to that of extant *Angiopteris* and leave no room for doubt about their relationships.

Linear synangia of the more completely fused type, found in the synangia of *Marattia*, are foreshadowed by the sessile synangia of the Upper Carboniferous *Acaulangium* (Millay, 1977). This circular synangium, with complete fusion of sporangia, is found in extant *Christensenia* and represents the retention of the primitive radial synangium characterized by *Ptychocarpus* of the Carboniferous.

Mesozoic Marattiales

We pick up the fossil record of the Marattiales in the Triassic rocks of Greenland. Here Harris (1931) discovered leaves assigned to *Marattiopsis*, which he indicates differ from extant *Marattia* only in age.

Figure 18.8. **A.** Restoration, abaxial surface of pinnule with *Scolecopteris*-type synangia. **B.** *Chorionopteris gleichenoides*, pinnule with marginal synangia on recurved stalk. (**B** redrawn from Kubart, 1916.) **C.** Diagrams (a–d) depicting the evolution of synangial position by "phyletic slide." **D.** Restoration of abaxial surface of pinnule with *Eoangiopteris*-type synangia. **B,C:** Carboniferous. **A,C,D:** Pennsylvanian. (**A,D** from Mamay, 1950; **C** redrawn from Mamay, 1950.)

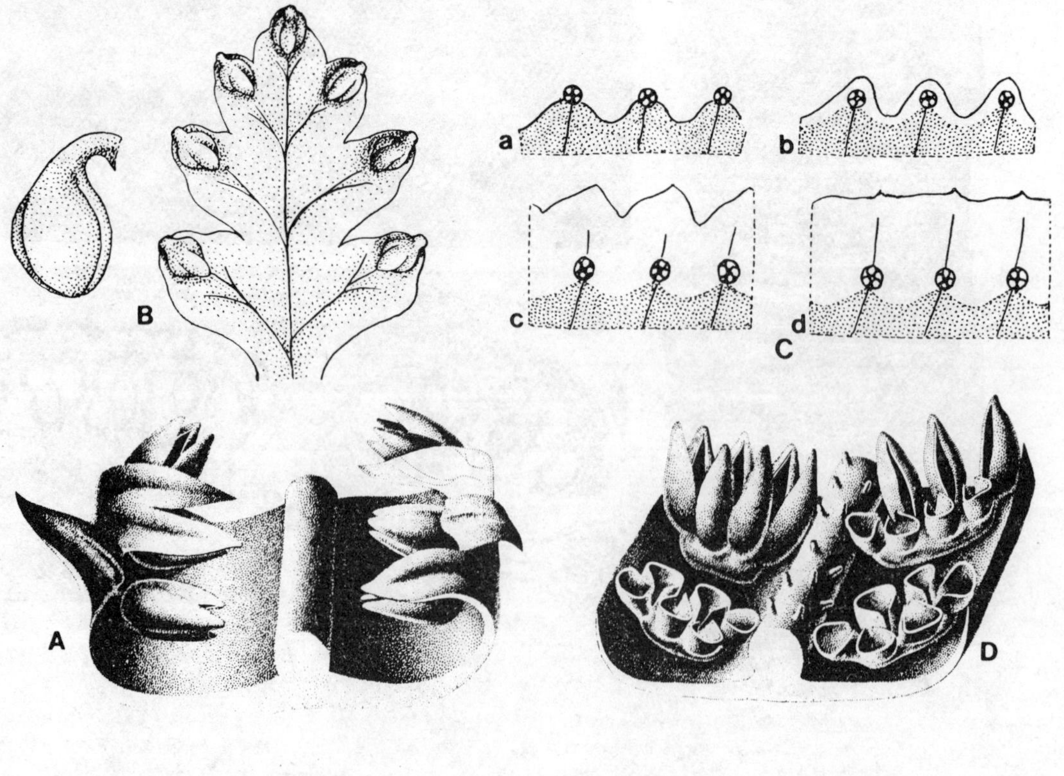

Other foliage specimens having linear synangia similar to those of extant *Danaea* have also been retrieved from the Triassic. From this time on to the end of the Jurassic, the Marattiales reached their climax. Compression–impression fossils of their foliage from these Mesozoic beds are identical to modern genera.

Ophioglossales

The Ophioglossales is a distinctive group of the small eusporangiate ferns, *Ophioglossum, Botrychium,* and *Helminthostachys,* which together have a worldwide distribution. Unlike the marattialean ferns, sporangia of the Ophioglossales are not fused into synangia, and the plants develop no sclerenchyma In *Ophioglossum* and *Botrychium* there is a short, subterranean rhizome, which produces adventitious roots and one frond member each year. The frond member is divided into a vegetative region called a trophophore and a fertile segment called a sporophore (Fig. 18.9C). In *Ophioglossum* the trophophore is usually undivided, and two rows of sporangia are sunken into the sporophore. Both the trophophore and sporophore of *Botrychium* are pinnately compound, and the sporangia occur at or near the margins of the highly dissected narrow pinnules. Internally, the stems are usually interpreted as having an endarch, ectophloic siphonostele, but there is some evidence for a eustelic protoxylem architecture in *Ophioglossum* and secondary xylem in *Botrychium.*

Figure 18.9. Habit sketch of *Botrychium,* showing frond divided into a vegetative segment (trophophore) and fertile segment (sporophore). Extant. (Redrawn from Gifford & Foster, 1989.)

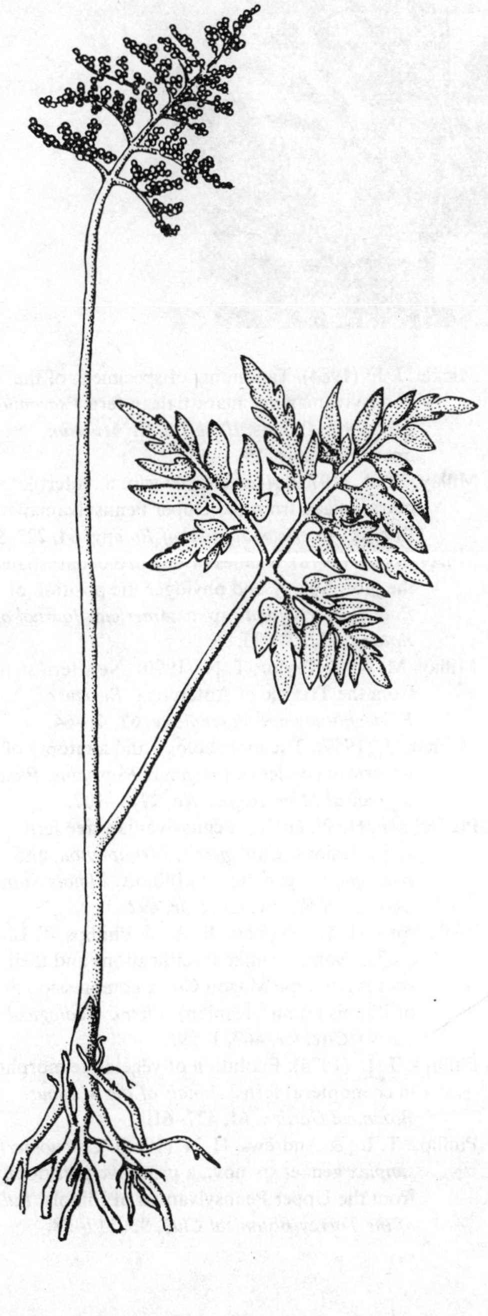

Until recently, there was no definite fossil record for the Ophioglossales, but Rothwell & Stockey (1989) have recently described *Botrychium wightonii* from the Paleocene of western Canada. The compressed frond parts of this species are pinnately compound (Fig. 18.10B) and remarkably similar to some living species such as *B. virginianum.* The sporophores are particularly striking with their large, globose sporangia borne at or near the margins of narrow pinnules (Fig. 18.10A). Trilete spores were produced that have ultrastructural features like those of the living species. *Botrychium wightonii* demonstrates that essentially modern species of the Ophioglossales had evolved by the beginning of the Tertiary, and strengthens the idea that the group is extremely old (Bierhorst, 1971).

Although the anatomy of the fossil *Botrychium* has not been preserved, there is some evidence for small ferns with endarch, ectophloic siphonosteles as early as the Upper Paleozoic. These features are present in the primitive filicalean genus, *Botryopteris* (Phillips, 1974), and in two genera of exquisitely preserved ferns (*Fremouwa* and *Schopfiopteris*) recently described from the Triassic of Antarctica. Millay & Taylor (1990) interpret these genera to be similar to *Ophioglossum* in size, stelar configuration, xylem maturation pattern, and phloem position, but the structures of the leaf traces and roots are more like those of the Filicales.

The previous lack of a fossil record and the possible occurrence of secondary xylem in *Botrychium* have fueled a lively discussion regarding relationships of the Ophioglossales. While most authors place the Ophioglossales among other ferns, some authors suggest that they have descended from the progymnosperms (Wagner, 1964; Bierhorst, 1971; Kato, 1988). None of the features of *Fremouwa* and *Schopfiopteris* are considered to be synapomorphies for either the

Figure 18.10. **A.** *Botrychium wightonii* showing compound sporophore with alternate rows of sporangia. **B.** *B. wightonii*, compression showing tripinnately compound trophophore. **A,B:** Paleocene. (From Rothwell & Stockey, 1989.)

Ophioglossales or the Filicales, so the relationships of the order remain uncertain.

References

Bierhorst, D. W. (1971). *Morphology of Vascular Plants*. New York: Macmillan.

DiMichele, W. A., & Phillips, T. L. (1977). Monocyclic *Psaronius* from the lower Pennsylvanian of the Illinois Basin. *Canadian Journal of Botany*, **55,** 2514–24.

Eames, A. J. (1936). *Morphology of Vascular Plants*. New York: McGraw-Hill.

Ehret, D. L., & Phillips, T. L. (1977). *Psaronius* root systems – morphology and development. *Palaeontographica B*, **161,** 147–64.

Gifford, E. M., & Foster, A. S. (1989). *Morphology and evolution of vascular plants*. 3rd. ed., New York: W. H. Freeman and Company.

Harris, T. M. (1931). The fossil flora of Scoresby Sound East, Greenland. Part I: Cryptogams (exclusive of Lycopodiales) *Meddelser om Gronland*, **85,** 1–100.

Janssen, R. E. (1957). *Leaves and Stems from Fossil Forests*, 2nd ed. Springfield: Illinois State Museum.

Kato, M. (1988). The phylogenetic relationship of Ophioglossaceae. *Taxon*, **37,** 381–86.

Kubart, B. (1916). Ein beitrag zur kenntnis von *Anachoropteris pulchra* Corda. *Denkschriften der Kaiserlichen Akademie der Wissenschaften in Wien Mathematische-Naturwissenschaftliche Klasse*, **93,** 1–34.

Mamay, S. H. (1950). Some American Carboniferous fern fructifications. *Annals of the Missouri Botanical Garden*, **37,** 409–75.

Mickle, J. E. (1984). Taxonomy of specimens of the Pennsylvanian age marattialean fern *Psaronius* from Ohio Illinois. *Illinois State Museum Scientific Paper*, **19,** 1–64.

Millay, M. A. (1977). *Acaulangium* gen. n., a fertile marattialean from the Upper Pennsylvanian of Illinois. *American Journal of Botany*, **64,** 223–9.

Millay, M. A. (1978). Studies of Paleozoic marattialeans: the morphology and phylogenetic position of *Eoangiopteris goodii* sp. n. *American Journal of Botany*, **65,** 577–83.

Millay, M. A., & Taylor, T. N. (1990). New fern stems from the Triassic of Antarctica. *Review of Palaeobotany and Palynology*, **62,** 41–64.

Morgan, J. (1959). The morphology and anatomy of American species of the genus *Psaronius*. *Illinois Biological Monographs, No.* **27,** 1–107.

Pfefferkorn, H. W. (1976). Pennsylvanian tree fern compressions, *Caulopteris, Megaphyton*, and *Artisophyton* gen. nov. in Illinois. *Illinois State Geological Survey, Circular*, **492.**

Pfefferkorn, H. W., Peppers, R. A., & Phillips, T. L. (1971). Some fernlike fructifications and their spores from the Mazon Creek compression flora of Illinois (Pennsylvanian). *Illinois Geological Survey Circular*, **463,** 1–29.

Phillips, T. L. (1974). Evolution of vegetative morphology in coenopterid ferns. *Annals of the Missouri Botanical Garden*, **61,** 427–61.

Phillips, T. L., & Andrews, H. N. (1966). *Catenopteris simplex* gen. et sp. nov., a primitive pteridophyte from the Upper Pennsylvanian of Illinois. *Bulletin of the Torrey Botanical Club*, **93,** 117–28.

Phillips, T. L., Peppers, R. A., & DiMichele, W. A. (1985). Stratigraphic and interregional changes in Pennsylvanian coal-swamp vegetation; environmental inferences. *International Journal of Coal Geology*, **5**, 43–109.

Rothwell, G. W., & Blickle, A. H. (1982). *Psaronius magnificus* n. comb., a marattialean fern from the Upper Pennsylvanian of North America. *Journal of Paleontology*, **56**, 459–68.

Rothwell, G. W., & Stockey, R. A. (1989). Fossil Ophioglossales in Paleocene of western North America. *American Journal of Botany*, **76**, 637–44.

Stidd, B. M. (1971). Morphology and anatomy of the frond of *Psaronius*. *Palaeontographica B*, **134**, 87–123.

Stidd, B. M. (1974). Evolutionary trends in the Marattiales. *Annals of the Missouri Botanical Garden*, **61**, 388–407.

Stidd, B. M., & Phillips, T. L. (1968). Basal stem anatomy of *Psaronius*. *American Journal of Botany*, **55**, 834–40.

Stubblefield, S. P. (1984). Taxonomic delimitation among Pennsylvanian marattialean fructifications. *Journal of Paleontology*, **58**, 793–803.

Takhtajan, A. L. (1956). *Telomophyta*. Moscow: Academiae Scientiarum.

Wagner, W. H. (1964). Evolutionary patterns of living ferns. *Memoirs of the Torrey Botanical Club*, **21**, 86–95.

19

Filicales of the Carboniferous

From Chapters 17 and 18 it is easy to understand why the various orders of ferns have traditionally been classified together. But as we have also seen, they constitute several distinct lineages that probably had independent origins from the Trimerophytopsida. The Filicales are another such lineage, and together with the heterosporous ferns they make up the largest clade of living pteridophytes. We interpret the heterosporous Marsileales and Salviniales to have separate origins from within the Filicales. In cladistic terms the three orders represent a monophyletic group in which the Marsileales and Salviniales are each monophyletic groups nested within the Filicales, and the Filicales alone are paraphyletic (Chart 3.1).

Next to the angiosperms (the flowering plants), the Filicales are the most diverse order of living land plants. Recent estimates place their diversity at about 10,000 species distributed through 300 genera. Over the past several years our concepts of primitive filicaleans have changed rapidly as a wealth of new information has become available. When the first edition of this book was published (Stewart, 1983), the Filicales were widely believed to have arisen from the Coenopteridales (Eggert, 1964), an admittedly artificial order of Paleozoic fernlike plants that traditionally has consisted of the families Stauropteridaceae, Zygopteridaceae, Botryopteridaceae, and Anachoropteridaceae (Delevoryas & Morgan, 1954b). Further studies by Phillips (1974), Cornet, Phillips, & Andrews (1976), Mickle (1980), and Taylor (1981) have made it increasingly clear that the order Coenopteridales, as originally visualized, actually comprises at least three orders of the class Filicopsida. These are the Stauropteridales, Zygopteridales, and Filicales. Mickle (1980) and Taylor (1981) have transferred four families, previously assigned to the Coenopteridales, to the Filicales. The families affected by the transfers are the Botryopteridaceae, Tedeleaceae, Anachopteridaceae, and Sermayaceae all of which have at least one genus with sporangial characteristics that indicate relationships with extant families of the Filicales. More recently, description of the family Psalixochlaenaceae (Holmes, 1981) and recognition of another new family of indusiate filicaleans (Rothwell, 1987) have contributed further to the growing amount of diversity that we find among true ferns from Paleozoic deposits. Because all species of the old order Coenopteridales have now been removed to other orders, the concept of this unnatural assemblage has become unnecessary.

At the present time it is most useful to consider the Filicales to have arisen from the Trimerophytopsida either directly, or through some intermediate group for which we currently have no evidence. Specializations that allow us to distinguish filicalean ferns from trimerophytes include a well-differentiated stem/leaf vegetative morphology with biseriate or planar fronds (Fig. 17.1), a sporangium that consists of a small capsule with a thin wall and a specialized dehiscence mechanism known as an annulus (Fig. 17.2E), and the production of relatively few spores within each sporangium (usually > 500).

The earliest Filicales

The first-known occurrence of the Filicales is in Lower Carboniferous (Tournaisian) sediments of France, where Galtier (1981) has discovered an isolated, permineralized sporangium of the filicalean type. The specimen is similar to Fig. 19.9D, and consists of a small capsule with a thin wall and a lateral annulus. Filicaleans become increasingly abundant in the Upper Carboniferous (Galtier & Scott, 1985), and by the end of the Carboniferous six families were represented. These families, the Tedeleaceae, Botryopteridaceae, Sermayaceae, Psalixochlaenaceae, Anachoropteridaceae, and a still-unnamed family are covered in this chapter. It is important to note that all of these families became extinct before the end of the Paleozoic, and were replaced by several families with living representatives by the beginning of the Triassic (Rothwell, 1987). More recent filicaleans and their heterosporous descendants, the Marsileales and Salviniales, are discussed in Chapter 20.

Tedeleaceae

Until 1966 it was customary to place *Ankyropteris* (Fig. 19.1) with the etapteroid members of the Coenopteridales. As studies of the North American species *A. glabra* proceeded, however, an increasing number of differences between it and other Zygopteridaceae became apparent, so that Eggert & Taylor (1966) placed this species in a new genus, *Tedelea,* and in its own family, the Tedeleaceae. A reinvestigation of all material recognized as *Ankyropteris* and *Tedelea* (Mickle, 1980) has resulted in the establishment of a single species, *A. brongniartii,* with all other species of *Ankyropteris* and *Tedelea* being placed in synonomy.

Figure 19.1. *Ankyropteris brongniartii.* **A.** Reconstruction of aerial portion of plant. Note axillary branching and aphlebiae. (Redrawn from Eggert, 1959a.) **B.** Portion of fertile frond. (From Eggert & Taylor, 1966.) **A,B:** Middle Pennsylvanian.

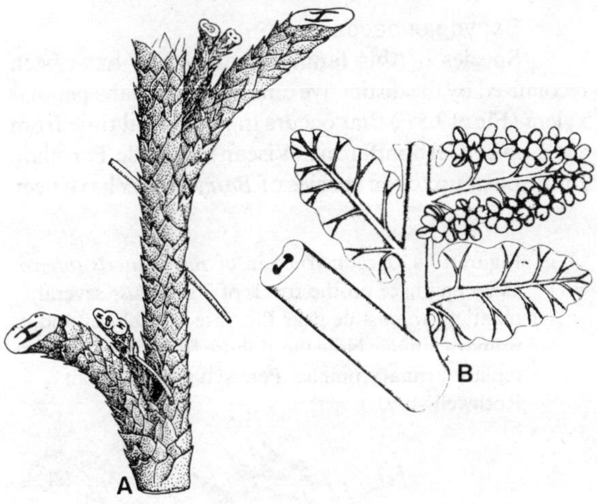

As a result of the efforts of Baxter (1951), Eggert (1959a, 1963), Eggert & Taylor (1966), and Mickle (1980) we can give the following description of *Ankyropteris brongniartii.* The internal anatomy of the stem, which reflects the external morphology, consists of a protostele that is usually a pentarch actinostele with a mixed pith (Fig. 19.2). Arising from the stem stele in a spiral sequence are frond traces (Fig. 19.2B) that assume the H-shaped configuration and have mesarch protoxylem in a position that is similar to the peripheral loops in *Etapteris* of the Zygopteridales (Chapter 17). Above each of the fronds there is an axillary branch (Fig. 19.2A) that has all of the characteristics of the stem except that it is smaller. The stem and branch steles also give rise to small traces that supply scalelike aphlebiae scattered on the stems and petioles (Fig. 19.1A). An investigation by Eggert (1963) reveals that traces are given off to the pinnae in two lateral rows and that the frond is biseriate and planated throughout, as in modern ferns. The frond which is bi- to tripinnate has pinnules with laminae and dichotomous venation. Sporangia have been discovered on the lower surface of pinnules either singly or in clusters near the margins (Fig. 19.1B). The sessile to short-stalked sporangia are pyriform with an annulus occupying the distal third. At maturity the sporangium wall is one cell layer thick. The spore output is approximately 120 per sporangium, and the isospores are like the *Raistrickia* dispersed-spore type.

Ankyropteris brongniartii shows at least two more or less unique characteristics that set it apart from other ferns. These are (1) the vascularized, scalelike aphlebiae and (2) axillary branching. Although we have seen some evidence of axillary branching in certain

Figure 19.2. *Ankyropteris brongniartii.* **A.** Transverse section of aerial axis above a node showing stem stele (ss); stele of axillary branch (ab); petiolar trace (pt). **B.** Transverse section of H-shaped petiolar trace. **A,B:** Pennsylvanian. (From Phillips, 1974.)

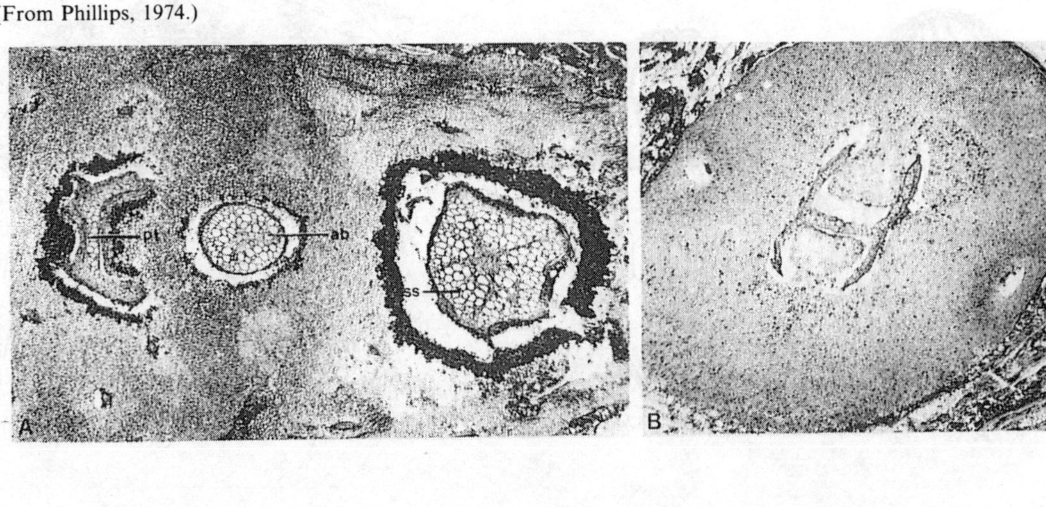

Trimerophytopsida, this is our first exposure to this kind of branching in ferns where axillary branching seldom occurs. The small scalelike aphlebiae with the trace of vascular tissue raises another interesting point. *A. brongniartii* is megaphyllous. At the same time, however, it developed structures (aphlebiae) that one might characterize as microphylls. Could these scattered structures have evolved from enations of the type found on certain of the trimerophytes? If this turns out to be the case, then the distinctions made between major groups of vascular plants based on microphylls and megaphylls fades dramatically.

Even though one might argue that ferns showing quadriseriate branching in their fronds are more specialized instead of primitive, the fact remains that the biseriate planated frond so characteristic of most modern ferns has been derived from the three-dimensional dichotomizing type through the processes of overtopping, reduction, planation, and webbing. Plants such as *Rhacophyton ceretangium* and *Zygopteris* give us some clues revealing the steps involved in the evolution of the biseriate frond from such a primitive branching system.

In the classification and identification of natural groups of ferns, much emphasis has been placed on the structure and development of sporangia. In looking for the origins of our modern eusporangiate and leptosporangiate ferns, paleobotanists search for and attempt to characterize the sporangia of fossil ferns in terms of those belonging to modern families. For example, the structure and position of sporangia of *Ankyropteris brongniartii* suggest affinities with members of the Schizaeaceae (Eggert & Taylor, 1966). It is this similarity that prompted Mickle (1980) to place the Tedeleaceae with the Filicales instead of leaving the genus in the Zygopteridales. In the primitive leptosporangiate Schizaeaceae, the sporangia are solitary or in small groups either at or near the margin of the pinnule. The sporangia are stalked or sessile and have a uniseriate annulus around a cap of cells at the distal end of the sporangium (Fig. 19.3A). In *A. brongniartii* the annulus has a cap plus two rows of annular cells in the distal end (Fig. 19.3B). We are cautioned, however, by Jennings & Eggert (1977) not to be too hasty in concluding that the affinities of *A. brongniartii* and *Senftenbergia,* both of which have similar sporangial types, are with the Schizaeaceae as suggested by Radforth (1938). One reason for this note of caution is the occurrence of sporangia with apical annuli in genera belonging to other extant families of ferns. Another is the vegetative structure and anatomy of *A. brongniartii* and *Senftenbergia,* which is unlike that of any known member of the Schizaeaceae. Until more is learned about members of the Tedeleaceae, it seems reasonable to conclude that the family shows a unique combination of characteristics within the Filicales.

Figure 19.3. **A.** Single sporangium of *Schizaea bifida*. Extant. (Redrawn from Smith, 1955.) **B.** A sporangium of *Ankyropteris brongniartii*. Middle Pennsylvanian. (Redrawn from Eggert & Taylor, 1966.)

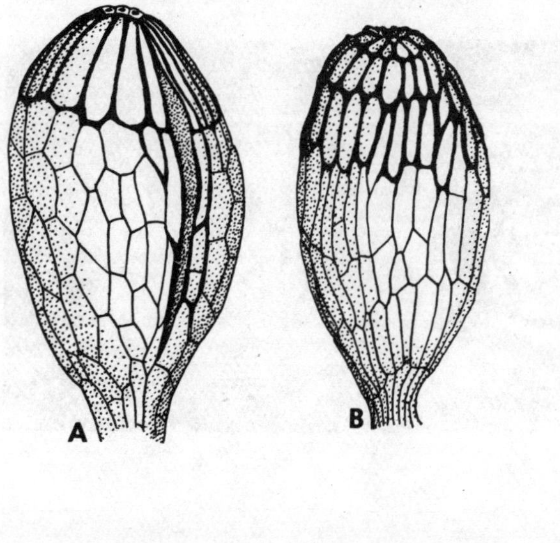

Botryopteridaceae

Species of this family traditionally have been recognized by the distinctive omega-shape of the petiolar xylem (Fig. 19.5F) that occurs in geological time from the Lower Carboniferous (Visean) into the Permian. In the past, up to ten species of *Botryopteris* have been

Figure 19.4. Reconstruction of *Botryopteris forensis* as an epiphyte on the trunk of *Psaronius;* several plantlets are visible near the base of rachises and primary pinnae. Note the globose fructifications that replace primary pinnae. Pennsylvanian. (From Rothwell, 1991.)

assigned to the family, but as we have learned more about the fertile parts of plants with such petioles, it has become clear that this feature is common to ferns that belong to at least two families of primitive Filicales (Good & Rothwell, 1988).

Currently, there are three or four species of *Botryopteris* assigned to the Botryopteridaceae, based on a relatively complete understanding of the whole plant (Rothwell, 1991). Several additional species with *Botryopteris* petioles may also belong to the family, but until their fertile structures are characterized, this familial assignment must remain tentative.

Vegetative structure

Botryopteris, like other genera of the Filicales, was an herbaceous plant. Some species may have had a prostrate rhizome with a semierect tip, as in the fern *Osmunda,* while others may have been scrambling plants supported by the surrounding vegetation, or vines. At least one species is known to have been an epiphyte growing on the trunk of the tree fern *Psaronius* (Fig. 19.4).

The rhizome bore helically arranged fronds (Fig. 19.5A,C) that, in at least one species (Delevoryas & Morgan, 1954b), are known to have biseriate tripinnate

Figure 19.5. **A.** Reconstruction of portion of petiole and frond of *Botryopteris forensis.* (From Delevoryas & Morgan, 1954b.) **B.** Diagram of vascular system of *Botryopteris* foliar member with plantlet (p). **C.** Reconstruction of *B. forensis* with prostrate rhizome bearing three petiolar traces with lateral shoots. (**B,C** from Phillips, 1974) **D.** *B. forensis* sporangia. (Redrawn from Galtier, 1971.) **E.** *Osmunda claytonia* sporangium. Extant. (Redrawn from Smith, 1955.) **F.** Transverse section of omega-shaped petiolar trace of *B. forensis.* **A–D,F**: Pennsylvanian.

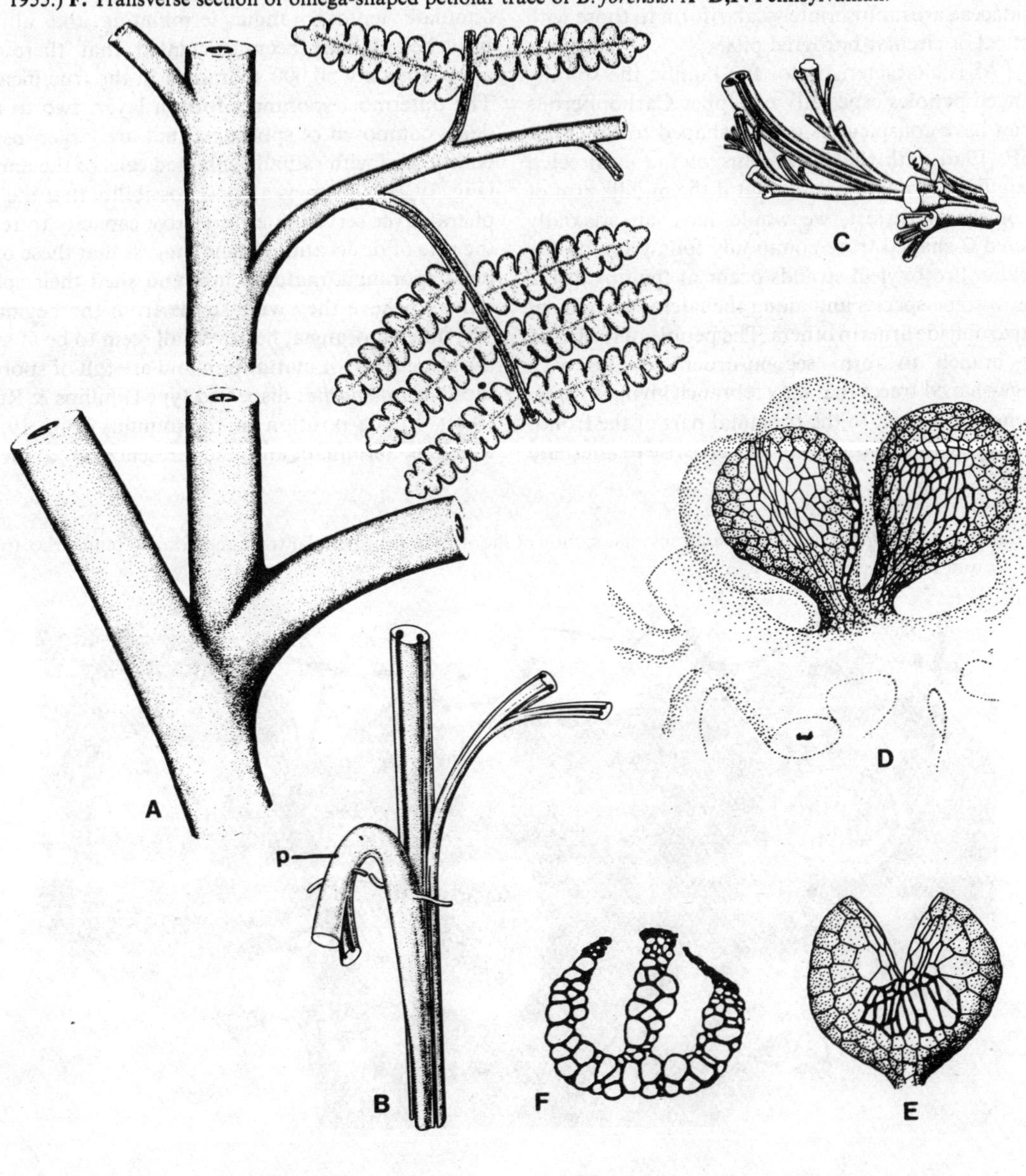

fern-type fronds with laminate pinnae. Where details of the pinnae have been worked out from permineralized specimens it appears that they resemble certain species of *Pecopteris* and *Sphenopteris* known from compression–impression fossils. The petioles of some *Botryopteris* species are known to have produced adventitious shoots (buds) or plantlets (Fig. 19.5B). Endogenous adventitious roots were borne on the rhizome, adventitious shoots, and bases of petioles. Most species had rhizomes covered with multicellular hairs.

In transverse section, the vascular system of *Botryopteris* is a somewhat lobed haplostele (Fig. 19.6) with mesarch to centrach primary xylem in some species and exarch in others. Cauline vascular systems of the foliar shoots are often siphonostelic and may show leaf gaps (Phillips, 1974). In general, the wall thickenings on the metaxylem tracheids of the Botryopteridaceae are multiseriate–scalariform to those with elliptical or circular bordered pits.

As is characteristic for the family, the spirally arranged petioles especially of Upper Carboniferous species have conspicuous omega-shaped traces (Figs. 19.5F, 19.6) with the three arms of the ω directed adaxially. It is worth noting that if the middle arm of the omega were lost, we would have an adaxially directed C-shaped trace commonly found among the Filicales. Protoxylem strands occur at the tips of the arms in some species and along the lateral margins of the two outside arms in others. The petiole of the frond may branch to form second-order petioles with omega-shaped traces that may rebranch giving a three-dimensional aspect to the proximal part of the frond. Distally, the third-order branches give rise to alternate biseriate, ω-shaped pinna traces to produce the planated portion of the frond.

Fertile frond parts

It is obvious that the massive globose sporangial aggregations of *Botryopteris globosa* (Fig. 19.7A) consist of the much-branched portion of fertile fronds. That this fructification belonged to *Botryopteris* was first determined by Darrah (1939). The relationship between the two was shown by the *Botryopteris*-like omega-shaped petiolar traces in *B. globosa*. The most complete accounts of *B. globosa*-type fructifications are by Murdy & Andrews (1957) and Galtier (1971). According to Murdy & Andrews, there was a central petiolelike branch with alternate biseriate pinnalike laterals. The latter branched repeatedly to produce a three-dimensional dichotomizing system with clusters of annulate leptosporangia terminating the ultimate branches. It has been calculated that there were approximately 50,000 sporangia in the fructification! The outermost sporangia form a layer, two to three deep, composed of sporangia that are larger, usually empty, and with radially enlarged cells of the annulus (Fig. 19.7B). There is a good possibility that the peripheral layer served in a protective capacity to reduce the rate of desiccation. Some suggest that these outermost sporangia matured first and shed their spores. Others believe they were sterile from the beginning. The inner sporangia, however, all seem to be at about the same stage of maturation and are full of spores of the *Punctatisporites* dispersed type (Phillips & Rosso, 1970). In the position of the annulus (Fig. 19.5D), which is terminal, and the presence of a median

Figure 19.6. *Botryopteris forensis*, transverse section of shoot (s) bearing petiolar trace and primary pinna trace (p). (From Phillips, 1974.)

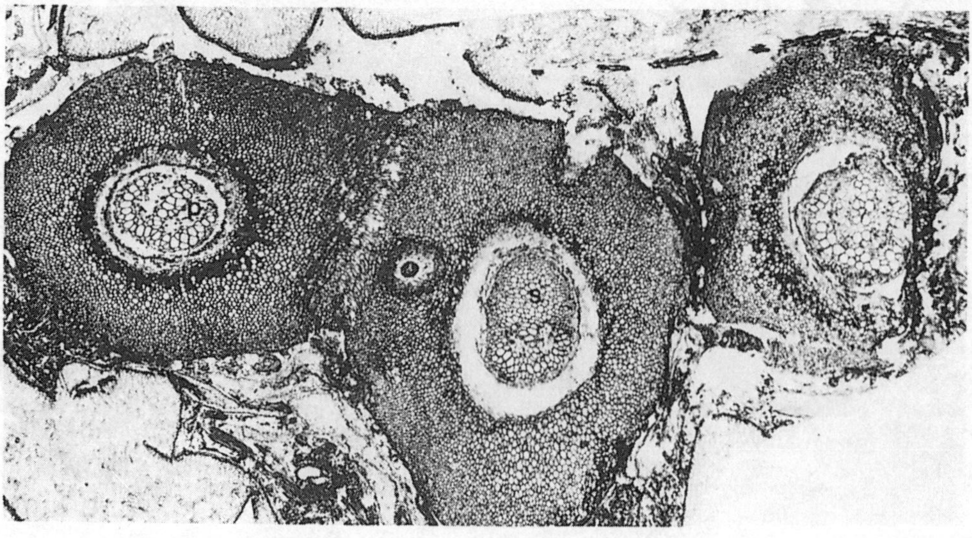

vertical zone of dehiscence these sporangia are similar to those of the Osmundaceae (Fig. 19.5E).

Botryopteris cratis, described by Millay & Taylor (1980), is similar to *B. globosa,* but is a looser aggregation of sporangia attached to small pinnae. The whole aggregation is, in turn, surrounded by larger sterile pinnae that may have partially enclosed and protected the sporangia.

Botryopteris forensis and other closely related species were epiphytes that grew on the surface of the marattiaceous tree fern, *Psaronius* (Chapter 18). Specimens of *B. forensis* were first described in 1875 by Renault, but a full reconstruction that reflects the structure, growth, and role of the plant in the swamp community (Fig. 19.4) has required the efforts of numerous workers over a period of more than 115 years! Most notable among these are the discovery by Mamay & Andrews (1950) that the rhizome branched profusely and bore helically arranged fronds and adventitious roots (Fig. 19.5C). Shortly thereafter Delevoryas & Morgan (1954b) demonstrated that the fronds were planar and pinnately compound with laminar pinnules (Fig. 19.5A). Phillips (1961) documented how plantlets with adventitious roots were produced by the fronds (Fig. 19.5B) and recognized that vegetative and fertile structures (including *B. globosa,* discussed above) known by several species names represented different parts of the shoot system of a single species (Phillips, 1966). Most recently, the discovery of rhizomes rooted on the trunks of the tree fern *Psaronius* (Rothwell, 1991) completes the evidence needed to document that the plant was indeed an epiphyte (Fig. 19.4).

Anachoropteridaceae

The family Anachoropteridaceae is similar to the Botryopteridaceae in that species traditionally have been recognized by a distinctive structure of the xylem in the petiole. Over the past 25 years as ancient ferns increasingly have become known as whole plants, we have discovered that species with *Anachoropteris*-type petioles represent a diverse assemblage that includes members of at least three additional families. These

Figure 19.7. **A.** Median section through the spherical fructification of *Botryopteris globosa.* **B.** Enlarged peripheral sporangia of *B. globosa.* Note outer sterile sporangia with large thick-walled cells. Inner sporangia are fertile and contain spores. **A,B:** Pennsylvanian. (**A,B** courtesy of Dr. T. L. Phillips.)

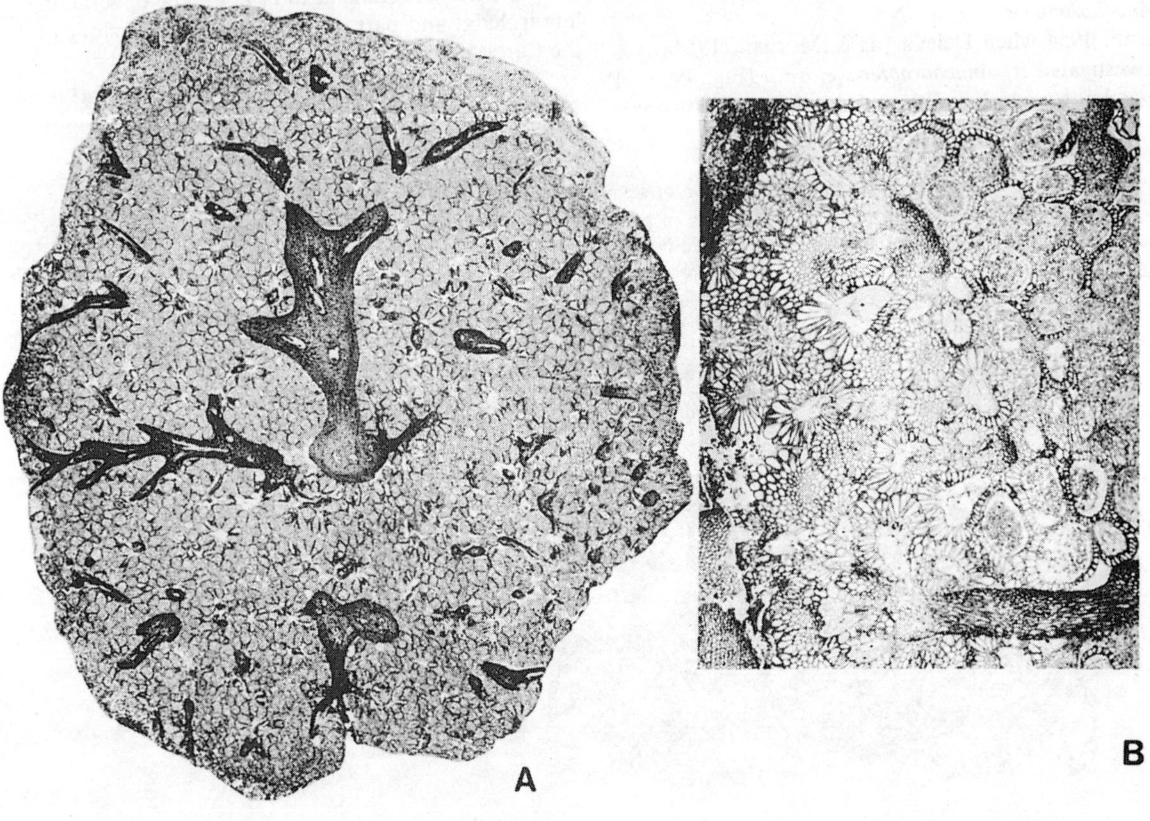

are the Sermayaceae, the Psalixochlaenaceae, and a new family with indusiate sori. Because we still recognize a large percentage of the species of *Anachoropteris* and similar ferns by vegetative structure only, their natural relationships are uncertain. In the future we may discover that all of these species belong in other families. If so, the concept of the family Anachoropteridaceae will no longer be useful. On the other hand, some of these species may represent a natural group that will allow us to more precisely characterize the family. In either case, however, the Anachoropteridaceae is presently a useful concept for grouping and understanding the vegetative remains of some of the most ancient filicalean ferns.

Taken as a group, the Anachoropteridaceae show us how branching systems, including axillary branching, evolved. Although these ancient plants have many characteristics of modern ferns, including spirally arranged biseriate fronds, circinate vernation, mesarch primary xylem, and sclerenchyma, they differ in one important feature. Unlike living ferns, where the leaf trace to a rachis is C-shaped with adaxial curvature (Fig. 17.2C), members of the Anachoropteridaceae have an abaxial orientation of the trace (Figs. 19.8A, 19.9A). The principal genera are *Anachoropteris, Apotropteris, Rhabdoxylon, Tubicaulis,* and possibly *Gramnopteris* (Eggert, 1964). All are known to occur in the Upper Carboniferous, with *Tubicaulis* and *Anachoropteris* continuing into the Lower Permian.

Anachoropteris

Until 1954 when Delevoryas & Morgan (1954a) reinvestigated it, *Anachoropteris clavata* (Fig. 19.10C) was known as a possible petiole of an unknown coenopterid. In cross section, the petiole trace appears as a C-shaped strand with clublike (clavate) arms. Like other *Anachoropteris* petiole traces, *A. clavata* has patches of protoxylem on what was apparently the adaxial surface. The actual position of protoxylem could not be confirmed, however, until the relationship of the petiole trace to a stem axis was determined.

In their specimens, Delevoryas & Morgan (1954a) discovered that small axes were produced from the lateral margins of the clavate petiole trace (Fig. 19.11D). The stem stele, which enlarged distally, is an exarch protostele. It in turn gave rise to adventitious roots and spirally arranged petiole traces. Where these become free from the stem stele, it is apparent that the curvature of the C-shaped trace and its opening is abaxial. This work also demonstrated, for the first time, the presence of a shoot bearing leaves and roots on a petiole of the Anachoropteridaceae, an observation further substantiated by Holmes (1979), who found similar foliar shoots on foliar members of *A. gillotii.* This propensity for the formation of adventitious shoots on foliar members is found in *Botryopteris* and a number of modern ferns.

Tubicaulis

The geological record of *Tubicaulis* extends from the Upper Carboniferous (Westphalian A, Middle Pennsylvanian) into the Permian. Like other anachoropterids, some species of *Tubicaulis* were erect to semierect plants, while others were thought to be epiphytic or scrambling in habit and part of the extensive understory of the Carboniferous swamps.

The vascular systems of *Tubicaulis* stems (Fig. 19.9A) vary from solid exarch protosteles to those

Figure 19.8. **A.** *Tubicaulis,* transverse section of stem stele and abaxially oriented C-shaped petiole traces. **B.** *Anachoropteris pulchra,* transverse section of petiole with a combined pinna and cauline trace at right. **A,B:** Pennsylvanian. (**A,B** from Phillips, 1974.)

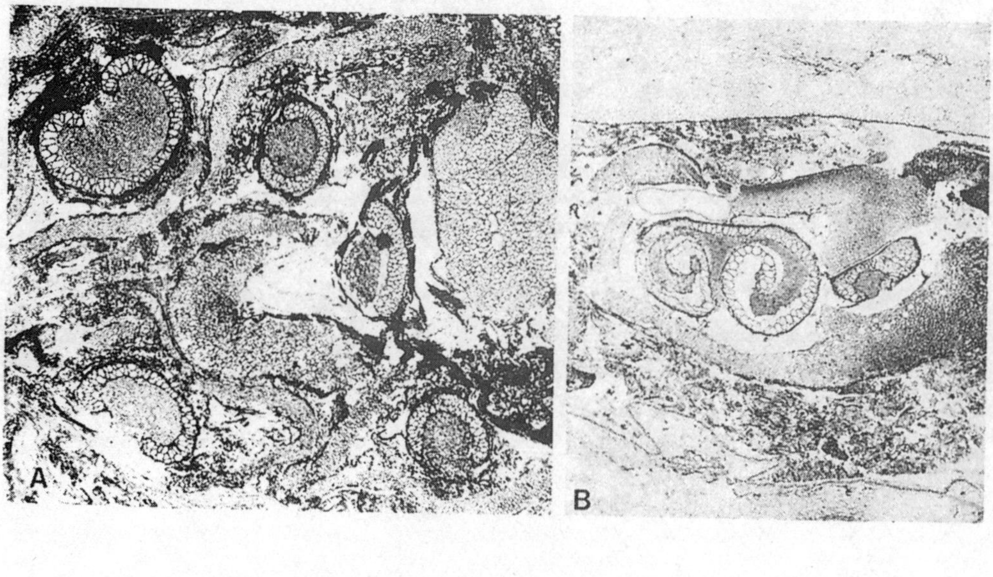

with a fair amount of parenchyma intermixed with the centrally located, metaxylem cells. It has been suggested by Eggert (1959b) that those species of *Tubicaulis* with "mixed" parenchymatous protosteles foreshadow the evolution of the fern siphonostele.

Abaxially curved petiole traces arise in a spiral sequence from the stem stele and have the characteristics of *Anachoropteris*. It was up to Hall (1961) to demonstrate attachment between a *Tubicaulis*-type stem (Fig. 19.8A) and petioles of *Anachoropteris* (Fig. 19.8B). Adventitious roots were produced from the stem and bases of the petioles. These roots, characteristic of coenopterid and other fern groups, are diarch.

Sermayaceae

Additional portions of sporangium-bearing fronds having the characteristics of *Anachoropteris* anatomy were found by Eggert & Delevoryas (1967) in permineralized specimens. It seems likely that the fertile fronds were borne by ferns with *Tubicaulis–Anachoropteris*-type vegetative anatomy. Until it is determined, however, that the three are congeneric, the authors have placed the fertile material in a new family, the Sermayaceae. This taxonomic procedure only emphasizes the belief that the Anachoropteridaceae is an artificial assemblage and that the Sermayaceae is a natural taxon with filicalean affinities. In addition to the Upper Pennsylvanian genus *Sermaya*, *Doneggia* and some other less well-known genera may be attributed to the family. The salient feature of *Sermaya* is the presence of leptosporangia, which have an obliquely horizontal two-rowed annulus. The sporangia are sessile or stalked and occur in sori on the abaxial surface of pinnules belonging to fronds of the *Sphenopteris* type (Fig. 19.9C). In the orientation and position of the annulus, the sporangia of the Sermayaceae are somewhat similar to those of the Gleicheniaceae of the Filicales. In the modern family the annulus is a single row of cells (Fig. 19.9D), not double. This difference along with differences in the anatomy of the fertile frond raises considerable doubt about a close relationship between the two families.

Another genus that may be at least partly referable to the Sermayaceae is *Oligocarpa* (Fig. 19.9B,C). Several species of *Oligocarpa* have been described from compression specimens collected in the Upper Carboniferous of Europe and the Pennsylvanian of North America. They consist of frond fragments that bear small sori on the abaxial surface of the pinnules. The sporangia of the various species have been described as having quite different features by various authors, but they apparently have a stalk with an oblique annulus. As interpreted by Abbott (1954) the annulus is uniseriate. but others interpret it to be bi- or multiseriate

Figure 19.9. **A.** Stereodiagram of the vascular system of *Tubicaulis*. Pennsylvanian. (From Phillips, 1974.) **B.** *Oligocarpia mixta*, sporangia showing features of the annulus. **C.** *O. mixta*, portion of a fertile frond. **B,C:** Pennsylvanian. (**B,C** redrawn from Abbott, 1954.) **D.** Sporangia of *Gleichenia pectinata* to show position of annulus. Extant. (Redrawn from Smith, 1955.)

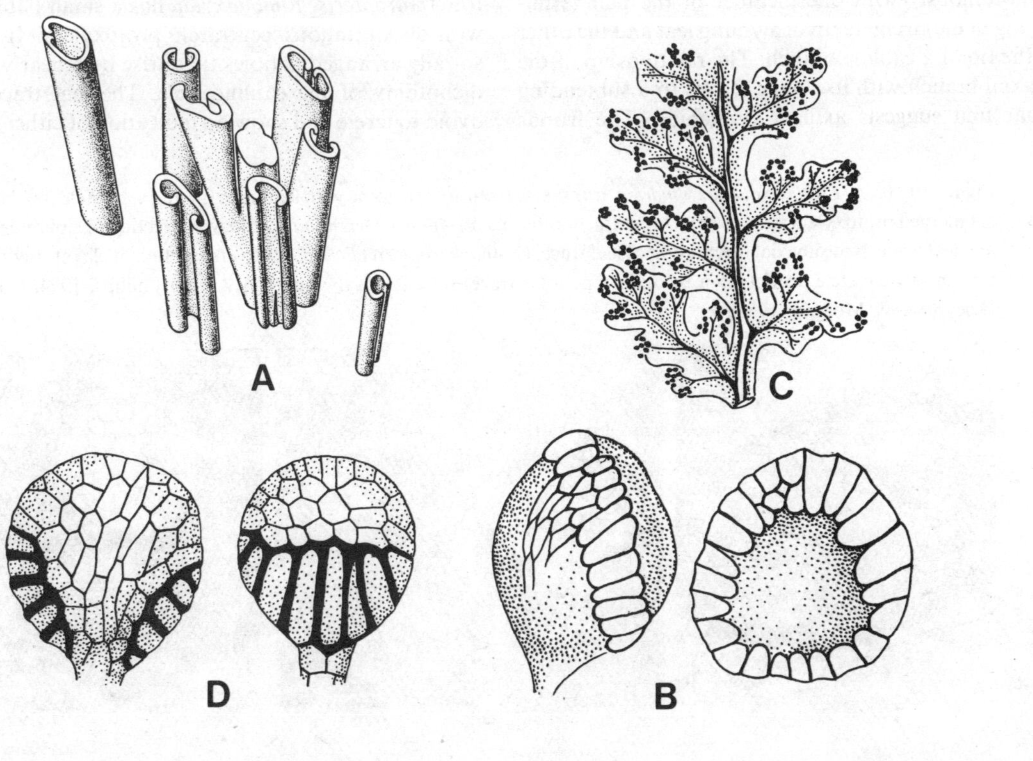

in some species. It is probable that future studies will establish that different species of *Oligocarpa* belong in two or more different families of Carboniferous Filicales.

Psalixochlaenaceae

Psalixochlaena

Psalixochlaena (Fig. 19.10A) was originally assigned to another genus of coenopterid ferns, *Botryopteris*. In 1960, Holden was able to demonstrate, however, that the petiole trace had an abaxial curvature, which became C-shaped at a higher level. These traces were given off in a spiral sequence from a terete, protostelic stem with mesarch groups of protoxylem. The protoxylem of the foliar trace is associated with its adaxial surface, and this distinguishes it from the stem (cauline) stele with its submerged protoxylem. With this distinction in mind, and working with large quantities of beautifully preserved permineralized specimens, Holmes (1977) was able to work out the general habit of *Psalixochlaena*.

The slender, rhizomelike stems of this plant are known to branch dichotomously or pseudomonopodially. The lateral branching from the stem is described as monopodial. It is clear, however, that the laterals are overtopped portions of a dichotomizing branching system. The vascular supply to a lateral branch (Fig. 19.10A) is a common trace that divides by an unequal dichotomy to produce an inferior petiolar strand of a leaf and a superior (adaxial) cauline branch. The petiole strand assumes an M-shape with three protoxylem ridges. This strand gives off small traces that supplied pinnae. The adaxial branch also branches dichotomously with one member of the pair terminating in the circinate tip of a young leaf and the other in the tip of a cauline branch. The relationship of the adaxial branch with its young leaves to a subtending foliar unit suggests axillary branching. The fronds of the plant are planated and alternately branched. Pinnules of the fronds are similar to those of the form genus *Sphenopteris*.

Apotropteris

In its anatomy, this tiny plant bears a resemblance to the *Psalixochlaena*. The suggestion has been made that *Apotropteris* may represent a fragment of the distal portion of a plant similar to *Psalixochlaena*. In transverse section, the stem stele of *Apotropteris* (Morgan & Delevoryas, 1954) is usually a terete protostele (Fig. 19.10B) with exarch primary xylem. Some specimens (Phillips, 1974) are siphonostelic and have leaf gaps. The petioles have no definite sequence in the way they are borne on the stem. When the trace to a petiole is produced there is a nearly equal dichotomy of the stem stele. That portion that vascularizes the petiole becomes C-shaped with abaxial orientation. Protoxylem of the trace, like other Anachoropteridaceae, lies along the adaxial convex surface of the petiole trace. Adventitious roots occur at random along the length of the stem.

According to the authors, this Upper Pennsylvanian plant shows certain primitive characteristics in the unequal dichotomy resulting in a foliar trace with a small amount of abaxial curvature.

Rhabdoxylon

This Upper Carboniferous (Upper Pennsylvanian) genus (Holden, 1960; Dennis, 1968) is another small and structurally simple plant that embodies the primitive characteristics suggested by Morgan & Delevoryas for *Apotropteris*. *Rhabdoxylon* has a small haplostele with discontinuous centrarch protoxylem. It bears spirally arranged petioles that arise by a nearly equal dichotomy of the cauline stele. The leaf traces are ovoid to terete and show no indication of either ab- or

Figure 19.10. **A.** *Psalixochlaena cylindrica,* transverse section of rhizome with leaf trace (lt) and stem stele (ss). Note submerged protoxylem strands (p). Upper Carboniferous. **B.** *Apotropteris minuta,* transverse section of siphonostele and leaf trace. Note abaxial orientation of leaf trace. **C.** *Anachoropteris clavata,* transverse section of clavate petiole trace and stem stele producing a secondary petiolar trace. **B,C:** Pennsylvanian. (**A,B** from Phillips, 1974; **C** from Delevoryas & Morgan, 1954a.)

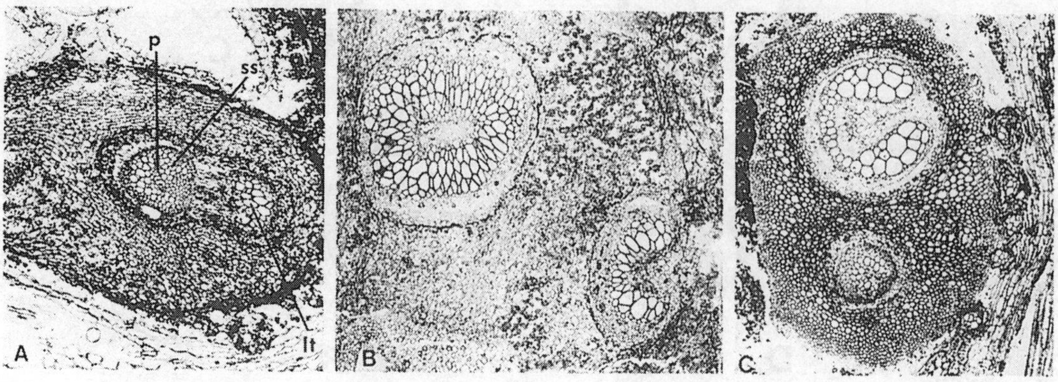

adaxial curvature. There is, however, a single protoxylem strand on the adaxial surface of the trace.

In his characterization of the Psalixochlaenaceae, Holmes (1981) described the sporangia of *Psalixochlaena* as occurring in small sori on the abaxial surface near the margins of truncated pinnules. Each sporangium has a short stalk and a small capsule with a lateral annulus located near the stalk (Good, 1981). In these features the sporangia of *Psalixochlaena* are similar to those of *Gleichenia* (Fig. 19.9D), except that the stalk of the fossils is generally shorter and broader, and the annulus of the Carboniferous ferns consists of two rows of thick-walled cells.

Sori of similar sporangia also occur on the abaxial surface of pinnules in the Middle Pennsylvanian fern *Norwoodia*, and species that previously had been assigned to the Botryopteridaceae on the basis of petiole trace configuration (Good, 1979). The latter ferns have now been transferred to the Psalixochlaenaceae as *P. antigua* from the Lower Carboniferous (Visean) of Europe and *Psalixochlaena* sp. from the Lower-Middle Pennsylvanian of Kentucky (Good & Rothwell, 1988). From the Psalixochlaenaceae it is clear that *Botryopteris* and *Anachoropteris*-type petiole traces were produced by species of several different families.

An indusiate filicalean

An additional segregate from the Anachoropteridaceae recently has been recognized as a new family of indusiate filicaleans from Upper Pennsylvanian deposits of Ohio (Rothwell, 1987). Vegetative parts of the plant were originally described as *Anachoropteris clavata* (Fig. 9.10C), and were shown by Delevoryas & Morgan (1954a) to produce foliar plantlets. The Ohio specimens have pinnately compound fronds with *Sphenopteris*-type pinnules, and are interpreted to have grown as vines like the living fern, *Lygodium japonicum* (Trivett & Rothwell, 1988). On the abaxial surface of the fertile pinnules are tiny, globose indusia like those of the extant tree fern *Cyathia*. Within each indusium is a large number of tightly packed sporangia with long, uniseriate stalks and a biseriate ringlike annulus. The sporangia are interpreted to have matured sequentially in a gradate developmental pattern to shed tiny, trilete spores.

Branching and growth habit in ferns

The general view of branching in vascular plants is that ferns and most other pteridophytes have dichotomous apical branching, while seed plants have lateral branching from the axils of leaves. In ferns, however, there is an array of branching types in both living and fossil species. Much attention has been given to axillary branching in *Psalixochlaena* because it supplies evidence for two supposedly different schemes explaining how axillary branching evolved. One model (Fig. 19.11A) depicts the branch starting from an adaxial position on the petiole and by "phyletic slide" migrating to the axillary position. The other envisages an overtopped branch arising in the same plane as the stem and petiole. In the first model a common trace

Figure 19.11. *Psalixochlaena*, models showing two types of axillary branching. **A.** Leaf and branch trace arising from a common trace. **B.** Leaf and branch trace arising separately from cauline vascular system. Petiolar axis stippled. (Redrawn from Holmes, 1977.) **C.** Stereodiagram, vascular system of *Apotropteris* sp. **D.** *Anachoropteris clavata*, stereodiagram, vascular system with plantlet arising from petiole trace. **C,D:** Pennsylvanian. (**C,D** from Phillips, 1974.)

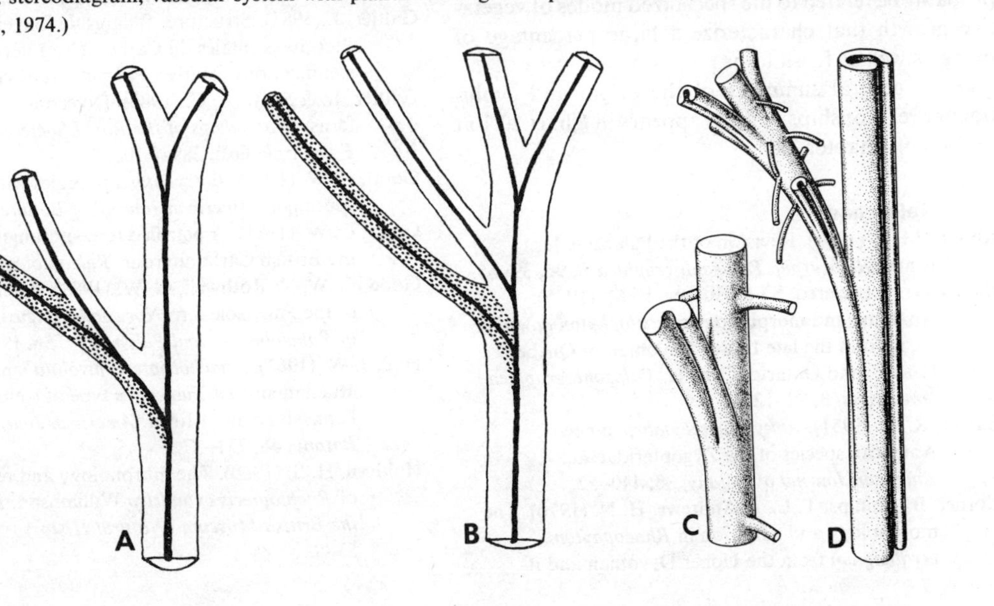

divides to supply the leaf and branch trace. In the second (Fig. 19.11B), the leaf and branch traces arise separately from the stelar vascular supply.

Both models are easily derived by the modification of three-dimensional telome trusses, where overtopping, reduction, and planation have occurred. You may recall (Chapter 13) that *Psilophyton crenulatum* of the Trimerophytopsida exhibits a variety of branching patterns, one of which appears to be axillary. Similar observations by Banks, Leclercq, & Hueber (1975) prompted them to conclude that *Psilophyton* in its branching seems to foreshadow axillary branching in the filicaleans.

More recently, Holmes (1989) has recognized a broad spectrum of branching types among Carboniferous filicalean ferns, and related them to specialized modes of vegetative growth that have been known to occur for many years (Mamay, 1952). In *Psalixochlaena* there is dichotomous, lateral, and extra-axillary branching of stems, as well as a variety of positions at which stems are borne on leaves (Holmes, 1989). Holmes equates the diverse patterns of branching in *Psalixochlaena cylindrica* with a growth form where the plants scrambled over the dry peat on the swamp floor, propagating by vegetative reproduction, to form a mat of ground cover.

As we emphasized earlier in this chapter, the prolific branching of stems and the production of foliar plantlets in *Botryopteris forensis* (Fig. 19.5B) are vegetative adaptations to the epiphytic habit of this species. Likewise, the distinctive branching that characterizes *Ankyropteris brongniartii* (Figs. 19.1 and 19.2) and the indusiate filicalean fern are undoubtedly related to the growth of these species as vines on the trunks of the tree fern *Psaronius*. As we continue to learn more about growth architecture in species of Carboniferous filicaleans, more branching patterns will probably be related to the specialized modes of vegetative growth that characterize a large percentage of living as well as fossil ferns.

A general summary of the origin and evolutionary relationships of ferns appears in Chart 20.1 at the end of Chapter 20.

References

Abbott, M. L. (1954). Revision of the Paleozoic fern genus *Oligocarpia*. *Palaeontographica B*, **96**, 39–65.

Banks, H. P., Leclercq, S., & Hueber, F. M. (1975). Anatomy and morphology of *Psilophyton dawsonii* sp. n. from the late Lower Devonian of Quebec (Gaspé) and Ontario, Canada. *Palaeontographica Americana*, **8**, 77–127.

Baxter, R. W. (1951). *Ankyropteris glabra*, a new American species of the Zygopteridaceae. *American Journal of Botany*, **38**, 440–52.

Cornet, B., Phillips, T. L., & Andrews, H. N. (1976). The morphology and variation in *Rhacophyton ceratangium* from the Upper Devonian and its bearing on frond evolution. *Palaeontographica B*, **158**, 105–29.

Darrah, W. C. (1939). The fossil flora of Iowa coal balls. II. The fructifications of *Botryopteris*. *Botanical Museum Leaflets, Harvard University*, **7**, 157–68.

Delevoryas, T., & Morgan, J. (1954a). A further investigation of the morphology of *Anachoropteris clavata*. *American Journal of Botany*, **41**, 192–203.

Delevoryas, T., & Morgan, J. (1954b). Observations on petiolar branching and foliage of an American *Botryopteris*. *American Midland Naturalist*, **52**, 374–87.

Dennis, R. L. (1968). *Rhabdoxylon americanum* sp. n., a structurally simple fern-like plant from the Upper Pennsylvanian of Illinois. *American Journal of Botany*, **55**, 989–95.

Eggert, D. A. (1959a). Studies of Paleozoic ferns. The morphology, anatomy, and taxonomy of *Ankyropteris glabra*. *American Journal of Botany*, **46**, 510–20.

Eggert, D. A. (1959b). Studies of Paleozoic ferns: *Tubicaulis stewartii* sp. nov. and evolutionary trends in the genus. *American Journal of Botany*, **46**, 594–602.

Eggert, D. A. (1963). Studies of Paleozoic ferns: the frond of *Ankyropteris glabra*. *American Journal of Botany*, **50**, 379–87.

Eggert, D. A. (1964). The question of the phylogenetic position of the Coenopteridales. *Bulletin of the Torrey Botanical Club*, **21**, 38–57.

Eggert, D. A., & Delevoryas, T. (1967). Studies of Paleozoic ferns: *Sermaya*, gen. nov. and its bearing on filicalean evolution in the Paleozoic. *Palaeontographica B*, **120**, 169–80.

Eggert, D. A., & Taylor, T. N. (1966). Studies of Paleozoic ferns: on the genus *Tedelea* gen. nov. *Palaeontographica B*, **118**, 52–73.

Galtier, J. (1971). La fructification de *Botryopteris forensis* Renault (Coenopteridales de Stéphanien Français): précisions sur la sporanges et les spores. *Naturalia Monspeliensia Botanique*, **22**, 145–55.

Galtier, J. (1981). Structures foliares de fougères et Pteridospermales du Carbonifère Inférieur et leur signification évolutive. *Paleontographica B*, **180**, 1–38.

Galtier, J., & Scott, A. C. (1985). Diversification of early ferns. *Proceedings of the Royal Society of Edinburgh*, **86B**, 289–301.

Good, C. W. (1979). *Botryopteris* pinnules with abaxial sporangia. *American Journal of Botany*, **66**, 19–25.

Good, C. W. (1981). A petrified fern sporangium from the British Carboniferous. *Paleontology*, **24**, 483–92.

Good, C. W., & Rothwell, G. W. (1988). A reinterpretation of the Paleozoic fern *Norwoodia angustum*. *Review of Palaeobotany and Palynology*, **56**, 199–204.

Hall, J. W. (1961). *Anachoropteris involuta* and its attachment to a *Tubicaulis* type of stem from the Pennsylvanian of Iowa. *American Journal of Botany*, **48**, 731–37.

Holdern, H. S. (1960). The morphology and relationships of *Rachiopteris cylindrica* Williamson. *Bulletin of the British Museum (Natural History), Geology*, **4**, 53–9.

Holmes, J. C. (1977). The Carboniferous fern *Psalixochlaena cylindrica* as found in Westphalian A coal balls from England. Part I. Structure and development of the cauline system. *Palaeontographica B*, **164**, 33–75.

Holmes, J. C. (1979). Foliar borne stems in *Anachoropteris gillotii* from the lower Westphalian of Belgium. *Canadian Journal of Botany*, **57**, 1518–27.

Holmes, J. C. (1981). The Carboniferous fern *Psalixochleana cylindrica* as found in Westphalian A coal balls from England. Part II. The frond and fertile parts. *Palaeontographica B*, **176**, 147–73.

Holmes, J. C. (1989). Anomalous branching patterns in some fossil Filicales: implications in the evolution of the megaphyll and the lateral branch, habit and growth pattern. *Plant Systematics and Evolution*, **165**, 137–58.

Jennings J. R., & Eggert, D. A. (1977). Preliminary report on permineralized *Senftenbergia* from the Chester Series of Illinois. *Review of Palaeobotany and Palynology*, **24**, 221–5.

Mamay, S. H. (1952). An epiphytic species of *Tubicaulis* Cotta. *Annals of Botany*, **62**, 145–63.

Mamay, S. H., & Andrews, H. N. (1950). A contribution to our knowledge of the anatomy of *Botryopteris*. *Bulletin of the Torrey Botanical Club*, **77**, 462–94.

Mickle, J. E. (1980). *Ankyropteris* from the Pennsylvanian of eastern Kentucky. *Botanical Gazette*, **141**, 230–43.

Millay, M. A., & Taylor, T. N. (1980). An unusual botryopterid sporangial aggregation from the Middle Pennsylvanian of North America. *American Journal of Botany*, **67**, 758–73.

Morgan, J., & Delevoryas, T. (1954). An anatomical study of a new coenopterid and its bearing on the morphology of coenopterid petioles. *American Journal of Botany*, **41**, 198–203.

Murdy, W. H., & Andrews, H. N. (1957). A study of *Botryopteris globosa* Darrah. *Bulletin of the Torrey Botanical Club*, **84**, 252–67.

Phillips, T. L. (1961). American species of *Botryopteris* from the Pennsylvanian. Unpublished Ph. D. dissertation, University of Illinois. 78 pp.

Phillips, T. L. (1966). Upper Carboniferous species of *Botryopteris*. *American Journal of Botany*, **53**, 630.

Phillips, T. L. (1974). Evolution of vegetative morphology in coenopterid ferns. *Annals of the Missouri Botanical Garden*, **61**, 427–61.

Phillips, T. L., & Rosso, S. W. (1970). Spores of *Botryopteris globosa* and *Botryopteris americana* from the Pennsylvanian. *American Journal of Botany*, **57**, 543–51.

Radforth, N. W. (1938). An analysis and comparison of the structural features of *Dactylotheca plumosa* Artis sp. and *Senftenbergia ophiodermatica* Göppert sp. *Transactions of the Royal Society of Edinburgh*, **59**, 385–96.

Renault, B. (1875). Recherches sur la végétaux silicifés d'Autun et de Saint-Etienne. Ètude du genre *Botryopteris*. *Annals de la Scientific Naturale, Botanique*, **1**, 220–40.

Rothwell, G. W. (1987). Complex Paleozoic filicaleans in the evolutionary radiation of ferns. *American Journal of Botany*, **74**, 458–61.

Rothwell, G. W. (1991). *Botryopteris forensis* (Botryopteriaceae), a trunk epiphyte of the tree fern *Psaronius*. *American Journal of Botany*, **78**, 782–88.

Smith, G. M. (1955). *Cryptogamic Botany*, 2nd ed. New York: McGraw-Hill, Vol. 2.

Stewart, W. N. (1983). *Paleobotany and the Evolution of Plants*. Cambridge: Cambridge University Press.

Taylor, T. N. (1981). *Paleobotany: An Introduction to Fossil Plant Biology*. New York: McGraw-Hill.

Trivett, M. L., & Rothwell, G. W. (1988). Modeling the growth architecture of fossil plants: a Paleozoic filicalean fern. *Evolutionary Trends in Plants*, **2**, 25–9.

20

The emergence of the modern Filicales, Salviniales, and Marsileales

Radiation of Mesozoic Filicales

The Filicales and their heterosporous derivatives the Marsileales and Salviniales are often described as leptosporangiate ferns. For paleobotanists, however, it is often difficult to distinguish between eusporangiate and leptosporangiate ferns based on the characteristics of the sporangia. Obviously, the developmental features of the two types are rarely available when studying fossil plants. As a result, most of the emphasis in characterizing ferns and distinguishing leptosporangiate from eusporangiate genera depends on mature, often dehisced sporangia. Among primitive fossil ferns – the Zygopteridaceae, for example – the presence of large fusiform sporangia with a high spore output suggests that they were eusporangiate until we see that they have an extensive annulus and have sporangial walls that appear at maturity to be one cell thick. The latter, however, is not a clean-cut distinction, for at maturity the multilayered wall of a developing eusporangium may be only one cell in thickness. Further, the annulate sporangia of the filicalean Osmundaceae show characteristics of eusporangia in their massive stalks and high spore output. With these ideas in mind, we think it is logical to interpret the sporangia of the Zygopteridaceae as a type in which the annulus has evolved, but the other characteristics of the primitive eusporangium are retained. A similar morphology has been described for microsporangia in some species of *Selaginella*, where a distinctive annulus affects dehiscence in large eusporangia (Koller & Scheckler, 1986). From this evidence it is clear that an annulus can not be interpreted as a synapomorphy that characterizes a clade incorporating both Zygopteridales and Filicales.

Osmundaceae

For reasons given above, the Osmundaceae are among the most primitive of the Filicales, and it is not surprising that they have a fossil record extending from the present into the Permian. We get our first clue of the origin of the *Osmunda*-type sporangium among the Carboniferous members of the Botryopteridaceae (Chapter 19). Recent discoveries of biseriate fronds with laminate pinnules bearing abaxial *Osmunda*-like sporangia make this origin more plausible. The only obstacle is the omega-shaped trace in the petiole of *Botryopteris*.

Based on structural features of the stem, we discover what approximates a phylogenetic "series" in evolution of *Osmunda* stems. When seen in transverse section, the sturdy rhizomes of these ferns exhibit an ectophloic dictyoxylic siphonostele (Fig. 20.1). This is essentially an ectophloic siphonostele (a siphonostele with phloem and endodermis forming a layer to the outside of the xylem) in which the xylem cylinder, but not the phloem or endodermis, is inter-

Figure 20.1. Transverse section, ectophloic dictyoxylic siphonostele of *Osmunda claytoniana*. Extant.

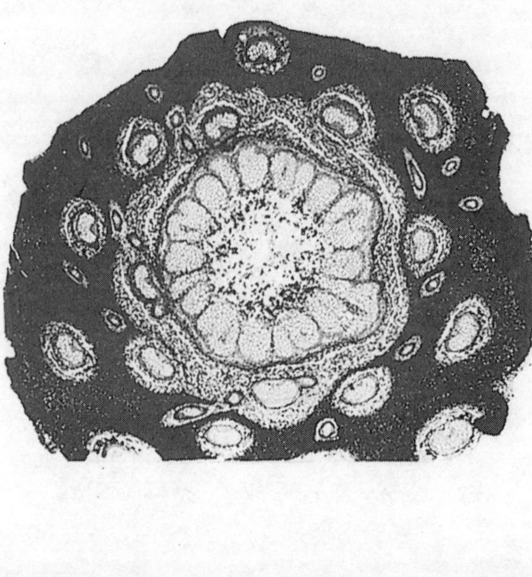

rupted by leaf gaps. Leaf traces arise from the vascular cylinder in a tight spiral so that the stem is clothed in a thick mantle of leaf bases with abundant sclerenchyma, which also occurs in the stem cortex. The leaf traces are C-shaped and adaxially directed. Diarch adventitious roots are borne on the rhizome.

Stems

We pick up the threads of *Osmunda* stelar evolution in the Permian with the stem genus *Grammatopteris* (Fig. 20.2A). In size of the protostele and spiral arrangement of leaf traces, *Grammatopteris* is similar to the Anachoropteridaceae. The leaf traces, however, are elliptical to bar-shaped without ab- or adaxial curvature. When leaf traces arise from the protostele of *G. baldaufi*, they leave an indentation in the xylem column, which has been interpreted as a rudimentary leaf gap (Arnold, 1947). In this same species, the central tracheids in the protostele are shorter than the peripheral ones. This feature is believed to foreshadow the evolution of a pith.

Thamnopteris (Fig. 20.2B) and *Zalesskya* (Fig. 20.2C) are two Permian stem genera that are considered precursors of the *Osmunda* stem type (Miller, 1971). Stems of *Thamnopteris* and *Zalesskya* may be protostelic or siphonostelic. The xylem cylinder is differentiated into an inner and outer region. In species of *Thamnopteris*, the differentiation of the metaxylem of the protostele into a central region of short tracheids and an outer zone of longer tracheids is especially apparent. Steles of *T. kidstoni* have small numbers of parenchyma cells mixed with the tracheids of the central xylem to form a mixed pith. Since there is no evidence of leaf gaps in this species, the only possible origin of the parenchyma was intrastelar by failure of cells to develop into metaxylem. The leaf traces indented the surface of the xylem cylinder at the point of their departure, but no leaf gaps were formed. The leaf traces become C-shaped with adaxial curvature as they enter the petioles.

A central pith surrounded by a ring of primary xylem showing the development of leaf gaps first appears in the Permian *Palaeosmunda* (Fig. 20.2D) (Gould, 1970). This genus is particularly interesting in the structure of the vascular cylinder. In some parts, the cylinder is essentially an ectophloic siphonostele giving off spirally arranged C-shaped leaf traces that form "incomplete gaps." These occur when a leaf trace departs from the vascular cylinder of the stem, leaving an indentation in the cylinder without actually forming a gap. Other parts of the vascular cylinder have very narrow leaf gaps of the "immediate" type (Fig. 20.3A) where the gap occurs directly above the point of leaf trace departure, and "delayed" gaps (Fig. 20.3B) where the gap occurs at some distance above the point of leaf trace departure. Thus, in one specimen, we see developmental stages in the formation of leaf gaps and the production of an ectophloic dictyoxylic siphonostele from a simple siphonostele essentially without leaf gaps. In its overall characteristics, *Palaeosmunda* is similar to the

Figure 20.3. **A.** Stereodiagram, "immediate" gap at the point of leaf trace (lt) departure. **B.** "Delayed" gap above the point of leaf trace departure. **C.** *Osmundacaulis kolbei*, Lower Cretaceous. Light stipple = mixed path. (**A,B** redrawn from Miller, 1971; **C** modified from Kidston & Gwynne-Vaughan, 1910.)

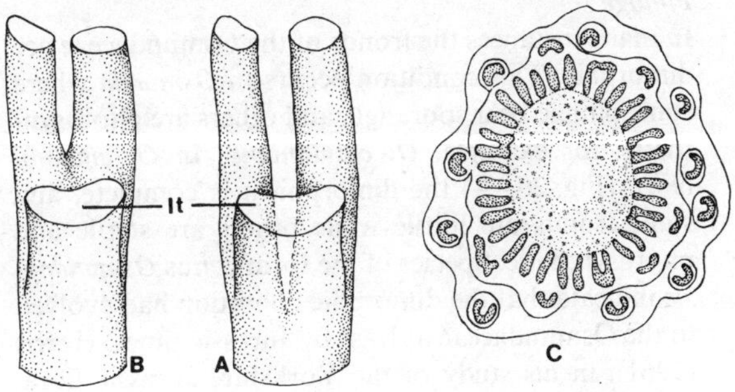

Figure 20.2. Series **A–D** depicting the evolution of the ectophloic siphonostele in the Osmundaceae. **A.** *Grammatopteris rigolloti*, Lower Permian. **B.** *Thamnopteris gwynnevaughni*, Upper Permian. **C.** *Zalesskya gracilis*, Upper Permian. **D.** *Palaeosmunda playfordii*. Dark stipple = primary xylem; light stipple = primary xylem and parenchyma; central white area = pith. (**C,D** modified from Gould, 1970.)

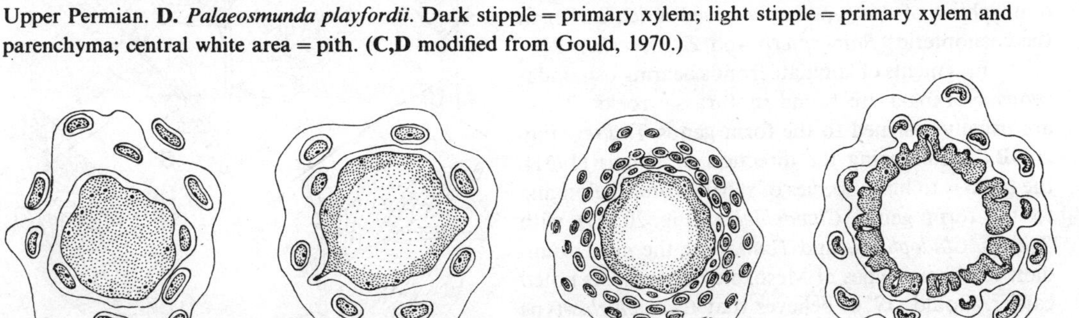

Jurassic *Osmundacaulis dunlopi* described by Kidston & Gwynne-Vaughn (1907–14) in their monumental work on the evolution of vascular systems of the Osmundaceae, based on the fossil record.

Two other species of *Osmundacaulis*, both from the Lower Cretaceous, have been selected because of their interesting and instructive characteristics. Like most species of *Osmundacaulis*, the arborescent *O. kolbei* has an ectophloic dictyoxylic siphonostele (Fig. 20.3C). The pith, however, has a considerable amount of tracheary tissue. The presence of a mixed pith is interpreted as a retention of a primitive characteristic showing how it evolved. Another of the arborescent osmundas is *O. skidegatensis*, which differs from other members of the family in having a dictyostele with immediate and delayed leaf gaps. In its dictyostele, the phloem and the endodermis are external and internal to the cylinder of xylem. This amphiphloic condition is reminiscent of the vasculature of *Psaronius* and many modern ferns. For the Osmundaceae, the amphiploic dictyostelic condition is considered derived from the ectophloic dictyoxylic siphonostele.

From the mid-Cretaceous onward, permineralized stems of the Osmundaceae have been found that have characteristics of modern *Osmunda, Todea,* and *Leptopteris*. Most of the fossil stems occur in rocks of Cenozoic age.

Foliage

In many instances the fronds of the Osmundaceae are dimorphic. This condition occurs in *Osmunda* where some pinnae bear sporangia and others are photosynthetic, for example, *O. claytoniana*. In *O. cinnamomea* (Fig. 20.4A) the dimorphism is complete, and some fronds are fertile while others are sterile and photosynthetic. Species of the form genus *Osmundopsis* indicate that the dimorphic condition had evolved in the Osmundaceae at least by Jurassic times. Harris (1961), in his study of the Yorkshire Jurassic flora, found specimens that lacked laminae in their pinnae and had pinnules with clusters of sporangia. Other specimens showing similar characteristics have been reported from the Cretaceous. It is worth noting that dimorphism of this type had evolved much earlier in the coenopterids *Botryopteris* and *Zygopteris*.

Fragments of laminate fronds bearing osmundaceous sporangia are found in Jurassic rocks. These are usually assigned to the form genus *Todites* (Fig. 20.4B,C). Following the direction of Harris (1961), the trend is to lump species of sterile fronds belonging to the form genus *Cladophlebis* (Fig. 20.4D) with *Todites*. *Cladophlebis* and *Todites* are the most abundant fern frond types of Mesozoic deposits. As noted earlier, Gould (1970) believes that *Cladophlebis*-type foliage was borne by his Permian *Palaeosmunda*. Frond fragments directly assignable to extant genera of the Osmundaceae do not appear prior to the Lower Cretaceous.

Although we have seen that the sporangia of the Carboniferous *Botryopteris* bear a striking resemblance to those of the Osmundaceae, it is not until the Jurassic that one finds sporangia that can confidently be assigned to the family. These belong to the genus *Todites* (Harris, 1961).

Osmundaceae and the origin of the siphonostele

In an earlier part of this chapter, we devoted considerable space to the description of the stelar structure of osmundaceous stems. The information from this account has a direct bearing on the answer to the question of the origin of the siphonostele in the Filicales and Marattiales. By 1914, as a result of the exhaustive studies by Kidston & Gwynne-Vaughn on stems of fossil Osmundaceae, it was demonstrated that the pith of these fern siphonosteles resulted from a phylogenetic parenchymatization of the central portion of a protostele. This has come to be called the intrastelar theory for the origin of the siphonostele. We have already seen (Chapter 11) that the siphonosteles of arborescent lycopods developed in this way.

This view of the origin of the pith was vigorously challenged by E. C. Jeffrey (1917), who claimed

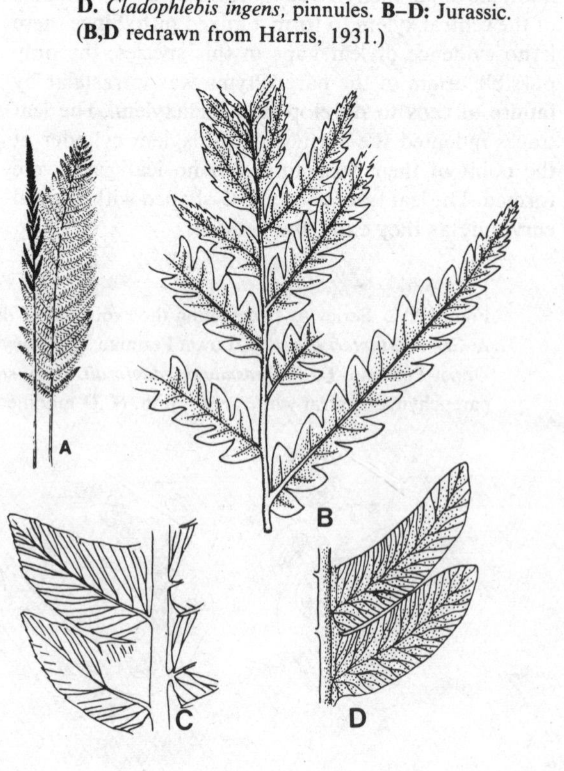

Figure 20.4. **A.** Leaf dimorphism, *Osmunda cinnamomea*. Extant. (From Takhtajan, 1956.) **B.** *Todites hartzi*, portion of frond. **C.** *T. hartzi*, pinnules. **D.** *Cladophlebis ingens*, pinnules. **B–D:** Jurassic. (**B,D** redrawn from Harris, 1931.)

that the pith had evolved by the "inrolling" of cortical parenchyma through "foliar and ramular lacunae" (leaf and branch gaps). Important support for this, the extrastelar theory, comes from the amphiphloic solenostele common among ferns – for example, the extant maidenhair fern *Adiantum* (Fig. 17.2C). The solenostele is a siphonostele with leaf gaps that do not overlap. If it is amphiphloic, then phloem and endodermis occur on inner and outer surfaces of the xylem cylinder. A particularly instructive diagram (Fig. 20.5) of the vascular system of an extant young *Gleichenia pectinata* has been prepared that shows the developmental stages of an amphiphloic solenostele.

Figure 20.5. Diagrams of vascular system of a young *Gleichenia pectinata*. Extant. Levels **a–d** of longitudinal section represented by corresponding transverse sections. **a** Protostele; **b** ectophloic siphonostele; **c** amphiphloic solenostele below a leaf gap; **d** amphiphloic solenostele at leaf gap. (Redrawn from Bower, 1935.)

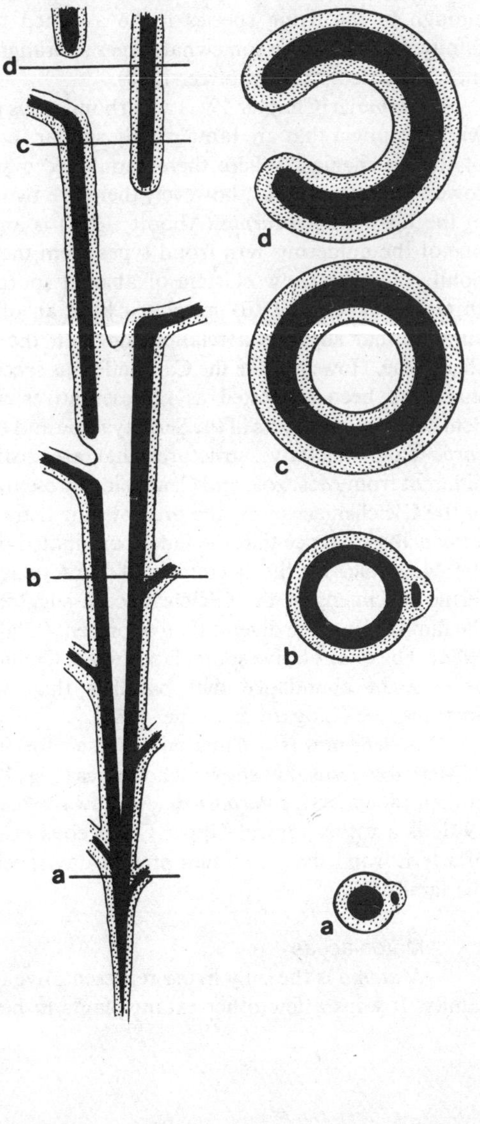

In the basal first-formed part of the plant, the vascular system is protostelic. Distally, this gives way to an ectophloic siphonostele prior to the level of the first leaf gap. The origin of the pith in this portion is clearly intrastelar until the second leaf gap is reached. Here there is evidence that internal phloem and endodermis have become continuous with these tissues to the outside of the xylem cylinder. Have these tissues, along with the cortical parenchyma, inrolled through the leaf gaps, as Jeffrey claimed?

In the fossil Osmundaceae all recent evidence points to an intrastelar origin of the pith, either prior to or concomitant with the evolution of leaf gaps. The studies of Gould (1970) and Miller (1971) are particularly supportive of Kidston & Gwynne-Vaughn in showing that phylogenetically it would have been impossible to have an extrastelar origin of the pith through leaf gaps when these did not exist as in siphonostelic species of *Thamnopteris, Zalesskya,* and other Permian members of the Osmundaceae. The presence of tracheary elements in the ectophloic dictyoxylic siphonostele of *Osmundacaulis kolbei* makes one wonder how these tracheary elements got intermingled with parenchyma cells of the pith if the origin of the latter was by invagination of cortical tissues through leaf gaps, which are present in this species. The only explanation is that the pith was intrastelar in origin and that we are seeing a reflection of a stage in its evolution.

The Lower Cretaceous position of *O. skidegatensis* with its amphiphloic dictyostele, along with abundant evidence for the intrastelar origin of the pith in this group, leaves little doubt that the amphiphloic condition evolved from more primitive Triassic and Jurassic osmundas with ectophloic dictyoxylic siphonosteles. It was Jeffrey's belief that the latter stelar type was derived from the amphiphloic siphonostele by loss of the internal phloem and endodermis. Evidence from the fossil record and ontogenetic studies have made this conclusion untenable. It is quite possible, however, that the amphiphloic dictyostele was derived from ancestors with ectophloic siphonosteles by "inrolling" of cortical tissues through leaf gaps as evidenced by the ontogeny of the shoots of some extant ferns.

Schizaeaceae

Radforth (1938) suggested that relatives of the Schizaeaceae were present in the Carboniferous. He discovered that the fern frond *Senftenbergia* bore sporangia similar to those of the Schizaeaceae. In genera belonging to this family, the annulus is a single row of cells at the distal end of the sporangium. The Lower Carboniferous *Senftenbergia* has sporangia with a distally placed annulus four to five cells deep (Fig. 20.6A). In Upper Carboniferous species, the annulus is reduced to two rows (Fig. 20.6B) and is similar in structure to sporangia of *Ankyropteris brongniartii.* It was Radforth's idea that reduction continued so that only a single terminal row of annular cells remained. This characteristic is shown by the widely distributed Jurassic genus *Klukia,* an undoubted representative of the Schizaeaceae (Harris, 1945). Jennings & Eggert (1977), however, have cautioned against considering *Senftenbergia* a Carboniferous representative of the family because the vegetative features of *Senftenbergia* are quite different from those of the Schizaeaceae. Further, there is no evidence from the fossil record to substantiate Radforth's hypothesized transition in sporangial structure. At the present time we consider *Senftenbergia* to be a member of the Tedeliaceae and similar to *Ankyropteris* in many features.

Figure 20.6. **A.** *Senftenbergia sturi,* sporangium. Lower Carboniferous. (Redrawn from Radforth, 1939.) **B.** *S. plumosa,* sporangium. Upper Carboniferous. (Redrawn from Radforth, 1938.) **C.** *Gleichenites coloradensis,* reconstruction showing branching of frond. Cretaceous. (From Andrews & Pearsall, 1941.)

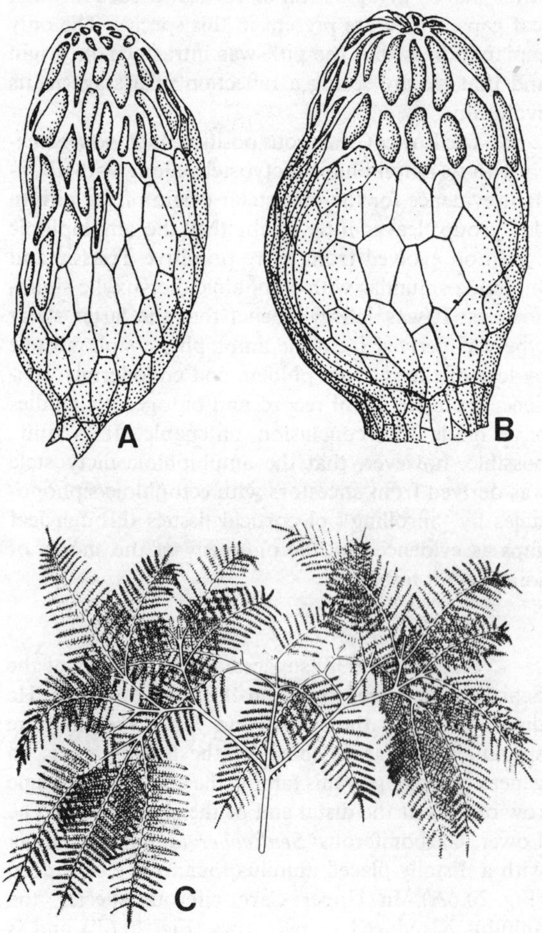

By the Jurassic, fern fronds with the characteristics of the extant Schizaeaceae *Lygodium* and *Anemia* had evolved. Beautiful compression specimens of *Anemia* (Fig. 20.7A) were found by Andrews & Pearsall (1941) in the Upper Cretaceous. These are almost identical with the living *A. adiantifolia.* The family appears to have increased in diversity into the Cretaceous (Crabtree, 1988), where recent quantitative spore data indicate that it underwent a dramatic decrease in abundance (Lidgard & Crane, 1988). *Lygodium* apparently retained a worldwide distribution into the Eocene, but today the principal distribution of the family is in tropical to subtropical environments.

Gleicheniaceae

Forking fronds with rounded pinnules like those of living members of the Gleicheniaceae (Fig. 20.6C) allow us to recognize the family in Jurassic and more recent deposits. Reports of even more ancient gleicheniaceous ferns are less convincing. Fertile specimens from the Late Jurassic onward are similar enough to the living species to be assigned to the family, but may have somewhat larger sporangia and more crowded sori (Crabtree, 1988).

Sermaya (Chapter 19) is a Carboniferous genus with sporangia that are tantalizingly similar to those of the gleichenias. Where there should be a single-rowed oblique annulus, however, there are two rows in the fossils. *Oligocarpia* (Abbott, 1954) is another one of the numerous fern frond types from the Carboniferous that show clusters of abaxial sporangia. In *O. mixta* (Fig. 19.9B) sporangia have an oblique annulus that suggests a relationship with the Gleicheniaceae. However, like the Carboniferous specimens that have been suggested as representatives of the Schizaeaceae, members of the Sermayaceae and *Oligocarpa* have vegetative structures that are distinctly different from Mesozoic and Cenozoic representatives of the Gleicheniaceae. At the present time there is no convincing evidence that the family originated before the Mesozoic. By the beginning of the Cretaceous, fertile specimens of the Gleicheniaceae suggest that the family was more diverse than at present (Crabtree, 1988). The quantitative spore data show a Cretaceous decrease in abundance that parallels that of the Schizaeaceae (Lidgard & Crane, 1988).

Gleichenites is a name widely used for fronds of Mesozoic ferns that show dichotomies (Fig. 20.6C) of their rachises. *G. coloradensis* (Andrews & Pearsall, 1941) is a rather typical Upper Cretaceous example of a fern frond that resembles present-day species of the family.

Matoniaceae

Matonia is the only living representative of the family. It joins a few other extant plants in being a "living fossil," because one of its extinct relatives, *Phlebopteris* (Fig. 20.8), was described long before the extant genus was discovered. *Matonia* has a highly restricted distribution in the Malayasian peninsula, in contrast to the worldwide distribution of its Mesozoic ancestors. The fronds belonging to genera of the Matoniaceae have a pedate arrangement of their pinnae; that is, they radiate from the tip of the petiole. Sporangia are in sori in a single row on each side of the pinnule midvein. In *Phlebopteris*, the sori lack indusia that are present in *Matonidium*. Structure of the sporangia is similar to that of the Gleicheniaceae with their oblique annuli.

Phlebopteris is a common compression–impression fossil making its first known appearence in the Triassic, while *Matonidium* has not been found prior to the Jurassic. Harris (1961) and others consider *Phlebopteris*, *Matonidium*, and *Matonia* to represent an evolutionary series in which there is increasing protection of the sporangia.

Dipteridaceae

The distribution of members of this family in time and space parallels that of the Matoniaceae, and the two are thought to be closely related. *Dipteris*, the only living genus, has fronds that are similar to those of *Matonia* except that the subdivisions of the leaf turn down rather than up. The sporangia are produced in sori with no regular pattern, and the annulus is slightly oblique.

As many as 6 genera and 60 species have been referred to fossil Dipteridaceae. The family is represented by many species in Triassic and Jurassic rocks of Europe, eastern Asia, North and South America, Australia, and Greenland. The Dipteridaceae reached their maximum abundance in the Upper Triassic–Lower Jurassic.

Dicksoniaceae

Establishing affinities with the Dicksoniaceae based solely on vegetative foliage remains is difficult at best because of the great variability presented by the leaf types of these arborescent ferns. The sorus that provides the most diagnostic characteristic is marginal to submarginal and protected by a bivalved structure where the outer valve is the reflexed lamina of the leaf margin and the inner one the indusium. The sporangia of the Dicksoniaceae, like those of the Gleicheniaceae, Matoniaceae, and Dipteridaceae, tend to have oblique annuli.

Of the fossil representatives, the genera *Coniopteris* (Fig. 20.7B) and *Dicksonia* are the best known.

Figure 20.7. **A.** *Anemia* sp., frond portion. Upper Cretaceous. (Courtesy of Dr. H. N. Andrews.) **B.** *Coniopteris weberi*, fertile and sterile pinnae. Jurassic. (Courtesy of Dr. T. Delevoryas.) **C.** *Dennstaedtiopsis aerenchymata*. Transverse section of rhizome and leaf base. Eocene. (From Taylor, 1981, with permission of Dr. R. Stockey.)

These genera appear quite suddenly in the Jurassic, with *Coniopteris* becoming widespread in the Jurassic and Lower Cretaceous (Crabtree, 1988).

Cyatheaceae

In some systems of classification, members of the Dicksoniaceae are placed in the Cyatheaceae. From the paleobotanical standpoint there is very little evidence, based on sporangium structure, suggesting a relationship between the two groups because the fossil record of known members of the family comes from casts and permineralized fragments of stems. These are recognized by their vermiform steles, which are interrupted by large leaf gaps, and characteristic leaf scars on the stems. Stems from arborescent ferns having these characteristics are *Cibotiocaulis, Cyathocaulis,* and *Protocyathea* from the late Jurassic and early Cretaceous of Japan and India. Another stem, *Cyathodendron,* was described from the Eocene.

Polypodiaceae

Of the homosporous leptosporangiate ferns, the Polypodiaceae have the largest number of derived (advanced) characteristics. It is the largest group (depending on the classification used) with the most species, and is the most widely distributed of extant ferns. Although it is customary to break the Polypodiaceae into several families (Bierhorst, 1971) based on the characteristics of extant species, paleobotanists must be content with the few generalized characteristics that have been used in the past to distinguish polypodiaceous ferns from the other major families of the Filicales. Because isolated compression–impression fossils of sterile fern foliage usually defy assignment to a natural group and because permineralized specimens are rare, the best that can be done in many cases is to look for fertile specimens in which the sporangia have a vertical annulus and transverse dehiscence.

Figure 20.8. *Phlebopteris* sp. Triassic member of the Matoniaceae. (From Bold, Alexopoulos, & Delevoryas, 1980, with the permission of Prof. R. C. Hope.)

The earliest known fossil showing these characteristic sporangia is *Aspidistes,* described by Harris (1961) from the Yorkshire Jurassic. Although Harris displayed an unusual reluctance in assigning the genus to the Polypodiaceae, the characteristics of the indusia and sporangia relate it to the polypodiaceous genus *Aspidium.* A few other leaf remains suggesting affinities with this family have been reported from the Jurassic (*Davallia*) and Cretaceous (*Asplenium* and *Onoclea*). It is not until the Tertiary, however, that the family is well represented. A few common genera that are similar to their extant counterparts are *Asplenium, Adiantum, Dennstaedtiopsis, Dryopteris, Onoclea, Polypodium, Pteridium,* and *Woodwardia.* Specimens assigned to one or more of these genera have been found in Tertiary rocks from around the world.

Beautifully preserved permineralized specimens have been obtained by Arnold & Daugherty (1963) and Basinger & Rothwell (1977) from Eocene cherts of western North America. These have been identified as the petioles of the *Acrostichum, Onoclea,* and *Aspidium* types. From at least two localities, stems assigned to *Dennstaedtiopsis* (Fig. 20.7C) have been found. These unusually fine specimens of fern rhizomes are often preserved cell for cell. The amphiphloic nature of the solenostele is clearly visible. Of particular interest is the arenchymatous tissue of the cortex and pith, which suggests that these plants grew in a marshy to semiaquatic environment. The cross-sectional configuration of the petiole trace provides the clue that the stem is dennstaedtioid. If more fossil fern material of this quality could be found, paleobotanists might be of assistance to those pteridologists who must depend on characteristics of living ferns in determining the relationships among those families of the Filicales that make their first known appearances in the Mesozoic and Tertiary.

Tempskyaceae

This family is represented by a single genus, *Tempskya* (Fig. 20.9A), from Cretaceous beds of

North America. The genus is extinct and known only from permineralized trunklike fragments of its stems. The trunks are unusual in their structure and are called false stems because the whole trunk is composed of many small ramifying axes, with amphiphloic solenosteles and leaf bases all bound together by masses of adventitious roots (Fig. 20.9B). Such false stems are found among the living ferns in *Todea barbara* (Osmundaceae) and *Hemitelia crenulata* (Cyatheaceae). The false stem is an obvious adaptation for support and elevation of leaves and fructifications. This is somewhat analogous to the supporting role of the adventitious roots of *Psaronius*.

Figure 20.9. **A.** Restoration of *Tempskya* sp. (Redrawn from Andrews, 1948.) **B.** Diagram, transverse section of false stem showing shoots and leaf traces embedded in a matrix of adventitious roots (stippled). (Redrawn from Read & Brown, 1937.) **A,B:** Cretaceous.

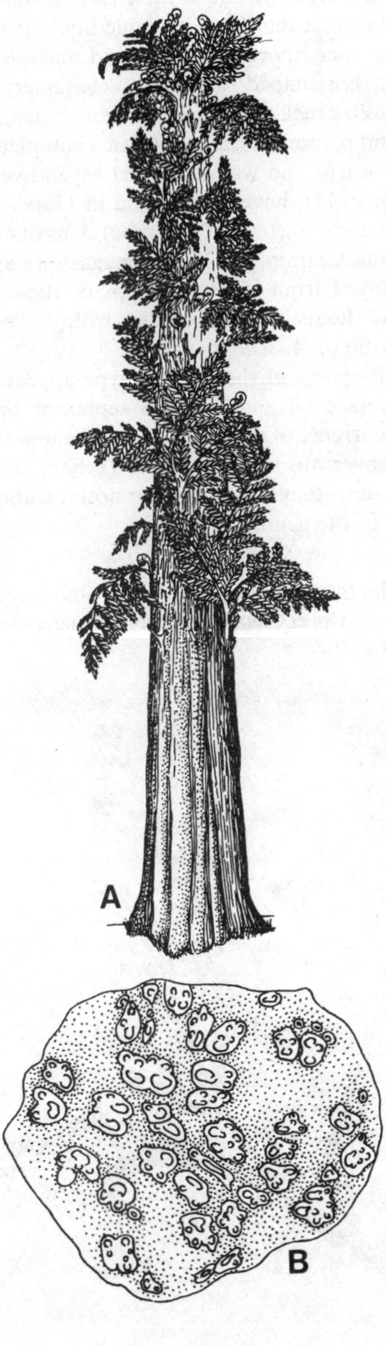

According to Andrews (1948), the trunks had a diameter of at least 40 cm and a height of 4.5 m. Again, the quality of preservation in the trunks is so superb that root hairs extending from the diarch roots can be seen. A detailed account of the structure of *Tempskya* has been prepared by Read & Brown (1937), who suggest that the affinities of this strange fern might be with the Schizaeaceae. More recently Ash & Read (1976) have prepared a detailed account of the species of *Tempskya* and their stratigraphic distribution in Cretaceous beds.

Paleoecology of filicalean ferns

The interpretation of several species of permineralized Carboniferous filicaleans as scrambling rhizomatous ground cover, as trees, as vines, and even as epiphytes has provided a window into interactions among extinct plants, and has played an important role in characterizing the community structure of extinct floras. Our understanding of compressed fossil filicaleans has recently also been enhanced by attention to paleoecology and improved collecting techniques. By careful evaluation of sedimentary features and floristic associations, fossilized ferns can sometimes be located and recovered as whole plants that remain attached to the soils in which they grew.

A good example of this occurs in Paleocene sediments of western Canada, where Rothwell & Stockey (1991) have identified whole plants of the living species *Onoclea sensibilis* rooted on the floor of a taxodiaceous marsh, along with *Equisetum* and overshadowed by *Metasequoia* trees. Using this approach in mid-Cretaceous sediments, Crabtree (1988) has identified a rhizome layer that clearly shows the interrelationships and growth habits of *Anemia fremontii* and *Gleicheniaceaphyllum falcarum*.

From palynological studies of the sediments that contained the fern rhizomes, Crabtree (1988) was also able to relate certain spores to megafossils of the ferns that produced them. Using these data, he interpreted the distribution of the various ferns in this horizon. In essence, this approach uses the relative abundance of spores of known affinities to assess the relative abundance of the plants that produced them. When fern spores (and pollen of seed plants) of known origin are evaluated in this fashion, one can interpret long-term changes in the geographic and stratigraphic abundance of plants as Lidgard & Crane (1988) have done to record the decline of the Schizaeaceae and Gleicheniaceae during the Cretaceous.

Salviniales and Marsileales

There are two orders of heterosporous aquatic ferns, the Salviniales and Marsileales. The fossil record of the latter is meager, represented by the Tertiary genus *Rodeites* (Surange, 1966), which is a sporocarp of the *Marsilea* type, containing mega- and microspores. The geological history of the Salviniales is much more impressive (Hall, 1974), with *Azollopsis, Parazolla,* and *Glomerisporites* first appearing in the Lower Cretaceous, giving way to *Azolla* and *Salvinia* in the Upper Cretaceous and Tertiary (Hall, 1969; Jain, 1971; Collinson, 1980). These genera are represented in fossiliferous beds by their mega- and microsporangiate reproductive parts. In living representatives of these genera, these structures are borne within sporocarps on the floating plants. The sporocarp may contain micro- or mega-sporangia. Within a microsporangium, the microspores are embedded in a foamy mass that forms three or more massulae or "floats." Each massula develops hairlike extensions (glochidia) from the surface that have anchor-shaped processes at the tips. When liberated into the water, the massulae with their glochidia are attached by chance to the megasporangiate apparatus (Fig. 20.10A). The latter consists of a functional megaspore and three apical massulae. In the Cretaceous genera, these microfossils usually have many massulae associated with the megaspore (Fig. 20.10B). By late Cretaceous and early Tertiary, the number of massulae seems to have been reduced and stabilized at 9. There are, however, many accounts of exceptions where 1 or 3 floats have been reported for Cretaceous azollas and multiple floats (18) for one Eocene species (Collinson, 1980). After reviewing some 60 species of fossil *Azolla* mega- and microspore reproductive structures thus far reported, Collinson is of the opinion that there were several "lines" to be recognized; those with 1, 3, 9, and those with more than 9 floats. This hypothesis is supported by Martin (1976), who demonstrated that different development pathways lead to the production of 3, 9, 15, and 24 floats. The fossil record indicates that the different lines have been most numerous at different times in the past (Fowler, 1975). Those with 15 or more floats were most numerous through the Paleocene; those with 9 being most diverse during the Oligocene and Miocene, and those with three being most numerous today. The microspore-containing massulae of these early forms usually have glochidia that have a simple hook at the tip. The 3-float megaspore apparatus and massulae with simple anchor-shaped glochidia, characteristic of modern azollas, make their appearance in the Oligocene.

Compression fossils of *Azolla,* complete with their sporocarps and well-preserved vegetative structures (Fig. 20.11), have been found in Upper Cretaceous and early Tertiary (Paleocene) deposits (Sweet & Chandrasekharam, 1973). The megaspore apparatuses removed from the sporocarps of these plants had many floats and glochidia with coiled tips characteristic of *A. schopfii.*

Megaspores of the *Salvinia* type appear in the Upper Cretaceous and seem to represent the first known occurrence of the genus. Compressions of their floating leaves are widespread in Tertiary deposits. Unfortunately, they tell us little or nothing about the evolution of the genus.

Figure 20.10. **A.** *Azolla filiculoides.* Optical section of megaspore (ms) with "floats" (f) above and attached massula (m) containing microspores. Extant. **B.** *Azolla schopfii,* megaspore complex with attached massula. Upper Cretaceous. (Drawn from Sweet & Chandrasekharam, 1973.)

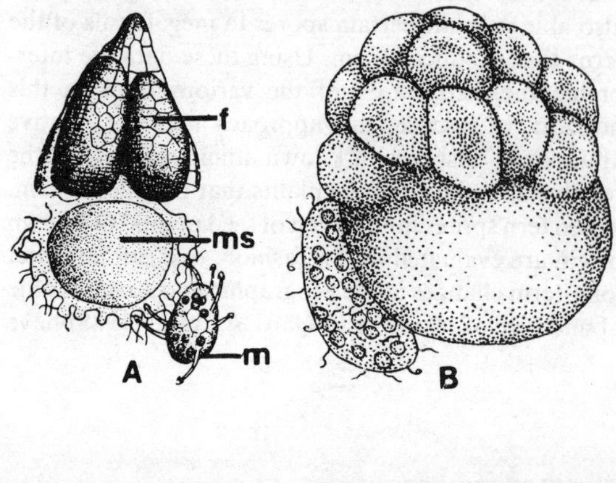

Figure 20.11. Compression vegetative portion, *Azolla schopfii.* Upper Cretaceous. (From Chandrasekharam, 1974.)

Thus far, we have considered the evolution of plant organs such as stems, leaves, and sporangia. It is obvious that *Azolla* provides us with an excellent example of evolution of their microscopic spores. The science of palynology hinges on the fact that spores and pollen grains are the product of evolution and have distinctive characteristics in size, symmetry, structure of wall layers, and ornamentation of the surface. As in the case of *Azolla* megaspores with their massulae, spores or pollen grain types can be recognized as occurring within certain geological boundaries and thus can be used to identify stratigraphic units. Thus, they serve as index fossils. Palynologists depend on assemblages of microfossils occurring within the stratigraphic units when making correlations with another unit or trying to define its limits. This assemblage may contain small parts of plants and animals as well as spores and/or pollen grain.

Summary of evolution and relationships: Filicopsida

The evolution of fern organs

When we prepare an overview of evolution within the class Filicopsida (Chart 20.1), we can see

Chart 20.1. Suggested origin and relationships of major groups of the Filicopsida and their distribution in geological time. Rhy. = Rhyniopsida; Trim. = Trimerophytopsida; Clad. = Cladoxylales; Ted. = Tedeleaceae; Ser. = Sermayaceae; Anach. = Anachoropteridaceae; Botry. = Botryopteridaceae; Psal. = Psalixochlaenaceae; Ophio. = Ophioglossales; Sal. = Salviniales; Mar. = Marsileales.

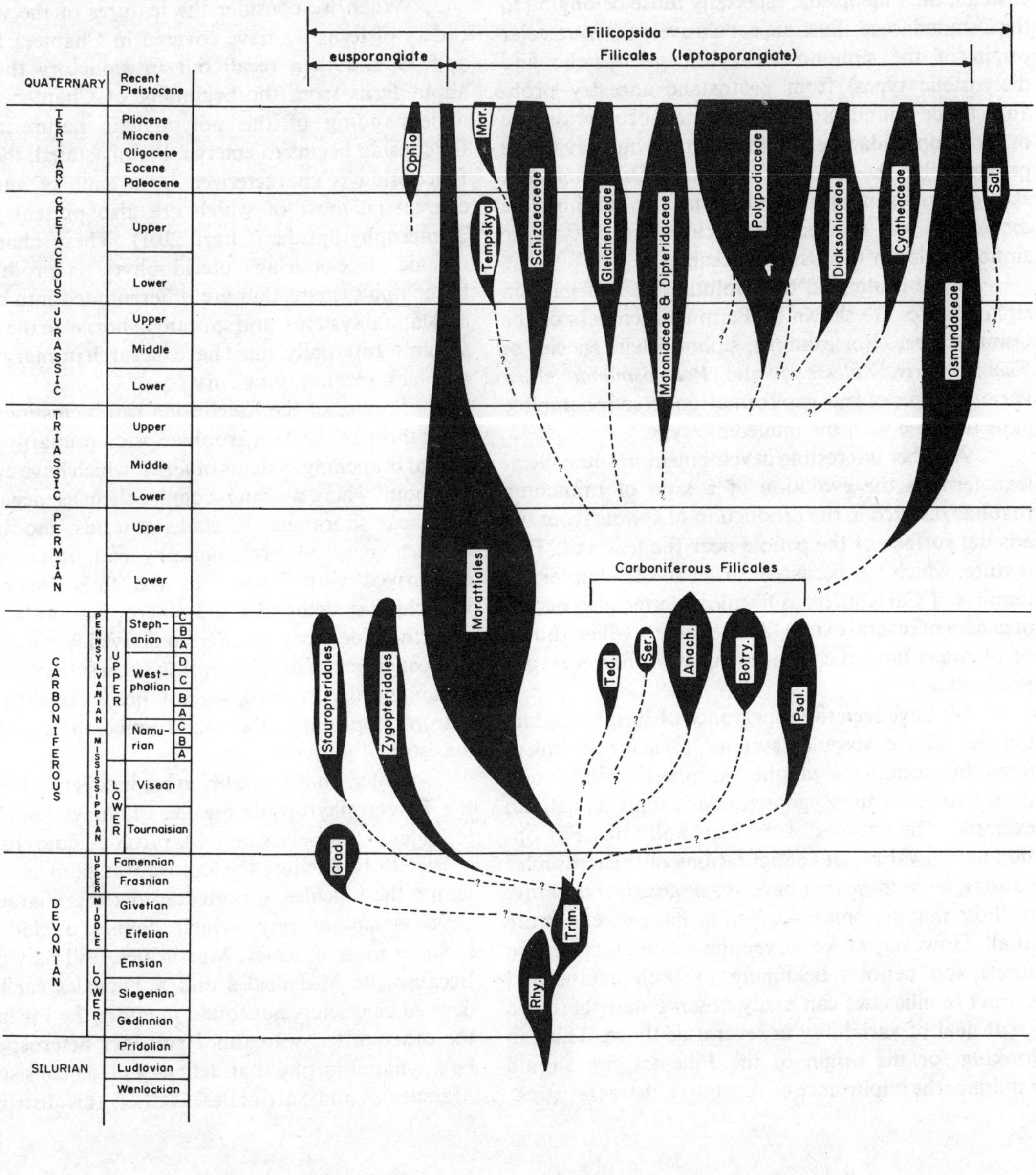

that the ferns, indeed, are a polyphyletic group of free-sporing pteridophytes with megaphylls and foliar-borne sporangia. Nevertheless, they show certain broad trends with respect to the appearance of those vegetative and reproductive structures that characterize ferns. Although we now know that these trends do not necessarily indicate phylogenetic relationships, they do give us insights into the evolution of the structural features that characterize modern forms. The derivation of the biseriate frond, for example, is foreshadowed in *Pseudosporochnus* of the Cladoxylales, *Rhacophyton* of the Zygopteridales, and Lower Carboniferous species of *Botryopteris* (*B. antiqua*). They provide us with unequivocal evidence that the fern frond (a megaphyll) evolved from the primitive three-dimensional branching systems so characteristic of trimerophytes and rhyniophytes.

Excluding the Cladoxylales that have evolved eusteles, the Filicopsida, especially those belonging to the Osmundaceae, illustrate conclusively the intrastelar origin of the siphonostele (both solenostelic and dictyostelic types) from protostelic ancestry probably to be found among the Anachoropteridaceae or Botryopteridaceae. In the same group, evidence provided by the Lower Cretaceous *Osmundacaulis skidegatensis* supports the idea that the amphiphloic dictyostele was derived from ectophloic dictyoxylic siphonosteles in the Osmundaceae.

Steps leading to the evolution of leaf gaps in siphonosteles are shown by Permian members of the Osmundaceae. For example, siphonostelic species of *Thamnopteris, Zalesskya,* and *Palaeosmunda* show various stages of leaf-gap formation from incomplete gaps to those with the immediate type.

Another interesting development in some filicalean ferns is the evolution of a kind of branching that has resulted in the production of shoots from the adaxial surface of the petiole near the leaf axil. This feature, which we discussed earlier in the chapter for families of Carboniferous filicalean ferns, also occurs in genera of several extant families of ferns. The shoots or plantlets have the obvious function of vegetative production.

We have seen the appearance of certain characteristics of the vascular systems of some primitive ferns that seem to be unique. Peripheral loops found in the fronds of the Zygopteridales supply us with an example. The petioles of the Carboniferous Filicopsida have a variety of configurations of their vascular strands, from those that have ab- or adaxial curvature to those that are omega-shaped or have no curvature at all. However, as we survey the vasculature of fern stems and petioles belonging to both extant and extinct families, we can easily observe that there is a great deal of variability in several of those. Thus, in looking for the origin of the Filicales, we should minimize the importance of vegetative characteristics.

With respect to sporangia, we find that the Marattiales are characterized by more or less fused clusters (synangia) of eusporangia from their first appearance in the fossil record to the present. The zygopteridaleans show a transition in their large, putative eusporangia, from the exannulate type of the Devonian *Rhacophyton* that is similar to that of primitive trimerophytes to the annulate forms that characterize Carboniferous and Lower Permian species. In contrast, both fossil and living filicaleans are characterized by sporangia with an annulus. However, there is a range of sporangial development, from clearly eusporangiate in the Osmundaceae to strictly leptosporangiate (that is, the entire sporangium develops from a single cell) in some of the more highly derived polypodiaceous lineages.

Fern phylogeny

When we consider the features of the various orders of ferns we have covered in Chapters 17, 18, and 19 and then recall our introductory thoughts about ferns from the beginning of Chapter 17, an understanding of the polyphyletic nature of the Filicopsida begins to emerge. Plainly stated, the class Filicopsida is characterized by a suite of ancestral characters, most of which are also present in the Trimerophytopsida (Chart 20.1). These characters include free-sporing pteridophytic reproduction, branching systems that are differentiated into central and lateral systems, and sporangia borne on the lateral systems. Internally, most have mesarch primary xylem and lack secondary xylem.

Species of the Filicopsida can be distinguished from those of the Trimerophytopsida primarily by the lateral branching systems of ferns, which have evolved to a point where we can recognize them as megaphylls that bear sporangia. In cladistic terms, shoots with distinct stem/leaf organography can be considered as derived with respect to the three-dimensional branching systems of the Trimerophytopsida. However, the wide array of different derived features found in most orders of the Filicopsida suggest that stem/leaf organography evolved several times from the Trimerophytopsida, rather than once in a common ancestor of all ferns.

Orders that probably arose independently from the Trimerophytopsida are the Cladoxylales, Zygopteridales, Ophioglassales, Marattiales, and Filicales (Chart 20.1). Because the leptosporangium arose only within the Filicales, we can consider this character to be a synapomorphy, which defines a clade that includes most Filicales, Marsileales, and Salvineales. Because the Marsileales and Salvineales each have derived characters not found in either the Filicales or the other order, we cannot consider heterospory to be a synapomorphy that defines a clade consisting of Marsileales and Salviniales. Rather, this distribution of characters suggests that heterospory evolved independently within the filicalean ancestors of each order. It is intriguing to consider how the results of a formal cladistic analysis of the ferns would compare to the interpretations of relationships that we have presented here.

Among the Filicales there appear to have been three major evolutionary radiations and a significant turnover in systematic diversity through time (Rothwell, 1987). From the fossil evidence we have considered in this chapter, the Filicales appear to have originated at the base of the Lower Carboniferous and have undergone an evolutionary radiation to establish several families including the Tedeleaceae, Botryopteridaceae, Anachoropteridaceae, Sermayaceae, and Psalixochlaenaceae. These families all became extinct by the Lower Permian. In the second major filicalean evolutionary radiation, the Carboniferous families were replaced with the most primitive filicalean families that have living representatives. These appear in the Permian, Triassic, and Jurassic, and include the Osmundaceae, Schizeaceae, Matoniaceae, Dipteridaceae, Dicksoniaceae, and Cyatheaceae.

The third major filicalean radiation was among the ferns that we place in the large, polyphyletic family Polypodiaceae. The first evidence for polypodiaceous ferns is from Jurassic sediments, but their evolutionary radiation apparently did not begin until the Upper Cretaceous after flowering plants had become dominant over much of the land surface.

References

Abbott, M. L. (1954). Revision of the Paleozoic fern genus *Oligocarpia*. *Palaeontographica B*, **96**, 39–65.

Andrews, H. N. (1948). Fossil tree ferns of Idaho. *Archaeology*, **1**, 190–5.

Andrews, H. N., & Pearsall, C. S. (1941). On the flora of the Frontier Formation of southwest Wyoming. *Annals of the Missouri Botanical Garden*, **28**, 165–92.

Arnold, C. A. (1947). *An Introduction to Paleobotany*. New York: McGraw-Hill.

Arnold, C. A., & Daugherty, L. H. (1963). The fern genus *Arcrostichum* in the Eocene Clarno Formation of Oregon. *Contributions from the Museum of Paleontology. University of Michigan*, **18**, 205–27.

Ash, S. R., & Read, C. B. (1976). North American species of *Tempskya* and their stratigraphic significance. U.S. *Geological Survey, Professional Paper*, **874**, 1–42.

Basinger, J. F., & Rothwell, G. W. (1977). Anatomically preserved plants from the Middle Eocene (Allenby Formation) of British Columbia. *Canadian Journal of Botany*, **55**, 1984–90.

Bierhorst, D. W. (1971). *Morphology of Vascular Plants*. New York: Macmillan.

Bold, H. J., Alexopoulos, C. J., & Delevoryas, T. (1980). *Morphology of Plants and Fungi*. 4th ed. New York: Harper & Row.

Bower, F. O. (1935). *Primitive Land Plants*. London: Macmillan. Reprint. New York: Macmillan (Hafner Press), 1959.

Chandrasekharam, A. (1974). Megafossil flora from the Genesee locality, Alberta, Canada, *Palaeontographica B*, **147**, 1–41.

Collinson, M. E. (1980). A new multiple-floated *Azolla* from the Eocene of Britain with a brief review of the genus, *Palaeontology*, **23**, 213–29.

Crabtree, D. R. (1988). Mid-Cretaceous ferns *in situ* from the Albino Member of the Mowry Shale, southwestern Montana. Palaeontographica B, **209**, 1–27.

Crane, P. R., & Lidgard, S. (1990). Angiosperm radiation and patterns of Cretaceous palynological diversity. In *Major Evolutionary Radiations*, eds. P. D. Tahlor & G. P. Larwood. Oxford: Clarendon Press.

Fowler, K. (1975). Megaspores and massulae of *Azolla prisca* from the Oligocene of the Isle of Wight. *Palaeontology*, **18**, 483–507.

Gould, R. E. (1970). *Palaeosmunda*, a new genus of siphonostelic osmundaceous trunks from the Upper Permian of Queensland. *Palaeontology*, **13**, 10–28.

Hall, J. W. (1969). Studies on fossil *Azolla:* primitive types of megaspores and massulae from the Cretaceous. *American Journal of Botany*, **56**, 1173–80.

Hall, J. W. (1974). Cretaceous Salviniaceae. *Annals of the Missouri Botanical Garden*, **61**, 354–67.

Harris, T. M. (1931). The fossil flora of Scoresby Sound East Greenland, Part I: Cryptograms (exclusive of Lycopodiales). *Meddeleser om Grønland*, **85**, 1–100.

Harris, T. M. (1945). Notes on the Jurassic flora of Yorkshire. 19. *Klukia exilis* (Phillips) Raciborski. *Annals of the Magazine of Natural History*, **12**, 257–65.

Harris, T. M. (1961). The Yorkshire Jurassic flora, I. *Thallophytes–Pteridophytes*. London: British Museum (Natural History).

Jain, R. K. (1971). Pre-Tertiary records of the Salviniaceae. *American Journal of Botany*, **58**, 487–96.

Jeffrey, E. C. (1917). *The Anatomy of Woody Plants*. Chicago: University of Chicago Press.

Jennings, J. R., & Eggert, D. A. (1977). Preliminary report on permineralized *Senftenbergia* from the Chester Series of Illinois. *Review of Palaeobotany and Palynology*, **24**, 221–5.

Kidston, R., & Gwynne-Vaughn, D. T. (1907–14). On the fossil Osmundaceae, Parts I–V. *Transactions of the Royal Society of Edinburgh*, **45**, 759–80; **46**, 213–32, 651–67; **47**, 455–77; **50**, 469–80.

Koller, A. L. & Scheckler, S. E. (1986). Variations in microsporangia and microspore dispersal in *Selaginella*. *American Journal of Botany*, **73**, 1274–88.

Lidgard, S., & Crane, P. R. (1988). Quantitative analyses of the early angiosperm radiation. *Nature*, **331**, 344–346.

Martin, A. R. H. (1976). Some structures in *Azolla* megaspores, and an anomalous form. *Review of Palaeobotany*, **21**, 141–69.

Miller, C. N. (1971). Evolution of the fern family Osmundaceae based on anatomical studies. *Contributions from the Museum of Paleontology, The University of Michigan*, **23**, 105–69.

Radforth, N. W. (1938). An analysis and comparison of the structural features of *Dactylotheca plumosa* Artis sp. and *Senftenbergia ophiodermatica* Göeppert sp.

Transactions of the Royal Society of Edinburgh, **59,** 385–96.

Radforth, N. W. (1939). Further contributions to our knowledge of the fossil Schizaeaceae; genus *Senftenbergia. Transactions of the Royal Society of Edinburgh*, **59,** 745–61.

Read, C. B., & Brown, R. W. (1937). American Cretaceous ferns of the genus *Tempskya. United States Geological Survey, Professional paper*, **186-F,** 105–29.

Rothwell, G. W. (1987). Complex Paleozoic Filicales in the evolutionary radiation of ferns. *American Journal of Botany*, **74,** 458–61.

Rothwell, G. W., & Stockey R. A. (1991). *Onoclea sensibilis* in the Paleocene of North America, a dramatic example of structural and ecological stasis. *Review of Palaeobotany and Palynology*, **709,** 113–24.

Surange, K. R. (1966). *Indian Fossil Pteriodophytes*. New Delhi: Botanical Monographs No. 4.

Sweet, A. R., & Chandrasekharam, A. (1973). Vegetative remains of *Azolla schopfii* Dijkstra from Genesee, Alberta. *Canadian Journal of Botany*, **51,** 1491.

Takhtajan, A. L. (1956). *Telomophyta*. Moscow: Academiae Scientiarum.

Taylor, T. N. (1981). *Paleobotany: An Introduction to Fossil Plant Biology*. New York: McGraw-Hill.

21

Free-sporing plants with gymnospermous secondary wood

Certain paleobotanical discoveries have assumed special significance in deciphering the relationships among vascular plants. One of these was the discovery by Beck (1960) of the connection between *Archaeopteris* and *Callixylon* that led to the establishment of the class Progymnospermopsida. As we shall see, this has had a profound effect on our interpretation of gymnosperm classification (Stewart, 1981).

To better understand the importance of Beck's discovery, we have to have some knowledge of the characteristics of *Archaeopteris* and *Callixylon* (Beck, 1981; Beck & Wight, 1988). Prior to 1960, *Archaeopteris* (Fig. 21.1) was believed to be the foliage of a late Devonian fernlike plant with large compound leaves. The idea that *Archaeopteris* was indeed a fern received support when it was discovered that it was free-sporing (Arnold, 1939). This information seemed to fit well with the interpretation that *Archaeopteris* was similar in structure to the bipinnate frond of a modern fern with opposite or subopposite fertile and sterile pinnae

Figure 21.1. *Archaeopteris macilenta,* portion of leafy shoot. Note opposite arrangement of branches on shoot. Upper Devonian. (From Beck, 1970.)

Figure 21.2. **A.** *Archaeopteris halliana*, sporangia attached along upper surface of dichotomously divided fertile leaves. **B.** *Archaeopteris* sp., single sporangium showing longitudinal dehiscence. **A,B:** Upper Devonian. (**A,B** from Phillips, Andrews, & Gensel, 1972.)

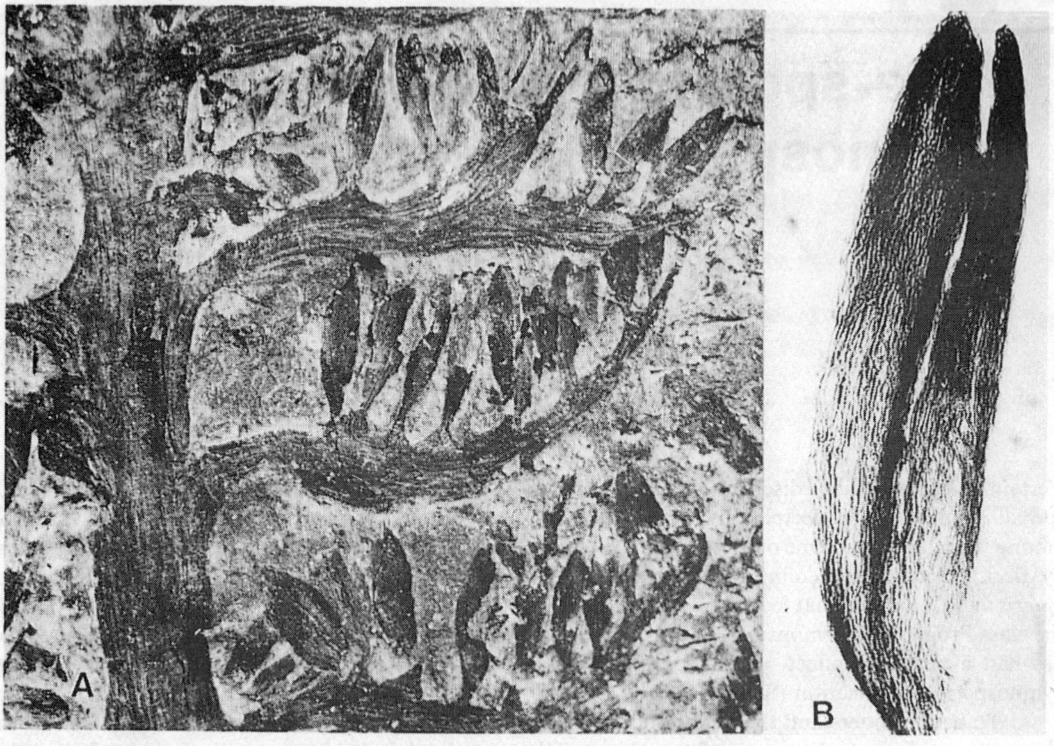

Figure 21.3. **A.** Transverse section, small branch of *Callixylon newberryi*. Note pycnoxylic secondary wood. **B.** *C. newberryi* showing a single sympodium with mesarch primary xylem. **A,B:** Upper Devonian. (**A,B** from Beck, 1970.)

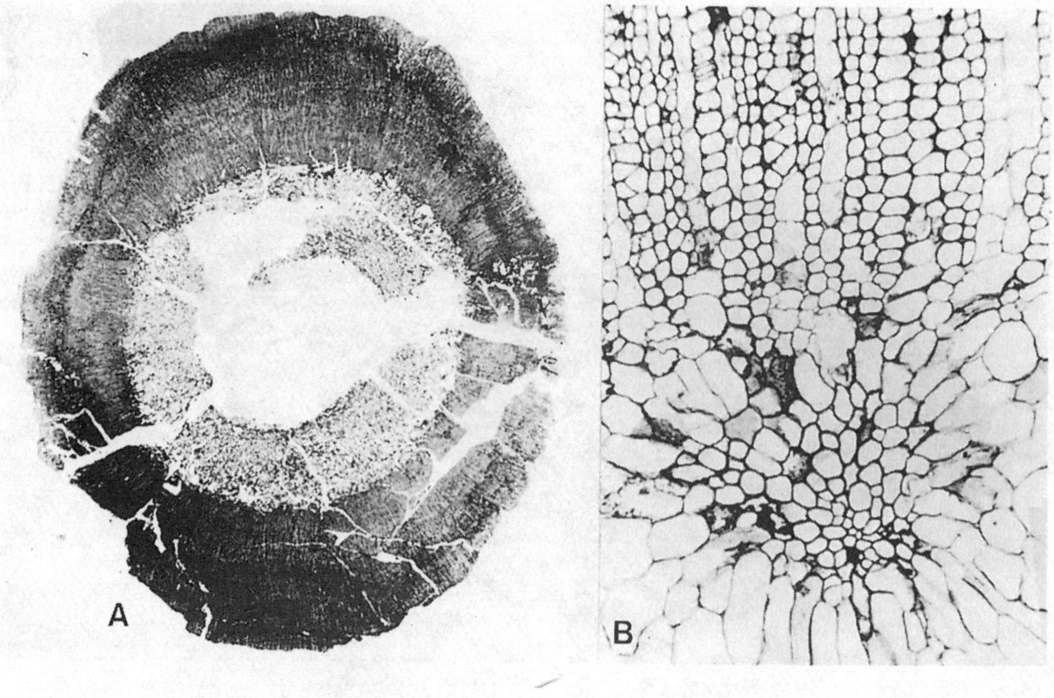

borne on a rachis. The pinnae in turn bore sterile and fertile pinnules (Fig. 21.2A). The sterile pinnules were planated and, depending on the species, were more or less webbed. The fertile pinnules bore one or two rows of adaxial fusiform sporangia with longitudinal dehiscence (Fig. 21.2B).

Callixylon also is a late Devonian genus often occurring in the same beds as *Archaeopteris*. Permineralized specimens of *Callixylon* range in size from fragments of small twigs and branches to trunks with a diameter of 1.5 m and a length up to 8.4 m. In transverse section (Fig. 21.3A), many of these specimens show a conspicuous pith of large thin-walled cells. At the periphery of the pith and in contact with the secondary xylem is a ring of mesarch primary xylem strands (Fig. 21.3B). The strands are constructed of spiral, reticulate, and pitted tracheids. The secondary xylem is compact, and is composed of radially arranged tracheids and narrow rays. Ray tracheids of the type found in the wood of many conifers also are present (Fig. 21.4A). Wood of this general structure is called pycnoxylic and is characteristic of modern conifers.

Figure 21.4. **A.** Highly magnified transverse section secondary wood of *Callixylon newberryi* showing ray tracheids (rt). **B.** Radial section, secondary wood of *C. newberryi* showing grouped pitting. **A,B:** Upper Devonian. (**A,B** from Beck, 1970.)

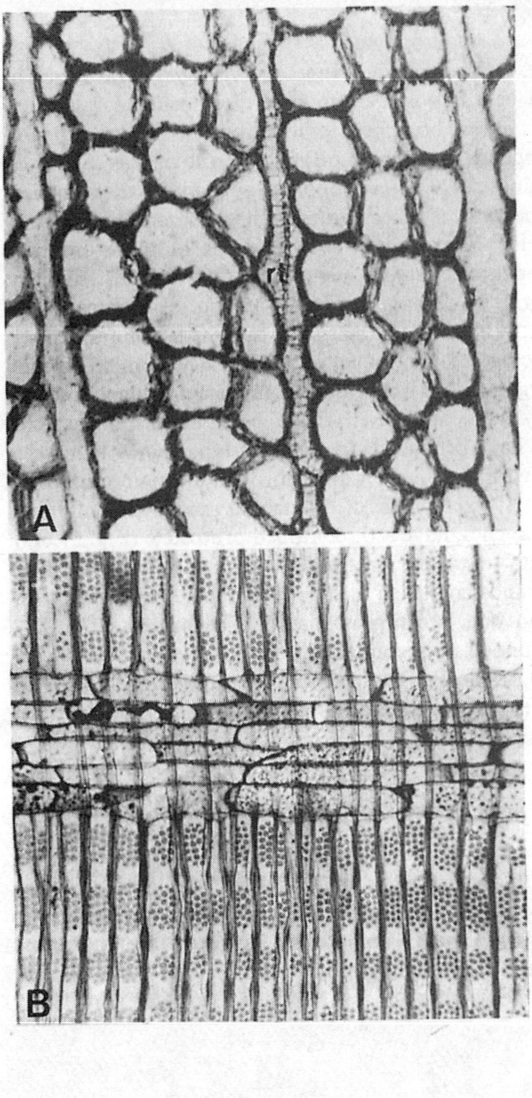

When observed in radial section (Fig. 21.4B), the tracheids of *Callixylon* wood (Beck, Coy, & Schmid, 1982) show a unique type of pitting, where the circular bordered pits occur in groups on the radial walls. The groups of 6 to 20 pits are arranged in two or three vertical rows; they are aligned horizontally and are separated by unpitted tracheid walls. Because of its coniferlike wood, *Callixylon* was widely assumed to have been an early member of the conifer alliance.

The specimen discovered by Beck and used to prove the connection between the fernlike *Archaeopteris* and the coniferlike *Callixylon* came from Upper Devonian beds in New York State. The specimen is 80 cm long and comprises compressed fertile and sterile foliage of the *Archaeopteris macilenta* type attached to partially permineralized stem and rachis axes. When examining sections of the permineralized axes, Beck made the exciting and almost unbelievable observation that these were axes of *Callixylon* showing all of the characteristics described above. The details of preservation were good enough to make it possible for him to assign the specimen to *C. zalesskyi*. The significance of his discovery was immediately apparent to Beck. He had found leaves of a free-sporing "fern" attached to a stem with gymnosperm characteristics, a situation previously unknown to occur among vascular plants! This set the stage for the development of the progymnosperm concept, stated by Beck (1962) as "plants having gymnospermic secondary wood and pteridophytic reproduction." Although the discovery of the specimen is another fine example of paleobotanical serendipity, the importance of the specimen would have gone unnoticed if it had not been examined by a paleobotanically oriented scientist.

Correctly using nomenclatural priority, the earlier name *Archaeopteris,* first used by Dawson in 1871, was selected by Beck for the whole plant. Based on the size of the trunk fragments and leaf remains of *Archaeopteris,* the plant was reconstructed (Beck, 1962) as a tree (Fig. 21.5) up to 18 m tall with a crown of spirally arranged bipinnate fronds.

Aneurophytales

After the reconstruction of *Archaeopteris* and the resulting establishment of the Progymnospermopsida, numerous Devonian plant fossils were assigned to the class and placed in one of three different orders: the Archaeopteridales, Aneurophytales, and

Protopityales (Beck & Wight, 1988). Those progymnosperms appearing first in the fossil record are Middle Devonian (Eifelian) in age and belong to the Aneurophytales. If we accept the suggestion that the Upper Devonian *Eospermatopteris* (Fig. 21.6A) and *Aneurophyton* are congeneric, and if the reconstruction of *Eospermatopteris* can be verified, then we can visualize a fernlike plant as high as 12 m with a crown of spirally arranged fronds up to 2.7 m long. The basal portion of the unbranched trunk formed a bulbous root-bearing base. Numerous such trunklike bases have been found in quarries near the town of Gilboa, New York.

Problems have been encountered by paleobotanists in their efforts to properly designate the orders of branches of the Aneurophytales. The problems arise because the fossils of their branches are fragmentary and not found attached to a structure that confidently can be called the main axis or "stem." At present, there is no known basic difference in the morphology of the branches comprising the branching system of aneurophytes that can be used to distinguish cauline from foliar organs. Of the various systems used to designate the orders of branching, that of Scheckler & Banks (1971a) seems best to reflect the morphological realities shown by internal structure and external form of the specimens that have been investigated. The designation starts (Fig. 21.8A) with the most proximal order of branching (usually of the greatest magnitude) and proceeds distally; the most proximal branch is a "first-order" branch, the next in succession distally is a "second-order" branch, and so on. The most distal order of branching usually bears ultimate vegetative appendages that, depending on the genus, may be bifurcate or pinnate, terete or planated. In those genera where the position of fertile organs on branching systems has been determined, it is clear that they are not the homologs of ultimate vegetative appendages, but replace branches of the most distal order. The vascular traces supplying both fertile organs and ultimate appendages are small terete strands, while the vascular systems of the various orders of branches are ribbed protosteles that may be immersed in secondary xylem.

Figure 21.5. Reconstruction of *Archaeopteris* sp. A tree about 4 m high. Upper Devonian. (Based on reconstruction by Beck, 1962.)

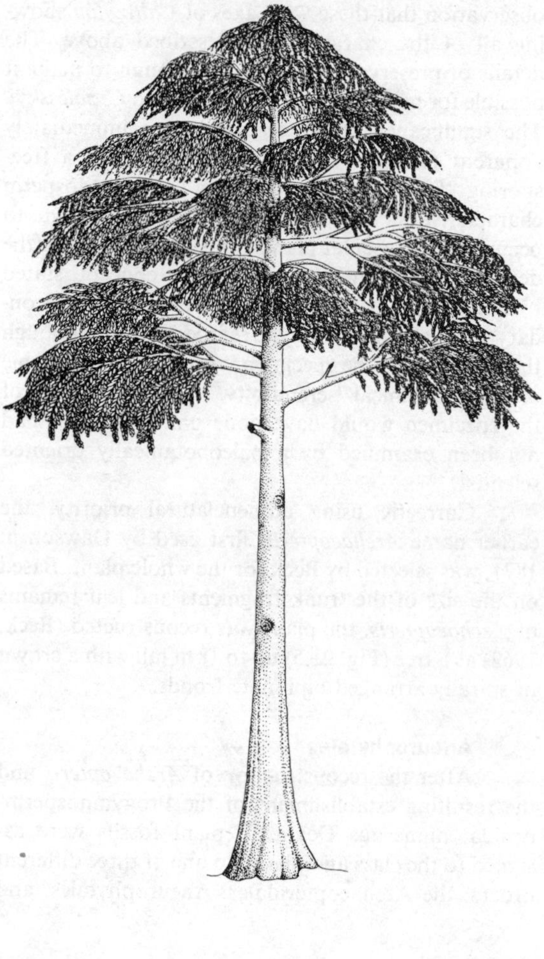

Aneurophyton

The branch systems of *Aneurophyton* are known to be composed of at least three orders of spirally arranged branches terminating in sterile ultimate appendages (Fig. 21.6B) or in fertile organs (Serlin & Banks, 1978). Sterile ultimate appendages dichotomize one to three times to form a three-dimensional telome truss that is unwebbed. The tips of these ultimate appendages are frequently slightly recurved. The fusiform to oblong sporangia are borne terminally on short branches and arise in two rows along the inner facing margins of an incurved, dichotomized fertile organ (Fig. 21.6C). Based on present evidence, *Aneurophyton* appears to have been homosporous and to have produced *Aneurospora*-type spores that were shed from the sporangia through a longitudinal slit.

All orders of branches, except the ultimate branches, have a three-ribbed protostele (Fig. 21.6D) with mesarch protoxylem strands. One of these strands is located in each of the three protostele ribs, and one is centrally placed. The protostele is surrounded by secondary xylem composed of tracheids and ray cells. The tracheids have one to five rows of elliptical bordered pits on their walls. The rays are uniseriate.

Rellimia

There has been a notable lack of agreement about the correct name for *Rellimia* (*Protopteridium*) with convincing arguments that *Protopteridium* is the proper name for the taxon (Matten & Schweitzer, 1982). Equally convincing, however, are the arguments of Bonamo (1983) that *Rellimia* is the correct name. We have chosen to follow Beck & Wight (1988) in using the name *Rellimia* for this primitive progymnosperm, which frequently is found in the Middle Devonian (Givetian) beds containing *Aneurophyton* in Europe and the USSR. The occurrence of *Rellimia* in the North American Middle Devonian, however, is rare (Bonamo, 1977). The genus shares many characteristics with *Aneurophyton*, the most important of these being gymnospermous secondary wood and free-sporing sporangia, the two essential characteristics of the Progymnospermopsida.

As characterized by Leclercq & Bonamo (1971), *Rellimia* probably was a bushy plant of fairly large size. All portions of the branching system of *Rellimia* show frequent dichotomies (Fig. 21.7A) that give rise to branches of at least three orders. Branches of all orders, including the ultimate appendages, are spirally arranged. The ultimate vegetative appendages are usually found on smaller branches of the highest numerical order, while fertile organs occur on larger branches of lower numerical order. Here they may be so abundant that they almost completely cover the axes that bear them. The sterile appendages dichotomize two to three times and may have a narrow lamina. There is no indication that the tips were recurved as in *Aneurophyton*. Each fertile organ dichotomizes once and then is branched pinnately two or three times (Fig. 21.7B). Clusters of fusiform sporangia terminate the pinnate branches. The whole fertile appendage tends to be adaxially recurved.

The primary xylem of *Rellimia* in all orders of branching, except ultimate branches, is a three-ribbed protostele (Fig. 21.7C) with mesarch development. The protoxylem strands occur near the tip of each of the radiating arms of the protostele and extend centripetally where some strands occupy a central position.

The secondary xylem, as characterized by Leclercq & Bonamo (1971), consists of tracheids and narrow rays. The tracheids have circular bordered pits on all walls. The extent of development of secondary

Figure 21.6. **A.** Reconstruction of *Eospermatopteris* sp. Upper Devonian. (After Goldring, in Andrews, 1961.) **B.** *Aneurophyton germanicum*, vegetative lateral branch system. Middle Devonian. (Redrawn from Arnold, 1947.) **C.** *A. germanicum*, lyre-shaped fertile branch. **D.** *A. germanicum*, transverse section of second-order branch showing three-ribbed mesarch protostele with secondary wood. Protoxylem, black; metaxylem, cellular; secondary xylem, hatched. **C,D:** Upper Devonian. (**C,D** drawn from photograph in Serlin & Banks, 1978.)

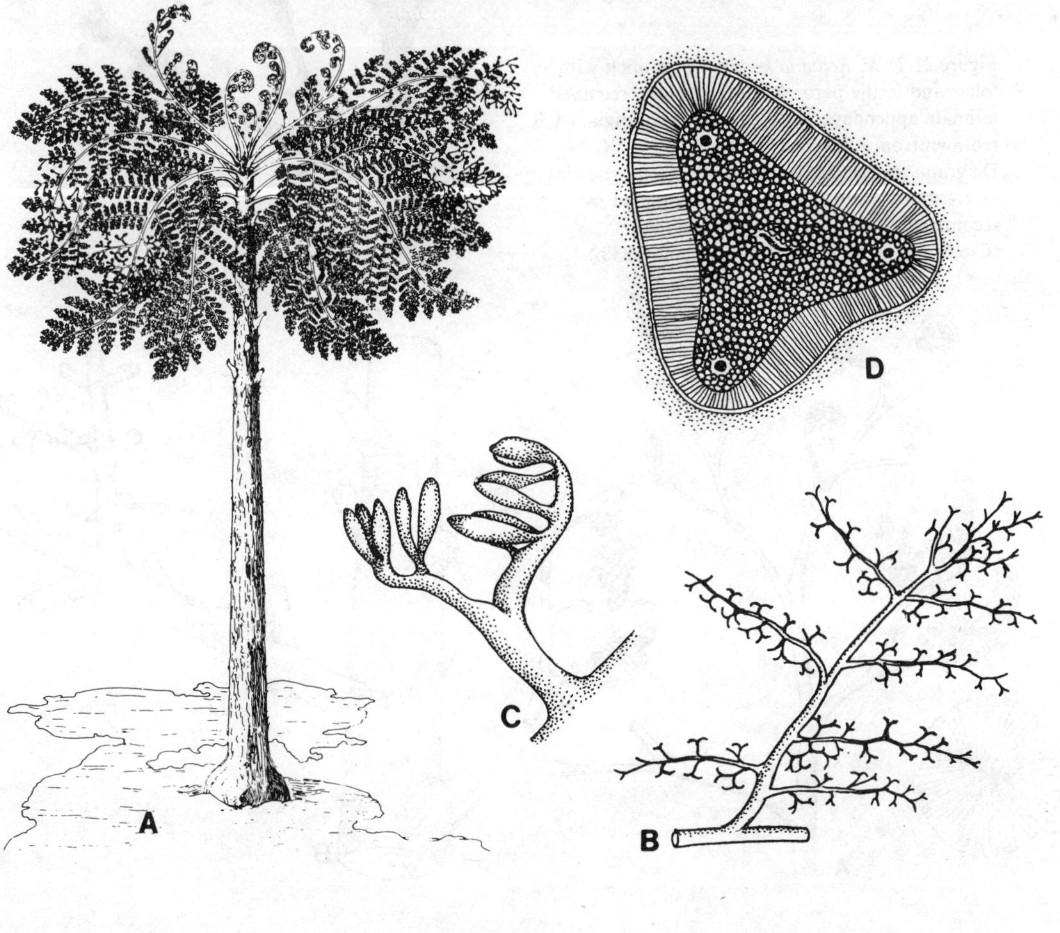

xylem in the various orders of branching is yet to be determined.

Tetraxylopteris

Of the presently known genera assigned to the Aneurophytales, the general morphology and detailed structure of *Tetraxylopteris* (Fig. 21.8B) is probably the most completely described (Beck, 1957; Bonamo & Banks, 1967; Scheckler, 1976). This Middle Devonian (Givetian)–Upper Devonian (Frasnian) progymnosperm is characterized by opposite-decussate branching and a four-ribbed actinostele.

Tetraxylopteris, like *Rellimia*, was probably a bushy shrub with stems that branched pseudomonopodially. The known vegetative axes of *Tetraxylopteris* comprise four orders of branches not including the ultimate appendages. The opposite–decussate arrangement of the known orders of branching produced a three-dimensional radiating pattern. The ultimate appendages borne on the fourth-order branches consist of short terete branches that dichotomize once or twice and are similar in morphology to the ultimate appendages of *Aneurophyton,* but dichotomize somewhat more. The arrangement of the ultimate appendages is decussate and thus repeats the branching pattern found in all other parts of the branch system.

The fertile organs of *Tetraxylopteris* (Fig. 21.9A) are strikingly similar to those of *Rellimia* except for their arrangement on the axis, which repeats the decussate arrangement of the vegetative branches. Each fertile appendage dichotomizes twice, and each resulting branch is tripinnate with the ultimate divisions bearing the sporangia. The sporangia are oblong–fusiform with some evidence of longitudinal dehiscence. The isospores are identical to the dispersed spore genus *Rhabdosporites*.

All vegetative axes and branches of *Tetraxylopteris,* except the ultimate appendages, contain a four-ribbed actinostele that, when seen in transverse section, is cruciform (Fig. 21.10A). The primary xylem of the actinostele is mesarch with numerous

Figure 21.8. **A.** Diagram of vascular system showing four orders of branching (1–4) and ultimate appendages (u) in *Tetraxylopteris*. (From Scheckler, 1976.) **B.** *T. schmidtii,* reconstruction of plant showing decussate branching of four orders and decussate ultimate appendages. **A,B:** Upper Devonian. (**B** redrawn from Scheckler & Banks, 1971a.)

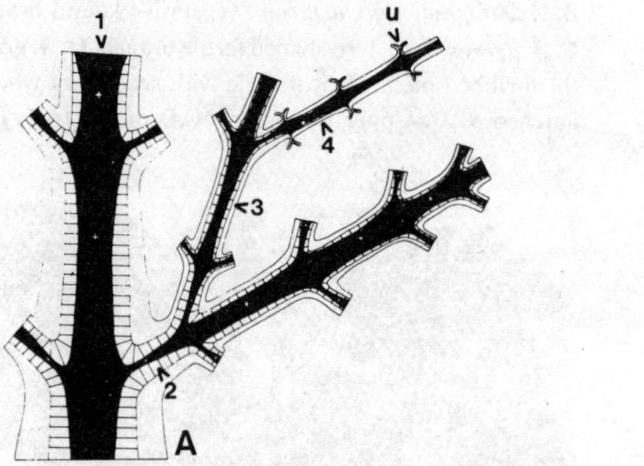

Figure 21.7. **A.** *Rellimia hostinense,* branch with foliar and fertile parts. **B.** *R. hostinense,* recurved ultimate appendages with terminal sporangia. **(A,B** redrawn from Kräusel & Weyland, 1933.) **C.** Diagram, transverse section *R. thomsonii* showing three-ribbed protostele (black) surrounded by secondary xylem. **A–C:** Middle Devonian. (**C** redrawn from Kräusel & Weyland, 1938.)

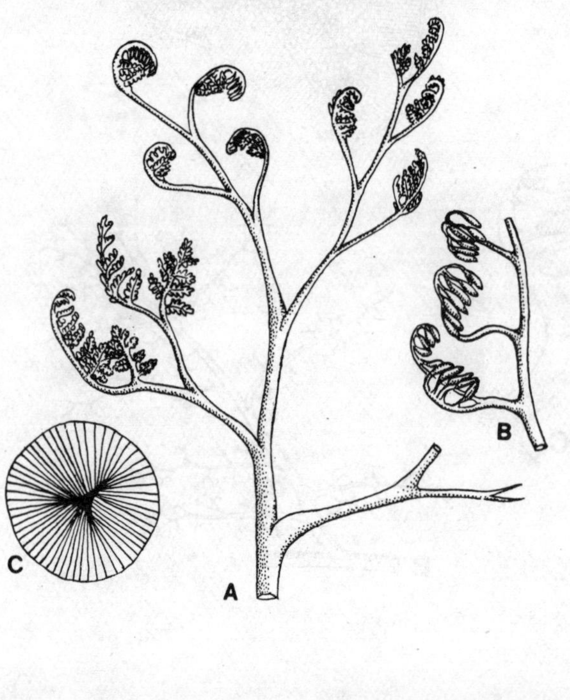

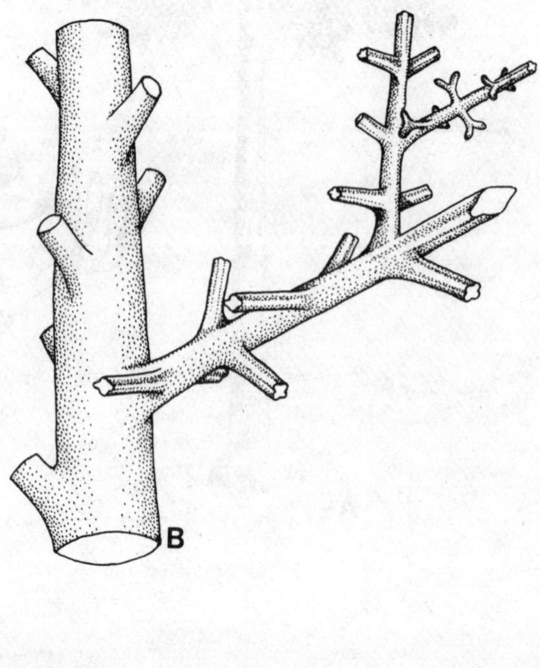

conspicuous protoxylem strands near the ends of the arms and in the center of the protostele. In addition, each arm of the protostele has at least one internal protoxylem strand more or less centrally located in the arm. Bordered pits occur on all walls of the metaxylem cells. Pitting of protoxylem is annular to helical (Scheckler & Banks, 1971a). The secondary xylem surrounding the protostele is composed of tracheids and rays and often shows some indication of growth rings. In its characteristics, the wood is pycnoxylic. Circular bordered pits are found on all walls of the tracheids and are often crowded and hexagonal in outline. They usually alternate on the tracheid wall and are similar to crowded araucarian-type pits. The uniseriate rays are high and contain some ray tracheids.

Unlike arborescent lycopsids, sphenopsids, and woody ferns, the vascular cambium of *Tetraxylopteris* and all other progymnosperms is bifacial, producing secondary xylem internally and abundant secondary phloem to the outside.

Triloboxylon and Proteokalon

Although sharing characteristics with other genera of the Aneurophytales, *Triloboxylon* (Fig. 21.11) and *Proteokalon* exhibit some distinctive features that will be important in our discussion of relationships of the Aneurophytales and leaf evolution. Both genera have been described (Matten & Banks, 1966; Scheckler & Banks, 1971a, 1971b; Scheckler, 1975) as having ultimate appendages that dichotomize several times in a single plane (Fig. 21.9B). These planated telome trusses are spirally arranged in *Triloboxylon*, but occur in pairs that alternate with a single abaxial appendage in *Proteokalon* (Fig. 21.9C). In *Proteokalon* second-order branches are decussate, a characteristic reminiscent of *Tetraxylopteris*, whereas all known orders of branching are spiral in *Triloboxylon*. Internally, both *Triloboxylon* (Fig. 21.10B) and *Proteokalon* are structurally similar to *Aneurophyton, Rellimia*, and *Tetraxylopteris*. In some specimens of *Triloboxylon* secondary wood, Scheckler & Banks (1971a) rather unexpectedly found aligned groups of pits on the radial walls, a characteristic believed to be diagnostic for the wood of *Archaeopteris* and *Tetraxylopteris*. Some specimens of *Triloboxylon, Proteokalon*, and *Tetraxylopteris* have outer cortical tissues composed of anastomosing strands of sclerenchyma, a conspicuous feature of many pteridosperm stems.

Fertile specimens of *Proteokalon* remain to be found. Scheckler (1975) was, however, able to demonstrate the position and structure of fertile appendages for *Triloboxylon*. Its fertile telome trusses (Fig. 21.11) occur in three vertical rows on first-order (penultimate) axes. Each dichotomizes twice, and the ultimate

Figure 21.9. **A.** *Tetraxylopteris schmidtii*, branches bearing terminal, fusiform sporangia. (Redrawn from Bonamo & Banks, 1967.) **B.** *Proteokalon petryi*, planated sterile appendage. **C.** *P. petryi*, shoot-bearing sterile appendages. **A–C:** Upper Devonian. (**B,C** redrawn from Scheckler & Banks, 1971b.)

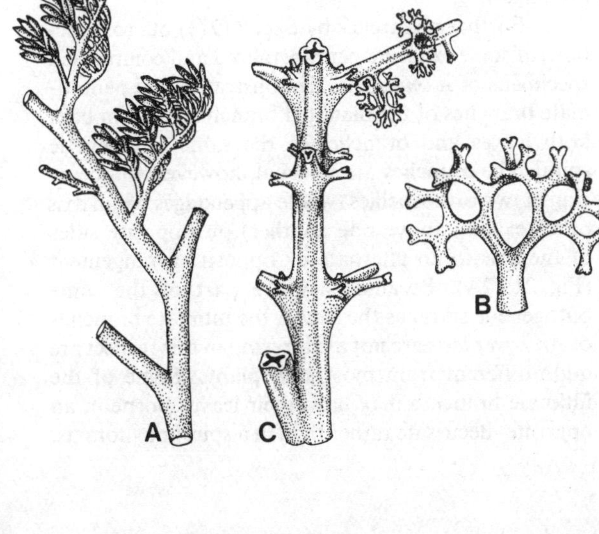

Figure 21.10. **A.** Transverse section of *Tetraxylopteris schmidtii* showing first-order branch with trace to second-order branch. Note cruciate configuration of primary xylem. **B.** *Triloboxylon ashlandicum*, transverse section, first-order branch showing three-ribbed protostele. **A,B:** Upper Devonian. (**A,B** from Scheckler & Banks, 1971a.)

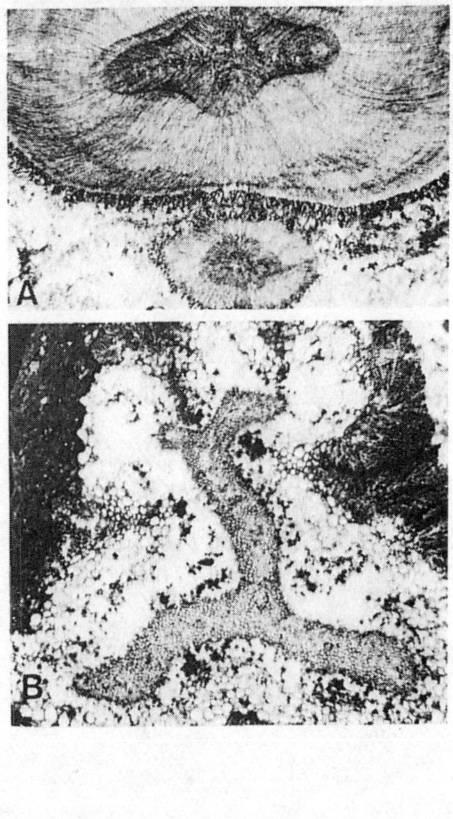

branches bear terminal fusiform sporangia with longitudinal dehiscence. These simple fertile appendages replace, and are apparently homologous with, the second-order vegetative branches.

Other genera have been assigned to the Aneurophytales, but these are generally not well enough known to be placed in the order with any degree of assurance. Analyses of these and other less well-known genera of the Aneurophytales have been prepared by Bonamo (1975) and Beck (1976). At present the only genera of the Aneurophytales that combine the reproductive and vegetative characteristics of the Progymnospermopsida are *Aneurophyton, Rellimia, Tetraxylopteris,* and *Triloboxylon.*

Archaeopteridales

In their first known appearance in the fossil record the Archaeopteridales are younger in age than the Aneurophytales. Fragments of *Archaeopteris*-like axes assigned to the genera *Svalbardia* and *Actinoxylon* have been discovered in beds of the Middle Devonian (Givetian) and are younger than the Eifelian

Figure 21.11. *Triloboxylon ashlandicum,* reconstruction showing fertile and sterile appendages. Upper Devonian. (From Scheckler, 1975.)

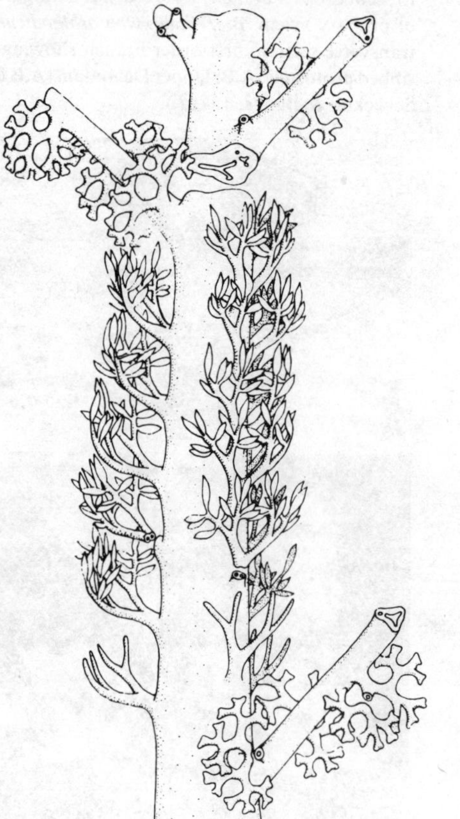

remains of *Rellimia.* The principal genus, *Archaeopteris,* first appears in the Frasnian and extends to the top of the Devonian.

After a comprehensive study revealing the ontogeny of *Archaeopteris* shoots, Scheckler (1978) has demonstrated that *Svalbardia, Actinopodium, Siderella,* and *Actinoxylon,* all genera assigned to the Archaeopteridales, are probably remains of axes representing various developmental stages of *Archaeopteris.* This leaves *Archaeopteris* as the only major representative of the Archaeopteridales exhibiting both vegetative and reproductive characteristics of the Progymnospermopsida.

Archaeopteris

In an earlier part of this chapter, we were introduced to the concept of the progymnosperms resulting from Beck's discovery in 1960 of the connection between *Archaeopteris* and *Callixylon.* At that time and until the appearance of a paper by Carluccio, Hueber, & Banks (1966), the leaves of *Archaeopteris* were thought to be homologous with the bipinnate fronds of ferns. This idea had been questioned, however, as early as 1894 by Schmalhausen, who suggested that the "pinnules" of the *Archaeopteris* frond were actually spirally arranged leaves. Obviously, little attention was paid to this suggestion for more than half a century. Using material of Upper Devonian *A. macilenta,* Carluccio et al. (1966) demonstrated conclusively that the "frond" of *Archaeopteris* is a planated branch system in which the ultimate appendages, thought to be pinnules, are in reality spirally arranged simple leaves. From permineralized segments of *Archaeopteris* these investigators correlated the internal structure with that of the *Callixylon* type and were able to show that leaf traces arise from the branches of *Archaeopteris* in a spiral sequence. On purely hypothetical grounds, Meeuse (1963) suggested that the vegetative leaves (pinnules) of *Archaeopteris* might be the homologs of coniferophyte leaves, a hypothesis that has since been supported by several investigators.

Further researches by Beck (1971) of cross sections of some exquisite permineralized and compressed specimens of *A. macilenta* demonstrated that penultimate branches of the planated branching system bear both leaves and branches in the same ontogenetic spiral. The branches are unusual, however, in occurring in two orthostichies (where appendages on an axis occur exactly above one another) on opposite sides of the axis in an alternate or opposite arrangement (Fig. 21.12A). Because they are part of the same ontogenetic spiral as the leaves, the ultimate branches of *Archaeopteris* are not axillary and in this respect are quite different from most seed plants. Some of the ultimate branches may have their leaves borne in an opposite–decussate rather than in a spiral phyllotaxis.

In spite of the morphological variability exhibited by the leaf- and branch-bearing units of *Archaeopteris*, some of which are quite un-coniferlike, Beck (1971) favors a relationship with the conifers because of pycnoxylic wood and the presence of simple decurrent leaves on the branches of *Archaeopteris*.

The leaves of various species of *Archaeopteris* show a spectrum of morphologies that suggest an origin from a planated and webbed telome truss. In their reconstruction of *A. fissilis* (Fig. 21.12B), Andrews, Phillips, & Radforth (1965) show its leaves to be planated terete telomes that dichotomize two to three times. Leaves of *A. macilenta* (Fig. 21.12C) show planation and various degrees of incomplete webbing so that the distal margins are deeply incised. In *A. halliana* and *A. obtusa* (Fig. 21.12D,E) the webbing is complete and the distal margins of the wedge-shaped leaves are smooth. In those species with webbed leaves (Fig. 21.12E), a single vein enters the decurrent base of the leaf and divides dichotomously several times in the lamina. The veins thus formed continue to the distal margin.

The fertile leaves of *Archaeopteris halliana* (Fig. 21.12F) have been described in detail by Phillips, Andrews, & Gensel (1972). These occur with sterile leaves on ultimate branches of the planated branching system (Fig. 21.13). The leaves are spirally arranged or four-ranked. Each fertile leaf consists of a terete, dichotomously branched telome truss with one to two rows of fusiform sporangia on the adaxial surface of the penultimate branches. The truss dichotomizes two to three times, and it is noteworthy that the terminal branches of the truss are sterile. An analysis of spores taken from the sporangia of *A. halliana* and *A. macilenta* (Phillips et al., 1972) conclusively shows that these two species were heterosporous. Of particular interest is their observation that sporangia containing megaspores were the same size or even a little smaller than those containg microspores. Heterospory in *Archaeopteris* was first discovered by Arnold (1939) in *A. halliana*. Here he found a marked difference in sporangium size, with megaspores (Fig. 21.14) in broad sporangia and microspores in slender ones. Megasporangia (Fig. 21.12G) contained 8 to 16 megaspores up to 300 μm in diameter. One hundred or more microspores (Fig. 21.12H) about 30 μm in diameter were removed from microsporangia. Megaspores of *Archaeopteris* are similar to the dispersed spore genus *Biharisporites* (Fig. 21.14).

Because heterospory has been demonstrated for at least two *Archaeopteris* species, Phillips et al. (1972) have suggested that when further studies are completed, all species may prove to be heterosporous.

Figure 21.12. **A.** *Archaeopteris macilenta,* transverse section of shoot-bearing leaf traces (lt), leaf bases (lb), and branch traces (bt). Note that branch traces are on opposite sides, but in the same spiral as the leaf traces. (Redrawn from Beck, 1971.) **B–E:** Stages in evolution of *Archaeopteris* leaf. **B.** *A. fissilis.* **C.** *A. macilenta.* **D.** *A. halliana.* **E.** *A. obtusa.* **F.** *A. halliana,* fertile appendages with micro- and megasporangia. **G.** *A. halliana,* cluster of megaspores macerated from megasporangium. **H.** *A. halliana,* cluster of microspores. **A–H:** Upper Devonian. (**F–H** redrawn from Arnold, 1939.)

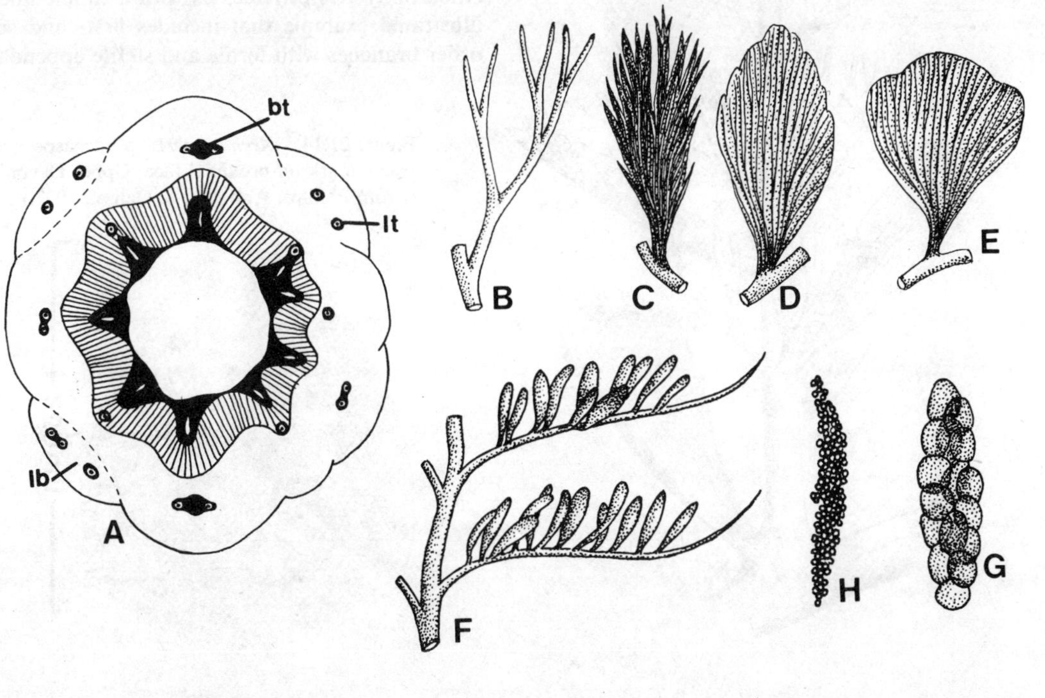

Cecropsis

While investigating anatomically preserved specimens from the Upper Pennsylvanian of Ohio, Stubblefield & Rothwell (1989) encountered a woody heterosporous pteridophyte whose characteristics are principally those of progymnosperms. This newly discovered fossil, which they describe as *Cecropsis* (Fig. 21.15), consists of a protostelic shoot with helically arranged leaves that compare favorably with leaves of *Archaeopteris*. The differentiation of simple leaves on a shoot again suggests an *Archaeopteris*-type morphology, unlike that of aneurophytes where aerial branching systems are not clearly differentiated into stems and leaves. In the construction of its protostele, *Cecropsis* is like that of aneurophytes; however, the cauline protoxylem forms a ring of sympodia as is found in the eusteles of the Protopityales and Archaeopteridales. In the architecture of its wood, *Cecropsis* shares features found in *Protopitys* and some aneurophytes.

Figure 21.13. *Archaeopteris halliana,* reconstruction of branch bearing spirally arranged four-ranked sterile and fertile leaves. Upper Devonian. (From Phillips, Andrews, & Gensel, 1972.)

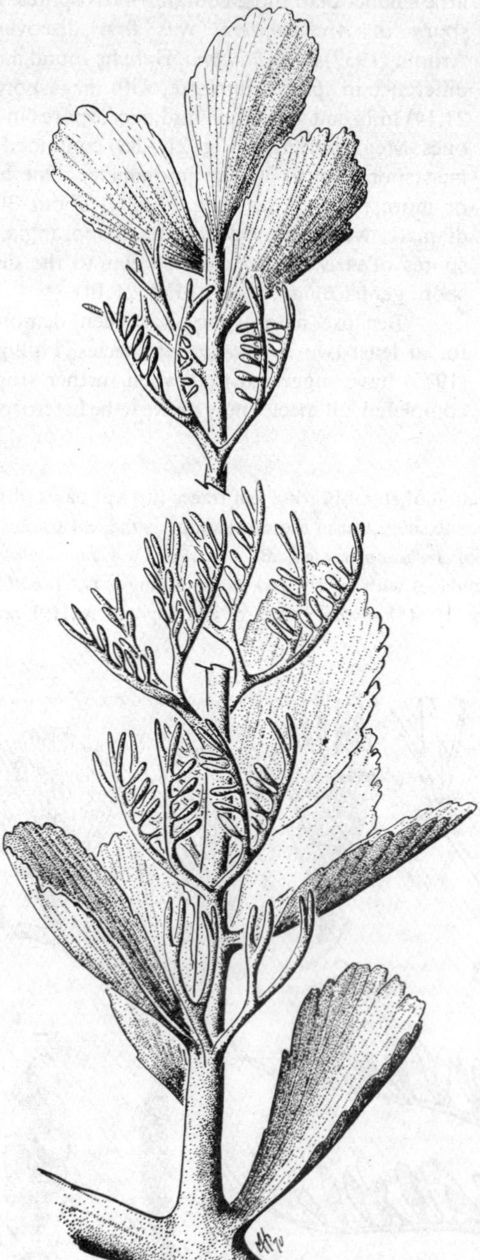

The eusporangia of heterosporous *Cecropsis* are on the adaxial surface of the leaves, a further indication of affinity with heterosporous archaeopterids rather than with the homosporous aneurophytes where sporangia are terminal on fronds. Although the question of relationships of *Cecropsis* remains open, Stubblefield & Rothwell conclude that plants with progymnosperm characteristics were present in the Upper Pennsylvanian of Euramerica.

Ontogenetic studies

In two provocative papers, Scheckler (1976, 1978) has undertaken ontogenetic studies of the Aneurophytales and Archaeopteridales. His studies reveal the changes in internal structure occurring in various orders of branches comprising the shoots of these progymnosperms. This information has been utilized to hypothesize the nature of the apical primordia in producing organs of indeterminate and determinate growth. *Triloboxylon* (Fig. 21.11) of the Aneurophytales provides us with a simple and well-illustrated example that includes first- and second-order branches with fertile and sterile appendages.

Figure 21.14. *Archaeopteris* sp. megaspore with trilete mark on proximal face. Upper Devonian. (From Phillips, Andrews, & Gensel, 1972.)

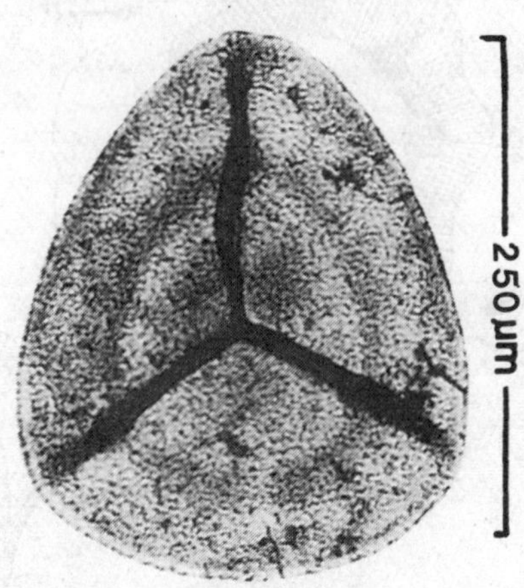

In both first- and second-order branches of *Triloboxylon* there is a clear indication of a change in the primary vascular system from the proximal to the distal portions. Proximally, the three-ribbed protostele tends to increase in size (epidogenesis), reflecting a stage of growth that was indeterminate. The increase in size of the primary vasculature was probably a reflection of an increase in volume of the apical meristem. As growth proceeded distally the maximum size of the primary vascular system was reached and maintained for some distance (menetogenesis) by an apical meristem that was in a more or less steady state of growth. With further growth the apical primordia decreased in volume, resulting in a decrease in the size of the axis and the amount of primary vascular tissue it formed (apoxogenesis). The onset of apoxogenesis marks the beginning of the determinate growth habit. Epidogenesis followed by menetogenesis and apoxogenesis occur in both first- and second-order branches of *Triloboxylon*.

Figure 21.15. *Cecropsis,* reconstruction of shoot segment showing central fertile region alternating with upper and lower vegetative regions. (From Stubblefield & Rothwell, 1989.)

Fertile organs on first-order branches and sterile appendages on second-order branches of *Triloboxylon* are supplied with small terete traces that decrease in size with each dichotomy of the appendage. The evident apoxogenesis in these structures is taken to indicate that their growth was determinate and that they are appendicular units resulting from the phylogenetic modification of fertile and sterile telome trusses. The change in vasculature from a three-ribbed protostele of the second-order branch to the terete strands of the sterile appendages they produce, plus the bilateral symmetry of the appendage, suggest that this structure is, by definition, a leaf (Bierhorst, 1971). The secondary xylem surrounding the three-ribbed protostele in the basal portion of first-order branches decreases in amount distally and disappears in second-order branches.

The various ontogenetic stages of the primary vasculature in *Archaeopteris–Callixylon* are shown in Figure 21.16B taken from Scheckler's 1978 paper. The primary vascular system with its ring of mesarch strands (sympodia) around a large central pith has the structure of a gymnosperm eustele. These primary structures with the accompanying secondary xylem change very little in pith size, number of sympodia, and amount of secondary xylem for relatively long distances in *Callixylon*-type axes and, in this way, conform with the concept of menetogenesis.

The penultimate branches of *Archaeopteris* that arise in a spiral sequence from sympodia of the *Callixylon*-type axes have a primary vasculature that is protostelic in the proximal portions. Distally this develops into a eustele with a ring of mesarch sympodia by medullation and subsequent ontogenetic dissection of the ring of primary vascular tissue. Toward the midregion of a penultimate branch we can find the maximum expression of epidogenesis. Here the pith has reached its greatest diameter and the sympodia are most numerous. In the distal portions of the penultimate branch, apoxogenesis occurs with sympodia decreasing in number and the pith of the eustele decreasing in volume. The same series of events occurs in the ultimate branches. Thus, ontogenetically both orders of branches were at first indeterminate but became determinate as growth and development proceeded.

Scheckler has demonstrated, because of changes in number of sympodia and the intercalation of branches into the ontogenetic spiral of penultimate branches, that the phyllotactic sequence often does not conform with the usual Fibonacci series of 1/3, 2/5, 3/8, 5/13... found in most vascular plants with spirally arranged leaves. This suggests that *Archaeopteris* had a rather archaic type of apical organization.

The leaves of *Archaeopteris,* like the ultimate sterile appendages of the Aneurophytales, are each supplied with a single terete trace that arises radially from one of the sympodia of a branch stele. The determinate nature of the *Archaeopteris* leaf is indicated by the decrease in size of the leaf traces as they dichotomize and enter the base of the leaf to become the veins. If a change in vasculature is a criterion indicating the appendicular nature of the leaf, then the ultimate appendages of *Archaeopteris* with their small terete traces are quite distinct from the branches that bear them where eusteles or protosteles with several mesarch strands occur. By position, vascularization, and ontogeny, the leaves of *Archaeopteris* seem to be the homologs of ultimate appendages of the Aneurophytales.

Secondary xylem is usually found in the proximal portions of the penultimate branches but decreases and then disappears in their distal portions. Ultimate branches lack secondary wood.

Cambial activity: first appearances

Although one might guess that cambial activity appeared relatively recently in vascular plant evolution, we have seen the products of the activity of a bifacial cambium in *Tetraxylopteris* (Beck, 1957) and *Triloboxylon* (Scheckler & Banks, 1971a) and the production of a phellogen in *Triloboxylon* and *Proteokalon* (Scheckler & Banks, 1971a, 1971b). All three genera first appear in the Middle Devonian (Givetian), approximately 360 m.y. ago. Cambial activity producing secondary xylem is first found in *Rellimia,* whose fossil record extends into the Eifelian, more than 370 m.y. ago.

Protopityales

A single Lower Carboniferous genus, *Protopitys* (Fig. 21.16C), represents the order. Because of the 45-cm diameter of one stem specimen and its abundant pycnoxylic secondary wood, it has been concluded that *Protopitys* was treelike (Walton, 1969). Pitting on the radial walls of secondary xylem tracheids consists of uni- to multiseriate circular bordered pits. The rays are uniseriate and vary in height from 2 to 15 cells depending on the species. The pith, which is elliptical in transverse section, is more or less enclosed by meta-

Figure 21.16. **A.** Reconstruction of *Triloboxylon ashlandicum* vascular system showing areas of epidogenesis (e), menetogenesis (m), and apoxogenesis (a). (From Scheckler, 1976.) **B.** *Archaeopteris/Callixylon,* diagrammatic reconstruction of vascular system. (From Scheckler, 1978.) **A,B:** Primary xylem, black; secondary xylem, hatched. **C.** *Protopitys scotica,* diagram of transverse section of shoot. Pith, stippled, secondary xylem, hatched; primary xylem, cellular. Lower Carboniferous. (Redrawn from Walton, 1957.)

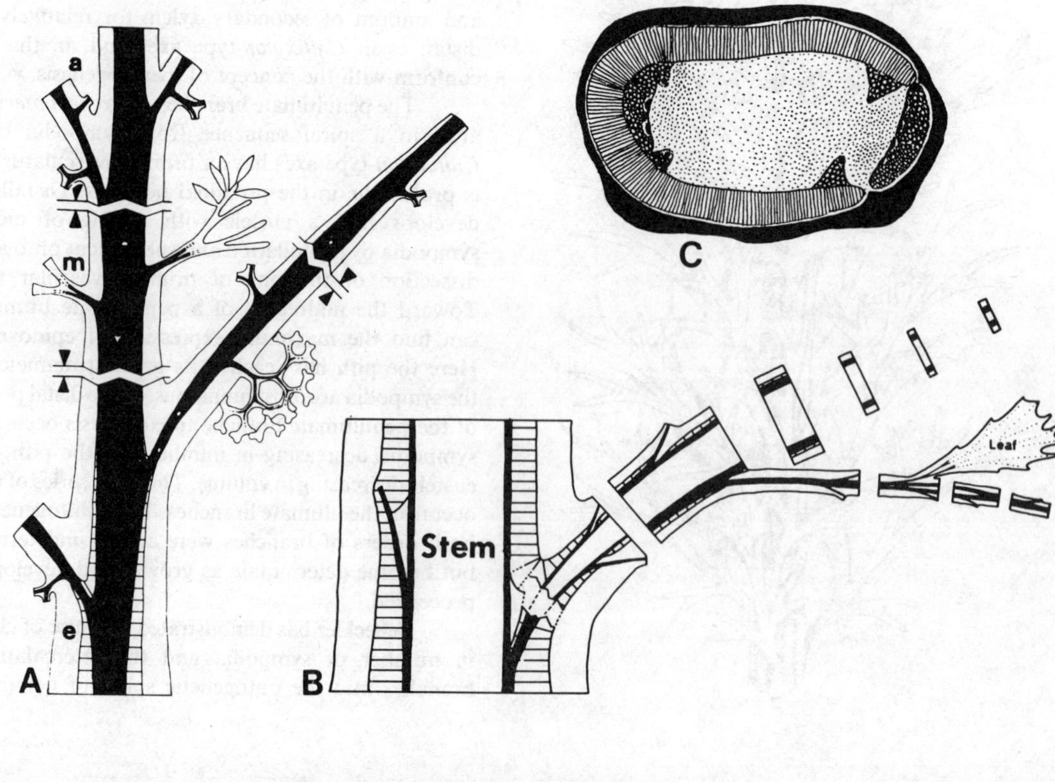

xylem that forms an irregular layer immediately to the inside of the secondary xylem. According to Walton (1957), a pair of protoxylem points occur at opposite ends of the elliptical pith and to the inside of the metaxylem. Each member of a pair of protoxylem strands divides below the node to form the two strands that are incorporated in an arc-shaped trace that is emitted from the side of the stele. A similar trace is emitted from the opposite side of the stele at a higher level, which tells us that the appendages of *Protopitys* were borne in an alternate and distichous manner on the stem.

Fertile organs, described by Walton (1957) and Smith (1962), consist of a dichotomous axis with each branch bearing pinnately arranged branchlets terminated by fusiform sporangia. These fertile organs are highly similar to those of *Rellimia* and *Tetraxylopteris.* The sporangia of *Protopitys* bear spores of two distinct sizes, and thus it is believed to have been heterosporous (Smith, 1962). Spores were shed through a longitudinal slit in the sporangial wall.

Although *Protopitys* is like the Aneurophytales in its fertile structures, it differs from both Aneurophytales and Archaeopteridales in its endarch primary xylem, the position of the primary vasculature in main axes, and the structure of its leaf traces.

The origin of the progymnosperms

We have looked to the Trimerophytopsida as a plexus of Lower (Emsian) and Middle Devonian plants, which at different times during their existence gave rise to the Cladoxylales, Coenopteridales, and some Sphenopsida. When we compare the characteristics of the Progymnospermopsida with those of trimerophytes, it seems highly probable that the progymnosperms also evolved from this ancient and primitive group of vascular plants. Generally, members of the two groups are alike in exhibiting dichotomous and pseudomonopodial branching. Ultimate and penultimate portions of their branching systems are modified (*Pertica* in the Trimerophytes, for example) into structures that are interpreted as prefronds. Internally, the protoxylem in axes of both groups is submerged in metaxylem so that the primary xylem is mesarch or centrarch. The trimerophytes, like the Aneurophytales of the progymnosperms, have massive protosteles. You may recall that sporangia of the Trimerophytopsida tend to be clustered and are borne terminally on ultimate branches of fertile telome trusses. The sporangia are fusiform, and spores are shed through a longitudinal slit in the sporangium wall. All of these reproductive characteristics are shared with progymnosperms. It has been noted that the only basic difference between the Trimerophytopsida and progymnosperms *Aneurophyton* and *Rellimia* is the presence of secondary vascular tissues produced by a bifacial cambium in the progymnosperms. Thus, it is generally agreed that the Progymnospermopsida have evolved from the Trimerophytopsida and that the Aneurophytales, which appear first in Middle Devonian, are the most primitive progymnosperms.

Andrews, Gensel, & Kasper (1975) have provided us with an early Middle Devonian (Eifelian) plant that they interpret as intermediate between the trimerophytes and progymnosperms. The plant, which they named *Oocampsa* (Fig. 21.17), comprises a main axis bearing spirally arranged branches of the first and second order. Second-order branches divide pseudomonopodially and dichotomously two to three times to form a fertile telome truss bearing erect oval sporangia at the tips of the ultimate branches. The sporangia have a longitudinal dehiscence mechanism. Because the fertile structure tends to have its erect sporangia distributed laterally and to have a somewhat pinnate organization, Andrews et al. (1975) speculate that *Oocampsa* is intermediate between a *Trimerophyton* type and *Tetraxylopteris.*

Interrelationships of the orders of progymnosperms

Several lines of evidence support the idea that Aneurophytales and Archaeopteridales are related. The

Figure 21.17. *Oocampsa catheta.* Restoration of portion of fertile shoot. Middle Devonian. (Redrawn from Andrews, Gensel, & Kasper, 1975.)

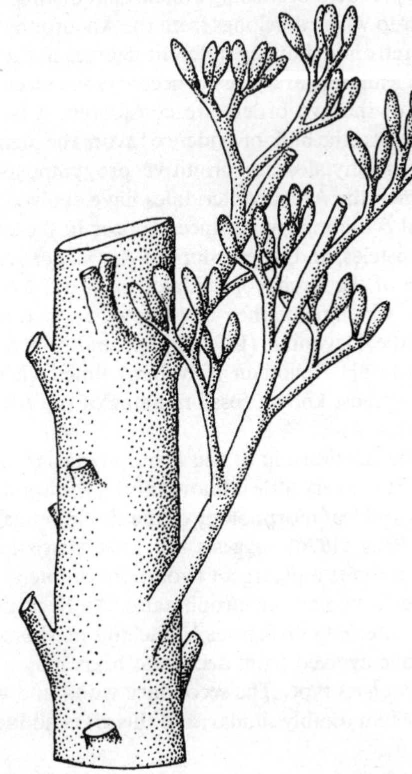

most obvious of these is the presence of a bifacial vascular cambium that produces abundant pycnoxylic secondary xylem and secondary phloem. The presence of ray tracheids and aligned and grouped bordered pits on the radial walls of some genera in both orders further support this relationship (Scheckler & Banks, 1971a). The primary xylem of genera belonging to aneurophytes and archaeopterids is highly uniform in having mesarch maturation. Although the Aneurophytales are exclusively protostelic, and *Callixylon*-type axes and some branches of *Archaeopteris* of the Archaeopteridales have eusteles, it is safe to conclude from the ontogenetic studies of Scheckler (1978) that the eustele with its ring of sympodia around a pith has been derived from a protostele by medullation and dissection. Thus, the distinction between the two orders, based on stelar morphology, is more apparent than real. Further evidence supporting this origin of the eustele will be discussed in Chapter 23.

With some minor variation in branching patterns, the fertile segments of the Aneurophytales and Archaeopteridales are remarkably similar. All have terminal sporangia that are fusiform or oval and shed their spores through a longitudinal slit in the sporangium wall.

In looking for differences, we must remember that the Aneurophytales appear to be homosporous, while the Archaeopteridales have several species that show marked heterospory. Further, it has been demonstrated that *Archaeopteris* bore simple leaves on an often planated branching system that distinguishes the order to which it belongs from the Aneurophytales, where prefronds showing different degrees in the evolution of leaflike characteristics occur. When all characteristics of the two orders are considered, it is quite apparent that the bulk of evidence favors the idea that the Aneurophytales are primitive progymnosperms from which the Archaeopteridales have evolved. The principal evolutionary advances appear in the simple leaves, eusteles, and heterospory of *Archaeopteris*. Divergence of the Archaeopteridales from the Aneurophytales, which made their first known appearance in the Middle Devonian (Eifelian), probably occurred during Middle Devonian (Givetian) times when we find the earliest known fossil remains of the Archaeopteridales.

The relationship of the Protopityales is still in doubt because very little is known about the branching patterns and leaf morphology of the single genus *Protopitys*. Beck (1976) suggests that this Mississippian plant represents a divergent evolutionary line arising from the Devonian Aneurophytales. As Beck points out, the alternate distichous branching of *Protopitys* could have evolved from decussate branching of the *Tetraxylopteris* type. The secondary wood and spore types are remarkably similar, and this gives additional support to the possible relationship between Aneurophytales and Protopityales.

Leaf evolution

The Aneurophytales, especially *Rellimia*, have often been cited as providing excellent examples illustrating early stages in the evolution of megaphylls in plants whose axes have not become differentiated into stem and leaf. That the compound leaves of cycads and ferns evolved from the three-dimensional dichotomized branching systems of primitive vascular plants was first proposed in 1884 by Bower and later incorporated into the telome concept by Zimmermann (1952). It is not difficult to visualize how, by overtopping, planation, and webbing of the branch systems of plants like *Rellimia, Aneurophyton,* or *Tetraxylopteris,* the fronds of cycadophytes evolved. We have seen a great deal of evidence that the megaphylls of ferns evolved in exactly the same way.

Meeuse (1963) recognizes the possibility that simple leaves of the coniferophytes and fronds of cycadophytes evolved from the primitive prefronds of those progymnosperms placed in the Aneurophytales. According to his hypothesis, Meeuse believes that the simple coniferophyte leaves evolved from ultimate appendages of a system composed of several orders of branches. This is exemplified by the simple leaves of *Archaeopteris* that have evolved from planated and webbed ultimate appendages on a system composed of at least two orders of branches. Cycadophyte fronds, on the other hand, evolved from the whole progymnosperm prefrond consisting of two to four orders of branching plus the ultimate appendages. This concept conforms with the conclusions of Beck (1957) and Bonamo & Banks (1967) after they completed studies of the aneurophyte *Tetraxylopteris*. Beck further suggested that because of the radial symmetry exhibited by the lateral branch systems of *Tetraxylopteris,* they may represent an early stage in the evolution of the leaf. Today the same conclusions could be applied to *Rellimia, Aneurophyton, Triloboxylon,* and *Proteokalon,* all of which show a degree of radial symmetry of the cauline portions of their prefronds. Only *Triloboxylon* and *Proteokalon* show any evidence of planation, and this is expressed in their sterile ultimate appendages. It may be of significance that the appendicular leaves and some ultimate branches of *Archaeopteris* are planated. If the cycadophyte frond evolved from radially symmetrical branching systems of the aneurophytes, then we have to assume that (1) planation as it appears in the ultimate appendages of *Triloboxylon* and *Proteokalon* extended proximally into second- and first-order branches to produce the bilateral symmetry characteristic of cycadophyte fronds; and (2) there was modification of the protostelic cauline vasculature in prefrond branches. Scheckler (1976,

1978) has demonstrated from his ontogenetic studies of the Archaeopteridales and Aneurophytales that the appendicular organs we call leaves are the products of apoxogeneis. With this interpretation in mind, we can explain the evolutionary origin of the compound cycadophyte leaf as the product of successive stages in apoxogenesis initiated in distal portions of the branching system and progressing proximally through one or more orders of a planated branching system, thus producing a compound appendicular structure with determinate growth and bilateral symmetry.

If we accept this as the mechanism for the evolution of the compound cycadophyte-type leaf, it adds credence to the idea that the Aneurophytales formed a plexus of Devonian plants from which the Archaeopteridales and seed plants diverged. However, as we discuss in more detail in the next chapter, there are several differing interpretations about how this evolution proceeded. According to the interpretation of Beck (1976, 1981) and others, the Aneurophytales gave rise to seed ferns and the other gymnosperms with cycadlike leaves (cycadophytes) and also to the Archaeopteridales. The Archaeopteridales later gave rise to cordaites and other gymnosperms with small leaves (coniferophytes). This interpretation implies that the entire lateral branch of an aneurophyte became the cycadophyte leaf, while only the ultimate segments of the lateral branch became a leaf in the coniferophyte line.

A second possible mechanism for the evolution of the small leaves of conifers and coniferlike plants is that they are derived from the cycadlike leaves by an alternation in the timing of developmental events (heterochrony) similar to the mechanism by which many leaves of cycads normally develop into small scales rather than into large foliar fronds (Rothwell, 1982). This interpretation implies that all seed plants may have been derived directly from the aneurophytes, with the Archaeopteridales being an evolutionary dead end.

The Progymnospermopsida and gymnosperm classification

One of the earliest classifications of the gymnosperms was proposed by Robert Brown prior to the middle of the nineteenth century. The basic criterion used in his classification was the production of "naked" seeds. This tended to place the focus on a single monophyletic origin for the gymnosperms. Although questioned from time to time, it was up to Arnold (1948) to interpret gymnosperm classification based on the evidence from the fossil record. Arnold concluded that the cycadophytes and coniferophytes had separate origins in the Devonian. Thus, he interpreted the class Gymnospermae as an artificial polyphyletic group in which the seed habit arose independently in cycadophytes and coniferophytes. He recommended that the class Gymnospermae be dropped from our systems of classification, an appropriate proposal because the common ancestor from which the two main lines of gymnosperms arose would not have been a gymnosperm, but a progymnosperm. If, on the other hand, the common ancestor of all gymnosperms was a gymnosperm, then a group of plants that includes gymnosperms plus angiosperms would be considered monophyletic. As we will see in Chapter 22, our uncertainty about the phylogenetic relationships of seed plants results as much from differences in our definition of a "gymnosperm" as from the lack of fossil evidence to clarify the question.

For a comprehensive review of the Progymnospermopsida, the student is referred to "Progymnosperms" by Beck & Wight (1988).

References

Andrews, H. N. (1961). *Studies in Paleobotany.* New York: Wiley.

Andrews, H. N., Gensel, P. G., & Kasper, A. E. (1975). A new fossil plant of probable intermediate affinities (Trimerophyte–Progymnosperm). *Canadian Journal of Botany,* **53,** 1719–28.

Andrews, H. N., Phillips, T. L., & Radforth, N. W. (1965). Paleobotanical studies in arctic Canada. I. *Archaeopteris* from Ellesmere Island. *Canadian Journal of Botany,* **43,** 545–56.

Arnold, C. A. (1939). Observations on fossil plants from the Devonian of Eastern North America. IV. Plant remains from the Catskill Delta deposits of Northern Pennsylvania and Southern New York. *Contributions from the Museum of Paleontology, University of Michigan,* **5,** 271–314.

Arnold, C. A. (1947). *An Introduction to Paleobotany.* New York: McGraw-Hill.

Arnold, C. A. (1948). Classification of gymnosperms from the viewpoint of paleobotany. *Botanical Gazette,* **110,** 2–12.

Beck, C. B. (1957). *Tetraxylopteris schmidtii* gen. et sp. nov., a probable pteridosperm precursor from the Devonian of New York. *American Journal of Botany,* **44,** 350–67.

Beck, C. B. (1960). The identity of *Archaeopteris* and *Callixylon. Brittonia,* **12,** 351–68.

Beck, C. B. (1962). Reconstruction of *Archaeopteris* and further consideration of its phylogenetic position. *American Journal of Botany,* **49,** 373–82.

Beck, C. B. (1970). The appearance of gymnospermous structure. *Biological Reviews,* **45,** 379–400.

Beck, C. B. (1971). On the anatomy and morphology of lateral branch systems of *Archaeopteris. American Journal of Botany,* **58,** 758–84.

Beck, C. B. (1976). Current status of the Progymnospermopsida. *Review of Palaeobotany and Palynology,* **21,** 5–23.

Beck, C. B. (1981). *Archaeopteris* and its role in vascular plant evolution. In *Paleobotany, Paleoecology, and Evolution,* ed. K. J. Niklas. pp. 193–229. New York: Praeger, Vol. I.

Beck, C. B., Coy, K., & Schmid, R. (1982). Observations on the fine structure of *Callixylon* wood. *American Journal of Botany*, **69**, 54–76.

Beck, C. B., & Wight, D. C. (1988). Progymnosperms. In *Origin and Evolution of Gymnosperms*, ed. C. B. Beck. pp. 1–84. New York: Columbia University Press.

Bierhorst, D. W. (1971). *Morphology of Vascular Plants*. New York: Macmillan.

Bonamo, P. M. (1975). The Progymnospermopsida: building a concept. *Taxon*, **24**, 569–79.

Bonamo, P. M. (1977). *Rellimia thompsonii* (Progymnospermopsida) from the Middle Devonian of New York State. *American Journal of Botany*, **64**, 1272–85.

Bonamo, P. M. (1983). *Rellimia thompsonii* (Dawson) Leclercq and Bonamo (1973): The only correct name for the aneurophytalean progymnosperm. *Taxon*, **32**, 449–54.

Bonamo, P. M., & Banks, H. P. (1967). *Tetraxylopteris schmidtii:* its fertile parts and its relationships within the Aneurophytales. *American Journal of Botany*, **54**, 755–68.

Carluccio, L. M., Hueber, F. M., & Banks, H. P. (1966). *Archaeopteris macilenta*, anatomy and morphology of its frond. *American Journal of Botany*, **53**, 719–30.

Kräusel, R., & Weyland, H. (1933). Die Flora des bömischen Mitteldevons. *Palaeontographica B*, **78**, 1–46.

Kräusel, R., & Weyland, H. (1938). Neue Pflanzenfunde im Mitteldevon von Elberfeld. *Palaeontographica B*, **83**, 172–95.

Leclercq, S., & Bonamo, P. M. (1971). A study of the fructification of *Milleria* (*Protopteridium*) *thompsonii* Lang from the Middle Devonian of Belgium. *Palaeontographica B*, **136**, 83–114.

Matten, L. C., & Banks, H. P. (1966). *Triloboxylon ashlandicum* gen. and sp. n. from the Upper Devonian of New York. *American Journal of Botany*, **53**, 1020–28.

Matten, L. C., & Schweitzer, H. J. (1982). On the correct name for *Protopteridium (Rellimia) thompsonii* (Fossil). *Taxon*, **31**, 322–6.

Meeuse, A. D. (1963). From ovule to ovary: a contribution to the phylogeny of the megasporangium. *Acta Biotheoretica*, **16**, 127–82.

Phillips, T. L., Andrews, H. N., & Gensel, P. G. (1972). Two heterosporous species of *Archaeopteris* from the Upper Devonian of West Virginia. *Palaeontographica B*, **139**, 47–71.

Rothwell, G. W. (1982). New interpretations of the earliest conifers. *Review of Palaeobotany and Palynology*, **37**, 7–28.

Scheckler, S. E. (1975). A fertile axis of *Triloboxylon ashlandicum*, a progymnosperm from the Upper Devonian of New York. *American Journal of Botany*, **62**, 923–34.

Scheckler, S. E. (1976). Ontogeny of progymnosperms I. Shoots of Upper Devonian Aneurophytales. *Canadian Journal of Botany*, **54**, 202–19.

Scheckler, S. E. (1978). Ontogeny of progymnosperms II. Shoots of Upper Devonian Archaeopteridales. *Canadian Journal of Botany*, **56**, 3136–70.

Scheckler, S. E., & Banks, H. P. (1971a). Anatomy and relationships of some Devonian Progymnosperms from New York. *American Journal of Botany*, **58**, 737–51.

Scheckler, S. E., & Banks, H. P. (1971b). *Proteokalon*, a new genus of progymnosperms from the Devonian of New York state and its bearing on phylogenetic trends in the group. *American Journal of Botany*, **58**, 874–84.

Schmalhausen, J. (1894). Ueber Devonische Pflanzen aus dem Donetz-Becken. *Mémoirs Comité Géologique*, **8**, 1–36.

Serlin, B. S., & Banks, H. P. (1978). Morphology and anatomy of *Aneurophyton*, a progymnosperm from the late Devonian of New York. *Palaeontographica Americana*, **8**, 343–53.

Smith, P. L. (1962). Three fructifications from the Scottish Lower Carboniferous. *Palaeontology*, **5**, 225–37.

Stewart, W. N. (1981). The Progymnospermopsida: the construction of a concept. *Canadian Journal of Botany*, **59**, 1539–42.

Stubblefield, S. P., & Rothwell, G. W. (1989). *Cecropsis luculentum* gen. et sp. nov.: Evidence for heterosporous progymnosperms in the Upper Pennsylvanian of North America. *American Journal of Botany*, **76**, 1415–28.

Walton, J. (1957). On *Protopitys* (Göppert): with a description of a fertile specimen *"Protopitys scotica"* sp. nov. from the Calciferous Sandstone series of Dunbartonshire. *Transactions of the Royal Society of Edinburgh*, **63**, 333–40.

Walton, J. (1969). On the structure of a silicified stem of *Protopitys* and roots associated with it from the Carboniferous (Mississippian) of Yorkshire, England. *American Journal of Botany*, **56**, 808–13.

Zimmermann, W. (1952). The main results of the "telome theory." *The Palaeobotanist*, **1**, 456–70.

22

Gymnosperm reproduction: early evolution

The evolution of the seed habit

The evolution of the seed habit ranks as one of the most important events in vascular plant evolution. The evidence from extinct and extant plants indicates that seed plants have become the most successful of land plants because of the selective advantage the formation of seeds gives these plants over all others. Textbook definitions tell us that a seed is a ripened ovule. We have to understand, of course, that ripening of the ovule is usually initiated by syngamy (fusion of egg and sperm nuclei) and is a postfertilization process. We should also know that the ovules of extant gymnosperms ready for fertilization (Fig. 22.1) consist of an envelope (the integument) with a micropyle and a megasporangium (the nucellus) inside of which there is a megagametophyte composed of nutritive tissue and several archegonia. After fertilization, to qualify as a seed, the structure should exhibit evidence of an embryo within the nutritive tissues of the megagametophyte. If we stick with this last criterion, then most of the seedlike structures that paleobotanists encounter should be called ovules, because embryos are seldom preserved. In actual practice, however, paleobotanists use the terms seed and ovule interchangeably since it is impossible in most cases to tell if fertilization has or has not occurred.

As described above, the ovule of an extant gymnosperm might be defined as an integumented megasporangium. In our opinion, however, one cannot divorce the evolution of the ovule from the evolution of pollen tubes that penetrate the nucellus and furnish the means by which sperm are deposited in proximity to eggs. Implicit here is the fact that in the ovules of modern plants the megasporangium remains indehiscent, and the only mechanism we know of that provides sperm access to female gametes through the indehiscent nucellus is the pollen tube. In short, it seems unlikely that the seed habit could have evolved without the evolution of pollen tubes or pollen tubelike structures as in cycads and *Ginkgo*. The two go hand in hand. If one agrees to this concept, then an ovule is better described as an integumented indehiscent megasporangium. By definition, such a reproductive structure as *Lepidocarpon* with its foliar integuments around a megasporangium must be excluded from the ovule category because it is known (Ramanujam & Stewart, 1969) that the megasporangium dehisced and probably admitted flagellated sperm directly to the archegonia. Free water was, no doubt, the medium for transport of sperms, as in all free-sporing plants.

We find it useful to recognize three levels of vascular plant evolution leading to and including the formation of ovules:

1. Free-sporing plants in which all spores, whether they be microspores or megaspores, are shed from dehiscent sporangia.
2. Dehiscent sporangia where microspores are shed, but megaspores (usually four megaspores one of which is functional) are retained in the megasporangium. Integuments are present or absent. Free water is the medium for transport of sperms to egg.
3. Indehiscent megasporangia and dehiscent microsporangia. Integuments with micropyles are present. The pollen tube provides the mechanism for transport of sperms to eggs.

These three levels of evolution are fairly easy to recognize in the life cycles of modern plants. We realize, as do most paleobotanists, that determining whether a fossil megasporangium is dehiscent or indehiscent and whether pollen tubes are present is difficult even with the best-preserved material. Thus, from this point on, all fossil reproductive structures of gymnosperms that have an integument, well-defined micropyle, and a nucellus surrounding a single functional megaspore, and that lack an embryo will be referred to as ovules unless it has been demonstrated that their megasporangia are dehiscent.

From isospores to microspores and megaspores

As concluded by Pettitt (1970) in an excellent review of the origin of the seed habit, one can hardly discuss ovule origin without understanding the role of homospory and heterospory. It is agreed by all concerned that homospory is the primitive condition and heterospory is derived. We also agree that heterospory preceded the evolution of seeds. When studying spores found in the sporangia of fossilized plants, paleobotanists frequently are not sure whether the plant they are studying is homosporous or heterosporous. The difficulty, of course, is the fragmentary nature of the fossil material. Not frequently a fossil plant – *Archaeopteris macilenta*, for example – was believed to be homosporous because the spores appeared to be similar in size and ornamentation. Later, a reinvestigation showed that some sporangia had spores that were larger and heterospory was proven. The difficulty for paleobotanists is compounded because they cannot determine the products of spore germination. It is quite conceivable that upon germination isospores may have given rise to bisexual (monoecious) exosporic gametophytes as found in the prothallus of a common fern; or sex differentiation might have occurred so that some of the isospores produced either exosporic male or female gametophytes. We will probably never know if such exosporic dioecism occurred in the past, but it might well have represented the first step in the evolution of heterospory.

A second possible step for which there is some evidence suggests that heterospory may have first appeared in plants where each sporangium contained both microspores and megaspores.

The third step for which there is abundant evidence in the fossil record occurs when some sporangia of the plant formed numbers of small spores (microspores), and other sporangia on either the same plant or another plant of the same species produced sporangia with a small number of larger spores (megaspores). We now know that fossil plants forming microspores produce endosporic micro- and megagametophytes, as do modern plants. At maturity the microspores with their endosporic gametophytes are shed (they are free-sporing). The megaspores with their megagametophytes may be free-sporing or retained in a dehisced or indehiscent megasporangium. There is a general tendency among heterosporous plants to reduce the number of megaspores from several to one functional megaspore. We have seen good examples of this tendency in the Lycopsida, Sphenopsida, and Filicopsida. This provides us with an excellent example of the independent and parallel evolution of heterospory in divergent groups of vascular plants.

Starting with the primitive free-sporing, homosporous, monoecious habit, the stages in evolution of heterospory can be summarized as follows:

1. Decrease in numbers of spores in some sporangia
2. Increase in size of remaining spores
3. Spore content of some sporangia remaining constant in size and number of spores
4. Change from monoecious to dioecious gametophytes
5. Change from exosporic to endosporic gametophytes.

As Pettitt (1970) emphasizes, there is no way of determining the order in which increase in spore size, sex determination, and endosporic development in the megasporangium occurred.

Figure 22.1. Diagram, longitudinal section of *Pinus* ovule. Megagametophyte (m); archegonium (a); integument (i); nucellus (n); pollen tube (pt). Extant.

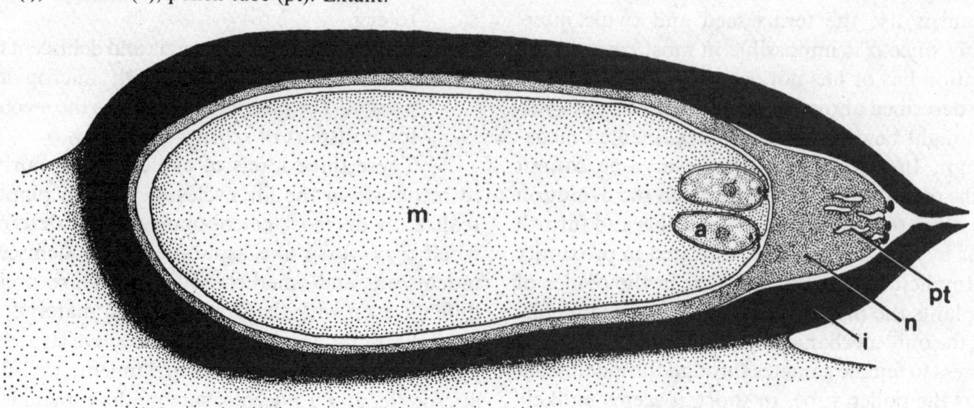

First evidences of free-sporing heterospory

Prior to the Middle Devonian all vascular plants were homosporous (Chaloner, 1970), and until the appearance of a paper by Andrews, Gensel, & Forbes (1974) describing heterospory in the Middle Devonian *Chauleria,* there was no concrete evidence of heterospory in strata older than the Upper Devonian (Frasnian), where heterosporous species of *Archaeopteris* occur.

Chauleria (Fig. 22.2) comprises an axis with spirally arranged pseudomonopodial branches. These in turn bear spirally arranged ultimate appendages that dichotomize one to two times. Fusiform sporangia with longitudinal dehiscence terminate some of the ultimate appendages. In its general morphology, *Chauleria* resembles a member of the Aneurophytales. Macerations of its sporangia yielded numerous spores of two sizes, those that are believed to be microspores in the 30–40-μm range and megaspores that are 60 to 156 μm in diameter. Some sporangia contain only microspores, while some seem to have a mixture. Another plant where a mixture of small and large spores occurs in the same sporangium is the enigmatic Upper Devonian *Barinophyton citrulliforme* (Brauer, 1980). Although Brauer raises the possibility that the small spores could be abortive megaspores, he is of the opinion that they

Figure 22.2. *Chauleria cirrosa* with spirally arranged branches terminated by sterile and fertile appendages. Middle Devonian. (From Andrews, Gensel, & Forbes, 1974.)

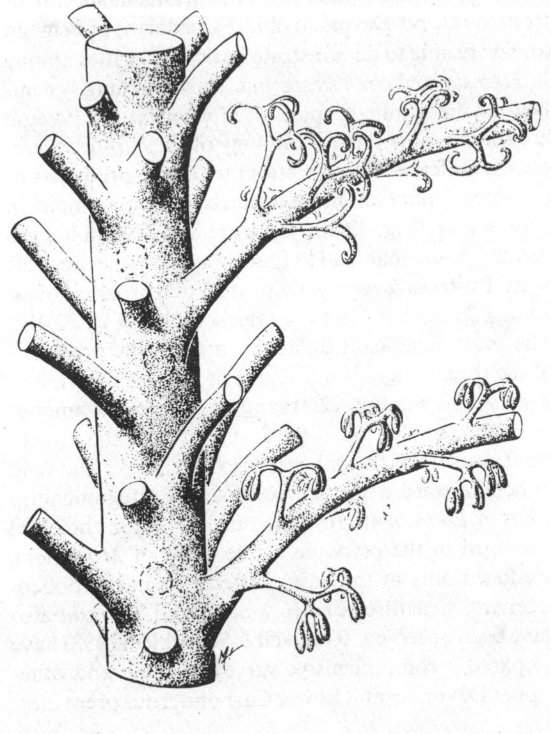

are microspores because of their well-developed spore coats. The microspores have a size range of 33 to 48.5 μm: the megaspores are very large, ranging from 700 to 900 μm. The presence of two spore types in a sporangium of these ancient vascular plants has been interpreted as "an evolutionary situation in heterospory which we would expect to find ... presaging a later and more definite distinction between microspore and megaspore" (Pettitt, 1970).

Beyond heterospory: the ovule

Before discussing the fossil evidence, it might help to present a list of those evolutionary events that took place in ovule evolution after the onset of heterospory:

1. Abortion of three megaspores and formation of a single functional megaspore in the megasporangium
2. Retention of the functional and abortive megaspores in the megasporangium (nucellus)
3. Formation of endosporic megagametophytes within an indehiscent megasporangium
4. Elaboration of the apex of the indehiscent nucellus for reception of pollen
5. Formation of an integument that delimited a micropyle
6. Formation of pollen tube or pollen tubelike structures from endosporic microgametophytes.

At present there is no way of telling for sure the exact order of these six events simply because the record is incomplete. What follows is a brief discussion of the first evidence in the fossil record for each of these events. To facilitate the discussion of the evidence, it is necessary at this point to introduce the term preovule. We envisage a preovule to be an ovulelike structure consisting of a megasporangium that either is naked or is invested by unfused or partially fused integumentary lobes, and thus lacks a well-defined micropyle. The term preovule obviously implies a primitive organization that preceded the evolution of the ovule as it has been defined earlier in the chapter.

A good place to start our discussion is with *Archaeosperma arnoldii* (Fig. 22.3), a cupulate, preovule-bearing organ from the Upper Devonian (Famennian). The interpretation of its structure by Pettitt & Beck (1968) is another landmark in paleobotany. Although we speculated that ovulelike structures would be found in the Devonian, it was not until Pettitt & Beck reinvestigated the specimens collected and described by Arnold (1935) that speculation became a reality.

The reconstruction of *Archaeosperma* shows the fossil to consist of a cupule that partially surrounds four preovules. The cupules are planated dichotomous sterile telome trusses. These are webbed in the proximal half and extend distally into long filiform tips. There are four such tips for each of the four segments comprising the cupule. Each orthotropous (upright) preovule is borne at the tip of a short stalk. The proximal portion of the integument of the preovule is spiny, while its distal portion is divided into a number of lobes forming a ring around a "micropyle." By macerating some preovules, Pettitt & Beck found a megaspore tetrad consisting of three abortive megaspores at the distal end of a large *Cystosporites*-type megaspore (Fig. 22.4A). The latter, often found as a dispersed spore, is clearly a functional megaspore. Unfortunately, the nucellus of the *Archaeosperma* specimens is not well preserved, but there are sufficient remains to demonstrate that the structure was present around the megaspore complex. Although some difficult questions remain concerning the organization and development of the nucellus, there is no doubt that *Archaeosperma* is well along the evolutionary road leading to the seed habit.

Figure 22.3. *Archaeosperma arnoldii,* restoration of cupule complex. Upper Devonian. (Redrawn from Pettitt & Beck, 1968.)

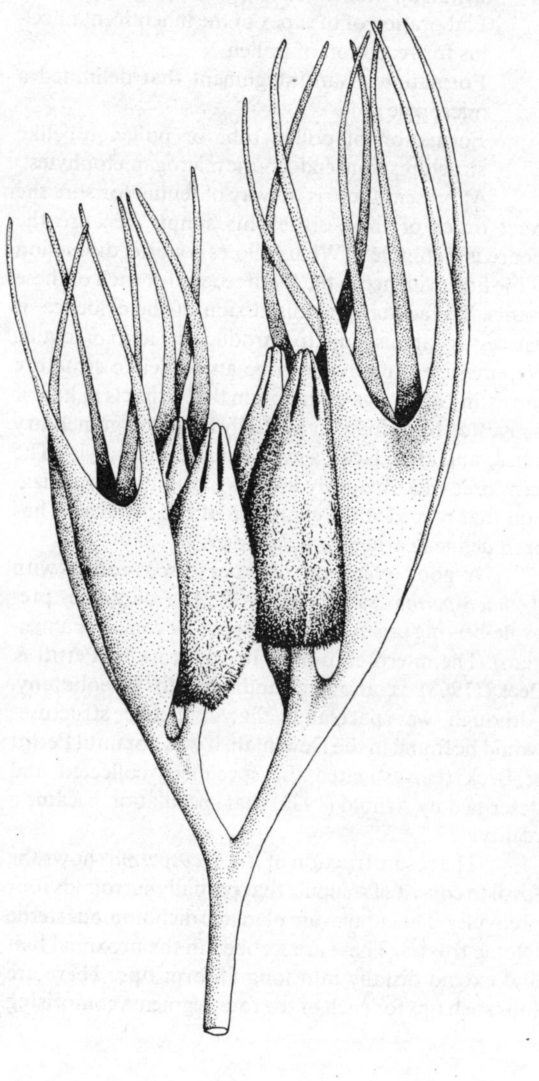

Until recently the preovules of *Archaeosperma* were the oldest to be reported from the fossil record. Gillespie, Rothwell, & Scheckler (1981), and Rothwell, Scheckler, & Gillespie (1989), however, described cupulate structures containing preovules from Famennian strata of the Upper Devonian that are somewhat older than the rocks containing *Archaeosperma.* The newly discovered preovules and their cupules have been assigned to a new genus *Elkinsia* (Fig. 22.4B). Although similar to *Archaeosperma* in general organization, *Elkinsia* differs in having little or no fusion between the four terete integumentary lobes that invest the nucellus. Fusion does, however, occur between the lobes and the nucellus in the basal one-third of the preovule. Distally, the integumentary lobes become free from the nucellus and separate into four or five lobes that curve inward at their tips to form a ring around the distal end of the nucellus. The latter is surmounted by a cup-shaped lagenostome (salpinx of some authors), which is a modification of the distal end of the nucellus adapted for more effective reception of pollen. The cupules containing the preovules are borne at the tips of dichotomously branched axes (Fig. 22.5A) where the branching is cruciate, i.e., each fork is in a plane at right angles to the last. The segments produced by each fork are unequal so that overtopping occurs not only in the branching axes but in the cupules as well.

Although *Moresnetia* (Fig. 22.5B) was originally described as a leaf-bearing system, it is now known that the "leaves" were in reality preovule-bearing cupules (Fairon-Demaret & Scheckler, 1986, 1987). These authors examined hundreds of *Moresnetia* specimens from the Upper Devonian (mid-Famennian) of Belgium and were able to demonstrate basic similarities among it, *Elkinsia,* and *Archaeosperma*. The branching systems bearing the terminal cupules of *Moresnetia* and *Elkinsia* are similar, as is the morphology of the cupules themselves. Differences in the structure of the preovules of the three genera are more distinctive. For example, in *Moresnetia* (Fig. 22.4C) there are 8 to 10 thin integumentary lobes that tend to flare away from the preovule apex. In *Archaeosperma* there are 5 to 6 broader lobes, while *Elkinsia* has 4 to 5 even thicker lobes (Fig. 22.4B). The most significant difference among the preovules of the three genera is the degree of fusion of the lobes. *Archaeosperma* (Fig. 22.3) exhibits the greatest amount of fusion, so that only about one-fifth of the integument consists of free lobes that extend above and tend to bend inward over the nucellus. The integumentary lobes of *Elkinsia* are fused to one another in the basal one-third of the preovule, while those of *Moresnetia* are fused only at the base. Differences in the pollen-receiving structures of *Moresnetia* and *Elkinsia* also have been observed. Rothwell & Scheckler (1988) have prepared a comprehensive survey of these and other Upper Devonian and Lower Carboniferous preovules.

The integumentary envelope

An overview of the ways in which green land plants have adopted to the terrestrial environment reveals a remarkable variety of structures that have evolved to protect their reproductive cells, tissues, and organs. A few examples that come to mind are thick-walled spores in multicellular sporangia, and multicellular sex organs that may be submerged in surrounding vegetative tissues, structures such as perianths, involucres, integuments, cupules, and ovary walls.

As far as the evolutionary origin of the seed habit is concerned we are, at this point, most interested in the origin of the integument. We have already seen how parts of sporophylls of *Lepidocarpon* and *Miadesmia* have been modified to form integuments around the megasporangia of these lycopsids. Available evidence indicates, however, that integuments of some gymnosperm ovules have evolved directly from fertile and sterile telome trusses. As early as 1935, Halle speculated, "In the light of the Psilophytales, both cupule and integument are easily accounted for as syntelomes." Nearly 20 years later Walton (1953) developed this hypothesis (Fig. 22.6) by incorporating the concepts of heterospory, heterangy, and reduction to a single functional megasporangium surrounded by sterile telomes. The latter, by tangential fusion (syngenesis), formed the integumentary envelope (Camp & Hubbard, 1963). In their investigations of *Archaeopteris,* Phillips, Andrews, & Gensel (1972) suggest that the integuments of preovules, such as those of *Archaeosperma,* could have evolved from the sporangium-bearing foliar units as they occur in some species of *Archaeopteris.* They state, "The telome-like morphology of fertile *Archaeopteris* leaves readily lends itself to enclosure of several separate megasporangia," and "the stalked nature of *Archaeopteris* . . . could result in extending a single megasporangium within flanking telomic lobes in a manner similar to the hypothetical stage illustrated by Andrews (1961)." Thus, it seems that free-sporing heterosporous species of *Archaeopteris* show us how the integument evolved from sterile and fertile telomes. Pettitt and Beck (1968) go one step further when they make the intriguing suggestion that *Archaeosperma* is the cupulate seed of a yet-to-be-determined species of *Archaeopteris.*

That the integuments of preovules have evolved by syngenesis of sterile telomes (Halle, 1935) is supported by the degree of fusion of the integumentary

Figure 22.4. **A.** *Archaeosperma arnoldii,* portion of preovule showing megaspore tetrad, nucellus (n?), and lobes of integument. (Redrawn from Pettitt & Beck, 1968.) **B.** *Elkinsia polymorpha,* diagrammatic representation of cupule containing preovules and showing forking branching pattern. **C.** *Moresnetia zalesskyi* reconstructed to show cupule containing preovules with one-quarter of cupule removed. **A–C:** Upper Devonian. (**B,C** redrawn from Rothwell & Scheckler, 1988.)

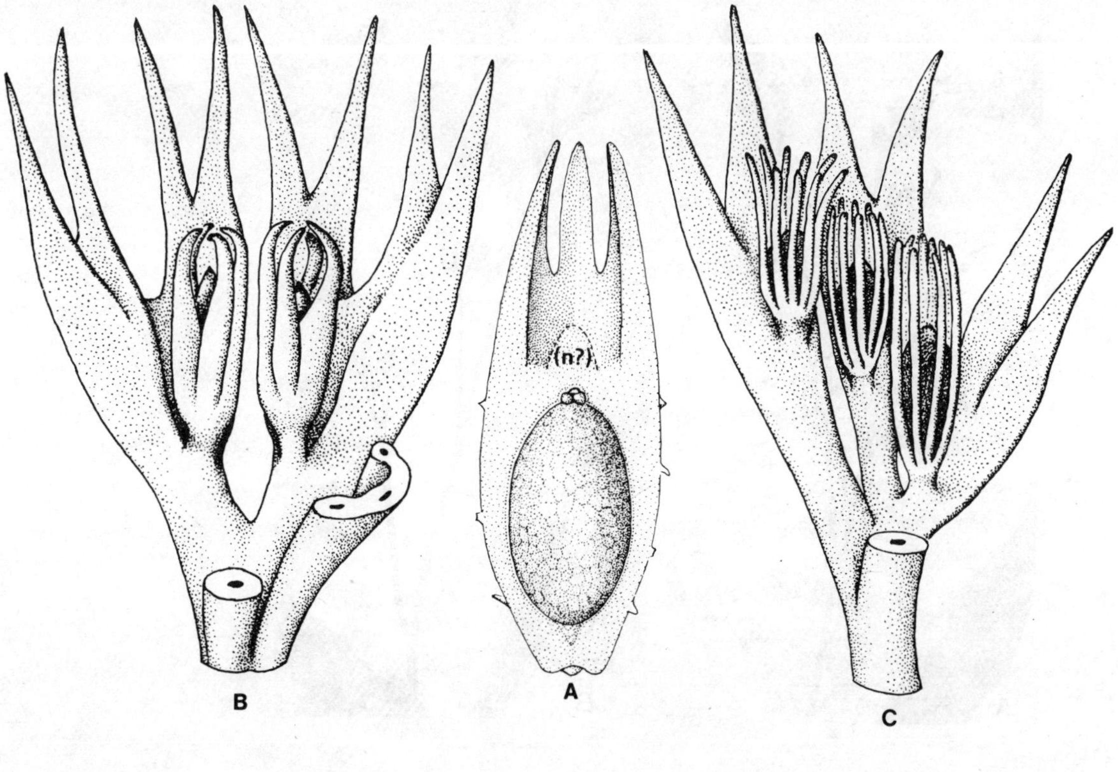

lobes in *Moresnetia, Elkinsia,* and *Archaeosperma.* Further evidence for the fusion of telomes in the evolution of integuments is well documented by Long (1966), who described a number of Lower Carboniferous preovules from Scotland. In his series of papers, which have now joined the ranks of paleobotanical classics, Long (1959, 1960a, 1960b, 1961a, 1965, 1966, 1975) has demonstrated for us how the integuments with its micropyle evolved from sterile telomes and how the apical end of the nucellus was modified for pollen reception and deposition. If as Pettitt (1970) suggests, an ovule is an integumented megasporangium where the integument forms a micropyle at the distal end, then the reproductive structures described by Long are best described as preovules showing early stages of integument and micropyle formation.

Although presented here as though they formed an evolutionary "series," the preovules described by Long actually show us stages in the evolution of the integument and micropyle. These stages have been clearly depicted and interpreted by Andrews (1963) in a fine review article on early seed plants. The least specialized in terms of integument and micropyle formation are the preovules of *Genomosperma latens* and *G. kidstoni* (Fig. 22.7A) (Long, 1959). In the latter, the integument is represented by a ring of eight telomes around a centrally placed megasporangium as in *Moresnetia.* The telomes flare outward and are fused only at their bases. There is no micropyle. *G. latens* (Fig. 22.7B) has an organization similar to *G. kidstoni* except that the eight syntelomes are more closely appressed to the megasporangium and are fused from the base distally for about one-third of their length. In another preovule, *Physostoma elegans,* the integumentary syntelomes are fused for about one-half of their length. *Archaeosperma* (Pettitt & Beck, 1968) and *Eurystoma angulare* (Long, 1960b) (Fig. 22.7C) show an even more advanced degree of fusion so that only four lobes represent the syntelomes at the distal end. It is a matter of opinion if the openings between the rings of lobes should be called a micropyle in these preovule structures. If the presence of a micropyle is a good criterion for an ovule, then *Stamnostoma huttonense* (Long, 1960a) seems to have reached the ovule level. Here (Fig. 22.7D) fusion of syntelomes is complete and a micropyle is formed.

All of the preovules and ovules described above are radiospermic (are more or less round in cross sections). Some preovules and ovules, however, appear to be flattened in transverse sections (platyspermic). At one time it was generally believed that radiospermic ovules were those of pteridosperms and

Figure 22.5. **A.** *Elkinsia polymorpha* with three cupules terminating an overtopped, cruciately forking system. (From Rothwell et al., 1989.) **B.** *Moresnetia zalesskyi* with several terminal raceme-like clusters of cupules. (**A,B:** Upper Devonian. (From Rothwell & Scheckler, 1988.)

Figure 22.6. Hypothetical stages in the evolution of the preovule. **A.** Telome truss consisting of fertile and sterile telomes. **B.** Heterangy and heterospory, micro- and megasporangia. **C.** Dioecious habit; megasporangium is enveloped in sterile telomes. **D.** Syngenesis of sterile telomes around megasporangium to form integument. (Redrawn from Andrews, 1961.)

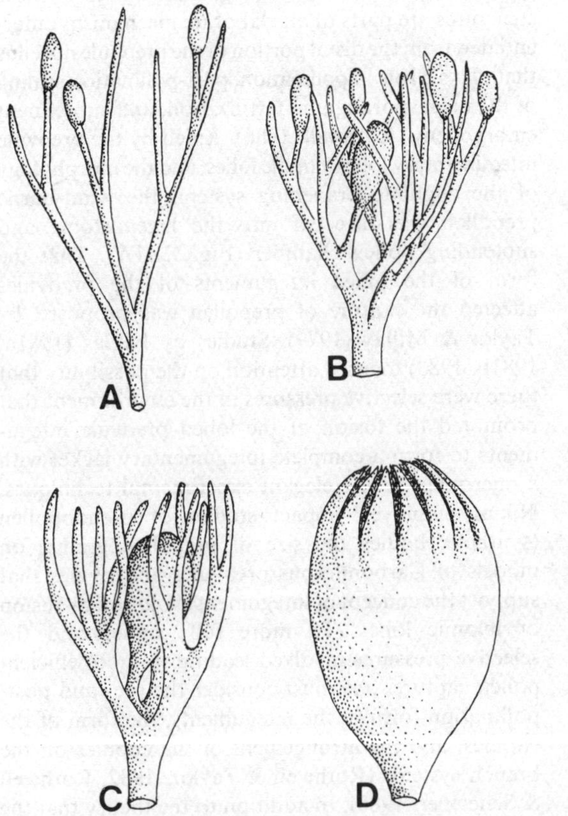

Figure 22.7. Stages in the evolution of the integument. **A.** *Genomosperma kidstoni,* unfused telomes. **B** *G. latens,* partial fusion of telomes, **C.** *Eurystoma angulare* fusion of telomes, except upper third of preovule. **D.** *Stamnostoma huttonense,* complete fusion of telomes to form integument and micropyle. **A,D:** Lower Carboniferous. (**A–D** redrawn from Long, 1959, 1960a,b.)

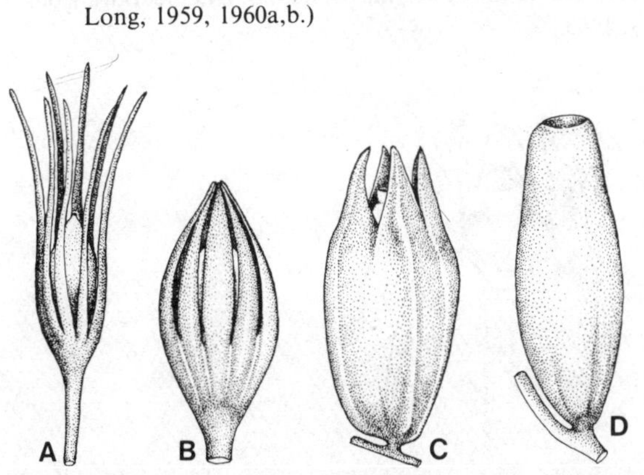

Figure 22.8. **A.** *Spermolithus devonicus,* a platysperm from the Upper Devonian. (Redrawn from Chaloner, Hill, & Lacey, 1977.) **B.** *Lyrasperma scotica.* Note lagenostome. Lower Carboniferous. (Redrawn from Long, 1960b.)

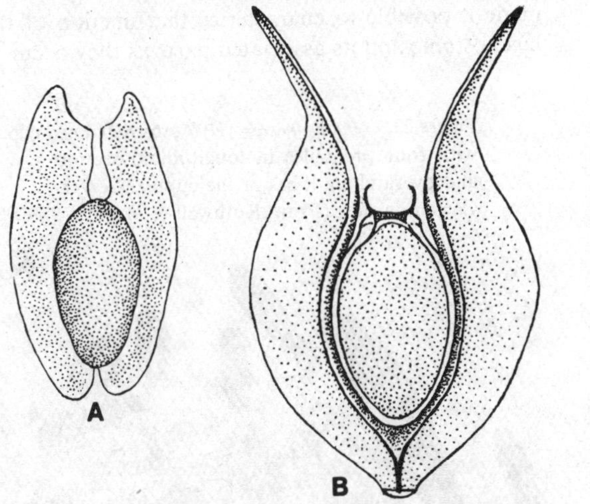

that platyspermic ones belonged to coniferophytes. As we shall see in Chapter 28, this generalization no longer applies. The oldest known flattened structure that may be an ovule or seed is *Spermolithus* (Fig. 22.8A), found as compression fossils in the Upper Devonian rocks of southern Ireland (Chaloner, Hill, & Lacey, 1977). In its general shape *Spermolithus* is cordate (heart-shaped) and has the appearance of ovules assigned to the Carboniferous form genus *Samaropsis* where the integument forms a winglike structure and a distinct micropyle is present.

Lyrasperma (Long, 1960b) is another preovulate structure that shows bilateral symmetry (Fig. 22.8B). These Lower Carboniferous preovules attain a length of 1.1 to 1.5 cm. The bilaterally winged integument is composed of two lobes, adnate in the basal two-thirds and continuing distally as free diverging horns. There is no micropyle, and the apical portion of the nucellus, which is differentiated into a lagenostome, is exposed. The unvascularized nucellus is fused with the integument up to the level of the lagenostome. There are two vascular bundles lying in the primary plane, each traversing a lobe of the integument. Although these preovules have not been found attached, Long is of the opinion, based on association and the well-differentiated lagenostome, that they belong to the pteridosperms.

Hydrasperman reproduction

Of particular importance to our understanding of pollen reception in preovules is the concept of hydrasperman reproduction (Rothwell, 1986). The preovules which have been utilized in developing this concept were first described by Long (1961b) from Lower Carboniferous strata. He assigned these specimens to a new genus, *Hydrasperma* (Fig. 22.9). Subsequent studies of these preovules by Matten, Lacy, & Lucas (1980), and Rothwell & Wight (1989) have made it possible to characterize the function of the lagenostome and its associated parts as they occur in *Hydrasperma* and most other archaic gymnosperm preovules that have been studied.

Figure 22.9. *Pullaritheca* = (*Hydrasperma*) *longii* cupule with four preovules in longitudinal sectior and overarching lobes of the cupule. Lower Carboniferous. (From Rothwell & Wight, 1989.)

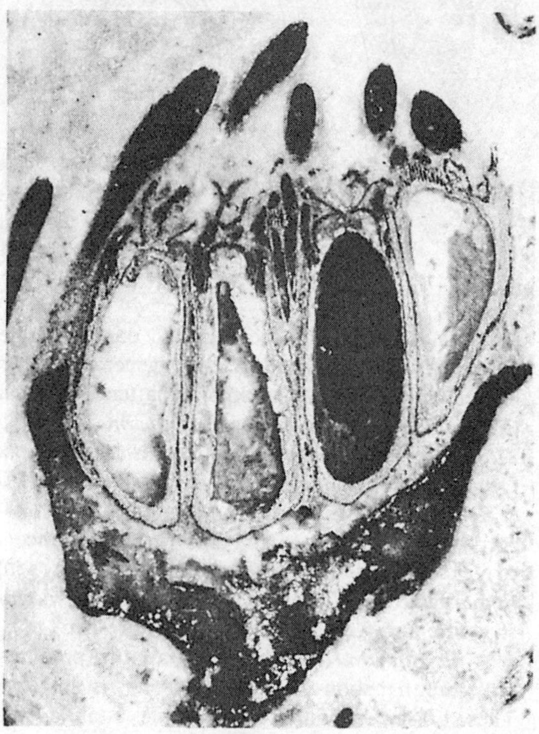

The lagenostome , which may have the form of a funnel (Fig. 22.10A), ring, or cup (Fig. 22.10C), rests on top of a hemispherical pollen chamber formed by the lysigenous disintegration of nucellar cells at the apex. The latter has a membranous floor with a prominent central column (Fig. 22.10B). All of these structures are parts of an elaborate mechanism differentiated from the distal portion of the preovule nucellus that plays a role in pollination, post-pollination sealing of the megasporangium, fertilization, and subsequent embryogeny (Rothwell, 1986). Aided by the preovule integumentary lobes, cupule lobes, and the morphology of the cupulate branching system, the wind-borne prepollen was directed into the lagenostome and subtending pollen chamber (Fig. 22.11A). That the form of the lobed integuments of the preovules affected the capture of prepollen was proposed by Taylor & Millay (1979). Studies by Niklas (1981a, 1981b, 1983) focused attention on the possibility that there were selective pressures in the environment that promoted the fusion of the lobed preovule integuments to form a complete integumentary jacket with a micropyle. Using elegant experimental techniques, Niklas employed impact studies of pseudopollen (synthetic bodies the size of prepollen grains) on models of Carboniferous preovules and ovules that support the concept of integument evolution by fusion of telomic lobes. To more fully understand the selective pressures involved leading to more efficient pollen capture, one must consider the pre- and post-pollination form of the integuments, the form of the cupules, and the arrangement of the cupules on the branch systems (Rothwell & Taylor, 1982; Rothwell & Scheckler, 1988). In addition to the theory that the integuments of early gymnosperms evolved in a way that increased the efficiency of pollination, there can be no doubt that the evolution of the integuments

Figure 22.10. Preovule pollen-receiving structures. **A.** *Physostoma elegans,* lagenostome an inverted funnel. **B.** Reconstruction of *P. elegans,* longitudinal section showing lagenostome (l), central column (c), and pollen chamber (p). **C.** *Eurystoma angulare,* cup-shaped lagenostome. **A–C:** Lower Carboniferous. (**A,C** redrawn from Niklas, 1981a; **B** redrawn and modified from Andrews, 1963.)

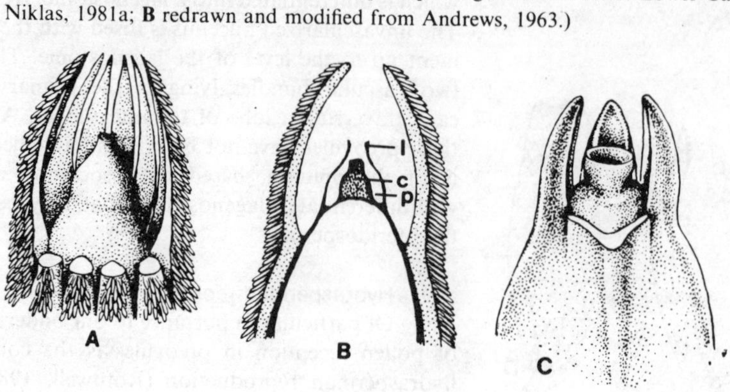

also provided an effective mechanism for the protection of the enclosed nucellus and developing megagametophyte (Andrews, 1963).

Following the introduction of the prepollen into the lagenostome and then into the pollen chamber (Fig. 22.11B), the subsequent growth of the megagametophyte pushed the central column into the base of the lagenostome, sealing the pollen chamber (Fig. 22.11C). Concurrently, the thin pollen-chamber floor was ruptured by the upward growing archegonium-bearing megagametophyte (Matten et al. 1984), allowing the latter structure to come into direct contact with the microgametophytes (Fig. 22.11A).

The final and equally important step in the evolution of hydrasperman reproduction, as outlined by Rothwell (1986) and Rothwell & Scheckler (1988), was the dissemination of propagules (preovules and cupules). This was probably initiated by abscission, with some propagules being transported to relatively dry habitats. Here their chances of survival were greatly enhanced because of the protection provided to the megagametophyte and presumed embryo by the nucellus, integuments, and surrounding lobes of the cupules. In the drier environments, the gymnosperm propagule had a better chance of germinating and producing a new sporophytic phase than the relatively unprotected propagules of their pteridophytic ancestors. As evolution of the integumentary system proceeded, it is apparent that the micropyle of the ovule replaced the lagenostome as the pollen-receiving and directing structure (Long, 1961b; Taylor & Millay, 1979). This function may have been promoted by the evolution of a pollination drop mechanism comparable to that of many gymnosperms. Here a noncellular viscous substance exudes from the micropyle. Pollen grains becoming adherent to the droplet are drawn by surface tension through the micropyle into the pollen chamber as the droplet dries out and shrinks. Such a droplet has been found by Rothwell (1979) in a pteridosperm ovule of the *Callospermarion* type (Fig. 22.12A). The droplet of this Middle Pennsylvanian seed contained palynomorphs of several types and confirms the anemophilous distribution of pollen and spores during the Carboniferous. Once in the pollen chamber of *Callospermarion* ovules, pollen grains of the *Vesicaspora* type germinated to form branched pollen tubes (Fig. 22.12B). This exciting discovery was made by Rothwell (1972) and is thus far the earliest record of a pollen tube. If, as suggested earlier, the formation of pollen tubes is a criterion delineating the seed habit in extant plants, then we can state unequivocally that the seed habit, as we know it to occur among modern gymnosperms, was achieved among some pteridosperms during the Carboniferous.

Figure 22.11. Schematic representation showing features of preovules with hydrasperman reproduction. **A.** Diagram of longitudinal section of mature generalized preovule showing vascularized integumentary lobes; lagenostome (l); central column (c); pollen chamber (p), and cellular megagametophyte (m). Note that the pollen chamber floor has ruptured allowing pollen grains access to the archegonia of the megagametophyte. **B.** Pollination stage where prepollen enters the pollen chamber. **C.** Growth of megagametophyte has pushed central column upward into the base of the lagenostome to seal the pollen chamber and rupture the pollen chamber floor (f). (Redrawn from Rothwell & Scheckler, 1988.)

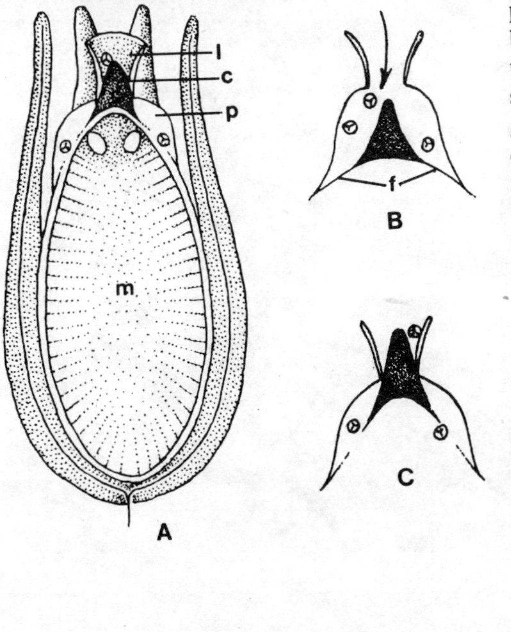

Pollen; micro- and megagametophytes

Clearly, pollen grains and microspores are homologous structures. Microspores produce endosporic gametophytes that liberate flagellated gametes from the proximal region of the spore. By contrast, pollen of many gymnosperms can be distinguished by distal germination and the production of a pollen tube. There are additional differences in the wall structure, especially the exine of spores and pollen grains. Trying to identify those characteristics that distinguish fossil pollen grains from microspores is another difficulty confronting the paleobotanist. Millay & Taylor (1975) and Taylor & Millay (1979) recognize four common pollen or prepollen types occurring in the Carboniferous. These are the small trilete forms of the lyginopterid seed ferns (Fig. 22.13A) that are morphologically identical to microspores; monosaccate types (Fig. 22.13B) that are known to belong to the callistophytalean seed ferns; the saccate grains (Fig. 22.13C) of the cordaites and conifers; and the large monolete grains (Fig. 22.13D) of the medullosan pteridosperms.

Although their discovery was more than one could hope for, endosporal microgametophytes have been described in three of the four pollen types. Millay & Eggert (1974) have reported stages of microgametophyte development in the pollen grains of two representatives of the Callistophytaceae (Fig. 22.13B, 22.14A). In some exquisitely preserved seed fern pollen grains, the authors were able to determine that three prothallial cells were produced in an axial row and that there was no antheridial jacket. These observations conform with those of Florin (1936) for the microgametophytes found in pollen of the cordaites. The interesting observation has been made that the microgametophytes of these extinct gymnosperms are very similar to those of modern gymnosperm families.

A stage in microgametophyte development was found by Stewart (1951) in the pollen chamber of the ovule *Pachytesta*. One of the monolete pollen grains, resting on the floor of the pollen chamber, contains two cells (Fig. 22.14B), which in their orientation and size are highly similar to the paired sperms of modern cycads. Of particular significance is the close proximity of this pollen grain and its microgametophyte to the multicellular, archegonium-containing megagametophyte (Fig. 2.11B). The two are separated by fragmented parts of nucellus and megaspore membrane. We would like to know if these fragmentary structures represent a stage in their dehiscence, thus exposing the archegonia to direct access by motile sperms, or if the nucellus remained indehiscent so that some mechanism, similar to a pollen tube, was required to accomplish fertilization.

After pollination, one of several mechanisms functioned to close the micropyle and thus seal the pollen chamber. In those ovules forming a resinous pollination drop, the drop itself sealed the micropyle. Proliferation of the integumentary tissue around the micropyle of other ovules effectively closed the micropyle, while growth of the megagametophyte tissue pushing the central column into the lower opening of the micropyle sealed the pollen chamber of yet others.

From remarkably well-preserved specimens of the Upper Devonian preovule *Hydrasperma*, Matten et al. (1984) describe stages in development of the megagametophyte from an apparent free nuclear stage (Matten, Lacey, & Lucas, 1980) to a cellular structure that was formed from the subdivision of centripetally developing alveoli. The thin-walled prothallial cells formed in this way may fill the megaspores of some ovules (Fig. 22.15B) and about three-quarters of the megaspores of large ovules. This pattern of development is highly similar to that of modern gymnosperm megagametophytes. In view of the observed similarity of micro- and megagametophyte development in extant and Upper Devonian gymnosperms, one can logically conclude that the rate of gametophyte evolution has been slow in this group for the last 300 m.y.

Figure 22.12. **A.** *Callospermarion*, a pteridosperm ovule with a pollination-drop mechanism (pd). (From Rothwell, 1979.) **B.** *Vesicaspora*-type pollen and pollen tube (pt) in pollen chamber of *Callospermarion*. **A,B:** Pennsylvanian. (From Rothwell, 1972.)

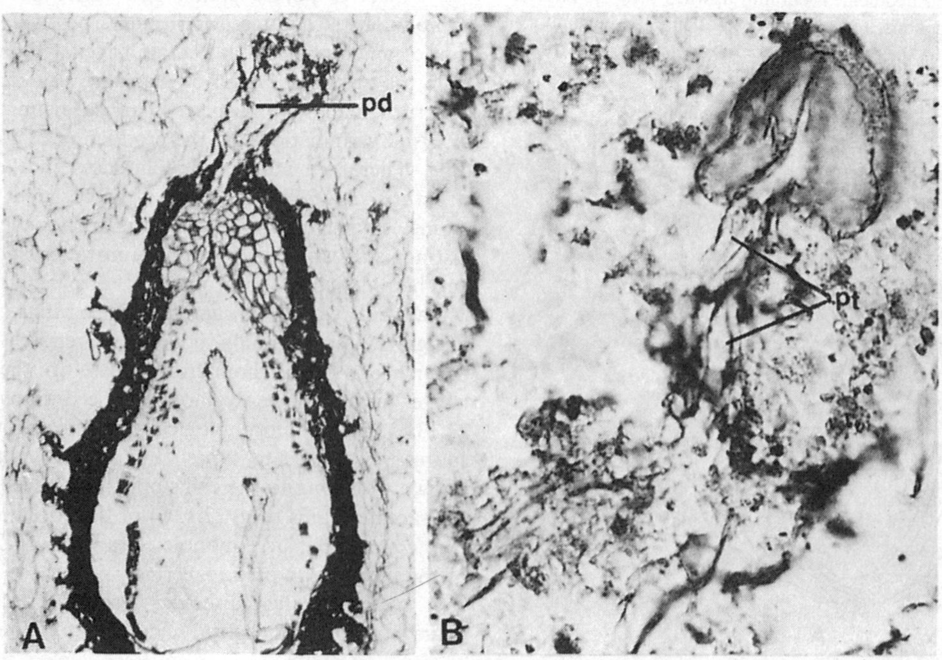

Figure 22.13. Prepollen and pollen types. **A.** Primitive *Crossotheca* type with proximal terilete suture. **B.** Monosaccate pollen grain of the Callistophytaceae. **C.** A saccate pollen grain of a cordaite. Note large central body. **D.** Proximal surface of *Monoletes* prepollen showing suture with median deflection. **A–D:** Pennsylvanian. (**A,D** from Millay, Eggert, & Dennis, 1978; **B** from Millay & Eggert, 1974; **C** from Taylor & Millay, 1979.)

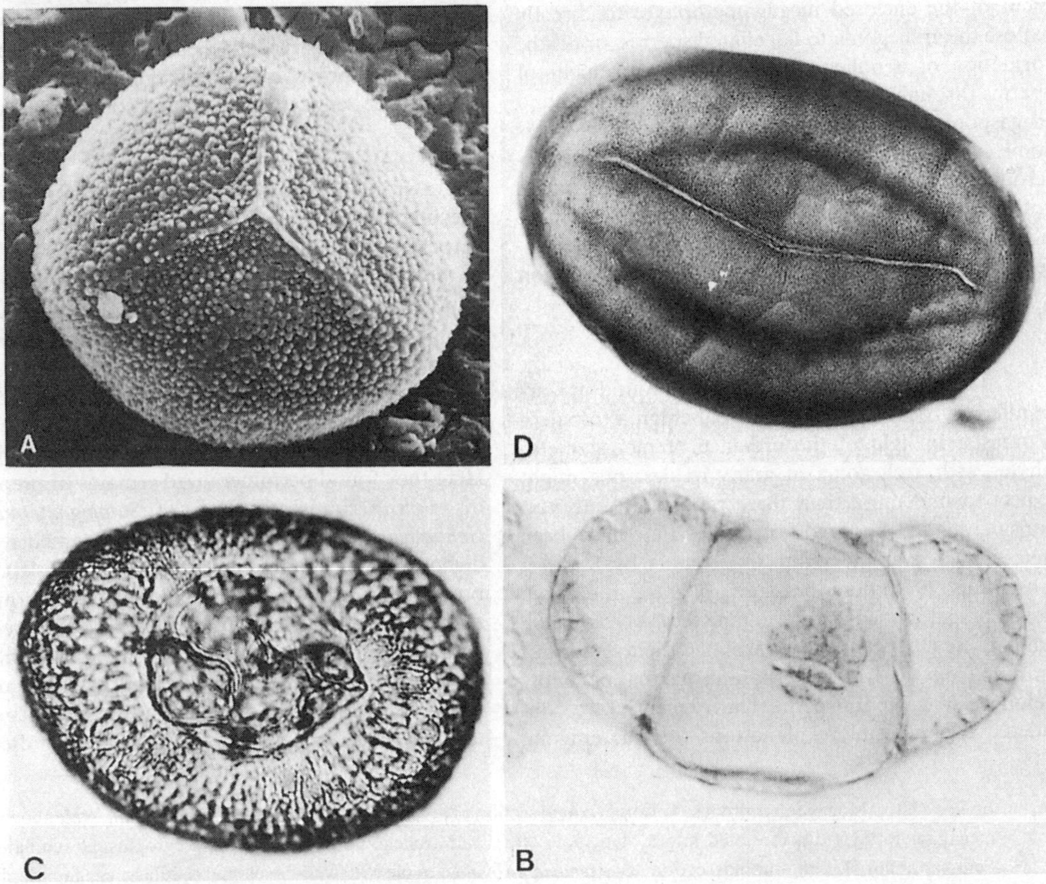

Figure 22.14. Microgametophytes. **A.** *Vesicaspora* pollen grain with axial row of four microgametophyte cells. (From Millay & Eggert, 1974.) **B.** Monolete pollen grain from pollen chamber of a pteridosperm ovule, *Pachytesta hexangulata.* Note two structures resembling sperms of extant cycads. (From Stewart, 1951.)

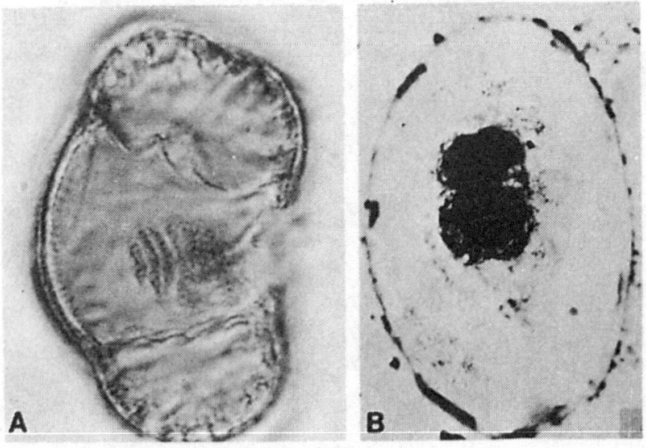

Archegonia occur in the distal end of the cellular megagametophyte. Here they have the appearance of an ovoid cavity, which may be partially filled with remains of vesiculate cytoplasm (Fig. 2.12B). Upper Devonian preovules are the first known to contain archegonia (Matten et al., 1984) with three archegonia being the most common number. Canals from archegonia open into a shallow ringlike depression around a low "tent pole" apparatus (Fig. 22.15A) at the apex of the megagametophyte. This structure with the basal ring-like depression forms an archegonial chamber similar to that found in the ovules of modern *Ginkgo* just prior to fertilization by swimming sperms. The cellular structure of archegonia is seldom preserved. Stidd & Cosentino (1976), however, were able to make out neck canal cells in the sunken archegonia of the Carboniferous ovule *Nucellangium.* In the same beautifully preserved material they found a funnel-shaped canal extending between the neck of an archegonium and the pollen chamber. They suggest that the canal is a pollen tube.

What information we have concerning the movement of sperms to archegonia of Upper Devonian and Carboniferous ovules or preovules suggests two possible mechanisms: (1) the rupture of the pollen chamber floor and megaspore membrane by enlargement of the enclosed megagametophyte to directly expose the archegonia to flagellated sperms; or (2) the formation of a pollen tube to allow movement of sperms through an indehiscent megasporangium and megaspore membrane to the archegonia. There is some evidence for both mechanisms in the Carboniferous, but they are not known to occur in any plants with hydrasperman reproduction. Therefore, it is most likely that pollen tubes evolved in the Upper Carboniferous. The importance of this step in the evolution of the seed habit cannot be overemphasized.

Embryos

The paucity of embryos in seeds found in Carboniferous rocks has become a real enigma to paleobotanists. In light of the presence of megagametophytes with archegonia in many ovules, one might expect to find more than three reports of embryos, considering the hundreds of ovules that have been investigated over the years. One report by Stidd & Cosentino (1976) describes an early stage in embryo development in the Pennsylvanian seed *Nucellangium,* while Long (1975) reports an isolated dicotyledonous embryo from the Lower Carboniferous that probably belonged to a pteridosperm. Until recently, we considered two plausible explanations for this enigma: (1) embryo development, once initiated, continued outside the integument; or (2) only a very small percentage of the ovules developed embryos. Lack of preservation cannot be claimed as the factor because the delicate tissues of the female gametophytes are frequently permineralized. That embryos can be and are preserved has recently been demonstrated by the discovery of polycotyledonous embryos in the seeds of an Upper Pennsylvanian conifer cone from Kansas.

The cupule

From an evolutionary point of view it has been speculated that the cupule was the precursor to the carpel (ovary wall) or became the second integument of flowering plant ovules. Because of its possible importance in understanding the evolutionary origin of angiospermy, we should have some comprehension of the origin of the cupule.

One of the oldest known cupules is described by Gillespie, Rothwell, & Scheckler (1981) from Upper Devonian (Famennian) rocks of West Virginia. Here they found permineralized remains of preovules of *Elkinsia* similar to those of *Salpingostoma* and *Genomosperma latens,* attached to segments of a dichotomizing branching system (Fig. 22.4B). The branching system forks twice at the base to form four segments, each of which bears an interior preovule and an exterior cupule lobe. The cupule lobe, in turn, forks twice. In the basal portion, the forking branches of the cupule lobe are fused. Distally they form four terete branches that are free from one another. Because

Figure 22.15. Megagametophytes. **A.** Longitudinal section of apical half of *Taxospermum undulatum* ovule showing megagametophyte differentiated into a "tent pole" (tp), and archegonium (a) containing cytoplasmic remnant. Pennsylvanian. **B.** Longitudinal section of *Lagenostoma ovoides* ovule with well-developed cellular megagametophyte (m) and archegonial cavity. Carboniferous. (**A,B** from Taylor & Millay, 1979.)

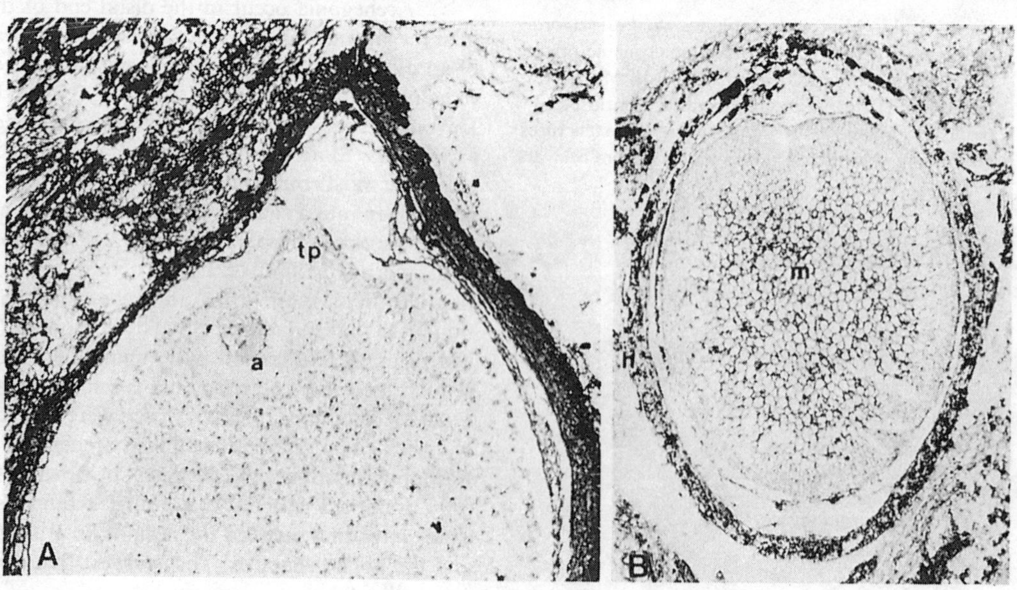

of the terete nature of the dichotomizing branches that comprise them, it is clear that the cupules of the *Elkinsia* reproductive structures are composed of telome trusses in which some planation and syngenesis have occurred.

In their description of the slightly younger *Archaeosperma*, Pettitt & Beck (1968) suggest that the cupule of this genus is foliar in origin because it is planated and partially webbed. Phillips, Andrews, & Gensel (1972) provided some evidence that the cupule of *A. arnoldii* was derived from basal sterile leaves on the fertile branches of *Archaeopteris*. Although the cupules of *A. arnoldii* are foliar in appearance, Long's (1960a, 1965, 1966) studies of cupulate structures from the Lower Carboniferous leave little doubt that they originated from sterile telome trusses. This is especially evident in the cupulate structure of *Eurystoma angulare* (Fig. 22.16), where the cupule is composed of a system of dichotomizing terete branches. The branches of the cupule dichotomize two to three times at right angles to form a three-dimensional system of branches. As many as 10 preovules occur in the cupule, each terminating an overtopped branch. In *Stamnostoma* (Long, 1960a, 1965), the telomic system is organized into two cupules each of which may have contained four preovules. By repeated dichotomies, each at right angles to the one below, a ring of 16 distally placed terete branches was produced. The cupules of *Archaeosperma arnoldii* are basically similar in structure to those of *Elkinsia*. As discussed earlier, it is reasonable to interpret *A. arnoldii* as bearing four preovules in a single cupule where syngenesis of the cupule lobes is incomplete. Irrespective of the interpretation, it is clear that the leaflike units of *A. arnoldii* cupules are homologous and derived from branching systems in the same way as leaves of megaphyllous vascular plants. We should keep this basic concept in mind, especially, when we consider the origin of the angiosperm carpel in a later chapter.

Figure 22.16. *Eurystoma angulare* showing telome truss enveloping preovules to form a poorly defined cupule. Carboniferous. (Redrawn from Long, 1965.)

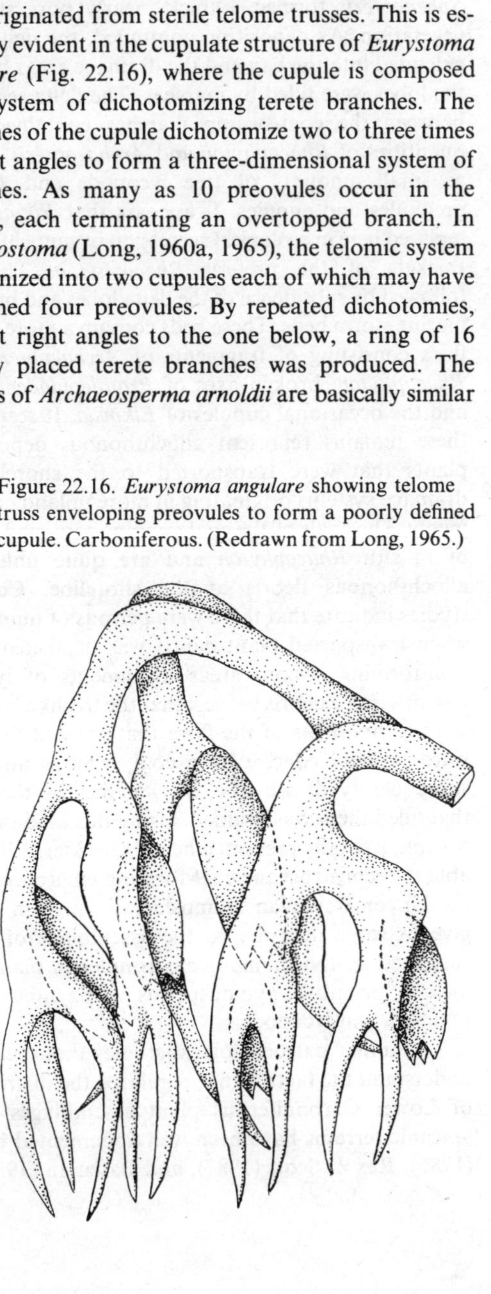

Preovule origin: monophyletic or polyphyletic?

Using the evidence for the hydrasperman type of reproduction found in the preovules of Upper Devonian and Lower Carboniferous gymnospermous plants, Rothwell (1986) and Rothwell & Scheckler (1988) have made a strong case for the monophyletic origin of this group. They visualize that hydrasperman reproduction, in which there is a special mechanism to facilitate pollination and fertilization in an undehisced megasporangium, evolved from heterosporous ancestors with pteridophytic reproduction. Here the magasporangium (nucellus) dehisced to allow flagellated, waterborne sperms to enter and move to the egg cells directly; or alternately by shedding the megagametophyte-containing megaspores into an aqueous environment where fertilization and subsequent embryo formation could occur away from the parent plant. To replace this pteridophytic type of reproduction in the ancestral gymnosperms with their single functional megaspore retained in a nucellus, a mechanism comprising a pollen chamber with a lagenostome and central column evolved for the capture of prepollen and their sperm-forming microgametophytes. If this scenerio has merit, then it can be assumed that the primitive gymnosperms with hydrasperman reproduction, where wind-blown prepollen was utilized, could complete their life cycles in relatively dry habitats even without the benefits of integuments.

Some support for the derivation of the hydrasperman preovule from a pteridophytic ancestor has been supplied by Galtier & Rowe (1989), who have described a single permineralized, integumented preovule from the early Carboniferous of France. Eight integumentary lobes, which are fused only at the base, envelop a massive megasporangium with an undifferentiated apex. One functional megaspore containing an endosporic, cellular megagametophyte is covered by an intact nucellus. This preovule corresponds in its general morphology to the hypothetical stage in preovule evolution shown in Figure 22.6C. Galtier & Rowe favor the interpretation that this does indeed represent a primitive stage in preovule evolution from a pteridophytic ancestor and that it is a stage in ovule evolution that did not require a hydrasperman precursor. If this latter concept has merit, we are obliged to consider the possibility that ovule evolution among gymnosperms was polyphyletic, not monophyletic. The additional possibility remains that the preovule described by Galtier & Rowe represents an immature stage in development prior to the differentiation of the prepollen-receiving mechanism so characteristic of hydrasperman reproduction. As previously indicated, we will see in later chapters that the monophyletic versus the polyphyletic origin of the seed habit is a topic of considerable debate.

Ecology of some early seed ferns

Upper Devonian (Famennian) coals and their associated beds belonging to the Hampshire Formation of West Virginia contain fossils of the hydrasperman seed fern *Elkinsia*. In hopes of discovering the environmental conditions and the assemblage of plants to which *Elkinsia* belonged, Scheckler (1986) undertook a massive study of 30 sites. The Hampshire formation is part of a clastic wedge formed to the west of the Acadian Mountains, which were the product of a disturbance called the Acadian orogeny. By late Devonian and Early Carboniferous times, the Acadian orogeny was subsiding but erosion of the mountains continued to produce broad, flat coastal plains. The clastics of the Hampshire beds, which comprise a part of the coastal plains, were deposited under alluvial and fluvial conditions. It was in this geological setting that the plants comprising the coal swamps grew and reproduced. Scheckler was "able to infer the original composition of the plant communities, their spatial distribution over a landscape and the partitioning of habitats," only after intense studies of a great number of rock exposures in his effort to analyze the floral and faunal content of the beds, the indirect traces of biogenetic activity, and the taphonomic histories of the fossils.

In most localities the fossils are preserved as coalified compressions–impressions with some parts anatomically preserved by pyrite, marcasite, or iron hydroxides. A few localities produced permineralized peat (coal balls) with good cellular detail. The techniques employed for determining the biomass contributed by each taxon depended on the type of preservation. For compression–impression remains a modification of a technique worked out by Scott (1977, 1979) was used, while procedures developed by Phillips, Kunz, & Mickish (1977) and modified by Pryor (1988) were employed for coal balls. Of the many sites where outcrops of the Hampshire Formation were found, Scheckler (1986) selected four to include in his study. The most productive of these was the Elkins locality, which yielded the greatest number of species as both compression–impression fossils and coal balls. Initial studies revealed that the Elkins locality represented an Upper Devonian, deltaic coastal marsh with a coal-forming environment. The principal genera at the locality are *Archaeopteris, Callixylon, Rhacophyton, Barinophyton, Sphenopteris* (a possible pteridosperm leaf type), *Elkinsia*, and a *Protolepidodendropsis*-like lycopod.

After studies of the distribution of the floral assemblage and stratigraphy at the Elkins locality were completed, comparisons were made with stratigraphically similar adjacent marine and upland localities at other sites. From these investigations Scheckler concluded that the in situ (autochthonous) concentrations of fossil plants occurred on the levee tops at the leading edges of small delta lobes. The latter were being deposited into a shallow water barrier-protected bay (Rothwell & Scheckler, 1988). In this part of the deltaic coastal marsh, *Elkinsia* accompanied by *Sphenopteris* formed a thick, nearly pure stand of vegetation. As deposition continued, the delta lobes extended into the bay and the drainage areas between the lobes were filled by marshes. The delta sediments between the coal-swamp marshes contained large quantities of *Rhacophyton* and *Archaeopteris*, as well as small amounts of tree lycopods and *Elkinsia* preovules and cupules. It appears that *Elkinsia* was replaced here in ecological succession by autochthonous thickets of the prefern *Rhacophyton*. This was followed by subsidence of the delta lobes and burial by marine storm beds. These beds contain a more diverse flora consisting of fragments of *Archaeopteris*, and *Rhacophyton*, broken axes of *Protolepidodendropsis*, and the occasional cupules of *Elkinsia*. It seems that these remains represent allochthonous deposits of plants that were transported to the shoreline by drainage systems originating in more upland environments. The coals of the coal swamps consisted mostly of in situ *Rhacophyton* and are quite unlike the allochthonous debris at the shoreline. Coal-ball studies indicate that there were periods of inundation when transported plant debris was deposited in the coal-forming swamp areas. Fragments of lycopod remains in the debris suggest that the treelike lycopods were components of the flora that grew at the landward, upland edges of the coal swamps and were subsequently drawn into the marshes as the water that filled them was drained. Using this kind of information, partially presented here, Scheckler (1986) was able to reconstruct the coal-forming environments of the Upper Devonian Hampshire Formation and to give us some insight into the assemblage of plants sustained by the changing environments, plants that include primitive gymnosperms, progymnosperms, preferns, and lycopods.

Similar paleoecological studies that help us to understand the factors that regulated the distribution of Lower Carboniferous plant assemblages within volcanic terrains have been well documented by Rex (1986), Rex & Scott (1987), and Bateman (1988).

References

Andrews, H. N. (1961). *Studies in Paleobotany*. New York: Wiley.

Andrews, H. N. (1963). Early seed plants. *Science*, **142**, 925–31.

Andrews, H. N., Gensel, P. G., & Forbes, W. H. (1974). An apparently heterosporous plant from the Middle Devonian of New Brunswick. *Palaeontology*, **17**, 387–408.

Arnold, C. A. (1935). On seedlike structures associated with *Archaeopteris*, from the Upper Devonian of northern Pennsylvania. *Contributions from the Museum of Paleontology, the University of Michigan*, **4**, 283–86.

Bateman, R. M. (1988). Palaeobotany and palaeoenvironments of Lower Carboniferous floras of two volcanigenic terrains in Scottish Midland Valley. *Unpublished Ph.D. thesis*. University of London.

Brauer, D. F. (1980). *Barinophyton citrulliforme* (Barinophytales *Incertae sedis*, Barinophytaceae) from the Upper Devonian of Pennsylvania. *American Journal of Botany*, **67**, 1186–206.

Camp, W. H., & Hubbard, M. M. (1963). On the origins of the ovule and cupule in lyginopterid pteridosperms. *American Journal of Botany*, **50**, 235–43.

Chaloner, W. G. (1970). The rise of the first land plants. *Biological Reviews*, **45**, 353–76.

Chaloner, W. G., Hill, A. J., & Lacey, W. S. (1977). First Devonian platyspermic seed and its implications in gymnosperm evolution. *Nature*, **265**, 233–5.

Fairon-Demaret, M., & Scheckler, S. E. (1986). A propos de *Moresnetia. L'evolution des gymnosperms, approche biologique et paléobiologique*. Colloque-Fondation Louis Emberger/Charles Sauvage. A l'Université du Languedoc, Montpellier, France. Resumés des communications, p. 19.

Fairon-Demaret, M., & Scheckler, S. E. (1987). Typification and redescription of *Moresnetia zalesskyi* Stockmans, 1948, an early seed plant from the Upper Famennian of Belgium. *Bulletin de L'Institut Royal des Sciences Naturelles de Belgique, Sciences de la Terre*, **57**, 183–99.

Florin, R. (1936). On the structure of pollen grains in the Cordaitales. *Svensk Botanisk Tidskrift*, **30**, 624–51.

Galtier, J., & Rowe, N. P. (1989). A primitive seed-like structure and its implications for early gymnosperm evolution. *Nature*, **340**, 225–7.

Gillespie, W. H., Rothwell, G. W., & Scheckler, S. E. (1981). The earliest seeds. *Nature*, **293**, 462–4.

Halle, T. G. (1935). The position and arrangement of the spore-bearing organs believed to belong to the Paleozoic pteridosperms. *Compte Rendu, Congress pour l'Avancement des Advance Etudes Stratragraphic et du Géologie du Carbonifere, Heerlen*, **1**, 227–35.

Long, A. G. (1959). On the structure of "*Calymmatotheca kidstoni*" Calder (emended) and "*Genomosperma latens*" gen. et sp. nov. from the Calciferous Sandstone series of Berwickshire. *Transactions of the Royal Society of Edinburgh*, **64**, 29–44.

Long, A. G. (1960a). "*Stamnostoma huttonense*" gen. et sp. nov. – pteridosperm seed and cupule from the Calciferous Sandstone series of Berwickshire. *Transactions of the Royal Society of Edinburgh*, **64**, 201–15.

Long, A. G. (1960b). On the structure of "*Samaropsis scotica*" Calder (emended) and "*Eurystoma angulare*" gen et sp. nov., petrified seeds from the Calciferous Sandstone series of Berwickshire. *Transactions of the Royal Society of Edinburgh*, **64**, 261–80.

Long, A. G. (1961a). On the structure of "*Deltasperma fouldenense*" gen. et sp. nov., and "*Camptosperma berniciense*" gen. et sp. nov., petrified seeds from the Calciferous Sandstone series of Berwickshire. *Transactions of the Royal Society of Edinburgh*, **64**, 282–95.

Long, A. G. (1961b). Some pteridosperm seeds from the Calciferous Sandstone Series of Berwickshire. *Transactions of the Royal Society of Edinburgh*, **64**, 401–19.

Long, A. G. (1965). On the cupule structure of *Eurystoma angulare*. *Transactions of the Royal Society of Edinburgh*, **66**, 111–27.

Long, A. G. (1966). Some Lower Carboniferous fructifications from Berwickshire, together with a theoretical account of the evolution of ovules, cupules, and carpels. *Transactions of the Royal Society of Edinburgh*, **66**, 345–75.

Long, A. G. (1975). Further observations on some Lower Carboniferous seeds and cupules. *Transactions of the Royal Society of Edinburgh*, **69**, 267–93.

Matten, L. C., Fine, T. I., Tanner, W. R., & Lacey, W. S. (1984). The megagametophyte of *Hydrasperma tenuis* Long from the uppermost Devonian of Ireland. *American Journal of Botany*, **71**, 1461–64.

Matten, L. C., Lacey, W. S., & Lucas, R. C. (1980). Studies on the cupulate seed genus *Hydrasperma* Long from Berwickshire and East Lothian in Scotland and County Kerry in Ireland. *Botanical Journal of the Linnean Society*, **81**, 249–73.

Millay, M. A., & Eggert, D. A. (1974). Microgametophyte development in the Paleozoic seed fern family Callistophytaceae. *American Journal of Botany*, **61**, 1067–75.

Millay, M. A., Eggert, D. A., & Dennis, R. L. (1978). Morphology and ultrastructure of four Pennsylvanian prepollen types, *Micropaleontology*, **24**, 303–15.

Millay, M. A., & Taylor, T. N. (1975). Evolutionary trends in fossil gymnosperm pollen. *Review of Palaeobotany and Palynology*, **21**, 65–91.

Niklas, K. J. (1981a). Airflow patterns around some early seed plant ovules and cupules: implications concerning efficiency in wind pollination. *American Journal of Botany*, **68**, 635–50.

Niklas, K. J. (1981b). Simulated wind pollination and airflow around ovules of some early seed plants. *Science*, **211**, 275–77.

Niklas, K. J. (1983). The influence of Paleozoic ovule and cupule morphologies on wind pollination. *Evolution*, **37**, 968–86.

Pettitt, J. (1970). Heterospory and the origin of the seed habit. *Biological Reviews*, **45**, 401–15.

Pettitt, J., & Beck, C. B. (1968). *Archaeosperma arnoldii* – a cupulate seed from the Upper Devonian of North

America. *Contributions from the Museum of Paleontology, the University of Michigan,* **22,** 139–54.

Phillips, T. L., Andrews, H. N., & Gensel, P. G. (1972). Two heterosporous species of *Archaeopteris* from the Upper Devonian of West Virginia. *Palaeontographica B,* **139,** 47–71.

Phillips, T. L., Kunz, A. B., & Mickish, D. J. (1977). Paleobotany of permineralized peat (coal balls) from the Herrin (No. 6) Coal Member of the Illinois Basin. In *Interdisciplinary Studies of Peat and Coal Origins,* eds. P. N. Given & A. B. Cohen. Geological Society of America Microform Publications, **7,** 18–49.

Pryor, J. S. (1988). Sampling methods for quantitative analysis of coal-ball plants. *Palaeogeography, Palaeoclimatology, Palaeoecology,* **63,** 313–26.

Ramanujam, C. G. K., & Stewart, W. N. (1969). A *Lepidocarpon* cone tip from the Pennsylvanian of Illinois. *Palaeontographica B,* **127,** 159–67.

Rex, G. M. (1986). The preservation and palaeoecology of Lower Carboniferous silicified plant deposits at Esnost, near Autun, France. *Geobios,* **19,** 773–800.

Rex, G. M., & Scott, A. C. (1987). The sedimentology, palaeoecology and preservation of Lower Carboniferous plant deposits at Pettycur, Fife, Scotland. *Geology Magazine,* **124,** 43–66.

Rothwell, G. W. (1972). Evidence of pollen tubes in Paleozoic pteridosperms. *Science,* **175,** 772–4.

Rothwell, G. W. (1979). Evidence for a pollination-drop mechanism in Paleozoic pteridosperms. *Science,* **198,** 1251–2.

Rothwell, G. W. (1986). Classifying the earliest gymnosperms. In *Systematic and Taxonomic Approaches in Palaeobotany,* eds. R. A. Spicer & B. A. Thomas, Systematics Association Special Vol. 31: Oxford: Oxford University Press.

Rothwell, G. W., & Scheckler, S. E. (1988). Biology of ancestral gymnosperms. In *Origin and Evolution of Gymnosperms,* ed. C. B. Beck. New York: Columbia University Press.

Rothwell, G. W., Scheckler, S. E., & Gillespie, W. H. (1989). *Elkinsia* gen. nov., a late Devonian gymnosperm with cupulate ovules. *Botanical Gazette,* **150,** 170–89.

Rothwell, G. W., & Taylor, T. N. (1982). Early seed plant wind pollination studies: A commentary. *Taxon,* **31,** 308–9.

Rothwell, G. W., & Wight, D. C. (1989). *Pullaritheca longii* gen. nov. and *Kerryia mattenii* gen. et sp. nov., Lower Carboniferous cupules with ovules of the *Hydrasperma tenuis*-type. *Review of Palaeobotany and Palynology,* **60,** 295–309.

Scheckler, S. E. (1986). Geology, floristics and paleoecology of late Devonian coal swamps from Applachian Laurentia (U.S.A.). *Annals de la Société Géologique de Belgique,* **109,** 209–22.

Scott, A. C. (1977). A review of the ecology of Upper Carboniferous plant assemblages, with new data from Strathclyde. *Palaeontology,* **20,** 447–73.

Scott, A. C. (1979). The ecology of coal measure floras from northern Britain. *Proceedings of the Geological Association,* **90,** 97–116.

Stewart, W. N. (1951). A new *Pachytesta* from the Berryville locality of Southeastern Illinois. *American Midland Naturalist,* **46,** 717–42.

Stidd, B. M., & Cosentino, K. (1976). *Nucellangium:* gametophytic structures and relationship to *Cordaites. Botanical Gazette,* **137,** 242–49.

Taylor, T. N., & Millay, M. A. (1979). Pollination biology and reproduction in early seed plants. *Review of Palaeobotany and Palynology,* **27,** 329–55.

Walton, J. (1953). The evolution of the ovule in pteridosperms. *British Association for the Advancement of Science,* **10,** 223–30.

23

Paleozoic gymnosperms with fernlike leaves

Prior to the introduction of cladistic analyses of seed plants (Crane, 1985, 1988; Doyle & Donoghue, 1986), it was customary to regard the gymnosperms as having evolved from the Progymnospermopsida as two phyletic lines (Beck, 1970). Today, however, the relationships among seed plants are less certain than previously thought (Rothwell, 1986), especially when the fossil evidence provided by archetypic gymnosperms is considered (Chapter 22). In the first edition of this book the Gymnospermopsida was reinstated as a class of vascular plants comprising at least two divergent phyletic lines, each of which evolved certain distinctive characteristics (Stewart, 1981). One of the lines is represented by those seed plants we commonly call the cycadophytes. We think of these plants as having large, frondlike leaves, stems with well-developed pith and cortex, and manoxylic secondary xylem. Xylem of this type is characteristically composed of very long tracheids that are relatively large in diameter and have several rows of circular bordered pits, mostly on the radial walls. The wood rays are multiseriate and very high.

The second line includes those gymnosperms we call the coniferophytes. Although there are many exceptions, the coniferophytes generally can be distinguished by their simple leaves and compact woody stems with pycnoxylic secondary wood. This type of wood first appears in the progymnosperms, where the tracheids of the secondary xylem are small in diameter and the rays are narrow.

While the recognition of cycadophytes and coniferophytes remains a useful concept for helping identify specimens of fossil gymnosperms, recent studies have demonstrated that these categories represent growth habits rather than evolutionary lineages (Rothwell, 1982; Crane, 1985; Doyle & Donoghue, 1986). We now recognize that there are several distinct groups included within both cycadophytes and coniferophytes.

At the present time, however, we do not know for sure if seed plants are monophyletic, biphyletic, or polyphyletic, the answer being at least partly determined by how we define "seed plants." We do recognize that the Class Gymnospermopsida is defined by the absence of flowering-plant characters, rather than by a uniquely derived character or set of characters. When considered in this light, the gymnosperms do not appear to be a natural lineage (that is, not a monophyletic group). It is anticipated that a new and more natural classification ultimately will be developed that recognizes a major clade of plants that is characterized by indehiscent megasporangia (i.e., plants with preovules plus all plants with seeds). However, such a classification has not yet been formulated. Therefore, as was done with the ferns in recognizing the artificial Class Filicopsida, we will continue to use the concept of Gymnospermopsida until a more natural classification emerges.

The first order of gymnosperms we will consider is the seed ferns or Pteridospermales. In the past, seed plants with dissected fernlike leaves that range in age from Devonian to Jurassic have been included here, resulting in an order that has become so broadly defined as to be systematically meaningless (Crane, 1985; Galtier, 1988). In this book we segregate many of the plants traditionally included in the Pteridospermales to their own orders. These include all of the Mesozoic forms as well as some Paleozoic groups with obviously derived characters. The plants we retain in the Pteridospermales are primarily the most ancient and presumably the most primitive of seed plants, as well as their immediate descendants. These include Upper Devonian plants with preovules (Chapter 22); Lower Carboniferous (Mississippian) plants with distinctly ribbed protosteles; and the families Lyginopteridaceae, Calamopityaceae, Medullosaceae, and Callistophytaceae.

Pteridospermales

The verification of seed ferns by Oliver & Scott (1904) (Chapter 3) represents another epoch-making discovery in the history of paleobotany. Prior to the publication where they demonstrated the probable attachment of ovules (*Lagenostoma*) to *Lyginopteris*-type stems, several other investigators had accumulated sufficient evidence suggesting the existence of this unusual group of Carboniferous plants that seemed to share vegetative characteristics of ferns and cycads. This bizarre combination of characteristics prompted the recognition of a new class, the Cycadofilices, which was later to be renamed the Pteridospermae by Oliver & Scott. The events leading to the establishment of the seed ferns are well summarized by Arnold (1947).

Some pteridosperms are known to have had the general appearance of modern tree ferns, while others had a more prostrate, scrambling, or vine-like habit (Fig. 23.1). Their large fronds were usually spirally arranged on the stem, and often the rachis or petiole dichotomized at least once proximal to the first basal pinnae. Ovules and microsporangia were borne on unmodified or only slightly modified foliage, and in this characteristic they are obviously different from coniferophytes, where the reproductive structures are borne in cones. Ovules of some pteridosperms are like those of modern cycads, but their eusporangiate

Figure 23.1. Restoration of seed ferns in a Carboniferous swamp setting. Note bifurcate fronds with large ovules and vinelike nature of specimen supported by the trunk of a lepidodendrid. (Photograph courtesy of the Field Museum of Natural History, Chicago.)

microsporangia differ in usually being fused to form a synangiate organ.

The outer cortical regions of pteridosperm stems characteristically show a supporting system of sclerenchyma strands that may anastomose to form a fibrous network. Protosteles are present in stems of many pteridosperms, but some are eustelic. The primary xylem is predominantly mesarch. The vascular cambium is bifacial, producing secondary phloem and secondary xylem. A large number of these plants have manoxylic wood of the cycad type, but others have pycnoxylic wood like that of conifers. The differences in wood appear to be related to the growth architecture of the plants, with manoxylic wood found on smaller, less-branched forms and pycnoxylic wood produced by those that grew into large much-branched trees (Long, 1979, 1987; Mapes, 1985).

Because of the predominance of their fernlike and cycadophytic characteristics outlined above, the pteridosperms were for many years thought to be an evolutionary link between ferns and the cycads. We have already examined evidence, however, which tells us that the origin of the Filicopsida was independent of the cycadophytes, and thus any similarities shared by the two groups must be interpreted as examples of parallel evolution.

There is reason to believe that the seed ferns originated from protostelic Aneurophytales of the Middle Devonian (Beck, 1970; Rothwell & Erwin, 1987). The Upper Devonian and Lower Carboniferous species are usually assigned to the Lyginopteridaceae or Calamopityaceae, depending primarily on features of the vascular tissues in the frond rachis (Galtier, 1988). Fronds with a ring of bundles in the rachis are placed in the genus *Kalymma* of the Calamopityaceae, whereas those with only one or two W-, U-, or V-shaped traces have been assigned to the Lyginopteridaceae as the genus *Lyginorachis*. In addition, two Lower Carboniferous genera with T-shaped rachial traces, *Triradioxylon* and *Buteoxylon*, have been placed in the Buteoxylaceae (Barnard & Long, 1973, 1975), a family of primitive pteridosperms with fronds that do not fork.

Despite the variations in vegetative structure among the Lyginopteridaceae, Calamopityaceae, Buteoxylaceae, and the most ancient seed ferns, we wish to stress that most or all of the species of these families had cupulate ovules with hydrasperman reproduction, and grouped or synangiate pollen organs that produced trilete prepollen (Meyer-Berthaud, 1989). Thus one can conclude that the species of these families probably represent fairly closely related plants, which reflect the initial evolutionary radiation of seed plants. It is also reasonable to suggest that the Medullosaceae and Callistophytaceae diverged separately from this complex of families with hydrasperman reproduction; the Medullosaceae first occurring at the base of the Upper Carboniferous and the Callistophytaceae appearing significantly later during the Middle Pennsylvanian (Rothwell, 1987).

The most ancient seed ferns

Over the past several years, studies by Barnard & Long (1973, 1975), Long (1961), Galtier (1977), Matten et al. (1980), May & Matten (1983), Matten, Tanner, & Lacey (1984), and Rothwell & Erwin (1987) indicate that those plants with archetypic preovules were probably small seed ferns with protostelic stems and simple petiole traces similar to those of the Lyginopteridaceae. These plants occur in Upper Devonian and Lower Carboniferous deposits of Europe and North America. The most ancient of these are associated with the preovulate cupules *Elkinsia polymorpha* in Upper Devonian deposits of West Virginia. The stems have a three-ribbed, mesarch protostele, and produce fronds with fluted traces in the petiole bases (Fig. 23.2). As is characteristic of other seed ferns, the stems produce secondary xylem and have sclerenchyma bundles in the outer cortex. Associated fronds have a rachis that dichotomizes and pinnules that are similar to the genus *Sphenopteris* (Fig. 23.4E).

Several basically similar genera found in the uppermost Devonian deposits of Ireland have been described by Matten and others. The best known of these are *Laceya hibernica* (May & Matten, 1983) and *Kerryoxylon hexalobatum* (Matten et al., 1984). *Laceya* stems were larger and produced more wood than the Elkins plant, and had distinctive scleren-

Figure 23.2. *Elkinsia*, transverse section of stem. Upper Devonian.

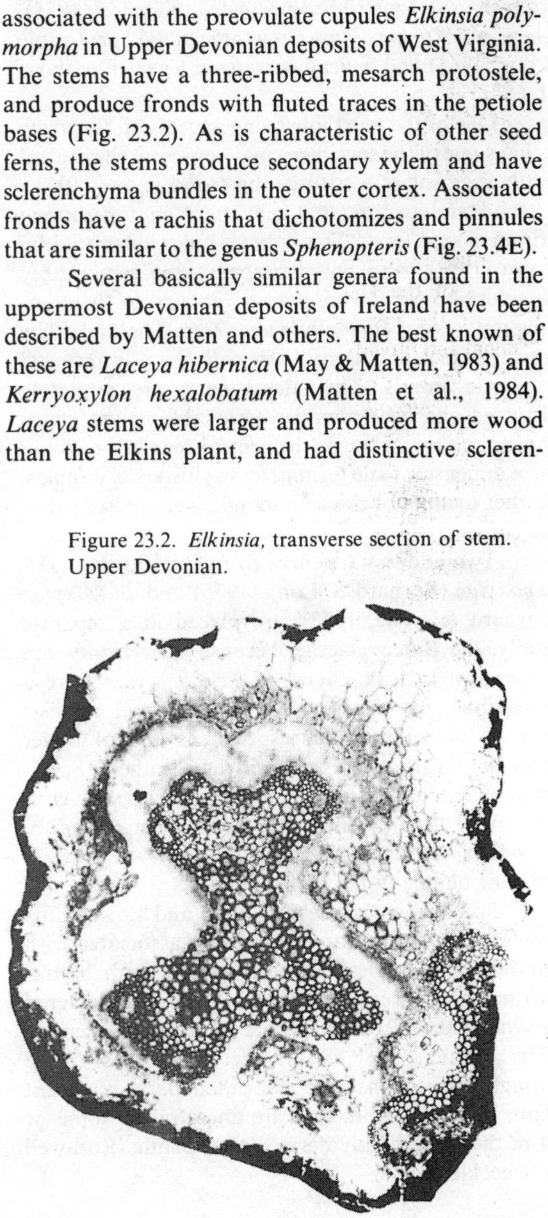

chyma strands near the periphery of the cortex. *Laceya* petioles often have a distinctly C-shaped bundle; they are known to dichotomize at some levels and to produce pinnate laterals at others. In *Kerryoxylon* the protostele typically has six ribs, and the petiole trace has a central rib that is directed away from the stele.

Additional genera that occur in the Irish flora are *Tristichia, Tetrastichia,* and *Buteoxylon,* all three of which were originally described from much more extensive specimens in a diverse flora from the Lower Carboniferous [Tournaisian (= lower Mississippian)] at Oxroad Bay in southern Scotland (Bateman & Rothwell, 1990). Of these, *Tetrastichia bupatides* is characterized by an opposite leaf arrangement and a cross-shaped, four-ribbed protostele (Gordon, 1938), whereas *Tristichia ovense* has a three-ribbed protostele (Long, 1961) and helically arranged leaves like those of *Elkinsia.*

One of the most intriguing aspects of *Tristichia* is that some of the specimens may represent the central fertile segment of a trifurcating frond (Long, 1961), a frond architecture that may be common among primitive seed ferns. This is beautifully illustrated by Rowe (1988), who has described some remarkably complete Lower Carboniferous remains of the compression genus *Diplopteridium* from England. The specimens include a narrow stem with helically arranged tripartate fronds. The lateral parts are pinnately dissected and bear pinnules assignable to the genus *Diplopteridium,* whereas the central segment forks in three dimensions and terminates in clusters of cupules. Neither ovules nor microsporangia were preserved in the cupules.

Two additional genera from Oxroad Bay, *Triradioxylon* (Barnard & Long, 1975) and *Buteoxylon* (Barnard & Long, 1973), are placed in a separate family, the Buteoxylaceae, because their fronds are not known to fork. Stems of *Triradioxylon* have a three-ribbed protostele, which is tiny and consists only of tracheids in some specimens and of larger protosteles with xylem parenchyma in others. Specimens of *Buteoxylon* have even larger steles with conspicuous xylem parenchyma. Both genera produce abundant secondary xylem and have a T-shaped vascular bundle in the rachis.

Specimens of these Devonian and Lower Carboniferous seed ferns are consistently associated with multiovulate cupules, dispersed ovules with hydrasperman reproduction (Rothwell, 1986), and clustered or synangiate microsporangiate structures (Bateman, 1988; Galtier, 1988; Meyer-Berthaud, 1989). We strongly suspect that the associated parts represent plants with preovules that are ancestral to some or all of the more highly derived seed plants (Rothwell & Scheckler, 1988).

Calamopityaceae

At one time paleobotanists treated the Calamopityaceae as a "catchall" group for stem and petiole fragments that displayed the general characteristics of pteridosperms. A review of the family by Read (1936) lists five different genera, all represented by permineralized specimens from the Upper Devonian and Lower Carboniferous of North America, England, Scotland, and Germany. Since Read's monographic work, several new genera have been added, and our understanding of the relationships of the Calamopityaceae has been greatly enhanced. A comprehensive review of the new information is presented by Galtier (1988).

Stenomyelon

Although Read (1936) was the first to suggest that *Stenomyelon* was a primitive member of the Calamopityaceae, it was not until the Aneurophytales of the Progymnospermopsida were established as ancestors of the seed ferns that the significance of this genus was fully realized (Namboodiri & Beck, 1968).

Stenomyelon differs from other genera of the Calamopityaceae in having a three-ribbed protostele that in its organization is highly similar to the three-lobed protosteles of the Aneurophytales. *S. primaevum* (Fig. 23.3A) (Long, 1964) has a protostele com-

Figure 23.3. **A.** *Stenomyelon primaevum,* transverse section of stem. **B.** *S. tuedianum,* transverse section of vascular system in stem. Note radiating plates of parenchyma (stippled) dissecting the three-ribbed protostele (cellular). Submerged protoxylem (black circles); secondary xylem (hatched). **A,B:** Lower Carboniferous. (**A,B** redrawn from Long, 1964). **C.** *Calamopitys* sp., transverse section of stem. Upper Devonian. (Drawn from photographs from Read, 1936, and Beck, 1970.)

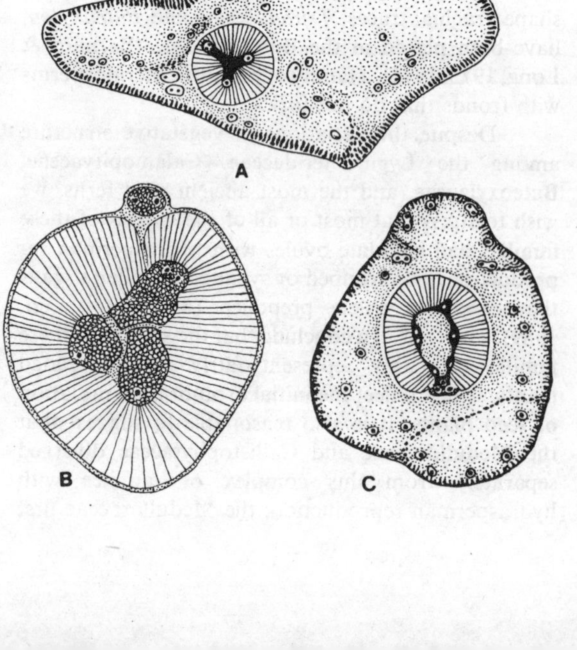

posed entirely of tracheids, with mesarch primary-xylem near the ends of the lobes. Leaf traces arise on the radii of the stem in a 2/5 spiral sequence (Beck, 1970) from the tips of the lobes. In their upward and outward course through the stem tissues to a leaf base, the leaf traces dichotomize three times to form eight traces that enter the leaf base. Another feature shared by these species is a well-developed hypodermal tissue consisting of radiating fibrous strands of sclerenchyma that anastomose. This characteristic, in addition to the mesarch primary xylem and manoxylic secondary wood, leaves little doubt that *Stenomyelon* is the stem of a pteridosperm.

The protosteles of *S. muratum* (Read, 1936) and *S. heterangioides* (Long, 1964) have "mixed" piths, while that of *S. tuedianum* (Fig. 23.3B) lacks the xylem parenchyma of a "mixed" pith, but has three radiating and interrupted bands of parenchyma instead. The bands have the effect of dividing the three-lobed protostele into three vertical columns of primary xylem. Except for the bases of petioles, nothing is known of the leaf structure of *Stenomyelon*. Reproductive structures have not been described.

Calamopitys

Calamopitys (Fig. 23.3C) is another stem genus probably belonging to Upper Devonian–Lower Carboniferous pteridosperms. The primary vasculature of their stems can vary from those that are essentially protostelic to those that have eusteles. Galtier (1975) was able to demonstrate this range of ontogenetic variability in stems belonging to *Calamopitys embergeri*. The eustelic stems usually have a "mixed" pith composed of parenchyma cells and scattered tracheids, with a ring of five sympodia at the periphery of the pith. The primary xylem comprising the axial sympodia is mesarch.

Depending upon the species, a single leaf trace arises by a radial or tangential division of a sympodium. The significance of the evolution of the eustele and method of leaf trace departure from axial sympodia in the Calamopityaceae are treated more fully later in this chapter.

In its upward and outward course through the cortex of the stem, a leaf trace divides dichotomously to form several bundles that assume a C-configuration in the petiole. Isolated petioles of the Calamopityaceae are placed in the genus *Kalymma*, which shares many characteristics with the medullosan pteridosperm petioles of the genus *Myeloxylon*. Like the fronds of many pteridosperms, those of *Calamopitys* have a rachis that dichotomizes. Internally the bundles of the rachis fuse tangentially prior to the dichotomy of the rachis and in this respect are similar to petioles of lyginopterid pteridosperms.

Other features of *Calamopitys* that indicate a close affinity with the pteridosperms are (1) the manoxylic secondary xylem that surrounds the primary body of the stem stele, and (2) the radiating strands of sclerenchyma found near the periphery of the massive cortex. The fronds of *Calamopitys* arise in a 2/5 phyllotactic sequence, a common leaf arrangement found among pteridosperms as well as other vascular plants. Axillary branching in *Calamopitys* has been discovered in permineralized specimens from the Lower Carboniferous (Galtier & Holmes, 1982). This discovery indicates that axillary branching was achieved earlier among the pteridosperms than in ferns where it first appears in the Upper Carboniferous.

Some other genera that are often ascribed to the Calamopityaceae have pycnoxylic wood. Among these are *Eristophyton* (Long, 1987), *Bilignia* (Bateman, 1988), and *Bostonia* (Stein & Beck, 1978) from the Lower Mississipian (Tournaisian). In transverse section the stem of *Bostonia* bears a superficial resemblance to the stem of the Carboniferous pteridosperm *Medullosa*. Like some stems of *Medullosa, Bostonia* has three vascular segments, each composed of a primary body surrounded by manoxylic secondary xylem. There are differences, however, in the composition of the primary body and position of the protoxylem, which suggest to the authors that we should not be too hasty to imply relationships between the Calamopityaceae and Medullosaceae. Additional studies of new specimens of *Bostonia* are required.

Chapelia, also from the Lower Mississippian, has been interpreted as a branching petiole with calamopityean or pteridosperm affinities. Only one specimen is described (Beck & Bailey, 1967). By two successive dichotomies the petiole of *Chapelia* forms four branches having a three-dimensional arrangement similar to the compression–impression fronds of *Diplothmema zeilleri.* In all preserved portions of the frond, the primary vasculature, which is a "mixed" protostele with mesarch primary xylem, is surrounded by secondary xylem of the manoxylic type. The proximal portions of the vascular system show a radial arrangement, while more distal segments display bilateral symmetry. This observation suggests to the authors that *Chapelia* is yet another example of a frond "caught in the act" of evolving from a primitive branching system. Affinity with lyginopterid pteridosperms is considered a possibility by Beck & Bailey because of the W-shaped (butterfly) configuration of the traces produced in the more distal portions of the petiolar system. This type of vascular configuration is described more fully in the section that follows.

Lyginopteridaceae

Although many authors shy away from characterizing the Lyginopteridaceae on the grounds that the family is poorly understood, there is a combination of characteristics that can be used to distinguish the Lyginopteridaceae from other families of pteridosperms (Stidd & Hall, 1970b; Rothwell & Erwin, 1987). These are as follows:

1. Cauline vasculature composed of a single vascular segment (monostelic)
2. Axillary branching
3. Small petioles with one V- or W-shaped trace resulting from fusion of several smaller traces
4. Fronds with a bifurcate rachis
5. Small ovules (3 to 5 mm in length) borne in cupules
6. Ovules with a hydrasperman pollen chamber
7. Nucellus fused with integument of ovule
8. Pollen organs small, laminar, or terminal in clusters on branches of fronds
9. Pollen or "prepollen" of the trilete type

As the study of Lyginopteridaceae has progressed it has become abundantly clear that there are many exceptions to the characteristics listed above and that those plants assigned to the family exhibit considerable morphological variability. Very useful summaries of this variability have been prepared by Taylor & Millay (1981a), and Galtier (1988).

Lyginopteris

Lyginopteris is known in the permineralized state in coal balls from the Lower and Middle Coal Measures (Westphalian A) of Yorkshire and Lancashire. It has also been reported from beds of similar age in continental Europe.

The stems of *Lyginopteris* do not exceed 3 to 4 cm in diameter, and for this reason the plant is depicted as vinelike and supported by surrounding trees and shrubs. The stems are known to have branches, with most branches arising in the axils of leaves. The leaves were large planated fronds that, when found as compression–impression fossils, are placed with form genera of those *Sphenopteris* or *Pecopteris* types with bifurcate rachises.

The spirally arranged fronds bore either cupulate ovules of the *Lagenostoma* type or synangiate clusters of microsporangia, presumably of the *Crossotheca* type on modified pinnules. *Crossotheca*-type microsporangia produce trilete prepollen of the dispersed type frequently found in *Lagenostoma* pollen chambers.

Large capitate glands (Fig. 23.4A) occur in the surfaces of stems, leaves, and cupules. These capitate glands, you may recall, are the structures that made it possible for Oliver & Scott (1940) to establish the existence of seed ferns (Chapter 3).

The stem of *Lyginopteris* (Fig. 23.4B) has a eustele with scattered nests of sclerotic cells in the pith. The eustele is surrounded by manoxylic secondary wood only a few millimeters thick. The tracheids of the wood are large with multiseriate bordered pits on the radial walls. The rays are numerous and some are quite wide. The vascular cambium producing the secondary xylem is bifacial and develops secondary phloem, to the outside of which is a "pericycle" with clusters of short cells forming sclerotic nests similar to those in the pith. A peridermlike tissue is often present in the outer tissues of the "pericycle." An inner cortex of uniform parenchyma cells is surrounded by an outer cortex with the characteristic structure of radially broadened fibrous strands that form an anastomosing network (Fig. 23.5) that is continuous into the base of the petioles.

After being formed by the division of an axial sympodium, the leaf trace divides into two bundles that bend outward and upward through the secondary wood, some of which may accompany the leaf traces as they traverse the cortex. In the cortex the leaf trace bundles reunite in the petiole to form a single V-, Y-, or W-shaped trace (Fig. 23.4C). When preserved as isolated fragments, petioles showing these "butterfly" traces are usually placed in the form genus *Lyginorachis*.

A highly interesting specimen of *Lyginorachis* was found by Long (1963) in association with stems of *Pitys primaeva*. At the point of bifurcation of the petiole, a branch arises with a three-ribbed protostele, highly similar in structure to protosteles of aneurophytes and *Stenomyelon* of the Calamopityaceae. The branches of *Lyginorachis* with three-ribbed protosteles are called the *Tristichia* type because of their resemblance to the presumed stem *Tristichia* (Long, 1961). A compression–impression fossil of *Sphenopteris*-type foliage (Fig. 23.4D) revealed a branch in the same position as the *Tristichia* branch (Long, 1963). The branch terminated in clusters of cupules having the characteristics of *Stamnostoma*. Long suggests that the stemlike rachises of the *Tristichia* type may have borne *Telangium* microsporangia. Galtier (1977) has confirmed the cauline nature of *Tristichia longii*, an observation that supports the idea that the fronds and cauline portions of the Lyginopteridaceae are homologous.

Cupules studded with capitate glands have been found attached to compression–impression fossils of *Sphenopteris hoeninghausi*, where the cupules terminate the ultimate naked branches of the frond. The frond, which is several times compound, has its petiole and rachises covered with capitate glands. Distal to the dichotomy of the petiole, the planated opposite pinnae are nearly at right angles to the rachis. Except where they are replaced by cupules, the ultimate pinnules of this *Sphenopteris* species (Fig. 23.4E) are small lobed structures, borne alternately on ultimate pinnae rachises.

The cupules have been described as tulip-shaped or like the husk of a hazelnut that more or less enclosed a single erect ovule attached by its base to the bottom of the cupule (Fig. 23.6A). The envelope formed by the cupule is deeply lobed in the distal half. Reconstructions show 8 to 10 lobes that apparently opened outward, allowing the maturing ovule to be shed. The cupule is supplied with a ring of 9 to 10 vascular bundles arising from the vascular system of the pedicel. The vascular bundles divide distally so that each lobe is supplied by two or more strands. Apparently the ovules abscised as they developed, so that the cupules are usually found empty and the ovules they produced are scattered.

Ovules of *Lyginopteris* (Fig. 23.6B) assigned to the genus *Lagenostoma* are small barrel-shaped structures about 4.25 mm in diameter and 5.5 mm in length. An ovule comprises a single vascularized integument and a nucellus that is fused with the integument except in the apical region. The ring of 8 to 9 vascular bundles traversing the integument suggests that it evolved from a similar number of telomes that have become completely fused to form a short micropyle (see Chapter 22). At its apex the nucellus is differentiated into a small pollen chamber with a prolonged flask-shaped lagenostome that projects into the micropyle. A central column of parenchyma tissue rises from the pollen-chamber floor into the cavity formed by the flask-shaped lagenostome. The space delineated by the outer surface of the central column and the inner face of the lagenostome is the pollen chamber. It is here that "prepollen" of the trilete type has been found, indicating that the lagenostome played a role in directing the grains into the pollen chamber, as it does in other ovules with hydrasperman reproduction.

Figure 23.4. **A.** Capitate gland on *Lyginopteris oldhamia.* (Drawing based on photographs in Scott, 1909.) **B.** *L. oldhamia,* transverse section of stem. Outer fibrous cortex (oc); sympodium (sy); secondary xylem (sx); pith with nests of sclereids (p); leaf trace (lt). (Redrawn from Andrews, 1961.) **C.** Transverse section of the petiole, *Lyginorachis* sp., showing "butterfly" configuration of vascular bundle, capitate glands, and fibrous outer cortex. **D.** *Lyginorachis-type* rachis terminating in *Stamnostoma* cupules. (Redrawn from Long, 1963.) **E.** Pinnules of *Sphenopteris* foliage. (Redrawn from Arnold, 1947.) **A–E:** Carboniferous.

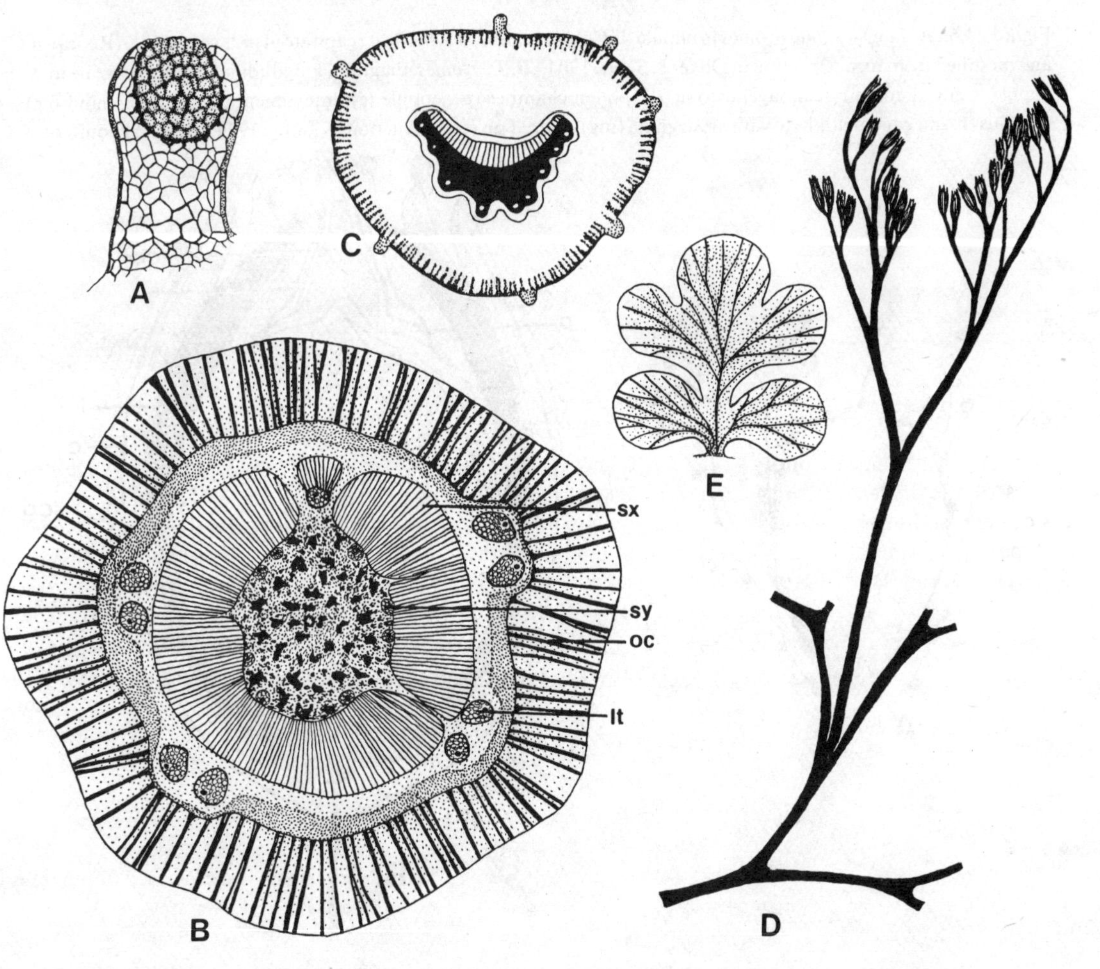

Figure 23.5. Tangential section through outer cortex of *Lyginopteris* showing anastomosing strands of fibrous cells. Carboniferous.

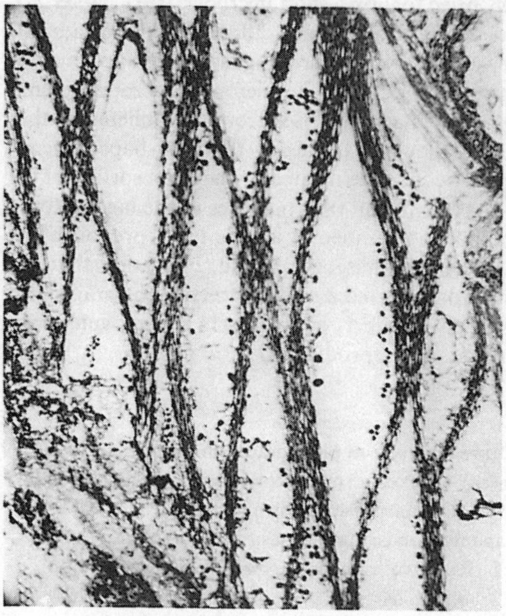

In his reconstruction of *Lagenostoma ovoides* containing megagametophytes, Long (1944) found that the central apical portion of the megagametophyte elongated into a "tent pole" apparatus that may have functioned in rupturing the megaspore membrane and floor of the pollen chamber to expose the archegonia to microgametophytes and their flagellated sperms. The "tent pole" also may have been important in pushing the central column into the distal end of the micropyle, effectively plugging it after pollination.

Although other parts of *Lyginopteris* have been well known for decades, there are still questions about the nature of the pollen-bearing structures. *Crossotheca,* a compression–impression fossil borne on *Sphenopteris-* or *Pecopteris*-type foliage, produces microspores with characteristics similar to those found in the pollen chambers of *Lagenostoma* ovules. The pollen-producing units of *Crossotheca* (Fig. 23.7) consist of flattened distal pinnules with fused sporangia pendant from the lower surface. These fertile pinnules bear a superficial resemblance to a miniature epaulet (Fig. 23.8A) and are associated with sterile pinnules on the same pinna. The partially or wholly fertile pinna may occur near the tip or at the midregion of the frond.

Figure 23.6. **A.** *Lagenostoma ovoides* in cupule of *Calymmatotheca* type. Note capitate glands on cupule. (Redrawn and modified from reconstruction in Oliver & Scott, 1904.) **B.** *L. ovoides*, diagram, longitudinal section of ovule in cupule. Central column (cc); lagenostome (l); pollen chamber (p); cupule (c); integument (i); capitate gland (cg); nucellus (n); megagametophyte with archegonia (mg). (Based on reconstruction by Long, 1944.) **A,B**: Carboniferous.

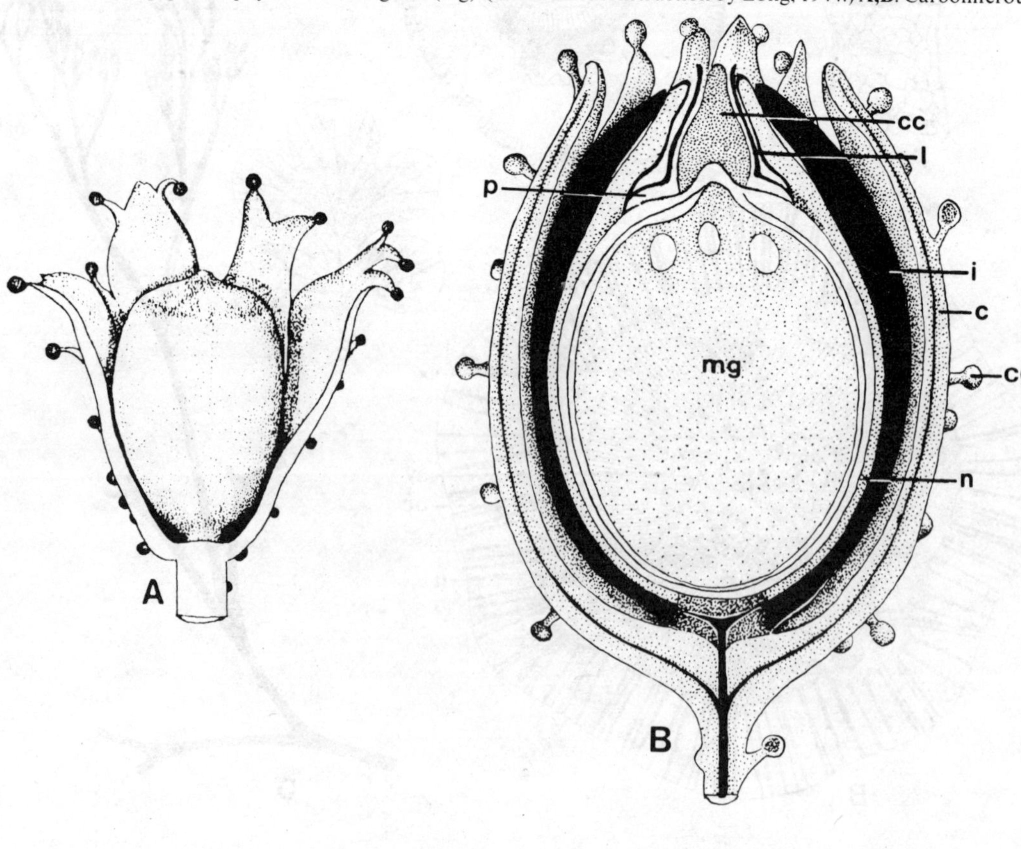

Other pollen-producing organs, presumed to have affinities with lyginopterid pteridosperms, are *Telangium, Telangiopsis,* and *Feraxotheca.* The latter (Fig. 23.8B), described by Millay & Taylor (1977), is known in the permineralized state and is quite similar in its characteristics to the compression–impression fossils of *Crossotheca.* The Pennsylvanian *Feraxotheca* has bilaterally symmetrical synangia attached to the abaxial surface of a vascularized planated pad of parenchyma. The sporangia of a synangium enclose a central space. Their inner walls facing the space are uniseriate as opposed to the thick outer walls. This suggests that dehiscence of the sporangia of a synangium was longitudinal along the inward-facing walls. Microspores (prepollen) isolated from the sporangia are of the *Cyclogranisporites,* dispersed type.

While *Crossotheca* and *Feraxotheca* bear their synangia on laminar structures, those of *Telangium* (Jennings, 1976) are borne terminally on fertile telome trusses (Fig. 23.8C) attached to an otherwise planated frond. The synangia are composed of sporangia arranged in a circle around a central space. In the Mississippian representatives, the sporangia walls are more or less uniform in thickness, while the younger Westphalian A species have inner walls with indications of longitudinal dehiscence along the inward facing walls. The synangium is vascularized in its base, but there is no indication of vascular tissue associated with individual sporangia.

Figure 23.7. *Crossotheca sagittata.* Note arrowhead-shaped fertile pinnules. Pennsylvanian.

Telangiopsis (Eggert & Taylor, 1971) is the name proposed for compression–impression fossils having the characteristics of the permineralized *Telangium.* Prepollen of the dispersed *Punctatisporites* type has been isolated from the compressed sporangia of this Mississippian genus.

Meyer–Berthaud (1989) has recently summarized our knowledge of primitive seed-fern pollen organs and explored the homologies of these fructifications more thoroughly. Pollen organs, and possibly ovulate cupules, occur on fronds with three types of morphologies. Some fronds have vegetative pinnules at the base, and are fertile distal to the fork of the rachis. Other plants appear to have forking fronds with fertile pinnules interspersed among the vegetative pinnules. In the third form the frond trifurcates, with the lateral pinnae being vegetative and the central unit being a three dimensionally branched fertile structure (Fig. 23.4D). A more thorough discussion of the fronds of this latter type is presented by Galtier (1988).

Heterangium

Paleobotanists are generally agreed that *Heterangium* is a close relative of *Lyginopteris.* The genus is fairly common in Upper Pennsylvanian coal balls of North America and the Coal Measures of England. Relative to some other genera of the pteridosperms little has been done in unraveling the life cycle of this genus.

The stem of *Heterangium* (Fig. 23.9) is small (0.5 to 5.0 cm in diameter) and protostelic. Its primary xylem consists of clusters of large metaxylem tracheids separated by parenchyma cells. The primary xylem is mesarch to exarch (Stidd, 1979), depending on the species. The striking resemblance of the *Heterangium* protostele and that of the extant fern *Gleichenia* was noted long ago, and suggested to early investigators that ferns and pteridosperms were related. If secondary xylem is present around the primary xylem, it is usually separated into segments by wide rays. Although smaller than the metaxylem tracheids, both secondary xylem tracheids and those of metaxylem have numerous circular bordered pits on the walls. Well-preserved phloem has been found exterior to and around the secondary xylem. The wide rays flare in the phloem, separating the phloem into wedgeshaped masses as in the phloem of the common linden (*Tilia*) when seen in cross section. In permineralized specimens, the phloem may be so well preserved that the sieve areas on the radial walls of the sieve cells can be observed (Hall, 1952).

Again depending on the species, the spirally

Figure 23.8. **A.** Single fertile pinnule of *Crossotheca sagittata* with pendant sporangia. **B.** *Feraxotheca culcitus*, branch bearing fertile pinnules. Single fertile pinnule enlarged. **A,B:** Pennsylvanian. (**A,B** redrawn from Millay & Taylor, 1977.) **C.** *Telangium* sp., portion of branching system with terminal synangia. Mississippian. (Based on Jennings, 1976, and Millay & Taylor, 1979.)

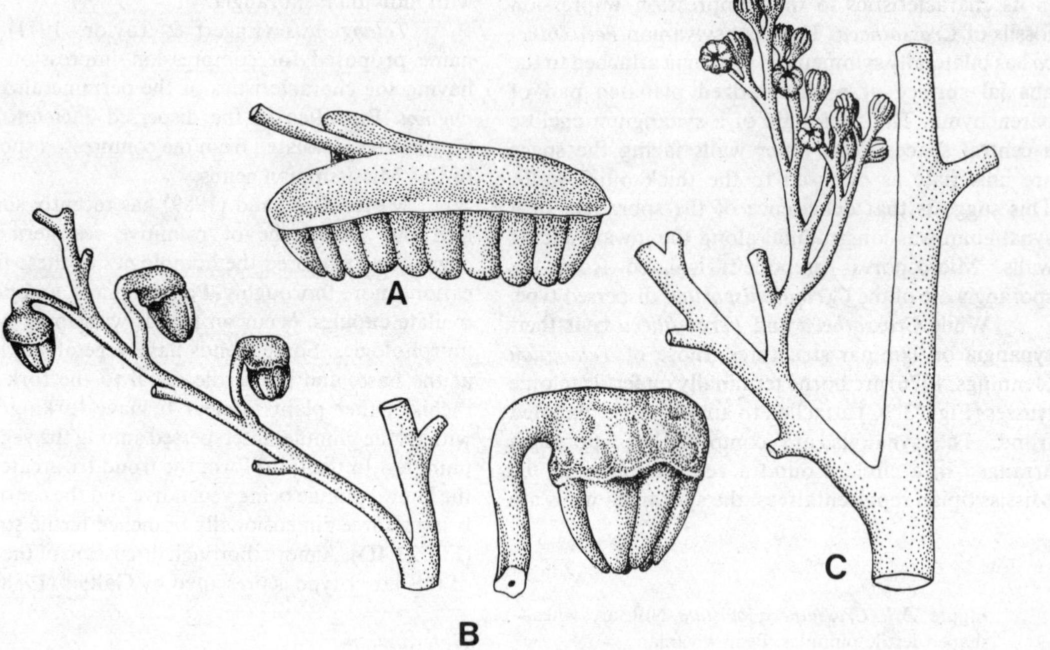

Figure 23.9. Diagram, transverse section of *Heterangium americanum* stem. Note "mixed" pith with large metaxylem cells (mx); "sparganum" cortex (sc); leaf trace (lt); secondary phloem (sp); secondary xylem (sx). Pennsylvanian.

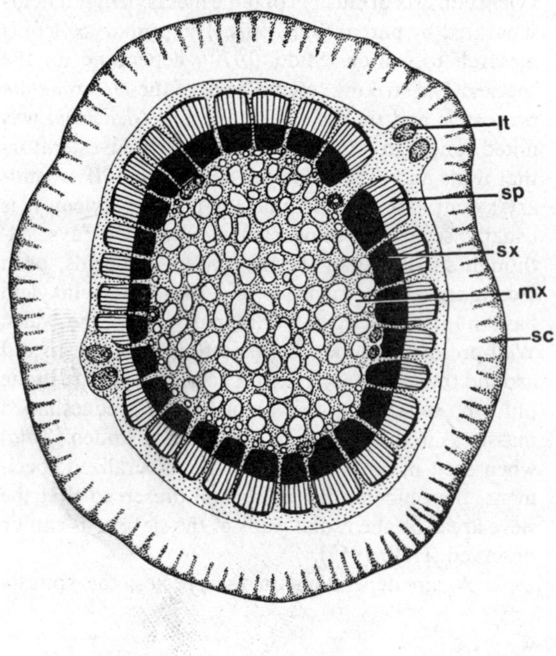

arranged leaf traces may arise singly and remain single until they enter a leaf base, or they may be double strands from the beginning, remain double, or divide further before the traces enter the leaf base (Pigg, Taylor, & Stockey, 1987). The inner cortex of *Heterangium* is characterized by horizontal plates of sclerotic cells. In the outer cortex, the sclerotic cells are arranged in vertical bundles to produce the "sparganum" or "dictyoxylon" configuration so characteristic of many pteridosperm stems.

Shadle & Stidd (1975), Jennings (1976), and Pigg, Taylor, & Stockey (1987) have ascertained that certain species of foliage belonging to the form genera *Rhodea* and *Sphenopteris* are the spirally arranged fronds of *Heterangium*. Specifically, Shadle & Stidd have demonstrated that stems of *H. americanum* bore fronds of the *S. obtusiloba* type (Fig. 23.10A). Typical of most pteridosperms, the petioles of *S. obtusiloba* dichotomize to produce a bipartite frond.

Although not found in actual connection, it is highly probable that ovules of the *Sphaerostoma* (Fig. 23.10B) and *Conostoma* (Fig. 23.10C) types were borne by *H. grievii* and *H. americanum*, respectively. In beds containing *Heterangium*, there are abundant remains of the small *Conostoma* ovules. They are notably absent in beds lacking *Heterangium* stems.

Sphaerostoma ovale (Fig. 23.10B) (Benson, 1914), from the Lower Carboniferous of Scotland, is a small radiospermic ovule borne singly in a delicate cupule. Although *Conostoma* has not been found in cupules, it shares many characteristics with *Sphaerostoma* in having a nucellus that is fused with the integument, except at the apex where the nucellus is modified into a pollen chamber, lagenostome, and central column (Fig. 23.11A). *Conostoma chappellicum* (Fig. 23.10C), a fairly representative species (Rothwell, 1971a; Stubblefield & Rothwell, 1980), is 3.6 to 5.5 mm long and 1.8 to 2.5 mm in diameter. The integument is composed of three layers: the outermost sarcotesta, consisting of a single layer of cells covered by a cuticle; a sclerotesta of radially aligned thick-walled cells; and inner fiberlike cells sandwiched between the sarcotesta and innermost endotesta. The latter is composed of several layers of thin-walled cells covered by an inner cuticle. Six traces of vascular tissue traverse the endotesta. Of the many species of *Conostoma*, most have four traces in the endotesta (Neely, 1951; Rothwell & Eggert, 1970; Rothwell, Taylor, & Clarkson, 1979). In some species, the integument in transverse section appears to be slightly six-angled; others are four-angled, and yet others are circular in outline.

Pollen chamber development in *Conostoma* has been worked out by several authors, including Rothwell (1971a), who used some exceptionally well-preserved permineralized specimens from the Lower Pennsylvanian. The sequence of events in the formation of the pollen chamber in *Conostoma* (Fig. 23.10D) is probably typical of species with hydrasperman re-

Figure 23.10. **A.** Diagram of *Heterangium* sp. frond with *Sphenopteris obtusiloba*-type pinnules. Pennsylvanian. (From Shadle & Stidd, 1975.) **B.** *Sphaerostoma ovale*. Note delicate cupule (c); integument (i); and nucellus (n). Lower Carboniferous. (From Andrews, 1961.) **C.** Diagram, *Conostoma chappellicum*, the probable ovule of *Heterangium*. Lower Pennsylvanian. (Redrawn from Rothwell, 1971a.) **D.** A series of diagrams showing the ontogeny of the apex of an ovule with a *Conostoma*-like lagenostome. See text for explanation. (Redrawn from Rothwell, 1971a.)

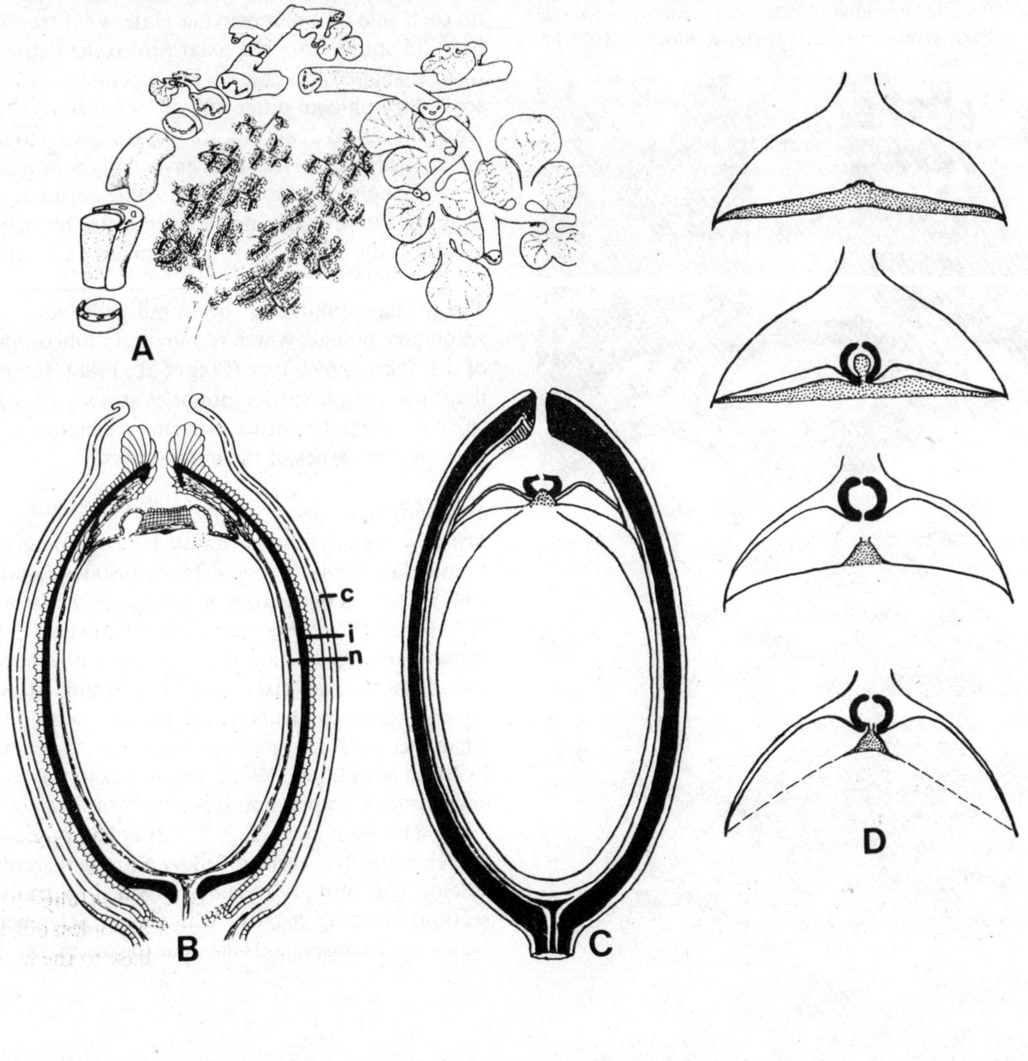

production. It starts with an apical pad of nucellar tissue, which then differentiates from the underlying tissue into a centrally placed lagenostome at the base of the micropyle and pollen chamber wall. As the pollen chamber expands the central column, which is a part of the pollen chamber floor, separates from the lagenostome, allowing pollen grains to enter the pollen chamber. If a pollination drop was present, the act of separation of the floor of the pollen chamber from the lagenostome probably provided the mechanism for pulling the pollination drop, with its included prepollen, into the pollen chamber. After pollination, the megagametophyte expanded, ruptured the pollen chamber floor, and pushed the central column into the inner opening of the lagenostome, effectively sealing the pollen chamber.

It is now known (Schabilion & Bortzman, 1979) that the functional megaspore within the nucellus (megasporagium) was derived from a tetrahedral tetrad of megaspores, three of which abort but are retained at the apex of the functional megaspore. The latter has a triradiate suture on its surface. This primitive tetrahedral arrangement of the one functional and three abortive megaspores finds a parallel in *Archaeosperma*, *Lepidocarpon*, and other lycopsids.

Based on frequency of association and anatomical similarities, it has been suggested by Jennings (1976) that *Telangium* and *Telangiopsis* represent the pollen organs of *Heterangium* borne on *Rhodea*-type foliage.

Figure 23.11. **A.** *Conostoma quadratum*, lagenostome at base of micropyle. Upper Pennsylvanian. (From Neely, 1951.) **B.** Transverse section of *Microspermopteris aphyllum* stem. Note protostele dissected by radiating plates of parenchyma. Middle Pennsylvanian. (From Taylor & Stockey, 1976.)

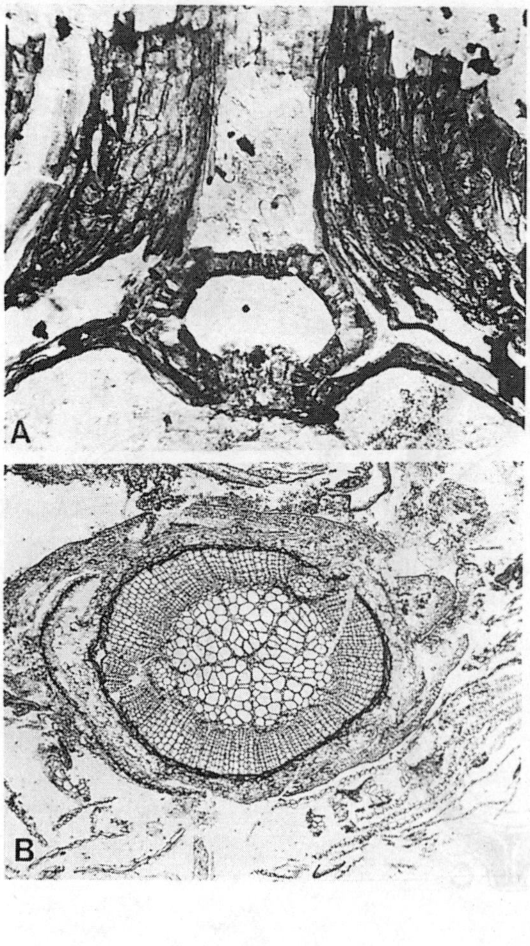

Microspermopteris

The vegetative structure of *Microspermopteris* (Fig. 23.11B) (Taylor & Stockey, 1976; Pigg, Stockey, & Taylor, 1986) approximates that of *Heterangium* in many ways. This stem genus from Middle and Lower Pennsylvanian deposits of the United States is characterized by a "mixed" exarch protostele with large metaxylem tracheids, and up to 10 exarch protoxylem strands. The primary xylem is pentagonal in cross section and is divided into 5 sections by longitudinal plates of parenchyma that radiate from the center of the stem. The protoxylem strands occur in pairs, one on each side of a parenchyma plate. Leaf traces arise in a 2/5 spiral from the axial protoxylem strands to form C-shaped leaf traces. The secondary xylem and secondary phloem differ only in detail from those of *Heterangium*. We can add *Microspermopteris* to the list of those pteridosperms that have axillary branching.

Longitudinal extensions of the cortex result in winglike lateral flaps in the more distal branches. In addition, the surfaces of the branches and stems are covered with irregularly shaped multicellular hairs. The primary pinnae of the frondlike leaves produce secondary pinnae, which terminate in lobed pinnules of the *Sphenopteris* type (Pigg et al., 1986). Internally, there are peripheral longitudinal strands of sclerenchyma in the outer cortex. Numerous secretory cavities occur in the tissues of the inner cortex.

Schopfiastrum

Schopfiastrum is another Middle Pennsylvanian genus, found thus far only in the United States (Andrews, 1945). The reconstruction of *Schopfiastrum* (Rothwell & Taylor, 1972) shows that the fronds, instead of being spirally arranged as are those of other lyginopterids, are alternate and distichous. The fronds show the typical bipartite structure of pteridosperm foliage. Characteristics of the pinna and pinnules, as determined by Stidd & Phillips, (1973), are similar to those of the compression–impression fronds of *Mariopteris*.

The stem stele (Fig. 23.12) is composed of an exarch protostele surrounded by abundant secondary xylem. The primary xylem is bilobed in transverse section, emitting massive leaf traces. Sclerenchyma bands and conspicuous lacunae occur in the outer cortex, while horizontal sclerotic plates have been reported in the inner cortex. Resin canals of the type frequently found in stems of pteridosperms occur in the cortex and all orders of the fronds. As many as five orders of branching have been observed in the planated bipartite fronds. From an evolutionary point of view, it is of considerable interest to note that the traces given off laterally from the petiolar trace to the primary pinnae have the same Y-configuration as the three-ribbed protosteles of aneurophytes, some members of the Calamopityaceae (*Stenomyelon* and *Tristichia*), and the frond segments of *Lyginorachis*. These observations suggest to Stidd & Phillips that *Schopfiastrum* is more closely related to the primitive pteridosperms.

You may have wondered how, in the absence of knowledge about their reproductive structures, we can confidently assign *Microspermopteris* and *Schopfiastrum* to the Pteridospermales. In our descriptions of the vegetative characteristics of those pteridosperms where the reproductive structures are known, we find that a combination of vegetative characteristics is consistently present. Some of these are (1) strands of sclerenchyma in the outer cortex (the "sparganum"-type outer cortex); (2) secretory cavities or resin ducts in the stem cortex and fronds; (3) bipartite fronds in which the rachis dichotomizes; (4) large metaxylem tracheids with circular multiseriate bordered pits on the walls; and (5) manoxylic secondary xylem.

Medullosaceae

The Medullosaceae has attracted the attention of many paleobotanists. As a result, a voluminous literature effectively summarized by Stidd (1981) has accumulated, which treats aspects of the vegetative and reproductive morphology of members of the family, their interrelationships, evolutionary orgins, and derivatives. As a group, the Medullosaceae are known to occur in the Mississippian (Lower Carboniferous), extending well into the Permian. Judging from their distribution in the fossil record, the family reached its zenith in the Upper Pennsylvanian. Currently four genera are recognized; they are *Quaestora, Sutcliffia, Medullosa,* and *Colpoxylon*.

Until the discovery of *Quaestora* (Mapes & Rothwell, 1980), we characterized the Medullosaceae as those Paleozoic pteridosperms with two or more vascular segments in their stems. Using the word in its descriptive sense, we describe those Medullosaceae where there is more than one vascular segment in the stem as polystelic. *Quaestora,* on the other hand, is like other pteridosperms in having a single vascular segment and is said to be monostelic. A vascular segment comprises a central primary body with its axial bundles and leaf traces surrounded by secondary xylem and phloem.

Other characteristics of the family, which in combination distinguish the Medullosaceae, are the following:

Figure 23.12. *Schopfiastrum decussatum,* transverse section of stem. Note well-developed "sparganum"-type outer cortex. Middle Pennsylvanian. (From Stidd & Phillips, 1973.)

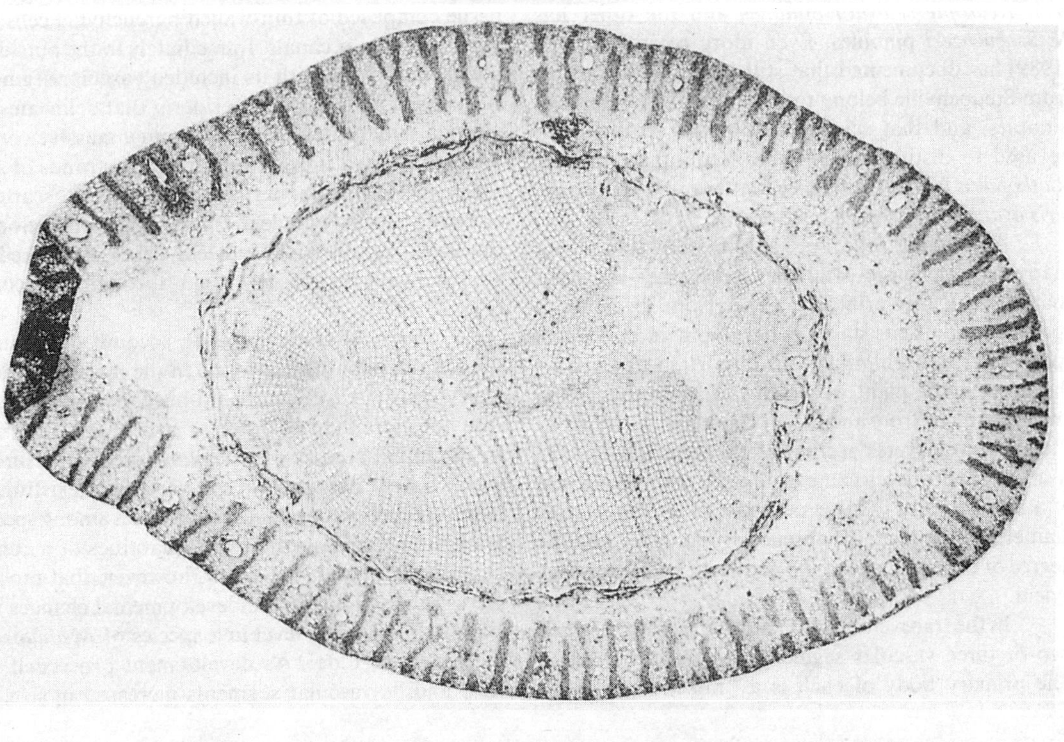

1. Large petioles with many, somewhat scattered traces
2. Fronds with a bifurcate rachis
3. Large three-angled, noncupulate ovules
4. Nucellus free from integument, except at base where it is attached by a stalk
5. A simple pollen chamber
6. Pollen organs that are synangiate structures with tubular sporangia
7. Pollen (prepollen) usually of the monolete type

Medullosa: vegetative structure

After examining hundreds of stem, leaf, and root specimens of *Medullosa,* it is possible to provide a reasonably accurate reconstruction (Fig. 23.13A) of *M. noei* (Stewart & Delevoryas, 1956). Here we see an erect plant having the general appearance of a tree fern 3.5 to 4.5 m high. The lower portion of the stem is covered by a periderm and adventitious roots. Higher on the stem are the remains of spirally arranged leaf bases, while the upper end supports large bipartite fronds. Although we know how the large ovules and pollen organs were borne on the fronds, we do not know whether the plants were monoecious or dioecious. For this reason the reproductive structures were omitted from the reconstruction.

As has been the case with many fossil plants, our increased understanding of *M. noei* has revealed that several species are represented by the fossils. In a recent study using hundreds of specimens from Upper Pennsylvanian coal balls collected near Steubenville, Ohio, Beeler (1983) reconstructed two species of plants that both have *M. noei* stems. One species has fronds with *Neuropteris ovata* pinnules and the other has *N. scheuchzeri* pinnules. Even more recently, Pryor (1989) has documented that still other *M. noei* stems from Steubenville belong to a plant with *Alethopteris* pinnules, and that several histological features can be used to distinguish *M. noei* stems that produce *Alethopteris* fronds from those that bear either *Neuropteris ovata* or *N. scheuchzeri* fronds.

As for some other species of fossil plants, where many specimens are available, it becomes apparent that much of the variability we see from specimen to specimen represents stages in development of a single species. With this in mind, it is important to consider the level in the plant at which the specimens being examined came from and the probable age of the plant. Thus, the transverse section of the stem of *M. noei* (Fig. 23.14) probably came from the proximal portion of a mature plant. This is determined by the relative diameter of the stem, the absence of leaf bases, and the degree of development of the periderm and secondary xylem.

In the transverse section of *M. noei* (Fig. 23.13B) two or three vascular segments are usually present. The primary body of each is a "mixed" protostele composed of clusters of large metaxylem cells, separated by abundant parenchyma (Fig. 23.16A). Axial bundles (sympodia) of primary xylem occur near the edge of the primary body that faces the outside of the stem. Each sympodium, of which there may be one to four per segment, is a strand of mesarch primary xylem. There may be a total of five to six sympodia for the two or three segments. Although isolated in separate vascular segments, the sympodia are arranged in a circle with respect to the transverse section of the stem (Fig. 23.15).

Leaf traces arise from the sympodia in an irregular sequence, and as they pass outward and upward through the cortex they divide dichotomously several times in the cortex and leaf base. The leaf traces derived from different levels of the same sympodium or different sympodia may supply the same leaf base (Basinger, Rothwell, & Stewart, 1974). The numerous collateral (phloem to one side of the bundle) leaf traces are often associated with a strand of sclerenchyma in the petiole where they occur in a ring as well as being scattered through the ground tissue of the petiole (Fig. 23.16D).

The secondary xylem around each of the two or three primary bodies is manoxylic (Fig. 23.16A). The vascular rays that traverse the secondary xylem and the surrounding secondary phloem are usually several cells wide. Pitting of the secondary xylem is multiseriate, circular bordered, and more or less confined to the radial walls (Fig. 23.16B). The development of the secondary wood is often endocentric (thickest on the side nearest the center of the stem).

The vascular segments are embedded in a ground tissue composed of thin-walled parenchyma cells and secretory cells or canals. Immediately to the outside of the ground tissue with its included vascular segments is a conspicuous band of periderm that delineates the internal structures of the stem from a massive cortex. The obvious features of the cortex are strands of sclerenchyma near the periphery, numerous scattered secretory ducts, and leaf traces. The separation of the large spirally arranged leaf bases is marked by sclerenchyma strands that cut through the cortex (Fig. 23.13C).

The most comprehensive account of medullosan stem development is found in the pioneer work by Delevoryas (1955), who based his conclusions on the study of more than 20 different specimens of *M. noei.* In the initial studies of *Medullosa* stem structure, it was believed that much of the variability existing between specimens represented differences among species. With the accumulation of vast quantities of material, Delevoryas was able to show, however, that most of the variability represented developmental changes that occurred at a given level in a species of *Medullosa* as its stem grew older. As development proceeded, the stem and its vascular segments increased in size, the

Figure 23.13. **A.** Reconstruction of *Medullosa noei*, a plant approximately 3.5 m high. Pennsylvanian. (From Stewart & Delevoryas, 1956.) **B.** Diagram, transverse section *M. noei*, stem showing two vascular segments. "Mixed" protostele (solid black); axial strands (circles); secondary xylem (hatched); ground tissue (stippled). Upper Pennsylvanian. (Redrawn from Basinger, Rothwell, & Stewart, 1974.) **C.** *M. primaeva*, transverse section of stem showing numerous vascular segments and large leaf bases. Middle Pennsylvanian. (Redrawn from Stewart & Delevoryas, 1952.)

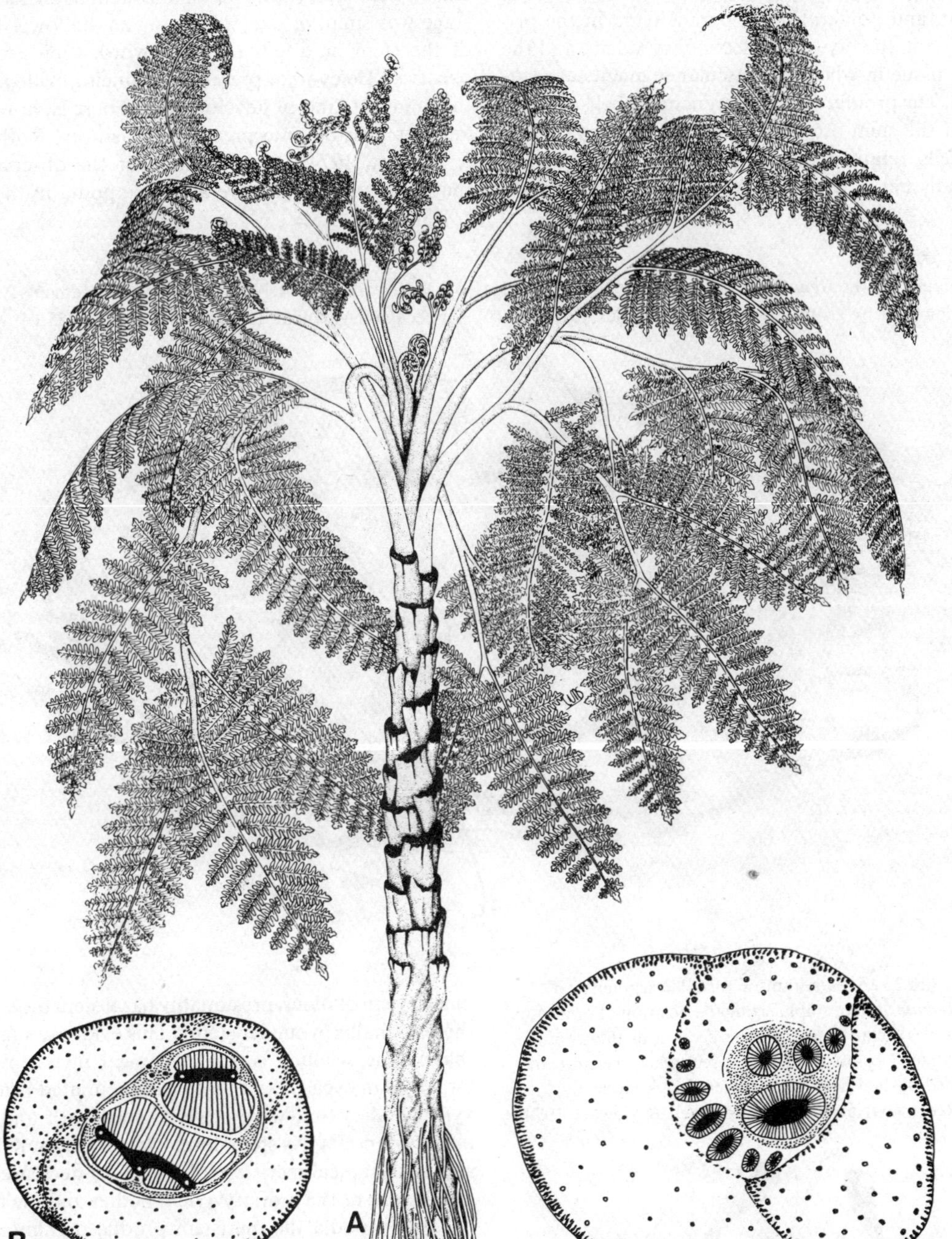

cortex with its leaf bases sloughed off, and the internal periderm, which increased in thickness, became the outer layer of stem tissue.

According to Delevoryas, the increase in size of the vascular segments was accomplished by an increase in size of the primary body of each segment, the formation of secondary vascular tissue around each segment, and poliferation of parenchyma in the primary xylem, the rays of the secondary xylem, and the ground tissue in which the vascular segments are embedded. The proliferation of parenchyma cells in older parts of the stem provides us with the evidence that these cells retained their capacity to divide in the various tissues of the stem, including the periderm.

If the interpretation by Delevoryas of the development of the primary xylem in a vascular segment is correct, then we have a most unusual situation where the primary body, after its formation by an apical meristem, continues to develop and increases in size. This interpretation presupposes that the primary vascular system of a young *Medullosa* stem in the seedling stage was small in size, increasing as the lower levels of the stem matured. In other words, epidogenesis occurred. Delevoryas presents convincing evidence for this unusual kind of development. There is, however, another possible interpretation (Basinger, Rothwell, & Stewart, 1974) that accounts for the observation that there are many more axial sympodia in the primary xylem of older, presumably basal stem fragments than in smaller, younger stems. This evidence suggests that in the seedling stage, the apical meristem was larger, as in cycads, and that it differentiated primary xylem bodies that were large to start with and had many sympodia. As stem development continued, the size of the apical meristem of the shoot decreased, as did the size of the primary xylem bodies and the number of sympodia the meristem produced. This interpretation invokes the process of apoxogenesis and suggests that *Medullosa* had a determinate growth habit. Irrespective of which explanation is the correct one for the development of the primary xylem of the vascular segments, we cannot avoid the conclusion that internal adjustments had to occur to accommodate the

Figure 23.14. *Medullosa noei,* transverse section of stem from older part of plant. Note three vascular segments (vs) separated by ground tissue (gt); and thick periderm (p). Upper Pennsylvanian. (From Delevoryas, 1955.)

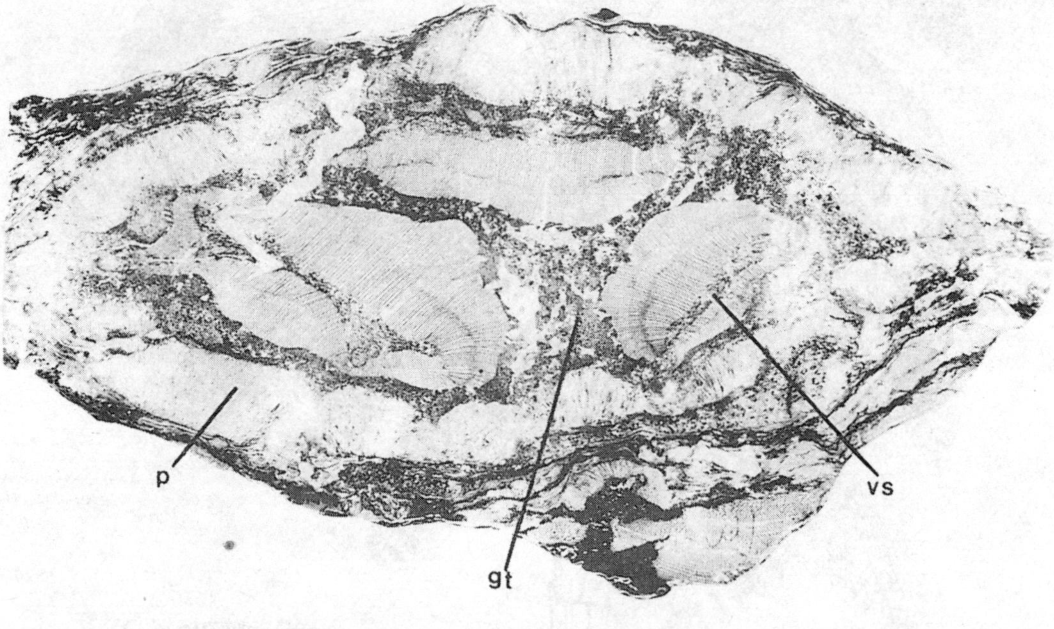

Figure 23.15. Diagram of vascular segments of *Medullosa* stem emphasizing the position of sympodia (circles) relative to the cross section of the stem. "Mixed" protostele (black); secondary xylem (outlined dashed lines); leaf traces (black circles). (Redrawn from Basinger, Rothwell, & Stewart, 1974.)

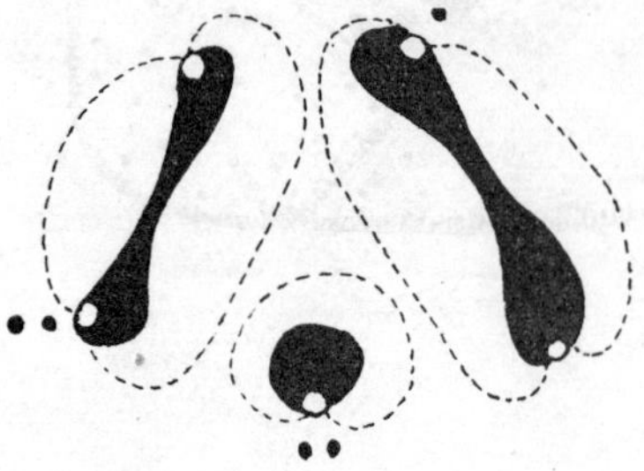

Figure 23.16. **A.** A vascular segment of *Medullosa* enlarged. Note "mixed" protostele surrounded by manoxylic secondary xylem, endocentric in development. **B.** Radial section of secondary xylem showing a single tracheid with alternate, circular bordered pits on the radial wall. **C.** *Myeloxylon* sp., transverse section of a medullosan petiole. **D.** Enlarged portion of *Myeloxylon* showing "sparganum" cortex and vascular bundles. **E.** Transverse section of *Alethopteris* pinnule showing revolute margins of the lamina. **A–D:** Upper Pennsylvanian. (**C,E** from Ramanujam, Rothwell, & Stewart, 1974.)

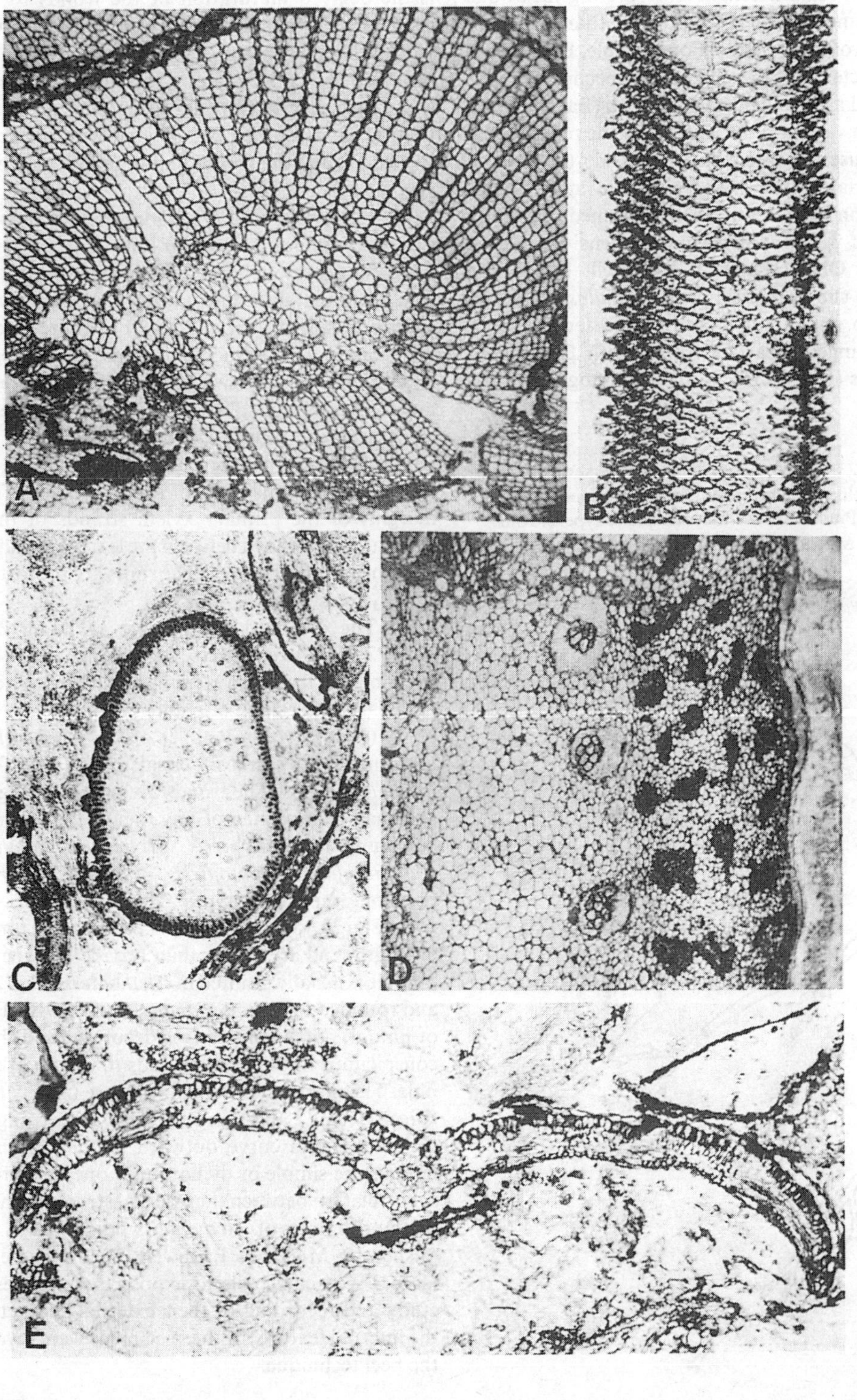

increase in size of the vascular segments produced by the subsequent formation of secondary tissues. It is quite possible that this increase in size stimulated the proliferation of parenchyma cells and periderm in the adjustment process.

The highly unusual mode of growth we have just described for *Medullosa* makes interpreting variations among specimens even more difficult than for most other groups of fossil plants. For example, Delevoryas (1955) suspected that the single stem specimen originally described as *Medullosa endocentrica* (Baxter, 1949) was a growth variant of *M. noei.* Later he retained the species because the specimen fell outside the normal range of variation for *M. noei.* This cautious taxonomic decision has recently been confirmed in a study of numerous *M. endocentrica* specimens from near Steubenville, Ohio (Hamer & Rothwell, 1988). The new material shows that stems of *Medullosa endocentrica* grew to only about 2 cm in diameter and bore helically arranged bipartate fronds with *Eusphenopteris* pinnules (Fig. 23.17). Both the configuration of the vascular tissues as two vascular segments and a distinctive pattern of leaf-trace production are characteristic for this species. *Medullosa endocentrica* also has axillary branching, the first known occurrence of this feature in the Medullosaceae.

Figure 23.17. *Medullosa endocentrica* showing slender vinelike stem bearing helically arranged bipartate fronds. Pinnules of the *Eusphenopteris* type. (From Hamer & Rothwell, 1988, with permission.)

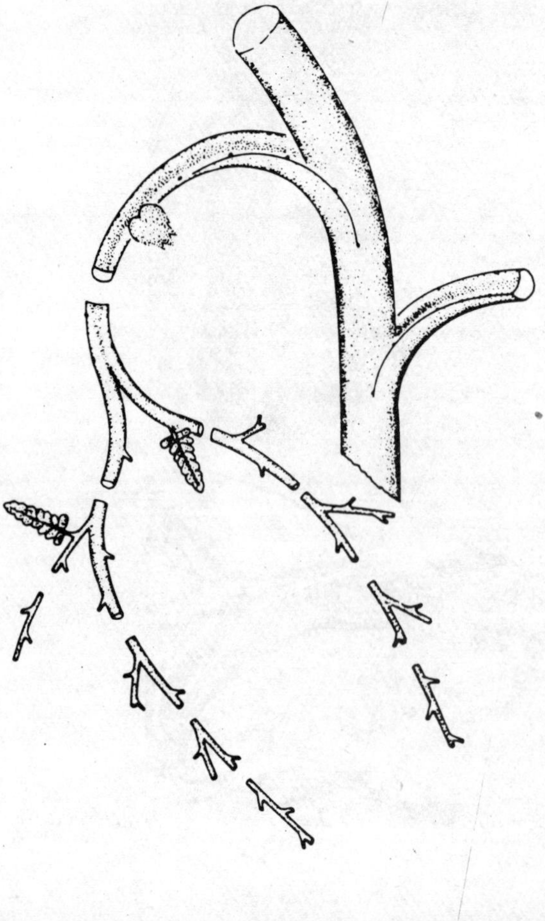

One of the unusual features of *M. endocentrica* is the delayed maturation of the fronds for several centimeters behind the apex. When considered together with the slender nature of the stems and the absence of adventitious roots along the stem, this pattern of shoot development provides strong evidence to indicate that *M. endocentrica* was a vine with an apex that twined around the trunks and branches of supporting vegetation.

Early studies of *Medullosa* stressed the number, size, and shape of vascular segments in a stem as important species characteristics (Stewart, 1951). We now know that the number of segments in a given stem specimen can vary from 2 to 20 or more (Stewart & Delevoryas, 1952). Such is the case for *M. primaeva* (Fig. 23.13C) where two segments proliferate to give rise to many "steles," some of which redivide to give leaf traces and others which continue in the stem to anastomose at higher levels.

Adventitious roots of *Medullosa* arise from the margins of the primary xylem strands of the stem. They turn downward between the leaf bases and emerge near the base of the stem. The roots are protostelic and usually tetrarch.

The detached petioles are called *Myeloxylon* (Fig. 23.16C). In transverse section they bear a striking resemblance to a corn stem in having many scattered vascular bundles and sclerenchyma fibers at the periphery (Fig. 23.16D). Petioles of the *Myeloxylon* type have numerous secretory canals of the kind found in *Medullosa* stems. The fronds are bipartite (Fig. 23.18) and may bear pinnae of any one of the form genera belonging to *Neuropteris, Mixoneura, Alethopteris, Callipteridium, Odontopteris,* or *Eusphenopteris* compression–impression types. The pinnules of *Alethopteris* (Fig. 23.18) and *Callipteridium* may be up to 5 cm in length, and are longer than broad. The apex of the pinnule is usually pointed in *Alethopteris* (Fig. 23.19A) and rounded in *Callipteridium* (Fig. 23.19B). The base of pinnules of *Alethopteris* is decurrent and there is a conspicuous midrib that extends to the tip of the pinnule. The margins of the pinnules are usually revolute (Fig. 23.16E). Lateral veins arise at a steep angle from the midrib and curve outward to the margins. The laterals are simple or dichotomize one or more times. A correlation between the internal structure and external morphology of *Alethopteris* pinnules was accomplished by Mickle & Rothwell (1982). These investigators split coal balls to expose the pinnules. After study and recording of their external characteristics, the internal features of the specimens were revealed by the peel technique.

Like *Alethopteris, Neuropteris* is widely distributed in the Upper Carboniferous. *Neuropteris* pinnules (Fig. 23.19C) have a rounded to cordate base and are attached to the pinna rachis by an inconspicuous stalk. A single vascular strand enters the base of the pinnule, which divides by repeated dichotomies to produce many veins that arch to the margins of the pinnule. *Mixoneura* superficially resembles *Neuropteris* foliage, except that the pinnules of *Mixoneura* are attached by the lower halves of their bases. Several veins enter the pinnule and dichotomize to produce a midrib composed of several strands. Other veins divide from one to five times as they arch toward the margins of the pinnules. *Odontopteris* has decurrent pinnules where the arched veins enter directly from the pinna rachis. *Odontopteris* ranges from the Upper Carboniferous into the Permian. *Linopteris* (Fig. 23.19D) resembles *Neuropteris,* but the veins of *Linopteris* anastomose to form a network.

Figure 23.18. Restoration of *Medullosa* shoot. Note bifurcations of petiole, position of attachment of pollen organ, and features of pinna and pinnules. Upper Pennsylvanian. (From Ramanujam, Rothwell, & Stewart, 1974.)

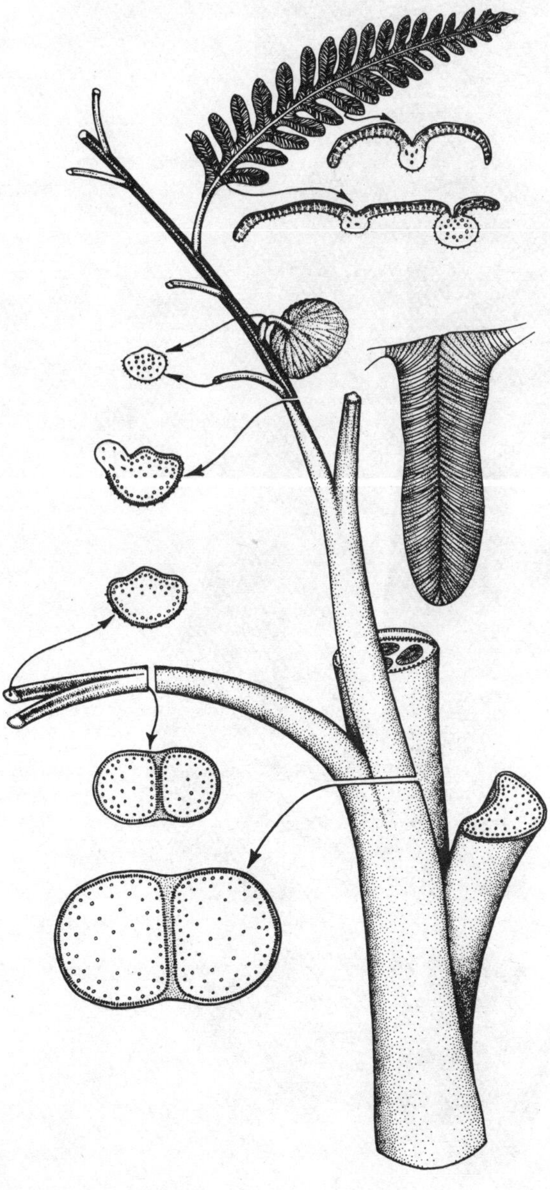

Several species of *Medullosa* have been described from the Permian of Europe. One of these, *M. leuckartii* (Fig. 23.20A), is similar to the Carboniferous *M. noei* in having three large vascular segments that are expanded tangentially in the periphery of the stem. In the centrally located ground tissue there are a number of smaller vascular segments. The leaf traces of this species are accompanied by secondary xylem. *Medullosa stellata* (Fig. 23.20B), another Permian species, is characterized by a cylindrical stele surrounding a central mass of parenchymatous ground tissue that occupies the position of a pith. A number of protostelic vascular segments, each surrounded by secondary xylem, are embedded in the ground tissue. The cylindrical stele of *M. stellata* comprises a usually continuous ring of primary xylem with both centripetal and centrifugal secondary xylem.

From the permineralized species of *Medullosa* there is good evidence for a wide variety of plant forms within the genus. As you will recall, *M. endocentrica* was a vine, and *M. noei* stems were produced by small trees that represent at least three species of plants. The growth forms of medullosans have been further elaborated from compressed remains from a Middle Pennsylvanian clastic swamp in Pennsylvania (Wunk & Pfefferkorn, 1984). The material includes what apparently are six different species of plants with *Neuropteris, Alethopteris,* and *Linopteris* foliage. Some, of the stems in this remarkable assemblage have been bent over and preserved where they grew. By uncovering and tracing the branching patterns of several plants on a single bedding plane, the authors have been able to demonstrate that some species had closely spaced leaves and were free standing. Other plants had narrow, flexuous stems with distantly spaced leaves, and are interpreted to have stood upright only by growing in clumps or by leaning on other vegetation.

It is interesting to note that unusual secondary tissues, remarkably similar to those of *Medullosa,* characterize numerous species of living flowering plants. The living plants that occur in highly stressed environments are vines or have flexuous stems (Mauseth, 1988). In this regard, the anatomical features of medullosan stems are consistent with the growth forms and environmental conditions that we have interpreted for many seed ferns.

Figure 23.19. Some medullosan foliage in Mazon Creek concretions. **A.** *Alethopteris*. **B.** *Callipteridium*. **C.** *Neuropteris*. **D.** *Linopteris*. Middle Pennsylvanian, Carbondale group. (**A,C,D** photographs by J. Wollin; **B** courtesy of the Illinois State Museum, Springfield.)

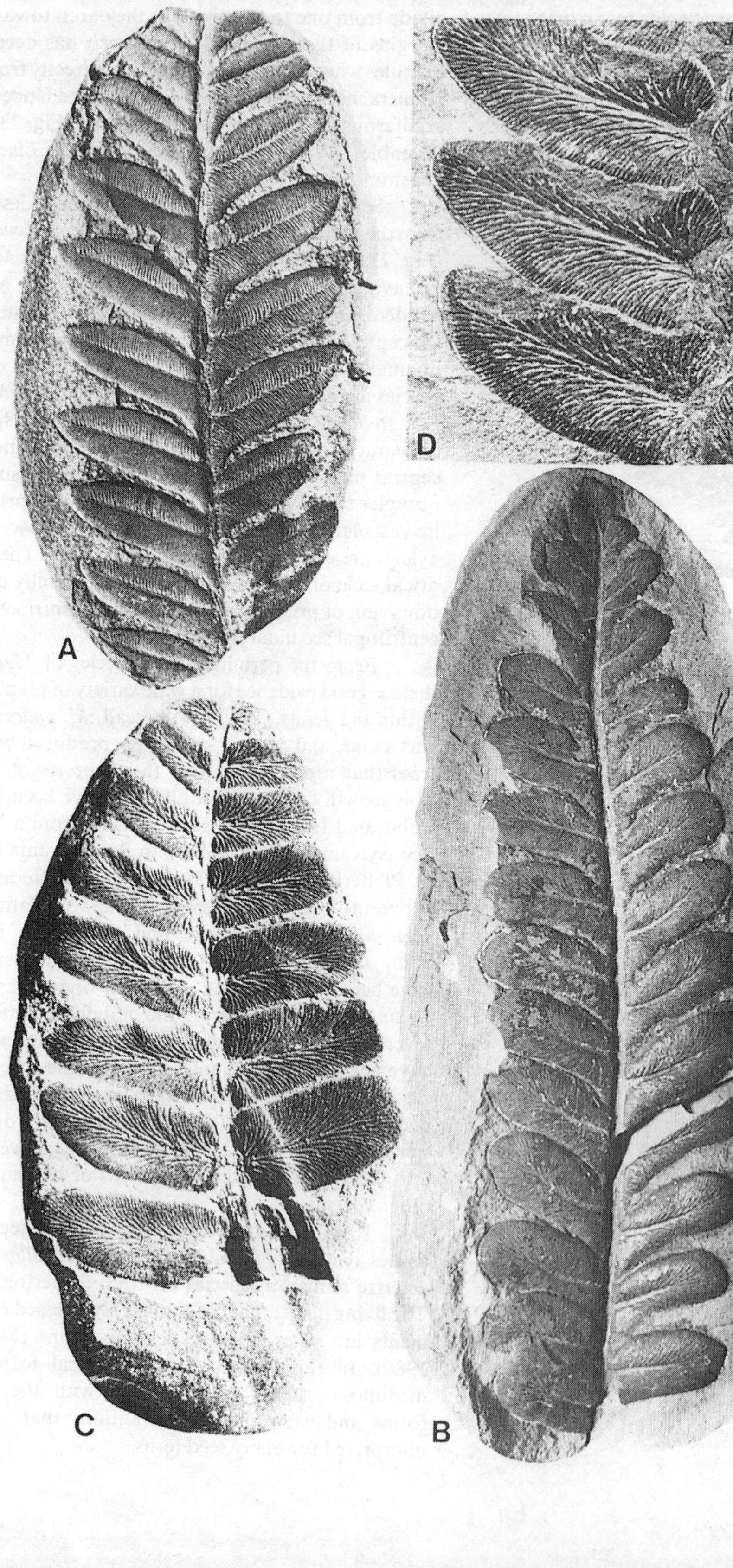

Colpoxylon

Colpoxylon (Fig. 23.20E), another medullosan stem genus from the Permian of France, is highly similar to *Medullosa leuckartii.* Like *M. leuckartii, Colpoxylon* has tangentially expanded vascular segments around a central parenchymatous ground tissue. The obvious difference between the two stems is the complete absence of central vascular strands in the ground tissue of *Colpoxylon.* The leaf traces are small and circular in cross section and possess secondary xylem.

Sutcliffia

Originally, the stem genus *Sutcliffia* (Fig. 23.20C) was known only from the Lower Coal Measures (Westphalian A) of England. It was not until 1963 that Phillips & Andrews reported its occurrence in Middle Pennsylvanian deposits of Illinois.

The stem is often described as having one or two large vascular segments surrounded by numerous smaller ones. Each large segment consists of a massive amount of primary xylem enveloped by weakly developed secondary xylem. The exarch primary xylem is composed of protoxylem and large metaxylem tracheids with interspersed parenchyma. The smaller surrounding segments arise from the large centrally located ones. They branch and rebranch, with some functioning as concentric leaf traces (traces where phloem surrounds the xylem). Some leaf traces have associated secondary xylem, while other vascular segments, produced by the profusely branched vascular system, anastomose at higher levels in the stem. The branching and anastomosing of the *Sutcliffia* cauline vascular system is highly similar to that of *Medullosa primaeva.* Unlike the medullosas, however, the vascular bundles in the petioles of *Sutcliffia* are concentric rather than collateral. Other than this distinctive characteristic, Stidd, Oestry, & Phillips (1975) describe the fronds of *Sutcliffia* as essentially similar to those of *Medullosa.* Pinnules having the characteristics of *Linopteris* compression–impression fossils have been found associated with axes of *Sutcliffia* fronds and are suggested (Stidd et al.) to be the pinnule type for *Sutcliffia.*

Some authors believe that the large and complex pollen organ of *Sutcliffia* is the *Potoniea* type (Stidd, 1978). This campanulate organ is 1 to 3 cm in diameter and consists of concentrically arranged tubular sporangia. These vertically oriented sporangia are further compartmentalized into groups of four to six radially arranged sporangia embedded in a ground tissue. Each group forms a cycle around a hollow center in the campanulum. The groups within a con-

Figure 23.20. Medullosan stem types in transverse section. **A.** *Medullosa leuckartii.* Note internal vascular segments. **B.** *M. stellata.* Note complete ring of vascular tissue with internal and external secondary xylem. **A,B:** Permian. **C.** *Sutcliffia insignis.* Carboniferous. (Based on Scott, 1909, and Phillips & Andrews, 1963.) **D.** *Quaestora amplecta.* Upper Mississippian. (Redrawn from Mapes & Rothwell, 1980.) **E.** *Colpoxylon aeduense.* Permian. (Redrawn from Delevoryas, 1955.)

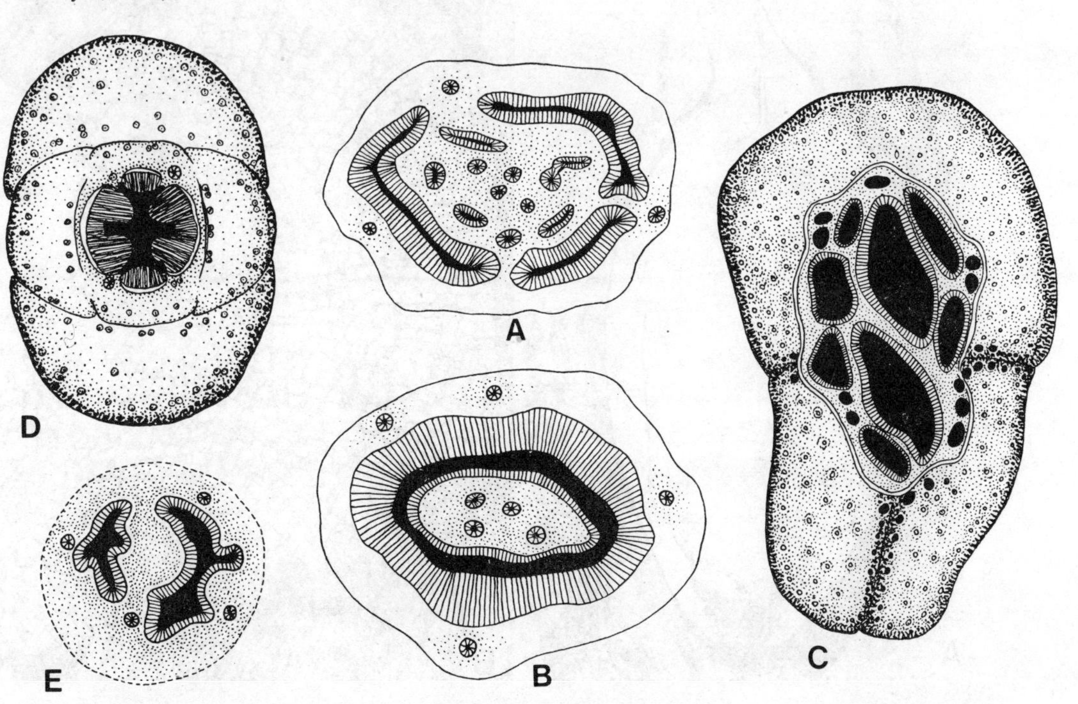

centric ring also are fused to one another, making the campanulum a compound structure. The ends of the sporangia of *Potoniea* protrude from the distal face of the campanulum so that their tips are free. *Potoniea* produced prepollen grains that are radially symmetrical with a trilete suture on the proximal surface. If found in the dispersed state, they would be identified as the *Punctatisporites* type. Based on frequency of association, Stidd suggests that *Potoniea* was borne on the fronds with *Linopteris*-type pinnules. Others suggest different relationships.

Quaestora

According to one phylogenetic interpretation of medullosan pteridosperm origin, it was anticipated that a monostelic representative would be found. Mapes & Rothwell (1980) have described such a monostelic stem, which they named *Quaestora* (Fig. 23.20D), from the Upper Mississippian of Arkansas. This may be the oldest known permineralized specimen of the Medullosaceae. Internally the stem displays a cruciform protostele surrounded by abundant secondary xylem, a vascular cambium, and secondary phloem of the medullosan type. The protostele is composed of metaxylem and interspersed parenchyma, with exarch protoxylem strands at the tips of the four arms. These strands, eight in number, form the axial sympodia that divide to produce leaf traces. The latter supply leaf bases that arise from the stem as decurrent opposite–decussate petioles. The internal periderm, a conspicuous feature of *Medullosa* stems, is lacking in *Quaestora*.

Although several features of the stem suggest affinities with the Calamopityaceae, Mapes & Rothwell place their new genus with the Medullosaceae because of similarities in leaf-trace production and petiolar structure. The petioles of *Quaestora* have secretory ducts, resin rodlets, vascular bundle number and distribution, and the "sparganum" cortex of the medullosan type. When all characteristics are considered, *Quaestora* seems to be transitional between the Calamopityaceae and Medullosaceae.

Medullosan ovules

Based on their proximity and frequency of association at the famous Berryville coal-ball locality of Illinois, it seems highly probable that the detached permineralized ovules of *Pachytesta illinoensis* (Fig. 23.21A) were borne on fronds of the *Alethopteris* type, which we know to be one of the detached foliage types of *Medullosa* – probably *M. noei* (Ramanujam, Rothwell, & Stewart, 1974). Although much effort has been expended in attempting to determine exactly where on

Figure 23.21. **A.** Composite of two photographs showing a median longitudinal section through *Pachytesta illinoensis*. Actual specimen 4.5 cm long. **B.** Paradermal section of *Bernaultia formosa*. Note radiating double rows of sporangia. Actual specimen 3.5 cm in diameter. **A,B:** Upper Pennsylvanian.

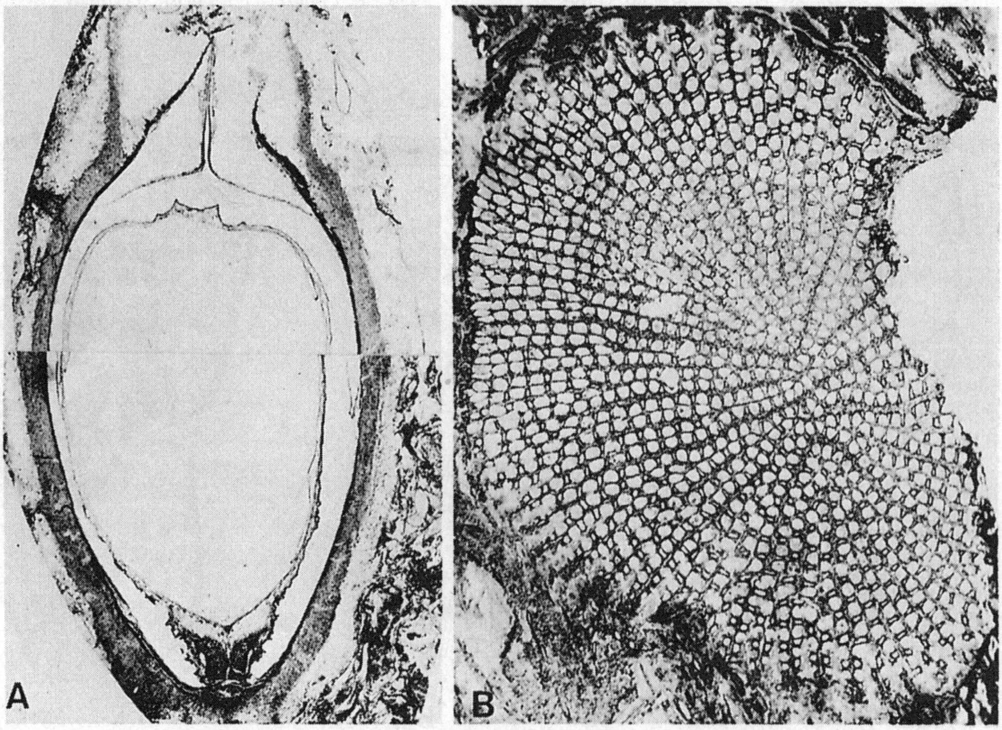

the frond *P. illinoensis* was borne, all ovules studied apparently had been abscised. We have some clues, however, provided by compression–impression fossils of frond parts from other localities that bear large ovules having characteristics of *Pachytesta.* These ovules were found on *Neuropteris* (Fig. 23.23A) and *Alethopteris* (Fig. 23.22A) pinnae, where they have the position of a lateral or terminal pinnule. Judging from observations of the vasculature in the base of *P. illinoensis,* where there is a ring of five to six bundles, it seems likely that this kind of ovule replaced a terminal pinnule on an *Alethopteris*-type frond. Terminal portions of a vegetative *Alethopteris* pinna rachis have approximately the same number of bundles arranged in a ring.

When preserved by authigenic cementation (Fig. 23.23B) or as compression–impression fossils, the large ovules of medullosan pteridosperms exhibiting characteristics of permineralized pachytestas are placed in the genus *Trigonocarpus*. The taxonomic problems introduced by using different generic names for different states of preservation of these large ovules has been thoroughly treated by Hoskins & Cross (1946).

The size of some species of *Pachytesta* is truly remarkable. The largest species, *P. incrassata* (Hoskins & Cross, 1946), attained a length of 11.0 cm and a diameter of 6.0 cm. The smallest member of the genus is *P. berryvillensis* (Taylor & Eggert, 1969), which has a maximum length of 6.8 mm and a diameter of 5.2 mm. This, the smallest *Pachytesta* known, is about the same size as the largest ovules of the Lyginopteridaceae.

In spite of their mode of preservation or variability in size, species of *Trigonocarpus* and *Pachytesta* show some indication of their basically trimerous (three-part) organization. This is particularly apparent in transverse sections of permineralized specimens (Fig. 23.22B) or in the casts of the internal cavity of the ovules (Fig. 23.23C). These casts, which are also assigned to *Trigonocarpus,* look very much like *Ginkgo* seeds with the outer fleshy testa removed. As we shall see, however, the ovules of medullosan pteridosperms are much more similar in their structure to ovules of modern cycads than to *Ginkgo*.

A fairly recent count of North American species of *Trigonocarpus* (Gastaldo & Matten, 1978) reveals

Figure 23.22. **A.** *Pachytesta*-type ovule attached to *Alethopteris* foliage in position of a pinnule. Carboniferous. (Redrawn from Halle, 1933.) **B.** *Pachytesta illinoensis,* transverse section at distal end of ovule showing trimerous nature of testa. Commissured rib (cr); endotesta (e); sclerotesta (sc); sarcotesta (sa). **C.** Diagram, longitudinal section, *P. illinoensis.* Note free nucellus and double vascular system. Micropyle (m); pollen chamber (pc); nucellus (n); integument (i). **B,C:** Upper Pennsylvanian. (**B,C** redrawn from Stewart, 1954.)

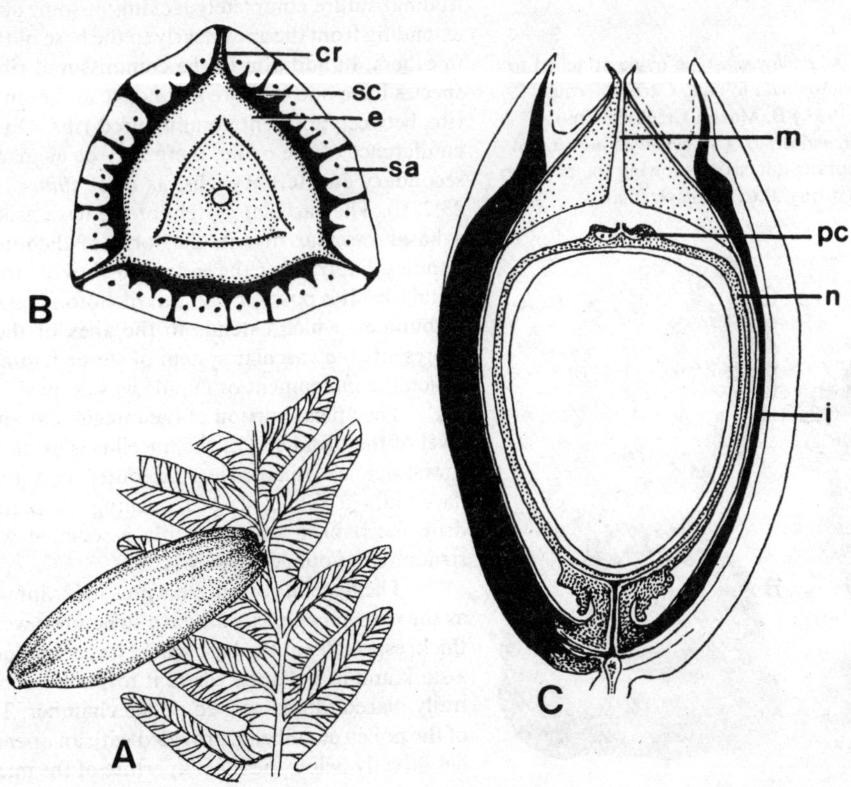

the rather staggering number of 43. It is almost certain that there is a great deal of synonymy here and that many specimens represent developmental or decorticated stages of a single taxon. In *Pachytesta,* where internal anatomy can be determined and many specimens are often available, the characterization of species is more reliable. In his monographic treatment of American species of *Pachytesta,* Taylor (1965) describes 13 species. To our knowledge, only one species has been added since then.

Other than its basic trimerous construction, one of the most striking structural features of *Pachytesta* (Fig. 23.22C) is its massive nucellus, which is free from the surrounding integument except for its point of attachment at the base. Examination of the vascularization of *Pachytesta* ovules reveals a double system where both nucellus and integument are supplied with vascular bundles. It is of more than passing interest that ovules of modern cycads also have double vascular systems and a nucellus that is free from the integument in the distal end.

When studying the structure of *Pachytesta* and related genera, one can hardly avoid wondering if the integument could be the homolog of a cupule as in *Lagenostoma* with its single ovule attached to the base of its cupule. This idea, expressed long ago, has been revived by Sporne (1974). If one adopts this idea, then that portion of the ovule we recognize here as a massive vascularized nucellus can be interpreted to be a compound structure composed of a nucellus that is completely fused with a vascularized inner integument. In view of its integumentary structure, that portion of the ovule derived from the cupule has been interpreted as an outer integument. There is no reason that we are aware of for suggesting that a cupulate structure could not have evolved into the covering layer of an ovule we call the integument. Until, however, we obtain fossil evidence that integument and nucellus have undergone phyletic fusion, we believe the wise choice is to stay with the more traditional interpretation of ovule structure for medullosan seeds. Since cupule and integument apparently evolved from sterile telome trusses that surrounded a megasporangium, we can interpret them as homologous structures. From a functional point of view, it makes little difference what we call the various layers that retain and protect the megagametophyte. For the sake of description, however, we see that the integument of *Pachytesta* is usually composed of three layers that frequently occur in the seed coats of modern seed plants and numerous fossil ones. There is an outer fleshy layer, the sarcotesta, covered by an epidermis; a thin inner layer, the endotesta, that lies adjacent to the nucellus; and a sclerotesta presumably composed of tough, fibrous cells sandwiched between the sarcotesta and endotesta. The trimerous nature of the ovule is clearly defined in the distal end of the ovule by the sarcotesta and endotesta, which in transverse section delineate the three segments of the inner testa separated by three commissured ribs (Fig. 23.22B). A commissured rib is a longitudinal suture completely lacking in some species or extending from the apex nearly to the base of the ovule in others. In addition to the commissured ribs, some species have one or more secondary and even tertiary ribs between adjacent commissured ribs. On the circumference of the ovule, there may be as many as 40 secondary and tertiary ribs, as in *P. illinoensis* (Fig. 23.22B). The vascular system of the testa arises from a basal vascular disc in the form of dichotomizing bundles that traverse the endotesta between the sarcotestal ribs. It is possible that this dichotomizing system of bundles, which extends to the apex of the ovule, represents the vascular system of sterile telomes from which the integument or cupule was derived.

The apical portion of the integument above the level of the distal end of the nucellus is extended into a well-defined micropyle, the outer end of which flares into a funnel-shaped opening. A pollination-drop mechanism was probably associated with this structure in younger ovules.

The nucellus or nucellus plus inner integument, as the case may be, is a massive structure several cells thick, especially at the base where it is constricted into a stalk and at the apex where it forms a hollow, centrally placed, dome-shaped pollen chamber. The wall of the pollen chamber is provided with an opening that lies directly below the inner aperture of the micropyle.

Figure 23.23. **A.** *Pachytesta*-type ovule attached to *Neuropteris heterophylla* foliage. Carboniferous. (From Halle, 1933.) **B.** Mazon Creek concretion containing *Trigonocarpus.* **C.** *Trigonocarpus,* a cast of the megasporangium cavity showing the position of commissured ribs. **B,C:** Pennsylvanian.

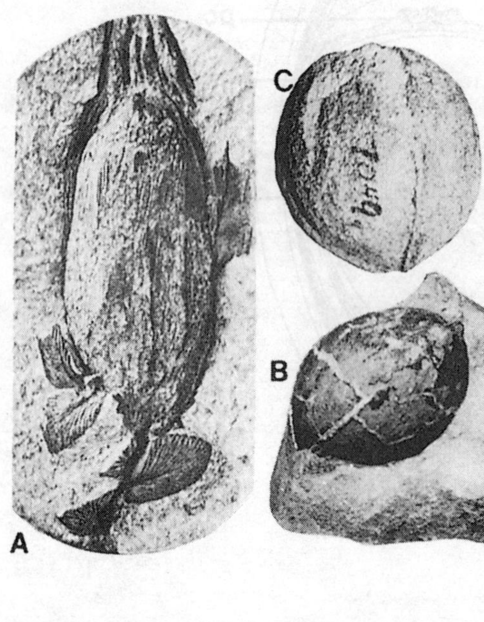

When an ovule developed, the opening into the pollen chamber was inserted into the base of the micropyle for pollen reception. In some species of *Pachytesta* and related genera, there is evidence of a central column that served to plug the pollen chamber after pollination, as in lyginopterid ovules.

Vascularization of the nucellus is from a disc of tracheary tissue in the distal end of the stalk. Vascular bundles radiate from the disc into the flanks of the nucellus, extend upward, and terminate at a level just below the pollen chamber.

Those students interested in the species variability displayed by members of the genus *Pachytesta* are referred to the monumental monograph by Taylor (1965), who also presents a tentative scheme depicting evolutionary trends within the genus. In general, Taylor portrays three such trends based on the structure and elaboration of the testa and its vascularization. One trend includes large ovules where the distinction between sarcotesta and sclerotesta is lost and the number of vascular bundles in the testa increases. Another, where the ovules are smaller, shows a series starting with the geologically oldest *P. shorensis*, where three commissured and three secondary ribs are present, to successively younger species with progressively more elaborate testas and increasing numbers of vascular strands. The third trend gives rise to geologically younger seeds that are small and with only three commissured ribs.

For additional information relating to the various genera of medullosan ovules, the papers by Hoskins & Cross (1946) and Combourieu & Galtier (1985) are most helpful. Hoskins & Cross give a well-illustrated comparison of all genera described up to that time. As far as we are aware, eight of the nine genera they list are still valid. The ninth genus *Rotodontiospermum*, is now known to be a *Pachytesta*. In addition to *Pachytesta*, the valid genera that share the characteristics of medullosan ovules are *Stephanospermum, Aetheotesta, Hexapterospermum, Polypterospermum, Ptychotesta, Polylophospermum, Colpospermum*, and *Codonospermum*. All are three-angled and have a simple pollen chamber without a lagenostome, a double vascular system, and a stalked or sessile nucellus that is free from the integument, except at its base.

Although large, possible medullosan ovules occur with *Quaestora* and *Medullosa* in Mississippian deposits of Arkansas (Taylor & Eggert, 1967; Mapes & Rothwell, 1980), the oldest permineralized ovules having all of these characteristics are *Pachytesta olivaeformis* and *P. shorensis*, both from the Lower Coal Measures (Westphalian A) of the Upper Carboniferous of Great Britain. The beautifully preserved silicified ovules from the Upper Carboniferous–Lower Permian coal fields of Saint Etienne, France, are geologically among the youngest known. They came from the Grand Croix beds, which have provided many specimens of Upper Carboniferous and Lower Permian medullosas. (Combourieu & Galtier, 1985).

Medullosan pollen organs

Numerous permineralized medullosan ovules have been sacrificed in order to study the contents of their pollen chambers. The objective here is to analyze the frequency with which pollen types occur in the chamber. Based on the observations of Stewart (1951) and others, there is no question that prepollen of the *Monoletes* type is congeneric with *Pachytesta illinoensis* and *P. hexangulata*, as this is the predominant pollen type found in the pollen chambers of these species. The pollen chamber contents of 10 specimens of *P. illinoensis* were investigated in one study. In 8 of these, *Monoletes* was the most abundant spore type. In the coal balls from the Berryville locality, where numerous specimens of *P. illinoensis* have been found, only three species of microsporangiate organs containing pollen of the *Monoletes* type have thus far been described. They are *Halletheca recticulatus* (Taylor, 1971) *Stewartiotheca warrenae* (Eggert & Rothwell, 1979), and *Bernaultia formosa* (Ramanujam, Rothwell, & Stewart, 1974). Of these only *B. formosa* is present in an abundance that corresponds to that of *Pachytesta illinoensis* ovules and *Medullosa noei* stems, and this leaves little doubt that these three organs are conspecific.

Based on studies of numerous permineralized specimens of *Alethopteris, Myeloxylon*, and *Bernaultia*, with particular attention being directed to understanding the primary vasculature of these genera, Ramanujam, Rothwell, & Stewart (1974) determined that campanula (bell-shaped structures) of *Bernaultia* replaced pinnae of the *Alethopteris–Myeloxylon*-type frond (Fig. 23.18). It has been suggested that a fertile frond may have many or all of its pinnae replaced by campanula.

One of the largest and certainly the most complex of pollen organs belonging to pteridosperms was originally described from small fragments as the genus *Dolerotheca* (Halle, 1933). For many years this name was used for all permineralized medullosan pollen organs that have numerous tubular pollen sacs embedded in a ground tissue. However, as so often happens in paleobotany, we have now discovered that at least three different medullosan pollen organs have features similar to the small fragments described as *Dolerotheca* (Rothwell & Eggert, 1986). Because we do not know which, if any, of these three are represented by the originally described specimens we have retained the name *Dolerotheca* for such fragments and use other generic names for pollen organs with structure that is completely known. The first detailed description of the campanulum we now know as *Bernaultia* was prepared by Schopf (1949). The campanulum of *B. formosa*

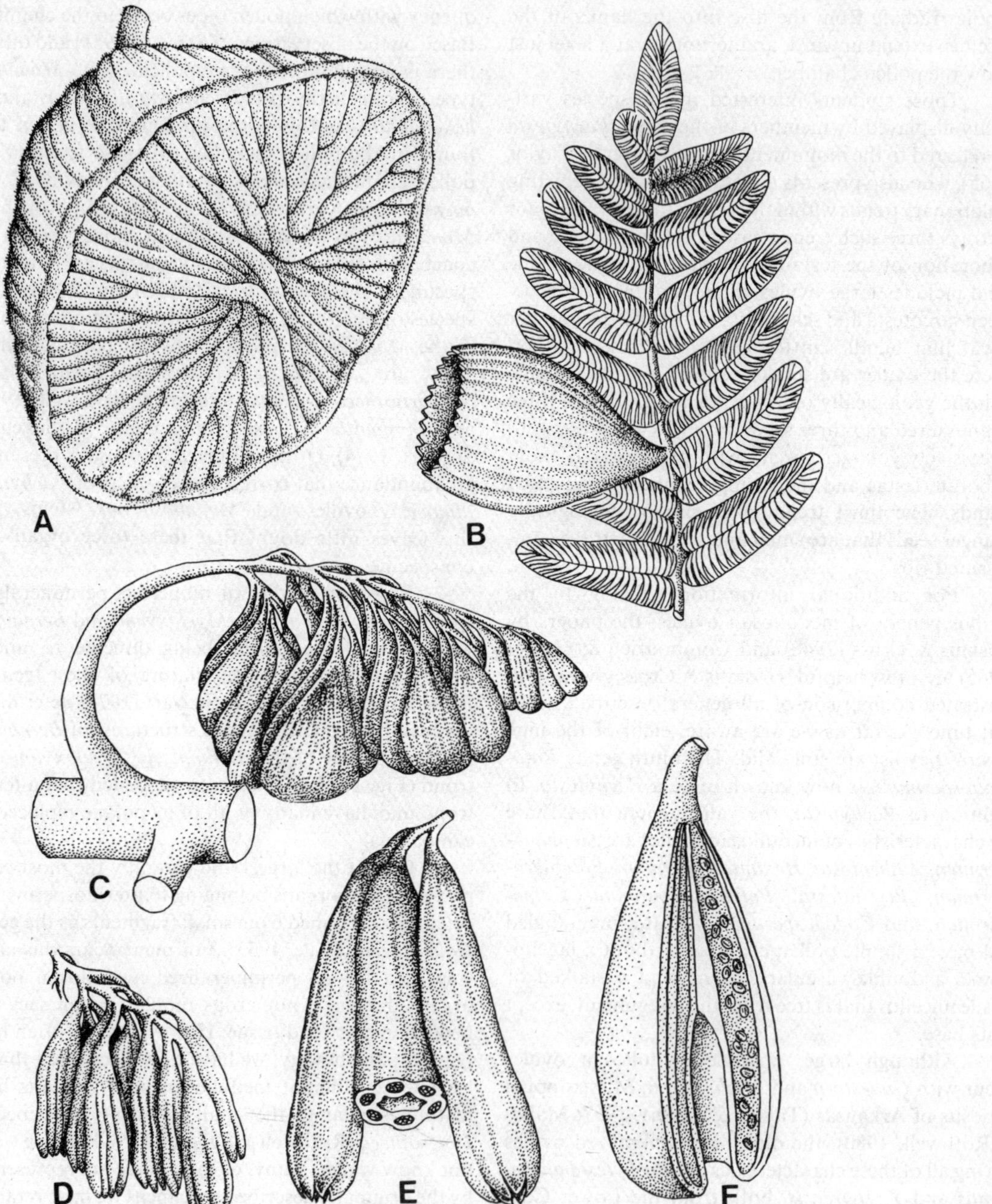

Figure 23.24. Some medullosan prepollen organs. **A.** *Bernaultia formosa,* campanulum prior to dehiscence. Upper Pennsylvanian. (Redrawn from Dennis & Eggert, 1978, and Millay & Taylor, 1979.) **B.** *Whittleseya media* attached to *Neuropteris*-type foliage. Carboniferous. (From Stewart & Delevoryas, 1956.) **C.** *Parasporotheca leismanii,* pinnate branching system bearing an aggregate of scoop-shaped synangia. (Redrawn from Dennis & Eggert, 1978.) **D.** *Codonotheca* sp. cluster of synangia. (Redrawn from Stidd, 1981.) **E.** *Aulacotheca* sp. (Reconstruction based on Millay & Taylor 1979, and Stidd, 1981.) **F.** *Halletheca reticulatus.* (Reconstruction modified from Millay & Taylor, 1979.) **C–F:** Pennsylvanian.

(Fig. 23.24A) may reach a diameter up to 4 cm. It is a hemispherical structure with an eccentrically placed pedicel. In paradermal section (Fig. 23.21B), the genus is easily recognized by its characteristic radiating rows of paired sporangia and multicellular hairs that cover the surface. Current interpretations (Dennis & Eggert, 1978; Eggert & Rothwell, 1979) of the campanulum describe it as a compound synangium composed of four radial synangia folded to form radiating plications. The folding of a synangium occurs in such a way that the dehiscence surfaces of opposite paired sporangia face one another. Dehiscence of sporangia is longitudinal. The longitudinal orientation of the dehiscence slits on essentially inward-facing walls of individual sporangia is similar to the arrangement previously described for many primitive vascular plants.

Vascularization of the campanulum is supplied from several veins arranged in a ring that enter the pedicel from the vascular system of the frond that bore it. Each vein or vascular bundle dichotomizes in the tissues that cover the campanulum. Delicate perpendicular vascular strands arise from the vascular system of the campanulum and are parallel to each sporangium on the side opposite from the longitudinal dehiscence slit. Thus, each microsporangium is supplied with vascular tissue. Some propose that the arrangement of the vascular tissue and sporangia reflect a derivation of the campanulum from several primitive fertile frond parts with pinnately arranged sporangia (Dufek & Stidd, 1981; Stidd, 1990) (Fig. 23.27) and not, as suggested by Dennis & Eggert (1978), the product of infolding of synangia, where the sporangia are essentially in a uniseriate ring with their dehiscence slits facing a hollow center (Fig. 23.26). The evidence for the interpretation by Dennis & Eggert is convincingly presented by its proponents, and it will help us to understand their ideas if we acquaint ourselves with the structure of a few other pteridosperm pollen organs that formed *Monoletes*-type pollen. We get considerable assistance from Millay & Taylor (1979) in their review article, "Paleozoic Seed Fern Pollen Organs."

Whittleseya (Halle, 1933), a compression–impression pollen organ (Fig. 23.24B), is suggested as the type of synangium that, by phyletic infolding of its uniseriate ring of sporangia, gave rise to synangia having the characteristics of *Bernaultia*. Theoretically, it would take four such structures to produce the compound synangium of *B. formosa*. There is now some doubt that *Whittleseya* was a campanulum. Instead, it is suggested that it was a planated structure with uniseriate or biseriate rows of sporangia. This is not unlike the planated synangium of *Parasporotheca* (Fig. 23.24C) described by Dennis & Eggert (1978), which is composed of a uniseriate row of sporangia that forms a curved scooplike structure. The authors suggest that such a structure could represent a stage in the evolution of a uniseriate ring of sporangia, similar to that described by Halle (1933) for *Whittleseya*, by recurvation and fusion along the lateral margins. In this way the radial symmetry typical of medullosan pollen organs may have evolved from a planated fertile telome truss.

The basic radial symmetry of pteridosperm pollen organs is portrayed by the compression–impression *Codonotheca* (Fig. 23.24D). Here there are usually 6 sporangia in a ring that are fused only at their bases. These sporangia extend from a vascularized receptacle. Nearly complete fusion of a ring of 6 sporangia is found in *Aulacotheca* (Fig. 23.24E), except at the tips of the sporangia, which are free. *Aulacotheca* is similar to the permineralized *Halletheca* (Fig. 23.24F), which has synangia composed of 5 to 12 sporangia around a central fibrous zone at the base and a central hollow in the distal end. There are several other genera that only serve to emphasize the basic radial symmetry of the simple synangium that forms a cycle around a partially or completely hollow center.

As early as 1933 Halle speculated that the primitive pteridosperm microsporangiate structure consisted of a ring of free microsporangia at the tip of a branch (Fig. 23.25A) and that syngenesis of the ring would produce simple synangia showing degrees of fusion, from partial fusion of the *Codonotheca* type to nearly complete fusion as in *Aulacotheca* (Fig. 23.25 B–D). It has now been determined (Taylor & Millay, 1981b) that the fusion of sporangia in the synangium of *Halletheca* is complete.

According to Eggert & Rothwell (1979), the two simple synangia *Sullitheca* (Stidd, Leisman, & Phillips 1977) and *Stewartiotheca* show evidence of infolding of a uniseriate cycle of fused sporangia (Fig. 23.26).

Figure 23.25. Steps in the evolution of some prepollen organs of medullosan pteridosperms. **A.** Trimerophyte ancestor with a ring of free sporangia at the tip of a branch. **B.** Partial syngenesis (tangential fusion) of sporangia in the proximal one-third; *Codonotheca* type. **C.** Nearly complete syngenesis of sporangia to produce a synangium of the *Whittleseya* type where there is a ring of sporangia around a hollow center. **D.** Complete syngenesis to form a synangium of the *Aulacotheca* type. (Redrawn from Halle, 1933.)

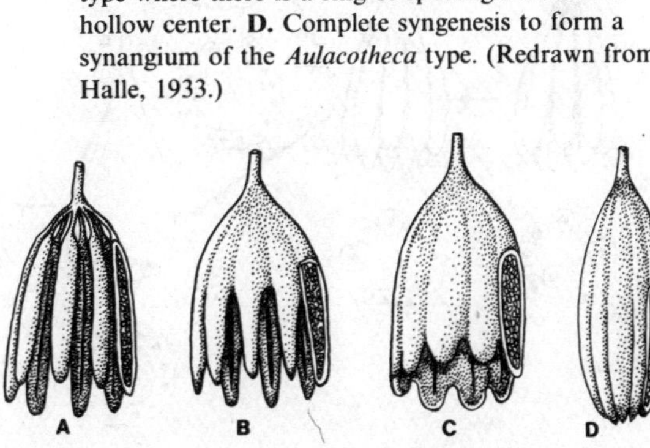

Both genera show a cavity around which the sporangia are arranged in double rows similar to those of *Bernaultia*. The characteristics of *Stewartiotheca* are highly similar to *Bernaultia*. The important difference is that *Stewartiotheca* is a simple synangium, while *Bernaultia* is often compound, most commonly being composed of the equivalent of four *Stewartiotheca* synangia (Fig. 23.26). Dufek & Stidd (1981) do not agree with the above interpretation of the evolution of *Bernaultia*, interpreting it instead as the fused remnants of an ancestral bifurcating pinna system (Fig. 23.27). Regardless of which interpretation proves to be most nearly correct, *Bernaultia* is the most specialized of all medullosan pollen organs and probably represents an evolutionary "dead end."

Medullosan prepollen

The dispersed prepollen genus *Monoletes* (= *Schopfipollenites*) is known to occur in situ in the vast majority of medullosan pollen organs thus far described. The grains, which have a smooth surface, are relatively large (100 to 500 μm in length), bilaterally symmetrical, and round to elliptical in shape. The distal surface of *Monoletes* (Fig. 23.28B) usually displays two longitudinal grooves separated by a convex umbo. The proximal surface (Fig. 23.28A) is marked by a suture that may show an angular deflection. The suture is probably a modification of the more primitive trilete suture characteristic of many isospores and other prepollen grains. It is believed that germination of *Monoletes* took place through the proximal suture (Millay, Eggert, & Dennis, 1978) as it occurs in ferns, for example. The absence of a preformed germination region on the distal surface of the grain, an essential feature of pollen, is the fundamental reason for calling the microspores of the *Monoletes* type prepollen instead of pollen. For the same reason, the microspores of *Crossotheca* and *Potoniea*, with their proximal trilete sutures, should also be called prepollen. Chaloner (1970) defines prepollen as any microspore, known to belong to a seed plant group, that germinates from the proximal surface of the grain.

Callistophytaceae

Up to this point we have stressed the idea that reconstruction of fossil plants, including information about their life cycles, is a very slow process taking many years of study. Our information about the Callistophytaceae, however, has accumulated in a relatively short time (approximately 20 years) and is surprisingly complete (Rothwell, 1981). The stem specimen of *Callistophyton* (Fig. 23.31B) that was later to become the type for the family was first seen in 1952 when it was found in a coal ball from the Upper Pennsylvanian deposits at the Berryville locality of Illinois. Our first impression of the specimen was that of a well-preserved stem of *Cordaites*. By 1954, however, Delevoryas & Morgan had demonstrated that

Figure 23.26. Suggested trends in the evolution of *Bernaultia* according to Eggert & Rothwell (1979). **A.** Cluster of terminal sporangia of an ancestral trimerophyte or progymnosperm. **B.** Sporangia arranged in a ring as in *Codonotheca*. **C.** Lateral fusion of some sporangia to form a bilaterally symmetrical organ similar to *Parasporotheca*. **D.** Laterally fused sporangia (synangium) to form a ring as in *Halletheca*. **E.** Simple synangium of sporangia showing plications as in *Stewartiotheca* or *Sullitheca*. **F.** Compound synangium of *Bernaultia* with radiating double rows of sporangia. (From Taylor, 1988, with permission.)

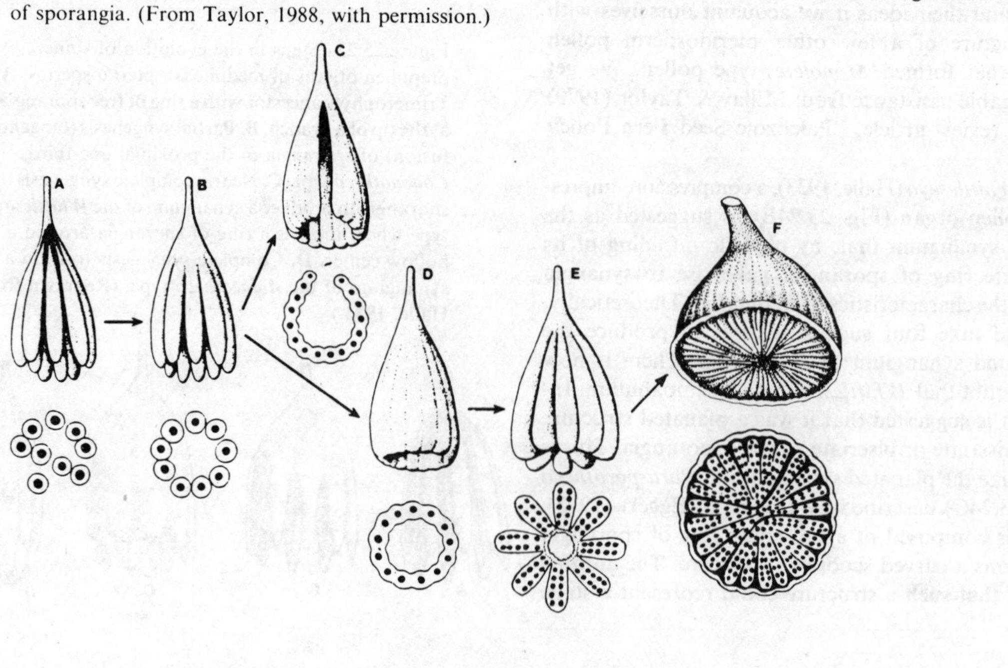

the stem had the characteristic anatomy of a pteridosperm most like that of *Lyginopteris*. Features that *Callistophyton poroxyloides* have in common with *Lyginopteris* are a pith that is surrounded by a eustele composed of sympodia where the axial strands are mesarch, well-developed secondary wood of the manoxylic type, secretory cells, a sparganum-type sclerenchymatous zone in the outer cortex, double leaf traces, capitate glands, and pinnately divided foliage.

By 1975, Rothwell was able to provide us with a monograph of the vegetative structures of the family as well as a reconstruction of *Callistophyton* (Fig. 23.29). Unlike *Lyginopteris*, which you will recall is portrayed as a vinelike plant, *Callistophyton* is reconstructed as a small shrubby plant, somewhat scrambling in habit, with stems up to 3 cm in diameter. This understory plant had spirally arranged fronds with axillary branches and adventitious roots at many of the nodes. The fronds are known to be at least bipinnate with pinnae that are similar to the compression–impression fossils of the *Medullopteris pluckeneti* type (Fig. 23.30). Circinate vernation characterizes the developing fronds. Pollen organs and ovules are now known to have been borne on the abaxial surfaces of unmodified pinnules where they are associated with veins. The pollen-producing structures are radially arranged synangia, and the ovules are small, platyspermic, and noncupulate. Pollen is monosaccate (with a bladder)

Figure 23.27. Suggested evolutionary origin of *Bernaultia* according to Dufek & Stidd (1981) and Stidd (1990). **A.** Hypothetical ancestral organ comprised of fertile dichotomizing pinnae. **B.** Enlarged segment of fertile pinna showing pinnate arrangement of sporangia. **C.** Enlarged portion of dichotomizing fertile pinnae with most sporangia removed to show the back-to-back fusion of sporangia forming the double rows of sporangia found in a *Bernaultia* campanulum. (Redrawn from Dufek & Stidd, 1981.)

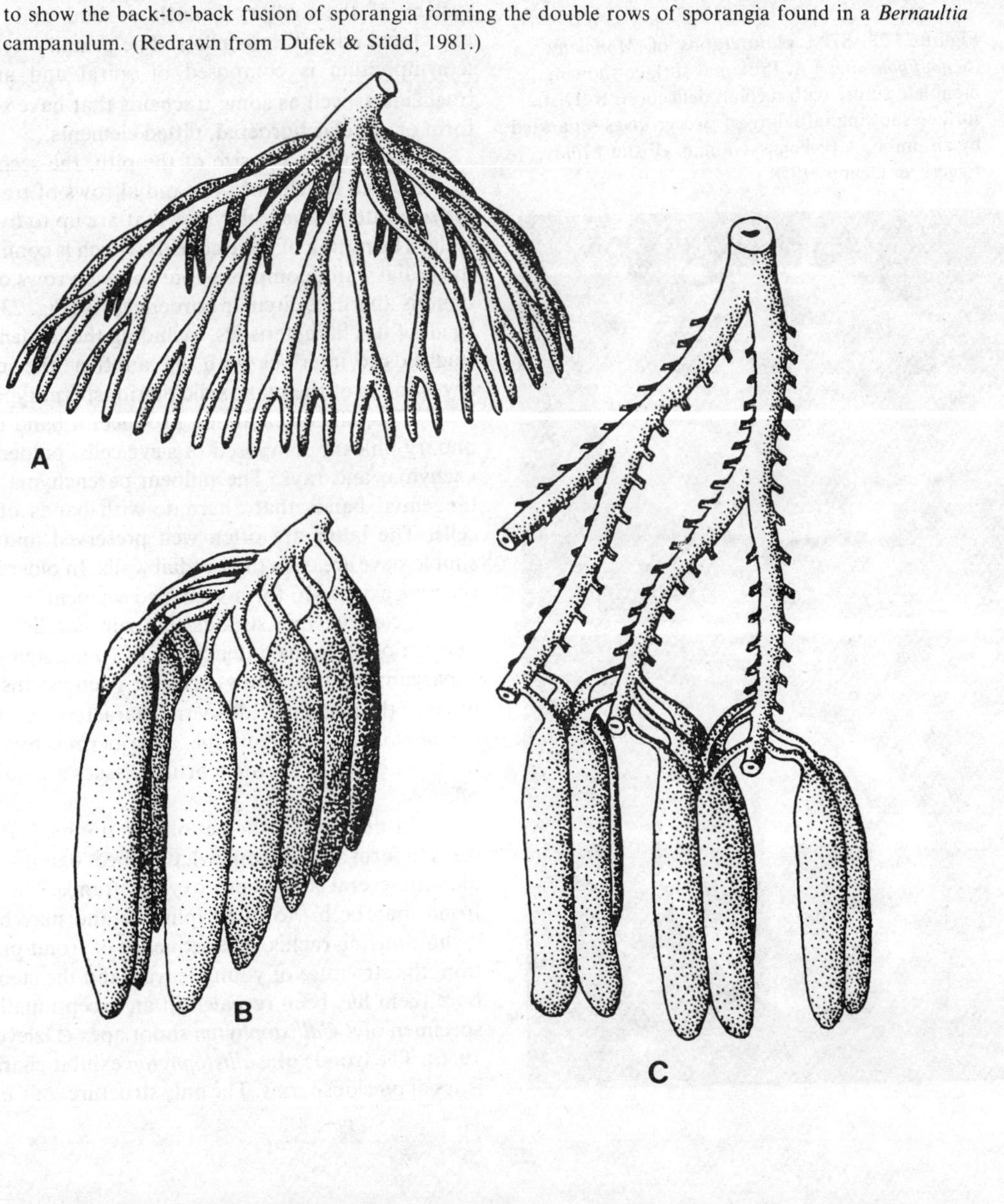

and simulates the pollen of fossil and modern conifers.

Even from the general description given above, it is obvious that *Callistophyton* shares several characteristics with both the pteridosperms and the conifers. We will amplify these shared characteristics in the description that follows and interpret them toward the end of the chapter where the origins and interrelationships of the pteridosperms are discussed.

Callistophyton: vegetative structure

A distinctive feature of *Callistophyton* is the secretory system, which is composed of spherical cavities lined with an epithelium (Fig. 23.31A). These structures occur in the pith and cortex of the stem and branches, all parts of the foliage, the pollen organs, and the testas of the ovules. Some are up to 300 μm in diameter. They are different in shape and structure from the rodlike secretory ducts of the medullosan pteridosperms and conifers with resin ducts. The spherical secretory cavities, along with the capitate glands, are characteristics used in reconstruction of the vegetative and reproductive parts of the plant, where actual attachment of organs has not been observed.

Figure 23.28. SEM photographs of *Monoletes* (*Schopfipollenites*.) **A.** Proximal surface showing monolete suture with median deflection. **B.** Distal surface showing infolding of two grooves separated by an umbo. **A,B:** Pennsylvanian. (From Millay, Eggert, & Dennis, 1978.)

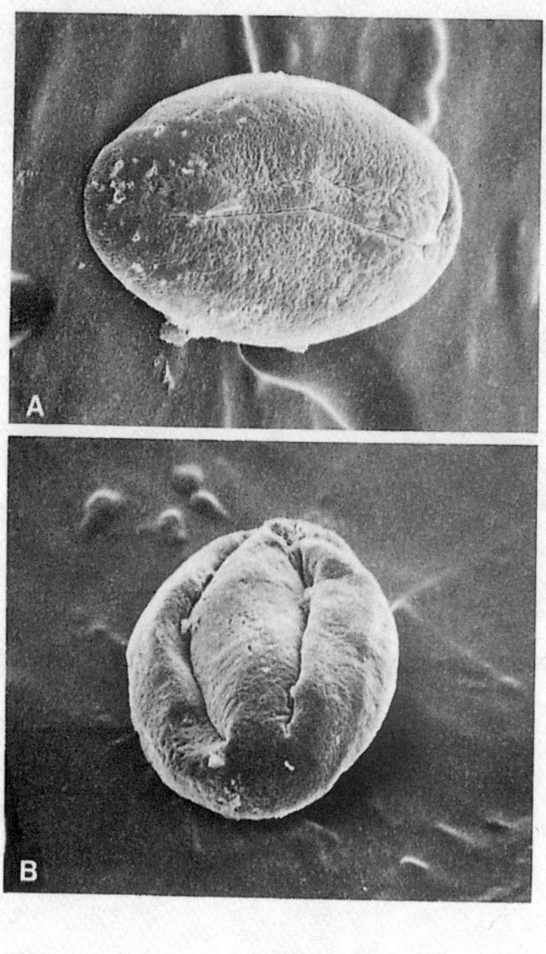

Similar to vascular plants such as *Archaeopteris, Calamopitys foerstei, Lyginopteris,* and many fossil conifers, *Callistophyton* (Fig. 23.31B) has a eustele composed of a ring of five sympodia at the periphery of the pith that give rise to leaf traces in a 2/5 spiral sequence. In *Callistophyton,* as in *Lyginopteris* and conifers, the division of the sympodium that gives rise to a leaf trace and an axial bundle is tangential. If we follow a leaf trace to the base of the leaf it supplies, we see it dichotomizes to form two to several bundles accompanied on the abaxial face by secondary xylem. In the stem this trace is bilobed, but in the petiole it is fused and the secondary xylem has disappeared. Adjacent sympodia divide to provide the primary vascular system of the axillary branch or bud and to the associated adventitious roots. The primary xylem of a sympodium is composed of spiral and annular tracheids as well as some tracheids that have scalariform or circular bordered, pitted elements.

Relative to the size of the pith, the secondary wood is well developed. The radial rows of tracheids are separated by vascular rays that are up to five cells wide. The pitting of the tracheids, which is confined to the radial walls, comprises four to seven rows of contiguous alternate circular bordered pits (Fig. 23.31C). In all of its characteristics, including the tendency for tracheids to interlock with one another, the secondary wood is of the manoxylic pteridosperm type.

The bifacial cambium produces a band of secondary phloem composed of sieve cells, phloem parenchyma, and rays. The phloem parenchyma forms tangential bands that alternate with bands of sieve cells. The latter are often well preserved and show simple sieve areas on their radial walls. In older stems, the rays expand to form V-shaped segments.

Sclerenchyma strands that run parallel to one another occur in the outer cortex of young stems. This "sparganum"-type outer cortex of young stems is another of the many pteridosperm characteristics of *Callistophyton.* As the stems age, a periderm is formed in the inner cortex and outer cortical tissues are sloughed off.

In the base of the petiole the double leaf trace fuses to form a single somewhat arched, bandlike bundle with several adaxial protoxylem strands. The whole frond may be bi- to quadripinnate and may have a dichotomizing rachis. The sequence of frond production, the structure of young leaves, and the stem that bore them has been revealed in an exceptionally fine specimen of a *Callistophyton* shoot apex (Delevoryas, 1956). The fronds of *Callistophyton* exhibit characteristics of pteridosperms. The only structures of *Callisto-*

phyton that approximate the simple leaves of conifers are the scales associated with its axillary buds.

The adventitious roots of *Callistophyton* have diarch exarch protosteles (Rothwell, 1975). Larger roots have secondary xylem of the same type found in the stem as well as a layer of periderm. Diarch protostelic roots occur in other groups of vascular plants, including ferns and conifers.

Callistophyton: reproductive structures

The isolated ovules of *Callistophyton* were first described by Eggert & Delevoryas (1960) and placed in a new genus of dispersed ovules called *Callospermarion* (Fig. 23.32A). The connection of this ovule type with stems and fronds of *Callistophyton* was demonstrated by Stidd & Hall (1970a) and was based on the presence of secretory cavities in the ovule integument similar to those in *Callistophyton* stems and fronds. Combining the descriptions of Eggert & Delevoryas and Stidd & Hall we can characterize *Callospermarion* as a small bilaterally symmetrical ovule up to 2.1 mm long, 1.8 mm wide in the primary plane, and 1.2 mm wide in the secondary plane. The integument is differentiated into sarcotesta, sclerotesta, and endotesta. There is a short micropyle in the apical region of the testa. Secretory cavities with their epithelial linings occur in the sarcotesta (Stidd & Hall, 1970a). The nucellus is fused basally with the inner integumentary layer, but is free from it for about four-fifths of its length. At the distal end the nucellus is differentiated into a hollow flask-shaped structure, terminating apically into a beak that is inserted into the inner aperture of the micropyle. Rothwell (1971b), in his ontogenetic studies of *Callospermarion,* shows us that the pollen chamber was formed by the disintegration of the apical porition of the nucellus, which may have been caused by growth of pollen tubes (Rothwell, 1972a) and by lysigenous breakdown of the tissue as in ovules of extant gymnosperms. All known stages of *Callospermarion* ontogeny, from the initiation of integuments to the formation of megagametophytes

Figure 23.29. Reconstruction of *Callistopnyton* showing scrambling habit. Note axillary branches, adventitious roots, and circinate vernation. Upper Pennsylvanian. (From Rothwell, 1975.)

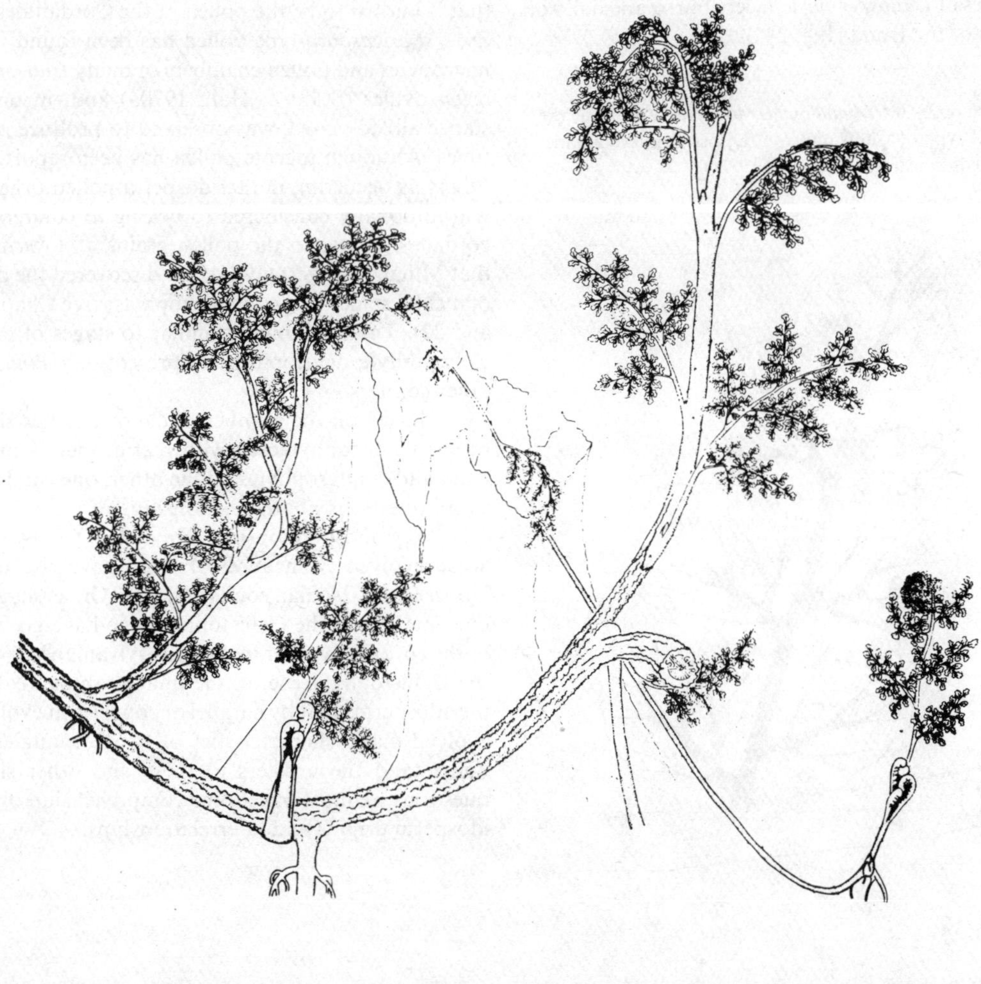

(Rothwell, 1980), parallel the stages known for ovules of present-day gymnosperms.

The vascular system in the base of the ovule consists of a single strand that expands into a disc of tracheids, from which arise two bundles in the primary plane. When seen in transverse section (Fig. 23.32B), the two bundles lie in the sarcotesta opposite the lateral ridges formed by the sclerotesta. These bundles terminate in the sarcotesta just below the level of the micropyle. There is reason to believe that the somewhat flattened ovules of *Callospermarion* evolved from a radiospermic type as evidenced by the presence of two minor bundles arising in the testa in the secondary plane. Ovules of the ancient conifers (Cordaitales) are known to be small and platyspermic, with a nucellus that is free from the integument except at the point of attachment at the base. They also have a pollen chamber similar to that of conifer ovules. This type of pollen chamber and the free nucellus are, however, characteristics shared with ovules of medullosan pteridosperms. We also know from lyginopterid pteridosperms that their ovules, which are characterized by lagenostomes, may be platyspermic or radiospermic. With this in mind, we should not rely on the symmetry of the ovule as a distinguishing characteristic. Although once believed to be stalked ovules, the evidence indicates that the ovules of *Callospermarion* were borne abaxially on pinnules of the frond (Fig. 23.32C).

Figure 23.30. *Medullopteris pluckeneti* pinnule; the leaf type of *Callistophyton*. Upper Pennsylvanian. (Redrawn from Rothwell, 1980.)

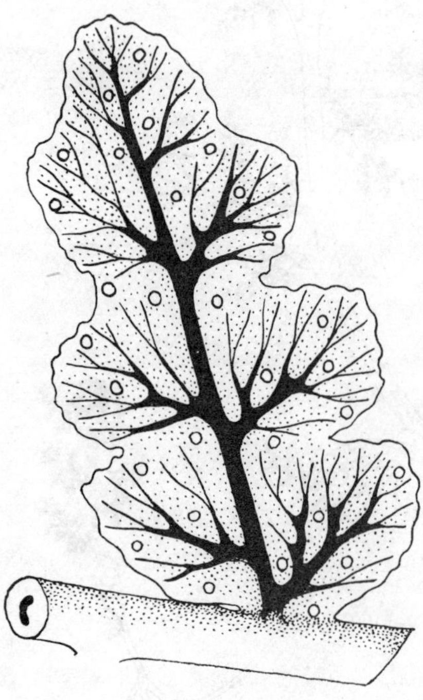

Idanothekion (Millay & Eggert, 1970) and *Callandrium* (Stidd & Hall, 1970a) have been identified as pollen organs of the Callistophytaceae (Rothwell, 1972b). As observed by Rothwell (1980), the two genera are so much alike that he has treated *Callandrium* as a synonym of *Idanothekion*. The pollen organs of *Idanothekion* are radially symmetrical synangiate structures with six to eight sporangia united at their bases to a central column (Fig. 23.33A). Dehiscence occurred along a longitudinal slit on the midline of the inner face of each sporangium. Once again we see a ring of microsporangia arranged around a hollow center, a rather consistent if not conservative characteristic of many pteridosperms. Another feature of *Idanothekion* is the presence of a median vascular strand in the outer wall of each sporangium. *Idanothekion* synangia have been found attached to the abaxial surface of *Callistophyton*-type pinnules near a vein ending (Fig. 23.33B).

The most surprising feature of these pollen organs is their pollen grains of the *Vesicaspora* dispersed type (Fig. 22.13B). This is a monosaccate pollen with a distal germination furrow (sulcus) and two lateral bladders. In its morphology the pollen is very similar to the bisaccate pollen of several extant and extinct coniferophytes and of the dispersed *Florinites* type that is known to be the pollen of the Cordaitales.

Vesicaspora-type pollen has been found in the micropyles and pollen chambers of many *Callospermarion* ovules (Stidd & Hall, 1970a) and, in one instance already cited, was observed to produce pollen tubes. Although saccate pollen has been reported by others as occurring in pteridosperm pollen organs, it is traditionally considered to belong to conifers and cordaites. It was in the pollen grains of *Vesicaspora* that Millay & Eggert (1974) first discovered the developmental stages of microgametophytes (see Chapters 2 and 22). These are highly similar to stages of microgametophyte development in present-day *Pinus* and other conifers.

Based on the number of characteristics shared by the Callistophytaceae with pteridosperms on one hand and coniferophytes on the other, one can hardly avoid questions about relationships of the family to other Gymnospermopsida. Have the Callistophytaceae evolved from ancestors that gave rise to the Upper Pennsylvanian coniferophytes? Or, as suggested by some, could the Callistophytaceae have given rise to the cordaites earlier in the Pennsylvanian? Or could the Callistophytaceae, in the final analysis, really be pteridosperms that by parallel or convergent evolution evolved many characteristics of the Cordaitales? To understand the answers to these and other similar questions we must have some comprehension of pteridosperm origins and interrelationships.

Figure 23.31. Anatomy of *Callistophyton* stems. **A.** Section through cortex showing epithelium-lined cavity at base of a capitate gland. (From Rothwell, 1975.) **B.** Transverse section. Note sympodia (s) at periphery of pith (p); and leaf trace (lt). **C.** Radial section of manoxylic secondary wood showing large tracheids with alternate bordered pits on the radial walls. **A–C:** Upper Pennsylvanian. (**B,C** from Delevoryas & Morgan, 1954.)

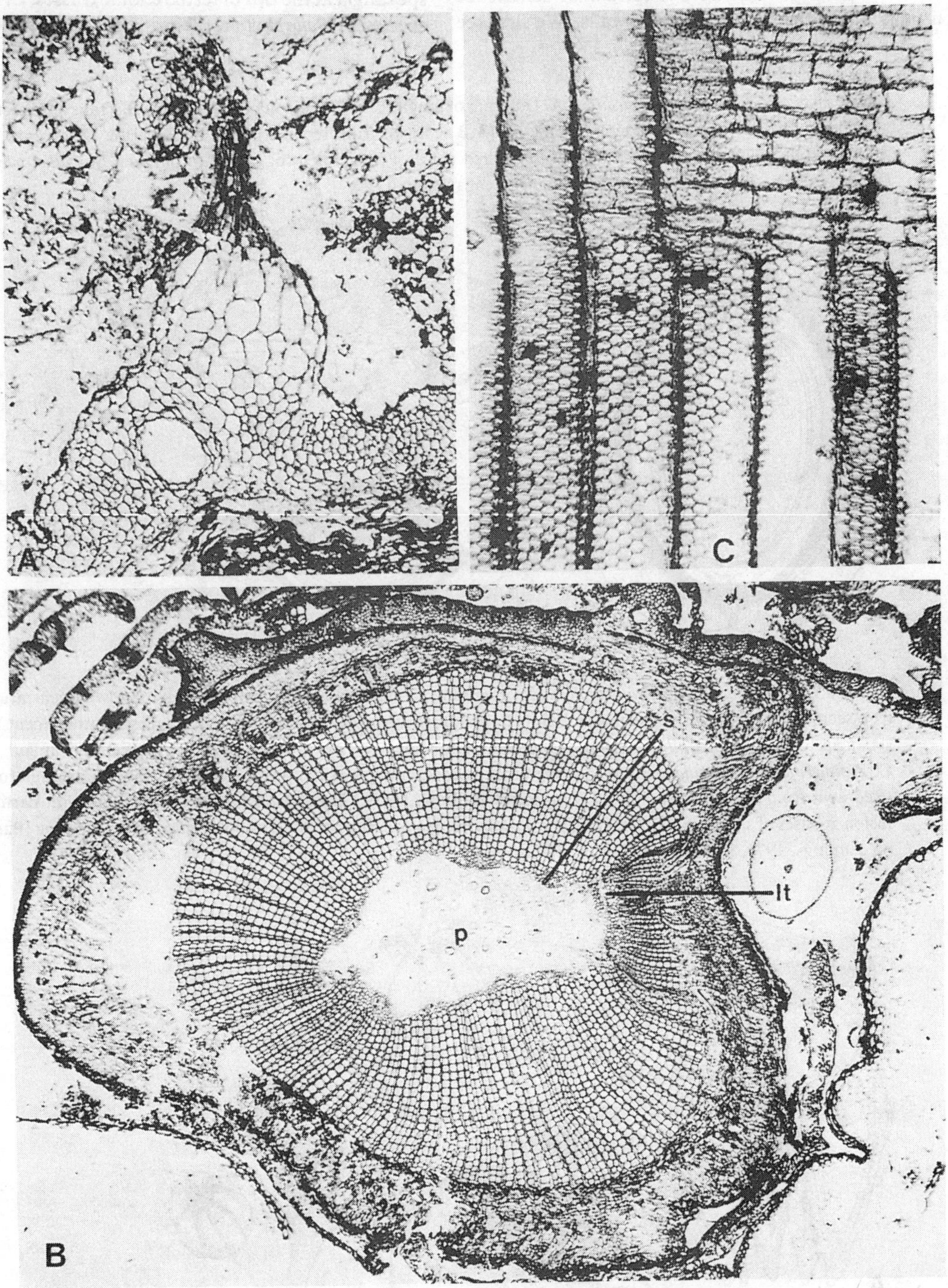

The origin of pteridosperm synangia

One of the more consistent features of the Paleozoic pteridosperms is their formation of synangiate microspore-producing organs composed of a cycle of linear eusporangia that have longitudinal dehiscence usually along their inward facing walls. We first see evidence of these characteristics among the Middle and Upper Devonian Trimerophytopsida (Chapter 13). One of the distinguishing characteristics of the trimerophytes is the formation of clusters of massive linear sporangia at the tips of fertile telome trusses. In several species of *Psilophyton*, the principal genus of the tri-

Figure 23.32. *Callospermarion pusillum*. **A.** Diagram of longitudinal section in primary plane. **B.** Transverse section at midpoint of ovule. Note bilateral symmetry. (**A,B** redrawn from reconstruction by Eggert & Delevoryas, 1960; Stidd & Hall, 1970b; Rothwell, 1981.) **C.** Suggested place of attachment of ovules to pinnules of *Callistophyton*. (Redrawn from Rothwell, 1981.) **A–C:** Upper Pennsylvanian.

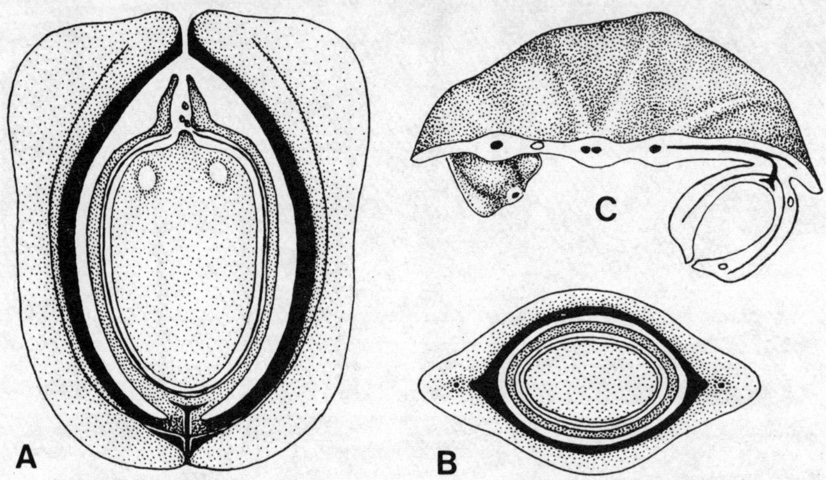

Figure 23.33. **A.** *Idanothekion callistophytoides*, diagram of longitudinal section of synangium showing central column (cc); secretory cavities (sc); vascular system (vs). (Redrawn from Rothwell, 1980.) **B.** Suggested attachment of synangia to pinnule of *Callistophyton*. (Modified from Millay & Taylor, 1979.) **A,B:** Upper Pennsylvanian. **C.** *Zimmermannitheca cupulaeformis*, a ring of unfused sporangia at the tip of a branch. Lower Carboniferous. (Redrawn from Millay & Taylor, 1979, in Remy & Remy, 1959.) **D,E.** *Telangium* sp., synangium with partial fusion at bases of sporangia. **E.** Diagram showing sporangia in a ring around a hollow center. Mississippian. (Based on Jennings, 1976, in Millay & Taylor, 1979.)

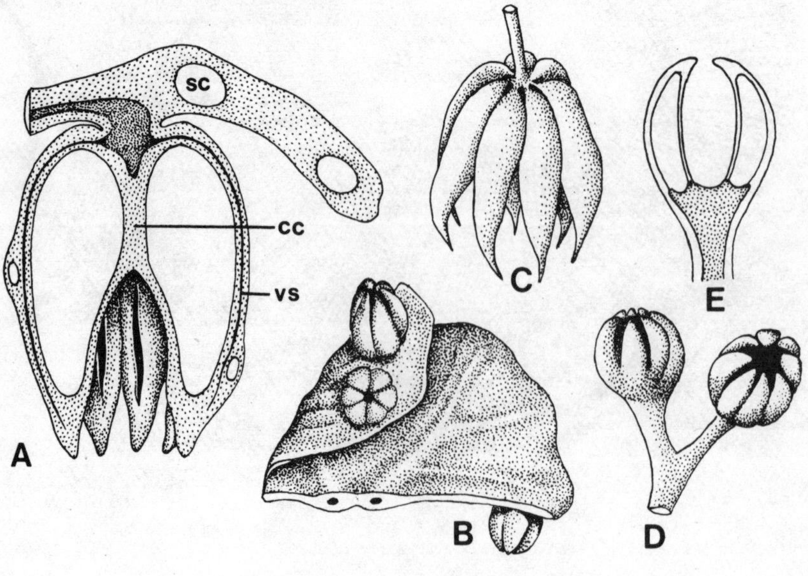

merophytes, the inward-facing walls of pairs of sporangia in a cluster dehisce longitudinally. It is worth noting that although the sterile telome trusses are subject to planation and overtopping, the fertile trusses tend to retain their three-dimensional dichotomous plan, and in this respect they can be interpreted as conservative structures. By slight reduction of those branches bearing paired sporangia in a cluster, a cycle or ring of sporangia would result (Millay & Taylor, 1979) that by tangential fusion (syngenesis) would produce a synangium.

That this represents the way in which the pteridosperm synangium evolved is evidenced by Carboniferous genera of presumed pteridosperm microsporangiate structures (Meyer–Berthaud, 1989). For example, in *Zimmermannitheca* (Remy & Remy, 1959) are clusters of two to seven linear sporangia borne at the tips of dichotomizing branch systems (Fig. 23.33C). There is no indication of fusion of the sporangia in the ringlike cluster. In *Telangium* (Fig. 23.33D, E), which is similar to *Zimmermannitheca,* the bases of the sporangia are fused and longitudinal dehiscence slits may occur along their inward-facing walls. The dichotomous branches of *Telangium* bearing the terminal synangia are three-dimensional telome trusses that occur alternately on a planated fertile frond (Jennings, 1976). *Telangium,* you may recall, is the pollen organ of a lyginopterid pteridosperm. It is known to occur in the Mississippian (Lower Carboniferous) and Westphalian A (Upper Carboniferous). By planation, overtopping, and webbing, the fertile telomes truss of the primitive type, accompanied by elaboration of the synangia, produced the many modifications of Paleozoic pteridosperm synangia already described.

It should be noted that sporangia of the Progymnospermopsida also are oblong–fusiform, usually have longitudinal dehiscence, and tend to be clustered at the tips of dichotomous branches having a pinnate arrangement. As far as we are aware, no presently known progymnosperm has synangia. It may be that the Lower Carboniferous *Geminitheca* is a progymnosperm with microsporangia that occur in clusters at the tips of dichotomies, as in *Zimmermannitheca*. Thus, there is some evidence of a morphology similar to *Zimmermannitheca* from which seed fern synangia were probably derived during the Lower Carboniferous.

The microspores of *Geminitheca* have characteristics of progymnosperms belonging to the Aneurophytales. The important characteristics are a trilete suture and an exine that tends to form a saccus separate from the central body of the microspore. In progymnosperms the girdling saccate exine is called a pseudosaccus, which in the aneurophyte *Rellimia* is attached proximally to the corpus. The microspores of the progymnosperms belonging to the Archaeopteridales and Protopityales are trilete, but without a pseudosaccus. Thus, among the progymnosperms we find microspores with characteristics of two of the four basic types of prepollen or pollen found among the Paleozoic pteridosperms. These two are the trilete prepollen types with a saccus and those without. Both types probably germinated through the proximal trilete suture. Monolete prepollen with proximal germination and monosaccate pollen with distal germination are clearly derived from the more primitive trilete types.

From the above explanation of the origin of pteridosperm synangia and pollen types, we can trace the evolution of seed ferns from the Middle Devonian Trimerophytes through the Progymnospermopsida of the Upper Devonian into the Lower Carboniferous (Chart 26.1). The evolution of a diverse assemblage of prepollen types in the Upper Devonian from trilete isospores correlates well with the evolution of heterospory and the seed habit. Until the discovery of heterospory in the Middle Devonian *Chauleria,* with its trilete spores of two distinct sizes, the earliest proven examples of heterospory were from the Upper Devonian *Archaeopteris.*

The origin of pteridosperm vegetative structure

Conclusive evidence for the origin of pteridosperms from progymnosperms is presented by Galtier (1977), who describes *Tristichia longii,* from the Lower Carboniferous of France, as being intermediate in structure between the Aneurophytales and Lyginopteridaceae. *Tristichia longii* has a three-ribbed protostele that is highly similar to that of *Aneurophyton, Rellimia,* and *Triloboxylon.* As in these genera, the maturation of the primary xylem of *Tristichia* is mesarch, with protoxylem strands near the tips of the radiating arms and extending centripetally to the center of the stele. Similar to *Triloboxylon* and *Proteokalon,* parenchyma cells occur among the metaxylem tracheids of the *Tristichia* stele and are interpreted as an early stage in the evolution of the eustele. Leaf traces are produced in pairs from the tips of the radiating arms of the protostele. The number, structure, and configuration of the leaf traces compare favorably with those of *Lyginorachis.* Other pteridosperm characteristics of *Tristichia longii* are the manoxylic secondary wood, metaxylem tracheids with multiseriate circular pits on the walls, secretory cells, and a "sparganum"-type outer cortex. The latter characteristic has been noted in *Triloboxylon* and *Proteokalon* of the Aneurophytales, as have metaxylem cells with circular bordered pits.

All the pteridosperm characteristics listed by Galtier (1977) for *Tristichia,* except for the *Lyginorachis*-type leaf traces, are also characteristics found among the various genera of the Calamopityaceae where some have three-ribbed protosteles composed entirely of tracheids (*Stenomyelon primaevum*), "mixed" protosteles (*S. heterangioides* and *Calamopitys americana*), or eusteles (*C. foerstei*). Some have secondary wood that is manoxylic, while others have the pycnoxylic type of the Aneurophytales. Considering all available characteristics, there is little doubt that most ancient seed ferns with ribbed protosteles, the Calamopityaceae, and Lyginopteridaceae, have evolved from the Aneurophytales during the Upper Devonian and Lower Carboniferous.

The origin of the Callistophytaceae

We can make the generalization, from the evidence presented earlier in the chapter, that the vegetative structure of the Callistophytaceae is most like that of lyginopterids and that their reproductive structures simulate those of the cordaites. The possibility has been suggested that the Callistophytaceae with saccate pollen and platyspermic ovules might have given rise to the cordaitalean conifers that make their first known appearance in the Lower Coal Measures (Westphalian A) of Great Britain or the Lower Pennsylvanian of North America. If the Middle Pennsylvanian specimens of *Callistophyton* represent their earliest known appearances in the fossil record, then it seems unlikely, from the stratigraphic evidence, that cordaites evolved from the Callistophytaceae. Another major obstacle in deriving cordaites from the Callistophytaceae is the way in which their reproductive structures are borne. You will recall that pollen organs and ovules of *Callistophyton* are on the abaxial surfaces of unmodified compound leaves; those of cordaites are spirally arranged in cones. Until there is some evidence that shows us how the cones of coniferophytes evolved from spirally arranged fronds (a possibility), or how the simple leaf of a cordaite evolved from plants with compound leaves (also a possibility), we think we should dismiss the idea of a callistophyte origin of the Cordaitales. It seems more likely that the Callistophytaceae originated from among primitive seed ferns with hydrasperman reproduction. That the Callistophytaceae is a group divergent from the Lyginopteridaceae is evidenced not only by their saccate pollen, but by ovules lacking cupules and lagenostomes.

The origin and evolution of the Medullosaceae

In this and previous chapters we have alluded to the important work of Beck (1970), who clearly set forth how the gymnosperm eustele evolved from protosteles. To understand how and from where the Medullosaceae originated, we should review the substance of the papers by Namboodiri & Beck (1968), Beck (1970), and Read (1936). In his monograph on the Calamopityaceae, Read suggested that species of *Stenomyelon* were primitive members of the family. As we have described in this chapter, some have three-ribbed protosteles that bear more than a superficial resemblance to protosteles of the progymnosperms belonging to the Aneurophytales (*Triloboxylon*, for example), not only in their mesarch primary xylem, but in the way their massive spirally arranged leaf traces arise on the radius from the tips of the arms of the protostele (Fig. 23.35B). With the protostelic *Stenomyelon primaevum* as a starting point, Beck has established a series illustrating the steps in the evolution of the eustele within the Calamopityaceae (Fig. 23.35A–C). The step that has attracted the attention of those interested in the origin of the medullosas is illustrated by *S. tuedianum*. Here you may recall from an earlier description that the three-ribbed protostele is dissected into three more or less equal segments by radially arranged longitudinal plates of parenchyma. Similar plates have been reported in protosteles of *Heterangium* and *Microspermopteris*. Of significance is the possibility that we see in *S. tuedianum* the way in which the segmented stele ("polystele") of the Medullosaceae evolved (Fig. 23.34A–C). According to Namboodiri & Beck, the bifacial cambium forming secondary xylem and phloem developed around each of the primary xylem segments. Among the Calamopityaceae, *Bostonia* displays three such vascular segments with secondary tissue. Unfortunately, the specimen of *Bostonia* is a short one, and thus it is impossible to tell whether it is a stem with a branching stele, each of which supplied a lateral branch at a higher level, or a "polystelic" structure.

It is important to understand that the protosteles of *Stenomyelon* have mesarch primary xylem with protoxylem strands that give rise to leaf traces near the tips of the arms to produce five orthostichies. In all species of *Stenomyelon* the leaf traces arise in a spiral 2/5 phyllotactic sequence. Even after phyletic dissection of the protostele, the spiral phyllotaxy is maintained, as are the protoxylem strands that give rise to the leaf traces. The latter lie in a ring, but are separated into vascular segments by the dissection of protostele, as in *S. tuedianum*. In a *Medullosa* stem with three vascular segments, the ring of protoxylem strands is maintained (Fig. 23.34C). Here, however, the strands are recognizable sympodia giving rise to leaf traces in vertical rows that supply the large spirally arranged leaf bases. The sympodia are submerged in a mixture of parenchyma cells and metaxylem tracheids along the outer-facing surfaces of the vascular segments. There are one to as many as five sympodia per vascular segment. A total of five sympodia per stem axis, however, is a common number. From the above description it should be clear that the primary vascular system of *Medullosa* has the basic structure of a eustele. According to this interpretation, the *Medullosa* stem is not "polystelic," but has been evolutionarily derived by dissection of a single stele (Basinger, Rothwell, & Stewart, 1974; Stewart, 1976).

Earlier, Delevoryas (1955) and Stewart &

Delevoryas (1956) proposed that the "polystelic" stems of the Medullosaceae were best explained as being derived from psilophyte ancestors by phyletic fusion (syngenesis) of the axes of these primitive plants but without fusion of their vascular systems (Fig. 23.34D–F). There are two problems that arise if one explains the origin of the medullosan "polystele" as a product of phyletic fusion. First, there is no evidence of ancestors that could be attributed to the Medullosaceae showing stages in fusion. Second, one would expect each vascular segment of a "polystelic" medullosan stem to have its own complete ring of sympodia with leaf traces arising on the circumference and not just from the outward-facing surfaces of each vascular segment.

Recent evidence from Mississippian specimens shows us that monostelic medullosan stems (*Quaestora*) coexisted with "polystelic" forms. The discoverers of *Quaestora* (Mapes & Rothwell, 1980) suggest that dissection of such a monostelic form could have given rise to the Upper Carboniferous *Sutcliffia* with its many vascular segments, some of which give rise to leaf traces and some that anastomose at a higher level in the axis. According to the interpretation of Delevoryas (1955), *Sutcliffia* represents a stage in syngenesis of axes where there is no distinction between cauline and foliar vasculature. Both have the same concentric organization with surrounding secondary tissues. This same lack of differentiation of cauline and foliar vasculature occurs in the Aneurophytales.

Regardless of which explanation you accept for the evolution of the "polystelic" stems of the Medullosaceae, by dissection or phyletic fusion, proponents of both suggest that *Sutcliffia* shows primitive features of the family. From the *Sutcliffia* type, Stewart & Delevoryas (1952) and Delevoryas (1955) envisage two main evolutionary trends within the Medullosaceae. One is described as a side branch in which ontogenetic fusion accompanied by phyletic fusion of numerous vascular segments occurs, so that the number of vascular segments becomes stabilized at two or three (*M. noei*). The side branch probably represents an evolutionary "dead end," a suggestion supported by the large size and complex nature of their pollen-producing structures. The second line of evolution is characterized by the tangential expansion and phyletic fusion of vascular segments, resulting in axes having many characteristics of monostelic cycad stems. To produce a monostelic axis from a medullosan stem with a ring of two or three vascular segments and several smaller internal ones (*M. leuckartii*) requires the phyletic tangential fusion of the large peripheral segments to form a continuous ring (*M. stellata*). This has the effect of producing a stem with concentric rings of secondary xylem, one ring with centrifugal and the other with centripetal development. Although *M. stellata* has small internal vascular segments, as do most Permian medullosas, one genus, *Colpoxylon* (Fig. 23.20E), in which tangential fusion of peripheral vas-

Figure 23.34. **A–C:** Steps in the evolution of medullosan stem structure according to phyletic dissection. **A.** Three-ribbed protostele with mesarch primary xylem (black); protoxylem strands (small circles). **B.** Dissection of protostele by radiating plates of parenchyma (stippled). **C.** Vascular segments invested with secondary xylem *(hatched)*. **D–F:** Steps in the evolution of medullosan stem structure according to the telome theory. **D.** Transverse section of protostelic telomes of the *Rhynia* type. Syngenesis of the telomes of the *Rhynia* type. **E.** Syngenesis of the telomes to form a "polystelic" axis. **F.** Development of cambial activity to produce secondary tissues around each vascular segment. (**A–C** redrawn from Stewart, 1976, **D–F** redrawn from Stewart, 1964.)

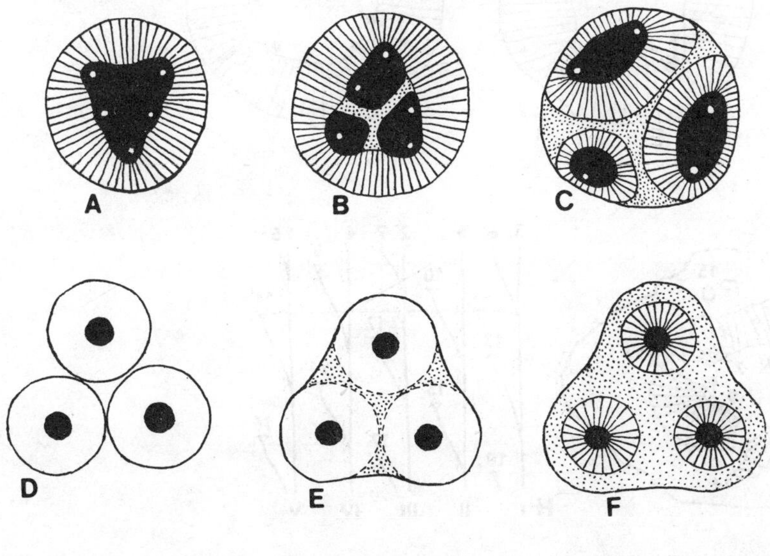

cular segments has occurred, has no small internal segments. The stem of *Colpoxylon* approximates the structure of the cycadean type more closely than any other pteridosperm (Delevoryas, 1955) and strongly suggests that the cycads, whose first known appearance is in the Permian, evolved from Upper Carboniferous–Lower Permian medullosas. The idea that cycads evolved from Medullosaceae was first developed by Worsdell (1906), who found many similarities in their stem anatomy, such as inverted zones of secondary xylem and phloem, the formation of concentric cycles of secondary vascular tissue around the main vascular segment, and the presence of cylindrical vascular strands in the cortex of some genera. The origin of the cycads from the medullosan pteridosperms is further supported by similarities in ovule structure, leaf characteristics, secondary wood, secretory systems, and so on. There are, however, some unanswered questions about the evolution of microsporangia in cycads from the basic medullosan type. These and other questions will be examined more fully in Chapter 24.

Figure 23.35. Steps in the evolution of the gymnosperm eustele as proposed by Beck (1970). **A–F:** Diagrams, transverse sections of stems assigned to the Calamopityaceae and Lyginopteridaceae, showing the evolution of axial sympodia from a protostele by medullation. Primary xylem (black); mixed pith (stippled); pith (white); secondary xylem (hatched). **A.** *Stenomyelon primaevum,* three-ribbed protostele. **B.** *S. tuedianum,* initiation of medullation. **C.** *Calamopitys americana,* increased medullation and delineation of sympodia. **D.** *Calamopitys* sp., completion of medullation and formation of a pith. Further delineation of five axial sympodia. **E.** *C. foerstei.* Note that as in previous stems leaf traces arise from sympodia on the radii of the stem. **F.** *Lyginopteris oldhamia.* Same as *C. foerstei* except that leaf traces arise by division of sympodia on a tangent, not a radius. **G.** Diagram of stem, *L. oldhamia,* showing the axial sympodia and their leaf traces numbered, showing the 2/5 spiral phyllotaxy. **H.** Diagram of stem stele in one plane. Roman numerals indicate numbers of axial sympodia. Arabic numerals indicate leaf trace numbers in each orthostichy. (A–F redrawn from Beck, 1970; G,H from Beck, 1970.)

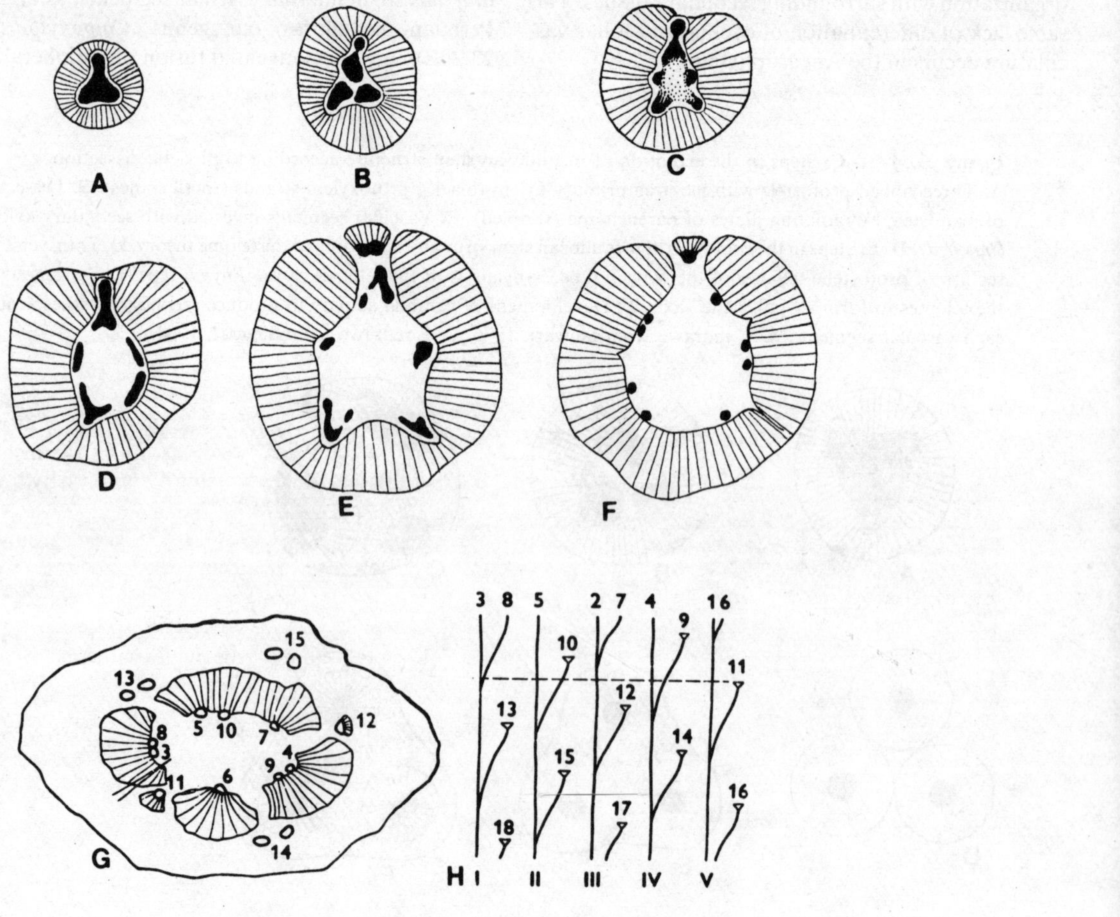

The evolution of the gymnosperm eustele

Most of the evidence supports an origin of the Calamopityceae, and possibly the Lyginopteridaceae, from the Aneurophytales. These two families have evolved eusteles, the basic arrangement of the primary vasculature of seed plants. As explained above, the early steps in the evolution of the eustele of the Calamopityaceae (Beck, 1970) start with the medullation of three-ribbed protosteles characteristic of species of *Stenomylon* (Fig. 23.35A,B). Additional medullation, resulting from failure of centrally located metaxylem cells to differentiate, occurs in *Calamopitys americana* (Fig. 23.35C). Like *Stenomyelon tuedianum, C. americana* has a "mixed" protostele. In *C. americana,* however, the five protoxylem strands appear as discrete axial bundles or sympodia from which leaf traces diverge on the radius of the stem. As in *Stenomyelon,* the leaf traces are produced in a 2/5 ontogenetic spiral. Further medullation results in the formation of a pith around which is a clearly defined ring of five axial bundles. This is apparent in *Calamopitys* sp. (Fig. 23.35D) and *C. foerstei* (Fig. 23.35E). In the latter species, the sympodia are spacially separated from one another. Prior to the departure of a leaf trace from the axial bundle, the bundle assumes a V-shape. One arm of the V separates to form an accessory bundle that diverges toward the axial bundle to the left in the ring, but does not fuse with it. Above the level of accessory bundle departure, the axial bundle divides on the radius of the stem to give rise to a leaf trace. These leaf traces also arise in a 2/5 ontogenetic spiral from the five axial bundles. Five leaf traces are produced in the spiral sequence before two leaf traces are formed from an axial trace in the same orthostichy. Beck (1970) shows us that, with one important difference, the primary vascular system of the *C. foerstei* eustele is the same as in *Lyginopteris.* In *Lyginopteris* (Fig. 23.35F–H), axial bundles divide on a tangent to produce leaf traces on the circumference of the stem rather than on the radius as in *Stenomyelon* and some species of *Calamopitys.* Eusteles of the *Lyginopteris* type (Fig. 23.36) are found in the Callistophytaceae, Cordaitales, and Coniferales. It has been suggested that angiosperm primary vascular systems may be of the same type. From the foregoing, it is clear that the seed plants do not have a primary vascular system where the departure of leaf traces from the stem stele produces leaf gaps as in the siphonostelic ferns (Fig. 10.3B). This evidence, along with evidence indicating an origin of pteridosperms and conifers from the progymnosperms, clearly indicates that the origin of the gymnosperms was quite independent of the ferns. There is no evidence supporting the idea that the pteridosperms formed an evolutionary bridge between the ferns and other groups of gymnosperms (Namboodiri & Beck, 1968; Stewart, 1976), an idea that persisted for 50 years.

Figure 23.36. Stereodiagram of primary vasculature of a gymnosperm eustele. Compare with the dictyoxylic siphonostele of a fern, Figure 10.3B. (Modified from Namboodiri & Beck, 1968.)

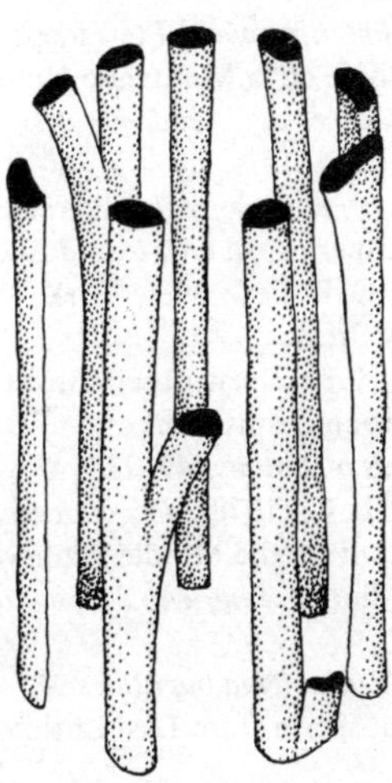

Summary: Major events in evolution of Paleozoic pteridosperms

The remarkable Paleozoic pteridosperms show many evolutionary adaptations and innovations that became established in groups of vascular plants derived from pteridosperms later in geological time. In summary, we can list the innovations discussed in this and preceding chapters.

1. The evolution of the planated bilaterally symmetrical frond from three-dimensional radially arranged telome trusses
2. The positioning of reproductive structures on the frond by modification of fertile telome trusses
3. The evolution of the eustele from protostelic ancestors
4. The evolution of manoxylic wood
5. The evolution of indehiscent megasporangia, of pollination, and of abscission of the megasporangium, thus producing plants with hydrasperman reproduction
6. The evolution of protective layers around the megasporangium from adjacent telomes to form integumentary layers and cupules, thus producing preovules
7. The evolution of true ovules with an integument that forms a functional micropyle, allowing for the independent evolution of the Medullosaceae, Callistophytaceae, and other groups of seed plants
8. The first known appearance of pollen tubes
9. The evolution of cyclic synangia from clusters of microsporangia terminating fertile telome trusses

10. The evolution of a diverse assemblage of pollen and prepollen types from trilete spores

In following chapters we will see how these and other characteristics evolved by the pteridosperms have played a role in the emergence of such groups as the glossopterids, Mesozoic pteridosperms, cycads, cycadeoids, and flowering plants (Chart 30.1).

References

Andrews, H. N. (1945). Contributions to our knowledge of American Carboniferous floras. *Annals of the Missouri Botanical Garden*, **32**, 323–60.

Andrews, H. N. (1961). *Studies in Paleobotany*. New York: Wiley.

Arnold, C. A. (1947). *An Introduction to Paleobotany*. New York: McGraw-Hill.

Barnard, P. D. W., & Long, A. G. (1973). On the structure of a petrified stem and some associated seeds from the Lower Carboniferous rocks of East Lothian, Scotland. *Transactions of the Royal Society of Edinburgh*, **69**, 91–108.

Barnard, P. D. W., & Long, A. G. (1975). *Triradioxylon* a new genus of Lower Carboniferous petrified stems and petioles together with a review of the classification of early Pterophytina. *Transactions of the Royal Society of Edinburgh*, **69**, 231–50.

Basinger, J. F., Rothwell, G. W., & Stewart, W. N. (1974). Cauline vasculature and leaf trace production in medullosan pteridosperms. *American Journal of Botany*, **61**, 1002–15.

Bateman, R. M., (1988). Palaeobotany and palaeoenvironments of Lower Carboniferous floras from two volcanigenic terrains in the Scottish Midland Valley. Unpublished Ph.D Thesis, University of London. 452 pp.

Bateman, R. M., & Rothwell, G. W. (1990). A reappraisal of the Dinantian floras at Oxroad Bay, East Lothian, Scotland. 1. Floristics and the development of whole-plant concepts. *Transactions of the Royal Society of Edinburgh*, **69**, 127–59.

Baxter, R. W. (1949). Some pteridospermous stems and fructifications with particular reference to the Medullosaceae. *Annals of the Missouri Botanical Garden*, **36**, 287–352.

Beck, C. B. (1970). The appearance of gymnospermous structure. *Biological Reviews*, **45**, 379–400.

Beck, C. B., & Bailey, R. E. (1967). Plants of the New Albany Shale. III: *Chapellia campbellii* gen. et sp. n. *American Journal of Botany*, **54**, 998–1007.

Beeler, H. E. (1983). Anatomy and frond architecture of *Neuropteris ovata* and *N. scheuchzeri* from the Upper Pennsylvanian of the Appalachian Basin. *Canadian Journal of Botany*, **61**, 2352–68.

Benson, M. J. (1914). *Sphaerostoma ovale* (*Conostoma oval et intermedium*, Williamson), a Lower Carboniferous ovule from Pettycur, Fifeshire, Scotland. *Transactions of the Royal Society of Edinburgh*, **50**, 1–15.

Chaloner, W. G. (1970). The evolution of microspore polarity. *Geoscience and Man*, **1**, 47–56.

Combourieu, N., & Galtier, J. (1985). Nouvelles observations sur *Polypterospermum, Polylophospermum, Colpospermum* et *Codonospermum*, ovules de Pteridospermales du Carbonifère Supérieur Francais. *Palaeontographica B*, **196**, 1–29.

Crane, P. R. (1985). Phylogenetic analysis of seed plants and the origin of angiosperms. *Annals of the Missouri Botanical Garden*, **72**, 716–93.

Crane, P. R. (1988). Major clades and relationships in the "higher" gymnosperms. In *Origin and Evolution of Gymnosperms*, ed. C. B. Beck. New York: Columbia University Press.

Delevoryas, T. (1955). The Medullosae–structure and relationships. *Palaeontographica B*, **97**, 114–67.

Delevoryas, T. (1956). The shoot apex of *Callistophyton poroxyloides*. *Contributions from the Museum of Paleontology, University of Michigan*, **12**, 285–99.

Delevoryas, T., & Morgan, J. (1954). A new pteridosperm from the Upper Pennsylvanian deposits of North America. *Palaeontographica B*, **96**, 12–23.

Dennis, R. L., & Eggert, D. A. (1978). *Parasporotheca*, gen. nov., and its bearing on the interpretation of the morphology of permineralized medullosan pollen organs. *Botanical Gazette*, **139**, 117–39.

Doyle, J. A., & Donoghue, M. J. (1986). Seed plant phylogeny and the origin of the angiosperms: An experimental cladistic approach. *Botanical Review*, **52**, 312–431.

Dufek, D., & Stidd, B. M. (1981). The vascular system of *Dolerotheca* and its phylogenetic significance. *American Journal of Botany*, **63**, 887–907.

Eggert, D. A., & Delevoryas, T. (1960). *Callospermarion* – a new seed genus from the Upper Pennsylvanian of Illinois. *Phytomorphology*, **10**, 323–60.

Eggert, D. A., & Rothwell, G. W. (1979). *Stewartiotheca* gen. n. and the nature and origin of complex permineralized medullosan pollen organs. *American Journal of Botany*, **66**, 851–66.

Eggert, D. A., & Taylor, T. N. (1971). *Telangiopsis* gen. nov., an Upper Mississippian pollen organ from Arkansas. *Botanical Gazette*, **132**, 30–7.

Galtier, J. (1975). Variabilité anatomique et ramifaction des tiges de *Calamopitys. Compte Rendu des Séances de l'Académie des Sciences, Paris*, **280**, 1967–70.

Galtier, J. (1977). *Tristichia longii*, nouvelle Ptéridospermale probable du Carbonifère de la Montagne Noire. *Compte Rendu des Séances de l'Académie des Sciences, Paris*, **284**, 2215–18.

Galtier, J. (1988). Morphology and phylogenetic relationships of early pteridosperms. In *Origin and Evolution of Gymnosperms*, ed. C. B. Beck. New York: Columbia University Press.

Galtier, J., & Holmes, J. C. (1982). New observations on the branching of Carboniferous ferns and pteridosperms. *Annals of Botany*, **49**, 737–46.

Gastaldo, R. A., & Matten, L. C. (1978). *Trigonocarpus leeanus*, a new species from the Middle Pennsylvanian of southern Illinois. *American Journal of Botany*, **65**, 882–90.

Gordon, W. T. (1938). On *Tetrastichia bupatides:* A Carboniferous pteridosperm from East Lothian.

Transactions of the Royal Society of Edinburgh, **59,** 351–70.

Hall, J. W. (1952). The phloem of *Heterangium americanum*. *American Midland Naturalist*, **47,** 763–8.

Halle, T. G. (1933). The structure of certain fossil spore-bearing organs believed to belong to pteridosperms. *Kungl. Svenska Vetenskapsakdemiens Handlingar*, **12,** 1–103.

Hamer, J. J., & Rothwell, G. W. (1988). The vegetative structure of *Medullosa endocentrica* (Pteridospermopsida). *Canadian Journal of Botany*, **66,** 375–87.

Hoskins, J. H., & Cross, A. T. (1946). Studies in the Trigonocarpales. Part II. Taxonomic problems and a revision of the genus *Pachytesta*. *American Midland Naturalist*, **36,** 331–61.

Jennings, J. R. (1976). The morphology and relationships of *Rhodea, Telangium, Telangiopsis*, and *Heterangium*. *American Journal of Botany*, **63,** 1119–33.

Long, A. G. (1944). On the prothallus of *Lagenostoma ovoides* Will. *Annals of Botany (N.S.)*, **8,** 105–17.

Long, A. G. (1961). *Tristichia ovensi* gen. et sp. nov., a protostelic Lower Carboniferous pteridosperm from Berwickshire and East Lothian, with an account of associated seeds and cupules. *Transactions of the Royal Society of Edinburgh*, **64,** 77–89.

Long, A. G. (1963). Some specimens of *Lyginorachis papilio* Kidston associated with stems of *Pitys*. *Transactions of the Royal Society of Edinburgh*, **65,** 211–24.

Long, A. G. (1964). Some specimens of *Stenomyelon* and *Kalymma* from the calciferous sandstone of Berwickshire. *Transactions of the Royal Society of Edinburgh*, **65,** 435–47.

Long, A. G. (1979). Observations on the Lower Carboniferous genus *Pitus* Witham. *Transactions of the Royal Society of Edinburgh*, **70,** 327–36.

Long, A. G. (1987). Observations on *Eristophyton* Zalessky, *Lyginorachis waltonii* Calder, and *Cladoxylon edromense* sp. nov. from the Lower Carboniferous Cementstone Group of Scotland. *Transactions of the Royal Society of Edinburgh*, **78,** 73–84.

Mapes, G. (1985). *Megaloxylon* in the midcontinent of North America. *Botanical Gazette*, **146,** 157–67.

Mapes, G., & Rothwell, G. W. (1980). *Quaestora amplecta* gen. et sp. n., a structurally simple medullosan stem from the Upper Mississippian of Arkansas. *American Journal of Botany*, **67,** 636–47.

Matten, L. C., Lacey, W. S., May, B. I., & Lucas, R. C. (1980). A megafossil flora from the uppermost Devonian near Ballyheigue, Ireland. *Review of Palaeobotany and Palynology*, **29,** 241–51.

Matten, L. C., Tanner, W. R., & Lacey, W. S. (1984). Additions to the silicified Upper Devonian/Lower Carboniferous flora from Ballyheigue, Ireland. *Review of Palaeobotany and Palynology*, **43,** 303–20.

Mauseth, J. D. (1988) *Plant Anatomy*. Menlo Park, Calif.: Benjamin/Cummings Publishing Company.

May, B. I., & Matten, L. C. (1983). A probable pteridosperm from uppermost Devonian near Ballyheigue, Co. Kerry, Ireland. *Botanical Journal of the Linnean Society*, **86,** 103–23.

Meyer–Berthaud, B. (1989). First gymnosperm fructifications with trilete prepollen. *Palaeontographica B*, **211,** 87–112.

Mickle, J. E., & Rothwell, G. W. (1982). Permineralized *Alethopteris* from the Upper Pennsylvanian of Ohio and Illinois. *Journal of Paleontology*, **56,** 392–402.

Millay, M. A., & Eggert, D. A. (1970). *Idanothekion* gen. n., a synangiate pollen organ with saccate pollen from the Middle Pennsylvanian of Illinois. *American Journal of Botany*, **57,** 50–61.

Millay, M. A., & Eggert, D. A. (1974). Microgametophyte development in the Paleozoic seed fern family, Callistophytaceae. *American Journal of Botany*, **61,** 1067–75.

Millay, M. A., Eggert, D. A., & Dennis, R. L. (1978). Morphology and ultrastructure of four Pennsylvanian prepollen types. *Micropaleontology*, **24,** 305–15.

Millay, M. A., & Taylor, T. N. (1977). *Feraxotheca* gen. n., a lyginopterid pollen organ from the Pennsylvanian of North America. *American Journal of Botany*, **64,** 177–85.

Millay, M. A., & Taylor, T. N. (1979). Paleozoic seed fern pollen organs. *Botanical Review*, **45,** 301–75.

Namboodiri, K. K., & Beck, C. B. (1968). A comparative study of the primary vascular system of conifers. III. Stelar evolution in gymnosperms. *American Journal of Botany*, **55,** 464–72.

Neely, F. E. (1951). Small petrified seeds from the Pennsylvanian of Illinois. *Botanical Gazette*, **113,** 165–79.

Oliver, F. W., & Scott, D. H. (1904). On the structure of the Paleozoic seed *Lagenostoma lomaxi*, with a statement of the evidence upon which it is referred to *Lyginodendron*. *Philosophical Transactions of the Royal Society of Edinburgh*, **197B,** 193–247.

Phillips, T. L., & Andrews, H. N. (1963). An occurrence of the medullosan seed-fern *Sutcliffia* in the American Carboniferous. *Annals of the Missouri Botanical Garden*, **50,** 29–51.

Pigg, K. B., Stockey, R. A., & Taylor, T. N. (1986). Studies of Paleozoic seed ferns: Additional studies of *Microspermopteris aphyllum* Baxter. *Botanical Gazette*, **147,** 124–36.

Pigg, K. B., Taylor, T. N., & Stockey, R. A. (1987). Paleozoic seed ferns: *Heterangium kentuckyensis* sp. nov., from the Upper Carboniferous of North America. *American Journal of Botany*, **74,** 1184–204.

Pryor, J. S. (1989). Delimiting species among permineralized medullosan pteridosperms: A plant bearing *Alethopteris* fronds from the Upper Pennsylvanian of the Appalachian Basin. *Canadian Journal of Botany*, **68,** 184–92.

Ramanujam, C. G. K., Rothwell, G. W., & Stewart, W. N. (1974). Probable attachment of the *Dolerotheca*. *American Journal of Botany*, **61,** 1057–66.

Read, C. B. (1936). The Flora of the New Albany Shale. Part 2. The Calamopityeae and their relationships. *United States Biological Survey, Professional Paper*, **186-E,** 81–91.

Remy, R., & Remy, W. (1959). *Zimmermannitheca cupulaeformis* n. gen. n. sp. *Monatsberichte der Deutschen Akademie Wissenchaft zu Berlin*, **1**, 767–76.

Rothwell, G. W. (1971a). Additional observation on *Conostoma anglogermanicum* and *C. oglongum* from the Lower Pennsylvanian of North America. *Palaeontographica B*, **131**, 167–78.

Rothwell, G. W. (1971b). Ontogeny of the Paleozoic ovule, *Callospermarion pusillum*. *American Journal of Botany*, **58**, 706–15.

Rothwell, G. W. (1972a). Evidence of pollen tubes in Paleozoic pteridosperms. *Science*, **175**, 772–4.

Rothwell, G. W. (1972b). Pollen organs of the Pennsylvanian Callistophytaceae (Pteridospermopsida). *American Journal of Botany*, **59**, 993–9.

Rothwell, G. W. (1975). The Callistophytaceae (Pteridospermopsida): I. Vegetative structures. *Paleontographica B*, **151**, 171–96.

Rothwell, G. W. (1980). The Callistophytaceae (Pteridospermopsida): II. Reproductive features. *Paleontographica, B.*, **173**, 85–106.

Rothwell, G. W. (1981). The Callistophytales (Pteridospermopsida). Reproductively sophisticated gymnosperms. *Review of Palaeobotany and Palynology*, **32**, 103–21.

Rothwell, G. W. (1982). New interpretations of the earliest conifers. *Review of Palaeobotany and Palynology*, **37**, 7–28.

Rothwell, G. W. (1986). Classifying the earliest gymnosperms. In *Systematic and Taxonomic Approaches in Palaeobotany*, eds. R. A. Spicer & B. A. Thomas. The Systematic Association of London Special Volume No. 31, London: Clarendon Press.

Rothwell, G. W. (1987). The role of development in plant phylogeny: A paleobotanical perspective. *Review of Palaeobotany and Palynology*, **50**, 96–114.

Rothwell, G. W., & Eggert, D. A. (1970). A *Conostoma* with a tentacular sarcotesta from the Upper Pennsylvanian of Illinois. *American Journal of Botany*. **131**, 359–66.

Rothwell, G. W., & Eggert, D. A. (1986). A monograph of *Dolerotheca* Halle, and related complex permineralized medullosan pollen organs. *Transactions of the Royal Society of Edinburgh*, **77**, 47–79.

Rothwell, G. W., & Erwin, D. M. (1987). Origin of seed plants: An aneurophyte/seed fern link elaborated. *American Journal of Botany*, **74**, 970–73.

Rothwell, G. W., & Scheckler, S. E. (1988). Biology of ancestral gymnosperms. In *Origin and Evolution of Gymnosperms*, ed. C. B. Beck. New York: Columbia University Press.

Rothwell, G. W., & Taylor, T. N. (1972). Carboniferous pteridosperm studies: morphology and anatomy of *Schopfiastrum decussatum*. *Canadian Journal of Botany*, **50**, 2649–58.

Rothwell, G. W., Taylor, T. N., & Clarkson, C. (1979). On the structural similarity of the ovule *Conostoma platyspermum* and *C. leptospermum*. *Journal of Paleontology*, **53**, 49–54.

Rowe, N. P. (1988). New observations on the Lower Carboniferous pteridosperm *Diplopteridium* Walton and an associated synangiate organ. *Botanical Journal of the Linnean Society*, **97**, 125–58.

Schabilion, J. T., & Brotzman, N. C. (1979). A tetrahedral megaspore arrrangement in a seed fern ovule of Pennsylvanian age. *American Journal of Botany*, **66**, 744–45.

Schopf, J. M. (1949). Pteridosperm male fructifications: American species of *Dolerotheca*, with notes regarding certain allied forms. *Report of Investigations*, **142.** State Geological Survey of Illinois, Urbana.

Scott, D. H. (1909). *Studies in Fossil Botany*. London: Black, Part II.

Shadle, G. L., & Stidd, B. M. (1975). The frond of *Heterangium*. *American Journal of Botany*, **62**, 67–75.

Sporne, K. R. (1974). *The Morphology of Gymnosperms. The Structure and Evolution of Primitive Seed Plants*. 2nd ed. London: Hutchinson University Library.

Stein, W. E., & Beck, C. B. (1978). *Bostonia perplexa* gen. et sp. nov., a calamopityan axis from the New Albany Shale of Kentucky. *American Journal of Botany*, **65**, 459–65.

Stewart, W. N. (1951). *Medullosa pandurata*, sp. nov. from the McLeansboro Group of Illinois. *American Journal of Botany*, **38**, 709–17.

Stewart, W. N. (1954). The structure and affinities of *Pachytesta illinoense* comb. nov. *American Journal of Botany*, **41**, 500–8.

Stewart, W. N. (1964). An upward outlook in plant morphology. *Phytomorphology*, **14**, 120–4.

Stewart, W. N. (1976). Polystely, primary xylem, and the Pteropsida. *Birbal Sahni Institute of Paleobotany*, Lucknow, 1–13.

Stewart, W. N. (1981). The Progymnospermopsida: the construction of a concept. *Canadian Journal of Botany*, **59**, 1539–42.

Stewart, W. N., & Delevoryas, T. (1952). Bases for determining relationships among the Medullosaceae. *American Journal of Botany*, **39**, 505–16.

Stewart, W. N., & Delevoryas, T. (1956). The medullosan pteridosperms. *Botanical Review*, **22**, 45–80.

Stidd, B. M. (1978). An anatomically preserved *Potoniea* with *in situ* spores from the Pennsylvanian of Illinois. *American Journal of Botany*, **65**, 677–83.

Stidd, B. M. (1979). A new species of *Heterangium* from the Illinois Basin of North America. *Review of Palaeobotany and Palynology*, **28**, 249–57.

Stidd, B. M. (1981). The current status of medullosan seed ferns. *Review of Palaeobotany and Palynology*, **32**, 63–101.

Stidd, B. M. (1990). Further documentation of the structure of *Dolerotheca* and a critique of other theories. *Palaeontographica B*, **217**, 51–86.

Stidd, B. M., & Hall, J. W. (1970a). *Callandrium callistophytoides* gen. et sp. nov., the probable pollen-bearing organ of the seed fern, *Callistophyton*. *American Journal of Botany*, **57**, 394–403.

Stidd, B. M., & Hall, J. W. (1970b). The natural affinity of the Carboniferous seed, *Callospermarion*. *American Journal of Botany*, **57**, 827–36.

Stidd, B. M., Leisman, G. A., & Phillips, T. L. (1977). *Sullitheca dactylifera* gen. et sp. n.: a new medullosan pollen organ and its evolutionary significance. *American Journal of Botany*, **64,** 994–1002.

Stidd, B. M., Oestry, L. L., & Phillips, T. L. (1975). On the frond of *Sutcliffia insignis* var. *tuberculata*. *Review of Palaeobotany and Palynology*, **20,** 55–66.

Stidd, B. M., & Phillips, T. L. (1973). The vegetative anatomy of *Schopfiastrum decussatum* from the middle Pennsylvanian of the Illinois Basin. *American Journal of Botany*, **60,** 463–74.

Stubblefield, S. P., & Rothwell, G. W. (1980). *Conostoma chappellicum* n. sp., lagenostomalean ovules from Kentucky. *Journal of Paleontology*, **54,** 1012–16.

Taylor, T. N. (1965). Paleozoic seed studies: a monograph of the American species of *Pachytesta*. *Palaeontographica B*, **117,** 1–46.

Taylor, T. N. (1971). *Halletheca reticulatus* gen. et sp. n.: synangiate Pennsylvanian pteridosperm pollen organ. *American Journal of Botany*, **58,** 300–8.

Taylor, T. N. (1988). Pollen and pollen organs of fossil gymnosperms: Phylogeny and reproductive biology. In *Origin and Evolution of Gymnosperms*, ed. C. B. Beck. New York: Columbia University Press.

Taylor, T. N., & Eggert, D. A. (1967). Petrified plants from the Upper Mississippian of North America. I: The seed *Rhynchosperma* gen. n. *American Journal of Botany*, **54,** 984–92.

Taylor, T. N., & Eggert, D. A. (1969). On the structure and relationships of a new Pennsylvanian species of the seed *Pachytesta*. *Palaeontology*, **12,** 382–8.

Taylor, T. N., & Millay, M. A. (1981a). Morphologic variability of Pennsylvanian lyginopterid seed ferns. *Review of Palaeobotany and Palynology*, **32,** 27–62.

Taylor, T. N., & Millay, J. A. (1981b). Additional information on the pollen organ *Halletheca* (Medullosales). *American Journal of Botany*, **68,** 1403–7.

Taylor, T. N., & Stockey, R. A. (1976). Studies of Paleozoic seed ferns: anatomy of *Microspermopteris aphyllum*. *American Journal of Botany*, **63,** 1302–10.

Wunk, C., & Pfefferkorn, H. W. (1984). The life habits and paleoecology of Middle Pennsylvanian medullosan pteridosperms based on an *in situ* assemblage from the Bernice Basin (Sullivan County, Pennsylvania, U.S.A.). *Review of Palaeobotany and Palynology*, **41,** 329–51.

Worsdell, W. C. (1906). The structure and origin of the Cycadaceae. *Annals of Botany*, **20,** 129–55.

24

Cycads: origins and relationships

The extant cycads comprise a small group of 10 genera that have a peculiar disjunct distribution in both Eastern and Western Hemispheres exemplified by *Stangeria* in Africa; *Microcyas* in Cuba; *Dioon* and *Ceratozamia* in Mexico; *Zamia* in Mexico and Cuba; *Macrozamia, Bowenia, Lepidozamia,* and *Cycas* in Australia; and species of *Cycas* in India, China, Japan, Madagascar, and East Africa. Attempts to explain their unusual distribution based on plant migration schemes and continental drift have failed. The best explanation thus far offered is by Arnold (1953), who suggests that the isolated genera are "leftovers" from much more widely distributed Mesozoic and early Cenozoic populations. As we shall see, the cycads reached their zenith during the Mesozoic.

Cycads: general features

We have already used the descriptive term cycadophyte to distinguish cycads and cycadlike plants from conifers and their relatives. This approach is useful as a first step in the classification of gymnospermous fossils. When other characters such as epidermal features and internal anatomy are also considered, specimens often can be identified with confidence as belonging to one of the various groups that share the cycadophyte growth form. Although there are many exceptions, it is possible to examine a cycad and come up with a fairly good set of distinguishing characteristics.

From a distance, large species of cycads look something like palm trees with stout stems. Including their crown of spirally arranged compound leaves, some cycads may be as much as 18 m high (*Macrozamia moorei*) without any branches arising from the trunk. This species, as with most cycads, has a trunk covered with an armor of persistent leaf bases. By contrast, however, some species of *Zamia* (Fig. 24.1) are small and have tuberous stems. Regardless of the species, the internal stem structure (Fig. 24.2A) of cycads is fairly uniform. All have a relatively massive pith with numerous secretory ducts. The primary xylem, which is obscure in older stems, is composed of clusters of cells that often merge with weakly developed radial files of secondary xylem at the periphery of the pith. Where it can be determined, the primary xylem is endarch. The secondary xylem is manoxylic with some rays that are very broad. The tracheids of the secondary xylem have alternate multiseriate bordered pits crowded on the radial walls. A bifacial vascular cambium is present. Relative to the development of the secondary tissues, stems of cycads have a broad parenchymatous cortex with secretory canals lined with epithelial cells that connect with those of the pith through the broad rays.

When we look at transections of cycad petioles we see many leaf traces (Fig. 24.2B) and secretory canals. In their general appearance they are similar to *Myeloxylon*, the detached petioles of medullosan pteridosperms. The leaf traces of cycads arise from the primary xylem strands in the stem. They are differentiated into numerous small radial traces as well as large traces that partially girdle the stem through the cortex before entering a leaf base. There are usually two such girdling traces in addition to a number of smaller radial traces that supply a petiole. The pair of girdling traces arise from primary xylem strands (sympodia) on the side of the stem opposite from the leaf they supply. In their 150° course to the leaf base, they traverse the cortex in opposite directions. Among the gymnosperms this girdling characteristic seems to be unique to the cycads. The petioles of the large compound leaves of cycads characteristically display numerous collateral bundles (Fig. 24.2B), often showing an inverted omega-shaped arrangement.

Early in development the leaflets of *Cycas* exhibit circinate vernation similar to the fronds of seed ferns, but the rachis of the frond has a more erect growth form. As the frond matures it becomes coriaceous (leathery), and the margins of the leaflets often become revolute (rolled downward). These characteristics seem to be shared with the pinnules of some medullosan fronds – *Alethopteris,* for example. In many species of cycads, several veins enter the base of a leaflet, increasing in number toward the distal end as a result of dichotomies. In *Cycas,* however, each leaflet has a single midvein and no laterals. *Stangeria* has a midvein from which lateral veins diverge that are dichotomously branched. Some veins in the leaflets of *Stangeria* anastomose and form loops near the margin. These venation patterns have proved to be useful in identification of fossil cycadophyte leaves as are stomatal characteristics that are of the haplocheilic type (Florin, 1933). This type (Fig. 24.2C) is characterized by a ring of subsidiary cells around the paired guard cells. The subsidiary cells and guard cells originate independently and directly from mother cells of the epidermis. In addition to the Cycadales, the haplocheilic stomatal apparatus is a consistent feature of pteridosperms, cordaites, and conifers, including *Ginkgo* and *Ephedra.*

Figure 24.1. *Zamia floridana* with ovulate cone. Extant. (Courtesy of the Field Museum of Natural History, Chicago.)

Seeds and microsporangia are produced in strobili (conelike structures where the sporophylls are loosely arranged) or cones, with pollen and seed cones borne on separate plants. In all cycads, the spirally arranged microsporophylls form compact cones. The abaxial microsporangia (Fig. 24.2D) occur in soral clusters (Fig. 24.2E) of three to six oval eusporangia placed in a ring around a papilla (Fig. 24.2F) from which they were formed. Each sporangium has a median longitudinal dehiscence mechanism. If the sporangia of a sorus were erect instead of being prostrate on the abaxial surface of the microsporophyll, then the dehiscence slits would be on inward-facing walls of the sporangia making up the ring. This, you may recall, is what has been suggested as the primitive arrangement of sporangia in the evolution of

Figure 24.2. **A.** Transverse section, stem of *Zamia floridana*. Note large pith (p); extensive cortex (c); poorly developed vascular system (vs); and girdling leaf traces (gt). (Redrawn from Chamberlain, 1935.) **B.** *Cycas revoluta*, transverse section of petiole showing inverted omega-shaped arrangement of the numerous collateral bundles. Xylem (black); phloem (white). **C.** Haplocheilie stoma of *Microcycas* sp. (Redrawn from Florin, 1933.) **D.** *Z. integrifolia*, microsporophyll with abaxial microsporangia. **E.** *C. circinalis*, with microsporangia in sori. **F.** *Z. integrifolia*, sorus with microsporangia attached to a papilla. **G.** *Cycas* sp. monocolpate pollen grain. **A–G:** Extant. (**B,D–F** redrawn from Pant & Mehra, 1962.)

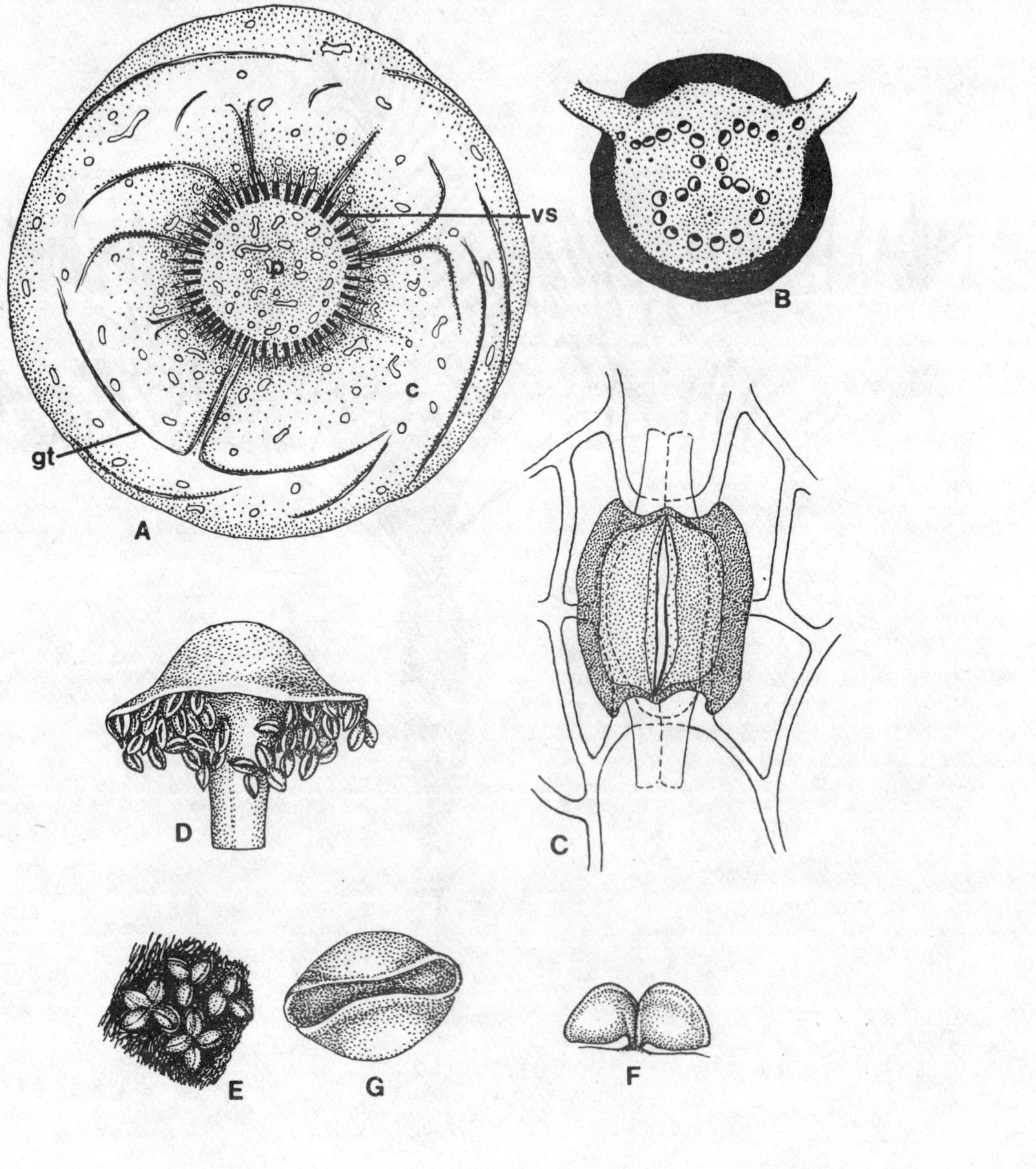

pteridosperm pollen organs. Although highly speculative, as far as we are aware this is the only explanation, with any supporting evidence from the fossil record, that accounts for the evolutionary origin of the soral clusters of extant cycads.

Although pollen grains of cycads (Fig. 24.2G) are boat-shaped and bilaterally symmetrical like the pre-pollen of *Monoletes*, they are much smaller monocolpate grains (with a single furrow on the distal surface). Germination occurs from the distal surface and is initiated by the formation of haustorial pollen tubes that become branched as they grow into the nucellus of the ovule. A microgametophyte develops in the proximal portion of the pollen grain, and this is pushed into the pollen chamber of the ovule by the elongation of the haustorial structures. At maturity each microgametophyte develops a pair of large ciliated sperm cells (up to 210 μm in *Cycas revoluta*).

The sporophylls of ovulate strobili and cones traditionally have been interpreted as an evolutionary series starting with the primitive type borne by *Cycas media* (Fig. 24.3A) and ending with the specialized megasporophylls of *Zamia* (Fig. 24.3C). *Cycas media* megasporophylls are leaflike pinnate structures that bear six to eight lateral ovules on the rachis with the micropyles of the ovules directed away from the axis of the sporophyll. *Macrozamia* (Fig. 24.3B), an intermediate member of the series, has a reduced blade with two ovules that have their micropyles directed toward the cone axis. *Zamia floridana*, with its pair of inverted ovules on scalelike, peltate megasporophylls, represents the most reduced member of the series. Whether the series is interpreted as presented here or in some other way, as suggested by the examples that follow, there is general agreement that the megasporophylls of cycads were derived from bilaterally symmetrical fronds that bore large ovules, as occurs among the medullosan pteridosperms.

As we have already indicated, the ovules of cycads (Fig. 24.3D) and medullosan pteridosperms share many characteristics. They may be large, up to 6 cm in length in cycads and as much as 10 cm in medullosan pteridosperms. Their massive integuments are differentiated into three layers, endotesta, sclerotesta, and sarcotesta. The well-developed nucellus of a cycad ovule is fused with the integument except

Figure 24.3. **A–C:** Megasporophylls of cycads. **A.** *Cycas media*. **B.** *Macrozamia*. **C.** *Zamia floridana*. **D.** Diagram of longitudinal section of *Dioon edule* ovule showing double vascular system (vs); nucellus (n); integument (i). **A–D:** extant. **E.** *Spermopteris coriaceae*. Upper Pennsylvanian. (**A** from Crane, 1988, with permission; **B** from Takhtajan, 1956; **D** redrawn from Chamberlain, 1935; **E** redrawn from Cridland & Morris, 1960.)

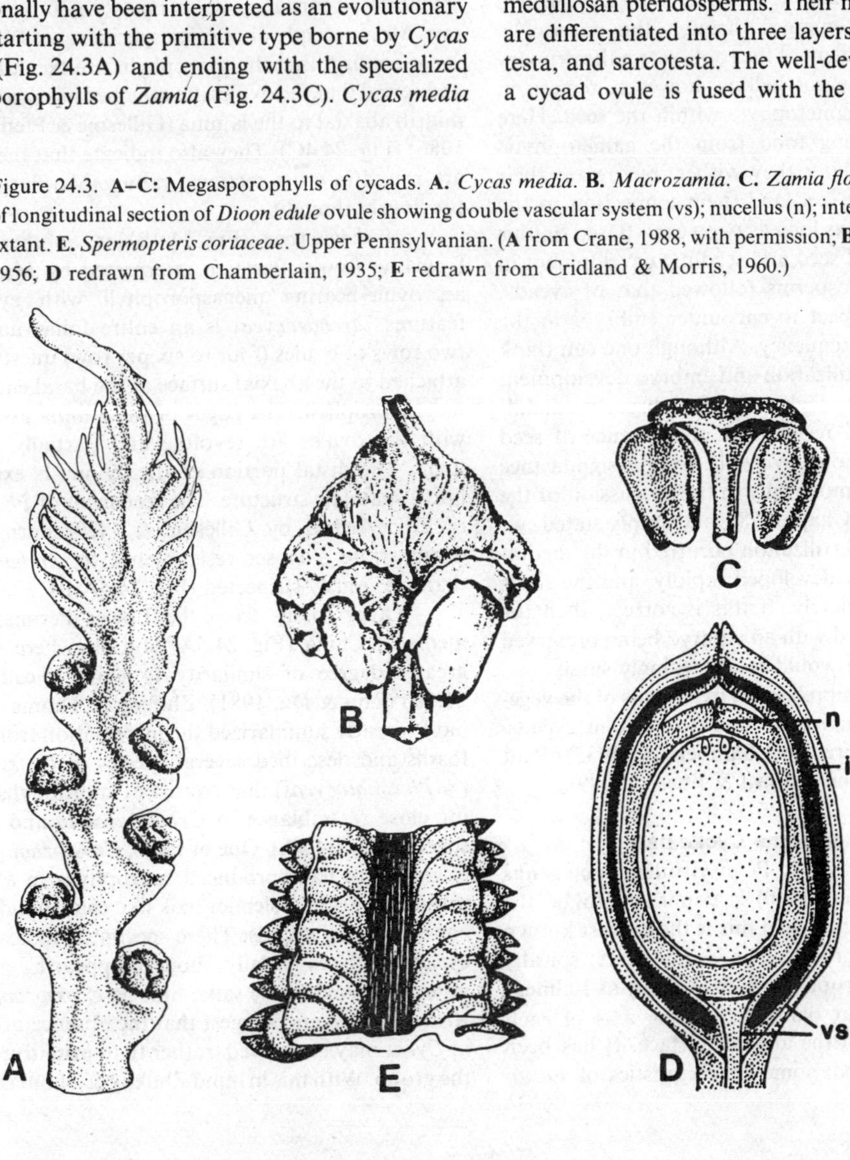

in the upper (micropylar) end of the ovule. In medullosan pteridosperm ovules, however, the two are unfused except at the base of the chalazal (vascularized) end of the ovule. Both integument and nucellus of cycad and medullosan ovules are vascularized. The megagametophytes of both also are alike in the position and number of archegonia (two to six) and development of the cellular structure.

We cannot leave this generalized presentation of the ways in which cycads and medullosan reproductive structures are alike without raising the question of why no embryos, in any stage of development, have been found in the medullosans. Literally hundreds of their ovules have been studied, and many of these contain megagametophytes. If we use the development of cycad embryos as the model for interpreting medullosan seed and embryo development, we should expect to find embryos in the fossil seeds. In cycads, the germination of the dicotyledonous embryo is initiated when the primary root, covered by a protective cap (coleorhiza), emerges through the micropyle. After the establishment of the root system the primary leaves emerge. The cotyledons, however, remain in the cellular megagametophyte within the seed. Here they stay, absorbing food from the gametophytic tissue. Sooner or later they wither, but retain their position inside the seed and their connection to the young seedling for as long as two years. Based on this observation, and if seed and seedling development in medullosan pteridosperms followed that of cycads, then we would expect to encounter embryos in the fossils with some frequency. Although one can think of reasons why fertilization and embryo development might not occur, the lack of fossil embryos in medullosan ovules could result from the absence of seed dormancy in this group. We will cover the significance of seed dormancy more fully in our discussion of the origin of conifers (Chapter 28), but simply stated, we suspect that once fertilization occurred in the medullosans the embryos developed rapidly, and the seeds germinated immediately. If this is correct, then the probability of a seed with an embryo being preserved before germination would be exceedingly small.

For more comprehensive treatments of the vegetative and reproductive structures of extant cycads, the student is referred to Chamberlain (1935), Pant & Mehra (1962), and Gifford & Foster (1989).

Early evidence of the Cycadales

If the problematical Pennsylvanian cone genus *Lasiostrobus* (Taylor, 1970) is confirmed to be the pollen cone of a cycad, then this is the earliest known evidence of the group. The cone bears spirally arranged microsporophylls, with as many as 10 linear microsporangia that parallel the long axis of each microsporophyll on the abaxial surface. It has been noted by Taylor that some characteristics of *Lasiostrobus* can, however, be interpreted as conifer- or ginkgo-like.

Of possible significance in our search for the first evidence of the cycads is *Spermopteris* (Fig. 24.3E), described by Cridland & Morris (1960) from the Upper Pennsylvanian of Kansas. *Spermopteris* is an ovulate phase of the leaf form genus *Taeniopteris* that is common in rocks of the Upper Paleozoic. The specimen comprises a fragment of a *Taeniopteris* frond with two rows of ovules attached to the abaxial surface, one row on each side of the midrib. The ovules are platyspermic with their micropylar ends projecting beyond the edge of the foliar lamina. *Phasmatocycas* (Fig. 24.4A), described by Mamay (1973, 1976) from the Lower Permian of Kansas, also has two rows of ovules attached laterally to the rachis of a taeniopterid frond. The ovules, which are 2.8 to 4 mm long, and, like those of *Spermopteris*, are erect with their micropyles directed away from the rachis on which they are borne. There may be as many as 30 ovules in a row. More recently discovered specimens of *Phasmatocycas* confirm that the sporophylls are taeniopteroid, and reveal that the ovules are attached along a swollen midrib abaxial to the lamina (Gillespie & Pfefferkorn, 1986) (Fig. 24.4C). They also indicate that the ovules are round in cross sections, rather than flattened as previously thought.

Archaeocycas (Fig. 24.4B) is another genus from the Lower Permian interpreted by Mamay as an ovule-bearing megasporophyll with cycadlike features. *Archaeocycas* is an entire foliar unit with two rows of ovules (four to six per row) questionably attached to the abaxial surface at the basal end of the megasporophyll. The edges of the lamina associated with the ovules are revolute and partially enclose them. The distal portion of the lamina is expanded into a spatulate structure. The specimens of *Phasmatocycas* described by Gillespie & Pfefferkorn (1986) demonstrate a closer resemblance to *Archaeocycas* than previously suspected (Fig. 24.4A,B).

Other fossils from the Lower Permian sediments of China (Fig. 24.5A) and elsewhere show a greater degree of similarity to living cycads (Fig. 24.5B) (Zhu & Du, 1981). Zhifeng & Thomas (1989) have recently summarized the information from these fossils and described several species of *Crossozamia* (= *Phasmatocycas*) that consist of sporophylls showing close resemblance to *Cycas revoluta* and similar species (Fig. 24.3A). One of these, *Crossozamia minor* is known to have produced sporophylls in a helical arrangement on a slender axis like that found in the cones of living cycads. These specimens indicate that cycads with essentially modern-appearing ovulate structures lived at the same time as *Crossozamia* and *Archaeocycas*, and suggest that the absence of a cone in *Cycas* may be derived, rather than primitive within the group. With this in mind Zhifeng & Thomas (1989)

Figure 24.4. **A.** *Phasmatocycas* sp. **B.** *Archaeocycas* sp. **C.** *Phasmatocycas*, showing details of sporophyll and attached ovules. **A–C:** Permian. (**A,B** redrawn from Mamay, 1976; **C** from Crane, 1988, with permission.)

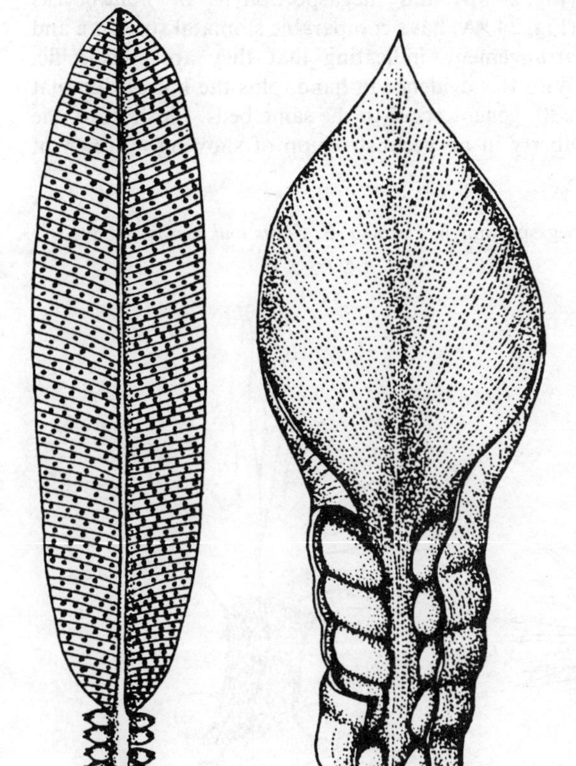

propose two possible evolutionary pathways from the *Crossozamia* type: (1) the reduction of the "strobilus" axis with little change in megasporophyll size and structure resulting in a *Cycas*-like terminal crown with lossely arranged spiral megasporophylls, and (2) the reduction of megasporophylls and their consolidation on a shortened axis as in *Zamia* (Fig. 24.1). There are several Triassic fossils with structural characteristics identical with those of extant genera. This supports the contention that the evolution of the Cycadales had started prior to the end of the Paleozoic. The most complete and well-preserved remains of the Cycadales come from the Upper Triassic and Jurassic.

Perhaps the most spectacular of these fossils occur in Lower/Middle Triassic peat deposits of the Transantarctic Mountains of Antarctica. From silicified peat deposits that are the equivalent of Pennsylvanian age coal balls, Smoot, Taylor, & Delevoryas (1985) have described nearly complete, anatomically preserved cycad stems as *Antarcticycas schopfii* (Fig. 24.6). The stems range from 1 to 4 cm in diameter, with a large pith and broad cortex. Mucilage canals like those found in modern cycads occur abundantly in the pith and cortex. Girdling leaf traces diverge from the stele, and a large number of apparently adventitious roots also are preserved in the cortex. The stele consists of a broad ring of endarch primary bundles and a narrow zone of manoxylic wood. Although the relationships of *Antarcticycas* to living genera of cycads remain obscure, the fossils show the greatest structural similarity to *Bowenia*.

A beautifully preserved permineralized specimen of *Lyssoxylon* is described by Gould (1971) from the Upper Triassic, Chinle Formation of Arizona. This stem genus is characterized by a large central pith, a cylinder of compact secondary wood, and an outer cortical region surrounded by persistent spirally arranged leaf bases. The stele of *Lyssoxylon* can be closely compared with that of the extant *Dioon spinulosum*, which has the same thick zone of wood, well-defined "growth rings," rays of two types, and tracheids with multiseriate alternate circular bordered pits of the araucarioid type on the radial walls. These characteristics, plus the girdling leaf traces in the cortex and the persistent leaf bases, leave little doubt that the affinities of *Lyssoxylon* are with the Cycadales.

In Upper Triassic beds of North Carolina, Delevoryas & Hope (1971, 1976) found cycad stems with leaves and a pollen cone attached. They established a new genus, *Leptocycas*, for these unusually complete compression–impression fossils. Fortunately, the cuticles of the pinnae of the compound leaves, rachises, and stem were preserved. The stomata of the cuticles are of the haplocheilic type with other features characteristic of cycad stomata. The attached fronds are comparable to the isolated cycad frond type *Pseudoctenis*. The reconstruction of *Leptocycas* prepared by Delevoryas & Hope (Fig. 24.7) depicts a cycad with a slender stem, about 1.5 m tall, bearing a crown of fronds some of which may have exceeded 30 cm in length. The stem of this cycad did not have persistent leaf bases. The general morphology of *Leptocycas* suggests that primitive cycads had slender stems and that the bulky, fleshy stem of the "typical" extant cycad is derived.

Other reconstructions of extinct cycads have been prepared by Florin (1933) and Harris (1961). Florin's reconstruction of an Upper Triassic cycad (Fig. 24.8A) was attempted only after an exhaustive study of epidermal and stomatal structures of isolated compression–impression fossils of leaves and reproductive structures. These studies showed that leaves of the dispersed *Taeniopteris* (*Doratophyllum*) type (Fig. 24.8B) and megasporophylls of *Palaeocycas* (Fig. 24.8A) have comparable stomatal structure and arrangement, indicating that they are conspecific. With this evidence in hand, plus the knowledge that both genera occur in the same beds. Florin took the liberty in his reconstruction of showing a crown of

Figure 24.5. **A.** *Crossozamia* (= *Primocycas*) *chinensis* megasporophyll, Permian. **B.** *Cycas tonkinensis*, extant. (**A,B** from Crane, 1988, with permission.)

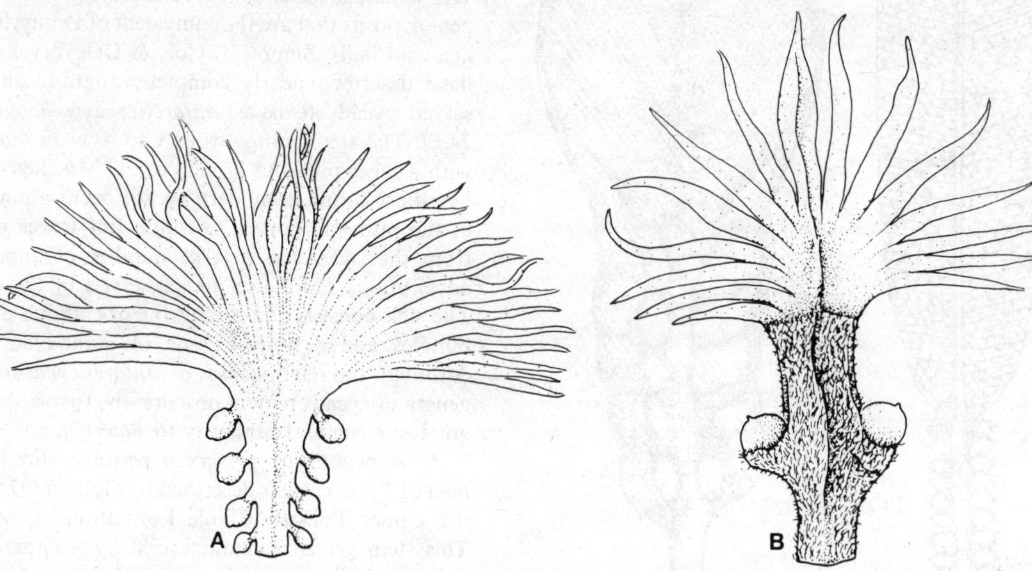

Figure 24.6. *Antarcticycas schopfii*. Transverse section showing ring of vascular bundles (vb); cortex (c); and mucilage canals (mc). Triassic. (From Smoot, Taylor, & Delevoryas, 1985, with permission.)

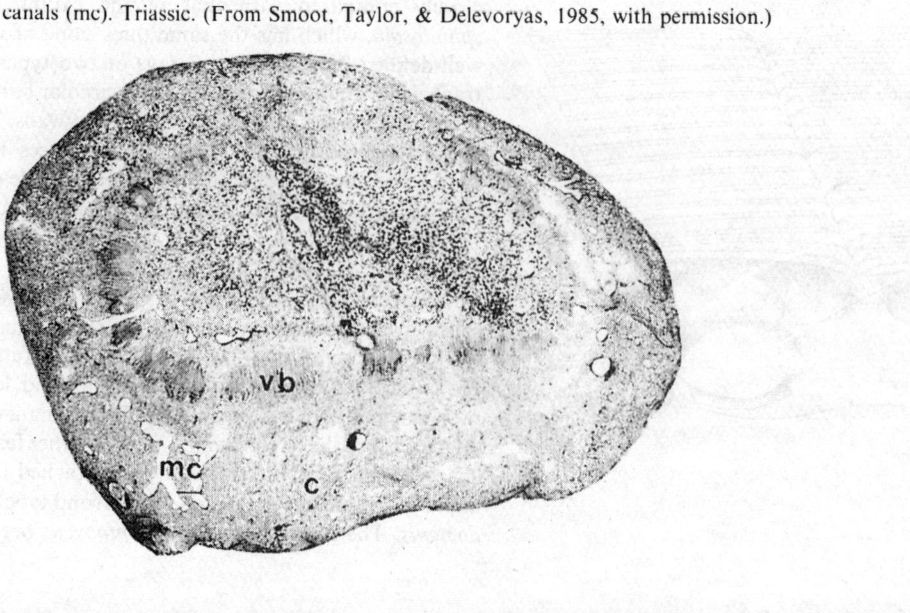

Doratophyllum leaves around a cluster of spirally arranged *Palaeocycas* megasporophylls at the apex of a stout unbranched trunk about 3 m high. The reconstructed plant, with an imaginary trunk, was named *Bjuvia simplex*.

In his reconstruction of a Jurassic cycad (Fig. 24.8C), Harris also took the liberty of placing reproductive organs and foliage on an imaginary stem. The latter is illustrated as a branched structure with an armor of spirally arranged leaf bases and a crown of *Nilssonia tenuinervis* leaves. Ovulate cones of the *Beania mamayi* type and *Androstrobus wonnacottii* pollen cones are believed to be conspecific with the *Nilssonia compta* foliage. All of these species occur in the Jurassic Yorkshire beds made famous by the prodigious works of Professor Harris.

Beania gracilis (Harris, 1964) ovulate cones (Fig. 24.9A) have an organization similar to seed cones of *Zamia*. The pendulous cone comprises megasporophylls in a loose spiral on the cone axis. Each

Figure 24.7. *Leptocycas gracilis*. Reconstruction of plant about 1.5 m tall. Upper Triassic. (From Delevoryas & Hope, 1971.)

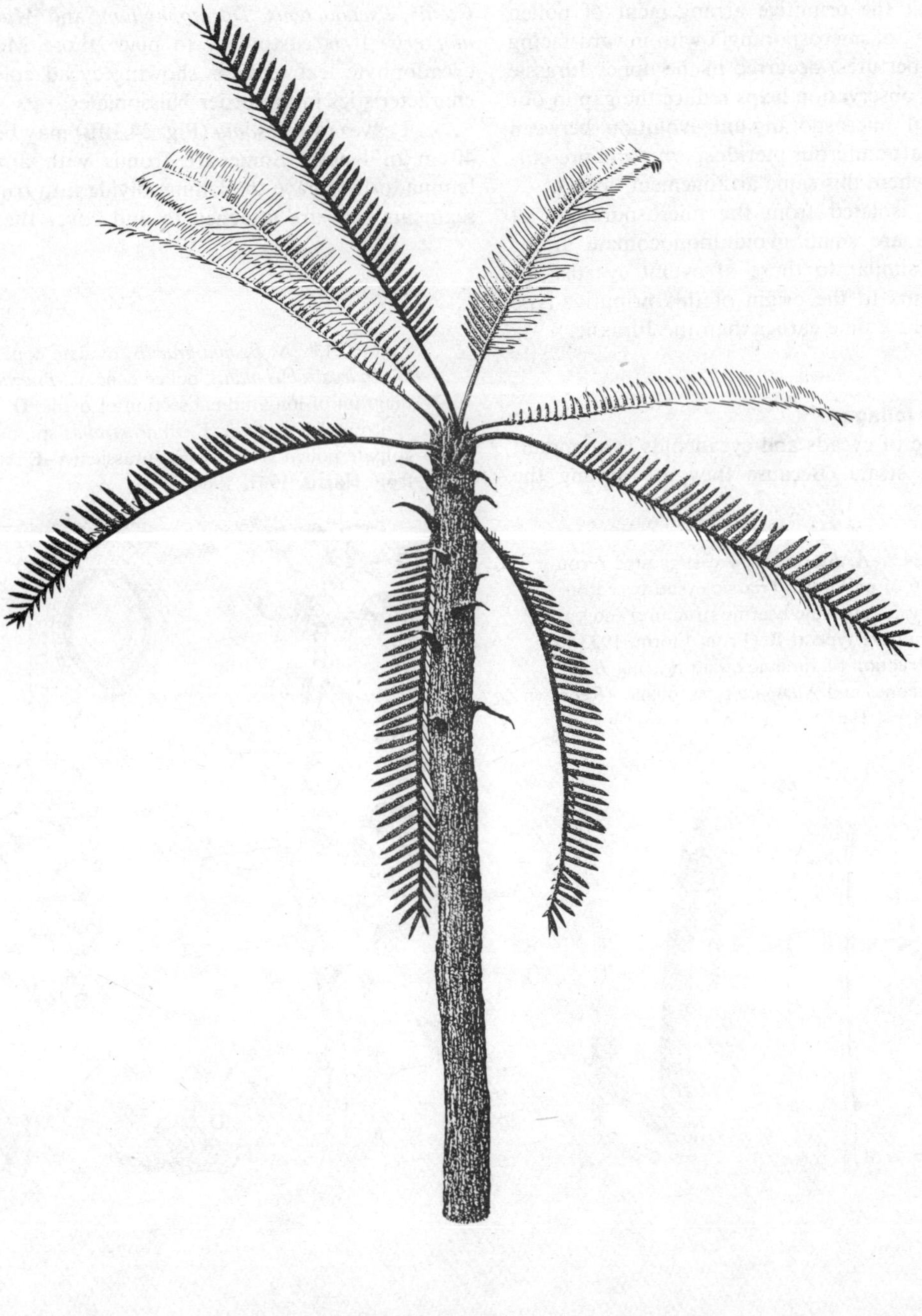

megasporophyll is peltate at the apex and bears two sessile ovules with their micropyles directed toward the cone axis. The cones may attain a length of 10 cm. The ovules (Fig. 24.9C) are about 1.5 cm in length and show an internal structure similar to that of extant cycads.

The features of the pollen cone *Androstrobus manis* (Harris, 1941) are particularly interesting. The cones (Fig. 24.9B), which are about 5 cm long, have spirally arranged microsporophylls each of which is terminated by a rhomboidal scale (Fig. 24.9D). The abaxial microsporangia are fairly large finger-shaped structures that appear to be arranged in sori with dehiscence slits along their inward-facing walls. This arrangement is of particular significance, because it indicates that the primitive arrangement of pollen organs (a ring of microsporangia with inward-facing dehiscence apertures) occurred in the upper Jurassic cycads. This observation helps reduce the gap in our knowledge of microsporangium evolution between the Lower Carboniferous pteridosperms and present-day cycads where this same arrangement occurs.

Pollen isolated from the microsporangia of *Androstrobus* are small ovoid monocolpate grains (Fig. 24.9E) similar to those of extant cycads. All evidence points to the origin of this primitive-type pollen grain at a time earlier than the Jurassic.

Cycad foliage

Foliage of cycads and cycadeoids is abundant in Mesozoic strata. Because they are among the commonest compression–impression fossils found, the Mesozoic has come to be knwon as the "age of the cycadophytes." In the previous section on the Cycadales, we have given an example illustrating how it is possible, based on the epidermal and stomatal characteristics, to distinguish cycad foliage from other cycadlike leaves that have a similar morphology. In addition to the haplocheilic organization (Fig. 24.2C), cycad stomata are characterized by relatively thin deposits of cuticle in the guard cells and irregular orientation of stomata. The epidermal cells have straight walls and are not oriented into rows. The genus *Nilssonia* is the only one that shows all of these characteristics. Other genera with haplocheilic stomata as well as some of the other characteristics listed are *Ctenis, Pseudoctenis, Doratophyllum,* and *Macrotaeniopteris*. It is customary to place those Mesozoic cycadophyte leaf remains showing cycad epidermal characteristics in the order Nilssoniales.

Leaves of *Nilssonia* (Fig. 24.10B) may be up to 40 cm in length. Some are fronds with an entire lamina, others have the lamina divide into truncated segments that are attached to and cover the upper

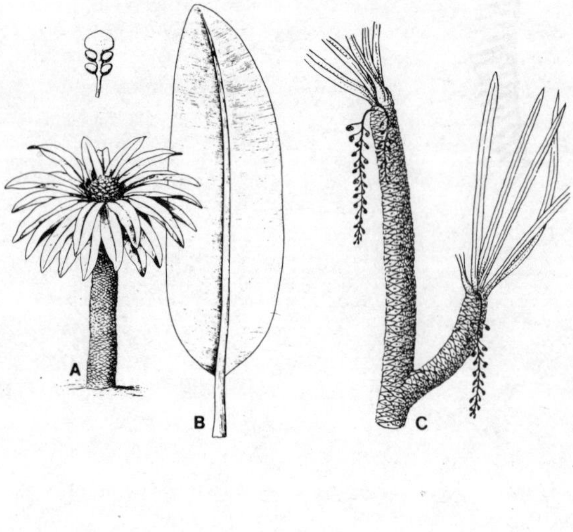

Figure 24.8. **A.** *Bjuvia simplex,* suggested reconstruction of an Upper Triassic cycad based on *Palaeocycas* (an ovule-bearing structure) and leaf of the *B. simplex* type at **B.** (From Florin, 1933.) **C.** Reconstruction of Jurassic cycad bearing *Beania* ovulate cones and *Nilssonia*-type foliage. (Redrawn from Harris, 1961.)

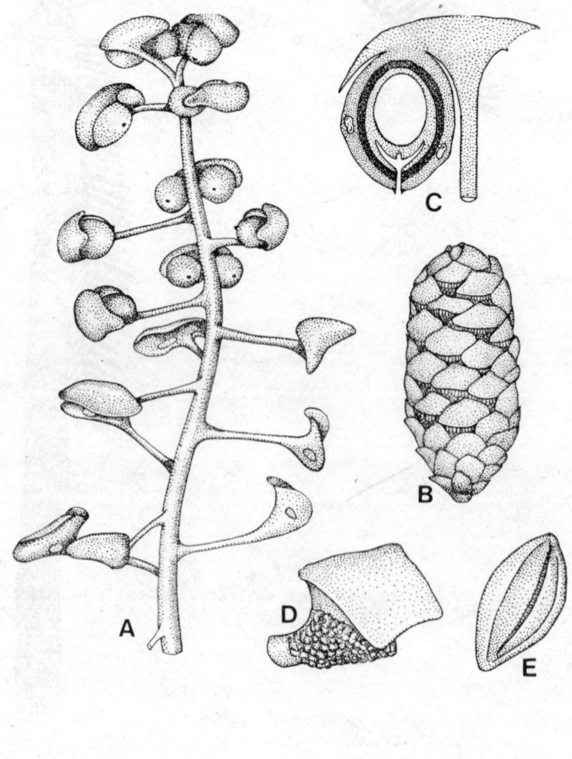

Figure 24.9. **A.** *Beania gracilis,* ovulate cone. **B.** *Androstrobus manis,* pollen cone. **C.** *B. gracilis,* diagram of longitudinal section of ovule. **D.** Microsporophyll, *A. manis.* **E.** *Androstrobus* sp., monocolpate pollen grain. **A–E:** Jurassic. (**A–E** redrawn from Harris, 1941, 1964.)

surface of the rachis. The parallel veins of the lamina arise from the rachis at right angles. There is no anastomosing of the veins. The leaves of *Ctenis* may attain a length of 2 m. They are compound with long, ascending pinnules that have several parallel veins entering their bases. The veins show interconnections, a feature lacking in *Pseudoctenis,* a genus which in all other respects is like *Ctenis*. The leaves of *Ctenis* and *Pseudoctenis* are highly similar to those of *Zamia*. *Doratophyllum* is a leaf with *Taeniopteris* morphology (Fig. 25.2C) and haplocheilic stomata. The oblong, lanceolate leaf has an entire lamina that arises from the sides of the midrib. The veins are at right angles to the midrib and occasionally fork but do not anastomose. *Macrotaeniopteris* is similar to *Doratophyllum,* but is much larger, attaining a length of 1 m and a maximum width of 33 cm. These large leaves must have had a superficial resemblance to the fronds of banana trees.

The evolutionary origin of Cycadales

The idea that the cycads originated from the pteridosperms has been supported over the years by such outstanding botanists as Chamberlain (1920) and Arnold (1953). In reviewing the evidence presented in this chapter, we think it is safe to conclude that an origin of cycads from pteridosperms is even more assured. We now have good reason to believe that cycads occurred in the Permian with possible ancestors in the Upper Carboniferous. This tells us that pteridosperms of the medullosan type and cycads coexisted during the Upper Paleozoic. Thus, a gap no longer exists between the fossil records of the two groups.

Figure 24.10. Portion of leaf, *Nilssonia compta,* foliage of a Jurassic cycad. (Redrawn from Andrews, 1961.)

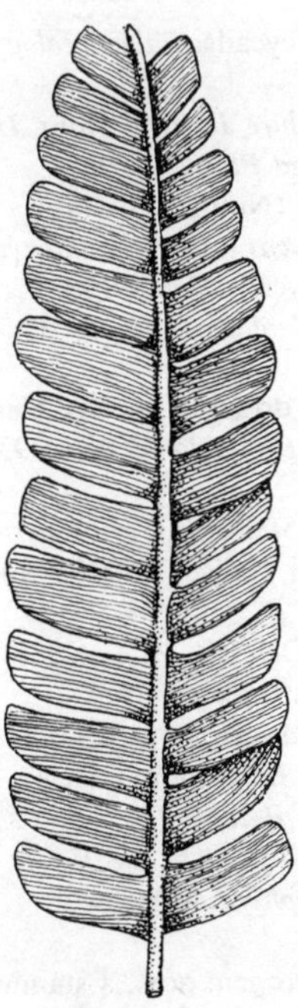

Vegetative structures

Following the arguments of Delevoryas (1955) we can reasonably conclude that the vegetative structure of cycads was derived with little modification from Permian monostelic medullosans. Both are characterized by their frondlike leaves, usually with haplocheilic stomata. They have large petioles with numerous vascular bundles and ample supply of supporting tissue and secretory ducts. The central portion of their stems is occupied by a large parenchymatous pith, also with secretory ducts. The pith is surrounded by usually poorly developed manoxylic secondary xylem with numerous, large rays. Anomalous development of the secondary xylem is another feature exhibited by both groups, as are tracheids with crowded circular bordered pits on the radial walls. Primary xylem of medullosan pteridosperms is mesarch, while the cycads have endarch primary xylem in their stems. Both groups have a broad parenchymatous cortex with secretory ducts and supporting tissues. The only fundamental difference in the vegetative structure of the two groups is the girdling of leaf traces in cycads not observed in the pteridosperms. Medullosan pteridosperms have a tendency to retain old leaf bases, as do may cycads. On the basis of this summary of vegetative characteristics one might jump to the conclusion that cycads are only slightly modified medullosas.

Reproductive structures

The most obvious difference between pteridosperms and cycads is the presence of specialized pollen and seed cones not found among any of the pteridosperms where microsporangia and ovules are borne on unmodified or slightly modified fronds. Medullosan pteridosperms have fertile and sterile fronds that are closely spaced on a stem that is usually unbranched. It seems that this tendency toward condensation of fronds, not apparent in other pteridosperms, foreshadows the consolidation and reduction of fronds required to produce the ovulate and pollen cones of cycads. Evidence from living and fossil cycads (*Crossozamia*) favors the origin of the megasporophyll from a pinnately compound frond. Other fossil evidence (*Archaeocycas*) suggests that the primitive frond was entire and became pinnate, that initially the lateral ovules on the frond were partially protected by the revolute margins of the laminae (Mamay, 1976). The reduction of megasporophylls and ovule number had been achieved by the Jurassic is shown by *Beania* (Harris, 1964). We have noted that the ovules of *Beania* and other cycads are similar in size and structure to those of medullosan pteridosperms (*Pachytesta*, for example).

The single largest "roadblock" in the derivation of cycads from medullosans is the origin of the cycad microsporophyll. The "roadblock" is removed, however, if we remember that the basic primitive arrangement of pteridosperma microsporangia is a synangium forming a ring around a hollow center. Each sporangium has a longitudinal dehiscence slit along its inward-facing wall. You will recall that Jurassic *Androstrobus* and present-day cycads also have their microsporangia arranged in rings (sori) with their bases fused to a short stalk (papilla) and longitudinal dehiscence along their inward-facing walls. This explanation obviates the impossible task of deriving the simple microsporangiate structure of cycads from complex structures such as *Bernaultia*. With minimal reduction of the frond and synangia, medullosan synangia of the *Aulacotheca* or *Hallatheca* types would produce rings of cycadlike microsporangia on the abaxial surface. The monolete prepollen of medullosan pteridosperms, although much larger than the monocolpate pollen of cycads, shares with the latter the characteristics of shape and modification of the distal surface of their grains.

Taking vegetative and reproductive characteristics into account, the case is indeed strong for the derivation of the Cycadales from the Upper Paleozoic Medullosaceae.

References

Andrews, H. N. (1961). *Studies in Paleobotany*. New York: Wiley.

Arnold, C. A. (1953). Origin and relationships of the cycads. *Phytomorphology*, **3,** 51–65.

Chamberlain, C. J. (1920). The living cycads and the phylogeny of seed plants. *American Journal of Botany*, **7,** 146–53.

Chamberlain, C. J. (1935). *Gymnosperms Structure and Evolution* Chicago: University of Chicago Press.

Crane, P. R. (1988). Major clades and relationships in the "higher" gymnosperms. In *Origin and Evolution of Gymnosperms*, ed. C. B. Beck. New York: Columbia University Press.

Cridland, A. A., & Morris, J. E. (1960). *Spermopteris*, a new genus of pteridosperms from the Upper Pennsylvanian series of Kansas. *American Journal of Botany*, **47,** 855–9.

Delevoryas, T. (1955). The Medullosae structure and relationships. *Palaeontographica B*, **97,** 114–67.

Delevoryas, T., & Hope, R. C. (1971). A new Triassic cycad and its phyletic implications, *Postilla*, **150,** 1–14.

Delevoryas, T., & Hope, R. C. (1976). More evidence for a slender growth habit in Mesozoic cycadophytes. *Review of Palaeobotany and Polynology*, **21,** 93–100.

Florin, R. (1933). Studien über die Cycadales des Mesozoikums. *Kungl. Svenska Vetenskapsakademiens Handlingar III*, **12,** 1–134.

Gifford, E. M., & Foster, A. S. (1989). *Morphology and Evolution of Vascular Plants*. 3rd ed., New York: W. H. Freeman and Company.

Gillespie, W. H., & Pfefferkorn, H. W. (1986). Taeniopterid lamina on *Phasmatocycas* megasporophylls (Cycadales) from the Lower Permian of Kansas, U.S.A. *Review of Palaeobotany and Palynology*, **49,** 99–116.

Gould, R. E. (1971). *Lyssoxylon grigsbyi*, a cycad trunk from the Upper. Triassic of Arizona and New Mexico. *American Journal of Botany*, **58,** 239–48.

Harris, T. M. (1941). Cones of extinct Cycadales from the Jurassic rocks of Yorkshire. *Philosophical Transactions of the Royal Society of London B*, **231,** 75–98.

Harris, T. M. (1961). The fossil cycads. *Palaeontology*, **4,** 313–23.

Harris, T. M. (1964). *The Yorkshire Jurassic Flora. II. Caytoniales, Cycadales and Pteridosperms.* London: British Museum (Natural History).

Mamay, S. H. (1973). *Archaeocycas* and *Phasmatocycas* – new genera of Permian cycads. *Journal of Research, United States Geological Survey*, **1,** 687–9.

Mamay, S. H. (1976). Paleozoic origin of cycads. *United States Geological Survey Professional Paper* 934, 1–48.

Pant, D. D., & Mehra, B. (1962). *Studies in Gymnospermous Plants – Cycas*. Allahabad, India: Central Book Deposit.

Smoot, E. L., Taylor, T. N., & Delevoryas, T. (1985). Structurally preserved fossil plants from Antarctica. I. *Antarcticycas* gen. nov., a Triassic cycad stem from the Beardmore Glacier area. *American Journal of Botany*, **72,** 1410–23.

Takhtajan, A. L. (1956). *Telomophyta*. Moscow: Academia Scientiarum.

Taylor, T. N. (1970). *Lasiostrobus* gen. nov., a staminate

strobilus of gymnosperm affinity from the Pennsylvanian of North America. *American Journal of Botany*, **57,** 670–90.

Zhifeng, G., & Thomas, B. A. (1989). A review of fossil cycad evidence of *Crossozamia* Pomel and its associated leaves from the Lower Permian of Taiyuan, China. *Review of Palaeobotany and Palynology*, **60,** 205–23.

Zhu, J., & Du, X. (1981). A new cycad – *Primocycas chineses* gen. et sp. nov. discovered from the Lower Permian in Shanxi, China and its significance. *Acta Botanica Sinica*, **23,** 401–4.

25

The enigmatic cycadeoids

The extinct order Cycadeoidales (the Bennettitales of some authors) is an enigmatic group of Mesozoic gymnosperms that disappeared from the fossil record during the Cretaceous. Members of this order have a growth habit reminiscent of genera of the Cycadales (Chapter 24). Like the cycads many cycadeoids have trunklike stems that are unbranched (Figs. 25.8, 25.9) or sparsely branched and clothed in spirally arranged persistent leaf bases (Fig. 25.6). The fronds borne by plants of both orders are highly similar in appearance, ranging from those with entire margins of the taeniopterid type to those that are pinnate.

Studies of seed- and pollen-producing structures of cycadeoids have demonstrated, however, that they are remarkably different from those of cycads. As explained in this and later chapters, the interpretations applied to the unusual reproductive organs of cycadeoids have assumed additional importance when we consider the origin of flowering plants.

Cycadeoid foliage

Until 1913, when Thomas & Bancroft's landmark work on the cuticles of extant and Mesozoic cycadophyte foliage was published, it was impossible to distinguish fossil fronds of cycads from those of cycadeoids. Where cuticles were preserved and available for study, they discovered that fronds of cycads have what we now call haplocheilic stomata and those of cycadeoids what are called syndetocheilic (Fig. 25.2A). Syndetocheilic differ from haplocheilic stomata in having a lateral subsidiary cell on each side of the two guard cells. The subsidiary cells are derived from the same mother cell that gives rise to the guard cells, independently of the surrounding epidermal cells. Cycadeoid guard cells also differ from those of cycads in having heavily cutinized outer and dorsal walls. The stomata of cycadeoids tend to be oriented at right angles and not parallel to the veins as in cycads. Cycadeoid epidermal cells are arranged in distinct rows and have walls with wavy outlines. Fronds with syndetocheilic stomata and at least some of the other characteristics given above are placed in the leaf order Bennettitales. It is not surprising, if we believe cycadeoids originated from pteridosperms, that some Carboniferous seed ferns would be found that have syndetocheilic stomata (Stidd & Stidd 1976). The extant gymnosperms *Gnetum* and *Welwitschia* also have syndetocheilic stomata, as do some flowering plants, an observation that has prompted the speculation that there is a relationship between the cycadeoids and these extant organisms. This idea, however, is tempered by the discovery of syndetocheilic stomata in the cuticles of the unrelated lycopod *Drepanophycus spinaeformis* (Stubblefield & Banks, 1978.)

Some common frond genera now known to belong to cycadeoids are *Pterophyllum, Ptilophyllum, Zamites, Otozamites, Dictyozamites, Anomozamites,* and *Nilssoniopteris*. Of these, *Pterophyllum* (Fig. 25.1A) is one of the commonest. It is characterized by long, slender pinnae with parallel margins. Each pinna is attached laterally to the rachis by the full width of its base (Fig. 25.1B). Several veins are supplied from the rachis to each pinna. The veins are parallel and dichotomize infrequently in the lamina. *Anomozamites* (Fig. 25.2B) shares the characteristics of *Pterophyllum,* except that the pinnae or segments of *Anomozamites* are usually short (less than twice as long as broad) while pinnae of *Pterophyllum* are long and slender.

Ptilophyllum (Fig. 25.1C), another common frond type of the Mesozoic, has linear to slightly sickle-shaped pinnae. The conspicuous characteristic of the genus is the attachment of the pinnae to the adaxial surface of the rachis. The broad, somewhat rounded bases of the pinnae nearly cover the upper surface of the rachis (Fig. 25.1D). *Zamites* has pinnae that are linear or lanceolate and attached to the upper surface of the rachis by their bases. Unlike the broad bases of *Ptilophyllum,* the constricted bases of *Zamites* do not cover the rachis. Another genus with leaflets attached to the adaxial surface of the rachis is *Otozamites* (Fig. 25.1E). The base of each pinna is rounded or almost lobed (Fig. 25.1F), with the distally directed lobe more pronounced than the proximal one. The pinnae are attached to the rachis by a constricted point. *Dictyozamites* is similar to *Otozamites* except that the veins form an anastomosing network.

In their general morphology the leaves of *Doratophyllum* and *Nilssoniopteris* (Fig. 25.2C) are alike. We have here, however, the unusual situation where one (*Doratophyllum*) has haplocheilic stomata and is placed in the Nilssoniales (cycads), and the other (*Nilssoniopteris*) with syndetocheilic stomata is assigned to the Bennettitales (cycadeoids). If epidermal structures are not preserved, the leaf is placed in the form genus *Taeniopteris*.

Mesozoic cycadophyte leaf remains have been recovered from many localities widely separated in time and space. In summary, the major collecting sites, which range in age from Triassic to Upper Cretaceous, occur in Western Europe, England, Greenland, many localities in North America including Canada and Alaska, southern Mexico, South Africa, India, and Australia. Although we know that the Cycadales persisted as a declining group from the Cretaceous to the present, their fossil record during the Tertiary is even more meager.

Figure 25.1. **A.** *Pterophyllum* sp. Abaxial surface of leaf. **B.** *Pterophyllum* sp., enlarged portion of leaf showing attachment of pinnae. **C.** *Ptilophyllum* sp., abaxial surface of leaf. **D.** *Ptilophyllum* sp., enlarged portion of leaf showing attachment of pinnae. **E.** *Otozamites* sp., portion of leaf. **F.** *Otozamites* sp., pinna enlarged. Note large distal lobe. **A–F:** Jurassic. (**A,B** redrawn from Harris, 1932, **C–F** redrawn from Harris, 1969.)

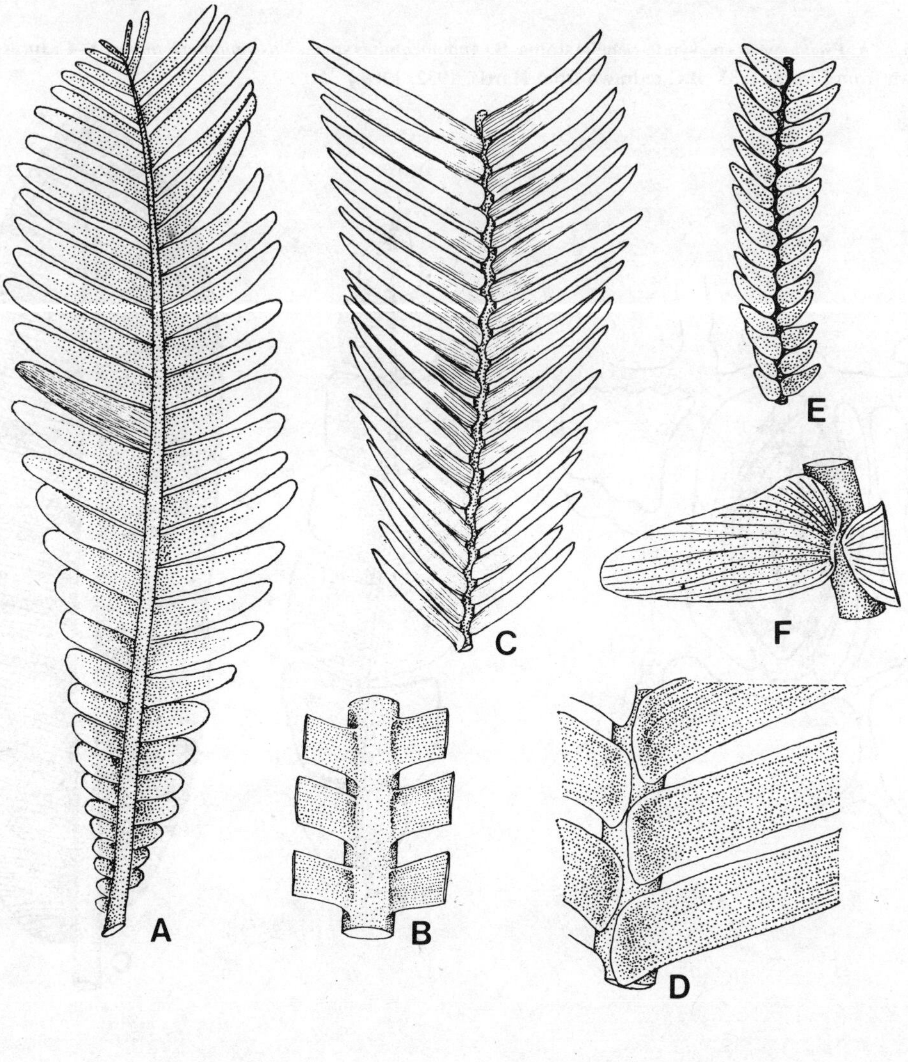

For the student interested in obtaining greater exposure to studies of Mesozoic cycadophyte foliage, the works of Florin (1933), Harris (1964), and Seward (1917) will provide something more than an introduction.

Structural features

Where the internal structure of the stem is known, there is a large pith surrounded by a cylinder of well-developed wood. The secondary xylem is usually compact and is reminiscent of conifer wood. Although some cycadeoids have scalariform pitting on the walls of secondary tracheids, others have multiseriate bordered pits on their radial walls. The primary xylem is endarch. Where leaf traces have been observed they arise from the primary xylem and pass directly through the secondary xylem into the cortex where they branch to form several traces that supply a leaf base. Cycadeoid stems lack the girdling traces that are so characteristic of cycad stems. Secretory canals, another of the characteristics of cycad and pteridosperm stems, occur in the pith and cortex.

The reproductive structures of cycadeoids have attracted much attention because of their flower-like characteristics (Crane, 1988). With at least one exception, the reproductive units of cycadeoids are bisporangiate and have microsporophylls and ovules occurring in the same cone. The ovules (Fig. 25.3A) are borne in large numbers on a dome-shaped receptacle. The integument of each ovule is extended and surrounds a long micropyle, a feature reminiscent of extant *Gnetum* and *Welwitschia*. The ovules are enveloped, except for the tips of their micropyles, by vascularized interseminal scales (Fig. 25.3A,B).

The microsporophylls have small saclike structures, each of which may contain a number of elongated sporangia fused into a synangium. Pollen grains are of the monocolpate type.

The order Cycadeoidales includes two families, the Williamsoniaceae and the Cycadeoidaceae. The Williamsoniaceae is the older, occurring in Upper Triassic and Jurassic strata.

Figure 25.2. **A.** *Pseudocycas* sp., syndetochelic stoma. **B.** *Anomozamites* sp. **C.** *Nilssoniopteris major*. **A–C:** Jurassic. (**A** redrawn from Florin, 1933; **B,C** redrawn from Harris, 1932, 1969.)

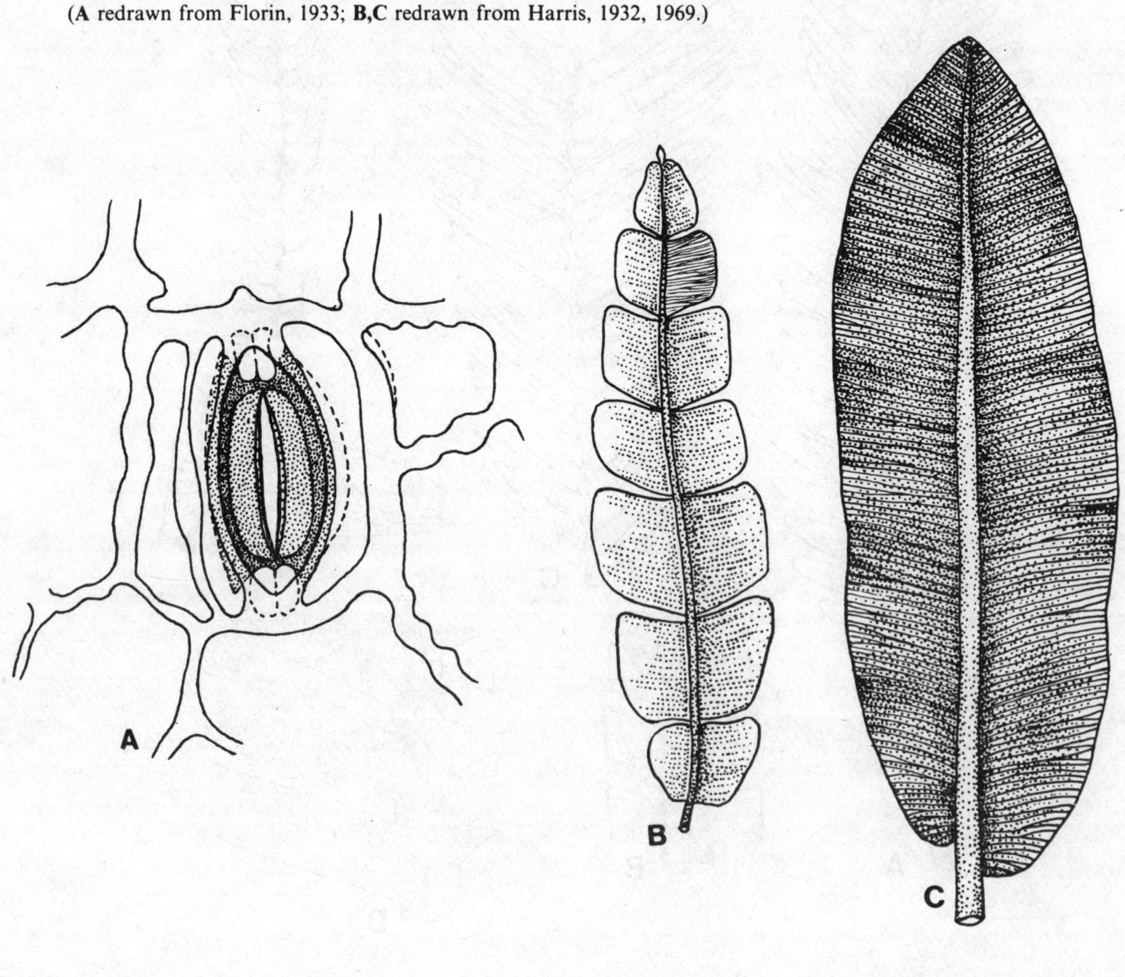

Williamsoniaceae

It has been suggested that those genera of the Williamsoniaceae with slender branched stems represent the primitive type of branching found among some pteridosperms (*Callistophyton*, for example).

Wielandiella

The first of the Williamsoniaceae to appear in the fossil record is the Upper Triassic (Rhaetian) *Wielandiella* (Fig. 25.4A) from Sweden. The stem axes, which seldom exceed 1.5 cm in diameter, exhibit a forked branching pattern that is probably a dichasium (false dichotomy). A dichasium appears to have formed where growth of the axis was terminated by a cone beneath which two lateral branches were formed. In the reconstruction the leaves, which are about 8 cm long, are concentrated around the dichasia. They are the *Anomozamites* type.

A single sessile cone is borne at each dichasium. Although not well preserved, the compression–impression fossils of cones show enough detail to assure the inclusion of *Wielandiella* in the Williamsoniaceae. The cone comprises a central ovulate receptacle with ovules and interseminal scales, subtended by a ring of rather broad linear microsporophylls. The latter probably produced sporangia on the adaxial surface. Bracts occurred around the base of each cone.

Figure 25.3. **A.** Diagram, longitudinal section of cycadeoid ovule with interseminal scales. Note long stalk (st); interseminal scale (is); ovule (o); projecting micropylar tube (m). Jurassic. (Redrawn from Sahni, 1932.) **B.** Paradermal view of a young cone of *Williamsonia harrisiana* showing ends of interseminal scales grouped about circular micropyles. Jurassic. (Redrawn from Bose, 1968.)

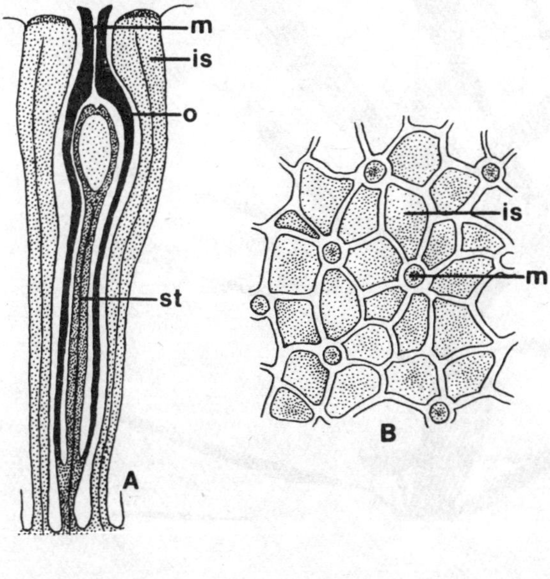

Williamsoniella

A much clearer picture of the Williamsoniaceae is presented by *Williamsoniella* from the Jurassic beds of Yorkshire (Thomas, 1915). As reconstructed by Zimmermann (1959), *Williamsoniella* (Fig. 25.4B) is a shrubby plant that bore foliage of the *Nilssoniopteris* type more or less scattered on the axes. The cones (Fig. 25.5.A), which are known to be bisporangiate, terminate slender peduncles borne in the axils of leaves. The centrally placed ovulate receptacle is covered with interseminal scales and small interspersed ovules with projecting micropyles. Approximately 300 ovules are borne on a receptable. Below the ovule-bearing portion of the receptacle there is a whorl of 12 or more microsporophylls (Fig. 25.5B) each of which has the shape of a segment from a tangerine. The narrow edge of the wedge-shaped microsporophyll is directed inward and bears the pollen sacs. A reinvestigation of microsporophylls by Harris (1944) revealed fingerlike processes, along the narrow inward-facing edge, which partially enclose pairs of sacs. Each sac has two compartments probably containing several sporangia. In their construction, the sacs are best described as synangia. The pollen grains obtained from the sporangia are monocolpate. Originally the cones were thought to lack subtending bracts. Harris, however, was able to demonstrate sets of hairy bracts that enclosed the reproductive parts and functioned as bud scales.

Williamsonia

Of all Williamsoniaceae, the genus *Williamsonia* has attracted the most attention, partially because of the beautiful although somewhat imaginary reconstruction prepared by Sahni (1932). The reconstruction has been rendered as a life-size model that stands as a monument to Dr. Sahni's paleobotanical works and to the Paleobotanical Institute he created in Lucknow, India. The often-reproduced reconstruction (Fig. 25.6) shows a small treelike plant, about 1.5 to 2 m high, having the appearance of a miniature *Cycas* with its armor of spirally arranged persistent leaf bases. The sparsely branched trunk is attributed to the dispersed stem *Bucklandia indica.* Fronds of the cycadeoid type, *Ptilophyllum cutchense,* were found attached to *B. indica* stems where they formed a crown of spirally arranged leaves. Bract-covered shoots bearing terminal cones are intercalated among the leaf bases. Based on the internal structure of permineralized peduncles of *Williamsonia scotica* cones, Sahni was able to deduce a connection with permineralized stem axes of *B. indica* that show a similar anatomy. After assembling the parts, Sahni named the reconstructed plant *Williamsonia sewardiana.*

Based on stems, leaves, and cones from the Yorkshire locality, Harris (1969) has prepared another

Figure 25.4. **A.** *Wielandiella,* reconstruction. Triassic. **B.** *Williamsoniella,* reconstruction. **A,B:** Jurassic. (Photographs courtesy of the Field Museum of Natural History, Chicago.)

reconstruction using *Bucklandia pustulosa*, *Williamsonia leckenbyi*, and *Ptilophyllum pecten*. The plant synthesized from these isolated species had axes not more than 10 cm in diameter bearing slender, highly branched laterals. Hughes (1976) suggests a plant that was small and inconspicuous, having a habit not unlike that of *Cycas rumphianus* of Malaysia with many lateral branches. Harris's reconstruction is also reminiscent of the bonsai form of *Cycas revoluta* with numerous lateral branches. The possibility that many Triassic and Jurassic cycadeoids had slender stems lacking an armor of persistent leaf bases is presented by Delevoryas & Hope (1976), who discovered just such a specimen in Upper Triassic rocks. The stem specimen bore several widely spaced spiral leaves with characteristics of *Otozamites* and *Zamites*. They named the leaf-bearing stem *Ischnopython*.

The stems of *Williamsonia sewardiana* (*Bucklandia indica*) exhibit many cycadophytic characteristics. Externally, the casts of *Bucklandia* show us that there was an armor of spirally arranged leaf bases. Where leaf traces are visible, there are six to eight present on the surface of the rhomboidal leaf scars. Internal preservation is rare, but where it has been found (Nishida, 1969) the characteristics are similar to those of cycad stems. Relative to the development of the secondary wood, the stem has a large pith with secretory ducts. The latter are also found in the cortex. The primary xylem is endarch, and the secondary xylem, although compact, is of the manoxylic type. The rays traversing the secondary xylem are one to three cells wide and very abundant. The pitting of secondary tracheids varies from scalariform on tracheids nearest the pith to circular bordered, with one to four rows of alternate pits crowded on the radial walls. Where they are contiguous, the pits tend to be hexagonal in outline and suggest the araucarioid pattern. The most obvious difference between *Bucklandia* and the stems of cycads is the absence of girdling leaf traces in *Bucklandia*.

That the cycads and cycadeoids are at best distantly related is dramatically emphasized by the differences in their reproductive structures as exemplified by their ovulate cones and pollen organs. The cones of *Williamsonia* are monosporangiate. Those shown terminating the fertile branches of the reconstruction of *W. sewardiana* are ovulate cones. It is not known how or if the dispersed pollen organs of *Weltrichia* were borne on the plant. Their attachment is inferred by association with other parts of *Williamsonia* and by their syndetochelic stomata. Sharma (1977) has prepared an illustrated review of those williamsonian reproductive structures obtained from the Upper Gondwana (Jurassic) rocks of India. A more comprehensive account of these fructifications from the Yorkshire Jurassic rocks has been prepared by Harris (1969).

We have had the privilege of examining some of the beautiful preserved permineralized ovulate

Figure 25.5. **A.** Diagram, longitudinal section of *Williamsoniella coronata* cone showing ovulate receptacle (or); microsporophyll (ms); bract (b). **B.** Microsporophyll enlarged to show embedded microsporangia. **A,B**: Jurassic. (**A,B** redrawn from Harris, 1964.)

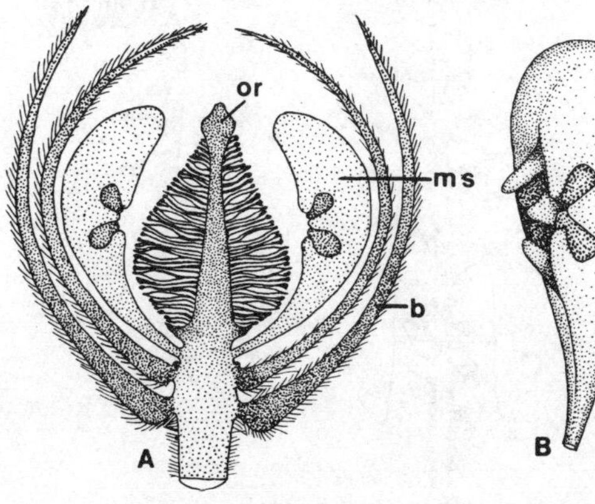

Figure 25.6. Reconstruction of *Williamsonia sewardiana* by Sahni (1932). Jurassic. (From Andrews, 1961.)

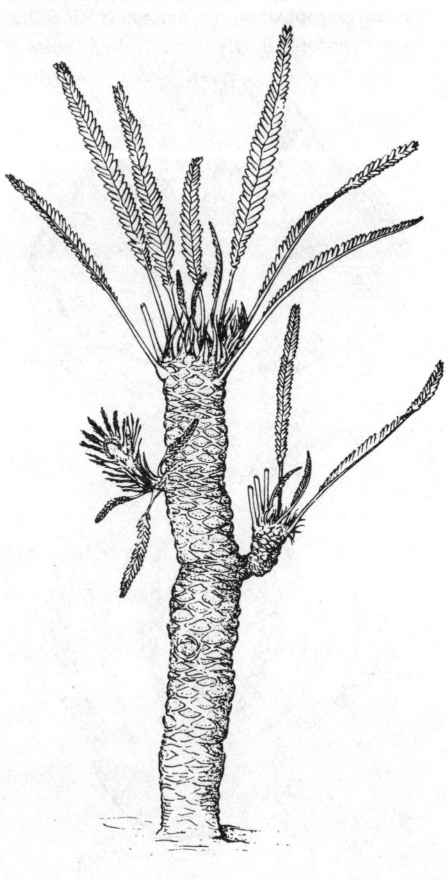

cones collected by Dr. M. N. Bose from the Upper Jurassic, Rajmahal Hills localities of India. In their structure, the cones described by Bose (1968) are quite similar to those of *W. sewardiana* (Sahni, 1932). In longitudinal section (Fig. 25.7A), the ovulate receptacle is nearly hemispherical or slightly conical. This is covered by a dense layer of stalked ovules and interseminal scales. The receptacle with its ovules and scales is subtended by spirally arranged bracts. As described and figured by Sahni (1932) and Sharma (1977), the interseminal scales are clublike structures each with a single trace of vascular tissue (Fig. 25.3A). Stomata have been observed on their exposed distal surfaces. The number of scales that surround and separate the ovules (Fig. 25.3B) far exceeds the number of ovules.

The orthotropous ovules have a tubular integument attached to the receptacle by a short stalk. The distal end of the integument, which surrounds a projecting micropyle, terminates in a funnel-shaped opening. Internal to the integument there is a long nucellar stalk that expands into an ovoid structure. Sahni described the nucellus as free from the integument, while Sharma believes the two were fused early in development. There is some evidence that the nucellus was vascularized by a single trace extending into the walls of the expanded distal end. Cellular megagametophytas and dicotyledonous embryos are reported in some nucellar cavities.

Because actual connection has not been demonstrated for presumed pollen-producing organs assigned to the Williamsoniaceae and because the name *Weltrichia* used for these organs has priority over *Williamsonia*, the name *Weltrichia* is used here as recommended by Harris (1969).

Specimens of *Weltrichia* are rare and, as far as we are aware, occur only as compression–impression fossils. Most species of *Weltrichia* have been found in the Jurassic Yorkshire beds of England and Rajmahal beds of India. These have been supplemented by Delevoryas (1991) who has described *W. ayuquilana* sp. nov. from the Jurassic beds of Oaxaca, Mexico. These pollen-producing organs have a cup-shaped base with as many as 20 fingerlike projections (microsporophylls) arising from the rim of the cup. Each microsporophyll has synangium-bearing branches along its upper (inner) face. The whole fructification may be as much as 10 cm in diameter.

Figure 25.7. **A.** *Williamsonia harrisiana*, longitudinal section of cone. (Based on photograph in Bose, 1968.) **B.** *Weltrichia spectabilis*, reconstruction of fructification. (Redrawn from Thomas, 1913.) **C.** *W. spectabilis*, detail of microsporophyll showing attached synangia. **D.** *W. whitbiensis*, reconstruction of fructification. **E.** *W. whitbiensis*, microsporophyll with synangia. (**C–E** redrawn from Nathorst, 1911.) **F.** *W. santalensis*, reconstruction of microsporophyll. (Redrawn from Sitholey & Bose, 1953.)

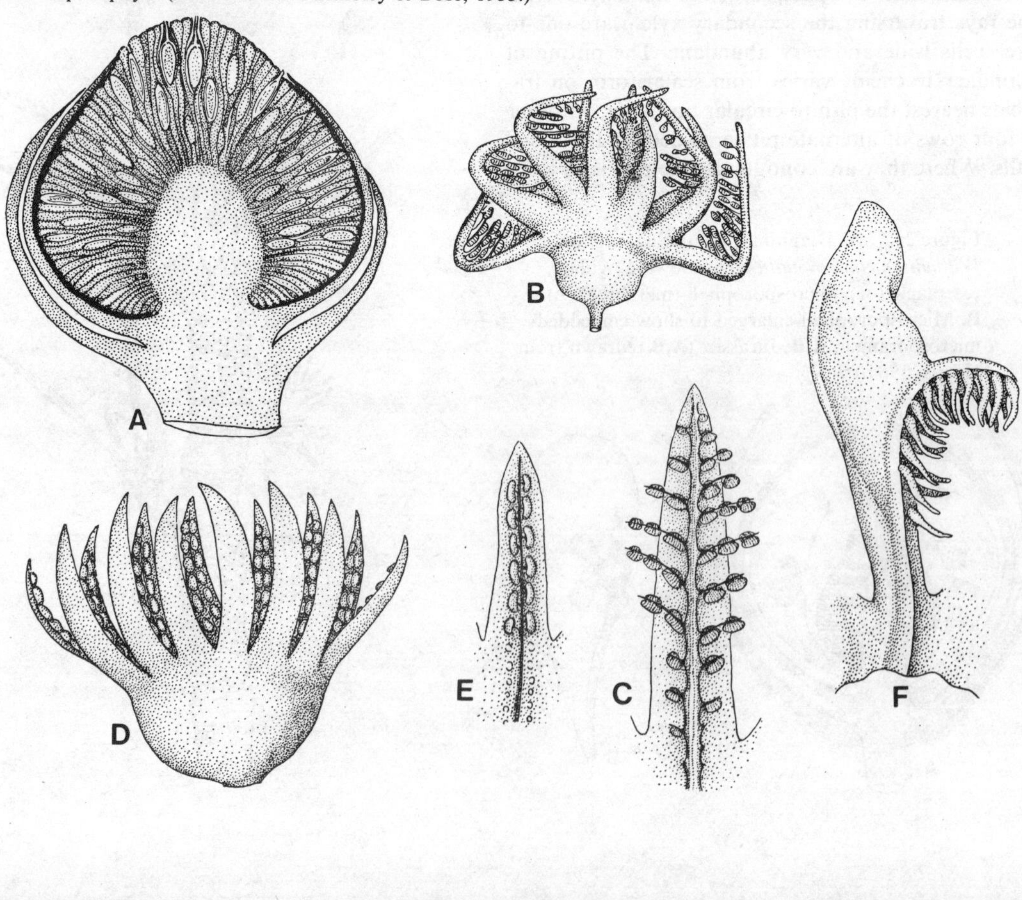

Weltrichia spectabilis (Fig. 25.7B) has approximately 12 microsporophylls that overarch the cup. The inner side of each has slender branches that, in turn, bear two rows of synangia (Fig. 25.7C). Although *W. whitbiensis* (Fig. 25.7D,E) has the basic structure of *W. spectabilis,* its fertile projections do not overarch the cup and it lacks a stalk. Two rows of toothlike structures are shown by Sitholey & Bose (1953) to occur on the inner faces of the twisted asymmetrical microsporophylls of *W. santalensis* (Fig. 25.7F). The suspected synangiate nature of the toothlike projections has since been confirmed by Sharma (1977). Although the synangia of several specimens of *Weltrichia* have been examined by others for pollen grains, the first convincing evidence that they are monocolpate was provided by Harris (1969).

Cycadeoidaceae

If one were to rely on growth habit alone, the relationships between the Cycadeoidaceae and Williamsoniaceae might seem obscure. While genera of the Williamsoniaceae may have slender branched axes (*Williamsoniella* and *Wielandiella*) or trunklike stems with lateral branches (*Williamsonia*), members of the Cycadeoidaceae have massive sparsely branched trunks (Fig. 25.8). Most are squat, not more than 50 cm high, by 40 cm wide, and ellipsoidal to globose in shape. However, an unusually large columnar specimen 1 m wide by 3 m high has been described. The resemblance of these bizarre fossils to old-fashioned beehives was noted long ago. As a matter of fact, many overly enthusiastic collectors thought that these interesting specimens actually were fossil beehives, wasp nests, corals, mushrooms, or various other structures.

Figure 25.8. *Cycadeoidea* sp., permineralized trunk with cones (c) and spirally arranged leaf scars (ls) separated by a thick ramentum (r). Cretaceous. (From Crepet, 1974.)

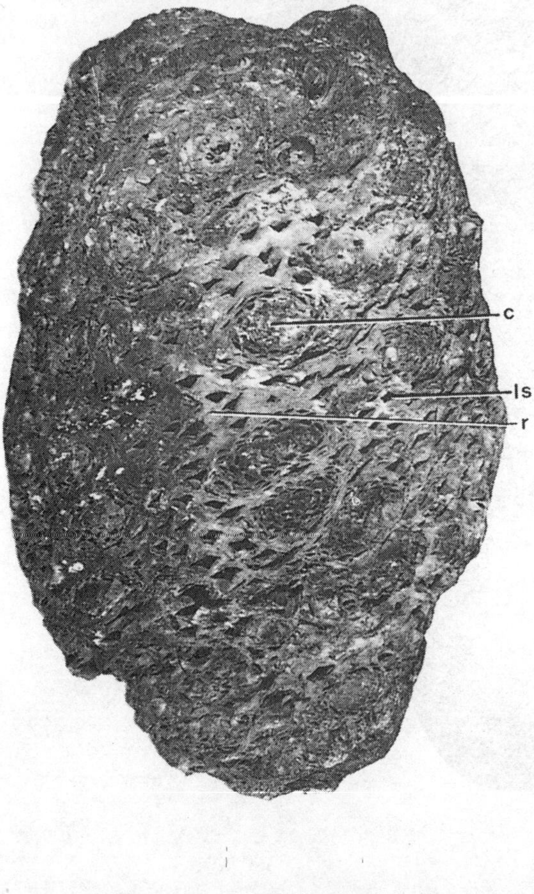

Permineralized trunks of *Cycadeoidea* occur in late Jurassic and Cretaceous strata and have a wide geographical distribution. The first *Cycadeoidea* specimens to be observed came from the Isle of Portland and the Isle of Wight over a century ago. Some three decades passed before the Lower Cretaceous, Black Hills cycadeoid localities of South Dakota were made famous by the work of MacBride (1893), Ward (1898), and Wieland (1906, 1916). Smaller numbers of cycadeoid trunks also have been found in Maryland, Wyoming, Kansas, Colorado, California, Texas, and New Mexico, while their presence in Jurassic and Lower Cretaceous rocks of Mexico, Italy, Belgium, France, Germany, Austria, Poland, and India attests to their wide distribution and abundance during the Mesozoic.

Collecting localities once known for an abundance of cycadeoid stems are now depleted, with most specimens residing in museum collections. One of the largest, with over 1,000 specimens, is part of the Peabody Museum collection at Yale University. Most of the specimens, which were collected and donated by interested amateurs, have been studied intensively by Wieland (1906, 1916), Delevoryas (1959, 1960, 1963, 1965, 1968a, 1968b), Crepet & Delevoryas (1972), and Crepet, (1972, 1974).

Cycadeoidea

Reconstruction of *Cycadeoidea* (Fig. 25.9) based on the early work of Wieland and others have appeared in countless publications. The plant is usually shown as having a stout, globose trunk clothed in an armor of spirally arranged leaf bases embedded in a ramentum of flat, tongue-shaped scales. The cones are borne at the apex of lateral axillary shoots that project only a short distance beyond the truncated stumps of the old leaf bases. In some species of *Cycadeoidea* the cones developed simultaneously and, according to Wieland's interpretation, matured into large flowerlike structures. Although never found attached, there is good evidence from undeveloped fronds found embedded in the ramentum of the stem that the plant had a crown of spirally arranged leaves of the *Ptilophyllum* or *Dictyozamites* types. An outstanding reconstruction of *Cycadeoidea* has been prepared by the curators and artists of the Field Museum of Natural

History, Chicago. As a result of the recent investigations by Delevoryas and Crepet, we will see, however, that the reconstruction had to be drastically modified, a situation that produced some trauma among the museum staff.

In transverse section the stele of a *Cycadeoidea* trunk (Fig. 25.10A) is cycadlike in having a large pith surrounded by a broad ring of vascular tissue composed of secondary phloem and manoxylic secondary xylem in about equal proportions. The ring is dissected by wide rays. Like the cycads, the cortex and pith of *Cycadeoidea* contains scattered secretory ducts. The primary xylem, which is located at the periphery of the pith at the inner edges of the wedges of secondary xylem, is endarch. Unlike the cycads, the preponderance of secondary xylem cells have scalariform thickenings, although circular bordered pits have been reported to occur on their radial walls. The rays of the rather compact secondary xylem are uniseriate and biseriate. When seen in transverse section the leaf traces traversing the broad cortex are composed of strands of primary and secondary vascular tissues that assumed a C-shaped configuration. In the leaf base the strands form a closed cylinder. Leaf traces, like those of other cycadeoids (*Bucklandia,* for example) show no indication of girdling as it occurs in cycads, but directly supply a leaf base. The formation of the cone vascular trace and its subtending leaf trace has been the object of investigations by Delevoryas (1959, 1960). The cone trace of *Cycadeoidea wyomingensis* is derived from the fusion of four leaf traces deep in the cortex. In its outward course the large cone trace gives rise to four leaf traces one of which supplies the leaf subtending the cone. The cone vascular supply of *Monanathesia,* where there is a cone in the axil of every leaf, is derived from the fusion of two cortical bundles that arise from leaf traces neither one of which supplies the leaf subtending the cone. These rather complicated and unexpected systems of leaf and cone vascularization serve to emphasize the variability in the vasculature of cycadophytes apparent in the cycads with their girdling traces

Figure 25.9. *Cycadeoidea* sp. reconstruction of plant. Cretaceous. (From Delevoryas, 1971.)

and medullosan pteridosperms where traces from different levels of a sympodium or from different sympodia supply the same leaf.

Although it was once thought that some species of *Cycadeoidea* had monosporangiate cones, it has been shown by Crepet (1972, 1974) that all cones studied were bisporangiate (Fig. 25.10C). The peduncle of the cone is terminated by a centrally located conical or dome-shaped receptacle that bears hundreds of stalked ovules (Fig. 25.11A,B) enveloped by club-shaped interseminal scales (Fig. 25.10B). These structures and their arrangement are highly similar to those already described for the Williamsoniaceae. As in *Williamsoniella,* the ovulate receptacle of *Cycadeoidea* is subtended by a whorl of microsporophylls. The reproductive parts of the cone are enclosed in spirally arranged bracts covered with a ramentum of scales. According to Wieland's 1906 interpretation, the microsporophylls were believed to be pinnate, frondlike structures with each pinna bearing two rows of kidney-shaped synangia. The microsporophylls and subtending bracts were supposed to open outward in much the same way as a flower bud. It was assumed that the microsporangia shed their monocolpate pollen and that wind was the vector in pollination. The similarity of the open cones of *Cycadeoidea* and flowers of the primitive ranalian type was soon recognized and was adopted as proof, in some systems of angiosperm classification, that flowering plants evolved from cycadeoid ancestors. As we shall see, however, there are major obstacles in the derivation of the angiosperm stamen from a compound microsporophyll and of the carpel from the ovulate gynoecium of a cycadeoid.

A further obstacle to the derivation of the angiosperms from cycadeoids was revealed by the researches of Delevoryas (1963, 1968a) on the ontogeny of the *Cycadeoidea* microsporophylls. His reinvestigation of Wieland's specimens led to the conclusion that *Cycadeoidea* microsporophylls could not expand into a flowerlike structure because of a massive dome of parenchymatous tissue that permanently enclosed them. Further, the microsporophylls bore their synangia along trabeculae (rodlike structures) that connected the inner faces of the parenchymatous covering around the microsporophylls, thus providing a structure that would physically prevent their opening out. The interpretation by Delevoryas raises the problem of polination mechanisms in *Cycadeoidea.* Wind pollination is eliminated; pollination by boring insects is a possibility (Crepet, 1972, 1974), but selfing seems to have been the chief mechanism. Because of this, outcrossing probably occurred with a low frequency. Later investigations by Delevoryas (1965, 1968a) led to the conclusion that the microsporangiate structure of a *Cycadeoidea* cone comprises individual microsporophylls, fused at their bases around the peduncle and ovulate receptacle, but free in their distal portions (Fig. 25.10C). The developmental stages of cones examined by Crepet (1947) clearly show that

Figure 25.10. **A.** Transverse section of trunk showing large pith (p); poorly developed secondary xylem (sx); thick ramentum (r) in which cone (c) and leaf bases (lb) are embedded. (Based on photograph in Wieland, 1906.) **B.** Diagram, longitudinal section of *Cycadeoidea morierei* ovule showing investing vascularized interseminal scale (is); micropyle (m); and integument (i). (Redrawn from Seward, 1917.) **C.** Reconstruction, *Cycadeoidea* cone. Note central ovulate receptacle (or), surrounded by whorl of microsporophylls (mi). **A,C:** Cretaceous. (Redrawn from Crepet, 1974.)

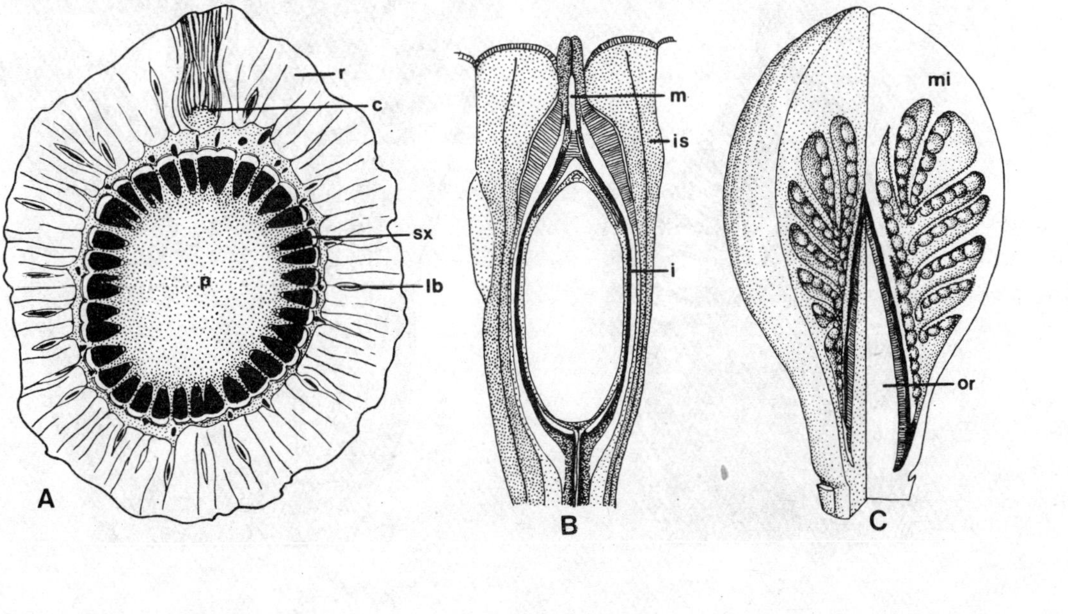

each microsporophyll in the whorl is a pinnate structure with each pinnule bearing two rows of synangia. A synangium (Fig. 25.11C) is a kidney-shaped structure containing 8 to 20 tubular sporangia within the periphery of the synangium. Monocolpate grains are found within the sporangia.

Because of the permineralized specimens where details are beautifully preserved, Crepet (1974) was able to demonstrate the early stages of microsporophyll development. A whorl of young microsporophylls appears around the apical meristem of the cone. As they matured, each pinnate microsporophyll became revolute, with the synangium-bearing pinnae folded inward parallel to the plane of the pinna rachis. The recurved microsporophyll rachises fused with one another above the ovulate receptacle, while the tips of the synangia-bearing pinnae become adnate (fused) as well.

In the same specimens showing developmental stages of the microsporophylls, Crepet & Delevoryas (1972) found ovules preserved in pregametophytic stages of nucellar ontogeny. Of particular interest is their discovery of remains of linear megaspore tetrads. You may recall that in other seed plants where megaspores are preserved (*Archaeosperma* and *Conostoma*), the three abortive megaspores and one functional megaspore have a tetrahedral arrangement. The discovery of a nucellus showing early stages of ontogeny in a Cretaceous gymnosperm is significant in allowing a better understanding of the life cycles of the fossil and as it compares with those of extant plants.

Although early stages in ontogeny of *Cycadeoidea* ovules are preserved, it is worth noting that stages in megagametophyte development have not been observed. This is most unusual in view of the discovery of well-preserved cycadeoid seeds with dicotyledonous embryos. Here the mysterious absence of preserved life cycle stages is the reverse of the situation in pteridosperms where ovules with mega-

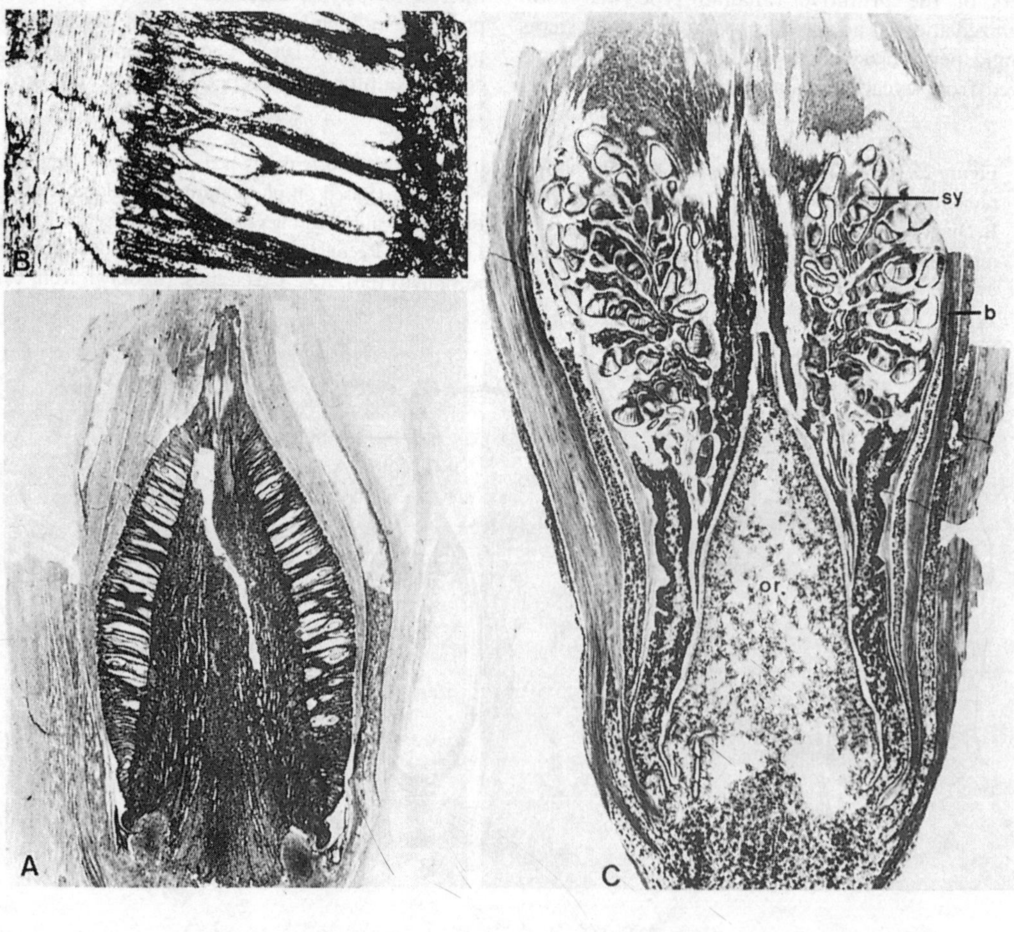

Figure 25.11. **A.** *Cycadeoidea* sp., longitudinal section through ovulate receptacle bearing stalked ovules separated by interseminal scales. **B.** Enlargement of stalked ovules and interseminal scales. **C.** *Cycadeoidea* sp., longitudinal section through cone showing ovulate receptacle (or), microsporophylls with synangia (sy), and investing bracts (b). **A–C:** Cretaceous. (**A–C** from Crepet, 1974.)

gametophytes are common but embryos have seldom been found.

The origin and relationships of the Cycadeoidales

Although the cycadeoids have many vegetative characteristics that are cycadlike, some of their reproductive structures, especially those bearing ovules, are unique. We know that cycadeoid fronds are so similar in their general morphology to the pinnate leaves of the cycads as to be indistinguishable without recourse to examining their stomatal structure. With few exceptions the internal structure of cycadeoid stems (*Bucklandia* and *Cycadeoidea*) is like that of cycads. The principal differences are the absence of girdling leaf traces, more abundant and compact secondary xylem in *Bucklandia*, and scalariform secondary xylem tracheids in *Cycadeoidea*.

The bisporangiate cones of *Williamsoniella* and *Cycadeoidea* distinguish these plants from cycads with their strictly monosporangiate cones. Although the cones of *Wielandiella* and *Williamsonia* appear to be dioecious, they may turn out to be monoecious when more is known about how and where their pollen-producing organs were borne. The structure of the stalked ovules surrounded by the enigmatic interseminal scales is unique to the Cycadeoidales, placing the members of the order well apart from the Cycadales. Further, the whorls of microsporophylls, with their pinnate synangium-bearing branches, represent another radical departure from the spirally arranged microsporophylls of cycads with their abaxial sporangia. Sporangia of both groups, however, produce monocolpate pollen.

If we were to make a judgment about the relationship between cycadeoids and cycads based on

Figure 25.12. **A,B.** Hypothetical stages in the evolution of the cycadeoid cone according to a foliar origin. **C.** Hypothetical origin of cycadeoid cone from an axillary shoot. **D.** Diagram showing origins of traces to cone and leaf bases. (A–D redrawn from Delevoryas, 1986b.)

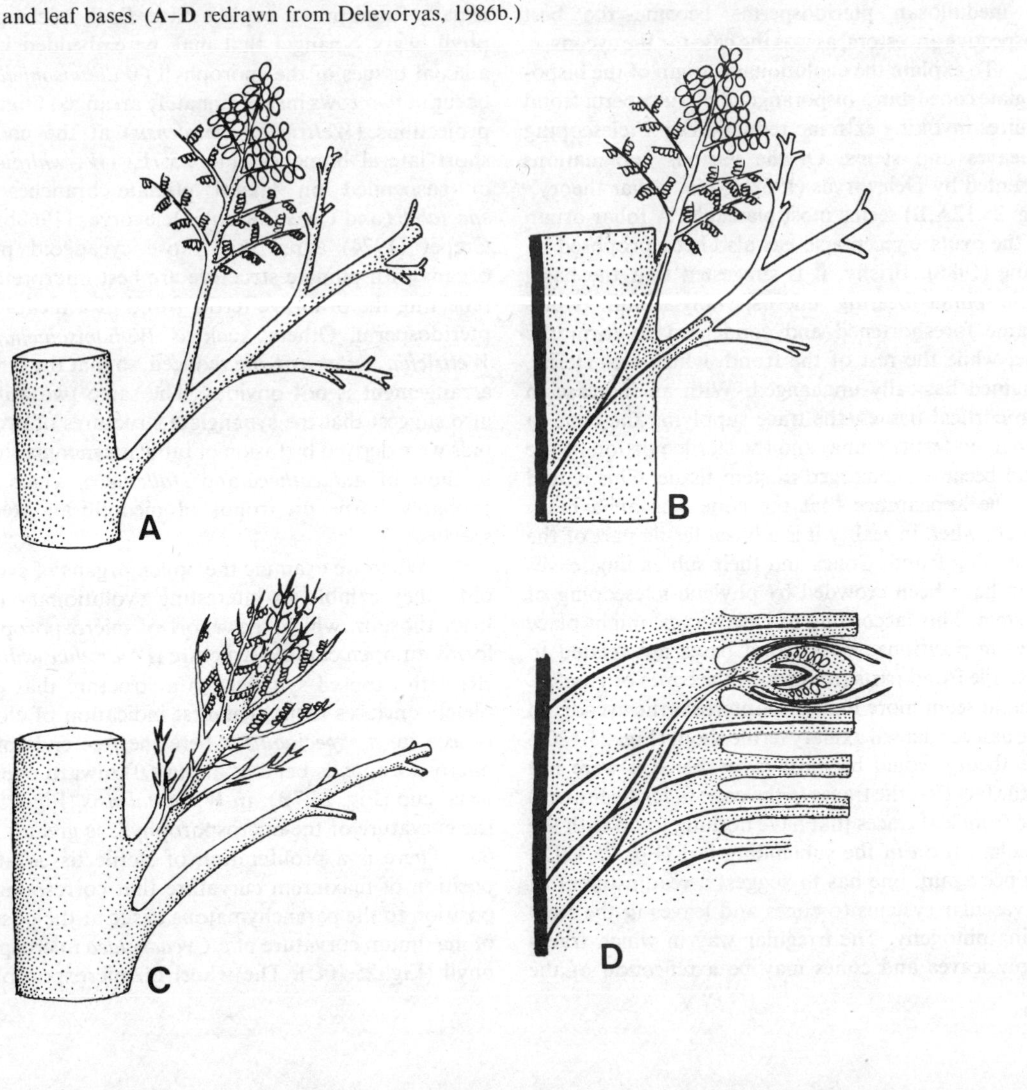

vegetative characteristics alone, one would be justified in suggesting a close relationship between the two groups. When, however, we consider the structure of cones, it becomes apparent that the relationship is a distant one at best. It seems that Chamberlain (1920) had the correct interpretation when he recognized the cycadeoids and cycads as two distinct groups with a possible common ancestry among the Paleozoic pteridosperms. What the ancestral pteridosperms were like becomes difficult to determine when all characteristics are considered. For example, the Triassic and Jurassic cycadeoids (*Williamsoniella* and *Wielandiella*) have slender, much-branched stems. The Yorkshire Jurassic *Williamsonia* may also have been highly branched with slender trunks. These and other observations have led Delevoryas (1968b) to suggest that cycadeoids with slender, much-branched stems are primitive. He also concludes that the primitive leaf type of cycadeoids is a pinnately compound one. If we look for pteridosperm ancestors with this kind of growth habit, the most likely candidates are the Lyginopteridaceae and Callistophytaceae. If, however, we consider internal stem structure of *Williamsonia* and *Cycadeoidea*, then the medullosan pteridosperms become the best prospective ancestors, as was the case for the cycads.

To explain the evolutionary origin of the bisporangiate cone from a bisporangiate pteridosperm frond requires invoking extreme reduction and telescoping of leaves and stems. Of the various explanations presented by Delevoryas (1968b), his "foliar theory" (Fig. 25.12A,B) seems most plausible. A foliar origin for the ovulate receptacle has also been proposed by Crane (1986). Briefly, it is suggested that the basal fertile pinna bearing microsporangia and ovules became foreshortened and condensed to form the cone, while the rest of the frond, which was sterile, remained basically unchanged. With an increase in stem cortical tissues, the trace supplying the cone (a condensed fertile pinna) and the sterile portion of the frond became submerged in stem tissue. This would give the appearance that the cone was an axillary branch, when in reality it is a basal fertile part of the subtending frond. Cones and their subtending leaves could have been crowded by phyletic telescoping of the stem. This, according to Delevoryas, might place cones in positions above frond bases not related to the sterile frond parts from which the cones originated. It might seem more logical to interpret the cycadeoid cone as a condensed axillary fertile shoot (Fig. 25.12C). This theory would be most favorable if it were not for the fact that the traces to the cone (a fertile branch) arise from leaf traces that have nothing to do with the vascularization of the subtending leaf (Fig. 25.12D). So once again, one has to suggest a readjustment of the vascular systems to cones and leaves in the stem during ontogeny. The irregular way in which traces supply leaves and cones may be a reflection of the unpredictable way leaf traces supply fronds of medullosan pteridosperms. We would like to suggest that an irregular disposition of leaf traces is characteristic of the vasculature of medullosan pteridosperms, cycads, and *Cycadeoidea*. This contrasts with leaf-trace vascular systems in conifers, Lyginopteridaceae, and Callistophytaceae where leaf traces are given off from sympodia and supply the spirally arranged leaves in an orderly and predictable way. It is quite possible that those gymnosperms with large fleshy stems and large leaf bases have evolved a system that supplies their leaves with large numbers of leaf traces whose origins from specific sympodia are unpredictable. In contrast, those gymnosperms with small leaf bases requiring only a few (one to eight) leaf traces are produced in a predictable way. With these ideas in mind it will be most interesting to find out what the leaf-trace pattern is like in such genera as *Williamsonia, Williamsoniella,* and *Wielandiella*.

The microsporangiate fructifications of cycadeoids compare favorably with one another and again suggest a common origin for the group. You may remember many consist of a whorl of microsporophylls usually fused into a cuplike base. Each microsporophyll bears synangia that may be embedded in the adaxial tissues of the sporophyll (*Williamsoniella*) or occur in two rows inside pinnately arranged fingerlike projections (*Weltrichia santalensis*) at the ends of short lateral branches (Fig. 25.7E) (*W. whitbiensis*) or suspended on lateral pinnate branches (*W. spectabilis* and *Cycadeoidea*). Delevoryas (1968b) and Crepet (1974) agree that those cycadeoid pollen organs with pinnate structure are best interpreted as reflecting the primitive fertile frond of a medullosan pteridosperm. Others, such as *Williamsoniella* and *Weltrichia santalensis*, are reduced, so that the pinnate arrangement is not obvious. The same two authors also suggest that the synangiate structures of cycadeoids were derived by fusion of tubular synangia similar to those of *Aulacotheca* and *Halletheca*, which were probably borne on fronds of medullosan pteridosperms.

When we examine the pollen organs of cycadeoids, they exhibit an interesting evolutionary trend from those in which the whorl of microsporophylls forms an open cuplike structure (*Weltrichia whitbiensis*) with exposed synangia to a structure that completely encloses them. The first indication of closure is seen in *W. spectabilis*, where the free ends of the microsporophylls bend at about 60° inward over the basal cup (Fig. 25.7B). In *W. santalensis* (Fig. 25.7F) the curvature of the microsporophylls is greater than 60°. There is a proliferation of sterile tissue at the position of maximum curvature that corresponds in position to the parenchymatous tissue at the position of maximum curvature of a *Cycadeoidea* microsporophyll (Fig. 25.10C). The whorl of microsporophylls of *W. santalensis* opened out, allowing exposure of their synangia and also allowing cross-pollination, whereas those of *Cycadeoidea* fused with one another to completely enclose the synangia and thus limit the possibilities of cross-pollination. Crepet (1974) suggests that self-pollination of *Cycadeoidea*, imposed by the phyletic closure of the pollen-producing fructification, was an important factor in the extinction of the group, which did not continue beyond the end of the Cretaceous.

The stalked ovules of cycadeoids (Fig. 25.10B) bear very little resemblance to ovules of the Medullosaceae. First, cycadeoid ovules are small (up to 3 mm long), unlike the much larger seeds of the medullosan pteridosperms. Although debatable, the nucellus of cycadeoid ovules is said to be fused with the integument except in the apical region. Unlike a *Pachytesta* of the medullosas, cycadeoid ovules have a single vascular system located in the basal portion of the nucellus. The integument has no vascular supply and is composed of only two relatively thin layers instead of the three layers found in the well-developed testas of medullosan ovules. Crane (1985) has made the novel suggestion that cycadeoid ovules have a thin, double integument like that of flowering plants. As elaborated by Crane (1986), structures that have been called a cutinized layer, a micropylar plate, or a cupule by previous authors occur in several cycadeoid species. It is this outer structure that has been interpreted as an integument. If this suggestion is accurate, then cycadeoids may be more closely related to both the flowering plants and to the Caytoniales (Chapter 26) than previously suspected.

The thin integument or integuments of cycadeoids presumably would have provided little protection for the nucellus. However, this poorly developed integument is more than made up for by the five to six surrounding interseminal scales. Each is expanded in the distal portion to form a club-shaped structure, and each contains at least one vascular strand. There is much speculation about the evolutionary origin of the stalked ovules and interseminal scales. If we follow Delevoryas (1968b), we must conclude that the scales represent either the products of reduction, derived from fertile and sterile parts of pteridosperm fronds, or ovules that have become sterile. It would be gratifying, however, if we had some intermediate stages at our disposal that would show us what part of the frond is represented by an interseminal scale, or how some 300 or more ovules with stalks in addition to five times as many interseminal scales assumed their position on a receptacle. It also would be most instructive to discover if and how the possible double integument of the ovules originated from a cupule or cupule-like structure. These deficiencies in our knowledge about cycadeoid evolution serve to emphasize the gap between the cycadeoids and their possible medullosan ancestors. If we simply ignored this gap, which we do not advocate, it would be easy to conclude a medullosan origin for cycadeoids as we have done for the cycads.

Some of the most controversial results from recent cladistic analyses have revised our understanding of the interrelationships among many groups of gymnosperms and their potentials as the ancestors of the flowering plants. These studies, which have been conducted primarily by Crane (1985, 1988) and Doyle & Donoghue (1986, 1987), have done much to alter our ideas about the relationships of the cycadeoids. As summarized by Crane (1988) we now suspect that the cycadeoids are much more closely related to the Gnetales (Chapter 26) and the flowering plants than they are to the cycads. This further emphasizes the unnatural status of the old group "cycadophytes."

Protection of sporangia

We believe it would be profitable at this point to undertake an overview of the ways the vascular plants we have studied thus far protect their sporangia against the vicissitudes of the environment. In the Lycopsida we have seen a shift from an exposed terminal position of sporangia to the adaxial surface of a microphyll, followed by condensation of microphylls into a cone. In some heterosporous lycopsids (*Lepidocarpon, Miadesmia*) the lateral laminae of the megasporophyll form an integument around the megasporangium. Recurvation of branches bearing terminal sporangia and condensation of these sporangiophores into cones seems to have provided the Sphenopsida with adequate protection of their sporangia. In some filicaleans (*Botryopteris globosa*), the terminal sporangia are borne in dense spherical clusters surrounded by sterile protective sporangia. Others, as in the Marattiales and Filicales, have accomplished a phyletic shift from terminal or marginal positions of their sporangia to a more protected abaxial one. Notably, the Marattiales have evolved synangia where the relative amount of desiccation from their clustered sporangia is reduced.

Primitive seed plants with their terminal mega- and microsporangia have evolved various innovative means of enclosing and protecting their reproductive structures. In preovules with hydrasperman reproduction, it is apparent that sterile telomes have fused around a megasporangium to form an integument as well as an additional covering, the cupule. The terminal microsporangia in the Lyginopteridaceae, Medullosaceae, and Callistophytaceae are fused into synangia, and those of the latter family have assumed an abaxial position on the sporophyll, a feature of the cycads. The cycads, however, afford additional protection, as do cycadeoids, by the condensation of their microsporophylls into cones. *Cycadeoidea* has achieved the ultimate in microsporangia protection by permanently enclosing its synangia, thus imposing self-pollination on the bisporangiate cones, which surely contributed to extinction of the group because of the imposition of ever-increasing homozygosity.

This brings us to the puzzling interseminal scale. The observation of abortive ovules terminating some of the scales suggests that the stalked ovules and scales are homologous structures, those with ovules representing megasporophylls, the latter sterile ones. When one gets to the "bottom line," it is quite clear that interseminal scales performed the same function for a cycadeoid plant as did integuments or cupules derived from the fusion of the telomes in pteridosperms. Thus the problematical interseminal scales around the ovules of cycadeoides provided additional protection to the ovules in much the same way as did the cone scales of conifers or the ovary walls of angiosperms. What seems important to us, however, is not what we call the structures, or even what their evolutionary origins might be, but that here we have one more example to add to our long list of ways in which green land plants (including the bryophytes) have provided mechanisms for the enclosure of their reproductive parts. Our next chapter will describe several different, often bizarre, and enigmatic innovations evolved by vascular plants. Some of these are designed to provide additional protection, especially for megasporangia, while others may also involve mechanisms for regulating outcrossing and natural selection before fertilization. Our ultimate goal is to evaluate these and other evolutionary experiments in terms of the possible origin of angiosperms.

References

Andrews, H. N. (1961). *Studies in Paleobotany*. New York: Wiley.

Bose, M. N. (1968). A new species of *Williamsonia* from the Rajmahal Hills, India. *Journal of the Linnean Society (Botany)*, **61**, 121–7.

Chamberlain, C. J. (1920). The living cycads and the phylogeny of seed plants. *American Journal of Botany*, **7**, 146–53.

Crane, P. D. (1985). Phylogenetic analysis of seed plants and the origin of angiosperms. *Annals of the Missouri Botanical Garden*, **72**, 716–93.

Crane, P. D. (1986). The morphology and relationships of the Bennettitales. In *Systematics and Taxonomic Approaches in Paleobotany*, eds. T. A. Spicer & B. A. Thomas. The Systematics Association Special Volume No. 31. Oxford: Clarendon Press.

Crane, P. D. (1988). Major clades and relationships in the "higher" gymnosperms. In *Origin and Evolution of Gymnosperms*, ed. C. B. Beck. New York: Columbia University Press.

Crepet, W. L. (1972). Investigations of North American cycadeoids: pollination mechanisms in *Cycadeoidea*. *American Journal of Botany*, **59**, 1048–56.

Crepet, W. L. (1974). Investigations of North American cycadeoids: the reproductive biology of *Cycadeoidea*. *Palaeontographica B*, **148**, 144–69.

Crepet, W. L., & Delevoryas, T. (1972). Investigations of North American cycadeoids: early ovule ontogeny. *American Journal of Botany*, **59**, 209–15.

Delevoryas, T. (1959). Investigations of North American cycadeoids: *Monanthesia*. *American Journal of Botany*, **46**, 657–66.

Delevoryas, T. (1960). Investigations of North American cycadeoids: trunks from Wyoming. *American Journal of Botany*, **47**, 778–86.

Delevoryas, T. (1963). Investigation of North American cycadeoids: cones of *Cycadeoidea*. *American Journal of Botany*, **50**, 45–52.

Delevoryas, T. (1965). Investigations of North American cycadeoids: microsporangiate structures and phylogenetic implications. *Palaeobotanist*, **14**, 89–93.

Delevoryas, T. (1968a). Investigations of North American cycadeoids: structure, ontogeny, and phylogenetic considerations of cones of *Cycadeoidea*. *Palaeontographica B*, **121**, 122–33.

Delevoryas, T. (1968b). Some aspects of cycadeoid evolution. *Journal of the Linnean Society (Botany)*. **61**, 137–46.

Delevoryas, T. (1971). Biotic provinces and the Jurassic – Cretaceous floral transition. *Proceedings of the North American Paleontological Convention, Part L*, 1660–74.

Delevoryas, T. (1991). Investigations of North American Cycadeoids: *Weltrichia* and *Williamsonia* from the Jurassic of Oaxaca, Mexico. *American Journal of Botany*, **78**, 177–82.

Delevoryas, T., & Hope, R. C. (1976). More evidence for a slender growth habit in Mesozoic cycadophytes. *Review of Palaeobotany and Palynology*, **21**, 93–100.

Doyle, J. A., & Donoghue, M. J. (1986). Seed plant phylogeny and the origin of angiosperms: An experimental cladistic approach. *The Botanical Review*, **52**, 321–431.

Doyle, J. A., & Donoghue, M. J. (1987). The origin of angiosperms: A cladistic approach. In *The Origins of Angiosperms and Their Biological Consequences*, eds. E. M. Friis, W. G. Chaloner, & P. R. Crane. Cambridge: Cambridge University Press.

Florin, R. (1933). Studien über die Cycadales des Mesozoikums. *Kungl. Svenska Vetenskapsakademiens Handlinger III*, **12**, 1–134.

Harris, T. M. (1932). The fossil flora of Scoresby Sound East Greenland. *Meddelelser om Grønland*, **85**, 1–133.

Harris, T. M. (1944). A revision of *Williamsoniella*. *Philosophical Transactions of the Royal Society of London B*, **231**, 313–28.

Harris, T. M. (1964). *The Yorkshire Jurassic Flora. II. Caytoniales, Cycadales and Pteridosperms*. London: British Museum (Natural History).

Harris, T. M. (1969). *The Yorkshire Jurassic Flora. III. Bennettitales*. London: British Museum (Natural History).

Hughes, N. F. (1976). *Paleobiology of Angiosperm Origins*, Cambridge: Cambridge University Press.

McBride, T. H. (1893). A new cycad. *American Geologist*, **12**, 248–50.

Nathorst, A. G. (1911). Poläobotanische Mitteilungen. *Kungl. Svenska Vetenskapsakademiens Handlingar*, **46**, 1–33.

Nishida, M. (1969). A petrified trunk of *Bucklandia choshiensis*, sp. nov. from the Cretaceous of Choshi, Chiba Prefecture, Japan. *Phytomorphology*, **19**, 28–34.

Sahni, B. (1932). A petrified *Williamsonia* (*W. sewardiana*, sp. nov.) from the Rajmahal Hills, India. *Palaeontologia Indica*, **20** (N. S.), 1–19.

Seward, A. C. (1917). *Fossil Plants*. Vol. III. Cambridge: Cambridge University Press. Reprint, New York: Macmillan (Hafner Press), 1969.

Sharma, B. D. (1977). Indian williamsonias – an illustrated review. *Acta Palaeobotanica*, **18**, 19–29.

Sitholey, R. V., & Bose, M. N. (1953). *Williamsonia santalensis* sp. nov. – a male fructification from the Rajmahal Series, with remarks on the structure of *Ontheanthus polyandra* Ganju. *The Paleobotanist*, **2**, 29–39.

Stidd, L. L. O., & Stidd, B. M. (1976). Paracytic (syndetocheilic) stomata in Carboniferous seed ferns. *Science*, **193**, 156–7.

Stubblefield, S., & Banks, H. P. (1978). The cuticle of *Drepanophycus spinaeformis*, long-ranging Devonian lycopod from New York and Eastern Canada. *American Journal of Botany*, **65**, 110–18.

Thomas, H. H. (1913). The fossil flora of the Cleveland District. *Quarterly Journal of the Geological Society, London*, **59**, 223–51.

Thomas, H. H. (1915). On *Williamsoniella*, a new type of bennettitalean flower. *Philosophical Transactions of the Royal Society of London B*, **207**, 113–48.

Thomas, H. H., & Bancroft, N. (1913). On the cuticles of some recent and fossil cycadean fronds. *Transactions of the Linnean Society (Botany)*, **8**, 155–204.

Ward, L. F. (1898). Descriptions of the species of *Cycadeoidea* of fossil cycadean trunks thus far determined from the Lower Cretaceous rim of the Black Hills. *Proceedings of the United States National Museum*, **21**, 195–229.

Wieland, G. R. (1906). *American Fossil Cycads*, Washington, D.C.: Carnegie Institute.

Wieland, G. R. (1916). *American Fossil Cycads*, Washington, D.C.: Carnegie Institute, Vol. II.

Zimmermann, W. (1959). *Die Phylogenie der Pflanzen*. Stuttgart: Fischer.

26

More innovation and diversification among gymnosperms

Relationships of the late Paleozoic and Mesozoic seed plants discussed in this chapter until recently have been either unknown or at best highly equivocal. We shall see some with characteristics that suggest origins from Paleozoic pteridosperms, others that are claimed to be preangiosperms, and some that present serious challenges to attempts at classification. The one thing all groups have in common, however, are novel mechanisms for the protection of ovules.

Caytoniales

To emphasize the difficulties in classification of these groups, we would like to use the example of the Caytoniales with the aid of a quotation from Harris (1964): "The affinities of the Caytoniales have from the first been regarded as open. Thomas (1925) emphasizes points of agreement with both the Pteridosperms and Angiosperms. On the whole, work since that date has made the Caytoniales seem even more isolated: the fruit in particular is strictly gymnospermous in pollination." In spite of this cautious evaluation of their possible relationships, several authors, notably Andrews (1961), see the Caytoniales as an offshoot of the pteridosperms that approximate the "angiospermous state." As you might guess, this group has attracted much attention in our search for the origin of flowering plants.

Authors of some classifications of the Mesozoic pteridosperms recognize three families belonging to the Caytoniales. They are the Caytoniaceae, Corystospermaceae, and Peltaspermaceae.

Caytoniaceae

Leaf remains of the Caytoniaceae, assigned to the genus *Sagenopteris,* have a wide geographic and stratigraphic distribution ranging from the Upper Triassic to Lower Cretaceous. The ovule-bearing organs *Caytonia* and pollen organs *Caytonanthus* have been found associated with leaves in Greenland, Sardinia, western Canada, the eastern USSR, England, and Siberia.

Sagenopteris

Sagenopteris phillipsi leaves (Fig. 26.1A) are petiolate, with two pairs of four leaflets in a palmate arrangement at the distal end of the petiole. The leaflets are lanceolate and have a more or less conspicuous midvein extending for most of their length. Lateral veins curve outward, dichotomizing and anastomosing to form a network of veins ending freely at the entire margins of the leaflets (Fig. 26.1B). The stomata are haplocheilic (Fig. 26.1C) and confined to the lower surface. The epidermal structure of *Sagenopteris* is distinctly angiospermous (Thomas, 1925).

Caytonia

As described by Thomas (1925), *Caytonia* (Fig. 26.2A) is a planated, bilaterally symmetrical megasporophyll. The rachis of the sporophyll bears lateral ovule-containing cupules in nearly opposite pairs. A cupule is globose and attached to the rachis by a short stalk. Each cupule is reflexed with a liplike projection above an opening that is contracted and curved back against the cupule stalk (Fig. 26.2B). At maturity, the opening becomes closed. Numerous small orthotropous ovules similar in size and structure to those of cycadeoids are borne in a curved row with their micropyles facing the opening (Fig. 26.2C). The ovules are platyspermic and have a single integument free to the base and enclosing the nucellus. The latter is differentiated into a minute chalaza at its base and a micropylar beak at the apex. Bisaccate pollen of the *Vitreisporites* dispersed type (Fig. 26.2D) has been found in the opening of the nucellar beak and in the micropyle (Fig. 26.2E).

Caytonanthus

Vitreisporites-type pollen is produced in antherlike synangia each composed of four elongated microsporangia (Fig. 26.2F). The synangium-bearing organ is called *Caytonanthus* (Harris, 1937). It is constructed of a main axis with short lateral branches that may rebranch (Fig. 26.2G). The ultimate branches bear terminal pendant tubular synangia that are similar in size and general morphology to *Aulacotheca* and *Halletheca* of the medullosan pteridosperms. The whole *Caytonanthus* organ is believed to be a modified pinnate microsporophyll (Harris, 1964), which is in keeping with the supposed pteridosperm origin of these Mesozoic plants.

Although the leaves, megasporophylls, and microsporophylls have not been found attached to a common axis, their frequency of association, similarity of epidermal features (especially stomatal structure), and presence of *Caytonanthus*-type pollen in the nucellar beak cavity of *Caytonia* leave little doubt that the three organ genera are congeneric.

Figure 26.1. *Sagenopteris phillipsi*. Jurassic. **A.** Palmate leaf. **B.** Enlarged portion of leaflet showing reticulate venation. (**A,B** redrawn from Thomas, 1925.) **C.** Stoma. (Redrawn from Harris, 1964.)

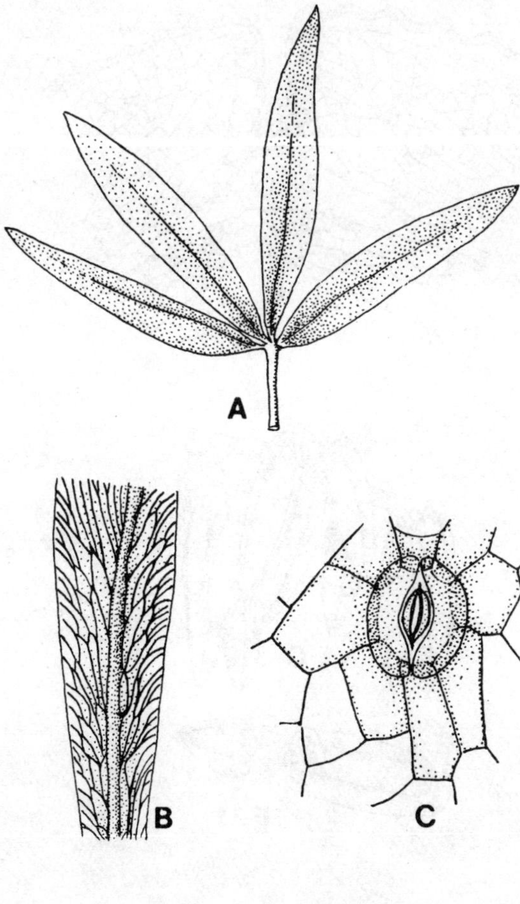

Corystospermaceae

The Corystospermaceae is a unique assemblage of fossil plants first described by Thomas (1933) from Triassic beds of Natal, South Africa. The principal and best known organ genera are those belonging to ovule-bearing megasporophylls *Umkomasia*, microsporophylls *Pteruchus*, and sterile foliage *Dicroidium*, *Pachypteris*, and *Stenopteris*.

Umkomasia

The megasporophyll of the corystosperm *Umkomasia* (Fig. 26.3A), is a planated branched axis with the branches borne in the axils of bracts. Opposite pairs of stalked cupules, reminiscent of *Caytonia*, occur on the branches. Each recurved cupule is partially divided to form a bivalved structure containing a single ovule. The detached ovule-containing cupules of *Pilophorosperma* (Fig. 26.3B) show clearly that the curved micropyle of the ovule, unlike those of *Caytonia*, projects beyond the helmetlike cupule. It has been suggested that *Umkomasia* and *Pilophorosperma* represent a stage in the evolution of the cupule intermediate between one that is open and one that becomes closed around the ovules (Delevoryas, 1962).

Pteruchus

Pteruchus africanus (Fig. 26.3C) is one of several species described by Thomas (1933) that is typical of the pollen-bearing organs. It is a small branching system that divides by unequal and equal dichotomies. Branches produced by the dichotomies terminate in a flattened circular to elliptical lamina, the lower side of which bore clusters of elongated pendant sporangia. The structure of *Pteruchus* is highly reminiscent of *Crossotheca*, which, you may remember, is a microsporangiate structure of certain Paleozoic pteridosperms. The pollen grains in the microsporangia of *Pteruchus* are bisaccate and nearly identical to those of *Caytonanthus*, differing only in having a well-defined colpus (Townrow, 1965).

Foliage

Although none has been found attached to axes bearing the reproductive structures, the associated frond genera *Dicroidium* (Fig. 26.3D), *Pachypteris* (Fig. 26.3E), and *Stenopteris* are believed to belong to corystosperms. The relationship is inferred by the similarity of epidermal features (Thomas, 1933; Townrow, 1965). All three genera have planated pinnate fronds, and some species of each genus are known to have open venation in their pinnules (Harris, 1964). In these characteristics the fronds of corystosperms are different from the palmately compound leaves of *Sagenopteris* with their anastomosing veins.

The recent description by Pigg (1988) of permineralized specimens of *Dicroidium* in association with ovulate cupules from Triassic deposits of Antarctica (Taylor & Taylor, 1987) provides information about the internal anatomy of the corystosperms. As is characteristic of compressed species of *Dicroidium*, the Antarctic leaves fork and bear pinnae with open dichotomous venation above and below the level of bifurcation. The Antarctic leaves are approximately 15 cm long, with pinnules that are bluntly elongate basely and lobed or bipinnatifid in the distal region. The rachis has 15 to 20 bundles arranged in a ring of seven or eight near the abaxial surface and a line of five to eight adaxially, a configuration that Pigg considers to be similar to that of cycads. Veins are surrounded by a bundle sheath, and some produce secondary xylem with scalariform–reticulate wall thickenings of the tracheids. Mesophyll of the pinnules is differentiated into palisade and spongy zones, and the epidermis has dicyclic stomata. Because corystosperm fronds are like the pinnately compound leaves of Paleozoic pteridosperms, it has been suggested that they are more primitive than the Caytoniales, which have palmately compound leaves.

A remarkable compression specimen from Argentina described by Taylor & Archangelsky (1985) documents that the Lower Cretaceous frond *Ruflorinia* bore cupules assignable to *Ktalenia*. This plant (Fig. 26.4) suggests that members of the Corystospermaceae extended through the Mesozoic into the Cretaceous, where they were contemporaneous with primitive flowering plants. Fronds are up to 8 cm long, with alternately arranged lanceolate pinnules that have wavy margins and open, dichotomous venation. The cupules are arranged (Fig. 26.4A) suboppositely to alternately on fertile pinnae, and are associated with clusters of linear, needlelike foliar structures. The cupules (Fig. 26.4B) are rounded, 3 to 4 mm long, and recurved with a terminal beaklike opening that is directed toward the base of the frond. Cuticular preparations show that each cupule contains one or two seeds that are rounded in cross section and have

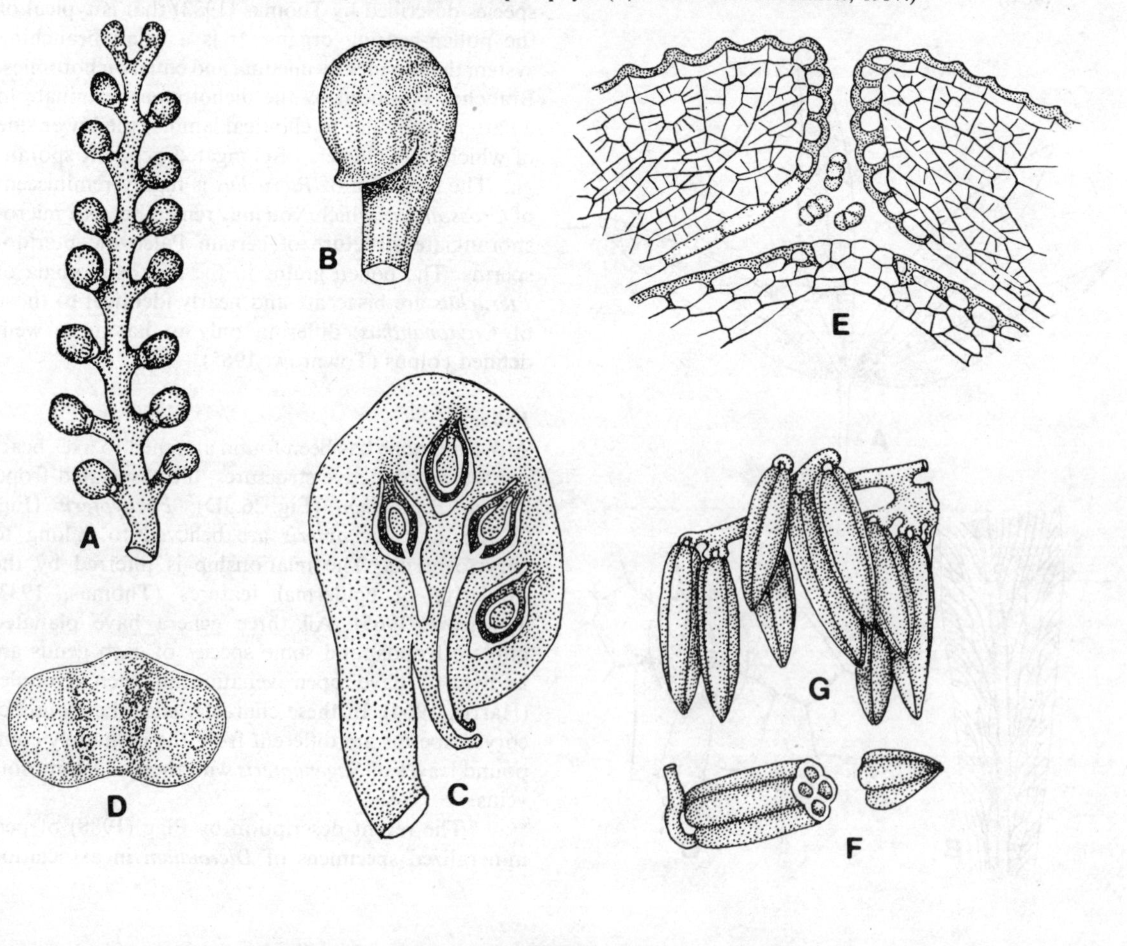

Figure 26.2. *Caytonia nathorsii*. Jurassic. **A.** Reconstruction of megasporophyll. (Redrawn from Thomas, 1925.) **B.** Young cupule showing position of lip and opening. (Redrawn from Harris, 1964.) **C.** *C. thomasi*. Diagram, longitudinal section through cupule showing position of ovules. (Redrawn from Harris, 1933.) **D.** *Caytonanthus*, bisaccate pollen grain. (Redrawn from Harris, 1964.) **E.** Bisaccate pollen grains of the *Caytonanthus* type in micropyle of *Caytonia* ovule. (Redrawn from Harris, 1951.) **F.** *Caytonanthus kochi*, sectioned synangium showing four microsporangia. **G.** *C. kochi*, portion of microsporophyll. (**F,G** redrawn from Harris, 1937.)

a distinct nucellar beak. The occurrence of this plant in Cretaceous sediments of South America emphasizes the global distribution of complex gymnosperms with enclosed seeds that characterize evolutionary radiations in the Mesozoic.

Peltaspermaceae

Upper Triassic beds of Natal, South Africa, and of Greenland have provided the vegetative and reproductive remains of this family. Instead of having several organ genera representing the parts of a plant, one genus, *Lepidopteris,* is designated for the whole plant. This is possible only because the separate parts, including stem fragments, leaves, ovulate discs, and

Figure 26.3. **A.** *Umkomasia macleani.* Megasporophyll with cupules. **B.** *Pilophorosperma* sp. Cupules with bifurcate micropyles projecting from ovule, and projecting seed. **C.** *Pteruchus africanus,* microsporophyll. **D.** Pinna of *Dicroidium lancifolia.* (**A–D** redrawn from Thomas, 1933.) **E.** *Pachypteris papillosa,* pinna. (Redrawn from Harris, 1964.) **A–E:** Jurassic.

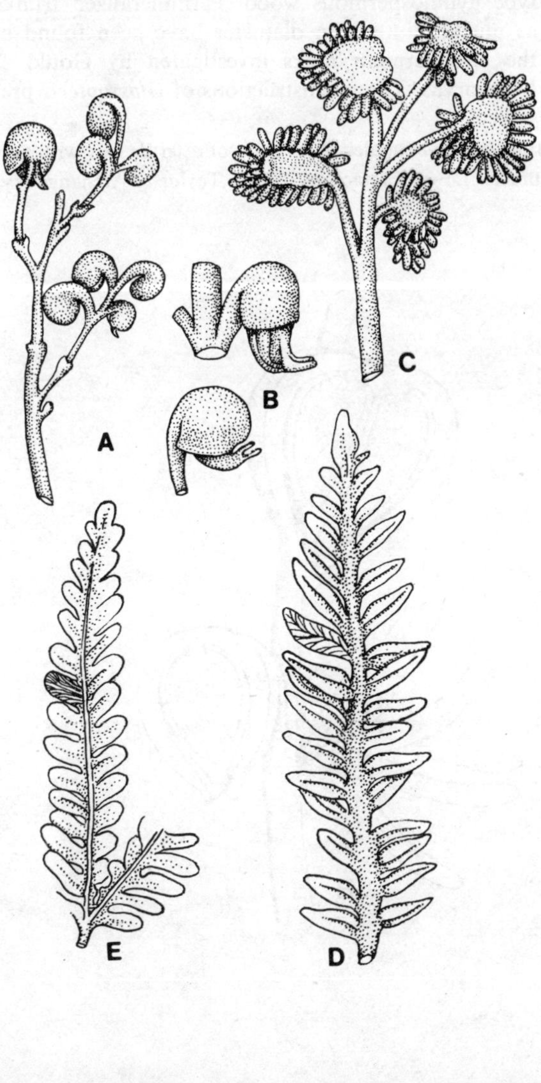

microsporangiate organs, have characteristic blisterlike swellings and similar stomatal structures.

Foliage

The frond of *Lepidopteris* (Fig. 26.5A) is a bipinnate structure about 30 cm long. The pinnules are attached by a broad base and the venation is the open type. They resemble those of *Callipteris* but are distinguished by the blisterlike swellings on the rachis and pinnules.

Microsporophylls

The largest specimen of a *Lepidopteris* microsporophyll found by Harris (1932) is 9 cm long. It is a pinnately divided axis with short branches at intervals of 1 to 2 cm. The branches subdivide into two or three ultimate branches that lie at right angles to the rest of the microsporophyll. Each ultimate branch bears two rows of pendant microsporangia. In general appearance the microsporophyll of *Lepidopteris* most resembles the pteridosperm *Crossotheca.*

Cupulate disc

The ovules are borne on stalked peltate discs, which have been found attached in a pinnate arrangement to a central axis (Fig. 26.5B). Each disc is about 1.5 cm in diameter and has a circle of about 20 cavities on the underside. The cavities represent scars where ovules were attached (Fig. 26.5C). The ovules are about 7 mm in length and, like ovules of the corystosperms, they have curved micropylar beaks. The internal structure of the ovule is not well enough preserved to make valid comparisons with ovules of other gymnosperms. Harris (1932) suggests, however, that the peltate disc of *Lepidopteris* "may be remotely comparable with a pteridosperm cupule with several seeds." If it is a primitive cupule (Delevoryas, 1962), it is conceivable that the cupules of corystosperms and *Caytonia* evolved from peltate ovulate structures like those of *Lepidopteris* by recurvation of its margins so as to enclose the ovule or ovules. It is equally plausible, however, that the ovulate structures of both corystosperms and peltasperms represent specializations of Paleozoic pteridosperm cupules. It is clear that the cupules of Paleozoic seed ferns were derived by the coalescence of telomes, whereas those of Mesozoic gymnosperms apparently were not. The latter cupules replace pinnae or pinnules on planar fronds. It is, therefore, likely that the cupules and cupulelike structures found in the Paleozoic and Mesozoic gymnosperms originated separately in response to the evolutionary advantages associated with enclosure of the ovules and seeds.

Glossopteridales

As another author observed, it would take a whole book to adequately cover the information that has accrued since the name *Glossopteris* was first applied to certain tongue-shaped leaves (Fig. 26.6A) by Brongniart in 1828. The entire linear leaves have a prominent midrib, especially near the base of the leaf, and a reticulate venation in that part of the lamina adjacent to the midrib. Leaves of this type are abundant in Permian and Triassic sedimentary rocks of the Southern Hemisphere. A notable exception is reported by Delevoryas (1969), who found leaves with all of the characteristics of *Glossopteris* in Jurassic deposits of Mexico. Other than this unusual Mexican find, *Glossopteris* occurs in India, Australia, South Africa, South America, and Antarctica (Cridland, 1963; Surange, 1966; Schopf, 1967) and is the primary constituent of the Permian Gondwanaland floras.

Presently over 50 species of *Glossopteris* have been described, many of which, no doubt, are variants of a single species. Surange (1966) makes the valid point, however, that we might except a fair degree of speciation during the 50 m.y. that glossopterids were on Earth.

Since the middle of this century there has been a revival of interest in the Glossopteridales stimulated by the discovery of ovule and pollen organs attached to leaves, and by the unexpected discovery of permineralized glossopterid remains in Antarctica (Schopf, 1970) and Australia (Gould & Delevoryas, 1977). These show structural details that could not be determined from observations of the usual compression–impression fossils. Based on the accumulated evidence we now have detailed reconstructions of whole *Glossopteris* plants prepared by Gould & Delevoryas (1977). More recent studies of the internal anatomy of two distinctly different species of *Glossopteris* from Antarctica (Pigg & Taylor, 1989) demonstrate greater anatomical variability than had been anticipated, and suggest that glossopterids may be as diverse as previously suspected. Another useful article summarizing the recent information about the *Glossopteris* plant was prepared by Pant (1977). Review articles by Surange & Chandra (1972) and Schopf (1976) are confined to descriptions, evaluations, and hypotheses about *Glossopteris* reproductive structures and their evolution.

Habit of the plant: Glossopteris

It has been suggested that *Glossopteris* was an herb, shrub, or small tree. Evidence accumulated by Pant (1977) and Gould & Delevoryas (1977) shows clearly, however, that *Glossopteris* was a large tree (Fig. 26.6B) with a substantial trunk of *Araucarioxylon*-type gymnospermous wood. Permineralized trunks as much as 40 cm in diameter have been found in the late Permian rocks investigated by Gould & Delevoryas. The reconstructions of *Glossopteris* pre-

Figure 26.4. **A.** Reconstruction of *Ruflorinia* frond bearing *Ktalenia* cupules. **B.** Portion of a fertile axis with two cupules containing one or two ovules and a cluster of bracts. Lower Cretaceous. (From Taylor & Archangelsky, 1985, with permission.)

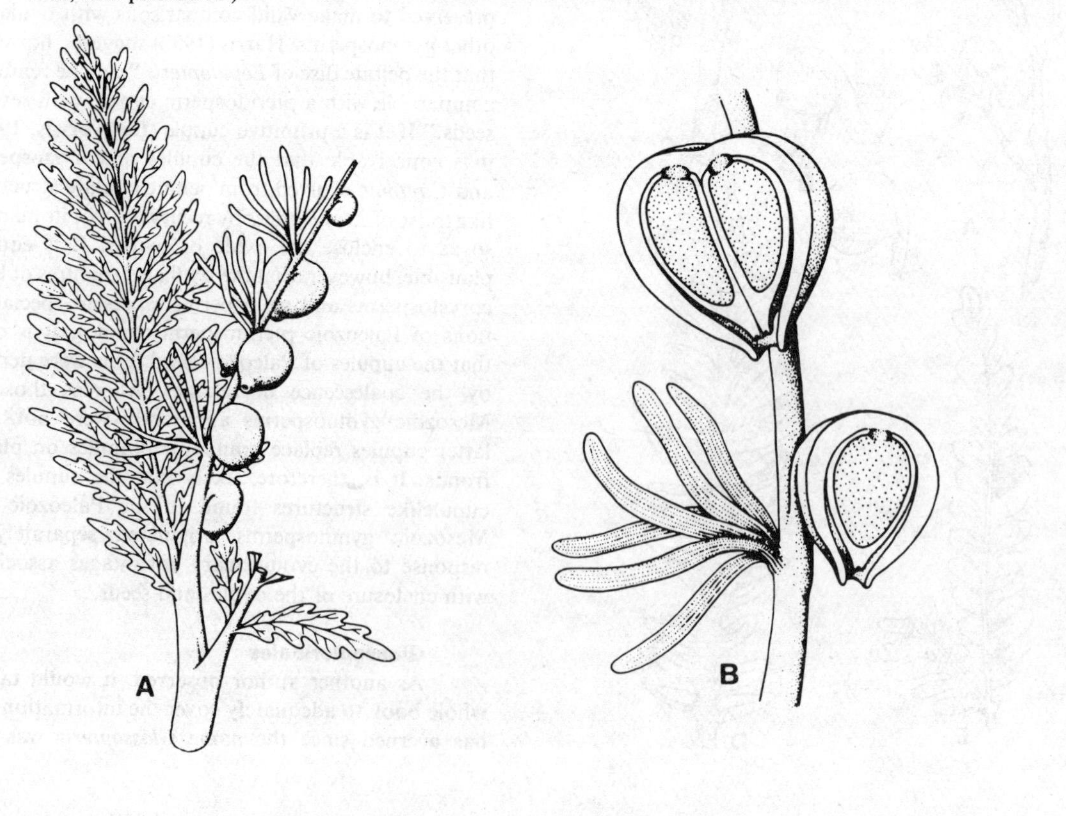

pared by these authors show trunks about 6 m tall supported by a root system of the *Vertebraria* type. Conspicuous growth rings are present in roots, trunks, and branches. Pant (1977), after reviewing the literature on leaf attachment, tells us that leaves were in spirals or whorls probably on short shoots.

Structural features: stems and roots

Following the discovery of permineralized plant remains from the Upper Permian containing the silicified remains of *Vertebraria,* Gould (1975) was able to verify that this enigmatic genus represents the detached roots of *Glossopteris* plants. Specimens of the *Vertebraria* are easy to recognize even as casts or compression–impression fossils because of the characteristic wedgelike sectors radiating from the center of the axis (Fig. 26.6C). It turns out that the wedges are composed of secondary xylem arranged around a central polyarch protostele. The protoxylem strands alternate with the radiating arms of secondary xylem. As far as Gould could determine, the cavities between the radiating arms were not filled with tissue at an earlier stage of development. This suggests that *Vertebraria* was the root system of a plant that grew in a semiaquatic environment. Some of the roots have pycnoxylic wood up to 9 cm thick developed around the outside of the centrally located cavities. That the roots branched frequently is indicated by numerous branch-root traces and an abundance of smaller roots in the rock specimens. Upon examining the secondary xylem of *Vertebraria,* Gould observed that it is exactly the same in structure as the pycnoxylic wood of *Araucarioxylon arberi* trunks. Both *Vertebraria* and *A. arberi* have secondary xylem tracheids with opposite to alternate circular bordered pits in up to five or six vertical rows. In some, the pits tend to be in groups of two to seven and often crowded on the radial walls so that the pits are hexagonal in outline. The rays are uniseriate, 1 to 20 cells high. These observations leave no doubt that the silicified roots of *Vertebraria* and stems of *A. arberi* are congeneric.

Figure 26.5. **A.** Portion of frond of *Lepidopteris* sp. **B.** Ovulate cone of *Lepidopteris* with spirally arranged cupulate discs. (**A,B** redrawn from Thomas, 1955.) **C.** *L. ottonis,* ovulate disc with pendant ovules. (Redrawn from Harris 1932.) **A–C:** Jurassic.

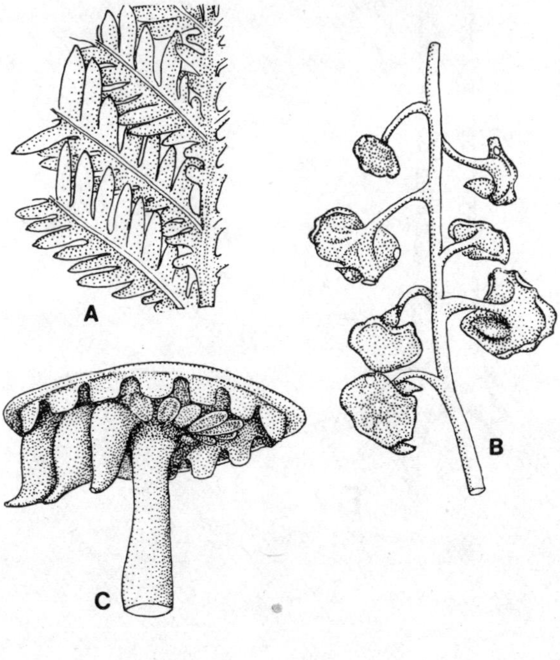

The discovery of silicified peat in Antarctica (Schopf, 1970) has provided us with an opportunity to see the anatomical details of Gondwana plants that rival those found in Carboniferous coal balls from North America and Europe. In her studies of permineralized *Glossopteris,* Pigg (1988) discovered that the stems bore both foliar leaves and small-scale leaves. The stems are typical of gymnosperms, with abundant pycnoxylic wood and a pith surrounded by a ring of primary bundles. The wood of these specimens is incompletely preserved, but is generally similar to the *Araucarioxylon* described from the Australian deposits.

Leaves

For many years species of *Glossopteris* leaves were identified on the basis of external form and venation pattern, but we now have internal anatomical information about species from both Antarctica and Australia. We all know that leaves of a given species, even from the same plant, can show considerable variation in these characteristics. It has been demonstrated that the various specimens of a species of *Glossopteris* can have the same venation pattern and form but different epidermal patterns. This means that investigators must consider cuticular structure as well as external morphology to fully identify and characterize glossopterid leaves. *Glossopteris* leaves are characteristically hypostomatic (stomata confined to the lower surface). The epidermal cells may be straight-walled or sinuous. The stomata are haplocheilic and irregularly placed between the veins.

Gould & Delevoryas (1977) and Pigg (1988) have demonstrated that internal leaf structure consisted of a heavy midrib composed of several vascular bundles from which laterals diverge. The laterals anastomose frequently, forming elongated reticulations termed meshes. In two Antarctic species the meshes are elongated longitudinally near the midrib while more peripherally, they are oriented toward the leaf margin (Fig. 26.6A). The Australian species shows differentiation of palisade and spongy mesophyll (Gould & Delevoryas, 1977). Palisade and spongy mesophyll are not differentiated in the Antarctic

Figure 26.6. **A.** *Glossopteris* sp., reconstruction of leaf showing conspicuous midrib and reticulate venation. **B.** Reconstruction of *Glossopteris* tree about 4 m tall. (Based on Gould & Delevoryas, 1977.) **C.** Diagram, transverse section *Vertebraria*. (Based on Gould, 1975.) A–C: Permian.

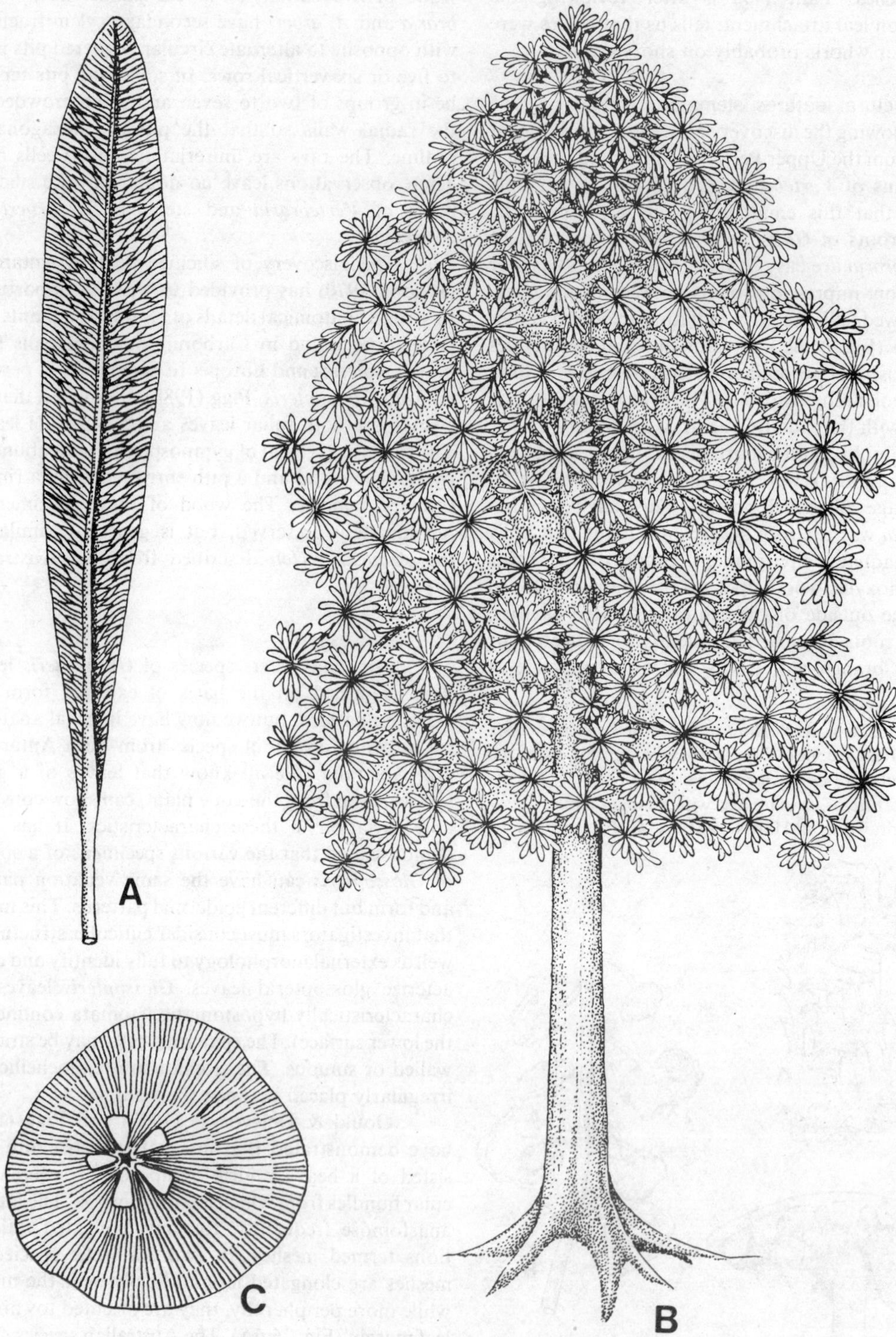

species, but there is a well-developed hypodermis, as well as bundle sheaths around the vascular tissues of the veins. Although there is a great deal of variation among species, the permineralized specimens are hypostomatic, with stomatal complexes randomly distributed within the meshes.

Pollen organs

What are now believed to be the pollen organs of *Glossopteris* (Surange & Maheshwari, 1970; Lacey, van Dijk, & Gordon-Gray, 1974, 1975; Gould & Delevoryas, 1977) have been recovered from Permian deposits of South Africa, India, and Australia. Some well-preserved specimens from Natal (Lacey et al.) confirm the reconstruction of the microsporophyll *Eretmonia* (Fig. 26.7A) by Surange & Maheshwari as a unit having an expanded, more-or-less triangular distal lamina on a stalk. Slightly beyond the midpoint on the adaxial surface of the stalk, two branches arise (Holmes, 1973) bearing whorls of *Arberiella*-type sporangia (Fig. 26.7B). The specimens examined by Holmes from Australia and by Lacey et al. from South Africa have the advantage of showing venation patterns in the distal lamina and some epidermal features that leave little doubt about their glossopterid affinities. Gould & Delevoryas (1977) reported an abundance of *Arberiella* sporangia and distinctive pollen grains (Fig. 26.7C) dispersed among the *Glossopteris* remains of their permineralized specimens.

Another pollen organ with clusters of *Arberiella* sporangia is called *Glossotheca* (Surange & Maheshwari, 1970). It is a spatulate foliar unit with a single branch arising from the vicinity of the midvein. According to the reconstruction, the branch bears paired (pinnate) laterals that terminate in clusters of sporangia.

Figure 26.7. **A.** *Eretmonia* sp., fertile leaf bearing two clusters of microsporangia. (Redrawn from Surnage & Maheshwari, 1970, and Pant, 1977.) **B.** *Arberiella,* dispersed microsporangia of glossopterids. **C.** *Striatites*-type dispersed pollen of *Glossopteris.* (**B,C** redrawn from Pant, 1977.) **A–C:** Permian.

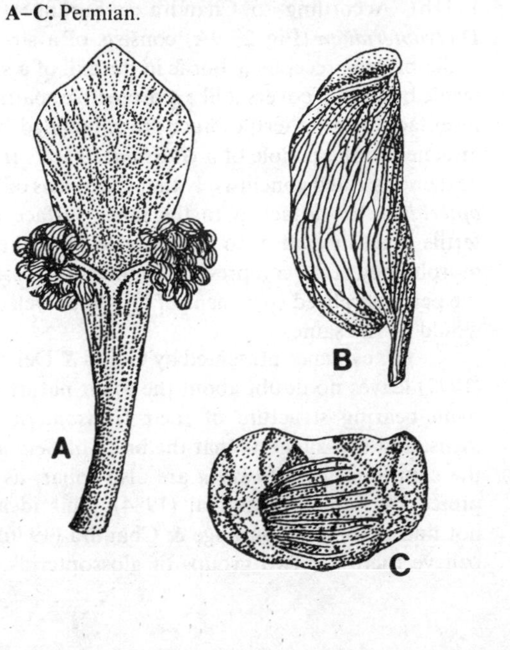

Ovulate structures

There has been more disagreement about how to interpret the ovulate structures of glossopterids than any other fossil plant organ that we know. This is, in part, the result of the very nature of compression–impression fossils where the position of cleavage through the specimen and the amount of coalified remains may determine what is observed on the negative (impression) or the positive (compression) surfaces. The controversy was initiated by the publication of two stimulating papers by Plumstead (1952, 1956). These represent early interpretations of the fructifications attached to leaves of *Glossopteris*. Briefly, Plumstead envisaged the detached organs she named *Scutum* as a bivalved cupule, not unlike the two halves of a mussel shell, attached by a pedicel to the midrib of the leaf. According to Plumstead, that half of the cupule nearest the adaxial surface of the subtending leaf bore carpels, while the other half was believed to

Figure 26.8. **A.** *Ottokaria bengalensis* showing capitulum and stalk adnate to subtending leaf. (Modified from Schopf, 1976.) **B.** *Glossopteris* (*Dictyopteridium*) showing relationship of ovuliferous capitulum to subtending leaf. **C.** Transverse section through ovule-bearing capitulum of *Glossopteris*. **D.** Longitudinal section of single ovule from *Glossopteris* capitulum. (**B–D** redrawn and modified from Gould & Delevoryas, 1977.) **A–D:** Permian.

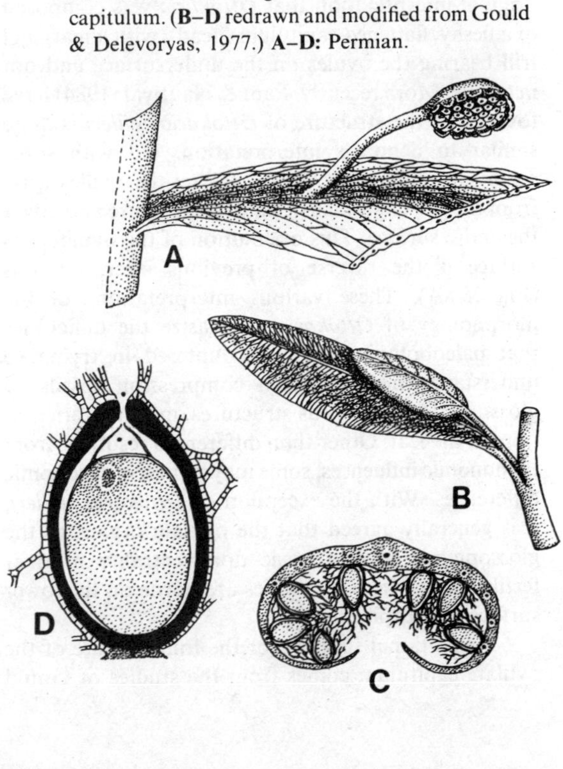

have borne microsporangia. The suggestion that carpels were borne by the reproductive structures of *Glossopteris* and that the affinities of the genus are with the angiosperms has been seriously questioned. A thorough discussion of counterproposals to Plumstead's interpretations is given by Schopf (1976) and will not be repeated here.

Agreement has been reached that the pedicel bearing the ovulate structure is axillary to a *Glossopteris* leaf (Fig.26.8B). In some cases, however, the axillary pedicel is adnate to the midrib of the subtending leaf for some distance before becoming free. Following the lead of earlier workers who evaluated Plumstead's interpretations, Surange & Maheshwari (1970) determined that the structure at the end of the pedicel of *Scutum* was a cone of spirally arranged ovules. Later, Surange & Chandra (1972) showed a species of *Scutum* having a bilateral receptacle bearing the ovules, covered on one side by a protective scale leaf with *Glossopteris*-type venation. Both receptacle and its subtending covering scale are borne on the pedicel.

Detached ovules with short stalks and bifid micropylar ends are assigned to the genus *Indocarpus*. Because of their association and abundance in the Permian beds of India that contain abundant *Glossopteris* remains, and because of their small size, these and other ovules (*Stephanostoma* and *Pterygospermum*) are thought to belong to the glossopterids.

The idea that *Glossopteris* bore ovules in cones has been extended by Surange & Chandra to the Lower Permian fructification *Ottokaria* (Fig. 26.8A), an ovulate fructification attached to the foliage of *Gangamopteris*. Schopf (1976), however, follows the original interpretation that *Ottokaria* was composed of a fleshy, flattened capitulum (head), with a marginal frill bearing the ovules on the undersurface and not in a cone. More recently, Pant & Nautiyal (1984) have found that the structure of *Ottokaria zeilleri* is quite similar to Schopf's interpretation, but with some surprising differences. In *O. zeilleri* the ovules arise from the upper surface of an oval structure and cover the entire surface. This orientation of the ovuliferous surface is the reverse of previous interpretations (Fig. 26.8A). These various interpretations of the morphology of *Ottokaria* emphasize the difficulties that paleobotanists have encountered in trying to understand the impression–compression fossils of glossopterid ovuliferous structures and their orientation on the leaf. Other than differences resulting from taphonomic influences, some may represent taxonomic differences. With the exception of *Ottokaria zeilleri*, it is generally agreed that the ovulate organs of the glossopterids have a basic dorsiventrality in their fertile heads where the ovules are borne on the lower surface of a capitulum.

Additional support for the foliar nature of the ovulate capitulum comes from the studies of Gould & Delevoryas (1977) using the permineralized specimens from the Permian of Australia. They established that the seed-bearing organs (Fig. 26.8B) found in their material have the same internal structure as vegetative leaves of *Glossopteris* and are essentially megasporophylls. The ovules are sessile and borne on the inner surface of the megasporophyll along the veins. The megasporophylls, which vary in width from 3 to 11 mm and in length from 10 to 42 mm, have revolute laminar margins that slightly overlap, thus partially enclosing the ovules (Fig. 26.8C) in an unsealed envelope. At maturity, the envelope opened out, allowing the seeds to be shed. The ovoid to pyriform ovules (Fig. 26.8D) are small and numerous, averaging a little over 1 mm in length and slightly more than 0.5 mm in diameter. They are erect, with their micropyles oriented toward the center of the envelope formed by the megasporophyll. The spaces between the ovules are filled with an unusual delicate network of cellular filaments that extends outward from the surface of the integuments. The integument of an ovule is slightly thicker around the micropyle. Its nucellus is free from the integument except for its attachment at the chalazal end. At the distal end, the nucellus is differentiated into a pollen chamber occasionally containing bisaccate pollen grains with transverse striations on the corpus, a characteristic of *Glossopteris* pollen (Fig. 26.7C). Some ovules have been found with megagametophytes containing a single archegonium. None of the ovules, whether shed or in the megasporophyll envelope, has been observed to contain an embryo.

When the general morphology of the permineralized ovulate organs described by Gould & Delevoryas (1977) is considered, one can interpret it as being similar to that of the compression–impression fossils of *Dictyopteridium* (Surange & Chandra, 1971b). According to Chandra & Surange (1975), *Dictyopteridium* (Fig. 26.9A) consists of a strobiloid ovule-bearing receptacle borne in the axil of a stalked fertile bract that covers it like a protective spathe. The receptacle and its fertile bract are subtended by and attached to the petiole of a *Glossopteris* leaf. If it can be demonstrated conclusively that the ovules of *Dictyopteridium* are attached to the inner surface of the fertile bract instead of to a receptacle, then the basic morphology of the compression–impression fossils and the permineralized specimens of Gould & Delevoryas would be the same.

The evidence presented by Gould & Delevoryas (1977) leaves no doubt about the foliar nature of the ovule-bearing structure of their *Glossopteris* specimens, and this suggests that the bract of *Scutum* and the capitulum of *Ottokaria* are also foliar, as interpreted by Pant & Nautiyal (1984). This idea does not find favor with Surange & Chandra (1976), who believe there are two groups of glossopterids, those with strobiloid receptacles and those with ovules borne on modified leaves. In the latter category, the genus *Lidgettonia* (Fig. 26.9B) is a good example, which all agree is a foliar ovule-bearing fructification.

The fertile leaves of *Lidgettonia* described by Thomas (1958) from South Africa are short, spatulate, and with a rounded apex. The veins spread from the base (petiole) of the leaf into the lamina where they fork and anastomose. The petiole of the leaf bears two rows of disclike capitula on the adaxial surface. Each

Figure 26.9. **A.** *Dictyopteridium* sp., seen from the underside showing ovules partially enveloped by the revolute margins of the capitulum. (Modified from Surnage & Chandra, 1975, and Gould & Delevoryas, 1977.) **B.** *Lidgettonia mucronata*. (Redrawn from Surange & Chandra, 1972.) **C.** *Denkania indica*. (Redrawn from Surange & Chandra, 1971a). A–C: Permian.

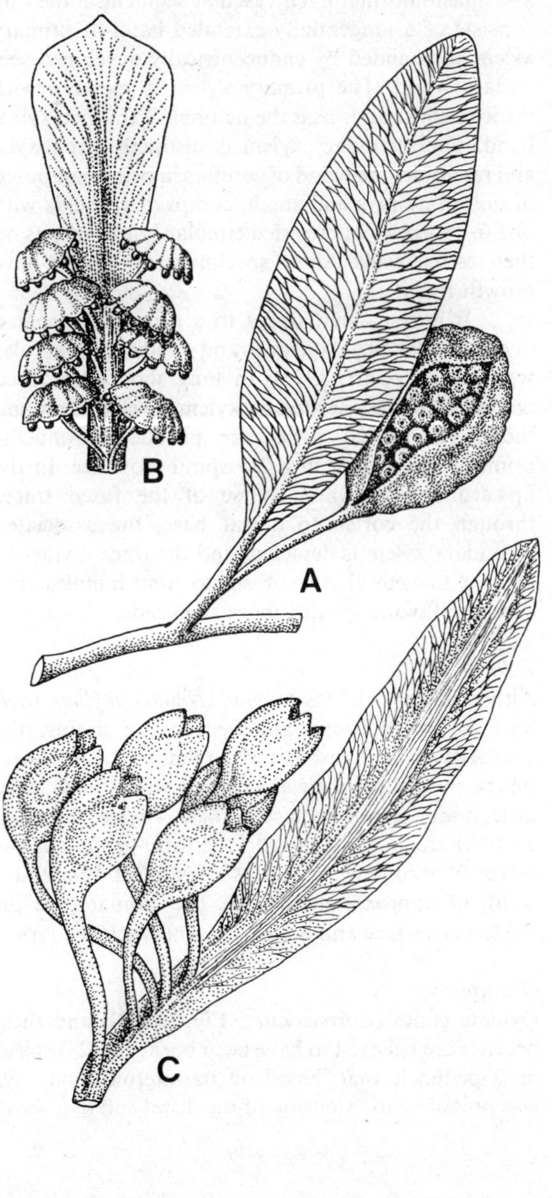

capitulum has a stalk arising from the side, which connects it with the leaf. There are three to four stalked capitula in each of the two rows. In its structure, the stalked capitulum of *Lidgettonia* is reminiscent of the homologous structure of *Ottokaria*. Instead of many ovules condensed on the capitulum, as in *Ottokaria*, the six to seven ovules of *Lidgettonia mucronata* (Surange & Chandra, 1972) are borne in a row on the surface of the capitulum near the undulating margin.

The paleobotanists of the Sahni Institute have described many other fructifications they attribute to the glossopterids, some of which appear to be strobiloid and others that are definitely foliar. One that deserves special attention, but seldom is mentioned in the literature, is *Denkania indica* (Surange & Chandra, 1971a). In its general appearance, *Denkania* (Fig. 26.9C) reminds one of *Lidgettonia indica* where there is a single row of ovules on the surface. *Denkania* also has pedicels arising in a row from the adaxial surface of the subtending glossopterid leaf. The most plausible interpretation of the capitulum, according to Surange & Chandra, is a cupule containing a single ovule. It seems unlikely that the structure terminating the pedicel is a naked ovule, as the impression of an ovule can be seen inside what appears to be a cupule. Another interpretation, suggested by material from South Africa, places the single ovule at the end of the pedicel associated with a bract or scale. From our observations, we are inclined to favor the idea that *Denkania* is cupulate. It takes very little imagination to believe that such a structure was derived from the *Lidgettonia* type capitulum. This idea finds a parallel in the possible evolution of the *Caytonia* cupule from the "cupulate disc" of *Lepidopteris* and partially developed cupules of the corystosperms. If these interpretations prove to be correct, then we have evidence that the structure we call the cupule has evolved independently three different times, first in the lyginopterid pteridosperms, then the Caytoniales (sensu lato), and the Glossopteridales. Based on the observable diversification of Mesozoic gymnosperms, parallel evolution of cupules and other reproductive organs is not unexpected.

A model of the evolution of the cupule in the Glossopteridales includes the following trends:

1. The tendency to cover a large number of ovules on the surface of a megasporophyll (capitulum) by inrolling of the margins (*Glossopteris, Ottokaria*)
2. Reduction of ovules from a large number more or less condensed on the surface of the megasporophyll (*Glossopteris, Ottokaria*), to a few ovules in a row (*Lidgettonia*).
3. Reduction of ovule number to one and enclosure by the revolute margins of the capitulum to form a cupule (*Denkania*).

If we can extrapolate from *Glossopteris,* where Gould & Delevoryas (1977) have given us unequivocal evidence of the foliar nature of the capitulum, then it appears that relationships between the Glossopteridales and Pteridospermopsida, where ovules are borne on leaves, is considerably strengthened. This is contrary to Schopf's hypothesis (1976) that the ovulate structures of glossopterids were derived from cordaitaleans where ovules are borne on axillary shoots, not on megasporophylls.

Pentoxylales

The permineralized specimens of *Pentoxylon,* which originated from the Lower and Middle Jurassic beds in the Rajmahal Hills of India, were first described by Srivastava (1935, 1937, 1946), who also established the cone genus *Carnoconites.* The organ genus for leaves of the Pentoxylales, *Nipaniophyllum* was made official by Sahni (1948), while Vishnu-Mittre (1952) proposed the name *Sahnia* for specimens of the pollen-producing organs. The best summarizing paper for the Pentoxylales, exclusive of the pollen organs, is that prepared by Sahni (1948).

Figure 26.10. *Pentoxylon sahnii.* Permian. **A.** Reconstruction of shoot with short shoot bearing leaves. (Modified from Sahni, 1948.) **B.** Reconstruction of short shoot bearing ovulate cones. **C.** Longitudinal section of *Carnoconites* ovule. (**C,D** redrawn from Sahni, 1948.)

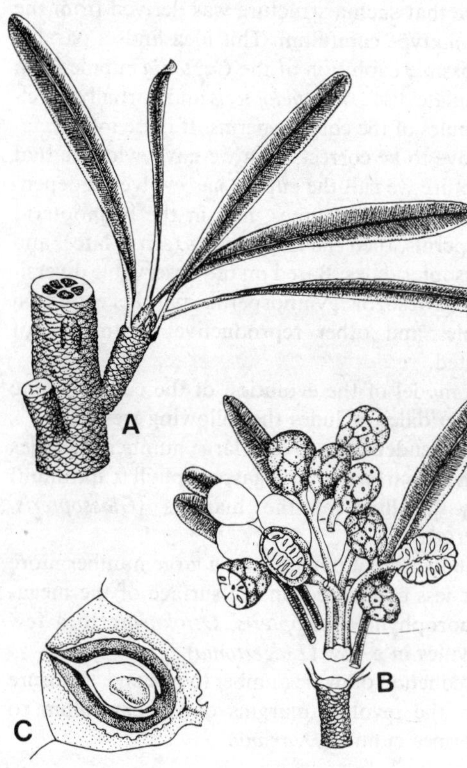

Pentoxylon

In his reconstruction of *Pentoxylon* (Fig. 26.10A), Sahni (1948) illustrated an axis about 1 cm in diameter bearing short shoots. The latter are terminated by a crown of spirally arranged *Nipaniophyllum*-type leaves. An armor of leaf bases, also spirally arranged, covers the short shoots and the long shoots as well (Fig. 26.11A). Some short shoots were terminated by clusters of ovulate cones (Sahni, 1948) or pollen organs (Vishnu-Mittre, 1952).

Stems

As suggests by the name of the organ genus *Pentoxylon,* the long shoots have an internal vascularization that frequently exhibits five segments of vascular tissue (Fig. 26.11B) arranged in a ring around a central ground tissue. The number of vascular segments varies, however, from 4 to 10 as a result of branching and anastomosing. Each vascular segment in the ring consists of a tangentially extended band of primary xylem surrounded by endocentrically developed secondary wood. The primary xylem is mesarch with protoxylem strands near the margins of the metaxylem band. The secondary xylem is distinctly pycnoxylic and resembles the wood of conifers in being composed of uniseriate rays and small, compact tracheids with one to two rows of crowded circular bordered pits on their radial walls. Some specimens appear to have growth rings.

It appears that traces to a leaf arise in pairs, one from each protoxylem strand of adjacent vascular segments (Stewart, 1976). In long shoots the fused pair of traces have secondary xylem developed around them. The paired traces are produced from the primary vasculature in a 2/5 spiral sequence. In the upward and outward course of the fused traces through the cortex to a leaf base, the associated secondary xylem is depleted, and the trace divides to form a tangential row of six to nine bundles that compare favorably with those of cycads.

Foliage

Although leaves of *Pentoxylon* (*Nipaniophyllum raoi*) have not been found attached to the shoots, the similarity of their vasculature to that of the leaf scars on the shoots and their association with shoots leave little doubt about their affinities. The leaves (Fig. 26.10A) are highly similar to the form genus *Taeniopteris. Nipaniophyllum* attains a length of 7 cm and a width of approximately 1 cm. The stomata are on the lower surface and are of the syndetocheilic type.

Ovulate cones

Ovulate cones (*Carnoconites,* Fig. 26.10B) and their pedicels are believed to have been borne in a 2/5 spiral on a peduncle that, based on its internal anatomy, was probably an extension of the distal end of a short shoot (Sahni, 1948). Although some pedicels remained unbranched and bore a single cone, some dichotomized with each branch bearing a cone. As described by Sahni, in life the cones might have resembled a cluster of fleshy "mulberry fruits." In the reconstruction (Fig. 26.10C) the ovules are shown to be spirally arranged and crowded on a central cone axis. The sessile ovules are erect, that is, their micropyles are directed outward. The integument is composed of a thick sarcotesta and an inner sclerotesta that in cross section gives the ovule a platyspermic appearance. The nucellus is free from the integument down to the base. Each cone axis has a ring of vascular bundles from which arise traces supplying the chalazal end of each ovule.

Microsporangiate organs

Sahnia, the microsporangiate organs of the Pentoxylales, bears some resemblance to the microsporangiate structure of cycadeoids, especially those species of *Weltrichia* that have their whorl of microsporophylls fused at the base to form a dish-shaped structure. *Sahnia* was found by Vishnu-Mittre (1952) attached to a broad conical receptacle at the tip of a dwarf shoot showing characteristics of *Pentoxylon*. The free ends of the microsporophylls comprising the whorl are filiform and bear short spirally arranged branches. Microsporangia are borne singly or in groups at the tips of these branches. Monocolpate pollen of the cycadophyte type is attributed to *Sahnia*.

Classification and relationships

If we agree that the reconstruction of *Pentoxylon*, based on the various parts described above, is accurate, then we are faced with a real dilemma when we attempt to establish the relationships of the plant. Consider that here we have a plant with coniferous wood organized into vascular segments; the leaves share characteristics with cycads and cycadeoids; the microsporangiate organs resemble cycadeoids in gross morphology; the pollen is cycadophytic; and the ovulate cones are unlike those of any gymnosperm. In their recent numerical cladistic analyses of seed plants, both Crane (1985, 1988) and Doyle & Donoghue (1986) consider the Pentoxylales to be closely related to the Cycadeoidales (Chapter 25), interpretations that are reasonable within the context of the large number of questions that remain to be answered about features of both groups. We agree with Andrews (1961), who, when confronted with the same problem of classifying *Pentoxylon*, concluded, "... it seems most appropriate to regard the Pentoxyleae as a wholly distinct group of plants." As for other groups of plants (cycadeoids, for example), *Pentoxylon* seems to represent another evolutionary experiment concluded by extinction.

As far as a possible origin of *Pentoxylon* is concerned, the Triassic and Jurassic genus *Rhexoxylon* deserves some attention. Fragments of trunlike stems of *Rhexoxylon* have been found that are as much as 3.3 m long and 12.5 cm in diameter. In most specimens of *Rhexoxylon* there is a ring of vascular segments with well-developed centrifugal secondary xylem to the outside of the primary xylem and a small amount of secondary xylem developed centripetally. The secondary wood is pycnoxylic and identical with that

Figure 26.11. *Pentoxylon sahnii.* Jurassic. **A.** Fragment of a long shoot showing spirally arranged leaf bases. **B.** Transverse section of long shoot showing five vascular segments (vs) alternating with leaf traces (lt). (**B** from Sahni, 1948.)

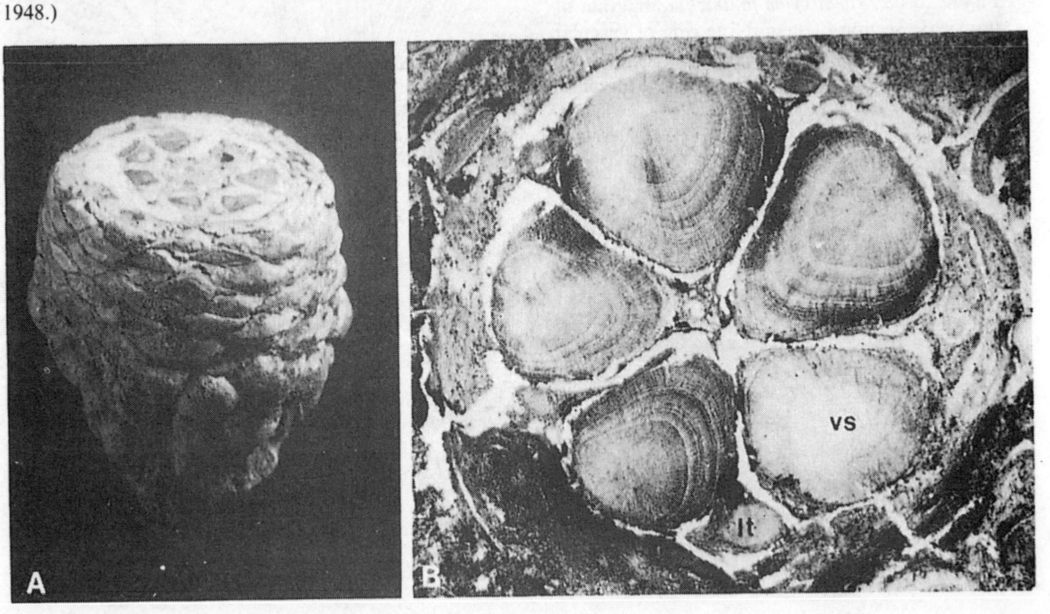

of *Pentoxylon*. The vascular segments of *Rhexoxylon piatnitzkyi* (Fig. 26.12), according to Archangelsky & Brett (1961), are highly similar to those of *Pentoxylon* in their organization. Their description of leaf trace origin in *R. piatnitzkyi* suggests that it is exactly the same as in *Pentoxylon*. Based on these observations of similarities of their vegetative structures, it seems plausible that some relationship can be assumed. The origin of *Rhexoxylon*, although once suggested to be from medullosan pteridosperms, is a moot question.

Czekanowskiales

Earlier classifications indicate an affinity of the Czekanowskiales with the Ginkgoales because it was believed that one of the two principal genera, *Czekanowskia* (Fig. 26.13A), had the characteristics of a primitive ginkgophyte leaf. This genus, which occurs in Jurassic and Cretaceous rocks, has leaves that are split into filiform segments. The leaves, each of which has only a single vein, are borne in clusters at the end of a short shoot with a scaly base. The whole unit must have looked like the short shoot of a pine with its needlelike leaves. Harris (1976) notes that, unlike *Ginkgo* where the leaves are shed singly, the short shoot with its attached leaves were shed intact. These and other differences suggest that *Czekanowskia* probably has nothing to do with the ginkgophytes.

In 1951, Harris, relying on the usual evidence from repeated association of parts and agreement in cuticular structure, showed that *Czekanowskia* and the ovulate structure *Leptostrobus* belonged to the same plant. *Leptostrobus* (Fig. 26.13B) has spirally arranged scales around the base of a long, slender axis bearing the ovulate structures, which Harris calls capsules. Each capsule is composed of two valves like mussel shells with each lobed valve containing a row of small ovules in a concavity (Fig. 26.13C). If the valves were too close, the ovules would be completely enveloped, an idea that suggests a bicarpellate ovary of an angiosperm. Pollen grains have been found on small hairlike cells along the suture between the valves. Although this implies an angiosperm affinity for the Czekanowskiales, Harris (1976) adopts a point of view with which we concur. He states, "Angiosperm affinity is indeed possible, but for my part I at present regard *Leptostrobus* as one of several Mesozoic gymnosperms that had approached the closed and protected ovary from different starting points; others are Bennettitales (cycadeoids), *Caytonia* and many conifers."

Gnetales

This order comprises three very different but apparently related genera of extant seed plants with uncertain affinities. These genera are *Ephedra, Gnetum,* and *Welwitschia*. Based on their general morphologies and habitats, one would hardly guess that the plants making up this disparate group are related. Many studies supported by cladistic analyses indicate that

Figure 26.12. *Rhexoxylon piatnizkyi*, diagram of transverse section of stem. Note ring of vascular segments. Protoxylem (circles); metaxylem (black); secondary xylem (hatched). Triassic. (Redrawn from Archangelsky & Brett, 1961.)

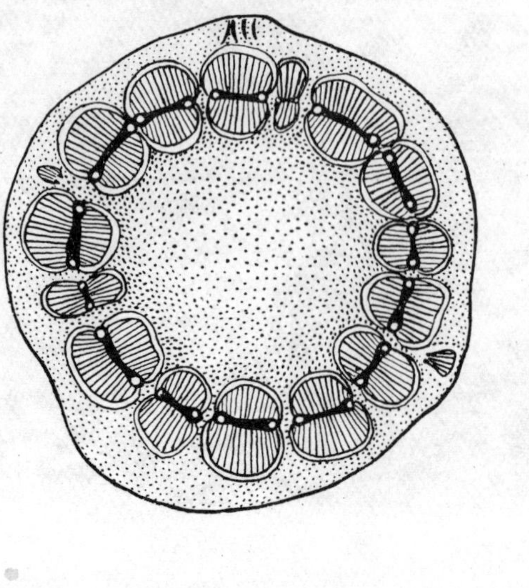

Figure 26.13. **A.** *Czekanowskia* sp., shoot with leaves. (Redrawn from Harris, 1935.) **B.** *Leptostrobus longus*, shoot with ovule-bearing "capsules." **C.** *L. longus*, section through capsule showing position of ovules. (**B,C** redrawn from Harris, 1951.) **A–C:** Jurassic–Cretaceous.

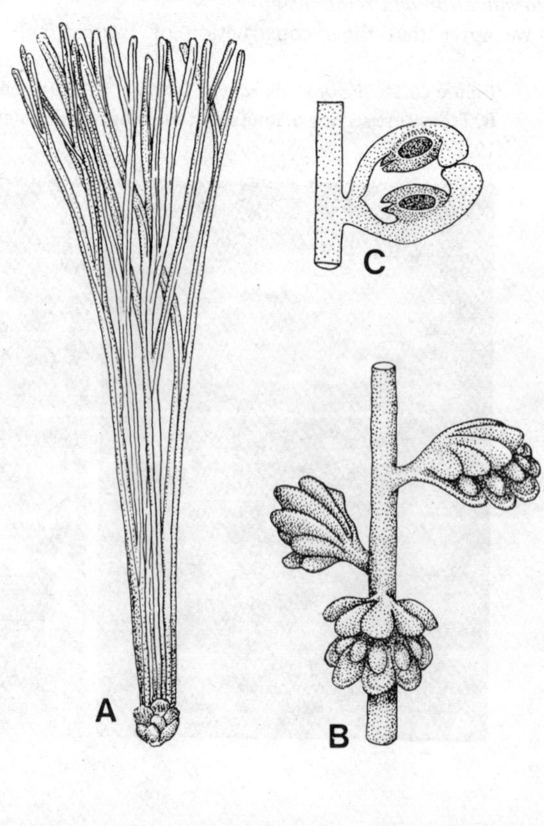

the three genera are indeed related. Of particular importance to our later discussion of angiosperm origins are the characteristics of the Gnetales, many of which are shared with the flowering plants. These characteristics are vessels in the wood; compound strobili made up of small flowerlike units with either a perianth and a whorl of more-or-less fused microsporophylls or a terminal ovule surrounded by one or more additional envelopes; reduction of the megagametophyte (no archegonia) and cellular embryogeny in *Welwitschia* and *Gnetum*; and dicotyledon-like leaves in *Gnetum*. Based on these and other similarities, such as siphonogamy and leaves with several orders of reticulate venation, many authors have suggested that some kind of relationship exists between the Gnetales and the angiosperms (Doyle & Donoghue, 1986; Crane, 1988; Friis & Endress, 1990).

Because of this possible relationship, the fossil record of the Gnetales has attracted much attention even though the accumulated evidence is sparse. Relative to the abundance of megafossils, microfossils in the form of ephedroid pollen grains occur with some frequency from the Upper Triassic to the Tertiary and are particularly abundant during the mid-Cretaceous. These striate, ribbed pollen grains (Fig. 26.14) are similar to those of *Ephedra* and *Welwitschia*. They are reported to first occur in the Middle Permian of North America (Crane, 1988); however, Friis & Endress (1990) maintain that the earliest record of striate ephedroid pollen comes from the Upper Triassic. Of the several fossil genera exhibiting the features of ephedroid pollen, *Ephedripites* and *Equisetosporites* (Fig. 26.14) have been most thoroughly investigated (Pocock & Vasanthy, 1988).

Ephedroid pollen of the *Equisetosporites* type has been found in the male cones of *Masculostrobus* from the Upper Triassic. The systematic position of this genus, however, is in question. *Masculostrobus* has been found in association with foliar shoots and ovulate organs of the genus *Dechellyia* (Ash, 1972), which was described as an unusual conifer of uncertain affinities. More recent studies of foliar and reproductive shoots of *Drewria* (Fig. 26.15) (Crane & Upchurch, 1987) reveal characteristics similar to those of *Dechellyia*, in which leaves are attached in opposite and decussate pairs, have flattened linear vegetative leaves with a thickened midrib, and have sporophylls with two longitudinal ribs. The foliar venation of *Drewria* compares favorably with that found in the cotyledons of *Welwitschia*. In addition, the structure of the seed-bearing, dichasially arranged inflorescences and their reproductive structures surrounded by pairs of bracts are similar to members of the Chloranthaceae and Gnetales (Crane & Upchurch, 1987). When all characteristics are considered, the authors conclude that *Drewria* is more closely related to the fossil and extant Gnetales than to any other group. It has been suggested that the vegetative and seed-bearing parts *Drewria* may be parts of the *Masculostrobus* plant with its probable gnetalean affinities.

Figure 26.14. *Equisetosporites chinleanthus* grain in longitudinal view. Note the obliquely oriented psilate bands. Upper Triassic. (From Pocock & Vasanthy, 1988, with permission.)

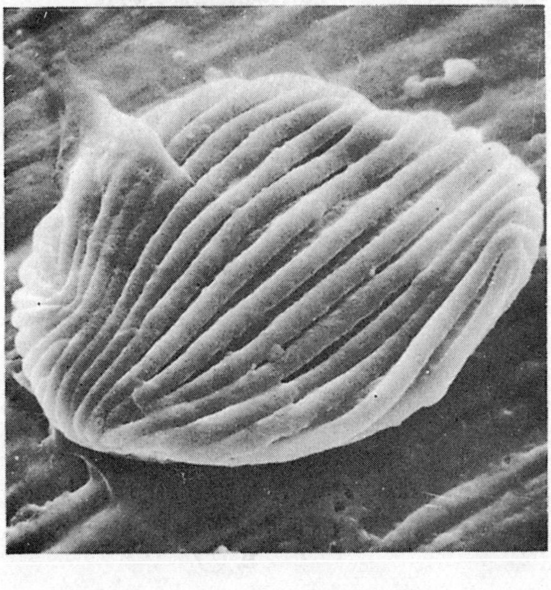

Figure 26.15. *Drewria patomacensis*. Reconstruction of vegetative and reproductive parts. The latter form a terminal dichasium on the shoot. Lower Cretaceous. (Redrawn from Crane & Upchurch, 1987.)

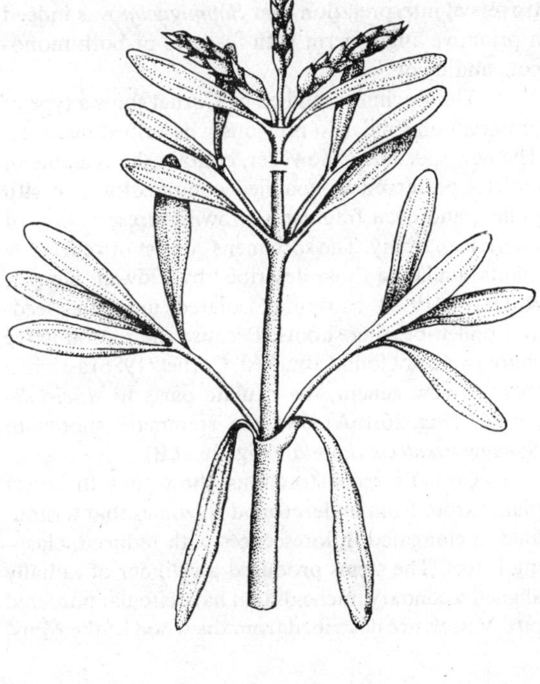

A minute Lower Cretaceous reproductive organ, *Eoanthus*, is described by Krassilov (1986) to have gnetalean characteristics. The fossil consists of four ovule-bearing structures in a whorl. Each structure contains a single orthotropus ovule, which in turn may contain ephedroid pollen. Whorls of bracts occur both above and below the ovule-bearing structures. These *Eoanthus* ovulate appendages have been compared with open angiosperm carpels, while the arrangement and position of the whorled bracts are similar to those of *Welwitschia*.

Although some progress has been made in understanding the gnetalean megafossil record, it is clear that much of what has been described remains to be clarified. If the variety of ephedroid pollen types is a criterion, then we can conclude that the Gnetales of the Mesozoic were much more diverse than our extant floras would indicate. The interpretation of the relationships between the Gnetales and angiosperms is elaborated in Chapter 30.

Sanmiguelia

One of the most intriguing and controversial Mesozoic plants with covered seeds is *Sanmiguelia lewisii* from the Upper Triassic of western North America. The similarity of its leaves to those of other gymnosperms with net venation and the absence of credible evidence for the existence of angiosperms prior to the Cretaceous led most workers to identify *Sanmiguelia* as a gymnosperm (Doyle, 1973; Hughes, 1976). In more recent studies, Cornet (1986, 1989) has described populations of *Sanmiguelia* from Upper Triassic deposits in Texas that show both vegetative and reproductive organs. His detailed studies of this new material have resulted in the surprising and controversial interpretation that *Sanmiguelia* was indeed a primitive angiosperm with features of both monocots and dicots.

The specimens studied by Cornet show a type of preservation similar to previously described material. The new specimens, however, contain the remains of cuticles preserved as coalified compressions, in situ pollen, and even fragments showing preservation of internal anatomy. The specimens consist of vegetative plants similar to those described by Tidwell, Simper, & Thayn (1977), as well as isolated, unisexual seed- and pollen-bearing shoots. Because the reproductive shoots were not found attached, Cornet (1986) assigned them to new genera; the ovulate parts to *Axelrodia burgeri* (Fig. 26.16A) and the staminate shoots to *Synangiospadixis tidwellii* (Fig. 26.16B).

Cornet's reconstructions show that the erect plants arose from underground rhizomes that terminated in elongated inflorescences with reduced, clasping leaves. The stems produced a cylinder of radially aligned secondary tracheids that have circular bordered pits. Vessels are described from the wood of the roots.

Figure 26.16. **A.** Reconstruction of *Axelrodia burgeri*, a cupulate, ovule-forming inflorescence emerging from folds of *Sanmiguelia lewisii* leaves. **B.** *Synangispadixis tidwellii*, a spadixlike inflorescence composed of crowded microsporophylls on secondary axes. Inflorescence not known to be attached to vegetative axes of *Sanmiguelia*. Upper Triassic. (From Cornet, 1986.)

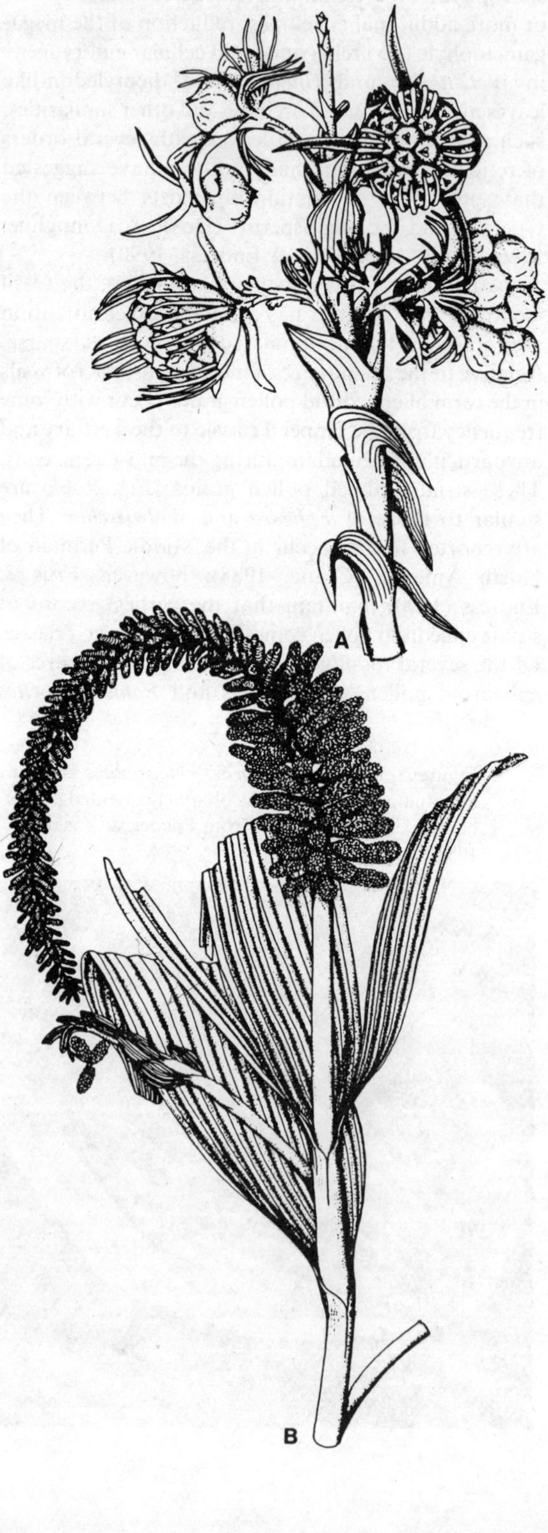

Staminate organs occur at the tip of the stem and lateral branches. Each inflorescence consists of helically arranged lateral axes that bear densely packed pairs of helically arranged pollen sacs with psilate, monosulcate pollen. No accessory organs are associated with the pollen sacs. Ovulate structures are reconstructed as partially closed cupules or carpels that are aggregated into heads at the tips of branches and also as groups of one or two in the axils of more proximal leaves (Fig. 26.16A). Each cupule or carpel is subtended by two or three types of bracts and is closed, except at the tip. A vertical suture located on one side opens toward the apex to separate two flaps that are interpreted to be stigmatic surfaces.

Although the preservation of the material is incomplete, and many of the features described by Cornet (1986, 1989) are difficult to recognize, he describes what appear to be two recurved ovules with a double integument near the base of each cupule/carpel. Cornet (1989) also interprets several features of development, fertilization, and embryology in *Sanmiguelia* as corresponding to those of living angiosperms, and concludes that this species demonstrates that flowering plants had evolved by Triassic time. We agree with Crane (1987), who states that "Cornet's conclusion, that *Sanmiguelia* is an angiosperm, is the most difficult to assess because this is a complex issue which hinges on the correct interpretation of specific details in the fossil material." Whether or not *Sanmiguelia* ultimately is recognized as a primitive angiosperm or a complex gymnosperm, this important species illustrates the rich diversity of complex seed plants that existed during the Mesozoic. Perhaps more significantly, fossils such as *Sanmiguelia* prompt us to clarify precisely which features of angiosperms are synapomorphies that distinguish them from gymnosperms. These features and their probability of preservation are discussed in detail in Chapter 30.

Protection of the ovule: a summary

We have come full circle to the start of this chapter, where it was suggested that we would find among the Mesozoic seed plants those that had evolved "novel" structures for protection of the ovules. In summary, you should recall such structures as the cupule of the Caytoniales. We may have glimpsed stages in the origin of the cupule of *Caytonia* revealed by the cupulate discs of *Lepidopteris* and the less well-developed cupules of *Umkomasia* and *Pilophorosperma*. Judging from the morphology of the structures that bear them, these cupules appear to have evolved as parts of pinnate fronds.

The enigma of the so-called cupule or capitulum of *Glossopteris* (*Scutum*) seems to have been solved by Gould & Delevoryas (1977). In providing protection to their ovules, these plants have evolved an enclosing lamina of a megasporophyll. Other ovulate structures of the Glossopteridales are probably modifications of this plan. If we accept the conclusion that the Glossopteridales evolved from pteridosperm ancestors, then the Callistophytaceae, which seem to combine conifer and pteridosperm characteristics, might be considered. Not only do both groups have bisaccate pollen, but their ovules are small and apparently were borne on the lower face of a leaf. If vegetative structure was the only consideration, one would have to look with favor on a cordaitalean origin of glossopterids, an idea favored by Schopf (1976). Regardless of its origin, we must be impressed by the way *Glossopteris* protects its ovules. It has been more than adequately demonstrated that the carpel of many angiosperms is a modified leaf whose margins have become revolute and fused to completely enclose the ovules. The partly closed cupule/carpel of *Sanmiguelia*, with its vertical suture and open apex, provides an excellent model for a morphological intermediate. One can hardly avoid the conclusion that the temporary enclosure of *Glossopteris* ovules by foliar lamina represents an evolutionary experiment well on the road to the condition of angiospermy. We should remember, however, that the product of angiospermy (where pollen grains germinate away from the nucellus, usually on a stigmatic surface) is only one of several characteristics of angiosperms that we will have an opportunity to consider at a later time.

Quite unlike other Mesozoic gymnosperms discussed thus far, the Pentoxylales have solved the problem of ovule protection by condensation of ovules with thick testas into a conelike structure where the ovules are borne directly on the cone axis. There are no structures such as bracts or interseminal scales associated with the cones that might give us clues as to their origin. This observation, along with the bizarre combination of other vegetative and reproductive features already explained, leaves us with but one possible conclusion – that is, the Pentoxylales represents another evolutionary experiment directed to the protection of ovules. That this also represents an experiment leading to angiospermy seems unlikely, but this has been suggested.

Like the Pentoxylales and Cycadeoidales, the evolutionary origin of the ovule-protecting structure of Czekanowskiales defies explanation. For the time being all are best considered as part of a diversified Mesozoic gymnosperm flora – a flora with some components becoming extinct and some, in ways which are not yet completely clear, having to do with the origins of flowering plants.

Chart 26.1 summarizes the suggested origins and relationships of the seed ferns and their presumed relatives, the Glossopteridales, Pentoxylales, and Czekanowskiales. The evidence and its evaluation appears in Chapters 23, 24, and 25, in addition to this one.

Chart 26.1. Suggested origin and relationships of major groups of the seed ferns and their presumed relatives, and their distribution in geological time. Trim. = Trimerophytopsida; Ane. = Aneurophytales; Arch. = Archaeopteridales; Calam. = Calamopityaceae; Pen. = Pentoxylales; Czek. = Czekanowskiales; Call. = Callistophytaceae.

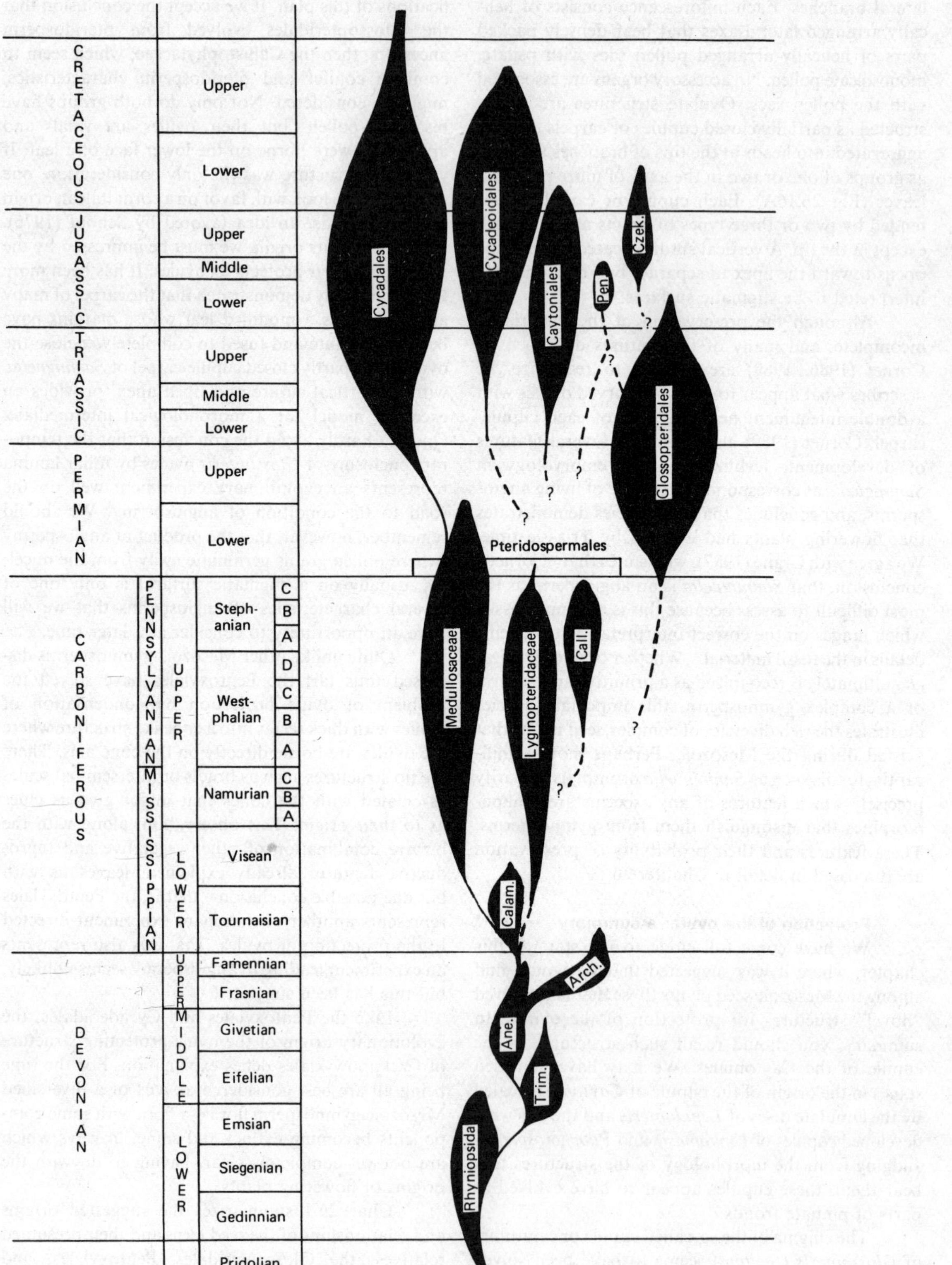

References

Andrews, H. N. (1961). *Studies in Paleobotany*. New York: Wiley.

Archangelsky, S., & Brett, D. W. (1961). Studies on Triassic fossil plants from Argentina. I. *Rhexoxylon* from the Ischigualasto Formation. *Philosophical Transaction of the Royal Society of London B*, **244,** 1–19.

Ash, S. R. (1972). Late Triassic plants from the Chinle Formation in northeastern Arizona. *Paleontology*, **15,** 598–618.

Cornet, B. (1986). The leaf venation and reproductive structures of a Late Triassic angiosperm, *Sanmiguelia lewisii. Evolutionary Theory*, **7,** 231–309.

Cornet, B. (1989). The reproductive morphology and biology of *Sanmiguelia lewisii*, and its bearing on angiosperm evolution in the Late Triassic. *Evolutionary Trends in Plants*, **3,** 25–51.

Crane P. R. (1985). Phylogenetic analysis of seed plants and the origin of angiosperms. *Annals of the Missouri Botanical Garden*, **72,** 716–93.

Crane, P. R. (1987). Review of Cornet, B. The leaf venation and reproductive structures of a Late Triassic angiosperm, *Sanmiguelia lewisii, Evolutionary Theory* **7**: 231–309 (1986). *Taxon*, **36,** 778–79.

Crane, P. R. (1988). Major clades and relationships in the "higher" gymnosperms. In *Origin and evolution of gymnosperms*, ed. C. B. Beck. New York: Columbia University Press.

Crane, P. R., & Upchurch, G. R., Jr. (1987). *Drewria potomacensis* gen. et sp. nov., an early Cretaceous member of the Gnetales from the Potomac Group of Virginia. *American Journal of Botany*, **74,** 1722–36.

Cridland, A. A. (1963). *A Glossopteris* flora from the Ohio Range, Antarctica. *American Journal of Botany*, **50,** 186–95.

Delevoryas, T. (1962). *Morphology and Evolution of Fossil Plants*. New York: Holt, Rinehart and Winston.

Delevoryas, T. (1969). Glossopterid leaves from the Middle Jurassic of Oaxaca, Mexico, *Science*, **165,** 895–96.

Doyle, J. A. (1973). Fossil evidence on early evolution of the monocotyledons. *Quarterly Review of Biology*, **48,** 399–413.

Doyle, J. A., & Donoghue, M. J. (1986). Seed plant phylogeny and the origin of angiosperms: An experimental cladistic approach. *Botanical Review*, **52,** 321–431.

Friis, E. M., & Endress, P. K. (1990). Origin and evolution of angiosperm flowers. *Advances in Botanical Research*, **17,** 99–162.

Gould, R. E. (1975). A preliminary report on petrified axes of *Vertebraria* from the Permian of Eastern Australia. In *Gondwana Geology: Papers Presented at the Third Gondwana Symposium, Canberra, Australia*, ed. K. S. W. Campbell. Canberra: Australian National University Press, pp. 109–15.

Gould, R. E., & Delevoryas, T. (1977). The biology of *Glossopteris:* evidence from petrified seed-bearing and pollen-bearing organs. *Alcheringa*, **1,** 387–99.

Harris, T. M. (1932). The fossil flora of Scoresby Sound, East Greenland, 2: Description of seed plants *Incerte Sedis* together with a discussion of certain cycadophyte cuticles. *Meddelelser om Grønland*, **85,** 1–112.

Harris, T. M. (1933). A new member of the *Caytoniales*. *New Phytologist*, **32,** 97–114.

Harris, T. M. (1935). The fossil flora of Scoresby Sound, East Greenland, 4: Ginkgoales, Coniferales; Lycopodiales, and isolated fructifications, *Meddelelser om Grønland*, **112,** 1–176.

Harris, T. M. (1937). The fossil flora of Scoresby Sound, East Greenland, 5: stratigraphic relations of the plant beds. *Meddelelser om Grønland*, **112,** 1–114.

Harris, T. M. (1951). The fructification of *Czekanowskia* and its allies. *Philosophical Transactions of the Royal Society of London B*, **235,** 483–508.

Harris, T. M. (1964). *The Yorkshire Jurassic Flora II. Caytoniales, Cycadales and Pteridosperms*. London: British Museum (Natural History).

Harris, T. M. (1976). The Mesozoic gymnosperms. *Review of Palaeobotany and Palynology*, **21,** 119–34.

Holmes, W. B. K. (1973). On some fructifications of the Glossopteridales from the Upper Permian of New South Wales. *Proceedings of the Linnean Society of New South Wales*, **98,** 132–41.

Hughes, N. F. (1976). *Palaeobiology of Angiosperm Origins*. Cambridge: Cambridge University Press.

Krassilov, V. A. (1986). New floral structure from the Lower Cretaceous of Lake Baikal area. *Review of Palaeobotany and Palynology*, **47,** 9–16.

Lacey, W. S., van Dijk, C. E., & Gordon-Gray, K. D. (1974). New Permian *Glossopteris* flora from Natal. *African Journal of Science*, **70,** 131–41.

Lacey, W. S., van Dijk, D. E., & Gordon-Gray, K. D. (1975). Fossils plants from the Upper Permian in the Mooi River district of Natal, South Africa. *Annals of the Natal Museum*, **22,** 349–420.

Pant, D. D. (1977). The plant of *Glossopteris*. *Journal of the Indian Botanical Society*, **56,** 1–23.

Pant, D. D., & Nautiyal, A. (1984). On the morphology and structure of *Ottokaria zeilleri* sp. nov. – a female fructification of *Glossopteris*. *Paleontographica B*, **193,** 127–52.

Pigg, K. B. (1988). Anatomically preserved *Glossopteris* and *Dicroidium* from the central Transantarctic Mountains. Unpublished Ph.D. Dissertation, Ohio State University, Columbus.

Pigg, K. B., & Taylor, T. N. (1989). Permineralized *Glossopteris* and *Dicroidium* from Antarctica. In *Antarctic Paleobiology*, eds. T. N. Taylor & E. L. Taylor. New York: Springer-Verlag.

Plumstead, E. P. (1952). Description of two new genera and six new species of fructifications borne on *Glossopteris* leaves. *Transactions of the Geological Society of South Africa*, **55,** 281–328.

Plumstead, E. P. (1956). Bisexual fructifications borne on *Glossopteris* leaves from South Africa. *Palaeontographica B*, **100,** 1–25.

Pocock, S. A. J., & Vasanthy, G. (1988). *Cornetipollis reticulata*, a new pollen with angiospermid features from the Upper Triassic (Carnian) sediments of Arizona (USA), with notes on *Equisetosporites. Review of Palaeobotany and Palynology*, **55,** 337–56.

Sahni, B. (1948). The Pentoxyleae: a new group of Jurassic gymnosperms from the Rajmahal Hills of India. *Botanical Gazette,* **110,** 47–80.

Schopf, J. M. (1967). Antarctic fossil plant collecting during the 1966–1967 season. *Antarctic Journal of the United States,* **2,** 114–16.

Schopf, J. M. (1970). Petrified peat from a Permian coal bed in Antarctica. *Science,* **169,** 274–77.

Schopf, J. M. (1976). Morphologic interpretations of fertile structures in *Glossopteris* gymnosperms. *Review of Palaeobotany and Palynology,* **21,** 25–64.

Srivastava, B. P. (1935). On silicified plant remains from the Rajmahal series of India. *Proceedings of the 22nd Indian Science Congress, Calcutta,* p. 285.

Srivastava, B. P. (1937). Studies on some silicified plant remains from the Rajmahal series. *Proceedings of the 24th Indian Congress, Hyderabad–Deccan,* pp. 273–4.

Srivastava, B. P. (1946). Silicified plant remains from the Rajmahal series of India. *Proceedings of the National Academy of Science (India),* **15,** 185–211.

Stewart, W. N. (1976). *Polystely, Primary Xylem, and the Pteropsida.* Lucknow, India: Birbal Sahni Institute of Palaeobotany.

Surange, K. R. (1966). The present position of the genus *Glossopteris. Proceedings of the Autumn School in Botany–Mahablalshwar,* pp. 316–27.

Surange, K. R., & Chandra, S. (1971a). *Denkania indica* gen. et sp. nov. – a glossopteridean fructification from the Lower Gondwana of India. *Palaeobotanist,* **20,** 264–68.

Surange, K. R. & Chandra, S. (1971b). *Dictyopteridium sporiferum* Feistmantel – female cone from the Lower Gondwana of India. *Palaeobotanist,* **20,** 127–36.

Surange, K. R., & Chandra, S. (1972). Fructification of Glossopteridae from India. *Palaeobotanist,* **21,** 1–17.

Surange, K. R., & Chandra, S. (1975). Morphology of the gymnospermous fructifications of the glossopteris flora. *Palaeontographica B,* **149,** 153–80.

Surange, K. R., & Chandra, S. (1976). Morphology and affinities of *Glossopteris. Palaeobotanist,* **25,** 509–24.

Surange, K. R., & Maheshwari, H. K. (1970). Some male and female fructifications of Glossopteridales from India. *Palaeontographica B,* **129,** 178–92.

Taylor, T. N., & Archangelsky, S. (1985). The Cretaceous pteridosperms *Ruflorinia* and *Ktalenia* and implications on cupule and carpel evolution. *American Journal of Botany,* **72,** 1842–53.

Taylor, T. N., & Taylor, E. L. (1987). An unusual gymnospermous reproductive organ of Triassic age. *Antarctic Journal of the United States,* **22**(5), 29–30.

Thomas, H. H. (1925). The Caytoniales, a new group of angiospermous plants from the Jurassic rocks of Yorkshire. *Philosophical Transactions of the Royal Society of London B,* **213,** 299–363.

Thomas, H. H. (1933). On some pteridospermous plants from the Mesozoic of South Africa. *Philosophical Transactions of the Royal Society of London B,* **222,** 193–265.

Thomas, H. H. (1955). Mesozoic pteridosperms. *Phytomorphology,* **5,** 177–84.

Thomas, H. H. (1958). *Lidgettonia,* a new type of fertile *Glossopteris. Bulletin of the British Museum (Natural History),* **3,** 179–89.

Tidwell, W. D., & Simper, A. D., & Thayn, G. F. (1977). Additional information concerning the controversial Triassic plant: *Sanmiguelia·B Palaeontographica B,* **163,** 143–51.

Townrow, J. A. (1965). A new member of the Corystospermaceae Thomas. *Annals of Botany (N. S.),* **29,** 295–311.

Vishnu-Mittre (1952). A male flower of the Pentoxyleae with remarks on the structure of the female cones of the group. *Palaeobotanist,* **2,** 75–84.

27

The record of a living fossil: Ginkgo

At a time when ecologists are predicting that 20 percent of known species will become extinct by the turn of the century because of man's activities, it is refreshing to know that the monotypic genus *Ginkgo biloba* seems to have been saved from extinction for the foreseeable future. *Ginkgo* was "rediscovered" long ago growing in temple gardens of China where the sacred trees were carefully tended and perpetuated. In 1956, a small population of *Ginkgo* trees was reported growing in the wild state in southeastern China. This is claimed to be the last natural refuge of a once prominent constituent of Mesozoic and early Tertiary floras of the Northern Hemisphere. As Seward (1919) observed, "*Ginkgo biloba* L. has a preeminent claim to be described in Darwin's words as a living fossil," a plant that has been reintroduced and cultivated in many parts of the world where it once flourished many millions of years ago.

Some features of extant Ginkgo

No effort is made here to give all of the information relating to the structure and reproduction of *Ginkgo biloba*. Such detailed accounts are found in several well-documented morphology texts. We will, however, focus our attention on those features that are germane to understanding the fossil record. This includes *Ginkgo* wood, a few reproductive structures, and countless leaves whose unequivocal remains first appear in the Upper Triassic.

The habit of *Ginkgo* resembles that of *Glossopteris* and many conifers, where there is a main trunk bearing branches with axillary long and short shoots (Fig. 27.1A). The characteristic often reniform to fan-shaped leaves of *Ginkgo* have long slender petioles and are scattered on the long shoots or crowded at the distal ends of the slow-growing short shoots, which are covered with leaf scars (Fig. 27.1B). *Ginkgo* shows a considerable variation in the size and form of its foliage from the "typical" bilobed leaf to those of rapidly growing long shoots where the lamina may be deeply divided into cuneate segments, or those of short shoots where the smaller leaves may have margins that are entire. Figure 27.2 shows silhouettes of *G. biloba* leaves to illustrate the degree of variability that can occur in leaves from a single tree. As you can imagine, if a paleobotanist were to find a fossil leaf from a short shoot and one from a long shoot, he or she would be inclined to place them in different species, especially if unaware of the range of variability within the species. Not all paleobotanists take cognizance of such possible species variability when describing and naming isolated fossil leaves, wood, and other structures.

Regardless of their size or form, foliage leaves of *G. biloba* (Fig. 27.3A) have two vascular bundles in the well-defined petiole where they diverge at the distal end and pass into the lower edge of the lamina. At the point of divergence each marginal vein gives rise to a succession of smaller dichotomizing veins that supply one-half the leaf lamina. The dichotomizing veins occasionally anastomose. The leaf is hypostomatic with stomata irregularly scattered between the veins. The stomata are haplocheilic, with four to six subsidiary cells each with a blunt papilla that tends to overarch the guard cells (Fig. 27.3B). When the characteristics of general leaf morphology, venation, epidermal cell patterns, and stomata are considered together, it simplifies the task of distinguishing leaves of *Ginkgo* from all others. Because many leaves found in Tertiary rocks display these characteristics and are indistinguishable from *G. biloba* leaves, some paleobotanists have assigned them to the same species. This of course implies that the fossil plant represented by the Tertiary leaves was the same, in all other structural and reproductive features, as an extant *G. biloba*. Since these features are not known, some object to assigning the leaves to the same species or even genus and use the generic name *Ginkgoites* instead.

It is not necessary to devote a great deal of space to the description of the internal structure of *Ginkgo* stems. Structurally, their short shoots differ from long shoots in the growth pattern of their pith and cortex, which in short shoots is somewhat reminiscent of a cycad stem. In older, more rapidly growing long shoots, older branches, and main trunk, the well-developed secondary xylem (Fig. 27.3C) is pycnoxylic with characteristics that are difficult to distinguish from those of conifer wood. The small tracheids have circular bordered pits in one to two rows mostly on the radial walls with bars of Sanio (crassulae) associated with the pits. The rays are uniseriate and do not exceed 11 cells in height. There are two to seven elliptical cross-field pits on the ray cells, and xylem parenchyma occurs here and there among the tracheids. The primary xylem is endarch, and it forms a ring of sympodia at the inner limits of the secondary xylem. Leaf traces rise in pairs from adjacent sympodia and become the paired traces that enter a leaf base. All of the stem characteristics given above for *Ginkgo* also are found among some conifers. Experts in gymnosperm wood anatomy believe that secondary xylem with characteristics similar to the wood of *G. biloba* makes its first known appearance in the Eocene. That there are, however, significant differences between extant and fossil *Ginkgo* woods has been demonstrated by Mastrogiuseppe, Cridland, & Bogyo (1979). Using multivariate statistical analyses of the wood of modern *G. biloba* and the Miocene *G. beckii*, the authors demonstrated that there were differences in numbers of pits on the radial walls of tracheids, ray height, and tangential diameter of late wood tracheids, differences that are not apparent when making comparisons without recourse to statistical analysis.

Like the cycads discussed earlier, *Ginkgo* is strictly dioecious. The pollen organs and ovules at the ends of stalks are restricted to short shoots on male and female trees. They arise in the axils of foliage leaves or inner bud scales. A pollen organ is a loose, catkinlike strobilus (Fig. 27.1A) consisting of an axis with several sporangiophores each bearing two pendant microsporangia. The pollen grains are similar to the monocolpate grains of cycads.

The erect ovules usually are borne in pairs at the end of stalks that arise on the short shoots in the axils of leaves (Fig. 27.1B), but there is a wide spectrum of variation on some very old trees (Fujii, 1896). Some

Figure 27.1. **A.** *Ginkgo biloba*, with catkinlike pollen organs and leaves on a short shoot. Note leaf scars on short shoot and the long shoot bearing the short shoot. **B.** *G. biloba*, short shoot of ovulate plant. Note pairs of ovules at tips of branches. (Photographs courtesy of General Biological Supply House, Chicago.)

ovules are borne at the margin of distorted leaves (Fig. 27.4A). All others occur at the tips of stalks that are borne in the axil of a leaf (Fig. 27.4B). Each stalk branches to produce from two to about ten ovules. There has been considerable discussion and confusion about the nature of these structures, but information needed to assess their homologies has been available for over 100 years. The ovules occur either on leaves or on axillary structures that are vascularized in the same way as branches, and the vascularization of each branch that terminates in an ovule is comparable to that of a leaf. All these features are in keeping with the interpretation that the ovulate stalks represent highly reduced and modified fertile branches that bear the ovules on modified leaves (Rothwell, 1987). All available evidence indicates that the ovulate stalks of ginkgoaleans originated within the group, and they may be regarded as a synapomorphy of the clade. However, in this respect they have had a parallel evolution with the ovulate cone scales of conifers, which also are regarded as highly modified fertile branches. The origin and homologies of the ovulate cone scales of conifers are subjects we will discuss in detail in the next chapter.

Figure 27.2. Silhouettes of five leaves from a *Ginkgo biloba* tree showing a range of variability in size and external morphology. Extant.

At maturity, the ovules are 1.5 to 2 mm in diameter, being much larger than the seeds of most conifers, but comparing favorably with those of cycads and medullosan pteridosperms. Around and partially enclosing the base of each ovule is an outgrowth called a "collar." The ovules have a three-layered integument composed of a fleshy sarcotesta, an inner flesh, and a stony sclerotesta between the two. The latter may be two-ribbed and bilaterally symmetrical or three-ribbed and radially symmetrical. The nucellus, which is more or less free from the surrounding integument except at its base, develops a pollen chamber at its apex. The vascular system is weakly developed and consists of a pair of anastomosing bundles in the inner fleshy layer of the integument. When the microgametophyte matures it produces haustorial pollen tubes and large motile sperms similar to those of cycads. Megagametophyte development also is cycadlike.

From the foregoing cursory description of *G. biloba* it must be apparent that it combines characteristics of conifers on one hand and cycads on the other. After our studies of the previous chapter of complex gymnospermous plants such as *Pentoxylon* and *Glossopteris,* we should not be surprised to find extant plants that do not fit into our preconceived ideas of certain plant taxa. Meyen (1987) has recently stressed possible relationships between *Ginkgo* and peltasperms (Chapter 26), and suggested that *G. biloba* may represent one of three major groups of seed plants that originated by the beginning of the Carboniferous. However, other authors do not agree, and recent cladistic analyses suggest that *Ginkgo* is more closely related to cordaites and conifers than to peltasperms. For these reasons taxonomists continue to place *Ginkgo,* extant and extinct, in the Ginkgoales, distinct from the Cycadales, Coniferales, Taxales, and other orders of Gymnospermopsida.

Early evidence of ginkgophytes

In an effort to solve the problems of origin and thus relationships of ginkgophytes, paleobotanists have scoured the fossil record, obtaining only a few clues as to when they first appeared and from what ancestral group they were derived. The most significant contributions to our understanding of the fossil record of ginkgophytes have been prepared by Seward (1919), Harris (1935, 1974), Florin (1936a, 1936b), Tralau (1968), and Krassilov (1970, 1972).

The generalization that leaves of primitive ginkgophytes tend to be linear and deeply dissected seems to be supported by finds of *Ginkgo*-like leaf fossils in the Upper Paleozoic. With some hesitation, which we share, Andrews (1941) suggests that the Upper Carboniferous *Dichophyllum moorei* is a "preginkgophyte" representing a stage in the early evolution of leaves.

The specimens, which are up to 10 cm in length, look like planated sterile telome trusses. The dichotomies terminate in slender linear subdivisions. It is not difficult to imagine that the trusses became partially webbed to form wedge-shaped leaves with dichotomous venation, not unlike those of ginkgophytes. Leaves of this type are characteristic of *Sphenobaiera* (Fig. 27.5A), which makes its first known appearance in the Lower Permian and extends into the Lower Cretaceous. *Sphenobaiera furcata* from the late Triassic bears the leaves in clusters on short shoots or long shoots. Some short shoots bore microsporangiate organs consisting of a central axis that branched two or three times and terminated in clusters of three to five sporangia. The leaves of some species of *Sphenobaiera* (Fig. 27.5B) are similar in appearance to those of *Czekanowskia*. The obvious difference, however, is the single vein that traverses the dichotomizing *Czekanowskia* leaves and the several dichotomizing veins in *Sphenobaiera*. There are, in addition, less obvious but no less important differences in epidermal characteristics. The leaves of *Arctobaiera flettii* (Fig. 27.5C,D) show the trend toward ginkgophyte leaves with entire margins from those that are deeply dissected.

Trichopitys (Fig. 27.6A) can be added to our short list of Paleozoic genera that may have affinities with the Ginkgoales. The specimens, which come from the Lower Permian, consist of axes about 8 mm in diameter bearing spirally arranged leaves. A single ovule-bearing branch is found in the axil of some of the leaves. Florin (1949) describes the ovule-bearing branch system as an overtopped fertile telome truss with an inverted ovule (Fig. 27.6B) at the tip of each ultimate branch. Four to six ovules are usually found per truss, but some specimens have as many as 20 (Fig. 27.6C).

There is no evidence of differentiation of *Trichopitys* axes into short shoots, an obvious feature of *Ginkgo biloba*. Florin (1949) speculates that as short shoots evolved the ovulate branches present on the long branches of *Trichopitys* were transferred to the short shoots when they evolved and that the number of ovules on a branch was reduced to one or two. Some support for this hypothesis is derived from abnormal specimens of *G. biloba* where several ovules are formed on an axillary branch system (Fig. 27.4B) very much like the fertile truss of *Trichopitys*. Utilizing this evidence, it is not difficult to suggest how the ovulate stalk in the axil of the leaf of *G. biloba* evolved from the sterile and fertile telome trusses of *Trichopitys*. It should be pointed out, however, that if *Trichopitys* represents an early stage in the evolution of ginkgophytes, then the relationship of the group seems to be with the conifer type. The reason for this assumption is that *Trichopitys* has a fertile shoot in the axil of a foliar unit as in the cordaites and thus is not a megasporophyll with ovules as in pteridosperms. On the other hand, evidence from the abnormal

Figure 27.3. **A.** *Ginkgo biloba*. Single leaf showing details of venation. Note two traces entering petiole. Extant. **B.** Stomatal apparatus of a *Ginkgoites lunzensis*. Note the papillae (p) arising from the subsidiary cells (sc) around the stoma (s). Triassic. (Redrawn from Andrews, 1961.) **C.** *G. biloba*. Radial section of secondary wood showing circular bordered pits (cp) with associated crassulae (c), and ray cells (rc) with cross-field pits (cf). Extant.

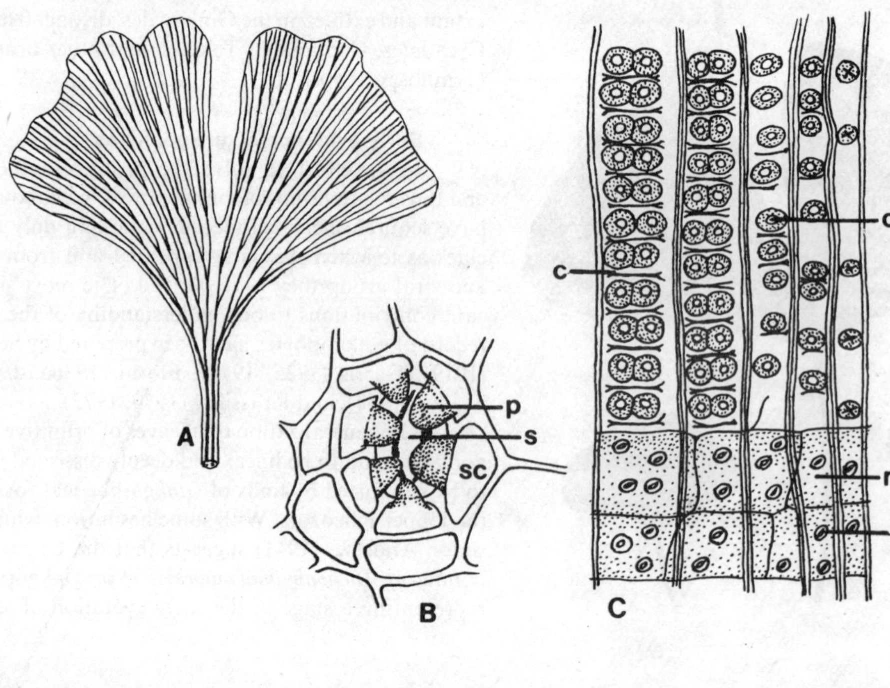

Figure 27.4. Abnormal development of *Ginkgo biloba* ovulate structures. **A.** Ovule borne on expanded lamina of a leaf. **B.** Highly branched stalk bearing a terminal ovule on each branch. (**A,B** redrawn from Chamberlain, 1935.) **A,B:** Extant.

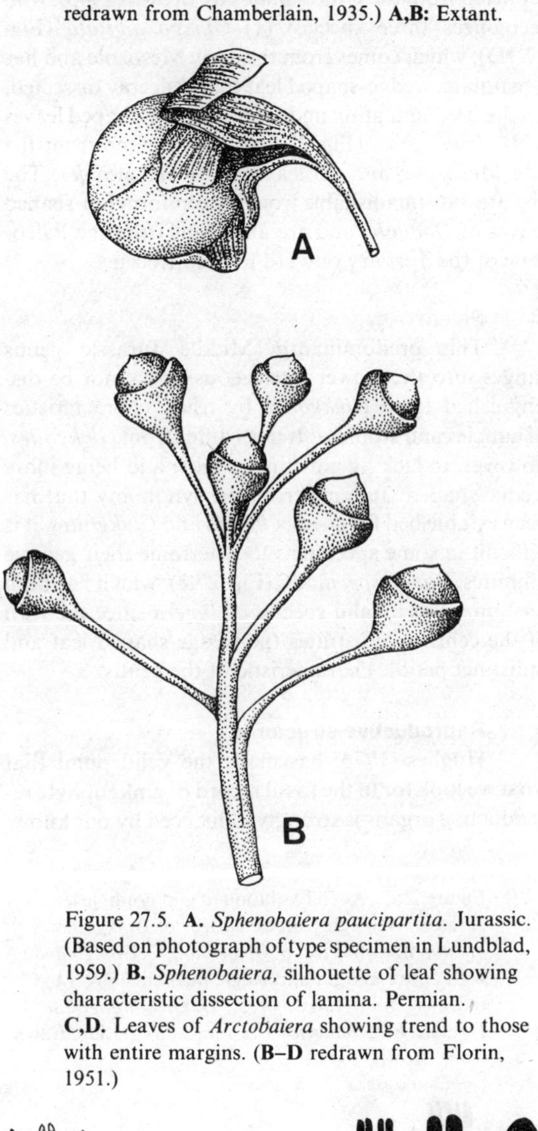

production of ovules on leaves of *G. biloba* does not support this view. This evidence, plus the numerous ways in which reproductive structures (pollen grains, microgametophytes, sperms, and megagametophyte development) simulate those of cycads, suggests to us that the origin of the Ginkgoales from the conifer type is far from settled. Until we learn more from the fossil record, it seems best to presume that the Ginkgoales had an origin with a Paleozoic ancestor, which also gave rise to conifers and pteridosperms.

The imprints of Mesozoic and Tertiary Ginkgoales

The abundance of compression–impression leaf remains available to the paleobotanist from Mesozoic and Tertiary deposits has presented many problems in identification and naming of species. In view of the known variability of leaf form and size in *G. biloba,* one author suggests that it might be proper to lump all "species" of ginkophyte leaves into two or three species instead of the many that have been described in the literature over the last several decades. Most agree this would be a poor solution and would not express the great diversity that occurred within the group, especially in the Jurassic. Since the introduction by Harris (1935) of studies of *Ginkgo*-like leaf compression–impressions utilizing the characteristics of epidermal and stomatal structures, it is possible, with well-preserved material, to distinguish what appear to be well-defined species.

The distinction of genera, however, offers some special problems of nomenclature. Many authors follow the system used by Seward (1919), who used the genus

Figure 27.5. **A.** *Sphenobaiera paucipartita.* Jurassic. (Based on photograph of type specimen in Lundblad, 1959.) **B.** *Sphenobaiera,* silhouette of leaf showing characteristic dissection of lamina. Permian. **C,D.** Leaves of *Arctobaiera* showing trend to those with entire margins. (**B–D** redrawn from Florin, 1951.)

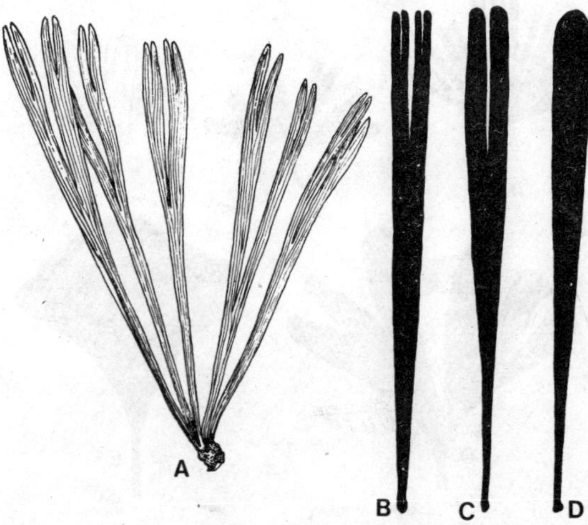

Figure 27.6. *Trichopitys heteromorpha.* Lower Permian. **A.** Portion of shoot-bearing sterile telome trusses (leaves) with axillary ovule-bearing shoots. **B.** A single anatropous ovule. **C.** Axillary shoot with numerous ovules. (Redrawn from Florin, 1949, 1951.)

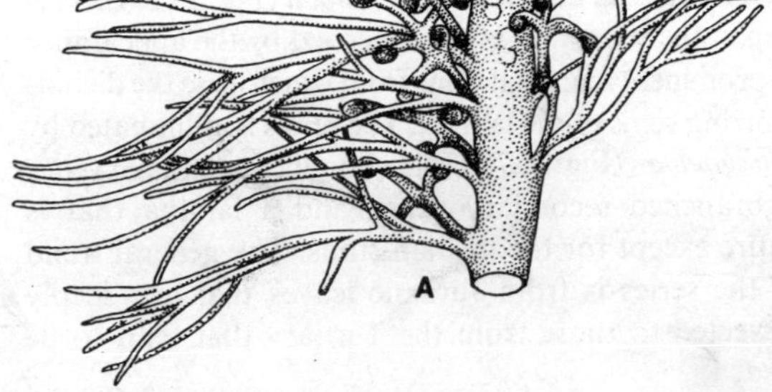

Ginkgoites for all fossil leaves resembling those of *G. biloba* even thought they are indistinguishable from leaves of this extant species. Adopting this system, however, implies that *Ginkgo* did not have a fossil record. To avoid this, some investigators have included those leaf fossils that are indistinguishable from leaves of the modern *Ginkgo* in the genus, but have used different species names. Those fossil *Ginkgo*-like leaves, which can be distinguished from *Ginkgo* by their morphological and anatomical characteristics, are placed by Harris (1935) and Florin (1936a) in the genus *Ginkgoites*.

Although both *Ginkgo* and *Ginkgoites* leaves are widely distributed in the Mesozoic, those of *Ginkgoites*, which appear in the late Triassic, are much more abundant in Mesozoic rocks. The fossil record of *Ginkgo* appears in the Jurassic but, as one might expect, it becomes much more extensive in the Tertiary. Of the seven or eight Mesozoic leaf genera included in the Ginkgoales (Harris, 1935, 1974), the two having the largest number of species are *Ginkgoites* and *Baiera*.

Ginkgoites and Ginkgo

Leaves of *Ginkgo* and *Gonkgoites* are usually bilobed. The two halves of the lamina give the leaves of these genera a semicircular appearance. Both have a distinct petiole with two traces that diverge into the basal edges of the lamina. To distinguish species of *Ginkgoites* Harris (1935) relies on characteristics of external morphology combined with structural features such as distribution of resin bodies, structure of mesophyll, size and shape of epidermal cells, and distribution of stomata and structure of their subsidiary cells. To describe a new species or identify an established one requires large quantities of well-preserved material, techniques for removing and studying cuticular remains, and a great deal of patience. Harris, as a result of painstaking studies, has provided us with a group of species that can be arranged in an interesting series showing variations in venation patterns and degrees of dissection (lobing) of lamina. In leaves of the *Ginkgoites minuta* type (Fig. 27.7A), the dissection of the leaf is pronounced and the branching of the leaf is by equal dichotomies. In the next member of the series, typified by *G. taeniata* (Fig. 27.7B), the dichotomies are unequal so that each half of the leaf appears to be divided into three lobes. This trend toward irregularity of lobing, from those where there are regular dichotomies, is amplified in *G. biloba* (Fig. 27.3A). This step in the series also is characterized by the appearance of prominent marginal veins from which arise the dichotomizing veins of the lamina. The series is culminated by *Ginkgoidium* (Fig. 27.7C) with prominent marginal veins, unbranched secondary veins, and a lamina that is entire except for the median sinus. The general trend of the series is from Jurassic leaves that are deeply dissected to those from the Tertiary that tend to be entire except for the median sinus (Tralau, 1968). That this series may have some evolutionary significance is illustrated by Mesozoic species of *Ginkgo* (*Ginkgoites*) reported from the Unites States by Brown (1943), who recognizes three species: (1) *Ginkgo digitata* (Fig. 27.7D), which comes from the early Mesozoic and has a petiolate, wedge-shaped leaf that is deeply dissected. (2) the predominantly undissected wedge-shaped leaves of *G. lamariensis* (Fig. 27.7E), which came from the late Mesozoic; and (3) leaves of *G. adiantoides*. The last are indistinguishable from the reniform fan-shaped leaves of *G. biloba* and are abundant from the Paleocene of the Tertiary onward to the Miocene.

Baiera

This predominantly Middle Jurassic genus ranges into the Lower Cretaceous. It cannot be distinguished from *Ginkgoites* by using characteristics of cuticles and stomata. It does differ from *Ginkgoites*, however, in lacking a distinct petiole and being more wedge-shaped. Judging from the synonomy that has been established for species *Baiera* and *Ginkgoites*, it is difficult in some specimens to determine their generic affinities. *Baiera spectabilis* (Fig. 27.8), which has been recognized as a valid species of *Baiera* since the turn of the century, illustrates the wedge-shaped leaf and indistinct petiole characteristic of the genus.

Reproductive structures

Hughes (1976) has made the valid point that what we look for in the fossil record of ginkgophyte reproductive organs is strongly influenced by our know-

Figure 27.7. **A–C:** Evolution of ginkgophyte leaves. Jurassic. **A.** Highly dissected leaf of *Ginkgoites minuta*. **B.** Lobed leaf of *G. taeniata*. **C.** *Ginkgoidium* with entire margin and single median sinus. (**A–C** redrawn from Harris, 1935.) **D.** *Ginkgo digitata*. Jurassic. **E.** *G. lamariensis*. Cretaceous. (**D,E** redrawn from Brown, 1943.)

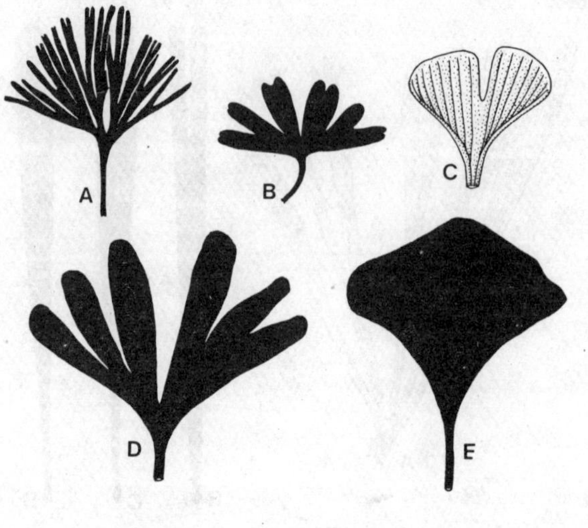

ledge of *Ginkgo biloba* fructifications. We forget that Jurassic and Cretaceous ginkgophytes were highly diverse and not represented by a single species. Thus, it is possible that fructifications not now interpreted as belonging to ginkgophytes have been found but not recognized as such. Perhaps the unusual ovulate structures of living *G. biloba* (Fig. 27.4) can provide a clue as to the morphology of some of the fossil forms.

Associated with the Jurassic and Cretaceous *Ginkgo*-like remains are numerous dispersed ovules (Fig. 27.9A) named *Allicospermum* (Harris, 1935) that are similar to ovules of *G. biloba* and cycads. *Karkenia,* an ovulate fructification described by Archangelsky (1965) from the Cretaceous of Argentina, consists of over a hundred or more small ovules crowded on a short stalk. Each ovule has a projecting microphyle and a short pedicel. The fructifications are associated with *Ginkgoites tigrensis* and are considered to belong to the same plant.

Material from the Yorkshire Jurassic beds has also contributed to our information about ginkgophyte fructifications. In 1976, Harris was able to report paired ovules joined by a pad of tissue with *Ginkgo*-like stomata. These occur in beds with abundant *Ginkgoites huttoni* where a search for pollen organs of the *Ginkgo* type was rewarded (van Konijnenburg-van Cittert, 1971). The organ is a little pollen-bearing catkin (Fig. 27.9B) with pairs of microsporangia at the tips of rather lax stalks attached to an axis about 5 mm long. The pollen obtained from the microsporangia is monocolpate (Fig. 27.9C) and like that of *G. biloba.* Very few other verified ginkgophyte reproductive structures have been reported.

Distribution in time and space

Although the "roots" of the Ginkgoales seems to be with Paleozoic ancestors (Chart 29.1), it was not until the end of the Triassic that they became an important part of the Mesozoic floras. During the Jurassic, especially by the Middle Jurassic, the Ginkgoales reached their zenith in numbers of species and distribution. A survey of Jurassic and Cretaceous localities shows us that the ginkgophytes were circumpolar, appearing in Alaska, Greenland, Scandinavia, Franz Joseph Land, Siberia, and Mongolia. Of these, the Siberian localities have been highly productive. Numerous localities in western Canada and the United States have produced leaf remains of *Ginkgoites* from Upper Mesozoic and Lower Tertiary deposits. Localities in Patagonia south to the tip of South America, South Africa, India, Australia, and New Zealand tell us that the Ginkgoales were not confined to those land areas that are now a part of the Northern Hemisphere. Many localities are known in Europe, including those in England, Scotland, Germany, Italy, Hungary, Turkestan, and Afghanistan. Many of these localities and others have been noted by Seward (1919) and Harris (1974). Toward the end of the Cretaceous the Ginkgoales were in an apparent decline, not only in their worldwide distribution, but in numbers of genera and species. The predominance of ginkgophytes in the high northern latitudes, starting with the early Cretaceous and their presence in southern latitudes in Argentina during the Jurassic, has suggested that the

Figure 27.8. *Baiera spectabilis.* Jurassic. (Redrawn from Harris, 1935.)

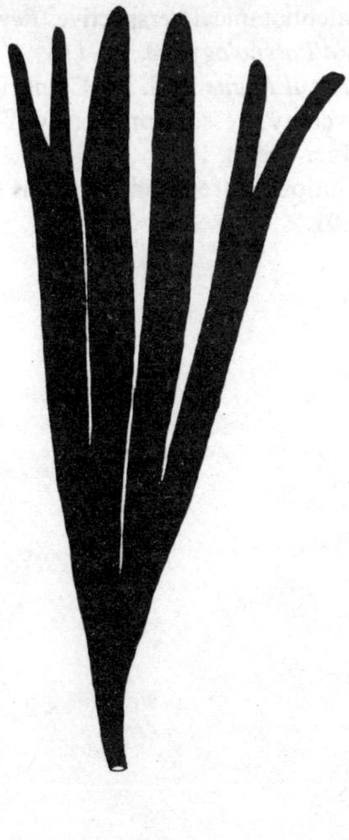

Figure 27.9. **A.** *Allicospermum xystrum,* diagram of longitudinal section of a *Ginkgo-like* ovule. (Redrawn from Harris, 1935.) **B.** Pollen organ associated with *Ginkgoites.* **C.** Monocolpate pollen grain from the fructification at B. **A–C:** Jurassic. (**B,C** redrawn from van Konijnenburg-van Cittert, 1971.)

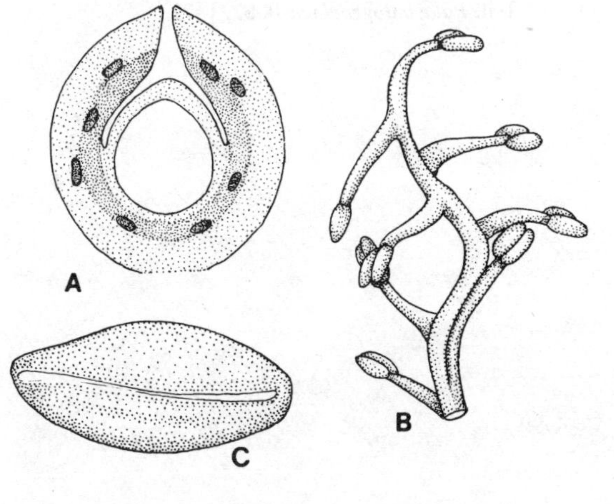

dispersal was from southern to northern latitudes during the Upper Mesozoic.

During the Tertiary, an interesting phenomenon in the decline of ginkgophytes has been recorded. By the start of the Oligocene epoch only 2 out of approximately 19 species remained. One of these is *Ginkgo adiantoides*, which, although not abundant, has been found in western Canada in the Upper Cretaceous–Tertiary (Paleocene) deposits. In the Eocene, *G. adiantoides* is more abundant, suggesting a cooling trend favorable to wider dispersal into areas that include Alaska, Alberta, British Columbia, Montana, South Dakota, and Wyoming. The paucity of *G. adiantoides* preserved during the Oligocene attests to a sharp decline that continued into the Miocene of western North America. During the Miocene the species disappeared from the fossil record in that part of the world. In Europe, however, it is reported to have continued a little longer into the Pliocene. Since then our present evidence indicates that the Ginkgoales has been represented by the extant monotypic "living fossil" – *Ginkgo biloba* of China.

References

Andrews, H. N. (1941). *Dichophyllum moorei* and certain associated seeds. *Annals of the Missouri Botanical Garden*, **28,** 375–84.

Andrews, H. N. (1961). *Studies in Paleobotany*. New York: Wiley.

Archangelsky, S. A. (1965). Fossil Ginkgoales from the Tico flora, Santa Cruz Province, Argentina. *Bulletin of the British Museum of Natural History, Geology*, **10,** 121–37.

Brown, R. W. (1943). Some prehistoric trees of the United States. *Journal of Forestry*, **41,** 861–8.

Chamberlain, C. J. (1935). *Gymnosperms Structure and Evolution*. Chicago: University of Chicago Press.

Florin, R. (1936a). Die fossilen Ginkgophyten von Franz-Joseph-Land nebst Eröterungen über vermeintliche Cordaitales mesozoischen Alters. I. Spezieller Teil, *Palaeontographica B*, **81,** 71–173.

Florin, R. (1936b). Die fossilen Ginkgophyten von Franz-Joseph-Land nebst Erörterungen über vermeintliche Cordaitales mesozoischen Alters. II. Allgemeiner Teil. *Palaeontographica B*, **82,** 1–72.

Florin, R. (1949). The morphology of *Trichopitys heteromorpha* Saporta, a seed plant of Paleozoic age, and the evolution of the female flowers in the Ginkgoinae. *Acta Horti Bergiani*, **15,** 79–109.

Florin, R. (1951). Evolution in cordaites and conifers. *Acta Horti Bergiani*, **15,** 285–388.

Fujii, K. (1896). On the different views hitherto proposed regarding the morphology of the flowers of *Ginkgo biloba* L. *Botanical Magazine Tokyo*, **10,** 7–8, 13–15, 104–10.

Harris, T. M. (1935). The fossil flora of Scoresby Sound, East Greenland 4: Ginkgoales, Coniferales, Lycopodales, and isolated fructifications. *Meddelelser om Grønland*, **112,** 1–176.

Harris, T. M. (1974). *The Yorkshire Jurassic Flora. IV. Ginkgoales and Czekanowskiales*. London: British Museum (Natural History).

Harris, T. M. (1976). The Mesozoic gymnosperms. *Review of Palaeobotany and Palynology*, **21,** 119–34.

Hughes, N. F. (1976). *Palaeobiology of Angiosperm Origins*. Cambridge: Cambridge University Press.

van Konijnenburg-van Cittert, J. H. A. (1971). In situ gymnosperm pollen from the Middle Jurassic of Yorkshire. *Botanica Neerlandica*, **20,** 1–96.

Krassilov, V. A. (1970). Approach to the classification of Mesozoic "Ginkgoalean" plants from Siberia. *Paleobotanist*, **18,** 12–19.

Krassilov, V. A. (1972). *Mesozoic flora of Burei* (*Ginkgoales and Czekanowskiales*). Moscow: Far-East Institute, Academy of Science.

Lundblad, A. B. (1959). Studies in the Rhaeto–Liassic floras of Sweden, II: 1. Ginkgophyta from a mining district of N W Scania. *Kungl. Svenska Vetenskapsakademiens Handlingar*, **6,** 3–38.

Mastrogiuseppe, J. D., Cridland, A. A., & Bogyo, T. P. (1970). Multivariate comparison of fossil and recent *Ginkgo wood. Lethaia*, **3,** 271–7.

Meyen, S. V. (1987). *Fundamentals of Palaeobotany*. London: Chapman and Hall.

Rothwell, G. S. (1987). The role of development in plant phylogeny: A paleobotanical perspective. *Review of Palaeobotany and Palynology*, **50,** 97–114.

Seward, A. C. (1919). *Fossil Plants*. Vol. IV. Cambridge: Cambridge University Press. Reprint. New York: Macmillan (Hafner Press), 1969.

Tralau, H. (1968). Evolutionary trends in the genus *Ginkgo*. *Lethaia*, **1,** 63–101.

28

The first coniferophytes

Unlike the Ginkgoales, which are represented in our present-day flora by a single species, the coniferophytes are highly diversified, with 51 genera and approximately 550 species, belonging to the Coniferales and Taxales. In addition, there are numerous extinct species belonging to the Cordaitales, Voltziales, and Coniferales. As the diversity within major groups of vascular plants increases, the task of presenting an adequate picture of their origins and relationships (Chart 29.1) becomes more difficult and we are forced, because of space limitations, to make more and more generalizations. Thus, when we try to characterize the coniferophytes we must be cognizant of the fact that there are many exceptions because of the great variability within the group.

Some characteristics of conifers

The conifers are woody plants that are highly branched trees or shrubs. The wood is pycnoxylic (Fig. 28.1) with circular bordered pits mostly confined to the radial walls of the tracheids. The rays are usually uniseriate, but may be multiseriate. In addition to ray parenchyma cells, rays may have ray tracheids. Wood parenchyma is also present in small amounts. The primary xylem associated with the eusteles of their stems is endarch. The pith and the cortex are relatively small compared with the development of the wood. The simple needlelike leaves are in whorls, opposite or spirally arranged. There are usually one or two veins in the leaf, but some have broad leaves with many veins. The stomata are haplocheilic.

Depending on the species, the plants are monoecious or dioecious and normally have monosporangiate cones. The ovulate cones are compound, except for the Taxales, where ovules terminate branches. Pollen cones are interpreted to be simple. Pollen grains are usually bisaccate (Fig. 28.2A). Germination of the pollen grain on the nucellus is distal and results in a microgametophyte consisting of a pollen tube with two nonflagellated sperms. In general, the stages in development of male and female gametophytes exhibit many features shared with cycads and *Ginkgo*. Polyembryony and polycotyledony are prevalent among the conifers.

From the preceding two paragraphs it should not be difficult to pick out those characteristics that, in a general way, distinguish the coniferophytes from the cycads (Chapter 24) and cycadeoids (Chapter 25). You will recall that most but not all cycads tend to be smaller and less branched than coniferophytes. Their secondary wood is manoxylic with, however, some notable exceptions. They have a relatively large pith and cortex in their stems. Their leaves are usually pinnate fronds with a large number of leaf traces supplying the rachis. Their cones are simple, composed of modified mega- and microsporophylls. Unlike the cycads where ovules and pollen organs are borne on modified leaves, ovulate and pollen cones of conifers are interpreted as fertile shoots usually borne in the axils of bracts or scale leaves.

The simple pollen cones of most conifers (Fig. 28.3A) are not more than a few centimeters in length and much smaller than those of cycads. With the exceptions of the Cupressaceae, pollen cones of conifers have their microsporophylls spirally arranged and borne directly on the cone axis. In the Cupressaceae they are cyclic in arrangement, and the pollen cones are borne terminally on specialized lateral shoots. Most microsporangia of living conifers develop on the abaxial surface of the sporophyll, which may be a planated leaflike structure or peltate. Members of the Pinaceae have two microsporangia per sporophyll, but the number varies from 2 to 15 for species of other families of the Coniferales.

The controversy about the morphology of the compound ovulate cones of conifers compares in intensity and duration with the leaf-gap controversy initiated by the work of Jeffrey. The problem is revealed by the structure of the ovulate cone of *Pinus* (Fig. 28.3B) where, by dissection or making longitudinal sections, one can see what appear to be simple megasporophylls attached in spirals to the cone axis. Closer examination reveals that these are, in reality, compound structures, each composed of a flattened ovuliferous scale in the axil of a subtending bract (Fig. 28.2B). In many conifers the ovuliferous scale bears a pair of adaxial inverted ovules near its base (Fig. 28.2C). The bract may be so reduced and adnate with the scale as to be difficult to see by dissection. The ovulate cones of the Douglas fir (Fig. 28.2D), however, have conspicuous forked bracts that extend well beyond the ovuliferous scales (Fig. 28.2E). A longitudinal section will show that in ovulate cones of pine the vascular systems to bract and scale are derived separately from the sympodia of the cone axis. Prior to the researches of Florin (1939, 1944, 1945, 1950a, 1950b, 1954) directed to the solution of the morphological nature of the compound cone, the literature was weighed down with every conceivable explanation of its structure (Florin, 1954). We recall that the ovuliferous scale was homologized with a ligule, a megasporophyll, and a modified branch. Later in this chapter we shall consider the researches of Florin that have led us to the final solution as to the nature of this structure.

Figure 28.1. *Pinus* sp., transverse section through branch showing pith (p); pycnoxylic secondary wood (sx); vascular cambium (vc); and cortex (c). (Courtesy of the General Biological Supply House, Chicago.)

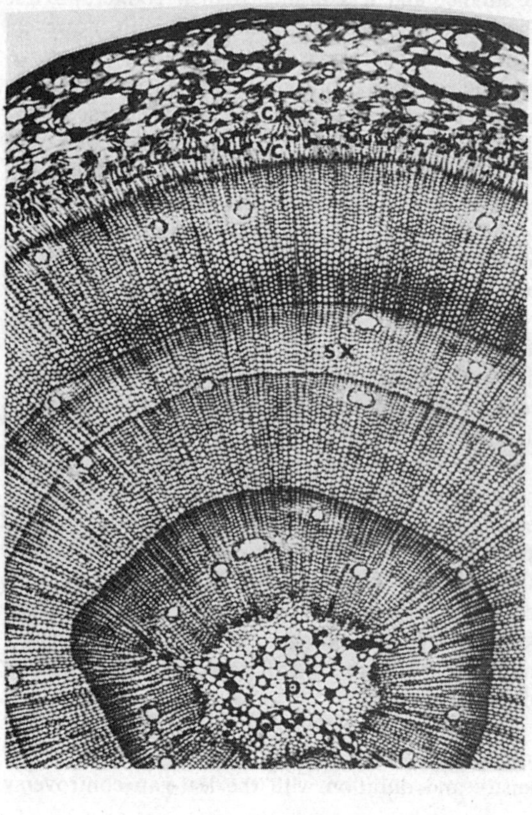

First evidence of conifer organization

For the last time we return to the Devonian for the "beginnings" of a group of vascular plants, this time the coniferophytes (Chart 29.1). It is appropriate that our attention focus on the Progymnospermopsida especially the Archaeopteridales of the Upper Devonian. You should recall (Chapter 21) that Namboodiri & Beck (1968), Beck (1970, 1981), Rothwell (1975b), Scheckler (1978), and several others have shown that *Archaeopteris* exhibits several characteristics found in the Cordaitales, Voltziales, Coniferales, and Taxales. In summary, these characteristics are (1) simple leaves that are spirally arranged; (2) pycnoxylic secondary xylem consisting solely of tracheids and narrow vascular rays that are derived from a bifacial vascular cambium that also produced secondary phloem; (3) a gymnosperm-type eustele, where the leaf traces diverge from sympodia; and (4) heterospory.

At about the time when the first edition of this book was published, there was general agreement that the Archaeopteridales represents a group of plants from which coniferophytes evolved. There is some question, however, whether *Archaeopteris* is on the direct "line" to the cordaites and conifers or a "side branch" (Rothwell, 1975b, 1982a). As we explained in Chapter 21, we continue to consider the Progymnospermopsida to be ancestral to cordaites and conifers, but it is unclear whether they arose from the Archaeopteridales or from some other group of progymnosperms with seed ferns as intermediates.

Cordaitales

Earlier classifications of the Cordaitales included three families – the Cordaitaceae, Pityaceae, and Poroxylaceae. The Lower Carboniferous Pityaceae, represented by the genus *Pitys,* was often thought to provide a connecting link between *Callixylon* of the Upper Devonian and Lower Carboniferous and the Cordaitaceae, first appearing in the Westphalian A (Lower Coal Measures of England and Lower Pennsylvanian of the United States). Researchers during the past three decades have shown that, as suspected, the affinity of *Pitys* is with the pteridosperms (Long, 1963, 1979) and that *Callixylon* is a progymnosperm (Beck, 1960). *Poroxylon,* the only genus assigned to the Poroxylaceae, is now determined to be a pteridosperm, having its affinities with the Callistophytaceae (Rothwell, 1975a). This leaves us with a single family, the Cordaitaceae representing the Cordaitales.

Cordaitaceae

The Cordaitaceae did not become an important element of the Carboniferous flora until Middle Pennsylvanian times, although they are recorded from

Lower Pennsylvanian deposits (Good & Taylor, 1970), which are roughly equivalent to Westphalian A of Great Britain, and may have arisen as early as the Namurian B. From that time on the cordaites flourished well into the Permian. Leaves having the general morphology of *Cordaites* have been reported from the Permo-Carboniferous of Siberia, China, India, Australia, South Africa, and South America. Their remains are particularly abundant at certain localities in the Middle Pennsylvanian, especially in Iowa and Kansas. Some coal balls obtained from these collecting sites contain nothing but leaves, stems, roots, cones, and seeds of these plants, suggesting that they were a predominant part of the forest.

Habit

The often-reproduced reconstruction of a cordaite (Fig. 28.4A) illustrates a tree that presumably attained a height of 30 m and a diameter of more than 1 m at the flared base. The estimate of the height of these trees was obtained from permineralized trunks with wood of the cordaitean type (Seward, 1919). The branches that formed the crown of the plant bore spirally arranged sessile leaves (Fig. 28.5A). Some species of *Cordaites* leaves (Fig. 28.5B) are straplike, bearing a striking resemblance to leaves of the flowering plant *Hippeastrum (Amaryllis)* or the conifer *Agathis*. Branches were formed in the axils of the leaves. The shoots bearing the pollen and ovule-forming fructifications were scattered among the leaves, although some may have been axillary. According to the early reconstructions, the root system is depicted as being shallow and extending laterally for several meters.

More recently, Cridland (1964) has reconstructed a cordaitean plant as a small tree about 5 m high (Fig. 28.4B). His highly speculative reconstruction

Figure 28.2. **A.** *Pinus strobus,* bisaccate pollen grain with microgametophyte. **B.** *P. strobus,* diagram of longitudinal section showing bract–scale complexes (b = bract, os = scale) in ovulate cone. Note ovule (o) with micropyle directed toward the cone axis. **C.** Single ovuliferous scale of *Pseudotsuga* sp. showing two ovules on upper surface. **D.** *P. taxifolia,* branch bearing first-year seed cone. **E.** A single bract–scale complex of *P. taxifolia.* Note the well-developed bract (b) subtending the ovuliferous scale (os). **A–E:** Extant.

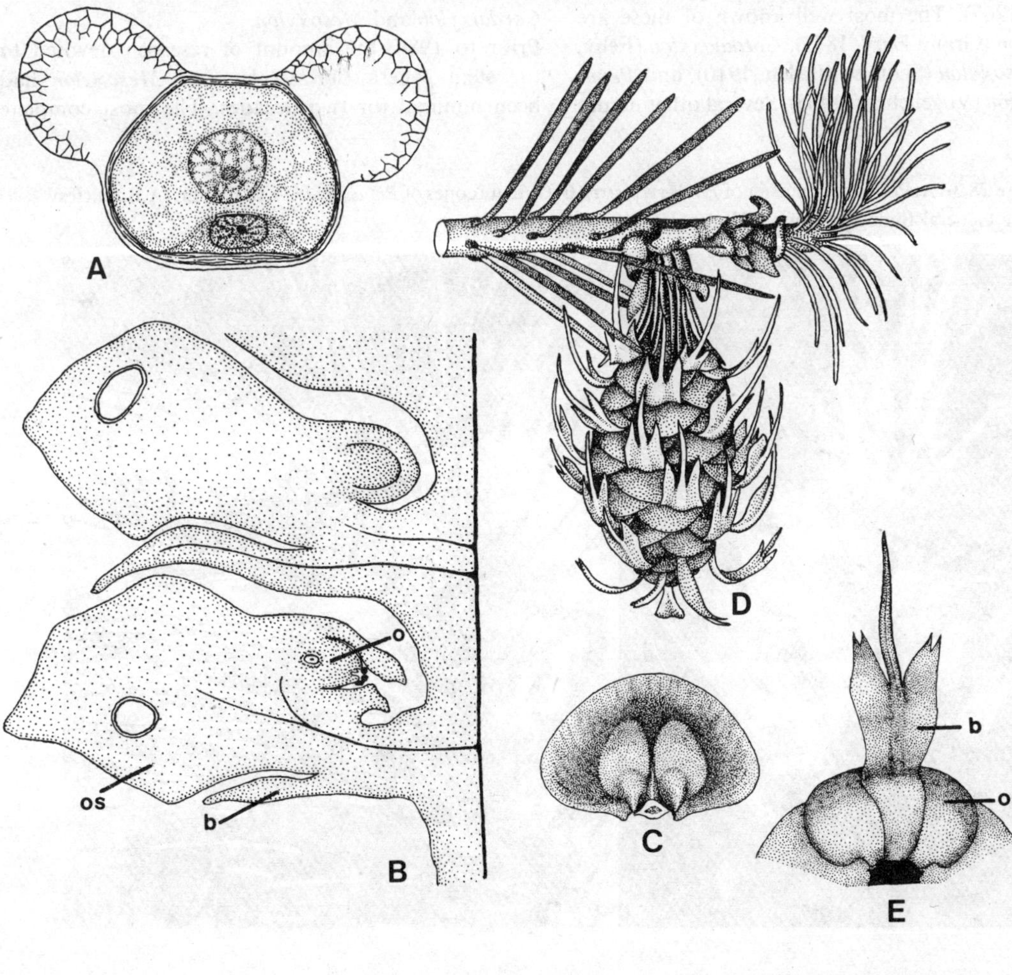

depicts the root system as being composed of stilt roots not unlike those of mangroves and suggests that like mangroves the habitat of the cordaites was in swamps along marine shores and estuaries. Ongoing studies of cordaitean remains suggest that at least one species was a scrambling shrublike plant (Fig. 28.4C).

Some problems of nomenclature
The genus *Cordaites* as we understand it today only applies to detached leaves of cordaitean plants (Arnold, 1967). This means that the name *Cordaites* should be abandoned as an organ genus for structurally preserved cordaitean stems. Researches conducted during the last four decades (Traverse, 1950; Cohen & Delevoryas, 1959; Baxter, 1959) suggest that there are two different types of anatomically preserved cordaitean stems. Recent studies confirm that the stem types represent separate genera of whole plants (Rothwell & Warner, 1984; Trivett & Rothwell, 1985, 1988; Trivett, 1991). Because the name *Cordaites* applies only to detached leaves, a valid name had to be substituted for *Cordaites* as it applied to anatomically preserved stems. To remedy the situation several workers have proposed adopting a variety of names for genera that are based on specimens with internal anatomy of the stems (Arnold, 1967). The most well known of these are *Cordaixylon* (Grand'Eury, 1877), *Cordaioxylon* (Felix, 1882), *Mesoxylon* (Scott & Maslen, 1910), and *Pennsylvanioxylon* (Vogellehner, 1965). Several other names have been used in the past (Scott, 1923). Although there is some disagreement, *Cordaixylon* is usually used for cordaitean stems showing pith and endarch primary vasculature including leaf and branch traces and wood, while *Mesoxylon* designates similar stems with mesarch xylem maturation.

Another problem is the Paleozoic form genus *Dadoxylon,* which is used for anatomically preserved fragments of pycnoxylic wood. Anatomically, *Dadoxylon* is the same in organization as the Mesozoic and Tertiary form genus *Araucarioxylon*. Some authors attempt to resolve the incongruity of using one generic name for Paleozoic wood fragments and a different name for those from the Mesozoic, but having the same characteristics, by proposing new names. As Andrews (1961) emphasizes, a great many such genera have been established designating such nebulous plant remains, and they contribute very little to our understanding of plant evolution.

Current studies of the anatomical structure of stems, leaves, roots, and pollen and ovulate cones belonging to cordaites have resulted in the reconstruction of whole cordaites plants (Trivett & Rothwell, 1991; Trivett, 1991).

Cordaixylon and *Mesoxylon*
Prior to 1980, the amount of research devoted to the stem genera *Cordaixylon* and *Mesoxylon* had been minimal for two decades. The most complete

Figure 28.3. **A.** Pollen cone cluster of *Pinus sylvestris*. **B.** Ovulate cones of *Pinus rigida*. **A,B:** Extant. (**A,B** courtesy of the General Biological Supply House, Chicago.)

Figure 28.4. **A.** Reconstruction of cordaitean plant. Carboniferous. (From Scott, 1909.) **B.** Reconstruction of cordaitean plant with stilt roots. Pennsylvanian. (From Cridland, 1964.) C. *Cordaixylon dumusum*. Reconstruction of plant up to 2 m tall. Note prostrate stem with adventitious roots and leaves that vary in size. (From Rothwell & Warner, 1984.)

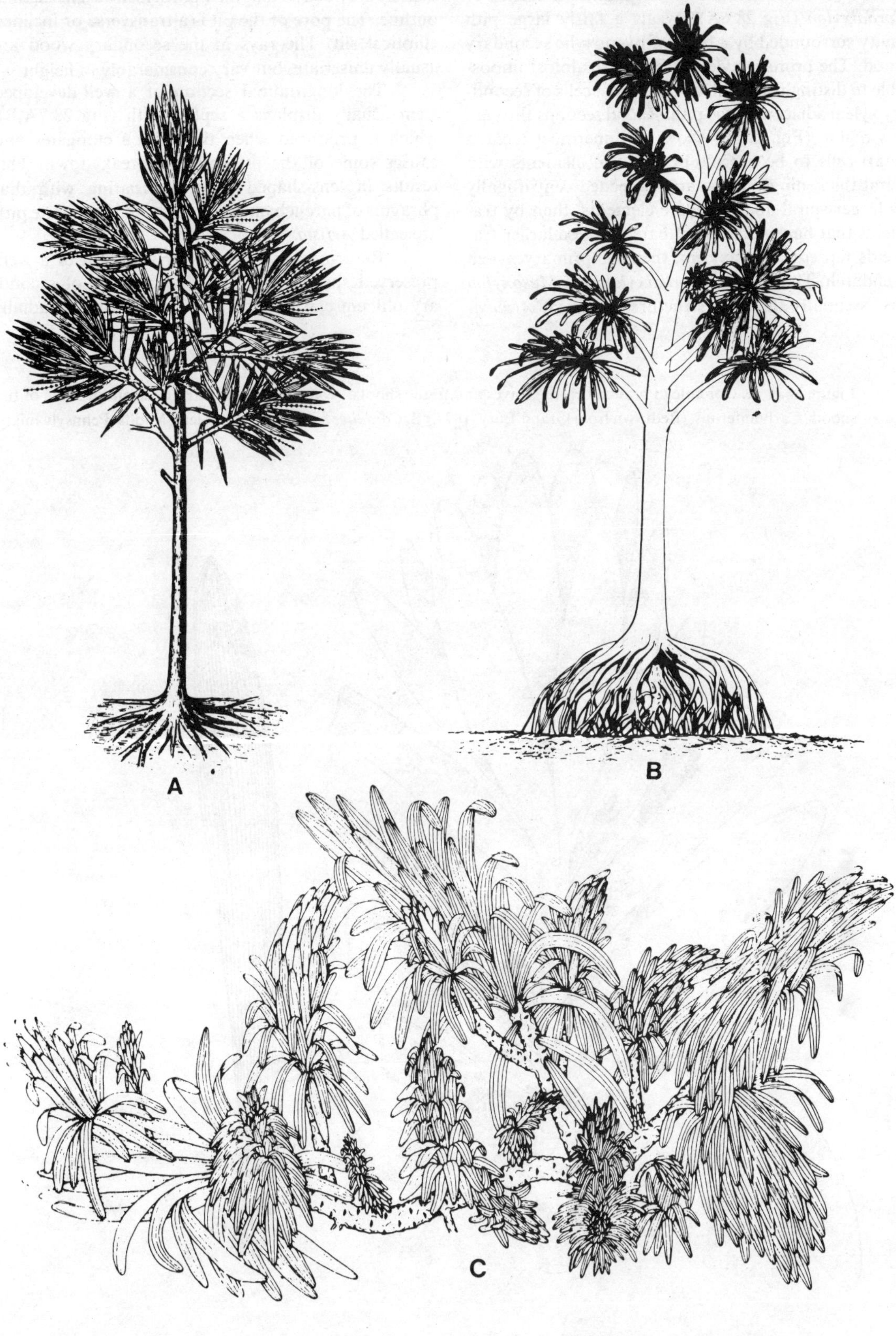

descriptions were provided by Grand'Eury (1877) and Renault (1879), and summarized by Scott (1909) and Seward (1919). The transverse section of a branch of *Cordaixylon* (Fig. 28.6A) reveals a fairly large pith cavity surrounded by a band of pycnoxylic secondary wood. The primary xylem strands are almost impossible to distinguish from the innermost cells of secondary xylem adjacent to the pith. Radial sections through this region (Fig. 28.6B) show the innermost (centripetal) cells to be narrow protoxylem elements with spiral thickenings. These are succeeded centrifugally by larger spiral and reticulate elements, then by tracheids that have scalariform bars. If these larger tracheids represent metaxylem, then the primary xylem is endarch. The wood of *Cordaixylon* and *Mesoxylon* has essentially the same structure as that of *Araucarioxylon*. Both have two or more rows of bordered pits restricted to the radial walls. They are alternate and densely crowded so that their borders have a hexagonal outline. The pore of the pit is a transverse or inclined elliptical slit. The rays in the secondary wood are usually uniseriate, but vary considerably in height.

The longitudinal section of a well-developed stem usually displays a septate pith (Fig. 28.7A,B), which is produced when the branch elongates and causes some of the pith cells to break down. This results in lens-shaped gaps, alternating with diaphragms of parenchyma cells. Casts of the septate pith are called *Artisia*.

Returning to the transverse section of a well-preserved specimen (Fig. 28.7C), a layer of secondary phloem can be identified. It consists of radially

Figure 28.5. **A.** Cordaitean branch bearing leaves and fertile shoots. Note the scattered arrangement of some of the shoots. Carboniferous. (Redrawn from Grand'Eury, 1877.) **B.** *Cordaites principalis*, single leaf. Middle Pennsylvanian.

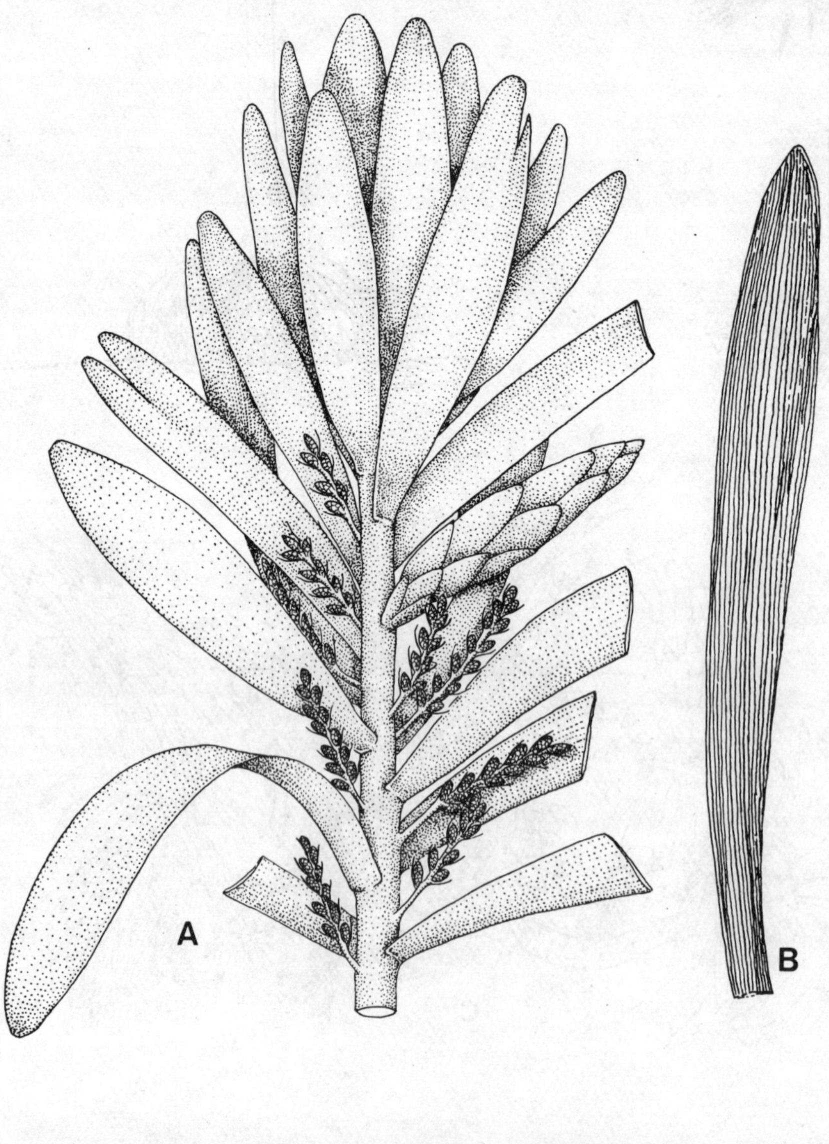

arranged sieve cells, parenchyma, and fibers. In young stems, where the cortex persists, secretory sacs can be found as well as vertical strands of anastomosing fibers that formed a hypodermal system. In older stems there is evidence of early formation of a periderm in the cortex that apparently replaced the cortical tissues as the stems increased in diameter.

The leaf and branch traces of *Cordaixylon* and *Mesoxylon* have been studied in detail by Traverse (1950), Baxter (1959), and Cohen & Delevoryas (1959). Their investigations confirm that the branching of these two genera is axillary. Using serial sections of *Mesoxylon thompsonii* branches, Traverse followed the course of leaf and branch traces from their inception at the periphery of the pith. At this position, the leaf traces appear to be double and are mesarch, as the generic name implies. Each double leaf trace of *M. thompsonii* is flanked by two branch traces. In *Cordaixylon* the primary xylem forms a sympodial system. The sympodia branch at intervals, as they do in conifers, with one strand from each branch becoming a leaf trace. In the axial bundles xylem maturation is either exarch or the protoxylem is indistinct, while leaf traces are exarch in some species and mesarch in others (Trivett, 1991). As they traverse the cortex of the stem and enter a leaf base, the leaf traces dichotomize several times to form 8 to 16 bundles. Branch trace formation varies from species to species, but in *Mesoxylon thompsonii,* the two branch traces retain their identity until they reach the cortex in the axillary position, where they fuse to form the single trace of the branch.

Whole plant concepts of Cordaixylon and Mesoxylon

During the past several years, there has been a concerted effort to develop whole plant concepts for cordaiteans so that now several species are well known. The characteristics used to distinguish *Cordaixylon* and *Mesoxylon* have changed somewhat from the time when these genera were recognized primarily by the characteristics of their stems. Some species have also been transferred from one genus to the other, but *Cordaixylon* and *Mesoxylon* have emerged as distinct genera. The systematics of cordaitean plants recently have been clarified by Trivett (1991), who summarizes the features of each genus. In *Cordaixylon* there is a sympodial system of primary xylem. The cones are small with a relatively round primary axis and are assigned to *Cordaitanthus* when isolated. The pollen sacs of *Cordaixylon* plants are arrranged in a ring at the tip of the fertile scale and produce *Florinites* pollen, while the ovules conform to the genus *Cardiocarpus.* In contrast, *Mesoxylon* stems have no axial bundles at the margin of the pith. They produce larger cones with a rectangular primary axis and are assigned to *Gothania* when found dispersed. The pollen sacs of *Mesoxylon* plants are arranged in a row at the tip of the fertile scale and produce either *Felixipollinites* or *Sullisaccites* prepollen. When isolated, ovules of *Mesoxylon* species with their membranous wings are assigned to the genus *Mitrospernum.*

Species of both genera inhabited a diversity of environments and had a variety of growth architectures. For example, some of the earliest descriptions

Figure 28.6. **A.** *Cordaixylon* sp., diagram of transverse section of young branch. Pith cavity (pc); secondary xylem (sx); axial sympodium (s); leaf trace (lt). Pennsylvanian **B.** Radial section through wood bordering pith. Note, from left to right, spiral elements of protoxylem, scalariform bordered and transitional bordered pits of metaxylem, and alternate multiseriate bordered pits on walls of secondary xylem cells. Carboniferous. (Redrawn from Takhtajan, 1956.)

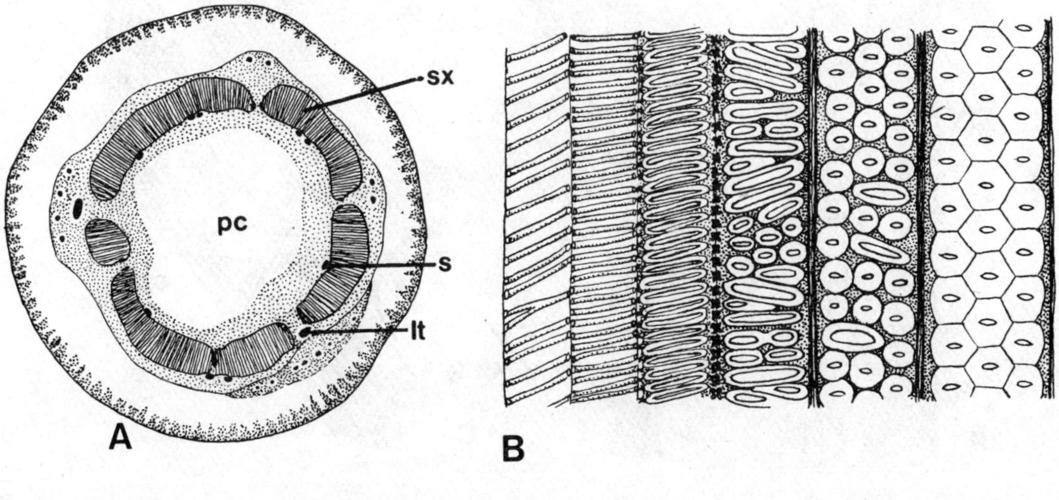

of *Cordaixylon* were from large logs that represent canopy trees from floodplain environments (Grand' Eury, 1877), whereas *Cordaixylon dumusum* (Fig. 28.4C) was a small shrub that grew in the peat-forming swamps (Rothwell & Warner, 1984). *Mesoxylon priapi* and *Cordaixylon iowensis* were both small trees similar to Figure 28.4A, which apparently grew on somewhat elevated ground near the edge of peat-forming swamps.

Cordaites

There is a wealth of information about the leaf genus *Cordaites* (Fig. 28.5B). Seward (1919) has reviewed the literature relating to compression–impression fossils of *Cordaites* up to that time. It reveals the great variety of morphologies to which species names have been assigned. We have collected specimens a few centimeters long, but some attain a length of 1 m and are as much as 15 cm in width. They are linear, but vary in form from lanceolate to spatulate or obovate. Some have pointed tips and others are rounded or blunt. In well-preserved specimens, one can see the parallel veins with occasional dichotomies. These specimens give the impression of a leaf that was leathery. The most reliable characteristics for distinguishing species of *Cordaites* are obtained by studying epidermal structures and internal anatomy (Harms & Leisman, 1961; Good & Taylor, 1970).

Transfer specimens of the relatively thick cuticles taken from *Cordaites* compression fossils show rather uniform rectangular cells in rows, with their long axes

Figure 28.7. **A.** Oblique section through young branch of *Cordaixylon* showing septations in pith region. **B.** Longitudinal section, *Cordaixylon* showing septation of pith into chambers. **C.** *Mesoxylon* sp., transverse section of an exceptional specimen showing large pith (p), surrounded by pycnoxylic secondary xylem (sx) containing leaf and branch traces (bt), secondary phloem (sp), and cortex (c). **A–C:** Pennsylvanian. (**A–C** courtesy of Dr. G. W. Rothwell.)

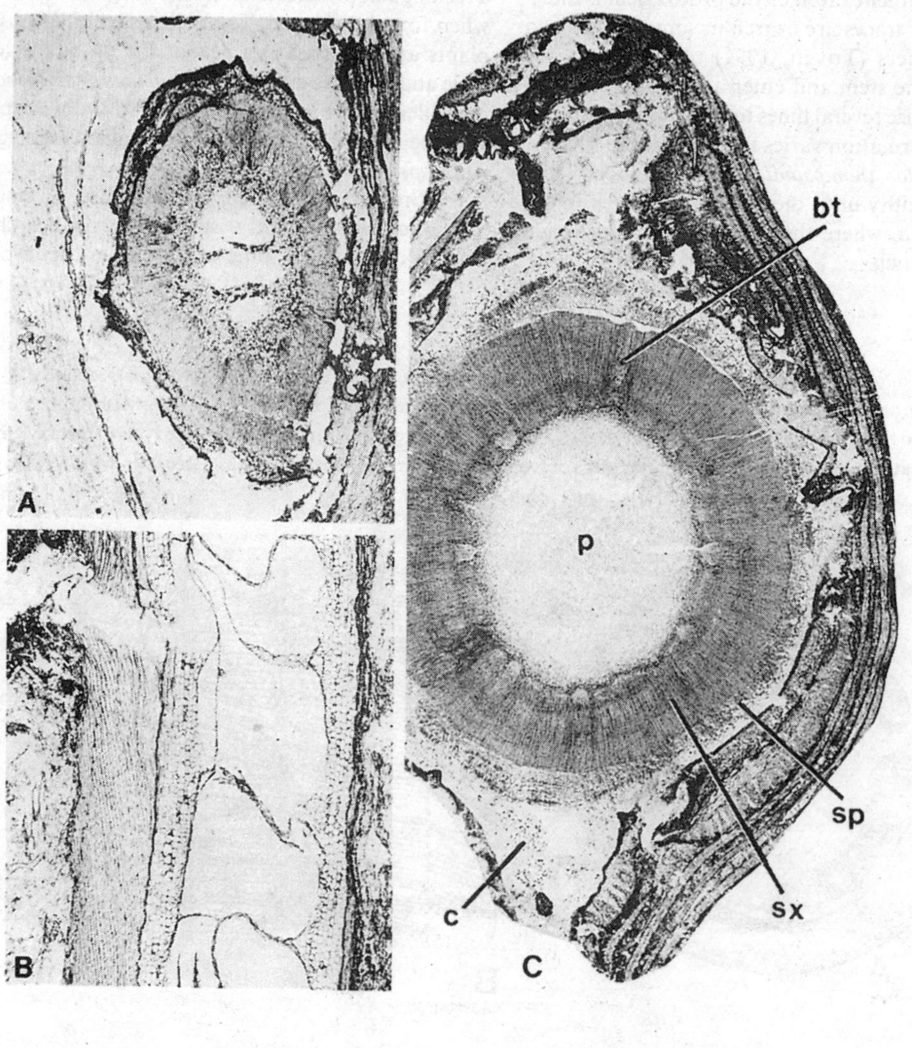

parallel to the long axis of the leaf. The cells have straight, thick walls. Most species have their stomata confined to the lower epidermis, but at least three are known with scattered stomata in the upper epidermis. Stomata of the lower epidermis are arranged in discrete bands between ribs formed by the veins (Fig. 28.8A). Each band is usually composed of several rows (two to four) of stomata consisting of a pair of bean-shaped guard cells elongated in a direction parallel to the veins (Fig. 28.8B). The guard cells are submerged in a pit and are surrounded by four to six subsidiary cells, two of which are polar. The stomatal apparatus is haplocheilic.

In their general internal anatomy (Fig. 28.8C) most species of *Cordaites* have a poorly differentiated palisade layer and a spongy mesophyll with a large number of lacunae between the veins. Hypodermal strands of sclerotic tissue are present in nearly all species. Their distribution and extent of development within the leaf are used to distinguish species. In some, the fibrous strands are confined to the area above and below a vein; others have the strands extending from the upper to the lower epidermis and enclosing the vein; some form an I-beam configuration between veins; and so on. The veins are surrounded by a vein sheath composed of two layers of cells. The xylem of the vein usually has ab- and adaxial development, but some lack centrifugal metaxylem and seem to vary in this characteristic. The phloem, if preserved, is located along the abaxial surface of the xylem.

Amyelon

The detached permineralized roots of cordaitean gymnosperms are placed in the genus *Amyelon*. One of the more comprehensive studies of their structure was prepared by Cridland (1964), who had well-preserved specimens. Because the roots are abundant in some coal-ball material and range from young to old roots. Cridland was able to work out some developmental stages. The cordaitean root system represented by *Amyelon* is highly branched and may have been fairly shallow, forming a pad of stilt roots that supported the stem.

A transverse section of *Amyelon* taken near the base of a branch root has the structure one might expect to find. There is a central exarch actinostele with two to four protoxylem strands, one at the tip of each arm of the stele. This is surrounded by a thick layer of secondary xylem having the same structure as that of the stem. Most roots, even those where there is little wood, have a well-developed periderm that initiates deep within the tissues of the cortex, which sloughs off early in development. As the periderm develops it produces an outer layer of phellem consisting of empty radially arranged cells. Lenticels are formed in these outer layers of the periderm. A broad zone of aerenchymatous phelloderm is formed to the inside, around the stele.

According to Cridland, as the root grew in length, the protostele gave way to a siphonostele by medullation of the metaxylem to form the pith. The protoxylem strands were retained at the margin of the pith where they lie adjacent to the secondary xylem.

The combination of aerenchyma, lenticels, periderm, and medullated protosteles is found in the stilt roots of plants growing in mangrove environments and suggests that some cordaites may have had such a root system (Fig. 28.4B).

Reproductive structures

Both pollen and ovulate fructifications of cordaites are assigned to either *Cordaitanthus* (Fig. 28.8D) or

Figure 28.8. *Cordaites affinis*. **A.** Lower epidermis showing disposition of stomatal furrows. **B.** Stomata in furrows enlarged. **C.** Transverse section of leaf. Note well-developed sclerenchyma (black) above and below veins. **A–C:** Pennsylvanian. (**A–C** redrawn from Reed & Sandoe, 1951.) **D.** *Cordaitanthus* sp. with fertile shoot (fs) in the axil of a leaf (l). The axillary position of the fertile shoot is not a feature of all cordaiteans.

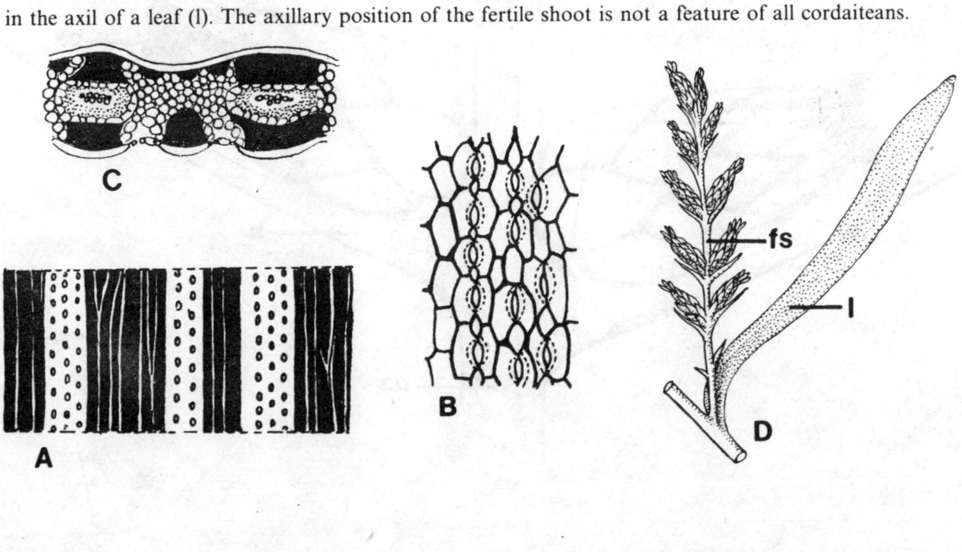

Gothania. These reproductive stuctures, referred to as strobili, inflorescences, or cones, are actually compound fructifications consisting of a primary axis that bears secondary shoots (cones) in the axils of modified leaves called bracts. The secondary shoots have determinate growth and bear spirally arranged modified leaves that we will refer to as scales. Most of the scales are sterile. A few, near or at the distal end of the secondary shoot, however, are fertile and terminate in pollen sacs or ovules. As far as is known, the reproductive structures are strictly monosporangiate.

Cordaitanthus concinnus. Although most species of *Cordaitanthus* are known from compression–impression fossils, abundant permineralized specimens have been discovered occurring in Middle Pennsylvanian coal-ball material (Delevoryas, 1953). From research by Baxter (1959), it seems that these fructifications were born by *Mesoxylon birame*, which occurs at the same localities as *C. concinnus* (Fig. 28.9). The primary axis arising from the axil of a foliage leaf is somewhat flattened and bilaterally symmetrical. It contains a medullated stele that gives off traces to four rows of secondary shoots and their subtending bracts (Rothwell, 1977). The trace to the bract diverges first from the stele and is followed by two shoot traces that fuse, forming a single trace that enters the shoot axis. This trace divides into sympodia that give rise to traces supplying the spirally arranged sterile and fertile scales. It is worth noting that the primary vasculature of the secondary shoot is clearly similar to the eustelic structure of many pteridosperms and all conifers.

There are about 25 to 40 scales on a secondary shoot of which the distal 5 to 10 are fertile (Fig. 28.10A). Each fertile scale usually is terminated with 6 microsporangia (pollen sacs), which are fused at the base (Fig. 28.10B). Pollen grains of the dispersed *Florinites* type (Fig. 28.10C) have been isolated from the microsporangia. *Florinites* is a monosaccate grain with reticulate ornamentation on the inner wall of the saccus (Fig. 22.13C). The latter is attached to the body of the pollen grain (corpus) on both its distal and proximal surfaces (Millay & Taylor, 1974). Pollen grains obtained from pollen sacs of *Gothania*, another Lower Pennsylvanian cordaitalean fructification (Daghlian & Taylor, 1979), are of the *Felixipollenites* dispersed type. These are monosaccate grains with the saccus attached distally to the corpus and have a suture that varies from monolete to trilete (Taylor & Daghlian, 1980). Grains preserved in the pollen cones of *Mesoxylon priapi* are assignable to *Sullisaccites* (Trivett & Rothwell, 1985). They are similar to *Felixipollinites*, but have less distinct ornamentation between arms of the proximal suture. It is quite possible the *Felixipollenites* and *Sullisaccites* represent an earlier stage in the evolution of bisaccate pollen.

Ovulate Cordaitanthus. Until recently, very few permineralized specimens of ovulate cordaitalean fructifications had been discovered and described. Most of our information came from compression–impression fossils originating from the Upper Carboniferous of England and Europe. We owe a debt of gratitude to the late Rudolf Florin (1939, 1950a, 1951) for his meticulous studies of these fossils. It turns out that the morphology and anatomy of the ovulate fructification are similar to those of the pollen organ. In *Cordaitanthus pseudofluitans*, the secondary shoots (dwarf shoots of

Figure 28.9. *Cordaitanthus concinnus*, portion of primary axis (pa) of fertile shoot with two secondary shoots subtended by bracts (b). Each secondary shoot consists of spirally arranged sterile scales (s) and fertile scales terminated by pollen sacs (ps). Pennsylvanian. (From Delevoryas, 1953.)

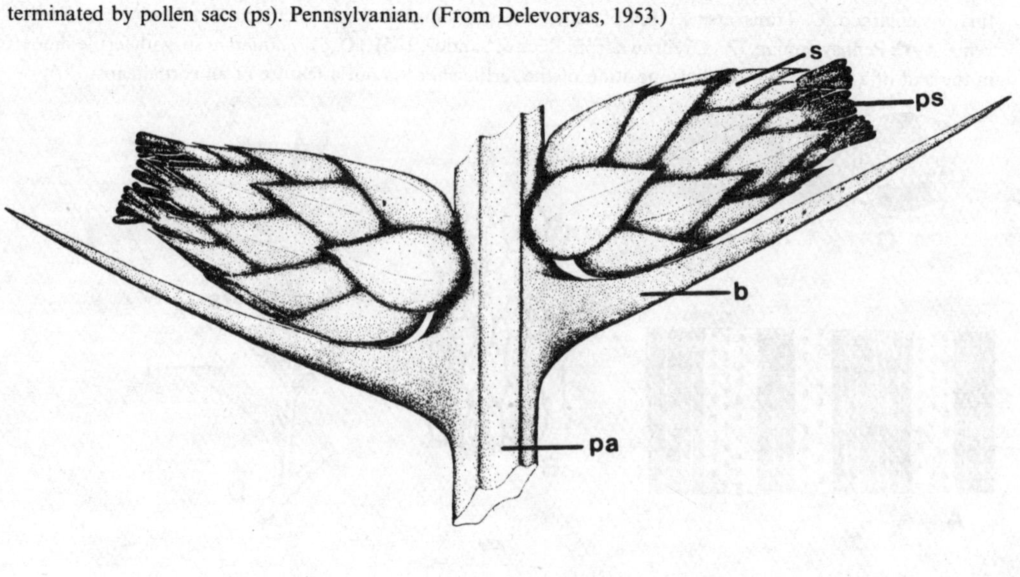

Florin) and their subtending bracts are distichous on the primary axis. Each secondary shoot has 16 to 20 sterile and fertile scales spirally arranged on the shoot axis. Four to six of the distal scales on the shoot are fertile. Each of these may have a single ovule at the tip (Fig. 28.11A), or the fertile scale may dichotomize so that there are two or more pendulous ovules at the tips of the branched scale. In *Cordaitanthus duquesnensis,* now known to be the ovulate cones of *Cordaixylon dumusum,* the ovules terminate fertile scales in the apical region of the cone. The cordate ovules are platyspermic and when found as dispersed compression–

Figure 28.10. *Cordaitanthus concinnus.* Pennsylvanian. **A.** Longitudinal section of secondary shoot showing sterile scales (ss) and pollen sacs (ps) at tips of fertile scales. **B.** Transverse section through six pollen sacs at the distal end of a secondary shoot. (**A,B** from Delevoryas, 1953.) **C.** SEM photograph proximal surface of *Florinites* pollen showing saccus (s) surrounding the corpus (c). (Courtesy of Dr. T. N. Taylor.) **D.** *Cardiocarpus spinatus,* oblique section through ovule showing details of sarcotesta (s), and spiny sclerotesta (sc).

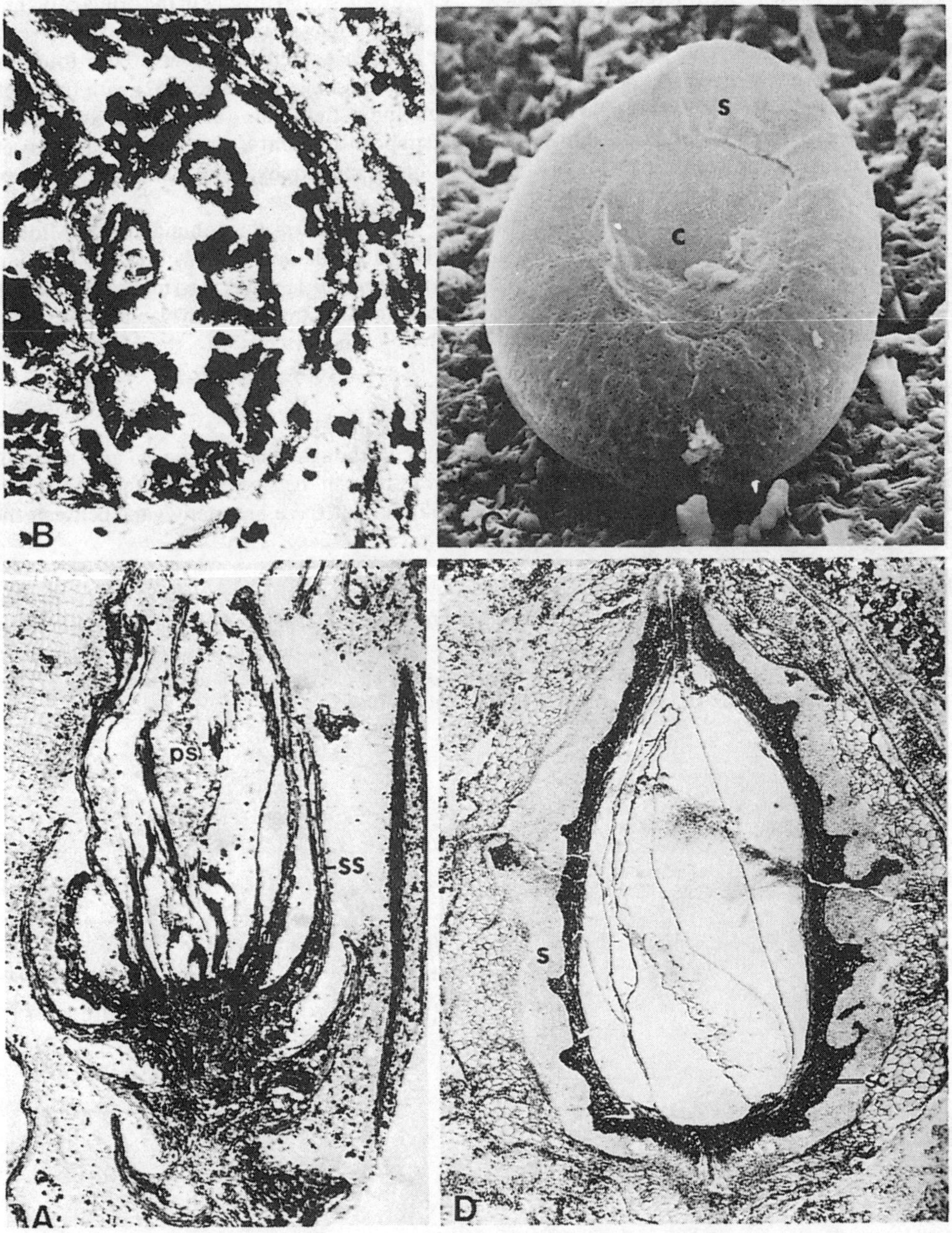

impression fossils they are the *Samaropsis* type. Structurally preserved, permineralized ovules are assigned to *Mitrospermum* and *Cardiocarpus.*

Cardiocarpus and Mitrospermum. On the surface of a single average-sized coal ball, we have counted as many as 20 specimens of *Cardiocarpus spinatus,* with twice this many revealed after sectioning. These ovules are about 15 mm long and have diameters of approximately 10 × 15 mm in the median transverse section. Both *Cardiocarpus* and *Mitrospermum* ovules are biconvex, with the primary plane of the ovule being parallel to the long axis when seen in transverse section and the secondary plane at right angles to it. Ovules sectioned longitudinally on the primary plane are circular to somewhat cordate. The most conspicuous structure of *C. spinatus* is the thick, elaborate testa (Andrews & Felix, 1952; Roth, 1955). The outer zone of the sarcotesta is composed of large thin-walled cells that grade rather abruptly into small cells of the inner sarcotesta. The sclerotesta, from which the specific epithet was derived, has spinelike projections into the sarcotesta (Fig. 28.10D). The spines are irregular and may branch. A thin endotesta composed of delicate cells lines the sclerotesta with its thick-walled sclerotic cells. In the primary plane the sarcotesta of *Cardiocarpus* and *Mitrospermum* is expanded into a flattened keel-like structure that in some species assumes the proportions of a "wing."

The nucellus of both genera is free from the integument except at the base. The distal portion of the nucellus just below the inner opening of the micropyle is differentiated into a beak and a well-developed pollen chamber (Fig. 28.11B). Several investigators have found pollen of the *Florinites* dispersed type in the micropyles and pollen chambers of these ovules.

A single vascular strand enters the base of the ovule and in *Cardiocarpus* it terminates at the base of the nucellus in a pad of tracheids. Two vascular strands arise from this central vascular supply and extend laterally, in the primary plane, into the sarcotesta. The lateral integumentary bundles terminate very near the apex of the ovule. In *Mitrospermum compressum* (Taylor & Stewart, 1964), the two lateral strands pass through the sclerotesta into the sarcotesta, where each strand divides in the primary plane to form several bundles (Fig. 28.11B).

Megagametophytes have been found in many ovules (Andrews & Felix, 1952) with two archegonia flanking a "tent pole" at the distal end. The unusual "tent pole" apparatus, you may recall, has been found in megagametophytes of some pteridosperms and *Ginkgo.*

Cardiocarpus is abundant in Middle Pennsylvanian beds and extends into the Permian, while *Mitrospermum* (Fig. 28.11B) makes its first known appearance in Lower Pennsylvanian (Westphalian A) deposits and extends into the Upper Pennsylvanian (Grove & Rothwell, 1980).

Voltziales

Although the Voltziales are referred to as "transitional" between the conifers and cordaites, we will see that there are many gaps between these two

Figure 28.11. **A.** *Cardiocarpus cordei,* primary axis with ovulate secondary shoots. Note *Samaropsis*-type pendant ovules. Carboniferous. (Based on Zimmermann, 1959.) **B.** *Mitrospermum compressum,* longitudinal section in primary plane. Micropyle (m); pollen chamber (pc); sclerotesta (sc); winglike extension of sarcotesta (s); vascular bundle (vb) entering sarcotesta. **C.** *M. compressum,* transverse section showing platyspermic nature of ovule with extended "wing" (w). Pennsylvanian. (Redrawn from Taylor & Stewart, 1964.)

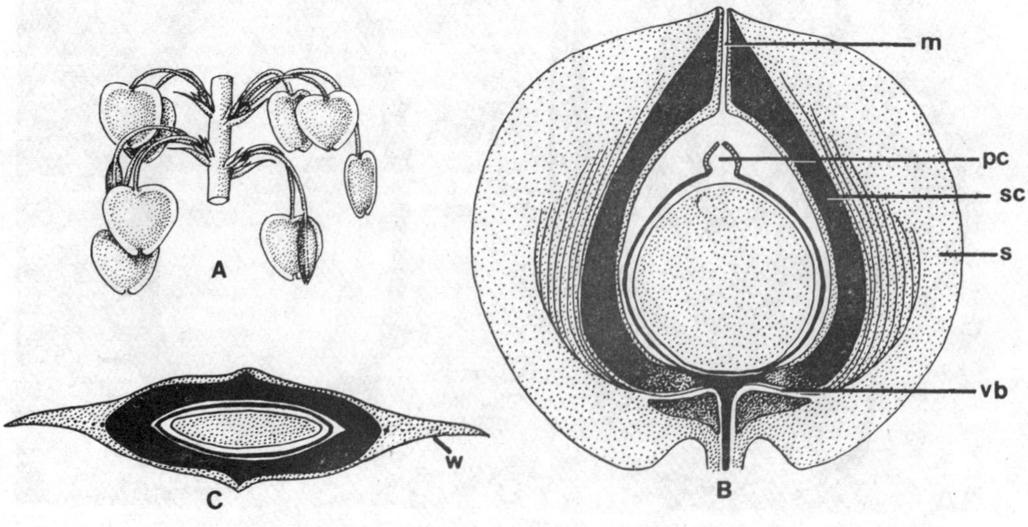

groups not filled by the Voltziales. The Upper Carboniferous and Permian genera of the Voltziales were the objects of intensive study by Florin (1939, 1944, 1945, 1950b, 1951). Of these articles, those published in 1950b and 1951 are excellent review articles written in English. Additional comprehensive reviews of the Upper Palaeozoic and Mesozoic conifers have been prepared by Miller (1977, 1982). It is obvious from these reviews that many of our concepts of most primitive conifers were either formulated by, or heavily influenced by Rudolf Florin. Only recently have new fossils been discovered that significantly expand our understanding of the structure and diversity among the Voltziales. These include an impressive array of compressed conifers from Europe that show cuticular features, and the first extensive permineralized remains of Paleozoic voltzialeans that have been discovered in North America (Clement-Westerhof, 1984, 1987, 1988; Mapes & Rothwell, 1984, 1991).

As we stressed in Chapter 3, the taxonomy and nomenclature for fossil plants can cause complex and confusing problems. A good example of such a problem has been discovered by Clement-Westerhof (1984) among Florin's names for the most primitive conifers. Florin recognized one family of Paleozoic voltzialeans, the Lebachiaceae, which contained four genera. The most well known of these is *Lebachia* for which the family is named. Florin also recognized several form genera that could not be assigned to a family. The most famous of these is *Walchia,* from which the informal category of "walchian conifers" has been derived.

Clement-Westerhof (1984) has argued that *Lebachia* is unfortunately an illegitimate name, and that new names are needed for both the genus and family. Also, Clement-Westerhof and her colleagues have discovered a greater diversity of walchian conifers than previously suspected, so that new genera and even families of Paleozoic conifers recently have been described (Clement-Westerhof, 1987, 1988; Visscher, Kerp, & Clement-Westerhof, 1986). The simplest and least confusing solution to the problem of legitimate names for Paleozoic voltzialeans recently has been suggested by Mapes & Rothwell, who propose that the name *Utrechtia* be used to replace the genus *Lebachia,* and that Utrechtiaceae replace Lebachiaceae. Other well-known Paleozoic voltzialeans are *Erenstiodendron, Ortiseia* (Clement-Westerhof, 1984), and *Otovicia* (Kerp et al., 1989) of the Utrechtiaceae; *Emporia* of the Emporiaceae (Mapes & Rothwell, 1988, 1991); and *Majonica* and *Dolomitia* of the Majonicaceae (Clement-Westerhof, 1987). *Walchia* is retained as a form genus for remains that can not be assigned to one of these families.

Vegetative morphology and anatomy

The external morphology of the Voltziales is quite different from that of the Cordaitales, from which they were previously thought to have been derived. While modern cladistic analyses place cordaites and conifers as closely related sister groups, an ancestor–descendant relationship is much less certain. According to Florin, those Voltziales recognized as the "walchias" were trees resembling the extant Norfolk Island Pine, *Araucaria heterophylla.* Like these trees, the walchias have spirally arranged, acicular (needlelike to bristlelike) leaves. Older plants had slender monopodial stems (Fig. 28.12A) with a regular arrangement of their branches in whorls. There are usually five to six branches in a whorl, as in *A. heterophylla.* In gross morphology the walchias differ from the cordaites, which lack regular whorled branching and the acicular vegetative leaves. Leaves of *Cordaites,* you will recall, usually are large and straplike and have many veins.

There is a considerable diversity of foliage produced by Upper Carboniferous and Permain Voltziales. For example, the leaves of *Utrechtia* and *Emporia* usually are entire, although ones with bifurcate tips

Figure 28.12. **A.** *Utrechtia Piniformis,* apex of shoot showing whorled branches. **B.** *Utrechtia* sp., branch with spirally arranged leaves. **C.** *Ernestiodendron,* leafy shoot. **A–C:** Permian. (**A–C** from Florin, 1951.)

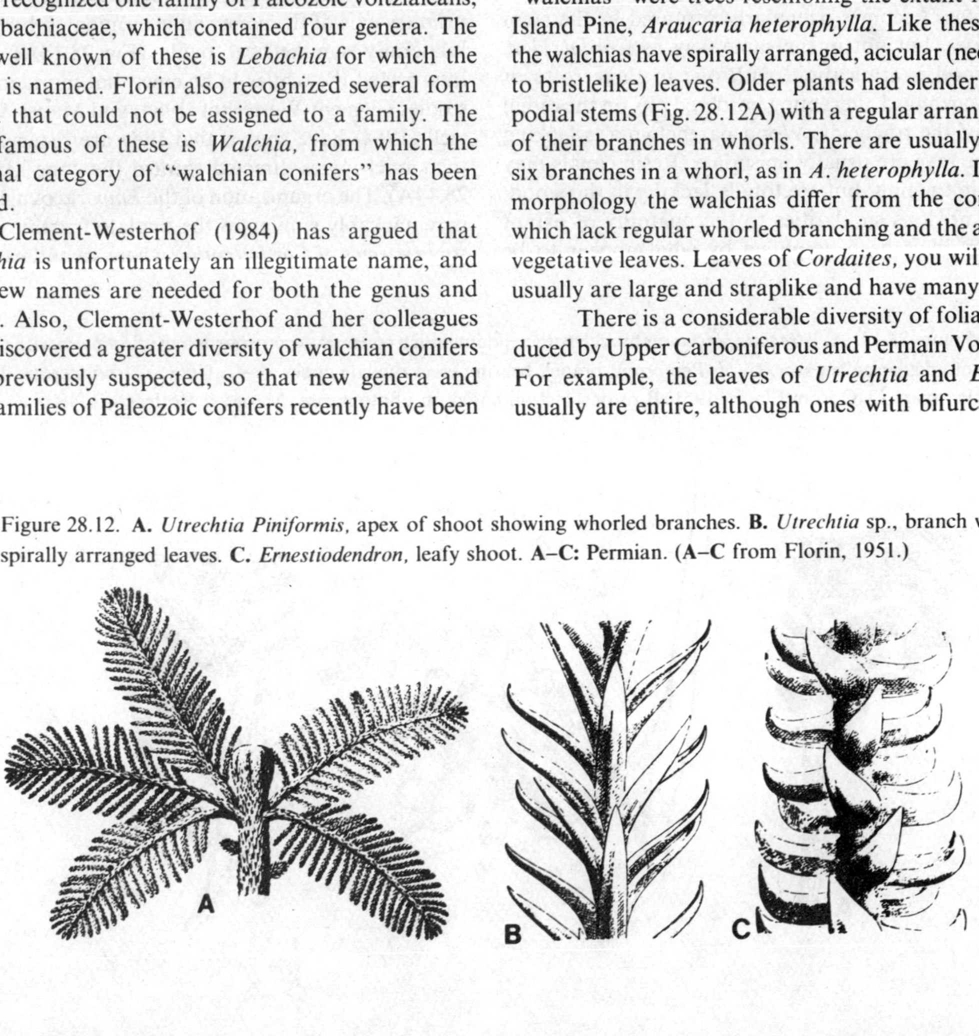

are found on penultimate shoots and bracts of ovulate cones. In *Ortiseia* the leaves and bracts are all entire, and those of all the genera are supplied by a single vein. The leaves of most genera (Fig. 28.12B) are decurrent along the branch and only slightly spreading, while those of *Ernestiodendron* (Fig. 28.12C) are borne at right angles to the branch. The leaves typically are amphistomatic (stomata on upper and lower surfaces). They are haplocheilic, with 4 to 10 subsidiary cells per stomatal apparatus. The stomata on the lower and upper surfaces of *Utrechtia* leaves tend to be longitudinally oriented in two bands. Those of *Ernestiodendron* are in isolated longitudinal rows and show less regular orientation. As indicated earlier, specimens that lack preserved cuticles and thus lack distinctive generic characteristics are placed in the form genus *Walchia*.

Until recently the anatomy of stems and branches of the Upper Paleozoic Voltziales was poorly known because of the lack of well-preserved permineralized specimens. Recently, however, well preserved floras containing permineralized voltzialeans have been described from North America (Rothwell, 1982b; Mapes & Rothwell, 1988; McComas, 1989). The stems have a eustele with endarch primary xylem strands, as in other conifers. The relatively large pith of the eustele sometimes becomes irregularly ruptured as the stem elongates, but not in the same way as *Cordaixylon*. The wood is araucarioid with one to three rows of closely arranged alternating bordered pits on the radial walls of the tracheids. Wood parenchyma is lacking and the rays are usually uniseriate. Resin canals may occur in the pith, but are totally lacking in the wood. These obvious similarities to the anatomy of extant conifers have been amplified by what appear to be resin ducts without epithelial cells in the pith and cortex of twigs, as well as mesophyll of leaves. The characteristics of leaves found attached to twigs also are those of extant conifers, except that the leaf traces in the fossils do not divide to form two or more bundles in a leaf.

Fructifications of the Carboniferous and Permian Voltziales

In their gross morphology, the fructifications of voltzialeans have the appearance of cones. The plants are monoecious with their monosporangiate cones borne at the tips of leafy branches where ovulate cones tend to be upright (Fig. 28.13C) and the pollen cones pendulous (Fig. 28.13A). The ovulate cone of *Utrechtia* (Fig. 28.13C) is compact and ellipsoidal to cylindrical. Its primary axis bears many spirally arranged bracts with bifurcate tips (Fig. 28.14A). Specimens of *Emporia* show that in the axil of each bract there is a tangentially flattened secondary shoot to which scales are attached. The scales are not helically arranged as Florin (1944) suggested, but appear to have an asymmetrical disposition (Mapes & Rothwell, 1984). In some species of *Utrechtia* all but one of the scales is sterile, and this – the fertile scale – faces the primary axis of the cone and is terminated by a single bilaterally symmetrical ovule (Fig. 28.14A). Florin interpreted the ovules to be erect, but in independent studies Clement-Westerhof (1984) and Mapes & Rothwell (1984) have shown that they are recurved with their micropyles directed toward the cone axis (Fig. 28.14A). The organization of the *Emporia* ovulate cone is remarkably similar to the ovulate fructification of some species of *Cordaitanthus* where secondary shoots

Figure 28.13. **A.** *Utrechtia piniformis* branch with pendant pollen cones. **B.** Microsporophyll of *Emporia* with cluster of adaxial pollen sacs. **C.** *U. Piniformis,* branch bearing erect ovulate cones. **A–C:** Upper Carboniferous–Lower Permian. (**A,C** from Florin, 1951; **B** reconstruction based on photographs, Mapes & Rothwell, 1988.)

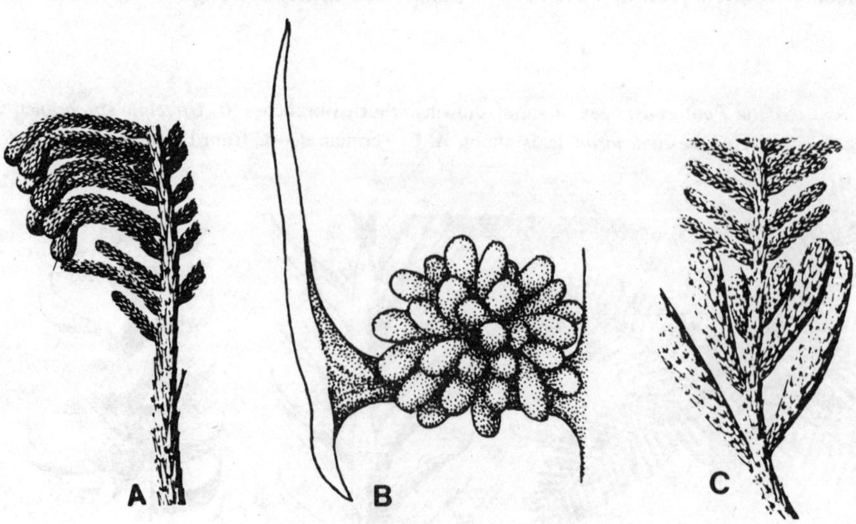

in the axils of bracts bear ovules on fertile scales. The primary differences are that secondary shoot–bract complexes of *Cordaitanthus* are lateral and distichous on the primary axis and spirally arranged in *Emporia*, and that the cordaitean ovules are erect while those of *Emporia* are inverted. Regardless of these differences, the ovulate fructifications of both genera are compound structures, and this is the single most important characteristic that relates the Cordaitales and Voltziales.

Other genera, for example *Ernestiodendron* (Fig. 28.14B) and *Walchiostrobus* (Fig. 28.14C), have ovulate cones with the same basic organization as that of *Emporia*. The secondary shoots of Upper Carboniferous and Lower Permian species of *Ernestiodendron* and *Walchiostrobus* have several ovule-bearing scales instead of one or two as is typical for *Emporia*. Their ovules may be erect or reflexed and, in *Ernestiodendron*, all of the scales on a secondary shoot are fertile. *Walchiostrobus* has sterile scales on the proximal portion of the secondary shoot and fertile ones distally.

The radially symmetrical cylindrical to ellipsoidal pollen cones of the walchians have a primary axis to which are attached spirally arranged dorsiventral scales or microsporophylls. Because there are no bracts subtending a secondary shoot in these cones, they are interpreted by Florin (1951) and many others as simple structures opposed to the ovulate cones which are compound. This interpretation recently has been supported by the anatomical structure of pollen cones found in association with the ovulate cones of *Emporia*. Each microsporophyll (Fig. 28.13B) has a narrow stalk-like base and a distal leaflike portion that is upturned. Florin interpreted the cones to have two microsporangia on the abaxial surface of each microsporophyll similar to microsporophylls of many conifers. However, the permineralized pollen cones show that there are several pollen sacs attached to the adaxial and lateral surfaces of the stalk (Fig. 28.13B) (Mapes & Rothwell, 1988). Pollen grains isolated from microsporangia of *Emporia* are monosaccate and similar to the *Potonieisporites* dispersed type. Some voltzialeans have bisaccate pollen not unlike that of the Pinaceae.

Ovulate fructifications of the Upper Permian Voltziales

No attempt will be made here to describe each genus of this order. All show some modification of the bract–scale complex, and some such as *Pseudovoltzia*, *Ullmannia*, and *Glyptolepis* may represent intermediate evolutionary stages between the Utrechtiaceae and modern conifers. Although we no longer use this information to infer direct ancestor–descendant relationships, these cones graphically illustrate the homologies described by Florin (1951) among ovuliferous cones of cordaites, voltzialean conifers, and modern conifers (Fig. 28.15). These Permian–Triassic genera of the Voltziaceae do not show the conelike charac-

Figure 28.14. **A.** *Emporia*, reconstruction of cone segment showing bract and fertile shoot. Note the reflexed ovule. Ovule sectioned to show integument (black), nucellus, and megaspore membrane. **B.** Bract–secondary shoot complex of *Ernestiodendron filiciforme*. Note that all scales are fertile with erect ovules. **C.** Secondary shoot of *Walchiostrobus*. Note reflexed ovules. **D.** *Ortiseia jonkeri*, reconstruction of bract and secondary shoot, adaxial view. Bract (b); sterile scale (ss); ovuliferous scale (os); attachment point of single ovule (o). **A–D:** Upper Carboniferous–Permian. (**A** redrawn from Mapes & Rothwell, 1984; **B,C** redrawn from FLorin, 1951; **D** redrawn from Clement-Westerhof, 1984.)

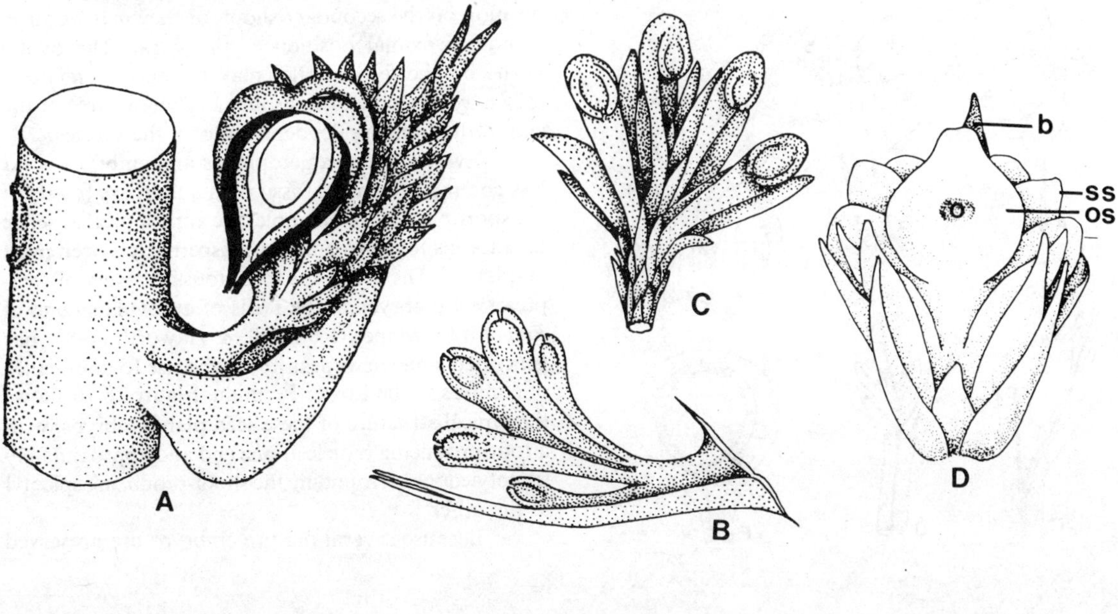

teristics of their ovulate secondary shoots so apparent in the Cordaitales. Instead of radially symmetrical secondary shoots, those of the Voltziaceae are distinctly planated and bilateral, a tendency apparent in the Utrechtiaceae (Clement-Westerhof, 1984) and the Emporiaceae (Mapes & Rothwell, 1984, 1991).

In *Ortesiea* of the Utrechtiaceae (Fig. 28.14D), the number of ovules is typically only one; the vegetative and fertile scales are more highly fused than in *Cordaitanthus* and *Emporia;* and the inverted ovule is fused to the upper surface of the shoot so that it appears to be abaxial in its placement (Clement-Westerhof, 1984).

The combined researches of Florin (1951) and Schweitzer (1963) show that the secondary shoot of *Pseudovoltzia* (Fig. 28.15A) consists of five lobes oriented in a plane. The central and two outer lobes each bear a reflexed ovule adnate at its base to the subtending lobe. The three ovule-bearing lobes and the two smaller sterile lobes are fused laterally to form an ovuliferous scale (Fig. 28.15B) that in turn is fused with a subtending bract at about its midpoint. Schweitzer was able to show that the vascular bundles to the bract and scale are not fused (Fig. 28.15C) and that the lobes and ovules are vascularized, each by a single strand, five of which arise from the end of the scale bundle (Fig. 15D). *Pseudovoltzia* shows clearly the homology of the ovuliferous cone scale and the ovuliferous shoots of cordaiteans, the Utrechtiaceae, and the Emporiaceae.

Glyptolepis (Fig. 28.15E) has ovulate fructifications that are elongated and lax and appear to have been borne in clusters. The secondary shoot is planated and bilaterally symmetrical. It consists of two inverted ovules that flank five to six sterile lobes. This unit is axillary to a bract. The bracts and their secondary shoots are spirally arranged on the lax primary axis.

Ullmannia (Fig. 28.15F) has a single inverted ovule on the adaxial surface of a large orbicular ovuliferous scale subtended by a bract. Whether the ovule is fused to the scale has not been determined. The bract–scale complex of *Ullmannia* is similar to that of *Araucaria* in general morphology and orientation of the single adaxial ovule.

Figure 28.15. *Pseudovoltzia liebeana*. **A.** Ovuliferous scale with three recurved ovules (o). **B.** Abaxial surface of bract–scale complex. Bract (b); ovuliferous scale (os). **C.** Diagram, bract–scale complex showing vascularization. **D.** Vascularization of ovuliferous scale. **E** *Glyptolepis longibracteata*. **F.** *Ullmannia bronnii*, adaxial surface of ovuliferous scale showing single ovule. **A–F:** Permian. (**A–D** redrawn from Schweitzer, 1963; **E,F** redrawn from Florin, 1951.)

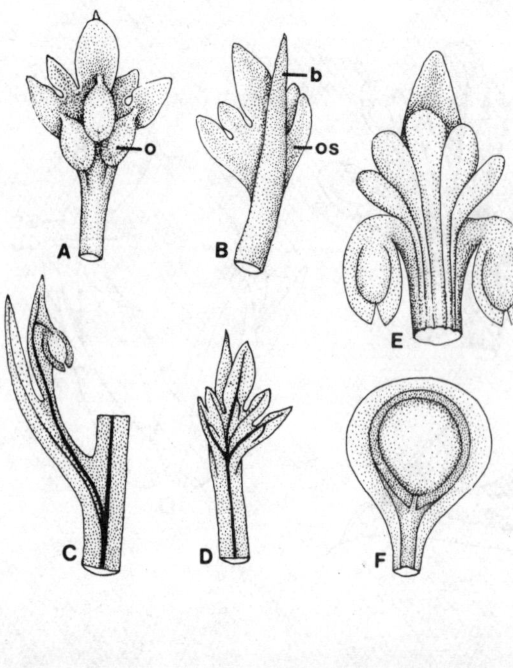

The origin of conifer cones

With the information presented above describing the ovulate fructifications of the Upper Carboniferous–Permian Cordaitales and Voltziales, we are able to interpret the major changes that occurred in the evolution of the compound cones of modern conifers from ancestors of the cordaite type.

Because we do not precisely know how or from what ancestral group the Cordaitales evolved, we are forced to start our interpretation of seed-cone evolution with cordaite ovulate fructifications. Basic to our understanding is that the secondary shoot in the axil of a bract of *Cordaitanthus* is homologous with the bract and axillary ovuliferous scale of a conifer seed cone. In *Cordaitanthus*, the secondary shoot bears (and this is important) spirally arranged sterile and fertile scales with the fertile scales borne in the same spiral sequence as the sterile ones. The fertile scales may occupy a distal position on the secondary shoot, or they may occur in a more proximal position of the spiral. The ovules borne on the fertile scales may be erect or reflexed; they may be single or in groups of two or three at the end of the fertile scale, depending on the species.

Evidence for gametophytes and embryos is far less common for most fossil plants than are fossils of the sporophyte. For example, the almost total absence of embryos for Paleozoic gymnosperms has been most perplexing. Therefore, the fortuitous discovery of well-preserved embryos in the seeds of an ovulate cone of *Emporia* by Mapes, Rothwell, & Haworth (1989) has generated some new ideas about the rise to dominance of conifers in the Lower Permian. Figure 28.16 shows the typical structure of these seed plant embryos, with a root-producing radicle at one end and approximately six cotyledons surrounding the shoot-producing epicotyl at the other.

Because several mature embryos are preserved

within seeds of the same cone, there is evidence that they had reached full size and then gone into a quiescent phase. If correctly interpreted, these embryos suggest that the origin of seed dormancy was associated with the ability of conifers to colonize and establish plant communities in areas that previously had been unforested. This ability may have played a major role in the rapid evolutionary radiation of conifers that occurred during the uppermost Pennsylvanian and Permian.

Utrechtia, Emporia, Ernestiodendron, and *Walchiostrobus* are like *Cordaitanthus* in having leafy secondary shoots composed of sterile and fertile scales. The secondary shoots of *Ernestiodendron* have predominantly fertile scales (four to seven) with erect or reflexed ovules, depending upon the species. *Emporia* is most similar to *Cordaitanthus* because both have sterile and fertile scales intermixed on the secondary shoot. However, in *Emporia* the secondary shoot axis is somewhat flattened, the scales are displaced from a strictly spiral arrangement, and the ovules are clearly inverted (Fig. 28.14A).

Without implying an evolutionary series, but looking to the Voltziales for evolutionary trends leading to the compound ovulate cone, we see (1) a tendency for the reduction of number of ovules on the secondary shoot *Emporia;* (2) the tendency for erect ovules to be inverted with their micropyles directed toward the primary axis of the fructification *Emporia* and *Utrechtia;* and (3) a tendency toward bilateral symmetry by flattening of the secondary shoot and ovule.

Pseudovoltzia, Ullmannia, Glyptolepis, and other Permian and Lower Triassic genera differ basically from their Upper Carboniferous and Lower Permian ancestors in having planated bilaterally symmetrical secondary shoots. *Pseudovoltzia* shows some tangential fusion of the planated sterile and fertile scales to form an ovuliferous scale. The three reflexed ovules have reduced stalks that are adnate to the ovuliferous scale. *Glyptolepis* shows reduction in number of ovules to two. Although the fertile and sterile scales lie in a single plane, they show very little tangential fusion. The maximum tangential fusion of sterile scales, to form an ovuliferous scale of the type found in many conifers, is shown by *Ullmannia.* Here, however, there is reduction to a single inverted ovule.

With new evidence provided by Mapes and Rothwell (1984) from their studies of permineralized specimens of *Emporia,* we now have a clue as to how the shift occurred from radial symmetry of the secondary shoot (Cordaitales) to bilateral symmetry as in the Voltziales. Once bilateral symmetry was established it was followed by tangential fusion of scales to form an ovuliferous scale having adaxial ovules with their micropyles directed toward the primary axis. These, you may recognize, are the characteristics of the compound ovulate cones of modern conifers. The evidence leading to this conclusion leaves no doubt that the ovuliferous scale of a conifer cone is a highly modified secondary shoot in the axil of a bract.

If we are to believe that the Voltziales owe their origin to the Cordaitales, a thesis strongly supported by the work of Florin, then we should have evidence showing us how the simple pollen cones of the Voltziales were derived from the compound pollen cones of the cordaites. Based on a study of the morphology of extant *Podocarpus* pollen cones, Wilde (1944) has offered an interesting and logical explanation that is supported by Banks (1970) and others. Primitive species of *Podocarpus* (Fig. 28.17A) have an organization of their pollen organs similar to that of the cordaites. A primary axis of *Podocarpus* bears secondary shoots (the "cones") in the axils of bracts. Each secondary shoot has spirally arranged sterile and fertile scales.

Figure 28.16. Reconstruction of seed with embryo. Megagametophyte (mg); embryo (e); micropyle (mi). Upper Carboniferous–Permian. (Redrawn from Mapes, Rothwell, & Haworth, 1989.)

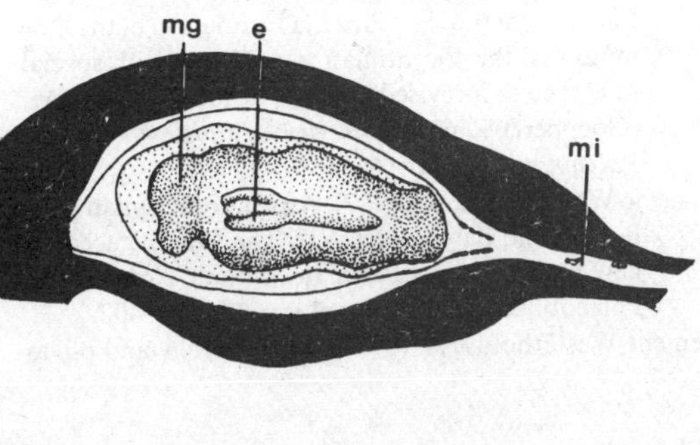

Figure 28.17. **A.** ***Podocarpus spicatus,*** axillary primary shoot bearing secondary shoots in the axils of bracts. Secondary shoots with pollen-producing scales. (Redrawn from Wilde, 1944.) **B.** *Pinus* pollen "cone" cluster with a bract subtending each cone. (From photograph in Banks, 1970.) **A,B:** Extant.

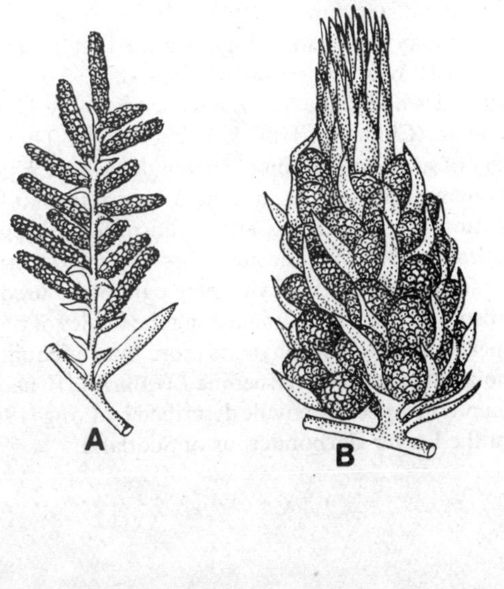

The fertile scales are organized into the "cone," and each scale bears a pair of abaxial pollen sacs. If the primary shoot became reduced so that the secondary shoots were condensed into a spiral, we would have a structure homologous with the cluster of "simple pollen cones" of modern pines. Banks (1970) shows such a cluster (Fig. 28.17B) where it is evident that each "cone" (the secondary shoot) is subtended by a bract. If this is the correct interpretation, then the pollen cone cluster is the basic pollen organ of modern conifers, not the simple pollen "cone." According to this idea, the pollen organ of modern conifers is a compound structure just as is the ovulate cone with its bract–scale complexes. Some support for this theory comes from fossil evidence of compound pollen cones belonging to *Darneya*, a Triassic member of the Voltziales (Grauvogel-Stamm, 1978). In *Darneya peltata*, the microsporangia are borne on small branches arising from the adaxial surface of a cone scale. Grauvogel-Stamm (1977) interprets the latter to be fused with a subtending bract to produce a pollen-forming bract–scale complex homologous with those of compound ovulate cones. It is difficult to confirm this interpretation from the features that are preserved in the compressed specimens of *Darneya*, but the more ancient permineralized pollen cones found in association with *Emporia* do provide such information. You will recall from our earlier description of these pollen cones that they are morphologically quite similar to *Darneya*, except that they have the pollen sacs arising directly from the sporophyll stalk. Mapes & Rothwell (1988) examined the base of the sporophylls and the vascular tissue of the cones to see if evidence of subtending bracts or vascular tissues like those in the ovulate cones was present. No such evidence was found. Rather, the sporophylls were vascularized by a simple trace as would be expected of a sporophyll. This evidence suggests that at least some members of the Voltziales represent two divergent groups, one with simple pollen cones and the other with compound ones. Other authors consider the evidence to be insufficient to determine whether any of the pollen cones of Paleozoic conifers could have had a compound structure.

It may be of some significance that flattened ovules with bilateral symmetry, characteristic of the Cordaitales and some Voltziales, occur in the Upper Devonian (Chaloner, Hill, & Lacey, 1976). The discovery of such platysperms (*Spermolithus*) in the Upper Devonian supports the idea held by some that the evolution of platysperms and radiosperms occurred independently among the ancestors of gymnosperms. The fact, however, that symmetry cannot be used as an absolute criterion distinguishing the ovules of pteridosperms and coniferophytes is more than adequately demonstrated by the platyspermic *Lyrasperma* (Chapter 22), a pteridosperm preovule described by Long (1960) from the Lower Carboniferous of Scotland.

The earliest record of the Voltziales shows that they co-existed with the Cordaitales as early the Westphalian B (Upper Carboniferous). Fragments of leaves having the morphology and cuticular characteristics of conifers have been found by Scott (1974) using bulk maceration techniques of the clays containing leaf remains.

The preceding, it is hoped, makes it clear that the origin of modern conifers directly from the Cordaitales via the "transition conifers" (Voltziales) (Chart 29.1) is far from settled. There are many information gaps that need to be filled, especially with respect to pollen cone morphology. Pant (1977) reflects these doubts about the origin of conifers from the Cordaitales when he presents an alternate hypothesis for their origin from elements of the Glossopteris (Permo-Carboniferous) flora called the Buriadiales. Whether this is a good alternate hypothesis remains to be seen. We envisage that ongoing cladistic analyses will help focus our attention on the most pertinent questions for the resolution of these relationships.

References

Andrews, H. N. (1961). *Studies in Paleobotany*. New York: Wiley.

Andrews, H. N., & Felix, C. J. (1952). The gametophyte of *Cardiocarpus spinatus* Graham. *Annals of the Missouri Botanical Garden*, **39**, 127–35.

Arnold, C. A. (1967). The proper designation of the foliage and stems of the Cordaitales. *Phytomorphology*, **17**, 346–50.

Banks, H. P. (1970). *Evolution and Plants of the Past*. Belmont, Calif.: Wadsworth.

Baxter, R. W. (1959). A new cordaitean stem with paired axillary branches. *American Journal of Botany*, **46**, 163–9.

Beck, C. B. (1960). The identity of *Archaeopteris* and *Callixylon*. *Brittonia*, **12**, 351–68.

Beck, C. B. (1970). The appearance of gymnospermous structure. *Biological Reviews*, **45**, 379–400.

Beck, C. B. (1981). *Archaeopteris* and its role in vascular plant evolution. In *Paleobotany, Paleoecology, and Evolution*, ed. K. J. Niklas. New York: Praeger, pp. 193–230.

Chaloner, W. G., Hill, A. J., & Lacey, W. S. (1976). First Devonian platyspermic seed and its implications in gymnospermous evolution. *Nature*, **265**, 233–5.

Clement-Westerhof, J. A. (1984). Aspects of Permian palaeobotany and palynology. IV. The conifer *Ortiseia* Florin from the Val Gardena Formation of the Dolomites and the Vicentinian Alps (Italy) with special reference to a revised concept of the Walchiaceae (Goeppert) Schimper. *Review of Palaeobotany and Palynology*, **41**, 51–166.

Clement-Westerhof, J. A. (1987). Aspects of Permian palaeobotany and palynology. VII. The Majonicaceae, a new family of Late Permian conifers. *Review of* Palaeobotany and Palynology, **52**, 375–402.

Clement-Westerhof, J. A. (1988). Morphology and phylogeny of Palaeozoic conifers. In *Origin and Evolution of Gymnosperms,* ed. C. B. Beck. New York: Columbia University Press.

Cohen, L., & Delevoryas, T. (1959). An occurrence of *Cordaites* in the Upper Pennsylvanian of Illinois. *American Journal of Botany,* **46,** 545–9.

Cridland, A. A. (1964). *Amyelon* in American coal-balls. *Palaeontology,* **7,** 186–209.

Daghilian, C. P., & Taylor, T. N. (1979). A new structurally preserved Pennsylvanian cordaitean pollen organ. *American Journal of Botany,* **66,** 290–300.

Delevoryas, T. (1953). A new male cordaitean fructification from the Kansas Carboniferous. *American Journal of Botany,* **40,** 144–50.

Felix, G. (1882). Ueber die Versteinerten Hölzer von Frankenberg. *Sitzungsberichte naturforschenden Ges.* Leipzig Neunte Jahrgang, 1882.: 5–9.

Florin, R. (1939). The morphology of the female fructifications in cordaites and conifers of Palaeozoic age. *Botaniska Notiser,* **36,** 547–65.

Florin, R. (1944). Die Koniferen des Oberkarbons und des Untern Perms. *Palaeontographica B,* **85,** 457–654.

Florin, R. (1945). Die Koniferen des Oberkarbons und des Unteren Perms. *Palaeontographica B,* **85,** 655–729.

Florin, R. (1950a). On female reproductive organs in the Cordaitinae. *Acta Horti Bergiani,* **15,** 111–134.

Florin, R. (1950b). Upper Carboniferous and Lower Permian conifers. *Botanical Review,* **16,** 258–82.

Florin, R. (1951). Evolution in Cordaites and conifers. *Acta Horti Bergiani,* **15,** 285–388.

Florin, R. (1954). The female reproductive organs of conifers and taxads, *Biological Reviews,* **29,** 367–89.

Good, C. W., & Taylor, T. N. (1970). On the structure of *Cordaites felicis* Benson from the Lower Pennsylvanian of North America. *Palaeontology* **13,** 29–39.

Grand'Eury, F. C. (1877). Flore Carbonifère du Department de la Loire et du centre de la France. *Memoire de l'Académie des Sciences, Paris,* **24,** 1–624.

Grauvogel-Stamm, L. (1978). La flore du grès à *Voltzia* (Buntsandstein Supérieur) de Vosges du Nord (France), interprétations phylogénetique paléogéographique. *Science Géologiques, Université Louis Pasteur de Strasbourg, Institut de Géologie, Mémoire,* **50,** 1–225.

Grove, G. G., & Rothwell, G. W. (1980). *Mitrospermum vinculum* sp. nov., a cardiocarpalean ovule from the Upper Pennsylvanian of Ohio. *American Journal of Botany,* **67,** 1051–8.

Harms, V. L., & Leisman, G. A. (1961). The anatomy and morphology of certain *Cordaites* leaves. *Journal of Paleontology,* **35,** 1041–64.

Kerp, J. H. J., Poort, H. A., Swinkels, J. M., & Verwer, R. (1989). Aspects of Permian Palaeobotany and palynology. IX. Conifer dominated Rotliegend floras from the Saar-Nahe Basin (? Upper Carboniferous-Lower Permian; SW-Germany) with special reference to the reproductive biology of early conifers. *Review of Palaeobotany and Palynology,* **62,** 205–48.

Long, A. G. (1960). On the structure of "*Samaropsis scotica*" Calder (emended) and "*Eurystoma angular*" gen. et. sp. nov., petrified seeds from the Calciferous Sandstone Series of Berwickshire. *Transactions of the Royal Society of Edinburgh,* **64,** 261–80.

Long, A. G. (1963). Some species of "*Lyginorachis papilio*" Kidston associated with stems of "Pitys." *Transactions of the Royal Society of Edinburgh,* **65,** 211–24.

Long, A. G. (1979). Observations on the Lower Carboniferous genus *Pitus* Witham. *Transactions of the Royal Society of Edinburgh,* **70,** 111–27.

Mapes, G., & Rothwell, G. W. (1984). Permineralized ovulate cones of *Lebachia* from Late Palaeozoic limestones of Kansas. *Palaeontology,* **27,** 69–94.

Mapes, G., & Rothwell, G. W. (1988). Diversity among Hamilton conifers. In *Regional Geology and Paleontology of Upper Paleozoic Hamilton Quarry Area in Southeastern Kansas,* eds. G. Mapes & R. H. Mapes. Guidebook Series 6, Kansas Geological Survey, Lawrence, Kansas.

Mapes, G., & Rothwell, G. W. (1991). Structure and relationships of primitive conifers. *Neues Jahrbuch für Geologie Palaeontollogie,* **183,** 269–87.

Mapes, G., Rothwell, G. W., & Haworth, M. T. (1989). Evolution of seed dormancy. *Nature,* **37,** 745–46.

McComas, M. A. (1989). Floristics and depositional ecology of two Upper Pennsylvanian (Lower Conemaugh floras from the 7–11 Mine in Northeastern Ohio), Unpublished thesis, Ohio University, Athens, Ohio.

Millay, M. A., & Taylor, T. N. (1974). Morphological studies of Paleozoic saccate pollen. *Palaeontographica B,* **147,** 75–99.

Miller, C. N. (1977). Mesozoic conifers. *Botanical Review,* **43,** 218–71.

Miller, C. N. (1982). Current status of Paleozoic and Mesozoic conifers. *Review of Palaeobotany and Palynology,* **37,** 99–114.

Namboodiri, K. K., & Beck, C. B. (1968). A comparative study of the primary vascular system of conifers. III. Stelar evolution in gymnosperms. *American Journal of Botany,* **55,** 464–72.

Pant, D. D. (1977). Early conifers and conifer allies. *Journal of the Indian Botanical Society,* **56,** 23–37.

Reed, F. D., & Sandoe, M. T. (1951). *Cordaites affinis:* a new species of cordaitean leaf from American coal fields. *Bulletin of the Torrey Botanical Club,* **78,** 449–57.

Renault, B. (1879). Structure comparèe de quelques tiges de la flora Carbonifère. *Nouvelles Archives du Museum, Ser. 2,* **2,** 213–348.

Roth, E. A. (1955). The anatomy and modes of preservation of the genus *Cardiocarpus spinatus* Graham. *University of Kansas Science Bulletin,* **37,** 151–74.

Rothwell, G. W. (1975a). The Callistophytaceae (Pteridospermopsida): I. Vegetative structure. *Palaeontographica B,* **151,** 171–96.

Rothwell, G. W. (1975b). Primary vasculature and gymnosperm systematics. *Review of Palaeobotany and Palynology,* **22,** 193–206.

Rothwell, G. W. (1977). The primary vasculature of *Cordaitanthus concinnus. American Journal of Botany,* **64,** 1235–41.

Rothwell, G. W. (1982a). New interpretation of the earliest conifers. *Review of Palaeobotany and Palynology,* **37,** 7–28.

Rothwell, G. W. (1982b). *Cordaitanthus duquesnensis* sp. nov., anatomically preserved ovulate cones from the

Upper Pennsylvanian of Ohio. *American Journal of Botany*, **69,** 239–47.

Rothwell, G. W. (1988). Cordaitales. In *Origin and Evolution of Gymnosperms*, ed. C. B. Beck. New York: Columbia University Press.

Rothwell, G. W., & Warner, S. (1984). *Cordaixylon dumusum* n. sp. (Cordaitales). I. Vegetative structures. *Botanical Gazette*, **145,** 275–91.

Scheckler, S. E. (1978). Ontogeny of progymnosperms. II. Shoots of Upper Devonian Archaeopteridales. *Canadian Journal of Botany*, **56,** 3136–70.

Schweitzer, H. J. (1963). Der weibliche Zapfen von *Pseudovoltzia Liebeana* und seine Bedeutung für die Phylogenie der Koniferen. *Palaeontographica B*, **113,** 1–29.

Scott, A. (1974). The earliest conifer. *Nature*, **252,** 707–8.

Scott, D. H. (1909). *Studies in Fossil Botany*, 2nd ed. London: Black, Vol. II.

Scott, D. H. (1923). *Studies in Fossil Botany*. 3rd. ed., London: Black, Vol. 2.

Scott, D. H., & Maslen, A. J. (1910). On *Mesoxylon*, a new genus of Cordaitales – preliminary note. *Annals of Botany*, n. s., **24,** 236–39.

Seward, A. C. (1919). *Fossil Plants*. New York, Vol. IV. Cambridge: Cambridge University Press. Reprint Macmillan (Hafner Press), 1969.

Takhtajan, A. L. (1956). *Telomophyta*. Moscow: Academiae Scientiarum.

Taylor, T. N., & Daghlian, C. P. (1980). The morphology and ultrastructure of *Gothania* (Cordaitales) pollen. *Review of Palaeobotany and Palynology*, **29,** 1–14.

Taylor, T. N., & Stewart, W. N. (1964). The Paleozoic seed *Mitrospermum* in American coal balls. *Palaeontographica, B*, **115,** 51–8.

Traverse, A. (1950). The primary vascular body of *Mesoxylon thompsonii*, a new American cordaitalean. *American Journal of Botany*, **37,** 318–25.

Trivett, M. L. (1991). Modeling the growth architecture of fossil plants. Unpublished Ph.D. dissertation, Ohio University, Athens, Ohio.

Trivett, M. L., & Rothwell, G. W. (1985). Morphology, systematics and paleoecology of Paleozoic fossil plants: *Mesoxylon priapi*, sp. nov. (Cordaitales). *Systematic Botany*, **10,** 205–23.

Trivett, M. L., & Rothwell, G. W. (1988). Diversity among Paleozoic Cordaitales: The vascular architecture of *Mesoxylon birame* Baxter. *Botanical Gazette*, **149,** 116–25.

Trivett, M. L., & Rothwell, G. W. (1991). Diversity among Paleozoic Cordaitales. *Neues Jahrbuch für Geologie. Paläontologie.*, **183,** 289–305.

Visscher, H., Kerp, J. H. J., & Clement-Westerhof, J. A. (1986). Aspects of Permian palaeobotany and palynology. VI. Towards a flexible system of naming Palaeozoic conifers. *Acta Bot. Neerl*, **35,** 87–99.

Vogellehner, D. (1965). Untersuchungen zur Anatomie und Systematik der verkieselten Hölzer aus dem frankischen und südthüringischen Keuper. *Erlanger Geologische Abhandlung*, **59,** 1–76.

Wilde, M. H. (1944). A new interpretation of coniferous cones: 1. Podocarpaceae (*Podocarpus*). *Annals of Botany* (N.S.), **8,** 1–41.

Zimmermann, W. (1959). *Die Phylogenie der Pflanzen*. Stuttgart: Fischer.

29

The diversification of conifers and taxads

Although the fossil record of conifers and taxads is difficult to interpret, it is clear that they reached their maximum diversification during the Mesozoic. Reviews of Mesozoic conifer and taxad remains prepared by Miller (1977, 1982) and Stockey (1981a, 1988) provide us with excellent summaries of the fossil evidence that records the first appearance of extinct and extant conifer and taxad families and their possible relationships.

Although some will disagree, we will continue to recognize the order Taxales as distinct from the Coniferales. The Taxales consists of a single family, the Taxaceae, while the Coniferales is comprised of six families, the Podocarpaceae, Araucariaceae, Cephalotaxaceae, Pinaceae, Cupressaceae, and Taxodiaceae. The placement of the taxads in an order separate from the Coniferales was first suggested by Sahni (1920) and supported by the researches of Florin (1951) into the fossil record of this group. These and other investigators (Sporne, 1965; Miller, 1977) point to the absence of clearly defined seed cones in the Taxales, which bear their seeds terminally on short lateral shoots. In their vegetative characteristics, genera of the Taxales such as *Taxus* and *Torreya* are similar to many conifers with their spirally arranged linear to needlelike leaves. Florin (1958) has demonstrated, however, that epidermal characteristics can be used to distinguish leaves of taxads from those of conifers. The secondary xylem of these shrubby to arborescent plants is pycnoxylic, with many coniferous features.

Wood of conifers and taxads

It has been recognized that the wood of conifers and taxads is simpler and more uniform in structure than that of angiosperms. The principal difference is the absence of vessels in the wood of conifers, taxads, and other gymnosperms except the Gnetales and their presence in most flowering plants. The study of coniferous fossil woods is a highly specialized subject that depends on a firsthand knowledge of the characteristics of modern conifer woods, an understanding of the variability of wood characteristics within species as well as a single plant, and an abundance of well-preserved material. Unfortunately, these criteria are seldom met in the hundreds of descriptions of conifer-type fossil woods. In almost every instance an attempt is made to relate the fossil wood fragment to the wood of an extant genus of conifer. This is indicated by using the name of the genus with the suffix *oxylon*. Thus, *Juniperoxylon, Sequoioxylon, Piceoxylon, Cedroxylon,* and so on. The list of such form genera is a long one, indeed, and only in the past three decades has there been a concerted attempt to decipher the natural affinities of these genera (Vogellehner, 1967, 1968). As we might expect, many of the geologically older fossil conifer woods have combinations of characteristics that do not "fit" those found in extant genera. It is quite possible that from these more primitive types, diversification of coniferous wood types occurred. Unfortunately there is seldom an opportunity to relate the diversified wood types to other conifer vegetative or reproductive structures that help immeasurably in determining natural affinities.

For those students interested in a greater exposure to studies of fossil coniferous and taxad woods, the works of Seward (1919) and Greguss (1967) provide keys and descriptions, while papers by Kräusel (1949), Grambast (1961), and Vogellehner (1967, 1968) consider the evolution of individual characteristics.

Some characteristics of conifer wood

The secondary xylem in stems and branches of conifers and taxads consists mostly of tracheids, rays, and some xylem parenchyma. In a growth ring, the late wood tracheids develop thicker walls and have pits with reduced borders. Such cells have the characteristics of fiber tracheids. Libriform wood fibers that frequently occur in the secondary xylem of woody angiosperms do not, however, occur in coniferous woods. As in most gymnosperms, the tracheids of Cordaitales, Voltziales, Ginkgoales, Coniferales, and Taxales are interconnected by circular or oval bordered pit pairs (Fig. 29.1A,B) in a single opposite or alternate arrangement (Fig. 29.1C–E). These usually are on the radial walls of the tracheids. Late wood tracheids may have some pits on their tangential walls. Multiseriate pitting, in which the pits alternate, is characteristic of many cycadophytes, Cordaitales, Voltziales, and Araucariaceae of the Coniferales.

The bordered pit pairs of *Ginkgo*, members of the Pinaceae, and some other conifers have tori present on their pit membranes. A torus (Fig. 29.1A) is a circular thickening of the pit membrane in the pit chamber that is somewhat larger in diameter than the pit aperture. Additional thickenings of intercellular and primary wall material may occur along the upper and lower margins of the pit pair (Fig. 29.1B). These thickenings, called crassulae, are also called the bars or rims of Sanio in the older literature. Other tracheary features used to characterize certain conifer woods are trabeculae, which are small bars extending across the lumina of tracheids, and spiral (helical) tertiary wall thickenings (Fig. 29.1F) that have been found in the tracheids of the Taxales, *Pseudotsuga*, some species of *Picea*, *Cephalotaxus*, and *Callitris*.

Xylem parenchyma, when it occurs among the Coniferales, is usually dispersed throughout the growth ring (Fig. 29.2A). In longitudinal section, the cells of xylem parenchyma occur in long vertical files. In some species of Podocarpaceae, Taxodiaceae, and Cupressaceae there is an abundance of xylem parenchyma. It is scarce, however, in the Pinaceae and absent in the Araucariaceae and Taxaceae.

The rays (Fig. 29.3A–D) of conifers and taxads are composed of parenchyma cells or of parenchyma cells and ray tracheids. The latter differ from ray parenchyma cells in having bordered pits and lacking protoplasts. Ray tracheids are a regular feature of the Pinaceae, except *Abies*, *Keteleeria*, and *Pseudolarix*. They occur in the Cupressaceae and have been reported in *Sequoia*. Their lignified walls often form toothlike projections into the lumen of the cell (Fig. 29.3A). Ray parenchyma cells in Araucariaceae, Taxaceae, Podocarpaceae, Cupressaceae, and Cephalotaxaceae only have primary walls.

Rays are usually 1 cell wide and 1 to 50 cells high. Those of the Pinaceae, with horizontal resin ducts, are regularly several cells wide above and below the duct and are called fusiform rays. The pitting of ray parenchyma cells (Fig. 29.3A–D), especially in those species with walls that have secondary thickening, is particularly distinctive. Their shape, number, and distribution on the radial walls, where a ray cell is in contact with a tracheid of the secondary xylem (the cross-field), are important characteristics in identification and classification of fossil coniferous woods.

Resin ducts are a constant feature of the secondary xylem of *Pinus* (Fig. 28.1), *Picea*, *Larix*, and *Pseudotsuga* of the Pinaceae. Other members of the family may lack resin ducts in the wood, but have them in leaves and roots. The Taxales and Araucariaceae are characterized by the lack of resin ducts. In many genera of conifers that normally lack resin ducts in the secondary xylem of their stems (*Cedrus*, *Pseudolarix*,

Figure 29.1. **A.** *Pinus*, circular bordered pit pair, section view. Torus (t). **B.** Face view. Crassula (c). (Redrawn and modified from Esau, 1977.) **C–E:** *Araucarioxylon* pitting on radial walls of tracheids. **C.** Note opposite pair of pits. **D.** Single row of flattened pits. **E.** Alternate, multiseriate pits with hexagonal outlines. **F.** Tracheid of *Taxus* with spiral, tertiary wall thickening. **A–F:** Extant.

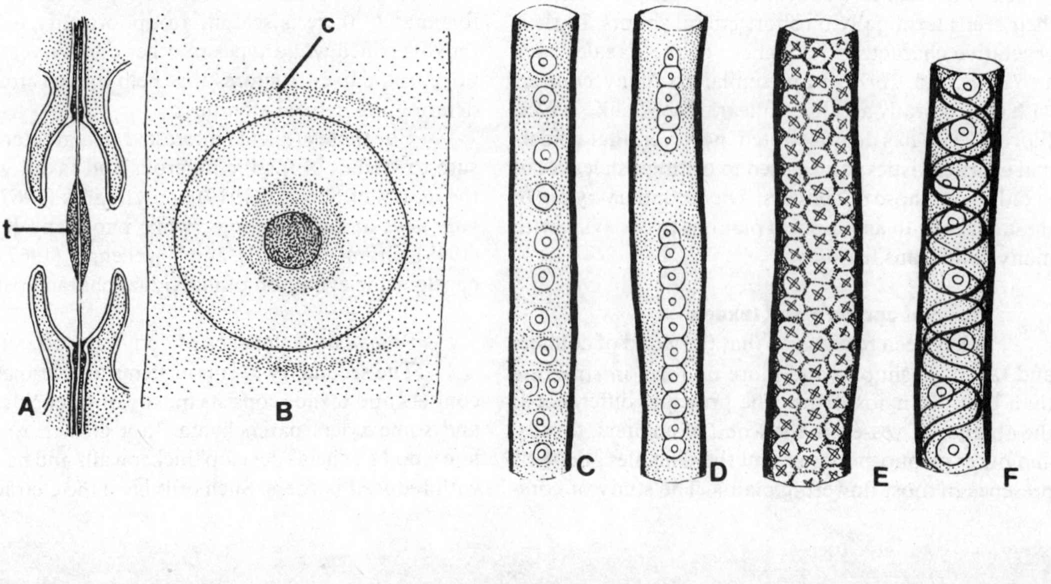

Figure 29.2. *Taxodioxylon* sp. **A.** Transverse section showing growth ring. Wood ray (r); xylem parenchyma (xp); scattered cells filled with black contents. **B.** Radial section. Note two rows of opposite pits on tracheid walls. **A,B:** Upper Cretaceous. (**A,B** courtesy of Dr. R. Stockey.)

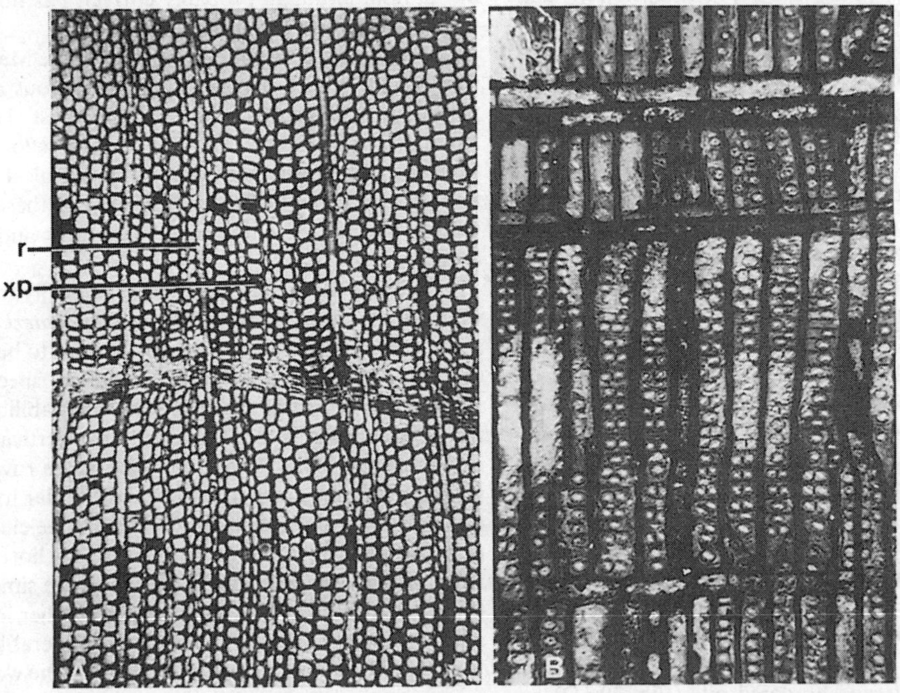

Figure 29.3. Types of cross-field pitting and ray cells (rp). **A.** *Pinus*, abeitineous cross-field pitting with dentate thickenings on ray tracheids (rt). **B.** *Pityoxylon*. **C.** *Podocarpoxylon*. **D.** *Taxodioxylon*. **A–D:** All similar to extant counterparts.

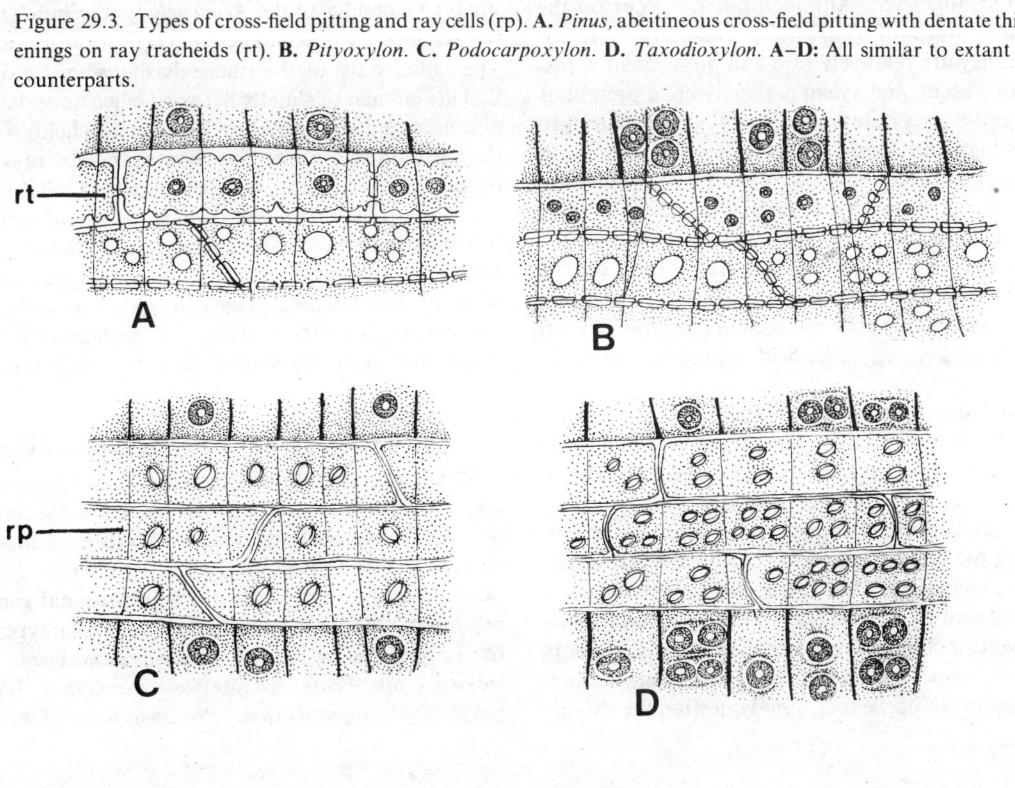

Tsuga, and members of the Taxodiaceae, for example), traumatic resin ducts may be induced by stress and physical injury. Resin ducts are schizogenous in origin, produced by the separation of parenchyma cells and the subsequent formation of a lining (epithelium) that excretes the resin. In some genera (*Abies* and *Tsuga*) the epithelial cells develop secondary walls. Sooner or later resin ducts may become closed by enlarged epithelial cells. These intrusions into the lumens of the ducts are called tylosoids.

This summary of the diagnostic characteristics of extant conifers and taxad woods is derived principally from the works of Seward (1919), Record (1934), Phillips (1941, 1948), Greguss (1955), and Esau (1977) and are those usually used by paleobotanists in the identification of fossil woods. No attempt has been made in the presentation of the following genera to give complete coverage to the various kinds of fossil conifer woods. Instead, we have selected a few that are considered representative of the principal families of conifers and taxads.

Dadoxylon–Araucarioxylon type

Of present-day genera only *Araucaria* and *Agathis* of the Araucariaceae display all the characteristics of the *Dadoxylon–Araucarioxylon* type. Here the pycnoxylic wood consists of tracheids with uniseriate, more or less flattened bordered pits (Fig. 29.1D) or two or more rows of polygonal pits on the radial walls (Fig. 29.1E) (Seward, 1919). Even though the pits of Paleozoic *Dadoxylon* may be circular rather than flattened or polygonal, they are characteristically alternate in arrangement. Although pits do occur on the tangential walls of *Dadoxylon–Araucarioxylon* tracheids, they are relatively rare and quite small. Crassulae are absent, and xylem parenchyma, if present, is scarce and poorly developed. The rays are usually 1 to 30 cells high and uniseriate. Occasional biseriate rays have been reported, however. The rays are composed exclusively of parenchyma cells. The ray cells have comparatively thin walls and lack pits in the horizontal and vertical tangential walls. Pitting on the radial walls of the cross-field may have 1 to 15 small pits with an oblique pore enclosed in a narrow border. In some araucarias, cross-field pitting seems to be absent.

Although the woods of *Dadoxylon* and *Araucarioxylon* cannot be distinguished, *Dadoxylon* is the name usually applied to the wood ascribed to Paleozoic cordaites and transition conifers, while *Araucarioxylon* is customarily applied to the Mesozoic woods showing the characteristics of *Dadoxylon*. However, we should remember *Dadoxylon–Araucarioxylon* wood was produced by a wide array of Paleozoic and Mesozoic plants including arborescent seed ferns, glossopterids, and possibly even some progymnosperms. Some investigators (Vogellehner, 1964) question the validity of *Cordaioxylon* and *Araucarioxylon* and recommend using the valid form genus *Dadoxylon* for all wood of the *Araucaria* type irrespective of age. This recommendation, although probably correct, has not been widely adopted in the current literature.

Among the specimens assigned to the Mesozoic form genus *Araucarioxylon*, the most famous are the permineralized trunks of the Petrified Forest National Monument of Arizona. Thousands of these logs have been discovered weathering from the Chinle Formation of the Upper Triassic. The largest of these spectacular specimens is 150 cm in diameter and 30 m long.

Gould (1975) has shown unequivocally that the wood of the Permo-Carboniferous *Glossopteris* is araucarioid. Fragments of wood now known to be those of *Glossopteris* were assigned to various species of *Araucarioxylon*. Some specimens show variability with respect to pitting, with pits in one to five vertical rows on the radial walls. When in two or more rows, the pits may be opposite or alternate and circular to polygonal. In some specimens the pits tend to be clustered in groups of two to seven in one or two horizontal rows. Other characteristics of the wood are similar to those of *Araucarioxylon*. As have many other investigators, Gould makes the point that considerable caution should be exercised when comparing the wood of *Glossopteris* with that of living araucarias because of known variations in their wood structure.

If one were to examine an isolated fragment of the wood from a stem of the Jurassic *Pentoxylon* (Sahni, 1948), there would be no hesitation in assigning it to a conifer of the *Araucaria* type. There are no ray tracheids, wood parenchyma cells, or resin canals. The radial walls of the tracheids show bordered pits that are circular or slightly flattened when uniseriate, or alternate and hexagonal on broader tracheids where they are multiseriate. The homogeneous rays are usually 2 to 7 cells high, with some up to 14 cells high.

We think most would agree, using characteristics other than those of secondary wood, that *Glossopteris* and *Pentoxylon* are not close relatives of the conifers. Whether the araucarioid type of wood evolved independently in these groups or originated from a common ancestral type remains to be determined.

Protopinaceae

The Protopinaceae, established by Kräusel (1949), is a family that now includes those Triassic and Jurassic genera, based entirely on wood specimens, that are transitional and combine features of modern conifer woods or possess characteristics that do not occur in any extant forms. Such transitional genera, presumably originating from the *Dadoxylon* type, are the Triassic and/or Jurassic *Protocupressinoxylon*, *Protocedroxylon*, *Prototaxodioxylon*, *Zenoxylon*, *Woodworthia*, *Protopiceoxylon*, *Protopolyporoxylon*, and

Protophyllocladoxylon. Of these, the Triassic genus *Woodworthia* provides us with a good example of a wood type exhibiting transitional characteristics. This genus, which is associated with the *Araucarioxylon* trunks in the Petrified Forest, has wood of the *Araucarioxylon* type. Studies of the internal structure of *Woodworthia* logs have shown, however, that they possess axillary short shoots subtended by leaf traces. Unlike the araucarias, the leaf traces of *Woodworthia* are not persistent in the secondary wood. This feature plus the presence of short shoots suggests that *Woodworthia* with its araucaroid features has evolved some characteristics of the abietinean (species belonging to the Pinaceae) conifers and is transitional between the two groups.

Podocarpoxylon (Mesembrioxylon)

The name *Podocarpoxylon* has been used for fossil wood agreeing more closely in structure with *Podocarpus* and *Dacrydium* than other conifers. Seward (1919), however, proposed the genus *Mesembrioxylon* for the reason that there is no adequate evidence for assuming affinities with the extant genera as their names imply. In spite of this, authors of current literature dealing with woods of this type (Greguss, 1955, 1967; Ramanujam, 1972) continue to use the name *Podocarpoxylon*. Generally, woods assigned to this genus are without resin canals and ray tracheids. Scattered xylem parenchyma is usually present. The rays, which are usually uniseriate and occasionally biseriate, are 2 to 35 cells high. Both tangential and horizontal walls of the ray cells are thin, smooth, and unpitted. Cross-field pitting on the radial walls of the ray cells is typically podocarpoid (Fig. 29.3C), where there are most frequently one or two, and occasionally more, pits per field. The pits appear to be bordered with an elliptical to linear pore that is oblique to more or less vertical. The radial walls of the tracheids have mostly uniseriate to locally biseriate bordered pits that are circular and separate. The pits are opposite to subopposite when biseriate. There is no evidence of a torus, but crassulae are faintly preserved.

Fossil woods having the characteristics described above have been reported from the Lower Jurassic into the Pleistocene. The fossil record tells us that the podocarps and the araucarias formed a predominant part of the Jurassic coniferous forests by the onset of the Jurassic.

Pityoxylon

The form genus *Pityoxylon* is generally used for wood fragments showing the anatomical features common to *Pinus, Picea, Larix, Pseudotsuga,* and a few members of the Pinaceae. Although the genera *Piceoxylon* and *Pinuxylon* have been used to indicate a more precise comparison with extant conifers, it is doubtful if such precision is possible in identification of fossil woods. Thus, the noncommittal *Pityoxylon* seems more reasonable.

Pityoxylon (Fig. 29.3B) is characterized by its resin canals and by the presence of ray tracheids in the rays. The walls of the ray tracheids may be smooth or have toothlike ingrowths. If the pits on the cross-field are large, they are usually simple; if small, they may have an inconspicuous border. There may be one to several pits on the cross-field. The epithelial cells lining the resin canals may have thin or thick walls. The pits on the radial walls of the tracheids are abietineous and accompanied by crassulae and tori. Although *Pityoxylon* tracheids usually lack helical wall thickenings, this characteristic is known to occur in some specimens of *Pseudotsuga*-like wood.

Wood fragments with the characteristics given above have been reported from deposits as old as the Carboniferous. The stratigraphic origin of these and other specimens, however, is very much in doubt. Authentic Lower Cretaceous specimens of *Pityoxylon* presently represent the earliest known wood remnants of the Pinaceae.

Protopiceoxylon (Pinoxylon)

Protopiceoxylon is one of several form genera belonging to the Protopinaceae that has been interpreted as transitional between the woods of the Araucariaceae and Pinaceae (Read, 1932; Vogellehner, 1967, 1968). The specimen of *Protopiceoxylon* (*Pinoxylon*) described by Read has all of the characteristics given for *Pityoxylon* except for the presence of partially araucarioid pits on the radial walls. The oldest known specimens of *Protopiceoxylon* are Jurassic. It is possible that they represent a primitive group of conifers that gave rise to Lower Cretaceous members of the Pinaceae.

Taxodioxylon

This form genus is assigned to fossil woods having the characteristics of the extant Taxodiaceae (*Athrotaxus, Cryptomeria, Cunninghamia, Glyptostrobus, Metasequoia, Sciadopitys, Sequoia, Sequoiadendron, Taiwania,* and *Taxodium*) and the Cupressaceae (*Cupressus, Juniperus, Libocedrus,* and *Thuja*). Fossil woods related to these modern genera also have been assigned to *Sequoioxylon, Cupressinoxylon,* and *Metasequoioxylon*. A study of exquisitely preserved wood of Eocene *Metasequoia milleri* (Basinger, 1981) showed that the only characteristic that distinguishes it from *Sequoia, Sequoiadendron,* and extant *Metasequoia* is ray height. Unlike these redwood genera, other genera of the Taxodiaceae can be distinguished only by such minute features as the presence of small dentate ingrowths on the horizontal walls of the wood parenchyma. We refer to this characteristic because it exemplifies the difficulty encountered in distinguishing woods of extant genera of conifers, to say nothing of those whose woods have been preserved and altered by fossilization. Although Greguss (1955) maintains that woods of the Cupressaceae can be distinguished from those of the Taxodiaceae by characteristics such as beadlike thickenings on the tangential walls of ray cells, there is no consistency with respect to this and other characteristics that are claimed to distinguish woods of the two families. With this in mind Seward (1919) uses the form genus *Cupressinoxylon* for all woods showing characteristics of the Taxodiaceae and Cupressaceae. Kräusel (1949), however, placed woods formerly assigned to *Sequoioxylon* in *Taxodioxylon*. After his careful researches of Eocene *Metasequoia* material Basinger (1981) proposes that *Metasequoioxylon* also should be placed in *Taxodioxylon*.

Taxodioxylon (Fig. 29.3D) possesses annual rings that are well defined. Vertical rows of wood parenchyma are abundant and scattered through the spring and summer wood (Fig. 29.2A). Although more easily seen in longitudinal sections, the wood parenchyma cells can be identified in transverse section by their thick walls, dark contents, and resin bodies. The circular bordered pits on the radial walls of the tracheids may be in one to four vertical rows (Fig. 29.2B). If they are in two or three rows, the pits are opposite. Crassulae associated with the pits are conspicuous even in fossilized specimens (Fig. 29.4B). The wood rays are homogeneous and range from 1 to 70 cells high. Most are uniseriate, but biseriate rays have been reported in some species (Ramanujam, 1972). Pitting of the rays is confined to the radial walls. All other walls of the ray cells are smooth and without pits. Pits on the cross-field are two to eight in number, often arranged in horizontal rows, and may be fairly large, simple, or bordered. Most of the pits tend to be horizontal on the cross-field, but some are strictly oblique (Fig. 29.4A). *Cupressinoxylon* cross-field pits are supposed to be oblique and simple or with narrow borders, but we have seen pits in wood known to be taxodioid with pitting of this type. It is clear that there is no good distinction between cupressoid and taxodioid cross-field pitting. Resin canals are absent except in regions of stress or wounding.

Based on the characteristics of their wood, it seems that the Taxodiaceae and Cupressaceae are more closely related to one another than to any other family of conifers. This close relationship is supported by Eckenwalder (1976), who proposes a merger between the two families because of similarities in morphology and phytochemistry. Wood of the *Taxodioxylon–Cupressinoxylon* type has been identified from the Upper Jurassic (Vaudois & Privé, 1971). As we shall see, leaf and reproductive remains assignable to the Cupressaceae and Taxodiaceae are also found as early as the Upper Triassic. This evidence suggests that the two families diverged from a common ancestral stock having the characteristics of *Pseudovoltzia* (Schweitzer, 1963).

Taxaceoxylon (Taxoxylon)

Extant members of the family Taxaceae include *Austrotaxus, Amentotaxus, Pseudotaxus, Torreya,* and *Taxus*. All species of these five genera have wood that is characterized by tertiary wall thickenings that form helical bands in their tracheids (Fig. 29.1F). Other features of their wood are similar to those of the *Taxodioxylon–Cupressinoxylon* type. Many wood specimens assigned to *Taxaceoxylon,* because of the apparent helical thickenings, are now known to belong to genera related to other families of conifers where some genera and species exhibit similar thickenings. We have already noted that *Pseudotsuga* and *Picea* of the Pinaceae have such helical bands in their tracheids. *Cephalotaxus* of the Cephalotaxaceae as well as *Callitris* of the Cupressaceae also are known to possess similar spiral thickenings. In the latter genus the spiral bands are flat and relatively broad. Fossil woods showing similar tertiary thickenings coupled with cupressoid wood characteristics are placed in the genus *Platyspiroxylon* known from the Upper Cretaceous (Ramanujam, 1972). From this example it

Figure 29.4. *Metasequoia milleri*. **A.** Taxodioid cross-field pit. **B.** Paired pits on radial walls of tracheids. Note crassulae. Eocene. (**A,B** courtesy of Dr. J. F. Basinger.)

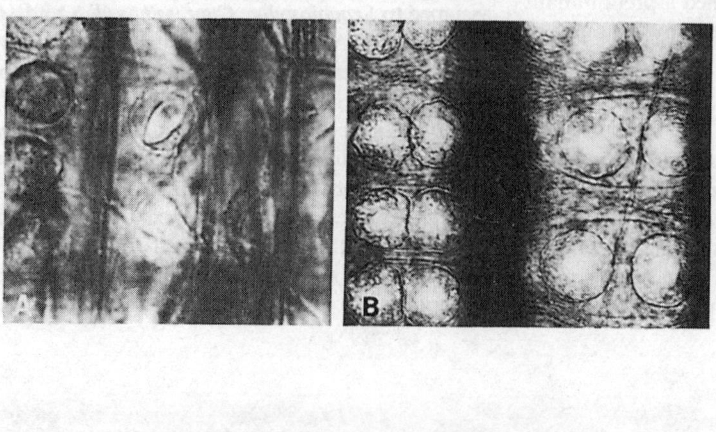

must be clear that every specimen showing helical thickenings should not automatically be placed in *Taxaceoxylon*. Another pitfall comes in the guise of false helical thickenings that occur in tracheids of all woods where physiological and mechanical changes have taken place in their walls to produce striations simulating the helical bands. Upon reexamination, many specimens thought to belong to *Taxaceoxylon* have been reassigned to the Pinaceae, Podocarpaceae, and other families. Woods that are like those of the extant Taxaceae in all characteristics are known from the Jurassic onward (Miller, 1977).

The evolution of coniferous wood structure

In our search for the ancestors of woody plants we can do no better than go back to the Middle Devonian Progymnospermopsida represented by *Aneurophyton* and *Rellimia* of the Aneurophytales (Chart 29.1). The wood of *Rellimia,* which is the earliest known for a vascular plant, is characterized by narrow vascular rays and tracheids with scalariform and reticulate thickenings on all walls (Leclercq & Bonamo, 1971). We have learned from ontogenetic and phylogenetic studies of tracheary elements that scalariform tracheids are the precursors of tracheids with circular pits and that tracheids with reticulate thickenings are often intermediate (*Mesoxylon,* for example). With this in mind, plus the great age of *Rellimia,* it is reasonable to conclude that this genus possesses, among other primitive characteristics, primitive secondary xylem from which multiseriate bordered pits of the type found in *Aneurophyton* have evolved. These pits are elliptical and occur on tangential as well as radial walls. The rays of *Aneurophyton* are uniseriate.

Moving to the Upper Devonian representatives of the Aneurophytales, we find that the secondary xylem of the three principal genera, *Triloboxylon, Proteokalon,* and *Tetraxylopteris,* shows further modification. In all three genera the circular bordered pits tend to be crowded and hexagonal in outline. They are usually multiseriate and alternate, occurring with greater frequency on the radial walls. Their pit apertures are horizontal or slightly inclined. The vast majority of the rays are uniseriate, but some in *Tetraxylopteris* are multiseriate (Beck, 1957). Rays with ray tracheids are described by Scheckler & Banks (1971a, 1971b), in *Proteokalon* and *Triloboxylon*. When observed in transverse section the secondary xylem appears to be pycnoxylic. The pitting on the walls of the tracheids is araucarioid.

In the classic, much-cited paper "The Appearance of Gymnospermous Structure," Beck (1970) states, "By Frasnian (early Upper Devonian) time, all of the major characteristic features of gymnosperm secondary xylem, except the torus, had evolved in the progymnosperms as illustrated by *Callixylon,* the wood of *Archaeopteris*." The secondary xylem of *Archaeopteris* (*Callixylon*), the principal genus of the Archaeopteridales, tends to be pycnoxylic and well developed in the trunks and branches. The circular bordered pits are confined to the radial walls except in the region of late wood in a growth ring, where they occur on the tangential walls. These pits are smaller and less abundant. This distribution of pits on tracheids in growth rings is a feature of conifer genera. Depending upon the species, the pits are uniseriate to multiseriate. If multiseriate, they are alternate and often crowded and hexagonal in outline (Fig. 29.5). The pits on radial walls of all species are grouped in radially aligned rows. Structures interpreted as crassulae are now believed to be artifacts of preservation (Beck, Coy, & Schmid, 1982). Trabeculae extending between the tangential walls of the tracheids also have been observed. Except for their grouping, the pits of *Archaeopteris* are araucarioid, as in the Aneurophytales.

Rays vary according to species from those that are very low and uniseriate to others that are high and multiseriate. Although the rays of conifers tend to be uniseriate, there are many examples, especially among the Taxodiaceae, where multiseriate rays occur in abundance. The rays of *Archaeopteris* are composed of ray tracheids and ray parenchyma. In one species, the ray tracheids have a helical wall superimposed on that containing the bordered pits, a structure noted by Beck to be similar to *Sequoia.* The ray tracheids may be marginal and discontinuous or interspersed in the ray and continuous. Cells of the ray tracheids are provided with bordered pit pairs in the walls of contiguous ray tracheids. The cross-fields of ray parenchyma cells have half-bordered pit pairs with oblique ellipti-

Figure 29.5. *Archaeopteris* (*Callixylon*), radial section of wood showing ray and tracheids with crowded bordered pits. Upper Devonian. (From Beck, 1970.)

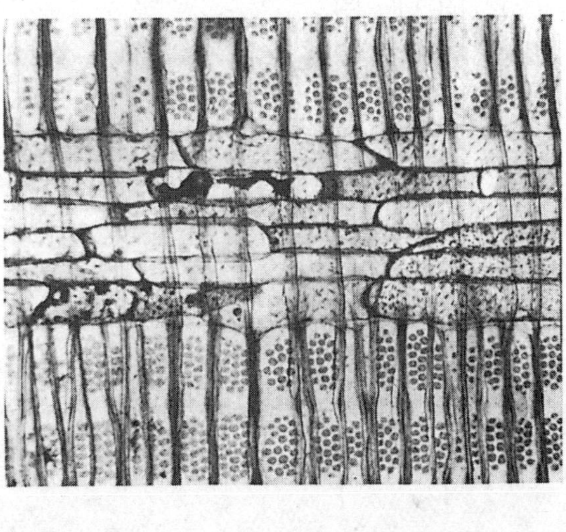

cal apertures reminiscent of those found in *Araucaria* and *Ginkgo*.

The only cells or structures commonly found among certain conifers that are lacking in the wood of *Archaeopteris* are wood parenchyma, resin ducts, and crassulae. Vertical resin ducts and well-defined tori do not appear in the fossil record until the Jurassic (Vogellehner, 1968; Schmid, 1967). Schmid suggests that the cycad type of pitting and pit membrane represent the primitive type. Most cycadophytes have secondary xylem tracheids with multiseriate bordered pits more or less confined to the radial walls. According to the studies of Schmid that utilize the electron microscope, the cycad type of pit membrane is a homogeneous structure that lacks openings or tori. The pit membranes of *Archaeopteris, Dadoxylon, Araucarioxylon,* and those other conifers lacking tori are usually of the homogeneous type. Studies of cross-field pitting

Chart 29.1. Suggested origins and relationships of Ginkgoales, Gnetales, and major groups of coniferophytes, and their distribution in geological time. Rhyn. = Rhyniopsida; Trim. = Trimerophytopsida; Ane. = Aneurophytales; Cala. = Calamopityaceae; Arch. = Archaeopteridales; Call. = Callistophytaceae.

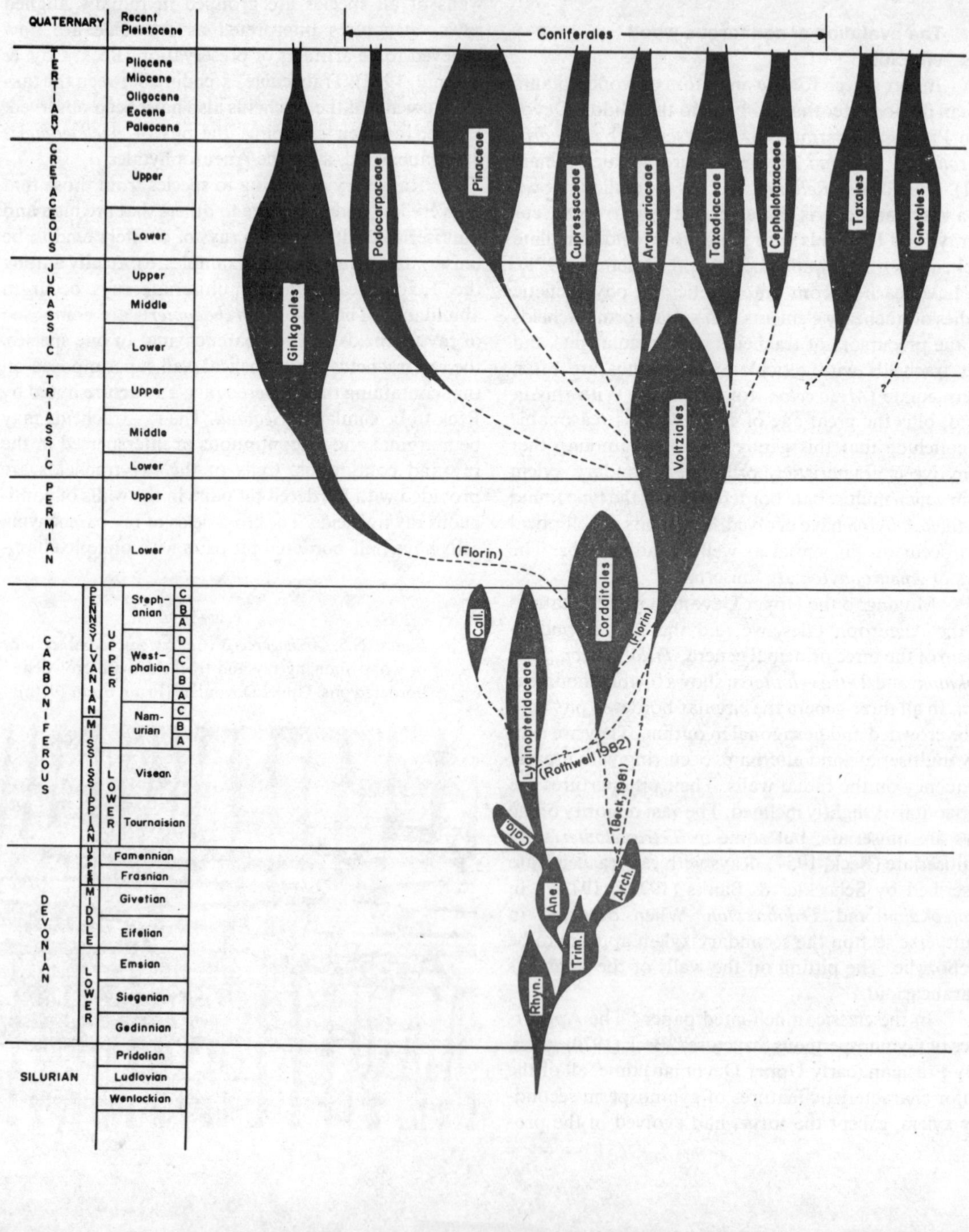

(Vogellehner, 1968) show that those of the taxodioid–cupressoid and pinoid types become more differentiated in the Jurassic.

All of the evidence from the study of fossil conifer-type woods indicates that the primitive type is to be found among the Middle and Upper Devonian Aneurophytales. By the Upper Devonian, the Archaeopteridales with their more specialized wood had diverged from the aneurophytes. Although *Archaeopteris* may not be on the evolutionary line leading to the cordaites, its wood characteristics indicate that it is part of a complex to which the Cordaitales and "transition" conifers, such as the Voltziales may also belong (Chart 29.1). Again, from studies of wood structure, it is now clear that diversification of modern Coniferales started in the Triassic and increased in the Jurassic and Lower Cretaceous, so that by the end of the Cretaceous all families and most genera of the Coniferales and Taxales were well established and formed a predominant part of the flora.

The fossil leaves of conifers

In the previous section on the identification of permineralized fossil conifer woods, we have tried to make it clear that paleobotanists have difficulty in assigning the specimens to families, to say nothing of genera, of present-day conifers and taxads. Even greater difficulty is encountered when one attempts to assign compression–impression leaf fossils with conifer and taxad characteristics to natural taxa. The reasons are many. Often the specimens are few in number and poorly preserved, lacking cuticles and epidermal features. If well preserved, the specimens available may not show the range of variability in leaf form that can occur in a single genus or species. For example, many conifer species exhibit both juvenile and mature leaves that in the fossil specimens may not be recognized as belonging to the same taxon. As a matter of fact, it would be very easy to assign fossils showing only juvenile leaves to one family of conifers and specimens with mature leaves to another, if epidermal characteristics are not available. Based on external morphology, DeLaubenfels (1953) recognizes only four main types of conifer and taxad leaves. These are not distinct, as there is a continuum of forms between the main types. He has demonstrated that as many as three of the four foliage types may occur in a single modern genus as mature foliage of the different species. A beautiful example of this leafy shoot polymorphism is described by Christophel (1976) for *Glyptostrobus* of the Taxodiaceae. Here the shoots are trimorphic, with some being cupressoid (Fig. 29.6A), cryptomeroid (Fig. 29.6B), or taxodioid (Fig. 29.6C). There are also intermediate morphologies between these three types. Based on their trimorphic nature, fossilized leafy shoots of *Glyptostrobus* might have the appearance of *Chamaecyparis* of the Cupressaceae or *Cryptomeria* and *Taxodium* of the Taxodiaceae. Confronted with this kind of evidence, it is little wonder that paleobotanists have given up trying to relate fossilized conifer leaf remains to those of modern conifers. This is especially true for compression–impression fossils of Mesozoic conifers where the number of genera is overwhelming and generic limits are poorly defined. There is no doubt that these factors have been a deterrent to many paleobotanists. In one instance (Hughes, 1976) it has been suggested that small leaves (conifers and taxads) of gymnosperms be separated into two groups – Brachyphylls and Linearphylls – without making any attempt at further classification until much more is known about their reproductive and vegetative structures. In a prodigious work, Florin (1958) did much to characterize the epidermal features of Mesozoic conifer and taxad shoots and relate them to genera and families of modern conifers. Another promising approach has been taken by Stockey and her colleagues (Stockey & Taylor, 1978; Stockey & Ko, 1989). These authors have initiated studies of the cuticular features of living conifers that provide a basis for accurately identifying and interpreting the relationships of conifer leaves. Unfortunately, there are very many specimens where epidermal features are lacking, and one must depend exclusively on external morphology. To accommodate these, Harris (1969, 1979) has adopted a system comprising eight form genera based primarily on external morphology of shoots and their leaves. Professor Harris has taken great pains to use only existing generic names, thus avoiding adding new ones to further burden the nomenclature. The Mesozoic genera are *Brachyphyllum, Pagiophyllum, Cyparissidium, Geinitzia, Elatocladus, Cupressinocladus, Pityocladus,* and *Podozamites*. A few other genera have been excluded for reasons given later. All genera except *Cupressinocladus* and *Podozamites* are represented in the Middle Jurassic beds of Yorkshire.

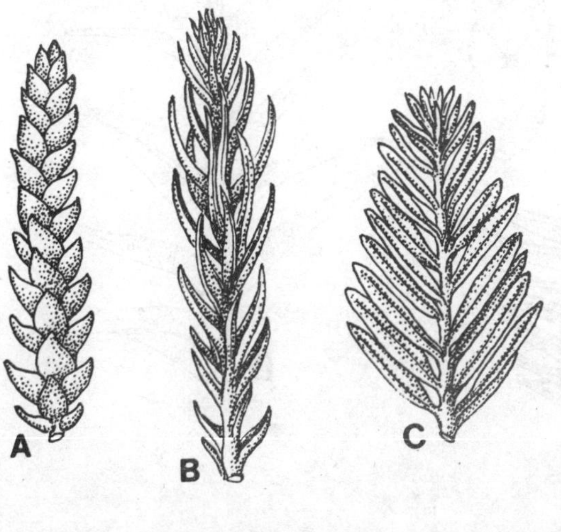

Figure 29.6. *Glyptostrobus nordenskioeldii* with three shoot types. **A.** Cupressoid type. **B.** Cryptomeroid type. **C.** Taxodioid type. Tertiary. (Modified from Christophel, 1976.)

Brachyphyllum

The shoots of *Brachyphyllum* have small helically arranged leaves. Each leaf is composed of a relatively broad basal cushion tapering into a minute free part. The total length of the free part is less than the width of the cushion, or the total height of the leaf (outward from the shoot) is less than the width of the cushion. The free part of a *B. mamillare* leaf (Fig. 29.7A) has the appearance of a short four-sided pyramid. Some specimens of *Brachyphyllum* have longer leaves, and these form a continuum overlapping the boundary of *Pagiophyllum* and *Cyparissidium*. *Dacrydium colensoe, Podocarpus ustus, Athrotaxus cupressoides* are examples of modern conifers with shoots similar to those of *Brachyphyllum*. It is known that *B. mamillare* has pollen cones and pollen of the *Araucaria* type. Harris (1979) is satisfied that the seed cone *Araucarites phillipsii* and the pollen cone bearing *B. mamillare* belong to the same plant. For what he describes as the "whole plant," Harris proposes the name *Araucaria phillipsii*.

Pagiophyllum

The shoots of *Pagiophyllum* have helically arranged leaves that contract gradually from a broad basal cushion. The free part of the leaf is broader than it is thick and is longer than the width of its cushion. *P. kurri* (Fig. 29.7B) has the general aspect of shoots described as *Araucaria* or *Araucarites* and extant *Araucaria bidwillii*. It is now known, however, that shoots of *P. kurri* have seed cones that are not araucarian, but are like those of *Hirmerella*.

Cyparissidium

Although the shoots of *Cyparissidium* (Fig. 29.7C) have leaves similar to those of *Pagiophyllum*, they differ in being somewhat narrower and appressed to the shoot axis. The leaves of both genera are spirally arranged and flattened. Fossils having this form have been placed in such genera as *Athrotaxites, Widdringtonites,* and *Glyptostrobus*. Present-day *Glyptostrobus pensilis, Taxodium ascendens,* and *Dacrydium cupressinum* have similar leafy shoots.

Geinitzia

The leafy shoots of *Geinitzia* (Fig. 29.7D) bear helically arranged leaves with the free part falcate or spreading. They are needlelike in form and of equal thickness in their horizontal and vertical dimensions. The proximal portion of the leaf merges into a basal cushion without contracting. Fossils similar in appearance have been placed in *Araucarites, Elatides, Sequoia, Pagiophyllum,* and *Cryptomerites*. Among living conifers, *Araucaria heterophylla* and *Cryptomeria* have similar shoots. Other conifers with entirely different mature foliage have juvenile leaves of this form. *Geinitzia*-like shoots of the Permian have been placed in the genus *Walchia* and those of the Triassic in *Voltzia*.

Elatocladus

The form genus *Elatocladus* (Fig. 29.7E) is characterized by shoots bearing leaves that are helically arranged or opposite. The leaves are elongated and dorsiventrally flattened, diverging from the shoot axis.

Figure 29.7. Principal form genera of fossil conifer foliage. **A.** *Brachyphyllum* sp. **B.** *Pagiophyllum* sp. **C.** *Cyparissidium* sp. **D.** *Geinitzia* sp. **E.** *Elatocladus* sp. **F.** *Cupressinocladus* sp. A–E: Mesozoic–Cenozoic. (A–E redrawn from Harris, 1979.)

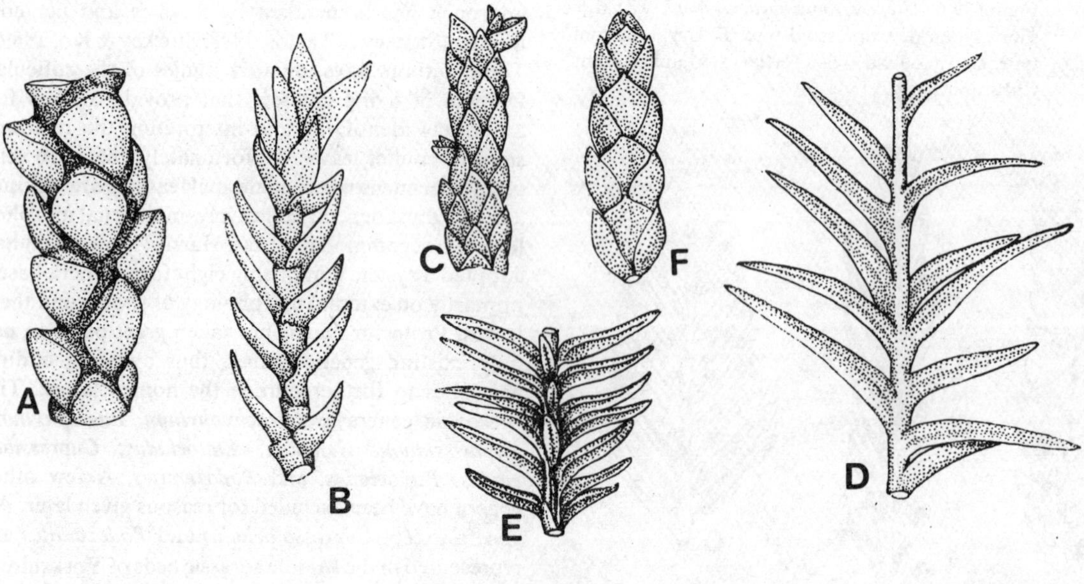

The base of the leaf is contracted into a short petiole attached to a basal cushion. There is a single vein in the leaf lamina. *Elatocladus* has been used for foliage types like those of *Taxus* and other conifers with helical spreading leaves such as *Taxodium distichum, Sequoia sempervirens,* and *Glyptostrobus*. Those with basically helical leaves that appear to be opposite occur in *Cephalotaxus, Amentotaxus,* and *Metasequoia*. The last is known to occur in the Upper Cretaceous, and its fossil record tells us that it became much more abundant and widely distributed in the early Tertiary.

Cupressinocladus

As the name implies, *Cupressinocladus* (Fig. 29.7F) is a genus to which are assigned fossil shoots with foliage similar to that of the Cupressaceae. The shoots have their leaves in decussate pairs or in alternating whorls. The leaves are small and scalelike or longer and appressed or spreading and dorsiventrally flattened. A count by Harris (1969) tells us that fossil foliage of this type has been described under at least 20 different generic names: *Thuja, Thuites, Cupressites,* and so on. Since it is difficult to distinguish foliage shoots of modern Cupressaceae at the generic level, the noncommittal genus *Cupressinocladus* for fossil remains, where identification is even more difficult, seems most appropriate.

Pityocladus

Specimens of *Pityocladus* usually consist of a branch bearing short shoots. The latter have clusters of spirally arranged needlelike leaves at the distal end. *Pseudolarix, Larix,* and *Cedrus* are members of the Pinaceae showing this characteristic. Cretaceous and Tertiary fossils with *Pinus*-like short shoots bearing two, three, four, or five needles (Fig. 29.8A) are retained in the genus *Pinus* or *Pinites* (Robison, 1977).

Podozamites

Podozamites (Fig. 29.8B) was at first thought to be the foliage of a Mesozoic cycad, but cuticular studies have revealed that the structure and arrangement of the stomata are coniferlike. The lanceolate leaves, which are quite large, were borne alternately or in a helix on long shoots that appparently had limited growth. Several veins enter the narrow base of the leaf where they branch to form a number of parallel veins at the midregion of the lamina. The veins converge at the apex. Leaves of this type are produced by *Agathis* and some species of *Podocarpus*. Although not found attached, leaves of *Podozamites distens* are frequently associated with seed cones and scales of *Cycadocarpidium* and may be congeneric (Harris, 1979).

Another leaf with the morphology of *Podozamites*, but different stomatal arrangement, is *Lindleycladus*. The latter agrees with the deciduous foliage of *Taxodium, Metasequoia,* and *Larix* in its cuticular and stomatal characteristics.

The origin of conifer leaves

We have no difficulty in finding abundant evi-

Figure 29.8. **A.** Five-leaved pine (*Pinus*). Eocene. **B.** *Podozamites lanceolatus*. Jurassic. (Redrawn from Seward, 1919.)

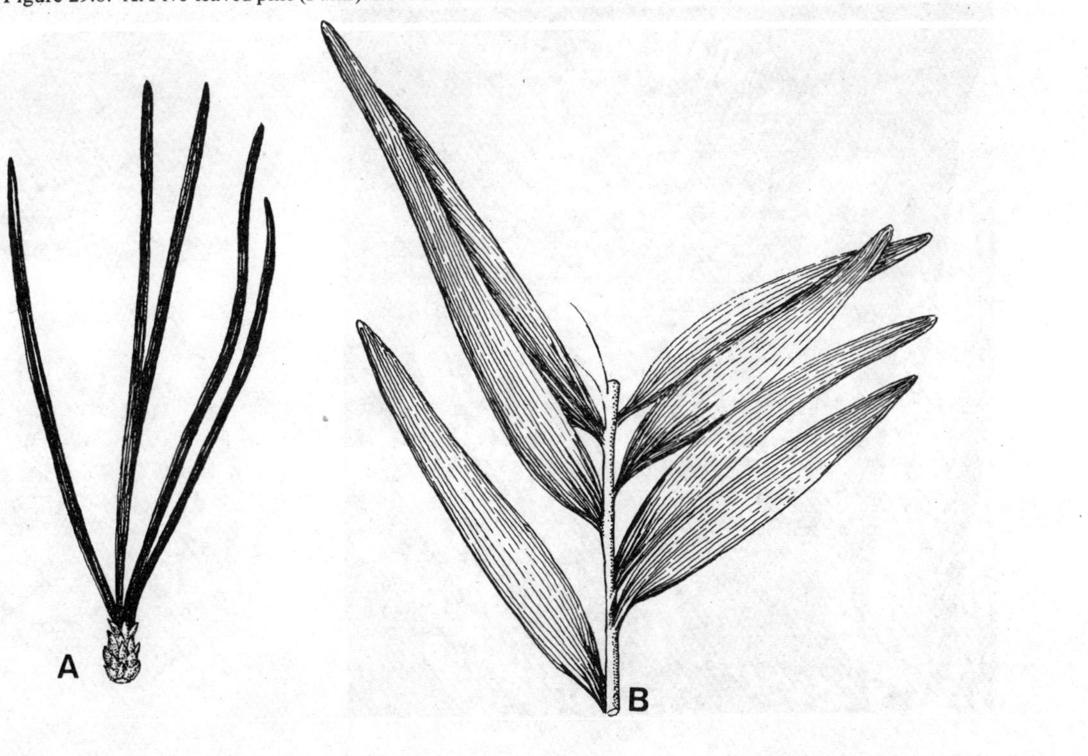

dence that shows us how the compound planated frond of a seed fern or cycadophyte evolved from telomic systems of the Aneurophytales (Progymnospermopsida). When we look for similar evidence revealing the origin of the simple spirally arranged leaves of conifers we turn again to the Progymnospermopsida, but this time to the Archaeopteridales for some answers. This would not have been possible prior to the researches of Carluccio, Hueber, & Banks in 1966 when it was verified that the leaves of *Archaeopteris* were simple and helically arranged. In Chapter 21 we explained that the simple planated and webbed leaf of the *A. macilenta* type evolved from the sterile telome trusses of the *A. fissilis* type. This suggests that the simple needlelike leaf of the conifers also may have evolved by reduction from the telome trusses of the *A. fissilis* type. This suggestion becomes even more plausible with the discovery in the Upper Carboniferous (Scott, 1974) of simple conifer leaves similar to those of *Lebachia* with bifurcate tips. However, the time gap between the Upper Carboniferous (Westphalian B) beds, from which the leaves were obtained, and those of the Upper Devonian–Lower Carboniferous containing *Archaeopteris,* is not as short as we once believed.

There is no difficulty in deriving shoots of the cordaitean type bearing broad helically arranged *Cordaites* leaves with many veins from the *A. macilenta* type. The evolution of planated multiveined leaves seems to have occurred not only in the Cordaitales, but independently in the Podocarpaceae and Araucariaceae of the Coniferales.

An alternative to this simple explanation for the origin of the conifer-type leaf has been proposed by Rothwell (1982). He notes that in addition to large compound fronds, there is a propensity for the production of simple scale leaves on such pteridosperms as *Callistophyton* and *Lyginopteris*. The scale leaves are characteristically produced at the bases of the branches borne by these genera. According to the hypothesis, a relatively minor change in the genome could suppress the development of fronds along the branches. The photosynthetic process would be taken over by development of simple leaves similar to the scale leaves at the bases of branches. The appearance of simple leaves along the branches could occur in a single generation. Such a dramatic and rapid change in morphology of a mature structure brought on by a "change in timing of integrated developmental events" within a single generation is called heterochrony. This is in contrast with allomorphosis where there are gradual changes in morphology of mature structures through

Figure 29.9. **A.** *Araucaria exselsa,* cone surrounded by branches. Compare with reconstruction of *Ernestiodendron* leafy shoot, Figure 28.12C. Extant. (Courtesy of General Biological Supply House, Chicago.) **B.** *A. mirabilis,* permineralized cone. **C.** Longitudinal section of *A. mirabilis* cone with ovules. **B,C:** Jurassic. (**B,C** courtesy of Dr. R. Stockey.)

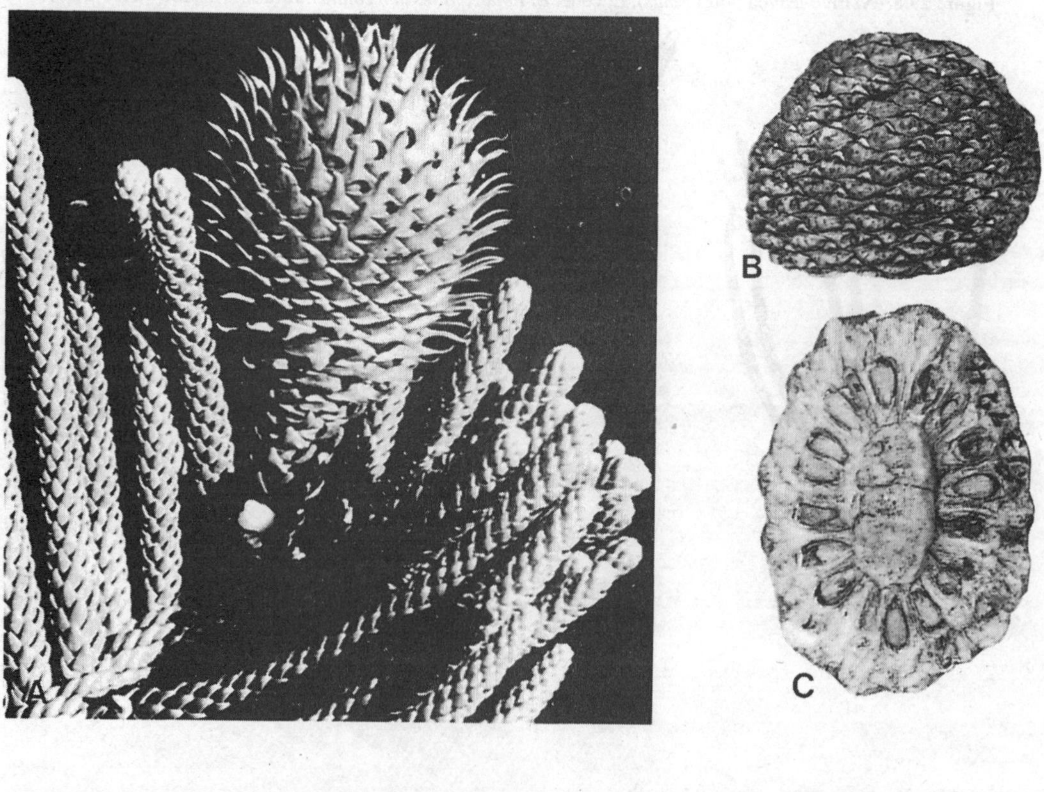

successive generations to produce a phylogenetic sequence. In the past few years the heterochronous hypothesis (Rothwell, 1982) for the origin of conifer leaves has received support both from fossil evidence and from the results of cladistic analyses.

Reproductive structures of Mesozoic coniferophytes

For the reason that reproductive structures of conifers and taxads with their attached or associated vegetative structures provide a much more reliable basis for determining the relationships of ancient conifers, we have chosen to present them as a separate unit of the chapter.

Voltziales

In Chapter 28 we described some Paleozoic members of the Voltziales and suggested that these "transition" conifers have a geological range that extends well into the Mesozoic. Among several genera prominent in this respect are representatives of the Voltziales such as *Voltziopsis, Glyptolepis,* and *Schizolepis* as summarized by Miller (1977).

As studies of Mesozoic conifers have proceeded, several genera have emerged that share characteristics with the Voltziales and families of the Coniferales. Obviously they cannot be assigned to either order. Based on our present information, these genera are in turn transitional and comprise a complex from which most families of modern conifers evolved during the Upper Cretaceous and early Tertiary. Some of these transitional genera have been assigned to families such as the Cheirolepidiaceae, the Pararaucariaceae, and the Palissyaceae, which are trated later in the chapter.

Araucariaceae

This family comprises two extant genera, *Agathis* and *Araucaria* (Fig. 29.9A), which are restricted to the Southern Hemisphere. These trees have helically arranged leaves and other appendages; the main branches, however, are usually whorled. Some species of *Araucaria* have falcate tetragonal leaves, while others have large flattened subsessile leaves with many parallel veins. The leaves of *Agathis* are of this latter type. The bract–scale complexes of the ovulate cones consist of a large woody bract and a smaller abaxial ovuliferous scale bearing a single submerged and inverted ovule. Bract and scale are fused, except in *Araucaria,* where the distal end of the scale is free and forms a "ligule." The pollen cones have bractlike microsporophylls with many pollen sacs. The pollen lacks wings.

For reasons given earlier, it is impossible to determine from wood and leaf fossils exactly when in geological time the Araucariaceae made its appearance. We get assistance, however, in making more accurate determinations from reproductive structures that can be assigned with some degree of assurance to families, even genera, of conifers. For example, cone scales similar to those of *Araucaria* have been found in beds as old as the Triassic. The idea of a Lower Triassic or Upper Permian origin of the family is consistent with the Permian occurrence of *Ullmannia frementaria,* whose foliage and wood are araucarioid and whose bract–scale complexes consist of a single inverted ovule on a large orbicular scale. It is still not known if the seed was fused with the ovuliferous scale or if it became detached as in mature *Araucaria bidwillii* cone scales (Schweitzer, 1963). Based on present evidence, it is reasonable to suggest that the Araucariaceae originated from *Ullmannia*-type representatives of the Permian Voltziales.

Araucarites

The cone–scale complexes assigned to *Araucarites* (Fig. 29.10A) are typically wedge-shaped with a broad base. As in living araucarias the complex is composed of a conspicuous bract with a beaklike apophysis. The bract is overlain by a smaller ovuliferous scale (Fig. 29.10C) with a distal free portion called the "ligule." An ovuliferous scale contains a single inverted ovule (Fig. 29.10B). When the cone matured the cone–scale complexes disarticulated. Thus, most specimens of *Araucarites* are incomplete and only show the bract with a median depression that indicates the former position of the seed. However, some specimens from Middle Jurassic beds (Kendall, 1952) show the seed and what appears to be the remains of the "ligule" in place (Fig. 29.10A). Cone–scale complexes of this type resemble those of living *Araucaria* belonging to the Section *Eutacta*.

Araucaria mirabilis

It has become customary to place compression–impression fossils of araucarioid seed–cone complexes

Figure 29.10. *Araucarites* sp. **A.** Bract–scale complex (b = bract). Ovuliferous scale with ovule (o) and ligule (l). **B.** Diagram longitudinal section ovule embedded in bract–scale complex. **C.** Transverse section. Jurassic. (**A** redrawn from Seward, 1919; **B,C** redrawn from Kendall, 1952.)

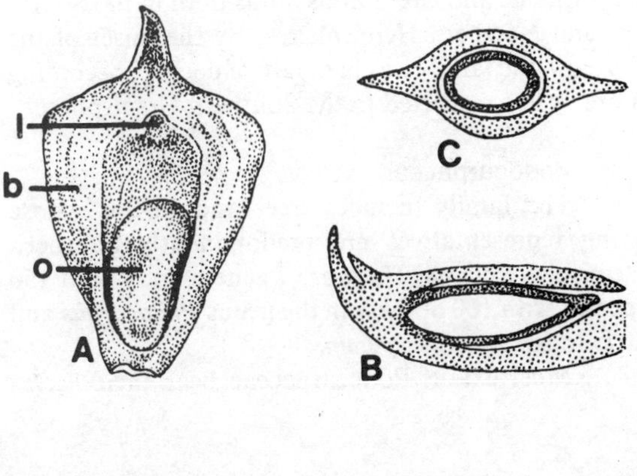

in the genus *Araucarites* and those of ovuliferous cones that are silicified in *Araucaria* (Fig. 29.9B). Stockey (1975, 1978, 1982) has made a detailed analysis of the permineralized cones from the famous Cerro Cuadrado petrified forest of Patagonia. Although the geological age of the petrified forest is in doubt, present evidence indicates that it is Middle to Late Jurassic. As a result of her studies of these beautifully preserved cones (Fig. 29.9C) Stockey was able to relate the structure of *A. mirabilis* cones to that of living *A. bidwillii*. Unlike most fossil ovules and seeds that are devoid of contents, those of *A. mirabilis* revealed an early free-nucleate stage of the megagametophyte and a later telo stage in the development of the embryo. At this stage the embryo, which is embedded in megagametophyte tissue, consists of two cotyledons, a shoot apex, root apex initials, a columella, and calypteroperiblem (a tissue that covers the root initials). The importance of this research is the demonstration that the telo stage of embryo formation had evolved by the Late Jurassic. Retention and development of the embryo prior to seed dispersal, as it occurs in *A. mirabilis*, is known to have evolved as early as the Upper Pennsylvanian or Lower Permian in seeds with the structure of the Cordaitales or some Voltziales (Miller & Brown, 1973). Retention and development of embryos within the nucellus and integuments of the developing seed seem to be characteristic of coniferous gymnosperms and Mesozoic cycadeoids as opposed to Paleozoic pteridosperms and cordaites. This feature suggests the origin of seed dormancy in primitive conifers, and it may have played a major role in allowing for the colonization and forestation of much of the land surface.

There are numerous accounts of araucarian cone-scale complexes and permineralized cones from many widely separated geographical locations that are Jurassic and Cretaceous in age. Many of the Jurassic localities are from India, Australia, and England, while those of the Cretaceous are from Europe, Africa, and North America. Other localities of Tertiary age are confined to the Southern Hemisphere. Based on their frequency of occurrence and wide geographical distribution, it is clear that the Araucariaceae became an abundant and important element of Upper Triassic, Jurassic, and Cretaceous floras both in the Southern and Northern Hemispheres. By the onset of the Tertiary, the family was in a state of decline, becoming more or less restricted to the Southern Hemisphere.

Podocarpaceae

The family includes trees and shrubs whose living representatives are predominantly Southern Hemisphere plants. There are 7 genera and about 150 species, with 100 of these in the genus *Podocarpus* and 20 belonging to *Dacrydium*.

Most extant Podocarpaceae bear their leaves and other parts in a helix, but some are opposite. Leaf morphology varies from species that have sessile, decurrent, and falcate leaves with a single vein to those that are large and flat and contain many parallel veins. The bract of the bract–scale complex is reduced and usually free from the ovuliferous scale. The latter bears a single ovule that may be erect, but is usually inverted in most species of *Podocarpus*. The ovuliferous scale partially or nearly completely envelops the ovule to form a fleshy epimatium (Fig. 29.11). The pollen cones produce bractlike microsporophylls each with two pollen sacs. The pollen grains have two to several wings.

The family has a fossil record, represented by reproductive shoots with leaves, that begins in the Upper Triassic and extends through the Mesozoic. They were particularly abundant in Antarctica during the Jurassic (Stockey, 1988). Compression–impression fossils of *Rissikia*, comprising leafy twigs, pollen cones, and seed cones, are the oldest remains that can be assigned with confidence to the Podocarpaceae. The specimens come from the Triassic of Africa and Australia. The leaves of *Rissikia* vary from scalelike to elongate leaves arranged in a helix on a spur shoot. Microsporophylls, which are borne in pollen cones,

Figure 29.11. *Podocarpus macrophyllus*. Ovule enclosed in epimatium (e) on receptacle (r). Extant. (Courtesy General Biological Supply House, Chicago.)

are peltate and have two pollen sacs. The pollen is striate and bisaccate.

The seed cones are of particular interest. They consist of an axis about 3 cm long bearing 15 to 25 bract–scale complexes (Fig. 29.12A) forming a loose helix. The subtending bract is trifid and about one-half the length of the axillary ovuliferous scale. The scale is divided into three lobes that are joined at their bases. Each lobe has one or two inverted stalked ovules, with the stalks fused basally to the lobe. When one examines the illustrations of *Rissikia,* one can see a striking similarity between its bract–scale complexes and those of the Voltziaceae. Miller (1977) makes the point that if it were not for the foliage, which is considered podocarpaceous, *Rissikia* would have been placed with the Voltziales. Regardless of how it is classified, *Rissikia* joins the ranks of "transition" conifers that seem to bridge the gap between Paleozoic and Mesozoic conifers.

Another genus that seems to be on the "line" leading to the Podocarpaceae is the Jurassic genus *Mataia* known from New Zealand and Australia. A seed cone of *Mataia,* which is about 3 cm long, has 8 to 12 bract–scale complexes (Fig. 29.12B). The bract is a small triangular structure subtending an expanded ovuliferous scale that has the distal one-third adaxially recurved in a way that partially covers the pair of inverted ovules on the adaxial surface of the scale. The recurved portion of the ovuliferous scale is interpreted as an early stage in the evolution of the podocarp epimatium. Other genera consisting of bract–scale complexes reflect a reduction of the lobed ovuliferous scale into a single unit with a trifid apex. This was accompanied by a reduction in the number of inverted ovules to one and the fusion of its stalk to the scale.

Figure 29.12. **A.** *Rissikia media.* Ovuliferous scale with paired, inverted ovules. Triassic. **B.** *Mataia podocarpoides,* ovuliferous scale partially enclosing a pair of ovules. **A,B:** Triassic. (**A,B** redrawn from Miller, 1977.) **C.** *Schizolepis braunii,* portion of cone. **D.** Ovuliferous scale. Jurassic. (Redrawn from Seward, 1919.)

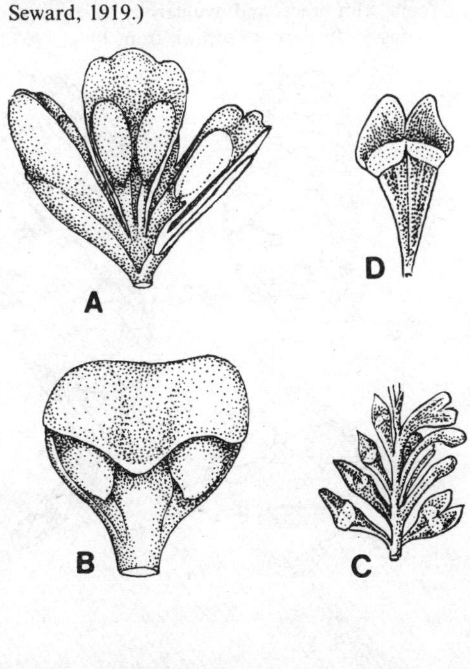

The Jurassic genus *Mehtaia* (Vishnu-Mittre, 1958) has cones consisting of spirally arranged bracts. Each bract bears a single erect ovule on the upper surface. There is no indication of an ovuliferous scale or epimatium. If the absence of these structures is an indication that the ovuliferous scale has undergone complete reduction, then we can place the origin of those extant podocarps showing this characteristic (*Pherosphaera*) as early as the Jurassic.

Pinaceae

Although the Pinaceae is the largest family of the modern conifers, it comprises only 10 genera, with about 200 species. The genera are *Pinus, Picea, Abies, Larix, Tsuga, Keteleeria, Pseudotsuga, Cedrus, Cathaya,* and *Pseudolarix.* All species belonging to these genera, with one exception, occur in the Northern Hemisphere.

The Pinaceae have in common needlelike leaves: leaves, cone bracts, and scales that are spirally arranged; woody ovuliferous scales that are distinct and free from the subtending bract; two abaxial pollen sacs on the cone scale, and pollen that is bisaccate.

When we look for the geologically oldest fossil that is anatomically comparable in structure with a modern conifer, we find that it is a cone from the Lower Cretaceous that has all the characteristics of cones of *Pinus.* As with other conifer families, however, there are many older genera that appear to be transitional in their reproductive characteristics. One of these is the Upper Triassic genus *Compsostrobus* (Delevoryas & Hope, 1973). The bract–scale complexes of this genus are borne in a loose helix on a cone axis. The bract is long and pointed; the axillary scale is spatulate and bears two inverted wingless seeds in the adaxial surface. *Elatocladus*-type foliage with cuticular characteristics similar to those of the cone were found in the same beds. Pollen cones, associated with the leaves and seed cones, have spirally arranged microsporophylls, each with two pollen sacs. The pollen is bisaccate and of the *Alisporites* dispersed type. Although the seed cones of *Compsostrobus* are imperfectly known, they exhibit a general morphology similar to that of modern pine cones (Miller, 1977). This suggests to Delevoryas & Hope that the Triassic period may represent a time when modern-type conifers were evolving (many modern families are recognized in the Jurassic period) with some ancestral and presumably more primitive forms persisting.

Other fossils that may be those of early Pinaceae are leafy shoots of *Pityocladus* and associated seed cones and scales of *Schizolepis* from the Upper Triassic (Seward, 1919) and Middle Jurassic (Harris, 1979).

The twigs bear dwarf shoots with spirally arranged persistent bud scales and simple linear single-veined leaves in a crowded helix. The similarity of these fossils and the foliage shoots of modern *Cedrus* and *Larix* has been noted by several authors. However, the only well-documented specimens that record the presence of *Cedrus* in the Mesozoic are those of the permineralized wood of *C. alaskensis* (Arnold, 1953).

The seed cone (Fig. 29.12C) and cone scales of *Schizolepis* (Fig. 29.12D) are reminiscent of those of the Pinaceae. The cone–scale complex consists of a small bract, probably united with the ovuliferous scale. The latter is a two-lobed structure with two abaxial inverted ovules, one near the base of each lobe.

A Mesozoic record for *Abies, Keteleeria, Picea, Larix, Pseudolarix, Pseudotsuga,* and *Tsuga* is presently lacking – or the evidence presented is questionable. Miller (1977) has provided a summary and evaluation of the evidence for the genera listed above.

Pine cones from the Cretaceous

This is an account of a few of the anatomically preserved seed cone genera whose affinities are with the Pinaceae. Seed cones belonging to this family are distinguishable by their helically arranged bract–scale complexes where each woody ovuliferous scale bears two inverted ovules. The bract is more or less free from and usually smaller than the ovuliferous scale it subtends. Many fossilized cones with these characteristics have been assigned to *Pityostrobus* or *Pseudoaraucaria,* both of which are genera combining characteristics not found in modern genera belonging to the Pinaceae (Robison & Miller, 1977; Stockey, 1981b). Without going into details, the characteristics of pinaceous cones that are most valuable in making generic assignments are size and shape; anatomy of the cone axis and its vascular cylinder, including bract and scale trace divergence; and bract and scale anatomy and morphology (Miller, 1976). Using these and other characteristics, Miller determined that of 15 cone specimens discovered in the Lower Cretaceous only 1 could be assigned to the genus *Pinus*. Of the remaining 14 specimens, 9 were best placed in *Pityostrobus* and 5 in *Pseudoaraucaria.* The paucity of Lower Cretaceous cones showing affinities with modern genera has led Miller to the conclusion that the single specimen assigned to *Pinus* represents the "nucleus" of an ancestral complex consisting of several different species of *Pityostrobus* from which other recent genera of the Pinaceae diverged during the Upper Cretaceous and Early Tertiary.

As its name implies, *Pseudoaraucaria* (Fig. 29.13A) was originally thought to be allied with the Araucariaceae until it was demonstrated by Alvin (1957) to be the cone of an extinct member of the Pinaceae. The cones are quite small (the largest is 6 cm in length). They have spirally arranged bract–scale complexes, where the thin scale apices terminate in a toothlike umbo. The ovuliferous scale consists of a perpendicular stalk and upturned distal lamina. The two seeds on the upper surface of each ovuliferous scale are separated by a median interseminal ridge (Fig. 29.13B) that partially overarches the ovules. This gives the impression of an *Araucaria* with two seeds that are partially embedded and explains its earlier designation as a member of the Araucariaceae.

The evidence provided by cone structures cited above shows us that the genus *Pinus* had evolved by the Lower Cretaceous. Although there is very little evidence that *Cedrus* and possibly *Larix* appeared prior to the Tertiary, there is abundant evidence that the remainder of the genera of the Pinaceae did not appear until the early Tertiary of later. The occurrence of leaves (*Pityocladus*) associated with seed cone scales (*Schizolepis*), both with characteristics of the Pinaceae, verifies the evidence from cones (*Pityostrobus, Pseudoaraucaria,* and *Pinus*) that the ancestors of Pinaceae were in place by the mid-Jurassic.

Some transitional taxodioids

Many Jurassic fossils, although exhibiting taxodiaceous characteristics, cannot be assigned to any one of the modern genera of the family. Once again we seem to have a complex of Jurassic and Cretaceous plants that combine characteristics of the Voltziales, genera of the Taxodiaceae, and other families of the Coniferales.

Figure 29.13. *Pseudoaraucaria heeri.* **A.** Reconstruction of cone with umbos at the tips of the bract–scale complex. **B.** Transverse section of scale showing two ovules embedded in adaxial surface. Cretaceous. (Redrawn from Alvin, 1957.) **C.** *Hirmerella muensteri,* cone with bracts and ovuliferous scales in place. Triassic–Jurassic. (Redrawn from Jung, 1968.)

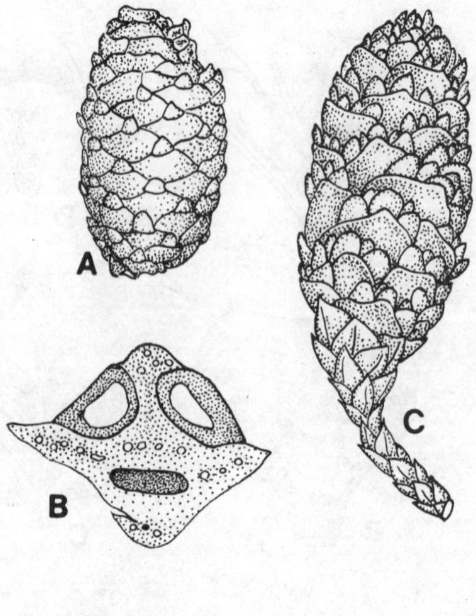

Cheirolepidiaceace

Our knowledge of the Cheirolepidiaceae has undergone a dramatic expansion in the last two decades, so that now it is possible to reconstruct the plants belonging to the family as large arborescent conifers. Alvin's (1983) reconstruction of *Pseudofrenelopsis parceramosa* from the Lower Cretaceous of the Isle of Wight provides a basis for such interpretations, and is an excellent example of the type of whole-plant reconstructions that are rapidly improving our understanding of fossil plant species. More recently, Watson (1988) has summarized our understanding of the Cheirolepidiaceae, and effectively emphasized that it was among the most diverse, reproductively sophisticated, and floristically important families of Mesozoic conifers. Vegetative and reproductive structures belonging to the Cheirolepidiaceae are known from the Upper Triassic through the Cretaceous of Europe and Argentina. Their remains constitute a predominant part of the Yorkshire Jurassic deposits of England, indicating that members of the family were an important part of the Jurassic coniferous forests.

Central to the development of our understanding of these conifers is dispersed pollen on the *Classopollis*-type (Harris, 1957) that extends from the Triassic to the mid-Cretaceous. Pollen grains of *Classopollis* (Fig. 29.14A) combine a unique set of characteristics. The grains, which are 20 to 30 μm in diameter, display a trilete mark on the proximal surface and a dishlike depression (cryptopore) on the distal surface. The

Figure 29.14. **A.** *Classopollis* pollen. Cryptopore (c); circumpolar groove (cg); triradiate scar (ts). **B.** *Hirmerella muensteri* with male cone above and two large female cones below. Female cone on left before shedding lobed ovuliferous scales, cone to right after shedding scales. **C.** *Frenelopsis ramosissima* showing terminal shoots with limited development of wood. **D.** *F. ramosissima,* suggested reconstruction of growth habit showing a succulent, shrubby xerophyte. **A:** Triassic–Cretaceous. **B:** Jurassic–Triassic. **C–D:** Lower Cretaceous. (From Watson 1988, with permission.)

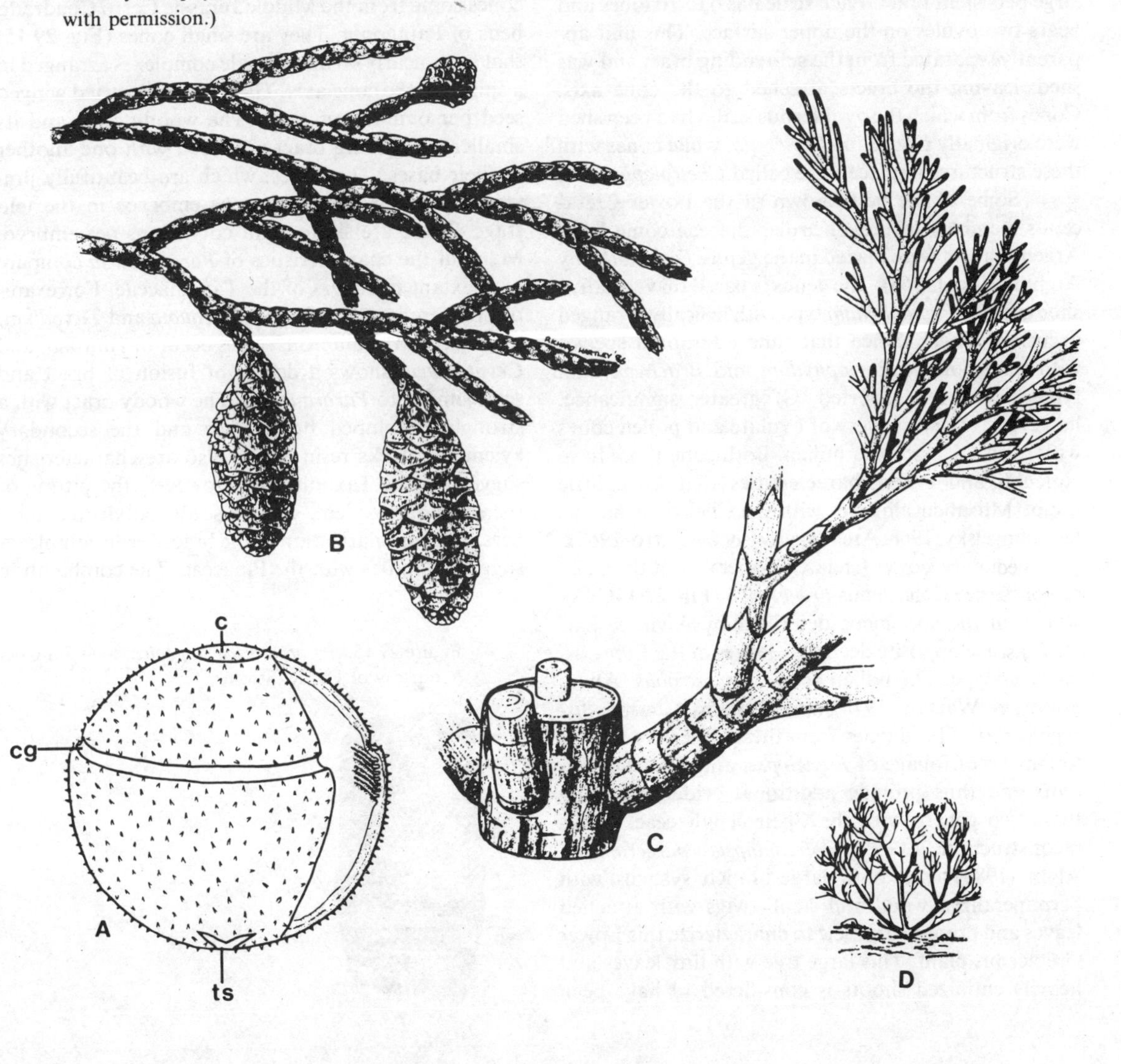

latter is thought to represent a germination pore. Approximately midway between the equator of the grain and the distal end is a circumpolar groove. Associated with the groove is a thickened band of sporoderm. The exine may be smooth to spiny, but most interesting is the tectate structure of the exine, a most unusual feature for pollen grains of conifers.

Of the megafossils, those assigned to the genus *Hirmerella* (Fig. 29.14B) are the best known. Stem fragments showing internal structure similar to that of conifers have been described by Jung (1968). The secondary wood is of the *Protocupressinoxylon* type. Leafy shoots with *Pagiophyllum-*, *Brachyphyllum-*, and *Geinitzia*-type foliage, known to be attached to stems, also bear pollen and ovulate cones. The pollen cones consist of microsporophylls surrounding an axis (Jung, 1968) with each microsporophyll probably with two pollen sacs from which *Classopollis*-type pollen has been isolated. The ovulate cones of *Hirmerella* (Fig. 29.13C), reinvestigated by Jung (1968), consist of spirally arranged ovuliferous scales each subtended by a large persistent bract. Each scale has 6 to 10 lobes and bears two ovules on the upper surface. This unit apparently separated from the subtending bract and was shed, leaving the bracts attached to the cone axis. Cones from which the ovuliferous scales had been shed were originally placed in *Hirmerella,* while cones with these structures in place were called *Cheirolepis*.

Some of the best known of the Lower Cretaceous members of the Cheirolepidiaceae come from Argentina and were placed in the genus *Tomaxellia* by Archangelsky (1963). The genus is based on vegetative shoots of the *Pagiophyllum* type with helically arranged decurrent leaves. Since that time additional species with leaves of the *Pagiophyllum* and *Brachyphyllum* types have been reported. Of greater significance, however, is the discovery of ovulate and pollen cones with in situ *Classopollis* pollen. Both cone types have cuticular and other characteristics that leave little doubt as to their affinities with the Cheirolepidiaceae (Archangelsky, 1968; Archangelsky & Gamerro, 1967).

Another Lower Cretaceous member of the Cheirolepidiaceae is the genus *Frenelopsis* (Fig. 29.14C,D), which, in the specimens described by Alvin & Pais (1978), have opposite decussate leaves of the *Cupressinocladus*-type. The pollen cone *Classostrobus* (Alvin, Spicer, & Watson, 1978) also contains *Classopollis*-type pollen. The cuticles from these cones are similar to cuticles of foliage of *Frenelopsis* and *Pseudofrenelopsis* and thus provide additional evidence linking these two genera with the Cheirolepidiaceae. In his reconstruction of *Pseudofrenelopsis parceramosa,* Alvin (1983) used both large-branch systems with permineralized wood and small twigs with attached leaves and preserved cuticle to characterize this Lower Cretaceous plant. This large tree with tiny leaves and heavily cutinized shoots is considered to have been evergreen, but to have shed ultimate branches in response to seasonal drying conditions.

Based on the characteristics of the ovulate cones, their structure, and the dehiscence of the seed-bearing scales, Jung (1968) suggests that members of the Cheirolepidiaceae are transitional between the Voltziales and the Taxodiaceae. However, cuticular preparations indicate that the ovules were surrounded by highly complex cone scales, which suggests that the group may be derived with respect to reproductive biology (Watson, 1988). This suggestion is supported by the complex structure of the *Classopollis* pollen, which may have served in a specialized pollen-recognition system similar to those of flowering plants (Chaloner, 1976; Watson, 1988).

Pararaucariaceae

Another good example that illustrates a Jurassic genus combining characteristics of various genera of the Taxodiaceae and other conifer families is the genus *Pararaucaria* (Stockey, 1977). These permineralized seed cones come from the Middle Jurassic Cerro Cuadrado beds of Patagonia. They are small cones (Fig. 29.15) that have nearly 40 bract–scale complexes arranged in a spiral on the cone axis. There is one inverted winged seed per ovuliferous scale. The woody scale and its smaller subtending bract are fused with one another at their bases. The cones, which are beautifully preserved, have polycotyledonous embryos in the telo stage. There are six to eight cotyledons per embryo. Many of the characteristics of *Pararaucaria* compare with extant members of the Taxodiaceae. For example, polyembryony is found in *Sequoia* and *Taxodium;* single-seeded ovuliferous scales occur in *Taiwana;* and *Cryptomeria* shows a degree of fusion of bract and scale similar to *Pararaucaria*. The woody bract with a strongly developed bract trace and the secondary xylem that lacks resin canals also are characteristics suggesting the Taxodiaceae. However, the pitting of the secondary xylem, winged seeds, cotyledon numbers, and vascularization of the bract–scale complexes suggest affinities with the Pinaceae. The combination

Figure 29.15. *Pararaucaria patagonica*, cone. Jurassic. (Courtesy of Dr. R. Stockey.)

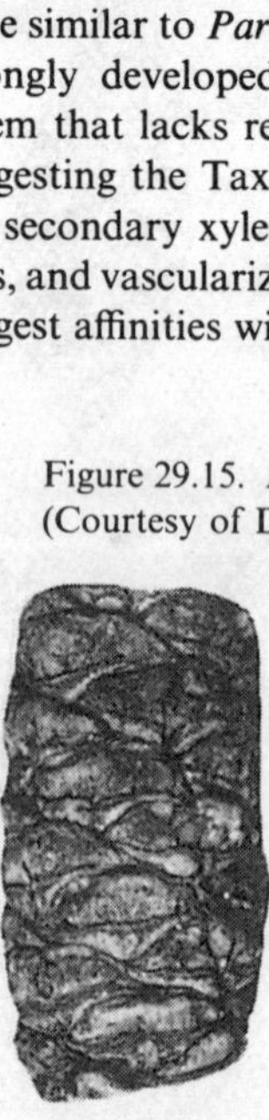

of taxodioid and pinoid characteristics in *Pararaucaria* has prompted some to suggest that the Taxodiaceae may have arisen from a primitive Triassic–Jurassic pinoid stock. As one might expect, the converse – that the pines have come from a primitive taxodioid stock – has been proposed. Regardless of which group came first, there is agreement that the two families are related, having a common ancestry in the Mesozoic.

Early evidence of extant families

The evidence of Mesozoic conifers that we have presented thus far in this chapter indicates the numerous difficulties paleobotanists face when attempting to classify extinct species. In the most rigorous approach employed to date, Miller (1988) has subjected fossil conifers to numerical cladistic analysis. Miller's study only partly resolves many important questions about the relationships among conifers. Nevertheless, it identifies which characters will be most useful for future study, and thus helps establish an agenda for the further resolution of conifer phylogeny.

Taxodiaceae

There are 10 genera assigned to the Taxodiaceae. These are *Arthrotaxis, Taxodium, Cunninghamia, Cryptomeria, Glyptostrobus, Metasequoia, Sciadopitys, Sequoia, Sequoiadendron*, and *Taiwania*. Fifteen modern species are recognized as belonging to these genera, none of which has species on more than one continent. Thus, the distribution of the genera is widespread and disjunct, which suggests that the modern representatives are relics of groups that were more abundant in the past.

With the exception of *Metasequoia* and *Sciadopitys*, the Taxodiaceae are characterized by leaves and cone parts that are spirally arranged; bract and ovuliferous scales with two to nine ovules on each scale; two to nine pollen sacs on each microsporophyll; and small nonsaccate pollen grains with a papilla.

Of the genera of the Taxodiaceae, the fossil record of *Cunninghamia* is probably the longest. Evidence for this genus appears as early as the Middle Jurassic in the form of the genus *Elatides*. Specimens of *Elatides* have accumulated for the past 150 years. Many of these have been used by T. M. Harris (1979) to reconstruct the whole plant where the internal structure of the wood (a *Cupressinoxylon*-type), stomatal characteristics, pollen cones, pollen, and seed cones with seeds are now known. With this much information at hand Harris is able to state that every characteristic of *Elatides* matches those found in the living genera of the Taxodiaceae. Except for its leaf characteristics, *Elatides* is most like the extant *Cunninghamia*. *Elatides williamsonii* is abundant from the Middle Jurassic to the Lower Cretaceous and clearly was a precursor to the Taxodiaceae.

Numerous genera have been described from the Upper Jurassic and Cretaceous that, like *E. williamsonii*, can be placed with the Taxodiaceae but cannot be assigned to a modern genus of the family. One of these that is particularly interesting is the Cretaceous genus *Parataxodium* (Arnold & Lowther, 1955). This genus was originally known from leaf remains, but later investigators have found pollen cones and seeds that they believed belong to the genus. The cones and leaves of *Parataxodium* are like those of *Metasequoia*. The arrangement of the leaves, however, is like that of *Taxodium*. Arnold & Lowther suggest that *Parataxodium* may represent a precursor of *Metasequoia* and *Taxodium*. The appearance of *Glyptostrobus* in the Upper Cretaceous conforms with that of *Metasequoia, Taxodium*, and *Sequoia*. Indentifications have been made using compression–impression fossils with seed and pollen cones. Isolated foliage shoots without reproductive parts and diagnostic cuticular characteristics cannot be made for *Glyptostrobus, Taxodium*, and *Sequoia*. Leaf fossils of these genera should be placed in the form genus *Elatocladus*.

Although permineralized wood, claimed to be that of *Sciadopitys*, has been reported from the Jurassic, this wood is nearly indistinguishable from that of *Podocarpus*. An Upper Cretaceous–Lower Tertiary origin for *Sciadopitys* is suggested, based on leafy shoots found by Christophel (1973) in Paleocene beds in Canada. The Upper Cretaceous diversification of the Taxodiaceae is further supported by *Austrosequoia*, a cone genus from Australia (Peters & Christophel, 1978). The latter combines characteristics of *Sequoia* and *Sequoiadendron*.

A Jurassic seed cone, *Swedenborgia*, has a structure similar to that of *Cryptomeria*. It is not until the Upper Cretaceous, however, that there is an abundance of *Cryptomeria*-like remains. This is usually in the form of foliage shoots that are now assigned to the form genus *Geinitzia*. As Miller (1977) indicates, the Mesozoic record of the *Cryptomeria* line is spotty at best.

Anatomically preserved cones with the characteristics of *Cunninghamia* are known to occur in Lower Cretaceous beds of California. The cone genus *Cunninghamiostrobus* has been assigned to these specimens (Miller, 1975). Other reports of *Cunninghamia* in the fossil record, and there are many, document its widespread occurrence in the Upper Cretaceous. Cone genera *Nephrostrobus* and *Rhombostrobus* (LaPasha & Miller, 1981), also from the Upper Cretaceous, are clearly those of the Taxodiaceae. The structure of the bract–scale complexes of *Nephrostrobus* approximates that of *Metasequoia*, while those of *Rhombostrobus* are similar to those of *Sequoia*. The authors conclude that *Nephrostrobus* and *Rhombostrobus* are elements of an ancestral complex from which the extant genera were derived. *Arthrotaxus* also appears in the Lower Cretaceous.

The above account of the distribution of the genera of the Taxodiaceae in the Mesozoic and Lower Tertiary leaves no doubt about the diversification of the family during the Cretaceous, especially the Upper Cretaceous. As a family, the Taxodiaceae, which shows an alliance with the Pinaceae, was well established by the Middle Jurassic.

Before leaving the Taxodiaceae, attention should be given to the remarkable genus *Metasequoia*. The history of its discovery, which makes a most interesting story, has been prepared by Li (1964). For nearly a century, paleobotanists had recognized fossil taxodiaceous foliage shoots that were assigned to *Sequoia langsdorfii*. It was thought to represent a forerunner to the modern redwoods, which during the Upper Cretaceous and Tertiary were widespread in the middle and high latitudes of Europe, Asia, and North America, including the Arctic. To Miki (1941) and other paleobotanists this northerly distribution of a typically warm-temperate evergreen needed explanation. In the course of his investigation Miki worked with specimens of *S. langsdorfii* from Pliocene deposits of Japan, Miki noticed, however, that many specimens had the striking characteristic of opposite leaves (Fig. 29.16A) on deciduous branches and seed cones with decussate scales and long slender stalks (Fig. 29.16B). Because these characteristics are most un-*Sequoia*-like, Miki established the genus *Metasequoia* for these specimens. In the same year that Miki described the fossil genus, a discovery was made in Szechuan Province in China of a large unknown conifer. After a period of years during which more collections were made it was determined by Hu & Cheng (1948) that the unknown extant gymnosperm was actually the same as the fossil *Metasequoia* earlier described by Miki. The living plants were called *M. glyptostroboides*. Until the researches of Rothwell & Basinger (1978), who established *M. milleri*, all fossil specimens were placed in a single species, *M. occidentalis* (Christophel, 1976). Here we have an example of compression–impression foliage shoots whose characteristics are well enough known to allow assignment to a natural instead of a form genus. The establishment of the new genus *Metasequoia* required a reexamination of *Sequoia langsdorfii* specimens and a reevaluation of the characteristics of fossil foliage. This was first attempted by Chaney (1951), with later revisions by Schweitzer (1974) and Christophel (1976).

Cupressaceae

The Cupressaceae, which is one of the three largest families of the conifers, has about 140 species distributed among 22 genera. The latter are about equally distributed between the Northern and Southern Hemispheres. Those genera that have a noteworthy Mesozoic fossil record are *Juniperus, Thuja,* and *Cupressus.*

Figure 29.16. *Metasequoia occidentalis.* **A.** Portion of leafy shoot showing opposite arrangement of leaves. **B.** Cone terminating long stalk. Upper Cretaceous. (From Chandrasekharam, 1974.)

Characteristics that have been used to identify fossils belonging to this family are the opposite and decussate or whorled arrangement of their parts; fusion between the bract and ovuliferous scale; erect ovules that are usually three or more in number; and a pollen sac number that varies from three to six. In their general construction, the cones of the Cupressaceae are similar to those of the Taxodiaceae.

One of the main problems in identification of presumed fossils of the Cupressaceae is determining if the leaf arrangement is decussate or helical. Trying to distinguish the Triassic remains of cupressoid and taxodioid remains also has proved to be difficult. It has been suggested that their similarity is evidence of a common origin for the two families. Unfortunately, the evidence demonstrating the Triassic occurrence of the Cupressaceae is based primarily on leafy shoots and wood specimens. However, bract–scale complexes that bear some resemblance to those of the Cupressaceae have been reported.

Leafy shoots assigned to *Cupressinocladus* have been described by Chaloner & Lorch (1960) from the Lower to Middle Jurassic. Although the construction of the epidermis is different from that of the Cupressaceae, there seems to be little doubt that the specimens belong to the family. Specimens similar to the foliage shoots, cones, and cone scales also have been described from the Upper Triassic.

We have to go to the Upper Cretaceous to find fossils that have been assigned to extant genera of the Cupressaceae. There are three species of *Thuja* reported and two species of *Juniperus*. Since foliage shoots provided the evidence of their affinities, there is some question of the validity of the identification. It is suggested that the fossils be placed in the form genus *Cupressinocladus* (Schweitzer, 1974; Christophel, 1976).

Many other Mesozoic genera with one or two characteristics suggesting affinities with extant genera of the Cupressaceae have been reported, but one has to look at the record of Tertiary fossils for those that can be assigned to such genera with some degree of assurance. The totality of the evidence indicates that the family began to differentiate in the Jurassic, but emergence of modern genera did not start until the Upper Cretaceous and Lower Tertiary.

Cephalotaxaceae

This family, comprising a single modern genus with six species, has a modest, unrevealing fossil record. Diagnostic seed cones of *Cephalotaxus* consist of opposite and decussate bracts, each subtending two erect ovules on a fertile short shoot. Usually one ovule develops into a large olivelike seed. Fossils that are claimed to show reproductive structure of this kind (*Cephalotaxospermum*) are reported from the Triassic to the Oligocene. Miller (1977) suggests that many of the specimens could be ovulate structures of *Podocarpus* or *Torreya*. *Cephalotaxopsis*, the presumed leafy shoots of *Cephalotaxus*, are reported from the Lower Cretaceous to the Miocene.

The most interesting part of the fossil record that may reflect the early evolution of the Cephalotaxaceae is found in the Lower Jurassic genera *Palissya* and *Stachyotaxus* of the Palissyaceae. The seed cones of both genera are long cylindrical structures consisting of a central axis with a number of spirally arranged bract–scale complexes. In *Palissya* (Fig. 29.17A) the bract is a large linear structure that has a decurrent base attached to the cone axis. The ovuliferous scale, which is in the axil of a bract, bears about 10 more or less erect ovules arranged in two rows. Each ovule is partially enclosed by an asymmetrical aril. The ovuliferous scale does not appear to be fused to the subtending bract. *Stachyotaxus* (Fig. 29.17B) has two ovules on a much shorter bract–scale complex and appears to be a reduced version of *Palissya* (Florin, 1951). The pollen cones of *Stachyotaxus* have the appearance of small catkins. The microsporophylls are bractlike. Each has two abaxial pollen sacs containing spherical wingless pollen. Leaves of the Palissyaceae are flattened needles with a single vein and a decurrent base. They are spirally arranged on the branches.

Figure 29.17. **A.** *Palissya sphenolepis*, bract and ovuliferous scale. **B.** *Stachyotaxus elegans* bract–scale complex. Jurassic, (Redrawn from Florin, 1951.) **C,D.** *Taxus baccata*, shoots showing arillate ovules at tips of lateral branches. Extant. **E.** *Palaeotaxus rediviva*, reconstruction showing arillate ovule surrounded by bracts. **F.** *Marskea thomasiana*, ovule. Jurassic. (**E** redrawn from Florin, 1951; **F** redrawn from Harris, 1976.)

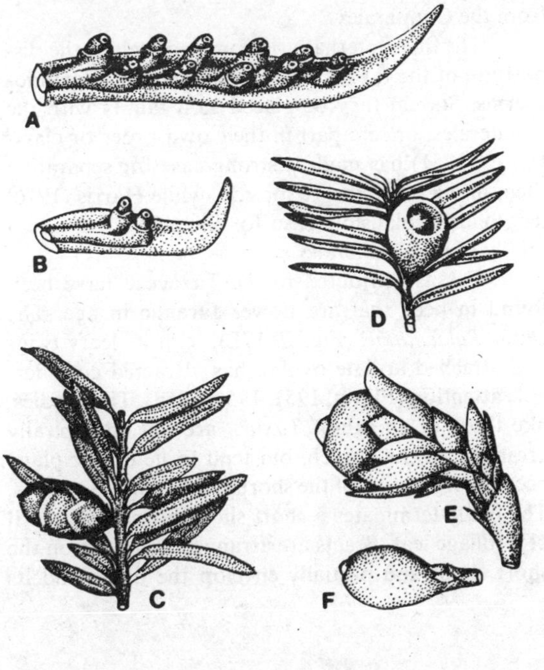

Although *Palissya* and *Stachyotaxus* cannot be placed in a modern family of conifers, their resemblance to *Dacrydium* of the Podocarpaceae and *Cephalotaxus* has been noted by various investigators (Florin, 1951, 1958; Schweitzer, 1963). They envisage the origin of the Palissyaceae bract–scale complex from genera of the Permian *Ernestiodendron* type where the secondary shoot bears several spirally arranged erect ovules. By reduction and planation, an ovuliferous scale in the axil of a bract similar to *Palissya* would be produced. By further reduction *Stachyotaxus* would evolve into the bract–scale complex characterized by *Cephalotaxus*, with its highly reduced ovuliferous scale that bears a pair of erect ovules.

Although attempts have been made to place *Cephalotaxus* with the Taxales because of their ovules that appear to be terminal on a lateral shoot, Florin and others have demonstrated that each "terminal" unit is a bract–scale complex with a highly reduced ovuliferous scale. If these interpretations are correct, then the origin of the Cephalotaxoceae is with the Permian Voltziales.

Taxales

This small group comprises only 5 living genera and about 20 species. The genera are *Taxus, Austrotaxus, Pseudotaxus, Torreya,* and *Amentotaxus*. Most are shrubs or small trees, but *Amentotaxus* and some species of *Taxus* and *Torreya* are large trees. As noted earlier in this chapter, the vegetative characteristics of the taxads are similar to those of the Coniferales. The pollen cones consist of several peltate microsporangiophores with two to eight pollen sacs. This remarkable un-coniferlike structure accompanied by the terminal position of the arillate ovules (Fig. 29.17C,D) comprise the characteristics used to separate the taxads from the Coniferales.

The question that arises immediately is the disposition of the taxads in the classification of gymnosperms. Should they be placed as a family with the Coniferales, or set apart in their own order or class? Florin (1951) has made a strong case for separating them into a class, the Taxopsida, while Harris (1976) has made a plausible case for including them as a family of the Coniferales.

Fossils attributed to the Taxaceae have been found in beds that are Lower Jurassic in age. The genus *Palaeotaxus* (Fig. 29.17E), with its leafy twigs and attached arillate ovules, has attracted considerable attention (Florin, 1951, 1954, 1958). The needles, like those of an extant *Taxus,* are flat and spirally arranged on the branch, but tend to lie in one plane because of twisting of the short petioles of the needles. The ovule terminates a short shoot borne in the axil of a foliage leaf. Bracts are arranged in a helix on the short shoot and partially envelop the ovule and its surrounding aril. With so many characteristics in common, it is rather surprising that the structure of the epidermis of *Palaeotaxus* is quite different from that of *Taxus*.

The Middle Jurassic beds of Yorkshire have provided a number of leafy shoots and ovulate and pollen organs whose affinities are unquestionably with the Taxaceae. Among these is the genus *Marskea* (Fig. 29.17F), which now includes the material described as *Taxus jurassica* by Florin (1958). After a painstaking reinvestigation of the Yorkshire material, Harris (1976, 1979) found that *T. jurassica* and *Marskea* were the same in having a highly variable cuticular structure and nonpetiolate leaves borne in a decussate, not spiral, manner. Taxaceae with these characteristics are *Amentotaxus* and *Torreya* rather than *Taxus*. The pollen organs are unique to *Marskea,* and the pollen is a nonsaccate type. Harris (1976) regards *Marskea* as another example of the many ancient genera of coniferophytes that combine characteristics of other genera belonging to a particular family.

Other seeds, leaves, and wood specimens attributed to the Taxaceae have been described from the Jurassic onward. Foliage remains are often assigned to the form genus *Elatocladus,* wood specimens to *Taxaceoxylon* and *Torreoxylon* (see earlier sections of this chapter).

Even though the fossil record of the taxads extends from the Lower Jurassic to the present, the record is devoid of clues that would help us to understand their origin and relationships. Florin (1954) holds strongly to the idea that the origin of the taxads can be traced to the Devonian psilophytes and that their origin is distinct from that of the conifers that stem from the Cordaitales. His argument is that the taxad ovules "are seated terminally on the strobilus [secondary shoot] axis itself and megasporophylls are accordingly absent. The strobili of these genera can therefore not be derived from those of Paleozoic cordaites and conifers." The latter have, as Florin has so carefully documented, compound cones with bract–ovuliferous scale complexes where the ovule represents a megasporophyll.

Harris (1976) is inclined to favor an origin of taxads from the Voltzialean type and believes that they are not fundamentally different from the conifers. He suggests that the ovuliferous short shoot of the taxads could be derived from fertile secondary shoots of *Utrechtia* by assuming a reduction of the shoot axis and its sterile scales above the ovule by suppression of the apical meristem – this followed by a shift of an ovule from the lateral to the terminal position. Although the fossil evidence for this origin of the terminal position of a taxad ovule is lacking, as Harris points out, it is a change that has occurred in plants in other groups. His recommendation is to place the family

Taxaceae in the Coniferales, a recommendation that can be supported when the vegetative structures of taxads and conifers are considered.

If Harris proves to be correct, then the generalization that the Paleozoic and Lower Mesozoic Voltziales provided the ancestors of the Coniferales would gain validity. We have seen good evidence that the Araucariaceae, Pinaceae, Podocarpaceae, Taxodiaceae, and Cephalotaxaceae can be traced to the Voltziales through intermediate families (Cheirolepidaceae, Palissyaceae, Pararaucariaceae, and Protopinaceae) and genera such as *Compsostrobus*. From these and other "intermediates" that combine characteristics of ancient and modern conifer genera, the modern families of conifers diverged and became delineated during the Upper Triassic, Jurassic, and Cretaceous. Most modern genera of the Coniferales were present by the onset of the Tertiary. The origin and subsequent evolution of those plants called coniferophytes is summarized in Chart 29.1.

References

Alvin, K. L. (1957). On *Pseudoaraucaria* Fliche emend., a genus of fossil pinaceous cones. *Annals of Botany* (N.S.), **21,** 33–51.

Alvin, K. L. (1983). Reconstruction of a Lower Cretaceous conifer. *Botanical Journal of the Linnean Society,* **86,** 169–76.

Alvin, K. L., & Pais, J. J. C. (1978). A *Frenelopsis* with opposite decussate leaves from the Lower Cretaceous of Portugal. *Palaeontology,* **21,** 873–9.

Alvin, K. L., Spicer, R. A., & Watson, J. (1978). A *Classopollis*-containing male cone associated with *Pseudofrenelopsis. Palaeontology,* **21,** 847–56.

Archangelsky, S. (1963). A new Mesozoic flora from Ticó, Santa Cruz Province, Argentina. *Bulleting of the British Museum (Natural History),* **8,** 4–92.

Archangelsky, S. (1968). On the genus *Tomaxellia* (Coniferae) from the Lower Cretaceous of Patagonia (Argentina) and its male and female cones. *Journal of the Linnean Society (Botany),* **61,** 153–65.

Archangelsky, S., & Gamerro, J. C. (1967). Pollen grains found in coniferous cones from the Lower Cretaceous of Patagonia (Argentina). *Review of Palaeobotany and Palynology,* **5,** 179–82.

Arnold, C. A. (1953). Silicified plant remains from the Mesozoic and Tertiary of North America. II. Some fossils from northern Alaska. *Michigan Academy of Sciences and Letters,* **38,** 9–20.

Arnold, C. A., & Lowther, J. C. (1955). A new Cretaceous conifer from northern Alaska. *American Journal of Botany,* **42,** 522–8.

Basinger, J. F. (1981). The vegetative body of *Metasequoia milleri* from the Middle Eocene of British Columbia. *Canadian Journal of Botany,* **59,** 2379–410.

Beck, C. B. (1957). *Tetraxylopteris schmidtii* gen. et sp. nov. A probable pteridosperm recursor from the Devonian of New York. *American Journal of Botany,* **44,** 350–67.

Beck, C. B. (1970). The appearance of gymnospermous structure. *Biological Reviews,* **45,** 379–400.

Beck, C. B., Coy, K., & Schmid, R. (1982). Observations on the fine structure of *Callixylon* wood. *American Journal of Botany,* **69,** 54–76.

Carluccio, L. M., Hueber, F. M., & Banks, H. P. (1966). *Archaeopteris macilenta,* anatomy and morphology of its frond. *American Journal of Botany,* **53,** 719–30.

Chaloner, W. G. (1976). The evolution of adaptive features in fossil exines. In *The Evolutionary Significance of the Exine,* eds. I. K. Ferguson & J. Muller. London: Academic Press.

Chaloner, W. G., & Lorch, J. (1960). An opposite-leaved conifer from the Jurassic of Israel. *Palaeontology,* **2,** 236–42.

Chandrasekharam, A. (1974). Megafossil flora from the Genessee locality, Alberta, Canada. *Palaeontographica B,* **147,** 1–41.

Chaney, R. W. (1951). A revision of fossil *Sequoia* and *Taxodium* in western North America based on the recent discovery of *Metasequoia. Transactions of the American Philosophical Society,* **40,** 171–263.

Christophel, D. C. (1973). *Sciadopitophyllum canadense* gen. et sp. nov.: a new conifer from western Alberta. *American Journal of Botany,* **60,** 61–6.

Christophel, D. C. (1976). Fossil floras of the Smoky Tower locality, Alberta, Canada. *Palaeontographica B,* **157,** 1–43.

DeLaubenfels, D. J. (1953). The external morphology of coniferous leaves. *Phytomorphology,* **3,** 1–20.

Delevoryas, T., & Hope, R. C. (1973). Fertile coniferophyte remains from the Late Triassic Deep River Basin, North Carolina. *American Journal of Botany,* **60,** 810–18.

Eckenwalder, J. E. (1976). Re-evaluation of Cupressaceae and Taxodiaceae: a proposed merger. *Madroño,* **23,** 237–56.

Esau, K. (1977). *The Anatomy of Seed Plants,* 2nd ed. New York: Wiley.

Florin, R. (1951). Evolution of cordaites and conifers. *Acta Horti Bergiani,* **15,** 285–388.

Florin, R. (1954). The female reproductive organs of conifers and taxads. *Biological Reviews,* **29,** 367–89.

Florin, R. (1958). On Jurassic taxads and conifers from northeastern Europe and eastern Greenland. *Acta Horti Bergiani,* **17,** 259–388.

Gould, R. E. (1975). A preliminary report on petrified axes of *Vertebraria* from the Permian of Eastern Australia. In *Gondwana Geology: Papers Presented at the Third Gondwana Symposium, Canberra, Australia,* ed. K. S. W. Campbell. Canberra: Australian National University Press, pp. 109–15.

Grambast, L. (1961). Evolution des structures ligneuses chez les Coniférophytes. *Bulletin de la Société Botanique de France, Memoirs,* **39,** 30–41.

Greguss, P. (1955). *Identification of Living Gymnosperms on the Basis of Xylotomy.* Budapest: Akadémiai Kiadó.

Greguss, P. (1967). *Fossil Gymnosperm Woods in Hungary from Permian to the Pliocene.* Budapest: Akadémiai Kiadó.

Harris, T. M. (1957). A Liasse–Rhaetic flora in South Wales.

Proceedings of the Royal Society, London B, **147,** 289–308.

Harris, T. M. (1969). Naming a fossil conifer. *Botanical Society of Bengal, J. Sen Memorial Volume,* 243–52.

Harris, T. M. (1976). The Mesozoic gymnosperms. *Review of Palaeobotany and Palynology,* **21,** 119–34.

Harris, T. M. (1979). *The Yorkshire Jurassic Flora V: Coniferales.* London: British Museum (Natural History).

Hu, H. H., & Cheng, W. C. (1948). On the new family Metasequoiaceae and *Metasequoia glyptostroboides,* a living species of *Metasequoia* found in Szechuan and Hupen. *Bulletin of the Fan Memorial Institue for Biology (N.S.),* **1,** 153–61.

Hughes, N. F. (1976). *Palaeobiology of Angiosperm Origins.* Cambridge: Cambridge University Press.

Jung, W. W. (1968). *Hirmerella muensteri* (Schenk) Jung nov. comb., eine bedeutsame Konifere des Mesozoikums. *Palaeontographica B,* **122,** 55–93.

Kendall, M. W. (1952). Some conifers from the Jurassic of England. *Annales, Magazine of Natural History,* **5,** 583–94.

Kräusel, R. (1949). Die fossilen Koniferen-Hölzer, *Palaeontographica B,* **89,** 83–203.

LaPasha, C. A., & Miller, C. N. (1981). New taxodiaceous seed cones from the Upper Cretaceous of New Jersey. *American Journal of Botany,* **68,** 1374–82.

Leclercq, S., & Bonamo, P. M. (1971). A study of the fructification of *Milleria* (*Protopteridium*) from the Middle Devonian of Belgium. *Palaeontographica B,* **136,** 83–114.

Li, H. L. (1964). *Metasequoia,* a living fossil. *American Scientist,* **52,** 93–109.

Miki, S. (1941). On the change of flora in eastern Asia since the Tertiary period. *Japanese Journal of Botany,* **11,** 237–303.

Miller, C. N. (1975). Petrified cones and needle-bearing twigs of a new taxodiaceous conifer from the Early Cretaceous of California. *American Journal of Botany,* **62,** 706–13.

Miller, C. N. (1976). Early evolution in the Pinaceae. *Review of Palaeobotany and Palynology,* **21,** 101–17.

Miller, C. N. (1977). Mesozoic conifers. *Botanical Review,* **43,** 217–80.

Miller, C. N. (1982). Current status of Paleozoic and Mesozoic conifers. *Review of Palaeobotany and Palynology,* **37,** 99–114.

Miller, C. N. (1988). The origin of modern conifer families. In *Origin and Evolution of Gymnosperms,* ed. C. B. Beck. New York: Columbia University Press.

Miller, C. N., & Brown, J. T. (1973). Paleozoic seeds with embryos. *Science,* **179,** 184–5.

Peters, M. D., & Christophel, D. C. (1978). *Austrosequoia wintonensis,* a new taxodiaceous cone from Queensland, Australia. *Canadian Journal of Botany,* **56,** 3119–28.

Phillips, E. W. J. (1941). The identification of coniferous woods by their wood structure. *Journal of Linnean Society, London (Botany),* **52,** 259–320.

Phillips, E. W. J. (1948). The identification of softwoods by their microscopic structure. *Forest Products Research Bulletin,* **22,** London: Department of Scientific and Industrial Research.

Ramanujam, C. G. K. (1972). Fossil coniferous woods from the Oldman Formation (Upper Cretaceous) of Alberta. *Canadian Journal of Botany,* **50,** 595–602.

Read, C. B. (1932). *Pinoxylon dakotense* Knowlton from the Cretaceous of the Black Hills. *Botanical Gazette,* **93,** 173–87.

Record, S. J. (1934). *Identification of the Timbers of Temperate North America.* New York: Wiley.

Robison, C. R. (1977). *Pinus triphylla* and *Pinus quinquefolia* from the Upper Cretaceous of Massachusetts. *American Journal of Botany,* **64,** 726–32.

Robison, C. R., & Miller, C. N. (1977). Anatomically preserved seed cones of the Pinaceae from the Early Cretaceous of Virginia. *American Journal of Botany,* **64,** 770–9.

Rothwell, G. W. (1982). New interpretation of the earliest conifers. *Review of Palaeobotany and Palynology,* **37,** 7–28.

Rothwell, G. W., & Basinger, J. F. (1978). *Metasequoia milleri* n. sp., anatomically preserved pollen cones from the Middle Eocene (Allenby Formation) of British Columbia. *Canadian Journal of Botany,* **57,** 958–70.

Sahni, B. (1920). On certain archaic features in the seed of *Taxus baccata,* with remarks on the antiquity of the Taxaceae. *Annals of Botany,* **34,** 117–33.

Sahni, B. (1948). The Pentoxyleae: A new group of Jurassic gymnosperms from the Rajmahal Hills of India. *Botanical Gazette,* **110,** 47–80.

Scheckler, S. E., & Banks, H. P. (1971a). *Proteokalon,* a new genus of progymnosperms from the Devonian of New York State and its bearing on phylogenetic trends in the group. *American Journal of Botany,* **58,** 875–84.

Scheckler, S. E., & Banks, H. P. (1971b). Anatomy and relationships of some Devonian progymnosperms from New York. *American Journal of Botany,* **58,** 737–51.

Schmid, T. (1967). Electron microscopy of wood of *Callixylon* and *Cordaites. American Journal of Botany,* **54,** 720–9.

Schweitzer, H. J. (1963). Der weibliche Zapfen von *Pseudovoltzia liebeana* und seine Bedeutung für die Phylogenie der Koniferen. *Palaeontographica B,* **113,** 1–29.

Schweitzer, H. J. (1974). Die "tertiaren" Koniferen Spitzbergens. *Palaeontographica B,* **149,** 1–89.

Scott, A. (1974). The earliest conifer. *Nature,* **251,** 707–8.

Seward, A. C. (1919). *Fossil Plants.* Vol. IV. Cambridge: Cambridge University Press. Reprint. New York: Macmillan (Hafner Press), 1969.

Sporne, K. R. (1965). *The Morphology of Gymnosperms.* London: Hutchinson University Library.

Stockey, R. A. (1975). Seeds and embryos of *Araucaria mirabilis. American Journal of Botany,* **62,** 856–68.

Stockey, R. A. (1977). Reproductive biology of the Cerro Cuadrado (Jurassic) fossil conifers: *Pararaucaria patagonica. American Journal of Botany,* **64,** 773–44.

Stockey, R. A. (1978). Reproductive biology of Cerro Cuadrado fossil conifers: ontogeny and reproductive strategies in *Araucaria mirabillis* (Spegazzini) Windhausen. *Palaeontographica B,* **166,** 1–15.

Stockey, R. A. (1981a). Some comments on the origin and

evolution of conifers. *Canadian Journal of Botany*, **59**, 1932–40.

Stockey, R. A. (1981b). *Pityostrobus mcmurrayensis* sp. nov., a permineralized pinaceous cone from the Cretaceous of Alberta. *Canadian Journal of Botany*, **59**, 75–82.

Stockey, R. A. (1982). The Araucariaceae: an evolutionary perspective. *Review of Palaeobotany and Palynology*, **37**, 133–54.

Stockey, R. A. (1988). Antarctic and Gondwana conifers. In *Antarctic Paleobiology*, eds. T. N. Taylor & E. L. Taylor. New York: Springer-Verlag.

Stockey, R. A., & Ko, H. (1989). Cuticle micromorphology of *Dacrydium* (Podocarpaceae) from New Caledonia. *Botanical Gazette*, **150**, 138–49.

Stockey, R. A., & Taylor, T. N. (1978). Cuticular features and epidermal patterns in the genus *Araucaria* de Jussieu. *Botanical Gazette*, **139**, 490–98.

Vaudois, N., & Privé, C. (1971). Révision des bois fossiles de Cupressaceae. *Palaeontographica B*, **134**, 61–86.

Vishnu-Mittre (1958). Studies on fossil flora of Nipania (Rajmahal) series, Bihar-coniferales. The *Paleobotanist*, **6**, 82–122.

Vogellehner, D. (1964). Zur Nomenklatur der fossilium Hölzgattung *Dadoxylon* Endlicher 1847. *Taxon*, **13**, 233–7.

Vogellehner, D. (1967). Zur Anatomie und Phylogenie Mesozoischer Gymnospermenhölzer, 5: Prodomus zu einer Monographie der Protopinaceae I. Die protopinoiden Hölzer der Trias. *Palaeontographica B*, **121**, 30–51.

Vogellehner, D. (1968). Zur Anatomie und Phylogenie Mesozoischer Gymnospermenhölzer, 7: Prodomus zu einer Monographie der Protopinaceae II. Die protopinoiden Hölzer der Jura. *Palaeontographica B*, **124**, 125–62.

Watson, J. (1988). The Cheirolepidiaceae. In *Origin and Evolution of Gymnosperms*, ed. C. B. Beck. New York: Columbia University Press.

30

The origin and early evolution of angiosperms

We approach the subject of this chapter with some trepidation, as have others who have tackled this topic. The magnitude and importance of the group; the plethora of conflicting concepts, hypotheses, theories, and ideas related to the origin of angiosperms; as well as our own limitations account for many of the difficulties. As students of paleobotany we have the advantage, however, of having examined the fossil records of chlorophyllous plants, starting with the algae and proceeding through the gymnosperms, prior to evaluating some of the multifaceted studies designed to provide insight into the "abominable mystery" of angiosperm origin(s). Our exposure to the fossil records of plants, prior to analyzing the origin of angiosperms, is the ultimate "upward outlook" in paleobotany. We believe that this approach is more productive than that used by many angiosperm morphologists who established sets of criteria for primitiveness, based on studies of extant flowering plants, and then looked "backward" to the fossil record for ancestors that seem to satisfy the criteria. The fossil record has been most disappointing to many who have attempted to discover the "missing links" in this way. As a partial result of this approach, ferns and most major groups of gymnosperms (Pteridospermales, Caytoniales, Coniferales, Cycadeoidales, Gnetales, Glossopteridales, and Pentoxylales) have been proposed at various times as ancestors of angiosperms.

Books and articles that have proved useful in obtaining a perspective on angiosperm origins and early evolution are by Just (1948); Eames (1959); Axelrod (1960); Melville (1960); Puri (1967); Cronquist (1968, 1981); Takhtajan (1969); Stebbins (1974, 1976); Beck (1976); Hughes (1976); Doyle (1978); Dilcher (1979); Meeuse (1979a, b); Crane (1985); Doyle & Donoghue (1986b, 1987); and Friis, Chaloner, & Crane (1987). These publications with their extensive lists of references will provide a basis for interested students who wish to become better acquainted with a spectrum of ideas about when angiosperms first appeared in the fossil record, the environment that supported them, whether the angiosperms were monophyletic or polyphyletic, the origin of flower parts, what characteristics are primitive, and so on. These are some of the topics we will briefly consider in this chapter.

Classification and characteristics of angiosperms

It is inevitable that those who devise "natural" systems of classification for plants, sooner or later find that the systems they construct are found to be incompatible in one or more parts with systems compiled by their colleagues. And so it is in this book with the classification established for the angiosperms. In Chapter 3 we gave the reasons why we believe the Division Tracheophyta to be a natural taxon that includes all plants having vascular tissue. Others disagree with this concept and separate vascular plants into many divisions for the reason that we can not be sure of the phylogenetic relationships of the major taxa (the rhyniophytes, lycopsids, sphenopsids, ferns, progymnosperms, gymnosperms, and angiosperms) that comprise them.

In his widely used classification of the flowering plants, Cronquist (1981) has adopted the latter point of view and assigned the angiosperms to the Division Magnoliophyta and subdivided it into two classes – Magnoliopsida (dicotyledons) and Liliopsida (monocotyledons). Because this classification cannot be reconciled with the concept of the Division Tracheophyta adopted here, we recommend a compromise that will allow the Cronquist classification for angiosperms to be utilized in this and the following chapter, and at the same time permit the recognition of the

Tracheophyta as a natural taxon. To do this we recognize the angiosperms as a subdivision of the Tracheophyta, which permits the adoption of the following classification:

Division – Tracheophyta
  Subdivision – Angiospermophytina
    Class – Magnoliopsida (dicots)
    Class – Liliopsida (monocots)

The designation Angiospermophytina, rather than Magnoliphytina, for the subdivision of flowering plants has the advantage of minimizing to a degree the widely accepted concept that magnoliid dicotyledons represent the ancestral plexus from which all other flowering plants were derived. Recent studies of Lower Cretaceous floras reveal fossil remains belonging to nonmagnoliid dicotyledons of the Hamamelidae. These plants were present at about the same time in the Lower Cretaceous as the magnoliid representatives and played a very important role in the subsequent derivation of a large and important group of nonmagnoliid dicots (Crane, 1989). The possibility remains, however, that the ancient hamamelids may have been derived in some way from as yet unknown magnoliid ancestors.

We regard this classification as one of accommodation until such a time when more has been resolved about relationships of major groups of vascular plants. For example, if it can be demonstrated that seed plants (including the angiosperms) and plants that have preovules and hydrasperman reproduction represent an inclusive assemblage – a clade forming an unnamed monophyletic group – then our classifications of "spermatophytes" will once more require major revisions.

Irrespective of what classification we use, whether it is one constructed by a "lumper" or a "splitter," the Subdivision Angiospermophytina comprises by far the largest number of species. The number varies according to the classification used, from 200,000 to 300,000 species assigned to 300 to 400 families. The numbers of angiosperm families and species are far greater than the numbers of families and species for all other plant groups combined. Taxonomists have agreed, since flowering plants were first classified, that the flowering plants can be subdivided into two natural groups, the monocotyledons and dicotyledons. In the classification by Cronquist (1981), the dicotyledons are divided into 64 orders and 318 families; the monocots into 17 orders and 65 families.

The distinction between the classes Liliopsida (monocotyledons) and Magnoliopsida (dicotyledons) is not as easy to make as their names imply. The number of cotyledons is not an absolute criterion. However, the structure and embryonic origin of their cotyledons are characteristics that taxonomists can depend on and use to distinguish members of the two groups (Bierhorst, 1971). The numbers of the cotyledons (one in monocots and two in dicots), the numbers of flower parts (threes in monocots and fours or fives in dicots), venation patterns ("parallel" in monocots and reticulate in dicots), cauline vascular bundle arrangement (scattered in monocots and a ring in dicots), and the presence (in dicots) or absence (in monocots) of a vascular cambium provide us with a set of characteristics by which one can distinguish the two subclasses. No one of these generalized characteristics, however, is exclusive to either subclass.

The single characteristic that seems to distinguish the Angiospermophytina is double fertilization and the consequent development of polyploid endosperm. In the vast majority of species, however, the possession of a carpel and the development of a fruit from it are distinctive. The evolution of the carpel precluded the germination of pollen on the nucellus or in the micropyle of the ovule, as it occurs in gymnosperms, shifting it instead to the stigmatic surface. Although there are exceptions, exemplified by those angiosperms with open stylar canals or incompletely fused carpels, the vast majority of flowering plants exhibit a feature of angiosperms where the germination of pollen grains occurs on a stigmatic surface.

Other characteristics not universally found in all angiosperms, and which to some degree may be shared with other groups of vascular plants, are the following:

1. Simple micro- and megagametophytes. The mature microgametophyte usually consists of two nonflagellated sperms, a tube cell, and the remains of a pollen grain (Fig. 30.1A). The mature megagametophyte (embryo sac) lacks archegonia. The female gametophytic tissue comprises an egg cell, synergids, antipodal cells, and polar nuclei (Fig. 30.1B). The numbers of cells and nuclei vary, but the "normal" embryo sac (monosporic eight-nucleate type) contains one cell that functions as the egg and two synergids at the micropylar end, three antipodals at the chalazal end, and two polar nuclei. The megagametophytes of the gymnosperms *Gnetum* and *Welwitschia* are like embryo sacs of angiosperms in lacking archegonia and in having certain developmental (free nucleate) stages.

2. Bitegmic ovules (ovules with double integuments). These occur in most monocots and many, but not all, groups of dicots (Fig. 30.1C). Members of the Sympetalae, for example, are unitegmic. Among the gymnosperms, the cupule of some Pteridospermales, the aril of the Taxales and *Cephalotaxus*, the "outer envelope" around the ovule of *Ephedra*, the "involucres" surrounding a *Gnetum* ovule, and the "bractioles" of *Welwitschia* have been interpreted as integumentary structures homologous with the outer integument of angiosperms.

3. Simple stamens. Most stamens are bilaterally symmetrical, each consisting of a filament

terminated by a two-lobed, four-loculate anther Fig. 30.1d).

4. Flowers with accessory reproductive structures such as sepals and petals. Some angiosperms lack one or both of these structures, as in many Amentiferae. In the Gnetales and Cycadeoidales of the gymnosperms, accessory reproductive structures in the form of bracts have prompted some to liken the cones of these plants to flowers.

5. Sporoderm (wall) of pollen grains with the exine differentiated into rodlike elements (columellae or baculae) covered by a tectum (Fig. 30.1E). Angiosperm pollen showing these structures is called tectate. Some ranalean families have an exine that is homogeneous (Walker, 1976). Most gymnosperms have exines that are alveolar instead of columellate, and in this characteristic they differ from the exine structure of angiosperm pollen.

6. Vessels in the xylem (Fig. 30.2A). Some angiosperms belonging to the Winteraceae lack vessels, while those of the ferns *Pteridium, Marsilea* and some other members of the Filicales, the lycopod *Selaginella,* the sphenopsid *Equisetum,* and the gymnosperm *Gnetum* have vessels with scalariform or reticulate perforation plates similar to those found in some angiosperms. Among other gymnosperms, vessel elements also occur in the wood of *Ephedra, Gnetum,* and *Welwitschia.* Vessels belonging to these genera usually have foraminate perforation plates (Fig. 30.2B) derived from circular bordered pits of the gymnosperm type. An exception, however, occurs in the wood of *Gnetum,* where Muhammad & Sattler (1982) found some vessel elements with scalriform perforation plates.

7. Libriform wood fibers, as well as sieve

Figure 30.1. **A.** Angiosperm microgametophyte. Pollen grain, pollen tube containing two sperms (s), and tube nucleus (tn). **B.** Embryo sac at the time of double fertilization. Egg (e); sperms (s); polar nucleus (pn). **C.** Longitudinal section (diagrammatic) of antropous, bitegmic (bi) ovule. **D.** Transverse section of angiosperm anther showing bilateral symmetry in the arrangement of the four microsporangia. **E.** Highly magnified portion of sporoderm of *Euphorbia.* Tectum (t); columella (c); foot layer (fl); intine (i). **A–E:** Extant.

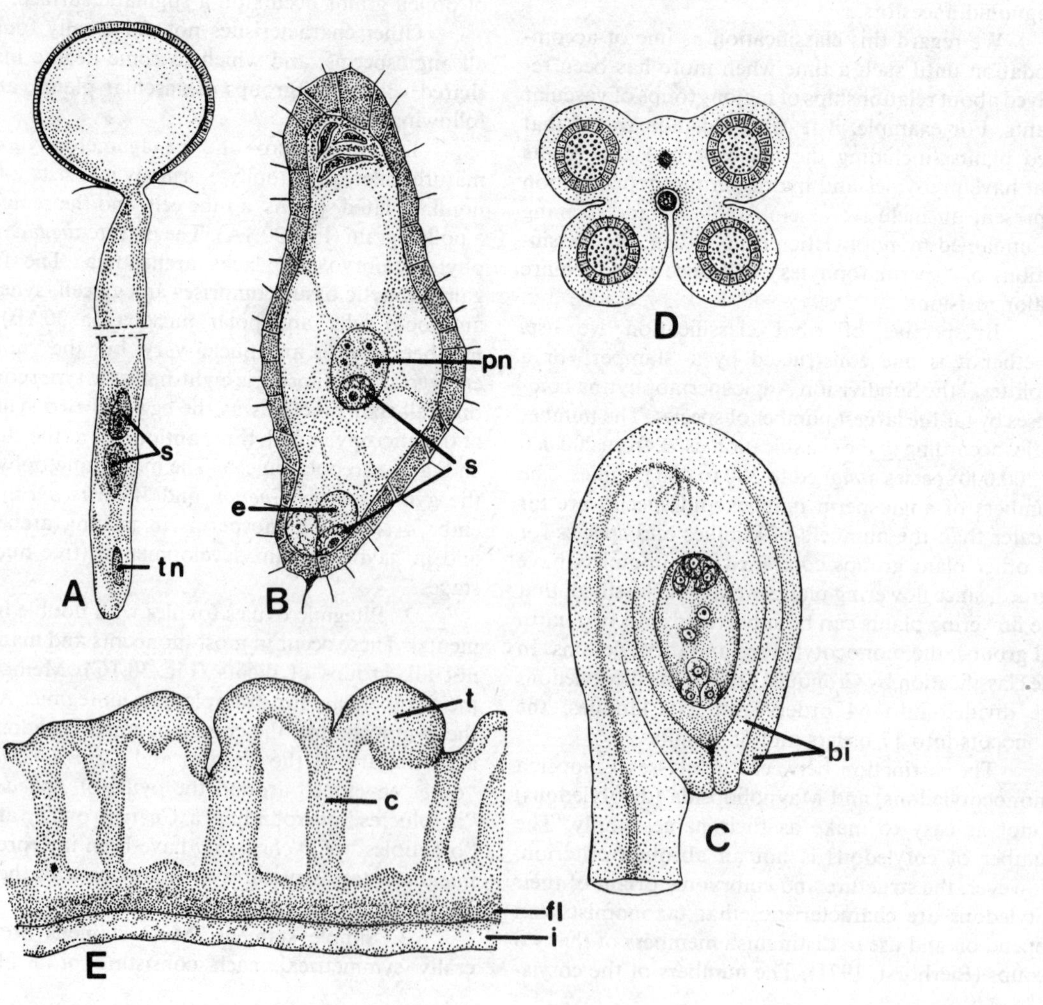

elements with companion cells in the phloem. The companion cells are derived from the same mother cell as the sieve tube elements.

8. Laminae of dicot leaves with reticulate venation (Fig. 30.2C) forming areoles and smaller veins traversing or terminating in the areoles (Fig. 30.2D). Laminae of monocot leaves with parallel major veins that are often arranged in sets of various sizes and interconnected by small veins (Fig. 30.2E). Among extant nonangiosperms, the leaves of *Gnetum* of the Gnetales have a venation pattern closely approximating that of a dicot in having several discrete vein orders with anastomoses between them to form areoles in which there are freely ending veinlets.

Figure 30.2. **A.** *Liquidamber* sp., vessel element with scalariform perforation plates. **B.** *Gnetum* sp., vessel element with perforation plates comprised of circular bordered pits. **C.** *Nicotiana* sp., reticulate venation with pinnate secondaries. **D.** Free vein endings in areoles of a dicot leaf. **E.** Venation in leaf of a monocot showing vein sets and cross veins. **A–E:** Extant. (**A–D** redrawn from Esau, 1977.)

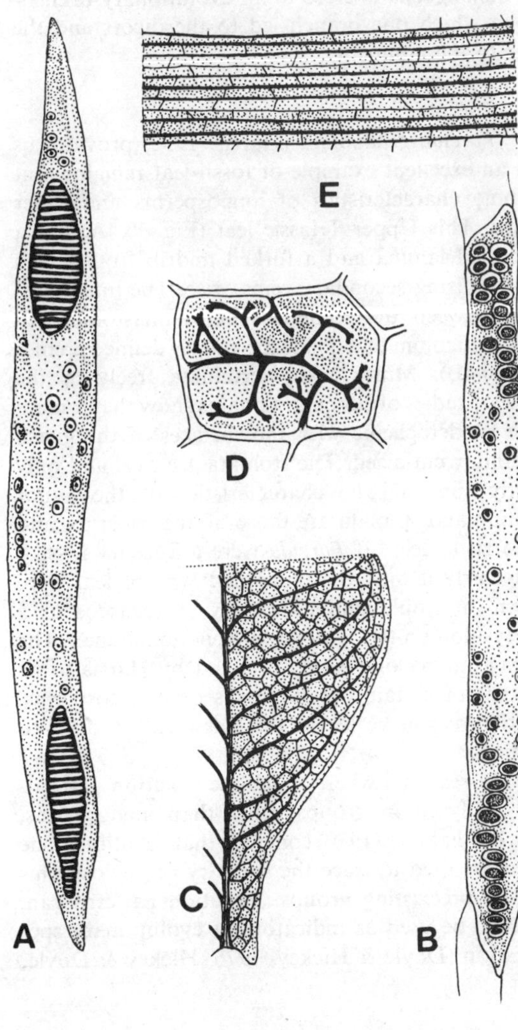

9. Alkaloids that differ in their chemical origin from those of non-angiosperms where their occurrence as secondary metabolites is uncommon.

It is a safe generalization, agreed to by most botanists interested in plant evolution, that angiosperms have evolved from gymnosperm ancestors. Exactly which one or ones remains a moot question. It is our opinion, however, that we are much closer to the answer than we were four decades ago, thanks to the efforts of paleobotanists who are using new techniques, new methods, and an accumulation of new material not available to earlier workers.

It is important to keep in mind, as we evaluate the gymnosperms as ancestors of angiosperms, that we should not expect to find all the above characteristics in a single species. Instead, from our experience with "transition" and "intermediate" conifers, the ancestral forms may show a combination of characteristics, some that we would interpret as angiospermous, some as gymnospermous, and some that do not fit the "recognized" characteristics of either group. We can expect some species to be "more angiospermous" in the totality of their characteristics, some others to be "more gymnospermous." The problem confronting paleobotanists is that the number of characteristics available may be few and may only pertain to a part of the life cycle and structure of the whole plant. Conclusions about relationships based on too few characteristics are always suspect. To partially alleviate the problems associated with trying to determine what group of gymnosperms presents the best possibilities as ancestors of angiosperms, Crane (1985) and Doyle & Donoghue (1986a,b, 1987) have adopted a cladistic approach in an effort to reduce the subjectivity attendant with earlier attempts to clarify this enigma.

The pre-Cretaceous record of presumed angiosperms

There is no good evidence in the fossil record of presumed angiosperm remains that suggests that

Figure 30.3. *Sanmiguelia lewisii.* Basal portion of plant, reconstructed. Triassic. (Redrawn from Tidwell, Simper, & Thayn, 1977.)

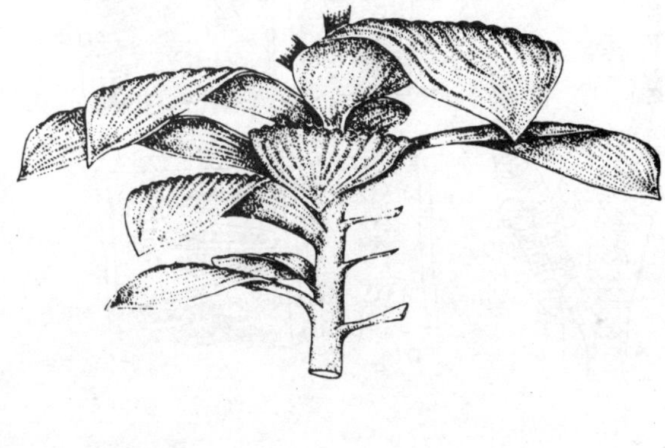

they had a late Paleozoic origin. Scott, Barghoorn, & Leopold (1960), Takhtajan (1969), Axelrod (1970), and others favor a Jurassic origin for flowering plants. After a careful evaluation of all Triassic and Jurassic fossils claimed to have affinities with angiosperms, Hughes (1976) has come to the conclusion that the first valid evidence does not appear until the Lower Cretaceous. Doyle (1978) concurs with this Cretaceous time of origin when he states, "Whatever the status of pre-Cretaceous angiosperm reports, the coherent fossil record of angiosperms begins in the early Cretaceous." This does not mean, however, that we should ignore those pre-Cretaceous plants that share one or more characteristics with angiosperms. It is reasonable to conclude that the "first" angiosperms did not spring fully formed as from a Medusa's head, although their evolution and subsequent diversification in the Cretaceous occurred in a relatively short time, geologically speaking. We have been exposed to similar examples of rapid evolution and diversification of land plants during the Devonian and Jurassic.

Sanmiguelia

This intriguing Upper Triassic plant (Fig. 30.3) was first brought to the attention of paleobotanists by Brown (1956), with later studies by Bock (1969), Becker (1972), and Tidwell, Simper, & Thayn (1977) attempting to determine the relationships of these enigmatic remains. In general morphology the leaves of *Sanmiguelia* resemble the pleated leaves of a palm, but they lack the midrib and petiole characteristics. Based on the shape of the leaves and their spiral arrangement on the stem, Tidwell et al. (1977) see a similarity between *Sanmiguelia* plants and those of the monocot *Veratrum* and related genera. Because of the stem surfaces with helical rhomboidal leaf scars and the internal stem structure, Bock has placed *Sanmiguelia* into synonymy with the form genus *Paloredoxites,* which he tentatively assigns to the cycadophytes. Others see affinities with broad-leaved presumed gymnosperms of the Upper Permian or *Schizoneura* of the Triassic. Further insight into this difficult material has been provided by Cornet (1986) in an extensive study of *Sanmiguelia* from productive localities in the Dockum Group of northwestern Texas. A description of the material with some conclusions about where this genus fits into the realm of seed plants is presented in Chapter 26. Cornet concludes that this enigmatic genus "may have been close to the morphology of an angiosperm ancestor" with some characteristics of a primitive dicot, below the level of the magnolias, and some of monocots and that *Sanmiguelia* is close to an evolutionary branching, in which one branch led to the dicots and the other to the monocots.

Figure 30.4. *Furcula granulifera.* **A.** Portion of leaf showing major veins. **B.** Fine venation forming a reticulum. **A,B:** Jurassic. (**A,B** redrawn from Harris, 1932.) **C.** *Trochodendron aralioidea,* longitudinal section of tracheids in secondary wood showing pitting. Extant. **D.** *Sahnioxylon rajmahalense,* pitting of secondary xylem tracheids. Triassic. (**C,D** redrawn from Sahni, 1932.)

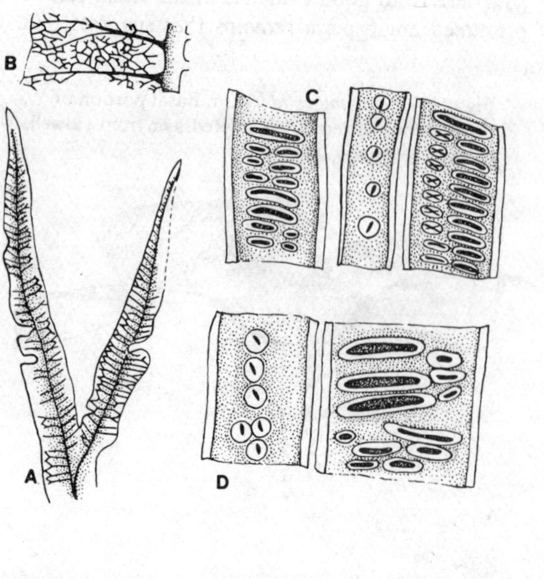

Furcula

Furcula granulifera (Harris, 1932) provides us with an excellent example of fossil leaf remains that combine characteristics of angiosperms and other groups. This Upper Triassic leaf (Fig. 30.4A) has a bifurcated lamina and a forked midrib from which dichotomizing secondary veins arise. The intercostal veins between the secondary veins anastomose to form a reticulum that delineates poorly defined areoles (Fig. 30.4B). Minor veins terminate freely in the areoles. Studies of cuticular remains show the stomata to be syndetocheilic with the surfaces of the guard cells thinly cutinized. The stomata are oval and only slightly sunken. The characteristics of the minor venation and stomata are those of angiosperms, and as Harris states, "If *Furcula* were a Tertiary fossil – particularly if only the undivided type of leaf were known – it would unquestionably be regarded as a Dicotyledon on the evidence of venation alone." The forking lamina of *Furcula* is noted by Harris to be like that of certain cycadophytes or pteridosperms. For this reason he leaves the classification open to question.

In research where reticulate venation patterns were analyzed in groups other than angiosperms, Alvin & Chaloner (1970) conclude that venation alone cannot be used to trace the ancestry of angiosperms back to preexisting groups. Venation patterns can, however, be used as indicators of evolutionary specialization (Doyle & Hickey, 1976; Hickey & Doyle, 1977).

Sahnioxylon (Homoxylon)

The families Winteraceae, Tetracentraceae, and Trochodendraceae of the dicotolyedonous angiosperms include several genera (*Trochodendron, Tetracentron, Tasmannia,* and others) with vesselless wood. Altogether there are more than 100 species that have "homoxylous" wood lacking vessels but have secondary xylem tracheids with bordered scalariform pits (Fig. 30.4C). These give way to circular bordered pits in the later-formed parts of the growth rings. Fossil woods exhibiting these characteristics (Fig. 30.4D) have been described from Triassic and Jurassic beds. Sahni (1932) was the first to suggest that his Jurassic *Homoxylon rajmahalense* was that of a vesselless angiosperm. He also noted, however, that the wood had characteristics very similar to the wood of cycadeoids (*Bucklandia,* for example). This has since been confirmed by others, and *Sahnioxylon* is now assigned to the Cycadeoidales.

It should be noted that there is a total absence of scalariform pitting in conifers, and *Ginkgo,* but that it is present in the secondary xylem of cycads, cycadeoids, angiosperms, and some Gnetales. Cordaiteans and some pteridosperms have both metaxylem and first formed secondary xylem composed of scalariform tracheids. These observations have led Takhtajan (1954, 1976) to suggest that scalariform pitting in cycads, cycadeoids, and angiosperms is an extrapolation of the juvenile feature (scalariform pitting in the metaxylem) into the secondary xylem. The resulting implication, supported by evidence from Jurassic woods such as *Sahnioxylon* of the cycadeoids, is that cycadophytes should be favored over the coniferophytes as ancestors of the angiosperms, a conclusion supported by evidence from other sources. The discovery of scalariform pitting in the vessels of *Gnetum* (Muhammad & Sattler, 1982), however, supports the thesis that the Gnetales must also be considered when searching for the ancestors of angiosperms. It is of some significance that the secondary wood of vesselless angiosperms and cycadeoids has achieved the same evolutionary level when compared with woods of other types.

Pollen types

More than a little interest was generated among angiosperm morphologists when the dispersed pollen genus *Eucommiidites* was first described from Jurassic deposits. For the better part of a quarter of a century different opinions about the relationships of *Eucommiidites* have been presented. Initially, these grains with their three germination furrows were thought to be similar to the tricolpate grains of nonmagnolioid dicots. Further investigations (Hughes, 1961, 1976) revealed that *Eucommiidites* occurs in the micropyles and pollen chambers of dispersed gymnosperm seeds and in pollen cones. Making use of transmission electron microscopy (TEM), Doyle, Van Campo, & Lugardon (1975) were able to show that the outer exine layer of the pollen grain wall has a granular structure found in either gymnosperms or angiosperms and that the inner layer of the exine is laminated, which is a characteristic of gymnosperm pollen.

These and other observations leave little doubt that *Eucommiidites* is the pollen of a gymnosperm and not of a dicotyledonous angiosperm. The tricolpate structure of these grains suggests, however, an evolutionary level approximating that found in many dicotyledonous angiosperms.

Based on extensive studies of the striate palynomorph *Equisetosporites chinleanus* Pocock & Vasanthy (1988) determined that the taxon included two distinctly different entities. The bands comprising the sporoderm of *E. chinleanus* are psilate (smooth), while others thought to belong to this species have obvious reticulate bands. Further investigation of the latter type, assigned to the new genus *Cornetipollis* (Fig. 30.5), showed it to be similar to pollen grains of some genera of the angiosperm family, Acanthaceae. Palynomorphs with *Cornetipollis* morphology have been found in the Upper Triassic of North and South America, and Australia into the Cretaceous (Barremian–Cenomanian) of North and South America.

Other candidates for pre-Cretaceous angiosperm pollen are monocolpate grains from Upper Triassic and Lower Jurassic beds of the eastern United States (Cornet, 1977). In their general morphology, these dispersed pollen grains are similar to monocolpate grains of many gymnosperms and monosulcate grains of some angiosperms. The grains discovered by Cornet have an outer exine that is differentiated into a well-defined tectum supported by columellae, as in angiosperms. There is no evidence of the alveolar structure found in the walls of gymnospermous pollen. All other previous accounts of pre-Cretaceous dispersed pollen type with supposed angiosperm affinities have been considered by Hughes (1976) and found to be contaminants, wanting in accuracy of stratigraphy, identification, and/or quality of preservation.

The Lower and Mid-Cretaceous record

From the research efforts of Doyle, Hickey, Hughes, and others it is increasingly clear that rapid diversification, if not the origin, of angiosperms occurred in the Lower Cretaceous (Chart 30.1). This conclusion is based in part on comprehensive studies of leaf compression–impression and dispersed pollen types from the Potomac Group of the Lower Cretaceous in the United States.

Leaves

The oldest megafossils that Hickey & Doyle (1977) acknowledge as those of bona fide angiosperms are remains of small pinnately veined simple leaves from the Neocomian equivalent in Siberia. The earliest evidence of angiosperm leaves in the slightly younger Potomac Group occurs in Zone I, established by Hickey & Doyle (1977), which is roughly equivalent to the Barremian–Aptian stages (Fig. 30.6). They report eight genera with such familiar names associated with dicots as *Ficophyllum* (Fig. 30.7A) *Proteaephyllum* (Fig. 30.7B), *Vitiphyllum* (Fig. 30.7C), and *Celastrophyllum* (Fig. 30.7D). Two of the eight genera are monocotlike leaves assigned to *Acaciaephyllum* (Fig. 30.7F) and *Plantaginopsis*. It is worth noting that angiosperm leaf remains make up less than 2 percent of the total number of leaf remains in Zone I.

As many others have discovered, there is a rapid increase in diversity of angiosperm leaf types higher in the Cretaceous. Although remains of cycads, cycadeoids, conifers, and ferns are abundant, the percentage of angiosperm remains increases dramatically to as much as 25 percent in the Patapsco Formation of the Lower Cretaceous. These tend to be dominated by members of the palmately lobed platanoid complex represented by *Araliaephyllum, Araliopsoides*, and others, as well as unlobed pinnately veined leaves assigned to the form genera *Betulites* and *Populites*. Pinnately compound leaves of *Sapindopsis* and leaves exhibiting a diversification of monocot and dicot types also make their appearance (Fig. 30.6).

Pollen

According to Hughes & McDougall (1987) the earliest acceptable record of angiosperm pollen is from the Lower Cretaceous, Hauterivian strata of England. Found here are grains with an angiospermous sporoderm comprising a tectum that covers columellae and a continuous inner layer of the exine. Similar pollen has been recovered from the slightly younger Zone I localities of the Lower Cretaceous by Hickey & Doyle (1977). Some of the grains are boat-shaped monosulcates of a type that could belong to either monocots or some magnoliaceous dicots. Such grains are assigned to the dispersed pollen genera *Clavatipollenites, Retimonocolpites,* and *Liliacidites*. The occurrence of these pollen types correlates well with the presence of monocot and dicot leaf types in Zone I (Fig. 30.6). Relying on the pollen record, Muller (1981, 1984) has also demonstrated that angiosperms with pollen types characteristic of the Magnoliidae were usually among the first to appear. More specifically, pollen similar to that of the extant magnoli-

Chart 30.1. Summary of suggested origins and relationships of angiosperms, cycadophytes, and glossopterids and their distribution in geological time. Medul. = Medullosaceae; Lyg. = Lyginopteridaceae; Call. = Callistophytaceae; Pen. = Pentoxylales; Czek. = Czekanowskiales; San. = *Sanmiguelia;* Coryst. = Corystospermaceae; Pelt. = Peltaspermaceae.

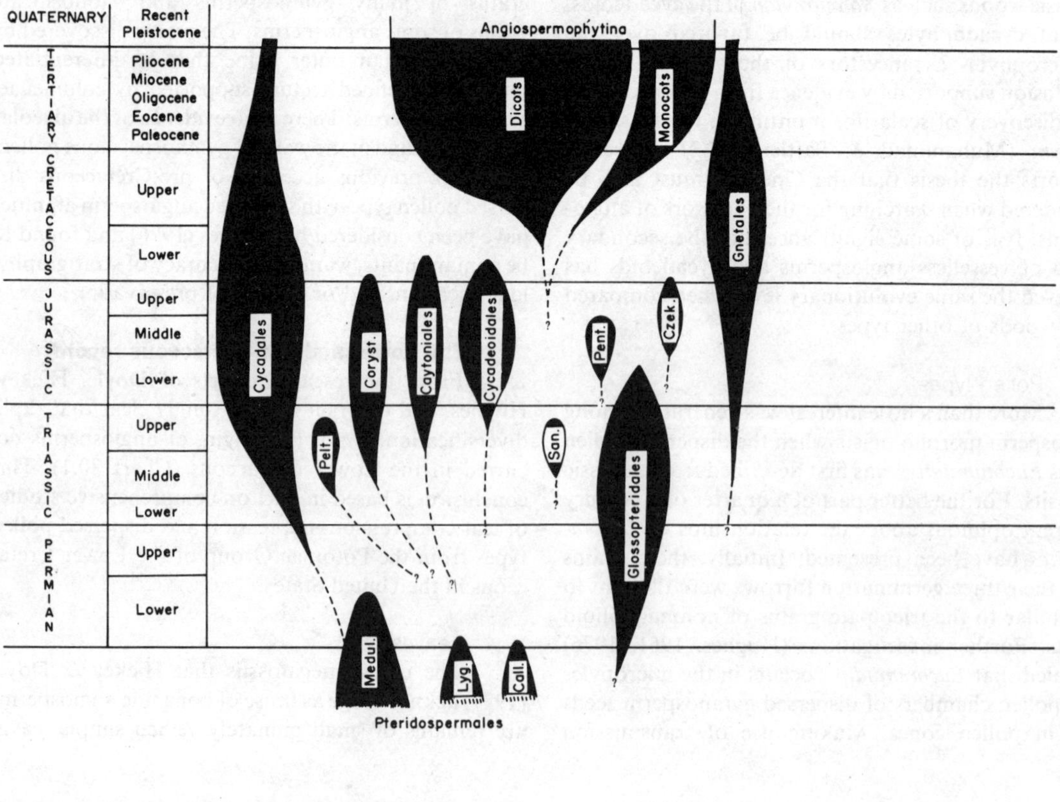

idean order Laurales occurs in the Aptian, followed by pollen of the Magnoliales in the Albian.

The diversification of pollen closely follows the rate of diversification of leaf types (Hickey & Doyle, 1977). The general trend, from Barremian to Cenomanian is from the tectate–columellar monosulcates (Fig. 30.8A) to even more elaborate smooth-walled tricolpates (Fig. 30.8B,C), triangular tricolporidates, and triporates (Fig. 30.8D). The evolution of triporate pollen is believed to be an adaptation facilitating germination on a stigmatic surface. The orientation of the triporate pollen grain on the stigma is of minimum importance because the grain will have at least one germination pore near or on the surface of the stigma, better insuring the growth of the pollen tube into the stigma.

Figure 30.5. *Cornetipollis reticulata.* Longitudinal view showing the reticulately perforated bands. Triassic–Lower Cretaceous. (From Pocock & Vasanthy, 1988, with permission.)

Wood

Unlike gymnosperm woods, Cretaceous angiosperm woods are relatively rare and poorly documented (Hughes, 1976). Only a few woods with angiosperm affinities have been reported from the Lower Cretaceous. One of these is a vesselless wood from the Aptian of Japan (Nishida, 1962). Woods with scalariform perforation plates in their vessels and other intermediate characteristics are known from Albian onward, increasing in variety from the Upper Cretaceous into the Tertiary.

Flowers and fruits

The oldest known remains reported to be fruits that bear some similarity to those of extant flowering plants come from the Barremian stage of the Lower Cretaceous. *Nyssidium,* as the name implies, resembles the strongly ribbed fruits of modern *Nyssa* and *Cercidiphyllum.* Other studies of *Nyssidium* suggest that the genus may represent the seeds of certain cycadeoids. They also bear a close resemblance to the follicles of

Figure 30.6. Summary of Cretaceous leaf and pollen sequences. Note pollen sequence from monosulcate to triporate types and primitive angiosperm leaf types in the Barremian–Aptian. See text for explanation. (From Hickey & Doyle, 1977.)

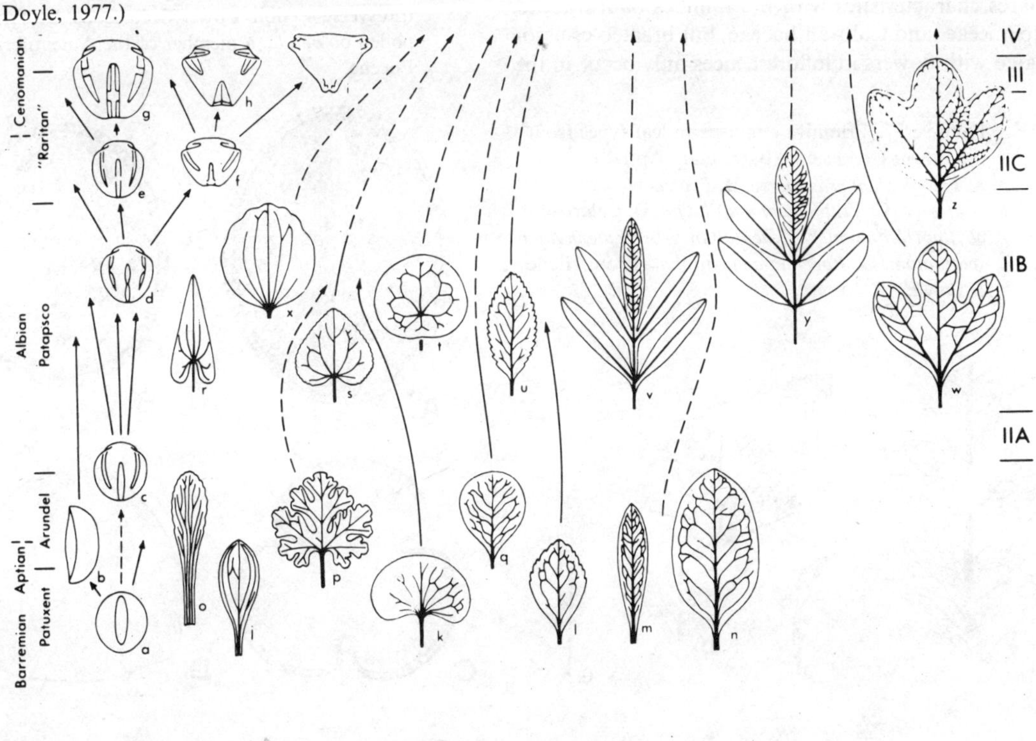

Cercidiphyllum. All of the dispersed seeds and follicles have been assigned to *Nyssidium* by Crane (1984).

Although there have been reports describing approximately 20 different reproductive organs ascribed to Lower Cretaceous (pre-Albian) angiosperms, except for that reported by Taylor & Hickey (1990), their angiosperm affinities have not been conclusively demonstrated (Friis & Crepet, 1987). The fossil described by Taylor & Hickey came from Aptian beds (Lower Cretaceous) at Koonwarra, Victoria, Australia. The fossil consists of a slender axis bearing small, fully expanded leaves with well-developed axillary inflorescences. The random reticulate venation pattern in the leaves is of a very low first rank, the lowest rank of any known for archaic angiosperms (Hickey & Doyle, 1977). With the evidence at hand that the leaves are unequivocally those of an early angiosperm, the characteristics of the small flowers attached to the axillary inflorescences assume particular importance in answering questions about what primitive angiosperm flowers were like.

The inflorescence has a peduncle and was probably a cyme. The ovaries of the flowers are associated with bracts that are in turn associated with bracts attached to a primary axis. There are two bracteoles that appear to be axillary to the ovate bract that partially encloses at least one ovary. The latter is small and oblong, terminated distally by a short stigma but no style. As we shall see, these diminutive apetalate flowers, subtended by bracts in a cymose inflorescence, do not fit our ideas of what primitive angiosperm flowers were like. The Aptian spicate inflorescence described by Taylor & Hickey (1990) shares characteristics with the families Saururaceae, Piperaceae, and Chloranthaceae, but bracteoles associated with flowers in inflorescences only occur in the Chloranthaceae. When all characteristics of the new fossil are considered, most are shared with various extant families of the Magnoliidae (*sensu* Cronquist, 1981) and archaic monocots.

Several unequivocal angiosperm fruits, seeds, and other floral parts have been found in Albian-age rocks of the Potomac group (Lower Cretaceous) (Crane, Friis, & Pedersen, 1986). Dispersed follicles and nutlets produced by taxa of magnoliidean affinities, as well as diminutive inflorescences with features similar to those of some Hamamelidae and Rosidae, are reported. Chloranthoid angiosperms are represented by androecia with fleshy cylindrical stamens (Fig. 30.9A), which occur in groups of three and are fused at their bases. The connectives, each bearing three stamens, are expanded at the distal end (Friis, Crane, & Pedersen, 1986).

Evidence of taxa belonging to the Platanaceae also appears in the Potomac flora. Several small platanoid inflorescences have been found attached to an axis (Fig. 30.9B). Each inflorescence is a sessile head consisting of tightly packed pistillate flowers each composed of five free carpels (Fig. 30.9C), a perianth, and bracts. The anthers are attached at their bases to short filaments. The discovery of these unisexual platanoid flowers confirms the early appearance of members of the Hamamelidae (*sensu* Cronquist, 1981) in the Lower Cretaceous.

Bisexual flowers have been found in Middle

Figure 30.7. Primitive angiosperm leaf types from the Lower Cretaceous (Barremian–Aptian). **A.** *Ficophyllum crassinerve*. **B.** *Proteaephyllum reniforme*. **C.** *Vitiphyllum multifidum*. **D.** *Celastrophyllum latifolium*. **E.** *P. dentatum*. **F.** *Acaciaephyllum spatulatum*. **G.** *Rogersia angustifolia*. (From Hickey & Doyle, 1977.)

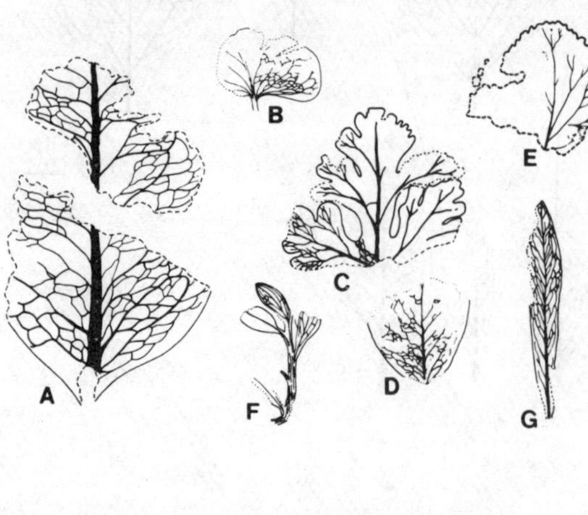

Figure 30.8. Some angiosperm pollen types. **A.** *Clavatipollenites* sp. Monosulcate, tectate pollen. **B.** Equatorial view of *Tricolpites* sp. **C.** *Tricolpites* sp. transverse section. Lower Cretaceous. **D.** Triporate pollen of *Betula*, a member of the Amentiferae. Eocene.

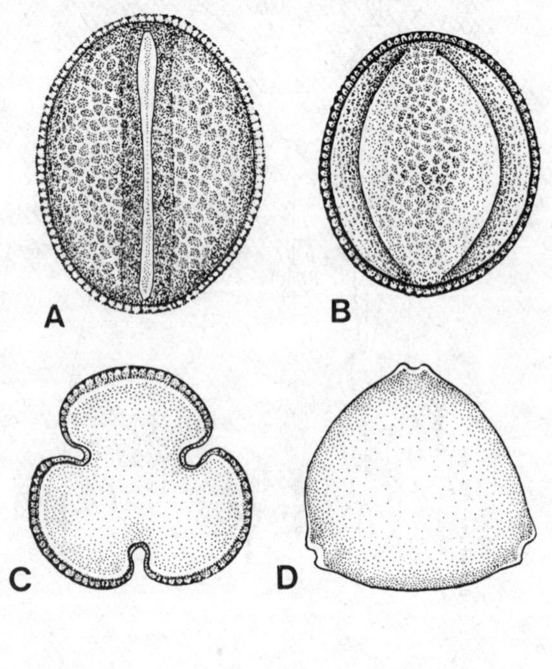

Albian deposits of the Soviet Union (Krassilov, Shilin, & Vakhrameev, 1983). The fossils consist of three to five elliptical follicles (Fig. 30.9D) borne on a flattened receptacle. Some specimens show the remains of a perianth and an androecium, suggesting that male and female parts occurred in the same flower. According to the authors, the fruits are similar to those of the Ranunculaceae and Paeoniaceae.

In addition to the growing number of angiosperm flower parts described from the Aptian and lower Albian of the Cretaceous, Dilcher (1979) and Dilcher & Crane (1985) have found the remains of magnoliidlike angiosperms in beds at the boundary between the Lower and Upper Cretaceous. The lower beds are placed in the uppermost Albian, which in turn transgresses the beds of the lowermost Cenomanian of the Upper Cretaceous. These beds belong to the Dakota Formation, which has been highly productive of fossil leaves and flowers. The parts of the flower *Lesqueria elocata* described by Crane & Dilcher (1984) come from these and other beds in the form of three-dimensional molds in sandstone. As reconstructed (Fig. 30.10), we can see that the gynoecium is composed of a large number of follicles (175–250), which are helically arranged on a swollen receptacle. Collectively, the follicles form a more-or-less spherical head, in which each follicle has a short stalk and two distal stylelike extensions. Within each follicle there are two longitudinal rows of 10 to 20 seeds, which were apparently shed through a single adaxial suture. In their morphology the follicles of *Lesqueria* are similar to the conduplicate carpels of many Magnoliidae. The receptacle below the fruit is cylindrical and elongated and bears many spirally arranged, laminar bractlike structures. Although it is apparent that *Lesqueria* shares many features with certain Magnoliidae, Crane & Dilcher have not been able to pinpoint a member of the subclass that shows the combination of characteristics exhibited by this genus.

The most complete of all the angiosperm fossils found in the rocks of the Dakota Formation is *Archaeanthus linnenbergeri,* another member of the

Figure 30.9. Fruits and flower parts from Lower Cretaceous. **A.** Chloranthoid androecium. **B.** Platanoid inflorescences. **C.** Pistillate platanoid flower. **D.** Fruits consisting of three follicles on a flattened receptacle. Lower Cretaceous. (From Friis & Crepet, 1987, with permission.)

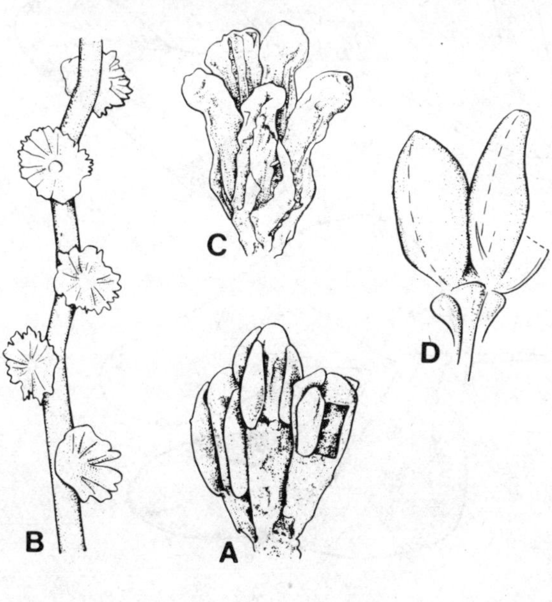

Figure 30.10. *Lesqueria elocata.* Reconstruction of fruiting axis. Mid-Cretaceous. (From Crane & Dilcher, 1984, with permission.)

Magnoliidae described and reconstructed (Figs. 30.11 and 30.12) by Dilcher (1979) and Dilcher & Crane (1985). As depicted, the plant consists of a stout woody branch with a large terminal flower (Fig. 30.11) or helically arranged stalked follicles (conduplicate carpels) (Fig. 30.12C) that dehisced along an adaxial suture. The 10 to 18 mature ovules inside a follicle were borne in two rows, one on each side of the suture (Fig. 30.12A,B). That portion of the receptacle below the follicles has three groups of scars; numerous small scars believed to represent the former position of stamens; 6 to 9 scars directly below the stamen scars, which probably represent the former position of perianth parts, and some distance below the perianth scars a single scar which represents the position of the floral bud scales. Dispersed perianth parts (*Archaepetala*), bud scales (*Kalymmanthus*), and leaves (*Liriophyllum kansense*) have been linked to *Archaeanthus*,

Figure 30.11. *Archaeanthus linnenbergeri*. Reconstruction of leafy twig and flower. Mid-Cretaceous. (From Dilcher & Crane, 1985, with permission. Original artwork by Megan Rohn.)

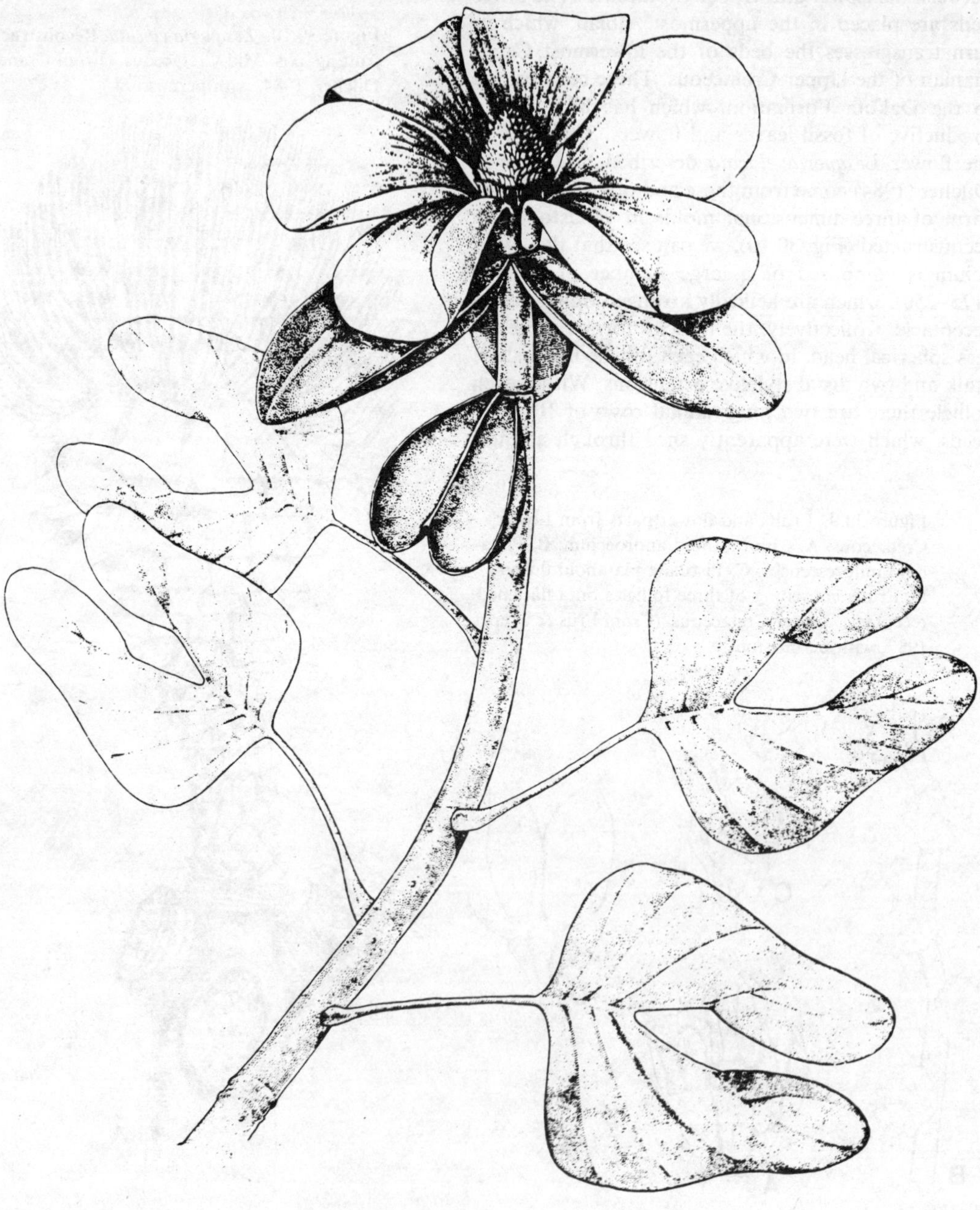

based on the frequency of association in the beds as well as evidence from structurally similar resin bodies found in all of these genera and species.

It is clear from its structure and morphology that *Archaeanthus* shares many characteristics with members of the Magnoliales (Dilcher & Crane, 1985) and clearly demonstrates the presence of magnolioid angiosperms in the Cenomanian. This evidence, in addition to Albian fossils showing magnoliidean affinities, strongly supports the theory that angiosperms sharing these floral characteristics represent an archetype from which insect-pollinated, bisexual flowers may have evolved.

The presence of unisexual staminate and pistillate platanoid inflorescences in the Albian and Cenomanian (Dilcher, 1979; Friis, Crane, & Pedersen, 1988) suggests that these diminutive unisexual flower types with inconspicuous perianths and small tricolpate pollen (Fig. 30.8B) were wind pollinated. The fact that the plants of this type coexisted with bisexual insect-pollinated magnolioids in this segment of the fossil record supports the concept that the differentiation of wind- and insect-pollinated angiosperms had already occurred at an early time in the evolution of flowering plants.

Additional evidence for the rapid radiation of angiosperms from the Albian into the Cenomanian is provided by *Prisca* (Fig. 30.14), *Caloda* (Fig. 30.13), and the Rose Creek flower (Fig. 30.15), all three of which were found in the upper Dakota Formation of the Cenomanian. Detailed observations have been made on the follicle-bearing racemes (Fig. 30.14A) of *Prisca* (Retallack & Dilcher, 1981b). These highly instructive impression fossils exhibit a set of characteristics that are unlike those of any extant group of angiosperms. That the fossils are those of angiosperms is evidenced by their conduplicate carpels, which are follicles (Fig. 30.14B) that contain what are interpreted to be bitegmic ovules (Fig. 30.14C). According to the authors, the inflorescences are simple racemes comprised of alternately arranged bracteate fruiting stalks or receptacles each of which bears the spirally arranged follicles. The ellipsoidal to globose follicles are attached to the receptacle by very short bases. Two to six orthotropous ovules are attached to the inside margins and have their micropyles directed

Figure 30.12. **A.** *Archaeanthus linnenbergeri*. Reconstruction of single follicle. **B.** Section through follicle; adaxial ridge (a); seed (s). **C.** Reconstruction of leafy twig bearing many helically arranged conduplicate carpels (follicles). Mid-Cretaceous. (From Dilcher & Crane, 1985, with permission. Original artwork by Sally Wolf.)

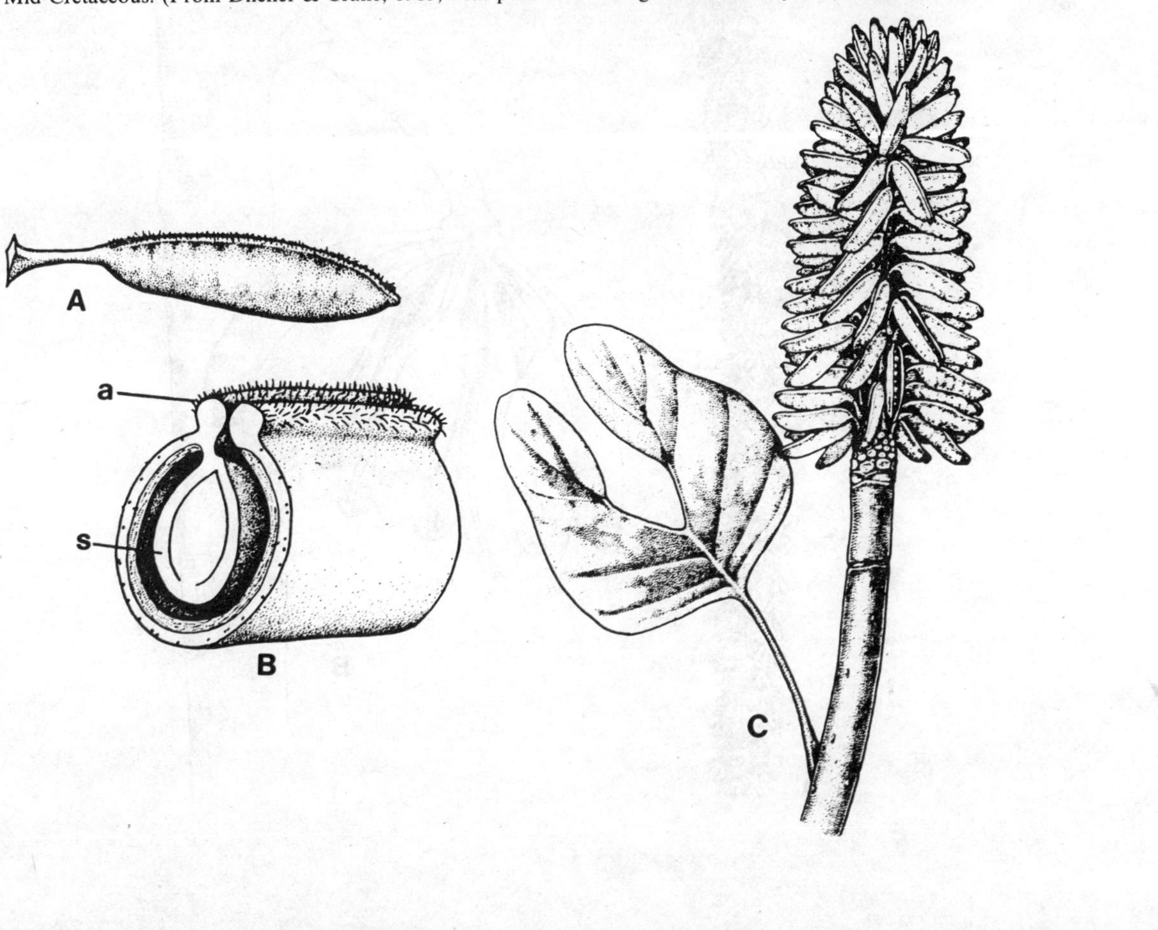

toward the interior of the follicle. The ovules are small ellipsoidal structures averaging only 0.8 mm long and 0.5 mm wide. There is no evidence of perianth parts or androecium. Based on the available evidence, Retallack & Dilcher (1981b) conclude that *Prisca* is unlike any known extant angiosperm and that it is presently best regarded as a member of an extinct group.

More recently Dilcher & Kovach (1986) described the angiosperm fructification *Caloda* (Fig. 30.13), which, like so many other angiosperm flowers and fruits from the Cenomanian, defy assignment to any known group of flowering plants. The inflorescence of *Caloda* consists of a main axis bearing many helically arranged secondary axes. Each of the latter terminates in a small receptacle with numerous con-

Figure 30.13. **A.** *Caloda delevoryana,* reconstruction in early fruiting stage prior to fruit abcission. **B.** *C. delevoryana,* reconstruction of secondary axis showing a cluster of carpels. Mid-Cretaceous. (From Dilcher & Kovach, 1986, with permission.)

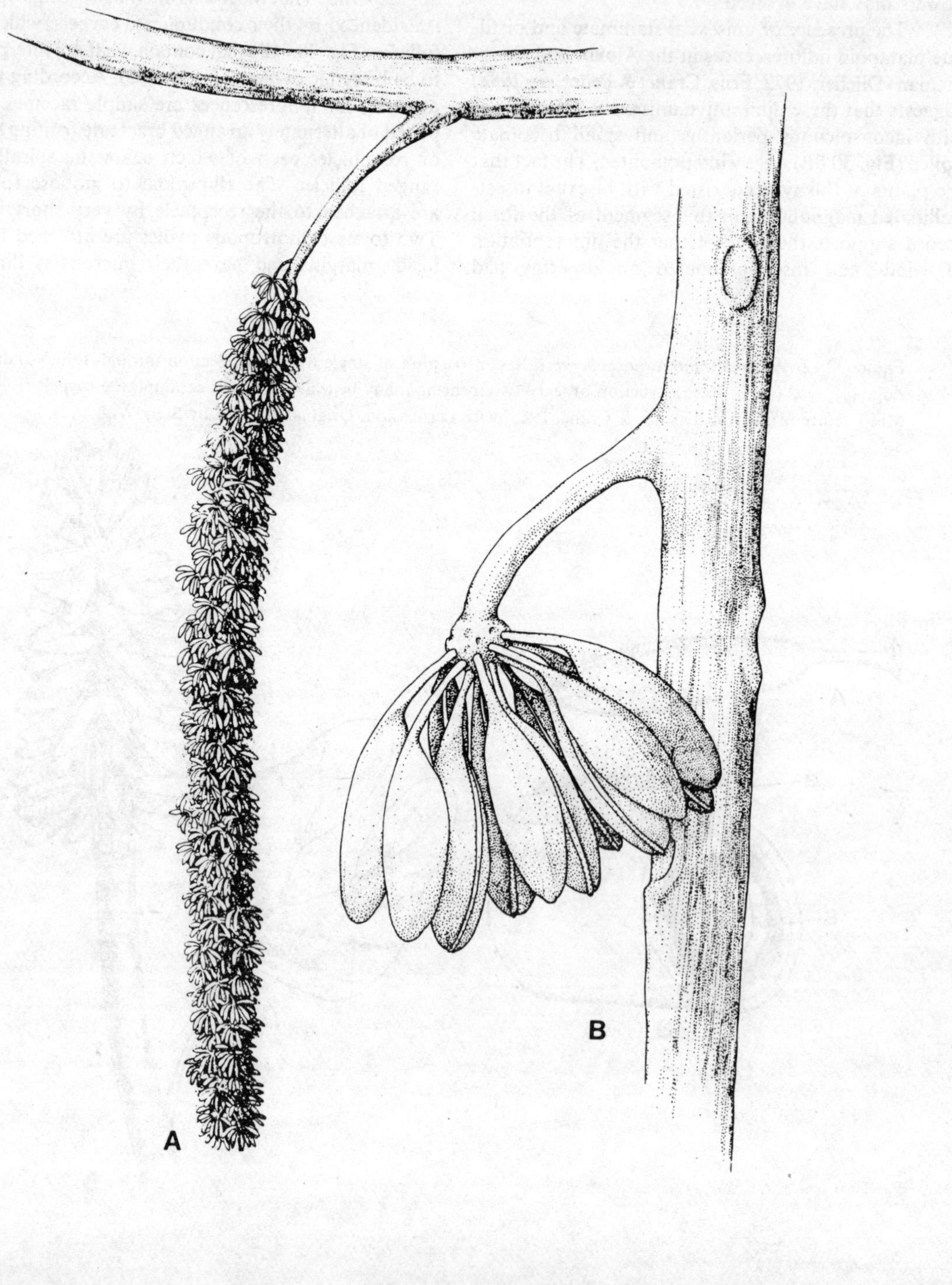

duplicate carpels (Fig. 30.13A) which are attached by a short stalk. The elliptical carpels are laterally flattened and exhibit an adaxial suture and abaxial vein that extends for the full length of the carpel. No other floral parts were found. The authors find some similarities between the characteristics of *Caloda* and members of the Platanaceae of the Hamamelidales and the Aponogetonaceae of the Najadales. Within the Magnoliidae the fossil can be envisioned as an early member of the Piperales where flowers are often unisexual and perianth parts are small or lacking and are arranged in densely packed spikes or racemes.

The third and most spectacular finds, however, are several more or less complete pentamerous (five-parted) flowers (Fig. 30.15) collected from the Rose Creek locality, Nebraska. As described by Basinger & Dilcher (1980, 1984), the Rose Creek flowers are radially symmetrical with five petals that alternate with five sepal lobes on the rim of a thick but shallow calyx cup. The stamens, five in number, are opposite the petals and have their filaments attached at the base to the margins of the calyx cup. These perfect flowers have a gynoecium composed of five closely appressed carpels in a ring. The configuration of the ring of carpels resembles the fruit casts of *Nordenskioldia* found in the Upper Cretaceous and Tertiary. These flowers, according to Basinger & Dilcher, have the generalized morphology of bisporangiate insect-pollinated flowers that incorporate characteristics of various orders (Saxifragales, Rosales, and Rhamnales) of the Rosidae.

From these preliminary observations of angiosperm flowers and fruits obtained from the mid-Cretaceous, it is clear that a considerable amount of diversification in flower structure had occurred by that time. Dilcher (1979) recognizes three main flower types: (1) the elongated axis of a catkin; (2) a globular head of the *Platanus* type; and (3) an enlarged receptacle terminated by a single flower. All of these and other floral types are represented in the Tertiary.

Although there are many other reports of Lower Cretaceous and early Upper Cretaceous fossil fruits and flowers, we can conclude from the evidence provided by the examples presented thus far that this was a period when many basic characteristics found among several groups of angiosperms first appeared. It has become clear that, although the fossil record of angiosperm leaves, pollen, and flowers is incomplete and does not provide an unequivocal basic archetype from which angiosperms evolved, it does show us that small simple flower types similar to those of some Chloranthaceae and Platanaceae appeared

Figure 30.14. *Prisca reynoldsii.* **A.** Carpellate inflorescence. **B.** Follicles showing position of ovules. **C.** Diagram, longitudinal section of ovule showing double integument. A–C: Dakota Formation, Cretaceous. (Redrawn from Retallack & Dilcher, 1981b.)

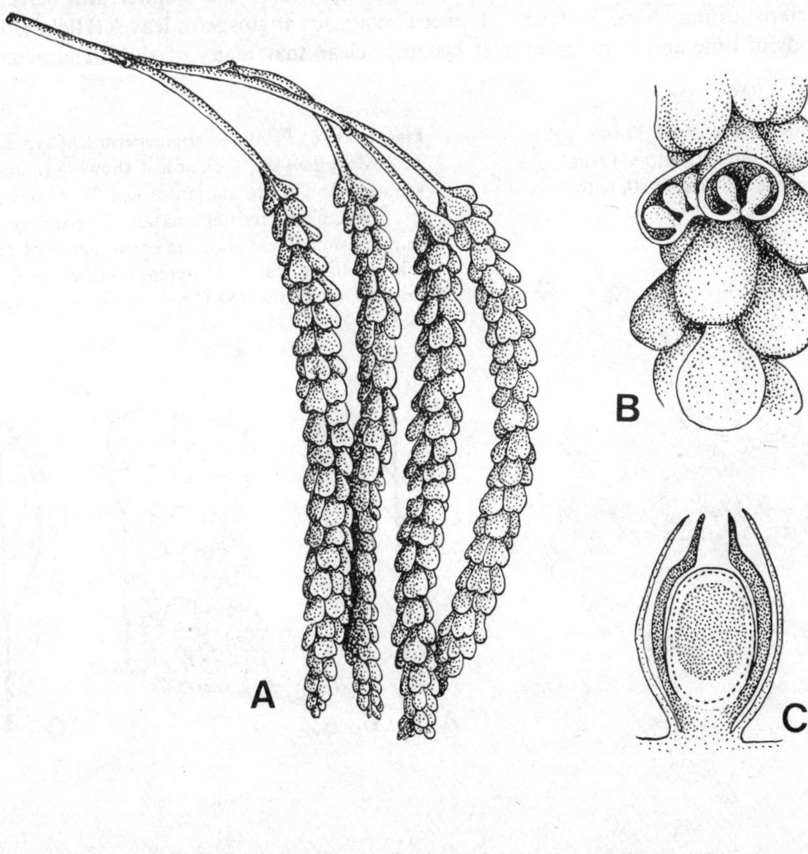

early in the history of angiosperms and coexisted in the Albian with more elaborate flowers of the magnolialean type (Friis & Crepet, 1987).

Some patterns of evolution

It is important to keep in mind that the characteristics by which we recognize angiosperms did not evolve simultaneously from a gymnosperm ancestor at the end of the Jurassic or Lower Cretaceous. Instead, many groups of gymnosperms evolved one or more characteristics that we associate with angiosperms prior to the Lower Cretaceous. Some of these groups became extinct, while others evolved new characteristics of particular selective advantage (the carpel and double fertilization, for example) in combination with already established characteristics such as reticulate venation or tectate pollen. Remember that an angiosperm, or any organism for that matter, is the product of natural selection operating on many characteristics. Natural selection, as we all know, can make many models obsolete, while others with different sets of characteristics continue to become the prototypes of new, successful groups that combine the "right" set of characteristics. It is probably correct to assume that at some time and some place in the semiarid environments of the Lower Cretaceous, there occurred plants having a synthesis of all those characteristics by which we recognize angiosperms. Examination of their leaf remains and reproductive parts should give us some insight into their origins as well as an understanding of what characteristics are primitive. Even here, however, care must be taken to keep in mind that some characteristics may have evolved for relatively longer periods of time and at different rates, so that some characteristics of early angiosperms may represent an advanced evolutionary level in combination with characteristics that mav be interpreted as intermediate or primitive.

Primitive dicot leaf characteristics

To prepare for the analysis of the characteristics of fossil angiosperm leaves, Hickey & Wolfe (1975) undertook a prodigious study of leaves belonging to most major groups of extant dicotyledonous angiosperms. The study resulted in the establishment of a set of characteristics that was later used by Doyle & Hickey (1976) to work out the phylogenies and systematics of Lower Cretaceous leaves. The characteristics are (1) organization of the leaf (simple versus compound); (2) the margin of the leaf (entire or toothed); (3) tooth shape, venation, and glandular processes; (4) major vein configurations; (5) intercostal venation; and (6) gland position. Based on their study, Hickey & Wolfe determined that the primitive leaf (Fig. 30.16A) of dicotyledonous angiosperms is simple and has pinnate venation, secondary veins that are camptodromous, a "first-rank" level of vein organization, and entire margins. The camptodromous leaf has secondary veins that do not intersect the leaf margins. In some, they form sets of loops and are called brochidodromous. First-rank level of organization is characterized by those leaves where all vein orders are poorly differentiated from one another and are irregular in their courses, manner of branching, and anastomoses (Fig. 30.16B).

When applied to the architectural features of Lower Cretaceous angiosperm leaves (Hickey, 1973), it becomes clear that many of those characteristics

Figure 30.15. Generalized Rose Creek flower from Dakota Formation, Cretaceous. (Redrawn from illustrations by Basinger & Dilcher, 1980, with permission.)

Figure 30.16. Primitive angiosperm leaf types. **A.** *Ficophyllum* sp., a dicot leaf showing pinnate venation and entire margin of leaf. **B.** *Ficophyllum* sp., first-rank intercostal venation. **C.** *Acaciaephyllum* sp., a monocot leaf showing convergence of main veins at tip of leaf and interconnecting veins. **A–C:** Lower Cretaceous. (**A–C** redrawn from Hickey & Doyle 1977.)

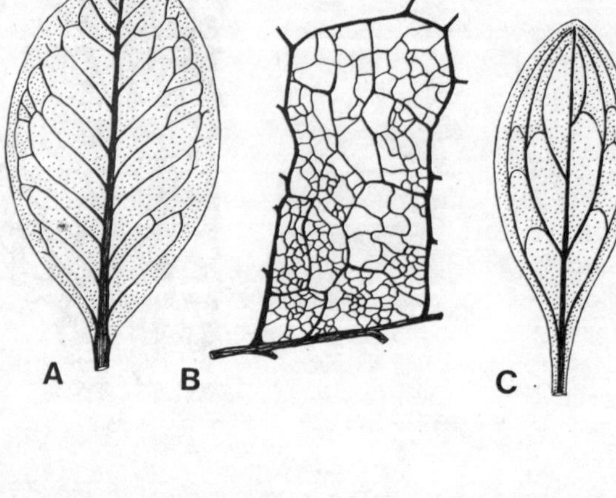

previously determined to be primitive for extant dicotyledonous angiosperms are those that occur in the oldest leaf fossils. For example, leaves from the Barremian–Aptian stages of the Lower Cretaceous are usually simple and symmetrical with entire margins. They have pinnate brochidodromous venation and minor vein orders that are poorly differentiated and random. The earliest leaves are elliptic and have an acute apex and a decurrent base lacking a petiole. Dicot leaves from the Lower Cretaceous that embody these characteristics are *Ficophyllum* (Fig. 30.7A). *Rogersia* (Fig. 30.7G), and *Celastrophyllum* (Fig. 30.7D). Among extant dicots those of the Magnoliales and some other orders of the subclass Magnoliidae also have leaves with these characteristics. The fossil evidence clearly indicates that such leaves are primitive and that dicotyledonous leaves with other morphologies and architecture are derived, many in the Albian and Cenomanian stage of the Cretaceous. As emphasized by Doyle & Hickey (1976), one should consider these ancient leaves of the first-rank condition as representing an early stage of evolution through which the leaves of other groups passed, as well as the level achieved by leaves of the modern Magnoliidae.

Primitive monocot leaf characteristics

The fossil record of leaves belonging to monocotyledonous angiosperms is relatively meager when compared with that of dicots. As a result, the evolution of their morphologies and architecture is poorly understood, as are those characteristics that may be primitive. There is some evidence (Doyle & Hickey, 1976), however, that plants having leaves with monocot characteristics were present in the Aptian stage of the Lower Cretaceous where they coexisted with dicotyledonous plants having leaves with first-rank characteristics. This implies that the basic separation of angiosperms into dicots and monocots had occurred in the Lower Cretaceous by Barremian–Aptian times. The leaf type, which departs radically in its architecture from that of a dicot, is *Acaciaephyllum* (Fig. 30.16C). Leaves of this genus are acrodromous and narrowly obovate and have long sheathing bases. Leaves with acrodomous venation have secondaries that form concentric arches that converge and appear to fuse toward the leaf apex. Chevronlike, somewhat finer crossveins occur between the secondary veins. The pattern of higher-order crossveins, elongated leaf bases, and essentially parallel venation suggests affinities with the monocots (Doyle & Hickey, 1976). Further support for the Lower Cretaceous origin of monocots is provided by Doyle (1973), who describes monosulcate pollen with monocotyledonous sculpturing in the same beds with *Acaciaephyllum*. If this evidence for the Lower Cretaceous presence of monocots stands the test of further investigation, then the primitive monocot leaf type is obovate, with acrodromous secondary veins that converge toward the apex, fine crossveins, and a sheathing leaf base.

Primitive pollen

If we base our conclusions on observations of pollen types found in the fossil record, then pollen of the *Clavatipollenites* (Fig. 30.8A) type must be considered as representing primitive angiosperm pollen (Hughes, 1976; Doyle, 1978). *Clavatipollenites* has a monosulcate aperture comparable with that found in pollen grains of many gymnosperms such as the Cycadales, Ginkgoales, Cycadeoidales, and Pentoxylales. The structure of the exine of *Clavatipollenites* with its tectum of clavate columellae distinguishes it from the gymnosperms and suggests that it is the pollen of an angiosperm. Some authors (Walker, 1976) maintain that because of the relatively small size of *Clavatipollenites*, it is a derived pollen type of a wind-pollinated angiosperm and that the large monosulcate grains of many extant Magnoliales with homogeneous exines (*Degeneria*, for example) are primitive and link the angiosperms to the gymnosperms. In this connection we should keep the possibility in mind that there could be more than one group of primitive angiosperms (Dilcher, 1979); wind-pollinated ones that have small monosulcate grains, and insect-pollinated ones with larger grains of the *Magnolia* type.

Accepting the evidence from the fossil record and from the studies of pollen of extant angiosperms and gymnosperms, we can conclude that monosulcate grains, with tectate or granular exines, found in Barremian and Aptian strata of the Lower Cretaceous represent a primitive type of angiosperm pollen. Not all agree, however, about how the wall structures of the grains evolved or how important grain size really is.

Primitive flowers

With the recent accumulation of evidence from the studies of fossil flowers we are presented for the first time with the opportunity of testing the "ranalian" concept of the primitive flower. This concept, emanating about 80 years ago from the work of C. E. Bessey on the phylogenetic classification of extant angiosperms, conceived the primitive flower to have the general characteristics of ranalian angiosperms belonging to the Magnoliales. These characteristics, as displayed by the flower of *Magnolia* (Fig. 30.17), include an indefinite and large number of carpels, stamens, and perianth parts all of which are helically arranged on a receptacle. The flowers, which are single and perfect, have the showy petals and well-developed calyx of insect-pollinated flowers. They are radially symmetrical and have carpels that are borne in the superior position on the receptacle. With the introduction of the Bessey system, the ranalian angiosperms replaced the amentiferous angiosperms of earlier classifications as the primitive representative of flowering plants. The Amentiferae of Engler & Prantl (Salicaceae, Juglandaceae, Fagaceae, Betulaceae, and many other families) were believed to be derived from coniferous ancestors, because of their conelike inflorescences and monosporangiate flowers. Like their supposed relatives, these flowers are wind-pollinated and lack conspicuous perianth parts and nectar. Subsequent to the advent of Bessey's classification, numerous studies of wood anatomy and floral structures of extant angiosperms have left little doubt that ranalian angiosperms having vesselless wood or wood with scalariform perforations (Fig. 30.2A) and pits on their vessels and flowers with conduplicate carpels (Fig. 30.20A) are primitive. This conclusion has been accepted for many decades. Paleobotanists, however, have had very little input into determining the primitive characteristics of flowers, primarily because of the paucity of an early angiosperm fossil record. Thus, until recently, those questions that one might expect paleobotanists to answer about primitive floral characteristics have remained mostly unanswered.

Figure 30.17. Flower of *Magnolia* sp. with some petals and sepals removed to show numerous stamens and spirally arranged carpels. Extant. (Courtesy of the General Biological Supply House, Chicago.)

Summary

As we examine the evidence provided by Dilcher (1979), Dilcher & Crane (1985), Friis & Crepet (1987), and others, it is clear that the conduplicate carpel and two basic types of flowers – diclinous (monosporangiate) and monoclinous (bisporangiate) –had evolved by the Lower Cretaceous. Diclinous flowers, which became more abundant in the Cenomanian stage of the Upper Cretaceous, are often spirally arranged and form spikelike or catkinlike inflorescences like those of *Prisca* and *Caloda*. These, along with heads of diclinous platanoid flowers from the Albian and Cenomanian, illustrate the early adaptation to wind pollination. The structure and arrangement of the floral parts of these flowers are obscure, but there is some evidence these were composed of undifferentiated parts arranged in whorls. By the Cenomanian, other major types of flowers adapted for insect pollination had evolved. Among these were the magnoliid types with indefinite numbers of radially arranged floral parts in spirals (*Lesqueria, Archaeanthus*) and those with definite numbers of parts (4s and 5s) in whorls (Rose Creek flowers). These insect-pollinated flowers probably had showy perianths comprised of a well-defined calyx and corolla that presaged the dominance of this floral type in the late Upper Cretaceous. From this summary we can conclude that flowers of the *Magnolia* type should no longer be thought of as epitomizing the primeval type (Dilcher, 1979). Rather they may represent one of several lines of early angiosperm evolution. In addition, it is now clear that independent lineages of some anemophilous (wind-pollinated) unisexual flowers had evolved early and perhaps independently of bisexual entomophilous (insect-pollinated) flowers. Dilcher (1979) speculates that both insect- and wind-pollinated types originated from a common monosporangiate stock. This stock may be represented by the Aptian bracteate ovary-bearing inflorescence described by Taylor & Hickey (1990).

Angiosperms: monophyletic or polyphyletic?

Based on the presence of small triporate pollen types characteristic of the amentiferous angiosperms and catkinlike inflorescences in the Upper Cretaceous, the suggestion is made that anemophily was firmly established by that time, if not before. Crepet's (1979, 1981) researches of Middle Eocene anemophilous flowers show us that the reproductive strategies and structures had achieved an evolutionary level very similar to many modern Amentiferae. Those with florets in catkins appear to be diclinous, with a perianth that is inconspicuous or absent, stamens that have elongated filaments, and abundant small pollen grains (20 to 30 μm) usually with a smooth exine. Walker (1976) has suggested that the small pollen grains of the Lower Cretaceous *Clavatipollenites hughesii* (12 to 25 μm) also are those of primitive wind-pollinated angiosperm types. The evidence interpreted in this way seems to support the idea that the angiosperms are polyphyletic and that the anemophilous, amentiferous angiosperms link wind-pollinated gymnosperms and angiosperms. Meeuse (1961, 1966, 1970, 1975, 1979a) has postulated, on hypothetical grounds, that angiosperms have evolved from a heterogeneous gymnosperm ancestry and that the amentiferous angiosperms are one of several different lines derived from the gymnosperm complex. Based on leaf and pollen studies, Hickey & Doyle (1977) support the more conventional monophyletic origin of angiosperms where the Amentiferae are interpreted as "a highly specialized group whose simple diclinous flowers are the result of reduction associated with reversion to wind pollination." They point to the derivation, during the Cretaceous, of triporate and other pollen types from monosulcate pollen of the Lower Cretaceous and the absence of triporate pollen in the early fossil record of angiosperms. Further, it is pointed out that the rigidly organized fourth-rank leaves of most amentiferous angiosperms do not appear until late in the Cretaceous. Hickey & Doyle do concede, however, based on their work with pollen and leaves of early angiosperms, that "the woody order Magnoliales should not necessarily be considered archetypic for the angiosperms as a whole, but represents only one of several early lines of specialization." You may recall that this conclusion is in keeping with that of Dilcher (1979) after his initial studies of Cretaceous flower types. These conclusions, *based on the fossil record of angiosperms,* indicate a shift away from the concept that the Magnoliales represent *the* ancestral type from which all other angiosperms evolved. The "door is now open" to other explanations, based on paleobotanical evidence, which are at variance with the classical concept originating from studies of extant flowering plants. Admittedly, much more critical study and documentation are required, making use of information from every part of the fossil record pertaining to the origin(s) of angiosperms from gymnosperms, as well as to relationships among groups of angiosperms, before we can hope for a resolution of their monophyletic or polyphyletic beginnings.

We cannot, however, conclude the topic of monophyletic versus the polyphyletic origin of angiosperms without being reminded of the very important fact that extant angiosperms – monocots and dicots – produce embryo sacs (megagametophytes) and microgametophytes that are very uniform in their basic structure and that they have double fertilization. Their megagametophytes lack archegonia, a condition only found in *Gnetum* and *Welwitschia* of the Gnetales. The uniformity in structure of their embryo sacs and double fertilization afford very strong arguments in favor of a monophyletic origin of angiosperms, arguments that are not likely to be subjected to analysis by paleobotanists. Nonetheless, paleobotanists are starting to establish an "upward outlook" in an effort to determine the origin and subsequent evolution of flowering plants.

The appearance of preangiosperm characteristics

The word preangiosperm as used here does not necessarily imply that there is a relationship between the organism whose angiospermlike characteristics are being described and flowering plants. What we are primarily interested in is describing levels of evolution with respect to characteristics that approximate those of angiosperms. In this way one might hope to zero in on those fossil gymnosperms that have the greatest potentialities as ancestors of flowering plants. However, many of these characteristics, such as the appearance of reticulate venation, are clearly the product of parallel evolution occurring in unrelated groups in much the same way that heterospory and reduction in the number of functional megaspores have evolved independently in lycopods, sphenopsids, ferns, and seed plants.

Preangiosperm growth habit

Two sources of evidence have had an appreciable influence on interpretations of preangiosperm growth habits. One is the observation that all Mesozoic gymnosperms produce secondary wood and are, for the most part, trees and shrubs. The herbaceous habit is not part of the gymnosperm syndrome. The second source is provided by the Magnoliales, the putative angiosperm archetype. Plants of this order are noted for their well-developed secondary xylem in plants that are trees or large shrubs. As we know, this feature is conspicuous among Coniferales, Taxales, and Glossopteridales. These observations have promoted the concept that preangiosperms and the angiosperm derivatives were arborescent components of upland mesic forest. A counterproposal by Stebbins (1974) suggests that early angiosperms and their ancestors were fast-growing pioneer shrubs inhabiting seasonally dry environments. Doyle & Hickey (1976), Hickey & Doyle (1977), Taylor & Hickey (1990), and others find evidence from their studies of Lower Cretaceous angiosperm leaves that support the Stebbins hypothesis. They note the association of the leaf fossils with coarse-grained stream margin lithofacies, the absence of fern- and gymnosperm-dominated back swamp deposits, the generally small leaf size, and the early occurrence of apparently herbaceous monocots. Hickey & Doyle conclude that the weedy immigrants from semiarid environs became adapted to locally disturbed areas such as stream banks and other riparian habitats. From these habitats, the leaf record suggests that they invaded the mesic conifer forests, first as understory plants, before competing with the well-established conifers that formed the canopy. If these interpretations are correct, one would not expect arborescent gymnosperms to have provided the growth habit of angiosperms during the Jurassic or Lower Cretaceous.

Of other Jurassic gymnosperms, where growth habits are now partially revealed, those of the Williamsoniaceae and Wielandiaceae, with slender, often highly branched stems, approximate the habit of the hypothetical angiosperm archetype as envisaged by Stebbins (1976). Unfortunately, very little is known about the growth habit of the Caytoniales. If, however, their leaves and reproductive parts were borne on trees with branching trunks, one would expect to find some fragmentary evidence of this. It is worth recalling at this point that the secondary wood of *Williamsonia sewardiana*, one of the cycadeoids, has many features of vesselless angiosperm wood. There is, however, no pre-Cretaceous evidence of wood with vessels.

Preangiosperm leaves

Following the evidence provided by Hickey & Doyle (1977), the fossil record indicates that we should look for the origin of the dicotyledonous leaf type among those gymnosperms that have leaves with pinnate secondary venation. This favors the pteridosperms, cycads, and cycadeoids over those gymnosperms with dichotomous and parallel venation (cordaites, conifers, Ginkgoales, and Czekanowskiales). According to Hughes (1976), evolution of the angiosperm leaf from its gymnospermous precursor involved the expansion of a petiole with parallel venation into a wedge-shaped or rounded lightweight lamina with a web of inconspicuous veins, a type of leaf exemplified by *Nelumbites minimus*. From this, secondary veins having a pinnate arrangement evolved. The reticulation of minor veins appeared later.

If we agree that the angiosperm syndrome of several ranks of anastomosing veins with freely ending veinlets provides us with a clue about the origin of angiosperm leaves, then we should turn our attention to those gymnosperms whose leaves show reticulate venation. Various groups of gymnosperms showing this pattern have been advanced as representing the archetype of the angiosperm leaf venation. Among these are leaves of Permian glossopterids and Mesozoic Caytoniaceae (*Sagenopteris*, for example). The problem here, according to Doyle (1978), is that the leaves of these groups show only one order of reticulate venation, a pattern also found in leaves of some Cycadales and Cycadeoidales. The possibility remains, however, that this type of reticulate venation represents a stage in the evolution of the angiosperm leaf venation. Of all leaf types belonging to gymnosperms, those of Recent *Gnetum* most closely approximate angiosperm leaf venation patterns. Some Permian gigantopterids and the Triassic genus *Furcula* are angiosperm-like in having more than one order of veins and veinlets that appear to end freely. If we agree that these leaf types belong to cycadophytes, then the propensity for reticulate venation among gymnosperms predominates among the cycadophytes, not coniferophytes. For this and other reasons Hickey & Doyle (1977) favor the pinnately compound cycadophyte leaf type as ancestral. They envisage a reduction from this type, in response to semixeric conditions, to produce a small simple leaf with pinnate secondary venation and entire margins. From these beginnings a secondary expansion of the lamina occurred, giving rise to Lower Cretaceous angiosperm leaves of the *Ficophyllum, Rogersia, Proteaephyllum*, and *Acaciaephyllum* types.

Preangiosperm perianths

Perianths of Cretaceous flowers are regularly radially symmetrical. Although the attachment of sepals and petals to receptacles in the fossils is often difficult to determine, there is good evidence that both helical and whorled arrangements occurred as early as the Lower Cretaceous. Examination of Upper Cretaceous perianth parts shows well-defined leaflike sepals and a corolla constructed in such a way as to suggest insect pollination. From this we can guess that the petals were showy. Bracts of female flowers belonging to Eocene members of the Juglandaceae (Dilcher, Potter, & Crepet, 1976) (Fig. 30.18A) clearly show leaflike venation patterns that also can be seen in detached corollas of Upper Cretaceous flowers. These and other observations substantiate the foliar nature of their floral parts.

Beyond this, the fossil record tells us little of the origin of perianth parts. The "flowers" (cones) of the Cycadeoidales (*Cycadeoidea, Wielandiella, Williamsonia*, and *Williamsoniella*) have bracts subtending the microsporophylls. These foliar structures are

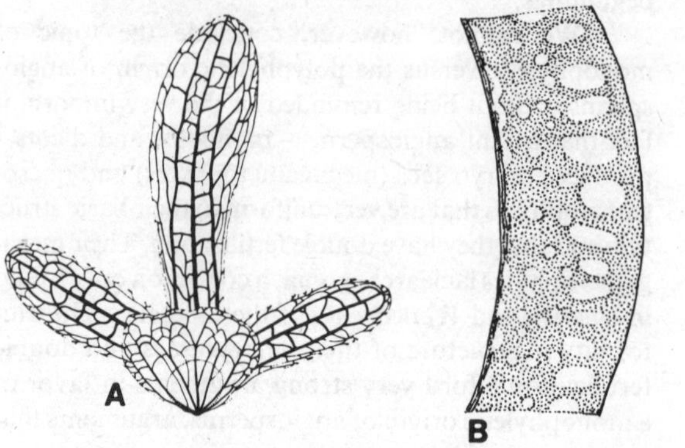

Figure 30.18. **A.** *Paraoreomunnea puryearensis*. Adaxial side of a single fruit. Note leaflike venation pattern. Eocene. (From Dilcher, Potter, & Crepet, 1976.) **B.** Portion of sporoderm of *Monoletes* showing alveolar structure. Pennsylvanian. (Drawn from photograph in Taylor, 1978.)

known to be spirally arranged in *Cycadeoidea* and *Williamsonia*. They clearly had the protective function of a calyx. According to Wieland's (1916) often illustrated interpretation, the "flower" of *Cycadeoidea* opened out to produce a large showy perianth that bore a resemblance to the flower of a magnolia. Much has been made of this, and it has been repeatedly suggested that *Cycadeoidea* is an ancestor of the angiosperms. Delevoryas (1968) and Crepet (1974) have demonstrated conclusively that the "flowers" of *Cycadeoidea* did not open as Wieland supposed. For reasons to be detailed later we do not seriously consider *Cycadeoidea* a potential angiosperm ancestor. As a group, however, they have evolved modified foliar structures (bracts) that had the same structure, origin, and protective function as perianth parts of angiosperms.

Preangiosperm stamens

The fossil record of flowers with preserved stamens are few indeed. The first well-preserved flowers with intact stamens come from the Upper Cretaceous Dakota Formation. These pentamerous flowers described by Basinger & Dilcher (1980, 1984) have five stamens, one attached to each of five petals. In general morphology the stamens of these ancient flowers are the same as stamens of extant angiosperms with a well-defined filament and an anther composed of four pollen sacs. The symmetry of the anther is bilateral. At present there is no published evidence from the fossil record of flowers where the stamen is an expanded foliar structure of the type indicated by some members of the extant Magnoliales (Canright, 1952). If we rely on what evidence there is, we might consider the possibility that the stamen is a modified microsporangiate telome truss, as proposed by Wilson (1942), where reduction of telomes has occurred followed by recurvation and syngenesis of microsporangia. Unfortunately, there is no evidence from the fossil record that supports this hypothesis.

Somewhat less hypothetical is the idea originated by Harris (1951) that pollen-producing synangia of *Caytonanthus,* each consisting of four elongated sporangia, are similar to anthers of angiosperms with their four embedded and elongate microsporangia. The basic difference between the two is the bilateral symmetry of angiosperm anthers and radial symmetry of *Caytonanthus*. We should also emphasize that stamens of angiosperms are simple structures, while the unit bearing the terminal clusters of pendant *Caytonanthus* synangia is compound and pinnate. More recently, Krassilov (1977) has found early stages in the development of *Caytonanthus* that display bilateral symmetry of the synangia. He concludes that these specimens are "the nearest approach to angiosperm stamens." If we agree that *Caytonanthus* does indeed represent a stage in the evolution of the stamen, then we must assume reduction from a compound pinnate structure to a simple one with foliar characteristics of the *Magnolia* type if the interpretations of Canright and others are correct. The possibility of more than one primitive type of angiosperm stamen remains, however.

The Cycadeoidales have pollen-producing organs that are leaflike microsporophylls. In *Cycadeoidea* and *Williamsonia* the whorled microsporophylls are compound bilaterally symmetrical structures that bear synangia. Nothing from the studies of fossil or extant flowers suggests that such organs were the precursors of stamens. *Williamsoniella* and possibly *Wielandiella,* however, have microsporophylls that are similar to the foliar stamens of some Magnoliales. All are simple structures with pollen sacs partially embedded in the tissue of the microsporophyll, usually on the adaxial surface. Each sac of the embedded pair contains several sporangia and is probably a synangium. Thus, of the Cycadeoidales where pollen organs are known, those of *Williamsoniella* most closely approximate foliar stamens of angiosperms.

Preangiosperm pollen

While examining the fossil record of angiosperms in search of evidence for primitive characteristics, we encountered considerable support for the idea that small tectate monosulcate pollen grains are characteristic of primitive angiosperms. In previous chapters we have reported that Pentoxylales, Cycadales, Cycadeoidales, Ginkgoales, and Peltaspermaceae have small monocolpate grains. These grains, however, are without a tectum. On the basis of pollen grain morphology alone, those groups of gymnosperms with mono- or bisaccate pollen (with wings or bladders) would seem to be excluded. You may recall that saccate pollen is characteristic of the Cordaitales, Voltziales, Callistophytaceae of the Pteridospermales, Glossopteridales, Corystospermaceae, Caytoniaceae, and some Coniferales (Pinaceae and Podocarpaceae, for example). Conifers with nonsaccate pollen (Araucariaceae and Taxodiaceae) are thought to be derived from those Permian–Triassic Voltziales with saccate pollen. This loss of the saccate condition in some Coniferales suggests that just because saccate pollen seems to typify the pteridosperm families Corystospermaceae and Caytoniaceae, we need not exclude them from consideration as angiosperm ancestors based on their pollen characteristics alone.

It is believed that the saccus originated by separation of the ectexine (= sexine, outer layer of the exine) at an early stage of development (Millay & Taylor, 1976). The internal corpus is bounded by a granular, sometimes laminated endexine (= nexine, inner layer of the exine). The ectexine of all saccate pollen belonging to the groups listed above is highly alveolar, a reflection of the finely alveolar structure of spores of *Archaeopteris* and the medullosan prepollen *Monoletes* (Fig. 30.18B). A honeycomb to spongy alveolar structure also is characteristic of monocolpate pollen grains of extant Cycadales. From this account, we are led to the unavoidable conclusion that pollen with an alveolar outer exine layer is primitive and that homogeneous to granular and tectate types have been derived (Doyle, 1978). This is at variance with Walker (1976), who concludes that the homogeneous and granular exines of some Magnoliales and other Magnoliidae represent the precolumellar tectate condition and are transitional between the nontectate pollen of gymnosperms and tectate pollen of other angiosperms. However, evidence provided by the studies of Cornet (1977) indicates that columellate tectate monocolpate pollen (the *Classopollis* type) had evolved by the Triassic long before the appearance of angiosperms. *Classopollis* is a pollen type assigned to the "intermediate" conifer group, the Cheirolepidiaceae (Chapter 29). This fossil evidence also supports the idea that the alveolar structure in the exine of gymnosperm pollen grains is primitive and that the homogeneous – granular and columellate – tectate types are derived.

The above account describing the origin of the fine structure of the angiosperm pollen-grain wall provides us with an excellent example illustrating how different conclusions about the origin of angiosperm characteristics can be reached when one studies the fossil record as opposed to studies of extant plants. It is apparent that evidence from all sources must be considered.

Preangiosperm ovules

Of the angiosperm characteristics given early in this chapter, those of the bitegmic and anatropous (reflexed) ovules are given special emphasis by Stebbins (1974, 1976). The widespread occurrence of ovules having these characteristics among the flowering plants leaves little doubt that they are indeed important. Bitegmic ovules occur in the majority of monocots and most dicots except the Rosidae, Asteridae, and Ericales. Anatropous ovules occur in the "core" orders of angiosperms, including many of the Magnoliales. However, a few members of the latter order belonging to the Chloranthaceae and Piperales have orthotropous (erect) ovules. It has been argued by Stebbins that with the evolution of the first integument of the gymnosperm ovule and subsequent evolution of the ovary wall, there would be little selective advantage to the later formation of a second integument, and for this reason it did not arise de novo among the angiosperms. Thus, the second integument as well as the first seem to represent primitive ovule characteristics retained from gymnospermous ancestors. This raises the obvious question about the origin of the integument among the gymnosperms.

We have seen unequivocal evidence that the initial envelope around the megasporangium of Paleozoic seed ferns was derived by the coalescence of telomes (Chapter 22). Although ovules of the lyginopterid pteridosperms and conifers provide little or no evidence of a double integument, Camp & Hubbard (1963) envisage the ovule wall of the cupulate *Lagenostoma* type to be composed of an inner integument fused with the nucellus. They claim the inner integument can be identified as that part of the ovule wall differentiated into the lagenostome. However, most paleobotanists believe this pollen-receiving structure to be part of the nucellus. The outer integument, according to the interpretation of Camp & Hubbard, is the one usually interpreted as the integument of *Lagenostoma* and other similar ovules.

Could it be that the cupule has become the second, outermost integument, and not the precursor of the ovary wall as frequently proposed by many paleobotanists for more than half a century? Based on evidence from lyginopterid pteridosperms, the scenario explaining this origin of the second integument from the cupule is as follows: There are three well-known genera of the Carboniferous Lyginopteridaceae that have cupules containing one or more orthotropous ovules. These are *Calathospermum* (Fig. 30.19A), *Gnetopsis* (Fig. 30.19B), and *Lagenostoma* (Chapter 23). *Calathospermum* has many (up to 60) stalked ovules arising from the base of the cupule. In *Gnetopsis* there are two ovules, and they are sessile in the base of the

Figure 30.19. **A.** *Calathospermum scoticum*, longitudinal section of cupule. (Redrawn and modified from Walton, 1949.) **B.** *Gnetopsis eliptica*, longitudinal section of cupule. **A,B:** Carboniferous. (Redrawn from Taylor & Millay, 1979.)

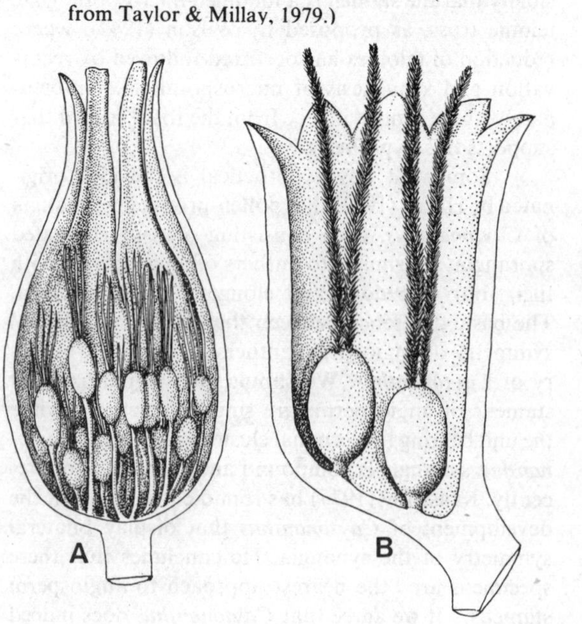

cupule. There is a single relatively large ovule with a short stalk that fills the cupule of *Lagenostoma*. It is suggested that these three genera represent a trend where there is reduction in the number of ovules in the cupule from many to one, accompanied by a reduction in the length of their stalks. The product of these reductions is a cupule that fits snugly around a single short-stalked ovule. Phyletic fusion between the unitegmic ovule and the surrounding cupule would produce an ovule with a double integument. This cupulate theory for the origin of the second integument among the Paleozoic pteridosperms is well argued by Smith (1964).

A reduction in number of ovules from several to one, as in the Paleozoic cupulate pteridosperms, is postulated for the Mesozoic pteridosperm families, the Caytoniaceae and Corystospermaceae. The cupule of *Caytonia,* which has the appearance of an anatropous ovule, contains several erect ovules each with a single integument. The anatropous cupules of *Umkomasia* and *Pilophorosperma,* on the other hand, each contain a single ovule with an exserted micropyle. Fusion between the surrounding cupule and integument of the enclosed ovule would produce a reproductive structure having the characteristics of an antropous bitegmic, angiosperm ovule. Unfortunately, there is very little evidence that the required fusion between cupule and ovary wall took place. Irrespective of which is the correct interpretation for the origin of the bitegmic, anatropous ovule, it is worth noting that all the supporting fossil evidence is derived from pteridosperms. Once again, the coniferophytes are eliminated.

Preangiosperm carpels

The just-concluded discussion of ideas relating to the evolution of the bitegmic ovule only serves to emphasize the tendency of land plants to cover their reproductive structures. In the case of the megasporangium, numerous mechanisms have been described, from the first appearance of Devonian preovules with their surrounding, fused, and planted telomes to the inferior ovary of a modern angiosperm embedded in the tissues of the receptacle. In addition to the integuments that ovules of all seed plants possess, we have seen examples of many different ways in which seeds are partially to completely enclosed. Here we can list structures such as the laminae of leaves that tend to cover or wrap around ovules (*Archaeocycas* with cycadean affinities, *Glossopteris,* and the pteridosperm *Spermopteris*); the interseminal scales of the Cycadeoidales; megasporophylls condensed into protective cones in the Cycadales, or ovules condensed into cones as in the Pentoxylales; woody well-developed cone scales of conifer seed cones; a great variety of cupules or cupulelike structures of the Paleozoic and Mesozoic pteridosperms (for example, *Lagenostoma, Caytonia, Umkomasia,* and *Lepidopteris*); bivalved structures of the Czekanowskiales; and finally the carpel. From the Devonian beginnings leading to the appearance of the carpel, there is a clear sequence with respect to the transfer of function related to the protection of the contents of the megasporangium (Stebbins, 1976). We need to stress, however, that the ovule enclosures that we find in many groups have almost certainly evolved independently in the different groups. Therefore, it would be a mistake to assume, for example, that the cupule of the Caytoniales evolved from that of the Lyginopteridaceae or that the angiosperm carpel evolved from the cupule of the Caytoniales. In Devonian plants with preovules, the function of protection was primarily accomplished by the megasporangium wall. This protective function was assumed by telomes that later evolved into an integumentary layer enveloping the megasporangium. Later this function was transferred to the cupule, which may have become the outer integument. Finally the protection of the ovules with their enclosed megasporangia became the function of the carpel or ovary wall.

Based on classical researches by Bailey & Swamy (1951) of extant angiosperms, placed at that time in the old order Ranales, we now have a clear picture of the conduplicate carpel (Fig. 30.20A) and its primitive characteristics. *Tasmannia* (*Drimys*), a ranalian genus, has a carpel that differs in many ways from carpels of most angiosperms, which are characterized by a closed ovary, stigma, and style. The carpel, which is considered the homolog of a megasporophyll, is folded together lengthwise; it is conduplicate. Early in development, the margins of the folded structure remain unfused and form a narrow cleft in the adaxial carpel wall that extends from the exterior to the inner cavity (locule) containing the ovules (Fig. 30.20B). The anatropous, bitegmic ovules form two rows attached to the inner walls of the carpel, about midway between the margins of the carpel and the center of the locule. There is an extensive development of epidermal hairs (Fig. 30.20C) from the flared margins of the carpel inward beyond the point of attachment of the ovules.The hairs and the trichomes lining the locule are thought collectively to represent the surface of a stigma. In order to germinate and produce functional pollen tubes, pollen grains must be deposited on the exposed epidermal hairs on the margins of the open carpel. There are three principal traces of vascular tissue in the carpel: a central dorsal trace and two ventral traces. The general morphology, vascularization, and method of marginal growth stongly support the conclusion that the carpel, similar to stamens, sepals, and petals, is a phyllome (modified leaf). Subsequent investigations of additional members of the ranalian angiosperms have shown how the extensive stigmatic surface of the conduplicate carpel has become localized at the tip of a style.

At least five different hypotheses and theories

have been advanced since 1925 that attempt to explain the origin of the carpel. Some are based on evidence from the fossil record, while others utilize observations of extant angiosperms. Primitive plants with telomic systems, ferns, and nearly every major group of gymnosperms have been proposed at one time or another as providing the precursors to carpels.

Since 1925, when the cupulate *Caytonia* was described by Thomas and thought to represent a primitive angiosperm carpel, numerous attempts have been made to homologize the cupule with the carpel. Later, however, Harris (1940) demonstrated the occurrence of bisaccate pollen in the micropyles and pollen chambers of ovules within the cupule of *Caytonia.* Thus, there was little doubt that in this characteristic, the plants bearing these organs were gymnospermous, not angiospermous. Subsequently, Thomas (1955, 1957) and many others have offered various models using cupules of Paleozoic and Mesozoic pteridosperms to show how the angiosperm carpel was derived. After the discovery of corystosperm cupules containing a single reflexed ovule, Thomas (1934) proposed a new model that involved the back-to-back fusion of two subopposite cupules of the *Pilophorosperma* type to produce a carpel with two anatropous ovules and a basal stigma. Andrews (1961) is inclined to favor the *Caytonia* cupule as showing us one way in which the carpel may have evolved.

The origin of the carpel from those Paleozoic pteridosperms forming cupules is supported by Delevoryas (1962), Andrews (1963), and Long (1966). Delevoryas favors lyginopterid pteridosperms of the *Calathospermum* and *Gnetopsis* type where the cupule bears several orthotropous ovules in the base. A closure of the cupule opening "should produce a structure closely resembling a carpel with basal placentation of ovules." If, as suspected, a cupule of *Calathospermum* is a modified frond, then the model suggested by Delevoryas provides us with a carpel that is foliar, and this is in keeping with the presumed foliar nature of the angiosperm carpel. Long (1966, 1977), on the other hand, favors those lyginopterid pteridosperms where the cupule is clearly only a segment of a frond. This offers difficulties because the angiosperm carpel is usually homologized with an entire leaf. Other problems are the orthotropous, unitegmic ovules of lyginopterid pteridosperms that are unlike the anatropous, bitegmic ovules of angiosperms. In homologizing the lyginopterid cupule with the ovary wall, Long is required to explain the origin of the second integument as an outgrowth of the first integument or the chalazal stalk of the ovule. Except for the model presented by Stebbins (1976), no others attempt to explain the origin of the anatropous, bitegmic ovule.

Wilson (1942) utilizes the telome theory in his explanation of carpel origin. He portrays the carpel as a planated webbed fertile telome truss with marginal ovules. The folding of this structure to enclose the

Figure 30.20. Conduplicate carpel of *Tasmannia (Drimys)*. **A.** Side view showing paired stigmatic crests. **B.** Transverse section showing position of attachment of two rows of anatropous ovules. **C.** Carpel opened out to show placentation of ovules, glandular hairs, and course of pollen tubes. Extant. (**A–C** redrawn from Bailey & Swamy, 1951.)

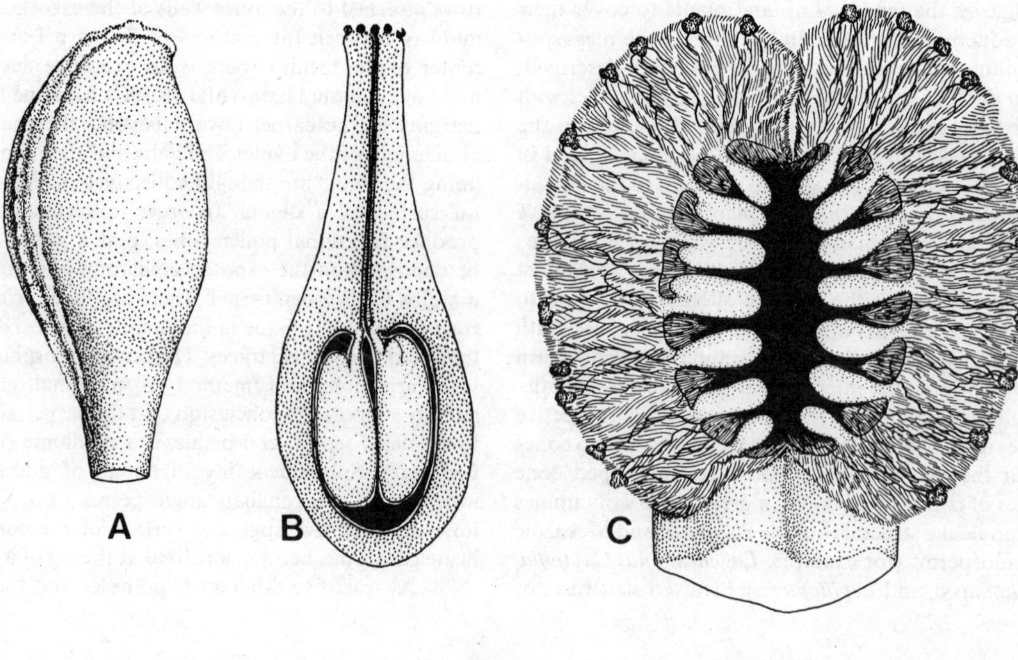

ovules would produce a conduplicate phyllome. The "gonophyll" hypothesis of Melville (1960, 1962) also is based on the telome theory. The fertile and sterile segments of the gonophyll are derived, according to Melville, from a planated telome truss composed of fertile segments that bore terminal ovules and a sterile webbed portion that became a leaf. According to his model, the fertile segment phyletically "moved" to the upper surface of the leaf where it became adnate and was finally covered by the inrolling of the margins of the subtending leaf. For fossil evidence supporting his elaborate theory, Melville cites the leaf of *Glossopteris* with its fertile ovulate structure (*Scutum*) adnate to the adaxial surface. The reticulate venation of the glossopterid leaf subtending the fertile branch is cited as additional evidence supporting the angiospermous nature of the gonophyll.

Recently, Krassilov (1977) has provided new information about the structure of *Caytonia,* which he contends suggests a closer relationship between the cupule of *Caytonia* and the angiosperm carpel. He and his colleagues find the ovules of a *Caytonia* cupule gathered into a cluster covered by a cutinized membrane. The ovules are connected with a pollen-conducting tube that is independent of the "mouth" in the cupule wall. The latter is envisaged as a modified gonophyll of the glossopterid type that has become folded over a fertile branch with its packet of ovules. According to this interpretation, each cupule of *Caytonia* is a modified leaf enclosing the ovules.

The "anthocorm" theory proposed by Meeuse (Meeuse, 1979a, 1979b), like the "gonophyll" theory, is an attempt to get away from the classical ranalian concept of the carpel. The anthocorm is a hypothetical seed plant unit comprising a main axis with ovuliferous cupules or microsporangiate synangia borne on lateral branches in the axils of bracts on the main axis. Meeuse (1966) makes the point that the cupule is not to be considered homologous with a leaf but with the aril of uniovulate arillate forms. He states that "a ranalean carpel is not a leaf homologue or a foliar carpel." Camp & Hubbard (1963) also see homology between the cupule and the aril of arillate angiosperms. They suggest that the origin of the cupule is from phyletic condensation of the three-dimensional telome truss that bore the ovules. Their model for such a primitive branching system is reflected in the dichotomous ovule-bearing branches of the Carboniferous *Eurystoma angulare*.

One of the simplest theories explaining the origin of the carpel is provided by Mamay (1969, 1976) after his studies of the Permian *Archaeocycas* (Fig. 30.21A). You may recall that this cycadophyte genus shows partial enclosure of the two rows of ovules by the lateral lamina of the sporophyll. By progressive expansion and fusion of the margins of the lamina, there would be complete enclosure of the ovules (Fig. 30.21B,C) to form an angiospermlike carpel.

Of all theories advanced to date attempting to explain the origin of a carpel with two rows of anatropous, bitegmic ovules, the proposals of Stebbins (1974, 1976) and Retallack & Dilcher (1981a) using the fossil record of the Permian Glossopteridales offer some of the most interesting possibilities. Their theories incorporate some of the ideas presented by Melville (1962). These ideas are only intended to show the ways that the carpel, with its included ovules, could have evolved. The possibility remains that carpel evolution followed more than one pathway, as proposed by Meeuse (1961). Because of the great time interval between the Permian glossopterids and the first appearance of the carpel in the Lower Cretaceous, there is a very good possibility that plants having primitive carpels and owing their origins to the glossopterids became extinct long before the advent of the angiosperms. When more is learned about the structure and life cycles of Jurassic cupulate pteridosperms, parallels with what is proposed here may become more apparent.

Turning to the evidence provided by the Glossopteridales, we can describe the following steps, some hypothetical, leading to the formation of conduplicate carpels basically similar to those of the mid-Cretaceous *Archaeanthus* (Dilcher & Crane, 1985).

1. Grouping of ovules. We have seen that among many cupulate pteridosperms, both Paleozoic and Mesozoic, their ovules tend to be grouped so that there are several ovules per cupule. Here we can identify such genera as *Calathospermum, Gnetopsis, Caytonia, Lepidopteris* with its cupulate discs, and the bivalved *Leptostrobus.* Among the Glossopteridales this tendency

Figure 30.21. Steps in the evolution of the carpel according to Mamay. **A.** *Archaeocycas,* partial covering of the adaxial ovules by reflexed margins of the lamina. **B,C.** Hypothetical stages in covering of ovules by lamina to form carpel. (Redrawn from Mamay, 1969.)

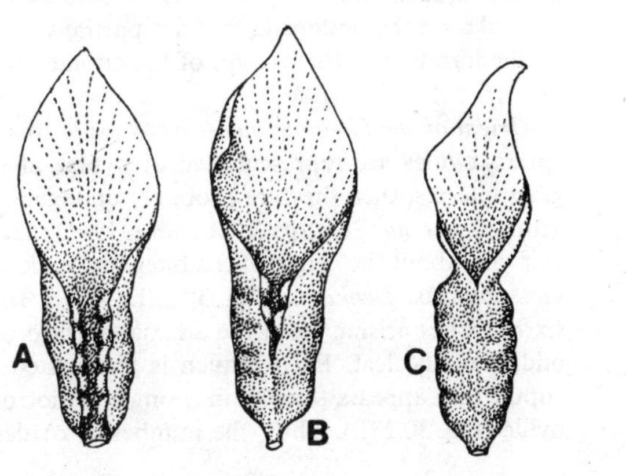

is well documented by the ovule-bearing structures of *Glossopteris* (*Dictyopteridium*), *Ottokaria*, and *Lidgettonia*, each with a cupulelike capitulum with inrolled margins that partially enclose a number of ovules.

2. Evolution of the fertile axis. In keeping with the gonophyll theory of Melville, the multiovulate cupules are arranged at the tips of branches (sporangiophores) of a pinnate fertile axis, not unlike that of *Caytonia*. The pinnate structure probably represents a reduced portion of a frond where the terminal pinnules have become the cupule wall. Similar uniovulate cupules terminate ultimate frond segments of several Carboniferous pteridosperms. The fronds may be entirely fertile, or fertile with sterile segments. Similar branched cupule-bearing structures are yet to be found among the glossopterids. Only unbranched sporangiophores terminated by the ovulate disc (cupule) are known. It is worth noting that sporangiophores bearing microsporangia are, however, highly branched.

3. The epiphyllous fertile branch. There is a juxtaposition of the sporangiophore and a foliage leaf so that the cupule-bearing sporangiophore is in the axil of the leaf (Fig. 30.22A). Specimens of *Glossopteris* (*Dictyopteridium*) show a degree of adnation between the proximal end of the sporangiophore and the proximal adaxial portion of the midrib of the leaf. Thus, the fertile branch (sporangiophore) is epiphyllous and simulates the gonophyll of Melville.

4. Adnation. Adnation between subtending leaf and an axillary fertile branch of the pinnate type, with cupules at the tips of the pinnae, is the next hypothetical step that finds some supporting evidence in the glossopterid genus *Jambadostrobus*. If adnation between the fertile branching system and leaf was complete, except for the distal ends of the cupule-bearing branches, then two rows of cupules with short stalks would be formed on either side of the apparent midrib of the leaf. *Lidgettonia mucronata* (Surange & Chandra, 1972) shows this kind of organization (Fig. 30.22B,C). There are three to five sporangiophores in each row, and the cupules have two rows of ovules on the underside that are partially covered by the inrolling of the margin of the cupule.

5. Origin of the bitegmic, anatropous ovule. If *Glossopteris* ovules are representative of ovules of other glossopterids, then they are probably unitegmic. The genus *Denkania* (Surange & Chandra, 1971) may give us a clue about the origin of the bitegmic ovule in the glossopterids. *Denkania* (Fig. 30.22D) has a row of six branches arising from the adaxial surface of the midrib of the leaf. Each branch is terminated by a cupule that appears to contain a single orthotropous ovule (Fig. 30.22E). Thus, the number of ovules per cupule is reduced from many (as in *Lidgettonia*) to one. We have noted a similar reduction in the lyginopterid pteridosperms of the Carboniferous and the Mesozoic Caytoniales. As in these groups, the cupule of *Denkania* with its single ovule provides the opportunity for fusion of the cupule and the integument of the ovule to form a double integument (Fig. 30.22F). Reflexing of the derived orthotropous, bitegmic ovule on its stalk or branch would result in an anatropous configuration.

6. Formation of the carpel. This involves the infolding of the leaf around two rows of bitegmic, anatropous ovules (Fig. 30.22G) to form the closed carpel. The distal end of the infolded leaf becomes transformed into the stigma (Fig. 30.22H). This critical step is hypothetical for the glossopterids, but is reflected in the follicles of mid-Cretaceous *Lesqueria, Prisca, Caloda,* and *Archaeanthus*.

Figure 30.22. Stages in the evolution of the carpel from glossopterid ancestors according to Stebbins, Retallack & Dilcher, and others. **A.** Modified "gonophyll" with pinnate ovulate "sporangiophore" in the axil of a leaf. **B.** Adnation of "sporangiophore" with adaxial surface of leaf leaving two rows of *Lidgettonia*-like capitula (**C**). **D,E.** Reduction in number of ovules per capitulum from several to one as in *Denkania*. **F.** Fusion of cupulelike capitulum with integument of ovule to form bitegmic ovule. **F,G.** Inversion of ovules to produce anatropous ovules. **H.** Reflexing of leaf margins to enclose two rows of ovules in a carpel. (Based on Stebbins, 1976; Retallack & Dilcher, 1981a.)

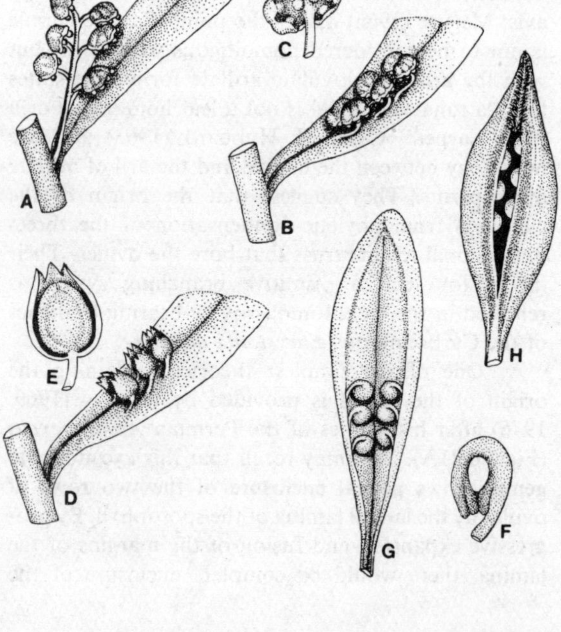

Conclusions

In presenting the chapters on gymnosperms and this chapter on angiosperms, we hope we have given you the correct impression that in the course of their evolution, seed plants have tended to evolve a variety of structures designed to protect ovules. Some of these provide insight about the origin of the carpel, while others represent evolutionary "experiments" ending in extinction. It is becoming increasingly clear that those gymnosperms loosely called the cycadophytes (especially the pteridosperms) have assumed a predominant place in our search for angiosperm origins. Collectively, they offer more potentialities as preangiosperms than any other group: in their leaf and epidermal structures, anatomy of their wood, growth habit, ovule structure and orientation, and flowerlike organization of their cones. No specific cycadophyte or group of cycadophytes is endowed with all these and other characteristics one might expect to find in a preangiosperm.

Recent researches of the fossil record have pinpointed the time of origin of angiosperms, thus limiting our studies to a more restricted part of the geological column. These same researches have improved our insight into what constitutes a primitive angiosperm. Based on flower structure, indications are that there may have been more than one type of primitive angiosperm; that those represented by the present-day Magnoliales, with their conduplicate carpels, single perfect flowers, and spirally arranged free parts, represent only one of two or three primitive types. Thus, based on the fossil record, a polyphyletic origin of angiosperms from gymnosperms seems more plausible than previously was thought. With this possibility in mind, we no longer need concentrate our efforts on looking for *the* precursor of angiosperms exhibiting all of those characteristics we have listed at the beginning of this chapter. To emphasize this last point, remember the Progymnospermopsida "with their gymnospermous wood and pteridophytic reproduction." By the same

Figure 30.23. Cladogram summarizing the relationships of major groups of seed plants. Key to abbreviations used for terminal taxa in cladogram: An. = Aneurophyton; Ar. = Archaeopteris; ML. = Early Carboniferous protostelic lyginopterids; HL. = higher lyginopterids (*Heterangium, Lyginopteris*); Md. = *Medullosa;* Go. = Ginkgoales; Cn. = coniferales; Cd. = Euramerican cordaites; Ca. = *Callistophyton;* Cs. = Corystospermaceae; Cy. = Caytoniales; Pl. = *Peltaspermum;* Gl. = Glossopteridales; Ct. = *Caytonia;* Ag. = angiosperms; Bn. = Bennettitales (Cycadeoidales); Pn. = *Pentoxylon;* Ep. = *Ephedra;* We. = *Welwitschia;* Gn. = *Gnetum.* (From Doyle & Donaghue, 1987, with permission.)

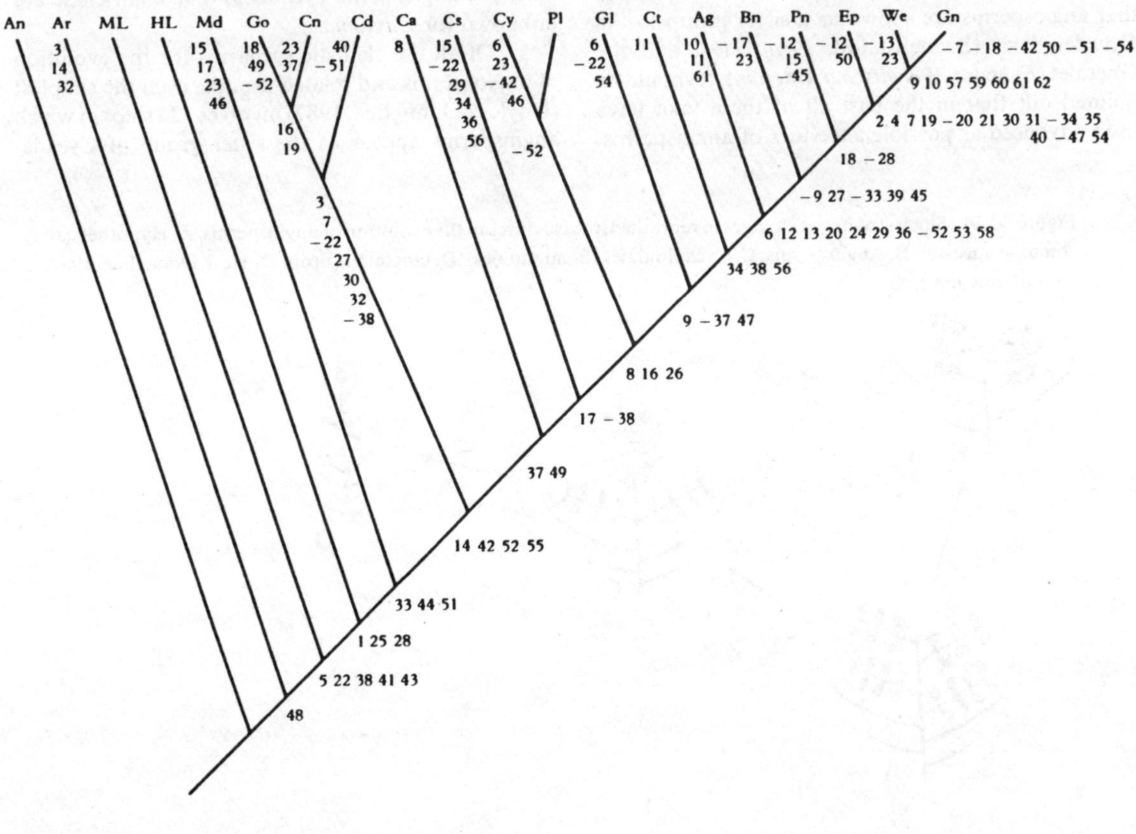

token, a preangiosperm could have carpels, gymnospermous wood, leaves, and pollen structure, or any one of a multitude of different combinations representing "a web of evolution with repeating themes of reproductive (and vegetative) adaptations" (Dilcher 1979).

In an effort to unravel the complex "web" of evolution that led to the origin of angiosperms prior to the Cretaceous, Hill & Crane (1982), Crane (1985), and Doyle & Donoghue (1986a, 1986b) employed for the first time an experimental cladistic approach. As explained in an earlier chapter, the function of this approach is to produce a natural system of classification in which the degree of subjectivity is reduced. In other words, it attempts to provide a more logical (more objective) framework within which the degrees of relationships among individuals or groups of individuals can be depicted in the form of a cladogram (Fig. 30.23). Rigorous testing and changing of working models of cladograms have been undertaken as new information is added or old information is reinterpreted.

In their 1987 cladistic study focusing on the origin of angiosperms, Doyle & Donoghue employed 20 terminal taxa, which include most major groups of vascular plants starting with the Aneurophytales, and 62 characteristics. Without going into the processes by which the characteristics are determined to be ancestral or the result of convergence or reversal when constructing the cladistic "tree," we can see that angiosperms are shown as a sister group to the Cycadeoidales (Bennettitales), *Pentoxylon,* and the Gnetales (*Ephedra, Welwitschia, Gnetum*). It should be pointed out that in the past all of these taxa have been advanced as possible ancestors of angiosperms. For example, Wettstein (1907) was among the first to seriously propose that angiosperms owed their origin to the Gnetales. He demonstrated that as a group the order exhibits more angiospermous features than any other group of gymnosperms. Many of these features are detailed earlier in this chapter. The compound strobili of the Gnetales were homologized with flowers of wind-pollinated Amentiferae. It is now known that the flowers of the Amentiferae are reduced and not primitive. Like the Gnetales, the Cycadeoidales were long ago proposed (Arber & Parkin, 1907) as putative ancestors of angiosperms. These authors originated the theory that the bisexual flowerlike structures of Mesozoic cycadeoids were homologous with flowers of the *Magnolia* type. It was this thesis that C. E. Bessey employed in his system for the classification of flowering plants by establishing *Magnolia* of the Ranales as a representative of the primitive type for angiosperms. Subsequent research has given substantial support to the idea that the extant Magnoliidae are indeed primitive. It is this group that provides the basis for most phylogenetic classifications of flowering plants (Cronquist, 1981).

In the cladistic study illustrated here (Fig. 30.23), it appears that *Caytonia* is also linked with the angiosperms, a proposal supported by Thomas (1925, 1955, 1957) and numerous other paleobotanists. Doyle & Donoghue (1987) point out, however, that in their cladistic studies most of the characteristics that unite the angiosperm–cycadeoid–gnetalian clade are unknown for *Caytonia*.

Of all the cladistic scenarios for the evolution of angiosperms and related groups, even the simplest (Doyle & Donoghue, 1987) involves 123 steps in which angiosperms appear as the sister group of Cycade-

Figure 30.24. Suggested major changes in reproductive structures in the evolution of angiosperms. **A.** Hypothetical common ancestor. **B.** Angiosperms. **C.** Cycadioidales (Bennettitales). **D.** Gnetales. (From Doyle & Donaghue, 1987, with permission.)

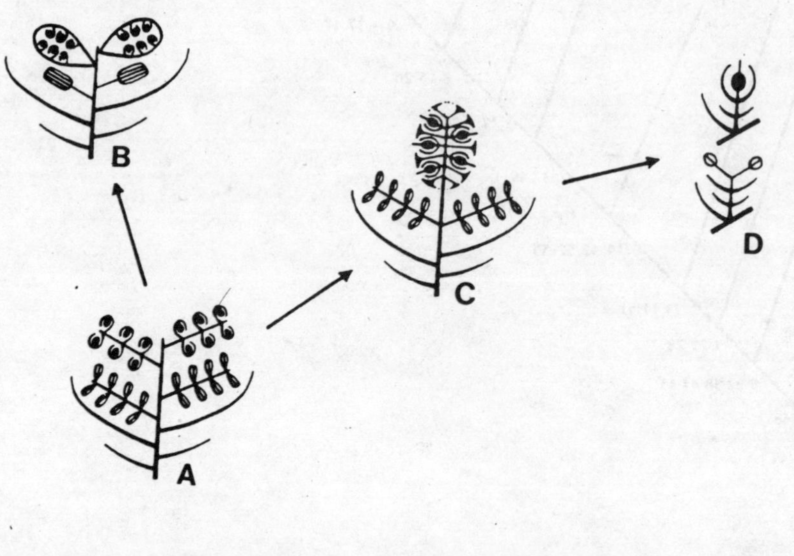

oidales and Gnetales (Fig. 30.23). In this scheme it is suggested that these three groups were derived from a common ancestor with bisexual flowers, anatropous ovules, and pinnate mega- and microsporophylls (Fig. 30.24A). The angiosperms (Fig. 30.24B) are primitive in retaining leaflike carpels, but are advanced in having simple stamens. The Cycadeoidales (Fig. 30.24C) have primitive pinnate microsporophylls, but highly reduced megasporophylls each with a single orthotropous ovule. The secondarily reduced unisexual flowers of the Gnetales (Fig. 30.24D) have simplified megasporophylls with orthotropous ovules and simple microsporophylls. Although the results of this cladistic study show the Gnetales to be the closest extant relatives of the angiosperms, this does not mean that the angiosperms originated from the Gnetales as proposed by Wettstein (1907). Rather, it supports the concept that there was a common hypothetical ancestor that gave rise to the angiosperms and cycadeoids (Fig. 30.24) and that the flowers of the gnetalean stock were subsequently reduced and aggregated into unisexual units in response to wind pollination as has occurred in the Amentiferae (Doyle & Donoghue 1987).

Even with the aid of cladistic analyses, it seems that all we can hope to find, when looking for the origin of angiosperms, are those gymnosperms that combine one or more angiosperm characteristics and form a complex from which the evolutionary lines leading to the flowering plants emerged. It is too much to expect to find that "moment" in the fossil record at which an evolutionary level was achieved where the organisms possess, *for the first time*, all of those characteristics we normally use to recognize a flowering plant. If we expect to discover this "moment," then as Darwin noted, the origin of the angiosperms will continue to be "an abominable mystery." If, on the other hand, we continue to determine the primitive characteristics of angiosperms, using the fossil record, then by further study of the potential gymnospermous ancestors and with the aid of cladistic techniques, we should be able to select the group or groups that offer the greatest potential as ancestors of the angiosperms. This approach could solve the "abominable mystery."

References

Alvin, K., & Chaloner, W. C. (1970). Parallel evolution in leaf venation: an alternative view of angiosperm origins. *Nature,* **226,** 662–3.

Andrews, H. N. (1961). *Studies in Paleobotany.* New York: Wiley.

Andrews, H. N. (1963). Early seed plants. *Science,* **142,** 925–31.

Arber, E. A. N., & Parkin, J. (1907). On the origin of angiosperms. *Journal of the Linnean Society., Botany,* **38,** 29–80.

Axelrod, D. I. (1960). The evolution of flowering plants. In *The Evolution of Life,* ed. S. Tax. Chicago: University of Chicago Press, pp. 227–305.

Axelrod, D. I. (1970). Mesozoic paleogeography and early angiosperm history. *Botanical Review,* **36,** 277–319.

Bailey, I. W., & Swamy, B. G. L. (1951). The conduplicate carpel of dicotyledons and its initial trend of specialization. *American Journal of Botany,* **38,** 373–9.

Basinger, J. F., & Dilcher, D. L. (1980). Bisexual flowers from the mid-Cretaceous of Nebraska. *Botanical Society of America, Miscellaneous Series,* **158,** 10.

Basinger, J. F., & Dilcher, D. L. (1984). Ancient bisexual flowers. *Science,* **224,** 511–13.

Beck, C. B. (1976). Origin and early evolution of angiosperms: a perspective. In *Origin and Early Evolution of Angiosperms,* ed. C. B. Beck. New York: Columbia University Press, pp. 1–10.

Becker, H. F. (1972). *Sanmiguelia* an enigma compounded. *Palaeontographica B,* **138,** 181–5.

Bierhorst, D. W. (1971). *Morphology of Vascular Plants.* New York: Macmillan.

Bock, W. (1969). The American Triassic flora and global distribution. *Geological Center Research Series,* **3,** 242–54.

Brown, R. W. (1956). Palmlike plants from the Dolores Formation (Triassic) in southwestern Colorado. *United States Geological Survey Professional Paper,* **274,** 205–9.

Camp, W. H., & Hubbard, M. M. (1963). On the origins of the ovule and cupule in lyginopterid pteridosperms. *American Journal of Botany,* **50,** 235–43.

Canright, J. E. (1952). The comparative morphology and relationships of the Magnoliaceae. I. Trends of specialization of the stamens. *American Journal of Botany,* **39,** 484–97.

Cornet, B. (1977). Angiosperm-like pollen with tectate-columellate wall structure from the Upper Triassic and Jurassic of the Newark Supergroup, USA. *American Association of Stratigraphic Palynologists, 10th Annual Meeting, Tulsa* (Abstract), pp. 8–9.

Cornet, B. (1986). The leaf venation and reproductive structures of a late Triassic angiosperm, *Sanmiguelia lewisii. Evolutionary Theory,* **7,** 231–309.

Crane, P. R. (1984). A re-evaluation of *Cercidiphyllum*-like plant fossils from the British early Tertiary. *Botanical Journal of the Linnean Society,* **89,** 199–230.

Crane, P. R. (1985). Phylogenetic analysis of seed plants and the origin of angiosperms. *Annals of the Missouri Botanical Garden,* **72,** 716–93.

Crane, P. R. (1989). Paleobotanical evidence on the early radiation of nonmagnoliid dicotyledons. *Plant Systematics and Evolution,* **162,** 165–91.

Crane, P. R., & Dilcher, D. L. (1984). *Lesqueria:* an early angiosperm fruiting axis from the mid-Cretaceous. *Annals of the Missouri Botanical Garden,* **71,** 384–402.

Crane, P. R., Friis, E. M., & Pedersen, K. R. (1986). Angiosperm flowers from the Lower Cretaceous: Fossil evidence on the early radiation of the dicotyledons. *Science,* **232,** 852–4.

Crepet, W. L. (1974). Investigations of North American cycadeoids: the reproductive biology of *Cycadeoidea. Palaeontographica B,* **148,** 144–69.

Crepet, W. L. (1979). Some aspects of the pollination

biology of Middle Eocene angiosperms. *Review of Palaeobotany and Palynology,* **27,** 213–38.

Crepet, W. L. (1981). The status of certain families of the Amentiferae during the Middle Eocene and some hypotheses regarding wind pollination in dicotyledonous angiosperms. In *Paleobotany, Paleoecology, and Evolution,* Vol. I., ed K. J. Niklas. New York: Praeger.

Cronquist, A. (1968). *The Evolution and Classification of Flowering Plants.* Boston: Houghton Mifflin.

Cronquist, A. (1981). *An Integrated System of Classification of Flowering Plants.* New York: Columbia University Press.

Delevoryas, T. (1962). *Morphology and Evolution of Fossil Plants.* New York: Holt, Rinehart and Winston.

Delevoryas, T. (1968). Investigations of North American cycadeoids: structure, ontogeny and phylogenetic considerations of cones of *Cycadeoidea. Palaeontographica B,* **121,** 122–33.

Dilcher, D. L. (1979). Early angiosperm reproduction: an introductory report. *Review of Palaeobotany and Palynology,* **27,** 291–328.

Dilcher, D. L., & Crane, P. R. (1985). *Archaeanthus:* an early angiosperm from the Cenomanian of the western interior of North America. *Annals of the Missouri Botanical Garden,* **71,** 351–83.

Dilcher, D. L., & Kovach, W. L. (1986). Early angiosperm reproduction: *Caloda delevoryana* gen. et sp. nov., a new fructification from the Dakota Formation (Cenomanian) of Kansas. *American Journal of Botany,* **73,** 1230–7.

Dilcher, D. L., Potter, F. W., & Crepet, W. L. (1976). Investigations of angiosperms from the Eocene of North America: juglandaceous winged fruits. *American Journal of Botany,* **63,** 532–44.

Doyle, J. A. (1973). Fossil evidence on early evolution of the monocotyledons. *Quarterly Review of Biology,* **48,** 339–413.

Doyle, J. A. (1978). Origin of angiosperms. *Annual Review of Ecology and Systematics,* **9,** 365–92.

Doyle, J. A., & Donoghue, M. J. (1986a). Relationships of angiosperms and Gnetales: a numerical cladistic analysis. In *Systematic and Taxonomic Approaches in Palaeobotany,* eds. B. A. Thomas & R. A. Spicer. Oxford: Oxford University Press, pp. 177–98.

Doyle, J. A., & Donoghue, M. J. (1986b). Seed plant phylogeny and the origin of angiosperms: an experimental cladistic approach. *The Botanical Review,* **52,** 321–421.

Doyle, J. A., & Donoghue, M. J. (1987). The origin of angiosperms: a cladistic approach. *In The Origin of Angiosperms and Their Biological Consequences,* eds. E. M. Friis, W. G. Chaloner, & P. R. Crane. Cambridge: Cambridge University Press.

Doyle, J. A., & Hickey, L. J. (1976). Pollen and leaves from the mid-Cretaceous Potomac Group and their bearing on early angiosperm evolution. In *Origin and Early Evolution of Angiosperms,* ed. C. B. Beck. New York: Columbia University Press, pp. 139–206.

Doyle, J. A., Van Campo, M., & Lugardon, B. (1975). Observations on exine structure of *Eucommiidites* and Lower Cretaceous angiosperm pollen. *Pollen et Spores.* **17,** 429–86.

Eames, A. J. (1959). The morphological basis for a Paleozoic origin of the angiosperms. *Recent Advances in Botany,* **1,** 721–5.

Esau, K. (1965). *Plant Anatomy,* 2nd ed. New York: Wiley.

Friis, E. M., Chaloner, W. G., & Crane, P. R. (1987). *The Origins of Angiosperms and Their Biological Consequences.* Cambridge: Cambridge University Press.

Friis, E. M., Crane, P. R., & Pedersen, K. R. (1986). Floral evidence for Cretaceous chloranthoid angiosperms, *Nature,* **320,** 163–4.

Friis, E. M., Crane, P. R., & Pedersen, K. R. (1988). Reproductive structures of Cretaceous Platanaceae. *Biologiske Skrifter,* **31,** 1–55.

Friis, E. M., & Crepet, W. L. (1987). Time of appearance of floral features. In *The Origin of Angiosperms and Their Biological Consequences,* eds. E. M. Friis, W. G. Chaloner, & P. R. Crane. Cambridge: Cambridge University Press.

Harris, T. M. (1932). The fossil flora of Scoresby Sound, East Greenland. 2: Description of seed plants *Incertae sedis* together with a discussion of certain cycadophytic cuticles. *Meddelelser om Grønland,* **85,** 4–7.

Harris, T. M. (1940). On *Caytonia* Thomas. *Annals of Botany* (N.S.), **4,** 713–34.

Harris, T. M. (1951). The relationships of the Caytoniales. *Phytomorphology,* **1,** 29–39.

Hickey, L. J. (1973). Classification of the architecture of dicotyledonous leaves. *American Journal of Botany,* **60,** 17–33.

Hickey, L. J., & Doyle, J. A. (1977). Early Cretaceous fossil evidence for angiosperm evolution. *Botanical Review,* **43,** 3–104.

Hickey, L. J., & Wolfe, J. A. (1975). The bases of angiosperm phylogeny; vegetative morphology. *Annals of the Missouri Botanical Garden,* **62,** 538–89.

Hill, C. R., & Crane, P. R. (1982). Evolutionary cladistics and the origin of angiosperms. In *Problems of Phylogenetic Reconstruction,* eds. K. A. Joysey & A. E. Friday. London: Academic Press.

Hughes, N. F. (1961). Further interpretation of *Eucommiidites* Erdtman 1948. *Paleontology,* **4,** 292–9.

Hughes, N. F. (1976). *Palaeobiology of Angiosperm Origins.* Cambridge: Cambridge University Press.

Hughes, N. F., & McDougall, A. B. (1987). Records of angiospermid pollen entry into the English Early Cretaceous succession. *Review of Palaeobotany and Palynology,* **50,** 255–72.

Just, T. (1948). Gymnosperms and the origin of angiosperms. *Botanical Gazette,* **110,** 91–103.

Krassilov, V. A. (1977). Contributions to the knowledge of the Caytoniales. *Review of Paleobotany and Palynology,* **24,** 115–78.

Krassilov, V. A., Shilin, P. V., & Vakhrameev, V. A. (1983). Cretaceous flowers from Kazakhstan. *Review of Palaeobotany and Palynology,* **40,** 91–113.

Long, A. G. (1966). Some Lower Carboniferous fructifications from Berwickshire, together with a theoretical account of the evolution of ovules, cupules, and carpels. *Transactions of the Royal Society of Edinburgh,* **66,** 345–75.

Long, A. G. (1977). Lower Carboniferous pteridosperm cupules and the origin of angiosperms. *Transactions of the Royal Academy of Edinburgh*, **70,** 13–35.

Mamay, S. H. (1969). Cycads: fossil evidence of late Paleozoic origin. *Science*, **164,** 295–6.

Mamay, S. H. (1976). Paleozoic origin of the cycads. *United States Geological Survey Professional Paper*, No. **934,** 1–47.

Meeuse, A. D. J. (1961). The Pentoxylales and the origin of the monocotyledons. *Proceedings of the Koninklijke Nederlandse, Akademie van Wentsenschappen c*, **64,** 543–59.

Meeuse, A. D. J. (1966). *Fundamentals of Phytomorphology*. New York: Ronald.

Meeuse, A. D. J. (1970). The descent of the flowering plants in the light of new evidence from phytochemistry and other sources. *Acta Botanica Neerlandica*, **19,** 61–72, 133–40.

Meeuse, A. D. J. (1975). Floral evolution in the Hammelididae. *Acta Botanica Neerlandica*, **24,** 155–79.

Meeuse, A. D. J. (1979a). Why were the early angiosperms so successful? A morphological, ecological and phylogenetic approach. *Proceedings of the Konicklijke Nederlandse, Akademie van Wentsenschappen c*, **82,** 343–69.

Meeuse, A. D. J. (1979b). *Anthocorm Theory*. Amsterdam: University of Amsterdam, H. deVries Laboratorium.

Melville, R. (1960). A new theory of the angiosperm flower. *Nature*, **188,** 14–18.

Melville, R. (1962). A new theory of the angiosperm flower. I. *Kew Bulletin*, **16,** 1–50.

Millay, M. A., & Taylor, T. N. (1976). Evolutionary trends in fossil gymnosperm pollen. *Review of Palaeobotany and Palynology*, **21,** 65–91.

Muhammad, A. F., & Sattler, R. (1982). Vessel structure of *Gnetum* and the origin of angiosperms. *American Journal of Botany*, **69,** 1004–21.

Muller, J. (1981). Fossil pollen records of extant angiosperms. *Botanical Review*, **47,** 1–142.

Muller, J. (1984). Significance of fossil pollen for angiosperm history. *Annals of the Missouri Botanical Garden*, **71,** 419–43.

Nishida, M. (1962). On some petrified plants from the Cretaceous of Choshi, Chiba Prefecture. *Japanese Journal of Botany*, **18,** 87–104.

Pocock, S. A. J., & Vasanthy, G. (1988). *Cornetipollis reticulata*, a new pollen with angiospermid features from Upper Triassic (Carnian) sediments of Arizona (U.S.A.),with notes on *Equisetosporites*. *Review of Palaeobotany and Palynology*, **55,** 337–56.

Puri, V. (1967). The origin and evolution of the angiosperms. *Journal of the Indian Botanical Society*, **56,** 1–14.

Retallack, G., & Dilcher, D. L. (1981a). Arguments for a glossopterid ancestry of angiosperms. *Paleobiology*, **7,** 54–67.

Retallack, G., & Dilcher, D. L. (1981b). Early angiosperm reproduction: *Prisca reynoldsii*, gen. et sp. nov. from mid-Cretaceous coastal deposits in Kansas, U.S.A. *Palaeontographica B*, **179,** 103–51.

Sahni, B. (1932). *Homoxylon rajmahalense* gen. et sp. nov., a fossil angiospermous wood, devoid of vessels, from the Rajmahal Hills, Behar. *Palaeontologia Indica*, **20,** 1–16.

Scott, R. A., Barghoorn, E. S., & Leopold, E. B. (1960). How old are the angiosperms? *American Journal of Science*, **258-A,** 284–9.

Smith, D. L. (1964). The evolution of the ovule. *Biological Reviews*, **39,** 137–59.

Stebbins, G. L. (1974). *Flowering Plants: Evolution Above the Species Level*. Cambridge: Harvard University Press.

Stebbins, G. L. (1976). Seeds, seedlings, and the origin of the angiosperms. In *Origin and Early Evolution of Angiosperms*, ed. C. B. Beck. New York: Columbia University Press, pp. 300–11.

Surange, K. R., & Chandra, S. (1971). *Denkania indica* gen. et sp. nov. – glossopteridean fructification from the Lower Gondwana of India. *Palaeobotanist*, **20,** 264–8.

Surange, K. R., & Chandra, S. (1972). *Lidgettonia mucoranata* sp. nov., a female fructification from the Lower Gondwana of India. *Palaeobotanist*, **21,** 212–26.

Takhtajan, A. L. (1954). *Essays on the Evolutionary Morphology of Plants*. Leningrad: Leningrad University.

Takhtajan, A. L. (1969). *Flowering Plants: Origin and Dispersal*. Edinburgh: Oliver.

Takhtajan, A. L. (1976). Neoteny and the Origin of Flowering Plants. In *Origin and Early Evolution of Angiosperms*, ed. C. B. Beck. New York: Columbia University Press, pp. 207–19.

Taylor, D. W., & Hickey, L. J. (1990). An Aptian plant with attached leaves and flowers: Implications for angiosperm origin. *Science*, **247,** 702–4.

Taylor, T. N. (1978). The ultrastructure and reproductive significance of *Monoletes* (Pteridospermales) pollen. *Canadian Journal of Botany*, **56,** 3105–18.

Taylor, T. N., & Millay, M. A. (1979). Pollination biology and reproduction in early seed plants. *Review of Palaeobotany and Palynology*, **27,** 329–55.

Thomas, H. H. (1925). The Caytoniales a new group of angiospermous plants from the Jurassic rocks of Yorkshire. *Philosophical Transactions of the Royal Society (London) B*, **213,** 299–363.

Thomas, H. H. (1934). The nature and origin of the stigma. *New Phytologist*, **33,** 173–98.

Thomas, H. H. (1955). Mesozoic pteridosperms. *Phytomorphology*, **15,** 177–85.

Thomas, H. H. (1957). Plant morphology and the evolution of flowering plants. *Proceedings of the Linnean Society (Botany)*, **168,** 125–33.

Tidwell, W. D., Simper, A. D., & Thayn, G. F. (1977). Additional information concerning the controversial Triassic plant *Sanmiguelia*. *Palaeontographica B*, **163,** 143–51.

Walker, W. J. (1976). Comparative pollen morphology and phylogeny of the ranalean complex. In *Origin and Early Evolution of Angiosperms*, ed. C. B. Beck. New York: Columbia University Press, pp. 241–99.

Walton, J. (1949). *Calathospermum scoticum* – an ovuliferous fructification of Lower Carboniferous age from Dunbartonshire. *Transactions of the Royal Society of Edinburgh*, **61,** 719–28.

Wettstein, R. R. von (1907). *Handbuch der systematischen Botanik*, 2nd ed. Leipzig: Franz Deuticke.

Wieland, G. R. (1916). *American Fossil Cycads*. Vol. II. Washington: Carnegie Institute.

Wilson, C. L. (1942). The telome theory and the origin of the stamen. *American Journal of Botany*, **29,** 759–64.

31

Angiosperms: diversification, radiation, and modernization

In Chapter 30 we have recorded the dramatic increase in variety and abundance of angiosperm fossils from their first appearance in the Hauterivian of the Lower Cretaceous to the Cenomanian (Chart 30.1). This is a record of rapid diversification and radiation of flowering plants so that by the mid-Cretaceous at least three of the six dicotyledonous subclasses, Magnoliidae, Hamamelidae, and Rosidae (*sensu* Cronquist, 1981) were established (Friis & Crepet, 1987; Crane, 1991). By the end of the Cenomanian, angiosperms had become the most diverse and floristically dominant group as evidenced by the composition of numerous macrofossils and pollen floras (Crabtree, 1987; Crane, 1987a; Muller, 1981; Tidwell, 1975; Wolfe, 1987a). Further increase in the radiation and modernization of angiosperms occurred during a period extending from the Upper Cretaceous to the end of the Eocene of the Tertiary (Chart 30.1). Much of the evidence for this rapid change comes from studies of fossil leaves, pollen, and flowers, studies that have been undertaken during the last three decades. Most of the late Lower Cretaceous angiosperm reproductive structures have affinities with the Magnoliidae, but it is not until the Cenomanian, with the discovery of *Archaeanthus* and *Lesqueria,* that reproductive and vegetative parts have been assigned to the order Magnoliales of the class Magnoliopsida.

Magnoliopsida (dicotyledons)

Magnoliidae

Magnoliales

New evidence in the form of a permineralized fructification (Nishida & Nishida, 1988) adds further support to the expected presence of the Magnoliales in the early Upper Cretaceous. The specimen, assigned to the new genus *Protomonimia* (Fig. 31.1A), is Turonian in age from Hokkaido, Japan. It consists of approximately 50 helically arranged follicles (Fig. 31.1B), each of which is derived from a conduplicate carpel attached to a concave receptacle. An adaxial longitudinal slit extends from near the base to the apex of the follicles. On either side of the slit there is a longitudinal ridge with clavate to spatulate hairs on the outside. It is apparent that these structures functioned as part of a stigmatic surface. Each follicle is provided with major vascular bundles, two ventral and one dorsal as in conduplicate carpels of other magnolioids. There are 12 to 15 bitegmic ovules per follicle (Fig. 31.1C), and these are arranged alternately on either side of the stigmatic ridges. The secondary wood of *Protomonimia* contains vessels with steeply inclined scalariform perforation plates and scalariform and bordered pits on the intervessel walls.

After comparing *Protomonimia* to other mid-Cretaceous magnoliids including *Archaeanthus* and *Lesqueria,* Nishida & Nishida (1982) conclude that these genera are part of a transitional group exhibiting a mosaic of primitive and more advanced features from which extant Magnoliales evolved.

Princetonia allenbyensis, first described by Stockey (1987) from the Middle Eocene of British Columbia, Canada, is thought to be an aquatic representative of the Magnoliidae allied with the Nymphaeales. Subsequent investigations (Stockey & Pigg, 1991) of permineralized *Princetonia* specimens have revealed fruits (Fig. 31.4), seeds, flowers, and inflorescences. The latter are racemes with up to 13 bisexual flowers in a helical arrangement. Each flower bud (Fig. 31.3) consists of two sepals and four or five petals that envelop 33 to 41 elongated tetrathecal anthers. These contain psilate pentacolpate pollen grains. The fruits are three to five loculate capsules (Fig. 31.4). Seed anatomy resembles that of the Nymphaeaceae, but the shape and absence of an operculum differ from seeds of most extant nymphaeaceous taxa. The combination of reproductive characteristics shown by *Princetonia*

Figure 31.1. *Protomonimia.* **A.** Lateral view of specimen showing follicles attached to receptacle. **B.** Transverse section of fructification with spirally arranged follicles. **C.** Longitudinal section showing follicles attached to concave receptacle. Note row of ovules in each follicle. Mid-Cretaceous. (From Nishida & Nishida, 1988, with permission.)

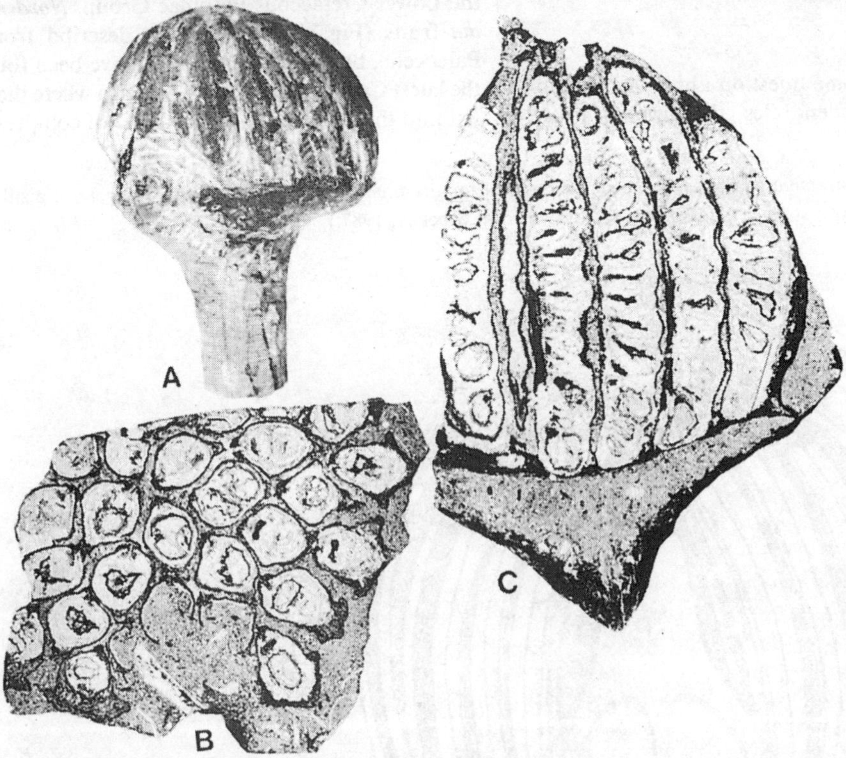

Figure 31.2. **A.** *Trochodendron araliodes.* Lateral view of young fruit with carpels. Extant. **B.** *Nordenskioldia borelais.* Transverse section through permineralized fruit containing 15 fruitlets. Paleocene. (From Crane, 1987b, with permission.)

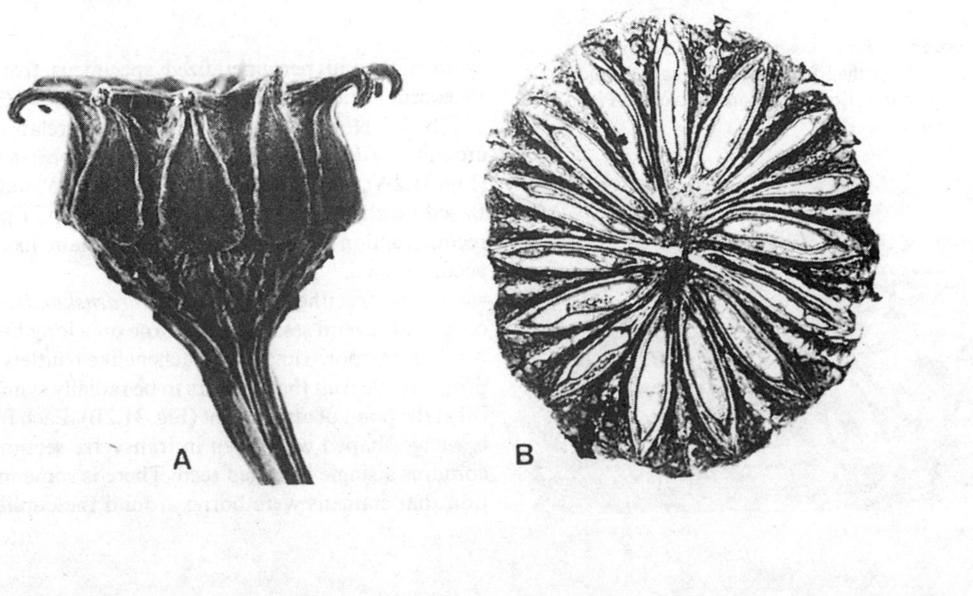

is not known to occur among present day angiosperms. It has been concluded that this genus represents an extinct family of the Magnoliidae.

Hamamelidae

Trochodendrales

Although there is some question about the relationships of the Trochodendrales, the order is usually placed with the Hamamelidae. In our earlier presentation of angiosperms (Chapter 30), we indicated that leaves of the order have a fossil record going back to the Lower Cretaceous Potomac Group. *Nordenskioldia* fruits (Fig. 31.2B) were first describd from the Paleocene, but subsequently they have been found in the latest Cretaceous of North America where they persist into the Miocene. Recent studies of compression–impression and permineralized specimens from the Paleocene (Crane, Dilcher, & Manchester, 1986; Crane, 1987b) of North America confirm the relationship among *Nordenskioldia* fructifications, *Trochodendron* (Fig. 31.2A), and *Tetracentron* of the Trochodendrales. Based on the evidence for this relationship, a partial reconstruction of the *Nordenskioldia* plant has been accomplished.

Figure 31.3. *Princetonia allenbyensis.* Reconstruction of longitudinal section of flower bud. Carpel (c); pistil (p); sepal (s); petal (pt); stamen (st). Eocene. (Redrawn from Stockey, 1987.)

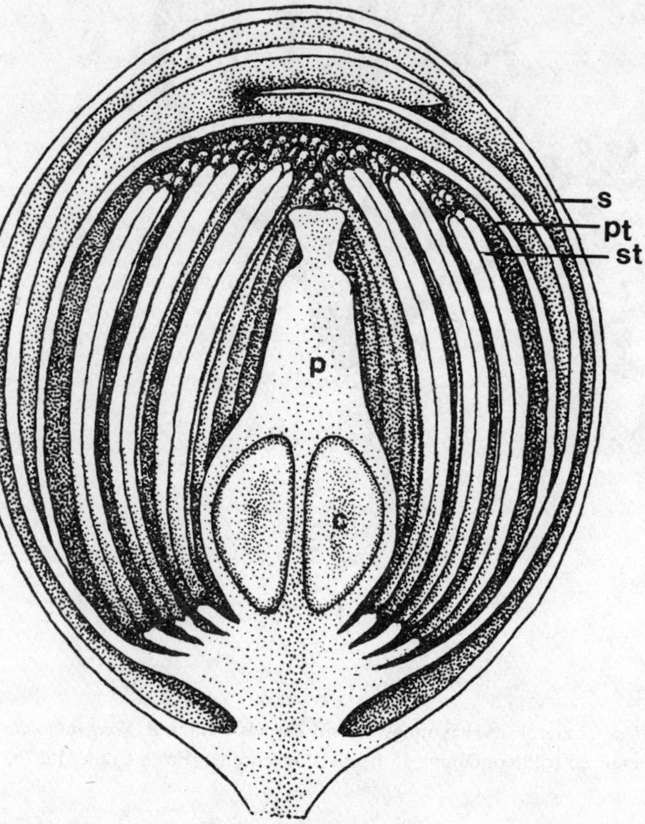

Figure 31.4. *Princetonia allenbyensis.* Transverse section of silicified fruit with four carpels and enclosed seeds. Eocene. (From Stockey & Pigg, 1990, with permission.)

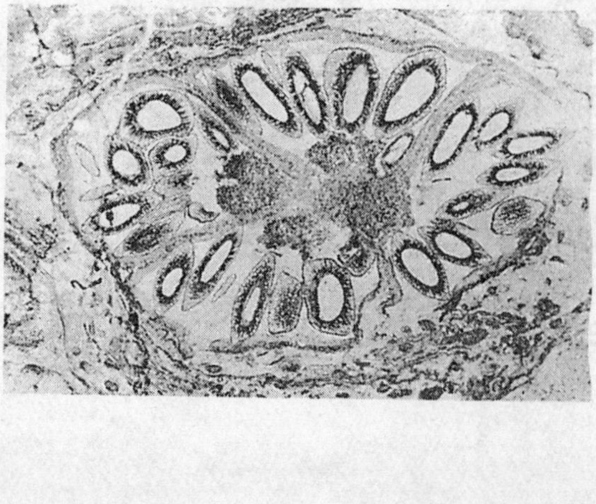

The fructifications of the *Nordenskioldia* plant consist of several sessile fruits borne on a long branch. A whorl of approximately 15 achenelike fruitlets comprise a single fruit that appears to be radially symmetrical at the point of attachment (Fig. 31.2B). Each fruitlet is wedge shaped when seen in transverse section and contains a single flattened seed. There is some indication that stamens were borne around the cuplike receptacle on which the fruits were borne. This would suggest that the flowers of the *Nordenskioldia* plant were bisexual. Nothing is known about its pollen or perianth.

Anatomical similarities between fructification axes and vegetative shoots of the *Nordenskioldia* plant have been used to establish the connection between the two. The vegetative axes consist of long and short shoots. Both have secondary wood that is vesselless. The leaves of the plant, plus the short shoots on which they probably were borne, are similar to extant *Tetracentron* and *Cercidiphyllum*. According to Crane (1987b), when all of its characteristics are considered, *Nordenskioldia* is more closely related to the Trochodendraceae than the Tetracentraceae.

Buxaceae

A new genus (*Spanomera*) has been established by Drinnan et al. (1991) as beautifully preserved, three-dimensional mummified flowers (Fig. 31.5A,B) from localities of early Cenomanian age of eastern North America. The floral units that were organized into an inflorescence have terminal pistillate and lateral staminate flowers. Each staminate flower (Fig. 31.5A,B) has five tepals, each of which is opposed to a stamen (Fig. 31.5C) with two pairs of pollen sacs. The stamens contain pollen comparable to the dispersed pollen *Striatopollis*. Pistillate flowers (Fig. 31.5A) have four tepals that are opposed to two carpels. The carpels are slightly fused at the base and are semicircular in outline. Each carpel has a long decurrent papillate ventral stigma (Fig. 31.5D).

From their study of *Spanomera*, Drinnan et al. suggest that these flowers provide additional evidence for the occurrence of unisexual, probably insect pollinated flowers among other mid-Cretaceous non-magnoliid dicotyledons. The details of flower structure leave little doubt that *Spanomera* is related to the Buxaceae. The position of this family, however, among the angiosperms is far from clear. At the generic level, the authors suggest that there is a close relationship among *Spanomera* and *Myrothamnus*, *Cercidiphyllum*, and *Tetracentron* of the "lower" Hamamelidae.

Hamamelidales

The family Cercidiphyllaceae of this order consists of one extant genus, *Cercidiphyllum*, and several extinct genera including *Joffrea* and *Nyssidium*. They comprise a small group now placed in the subclass Hamamelidae. Fossil leaves assigned to *Cercidiphyllum* occur with some frequency in the Upper Cretaceous. They become very abundant in the Paleocene of the lower Tertiary and persist until the late Eocene at middle and high paleolatitudes of North America. Recent investigations of *Cercidiphyllum*-like leaves and associated development and reproductive stages (Stockey & Crane, 1983; Crane & Stockey, 1985, 1986a) have resulted in one of the most completely understood and reconstructed of all Tertiary plants (Fig. 31.6). The fossil material was obtained from the Paskapoo Formation (late Paleocene) at Joffre Bridge near Red Deer, Alberta, Canada. The generic name *Joffrea* selected for the "whole plant" was derived from the bridge site. The assignment of the *Cercidiphyllum*-like leaves to the genus *Joffrea* rather than to the extant genus *Cercidiphyllum* is in accordance with the generally adopted concept that fossil leaves should not be assigned to extant genera, because this assumes that all characteristics of both fossil and extant plants represented by their leaves will be the same at the generic level. Such has not proved to be the case where "whole plants" have been reconstructed from the fossil record and compared to their supposed extant counterparts.

Joffrea speirsii is known from seedlings (Fig. 31.7A), shoots, leaves, pistillate inflorescences with carpels (Fig. 31.7B), follicles (Fig. 31.7C), seeds (Fig. 31.7D), and possibly staminate inflorescences. The evidence for including the different organs in a single binomial includes attachment of inflorescences and leaf petioles to long and short shoots, developmental intermediates between carpels and follicles, expulsion of seeds from follicles, preserved germinating seeds, stages in seedling development, similarities between adult foliage and seedlings, and association of dispersed parts at the site (Crane & Stockey, 1985).

Similarities between *Joffrea* and extant *Cercidiphyllum* include (1) opposite phyllotaxy; (2) leaf architecture with major venation of the actinodromous type (Fig. 2.7A) and minute glands at the apices of crenations on the leaf margins; (3) winged seeds with a distinctive looped raphe (Fig. 31.7D); (4) well-developed long and short shoots (Fig. 31.8A,B); and (5) the general organization of the pistillate inflorescence (Fig. 31.7B) which is a raceme.

The major difference between the two genera is in their short shoots and the inflorescences they produce. In the fossil *Joffrea* the short shoots were monopodial and had a terminal bud with the inflorescence (raceme) in the leaf axil. The raceme axis was as much as 130 mm long and produced up to 40 follicles per axis (Fig. 31.9B). By comparison *Cercidiphyllum* produces its fructifications terminally on a sympodial short shoot (Fig. 31.9A). The raceme axis has a maximum length of 25 mm and produces 2 to 8 follicles. It is apparent that the inflorescence of *Cercidiphyllum* evolved by reduction in length of the raceme axis from an elongated form of the *Joffrea* type (Crane & Stockey, 1986a). Comparison of *Joffrea* with other fossil plants of similar types indicates that there was considerable diversity among the extinct members of the Cercidiphyllaceae.

The presence of large numbers of leaves, seeds, and stages in their development in fluvial depositional environments leads to the conclusion that the *Cercidiphyllum*-like plants were early colonizers of disturbed habitats and were probably important Tertiary weeds (Crane, 1987a).

Platanaceae
From the time when evidence of the Platanaceae was first discovered in the Albian of the Lower Cretaceous, the family has become an important part of the fossil record of flowering plants in the Upper Cretaceous and Tertiary. By the Eocene the modernization of the family had become more-or-less complete as evidenced by the "whole plant" reconstruction (Fig. 31.10) of the extinct "Clarno plane tree" (Manchester, 1986a). This tree is similar in many respects to the extant plane tree, *Platanus* (sycamore), but differs in several features that warrant assigning it to the new genus and species, *Macginitiea angustiloba*. The "whole plant" *M. angusti-*

Figure 31.5. *Spanomera mauldinensis*. **A.** Inflorescence unit with terminal pistillate flower with two carpels (cp). **B.** Inflorescence unit with staminate flower (st) in the axil of a bract (b). **C.** Ventral view of stamen with two pollen sacs and short filament. **D.** Oblique view of carpel showing the ventral suture (vs) flanked by long stigmatic surfaces. Mid-Cretaceous. (From Drinnan et al., 1991, with permission.)

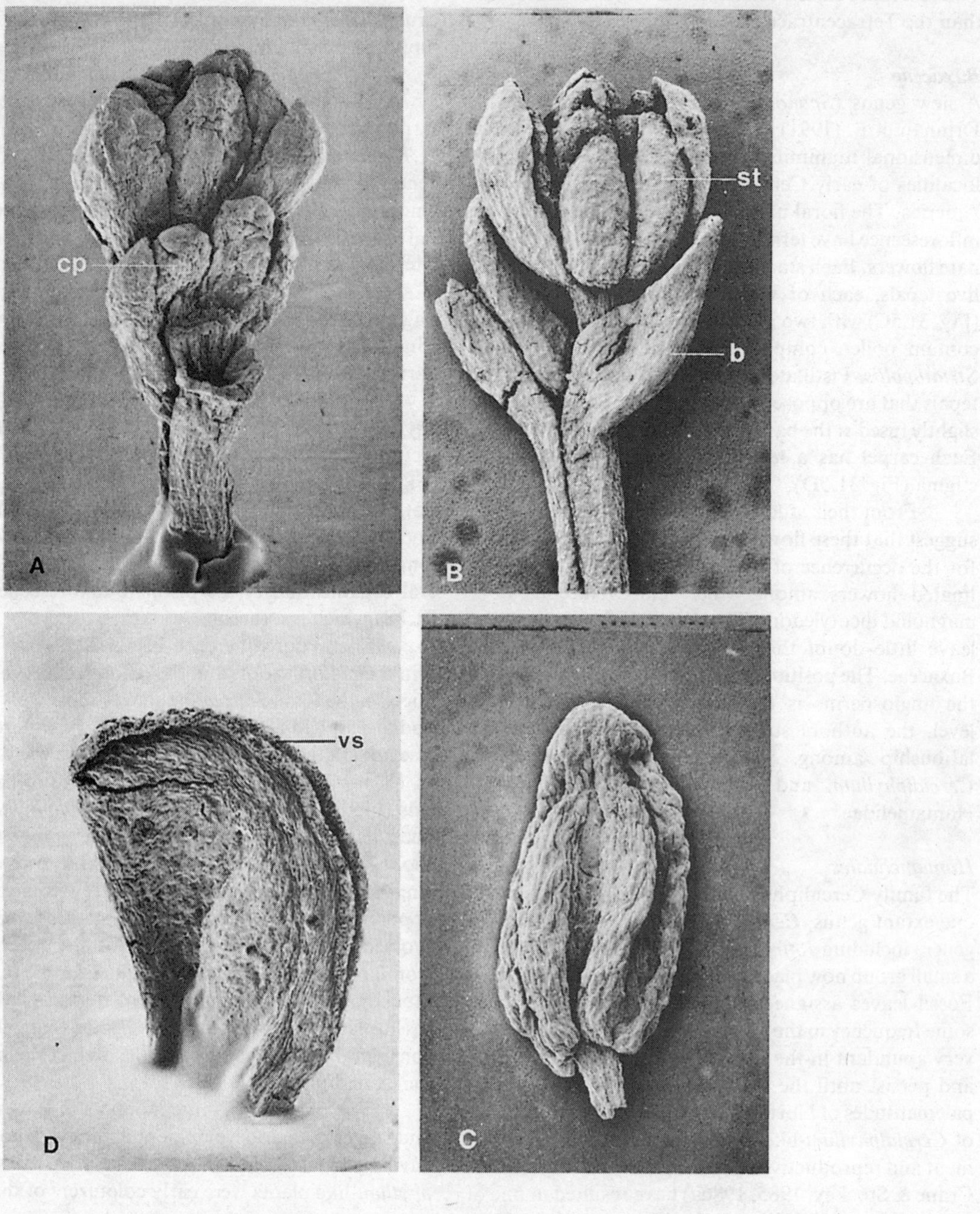

Figure 31.6. *Joffrea speirsii*. **A.** Reconstruction of pistillate inflorescences and developing leaves attached to a short shoot. **B.** Reconstruction of a long shoot with short shoots, leaves, and follicles. Paleocene. (From Crane & Stockey, 1985, with permission; Polyanna Quasthoff, artist.)

loba is known from leaves (Fig. 31.11), petiole and stem anatomy, pistillate inflorescences (*Macginicarpa*), fruits, staminate inflorescences (*Platananthus*), and pollen. The assembly of parts for the reconstruction was based on shared platanaceous characteristics and degree of association of dispersed parts in the field.

Studies of the reproductive features of the *Macginitiea* plant and those of extant *Platanus* show that the fossil combines both archaic and specialized features. The inflorescences, which are globose heads (Fig. 31.10B,F) borne on separate staminate and pistillate axes, have florets with well-developed perianths and whorls of five carpels or stamens (Fig. 31.12A,B). In these features the *Macginitiea* plant differs from *Platanus*, which has poorly developed perianths with a variable number of stamens and carpels. Another conspicuous difference is the presence of fruit dispersal hairs associated with the fruits of *Platanus* and their absence in the fossil. It has been discovered that the pollen of the *Macginitiea* plant is smaller than that of *Platanus*, suggesting that the *Macginitiea* plant was not well adapted for wind pollination and that insects may have been the vector. The pollen is tricolpate and reticulate (Fig. 31.10D).

Figure 31.7. *Joffrea speirsii*. **A.** Seedling with first and second leaves. **B.** Inflorescence with carpels. Note the curved tips of stigma and style. **C.** Mature inflorescence with widely spaced follicles. **D.** Seed showing characteristic raphe (r). Paleocene. (From Crane & Stockey, 1985, with permission.)

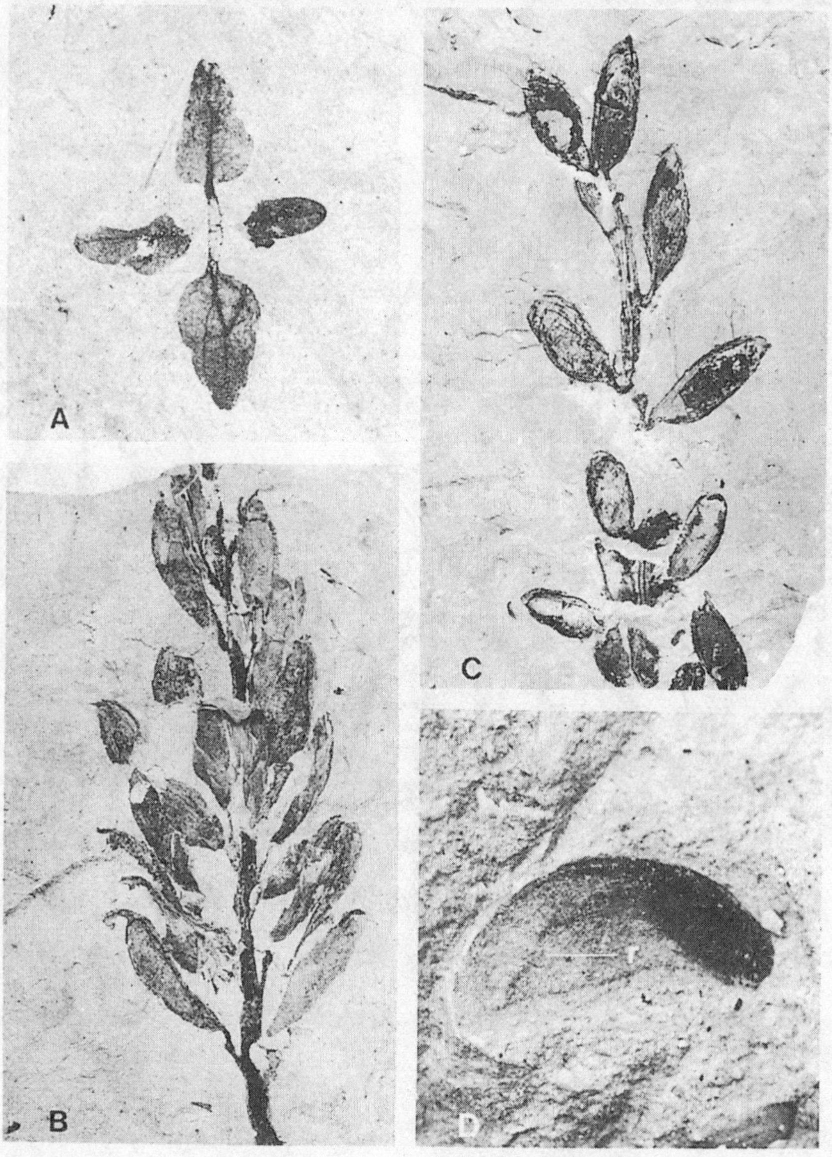

A more recent comprehensive investigation describing a new platanoid plant (Pigg & Stockey, 1991) has further expanded our understanding of fossil representatives of the Hamamelidales. The material comprises a suite of compression–impression fossils from the Paleocene age locality at Joffre Bridge, Alberta, Canada. Included here are leaves similar to *Platanus nobilis,* pistillate inflorescences (Fig. 31.13A), and infructescences (Fig. 31.13C) placed in a new taxon *Maginicarpa manchesteri,* staminate inflorescences assigned to another new taxon *Platananthus speirsae* (Fig. 31.13B), isolated stamens containing platanoid pollen, and young seedlings in several stages of development (Fig. 31.13D). These disarticulated organs are believed to represent the same platanoid plant because of their exclusive co-occurrence at the Joffre Bridge site. Including the *Cercidiphyllum*-like *Joffrea speirsii* (Crane & Stockey, 1985a) the new *Joffrea* is now one of the best known of Paleocene angiosperms. As emphasized by Pigg & Stockey (1991), the establishment of multiorgan reconstructions clearly increases the opportunity for understanding the evolution of the Platanaceae.

Figure 31.8. *Joffrea speirsii.* **A.** Long shoot with probable young staminate inflorescence. **B.** Twig with a pair of short shoots. Paleocene. (From Crane & Stockey, 1985, with permission.)

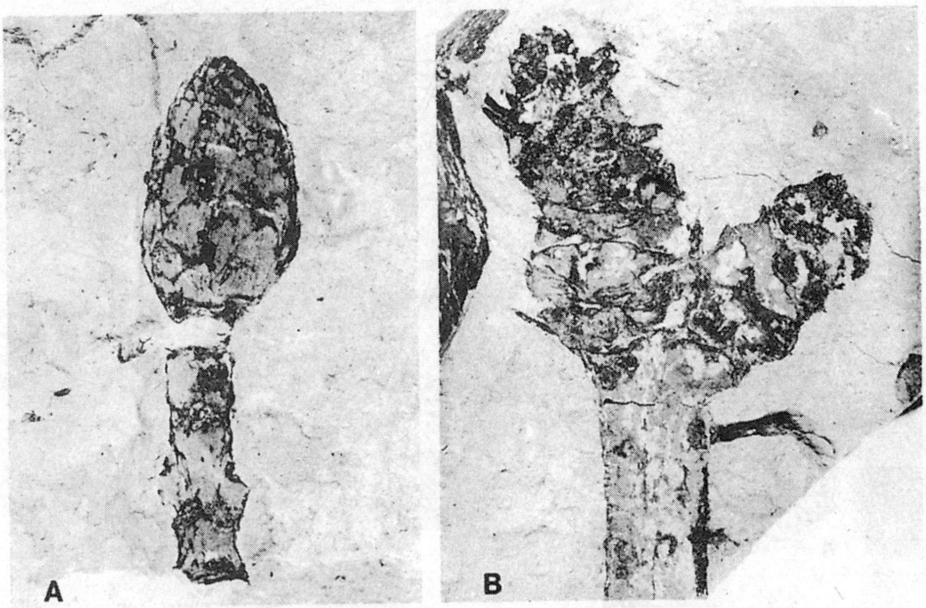

Figure 31.9. Diagrams showing shoot morphology of extant and fossil Cercidiphyllaceae. **A.** *Cercidiphyllum japonicum* (extant) with sympodial short shoot, single leaf, axillary bud, and terminal infructescence. **B.** *Joffrea speirsii* (Paleocene) with sympodial short shoot, opposite and decussate leaves, terminal bud, and axillary infructescence. (From Crane & Stockey, 1985, with permission.)

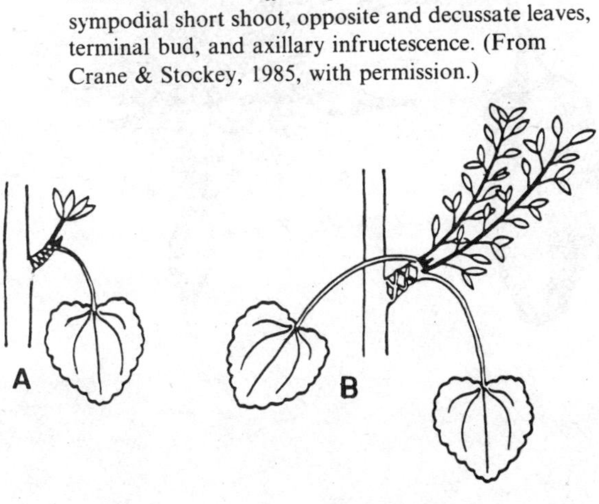

As the evolution of the Platanaceae proceeded from their first appearance in the Albian of the Lower Cretaceous, their leaf morphotypes became quite diverse (Crabtree, 1987) so that there are some that are pinnately compound, others that have pinnate instead of palmate venation, and several that differ in their minor venation patterns. General stratigraphic trends have been utilized to reveal the evolution of reproductive structures. Here, there is a general decrease in perianth and pollen size, an increase in inflorescence size, and a reorganization of branches bearing the reproductive structures. To date and including previous investigations by other authors (Manchester, 1986a), the family Platanaceae is now one of the best known groups of dicotyledonous

Figure 31.10. Diagrammatic reconstruction of the Clarno plane tree (*Macginitiea angustiloba* plant). **A.** Branch with leaf of *M. angustiloba* and pistillate and staminate flowering axes. **B.** Staminate inflorescence of *Platananthus synandrus* with infructescences shedding stamens. **C.** Group of stamens. **D.** Tricolpate, reticulate pollen grain from *Platananthus*. **E.** Branch with leaf and mature pistillate infloreseene. **F.** *Macginicarpa glabra*, showing fruit dispersal. **G.** Single dispersed fruit (achene). **H.** Transverse section of fruit containing a single seed. **A–H:** Eocene. (From Manchester, 1986a, with permission.)

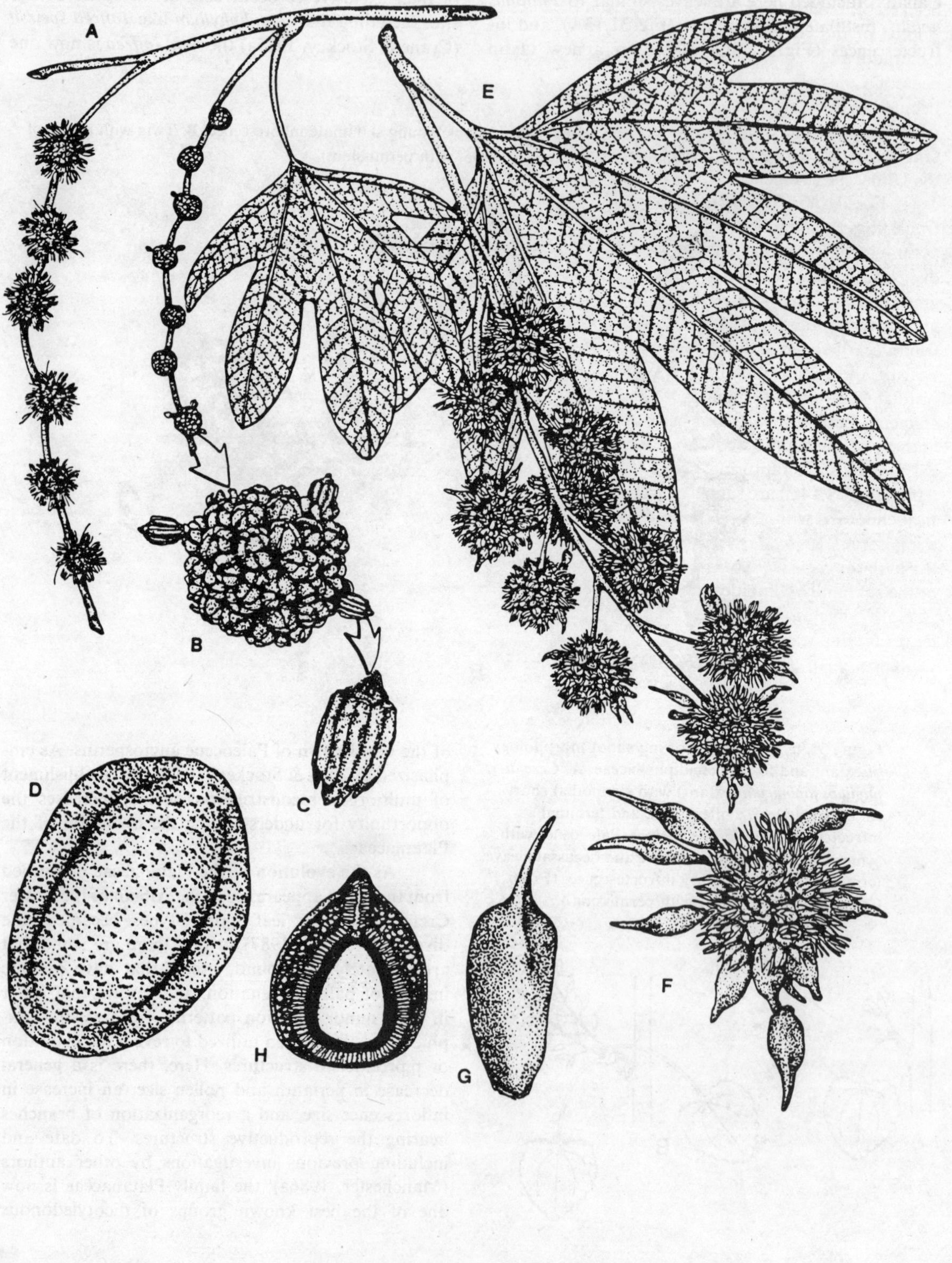

angiosperms in the fossil record (Pigg & Stockey, 1991).

Urticales
Exceedingly well-preserved compression–impression fossils of the genus *Cedrelospermum* (Fig. 31.14A) have been recovered by Manchester (1989) from the middle Eocene of the Green River Formation in Utah and Colorado and from the early Oligocene Florissant beds of Colorado. This extinct genus is widely distributed in Tertiary beds of Europe as well as North America. The fruits assigned to *Cedrelospermum* represent an extinct genus in the subfamily Ulmoideae of the Ulmaceae (Manchester, 1986b).

Fossil branches (Fig. 31.14A) with attached fruits, flowers, and foliage have made possible a "whole plant" reconstruction. The leaves (Fig. 31.14B) are distichously arranged and pinnately veined with margins that are serrate to entire. The axillary flowers appear to be unisexual, with one specimen showing fruits in early stages of development. Another specimen bearing the same kind of leaves as the twigs with pistillate flowers has staminate flowers with 4 to 6 stamens. Perianth parts are poorly preserved. The anthers of the stamens contain 3 to 5 porate pollen with verrucate sculpturing. The pedicellate axillary fruits (Fig. 31.14C) are samaras composed of an elliptical endocarp with a large laterally placed wing and a small secondary wing. An analysis of the characteristics of the extinct *Cedrelospermum* and extant members of the subfamily Ulmoideae places this genus close to *Zelkova,* which is represented in the fossil record by leaves first appearing in the Paleocene. The first appearance of an extant genus of the Ulmoideae that has been verified by both fruits and foliage, belongs to *Ulmus* (elm) from the early Eocene.

Figure 31.11. *Macginitiea angustiloba,* compression fossil of complete leaf. Eocene. (From Manchester, 1986a, with permission.)

Amentiferae
The Amentiferae is an informal taxon of some historical interest which includes some orders of the Hamamlidae with small, unisexual, simple wind-pollinated flowers often borne in inflorescences called catkins (aments). The orders usually included in the Amentiferae are the Casuarinales, Fagales, Juglandales, and Myricales. Catkinlike inflorescences with pollen-producing florets have been reported by Dilcher (1979) from the Cenomanian of the mid-Cretaceous. This is one of the first floral structures to be described that

Figure 31.12. **A.** *Macginicarpa glabra,* longitudinal section of silicified head showing single-seeded carpels (c) in florets with enveloping perianths (p). **B.** *Platananthus synandrus* showing florets with stamens and enveloping perianths (p). **A,B:** Eocene. (From Manchester, 1986a, with permission.)

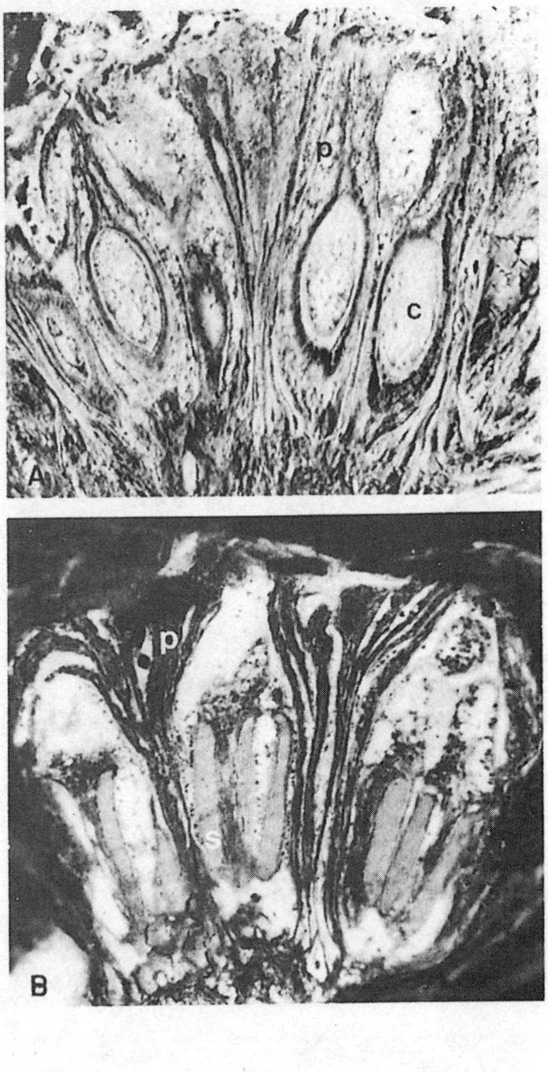

was suited for wind pollination as it occurs among the Amentiferae.

Juglandales

A report by Crepet, Dilcher, & Potter (1974) reveals a variety of mid-Eocene floral types, including catkins with reduced perianths some of which show characteristics of the Juglandaceae (walnut family) (Crepet, Dilcher, & Potter, 1985). The catkins assigned to the genus *Eokachyra* (Fig. 31.15) have helically arranged staminate flowers. Each flower is bilaterally symmetrical and has a conspicuous perianth and a three-lobed bract. Triporate pollen has been isolated from the stamens. The discovery of these and other flowers adapted for wind pollination supports the idea that wind-pollination strategies were well developed at least by the mid-Eocene and may have evolved as early as the Upper Cretaceous (Dilcher, 1979).

The distinctive winged fruits of some Juglandaceae are also present in the Eocene and other younger Tertiary floras. Those described by Dilcher et al. (1976) were similar to the extant juglandaceous genera, *Engelhardtia* and *Orromunnea*. The latter prompted the generic name *Paraoreomunnea* (Fig. 30.18A) for the fossil specimens. The wing of the fruit is a three-lobed structure with the central lobe larger than the two laterals. An abaxial bract attached to the base of the winged fruit covers the nut with its stigma and style.

Although first appearing in the mid-Cenomanian, a pollen complex called the Normapolles became

Figure 31.13. **A.** *Macginicarpa manchesteri.* Reproductive axis bearing globose pistillate infructescences. **B.** *Platananthus speirsae.* Reproductive axis with staminate infructescences. **C.** *Macginicarpa* pistillate infructescence enlarged. Note the curved persistent styles projecting from the head. **D.** Young seedling with two cotyledons showing petioles and venation. **A–D:** Paleocene. (From Pigg & Stockey, 1991, with permission.)

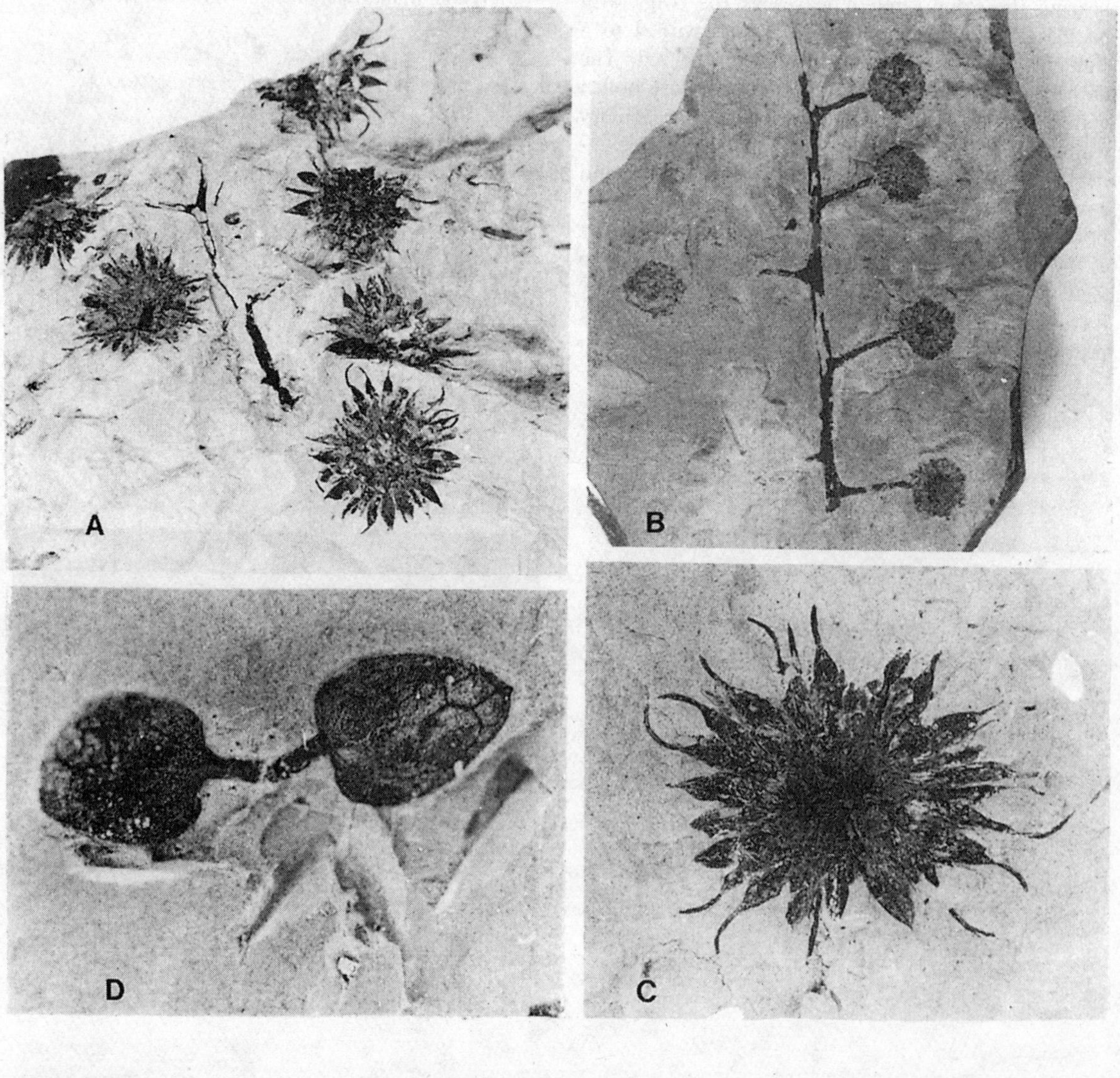

more abundant during the Turonian–Campanian interval attaining maximum diversity during the Santonian. Typical Normapolles pollen grains are triangular, and triporate with complex pore and wall structures (Fig. 31.16A). This pollen is a predominant feature of North American and European palynofloras and is part of a province that is delimited on the north and south by climatic boundaries. In spite of its great abundance during the Turonian–Campanian interval, very little was known about its affinities until 1983 when Friis found three-dimensional preserved mummified flowers containing Normapolles pollen (Fig. 31.16B) from the Upper Cretaceous of southern Sweden. The flowers with which the pollen was associated compared favorably with members of the Hamamelidae, especially the Myricales and Juglandales. The suggestion has been made that the extant hamamelids that have, for the most part, imperfect wind-pollinated flowers were derived from a complex producing Normapolles pollen.

Fagales

The Fagales, which includes the families Betulaceae and Fagaceae, is well represented by fossils from the Paleocene to the present. For the Betulaceae (birch and alder) the earliest-known, now-extinct genus is *Palaeocarpinus* from the Paleocene (Crane, 1984). The most fully documented record of *Betula* is from the Middle Eocene of British Columbia, Canada (Crane & Stockey, 1986b). Compression–impression fossils from the Allenby Formation comprise pistillate fructification, fruits, staminate inflorescences, pollen, and leaves. Based on association of these structures at the site and independently determined relationships, the authors suggest that these vegetative and reproductive structures belong to the fossil *Betula leopoldae*.

Difficulties are encountered in the assignment of leaves belonging to extant and extinct genera of the Betulaceae because of their great variability. The leaves of *Betula* (Fig. 31.17A,B) at the locality also exhibit great diversity in size, shape, and morphology

Figure 31.14. **A.** *Cedrelospermum nervosum*. Twig with serrate leaves and well-preserved axillary samaras. **B.** *C. nervosum* leaf with blunt, simple teeth and prominent midvein. **C.** *C. nervosum* fruit (samara) impression fossil showing striated endocarp (e) and primary (p) and small secondary (s) wings. **A–C:** Eocene. (From Manchester, 1989, with permission.)

of the teeth. In spite of this, Crane & Stockey believe this variability is intraspecific and that these are the leaves of *B. leopoldae*. The pistillate catkins are composed of spirally arranged trilobed bracts (Fig. 31.17C) and small, flattened fruits with a lateral flange and

Figure 31.15. *Eokachyra aeolia,* reconstruction of portion of catkin with staminate flowers. Eocene. (Redrawn From Crepet, Dilcher, & Potter, 1985.)

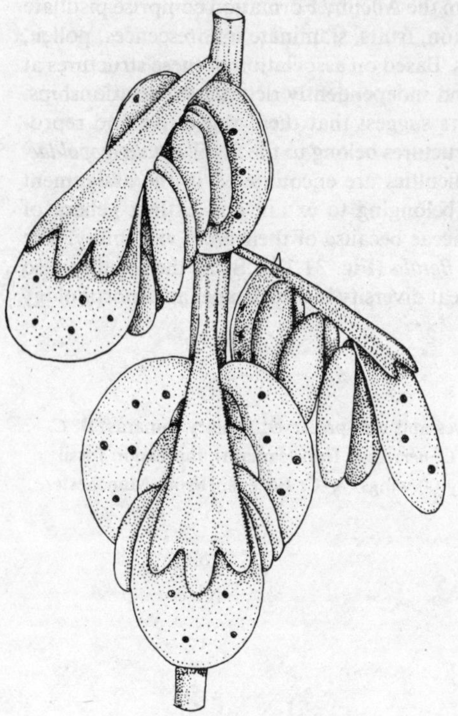

two short styles (Fig. 31.17D). The trilobed bracts are diagnostic for *Betula,* but fruit morphology is found in extant species of *Alnus* and *Betula*. The staminate catkins (Fig. 31.17E) are composed of 30 to 40 staminate flowers, each consisting of a triangular primary bract, secondary bracteoles, and stamens. Pollen isolated from the stamens is similar to extant species of *Betula* (Fig. 30.8E).

The information gained from *B. leopoldae,* supplemented by previous macrofossils and palynological studies, suggests that the tribe Betulae may have differentiated in the Upper Cretaceous. Both *Betula* and *Alnus* were certainly present by the Middle Eocene.

Further evidence of the Fagales is provided by staminate catkins (Fig. 31.18) with characteristics of extant *Castanea* (chestnut) of the Fagaceae (Crepet & Daghlian, 1980). These catkins are large, up to 9 cm in length, with spirally arranged dichasia each consisting of three florets subtended by bracts. The individual florets have a floral envelope terminating in five to six lobes and up to ten stamens. These Middle Eocene catkins, named *Castaneoidea,* were probably insect pollinated as are the flowers of the modern chestnut.

The fossil record of leaves and pollen of *Fagus* (beech) and *Quercus* (oak) are not adequate to reveal the time of origin and their relationships within the Fagaceae. However, a spectacular display of exceptionally well-preserved flowers of *Quercus, Fagus,* and *Castanea* (Fig. 31.19) have been found in Baltic amber. The deposits containing the greatest number of floral parts are probably of Late Eocene or Oligocene age. *Quercus* acorns and *Fagus* fruits have also been found

Figure 31.16. **A.** Three types of Normapolles pollen. Mid-Cretaceous to Eocene. (Redrawn from Thomas & Spicer, 1986.) **B.** *Antiquocarya,* reconstruction of a flower type producing Normapolles pollen. Upper Cretaceous. (Redrawn, based on Friis 1983; Friis & Crepet, 1987.)

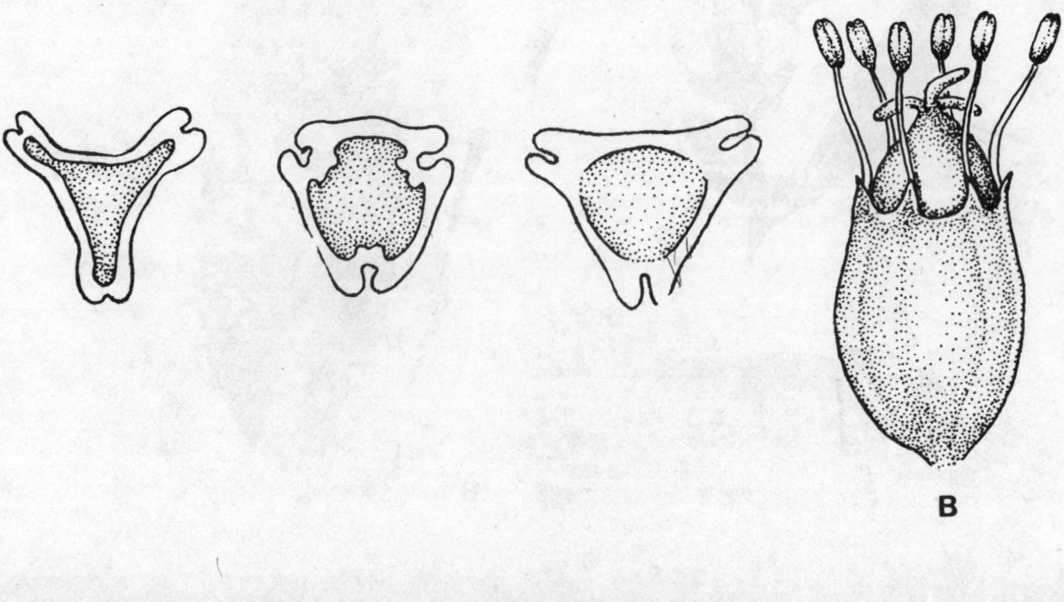

Figure 31.17. **A,B.** leaves of *Betula leopoldae* showing variation in physiognomy. **C.** Single dispersed trilobed bract from psillate infructescence. **D.** Dispersed fruit showing two apical styles. Note narrow marginal wing (w). **E.** Staminate inflorescence showing pedicel. **A–D:** Middle Eocene. (From Crane & Stockey, 1986b, with permission.)

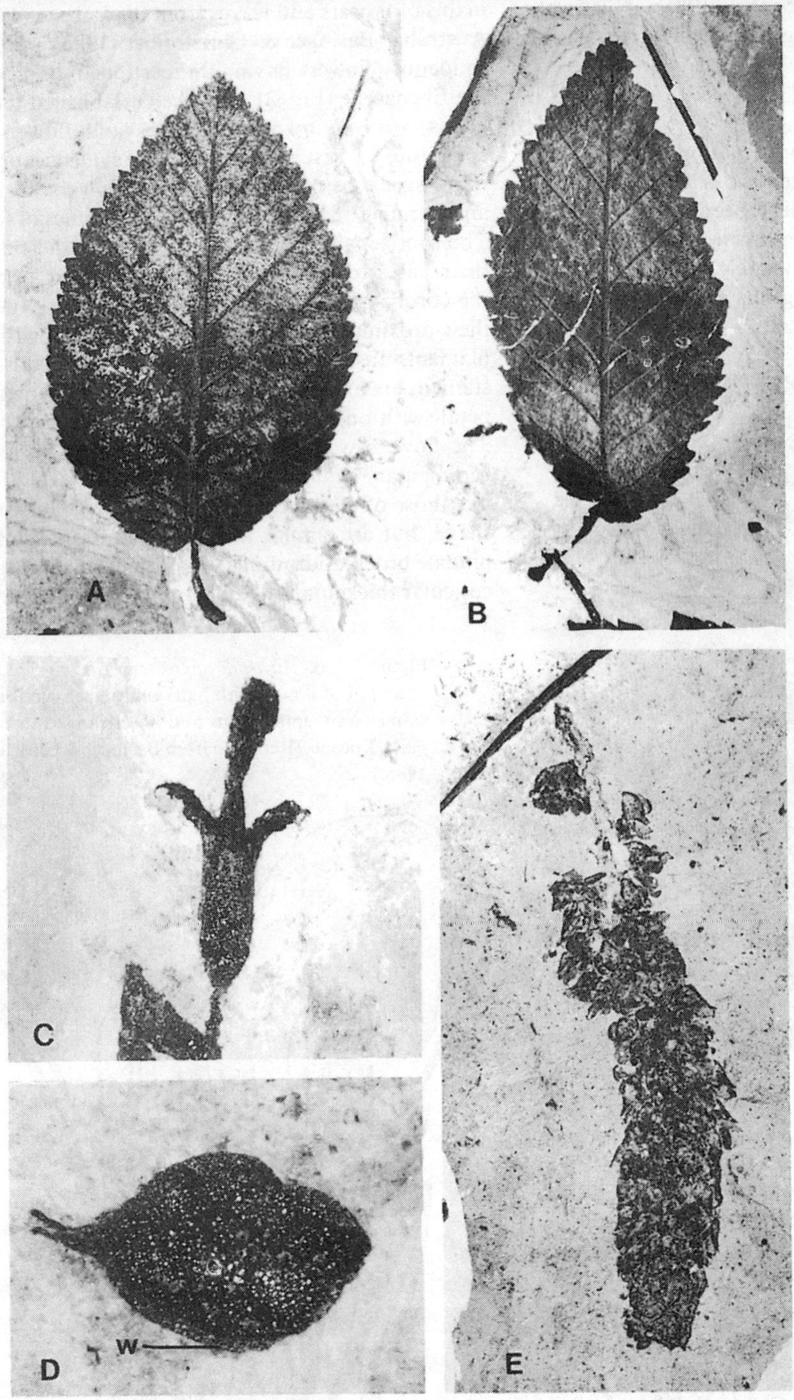

in other Eocene localities. Mummified leaves, cupules, and seeds of the Southern Hemisphere *Nothofagus* have been recovered from a Miocene age locality in Victoria, Australia (Christophel, 1985). This is the first report of well-preserved *Nothofagus* megafossils from Australia.

Dilleniidae

Perhaps the best representation of fossil members of the subclass Dilleniidae is from the record of the Ebenaceae. There have been many reports of fossil leaves belonging to members of this family, for example *Diospyros*. Unfortunately, the characteristics of the leaves of this large genus are of a generalized nature making the unequivocal assignment of fossil leaves to the Ebenaceae almost impossible. In a recent investigation of beautifully preserved mummified remains of flowers and leaves from the Eocene of South Australia, Basinger & Christophel (1985) were able to identify flowers having unquestioned affinity with the Ebenaceae (Fig. 31.20). They established the new genus *Austrodiospyros* for leaves and flowers that constitute the first well-documented evidence for early floral structure of the family. The flowers are small and tetramerous (parts in fours or multiples of fours). The four sepals are triangular in shape and fused at their bases to form a calyx cup. The four petals of the corolla alternate with the sepals and are fused in their proximal third. There are 16 stamens with their filaments fused at their bases with the petals. The stamens are arranged in radial pairs, four stamens per petal, with one short and one long stamen per pair. The flowers were all staminate with the occasional rudimentary ovary in the superior position. The leaves like those of *Diospyros* are highly variable in size and shape, but are similar in having entire margins and pinnate brochidodromous venation. Peculiar ridgelike cuticular thickenings of epidermal cells made identifi-

Figure 31.18. *Fagus*-type staminate catkin. Eocene. (From Crepet, 1979.)

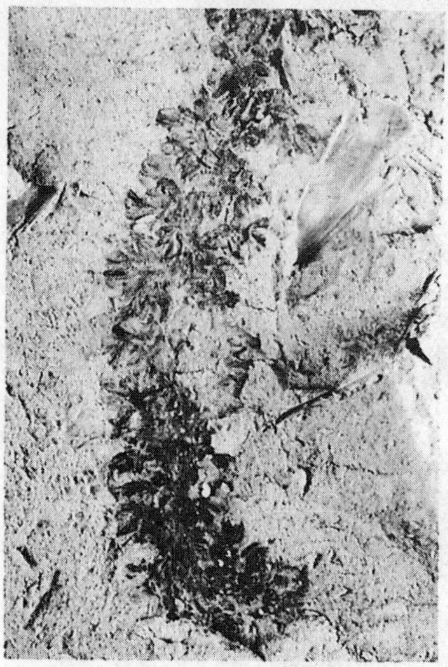

Figure 31.19. **A.** *Quercus meyeriana*, staminate flower. **B.** *Quercus taeniatopilosa*, staminate flower. Late Eocene Baltic amber. (After Conwentz in Friis & Crepet, 1987, with permission.)

Figure 31.20. *Austrodiospyros cryptostoma*, reconstruction of flower with hairs omitted. Note that stamens are epipetalous and are arranged in radial pairs. Eocene. (Redrawn from Basinger & Christophel, 1985.)

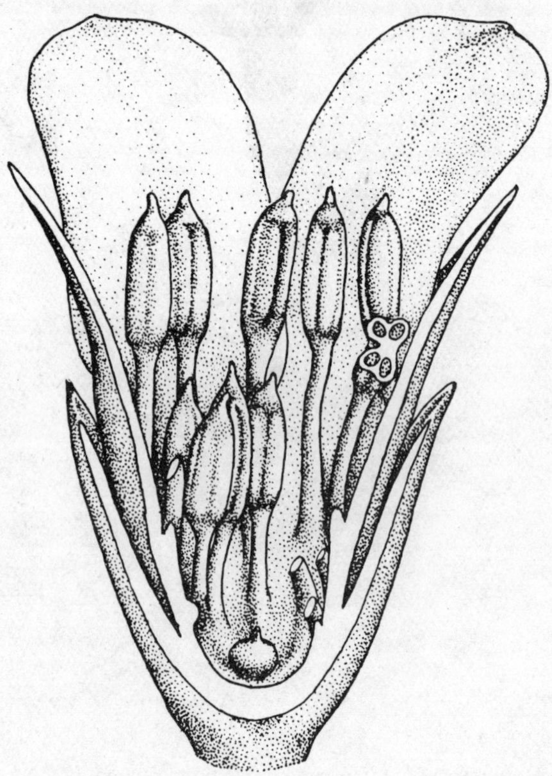

cation of leaves of *Austrodiospyros* relatively easy in spite of their variable physiognomy. Similar cuticular thickenings occur on the sepals and petals of some flowers. This provides us with another good example of how "whole plant" reconstructions are possible from dispersed fossil plant organs.

Other species of the Dilleniidae with preserved floral structures are the Eocene-age *Myrsinopsis* (Fig. 31.21A) and *Berendtia* (Fig. 31.21B) found in Baltic amber and assigned to the Primulales, the late Cretaceous *Actinocalyx* (Fig. 31.21C) of the Ericales, and the Paleocene *Sezannella* of the Malvales. The Euphorbiales are represented by Paleocene inflorescences with pseudanthia as well as inflorescences from the Eocene (Crepet & Daghlian, 1981).

Rosidae

The fossil flora from Scania, southern Sweden, has produced the most diverse and best-preserved assemblage of fossil flowers known from the Santonian–Campanian (Friis & Skarby, 1981). The fossils are three-dimensional mummified coalifications that include more than 100 different taxa of angiosperm reproductive structures. Among these are members of the Hamamelidales, Fagales, Myricales, and Juglandales of the Hamamelidae. Many others have affinities with the Rosidae, especially the Saxifragales. This order is represented by *Scandianthus* (Friis & Skarby, 1982) (Fig. 31.21D). The genus is a typical example of angiosperms with insect-pollinated flowers, unlike the Hamamelidae with predominantly wind-pollinated genera. The flowers are small with a dish-shaped perianth composed of a calyx and corolla. These parts and the androecium are radially arranged in whorls of fives. The gynoecium consists of two carpels that develop into unilocular, inferior ovaries containing small ovules. The fruit is a capsule that opened apically between the styles. A ten-lobed nectary is placed between the androecium and gynoecium.

Among the other saxifragalean remains from the Scania beds are the faithfully preserved remains of *Silvianthemum* (Friis, 1990) (Fig. 31.22), which shares many of its characteristics with *Scandianthus*. *Silvianthemum* differs (Fig. 31.23A), however, in having a gynoecium composed of three carpels fused at the base to form a unilocular ovary. The ovary contains three stalked placentae bearing parietal anatropous ovules. At maturity the ovary forms what appears to be a capsule containing numerous minute seeds. Pollen grains are usually tricolpate with a tectate wall that is perforate. The flowers of *Silvianthemum* compare most favorably with those of the woody members of the Saxifragales typical of the Southern Hemisphere.

Rosales

For the Rosales, the genus *Paleorosa* (Basinger, 1976b) is the oldest known flower of the family Rosaceae. The superbly preserved permineralized specimens of this flower came from the productive Middle Eocene outcrops of the Allenby formation near Princeton, British Columbia. *Paleorosa* flowers (Fig. 31.24) are small (about 2 mm in diameter) and pentamerous. The five sepals and petals are inserted alternately and fused to form a floral cup. There are 13 to 19 stamens inserted in a ring on the inner face of the cup. The gynoecium (Fig. 31.23B) consists of five free carpels each containing two collateral erect ovules. These structural features of an insect-pollinated flower suggest that affinities of *Paleorosa* are with those Rosaceae having primitive characteristics. Although the flower does not fit into any extant group, it is

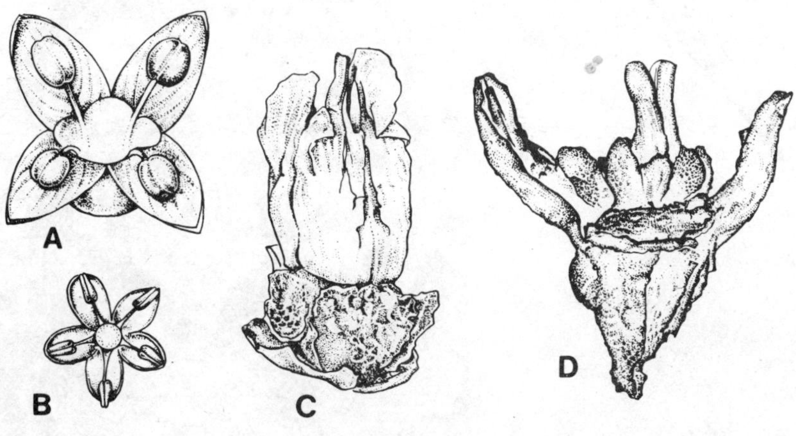

Figure 31.21. **A.** *Myrsinopsis succinea,* corolla with attached stamens. **B.** *Berendtia pimuloides,* corolla with attached stamens. **A,B:** Eocene. (After Conwentz in Friis & Crepet, 1987, with permission.) **C.** Sympetalous flower of *Actinocalyx bohrii.* **D.** *Scandianthus costatus* with ten-lobed nectary. **C,D:** Late Cretaceous. (After Friis & Crepet, 1987, with permission.)

possible that *Paleorosa* represents an early stage in the evolution of the family.

Fabales

The order Fabales is divided into three families: Caesalpiniaceae, Mimosaceae, and Fabaceae. Flowers belonging to the latter family are known from the late Paleocene (Crepet & Taylor, 1985). They are noteworthy because they represent the first occurrence of fossil flowers to exhibit well-defined zygomorphic (bilateral) symmetry. The flowers have a keel, wing petals, and standard typical of flowers of legumes.

Representatives of the Mimosoideae also make their first appearance in the late Paleocene and continue into the uppermost Eocene (Crepet & Taylor, 1986). The late Paleocene genus *Protomimosidea* (Fig. 31.25) shows an assemblage of characteristics that are believed to be primitive for the subfamily. Among these are inflorescences that are racemes, flowers with 5 petals and 10 exserted stamens, hairy carpels with filiform styles, and short tubular stigmas. Pollen occurs as monads and is tricolporate. The inflorescence of the Eocene–Oligocene genus *Eomimosoidea* (Crepet & Dilcher, 1977) is a delicate spike with alternately arranged flowers (Fig. 31.26) that are sessile and perfect, each with a four-part calyx and corolla. A

Figure 31.22. Reconstruction of *Silvianthemum suecicum*. Upper Cretaceous. (From Friis, 1990, with permission.)

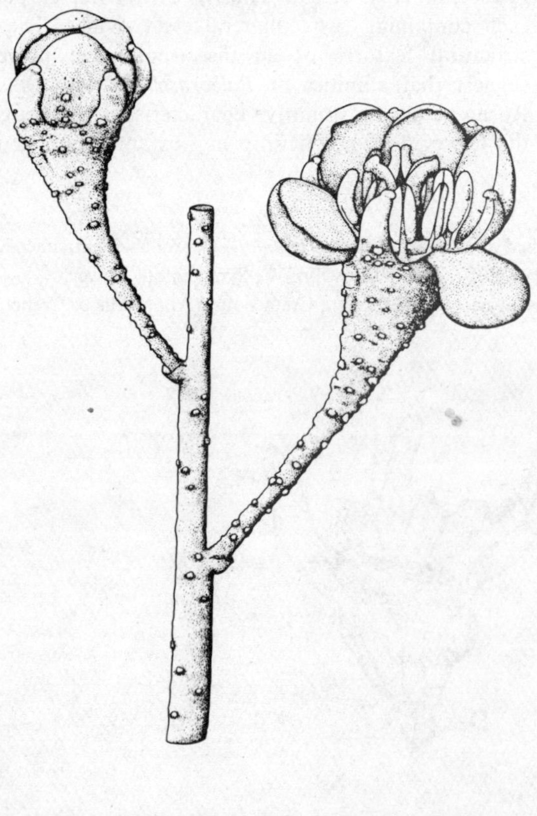

Figure 31.23. **A.** *Silvianthemum suecicum* flower with abscissed perianth showing three conspicuous styles. Upper Cretaceous. (From Friis, 1990, with permission) **B.** *Paleorosa similkameenensis*, transverse section of gynoecium showing five carpels each with two ovules. Eocene. (From Basinger, 1976b, with permission.)

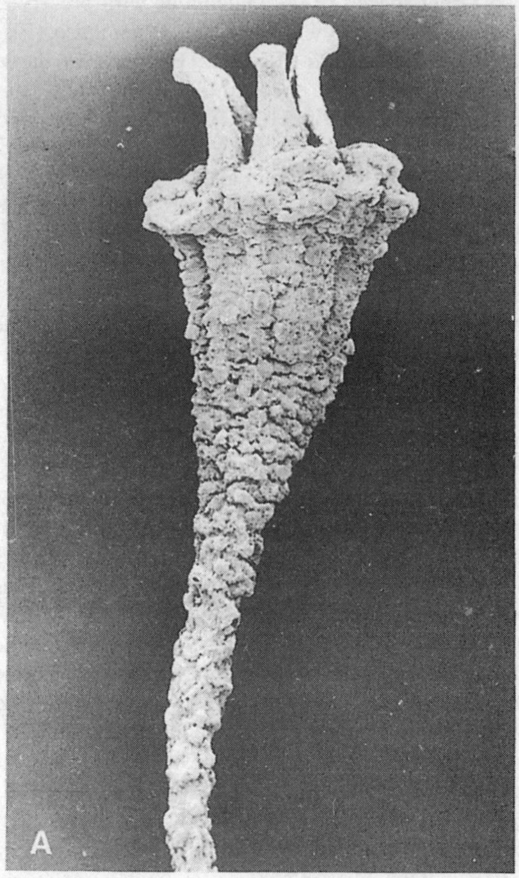

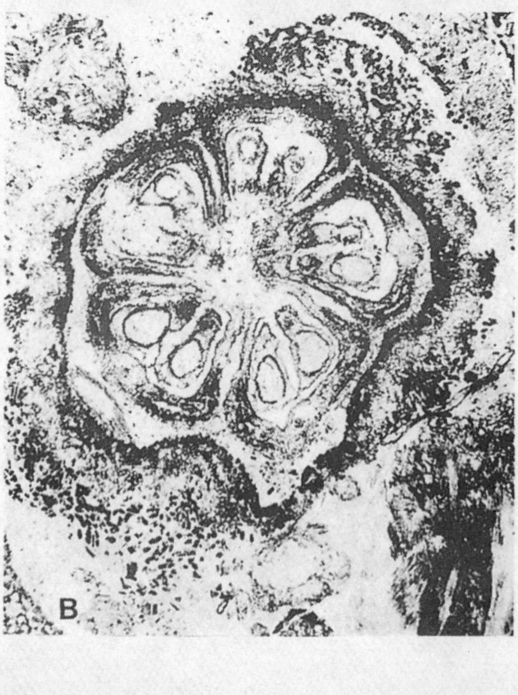

Figure 31.24. Reconstruction of *Paleorosa similkameenensis*. Eocene. (Redrawn from Basinger, 1976b.)

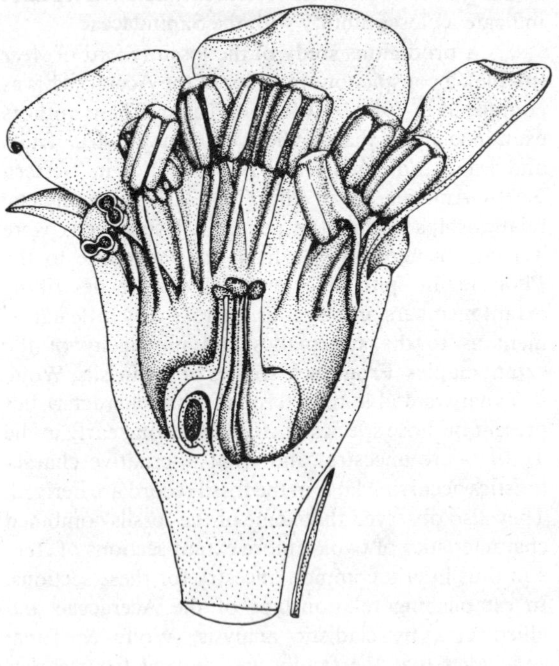

Figure 31.25. *Protomimosidea buchananensis*, the earliest evidence of the Mimosoidae. Paleocene–Eocene. (From Friis & Crepet, 1987.)

conspicuous hairy style protrudes beyond the floral envelope and terminates in a rounded stigma. Anthers are at the tips of long slender filaments, which extend as much as 5 mm beyond the floral envelope. There are 8 to 10 stamens for each flower with anthers containing tricolporate pollen grains which occur in permanent tetrahedral tetrads. A statistical analysis of *Protomimosoidea* occurring in the late Paleocene–Upper Eocene interval showed no significant differences except for the length of the floral envelope. The lack of notable differences during this long interval of at least 8 m.y. suggests considerable morphological uniformity in floral structure. *Eomimosoidea* shows a similar stasis in the evolution of its flowers from the Middle Eocene to the Oligocene (Daghlian, Crepet, & Delevoryas, 1980).

Permineralized flowers and fruits that compare favorably with the Myrtales have been discovered in the Deccan Intertrappean beds of central India. Although still debated, the age of the beds is thought to be Paleocene to early Eocene (Nambudiri, Tidwell, & Chitaley, 1987). The fossil flowers (*Sahnianthus*) and associated fruits (*Enigmocarpon*) are fairly common components of the Intertrappean beds. The flowers have radially arranged more-or-less undifferentiated floral parts in a single whorl of eight tepals. The bisexual flowers have eight stamens and an ovary with six to eight locules. Two rows of ovules showing axile placentation occur in each locule. Other Intertrappean flowers assigned to the Myrtales show variations in number of perianth parts, stamens, and

Figure 31.26. *Eomimosoidea plumosa* inflorescence. Note stigma and style (s); stamens (st); floral envelope (fe). Eocene. (Redrawn from Crepet & Dilcher, 1977.)

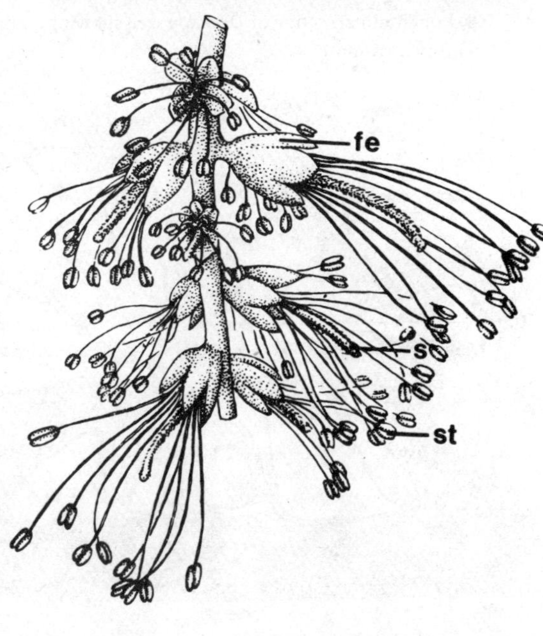

carpels; these are *Sahnipushpam*, *Chitaleypushpam*, and *Raoanthus*.

Exceptionally well-preserved fruits and seeds assigned to *Decodon allenbyensis* (Fig. 31.27A) of the myrtalean family Lythraceae recently have been described from the Allenby Formation cherts of Princeton, British Columbia (Cevallos-Ferriz & Stockey, 1988a). Embryos have been found in the seeds (Fig. 31.27B) one of which shows two cotyledons and a possible suspensor. The detailed comparisons of the fossil fruits and seeds with those of extant members of the Lythraceae have shown the fossil species to be similar to the swamp willow *D. verticillatus*. When the fossil assemblages comprising the Eocene floras containing *Decodon* are evaluated, it is clear that they grew in riparian or lacustrine habitats of a subtropical paleoenvironment.

The fossil record of the Sapindales, especially those belonging to the family Sapindaceae, is represented in the fossil record by pollen, leaves, fruits, and seeds. The new genus *Wehrwolfea* (Erwin & Stockey, 1990) is the earliest record of a sapindaceous flower from western North America. These small unisexual flowers (Fig. 31.28) were found in the Middle Eocene Princeton chert locality of southern British Columbia, Canada. All of the flowers observed are staminate; some are in the form of buds, while others are mature pedicellate flowers with a perianth of at least three sepals and as many as four petals. Some flowers contain a centrally located rudimentary pistil surrounded by a conspicuous nectary disc. There are 10 relatively large stamens with long filaments attached by a slender connective to large dithecal anthers. The *in situ* pollen is tricolporate, prominently striated with equatorially bridged colpi. Although the floral and pollen morphology of these diminutive flowers are similar to the Aceraceae, detailed features indicate a closer affinity with the Sapindaceae.

A prodigious study of the fossil record of *Acer* (maple), a conspicuous member of the Aceraceae, was completed in 1987 by Wolfe & Tanai. These authors examined more than 2,000 specimens of *Acer* fruits and leaves obtained from 170 localities in western North America in an effort to clarify the phyletic relationships within the genus. All specimens were Tertiary in age, ranging from the Paleocene to the Pliocene. In spite of extensive literature describing extant members of the genus, there is very little agreement as to the systematics and phylogeny of the extant maples. From their studies of the fossils, Wolfe & Tanai were able to determine what characteristics present in those species of *Acer* appearing early in the Tertiary are ancestral and what alternative characteristics occurring later in the fossil record are derived. They also observed that some of the fossils combined characteristics of two or more extant sections of *Acer* and thus infer a common ancestry for these sections. In establishing relationships of the Aceraceae and allied taxa by cladistic analysis, Wolfe & Tanai concluded that the family was derived from earlier members of the Sapindaceae, such as *Bohlenia*, by loss of a stipule, loss of a locule, and a change from alternate to opposite leaves. The divergence of the *Dipteronia* line (Fig. 31.29) of the Aceraceae involved the maintenance of pinnately compound leaves of the bohlenoid line, but a change in the secondary venation pattern of that line. Continuing on into the "*Acer*" *Arcticum* line involved the fusion of at least three

Figure 31.27. **A.** *Decodon allenbyensis*, transverse section of fruit showing several seeds (s) and septation (sp). **B.** Longitudinal section of *Decodon* seed showing embryo (e). **A,B:** Eocene. (From Cevallos-Ferriz & Stockey, 1988a, with permission.)

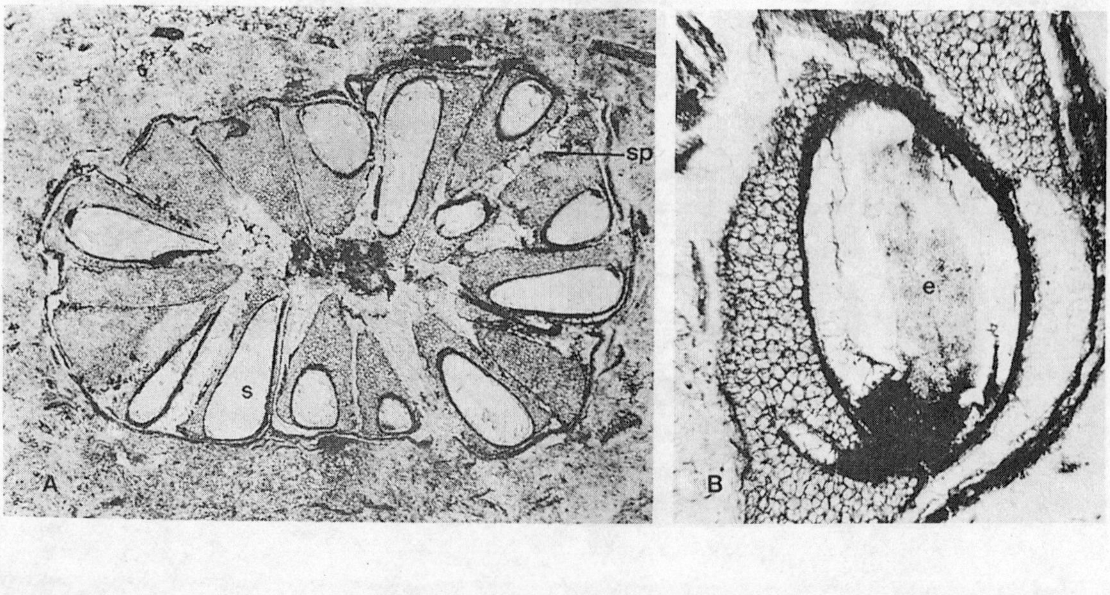

Figure 31.28. *Wehrwolfea striata,* reconstruction showing five stamens, perianth, rudimentary pistil, and interstaminal nectary disc (n). Eocene. (From Erwin & Stockey, 1990, with permission.)

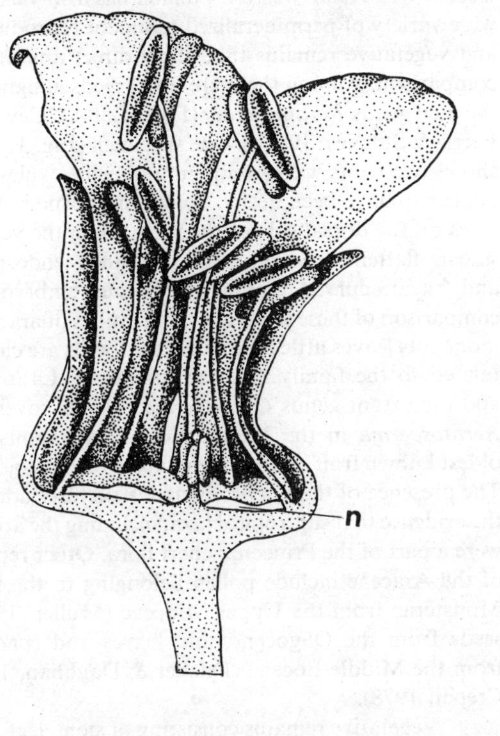

leaflets into an actinodromous maple leaf. The fruit of the *Dipteronia* line would be similar to that of *Bohlenia* (Fig. 31.29) so that the wing surrounds the nutlet of the fruit except along the base at the point of its attachment. The divergence in evolution of the fruit of the *Acer* and "*Acer*" *arcticum* lines is marked by a coalescence of veins along the proximal margin of the wing in *Acer* (Fig. 31.29) and along the distal margin of the wing in the "*Acer*" *arcticum* complex.

The comprehensive cladistic study by Wolfe & Tanai of the fossil and extant members of the Aceraceae provides us with a graphic example of how the use of cladistic and phenetic analyses of large numbers of fossil specimens representing a single genus can make an important contribution to our understanding of the systematics and phylogeny of both fossil and extant members of the taxon.

Liliopsida (monocotyledons)

The fossil record of the angiosperms shows us that a very high percentage of their remains belong to the families of the Magnoliopsida (dicotyledons). The Liliopsida, on the other hand, are represented by approximately 20 families (Daghlian, 1981), only four of which extend back into the Cretaceous. This paucity of monocots may reflect the fact that there are many more genera of dicots than monocots, as well as the possibility that monocots are mostly

Figure 31.29. Suggested evolution of fruit in the Aceraceae and allied taxa. Cenozoic. (From Wolfe & Tanai, 1987, with permission.)

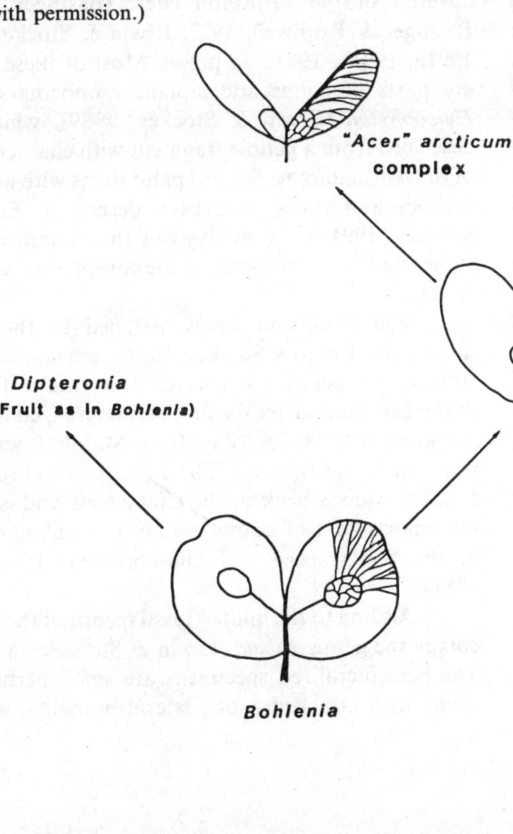

herbaceous plants that would not be as readily preserved as are woody dicots (Erwin & Stockey, 1991a).

The earliest fossils with possible monocot affinities are recorded from the Aptian and Albian of the Early Cretaceous (Doyle, 1973; Walker & Walker, 1984). Pollen grains with characteristics similar to those described by these investigators are known to occur in extant entomophilous groups of monocots. The monoporate pollen grains typical of anemophilous monocots are not recorded until the Late Cretaceous (Friis, Chaloner, & Crane, 1987). Leaves having apically fused veins and sheathing bases characteristic of monocot leaves also occur in the Aptian (Doyle & Hickey, 1976).

From the Lower Cretaceous to the Maastrichtian of the Upper Cretaceous, the fossil record of monocots thus far described is very poor. Much of the evidence for Upper Cretaceous monocots has been obtained from the Intertrappean beds of India, which are now believed to be Maastrichtian in age. Among the fossils are *Musa*-like of the Zingiberaceae (Jain, 1963) of the Liliidae. Other flowers and fruits described as monocots from the Intertrappean beds are *Deccananthus, Nipa, Tricoccites, Palmocarpon* and *Variocarpon,* all exhibiting a trimerous organization of monocots that places them with the palms of the Arecaceae, leaf remains of palms and the Zingiberaceae (ginger family) occur in the Upper Cretaceous and Paleocene of the Tertiary. *Zingiberopsis* leaves (Hickey & Peterson, 1978) show unequivocal monocot parallelodromous venation patterns in their elliptical to ovate leaves, which have a distinct midrib. The secondary veins tend to converge toward the leaf apex, while the fine veins run between adjacent parallel veins. All are features common to leaves of many monocots.

Beginning in the Early Eocene, the Liliopsida as in the Magnoliopsida show trends leading to diversification and modernization so that by the Late Eocene the monocotyledonous components of these Early Tertiary floras were similar to extant genera. Among these, reported by Collinson & Hooker (1987) from the Early Tertiary of southern England, are *Stratiotes* of the Hydrocharitaceae, *Potamogeton* of the Potamogetonaceae, *Scirpis* and *Mariscus* of the Cyperaceae, *Manicaria* and *Nipa* of the Arecaceae, and *Typha* of the Typhaceae. Many of the specimens assigned to the Typhaceae comprise fossil seeds, fruits, and megaspores separated from the sands and clays. Perhaps the most well-known collections of these specimens come from the Early Eocene London Clay flora of southern England made famous by the pioneering work of Reid & Chandler (1933) and expanded later in numerous researches by Chandler and more recently by Collinson and others. Their specimens come from Tertiary beds covering a period of 25 m.y. – Paleocene to Oligocene.

In addition to substantially increasing our information about Eocene dicots, the Princeton Middle Eocene chert from western Canada has also yielded a wide variety of permineralized monocot reproductive and vegetative remains that allow direct anatomical comparison. Among these are fossil seeds assigned to the new genus *Keratosperma* (Fig. 31.30A) (Cevallos-Ferriz & Stockey, 1988b). One specimen (Fig. 31.30B) shows two fruits with a minimum of eight ovules that appear to have been fleshy. Spines are borne in three rows on the dorsal side of the seed, while the ventral side is flattened. Some seeds contained endosperm and linear, curved monocotyledonous embryos. A comparison of these seeds with subfamilies of araceous monocots leaves little doubt that the fossils are closely related to the family Araceae, subfamily Lasioidae, and the extant genus *Cyrtosperma*. The discovery of *Keratosperma* in the Middle Eocene represents the oldest known fruit and seed remains of the Lasioideae. The presence of these reproductive structures adds to the evidence that subtropical taxa, including the aroids, were a part of the Princeton chert flora. Other reports of the Araceae include pollen belonging to the tribe Monsterae from the Upper Miocene (Muller, 1981), seeds from the Oligocene, and leaves and spadices from the Middle Eocene (Dilcher & Daghlian, 1977; Crepet, 1978).

Vegetative remains consisting of stem, leaf, and root fossils belonging to monocots have been discovered in the Princeton chert (Basinger, 1976a; Basinger & Rothwell, 1977; Erwin & Stockey 1989, 1991a, 1991b, 1991c, in press). Most of these organs are parts of palms and aquatic monocots such as *Heleophyton* (Erwin & Stockey, 1989), which was described from a petiole fragment with characteristics of the Alismataceae. Several palm stems with attached petioles and roots have been described (Erwin & Stockey, 1991a). An analysis of their characteristics shows that they are related to the coryphoid (sabaloid) Arecaceae.

Rhizomes and stems assigned to the genus *Soleredera* (Erwin & Stockey, 1991b) are anatomically similar to vegetative structures occurring in families of the Liliales and are the first permineralized lilialean megafossils to be described from Middle Eocene deposits in North America. The megafossil record of the Liliales extends back to the Cretaceous and is based on reports of leaf compressions resembling species of the Smilacaceae and Dioscoreaceae (Daghlian, 1981).

Adding to the limited fossil record of the monocots is the genus *Ethela* (Erwin & Stockey, in press). The permineralized specimens are small herbaceous stems with attached roots, lateral branches, and remains of sheathing leaves. They occur in the Middle Eocene Princeton chert and have been identified as belonging to the Subclass Commeliniae, which includes the Cyperales and Juncales.

Moving to the Middle Miocene, Tidwell & Parker (1990) record the discovery of a permineralized *Yucca* of the Agavaceae. This rare specimen called *Protoyucca* is the first report of a permineralized monocotyledon with secondary growth.

The limited amount of literature describing fossil grasses (Poaceae) attests to the paucity of fossils belonging to this group. A partial explanation for their infrequent preservation may be related to the fact that they grew, as do most extant grasses, in arid to semiarid upland environments where chances of their preservation were severely limited. The first records of grass pollen and megafossils are from the Paleocene–Eocene (Muller, 1981; Crepet & Feldman, 1991). Megafossils of grasses, however, occur with greater frequency in the Oligocene, Miocene, and Pleistocene (Thomasson 1978; Tidwell & Nambudiri, 1989).

Pollination syndromes

Although anemophily (wind pollination) is known to occur among the vast majority of extinct and extant gymnosperms, the possibility that anthophilous (flower-loving) insects were pollen vectors for Late Jurassic and Early Cretaceous Cycadeoidales was first proposed by Crepet (1974). He provides us with good evidence that members of the Coleoptera (beetles) with chewing mouth parts were probably the insects that invaded the closed microsporophylls of *Cycadeoidea* to obtain pollen. In addition to beetles, which had become diverse during the Mesozoic, there were members of the Diptera (flies) that may have functioned as insect pollen vectors for other cycadeoids (*Wielandiella* and *Williamsoniella*). In these genera the microsporophylls open out, and microsporangia containing pollen would be exposed to both beetles and flies. After examining the fossil record of cycadeoid fructifications and insects available for pollination, Crepet & Friis (1987) suggest that the insects, in conjunction with the Jurassic-Lower Cretaceous hermaphroditic cycadeoid fructifications, provided the "archetypal syndrome of insect pollination." It has been accepted for many years that the fructifications of *Cycadeoidea* were the functional analogs of magnolialean flowers and that these insect-pollinated fructifications represent the primitive type from which wind-pollinated flowers evolved. This concept was generally supported until the recent discovery of anemophilous flowers belonging to the Lower Cretaceous (Albian) Platanaceae coexisting with taxa

Figure 31.30. **A.** *Keratosperma allenbyensis*. Oblique, longitudinal section showing micropyle (m). **B.** Sectioned fruits containing seeds. Eocene. (From Cevallos-Ferriz & Stockey, 1988b, with permission.)

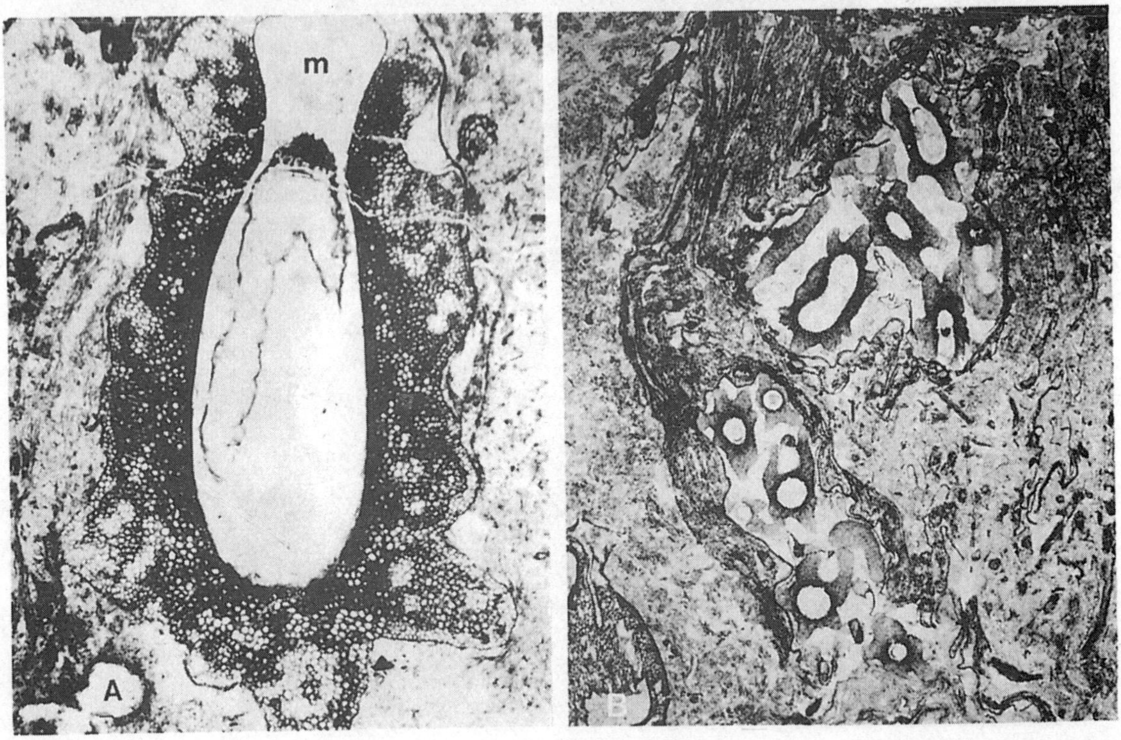

of the Magnoliidae. With this new information at hand, Crepet & Friis (1987) suggest that archetype angiosperms with simple anemophilous flowers possessed the requisite evolutionary flexibility to participate in a coevolutionary relationship with more advanced "faithful" insect pollinators. These same simple flowers would also be the logical ancestors of other anemophilous flowering plants similar to those of the Hamamelidales, e.g., the Platanaceae. If this concept is accepted, then those members of the Magnoliales with large complex flowers (e.g., *Magnolia*) may not represent the archetype for the class Magnoliopsida.

In the establishment of entomophily insects, most probably beetles, feeding on other parts of the plants fortuitously encountered and fed upon the protein-rich pollen grains and the sticky pollination drop. When a preference for pollen as a food supply was established and visits to the ovules became frequent, the beetles inadvertently established a haphazard type of insect pollination that supplemented the normal anemophily. From such a beginning, mutations occurred in plants that provided food and attraction to insects. In the attraction of insects to the reproductive parts, however, the consumption of ovules by the insects presented a problem that was solved by the evolution of the carpel or similar enveloping structures. Other mutations that must have

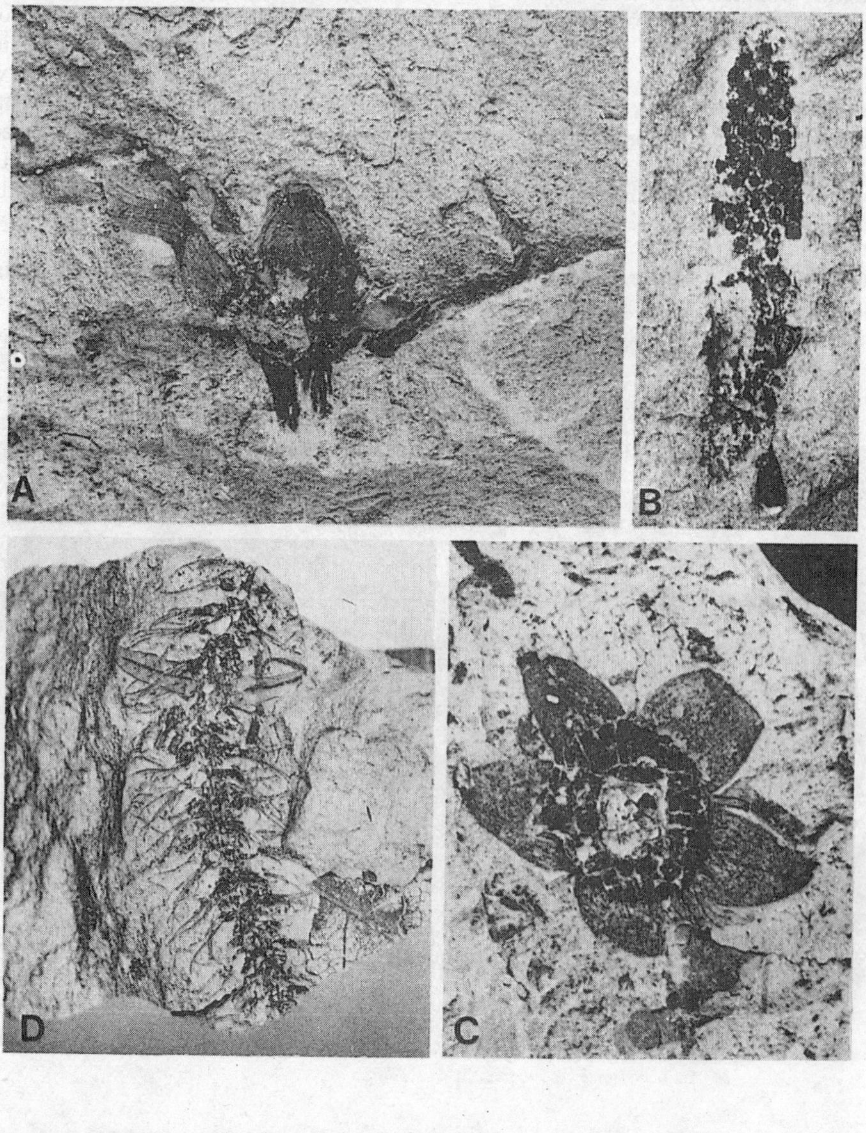

Figure 31.31. Eocene flower and inflorescence types modified for insect pollination. **A.** Large flat flower type adapted for pollination by beetles. **B.** Spadix of an aroid probably by flies. **C.** Unspecialized small flowers where flies may be the pollinators. **D.** Mimosoid inflorescence of florets with small perianths and exserted stamens, possibly pollinated by bees. (**A–D** from Crepet, 1979.)

been of great selective advantage in the attraction of insects were production of distinctive perianth parts and nectaries with their secretions high in sugar content. Concomitant with the further specialization and diversification of insect pollinators, flowers became more diversified, evolving bisporangiate flowers that brought stamens and carpels into closer proximity, further facilitating pollination.

First occurrences of insects in the fossil record show us that representatives of all four major groups involved in insect pollination were present at least by the Early Cretaceous. In addition, to the Coleoptera and Diptera, both present in the Mesozoic, are members of the Hymenoptera (bees and wasps). The Lepidoptera (butterflies and moths) first appear in the Early Cretaceous. Thus, the insect vectors that participate in pollination were in place prior to or at the time of the early evolution of flowering plants.

In the earlier part of this chapter, space has been devoted to the description of several of the more complete and better preserved types of flowers belonging to the major orders of the class Magnoliopsida, many of which are entomophilous angiosperms. Many taxa with flowers adapted to insect pollination had evolved by the Middle Eocene (Fig. 31.31) (Crepet & Friis, 1987; Friis & Crepet, 1987). Among these are small simple flowers with few floral parts. These chloranthoid flowers, appearing first in the Lower Cretaceous, may be bisexual or unisexual. In addition to wind pollination, possible insect pollinators include beetles, certain moths, wasps, and sawflies, all present in the Early Cretaceous.

Large, flat bisexual flowers with numerous helically arranged parts, for example *Archaeanthus* and other large (6 cm in diameter) magnolialean flower types with bowl-shaped perianths (Fig. 31.31A), are generally pollinated by beetles. Their structure provides easy access to the numerous stamens found in these flowers. The small unisexual flowers with more or less undifferentiated perianth parts of the *Platanus*-type provide evidence that wind pollination was well established in the Early Cretaceous. By the Late Cretaceous (Cenomanian) flowers having a low number of floral parts in a cyclic (whorled) arrangement, for example the Rose Creek flowers, had evolved. These and similar flowers from the Tertiary may have had nectaries (Fig. 31.28). Their morphologies suggest that they were well adapted to pollination by flies and possibly wasps and beetles. Fly-pollinated flowers comprising the spadices of members of the Araceae occur in the Middle Eocene and elsewhere in the Tertiary.

The radiation of primitive members of the Rosidae during the Cenomanian and culminating in the Santonian-Campanian is believed to have been in response to the evolution of additional groups of nectar feeding Hymenoptera and Lepidoptera. During the Santonian–Campanian interval floral characteristics associated with advanced (faithful) constant pollintors began to appear (Crepet, 1985). These include fusion of floral parts as in *Actinocalyx*, bilateral symmetry, and the appearance of the common style. The latter suggests more efficient pollination and increased gametophytic competition.

Flowers with bilateral symmetry orient insect vectors (bees) in their approach to the flowers, enabling the insects to find their way to nectar or to operate pollination mechanisms. The first floral evidence of zygomorphy is a weakly developed bilateral symmetry in flowers and fruits of Maastrichtian (*Raoanthus* and *Musa*-like fruits) Zingiberales. It is not until the Paleocene that the first distinctly zygomorphic flowers (Fig. 31.32A) appear. These papilionoid legume flowers, which are pollinated by bees, occur at the same time as the first fossil evidence for the presence of bees in the Early Tertiary. Indirect evidence indicates that bees had evolved before the end of the Cretaceous.

Figure 31.32. **A.** Papilionoid legume flower. Note the two planes (bilateral) of symmetry. Keel petals (k); wing (w); standard (s). **B.** Flower with narrow tubular corolla. **A,B:** Paleocene–Eocene. (**A** from Friis & Crepet, 1987, with permission; **B** from Crepet, 1985.)

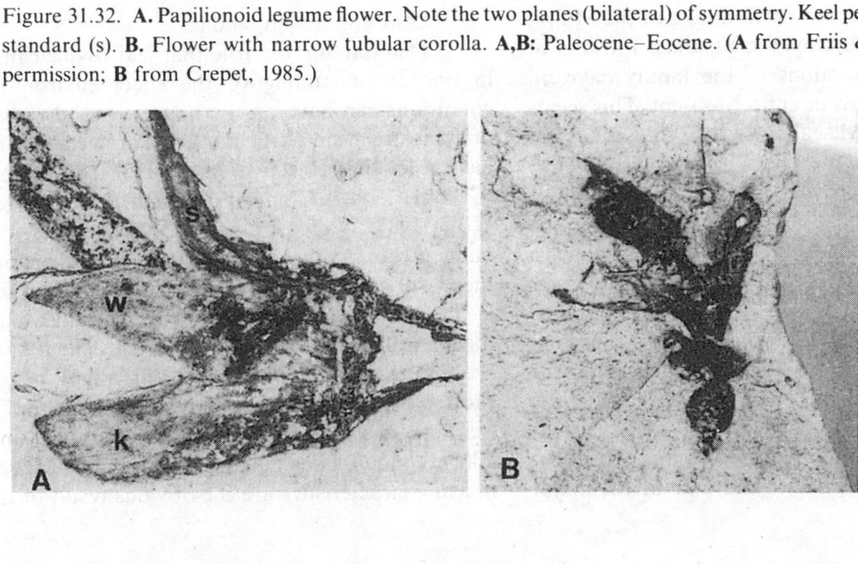

The Maastrichtian–Eocene interval marks the time of the second radiation and modernization of flowering plants and involved advanced (constant) pollinators, especially bees and haustellate Lepidoptera (butterflies and moths). In addition to the papilionoid legume flowers, genera of the diversified Fabales include the Paleocene–Middle Eocene brush-type inflorescences of the Mimosaceae (Fig. 31.25). All of these floral types are highly adapted to their bee pollinators. During the same time interval, flowers with funnelform corollas and narrow corolla tubes (Fig. 31.32B) formed from fused petals also appeared. This suggests that advanced lepidopterans with elongate mouthparts were present. Flowers with shorter and wider corolla tubes also occur and may have been pollinated by long-tongued bees.

An important discovery of flowers with affinities to the Malpighiaceae has been reported by Taylor & Crepet (1987). The flowers they describe, which come from the Middle Eocene of the Claiborne formation, Tennessee, provide us with a specific example of the relationship between floral structure and a constant insect pollinator. The flowers are distinctive in having sepals with large paired abaxial oil glands and petals with clawlike edges. These fossil flowers, assigned to the genus *Eoglandulosa,* have ten stamens in two whorls that produce tricolporate pollen with a supertectal reticulate structure. These features are found among extant genera of the Malpighiaceae, a predominantly South American family with some representatives occurring in the Old World tropics. It has been suggested (Crepet & Friis, 1987; Taylor & Crepet, 1987) that the close relationship existing between the Malpighiaceae and certain tribes of bees may have resulted in the highly uniform floral structure found in the family. At the time of pollination, it has been observed that the bees reach around the clawed bases of the petals to obtain oil from the oil glands on the abaxial surfaces of the sepals. The bee pollinators from the tribes involved are confined to the South American tropics where all but a few derived taxa of the Malpighiaceae have well-developed oil glands. Flowers of the Old World populations of the family have oil glands that are reduced or entirely absent. This seems to be related to the absence of the tribes of oil-collecting bees in the Old World environments. This fossil evidence provides us with an example of a specific plant-pollinator relationship during the Eocene (Taylor & Crepet, 1987), as well as evidence that pollination by bees occurred at least by the Early Tertiary.

From the examples given above and the accumulation of other paleontological evidence for the appearance of constant insect pollinators, Crepet (1985) has demonstrated that the inception of advanced, constant insect-pollination syndromes was coincident with or occurred just prior to the time of the major diversification and radiation of flowering plants during the Late Cretaceous and Early Tertiary. With this paleontological evidence at hand, one can no longer exclude advanced, constant pollinators as participants in the diversification of Tertiary angiosperms. This evidence from the fossil record adds considerably to the neontological evidence indicating that advanced insect pollinators played an important role in establishing much of the diversity found among contemporary flowering plants.

Cretaceous paleoclimates and paleogeography

During the Cretaceous of North America, the potential barriers restricting the migration of angiosperm floras were temperature and oceans (Parrish, 1987). The temperature barriers, however, were probably not of great importance at this time as indicated by the apparent warmth of the entire planet. In addition, it has been postulated that the warm global temperatures were accompanied, in Laurasia, by a general high degree of humidity that was initiated in the Jurassic and extended into the Cretaceous. Continued humidification has been related to widespread flooding of the continental crust by epeiric seas and the opening of the Atlantic Ocean. The first major expansion of the Cretaceous epicontinental seas began in the late Aptian, continuing into the mid-Albian when the sea extended unbroken from the Arctic Ocean to the Gulf of Mexico. In the regions now occupied to the east by the Atlantic Coastal Plain and to the west by the Rocky Mountains, the rainfall has been mapped as humid in the east and low to moderately low in the west, with areas of heavy precipitation in the Upper Cretaceous. It now seems reasonably clear that the early evolution and radiation of angiosperms took place on a planet with an equable climate (temperature and rainfall) and at a time that was relatively stable in its tectonic history.

Leaf and pollen studies

Determining the role that leaf fossils can play in our understanding of the effect environmental conditions can have on the migration and radiation of angiosperms in particular areas at particular times has been the objective of several paleobotanists (Crabtree, 1987; Crane, 1987a, 1987b; Upchurch & Wolfe 1987; Wolfe & Upchurch, Wolfe, 1987a) and many others. As these investigators and others before them will agree, the study of angiosperm leaf fossils is fraught with difficulties that can produce bias in their results. The vast majority of specimens studied are impression–compression fossils, many of which show the effects of various taphonomic factors. For example, the leaf remains are commonly found in allochthonous depositional sites where drastic changes in leaf characteristics are the obvious result of transportation and fragmentation. Mistaken conclusions about leaf size can result from sorting of leaves by wind and water prior to deposition. Critical characteristics of minor venation patterns and epidermal features can be altered by various steps in diagenesis, and so on. These, and many other physical and chemical factors associated with the depositional site, must be taken into consideration in an effort to eliminate bias from that quarter.

Paleoecologists are dependent on the proper identification of the leaf fossils and pollen grains they work with. In the past, as late as the last decade, the usual method of identification was to match the gross morphological characteristics of the leaf fossil as nearly as possible with the leaf type of a genus belonging to an extant flowering plant and then assign that genus to the fossil. Concluding that extant and extinct floras are alike, however, is risky if one depends on compression–impression leaf fossils. The difficulty is exemplified by the form genera *Zizyphus, Grewia, Populus, Cissus,* and a number of others that are now known to belong to *Cercidiphyllum*-like plants. Thus, genera belonging to different families of flowering plants are not readily distinguished by the morphology of their leaves. Leaf polymorphism also presents problems, as in the form genus *Ficus,* with over 150 species. Many of these, without doubt, represent variants of a single species. Further, it has been estimated that less than 19 percent of all species of fossil *Ficus* leaves actually belong to the genus.

The importance of fine venation patterns and epidermal features, when assigning leaf fossils to supposed natural genera and families, has been emphasized by the works of Hickey (1973), Dilcher (1974), Hickey & Wolfe (1975), Crabtree (1987), and others. The evidence provided by their works suggests that as much as 60 percent or more of these Cretaceous and Tertiary angiosperm leaves have been incorrectly identified and thus placed in wrong genera, families, or even orders. It is now recognized that many leaf types represent taxa that have become extinct and thus cannot be assigned to an extant group.

In an effort to avoid the pitfalls introduced by faulty indentification of fossil dicot leaf genera, several authors (Dilcher, 1974; Hickey & Wolfe, 1975; Doyle & Hickey 1976; Hickey, 1984; Crabtree, 1987; Upchurch & Wolfe, 1987) have established morphotypes for the fossil leaves. Here, informal descriptive names have been proposed that unite species and genera having similar morphologies. Morphotypes are based on their distinctive venation patterns, leaf shapes and sizes, leaf organizations, and leaf margins. Using these criteria, Crabtree (1987) has illustrated his proposed Cretaceous dicotyledonous leaf morphotypes (Fig. 31.23) and listed the botanical affinities and important genera assigned to the morphotypes. This list emphasizes the number of extant genera mistakenly identified as belonging to Cretaceous floras.

The modernization of Cretaceous floras

In summarizing the results of his studies based on 23 Cretaceous age localities in the northern Rocky Mountains, Crabtree concludes that angiosperms entered this region during the Middle Albian, or about 8 m.y. later than they appeared in the Lower Albian Potomac Group of the Atlantic Coastal Plain. Pollen and megafossils have been found throughout the northern Rocky Mountain region of North America. The palynological and angiosperm leaf fossil evidence indicates a close similarity between floras of this region and the Potomac Group of the Atlantic coast. Similarities include monosulcate pollen grains of the angiosperm type, associated with sapindophyll and platanophyll morphotypes (Fig. 31.33). Other leaf morphotypes also indicate that early members of extant angiosperm taxa were present in the Albian of the northern Rocky Mountains. For example, sapindophylls appear to have an alliance with the Rosidae; the platanophylls, protophylls, and trochodendrophylls with the Hamamelidae. Probable early Magnoliidae include Laurales, Magnoliales, Chloranthales, and Nymphaeales. Most all of these taxa also occur in the Albian of the Potomac Group and the lower Cenomanian of the Atlantic Coastal Plain. In addition to these, cinnamomophylls became abundant during the Cenomanian, like to many other leaf morphotypes common to the Albian but showing more modern aspects in their high-rank venation patterns. Diversification and modernization of angiosperm floras continued gradually after the Cenomanian through the Campanian with the addition of members of the Menispermaceae, Cercidiphyllales, Fagales, Dilleniidae, and pinnate palms.

This increase in diversity has been confirmed by the palynological studies by Muller (1981) of Cretaceous–Tertiary dispersed pollen. He shows that during the Turonian–Campanian interval, there was a considerable increase in the number of angiosperms related to extant taxa at the ordinal level or below. By late Campanian age, Muller was able to identify pollen belonging to 13 orders and 12 families of modern flowering plants. Again based on palynological studies, it is estimated that, during the Maastrichtian, angiosperms underwent another period of diversification and modernization as evidenced by approximately twice the number of recognizable extant families and orders. Some of the orders appearing for the first time in the Maastrichtian are the Illiciales, Caryophyllales, Ericales, Ebenales, Malvales, Fabales, Geraniales, Santalales, Proteales, Arecales, and Pandanales. The evidence from combined studies of angiosperm megafossils and pollen reveals that all but one (Asteridae) of the dicotyledonous subclasses had representative genera by the end of the Cretaceous. By this time monocotyledonous pollen types belonging to palms were present in abundance.

Leaves as indicators of change

Analyses of leaf physiognomy (characteristic external features of extant vegetation) have been used to obtain more precise information, which in turn can be extrapolated when preparing reconstructions of climatic and vegetational changes of the past (Upchurch & Wolfe, 1987). Using studies of the temperature parameters of humid and mesic forests of eastern Asia, Wolfe (1979) established a classification of vegetational types based on the physiognomic characteristics of extant angiosperm leaves that occur within the temperature parameters. The latter comprise three categories: (1) microthermal vegetation growing within a range of −5°C to 13°C; (2) mesothermal vegetation, 13°C to 20°C; and (3) megathermal vegetation, 20°C to 30°C (Wolfe, 1985). With these parameters in mind, it is possible to evaluate the physiognomic leaf characteristics of extant angiosperms that have proved to be useful in determining the vegetative and climatic conditions that supported extinct floras within each of the Cretaceous temperature parameters (Upchurch & Wolfe, 1987).

1. A correlation exists between leaf size and decreasing temperatures and/or precipitation. Thus, large leaf size is found to occur in understories of evergreen multistratal tropical rain forests. Here, the mesophyllous leaves are more than 12 cm in length. Notophyllous broad-leaved evergreen forests have leaves 8 to 12 cm in length and reflect lower temperatures and reduced precipitation. The smallest (microphyllous) leaves, less than 8 cm, occur in xerophyllous scrub and Arctic vegetation.
2. Serrate (toothed) leaf margins are typical of humid microthermal (mean annual temperature, < 13°C) vegetation. They show a marked decrease in abundance in forests where higher temperatures prevail and where they are replaced by entire margined species. At a mean annual temperature of 20°C, the entire margined species make up 60 to 70 percent of the forest.
3. Thick leaf texture is typical of evergreen angiosperms that predominate in megathermal and mesothermal climax vegetation where precipitation may be limited. Plants with leaves that are thin and usually deciduous predominate in microthermal climax vegetation or successional mesothermal vegetation.
4. Evergreen leaves in humid environments usually have attenuated apices (drip-tips), especially in understories of multistratal forests.
5. Compound leaves tend to occur on plants that

Figure 31.33. Cretaceous dicotyledonous leaf morphotypes. (From Crabtree, 1987, with permission.)

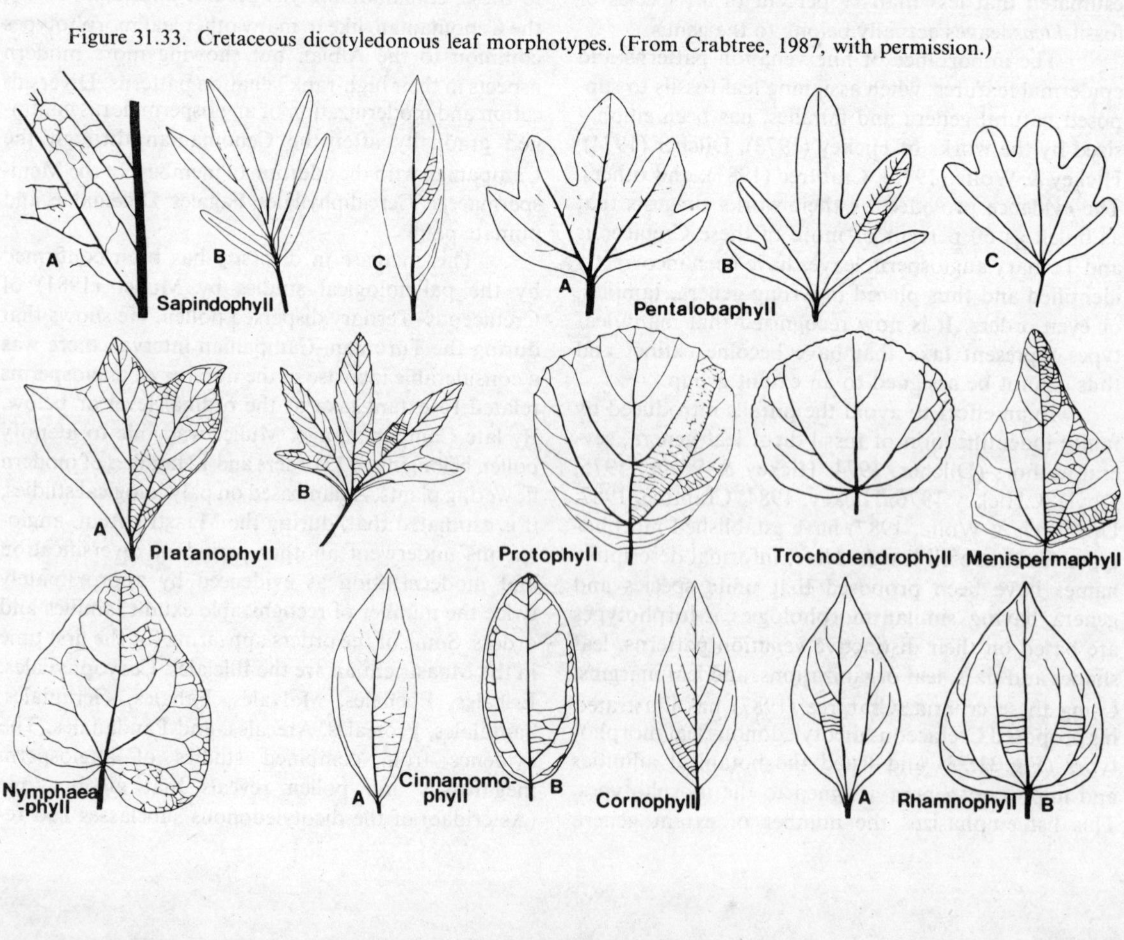

are frequently associated with deciduous plants.

6. Narrow leaves are often found on plants inhabiting stream banks where the substrate is disturbed. Broader heart-shaped leaves, usually with palmate venation, are produced by lianas, sprawling shrubs in successional vegetation, or colonizers in open areas of the forest. Lobed leaves are often found on successional or understory species.

Applying these physiognomic characteristics when studying fossil angiosperm leaves has the advantage of providing climatic and vegetational interpretations directly to the flora quite independently of the questionable affinities and presumed climatic tolerances of extant species (Wolfe, 1985).

Time and place of origin and radiation

Based on many studies of extant angiosperm leaves and pollen, various ideas have been formulated predicting the time and place of origin and subsequent radiation of flowering plants and the climates that supported them.

Because of the predominance of primitive extant ranalean species in the southwestern Pacific and southeastern Asia, Bailey (1949), Axelrod (1952, 1960), Takhtajan (1969), and others have promoted the widely held view that angiosperms originated in Mesozoic tropics during the Jurassic or earlier. Acceptance of this hypothesis presupposes that the tropical, equatorial zone has remained in about the same position during the time since the initial radiation and that the angiosperms were monophyletic. Doyle (1969) tells us that there is no conclusive evidence from palynology that supports this early time of origin of angiosperms in the tropics. Rather, based on fossil evidence, Doyle & Hickey (1977) and subsequent authors favor a primary adaptive radiation of angiosperms close to the base of the Lower Cretaceous.

The early radiation of angiosperms into the Lower Cretaceous of the Southern and Northern Hemispheres has been well documented by Axelrod (1959). Using first occurrences from palynological records at different latitudes, Hickey & Doyle (1977) have clearly shown (Fig. 31.34) a poleward spread of angiosperms, initiated in the Barremian and extending through the Cenomanian of the Lower to mid-Cretaceous. With the subsequent discovery of unequivocal angiosperm pollen, this record has been extended by Hughes & McDougall (1987) into the older Hauterivian.

The Cretaceous environment that supported the initial radiation of angiosperms in the Aptain–early Albian of the Atlantic Coastal Plain was an equable, mesic to humid environment with a mean annual temperature estimated to be 20°C. Whether or not precipitation was seasonal has not been determined. During this interval, angiosperms were an unimportant early successional part of the flora with a relative abundance of 1 percent or less in the palynoflora. They formed only a sporadic element in the megafloras of stream-bank environments. Approximately 75 percent of the leaf species from this part of the Cretaceous have entire margins and are microphyllous. Most of these are unlobed and pinnately veined, features that

Figure 31.34. Diagram showing the latitudinal distribution of first occurrences of angiosperm pollen. Open circles, monosulcate grains; closed circles, tricolpate grains; split circles, monosulcate and tricolpate grains. (After Hickey & Doyle, 1977.)

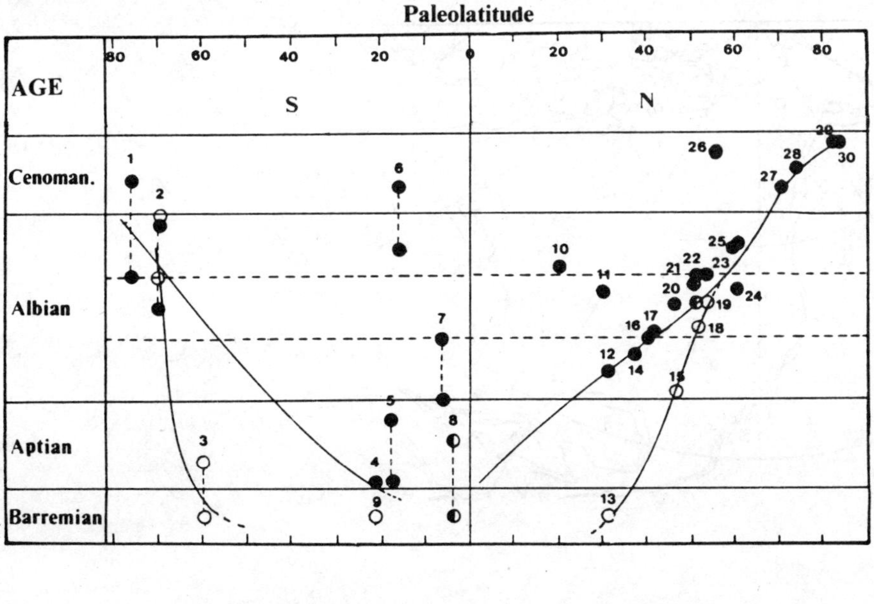

often are associated with late successional vegetation. Many, however, are stenophylls (narrow and elongated), typical of extant stream-side species. Other species are mesophyllous, unlobed with pinnate venation, a physiognomy associated with understory plants. The physiognomy of other leaf types suggests plants that grew in semiaquatic environments or were herbaceous with a lianalike habit (Upchurch & Wolfe, 1987).

By the middle to late Albian of the Atlantic Coastal Plain region, the relative abundance of angiosperms had increased to as much as 20 percent of the palynoflora. Leaf fossils occur in facies associated with stream margins that show evidence of environmental disturbance. The physiognomic diversity of these leaf fossils is greater than that of Aptian-early Albian angiosperms. All of the new types of leaves have characteristics of extant successional plants. Leaves from woody plants (Fig. 31.33) include pinnatifid and pinnately compound leaves (sapindophylls), platanophylls with palmately lobed leaves, and trochodendrophylls with shallow cordate leaves having serrate margins and palmate venation.

Although the late Cretaceous (Cenomanian-Maastrichtian) climate exhibited an equator-to-pole temperature gradient and was warmer on a global basis than it is today, the floras were not uniform but showed a latitudinal vegetational zonation (Fig. 31.35). For the Northern Hemisphere, Upchurch & Wolfe (1987) recognize three zones; (1) low-middle paleolatitudes (30° to 45°N), (2) high-middle paleolatitudes (45° to 65°N), and (3) high paleolatitudes (greater than 65°N).

During the early Cenomanian of the Northern Hemisphere, megafloras are diverse at lower-middle paleolatitudes and are comprised of an overwhelming number of angiosperm leaf fossils. Based on palynofloras the relative abundance of angiosperms reaches 40 percent. On the basis of their leaf physiognomy, these angiosperms were thought to have been successional plants during the Aptian and to have become the canopy of later successional forests where ferns and some gymnosperms (cycadophytes) had gradually been replaced by a variety of angiosperm successional components (Crane, 1987a). At middle paleolatitudes the physiognomy of most leaf assemblages, with their small leaf size, some with drip-tips and probable vine habit, reflect the vegetation of subhumid megathermal regions of the tropics with sandy porous soils. These regions typically have an open canopy comprised of plants of relatively short stature. A few assemblages at this paleolatitude show the characteristics of megathermal, closed canopy, multistratal forests that document wetter and warmer tropical environments. The proportion of entire margined species in these assemblages is usually greater than 70 percent, and most of the leaves are thick textured, indicating that megathermal vegetation was evergreen and paratropical to tropical, with mean annual temperatures of 20° to 25°C.

In the Upper Cretaceous between the Turonian and Campanian, palynological studies show that angiosperms came into dominance at high-middle paleolatitudes. By the Santonian, angiosperm leaves became the most abundant elements of the megafossil record. Based on leaf size and percentages of entire

Figure 31.35. Late Cretaceous vegetational map for North America, based on the Late Maastrichtian. Tropical rain forest (T); paratropical forest (P); notophyllous broad-leaved evergreen forest (N); polar broad-leaved deciduous forest (D). (From Upchruch & Wolfe, 1987, with permission.)

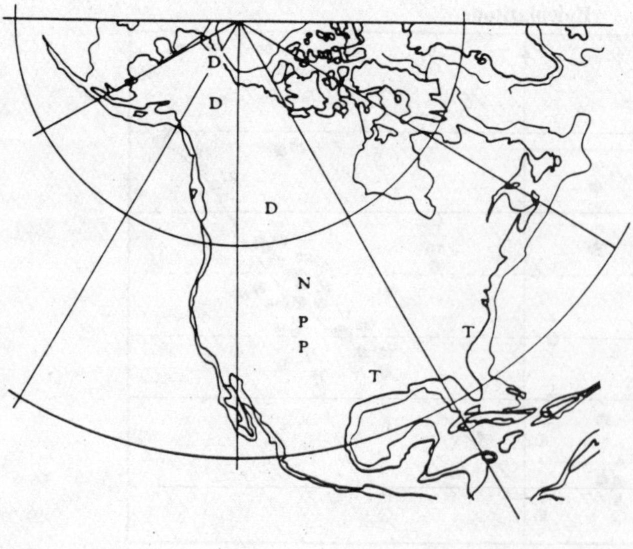

Figure 31.36. Diagram summarizing the distribution of major groups of plants present in the floras of the Neocomian, Aptian, Cenomanian; Campanian, and recent mid-latitude vegetation. Araucarian conifer (a); taxodiaceous conifer (b); cycadophyte (c); cycadeoid (d); herbaceous lycopod (e); fern (f); angiosperm shrub (g); angiosperm herb (h); gnetalean herb or small shrub (i); angiosperm tree (j). (From Crane, 1987a, with permission.)

margined leaves that are thick textured, a mesothermal evergreen vegetation growing in humid conditions is indicated. Leaf-size evaluations also indicate more favorable growth at high-middle paleolatitudes than at low-middle paleolatitudes. These high-middle mesothermal evergreen forests have leaves whose physiognomic features compare favorably with present-day notophyllous broad-leaved evergreen forests growing in environments with seasonal temperatures (warm month mean less than 20°C, cold month mean greater than 1°C). During the Turonian–Campanian interval, the number of angiosperms related to present-day groups increases at the ordinal and family levels. The palynological studies made during this interval (Muller, 1981) list 13 orders and 12 families as appearing by the Campanian. Most of these, as in the Cenomanian, belong to the subclasses Magnoliidae, Hamamelidae, and Rosidae. At low paleolatitudes during this interval, palms became an important part of the vegetation.

High-latitude vegetation of the Cenomanian–Maastrichtian interval is known in North America to occur in megafossil sequences from Alaska to Alberta that range in age from Cenomanian to Campanian. The angiosperm leaves from these latitudes are often associated with deciduous gymnosperms (*Metasequoia* and *Gingko*) and are characterized by small numbers of entire margined species with large average leaf size and thin texture. According to Wolfe (1985) and Upchurch & Wolfe (1987), these are characteristics of an extinct vegetational type called broad-leaved deciduous forest (Fig. 31.35). Here it is estimated that the temperature range was from mesothermal to microthermal. The leaf fall was apparently controlled by the onset of the Arctic night. Variability in leaf characteristics during this interval at this latitude is low but includes platanophylls, trochodendrophylls, and protophylls. The large size of the leaves conforms with other evidence indicating high precipitation.

As with other physiognomic features of angiosperm leaves, those of the polar broad-leaved deciduous forests can be accounted for by adaptations to various regimens of temperature, light, and precipitation.

Summary of succession

In previous paragraphs considerable emphasis has been placed on the radiation and diversity of angiosperms and their succession into new environments. This information has been summarized by Crane (1987a), who illustrates (Fig. 31.36) the predicted changes in terrestrial communities during the Cretaceous. We start with those early angiosperms of the Lower Cretaceous (Aptian–early Albian) that became established as early successional flowering plants. These were weedy herbs or shrubs that colonized more-or-less open environments previously occupied by ferns and lycopods. It is also postulated that small angiospermous trees and shrubs that appeared in the early Upper Cretaceous became efficient colonizers. In time they dominated environments formerly occupied by ferns and cycadophytes. Conifers, on the other hand, were only slightly affected by radiation of angiosperms. It is clear that the conifers represent the only group of gymnosperms that has retained dominance over large areas of Earth in spite of competition from the flowering plants. The large canopy-forming trees that make up conspicuous parts of our present-day angiosperm hardwood forests apparently did not become widespread until the latest Upper Cretaceous or early Tertiary. The radiation of herbaceous angiosperms, which have come to account for 50 percent of extant flowering plant species, did not take place until later in the Tertiary.

The Cretaceous–Tertiary boundary

The Cretaceous–Tertiary boundary (K/T boundary) is usually associated with mass extinction of dinosaurs. This extinction has been described as sudden and simultaneous on one hand, or possibly the product of a gradual decline involving a broad spectrum of organisms on the other. Explanations that have received the most attention are a late Cretaceous cooling of climate (Hickey, 1981) and/or a catastrophic impact by an extraterrestrial body (Alvarez et al., 1980). These authors report an iridium-rich horizon at the K/T boundary. Because iridium is fairly common in meteorites but rare on Earth, they postulate that a large asteroid estimated to have been 12 km in diameter collided with Earth 65 m.y. ago. The collision resulted in a dust cloud that eliminated sunlight from Earth's surface. Such an event would have greatly reduced surface temperature during the dark period, which could have resulted in sudden mass extinctions. When the dust cloud settled an iridium-rich layer resulted which marks the K/T boundary at the end of the Cretaceous.

Hickey (1984) makes the point, after studying the fossil plant record of horizons associated with the K/T boundary, that the number of plant extinctions was not unusual when compared with the number of extinctions at the Paleocene–Eocene boundary where there is no evidence of a universal catastrophe.

Subsequent investigation of the vegetation at the K/T boundary by Wolfe & Upchurch (1987b) supports the thesis that some kind of global catastrophe occurred, resulting in extensive and severe climatic change, extinctions, and ecological disturbance. The studies focus on the K/T boundary as it occurs in the Raton Basin of New Mexico and Colorado. Intact sequences containing leaf fossils have been found in iridium-rich boundary clays that include pollen assemblages. After a detailed analysis of the vegetational sequences below, at, and above the K/T boundary, Wolfe & Upchurch (1987b) (Fig. 31.37) recognize five floristic phases. The first of these is represented by fossil leaves from the upper part of the Vermejo Formation and the base of the Raton Formation. The phase 1 flora is diverse, containing a wide variety of dicots and some palms. Evergreen conifers occur with some frequency and show affinities with extant Cupressaceae of Australia and New Caledonia. Study of the physiognomy and cuticles of phase 1 vegetation indicates a climate with megathermal temperatures and moderate precipitation in dry sunny habitats. Palynological studies show that fern spores make up less than 25 percent of the pollen flora at this time.

The K/T boundary is marked by the iridium-rich clay between phases 1 and 2. Immediately above the boundary in phase 2, fern spores reach an abundance of 96 to 99 percent. This so-called "fern spike," which was of short duration, is interpreted as representing the vegetation following a mass kill similar to that observed after the eruption of Krakatoa. The point should be made here that a mass kill does not necessarily mean there was a mass extinction. The fern fronds found in phase 2 just above the clay represent a genus of extant ferns that are known to be primary colonizers of vegetation in Africa and Indomalaya.

When phase 1 is compared with phase 3, a marked decrease in diversity of angiosperms is evident. The physiognomy of phase 3 leaves indicates that there were early successional stream-side plants growing in a megathermal, wet climate where recolonization was occurring following the K/T event.

During phases 4 and 5 a gradual increase in

Figure 31.37. Changes in leaf physiognomy from the latest Cretaceous to the early Paleocene, including the K/T boundary, in the Raton basin of New Mexico and Colorado. See text for explanation. (From Wolfe & Upchurch, 1987b, with permission.)

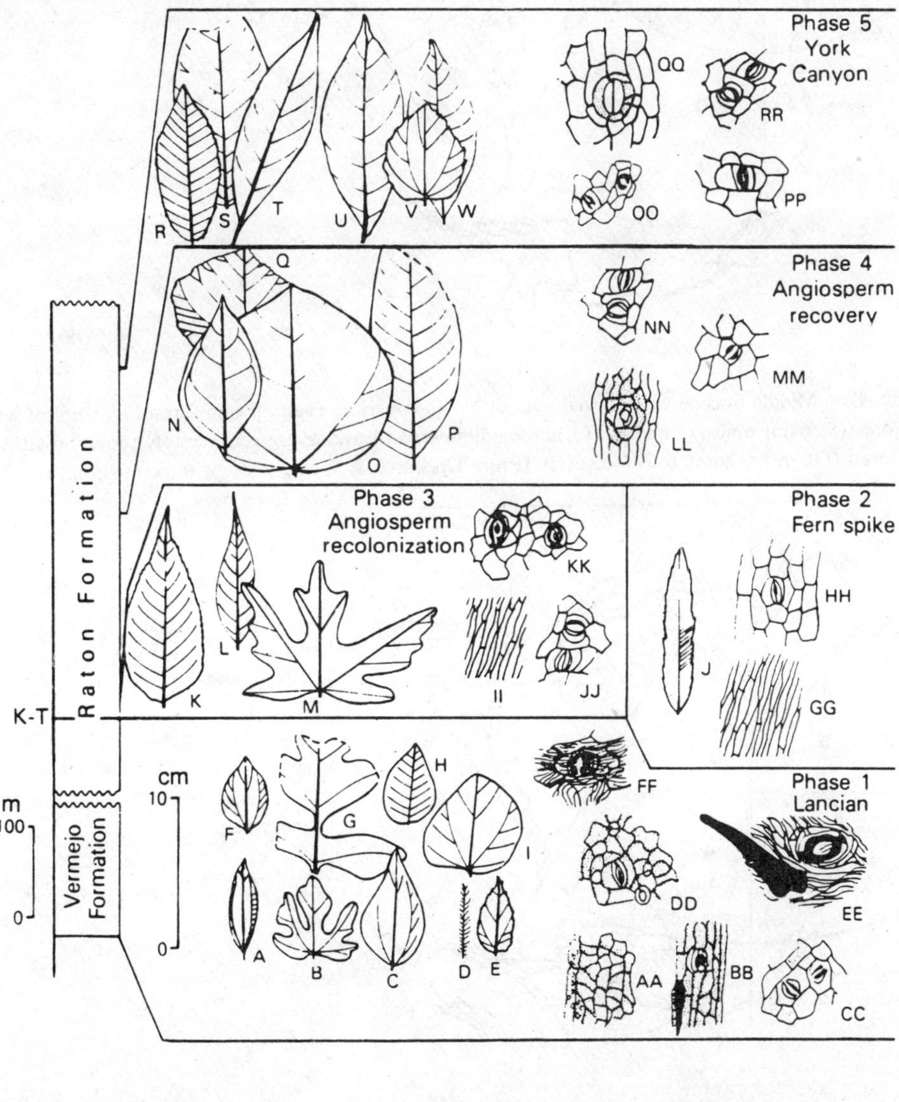

diversity is observed. However, following a period of 115 m.y. after the K/T event, the diversity of phase 5 is not as great as it was in phase 1. Nonetheless, phase 4 and 5 dicots are more diverse than in phase 3. The dominant plants of these phases are palms and evergreen dicots with leaves having drip-tips, a high leaf index, and cuticle types that occur on leaves of plants in megathermal rain forests.

In general, the post-K/T changes in vegetation reflect a normal ecological succession and recovery after a mass kill in a gradually changing environment. However, the magnitude of the mass kill and its relation to the amount of extinction is not clear. Estimates of the amount of extinction across the K/T boundary, based on megafossils, vary from 40 to 75 percent.

Figure 31.38. Early Paleocene vegetational map for North America. Note the absence of notophyllous broad-leaved evergreen vegetation and the close proximity of paratropical rain forest and broad-leaved deciduous forest. Tropical rain forest (T); paratropical rain forest (P); broad-leaved deciduous forest (D). (From Upchurch & Wolfe, 1987, with permission.)

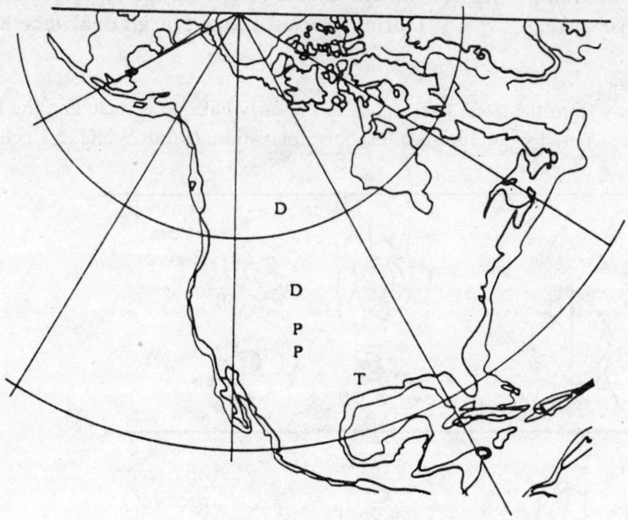

Figure 31.39. Late Middle Eocene vegetational map of North America. Tropical rain forest (T); tropical semi-deciduous forest (S); paratropical rain forest (P); notophyllous broad-leaved evergreen forest (N); polar broad-leaved deciduous forest (D); mixed coniferous forest (C). (From Upchurch & Wolfe, 1987, with permission.)

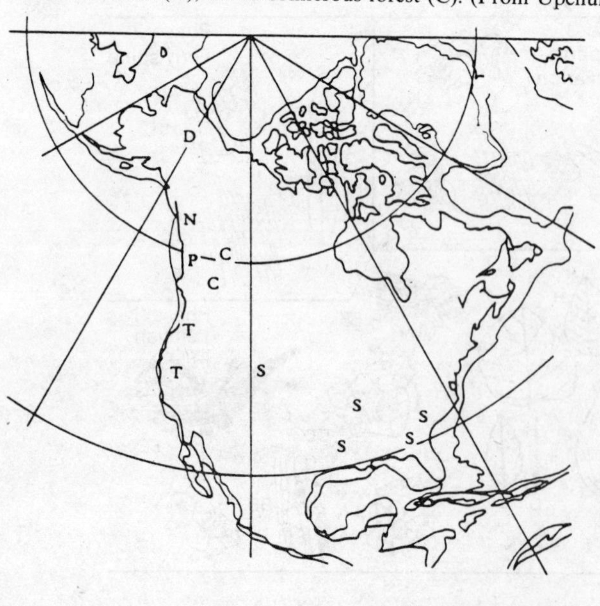

Deciduousness. Although deciduousness is known to have been a feature of some Cretaceous floras, it was not until the late Cretaceous that polar broad-leaved deciduous forests became important components of North American vegetation. The deciduous characteristic of angiosperm vegetation may have evolved early in their radiation as an adaptation to xeric climatic conditions. In the early Cretaceous, the evolving angiosperms apparently were subjected to fluctuations of water-table levels along stream margins and thus periodic drought conditions. In such an environment deciduousness, followed by a period of dormancy, would have promoted survival. The terminal event at the K/T boundary may have resulted in widespread selection for angiosperms with the deciduous habit leading to their diversification in the Tertiary (Wolfe, 1987b).

Tertiary climate and vegetation

Paleocene–Eocene

The period of vegetational recovery in the Paleocene that followed the K/T event in many ways exhibits a modern secondary succession extending over a long period of time. Although there is no evidence for a decrease in temperature, the amount of precipitation increased considerably at low and middle paleolatitudes. This increase initiated the development of tropical and paratropical rain forests in megathermal regions (Fig. 31.38). The megafossils in these regions are characterized by large leaves and some species with drip-tips. In the high-middle latitudes extensive deciduous vegetation replaced much of the broad-leaved evergreen taxa, which became extinct at the time of the K/T event (Upchurch & Wolfe, 1987). During the Paleocene, deciduous vegetation became an important component of the northern Rocky Mountain and Great Plains regions. The floras of these regions are characterized by deciduous genera belonging to families that are important members of our extant vegetation. Among these families are the Betulaceae and Juglandaceae, which became highly diversified during the Paleocene. Throughout the Paleocene and Eocene the floras became increasingly modern in composition. Increase in warming of the climate continued into the early Eocene so that the global climate became warmer than at present. Precipitation also increased dramatically, making it possible for floras with tropical and paratropical aspects to extend to latitudes approximating 50° to 60°N. With the changes in temperature and precipitation, the broad-leaved deciduous and coniferous forests were restricted to even higher latitudes.

During the Middle Eocene (Fig. 31.39), the megathermal vegetation of southeastern North America and Europe was exposed to seasonally dry climates and is characterized by semideciduous forests. A major decline in the temperature at the end of the Eocene greatly reduced the northerly latitudinal extent of the megathermal vegetation, allowing the development of large areas of microthermal broad-leaved deciduous forests to expand southward for the first time. This change was accompanied by a decline in floral diversity at middle and high paleolatitudes. Although minor changes in temperature and precipitation have occurred since the end of the Eocene, the general composition and disposition of the floras of North America are similar to those of the present day. One notable exception is the expansion of open grasslands at middle and high paleolatitudes during the Miocene. Unequivocal grass pollen is first recorded from the Paleocene with a record of permineralized grass from the Upper Miocene of California (Tidwell & Nambudiri, 1989). This discovery suggests that at this time a chaparral-type grassland had been established in the Mojave desert region. More recently Crepet & Feldman (1991) have described compression–impression fossils of grasses from the Paleocene–Eocene Wilcox Formation of Tennessee. These megafossils exhibit a suite of characters belonging to members of the family Poaceae and represent the earliest known remains of grasses in the fossil record.

References

Alvarez, L. W., Alvarez, W., Asaro, F., & Michel, H. V. (1980). Extraterrestrial cause for the Cretaceous/Tertiary extinction. *Science*, **208**, 1095.

Axelrod, D. I. (1952). A theory of angiosperm evolution. *Evolution*, **6**, 29–60.

Axelrod, D. I. (1959). Poleward migration of early angiosperm floras. *Science*, **130**, 203–7.

Axelrod, D. I. (1960). The evolution of flowering plants. *Evolution after Darwin*. **1**, 277–305.

Bailey, I. W. (1949). Origin of the angiosperms: Need for a broadened outlook. *Journal of the Arnold Arboretum*, **30**, 64–70.

Basinger, J. F. (1976a). Permineralized plants from the Eocene, Allenby Formation of southern British Columbia. M.Sc. thesis, University of Alberta, Edmonton, Canada.

Basinger, J. F. (1976b). *Paleorosa similkameenensis* gen. et sp. nov. permineralized flowers (Rosaceae) from the Eocene of British Columbia. *Canadian Journal of Botany*, **54**, 2293–305.

Basinger, J. F., & Christophel, D. C. (1985). Fossil flowers and leaves of the Ebenaceae from the Eocene of southern Australia. *Canadian Journal of Botany*, **63**, 1825–43.

Basinger, J. F., & Rothwell, G. W. (1977). Anatomically preserved plants form the Middle Eocene (Allenby) Formation of British Columbia. *Canadian Journal of Botany*, **55**, 1984–90.

Cevallos-Ferriz, S. R. S., & Stockey, R. A. (1988a). Permineralized fruits and seeds from the Princeton chert (Middle Eocene) of British Columbia: Lythraceae. *Canadian Journal of Botany*, **66**, 303–12.

Cevallos-Ferriz, S. R. S., & Stockey, R. A. (1988b). Permineralized fruits and seeds from the Princeton chert (Middle Eocene) of British Columbia: Araceae. *American Journal of Botany,* **75,** 1099–113.

Christophel, D. C. (1985). First record of well-preserved megafossils of *Nothofagus* from mainland Australia. *Proceedings of the Royal Society of Victoria,* **97,** 175–78.

Collinson, M. E., & Hooker, J. J. (1987). Vegetational and mammalian faunal changes in the Early Tertiary of southern England. In *The Origins of Angiosperms and Their Biological Consequences,* eds. E. M. Friis, W. G. Chaloner, & P. R. Crane. New York: Cambridge University Press.

Crabtree, D. R. (1987). Angiosperms of the northern Rocky Mountains: Albian to Campanian (Cretaceous) megafossil floras. *Annals of the Missouri Botanical Garden,* **74,** 707–47.

Crane, P. R. (1984). Early fossil history of the Betulaceae: A preliminary report. *American Journal of Botany,* **71,** 109 (Abstract).

Crane, P. R. (1987a). Vegetational consequences of angiosperm diversification. In *The Origins of Angiosperms and Their Biological Consequences,* eds. E. M. Friis, W. G. Chaloner, & P. R. Crane. New York: Cambridge University Press.

Crane, P. R. (1987b). Paleobotanical evidence on the early radiation of nonmagnoliid dicotyledons. *Plant Systematics and Evolution.* **162,** 165–91.

Crane, P. R., Dilcher, D. L., & Manchester, S. R. (1986). *Nordenskioldia:* a vesselless angiosperm from the Paleocene of North America. *American Journal of Botany,* **73,** 697. (Abstract)

Crane, P. R., & Stockey, R. A. (1985). Growth and reproductive biology of *Joffrea speirsii* gen. et sp. nov., a *Cercidiphyllum*-like plant from the Late Paleocene of Alberta, Canada. *Canadian Journal of Botany,* **63,** 340–64.

Crane, P. R., & Stockey, R. A. (1986a). Morphology and development of pistillate inflorescences in extant and fossil Cercidiphyllaceae. *Annals of the Missouri Botanical Garden,* **73,** 382–93.

Crane, P. R., & Stockey, R. A. (1986b). *Betula* leaves and reproductive structures from the Middle Eocene of British Columbia, Canada. *Canadian Journal of Botany,* **65,** 2490–500.

Crepet, W. L. (1974). Investigations of North American cycadeoids: The reproductive biology of Cycadeoidea. *Palaeontographica B,* **148,** 144–69.

Crepet, W. L. (1978). Investigations of angiosperms from the Eocene of North America: an aroid inflorescence. *Review of Palaeobotany and Palynology,* **25,** 241–52.

Crepet, W. L. (1979). Some aspects of pollination biology of Middle Eocene angiosperms. *Review of Palaeobotany and Palynology,* **27,** 213–38.

Crepet, W. L. (1985). Advanced (constant) insect pollination mechanisms: patterns and implications vis-a-vis angiosperm diversity. *Annals of the Missouri Botanical Garden,* **71,** 607–30.

Crepet, W. L., & Daghlian, C. P. (1980). Castaneoid inflorescences from the Middle Eocene of Tennessee and the diagnostic value of pollen (at the subfamily level) in the Fagaceae. *American Journal of Botany,* **67,** 739–57.

Crepet, W. L., & Daghlian, C. P. (1981). Euphorbioid inflorescences from the Middle Eocene of the Claiborne Formation. *American Journal of Botany,* **69,** 258–66.

Crepet, W. L., & Dilcher, D. L. (1977). Investigations of angiosperms from the Eocene of North America: a mimosoid inflorescence. *American Journal of Botany,* **64,** 714–25.

Crepet, W. L., Dilcher, D. L., & Potter, F. W. (1974). Eocene angiosperm flowers. *Science,* **185,** 781–2.

Crepet, W. L., Dilcher, D. L., & Potter, F. W. (1985). Investigations of angiosperms from the Eocene of North America: A catkin with juglandaceous affinities. *American Journal of Botany,* **62,** 813–23.

Crepet, W. L., & Feldman, G. D. (1991). The earliest remains of grasses in the fossil record. *American Journal of Botany,* **78,** 1010–14.

Crepet, W. L., & Friis, E. M. (1987). The evolution of insect pollination. In *The Origins of Angiosperms and Their Biological Consequences,* eds. E. M. Friis, W. G. Chaloner, & P. R. Crane. New York: Cambridge University Press.

Crepet, W. L., & Taylor, D. W. (1985). The diversification of the Leguminosae: first fossil evidence of the Mimosoideae and Papilionoideae. *Science,* **228,** 1087–9.

Crepet, W. L., & Taylor, D. W. (1986). Primitive mimosoid flowers from the Paleocene/Eocene and their systematic and evolutionary implications. *American Journal of Botany,* **73,** 548–63.

Cronquist, A. (1981). *An Integrated System of Classification of Flowering Plants.* New York: Columbia University Press.

Daghlian, C. P. (1981). A review of the fossil record of monocotyledons. *The Botanical Review,* **47,** 517–55.

Daghlian, C. P., Crepet, W. L., & Delevoryas, T. (1980). Investigations of Tertiary angiosperms: a new flora including *Eomimosoidea plumosa* from the Oligocene of eastern Texas. *American Journal of Botany,* **67,** 309–20.

Dilcher, D. L. (1974). Approaches to the identification of angiosperm leaf remains. *Botanical Review,* **40,** 1–157.

Dilcher, D. L. (1979). Early angiosperm reproduction: an introductory report. *Review of Palaeobotany and Palynology,* **27,** 291–328.

Dilcher, D. L., & Daghlian, C. P. (1977). Investigations of angiosperms from the Eocene of North America: *Philodendron* leaf remains. *American Journal of Botany,* **64,** 526–34.

Dilcher, D. L., Potter, F. W., & Crepet, W. L. (1976). Investigations of angiosperms from the Eocene of North America: Juglandaceous winged fruits. *American Journal of Botany,* **63,** 532–44.

Doyle, J. A. (1969). Cretaceous angiosperm pollen of the Atlantic Coastal plain and its evolutionary significance. *Journal of the Arnold Arboretum,* **50,** 1–35.

Doyle, J. A. (1973). The monocotyledons: their evolution and comparative biology. V. Fossil evidence on early evolution of monocotyledons. *Quarterly Review of Biology,* **48,** 399–413.

Doyle, J. A., & Hickey, L. J. (1976). Pollen and leaves from the mid-Cretaceous Potomac Group and their bearing on angiosperm evolution. In *Origin and Early Evolution of Angiosperms,* ed. C. B. Beck. New York: Columbia University Press.

Drinnan, A. H., Crane, P. R., Friis, E. M., & Pedersen, K. R. (1991). Angiosperm flowers and tricolpate pollen of buxaceous affinity from the Patomac Group (mid-Cretaceous) of eastern North America. *American Journal of Botany,* **78,** 153–76.

Erwin, D. M., & Stockey, R. A. (1989). Permineralized monocotyledons from the Middle Eocene Princeton chert (Allenby Formation) of British Columbia: Alismataceae. *Canadian Journal of Botany,* **67,** 2636–45.

Erwin, D. M., & Stockey, R. A. (1990). Sapindaceous flowers from the Middle Eocene Princeton chert (Allenby Formation) of British Columbia, Canada. *Canadian Journal of Botany,* **68,** 2025–34.

Erwin, D. M., & Stockey, R. A. (1991a). Silicified monocotyledons from the Middle Eocene Princeton chert (Allenby Formation) of British Columbia, Canada. *Review of Palaeobotany and Palynology,* **70,** 147–62.

Erwin, D. M., & Stockey, R. A. (1991b). *Soleredera rhizomorpha* gen. et sp. nov., a permineralized monocotyledon from the Middle Eocene Princeton chert of British Columbia, Canada. *Botanical Gazette* **152,** 231–47.

Erwin, D. M., & Stockey, R. A. (1991c). Silicified monocotyledons from the Middle Eocene Princeton chert (Allenby Formation) of British Columbia. *Review of Palaeobotany and Palynology,* **70,** 147–62.

Erwin, D. M., & Stockey, R. A. (In press). Vegetative body of a permineralized monocotyledon from the Middle Eocene Princeton chert of British Columbia, Canada. *Couriėr Forshungsinstitut Senckenberg.*

Friis, E. M. (1983). Upper Cretaceous (Senonian) floral structures of juglandalean affinity containing Normapolles pollen. *Review of Palaeobotany and Palynology,* **39,** 161–88.

Friis, E. M. (1990). *Silvianthemum suecicum* gen. et sp. nov., a new saxifragalean flower from the Late Cretaceous of Sweden. *Biologiske Skrifter,* **36,** 5–21.

Friis, E. M., Chaloner, W. G., & Crane, P. R. (1987). Introduction to the angiosperms. In *The Origins of Angiosperms and Their Biological Consequences,* eds. E. M. Friis, W. G. Chaloner, & P. R. Crane. New York: Cambridge University Press.

Friis, E. M., & Crepet, W. L. (1987). Time of appearance of floral features. In *The Origins of Angiosperms and Their Biological Consequences,* eds. E. M. Friis, W. G. Chaloner, & P. R. Crane. New York: Cambridge University Press.

Friis, E. M., & Skarby, A. (1981). Structurally preserved angiosperm flowers from the Upper Cretaceous of southern Sweden. *Nature,* **291,** 485–6.

Friis, E. M., & Skarby, A. (1982). *Scandianthus* gen. nov., angiosperm flowers of saxifragalean affinity from the Upper Cretaceous of southern Sweden. *Annals of Botany,* **50,** 569–83.

Hickey, L. J. (1973). Classification of the architecture of dicotyledonous leaves. *American Journal of Botany,* **60,** 17–33.

Hickey, L. J. (1981). Land plant evidence compatible with gradual, not catastrophic change at the end of the Cretaceous. *Nature,* **292,** 529–31.

Hickey, L. J. (1984). Changes in the angiosperm flora across the Cretaceous-Tertiary boundary. In *Catastrophies and Earth History,* eds. W. A. Berggren & J. A. VanCouvering. Princeton: Princeton University Press.

Hickey, L. J., & Doyle, J. A. (1977). Early Cretaceous fossil evidence for angiosperm evolution. *Botanical Review,* **43,** 3–104.

Hickey, L. J., & Peterson, R. K. (1978). *Zingiberopsis,* a fossil genus of the ginger family from the Late Cretaceous to early Eocene sediments of western interior North America. *Canadian Journal of Botany,* **56,** 1136–52.

Hickey, L. J., & Wolfe, J. A. (1975). The bases of angiosperm phylogeny: vegetative morphology. *Annals of the Missouri Botanical Garden,* **62,** 538–89.

Hughes, N. F., & McDougall, A. B. (1987). Records of angiosperm pollen entry into English Early Cretaceous succession. *Review of Palaeobotany and Palynology,* **50,** 255–72.

Jain, R. K. (1963). Studies in the Musaceae. 1. *Musa cardiosperma* sp. nov., a fossil banana fruit from the Deccan Intertrappean Series. *The Palaeobotanist* **12,** 45–8.

Manchester, S. R. (1986a). Vegetative and reproductive morphology of an extinct plane tree (Platanaceae) from the Eocene of western North America. *Botanical Gazette,* **147,** 200–26.

Manchester, S. R. (1986b). Extinct ulmaceous fruits from the Tertiary of Europe and western North America. *Review of Palaeobotany and Palynology,* **52,** 119–29.

Manchester, S. R. (1989). Attached reproductive and vegetative remains of the extinct American-European genus *Cedrelospermum* (Ulmaceae) from the early Tertiary of Utah and Colorado. *American Journal of Botany,* **76,** 256–76.

Muller, J. (1981). Fossil pollen records of extant angiosperms. *The Botanical Review,* **47,** 1–142.

Nambudiri, E. M. V., Tidwell, W. D., & Chitaley, S. (1987). Revision of *Sahniocarpon harrisii* Chitaley & Patil based on new specimens from the Deccan Intertrappean beds of India. *Great Basin Naturalist,* **47,** 527–35.

Nishida, H., & Nishida, M. (1988). *Protomonimia kasai-nakajhongii* gen. et sp. nov.: a permineralized magnolialean fructification from the mid-Cretaceous of Japan. *The Botanical Magazine, Tokyo,* **101,** 397–426.

Parrish, J. T. (1987). Global palaeogeography and palaeoclimate of the Late Cretaceous and Early Tertiary. In *The Origins of Angiosperms and Their Biological Consequences,* eds. E. M. Friis, W. G. Chaloner, & P. R. Crane. New York: Cambridge University Press.

Pigg, K. B., & Stockey, R. A. (1991). Platanaceous plants from the Paleocene of Alberta, Canada. *Review of Palaeobotany and Palynology,* **70,** 125–46.

Reid, E. M., & Chandler, M. E. (1933). *The London Clay Flora.* London: British Museum (Natural History).

Stockey, R. A. (1987). A permineralized flower from the Middle Eocene of British Columbia. *American Journal of Botany,* **74,** 1878–87.

Stockey, R. A., & Crane, P. R. (1983). In situ *Cercidiphyllum*-like seedlings from the Paleocene of Alberta, Canada. *American Journal of Botany*, **70,** 1564–68.

Stockey, R. A., & Pigg, K. B. (1991). Flowers and fruits of *Princetonia allenbyenesis* from the Middle Eocene chert of British Columbia. *Review of Palaeobotany and Palynology*, **70,** 163–72.

Takhtajan, A. (1969). *Flowering Plants: Origin and Dispersal* Edinburgh: Oliver & Boyd.

Taylor, D. W., & Crepet, W. L. (1987). Fossil floral evidence of Malpighiaceae and an early plant-pollinator relationship. *American Journal of Botany*, **74,** 274–86.

Thomas, B. A., & Spicer, R. A. (1986). *The Evolution and Palaeobiology of Land Plants*. Portland, Oregon: Dioscorides Press.

Thomasson, J. R. (1978). Observations on the characteristics of the lemma and palea of the late Cenozoic grass *Panicum elegans*. *American Journal of Botany*, **65,** 34–9.

Tidwell, W. D. (1975). *Common Fossil Plants of Western North America*. Provo, Utah: Brigham Young University Press.

Tidwell, W. D., & Nambudiri, E. M. V. (1989). *Tomlinsonia thomassonii*, gen. et sp. nov., a permineralized grass from the Upper Miocene Ricardo Formation, California. *Review of Palaeobotany and Palynology*, **60,** 165–177.

Tidwell, W. D., & Parker, L. R. (1990). *Protoyucca shadishii* gen. et sp. nov., an arborescent monocotyledon with secondary growth from the Middle Miocene of northwestern Nevada, U.S.A. *Review of Palaeobotany and Palynology*, **62,** 79–95.

Upchurch, G. R., & Wolfe, J. A. (1987). Mid-Cretaceous to Early Tertiary vegetation and climate: evidence from fossil leaves and woods. In *The Origins of Angiosperms and Their Biological Consequences*, eds. E. M. Friis, W. G. Chaloner, & P. R. Crane. New York: Cambridge University Press.

Walker, J. W., & Walker, A. G. (1984). Ultrastructure of Lower Cretaceous angiosperm pollen and the origin and early evolution of flowering plants. *Annals of the Missouri Botanical Garden*, **71,** 464–521.

Wolfe, J. A. (1979). Temperature parameters of humid to mesic forests of eastern Asia and relation to forests of other regions of the Northern Hemisphere and Australasia. *U. S. Geological Survey Professional Paper*, **1106,** 37 pp.

Wolfe, J. A. (1985). Distribution of major vegetational types during the Tertiary. In *The Carbon Cycle and Atmospheric CO_2: Natural Variations Archean to Present. Geophysical Monograph*, **32,** eds. E. T. Sundquist & W. S. Broecker. Washington, D.C.: American Geophysical Union.

Wolfe, J. A. (1987a). An overview of the origins of the modern vegetation of the northern Rocky Mountains. *Annals of the Missouri Botanical Garden*, **74,** 785–803.

Wolfe, J. A. (1987b). Cretaceous-Cenozoic history of deciduousness and the terminal Cretaceous event. *Paleobiology*, **13,** 215–226.

Wolfe, J. A., & Tanai, T. (1987). Systematics, phylogeny and distribution of *Acer* (maples) in the Cenozoic of western North America. *Journal of the Faculty of Science, Hokkaido University*, **22,** I. 1–246.

Wolfe, J. A., & Upchurch, G. R. (1987a). North American nonmarine climates and vegetation during the Late Cretaceous. *Palaeogeography, Plaeoclimatology, Palaeoecology*, **61,** 33–77.

Wolfe, J. A., & Upchurch, G. R. (1987b). Leaf assemblages across the Cretaceous-Tertiary boundary in the Raton Basin, New Mexico and Colorado. *Proceedings of the National Academy of Science, U.S.A.*, **84,** 5096–100.

32

Major evolutionary events and trends – in retrospect

In the time it takes to cover the information from the first evidence of living organisms in the Precambrian to the appearance of bilateral symmetry in an Eocene flower, one tends to lose sight of the many important evolutionary trends and events that have occurred in between. Thus, the purpose of this short chapter is to present a brief and concise review of these trends and events and to relate them, in a general way, to the geological record. Similar useful summaries have been prepared by Banks (1970), Chaloner (1970), and Chaloner & Sheerin (1979, 1981). Documentation obtained from the works of a multitude of researches has been presented in the preceding chapters and will not be repeated. To make the material that follows more useful, words and phrases describing the first known appearances of groups of organisms, structures, or conditions are in boldface. All dates given for each segment of geological time are in terms of "years ago." Dates used are those of Harland, Smith, & Wilcock (1964) and van Eysinga (1975).

1. *Precambrian, about 3.5 billion years.* The first evidence of cellular organization. Both **unicells** and **filaments of cells** having the size, shape, and organization of **prokaryotic bacteria** and **blue-green algae.** Some kind of **photosynthetic mechanism** probably accompanied the evolution of some of these organisms, whereas some were no doubt **heterotrophs.**

2. *Precambrian, about 2.1 billion years.* Organisms recognizable as blue-green algae because of the organization of their **filamentous colonies. Coccoid-** and **bacillus-**type cells comparing favorably with those of extant bacteria are present.

3. *Precambrian, about 1400 million years.* Diversification of blue-green algae to produce several morphologies identical with modern blue-greens. Unicellular organisms resembling extant **green** and **red algae** have been found. Many cells of these organisms contain a structure that has the size and shape of a **nucleus.** The evidence is equivocal. If proven, we can conclude that the process of **mitosis** had evolved in **eukaryotic** organisms. Branching nonseptate filaments suggest that **fungi** similar to **phycomycetes** had evolved.

4. *Cambrian and Ordovician, between 570 and 435 million years.* Great diversification of algal types. In addition to blue-green algae, with many types identical to extant genera, there are morphologies similar to green, red, and **brown algae.** These organisms and the occurrence of invertebrates leave no doubt that the eukaryotic condition had evolved, as had **multicellular organization. Spores** with **triradiate ornamentations** provide evidence that **tetrahedral tetrads** were produced by **meiosis.** This leads to the inevitable conclusion that there were plant life cycles incorporating **sexual reproduction.**

5. *Mid-Silurian (Wenlockian), about 420 million years.* The first evidence of **vascular land plants.** These small plants had **naked dichotomizing axes** with **terminal eusporangia.** The **parenchymatous axes** contain a slender strand of putative **primary xylem** in the center. The **sporangia** are **multicellular** and produce meiospores with a thick **sporoderm.**

6. *Lower Devonian, between 395 and 374 million years.* **Rhyniophytes** and **zosterophylls** with diverse morphologies. First demonstration of simple **stomata** and **cuticlelike** covering on aerial axes. **Rhizoids** extended from basal portions. **Pseudomonopodial branching** derived by overtopping of dichotomous axes. Some axes with **enations** or **microphylls** with a single vein. Enations with an irregular arrangement on the axes; microphylls usually **spirally arranged.** Evolution of **protosteles**

that are circular, **centrarch,** elliptical or stellate, and **exarch.** In addition to terminal sporangia on main axes (rhyniophytes), **sporangia** are at the **tips** of **short lateral branches** (zosterophylls). These are mostly **reniform sporangia** with **transverse dehiscence.** Phycomycetes present. **Trimerophytes** appear and diversify in later Lower Devonian.

7. Middle Devonian, between 374 and 359 million years. Other trimerophyte derivatives and **aneurophytes** make their appearance. Increase in plant size, including **arborescence** in some **progymnosperms, lycopods,** and **Cladoxylales** and a variety of branching patterns. Indication of **axillary branching** in trimerophytes. Aneurophytes with modified, **planated,** and partially **webbed** branches considered to be **prefronds** with **determinate growth.** First evidence of a **bifacial cambium, secondary xylem,** and **secondary phloem** in aneurophytes. **Fusiform sporangia** in clusters. **Longitudinal dehiscence** of sporangial wall.

8. Upper Devonian, between 359 and 353 million years. **Archaeopterids,** aneurophytes, **lycopsids, sphenopsids,** and **ferns** indicate diversification from the primitive types. Microphylls and spirally arranged **simple** and **compound megaphylls.** Evolution of the **eustle** (*Archaeopteris*) from protostele, the eustele with a pith surrounded by a ring of sympodia. Secondary xylem of gymnosperm types. Well-defined **free-sporing heterospory.** First evidence of **liverworts.**

9. Upper Devonian, between 353 and 345 million years. **Pteridosperms** and **arborescent sphenopsids.** First known appearance of **preovules** in **cupule**like structures. These associated with **fronds** of pteridosperms. Unifacial vascular cambium in arborescent lycopods.

10. Mississippian and Pennsylvanian (Carboniferous), between 345 and 280 million years. Great diversification to produce the Carboniferous floras consisting of **cordaites,** conifers, seed ferns in great variety, herbaceous and arborescent lycopsids and sphenopsids, and ferns including **leptosporangiate** and **eusporangiate** types that share some characteristics of the **Filicales** and **Marattiales.** Plants with life cycles in which well-developed **ovules** and **cupules** are produced. In the life cycles of these plants tetrahedral tetrads are produced with a **single functional megaspore, endosporic megagametophytes** with **archegonia** and **microgametophytes** with **pollen tubes,** several **prothallial cells,** and **paired sperms.** Some showing evidence of **dormancy.** Embryos are associated with megagametophytes. Evolution of conifer **compound seed cones. Ascomycetes** and **basidiomycetes** make their appearance, as do the **mosses.**

11. Permian and Triassic, between 280 and 195 million years. Permian extinction of Carboniferous pteridosperms, arborescent lycopsids, and sphenopsids. Diversification of **Voltziales** (transition conifers) and ferns. First evidence of **cycads** and **cycadeoids. Glossopterids** become a conspicuous part of the Southern Hemisphere Permian floras; **ginkgophytes** appear in the Permian and increase in the Triassic.

12. Jurassic, between 195 and 141 million years. Complex cupulate pteridosperms in some variety. Rapid diversification of ferns, cycads, cycadeoids, conifers, and ginkgophytes, with all of these groups reaching their maximum abundance during the Jurassic. Sphenopsids and lycopsids a less conspicuous part of the flora. Extinction of glossopterids. Some isolated examples of plants with angiospermlike leaf and wood characteristics. **Tectate pollen** first known to occur.

13. Lower Cretaceous, between 141 and 100 million years. Unequivocal evidence of **angiosperm leaves, pollen,** and **flowers.** Both **monocots** and **dicots** present, as are wind- and insect-pollinated flowers. **Carpels, perianths, stamens,** and **tectate columellate pollen** of the **monosulcate, tricolpate,** and **triporate** types. Primitive leaf types with **reticulate venation** of the "first-rank" angiosperm type. Extremely rapid diversification of angiosperms that began replacement of the declining cycads, cycadeoids, and ginkgophytes. Mesozoic pteridosperms and other minor groups become extinct.

14. Upper Cretaceous, between 100 and 65 million years. Many angiosperms with characteristics of extant families in addition to those that have no living counterparts. Generalized flower types present that may represent precursors of extant groups. Great diversification of angiosperm pollen and leaf types. Most families of ferns and conifers continue to the present. The variety of ginkgophtyes and cycads declines dramatically. Cycadeoids become extinct.

15. Tertiary, between 65 million and present. Steady evolution of new angiosperm types including those with bilaterally symmetrical flowers, herbaceous growth habits such as grasses, and a multitude of seed, fruit, leaf, and pollen types. Conifers decline in diversity, but ferns expand.

The evidence summarized above makes it clear that each major group of vascular plants that we recognize today and some that have become extinct originated by radiations from simple generalized forms that produced, at first, a variety of organisms having different sets of characteristics than those we associate with extant plants. Subsequent selection and radiation have produced those organisms, mostly extant, with which we are familiar. Because of different rates of evolution, new variants in some groups, such as the angiosperms, appeared in a relatively short time, while in others where there are slower rates, as in the conifers, new species appeared over longer periods of time. From these examples it is apparent that there are different rates for the diversification and evolution of vascular plants.

Patterns of diversification: vascular plants

According to a retrospective analysis of fossils at the species level, Niklas, Tiffney, & Knoll (1983, 1985) have identified an overall pattern for the diversification of vascular plants. In preparing the analysis, Niklas et al. (1983, 1985) compiled approximately 18,000 citations for fossil plant species which were computerized according to affinity, age, and location. Using this information, it was possible to arrive at rough approximations of the evolutionary rates of the plants based on the values obtained for rates of species origination and longevity. Figures 32.1, 32.2, and 32.3) show the rates of changes in diversity (at the species level) for suprageneric taxa through geological time. Collectively, the graphs show that four phases have successively dominated the vascular plant floras. The dominance of the first phase is marked by the time when there was a maximum number of species of primitive vascular plants. Concomitant with the radiation and diversification of derived taxa comprising the second phase, the number of species of phase one was gradually reduced and finally terminated by extinction. The diversifying species of the third phase, in a similar way, replaced many of those of the second phase and so on to the fourth phase where the dominant number of species is that of the flowering plants that are with us today. Thus, Niklas et al. (1983, 1985) have demonstrated that shortly after the initial appearance of members of each major taxon, there was a time of rapid speciation. With the passage of time, however, the rate of speciation was reduced, with those species characterized by longevity remaining often to be replaced by rapidly radiating and diversifying species of the next phase. This obviously does not apply to the fourth phase where the angiosperms are still undergoing rapid diversification with no indication of their replacement.

As shown in Figure 32.1, the phase one primitive early vascular plants (rhyniophytes, trimerophytes, zosterophylls), after reaching their peaks of diversifi-

Figure 32.1. Changes in diversity, at the species level, of suprageneric taxa represented in the Silurian and Devonian Periods. For explanation, see text. (From Niklas, Tiffney, & Knoll, 1985, with permission.)

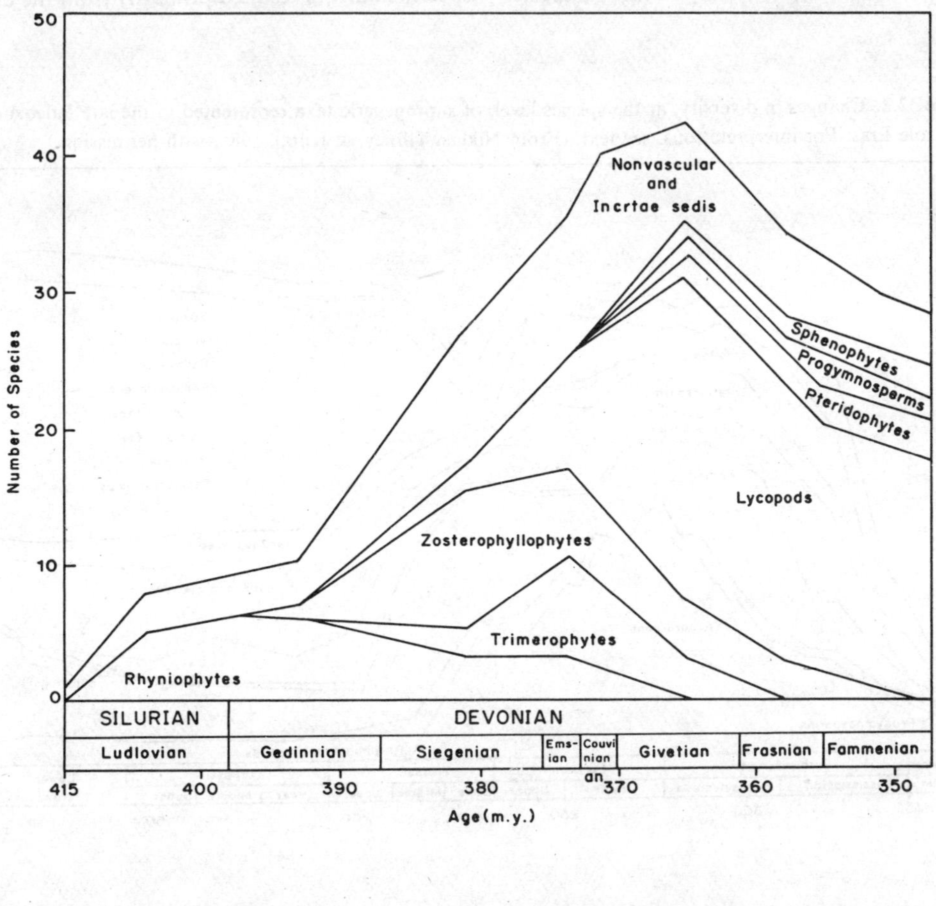

cation, were gradually replaced in the Upper Devonian by the second-phase lycopods and other derived taxa including ferns, progymnosperms, and sphenophylls. These Devonian genera, derived from their first-phase primitive ancestors, continued on into the Carboniferous to become dominant constituents of phase two including the coal-swamp floras (Fig. 32.2).

During the Carboniferous, those taxa evolving during the second phase of radiation and diversification were augmented by the appearance of the third-phase Upper Devonian seed plants. We have seen that diversification of these plants resulted in the subsequent evolution of cordaites, coniferophytes, pteridosperms, cycads, cycadeoids, and ginkgophytes. All are gymnospermous seed plants that replaced many of the pteridophytes, especially the arborescent lycopods and sphenopsids. It is clear that the appearance of the seed habit, which involved retention of the megagametophyte within sporophytic tissues and the formation of pollen tubes which eliminated the requirement of free water for fertilization, provided the basis for their adaptive radiation into uplands during the Upper Paleozoic and on into the Mesozoic. The latter shows abundant evidence of drastic changes in the terrestrial environments from the humid conditions of the great coal swamps to much drier, extensive uplands. Because of this the early Mesozoic is characterized by a major shift in the composition of floras so that by the Middle Triassic over 60 percent of the known species are gymnosperms, a number that rose to 80 percent by the Middle Jurassic. This rise in the number of gymnosperm species is associated with some decrease in the number of ferns and the final extinction of arborescent lycopods and cordaites during the Permian (Fig. 32.2).

Derived taxa showing increased rates of species origination and radiation during the Mesozoic include the voltzias, conifers, taxads, ginkgophytes, cycads, and cycadeoids. During this time the cycads, ferns, sphenopsids, and herbaceous lycopods show some decline in numbers of new species with some species exhibiting greater longevity and continuing on into the Tertiary. The pteridosperms and cycadeoids that became extinct in the Cretaceous were replaced during phase four (Fig. 32.3) by the rapidly diversifying and radiating angiosperms in the Lower Cretaceous.

Three of the four phases in vascular plant evolution outlined above show important increases in the rates of species originations. During the initial phase, which lasted approximately 60 m.y., the diversity of primitive land plants, represented by the maximum number of species, increased four times. During the second phase, when there was an adaptive radiation of the derived Upper Devonian pteridophytes into the Carboniferous (progymnosperms, arborescent lycopods, ferns, and sphenopsids), there was another fourfold increase (Fig. 32.2). The last major increase in species numbers resulted primarily from the diversifi-

Figure 32.2. Changes in diversity, at the species level, of suprageneric taxa represented in the late Paleozoic and Mesozoic Eras. For interpretations, see text. (From Niklas, Tiffney, & Knoll, 1985, with permission.)

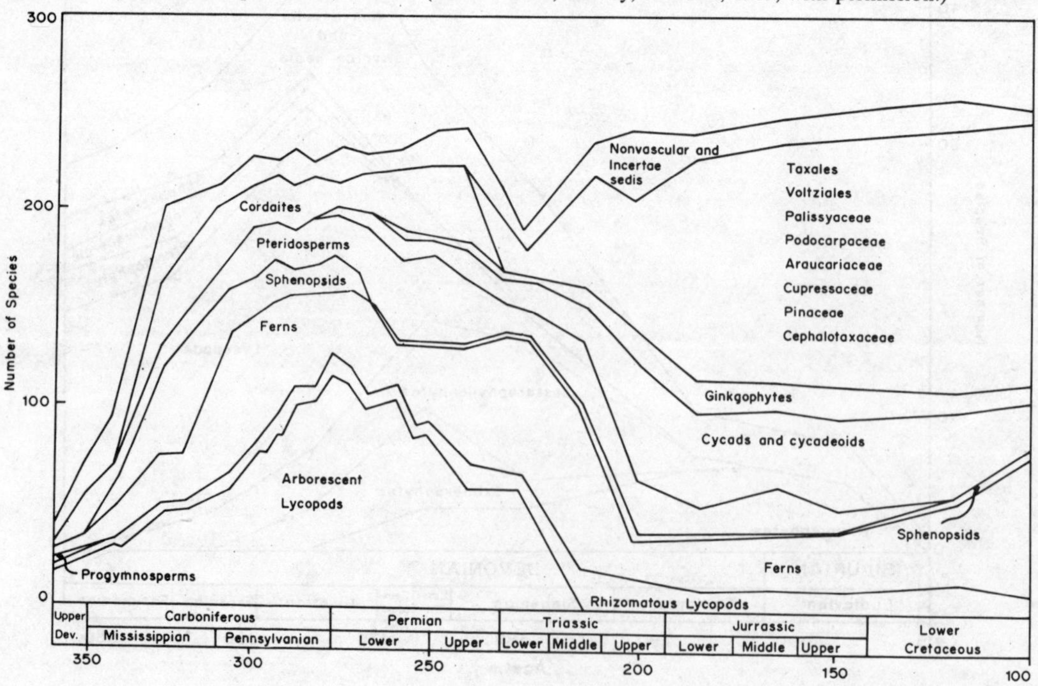

cation of the flowering plants during the Tertiary where they comprise almost 80 percent of all vascular plants (Fig. 32.3). Here they have replaced all but a few species of cycadophytes and ginkgophytes and to some extent ferns and coniferophytes. This threefold increase in species originations by the flowering plants is reflected by the number of present-day species, estimated to be from 200,000 to 300,000. It is clear that this spectacular increase by the angiosperms and their subsequent rapid radiation are related to the evolution of protective and selective function, provided by the carpel, changes in the life cycles, and insect-pollination mechanisms.

The species origination rate for the angiosperms is the greatest of any major taxon, while the duration of their species is the lowest, from 4 to 9.4 m.y. This is in marked contrast with the primitive vascular plants of the Silurian–Devonian (rhyniophytes, trimerophytes, and zosterophylls) where species originations are the lowest of all, but where species durations are the longest, from 12 to 14 m.y. Environmental changes, differences in life cycles, reproductive structures (e.g., seeds, cupules, carpels, flowers, and vegetative reproduction) are interpreted to be important, if not the ultimate, factors that determine the evolution rates as expressed by species originations and longevity (Niklas et al., 1985). That these rates can show marked differences is clearly demonstrated by such taxa as those lycopods, ferns, and sphenopsids with rhizomatous habits where there are low rates of species origination but greater species longevity. Of additional interest, it has long been noted that many of the characteristics of species belonging to these rhizome-forming groups first appearing in the Devonian have morphologies that appear to have remained more-or-less unchanged for a period of time in excess of 350 m.y. One can conclude that the longevity rates of their species are indicative of a capacity for survival not exhibited in other groups that were marked for extinction with radical changes in the environment. As far as the origination and extinction rates of their species are concerned, the stasis exhibited by the lineages of lycopods, sphenopsids, and ferns is more apparent than real. When one examines the total fossil record of these taxa, it appears that many new species have originated, while many have become extinct. Thus the more-or-less static cumulative species numbers for these groups actually represent dramatically changing groups where there is an approximate balance between the rates of originations and extinctions of species.

The distribution of the major taxa described in the previous chapters and in this retrospective analysis are diagrammatically represented by (Chart 32.1) as they occur in geological time.

Figure 32.3. Changes in diversity, at the species level, of suprageneric taxa represented in the late Mesozoic and Cenozoic Eras. The data used in the graphs for Figure 32.1, 32.2, and 32.3 are predominantly from the Northern Hemisphere. For interpretations of the diversity depicted in the graph, see text. (From Niklas, Tiffney, & Knoll, 1985, with permission.)

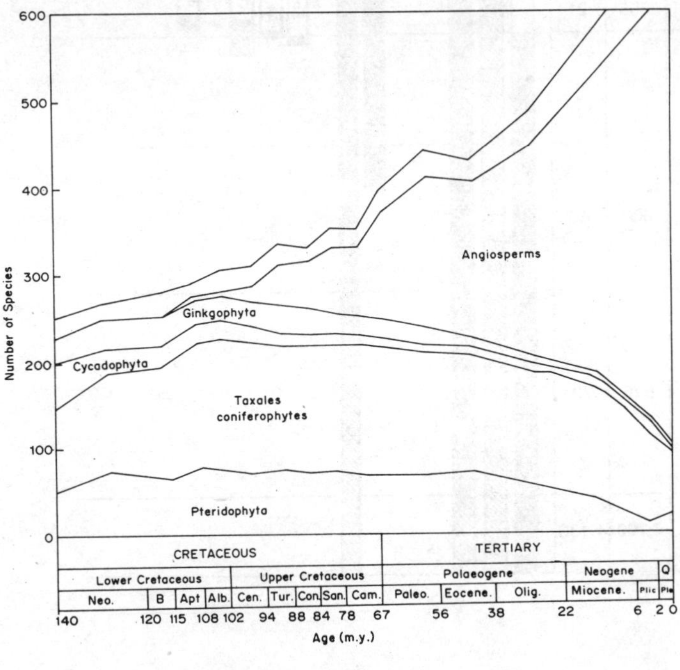

ERA	PERIOD		BEGAN M.Y. AGO	DURATION M.Y.
Cenozoic	Quaternary			2.5
	Tertiary		65	62.5
Mesozoic	Cretaceous	Upper	100	35
		Lower	141	41
	Jurassic		195	54
	Triassic		225	30
Paleozoic	Permian		280	55
	Pennsylvanian	Carboniferous	325	45
	Mississippian		345	20
	Devonian	Upper	359	14
		Middle	374	15
		Lower	395	21
	Silurian		435	40
	Ordovician		500	65
	Cambrian		570	70
Precambrian			4,700	4,130

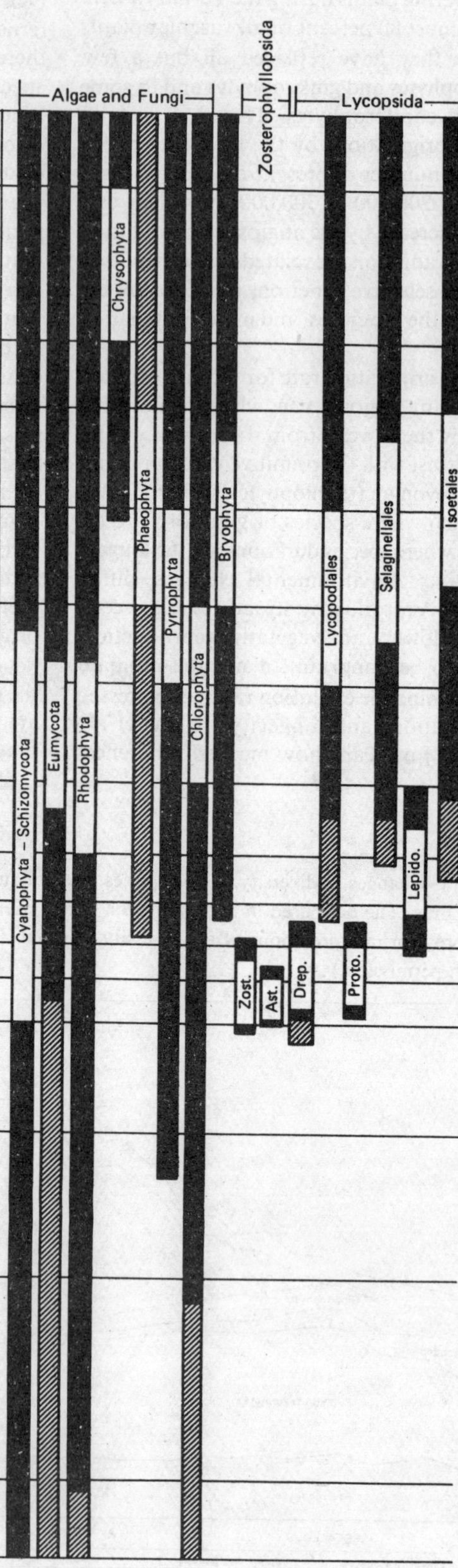

Chart 32.1. General summary of the distribution of major plant groups in geological time. Zost. = Zosterophylallales; Ast. = Asteroxylales; Drep. = Drepanophycales; Proto. = Protolepidodendrales; Lepido. = Lepidodendrales; Sphen. = Sphenophyllales; Hye. = Hyeniales; Arc. = Archaeocalamitaceae; Calam. = Calamitaceae; Koretro. = *Koretrophyllites;* Phyllo. = *Phyllotheca;* Neo. = Neocalamitaceae; Ps. = Pseudoborniales; Rhy. = Rhyniopsida; Trim. = Trimerophytopsida; Clad. = Cladoxylales; Stau. = Stauropteridales; Zygop. = Zygopteridales; Ophio. = Ophioglossales; Pterido. = Pteridospermales; Ane. = Aneurophytales; Cayt. = Caytoniales; Glossop. = Glossopteridales; Pen. = Pentoxylales; Czek. = Czekanowskiales; Arch. = Archaeopteridales; Cord. = Cordaitales.

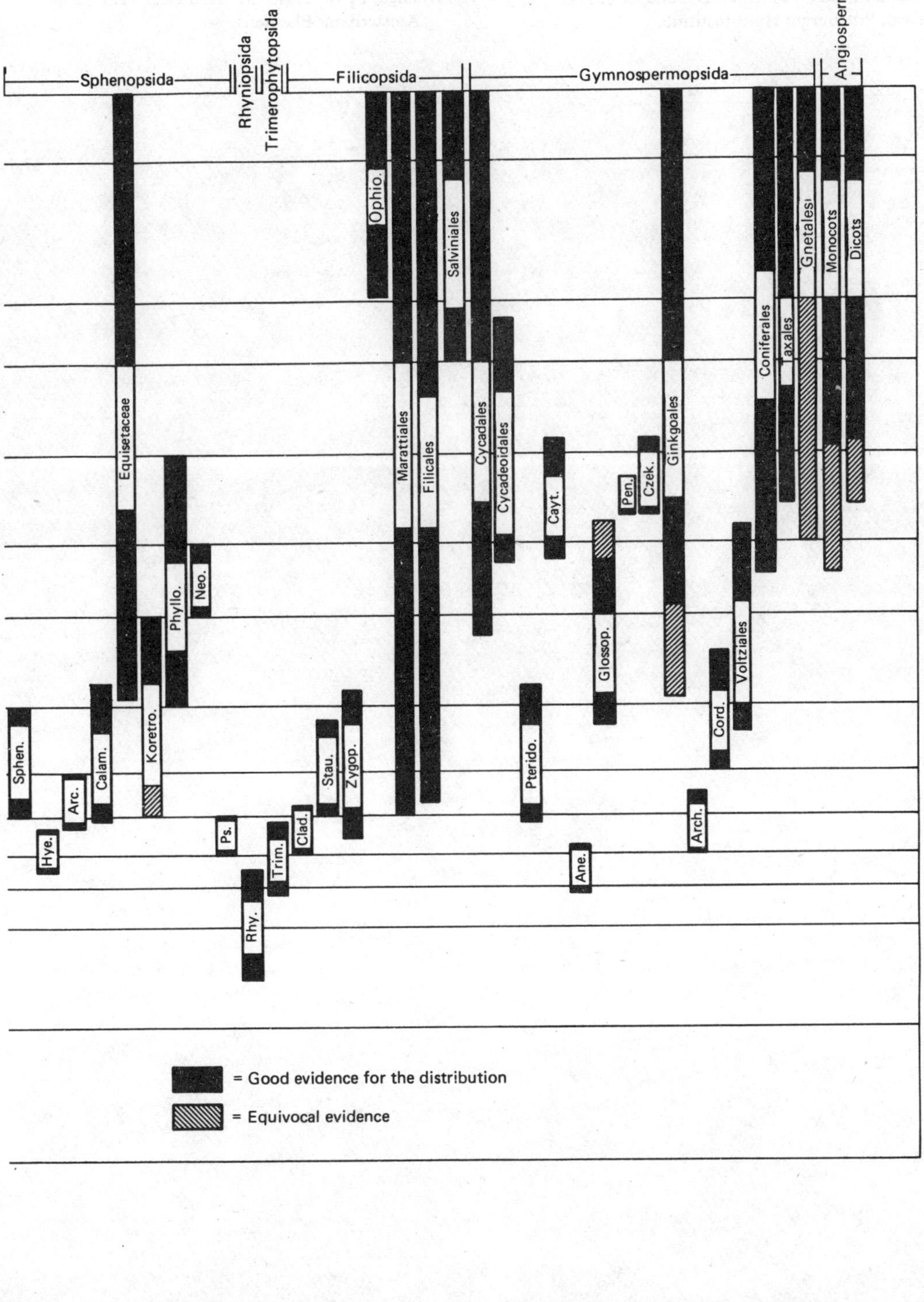

References

Banks, H. P. (1970). Major evolutionary events and the geological record of plants. *Biological Reviews*, **47,** 451–4.

Chaloner, W. G. (1970). The rise of the first land plants. *Biological Reviews,* **45,** 353–76.

Chaloner, W. G., & Sheerin, A. (1979). Devonian Macrofloras. In *The Devonian System, Special Papers. Palaeontology,* **23,** 145–61.

Chaloner, W. G., & Sheerin, A. (1981). The Evolution of Reproductive Strategies in Early Land Plants. In *Evolution Today,* eds. G. G. E. Scudder & J. L. Reveal. Pittsburgh; Hunt Institute.

Harland, W. B., Smith, A. G., & Wilcock, B. (1964). The phanerozoic time scale. *Quarterly Journal of the Geological Society, London,* **120**(S), 260–2.

Niklas, K. J., Tiffney, B. H., & Knoll, A. H. (1983). Patterns in vascular land plant diversification. *Nature,* **303,** 614–16.

Niklas, K. J., Tiffney, B. H., & Knoll, A. H. (1985). Patterns in vascular plant diversification: an analysis at the species level. In *Phanerozoic Diversity Patterns: Profiles in Macroevolution,* ed. J. W. Valentine. Princeton; Princeton University Press.

van Eysinga, F. W. B. (1975). *Geological Time Table.* Amsterdam: Elsevier.

Index

Letter f following page number indicates illustration. Letter c following page number indicates chart.